INTEGRATED SCIENCE

INTEGRATED SCIENCE

BILL W. TILLERY
Arizona State University

ELDON D. ENGER
Delta College

FREDERICK C. ROSS
Delta College

Boston   Burr Ridge, IL   Dubuque, IA   Madison, WI   New York   San Francisco   St. Louis
Bangkok   Bogotá   Caracas   Lisbon   London   Madrid
Mexico City   Milan   New Delhi   Seoul   Singapore   Sydney   Taipei   Toronto

McGraw-Hill Higher Education

A Division of The McGraw-Hill Companies

INTEGRATED SCIENCE

Published by McGraw-Hill, an imprint of The McGraw-Hill Companies, Inc., 1221 Avenue of the Americas, New York, NY 10020. Copyright © 2001 by The McGraw-Hill Companies, Inc. All rights reserved. No part of this publication may be reproduced or distributed in any form or by any means, or stored in a database or retrieval system, without the prior written consent of The McGraw-Hill Companies, Inc., including, but not limited to, in any network or other electronic storage or transmission, or broadcast for distance learning.

Some ancillaries, including electronic and print components, may not be available to customers outside the United States.

This book is printed on acid-free paper.

1 2 3 4 5 6 7 8 9 0 VNH/VNH 0 9 8 7 6 5 4 3 2 1 0

ISBN 0–07–229766–2

Vice president and editor-in-chief: *Kevin T. Kane*
Publisher: *JP Lenney*
Sponsoring editor: *Daryl Bruflodt*
Developmental editor: *Lori A. Sheil*
Editorial assistant: *Jenni Lang*
Marketing managers: *Mary K. Kittell/Debra A. Besler*
Senior project manager: *Gloria G. Schiesl*
Lead media producer: *Steve Metz*
Senior production supervisor: *Sandra Hahn*
Designer: *K. Wayne Harms*
Cover/interior designer: *Elise Lansdon*
Cover image: *Darryl Torckler/Tony Stone Images*
Photo research coordinator: *John C. Leland*
Photo research: *Mary Reeg Photo Research*
Senior supplement coordinator: *Audrey A. Reiter*
Compositor: *Carlisle Communications, Ltd.*
Typeface: *10/12 Minion*
Printer: *Von Hoffmann Press, Inc.*

The credits section for this book begins on page 850 and is considered an extension of the copyright page.

Library of Congress Cataloging-in-Publication Data

Tillery, Bill W.
    Integrated science / Bill W. Tillery, Eldon D. Enger, Frederick C. Ross. — 1st ed.
        p.   cm.
    Includes index.
    ISBN 0–07–229766–2
    1. Science. I. Enger, Eldon D. II. Ross, Frederick C. III. Title.

Q161.2 .T54 2001
500—dc21   00–033935
                                                          CIP

CONTENTS

CHAPTER | Fourteen
Organic Chemistry 321

CHAPTER | Fifteen
Nuclear Reactions 351

CHAPTER | Sixteen
The Universe 377

CHAPTER | Seventeen
The Solar System 405

CHAPTER | Eighteen
Earth in Space 435

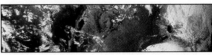

CHAPTER | Nineteen
The Earth 461

CHAPTER | Twenty
The Earth's Surface   487

CHAPTER | Twenty-One
Earth's Weather   527

CHAPTER | Twenty-Two
The Earth's Waters   565

CHAPTER | Twenty-Three
What Is Life?   587

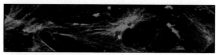

CHAPTER | Twenty-Four
The Origin and Evolution of Life   617

CHAPTER | Twenty-Seven
Mendelian and Molecular Genetics 743

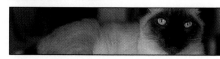

PREFACE

Introduction

Integrated Science is a straightforward, easy-to-read, but substantial introduction to the fundamental behavior of matter and energy in living and nonliving systems. It is intended to serve the needs of nonscience majors who are required to complete one or more science courses as part of a general or basic studies requirement. It introduces basic concepts and key ideas while providing opportunities for students to learn reasoning skills and a new way of thinking about their environment. No prior work in science is assumed. The language, as well as the mathematics, is as simple as can be practical for a college-level science course.

The *Integrated Science* sequence of chapters is flexible, and the instructor can determine topic sequence and depth of coverage as needed. The materials are also designed to support a conceptual approach or a combined conceptual and problem-solving approach. With laboratory studies, the text contains enough material for the instructor to select a sequence for a one- or two-semester course.

The Goals of Integrated Science

1. Create an introductory science course aimed at the nonscience major. The origin of this book is rooted in our concern for the education of introductory-level students in the field of science. Historically, nonscience majors had to enroll in courses intended for science or science-related majors such as premeds, architects, or engineers. Such courses are important for these majors, but are mostly inappropriate for introductory-level nonscience students who are simply not interested in science, and perhaps anxious about taking a science course. To put a nonscience student into such a course is a mistake. Few students will have the time or background to move through the facts, equations, and specialized language to gain any significant insights into the logic or fundamental understandings; instead, they will leave the course with a distaste for

science. Today, society has a great need for a few technically trained people, but a much larger need for individuals who understand science.

2. Introduce a course that presents a coherent and clear picture of all science disciplines—an interdisciplinary approach—which helps students confirm and calibrate the big picture with the real world. Recent studies and position papers have called for an interdisciplinary approach to teaching science to nonmajors. For example, the need is discussed in *Science for All Americans—Project 2061* (American Association for the Advancement of Science), *National Science Education Standards* (National Research Council, 1994), and *Science in the National Interest* (The White House, 1994). Interdisciplinary science is an attempt to broaden and humanize science education by reducing and breaking down the barriers that enclose traditional science disciplines as distinct subjects.

3. Help instructors build their own mix of descriptive and analytical aspects of science, arousing student interest and feelings as they help students reach the educational goals of their particular course. The spirit of interdisciplinary science is sometimes found in courses called "General Science," "Combined Science," or "Integrated Science." These courses draw concepts from a wide range of the traditional fields of science, but are not concentrated around certain problems or questions. For example, rather than just dealing with the physics of energy, an interdisciplinary approach might consider broad aspects of energy—dealing with potential problems of an energy crisis—including social and ethical issues. There are a number of approaches that can be used in interdisciplinary science, including the teaching of science in a *social, historical, philosophical,* or *problem-solving* context, but there is no single best approach. One of the characteristics of interdisciplinary science is that it is not constrained by the necessity of

teaching certain facts or by traditions. It likewise cannot be imposed as a formal discipline, with certain facts to be learned. It is justified by its success in attracting and holding the attention and interest of students, making them a little wiser as they make their way toward various careers and callings.

4. Humanize science for nonscience majors. Each chapter presents historical background where appropriate, uses everyday examples in developing concepts, and follows a logical flow of presentation. The historical chronology, of special interest to the humanistically inclined nonscience major, serves to humanize the science being presented. The use of everyday examples appeals to the nonscience major, typically accustomed to reading narration, not scientific technical writing, and also tends to bring relevancy to the material being presented. The logical flow of presentation is helpful to students not accustomed to thinking about relationships between what is being read and previous knowledge learned, a useful skill in understanding the sciences.

Features

To achieve the goals stated, this text includes a variety of features that should make your study of *Integrated Science* more effective and enjoyable. These aids are included to help you clearly understand the concepts and principles that serve as the foundation of the integrated sciences.

Overview

Chapter 1 provides an overview or orientation to integrated science in general, and this text in particular. It also describes the fundamental methods and techniques used by scientists to study and understand the world around us.

Introductory Overviews

Each chapter begins with an introductory overview. The overview previews the chapter's contents and what you can expect to learn from reading the chapter. After reading the introduction, browse through the chapter, paying particular attention to the topic headings and illustrations so that you get a feel for the kinds of ideas included within the chapter.

Chapter Outlines

The chapter outline includes all the major topic headings and subheadings within the body of the chapter. It gives you a quick glimpse of the chapter's contents and helps you locate sections dealing with particular topics.

Bold-Faced/Italicized Terms

As you read each chapter you will notice that various words appear darker than the rest of the text, and others appear in italics. The darkened words, or bold-faced terms, signify key terms that you will need to understand and remember to fully comprehend the material in which they appear. These important terms are defined in context the first time they are used. Italicized words are meant to emphasize their importance in understanding explanations of ideas and concepts discussed.

Activities

As you look through each chapter you will find one or more activities. These activities are simple investigative exercises that you can perform at home or in the classroom to demonstrate important concepts and reinforce your understanding of them.

Closer Look and Connections

Each chapter of *Integrated Science* also includes one or more **Closer Look** readings that discuss topics of special human or environmental concern, topics concerning interesting technological applications, or topics on the cutting edge of scientific research. All boxed features are informative materials that are supplementary in nature. In addition to the **Closer Look** readings, each chapter contains concrete interdisciplinary **Connections** that are set aside and highlighted. **Connections** will help you better appreciate the interdisciplinary nature of the sciences. The **Closer Look** and **Connections** serve to underscore the relevance of integrated science in confronting the many issues we face in our day-to-day lives. They are identified with the following icons:

General: This icon identifies interdisciplinary topics that cross over several categories; for example, life sciences and technology.

Life: This icon identifies interdisciplinary life science topics, meaning connections concerning all living organisms collectively: plant life, animal life, marine life, and any other classification of life.

Technology: This icon identifies interdisciplinary technology topics, that is, connections concerned with the application of science for the comfort and well being of people, especially through industrial and commercial means.

Measurement, Thinking, Scientific Methods: This icon identifies interdisciplinary concepts and understandings concerned with people trying to make sense out of their surroundings by making observations, measuring, thinking, developing explanations for what is observed, and experimenting to test those explanations.

Environmental Science: This icon identifies interdisciplinary concepts and understandings about the problems caused by human use of the natural world and remedies for those problems.

End-of-Chapter Features

At the end of each chapter you will find the following materials:

- *Summary:* highlights the key elements of the chapter
- *Summary of Equations* (Chapters 1–9, 11–13, 15): to reinforce your retention of them
- *Key Terms:* page-referenced where you will find the terms defined in context
- *Applying the Concepts:* a multiple choice quiz to test your comprehension of the material covered
- *Questions for Thought:* designed to challenge you to demonstrate your understandings of the topic

- *Parallel Exercises* (Chapters 1–15): There are two groups of parallel exercises, Group A and Group B. The Group A parallel exercises have complete solutions worked out, along with useful comments in appendix D. The Group B parallel exercises are similar to those in Group A but do not contain answers in the text. By working through the Group A parallel exercises and checking the solution in appendix D you will gain confidence in tackling the parallel exercises in Group B, and thus reinforce your problem-solving skills.

End-of-Text Material

At the back of the text you will find appendices that will give you additional background details, charts, and answers to chapter exercises. There is also a glossary of all key terms, an index organized alphabetically by subject matter, and special tables printed on the inside covers for reference use.

Supplementary Materials

Integrated Science is accompanied by a variety of supplementary materials, including an instructor's manual for the text, a laboratory manual, an instructor's edition of the laboratory manual, a test bank containing multiple choice test items, and a fully interactive website.

Laboratory Manual

The laboratory manual, written and classroom tested by the authors, presents a selection of laboratory exercises specifically written for the interest and abilities of nonscience majors. There are laboratory exercises that require measurement, data analysis, and thinking in a more structured learning environment. Alternative exercises that are open-ended "Invitations to Inquiry" are provided for instructors who would like a less-structured approach. When the laboratory manual is used with *Integrated Science,* students will have an opportunity to master basic scientific principles and concepts, learn new problem-solving and thinking skills, and understand the nature of scientific inquiry from the perspective of hands-on experiences. There is also an **instructor's edition lab manual** available for professors upon request.

Instructor's Manual/Test Item File

The instructor's manual, also written by the text authors, provides a chapter outline, an introduction/summary of each chapter, suggestions for discussion and demonstrations, and multiple choice questions (with answers) that can be used as resources for cooperative teaching. It also includes answers and solutions to all end-of-chapter questions and exercises not provided in the text.

Microtest

This computerized test bank is available in both Windows and Macintosh formats.

Interactive Website—Found at http://www.mhhe.com/

For Instructors: This text-specific website includes the fully downloadable Instructor's Manual/Test Item File, a powerpoint presentation of figures from the text that can be integrated into your own lecture, web links, an "Ask the Author" message board, and many other features. In addition, instructors can gain access to the Online Learning Center, which contains many other features that can be pulled into PageOut™, McGraw-Hill's solution for helping instructors create their own web pages. PageOut™ offers a series of templates. Simply fill them with your course information and click on one of 16 designs. The process takes under an hour and leaves you with a professionally designed website. PageOut™ is so easy and intuitive, it's little wonder why over 5,000 of your colleagues are using it.

For Students: Students can use our website to study! It contains scorable practice quizzes and crossword puzzles that use key terms and definitions from the text, as well as a career center and web links. Accessing the Online Learning Center will allow the student to use flashcards, take additional self-assessment quizzes, and utilize the online glossary. For students wanting additional help, they can post a message to the "Ask the Author" message board, which is mediated by the authors. Check it out today—new features are always being added!

Reviewers of Integrated Science

Many constructive suggestions, new ideas, and invaluable advice were provided by reviewers through several stages of manuscript development. Special thanks and appreciation to those who reviewed all or parts of the manuscript.

Douglas Allchin   *University of Texas, El Paso*
B. J. Bateman   *Troy State University*
William Blaker   *Furman University*
Lauretta Buschar   *Beaver College*
Tracey Cascadden   *University of New Mexico*
Tim Champion   *J. C. Smith University*
James Courtright   *Marquette University*
David Fawcett   *Columbia State Community College*
David Grainger   *Colorado State University*
James J. Grant   *St. Peter's College*
William Keller   *St. Petersburg Junior College*
Mary Hurn Korte   *Concordia University*
Theo Koupelis   *University of Wisconsin, Marathon*
Bryan Long   *Columbia State Community College*
Bruce MacLaren   *Eastern Kentucky University*
Donald Miller   *University of Michigan, Dearborn*
Scott Mohr   *Boston University*
Joaquin Ruiz   *University of Arizona*
John Oakes   *Marian College*
C. Dianne Phillips   *University of Arkansas*
Richard Sapanaro   *Contra Costa College*
Herbert Stewart   *Florida Atlantic University*
Deborah Tull   *University of New Mexico*
Andrew Wallace   *Angelo State University*
Jim Westgate   *Lamar University*
Robert Wingfield, Jr.   *Fisk University*
Yen-Yuan James Wu   *Texas A & M University—Commerce*

Science is concerned with your surroundings and your concepts and understanding of these surroundings.

CHAPTER | One

The World Around You

Have you ever thought about your thinking and what you know? On a very simplified level, you could say that everything you know came to you through your senses. You see, hear, and touch things of your choosing and you can also smell and taste things in your surroundings. Information is gathered and sent to your brain by your sense organs. Somehow, your brain processes all this information in an attempt to find order and make sense of it all. Finding order helps you understand the world and what may be happening at a particular place and time. Finding order also helps you predict what may happen next, which can be very important in a lot of situations.

This is a book on thinking about and understanding your surroundings. These surroundings range from the obvious, such as the landscape and the day-to-day weather, to the not so obvious, such as how atoms are put together. Your surroundings include natural things as well as things that people have made and used (figure 1.1). You will learn how to think about your surroundings, whatever your previous experience with thought-demanding situations. This first chapter is about "tools and rules" that you will use in the thinking process.

OBJECTS AND PROPERTIES

Science is concerned with making sense out of the environment. The early stages of this "search for sense" usually involve objects in the environment, things that can be seen or touched. These could be *objects* you see every day, such as a glass of water, a moving automobile, or a running dog. They could be quite large, such as the sun, the moon, or even the solar system, or invisible to the unaided human eye. Objects can be any size, but people are usually concerned with objects that are larger than a pinhead and smaller than a house. Outside these limits, the actual size of an object is difficult for most people to comprehend.

As you were growing up, you learned to form a generalized mental image of objects called a *concept.* Your concept of an object is an idea of what it is, in general, or what it should be according to your idea (figure 1.2). You usually have a word stored away in your mind that represents a concept. The word "chair," for example, probably evokes an idea of "something to sit on." Your generalized mental image for the concept that goes with the word "chair" probably includes a four-legged object with a backrest. Upon close inspection, most of your (and everyone else's) concepts are found to be somewhat vague. For example, if the word "chair" brings forth a mental image of something with four legs and a backrest (the concept), what is the difference between a "high chair" and a "bar stool"? When is a chair a chair and not a stool? Thinking about this question is troublesome for most people.

Not all of your concepts are about material objects. You also have concepts about intangibles such as time, motion, and relationships between events. As was the case with concepts of material objects, words represent the existence of intangible concepts. For example, the words "second," "hour," "day," and "month" represent concepts of time. A concept of the pushes and pulls that come with changes of motion during an airplane flight might be represented with such words as "accelerate" and "falling." Intangible concepts might seem to be more abstract since they do not represent material objects.

By the time you reach adulthood you have literally thousands of words to represent thousands of concepts. But most, you would find on inspection, are somewhat ambiguous and not at all clear-cut. That is why you find it necessary to talk about certain concepts for a minute or two to see if the other person has the same "concept" for words as you do. That is why when one person says, "Boy, was it hot!" the other person may respond, "How hot was it?" The meaning of "hot" can be quite different for two people, especially if one is from Arizona and the other from Alaska!

The problem with words, concepts, and mental images can be illustrated by imagining a situation involving you and another person. Suppose that you have found a rock that you believe would make a great bookend. Suppose further that you are talking to the other person on the telephone, and you want to discuss the suitability of the rock as a bookend, but you do not know the name of the rock. If you knew the name, you would simply state that you found a "_____." Then you would probably discuss the rock for a minute or so to see if the other person really understood what you were talking about. But not knowing the name of the rock, and wanting to communicate about the suitability of the object as a bookend, what would you do? You would probably describe the characteristics, or **properties,** of the rock. Properties are the qualities or attributes that, taken together, are usually peculiar to an object. Since you commonly determine properties with your senses (smell, sight, hearing, touch, and taste), you could say that the properties of an object are the effect the object has on your senses. For example, you might say that the rock is a "big, yellow, smooth rock with shiny gold cubes." But consider the mental image that the other person on the telephone forms when you describe these properties. It is entirely possible that the other person is thinking of something very different from what you are describing (figure 1.3)!

As you can see, the example of describing a proposed bookend by listing its properties in everyday language leaves much to be desired. The description does not really help the other person form an accurate mental image of the rock. One problem with

FIGURE 1.2

What is your concept of a chair? Are all of these pieces of furniture chairs? Most people have concepts, or ideas of what things in general should be, that are loosely defined. The concept of a chair is one example of a loosely defined concept.

FIGURE 1.1

Your surroundings include naturally occurring objects and manufactured objects such as sidewalks and buildings.

the attempted communication is that the description of any property implies some kind of *referent.* The word **referent** means that you *refer to,* or think of, a given property in terms of another, more familiar object. Colors, for example, are sometimes stated with a referent. Examples are "sky blue," "grass green," or "lemon yellow." The referents for the colors blue, green, and yellow are, respectively, the sky, living grass, and a ripe lemon.

Referents for properties are not always as explicit as they are with colors, but a comparison is always implied. Since the comparison is implied, it often goes unspoken and leads to assumptions in communications. For example, when you stated that the rock was "big," you assumed that the other person knew that you did not mean as big as a house or even as big as a bicycle. You assumed that the other person knew that you meant that the rock was about as large as a book, perhaps a bit larger.

Another problem with the listed properties of the rock is the use of the word "smooth." The other person would not know if you meant that the rock *looked* smooth or *felt* smooth. After all, some objects can look smooth and feel rough. Other objects

FIGURE 1.3

Could you describe this rock to another person over the telephone so that the other person would know *exactly* what you see? This is not likely with everyday language, which is full of implied comparisons, assumptions, and inaccurate descriptions.

can look rough and feel smooth. Thus, here is another assumption, and probably all of the properties lead to implied comparisons, assumptions, and a not very accurate communication. This is the nature of your everyday language and the nature of most attempts at communication.

1. Find out how people communicate about the properties of objects. Ask several friends to describe a paper clip while their hands are behind their backs. Perhaps they can do better describing a goatee? Try to make a sketch that represents each description.
2. Ask two classmates to sit back to back. Give one of them a sketch or photograph that shows an object in some detail, perhaps a guitar or airplane. This person is to describe the properties of the object *without naming it*. The other person is to make a scaled sketch from the description. Compare the sketch to the description; then see how the use of measurement would improve the communication.

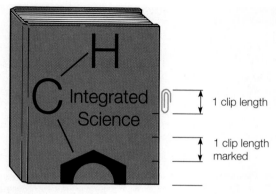

FIGURE 1.4

As an example of the measurement process, a standard paper-clip length is selected as a referent unit. The unit is compared to the property that is being described. In this example, the property of the book length is measured by counting how many clip lengths describe the length.

QUANTIFYING PROPERTIES

Typical day-to-day communications are often vague and leave much to be assumed. A communication between two people, for example, could involve one person describing some person, object, or event to a second person. The description is made by using referents and comparisons that the second person may or may not have in mind. Thus, such attributes as "long" fingernails or "short" hair may have entirely different meanings to different people involved in a conversation. Assumptions and vagueness can be avoided by using **measurement** in a description. Measurement is a process of comparing a property to a well-defined and agreed-upon referent. The well-defined and agreed-upon referent is used as a standard called a **unit.** The measurement process involves three steps: (1) *comparing* the referent unit to the property being described, (2) following a *procedure,* or operation, which specifies how the comparison is made, and (3) *counting* how many standard units describe the property being considered.

As an example of how the measurement process works, consider the property of *length*. Most people are familiar with the concept of the length of something (long or short), the use of length to describe distances (close or far), and the use of length to describe heights (tall or short). The referent units used for measuring length are the familiar inch, foot, and mile from the English system and the centimeter, meter, and kilometer of the metric system. These systems and specific units will be discussed later. For now, imagine that these units do not exist but that you need to measure the length and width of this book. This imaginary exercise will illustrate how the measurement process eliminates vagueness and assumption in communication.

The first requirement in the measurement process is to choose some referent unit of length. You could arbitrarily choose something that is handy, such as the length of a standard paper clip, and you could call this length a "clip." Now you must decide on a procedure to specify how you will use the clip unit. You could define some specific procedures. For example:

1. Place a clip parallel to and on the long edge, or length, of the book so the end of the referent clip is lined up with the bottom edge of the book. Make a small pencil mark at the other end of the clip, as shown in figure 1.4.
2. Move the outside end of the clip to the mark and make a second mark at the other end. Continue doing this until you reach the top edge of the book.
3. Compare how many clip replications are in the book length by counting.
4. Record the length measurements by writing (a) how many clip replications were made and (b) the name of the clip length.

If the book length did not measure to a whole number of clips, you might need to divide the clip length into smaller subunits to be more precise. You could develop a *scale* of the basic clip unit and subunits. In fact, you could use multiples of the basic clip unit for an extended scale, using the scale for measurement rather than moving an individual clip unit. You could call the scale a "clipstick" (as in yardstick or meterstick).

The measurement process thus uses a defined referent unit, which is compared to a property being measured. The *value* of the property is determined by counting the number of referent units. The name of the unit implies the procedure that results in the number. A measurement statement always contains a *number* and *name* for the referent unit. The number answers the question "How much?" and the name answers the question "Of what?" Thus a measurement always tells you "how much of what." You will find that using measurements will sharpen your communications. You will also find that using measurements is one of the first steps in understanding your physical environment.

MEASUREMENT SYSTEMS

Measurement is a process that brings precision to a description by specifying the "how much" and "of what" of a property in a particular situation. A number expresses the value of the property, and the name of a unit tells you what the referent is as well as implying the procedure for obtaining the number. Referent units must be defined and established, however, if others are to

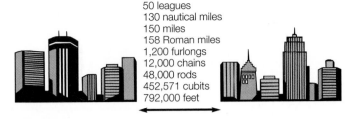

50 leagues
130 nautical miles
150 miles
158 Roman miles
1,200 furlongs
12,000 chains
48,000 rods
452,571 cubits
792,000 feet

FIGURE 1.5

Any of these units and values could have been used at some time or another to describe the same distance between these hypothetical towns. Any unit could be used for this purpose, but when one particular unit is officially adopted, it becomes known as the *standard unit*.

understand and reproduce a measurement. It would be meaningless, for example, for you to talk about a length in "clips" if other people did not know what you meant by a "clip" unit. When standards are established the referent unit is called a **standard unit** (figure 1.5). The use of standard units makes it possible to communicate and duplicate measurements. Standard units are usually defined and established by governments and their agencies that are created for that purpose. In the United States, the agency concerned with measurement standards is appropriately named the National Institute of Standards and Technology.

There are two major *systems* of standard units in use today, the English system and the metric system. The metric system is used throughout the world except in the United States, where both systems are in use. The continued use of the English system in the United States presents problems in international trade, so there is pressure for a complete conversion to the metric system. More and more metric units are being used in everyday measurements, but a complete conversion will involve an enormous cost. Appendix A contains a method for converting from one system to the other easily. Consult this section if you need to convert from one metric unit to another metric unit or to convert from English to metric units or vice versa. Conversion factors are listed inside the front cover.

People have used referents to communicate about properties of things throughout human history. The ancient Greek civilization, for example, used units of *stadia* to communicate about distances and elevations. The "stadium" was a referent unit based on the length of the racetrack at the local stadium ("stadia" is the plural of stadium). Later civilizations, such as the ancient Romans, adopted the stadia and other referent units from the ancient Greeks. Some of these same referent units were later adopted by the early English civilization, which eventually led to the *English system* of measurement. Some adopted units of the English system were originally based on parts of the human body, presumably because you always had these referents with you (figure 1.6). The inch, for example, used the end joint of the thumb for a referent. A foot, naturally, was the length of a foot, and a yard was the distance from the tip of the nose to the end of the fingers on an arm held straight out. A cubit was the distance from the end of an elbow to the fingertip, and a fathom

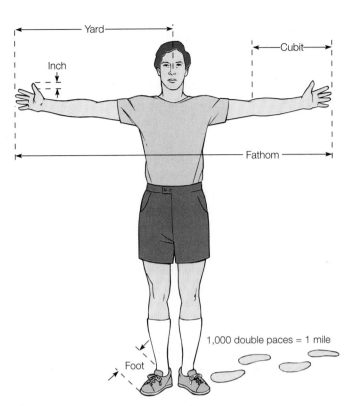

1,000 double paces = 1 mile

FIGURE 1.6

Many early units for measurement were originally based on the human body. Some of the units were later standardized by governments to become the basis of the English system of measurement.

TABLE 1.1	Early conversion table for English units of volume
Two Quantities	**Equivalent Quantity**
2 mouthfuls	= 1 jigger
2 jiggers	= 1 jack
2 jacks	= 1 jill
2 jills	= 1 cup
2 cups	= 1 pint
2 pints	= 1 quart
2 quarts	= 1 pottle
2 pottles	= 1 gallon
2 gallons	= 1 pail
2 pails	= 1 peck
2 pecks	= 1 bushel

was the distance between the fingertips of two arms held straight out. As you can imagine, there were problems with these early units because everyone was not the same size. Beginning in the 1300s, the sizes of the units were gradually standardized by various English kings (table 1.1). In 1879, the United States, along with sixteen other countries, signed the *Treaty of the Meter*, defining the English units in terms of the metric system. The

TABLE 1.2	The SI standard units	
Property	**Unit**	**Symbol**
Length	meter	m
Mass	kilogram	kg
Time	second	s
Electric current	ampere	A
Temperature	kelvin	K
Amount of substance	mole	mol
Luminous intensity	candela	cd

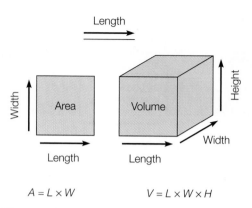

$A = L \times W$ $V = L \times W \times H$

FIGURE 1.7

Area, or the extent of a surface, can be described by two length measurements. Volume, or the space that an object occupies, can be described by three length measurements. Length, however, can be described only in terms of how it is measured, so it is called a *fundamental property.*

United States thus became officially metric, but not entirely metric in everyday practice.

The *metric system* was established by the French Academy of Sciences in 1791. The academy created a measurement system that was based on invariable referents in nature, not human body parts. These referents have been redefined over time to make the standard units more reproducible. In 1960, six standard metric units were established by international agreement. The *International System of Units,* abbreviated *SI,* is a modernized version of the metric system. Today, the SI system has seven units that define standards for the properties of length, mass, time, electric current, temperature, amount of substance, and light intensity (table 1.2). The standard units for the properties of length, mass, and time are introduced in this chapter. The remaining units will be introduced in later chapters as the properties they measure are discussed.

STANDARD UNITS FOR THE METRIC SYSTEM

If you consider all the properties of all the objects and events in your surroundings, the number seems overwhelming. Yet, close inspection of how properties are measured reveals that some properties are combinations of other properties (figure 1.7). Volume, for example, is described by the three length measurements of length, width, and height. Area, on the other hand, is described by just the two length measurements of length and width. Length, however, cannot be defined in simpler terms of any other property. There are four properties that cannot be described in simpler terms, and all other properties are combinations of these four. For this reason they are called the *fundamental properties.* A fundamental property cannot be defined in simpler terms other than to describe how it is measured. These four fundamental properties are (1) *length,* (2) *mass,* (3) *time,* and (4) *charge.* Used individually or in combinations, these four properties will describe or measure what you observe in nature. Metric units for measuring the fundamental properties of length, mass, and time will be described next. The fourth fundamental property, charge, is associated with electricity, and a unit for this property will be discussed in a future chapter.

Length

The standard unit for length in the metric system is the *meter* (the symbol or abbreviation is m). The meter was originally defined in 1793 as one ten-millionth of the distance between the geographic North Pole and the equator of the earth. In order to make this standard accessible, the length of a meter was determined and a one-meter metal bar was made as a prototype. This prototype was used to make copies for the countries of the world. The United States received its prototype meter in 1890. Beginning in 1893, the yard was legally defined in terms of the meter. Metal bars, however, tend to expand and contract with changes in temperature, so every precise measurement with the bar required a correction for the temperature. In 1960 the definition of a meter was changed to one that used the wavelength of a certain color of light given off from a particular element. In 1983 the definition was again changed, this time in terms of the distance that light travels in a vacuum during a certain time period, 1/299,792,458 second. The important thing to remember, however, is that the meter is the metric *standard unit* for length. A meter is slightly longer than a yard, 39.3 inches. It is approximately the distance from your left shoulder to the tip of your right hand when your arm is held straight out. Many doorknobs are about one meter above the floor. Think about these distances when you are trying to visualize a meter length.

Mass

The standard unit for mass in the metric system is the *kilogram* (kg). The kilogram is defined as the mass of a certain metal cylinder kept by the International Bureau of Weights and Measures in France. This is the only standard unit that is still defined in terms of an object. The property of mass is sometimes confused with the property of weight since they are directly proportional to each other at a given location on the surface of the earth. They are, however, two completely different properties

A CLOSER LOOK | The Leap Second

Most people have heard of a leap year, but not a leap second. A *leap year* is needed because the earth does not complete an exact number of turns on its axis while completing one trip around the sun. Our calendar system was designed to stay in step with the seasons with 365-day years and a 366-day year (leap year) every fourth year.

Likewise, our clocks are occasionally adjusted by a one-second increment known as a *leap second*. The leap second is needed because the earth does not have a constant spin. Coordinated Universal Time is the worldwide scientific standard of timekeeping. It is based upon the earth's rotation and is kept accurate to within microseconds with carefully maintained atomic clocks. A leap second is a second added to Coordinated Universal Time to make it agree with astronomical time to within 0.9 second.

In 1955 astronomers at the U.S. Naval Observatory and the National Physical Laboratory in England measured the relationship between the frequency of the cesium atom (the standard of time) and the rotation of the earth at a particular period of time. The standard atomic clock second was defined to be equivalent to the fraction 1/31,556,925.9747 of the year 1900—or, an average second for that year. This turned out to be the time required for 9,192,631,770 vibrations of the cesium 133 atom. The second was defined in 1967 in terms of the length of time required for 9,192,631,770 vibrations of the cesium 133 atom. So, the atomic second was set equal to an average second of the earth rotation time near the turn of the twentieth century, but defined in terms of the frequency of a cesium atom.

The earth is constantly slowing from the frictional effects of the tides. Evidence of this slowing can be found in records of ancient observations of eclipses. From these records it is possible to determine the slowing of the earth to be roughly 1–3 milliseconds per day per century. This causes the earth's rotational time to slow with respect to the atomic clock time. It has been nearly a century since the referent year used for the definition of a second, and the difference is now roughly 2 milliseconds per day. Other factors also affect the earth's spin, such as the wind from hurricanes, so that it is necessary to monitor the earth's rotation continuously and add or subtract leap seconds when needed.

---

and are measured with different units. All objects tend to maintain their state of rest or straight-line motion, and this property is called "inertia." The *mass* of an object is a measure of the inertia of an object. The *weight* of the object is a measure of the force of gravity on it. This distinction between weight and mass will be discussed in detail in chapter 3. For now, remember that weight and mass are not the same property.

Time

The standard unit for time is the *second* (s). The second was originally defined as 1/86,400 of a solar day (1/60 × 1/60 × 1/24). The earth's spin was found not to be as constant as thought, so the second was redefined in 1967 to be the duration required for a certain number of vibrations of a certain cesium atom. A special spectrometer called an "atomic clock" measures these vibrations and keeps time with an accuracy of several millionths of a second per year.

METRIC PREFIXES

The metric system uses prefixes to represent larger or smaller amounts by factors of 10. Some of the more commonly used prefixes, their abbreviations, and their meanings are listed in table 1.3. Figure 1.8 illustrates how these prefixes are used. Suppose you wish to measure something smaller than the standard unit of length, the meter. The meter is subdivided into ten equal-sized subunits called *decimeters*. The prefix *deci-* has a meaning of "one-tenth of," and it

TABLE 1.3	Some metric prefixes	
Prefix	**Symbol**	**Meaning**
Giga-	G	1,000,000,000 times the unit
Mega-	M	1,000,000 times the unit
Kilo-	k	1,000 times the unit
Hecto-	h	100 times the unit
Deka-	da	10 times the unit
Unit		
Deci-	d	0.1 of the unit
Centi-	c	0.01 of the unit
Milli-	m	0.001 of the unit
Micro-	μ	0.000001 of the unit
Nano-	n	0.000000001 of the unit

takes 10 decimeters to equal the length of 1 meter. For even smaller measurements, each decimeter is divided into ten equal-sized subunits called *centimeters*. It takes 10 centimeters to equal 1 decimeter and 100 to equal 1 meter. In a similar fashion, each prefix up or down the metric ladder represents a simple increase or decrease by a factor of 10.

When the metric system was established in 1791, the standard unit of mass was defined in terms of the mass of a certain volume of water. A cubic decimeter (dm³) of pure water at 4°C was *defined* to have a mass of 1 kilogram (kg). This definition

1 meter
1 decimeter
1 centimeter
1 millimeter

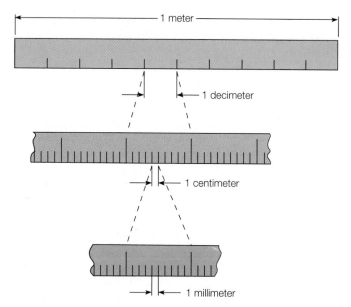

FIGURE 1.8

Prefixes are used with the standard units of the metric system to represent larger or smaller amounts by factors of 10. Measurements somewhat smaller than the standard unit of the meter, for example, are measured in decimeters. The prefix "deci-" means "one-tenth of," and it takes 10 decimeters to equal the length of 1 meter. For even smaller measurements, the decimeter is divided into 10 centimeters. Continuing to even smaller measurements, the centimeter is divided into 10 millimeters. There are many prefixes that can be used (table 1.3), but all are related by multiples of 10.

1 decimeter
(10 centimeters)
1 cm^3
1 mL H_2O
1 g H_2O
1 dm^3 (1,000 cm^3)
By definition:
1 kg H_2O
1L H_2O
1 decimeter
(10 centimeters)

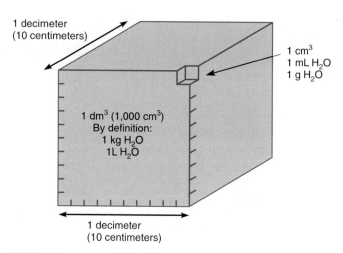

FIGURE 1.9

A cubic decimeter of water (1,000 cm^3) has a liquid volume of 1 L (1,000 mL) and a mass of 1 kg (1,000 g). Therefore, 1 cm^3 of water has a liquid volume of 1 mL and a mass of 1 g.

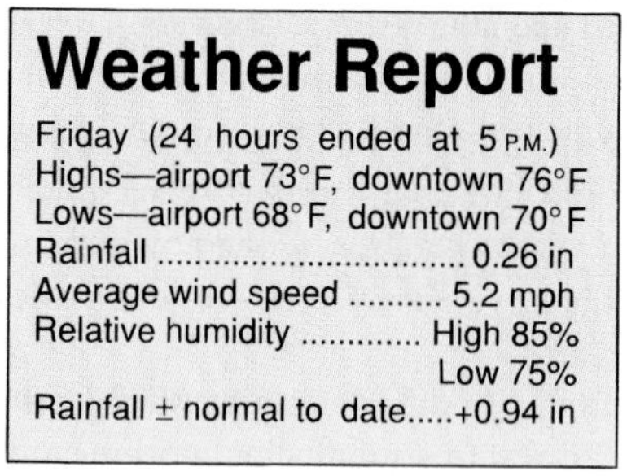

Weather Report
Friday (24 hours ended at 5 P.M.)
Highs—airport 73°F, downtown 76°F
Lows—airport 68°F, downtown 70°F
Rainfall .... 0.26 in
Average wind speed .... 5.2 mph
Relative humidity .... High 85%
Low 75%
Rainfall ± normal to date.....+0.94 in

FIGURE 1.10

A weather report gives exact information, data that describe the weather by reporting numerically specified units for each condition being described.

was convenient because it created a relationship between length, mass, and volume. As illustrated in figure 1.9, a cubic decimeter is 10 cm on each side. The volume of this cube is therefore 10 cm × 10 cm × 10 cm, or 1,000 cubic centimeters (abbreviated as cc or cm^3). Thus, a volume of 1,000 cm^3 of water has a mass of 1 kg. Since 1 kg is 1,000 g, 1 cm^3 of water has a mass of 1 g.

The volume of 1,000 cm^3 also defines a metric unit that is commonly used to measure liquid volume, the *liter* (L). For smaller amounts of liquid volume, the milliliter (mL) is used. The relationship between liquid volume, volume, and mass of water is therefore

$$1.0 \text{ L} \equiv 1.0 \text{ dm}^3 \text{ and has a mass of } 1.0 \text{ kg}$$

or, for smaller amounts,

$$1.0 \text{ mL} \equiv 1.0 \text{ cm}^3 \text{ and has a mass of } 1.0 \text{ g}$$

UNDERSTANDINGS FROM MEASUREMENTS

One of the more basic uses of measurement is to *describe* something in an exact way that everyone can understand. For example, if a friend in another city tells you that the weather has been "warm," you might not understand what temperature is being described. A statement that the air temperature is 70°F carries more exact information than a statement about "warm weather."

The statement that the air temperature is 70°F contains two important concepts: (1) the numerical value of 70 and (2) the referent unit of degrees Fahrenheit. Note that both a numerical value and a unit are necessary to communicate a measurement correctly. Thus, weather reports describe weather conditions with numerically specified units; for example, 70° Fahrenheit for air temperature, 5 miles per hour for wind speed, and 0.5 inch for rainfall (figure 1.10). When such numerically specified units are used in a description, or a weather report, everyone understands *exactly* the condition being described.

Data

Measurement information used to describe something is called **data.** Data can be used to describe objects, conditions, events, or changes that might be occurring. You really do not know if the weather is changing much from year to year until you compare the yearly weather data. The data will tell you, for example, if the weather is becoming hotter or dryer or is staying about the same from year to year.

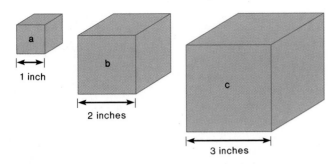

FIGURE 1.11

Cube *a* is 1 inch on each side, cube *b* is 2 inches on each side, and cube *c* is 3 inches on each side. These three cubes can be described and compared with data, or measurement information, but some form of analysis is needed to find patterns or meaning in the data.

Let's see how data can be used to describe something and how the data can be analyzed for further understanding. The cubes illustrated in figure 1.11 will serve as an example. Each cube can be described by measuring the properties of size and surface area.

First, consider the size of each cube. Size can be described by *volume*, which means *how much space something occupies.* The volume of a cube can be obtained by measuring and multiplying the length, width, and height. The smallest cube in figure 1.11, cube *a*, is 1 inch on each side, so the volume of this cube is 1 in $\times$ 1 in $\times$ 1 in, or 1 in^3. Note that both the numbers and the units were treated mathematically and $1 \times 1 \times 1 = 1$ and in $\times$ in $\times$ in = in^3. The cubic unit is thus a result of multiplying units, and *cubic inch* does not necessarily mean that the volume is in the shape of a cube. The middle-sized cube in figure 1.11, cube *b*, is 2 inches on each side, so the volume of this cube is 2 in $\times$ 2 in $\times$ 2 in, or 8 in^3. The largest cube, cube *c*, is 3 inches on each side, so the volume of this cube is 27 in^3. The data so far is

volume of cube *a*	1 in^3
volume of cube *b*	8 in^3
volume of cube *c*	27 in^3

Now consider the surface area of each cube. *Area* means *the extent of a surface,* and each cube has six surfaces, or faces (top, bottom, and four sides). The area of any face can be obtained by measuring and multiplying length and width. The smallest cube in figure 1.11, cube *a*, is 1 inch on each side, so the area of one face of this cube is 1 in $\times$ 1 in, or 1 in^2. Again, note that the numbers and units were treated mathematically: in $\times$ in = in^2. The square unit is a result of the multiplication and does not necessarily mean that the area is in the shape of a square. The area of one face of the smallest cube is 1 in^2, and there are six faces, so the total surface area of the smallest cube is 6×1 in^2, or 6 in^2. The middle-sized cube, cube *b*, is 2 inches on each edge, so the area of one face on this cube is 2 in $\times$ 2 in = 4 in^2. The total surface area is 6×4 in^2, or 24 in^2. The largest cube, cube *c*, is 3 inches on each edge, so each face has an area of 9 in^2, and

the total surface area is 6×9 in^2, or 54 in^2. The data for the three cubes thus describes them as follows:

	Volume	Surface Area
cube *a*	1 in^3	6 in^2
cube *b*	8 in^3	24 in^2
cube *c*	27 in^3	54 in^2

Ratios and Generalizations

Data on the volume and surface area of the three cubes in figure 1.11 describes the cubes, but whether it says anything about a relationship between the volume and surface area of a cube is difficult to tell. Nature seems to have a tendency to camouflage relationships, making it difficult to extract meaning from raw data. Seeing through the camouflage requires the use of mathematical techniques to expose patterns. You have spent your time in mathematics classes "doing" mathematics, but here is a chance to *use* it in the real world to understand something. The key is to reduce the data to something manageable that permits you to make comparisons. Using mathematics as a vehicle to understand your surroundings can be very exciting and satisfying. Let's see how such operations can be applied to the data on the three cubes and what the pattern means.

One mathematical technique for reducing data to a more manageable form is to expose patterns through a *ratio.* A ratio is a relationship between two numbers. You could think of a ratio as a *rate* obtained when one number is divided by another number. Suppose, for example, that an instructor has 50 sheets of graph paper for a laboratory group of 25 students. The relationship, or ratio, between the number of sheets and the number of students is 50 papers to 25 students, and this can be written as 50 papers/25 students. This ratio is *simplified* by dividing 25 into 50, and the ratio becomes 2 papers/1 student. The 1 is usually understood (not stated), and the ratio is written as simply 2 papers/student. It is read as 2 papers "for each" student, or 2 papers "per" student. The concept of simplifying with a ratio is an important one, and you will see it time and time again throughout science. It is important that you understand the meaning of "per" and "for each" when used with numbers and units.

Applying the ratio concept to the three cubes in figure 1.11, the ratio of surface area to volume for the smallest cube, cube *a*, is 6 in^2 to 1 in^3, or

$$\frac{6 \text{ in}^2}{1 \text{ in}^3} = 6 \frac{\text{in}^2}{\text{in}^3}$$

meaning there are 6 square inches of area *for each* cubic inch of volume.

The middle-sized cube, cube *b*, had a surface area of 24 in^2 and a volume of 8 in^3. The ratio of surface area to volume for this cube is therefore

$$\frac{24 \text{ in}^2}{8 \text{ in}^3} = 3 \frac{\text{in}^2}{\text{in}^3}$$

meaning there are 3 square inches of area *for each* cubic inch of volume.

The largest cube, cube *c*, had a surface area of 54 in^2 and a volume of 27 in^3. The ratio is

$$\frac{54 \text{ in}^2}{27 \text{ in}^3} = 2 \frac{\text{in}^2}{\text{in}^3}$$

or 2 square inches of area *for each* cubic inch of volume. Summarizing the ratio of surface area to volume for all three cubes, you have

small cube	*a*–6:1
middle cube	*b*–3:1
large cube	*c*–2:1

Now that you have simplified the data through ratios, you are ready to generalize about what the information means. You can generalize that the surface-area-to-volume ratio of a cube *decreases* as the volume of a cube becomes larger. Reasoning from this generalization will provide an explanation for a number of related observations. For example, why does crushed ice melt faster than a single large block of ice with the same volume? The explanation is that the crushed ice has a larger surface-area-to-volume ratio than the large block, so more surface is exposed to warm air. If the generalization is found to be true for shapes other than cubes, you could explain why a log chopped into small chunks burns faster than the whole log. Further generalizing might enable you to predict if 10 lb of large potatoes would require more or less peeling than 10 lb of small potatoes. When generalized explanations result in predictions that can be verified by experience, you gain confidence in the explanation. Finding patterns of relationships is a satisfying intellectual adventure that leads to understanding and generalizations that are frequently practical.

A Ratio Called Density

The power of using a ratio to simplify things, making explanations more accessible, is evident when you compare the simplified ratio 6 to 3 to 2 with the hodgepodge of numbers that you would have to consider without using ratios. The power of using the ratio technique is also evident when considering other properties of matter. Volume is a property that is sometimes confused with mass. Larger objects do not necessarily contain more matter than smaller objects. A large balloon, for example, is much larger than this book, but the book is much more massive than the balloon. The simplified way of comparing the mass of a particular volume is to find the ratio of mass to volume. This ratio is called mass **density,** which is defined as *mass per unit volume.* The "per" means "for each" as previously discussed, and "unit" means one, or each. Thus "mass per unit volume" literally means the "mass of one volume" (figure 1.12). The relationship can be written as

$$\text{mass density} = \frac{\text{mass}}{\text{volume}}$$

or

$$\rho = \frac{m}{V}$$

equation 1.1

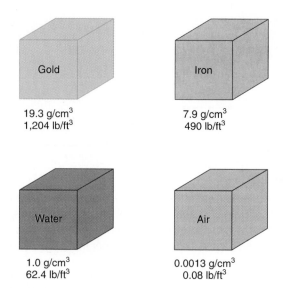

FIGURE 1.12

Equal volumes of different substances do not have the same mass. The ratio of mass to volume is defined as a property called *mass density,* which is identified with the Greek symbol ρ. The mass density of these substances is given in g/cm^3. The weight density (*D*) is given in lb/ft^3.

As with other ratios, density is obtained by dividing one number and unit by another number and unit. Thus, the density of an object with a volume of 5 cm^3 and a mass of 10 g is

$$\text{density} = \frac{10 \text{ g}}{5 \text{ cm}^3} = 2 \frac{\text{g}}{\text{cm}^3}$$

The density in this example is the ratio of 10 g to 5 cm^3, or 10 g/5 cm^3, or 2 g to 1 cm^3. Thus, the density of the example object is the mass of *one* volume (a unit volume), or 2 g *for each* cm^3.

Any unit of mass and any unit of volume may be used to express density. The densities of solids, liquids, and gases are usually expressed in grams per cubic centimeter (g/cm^3), but the densities of liquids are sometimes expressed in grams per milliliter (g/mL). Using SI standard units, densities are expressed as kg/m^3. When density is expressed in terms of mass, it is known as *mass density* and is given the symbol ρ (which is the Greek symbol for the letter rho). Density expressed in terms of weight is known as *weight density* and is given the symbol *D* (table 1.4). The units for weight and mass will be discussed fully in chapter 3. The "weight per unit volume" relationship can be written as

$$\text{weight density} = \frac{\text{weight}}{\text{volume}}$$

or

$$D = \frac{w}{V}$$

equation 1.2

If matter is distributed the same throughout a volume, the *ratio* of mass to volume will remain the same no matter what mass and volume are being measured. Thus, a teaspoonful, a cup, or a lake full of freshwater at the same temperature will all have a density of about 1 g/cm^3 or 1 kg/L.

CONNECTIONS

Density Matters—Fish, Nuclear Fusion, and Cola

Sharks and rays are marine animals that have an internal skeleton made entirely of cartilage. These animals have no swim bladder to adjust their body density in order to maintain their position in the water; therefore, they must constantly swim or they will sink. The bony fish, on the other hand, have a skeleton composed of bone and most also have a swim bladder. These fish can regulate the amount of gas in the bladder to control their density. Thus, the fish can remain at a given level in the water without expending large amounts of energy.

Scientists plan to study nuclear fusion involving extremely dense plasmas, up to six times denser than the center of the sun. The proposed National Ignition Facility (NIF) is expected to be funded and completed by 2010. This research facility will use high-power lasers to achieve self-sustaining fusion reactions, with densities reaching up to 1,000 g/cm³. Existing research facilities today are capable of creating plasma densities up to about 50 g/cm³. Compare these densities with what you might find at the center of the sun, about 160 g/cm³.

Finally, have you ever noticed the different floating characteristics of cans of the normal version of a carbonated cola beverage and a diet version? The surprising result is that the normal version usually sinks and the diet version usually floats. This has nothing to do with the amount of carbon dioxide in the two drinks. It is a result of the increase in density from the sugar added to the normal version, while the diet version has much less of an artificial sweetener that is much sweeter than sugar.

TABLE 1.4	Densities of some common substances	
Substance	**Mass Density** (ρ)	**Weight Density** (D)
	(g/cm³)	(lb/ft³)
Aluminum	2.70	169
Copper	8.96	555
Iron	7.87	490
Lead	11.4	705
Water	1.00	62.4
Seawater	1.03	64.0
Mercury	13.6	850
Gasoline	0.680	42.0

Activities

1. What is the mass density of this book? Measure the length, width, and height of this book in centimeters, then multiply to find the volume in cm³. Use a balance to find the mass of this book in grams. Compute the density of the book by dividing the mass by the volume. Compare the density in g/cm³ with other substances listed in table 1.4.
2. Compare the densities of some common liquids. Pour a cup of vinegar in a large bottle. Carefully add a cup of corn syrup, then a cup of cooking oil. Drop a coin, tightly folded pieces of aluminum foil, and toothpicks into the bottle. Explain what you observe in terms of density. Take care not to confuse the property of *density*, which describes the compactness of matter, with *viscosity*, which describes how much fluid resists flowing under normal conditions. (Corn syrup has a greater viscosity than water—is this true of density, too?)

THE NATURE OF SCIENCE

Most humans are curious, at least when they are young, and are motivated to understand their surroundings. These traits have existed since antiquity and have proven to be a powerful motivation. In recent times the need to find out has motivated the launching of space probes to learn what is "out there," and humans have visited the moon to satisfy their curiosity. Curiosity and the motivation to understand nature were no less powerful in the past than today. Over two thousand years ago the ancient Greeks lacked the tools and technology of today and could only make conjectures about the workings of nature. These early seekers of understanding are known as *natural philosophers*, and they thought and wrote about the workings of all of nature. They are called philosophers because their understandings come from reasoning only, without experimental evidence. Nonetheless, some of their ideas were essentially correct and are still in use today. For example, the idea of matter being composed of *atoms* was first reasoned by certain ancient Greeks in the fifth century B.C. The idea of *elements*, basic components that make up matter, was developed much earlier but refined by the ancient Greeks in the fourth century B.C. The concept of what the elements are and the concept of the nature of atoms have changed over time, but the idea first came from ancient natural philosophers.

The Scientific Method

Some historians identify the time of Galileo and Newton, approximately three hundred years ago, as the beginning of modern science. Like the ancient Greeks, Galileo and Newton were interested in studying all of nature. Since the time of Galileo and Newton, the content of physical science has increased in scope and specialization, but the basic means of acquiring understanding, the scientific investigation, has changed little. A *scientific investigation* provides understanding through *experimental evidence,* as opposed to the conjectures based on thinking only of the ancient natural philosophers. In the next chapter, for example, you will learn how certain ancient Greeks described how objects fall toward the earth with a thought-out, or reasoned, explanation. Galileo, on the other hand, changed how people thought of falling objects by developing explanations from both creative thinking and precise measurement of physical quantities, providing experimental evidence for his explanations. Experimental evidence provides explanations today, much as it did for Galileo, as relationships are found from precise measurements of physical quantities. Thus, scientific knowledge about nature has grown as measurements and investigations have led to understandings that lead to further measurements and investigations.

What is a scientific investigation and what methods are used to conduct one? Attempts have been made to describe scientific methods in a series of steps (define problem, gather data, make hypothesis, test, make conclusion), but no single description has ever been satisfactory to all concerned. Scientists do similar things in investigations but there are different approaches and different ways to evaluate what they find. Overall, the similar things might look like this:

1. Observe some aspect of nature.
2. Invent an explanation for something observed.
3. Use the explanation to make predictions.
4. Test predictions by doing an experiment or by making more observations.
5. Modify explanation as needed.
6. Return to step 3.

The exact approach a scientist uses depends on the individual doing the investigation as well as the particular field of science being studied.

Another way to describe what goes on during a scientific investigation is to consider what can be generalized. There are at least three separate activities that seem to be common to scientists in different fields as they conduct scientific investigations, and these generalized activities are:

- Collecting observations.
- Developing explanations.
- Testing explanations.

No particular order or routine can be generalized about these common elements. In fact, individual scientists might not even be involved in all three activities. Some, for example, might spend all of their time out in nature, "in the field" collecting data and generalizing about their findings. This is an acceptable means of scientific investigation in some fields of science. Yet, other scientists might spend all of their time indoors, at computer terminals, developing theoretical equations that offer explanations for generalizations made by others. Again, the work at a computer terminal is an acceptable means of scientific investigation. Thus, there is not an order of five steps that are followed, particularly by today's specialized scientists. This is one reason why many philosophers of science argue there is no such thing as *the* scientific method. There are common activities of observing, explaining, and testing in scientific investigations in different fields, and these activities will be discussed next.

Explanations and Investigations

Explanations in the natural sciences are concerned with things or events observed, and there can be several different means of developing or creating explanations. In general, explanations can come from the results of experiments, from an educated guess, or just from imaginative thinking. In fact, there are several examples in the history of science of valid explanations being developed even from dreams. Explanations go by various names, each depending on intended use or stage of development. For example, an explanation in an early stage of development is sometimes called a *hypothesis.* A **hypothesis** is a tentative thought- or experiment-derived explanation. It must be compatible with all observations and provide understanding of some aspect of nature, but the key word here is "tentative." A hypothesis is tested by experiment and is rejected, or modified, if a single observation or test does not fit. The successful testing of a hypothesis may lead to the design of experiments, or it could lead to the development of another hypothesis, which could, in turn, lead to the design of yet more experiments, which could lead to. . . . As you can see, this is a branching, ongoing process that is very difficult to describe in specific terms. In addition, it can be difficult to identify a conclusion, an endpoint in the process. The search for new concepts to explain experimental evidence may lead from a hypothesis to a new theory, which results in more new hypotheses. This is why one of the best ways to understand scientific methods is to study the history of science. Or, you can conduct a scientific investigation yourself.

In some cases a hypothesis may be tested by simply making additional observations. For example, if you hypothesize that a certain species of bird uses cavities in trees as places to build nests, you could observe several birds of the species and record the kinds of nests they build and where they are built.

Another common method for testing a hypothesis involves devising an experiment. An experiment is a re-creation of an event or occurrence in a way that enables a scientist to support or disprove a hypothesis. This can be difficult since a particular event may be influenced by a great many separate things. For example, the production of a song by a bird involves many activities of the bird's nervous and muscular systems and is influenced by a wide variety of environmental factors. It might seem that developing an understanding of the factors involved in birdsong production is an impossible task. To help unclutter such situations, scientists have devised what is known as a controlled experiment. A **controlled experiment** compares two situations that have all the influencing factors identical except one. The situation used as the basis of comparison is called the *control group* and the other is called the *experimental group.* The single influencing factor that is allowed to be different in the experimental group is called the *experimental variable.*

The situation involving birdsong production would have to be broken down into a large number of simple questions, as previously mentioned. Each question would provide the basis on which experimentation would occur. Each experiment would provide information about a small part of the total process of birdsong production. For example, in order to test the hypothesis that male sex hormones are involved in stimulating male birds to sing, an experiment could be performed in which one group of male birds had their testes removed (the experimental group), while the control group was allowed to develop normally. After the experiment, the new data (facts) gathered would be analyzed. If there were no differences between the two groups, scientists could conclude that the variable evidently did not have a cause-and-effect relationship (i.e., was not responsible for the event). However, if there were a difference, it would be likely that the variable was responsible for the difference between the control and experimental groups. In the case of songbirds, removal of the testes does change their singing behavior.

Scientists are not likely to accept the results of a single experiment, since it is possible that a random event that had nothing to do with the experiment could have affected the results and caused people to think there was a cause-and-effect relationship when none existed. For example, the operation necessary to remove the testes of male birds might cause illness or discomfort in some birds, resulting in less singing. A way to overcome this difficulty would be to subject all birds to the same surgery but to remove the testes of only half of them. (The control birds would still have their testes.) The results of the experiment are considered convincing only when there is one variable, many replicates (copies) of the same experiment have been conducted, and the results are consistent.

Furthermore, scientists often apply statistical tests to the results to help decide in an impartial manner if the results obtained are *valid* (meaningful; fit with other knowledge), *reliable* (give the same results repeatedly), and show cause-and-effect, or if they are just the result of random events.

During experimentation, scientists learn new information and formulate new questions that can lead to yet more experiments. One good experiment can result in a hundred new questions and experiments. The discovery of the structure of the DNA molecule by Watson and Crick resulted in thousands of experiments and stimulated the development of the entire field of molecular biology (figure 1.13). Similarly, the discovery of molecules that regulate the growth of plants resulted in much research about how the molecules work and which molecules might be used for agricultural purposes.

If the processes of questioning and experimentation continue, and evidence continually and consistently supports the original hypothesis and other closely related hypotheses, the scientific community will begin to see how these hypotheses and facts fit together into a broad pattern.

Laws and Principles

Sometimes you can observe a series of relationships that seem to happen over and over again. There is a popular saying, for example, that "if anything can go wrong, it will." This is called Murphy's law. It is called a *law* because it describes a relationship between events that seems to happen time after time. If you drop a slice of buttered bread, for example, it can land two ways, butter side up or butter side down. According to Murphy's law, it will land butter side down. With this example, you know at least one way of testing the validity of Murphy's law.

Another "popular saying" type of relationship seems to exist between the cost of a houseplant and how long it lives. You could call it the "law of houseplant longevity." The relationship is that the life of a houseplant is inversely proportional to its purchase price. This "law" predicts that a $10 houseplant will wilt and die within a month, but a 50¢ houseplant will live for years. The inverse relationship is between the variables of (1) cost and (2) life span, meaning the more you pay for a plant the shorter the time it will live. This would also mean that inexpensive plants will live for a long time. Since the relationship seems to occur time after time, it is called a law.

A **scientific law** describes an important relationship that is observed in nature to occur consistently time after time. The law is often identified with the name of a person associated with the formulation of the law. For example, with all other factors being equal, an increase in the temperature of the air in a balloon results in an increase in its volume. Likewise, a decrease in the temperature results in a decrease in the total volume of the balloon. The volume of the balloon varies directly with the temperature of the air in the balloon and this can be observed to occur consistently time after time. This relationship was first discovered in the later part of the eighteenth century by two French scientists, A. C. Charles and Joseph Gay-Lussac. Today, the relationship is sometimes called *Charles' law.* When you read about a scientific *law,* you should remember that a law is a statement that means something about a relationship that you can observe time after time in nature.

Have you ever heard someone state that something behaved a certain way *because* of a scientific principle or law? For example, a big truck accelerated slowly *because* of Newton's laws of motion. Perhaps this person misunderstands the nature of scientific principles and laws. Scientific principles and laws do not dictate the behavior of objects, they simply describe it. They do not say how things ought to act but rather how things *do* act. A scientific principle or law is *descriptive;* it describes how things act.

A **scientific principle** is a relationship that describes a more specific set of relationships than is usually identified in a law. The difference between a scientific principle and a scientific law is usually one of the extent of the phenomena covered by the explanation,

A

B

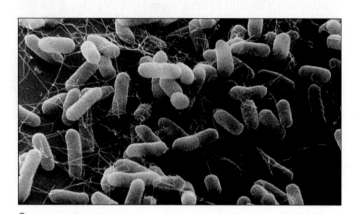

C

FIGURE 1.13

James D. Watson and Francis W. Crick are theoretical scientists who, in 1953, determined the structure of the DNA molecule, which contains the genetic information of a cell. (*A*) This photograph shows the model of DNA they constructed. The discovery of the structure of the DNA molecule was followed by much research into how the molecule codes information, how it makes copies of itself, and how the information is put into action. Ultimately, these lines of research have led to altering the DNA of bacteria so that they produce useful materials such as vitamins, proteins, and antibiotics. (*B*) Genetically altered bacteria are grown in special vats and the useful materials are harvested. (*C*) The bacterium *E. coli* is commonly found in the human intestine.

but there is not always a clear distinction between the two. As an example of a scientific principle, consider Archimedes' principle. This principle is concerned with the relationship between an object, a fluid, and buoyancy, which is a specific phenomenon.

Models and Theories

Often the part of nature being considered is too small or too large to be visible to the human eye and the use of a *model* is needed. A **model** (figure 1.14) is a description of a theory or idea that accounts for all known properties. The description can come in many different forms, such as an actual physical model, a computer model, a sketch, an analogy, or an equation. No one has ever seen the whole solar system, for example, and all you can see in the real world is the movement of the sun, moon, and planets against a background of stars. A physical model or sketch of the solar system, however, will give you a pretty good idea of what the solar system might look like. The physical model and the sketch are both models since they give you a mental picture of the solar system.

At the other end of the size scale, models of atoms and molecules are often used to help us understand what is happening in this otherwise invisible world. Also, a container of small, bounc-

ing rubber balls can be used as a model to explain the relationships of Charles' law. This model helps you see what happens to invisible particles of air as the temperature, volume, and pressure of the gas change. Some models are better than others, and models constantly change along with our understanding about nature. Early models of atoms, for example, were based on a "planetary model," which had electrons in the role of planets moving around the nucleus, which played the role of the sun. Today, the model has changed as our understandings about the nature of the atom have changed. Electrons are now pictured as vibrating with certain wavelengths, which can make standing waves only at certain distances from the nucleus. Thus the model of the atom changed from one with electrons viewed as solid particles to one that views them as vibrations on a string.

The most recently developed scientific theory was refined and expanded during the 1970s. This theory concerns the surface of the earth, and it has changed our model of what the earth is like. At first, however, the basic idea of today's accepted theory was pure and simple conjecture. The term "conjecture" usually means an explanation or idea based on speculation, or one based on trivial grounds without any real evidence. Scientists would look at a map of Africa and South America, for example,

A

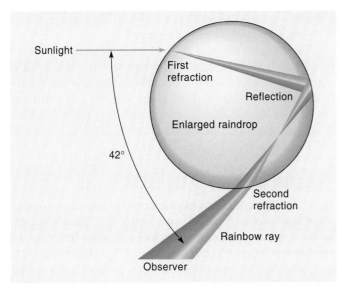

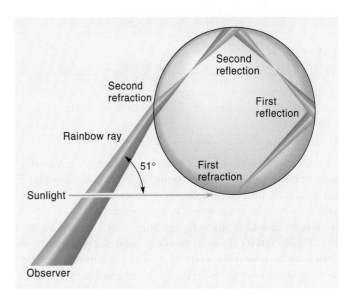

B

FIGURE 1.14

A model helps you visualize something that cannot be observed. You cannot observe what is making a double rainbow, for example, but models of light entering the upper and lower surface of a raindrop help you visualize what is happening.

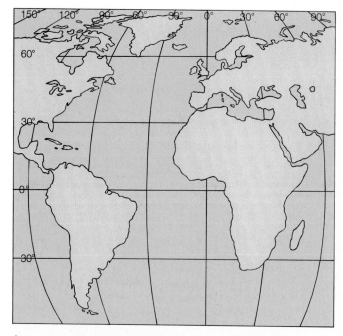

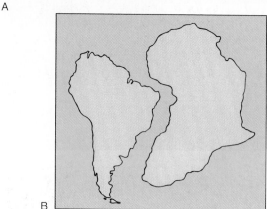

FIGURE 1.15

(*A*) Normal position of the continents on a world map. (*B*) A sketch of South America and Africa, suggesting that they once might have been joined together and subsequently separated by a continental drift.

and mull over how the two continents seem to be as pieces of a picture puzzle that had moved apart (figure 1.15). Any talk of moving continents was considered conjecture because it was not based on anything acceptable as real evidence.

Many years after the early musings about moving continents, evidence was collected from deep-sea drilling rigs that the ocean floor becomes progressively older toward the African and South American continents. This was good enough evidence to establish the "seafloor spreading hypothesis" that described the two continents moving apart.

If a hypothesis survives much experimental testing and leads, in turn, to the design of new experiments with the generation of new hypotheses that can be tested, you now have a working *theory*. A **theory** is defined as a broad, working hypothesis that is based on extensive experimental evidence. For example, the "seafloor spread-

ing hypothesis" did survive requisite experimental testing, and together with other working hypotheses is today found as part of the *plate tectonic theory.* The plate tectonic theory describes how the continents have moved apart, just like pieces of a picture puzzle. Is this the same idea that was once considered conjecture? Sort of, but this time it is supported by experimental evidence.

The term *scientific theory* is reserved for historic schemes of thought that have survived the test of detailed examination for long periods of time. The *atomic theory,* for example, was developed in the late 1800s and has been the subject of extensive investigation and experimentation over the last century. The atomic theory and other scientific theories form the framework of scientific thought and experimentation today. Scientific theories point to new ideas about the behavior of nature and these ideas result in more experiments, more data to collect, and more explanations to develop. All of this may lead to a slight modification of an existing theory, a major modification, or perhaps the creation of an entirely new one. These activities continue in an ongoing attempt to satisfy the curiosity of people by understanding nature.

SCIENCE, NONSCIENCE, AND PSEUDOSCIENCE

As you can see from the discussion of the nature of science, a scientific approach to the world requires a certain way of thinking. There is an insistence on ample supporting evidence by numerous studies rather than easy acceptance of strongly stated opinions. Scientists must separate opinions from statements of fact. A scientist is a healthy skeptic.

Careful attention to detail is also important. Since scientists publish their findings and their colleagues examine their work, there is a strong desire to produce careful work that can be easily defended. This does not mean that scientists do not speculate and state opinions. When they do, however, they take great care to clearly distinguish fact from opinion.

There is also a strong ethic of honesty. Scientists are not saints, but the fact that science is conducted out in the open in front of one's peers tends to reduce the incidence of dishonesty. In addition, the scientific community strongly condemns and severely penalizes those who steal the ideas of others, perform shoddy science, or falsify data. Any of these infractions could lead to the loss of one's job and reputation.

From Experimentation to Application

The scientific method has helped us to understand and control many aspects of our natural world (table 1.5). Some information is extremely important in understanding the structure and functioning of things in our world but at first glance appears to have little practical value. For example, understanding the life cycle of a star may be important for people who are trying to answer questions about how the universe is changing, but it seems of little practical value to the average citizen. However, as our knowledge has increased, the time between first discovery to practical application has decreased significantly.

For example, scientists known as *genetic engineers* have altered the chemical code system of small organisms (microorganisms) so

TABLE 1.5	Professional science degrees and titles	
Degree	**Education**	**Title**
A.S. (Associate of Science)	2 years community or junior college; focus on introductory courses in science in general (e.g., biology, chemistry, physics)	Mr. Mrs. Ms.
A.A.S. (Associate Degree in Applied Science)	2 years community or junior college; focus on courses of study such as nursing, dental hygiene, radiography, and respiratory therapy; requires biology, chemistry, and physics	Mr. Mrs. Ms.
B.S. (Bachelor of Science)	4–5 years college or university; major course work in a specific area (e.g., biology) with minor course work in related field (e.g., chemistry) along with general education courses (e.g., English, math, government)	Mr. Mrs. Ms.
M.S. (Master of Science)	2–3 years past bachelor's degree, college or university; major course work in a specific area (e.g., bacteriology) with minor course work in related field (e.g., biochemistry); lab research project may be required, thesis research paper required	Mr. Mrs. Ms.
Ph.D. (Doctor of Philosophy)	4–5 years past bachelor's degree, university; major course work in a specific area (e.g., exotoxin production of the *Streptococcacea*) with major research project and doctoral dissertation paper required	Dr.
M.D. (Medical Doctor)	4–5 years past bachelor's degree, university; concentrated course work and clinical experience in the practice of medicine	Dr.
D.D.S. (Doctor of Dental Science)	4–5 years past bachelor's degree, university; concentrated course work and clinical experience in the practice of dentistry	Dr.
Ed.D. (Doctor of Education)	4–5 years past bachelor's degree, university; major course work in a specific area (e.g., learning theory as applied to adult learners) with major research project and doctoral dissertation paper required	Dr.

that they may produce many new drugs such as antibiotics, hormones, and enzymes. The ease with which these complex chemicals are produced would not have been possible had it not been for the information gained from the basic, theoretical sciences of microbiology, molecular biology, and genetics (see figure 1.13). Our understanding of how organisms genetically control the manufacture of proteins has led to the large-scale production of enzymes. Some of these chemicals can remove stains from clothing, deodorize, clean contact lenses, remove damaged skin from burn patients, and "stone wash" denim for clothing.

Another example is Louis Pasteur, a French chemist and microbiologist. Pasteur was interested in the theoretical problem of whether life could be generated from nonliving material. Much of his theoretical work led to practical applications in disease control. His theory that there are microorganisms that cause diseases and decay led to the development of vaccinations against rabies and the development of pasteurization for the preservation of foods (figure 1.16).

Science and Nonscience

The differences between science and nonscience are often based on the assumptions and methods used to gather and organize information and, most important, the testing of these assumptions. The difference between a scientist and a nonscientist is that a scientist continually challenges and tests principles and assumptions to determine a cause-and-effect relationship, whereas a nonscientist may not feel that this is important.

FIGURE 1.16

Louis Pasteur (1822–1895) performed many experiments while he studied the question of the origin of life, one of which led directly to the food-preservation method now known as pasteurization.

Once you understand the nature of science, you will not have any trouble identifying astronomy, chemistry, physics, and biology as sciences. But what about economics, sociology, anthropology, history, philosophy, and literature? All of these fields may make use of certain central ideas that are derived in a logical way, but they are also nonscientific in some ways. Some

FIGURE 1.17

Not many people realize that electronic music was first created in the late 1800s. Some of those early instruments are still used in the production of movie sound tracks, especially science fiction and horror movies. Although scientists are not in the business of evaluating whether such music is "good," they have certainly contributed to the understanding of (1) the physics of how the sounds are created, (2) the anatomy and physiology of how the sound stimulates nerve endings in the ear, and (3) the biochemistry of how the nerve impulses travel through the body and affect the various body structures such as tear glands (crying), muscles (tension during a fear response), hormone levels (that pounding in your chest), and vocalizations (screaming!).

things are beyond science and cannot be approached using the scientific method. Art, literature, theology, and philosophy are rarely thought of as sciences. They are concerned with beauty, human emotion, and speculative thought rather than with facts and verifiable laws. On the other hand, physics, chemistry, geology, and biology are almost always considered sciences.

Music is an area of study in a middle ground where scientific approaches may be used to some extent. "Good" music is certainly unrelated to science, but the study of how the human larynx generates the sound of a song is based on scientific principles (figure 1.17). Any serious student of music will study the physics of sound and how the vocal cords vibrate to generate sound waves. Similarly, economists use mathematical models and established economic laws to make predictions about future economic conditions. However, the regular occurrence of unpredicted economic changes indicates that economics is far from scientific, since the reliability of predictions is a central criterion of science. Anthropology and sociology are also scientific in nature in many respects, but they cannot be considered true sciences because many of the generalizations they have developed cannot be tested by repeated experimentation. They also do not show a significantly high degree of cause-and-effect, or they have poor predictive value.

Pseudoscience

Pseudoscience ("pseudo" means "false") takes on the flavor of science but is not supportable as valid or reliable. Often, the purpose of pseudoscience is to confuse or mislead. The area of nutrition is flooded with pseudoscience (figure 1.18). We all know that we must obtain certain nutrients like amino acids,

FIGURE 1.18

"Nine out of ten doctors surveyed recommend Brand X." It is obvious that there are many things wrong with this statement. First of all, is the person in the white coat a physician? Second, if only 10 doctors were asked, the sample size is too small. Third, only selected doctors might have been asked to participate. Finally, the question could have been asked in such a way as to obtain the desired answer: "Would you recommend brand X over Dr. Pete's snake oil?"

vitamins, and minerals from the food that we eat or we may become ill. Many scientific experiments reliably demonstrate the validity of this information. However, in most cases, it has not been proven that the nutritional supplements so vigorously promoted are as useful or desirable as advertised. Rather, selected bits of scientific information (amino acids, vitamins, and minerals are essential to good health) have been used to create the feeling that additional amounts of these nutritional supplements are necessary or that they can improve your health. In reality, the average person eating a varied diet will obtain all of these nutrients in adequate amounts, and nutritional supplements are not required.

In addition, many of these products are labeled as organic or natural, with the implication that they have greater nutritive value because they are organically grown (grown without pesticides or synthetic fertilizers) or because they come from nature. The poisons curare, strychnine, and nicotine are all organic molecules that are produced in nature by plants that could be grown organically, but we would not want to include them in our diet.

Limitations of Science

By definition, science is a way of thinking and seeking information to solve problems. Therefore, scientific methods can be applied only to questions that have a factual basis. Questions concerning morals, value judgments, social issues, and attitudes cannot be answered using the scientific methods. What makes a painting great? What is the best type of music? Which wine is best? What color should I paint my car? These questions are related to values, beliefs, and tastes; therefore, scientific methods cannot be used to answer them.

Science is also limited by the ability of people to pry understanding from the natural world. People are fallible and do not always come to the right conclusions, because information is lacking or misinterpreted, but science is self-correcting. As new information is gathered, old incorrect ways of thinking must be changed or discarded. For example, at one time people were sure that the sun went around the earth. They observed that the sun rose in the east and traveled across the sky to set in the west. Since they could not feel the earth moving, it seemed perfectly logical that the sun traveled around the earth. Once they understood that the earth rotated on its axis, people began to understand that the rising and setting of the sun could be explained in other ways. A completely new concept of the relationship between the sun and the earth developed.

Although this kind of study seems rather primitive to us today, this change in thinking about the sun and the earth was a very important step in understanding the universe and how the various parts are related to one another. This background information was built upon by many generations of astronomers and space scientists, and finally led to space exploration.

People also need to understand that science cannot answer all the problems of our time. Although science is a powerful tool, there are many questions it cannot answer and many problems it cannot solve. The behavior and desires of people generate most of the problems societies face. Famine, drug abuse, and pollution are human-caused and must be resolved by humans. Science may provide some tools for social planners, politicians, and ethical thinkers, but science does not have, nor does it attempt to provide, all the answers to the problems of the human race. Science is merely one of the tools at our disposal.

SUMMARY

Science is a search for order in our surroundings. People have *concepts,* or mental images, about material *objects* and intangible *events* in their surroundings. Concepts are used for thinking and communicating. Concepts are based on *properties,* or attributes that describe a thing or event. Every property implies a *referent* that describes the property. Referents are not always explicit, and most communications require assumptions.

Measurement is a process that uses a well-defined and agreed-upon *referent* to describe a *standard unit.* The unit is compared to the property being defined by an *operation* that determines the *value* of the unit by *counting.* Measurements are always reported with a *number,* or value, and a *name* for the unit.

The two major *systems* of standard units are the *English system* and the *metric system.* The English system uses standard units that were originally based on human body parts, and the metric system uses standard units based on referents found in nature. The metric system also uses a system of prefixes to express larger or smaller amounts of units. The metric standard units for length, mass, and time are the *meter, kilogram,* and *second.*

Measurement information used to describe something is called *data.* One way to extract meanings and generalizations from data is to use a *ratio,* a simplified relationship between two numbers. Density is a ratio of mass to volume, or $\rho = m/V$.

Modern science began about three hundred years ago during the time of Galileo and Newton. Since that time, *scientific investigation* has been used to provide *experimental evidence* about nature. The investigations provide *accurate, specific,* and *reliable* data that are used to develop and test *explanations.* A *hypothesis* is a tentative explanation that is accepted or rejected from experimental data. An accepted hypothesis may result in a *principle,* an explanation concerned with a specific range of phenomena, or a *scientific law,* an explanation concerned with important, wider-ranging phenomena. Laws are sometimes identified with the name of a scientist and can be expressed verbally, with an equation, or with a graph.

A *model* is used to help understand something that cannot be observed directly, explaining the unknown in terms of things already understood. Physical models, mental models, and equations are all examples of models that explain how nature behaves. A *theory* is a broad, detailed explanation that guides development and interpretations of experiments in a field of study.

Science and *nonscience* can be distinguished by the kinds of laws and rules that are constructed to unify the body of knowledge. Science involves the continuous *testing* of rules and principles by the collection of new facts. If the rules are not testable, or if no rules are used, it is not science. *Pseudoscience* uses scientific appearances to mislead.

Summary of Equations

1.1

$$\text{mass density} = \frac{\text{mass}}{\text{volume}}$$

$$\rho = \frac{m}{V}$$

1.2

$$\text{weight density} = \frac{\text{weight}}{\text{volume}}$$

$$D = \frac{w}{V}$$

KEY TERMS

controlled experiment (p. **13**)

data (p. **8**)

density (p. **10**)

hypothesis (p. **12**)

measurement (p. **4**)

model (p. **14**)

properties (p. **2**)

referent (p. **3**)

scientific law (p. **13**)

scientific principle (p. **13**)

standard unit (p. **5**)

theory (p. **16**)

unit (p. **4**)

APPLYING THE CONCEPTS

1. The process of comparing a property of an object to a well-defined and agreed-upon referent is called the process of
   a. generalizing.
   b. measurement.
   c. graphing.
   d. scientific investigation.

2. The height of an average person is closest to
   a. 1.0 m.
   b. 1.5 m.
   c. 2.5 m.
   d. 3.5 m.

3. Which of the following standard units is defined in terms of an object as opposed to an event?
   a. kilogram
   b. meter
   c. second
   d. none of the above

4. One-half liter of water has a mass of
   a. 0.5 g.
   b. 5 g.
   c. 50 g.
   d. 500 g.

5. A cubic centimeter (cm^3) of water has a mass of about 1
   a. mL.
   b. kg.
   c. g.
   d. dm.

6. Measurement information that is used to describe something is called
   a. referents.
   b. properties.
   c. data.
   d. a scientific investigation.

7. The property of volume is a measure of
   a. how much matter an object contains.
   b. how much space an object occupies.
   c. the compactness of matter in a certain size.
   d. the area on the outside surface.

8. As the volume of a cube becomes larger and larger, the surface-area-to-volume ratio
   a. increases.
   b. decreases.
   c. remains the same.
   d. sometimes increases and sometimes decreases.

9. If you consider a very small portion of a material that is the same throughout, the density of the small sample will be
   a. much less.
   b. slightly less.
   c. the same.
   d. greater.

10. A hypothesis concerned with a specific phenomenon is found to be acceptable through many experiments over a long period of time. This hypothesis usually becomes known as a
    a. scientific law.
    b. scientific principle.
    c. theory.
    d. model.

11. A scientific law can be expressed as
    a. a written concept.
    b. an equation.
    c. a graph.
    d. any of the above.

Answers

1. b 2. b 3. a 4. d 5. c 6. c 7. b 8. b 9. c 10. b 11. d

QUESTIONS FOR THOUGHT

1. What is a concept?
2. What two things does a measurement statement always contain? What do the two things tell you?
3. Other than familiarity, what are the advantages of the English system of measurement?
4. Describe the metric standard units for length, mass, and time.
5. Does the density of a liquid change with the shape of a container? Explain.
6. Does a flattened pancake of clay have the same density as the same clay rolled into a ball? Explain.
7. Compare and contrast a scientific principle and a scientific law.
8. What is a model? How are models used?
9. Are all theories always completely accepted or completely rejected? Explain.
10. What is pseudoscience and how can you always recognize it?

PARALLEL EXERCISES

The exercises in groups A and B cover the same concepts. Solutions to group A exercises are located in appendix D.

Group A

Note: *You will need to refer to table 1.4 to complete some of the following exercises.*

1. What is your height in meters? In centimeters?
2. What is the mass density of mercury if 20.0 cm³ has a mass of 272 g?
3. What is the mass of a 10.0 cm³ cube of lead?
4. What is the volume of a rock with a mass density of 3.00 g/cm³ and a mass of 600 g?
5. If you have 34.0 g of a 50.0 cm³ volume of one of the substances listed in table 1.4, which one is it?
6. What is the mass of water in a 40 L aquarium?
7. A 2.1 kg pile of aluminum cans is melted, then cooled into a solid cube. What is the volume of the cube?
8. A cubic box contains 1,000 g of water. What is the length of one side of the box in meters? Explain your reasoning.
9. A loaf of bread (volume 3,000 cm³) with a density of 0.2 g/cm³ is crushed in the bottom of a grocery bag into a volume of 1,500 cm³. What is the density of the mashed bread?
10. According to table 1.4, what volume of copper would be needed to balance a 1.00 cm³ sample of lead on a two-pan laboratory balance?

Group B

Note: *You will need to refer to table 1.4 to complete some of the following exercises.*

1. What is your mass in kilograms? In grams?
2. What is the mass density of iron if 5.0 cm³ has a mass of 39.5 g?
3. What is the mass of a 10.0 cm³ cube of copper?
4. If ice has a mass density of 0.92 g/cm³, what is the volume of 5,000 g of ice?
5. If you have 51.5 g of a 50.0 cm³ volume of one of the substances listed in table 1.4, which one is it?
6. What is the mass of gasoline ($\rho = 0.680$ g/cm³) in a 94.6 L gasoline tank?
7. What is the volume of a 2.00 kg pile of iron cans that are melted, then cooled into a solid cube?
8. A cubic tank holds 1,000.0 kg of water. What are the dimensions of the tank in meters? Explain your reasoning.
9. A hot dog bun (volume 240 cm³) with a density of 0.15 g/cm³ is crushed in a picnic cooler into a volume of 195 cm³. What is the new density of the bun?
10. According to table 1.4, what volume of iron would be needed to balance a 1.00 cm³ sample of lead on a two-pan laboratory balance?

The whirlpool is but one example of circular motion in nature.

CHAPTER | Two

Motion

In chapter 1, you learned some "tools and rules" and some techniques for finding order in your surroundings. Order is often found in the form of patterns, or relationships between quantities that are expressed as equations. Equations can be used to (1) describe properties, (2) define concepts, and (3) describe how quantities change together. In all three uses, patterns are quantified, conceptualized, and used to gain a general understanding about what is happening in nature.

In the study of science, certain parts of nature are often considered and studied together for convenience. One of the more obvious groupings involves *movement*. Most objects around you appear to spend a great deal of time sitting quietly without motion. Buildings, rocks, utility poles, and trees rarely, if ever, move from one place to another. Even things that do move from time to time sit still for a great deal of time. This includes you, automobiles, and bicycles (figure 2.1). On the other hand, the sun, the moon, and starry heavens always seem to move, never standing still. Why do things stand still? Why do things move?

Questions about motion have captured the attention of people for thousands of years. But the ancient people answered questions about motion with stories of mysticism and spirits that lived in objects. It was during the classic Greek culture, between 600 B.C. and 300 B.C., that people began to look beyond magic and spirits. One particular Greek philosopher, Aristotle, wrote a theory about the universe that offered not only explanations about things such as motion but also offered a sense of beauty, order, and perfection. The theory seemed to fit with other ideas that people had and was held to be correct for nearly two thousand years after it was written. It was not until the work of Galileo and Newton during the 1600s that a new, correct understanding about motion was developed. The development of ideas about motion is an amazing and absorbing story. You will learn in this chapter how equations are used to (1) define the properties of motion and (2) describe how quantities of motion change together. These are basic understandings that will be used to define some concepts of motion in the next chapter.

DESCRIBING MOTION

Motion is one of the more common events in your surroundings. You can see motion in natural events such as clouds moving, rain and snow falling, and streams of water moving in a never-ending cycle. Motion can also be seen in the activities of people who walk, jog, or drive various machines from place to place. Motion is so common that you would think everyone would intuitively understand the concepts of motion, but history indicates that it was only during the past three hundred years or so that people began to understand motion correctly. Perhaps the correct concepts are subtle and contrary to common sense, requiring a search for simple, clear concepts in an otherwise complex situation. The process of finding such order in a multitude of sensory impressions by taking measurable data, and then inventing a concept to describe what is happening, is the activity called *science*. We will now apply this process to motion.

What is motion? Consider a ball that you notice one morning in the middle of a lawn. Later in the afternoon, you notice that the ball is at the edge of the lawn, against a fence, and you wonder if the wind or some person moved the ball. You do not know if the wind blew it at a steady rate, if many gusts of wind moved it, or even if some children kicked it all over the yard. All you know for sure is that the ball has been moved because it is in a different position after some time passed. These are the two important aspects of motion: (1) a change of position and (2) the passage of time.

If you did happen to see the ball rolling across the lawn in the wind, you would see more than the ball at just two locations. You would see the ball moving continuously. You could consider, however, the ball in continuous motion to be a series of individual locations with very small time intervals. Moving involves a change of position during some time period. Motion is the act or process of something changing position.

The motion of an object is usually described with respect to something else that is considered to be not moving. (Such a stationary object is said to be "at rest.") Imagine that you are traveling in an automobile with another person. You know that you are moving across the land outside the car since your location on the highway changes from one moment to another. Observing your fellow passenger, however, reveals no change of position. You are in motion relative to the highway outside the car. You are not in motion relative to your fellow passenger. Your motion, and the motion of any other object or body, is the process of a change in position *relative* to some reference object or location. Thus *motion* can be defined as the act or process of changing position relative to some reference during a period of time.

MEASURING MOTION

You have learned that objects can be described by measuring certain fundamental properties such as mass and length. Since motion involves (1) a change of position, or *displacement*, and

(2) the passage of *time*, the motion of objects can be described by using combinations of the fundamental properties of length and time. Combinations of these measurements describe the motion properties of *speed*, *velocity*, and *acceleration*.

Speed

Suppose you are in a car that is moving over a straight road. How could you describe your motion? Motion was defined as the process of a change of position. You need at least two measurements to describe the motion. These are (1) how much the position was changed, or the *displacement* between the two positions, and (2) how long it took for the change to take place, or the *time* that elapsed while the object moved between the two positions. Displacement and time can be combined as a ratio to define the *rate* at which a distance is covered. This rate is a property of motion called **speed,** a measurement of how fast you are moving. Speed is defined as distance per unit time, or

$$\text{speed} = \frac{\text{distance (how much the position changed)}}{\text{time (elapsed time during change)}}$$

Suppose your car is moving over a straight highway and you are covering equal distances in equal periods of time (figure 2.2). If you use a stopwatch to measure the time required to cover the distance between highway mile markers (those little signs with numbers along major highways), the time intervals will all be equal. You might find, for example, that one minute lapses between each mile marker. Such a uniform straight-line motion that covers equal distances in equal periods of time is the simplest kind of motion.

If your car were moving over equal distances in equal periods of time, it would have a *constant speed*. This means that the car is neither speeding up nor slowing down. It is usually difficult to maintain a constant speed. Other cars and distractions such as interesting scenery cause you to reduce your speed. At other times you increase your speed. If you calculate your speed

FIGURE 2.1

The motion of this windsurfer, and of other moving objects, can be described in terms of the distance covered during a certain time period.

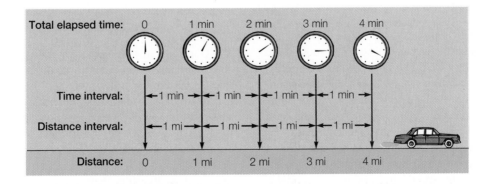

FIGURE 2.2

This car is moving in a straight line over a distance of 1 mi each minute. The speed of the car, therefore, is 60 mi each 60 min, or 60 mi/hr.

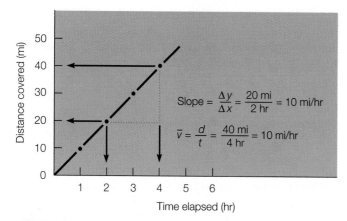

FIGURE 2.3

Speed is defined as a ratio of the displacement, or the distance covered, in straight-line motion for the time elapsed during the change, or $v = d/t$. This ratio is the same as that found by calculating the slope when time is placed on the x-axis. The answer is the same as shown on this graph because both the speed and the slope are ratios of distance per unit of time. Thus you can find a speed by calculating the slope from the straight line, or "picture," of how fast distance changes with time.

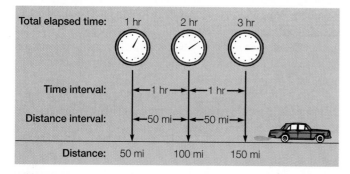

FIGURE 2.4

If you know the value of any two of the three variables of distance, time, and speed, you can find the third. What is the average speed of this car?

over an entire trip, you are considering a large distance between two places and the total time that elapsed. The increases and decreases in speed would be averaged. Therefore, most speed calculations are for an *average speed*. The speed at any specific instant is called the *instantaneous speed*. To calculate the instantaneous speed, you would need to consider a very short time interval—one that approaches zero. An easier way would be to use the speedometer, which shows the speed at any instant.

It is easier to study the relationships between quantities if you use symbols instead of writing out the whole word. The letter v can be used to stand for speed when dealing with straight-line motion, which is the only kind of motion that will be considered in the problems in this text. The letter d can be used to

stand for distance and the letter t to stand for time. The relationship between average speed, distance, and time is therefore

$$\bar{v} = \frac{d}{t}$$

equation 2.1

This is one of the three types of equations that were discussed earlier, and in this case the equation defines a motion property. The bar over the v ($\bar{v}$) is a symbol that means *average* (it is read "v-bar" or "v-average"). You can use this relationship to find average speed (figure 2.3). For example, suppose a car travels 150 mi in 3 hr. What was the average speed? Since $d = 150$ mi, and $t = 3$ hr, then

$$\bar{v} = \frac{150 \text{ mi}}{3 \text{ hr}}$$

$$= 50 \frac{\text{mi}}{\text{hr}}$$

As with other equations, you can mathematically solve the equation for any term as long as two variables are known (figure 2.4). For example, suppose you know the speed and the time but want to find the distance traveled. You can solve this by first writing the relationship

$$\bar{v} = \frac{d}{t}$$

and then multiplying both sides of the equation by t (to get d on one side by itself),

$$(\bar{v})(t) = \frac{(d)(t)}{t}$$

and the t's on the right cancel, leaving

$$\bar{v}t = d \text{ or } d = \bar{v}t$$

If the $\bar{v}$ is 50 mi/hr and the time traveled is 2 hr, then

$$d = \left(50 \frac{\text{mi}}{\text{hr}}\right)(2 \text{ hr})$$

$$= (50)(2)\left(\frac{\text{mi}}{\text{hr}}\right)(\text{hr})$$

$$= 100 \frac{(\text{mi})(\text{hr})}{\text{hr}}$$

$$= 100 \text{ mi}$$

Constant, instantaneous, or average speeds can be measured with any distance and time units. Common units in the English system are miles/hour and feet/second. Metric units for speed are commonly kilometers/hour and meters/second. The ratio of any distance/time is usually read as distance per time, such as miles per hour. The "per" means "for each." Sometimes you will hear or read about the **magnitude** of speed. Magnitude is defined as the numerical value and the unit, such as 30 mi/hr. It simply means the size or extent of a quantity. For example, a speed of 60 mi/hr is twice the magnitude of a speed of 30 mi/hr.

C O N N E C T I O N S

Speed and Predators

Predation occurs when one animal captures, kills, and eats another animal. The organism that is killed is called the *prey*, and the one that does the killing is called the *predator*. Most predators are relatively large compared to their prey and have specific adaptations that aid them in catching prey. Predators such as leopards, lions, and cheetahs use *speed* to run down their prey, while others such as frogs, toads, and many kinds of lizards strike quickly when a prey organism happens by. For example, the cheetah can reach estimated speeds of 101 km/hr (63 mi/hr) during sprints to capture its prey and the peregrine falcon reaches a speed of 349 km/hr (217 mi/hr) when diving at an angle of 45 degrees. The dragonfly, a predator insect, reaches speeds up to 58 km/hr (36 mi/hr). How does the human compare to all these speed records? The fastest speed for a human has been about 44 km/hr (27 mi/hr) for a short distance.

Many kinds of predators are useful to us because they control the populations of organisms that do us harm. For example, snakes eat many kinds of rodents that eat stored grain and other agricultural products. Many birds and bats eat insects that are agricultural pests. It is even possible to think of a predator as having a beneficial effect on the prey species. Certainly the individual organism that is killed is harmed, but the population can benefit. Predators act as selecting agents and the individuals who fall to them as prey are likely to be less well adapted. Predators usually kill slow, unwary, sick, or injured individuals. Thus, the genes that may have contributed to slowness, inattention, illness, or the likelihood of being injured are removed and a better-adapted population remains. Because predators eliminate poorly adapted individuals, the species benefits. What is bad for the individual can be good for the species.

Velocity

The word "velocity" is sometimes used interchangeably with the word "speed," but there is a difference. **Velocity** describes the *speed and direction* of a moving object. For example, a speed might be described as 60 mi/hr. A velocity might be described as 60 mi/hr to the west. To produce a change in velocity, either the speed or the direction is changed (or both are changed). A satellite moving with a constant speed in a circular orbit around the earth does not have a constant velocity since its direction of movement is constantly changing. Measurements that have magnitude only, such as speed, are called *scalar quantities*, or *scalars*. Measurements that have both magnitude and direction, such as velocity, are called *vector quantities*, or *vectors*. Speed is a scalar quantity, and velocity is a vector quantity. Vector quantities can be represented graphically with arrows. The lengths of the arrows are proportional to the magnitude, and the arrowheads indicate the direction (figure 2.5).

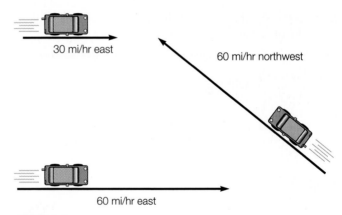

30 mi/hr east

60 mi/hr northwest

60 mi/hr east

FIGURE 2.5

Velocity is a vector that we can represent graphically with arrows. Here are three different velocities represented by three different arrows. The length of each arrow is proportional to the speed, and the arrowhead shows the direction of travel.

Acceleration

Motion can be changed in three different ways: (1) by changing the speed, (2) by changing the direction of travel, or (3) by changing both the speed and direction of travel. Since velocity describes both the speed and the direction of travel, any of these three changes will result in a change of velocity. You need at least one additional measurement to describe a change of motion, which is how much time elapsed while the change was taking place. The change of velocity and time can be combined to define the *rate* at which the motion was changed. This rate is called **acceleration.** Acceleration is defined as a change of velocity per unit time, or

$$\text{acceleration} = \frac{\text{change of velocity}}{\text{time elapsed}}$$

Another way of saying "change in velocity" is the final velocity minus the initial velocity, so the relationship can also be written as

$$\text{acceleration} = \frac{\text{final velocity} - \text{initial velocity}}{\text{time}}$$

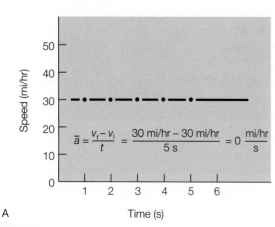

A

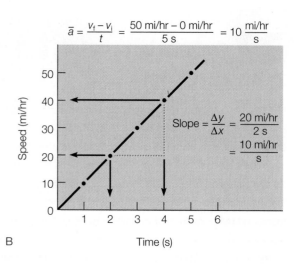

B

FIGURE 2.6

(A) This graph shows how the speed changes per unit of time while driving at a constant 30 mi/hr in a straight line. As you can see, the speed is constant, and for straight-line motion, the acceleration is 0. (B) This graph shows the speed increasing 50 mi/hr when moving in a straight line for 5 s. The acceleration, or change of velocity per unit of time, can be calculated from either the equation for acceleration or by calculating the slope of the straight-line graph. Both will tell you how fast the motion is changing with time.

Acceleration due to a change in speed only can be calculated as follows. Consider a car that is moving with a constant, straight-line velocity of 60 km/hr when the driver accelerates to 80 km/hr. Suppose it takes 4 s to increase the velocity of 60 km/hr to 80 km/hr. The change in velocity is therefore 80 km/hr minus 60 km/hr, or 20 km/hr. The acceleration was

$$\text{acceleration} = \frac{80\ \dfrac{\text{km}}{\text{hr}} - 60\ \dfrac{\text{km}}{\text{hr}}}{4\ \text{s}}$$

$$= \frac{20\ \dfrac{\text{km}}{\text{hr}}}{4\ \text{s}}$$

$$= 5\ \frac{\text{km/hr}}{\text{s}}\ \text{or,}$$

$$= 5\ \text{km/hr/s}$$

The average acceleration of the car was 5 km/hr for each ("per") second. This is another way of saying that the velocity increases an average of 5 km/hr in each second. The velocity of the car was 60 km/hr when the acceleration began (initial velocity). At the end of 1 s, the velocity was 65 km/hr. At the end of 2 s, it was 70 km/hr; at the end of 3 s, 75 km/hr; and at the end of 4 s (total time elapsed), the velocity was 80 km/hr (final velocity). Note how fast the velocity is changing with time. In summary,

start (initial velocity)	60 km/hr
first second	65 km/hr
second second	70 km/hr
third second	75 km/hr
fourth second (final velocity)	80 km/hr

As you can see, acceleration is really a description of how fast the speed is changing (figure 2.6); in this case, it is increasing 5 km/hr each second.

Usually, you would want all the units to be the same, so you would convert km/hr to m/s. You convert to m/s because, as you will see later, many other quantities are defined in terms of meters in the form of the metric system you will be using (meter-kilogram-second). The other form of the metric system (centimeter-gram-second) will not be used. If you were given English units of mi/hr for velocity and s for time, you would convert the mi/hr to ft/s.

A change in velocity of 5.0 km/hr converts to 1.4 m/s and the acceleration would be 1.4 m/s/s. The units m/s per s mean what change of velocity (1.4 m/s) is occurring every second. The combination m/s/s is rather cumbersome, so it is typically treated mathematically to simplify the expression (to simplify a fraction, invert the divisor and multiply, or m/s × 1/s = m/s²). Remember that the expression 1.4 m/s² means the same as 1.4 m/s per s, a change of velocity in a given time period.

The relationship among the quantities involved in acceleration can be represented with the symbols a for average acceleration, v_f for final velocity, v_i for initial velocity, and t for time. The relationship is

$$a = \frac{v_f - v_i}{t}$$

equation 2.2

As in other equations, any one of these quantities can be found if the others are known. For example, solving the equation for the final velocity, v_f, yields:

$$v_f = at + v_i$$

In problems where the initial velocity is equal to zero (starting from rest), the equation simplifies to

$$v_f = at$$

Connections

SuperTrain

The super-speed magnetic levitation (maglev) train is a completely new technology based on magnetically suspending a train 3 to 10 cm (about 1 to 4 in) above a monorail, then moving it along with a magnetic field that travels along the monorail guides. Very short acceleration and braking distances are possible, thanks to the easily manipulated magnetic fields and the lack of rolling resistance. For example, a German maglev train can accelerate from 0 to 300 km/hr (about 185 mi/hr) over a distance of just 5 km (about 3 mi). A conventional train with wheels requires about 30 km (about 19 mi) in order to reach the same speed from a standing start. The maglev is attractive for short runs because of its superior acceleration and breaking abilities. It is also attractive for longer runs because of its high top speed—up to about 500 km/hr (about 310 mi/hr). Today, only an aircraft can match such a speed.

So far, you have learned only about straight-line, uniform acceleration that results in an increased velocity. There are also other changes in the motion of an object that are associated with acceleration. One of the more obvious is a change that results in a decreased velocity. Your car's brakes, for example, can slow your car or bring it to a complete stop. This is sometimes called *negative acceleration,* or *deceleration.* Another change in the motion of an object is a change of direction. Velocity encompasses both the rate of motion as well as direction, so a change of direction is an acceleration. The satellite moving with a constant speed in a circular orbit around the earth is constantly changing its direction of movement. It is therefore constantly accelerating because of this constant change in its motion. Your automobile has three devices that could change the state of its motion. Your automobile therefore has three accelerators—the gas pedal (which can increase magnitude of velocity), the brakes (which can decrease magnitude of velocity), and the steering wheel (which can change direction of velocity). (See figure 2.7.) The important thing to remember is that acceleration results from any *change* in the motion of an object.

The final velocity (v_f) and the initial velocity (v_i) are different variables than the average velocity ($\bar{v}$). You cannot use an initial or final velocity for an average velocity. You may,

Connections

Storm Speeds

What is the difference between a tropical depression, tropical storm, and a hurricane? In general, they are all storms with strong upward atmospheric motion and a cyclonic surface wind circulation. They occur over tropical or subtropical waters and are not associated with a weather front. The varieties of storm intensities are classified according to the *speed* of the maximum sustained surface winds. In the United States the maximum sustained surface wind is measured by averaging the wind speed over a 1-minute period. Here is the classification scheme:

Tropical Depression. This is a center of low pressure around which the winds are generally moving 55 km/hr (about 35 mi/hr) or less. The tropical depression might dissolve into nothing or it might develop into a more intense disturbance.

Tropical Storm. This is a more intense, highly organized center of low pressure with winds between 56 and 120 km/hr (about 35 to 75 mi/hr).

Hurricane. This is an intense low-pressure center with winds greater than 120 km/hr (about 75 mi/hr). A strong storm of this type is called a "hurricane" if it occurs over the Atlantic Ocean or the Pacific Ocean east of the international date line. It is called a "typhoon" if it occurs over the North Pacific Ocean west of the international date line. Hurricanes are further classified according to category and damage to be expected. Here is the classification scheme:

Category	Damage	Winds
1	minimal	120–153 km/hr (75–95 mi/hr)
2	moderate	154–177 km/hr (96–110 mi/hr)
3	extensive	178–210 km/hr (111–130 mi/hr)
4	extreme	211–250 km/hr (131–155 mi/hr)
5	catastrophic	>250 km/hr (>155 mi/hr)

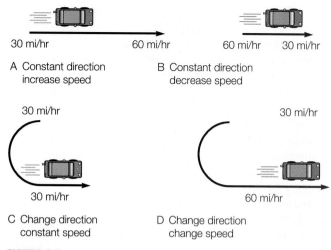

FIGURE 2.7

Four different ways (A–D) to accelerate a car.

however, calculate an average velocity ($\bar{v}$) from the other two variables as long as the acceleration taking place between the initial and final velocities is uniform. An example of such a uniform change would be an automobile during a constant, straight-line acceleration. This is done just as in finding other averages, such as the average weight of your class. The average class weight is the sum of all the weights divided by the number of people. To find an average velocity *during* a uniform acceleration, you add the initial velocity and the final velocity and divide by 2. This averaging can be done for a uniform acceleration that is increasing the velocity or for one that is decreasing the velocity. In symbols,

$$\bar{v} = \frac{v_f + v_i}{2}$$

equation 2.3

Activities

1. Is the speedometer in your car accurate? Take a friend and a stopwatch to a highway with mile markers. Drive at a constant speed while your friend measures the time required to drive 1 mile. Do this several times, and then compare the indicated speed with the calculated speed. Which do you believe is more accurate, the speedometer or the calculated speed?

2. As long as you are driving with a friend and a stopwatch, measure the time required for your car to accelerate from one speed to another. Calculate the average acceleration. Compare your finding to the acceleration of other cars.

ARISTOTLE'S THEORY OF MOTION

Some of the first ideas about the causes of motion were recorded back in the fourth century B.C. by the Greek philosopher Aristotle (figure 2.8). Aristotle was a former student of Plato and founded a new school, the Lyceum, in Athens in 337 B.C. As head of the Lyceum, Aristotle taught his students while walking within the school grounds. He also wrote an encyclopedia of classified knowledge that organized all knowledge known at that time as a unified theory of nature. He produced a philosophic theory of the physical world that the Greeks used to interpret and understand nature. The theory is said to be philosophic because it was developed from reasoning alone.

Aristotle's theory presented a grand scheme of the universe, which was considered to have two main parts: (1) the *sphere of change* within the orbit of the moon and (2) the *sphere of perfection* outside the orbit of the moon. The earth was seen to be fixed and unmoving at the center of the sphere of change. All other heavenly bodies (the moon, the sun, planets, stars, etc.) were seen to be in the sphere of perfection. Everything in the sphere of perfection was thought to be perfect in every respect. Everything was perfectly round and smooth and moved in perfect circles at perfectly uniform speeds. These perfect and unchanging bodies moved in "ether," a rarefied material believed to fill the sphere of perfection. The ether was a necessary part of the theory because Aristotle did not consider a vacuum possible.

The part of the universe inside the sphere of change was considered to be made of four elements: earth, air, fire, and water. These were not your ordinary earth, air, fire, and water; they were idealized elements that did not exist in their pure forms. They always existed in combination with each other, and the different, varying combinations made up the materials of the earth (figure 2.9). Rocks, for example, were thought to be mostly the element earth in combination with lesser amounts of air, water, and fire. The various amounts of the other elements accounted for the different kinds of rocks. Other materials, such as steam, were considered combinations of two elements, perhaps fire and water. Wood was a combination of all four elements of earth, air, fire, and water.

Natural Motion

Ideally, the four elements would exist alone in their own spheres. The center sphere would consist of the heaviest element, earth. This would be surrounded by a sphere of water, which in turn would be surrounded by a sphere of air. Fire, the lightest element, would surround the sphere of air (see figure 2.9). But this region of the universe, the sphere of change, was thought not to be perfect, and mixing did occur. When the mixing occurred, each element would seek out its ideal, or natural, place. Thus a rock, being made up of mostly the heavy element earth, would tumble down a hill through the air because the element earth was seeking its natural home. Water would flow over the rock, and bubbles of air would rise through the water as water and air moved toward their natural homes. Aristotle called this spontaneous movement of materials seeking their natural place a *natural motion*. In natural

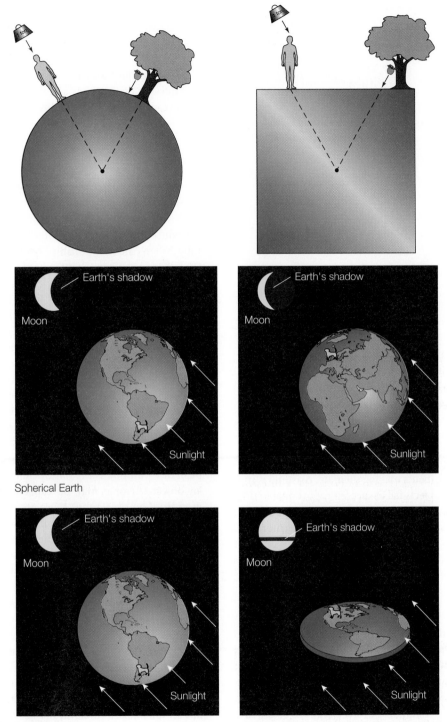

A   Spherical Earth

B   Flat Earth

FIGURE 2.8

One of Aristotle's proofs that Earth is a sphere. Aristotle argued that falling bodies move toward the center of Earth. Only if Earth is a sphere can that motion always be straight downward, perpendicular to Earth's surface. Another of Aristotle's proofs that Earth is a sphere: (*A*) Earth's shadow on the Moon during lunar eclipses is always round. (*B*) If Earth were a flat cylinder, there would be some eclipses in which Earth would cast a flat shadow on the Moon.

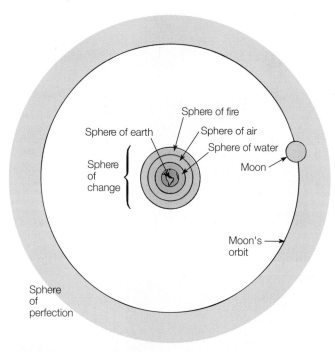

FIGURE 2.9

The shape of a sphere represented an early Greek idea of perfection. Aristotle's grand theory of the universe was made up of spheres, which explained the "natural motion" of objects.

motion, heavy bodies moved down and light bodies moved up according to the four elements making them up. Aristotle is supposed to have said that objects fall at speeds proportional to their weight but at a constant speed. In other words, a ten-pound ball would fall twice as fast as a five-pound ball. Furthermore, the speed was seen to be acquired immediately and remain constant during the fall. This belief was never checked by measurement, as reasoning, not observation, was used to develop the concept.

Forced Motion

Aristotle recognized that not all motion was in response to objects seeking their "natural places." People could throw rocks and horses could move carts, and the motion of the rocks and the carts was not from these objects seeking their natural places. This motion was imposed on the rocks and the carts. He called imposed motion a *forced motion*. A forced motion required a push or pull, such as a person pushing on a rock or a horse pulling a cart. But Aristotle believed that the push or pull must be continuously acting or else the motion would stop. If a horse stops pulling on a cart, for example, the cart will come to rest. A rock thrown through the air continues to move, he believed, because air moved in behind the moving rock to force it along. If it were not for the air forcing the rock to move, it would stop and fall straight down according to its natural motion.

Aristotle's work did offer explanations for the motion of falling objects and for horizontal motion, but they were disastrously wrong. His physical science theory was based on think-

ing, not measurement, and no one checked to see if his beliefs were accurate. He should not be blamed, however, for the fact that his writings later became part of Middle Ages theology, acquiring the status of priestly authority. With such authority, a person who questioned Aristotle's reasoned beliefs was viewed as a devious devil's advocate. It therefore took about two thousand years to correct his beliefs about motion.

FORCES

Aristotle recognized the need for a force to produce forced motion. He was partly correct, because a force is closely associated with *any* change of motion, as you will see. This section introduces the concept of a force, which will be developed more fully when the relationship between forces and motion is explained in the next chapter.

A **force** is usually considered as a push or a pull. These pushes and pulls can result from two kinds of interaction, (1) *contact interaction* and (2) *interaction at a distance*. An example of a contact interaction is people exerting a force by contact with a stalled automobile. They push on it directly until it moves. Likewise, a tow truck pulls through the contact of a tow cable on the automobile until it moves. A common example of interaction at a distance is gravitational attraction. A meteor is pulled toward the surface of the earth by this gravitational force. Raindrops and autumn leaves are also pulled by interaction at a distance, the gravitational force.

Through contact or interaction at a distance, a force is a push or a pull that is *capable* of changing the state of motion of an object. A force may be capable of changing the state of motion of an object, but there are other considerations. Consider, for example, a stalled automobile. Suppose you find that two people can push with sufficient force to move the car. Does it matter where on the car they push? It should be obvious that the people can push with a certain strength and determine the direction that the force is directed. Since a force has magnitude, or strength, as well as direction, it is a vector (figure 2.10).

What effect does direction have on two force vectors acting on one object at the same point? If they act in exactly opposite directions, the overall force on the object is the difference between the strength of the two forces. If they have the same magnitude, the overall effect is to cancel each other without producing any motion. The forces balance each other, and there is not an *unbalanced* force, so the *net force* is said to be zero. The **net force** is the sum of all the forces acting on the object. *Net* means "final," after the vector forces are added (figure 2.11).

When two parallel forces act in the same direction, they can be simply added together. In this case, there is a net force that is equivalent to the vector sum of the two forces (see figure 2.11). The vector sum is the addition of all the magnitude-of-displacement values and the combining of the direction values.

When two vector forces act in a way that is neither exactly together nor exactly opposite each other, the result will be like a new, different net force having a new direction and magnitude. The new net force is called a *resultant* and can be calculated by

FIGURE 2.10

The rate of movement and the direction of movement of this ship are determined by a combination of direction and magnitude of force from each of the tugboats. A force is a vector, since it has direction as well as magnitude. Which direction are the two tugboats pushing? What evidence would indicate that one tugboat is pushing with a greater magnitude of force? If the tugboat by the numbers is pushing with a greater force and the back tugboat is keeping the ship from moving, what will happen?

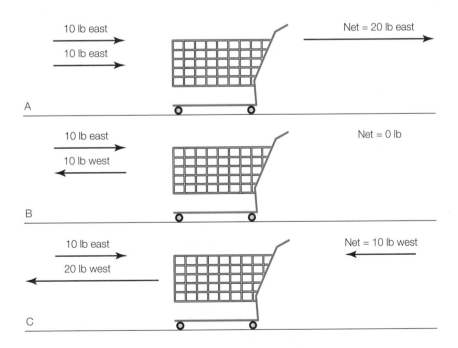

FIGURE 2.11

(A) When two parallel forces are acting on the cart in the same direction, the net force is the two forces added together. (B) When two forces are opposite and of equal magnitude, the net force is zero. (C) When two parallel forces are not of equal magnitude, the net force is the difference in the direction of the larger force.

FIGURE 2.12

Galileo (left) challenged the Aristotelian view of motion and focused attention on the concepts of distance, time, velocity, and acceleration.

drawing vector arrows to scale (length and direction) and using geometry. The tail of the vector arrow is placed on the object that feels the force, and the arrowhead points in the direction in which the force is exerted. The length of the arrow is proportional to the magnitude of the force. In summary, there are four things to understand and remember about vector arrows that represent forces:

1. The tail of the arrow is placed on the object that *feels* the force.
2. The arrowhead points in the *direction* in which the force is applied.
3. The *length* of the arrow is proportional to the magnitude of the force.
4. The *net force* is the sum of the vector forces.

HORIZONTAL MOTION ON LAND

Everyday experience seems to indicate that Aristotle's idea about horizontal motion on the earth's surface is correct. After all, moving objects that are not pushed or pulled do come to rest in

FIGURE 2.13

According to a widespread story, Galileo dropped two objects with different weights from the Leaning Tower of Pisa. They were supposed to have hit the ground at about the same time, discrediting Aristotle's view that the speed during the fall is proportional to weight.

a short period of time. It would seem that an object keeps moving only if a force continues to push it. A moving automobile will slow and come to rest if you turn off the ignition. Likewise, a ball that you roll along the floor will slow until it comes to rest. Is the natural state of an object to be at rest, and is a force necessary to keep an object in motion? This is exactly what people thought until Galileo (figure 2.12) published his book *Two New Sciences* in 1638, which described his findings about motion.

Consider the era when Galileo lived, from 1564 to 1642. Aristotle's ideas about motion had become joined with theology some three hundred years earlier, and to challenge Aristotle's ideas was considered to be heresy. Indeed, people were burned at the stake at that time for questioning the current theologic dogma. But this was a time of change. Europe was in the middle of the Renaissance movement, and it was the time of Columbus, da Vinci, Shakespeare, and Michelangelo, among others. Galileo was a key figure in this change as he challenged the Aristotelian views and focused attention on the concepts of distance, time, velocity, and acceleration.

Galileo was a professor at the University of Pisa and studied a variety of physical science topics. The telescope was invented about 1608 and Galileo built his own in 1609, focusing on many things that had not been previously observed. For the first time he could see mountains on the Moon, spots on the Sun, and a bulging Saturn. These were not the perfectly round and perfectly smooth celestial objects that Aristotle's theory called for. Furthermore, Galileo observed moons revolving around Jupiter, an impossibility if Earth was the center of rotation for the universe. Galileo was probably best known, however, for the widespread story of his experiments of dropping objects from the Leaning Tower of Pisa to study the motion of falling objects (figure 2.13).

Galileo's constant challenge of the ancient authority of Aristotle created many enemies and eventually led to a trial by the Inquisition. Galileo was a prisoner of the Inquisition when he wrote *Two*

A CLOSER LOOK | Transportation and the Environment

Environmental science is an interdisciplinary study of the earth's environment. The concern of this study is the overall problem of human degradation of the environment and remedies for that damage. As an example of an environmental science topic of study, consider the damage that results from current human activities concerning the use of transportation. Researchers estimate that overall transportation activities are responsible for about one-third of the total U.S. carbon emissions that are added to the air every day. In some areas of the country, transportation contributes up to 50 percent of the carbon emissions. Carbon emissions are a problem because they are directly harmful in the form of carbon monoxide and indirectly harmful because of the contribution of carbon dioxide to global warming and the consequences of climate change.

Carbon emission from transportation is a problem, but what remedies might exist? First, people must understand that the operation of cars, minivans, sport utility vehicles, and pickup trucks (small and large) cause the greatest amount of environmental damage when compared to the damage resulting from any other human activity. The act of driving does damage to the environment, and this includes a short trip to the grocery store, commuting to work or school, or a drive to the mall. With this understanding you can see that ways to reduce the damage caused by transportation can be found in planning and making better consumer choices.

The need for better consumer choices is evident when you examine the trends of more people driving more miles coupled with the use of more vehicles with lower fuel economy. About half the new vehicles sold are pickup trucks, vans, and sport utility vehicles, which do much more environmental damage than cars. One solution to this part of the problem is simply a consumer decision not to purchase pickup trucks or sport utility vehicles—or require higher emission standards for all vehicles.

Planning for environmental consideration could also help lower the damage. Environmental consideration means doing such things as planning for more compact city development rather than urban sprawl. This planning would include the development of bus or rail transit, bikeways, and walkways to help reduce carbon emissions by making the use of cars less necessary and less desirable. Access to other forms of transportation makes it possible to do less damage to the environment.

With or without such developments, individual consumer decisions are most important in reducing carbon emissions. This includes a decision by individuals to leave the car parked whenever and wherever possible. Consider the use of bike and pedestrian facilities, carpooling, or bus or rail transit. As they become more available and affordable, individuals will also be better able to consider electric cars and alternative-fuel vehicles.

CONNECTIONS

Move It!

Many animals do not use legs or wings for locomotion. For example, snakes, worms, clams, and slugs move by waves of muscle contractions down the length of their bodies. Though these animals have no limbs, they can move easily through their environments. An earthworm moves through soil by extending its body to push its head forward. Once inched forward, it expands its front end, creating a larger space in the ground into which the rest of its body moves. Small hairlike projections on its lower surface prevent the worm from slipping backward. Snakes have powerful muscles that ring and run the length of their bodies. These muscles allow snakes to skillfully move across land as well as climb and swim. To move forward, a snake forms S-shaped curves that proceed from its anterior to its posterior end. As the wave passes along, the animal pushes against objects in its environment and glides along the surface.

The movement of land snails or slugs is very different. These animals lay down a layer of slime or mucus upon which they glide by waves of muscle contractions in their lower flat structure, or foot. Anything that prevents them from contacting the mucus will prevent them from moving. If you are having problems with snails or slugs in your garden, one way to trap them is to set a shallow pan of beer into the ground. The slugs will move into the dish to feed but will not be able to leave because the beer prevents them from making good contact with their slime layer.

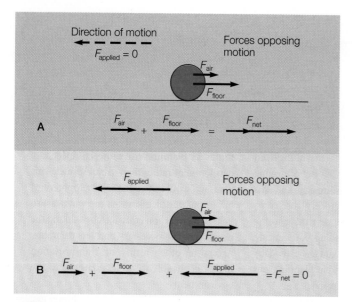

Direction of motion
$F_{applied} = 0$

Forces opposing motion

F_{air}
F_{floor}

A

F_{air} + F_{floor} = F_{net}

Forces opposing motion

$F_{applied}$

F_{air}
F_{floor}

B

F_{air} + F_{floor} + $F_{applied}$ = $F_{net} = 0$

FIGURE 2.14

(A) This ball is rolling to your left with no forces in the direction of motion. The vector sum of the force of floor friction (F_{floor}) and the force of air friction (F_{air}) result in a net force opposing the motion, so the ball slows to a stop. (B) A force is applied to the moving ball, perhaps by a hand that moves along with the ball. The force applied ($F_{applied}$) equals the vector sum of the forces opposing the motion, so the ball continues to move with a constant velocity.

New Sciences, which was secretly published in Holland in 1638. The book had three parts that dealt with uniform motion, accelerated motion, and projectile motion. Galileo described details of simple experiments, measurements, calculations, and thought experiments as he developed definitions and concepts of motion. In one of his thought experiments, Galileo presented an argument against Aristotle's view that a force is needed to keep an object in motion. Galileo imagined an object (such as a ball) moving over a horizontal surface without the force of friction. He concluded that the object would move forever with a constant velocity as long as there was no unbalanced force acting to change the motion.

Why does a rolling ball slow to a stop? You know that a ball will roll farther across a smooth, waxed floor such as a bowling lane than it will across floor covered with carpet. The rough carpet offers more resistance to the rolling ball. The resistance of the floor friction is shown by a force vector in figure 2.14, as is the resistance produced by air friction. Now imagine what force you would need to exert by pushing with your hand, moving along with the ball to keep it rolling at a uniform rate. The answer is the same force as the combined force from the floor and air resistance. Therefore, the ball continues to roll at a uniform rate when you *balance* the force opposing its motion. The force that you applied with your hand simply balanced the sum of the two opposing forces. It is reasonable, then, to imagine that if the opposing forces were removed you would not need to apply *any* force to the ball to keep it moving; it would roll forever on the frictionless floor. This was the kind of reasoning that Galileo did when he discredited the Aristotelian view that a force was necessary to keep an

object moving. Galileo concluded that a moving object would continue moving with a constant velocity if no unbalanced forces were applied, that is, if the net force were zero.

It could be argued that the difference in Aristotle's and Galileo's views of forced motion is really a degree of analysis. After all, moving objects on the earth do come to rest unless continuously pushed or pulled. But Galileo's conclusion describes *why* they must be pushed or pulled and reveals the true nature of the motion of objects. Aristotle argued that the natural state of objects is to be at rest and attempted to explain why objects move. Galileo, on the other hand, argued that it is just as natural for an object to be moving and attempted to explain why they come to rest. Galileo called the behavior of matter to persist in its state of motion **inertia.** Inertia is the *tendency of an object to remain in unchanging motion or at rest in the absence of an unbalanced force* (friction, gravity, or whatever). The development of this concept changed the way people viewed the natural state of an object and opened the way for further understandings about motion. Today, it is understood that a satellite moving through free space will continue to do so with no unbalanced forces acting on it (figure 2.15A). An unbalanced force is needed to slow the satellite (figure 2.15B), increase its speed (figure 2.15C), or change its direction of travel (figure 2.15D).

FALLING OBJECTS

Return now to the other kind of motion described by Aristotle, natural motion. Recall that Aristotle viewed natural motion as the motion produced when objects return to their natural places. For example, if you were to hold a rock above the ground and let it go, the rock would fall toward the ground, its natural place. Today, you know that the rock falls because the gravitational force of the earth pulls the rock downward. But the concern here is what happens during the fall. Aristotle is supposed to have said that objects fall at speeds proportional to their weight. In other words, suppose you drop a 100-pound iron ball at the same time a 1-pound iron ball is dropped from a height of 100 feet. According to Aristotle, the 100-pound ball would reach the ground before the 1-pound ball had fallen 1 foot. As stated in a popular story, Galileo discredited Aristotle's conclusion by dropping a solid iron ball and a solid wooden ball simultaneously from the top of the Leaning Tower of Pisa (see figure 2.13). Both balls, according to the story, hit the ground nearly at the same time. To do this, they would have to fall with the same velocity. In other words, the velocity of a falling object does not depend on its weight. Any difference in freely falling bodies is explainable by air resistance. Soon after the time of Galileo the air pump was invented. The air pump could be used to remove the air from a glass tube. The effect of air resistance on falling objects could then be demonstrated by comparing how objects fall in the air with how they fall in an evacuated glass tube. You know that a coin falls faster when dropped with a feather in the air. A feather and heavy coin will fall together in the near vacuum of an evacuated glass tube because the effect of air resistance on the feather has been removed. When objects fall toward the earth without considering air resistance, they are said to be in **free fall.** Free fall considers only gravity and neglects air resistance.

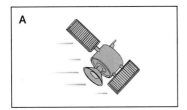

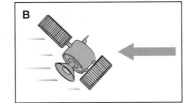

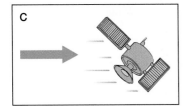

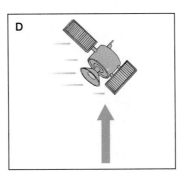

FIGURE 2.15

Galileo concluded that objects persist in their state of motion in the absence of an unbalanced force, a property of matter called *inertia.* Thus, an object moving through space without any opposing friction (*A*) continues in a straight-line path at a constant speed. The application of an unbalanced force in the direction of the change, as shown by the large arrow, is needed to (*B*) slow down, (*C*) speed up, or (*D*) change the direction of travel.

A CLOSER LOOK | A Bicycle Racer's Edge

Galileo was one of the first to recognize the role of friction in opposing motion. As shown in figure 2.14, friction with the surface and air friction combine to produce a net force that works against anything that is moving on the surface. This article is about air friction and some techniques that bike riders use to reduce that opposing force—perhaps giving them an edge in a close race.

The bike riders in box figure 2.1 are forming a single-file line, called a *paceline,* because the slipstream reduces the air resistance for a closely trailing rider. Cyclists say that riding in the slipstream of another cyclist will save as much as 25 percent of their energy, and they can move up to 5 mi/hr faster than they could riding alone.

In a sense, riding in a slipstream means that you do not have to push as much air out of your way. It has been estimated that at 20 mi/hr a cyclist must move a little less than half a ton of air out of the way every minute. One of the earliest demonstrations of how a slipstream can help a cyclist was done back about the turn of the century. Charles Murphy had a special bicycle trail built down the middle of a railroad track. Riding very close behind a special train caboose, Murphy was

BOX FIGURE 2.1

The object of the race is to be in the front, to finish first. If this is true, why are these racers forming a single-file line?

able to reach a speed of over 60 mi/hr for a one mile course. More recently, cyclists have reached over 125 mi/hr by following close, in the slipstream of a race car.

Along with the problem of moving about a half-ton of air out of the way every minute, there are two basic factors related to air resistance. These are (1) a turbulent versus a smooth flow of air, and (2) the problem of frictional drag. A turbulent flow of air contributes to air resistance because it causes the air to separate slightly on the back side, which increases the pressure on

the front of the moving object. This is why racing cars, airplanes, boats, and other racing vehicles are streamlined to a teardrop-like shape. This shape is not as likely to have the lower-pressure-producing air turbulence behind (and resulting greater pressure in front) because it smoothes or streamlines the air flow.

The frictional drag of air is similar to the frictional drag that occurs when you push a book across a rough tabletop. You know that smoothing the rough tabletop will reduce the frictional drag on the book. Likewise, the smoothing of a surface exposed to moving air will reduce air friction. Cyclists accomplish this "smoothing" by wearing smooth lycra clothing, and by shaving hair from arm and leg surfaces that are exposed to moving air. Each hair contributes to the overall frictional drag, and removal of the arm and leg hair can thus result in seconds saved. This might provide enough of an edge to win a close race. Shaving legs and arms, together with the wearing of lycra or some other tight, smooth-fitting garments, are just a few of the things a cyclist can do to gain an edge. Perhaps you will be able to think of more ways to reduce the forces that oppose motion.

Galileo versus Aristotle's Natural Motion

Galileo concluded that light and heavy objects fall together in free fall, but he also wanted to know the details of what was going on while they fell. He now knew that the velocity of an object in free fall was *not* proportional to the weight of the object. He observed that the velocity of an object in free fall *increased* as the object fell and reasoned from this that the velocity of the falling object would have to be (1) somehow proportional to the *time* of fall and (2) somehow proportional to the *distance* the object fell. If the time and distance were both related to the velocity of a falling object, how were they related to one another? To answer this question, Galileo made calculations involving distance, velocity, and time and, in fact, introduced the concept of acceleration. The relationships between these variables are found in the same three equations that you have already learned. Let's see how the equations can be rearranged to incorporate acceleration, distance, and time for an object in free fall.

Step 1:   Equation 2.1 gives a relationship between average velocity ($\bar{v}$), distance (d), and time (t). Solving this equation for distance gives

$$d = \bar{v}t$$

Step 2:   An object in free fall should have uniformly accelerated motion, so the average velocity could be calculated from equation 2.3,

$$\bar{v} = \frac{v_f + v_i}{2}$$

Substituting this equation in the rearranged equation 2.1, the distance relationship becomes

$$d = \left(\frac{v_f + v_i}{2}\right)(t)$$

Step 3:   The initial velocity of a falling object is always zero just as it is dropped, so the v_i can be eliminated,

$$d = \left(\frac{v_f}{2}\right)(t)$$

Step 4:   Now you want to get acceleration into the equation in place of velocity. This can be done by solving equation 2.2 for the final velocity (v_f), then substituting. The initial velocity (v_i) is again eliminated because it equals zero.

$$a = \frac{v_f - v_i}{t}$$

$$v_f = at$$

$$d = \left(\frac{at}{2}\right)(t)$$

Step 5:   Simplifying, the equation becomes

$$d = \frac{1}{2}at^2$$

equation 2.4

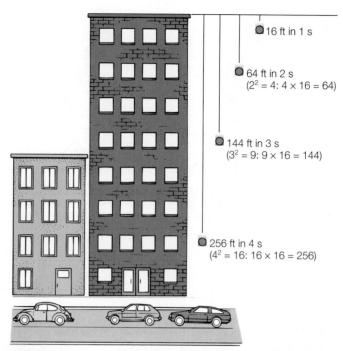

16 ft in 1 s

64 ft in 2 s
($2^2 = 4$: $4 \times 16 = 64$)

144 ft in 3 s
($3^2 = 9$: $9 \times 16 = 144$)

256 ft in 4 s
($4^2 = 16$: $16 \times 16 = 256$)

FIGURE 2.16

An object dropped from a tall building covers increasing distances with every successive second of falling. The distance covered is proportional to the square of the time of falling ($d \propto t^2$).

Thus, Galileo reasoned that a freely falling object should cover a distance *proportional to the square of the time of the fall* ($d \propto t^2$). In other words, the object should fall 4 times as far in 2 s as in 1 s ($2^2 = 4$), 9 times as far in 3 s ($3^2 = 9$), and so on. Compare this prediction with figure 2.16.

Galileo checked this calculation by rolling balls on an inclined board with a smooth groove in it. He used the inclined board to slow the motion of descent in order to measure the distance and time relationships, a necessary requirement since he lacked the accurate timing devices that exist today. He found, as predicted, that the falling balls moved through a distance proportional to the square of the time of falling. This also means that the *velocity of the falling object increased at a constant rate,* as shown in figure 2.17. Recall that a change of velocity during some time period is called *acceleration.* In other words, a falling object *accelerates* toward the surface of the earth.

Acceleration Due to Gravity

Since the velocity of a falling object increases at a constant rate, this must mean that falling objects are *uniformly accelerated* by the force of gravity. *All objects in free fall experience a constant acceleration.* During each second of fall, the object gains 9.8 m/s (32 ft/s) in velocity. This gain is the acceleration of the falling object, 9.8 m/s^2 (32 ft/s^2).

The acceleration of objects falling toward the earth varies slightly from place to place on the earth's surface because of the

C O N N E C T I O N S

Free Fall

There are two different meanings for the term "free fall." In physics, "free fall" means the unconstrained motion of a body in a gravitational field, without considering air resistance. Without air resistance all objects are assumed to accelerate toward the surface at 9.8 m/s².

In the sport of skydiving, "free fall" means falling within the atmosphere without a drag-producing device such as a parachute. Air provides a resisting force that opposes the motion of a falling object, and the net force is the difference between the downward force (weight) and the upward force of air resistance. The weight of the falling object depends on the mass and acceleration from gravity, and this is the force downward. The resisting force is determined by at least two variables, (1) the area of the object exposed to the airstream, and (2) the speed of the falling object. Other variables such as streamlining, air temperature, and turbulence play a role, but the greatest effect seems to be from exposed area and the increased resistance as speed increases.

A skydiver's weight is constant, so the downward force is constant. Modern skydivers typically free-fall from about 3,650 m (about 12,000 ft) above the ground until about 750 m (about 2,500 ft), where they open their parachutes. After jumping from the plane, the diver at first accelerates toward the surface, reaching speeds up to about 185–210 km/hr (about 115–130 mi/hr). The air resistance increases with increased speed and the net force becomes less and less. Eventually, the downward weight force will be balanced by the upward air resistance force, and the net force becomes zero. The person now falls at a constant speed and we say the terminal velocity has been reached. It is possible to change your body position to vary your rate of fall up or down to 32 km/hr (about 20 mi/hr). However, by diving or "standing up" in free fall, experienced skydivers can learn to reach speeds of up to 290 km/hr (about 180 mi/hr). The record free fall speed, done without any special equipment, is 517 km/hr (about 321 mi/hr). Once the parachute opens, a decent rate of about 16 km/hr (about 10 mi/hr) is typical.

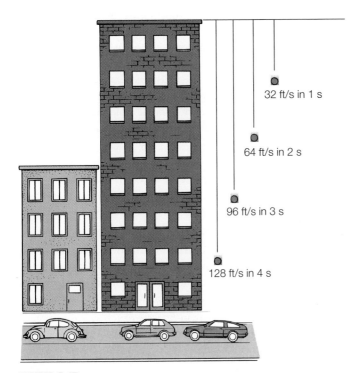

FIGURE 2.17
The velocity of a falling object increases at a constant rate, 32 ft/s².

Activities

1. Place three or four coins of different masses such as a dime, nickel, and quarter flat on the edge of a table. Place a notebook or book flat on the table behind the coins so it is parallel to the edge of the table. Keeping the book parallel to the table, push the coins over the edge at the same time. Do this several times. Does the way they hit the floor support Aristotle's view?

2. What is your reaction time? Have a friend hold a meterstick vertically from the top while you position your thumb and index finger at the 50 cm mark. Your friend will drop the stick (unannounced), and you will catch it with your thumb and finger. Measure how far the stick drops. Use this distance in the equation,

$$t = \sqrt{\frac{2d}{g}}$$

to calculate your reaction time. (This is equation 2.4 solved for t.)

A CLOSER LOOK | Deceleration and Seat Belts

Do you "buckle up" when you ride in a car? As you know, seat belts and shoulder harnesses are designed to protect people if an accident occurs. Clearly, many accidents are survivable if such restraints are worn (box figure 2.2).

Car collisions range from the relatively simple, straightforward fender-bender to the complex interactions that occur when a car is crunched into a lump of deformed metal. It is not the deceleration that hurts people. Injuries occur from the tendency of the passengers to retain their states of motion as the car rapidly decelerates. As a result, passengers hit the steering wheel, the dash, or the windshield because of their inertia.

Deceleration is measured the same as acceleration, except it is a decrease of velocity. The acceleration due to gravity, g, is 9.8 m/s² or 32 ft/s². This value of g is used as a base unit when discussing the acceleration or deceleration of cars, aircraft, and rockets. A normally braking car might decelerate with a velocity reduction of 16 ft/s every second. Since this is one-half of 32 ft/s², this car is decelerating at 0.5 g. A hard-braking car might decelerate at 1 g, or 32 ft/s². A 1 g deceleration is equivalent to reducing the velocity about 22 mi/hr each second.

The human body is surprisingly hardy when dealing with deceleration. People have survived a deceleration of 35 g's without serious injury and have a good chance of surviv-

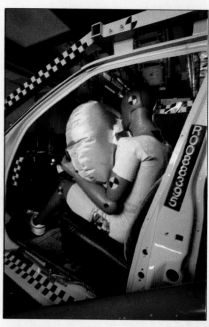

BOX FIGURE 2.2

General Motors' driver- and passenger-side air bags provide supplemental protection in testing equivalent to a 70 mi/hr frontal car-to-parked-car crash.

ing a car collision without injury if the deceleration does not exceed 25 g's. Compare this level of acceptable deceleration—25 g's—to the deceleration experienced when coming to a dead stop in a quarter of a second from the velocities shown in box table 2.1.

BOX TABLE 2.1

Deceleration to a complete stop in a quarter of a second

Speed		Deceleration	
mi/hr	*ft/s*	*ft/s²*	*g's*
10	15	59	1.8
20	29	117	3.7
30	44	176	5.5
40	59	235	7.3
50	73	293	9.1
60	88	352	11
70	103	411	12.8

As you can see in box table 2.1, you can survive a deceleration of even 70 mi/hr in a quarter of a second. In real crashes a number of factors influence the time involved in the deceleration. The front end of most cars is designed to collapse, extending the time variable. An air bag is a safety feature designed to inflate in the passenger compartment at the instant of a collision. The air bag keeps passengers from hitting the steering wheel, windshield, and so forth, but also increases the stopping time. This reduces the deceleration as it helps prevent injury. Most injuries that occur in a car collision are not from the rapid deceleration but occur because people fail to wear seat belts.

---

earth's shape and spin. The acceleration of falling objects decreases from the poles to the equator and also varies from place to place because the earth's mass is not distributed equally. The value of 9.8 m/s² (32 ft/s²) is an average value that is fairly close to, but not exactly, the acceleration due to gravity in any particular location. The acceleration due to gravity is important in a number of situations, so the acceleration from this force is given a special symbol, **g**.

COMPOUND MOTION

So far we have considered two types of motion: (1) the horizontal, straight-line motion of objects moving on the surface of the earth and (2) the vertical motion of dropped objects that accelerate toward the surface of the earth. A third type of motion occurs when an object is thrown, or projected, into the air.

Essentially, such a projectile (rock, football, bullet, golf ball, or whatever) could be directed straight upward as a vertical projection, directed straight out as a horizontal projection, or directed at some angle between the vertical and the horizontal. Basic to understanding such compound motion is the observation that (1) gravity acts on objects *at all times,* no matter where they are, and (2) the acceleration due to gravity (*g*) is *independent of any motion* that an object may have.

Vertical Projectiles

Consider first a ball that you throw straight upward, a vertical projection. The ball has an initial velocity but then reaches a maximum height, stops for an instant, then accelerates back toward the earth. Gravity is acting on the ball throughout its climb, stop, and fall. As it is climbing, the force of gravity is accelerating it back to the earth. The overall effect during the climb is deceleration, which continues to slow the ball until the instantaneous stop. The ball then accelerates back to the surface just like a ball that has been dropped. If it were not for air resistance, the ball would return with the same velocity that it had initially. The velocity vectors for a ball thrown straight up are shown in figure 2.18.

Horizontal Projectiles

Horizontal projections are easier to understand if you split the complete motion into vertical and horizontal parts. Consider, for example, a bullet that is fired horizontally from a rifle over perfectly level ground. The force of gravity accelerates the bullet downward, giving it an increasing velocity as it falls vertically toward the earth. This increasing downward velocity, illustrated with an arrow in figure 2.19, is the same as that of a dropped bullet and is represented by the vector arrows. There are no forces in the horizontal direction (if you ignore air resistance), so the horizontal velocity remains the same as shown by the v_h arrows. The combination of the vertical (v_v) motion and the horizontal (v_h) motion causes the bullet to follow a curved path until it hits the ground. An interesting prediction that can be made from this analysis is that a second bullet dropped from the same height as the horizontal rifle at the same time the rifle is fired will hit the ground at the same time as the bullet fired from the rifle.

Golf balls, footballs, and baseballs are usually projected upward at some angle to the horizon. The horizontal motion of these projectiles is constant as before because there are no horizontal forces involved. The vertical motion is the same as that of a ball projected directly upward. The combination of these two motions causes the projectile to follow a curved path called a

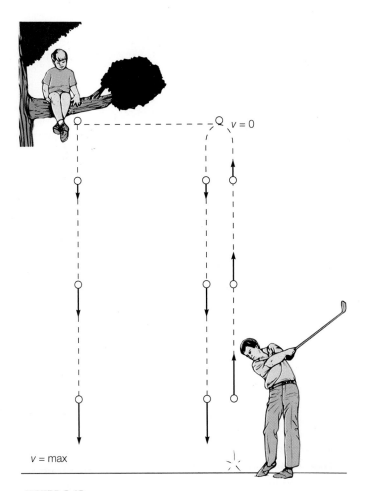

$v = 0$

$v = $ max

FIGURE 2.18

On its way up, a vertical projectile such as this misdirected golf ball is slowed by the force of gravity until an instantaneous stop; then it accelerates back to the surface, just as another golf ball does when dropped from the same height. The straight up and down moving golf ball has been moved to the side in the sketch so we can see more clearly what is happening.

parabola, as shown in figure 2.20. The next time you have the opportunity, observe the path of a ball that has been projected at some angle (figure 2.21). Note that the second half of the path is almost a reverse copy of the first half. If it were not for air resistance, the two values of the path would be exactly the same. Also note the distance that the ball travels as compared to the angle of projection. An angle of projection of 45° results in the maximum distance of travel.

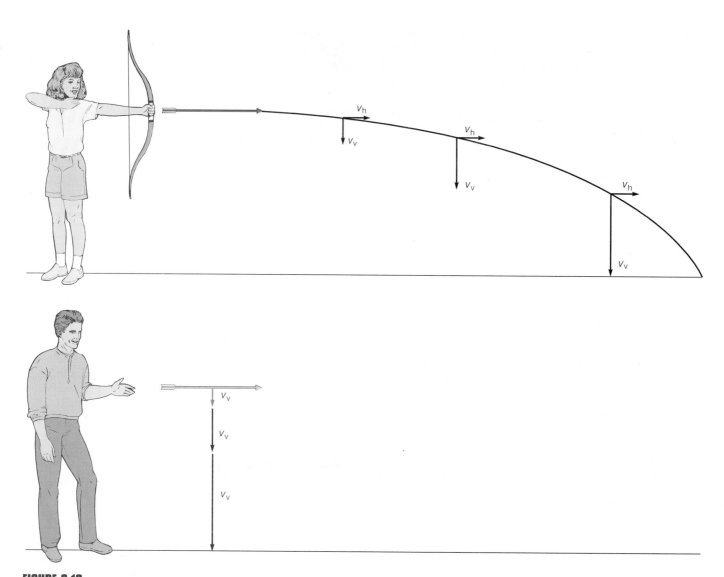

FIGURE 2.19

A horizontal projectile has the same horizontal velocity throughout the fall as it accelerates toward the surface, with the combined effect resulting in a curved path. Neglecting air resistance, an arrow shot horizontally will strike the ground at the same time as one dropped from the same height above the ground, as shown here by the increasing vertical velocity arrows.

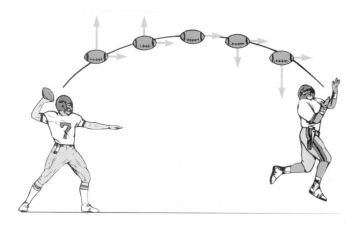

FIGURE 2.20

A football is thrown at some angle to the horizon when it is passed downfield. Neglecting air resistance, the horizontal velocity is a constant, and the vertical velocity decreases, then increases, just as in the case of a vertical projectile. The combined motion produces a parabolic path. Contrary to statements by sportscasters about the abilities of certain professional quarterbacks, it is impossible to throw a football with a "flat trajectory" because it begins to accelerate toward the surface as soon as it leaves the quarterback's hand.

FIGURE 2.21
Without a doubt, this baseball player is aware of the relationship between the projection angle and the maximum distance acquired for a given projection velocity.

SUMMARY

Motion is defined as the act or process of changing position. The change is usually described by comparing the moving object to something that is not moving. Motion can be measured by speed, velocity, and acceleration. *Speed* is a measure of how fast something is moving. It is a ratio of the distance covered between two locations to the time that elapsed while moving between the two locations. The *average speed* considers the distance covered during some period of time, while the *instantaneous speed* is the speed at some specific instant. *Velocity* is a measure of the speed and direction of a moving object. *Acceleration* is a change of velocity per unit of time. The *average acceleration* is the change of velocity during some period of time, while the *instantaneous acceleration* is the acceleration at some specific instant. Speed has magnitude only and is a *scalar*. Velocity and acceleration have both magnitude and direction and are *vectors*.

Early ideas about motion were recorded in the fourth century B.C. by Aristotle. Aristotle's theory of the universe considered two types of motion: *natural motion* and *forced motion*. Natural motion was the motion of falling objects. Aristotle's theory explained the motion of falling objects as the result of objects seeking their natural places. During the fall, an object was seen to acquire an immediate and constant velocity that was proportional to its weight. Forced motion required an imposed force to move an object. The motion was understood to stop unless the force was continuously acting.

A *force* is a push or a pull that can change the motion of an object. A force results from *contact interactions* or *interactions at a distance*. Force is a vector, and *net force* is the vector sum of all the forces acting on an object.

Galileo challenged the authority of Aristotle's views, which had become a part of Middle Ages theology. Galileo determined that a continuously applied force is unnecessary for motion and defined the concept of *inertia*: an object remains in unchanging motion in the absence of an unbalanced force. Galileo also determined that falling objects accelerate toward the earth's surface and that the acceleration is independent of the weight of the object. He found the acceleration due to gravity, g, to be 9.8 m/s^2 (32 ft/s^2), and the distance an object falls is proportional to the square of the time of free fall ($d \propto t^2$).

Compound motion occurs when an object is projected into the air. Compound motion can be described by splitting the motion into vertical and horizontal parts. The acceleration due to gravity, g, is a constant that is acting at all times and acts independently of any motion that an object has. The path of an object that is projected at some angle to the horizon is therefore a parabola.

Summary of Equations

2.1

$$\text{average speed or magnitude of velocity} = \frac{\text{distance}}{\text{time}}$$

$$\bar{v} = \frac{d}{t}$$

2.2

$$\text{acceleration} = \frac{\text{change of velocity}}{\text{time}}$$

$$= \frac{\text{final velocity} - \text{initial velocity}}{\text{time}}$$

$$a = \frac{v_f - v_i}{t}$$

2.3

$$\text{average velocity} = \frac{\text{final velocity} + \text{initial velocity}}{2}$$

$$\bar{v} = \frac{v_f + v_i}{2}$$

2.4

$$\text{distance} = \frac{1}{2}(\text{acceleration})(\text{time})^2$$

$$d = \frac{1}{2}at^2$$

KEY TERMS

acceleration (p. **27**)   magnitude (p. **26**)

force (p. **32**)   net force (p. **32**)

free fall (p. **36**)   speed (p. **25**)

g (p. **40**)   velocity (p. **27**)

inertia (p. **36**)

APPLYING THE CONCEPTS

1. The speedometer of a car gives you a reading of the car's
   a. constant speed.
   b. average speed.
   c. instantaneous speed.
   d. total speed.

2. For a trip of several hours duration, a ratio of the total distance traveled to the total time elapsed is a(an)
   a. instantaneous speed.
   b. average speed.
   c. constant speed.
   d. total speed.

3. The symbol $\bar{v}$ is used to represent
   a. instantaneous speed.
   b. average speed.
   c. constant speed.
   d. total speed.

4. If $\bar{v} = d/t$, then $d = ?$
   a. $\bar{v}/t$
   b. $t/\bar{v}$
   c. $\bar{v}t$
   d. $(\bar{v})(1/t)$

5. The size of a quantity is called
   a. vector.
   b. scalar.
   c. ratio.
   d. magnitude.

6. A quantity of 15 m/s to the north is a measure of
   a. velocity.
   b. acceleration.
   c. speed.
   d. directional distance in the metric system.

7. A quantity of 60 km/hr describes a(an)
   a. vector.
   b. acceleration.
   c. speed.
   d. metric distance.

8. A quantity of 5 m/s^2 is a measure of
   a. metric area.
   b. acceleration.
   c. speed.
   d. velocity.

9. Measurements that have both magnitude and direction are
   a. scalar quantities.
   b. vector quantities.
   c. dual quantities.
   d. ratios.

10. A ratio of $\Delta v/\Delta t$ is a measure of motion that is known as (recall that the symbol Δ means "a change in")
    a. speed.
    b. velocity.
    c. acceleration.
    d. none of the above.

11. A ratio of $\Delta d/\Delta t$ is a measure of motion that is known as
    a. speed.
    b. acceleration.
    c. mass density.
    d. none of the above.

12. Which one of the following is *not* the same as the expression of m/s^2?
    a. m/s/s
    b. (m/s)(s)
    c. (m/s)(1/s)
    d. none of the above

13. An automobile has how many different devices that can cause it to undergo acceleration?
    a. none
    b. one
    c. two
    d. three or more

14. You can find an average velocity by adding the initial and final velocity and dividing by 2 only when the acceleration is
    a. changing.
    b. uniform.
    c. increasing.
    d. decreasing.

15. Ignoring air resistance, an object falling toward the surface of the earth has a *velocity* that is
    a. constant.
    b. increasing.
    c. decreasing.
    d. acquired instantaneously, but dependent on the weight of the object.

16. Ignoring air resistance, an object falling near the surface of the earth has an *acceleration* that is
    a. constant.
    b. increasing.
    c. decreasing.
    d. dependent on the weight of the object.

17. Two objects are released from the same height at the same time, and one has twice the weight of the other. Ignoring air resistance,
    a. the heavier object hits the ground first.
    b. the lighter object hits the ground first.
    c. they both hit at the same time.
    d. whichever hits first depends on the distance dropped.

18. A ball rolling across the floor slows to a stop because
    a. there are unbalanced forces acting on it.
    b. the force that started it moving wears out.
    c. the forces are balanced.
    d. the net force equals zero.
19. If there are no unbalanced forces acting on a moving object it will
    a. slow to a stop.
    b. continue to move.
20. The distance that an object in free fall has covered is proportional to the square of the
    a. weight of the object.
    b. mass of the object.
    c. time of fall.
    d. distance of the fall.
21. When you throw a ball directly upward, it is accelerated
    a. only as it falls.
    b. during the instantaneous stop and during the fall.
    c. at all times.
    d. none of the time.
22. After leaving the rifle, a bullet fired horizontally has how many forces acting on it (ignoring air resistance)?
    a. one, from the gunpowder explosion
    b. one, from the pull of gravity
    c. two, one from the gunpowder explosion and one from gravity
    d. none
23. Which of the following best describes the path that a bullet follows after it leaves a rifle?

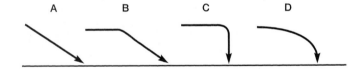

Answers

1. c 2. b 3. b 4. c 5. d 6. a 7. c 8. b 9. b 10. c 11. a 12. b 13. d 14. b 15. b
16. a 17. c 18. a 19. b 20. c 21. c 22. b 23. d

QUESTIONS FOR THOUGHT

1. What is the difference between speed and velocity?
2. What is acceleration?
3. An insect inside a bus flies from the back toward the front at 5.0 mi/hr. The bus is moving in a straight line at 50.0 mi/hr. What is the speed of the insect?
4. Explain what happens to the velocity and the acceleration of an object in free fall.
5. In the equation $d = 1/2at^2$, if g is 9.8 m/s^2 and t is in s, what is the unit for d?
6. What is inertia?
7. In the unit of acceleration 9.8 m/s^2, what is the meaning of a "square second"?
8. Disregarding air friction, describe all the forces acting on a bullet shot from a rifle into the air.
9. Can gravity act in a vacuum? Explain.
10. Does the force of gravity acting on a ball change as the ball is thrown straight up from the surface of the earth? Explain.
11. How does the velocity of a ball thrown straight up compare to the velocity the ball has when it falls to the same place on the way down? Explain.
12. Are there any unbalanced forces acting on an object that falls a short distance? Explain.

PARALLEL EXERCISES

The exercises in groups A and B cover the same concepts. Solutions to group A exercises are located in appendix D.

Note: Neglect all frictional forces in all exercises.

Group A

1. What is the average speed in km/hr for a car that travels 22 km in exactly 15 min?
2. Suppose a radio signal travels from the earth and through space at a speed of 3.0×10^8 m/s. How far into space did the signal travel during the first 20.0 minutes?
3. How far away was a lightning strike if thunder is heard 5.00 seconds after seeing the flash? Assume that sound traveled at 350.0 m/s during the storm.
4. A car is driven at an average speed of 100.0 km/hr for two hours, then at an average speed of 50.0 km/hr for the next hour. What was the average speed for the three-hour trip?
5. What is the acceleration of a car that moves from rest to 15.0 m/s in 10.0 s?

Group B

1. A boat moves 15.0 km across a lake in 45 min. (a) What was the average speed in meters per second? (b) What was the average speed in kilometers per hour?
2. If the Sun is a distance of 1.5×10^8 km from Earth, how long does it take sunlight to reach Earth if it moves at 3.0×10^8 m/s?
3. How many meters away is a cliff if an echo is heard one-half second after the original sound? Assume that sound traveled at 343 m/s on that day.
4. A car has an average speed of 80.0 km/hr for one hour, then an average speed of 90.0 km/hr for two hours during a three-hour trip. (a) How far did the car travel during the trip? (b) What was the average speed for the trip?

6. How long will be required for a car to go from a speed of 20.0 m/s to a speed of 25.0 m/s if the acceleration is 3.0 m/s²?

7. A car moving at 80.0 km/hr is brought to a stop in 5.00 seconds. How far did the car travel while stopping?

8. What is the average speed of a truck that makes a 285 mile trip in 5.0 hours?

9. A sprinter runs the 200.0 m dash in 21.4 s. (a) What was the sprinter's speed in m/s? (b) If the sprinter were able to maintain this pace, how much time (in hours) would be needed to run the 420.0 km from St. Louis to Chicago?

10. What average speed must you maintain to make (a) a 400.0 mile trip in 8.00 hours? (b) the same 400.0 mile trip in 7.00 hours?

11. A bullet leaves a rifle with a speed of 2,360 ft/s. How much time elapses before it strikes a target 1 mile (5,280 ft) away?

12. A pitcher throws a ball at 40.0 m/s, and the ball is electronically timed to arrive at home plate 0.4625 s later. What is the distance from the pitcher to the home plate?

13. The Sun is 1.50×10^8 km from Earth, and the speed of light is 3.00×10^8 m/s. How many minutes elapse as light travels from the Sun to Earth?

14. An archer shoots an arrow straight up with an initial velocity magnitude of 100.0 m/s. After 5.00 s, the velocity is 51.0 m/s. At what rate is the arrow decelerated?

15. An airplane starts from rest and accelerates for 5.0 s to 235 ft/s before taking off. What is the acceleration of the plane?

16. A racing car accelerates from 145 ft/s to 220 ft/s in 11 s. What is the acceleration?

17. What is the velocity of a car that accelerates from rest at 9.0 ft/s² for 8.0 s?

18. A sports car moving at 88.0 ft/s is able to stop in 100.0 feet. (a) How much time (in seconds) was required for the stop? (b) What was the deceleration in *g*'s?

19. A ball thrown straight up climbs for 3.0 s before falling. Neglecting air resistance, with what velocity was the ball thrown?

20. A ball dropped from a building falls for 4.00 s before it hits the ground. (a) What was its final velocity just as it hit the ground? (b) What was the average velocity during the fall? (c) How high was the building?

21. You drop a rock from a cliff, and 5.00 s later you see it hit the ground. How high is the cliff?

22. How long must a car accelerate at 4.00 m/s² to go from 10.0 m/s to 50.0 m/s?

23. What is the velocity of a rock in free fall 3.0 s after it is released from rest?

5. What is the acceleration of a car that moves from a speed of 5.0 m/s to a speed of 15 m/s during a time of 6.0 s?

6. How much time is needed for a car to accelerate from 8.0 m/s to a speed of 22 m/s if the acceleration is 3.0 m/s²?

7. An arrow accelerates to 36 km/hr in one-half second. How far did the arrow travel in one-half second?

8. The Empire State Building is 450 m high. A penny dropped from the top would have what speed when it hits the sidewalk?

9. A baseball player pops a fly ball 20.0 m straight up and the catcher tosses his mask to catch the ball. How long will the catcher have to wait after the ball starts back down?

10. What was the initial speed of a car that required 13.7 m to stop in 4.00 s?

11. What is the average speed of a car that travels 270.0 miles in 4.50 hours?

12. A ferryboat requires 45.0 min to travel 15.0 km across a bay. (a) What was the average speed in m/s? (b) At this speed, how much time would be required for the boat to take a 420 km trip up the river?

13. A car with an average speed of 60.0 mi/hr (a) will require how much time to cover 300.0 miles? (b) will travel how far in 8.0 hours?

14. A rocket moves through outer space at 11,000 m/s. At this rate, how much time would be required to travel the distance from Earth to the Moon, which is 380,000 km?

15. Sound travels at 1,140 ft/s in the warm air surrounding a thunderstorm. How far away was the place of discharge if thunder is heard 4.63 s after a lightning flash?

16. How many hours are required for a radio signal from a space probe near the planet Pluto, 6.00×10^9 km away, to reach Earth? Assume that the radio signal travels at the speed of light, 3.00×10^8 m/s.

17. A rifle is fired straight up, and the bullet leaves the rifle with an initial velocity magnitude of 724 m/s. After 5.00 s the velocity is 675 m/s. At what rate is the bullet decelerated?

18. An airplane is accelerated uniformly from rest to 95.0 m/s in 6.33 s. What is the acceleration of the plane?

19. A racing car accelerates from 40.0 m/s to 80.0 m/s in 10.0 s. What is the acceleration?

20. What is the velocity of a car that accelerates from rest at 6.0 m/s² for 5.0 s?

21. A car with an initial velocity of 30.0 m/s is able to come to a stop over a distance of 100.0 m when the brakes are applied. (a) How much time was required for the stopping process? (b) How many *g*'s did the driver experience?

22. A rock thrown straight up climbs for 2.50 s, then falls to the ground. Neglecting air resistance, with what velocity did the rock strike the ground?

23. An object is observed to fall from a bridge, striking the water below 2.50 s later. (a) With what velocity did it strike the water? (b) What was its average velocity during the fall? (c) How high is the bridge?

24. A ball dropped from a window strikes the ground 2.00 s later. How high is the window above the ground?

25. What is the resulting velocity if a car moving at 10.0 m/s accelerates at 4.00 m/s² for 10.0 s?

26. If a rock in free fall is moving at 96.0 ft/s, how long has it been falling?

Information about the mass of a hot air balloon and forces on the balloon will enable you to predict if it is going to move up, down, or drift across the river. This chapter is about such relationships between force, mass, and changes in motion.

CHAPTER | Three

Patterns of Motion

In the previous chapter you learned how to describe motion in terms of distance, time, velocity, and acceleration. In addition, you learned about different kinds of motion, such as straight-line motion, the motion of falling objects, and the compound motion of objects projected up from the surface of the earth. You were also introduced, in general, to two concepts closely associated with motion: (1) that objects have inertia, a tendency to resist a change in motion, and (2) that forces are involved in a change of motion.

The relationship between forces and a change of motion is obvious in many everyday situations (figure 3.1). When a car, bus, or plane starts moving, you feel a force on your back. Likewise, you feel a force on the bottoms of your feet when an elevator starts moving upward. On the other hand, you seem to be forced toward the dashboard if a car stops quickly, and it feels as if the floor pulls away from your feet when an elevator drops rapidly. These examples all involve patterns between forces and motion, patterns that can be quantified, conceptualized, and used to answer questions about why things move or stand still. These patterns are the subject of this chapter.

LAWS OF MOTION

Isaac Newton was born on Christmas Day in 1642, the same year that Galileo died. Newton was a quiet farm boy who seemed more interested in mathematics and tinkering than farming. He entered Trinity College of Cambridge University at the age of eighteen, where he enrolled in mathematics. He graduated four years later, the same year that the university was closed because the bubonic plague, or Black Death, was ravaging Europe. During this time, Newton returned to his boyhood home, where he thought out most of the ideas that would later make him famous. Here, between the ages of twenty-three and twenty-four, he invented the field of mathematics called *calculus* and clarified his ideas on motion and gravitation. After the plague he returned to Cambridge, where he was appointed professor of mathematics at the age of twenty-six. He lectured and presented papers on optics. One paper on his theory about light and colors caused such a controversy that Newton resolved never to publish another line. Newton was a shy, introspective person who was too absorbed in his work for such controversy. In 1684, Edmund Halley (of Halley's comet fame) asked Newton to resolve a dispute involving planetary motions. Newton had already worked out the solution to this problem, in addition to other problems on gravity and motion. Halley persuaded the reluctant Newton to publish the material. Two years later, in 1687, Newton published *Principia*, which was paid for by Halley. Although he feared controversy, the book was accepted almost at once and established Newton as one of the greatest thinkers who ever lived.

Newton built his theory of motion on the previous work of Galileo and others. In fact, Newton's first law is similar to the concept of inertia presented earlier by Galileo. Newton acknowledged the contribution of Galileo and others to his work, stating that if he had seen further than others "it was by standing upon the shoulders of giants."

Newton's First Law of Motion

Newton's first law of motion is also known as the *law of inertia* and is very similar to one of Galileo's findings about motion. Recall that Galileo used the term *inertia* to describe the tendency

FIGURE 3.1

In a moving airplane, you feel forces in many directions when the plane changes its motion. You cannot help but notice the forces involved when there is a change of motion.

of an object to resist changes in motion. Newton's first law describes this tendency more directly. In modern terms (not Newton's words), the **first law of motion** is as follows:

> Every object retains its state of rest or its state of uniform straight-line motion unless acted upon by an unbalanced force.

This means that an object at rest will remain at rest unless it is put into motion by an unbalanced force; that is, the vector sum of the forces must be greater than zero if more than one force is involved. Likewise, an object moving with uniform straight-line motion will retain that motion unless an unbalanced force causes it to speed up, slow down, or change its direction of travel. Thus, Newton's first law describes the tendency of an object to resist *any* change in its state of motion, a property defined by inertia.

Some objects have greater inertia than other objects. For example, it is easier to push a small car into motion than to push a heavy truck into motion. The truck has greater inertia than the

CONNECTIONS

Bubonic Plague—The Black Death

The bacterium *Yersinia pestis* is responsible for the disease known as the plague. This disease probably originated in Asia or central Asia and spread throughout the world as humans became increasingly mobile. One of the earliest recorded epidemics of plague occurred in 542 B.C. during the reign of the Byzantine emperor Justinian. The Great Plague that swept Europe in the fourteenth century claimed an estimated 25 million lives (one-quarter of Europe's population) when living conditions were poor and rats and fleas were common. This disease killed 50 percent of the population in some countries in Western Europe. Another wave swept through London in 1665, killing more than 70 thousand. Since that time, epidemics have been reported in Hong Kong and Bombay (1876); Santos, Brazil (1899); San Francisco (1900); Manchuria, China (1910); and Vietnam (1965).

The bacteria are harbored in wild rodents. It does little harm to the rat but can cause disease if it is transmitted to humans. If a flea sucks blood from an infected rat and then bites a human, the bacteria may enter the human bloodstream and cause disease. Once inside the body, the bacteria enter the lymph system and spread through the body. The bacteria lodge in the lymph nodes of the groin and arm pit, and release toxin molecules that cause enlarged, hemorrhagic, inflamed nodes called buboes. Once they move into the bloodstream, they can invade any organ.

If in the skin, the hemorrhaging capillaries will form dark patches that join together and appear as large, blackened areas. That symptom was quite predominant in the Great Plague, and thus the disease became known as the "Black Death."

small car. It is also more difficult to stop the heavy truck from moving than it is to stop a small car. Again the heavy truck has greater inertia. The amount of inertia an object has describes the **mass** of the object. Mass is a *measure of inertia.* The more inertia an object has, the greater its mass. Thus the heavy truck has more mass than the compact car. You know this because the truck has greater inertia. Newton originally defined mass as the "quantity of matter" in an object, and this definition is intuitively appealing. However, Newton needed to measure inertia because of its obvious role in motion and redefined mass as a measure of inertia. Thinking of mass in terms of a resistance to a change of motion may seem strange at first, but it will begin to make more sense as you explore the relationships between mass, forces, and acceleration.

Think of Newton's first law of motion when you ride standing in the aisle of a bus. The bus begins to move, and you, being an independent mass, tend to remain at rest. You take a few steps back as you tend to maintain your position relative to the ground outside. You reach for a seat back or some part of the bus. Once you have a hold on some part of the bus it supplies the forces needed to give you the same motion as the bus and you no longer find it necessary to step backward. You now have the same motion as the bus, and no forces are involved, at least until the bus goes around a curve. You now feel a tendency to move to the side of the bus. The bus has changed its straight-line motion, but you, again being an independent mass, tend to move straight ahead. The side of the seat forces you into following the curved motion of the bus. The forces you feel when the bus starts moving or turning are a result of your inertia. You tend to remain at rest or follow a straight path until forces correct your motion so that it is the same as that of the bus (figure 3.2).

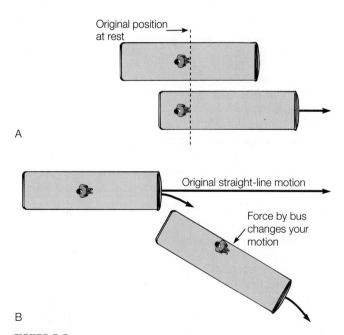

A

B

FIGURE 3.2

Top view of a person standing in the aisle of a bus. (*A*) The bus is at rest, and then starts to move forward. Inertia causes the person to remain in the original position, appearing to fall backward. (*B*) The bus turns to the right, but inertia causes the person to retain the original straight-line motion until forced in a new direction by the side of the bus.

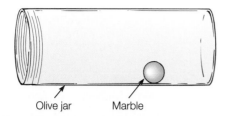

FIGURE 3.3

This marble can be used to demonstrate inertia.

Activities

1. Place a marble inside any cylindrical container, such as an olive bottle or pill vial. Place the container on its side on top of a table and push it along with the mouth at the front end (figure 3.3). Note the position of the marble inside the container. Stop the container immediately. Explain the reaction of the marble.

2. Place the marble inside the container and push it along as before. This time push the container over a carpeted floor, a smooth floor, a waxed tabletop, and so forth. Notice the distance that the marble rolls over the different surfaces. Repeat the procedure with large and small marbles of various masses as you use the same velocity each time. Explain the differences you observe.

Consider a second person standing next to you in the aisle of the moving bus. When the bus stops, the two of you tend to retain your straight-line motion, and you move forward in the aisle. If it is more difficult for the other person to stop moving, you know that the other person has greater inertia. Since mass is a measure of inertia, you know that the other person has more mass than you. But do not confuse this with the other person's weight. Weight is a different property than mass and is explained by Newton's second law of motion.

Newton's Second Law of Motion

Newton had successfully used Galileo's ideas to describe the nature of motion. Newton's first law of motion explains that any object, once started in motion, will continue with a constant velocity in a straight line unless a force acts on the moving object. This law not only describes motion but establishes the role of a force as well. A change of motion is therefore *evidence* of the action of some unbalanced force (net force). The association of forces and a change of motion is common in your everyday experience. You have felt forces on your back in an accelerating automobile, and you have felt other forces as the automobile turns or stops. You have also learned about gravitational forces that accelerate objects toward the surface of the earth. Unbalanced forces

FIGURE 3.4

This bicycle rider knows about the relationship between force, acceleration, and mass.

and acceleration are involved in any change of motion. The amount of inertia, or mass, is also involved, since inertia is a resistance to a change of motion. Newton's second law of motion is a relationship between *net force, acceleration,* and *mass* that describes the cause of a change of motion (figure 3.4).

Consider the motion of you and a bicycle you are riding. Suppose you are riding your bicycle over level ground in a straight line at 3 miles per hour. Newton's first law tells you that you will continue with a constant velocity in a straight line as long as no external, unbalanced force acts on you and the bicycle. The force that you *are* exerting on the pedals seems to equal some external force that moves you and the bicycle along (more on this later). The force exerted as you move along is needed to *balance* the resisting forces of tire friction and air resistance. If these resisting forces were removed, you would not need to exert any force at all to continue moving at a constant velocity. The net force is thus the force you are applying minus the forces from tire friction and air resistance. The *net force* is therefore zero when you move at a constant speed in a straight line (figure 3.5). When the net force on an object is zero, no unbalanced force acts on it.

If you now apply a greater force on the pedals the *extra* force you apply is unbalanced by friction and air resistance. Hence there will be a net force greater than zero, and you will accelerate. You will accelerate during, and *only* during, the time that the (unbalanced) net force is greater than zero. Likewise, you will slow down if you apply a force to the brakes, another kind of resisting friction. A third way to change your velocity is to apply a force on the handlebars, changing the direction of your velocity. Thus, *unbalanced forces* on you and your bicycle produce an *acceleration.*

Starting a bicycle from rest suggests a relationship between force and acceleration. You observe that the harder you push on the pedals, the greater your acceleration. If, for a time, you exert a force on the pedals greater than the combined forces of tire

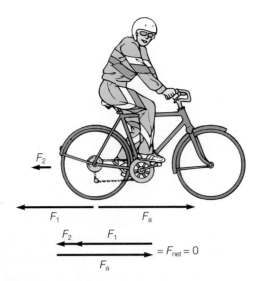

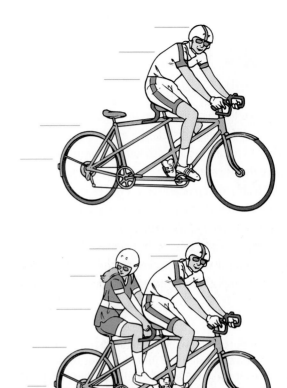

FIGURE 3.5

At a constant velocity the force of tire friction (F_1) and the force of air resistance (F_2) have a vector sum that equals the force applied (F_a). The net force is therefore zero.

friction and air resistance, then you will create an unbalanced force and accelerate to a certain speed. If, by exerting an even greater force on the pedals, you double the net force in the direction you are moving, then you will also double the acceleration, reaching the same velocity in half the time. Likewise, if you triple the unbalanced force you will increase the acceleration three-fold. Recall that when quantities increase or decrease together in the same ratio, they are said to be *directly proportional*. The acceleration is therefore directly proportional to the unbalanced force applied. Recall also that the symbol $\propto$ means "is proportional to." The relationship between acceleration (a) and the unbalanced force (F) can thus be abbreviated as

$$\text{acceleration} \propto \text{force}$$

or in symbols,

$$a \propto F$$

Suppose that your bicycle has two seats, and you have a friend who will ride with you. Suppose also that the addition of your friend on the bicycle will double the mass of the bike and riders. If you use the same extra unbalanced force as before, the bicycle will undergo a much smaller acceleration. In fact, with all other factors equal, doubling the mass and applying the same extra force will produce an acceleration of only half as much (figure 3.6). An even more massive friend would reduce the acceleration even more. If you triple the mass and apply the same extra force, the acceleration will be one-third as much. Recall that when a relationship between two quantities shows that one quantity increases as another decreases, in the same ratio, the quantities are said to be *inversely proportional*. The acceleration (a) of an object is therefore inversely proportional to its mass (m). This relationship can be abbreviated as

FIGURE 3.6

More mass results in less acceleration when the same force is applied. With the same force applied, the riders and bike with twice the mass will have half the acceleration, with all other factors constant. Note that the second rider is not pedaling.

$$\text{acceleration} \propto \frac{1}{\text{mass}}$$

or in symbols,

$$a \propto \frac{1}{m}$$

Since the mass (m) is in the denominator, doubling the mass will result in one-half the acceleration, tripling the mass will result in one-third the acceleration, and so forth.

Now the relationships can be combined to give

$$\text{acceleration} \propto \frac{\text{force}}{\text{mass}}$$

or

$$a \propto \frac{F}{m}$$

The acceleration of an object therefore depends on *both* the *net force applied* and the *mass* of the object. The **second law of motion** is as follows:

The acceleration of an object is directly proportional to the net force acting on it and inversely proportional to the mass of the object.

Swimming Scallop

A scallop is the shell often seen as a logo for a certain petroleum company, a fan-shaped shell with a radiating fluted pattern. The scallop is a marine mollusk that is most unusual since it is the only bivalve that is capable of swimming. By opening and closing its shell it is able to jet propel itself in a way by forcing water from the interior of the shell in a jetlike action. The popular seafood called "scallops" is the edible muscle that the scallop uses to close its shell.

Scallops are able to swim by orienting their shell at a proper angle and maintaining a minimum acceleration to prevent sinking. For example, investigations have found that one particular species of scallop must force enough water backward to move about 6 body lengths per second with a 10 degree angle of attack to maintain level swimming. Such a swimming effort can be maintained for up to about 20 seconds, enabling the scallop to escape predation or some other disturbing condition.

A more massive body limits the swimming ability of the scallop, as a greater force is needed to give a greater mass the same acceleration (as you would expect from Newton's second law of motion). This problem becomes worse as the scallop grows larger and larger without developing a greater and greater jet force.

The proportion between acceleration, force, and mass is not yet an equation, because the units have not been defined. The metric unit of force is *defined* as a force that will produce an acceleration of 1.0 m/s² when applied to an object with a mass of 1.0 kg. This unit of force is called the **newton** (N) in honor of Isaac Newton. Now that a unit of force has been defined, the equation for Newton's second law of motion can be written. First, rearrange

$$a \propto \frac{F}{m} \text{ to } F \propto ma$$

Then replace the proportionality with the more explicit equal sign, and Newton's second law becomes

$$F = ma$$

equation 3.1

The newton unit of force is a derived unit that is based on the three fundamental units of mass, length, and time. This is readily observed if the units are placed in the second-law equation

$$F = ma$$

$$1\text{ N} = (1\text{ kg})\left(1\frac{\text{m}}{\text{s}^2}\right)$$

or

$$1\text{ N} = 1\frac{\text{kg·m}}{\text{s}^2}$$

Until now, equations were used to *describe properties* of matter such as density, velocity, and acceleration. This is your first example of an equation that is used to *define a concept*, specifically the concept of what is meant by a force. Since the concept is defined by specifying a measurement procedure, it is also an example of an *operational definition*. You are not only told what a newton of force is but also how to go about measuring it. Notice that the newton is defined in terms of mass measured in kg and acceleration measured in m/s². Any other units must be

converted to kg and m/s² before a problem can be solved for newtons of force.

The difference between mass and weight can be confusing, since they are proportional to one another. If you double the mass of an object, for example, its weight is also doubled. But there is an important distinction between mass and weight and, in fact, they are different concepts. *Weight is a downward force*, the gravitational force acting on an object. Mass, on the other hand, refers to the amount of matter in the object and is *independent* of the force of gravity. Mass is a measure of inertia, the extent to which the object resists a change of motion. The force of gravity varies from place to place on the surface of the earth. It is slightly less, for example, in Colorado than in Florida. Imagine that you could be instantly transported from Florida to Colorado. You would find that you weigh less in Colorado than you did in Florida, even though the amount of matter in you has not changed. Now imagine that you could be instantly transported to the moon. You would weigh one-sixth as much on the moon because the force of gravity on the moon is one-sixth of that on the earth. Yet, your mass would be the same in both locations.

Weight is a measure of the force of gravity acting on an object, and this force can be calculated from Newton's second law of motion,

$$F = ma$$

or

downward force = (mass)(acceleration due to gravity)

or

$$\text{weight} = (\text{mass})(g)$$

$$\text{or } w = mg$$

equation 3.2

You learned in the previous chapter that g is the symbol used to represent acceleration due to gravity. Near the earth's

TABLE 3.1	Units of mass and weight in the metric and English systems of measurement				
	Mass	**×**	**Acceleration**	**=**	**Force**
Metric system	kg	×	$\dfrac{m}{s^2}$	=	$\dfrac{kg \cdot m}{s^2}$ (newton)
English system	$\left(\dfrac{lb}{ft/s^2}\right)$	×	$\dfrac{ft}{s^2}$	=	lb (pound)

surface, g has an average value of 9.8 m/s^2. To understand how g is applied to an object not moving, consider a ball you are holding in your hand. By supporting the weight of the ball you hold it stationary, so the upward force of your hand and the downward force of the ball (its weight) must add to a net force of zero. When you let go of the ball the gravitational force is the only force acting on the ball. The ball's weight is then the net force that accelerates it at g, the acceleration due to gravity. Thus, $F_{net} = w = ma = mg$. The weight of the ball never changes when near the surface of the earth, so its weight is always equal to $w = mg$, even if the ball is not accelerating.

In the metric system, *mass* is measured in kilograms. The acceleration due to gravity, g, is 9.8 m/s^2. According to equation 3.2, weight is mass times acceleration. A kilogram multiplied by an acceleration measured in m/s^2 results in kg·m/s^2, a unit you now recognize as a force called a newton. The *unit of weight* in the metric system is therefore the *newton* (N).

In the English system, the pound is the unit of *force*. The acceleration due to gravity, g, is 32 ft/s^2. The force unit of a pound is defined as the force required to accelerate a unit of mass called the *slug*. Specifically, a force of 1.0 lb will give a 1.0 slug mass an acceleration of 1.0 ft/s^2.

The important thing to remember is that *pounds* and *newtons* are units of *force*. A *kilogram*, on the other hand, is a measure of *mass*. Thus the English unit of 1.0 lb is comparable to the metric unit of 4.5 N (or 0.22 lb is equivalent to 1.0 N). Conversion tables sometimes show how to convert from pounds (a unit of weight) to kilograms (a unit of mass). This is possible because weight and mass are proportional on the surface of the earth. It can also be confusing, since some variables depend on weight and others depend on mass. To avoid confusion, it is important to remember the distinction between weight and mass and that a kilogram is a unit of mass. Newtons and pounds are units of force that can be used to measure weight (table 3.1).

Newton's Third Law of Motion

Newton's first law of motion states that an object retains its state of motion when the net force is zero. The second law states what happens when the net force is *not* zero, describing how an object with a known mass moves when a given force is applied. The two laws give one aspect of the concept of a force; that is, if you observe that an object starts moving, speeds up, slows down, or changes its direction of travel, you can conclude that an unbalanced force is acting on the object. Thus, any change in the state of motion of an object is *evidence* that an unbalanced force has been applied.

Newton's third law of motion is also concerned with forces and considers how a force is produced. First, consider where a force comes from. A force is always produced by the interaction of two or more objects. There is always a second object pushing or pulling on the first object to produce a force. To simplify the many interactions that occur on the earth, consider a satellite freely floating in space. According to Newton's second law ($F = ma$), a force must be applied to change the state of motion of the satellite. What is a possible source of such a force? Perhaps an astronaut pushes on the satellite for 1 second. The satellite would accelerate *during* the application of the force, then move away from the original position at some constant velocity. The astronaut would also move away from the original position, but in the opposite direction (figure 3.7). A *single* force *does not exist* by itself. There is always a matched and opposite force that occurs at the same time. Thus, the astronaut exerted a momentary force on the satellite, but the satellite evidently exerted a

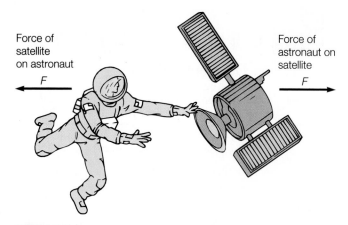

Force of satellite on astronaut

F

Force of astronaut on satellite

F

FIGURE 3.7

Forces occur in matched pairs that are equal in magnitude and opposite in direction.

momentary force back on the astronaut as well, for the astronaut moved away from the original position in the opposite direction. Newton did not have astronauts and satellites to think about, but this is the kind of reasoning he did when he concluded that forces always occur in matched pairs that are equal and opposite. Thus the **third law of motion** is as follows:

> **Whenever two objects interact, the force exerted on one object is equal in size and opposite in direction to the force exerted on the other object.**

The third law states that forces always occur in matched pairs that act in opposite directions and on two *different* bodies. You could express this law with symbols as

$$F_{\text{A due to B}} = F_{\text{B due to A}}$$

equation 3.3

where the force on the astronaut, for example, would be "A due to B," and the force on the satellite would be "B due to A."

Sometimes the third law of motion is expressed as follows: "For every action there is an equal and opposite reaction," but this can be misleading. Neither force is the cause of the other. The forces are at every instant the cause of each other and they appear and disappear at the same time. If you are going to describe the force exerted on a satellite by an astronaut, then you must realize that there is a simultaneous force exerted on the astronaut by the satellite. The forces (astronaut on satellite and satellite on astronaut) are equal in magnitude but opposite in direction.

Perhaps it would be more common to move a satellite with a small rocket. A satellite is maneuvered in space by firing a rocket in the direction opposite to the direction someone wants to move the satellite. Exhaust gases (or compressed gases) are accelerated in one direction, and the expelled gases exert an equal but opposite force on the satellite that accelerates it in the opposite direction. This is another example of the third law.

Consider how the pairs of forces work on the earth's surface. You walk by pushing your feet against the ground (figure 3.8). Of

FIGURE 3.8

The football player's foot is pushing against the ground, but it is the ground pushing against the foot that accelerates the player forward to catch a pass.

course you could not do this if it were not for friction. You would slide as on slippery ice without friction. But since friction does exist, you exert a backward horizontal force on the ground, and, as the third law explains, the ground exerts an equal and opposite force on you. You accelerate forward from the unbalanced force as explained by the second law. If the earth had the same mass as you, however, it would accelerate backward at the same rate that you were accelerated forward. The earth is much more massive than you, however, so any acceleration of the earth is a vanishingly small amount. The overall effect is that you are accelerated forward by the force the ground exerts on you.

Return now to the example of riding a bicycle that was discussed previously. What is the source of the *external* force that accelerates you and the bike? Pushing against the pedals is not external to you and the bike, so that force will *not* accelerate you and the bicycle forward. This force is transmitted through the bike mechanism to the rear tire, which pushes against the ground. It is the ground exerting an equal and opposite force against the system of you and the bike that accelerates you forward. You must

A CLOSER LOOK | The Greenhouse Effect or the Ozone Hole?

What is the greenhouse effect? Is it the same thing as the ozone hole? The greenhouse effect is the warming of the *lower part* of the atmosphere that is caused by certain gases that are sometimes referred to as the "greenhouse gases" (carbon dioxide, methane, nitrous oxide, and others). These gases absorb infrared radiation from the earth's surface, then reradiate it in all directions. The overall effect is that a significant part of the infrared energy is redirected back to the earth's surface, which increases the average temperatures.

The ozone hole problem occurs in the *upper part* of the atmosphere and concerns the destruction of ozone by certain hydrocarbon molecules that have the element chlorine and/or fluorine. Destruction of ozone is a concern because it normally absorbs part of the incoming ultraviolet radiation from the sun. Without the ozone layer, increased ultraviolet radiation can have adverse effects on the people, plants, and animals that are exposed to sunlight.

What can you do about the greenhouse effect and the ozone hole problem? Several clues can be found by comparing the yearly pollution emissions for an "average" passenger car with the emissions from an "average" light truck (including pickups, vans, minivans, and sport utility vehicles). How many solutions can you think of after comparing the following two tables?

Average passenger car

Pollutant Problem	Amount/Mile	Miles/Year	Pollution Contribution/Year
Hydrocarbons— urban ozone (smog) and air toxins	2.9 g/mi (about 0.0064 lb/mi)	12,500	36.3 kg of HC (about 80 lb)
Carbon monoxide— poisonous gas	22 g/mi (about 0.0485 lb/mi)	12,500	275 kg of CO (about 606 lb)
Nitrogen oxides— urban ozone (smog) and acid rain	1.5 g/mi (about 0.003 lb/mi)	12,500	18.8 kg of NO_x (about 41 lb)
Carbon dioxide— global warming	360 g/mi (about 0.8 lb/mi)	12,500	4,536 kg of CO_2 (about 10,000 lb)

Average light truck (pickups, minivans, sport utility vehicles)

Pollutant Problem	Amount/Mile	Miles/Year	Pollution Contribution/Year
Hydrocarbons— urban ozone (smog) and air toxins	3.7 g/mi (about 0.0082 lb/mi)	14,000	51.8 kg of HC (about 114 lb)
Carbon monoxide— poisonous gas	29 g/mi (about 0.0639 lb/mi)	14,000	406 kg of CO (about 895 lb)
Nitrogen oxides— urban ozone (smog) and acid rain	1.9 g/mi (about 0.0042 lb/mi)	14,000	26.6 kg of NO_x (about 59 lb)
Carbon dioxide— global warming	544 g/mi (about 1.2 lb/mi)	14,000	7,620 kg of CO_2 (about 16,800 lb)

Data from U.S. Environmental Protection Agency, National Vehicle and Fuel Emissions Laboratory, EPA420-F-97-037.

Notes:

1. These values are averages. Individual vehicles may travel more or less miles and may emit more or less pollution per mile than indicated here. Emission factors and pollution/fuel consumption totals may differ slightly from original sources due to rounding.

2. The emission factors used here come from standard EPA emission models. They assume an "average," properly maintained car or truck on the road in 1997, operating on typical gasoline on a summer day (72°–96°F). Emissions may be higher in very hot or very cold weather.

3. Average annual mileage source: EPA emissions model MOBILE5.

4. Fuel consumption is based on average in-use passenger car fuel economy of 22.5 miles per gallon and average in-use light truck fuel economy of 15.3 miles per gallon. Source: DOT/FHA, Highway Statistics 1995.

consider the forces that act on the system of you and the bike before you can apply $F = ma$. The only forces that will affect the forward motion of the bike system are the force of the ground pushing it forward and the frictional forces that oppose the forward motion. This is another example of the third law.

MOMENTUM

Sportscasters often refer to the *momentum* of a team, and newscasters sometimes refer to an election where one of the candidates has *momentum*. Both situations describe a competition where one side is moving toward victory and it is difficult to stop. It seems appropriate to borrow this term from the physical sciences because momentum is a property of movement. It takes a longer time to stop something from moving when it has a lot of momentum. The physical science concept of momentum is closely related to Newton's laws of motion. **Momentum** (p) is defined as the product of the mass (m) of an object and its velocity (v),

$$\text{momentum} = \text{mass} \times \text{velocity}$$

or

$$p = mv$$

equation 3.4

The astronaut in figure 3.9 has a mass of 60.0 kg and a velocity of 0.750 m/s as a result of the interaction with the satellite. The resulting momentum is therefore (60.0 kg) (0.750 m/s), or 45.0 kg·m/s. As you can see, the momentum would be greater if the astronaut had acquired a greater velocity or if the astronaut had a greater mass and acquired the same velocity. Momentum involves both the inertia and the velocity of a moving object.

Notice that the momentum acquired by the satellite in figure 3.9 is *also* 45.0 kg·m/s. The astronaut gained a certain momentum in one direction, and the satellite gained the *very same momentum in the opposite direction*. Newton originally defined the second law in terms of a rate of change of momentum being proportional to the net force acting on an object. Since the third law explains that the forces exerted on both the astronaut and satellite were equal and opposite, you would expect both objects to acquire equal momentum in the opposite direction. This result is observed any time objects in a system interact and the only forces involved are those between the interacting objects (figure 3.9). This statement leads to a particular kind of relationship called a *law of conservation*. In this case, the law applies to momentum and is called the **law of conservation of momentum:**

> **The total momentum of a group of interacting objects remains the same in the absence of external forces.**

Conservation of momentum, energy, and charge are among examples of conservation laws that apply to everyday situations. These situations always illustrate two understandings, that (1) each conservation law is an expression of symmetry that describes a physical principle that can be observed; and (2) each law holds regardless of the details of an interaction or how it

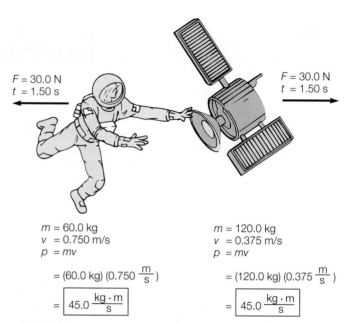

$$F = 30.0 \text{ N} \qquad F = 30.0 \text{ N}$$
$$t = 1.50 \text{ s} \qquad t = 1.50 \text{ s}$$

$$m = 60.0 \text{ kg} \qquad m = 120.0 \text{ kg}$$
$$v = 0.750 \text{ m/s} \qquad v = 0.375 \text{ m/s}$$
$$p = mv \qquad p = mv$$

$$= (60.0 \text{ kg}) (0.750 \tfrac{\text{m}}{\text{s}}) \qquad = (120.0 \text{ kg}) (0.375 \tfrac{\text{m}}{\text{s}})$$

$$= \boxed{45.0 \tfrac{\text{kg} \cdot \text{m}}{\text{s}}} \qquad = \boxed{45.0 \tfrac{\text{kg} \cdot \text{m}}{\text{s}}}$$

FIGURE 3.9

Both the astronaut and the satellite received a force of 30.0 N for 1.50 s when they pushed on each other. Both then have a momentum of 45.0 kg·m/s in the opposite direction. This is an example of the law of conservation of momentum.

took place. Since the conservation laws express symmetry that always occurs, they tell us what might be expected to happen, and what might be expected not to happen in a given situation. The symmetry also allows unknown quantities to be found by analysis. The law of conservation of momentum, for example, is useful in analyzing motion in simple systems of collisions such as those of billiard balls, automobiles, or railroad cars. It is also useful in measuring action and reaction interactions, as in rocket propulsion, where the backward momentum of the expelled gases equals the momentum given to the rocket in the opposite direction (figure 3.10). When this is done, momentum is always found to be conserved.

Compared to the other concepts of motion, there are two aspects of momentum that are unusual: (1) the symbol for momentum (p) does not give a clue about the quantity it represents, and (2) the combination of metric units that results from a momentum calculation (kg·m/s) does not have a name of its own. So far, you have been introduced to one combination of units with a name. The units kg·m/s^2 are known as a unit of force called the *newton* (N). More combinations of metric units, all with names of their own, will be introduced later.

FORCES AND CIRCULAR MOTION

Consider a communications satellite that is moving at a uniform speed around the earth in a circular orbit. According to the first law of motion there *must be* forces acting on the satellite, since it does *not* move off in a straight line. The second law of motion

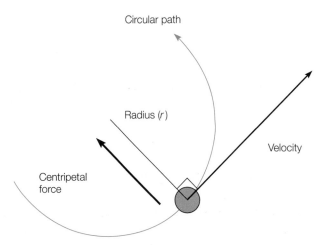

FIGURE 3.11

Centripetal force on the ball causes it to change direction continuously, or accelerate into a circular path. Without the unbalanced force acting on it, the ball would continue in a straight line.

FIGURE 3.10

According to the law of conservation of momentum, the momentum of the expelled gases in one direction equals the momentum of the rocket in the other direction in the absence of external forces.

also indicates forces, since an unbalanced force is required to change the motion of an object.

Recall that acceleration is defined as a change in velocity, and that velocity is a vector quantity, having both magnitude and direction. The vector quantity velocity is changed by a change in speed, direction, or both speed and direction. The satellite in a circular orbit is continuously being accelerated. This means that there is a continuously acting unbalanced force on the satellite that pulls it out of a straight-line path.

The force that pulls an object out of its straight-line path and into a circular path is called a **centripetal** (center-seeking) **force.** Perhaps you have swung a ball on the end of a string in a horizontal circle over your head. Once you have the ball moving, the only unbalanced force (other than gravity) acting on the ball is the centripetal force your hand exerts on the ball through the string. This centripetal force pulls the ball from its natural straight-line path into a circular path. There are no outward forces acting on the ball. The force that you feel on the string is a consequence of the third law; the ball exerts an equal and opposite force on your hand. If you were to release the string, the ball would move away from the circular path in a *straight line* that has a right angle to the radius at the point of release (figure 3.11). When you release the string, the centripetal force ceases, and the

ball then follows its natural straight-line motion. If other forces were involved, it would follow some other path. Nonetheless, the apparent outward force has been given a name just as if it were a real force. The outward tug is called a *centrifugal force.*

The magnitude of the centripetal force required to keep an object in a circular path depends on the inertia, or mass, of the object and the acceleration of the object, just as you learned in the second law of motion. The acceleration of an object moving in a circle can be shown by geometry or calculus to be directly proportional to the square of the speed around the circle (v^2) and inversely proportional to the radius of the circle (r). (A smaller radius requires a greater acceleration.) Therefore, the acceleration of an object moving in uniform circular motion (a_c) is

$$a_c = \frac{v^2}{r}$$

equation 3.5

The magnitude of the centripetal force of an object with a mass (m) that is moving with a velocity (v) in a circular orbit of a radius (r) can be found by substituting equation 3.5 in $F = ma$, or

$$F = \frac{mv^2}{r}$$

equation 3.6

NEWTON'S LAW OF GRAVITATION

You know that if you drop an object, it always falls to the floor. You define *down* as the direction of the object's movement and *up* as the opposite direction. Objects fall because of the force of gravity, which accelerates objects at $g = 9.8$ m/s^2 (32 ft/s^2) and gives them weight, $w = mg$.

CONNECTIONS

Circular Fun

Amusement park rides are designed to accelerate your body, sometimes producing changes in the acceleration (jerk) as well. This is done by changes in speed, changes in the direction of travel, or changes in both direction and speed. Many rides move in a circular path, since such movement is a constant acceleration.

Why do people enjoy amusement park rides? It is not the high speed, since your body is not very sensitive to moving at a constant speed. Moving at a steady 600 mi/hr in an airplane, for example, provides little sensation when you are seated in an aisle seat in the central cabin. You really need a visual comparison to realize you are traveling at a high speed.

Your body is not sensitive to high-speed traveling, but it is sensitive to acceleration and changes of acceleration. Accelera-

tion seems to produce an "adrenaline rush," whereas constant speed produces no response. Acceleration affects the fluid in your inner ear, which controls your sense of balance. In most people, acceleration also produces a reaction that results in the release of the hormones epinephrine and norepinephrine from the adrenal medulla, located near the kidney. The heart rate increases, blood pressure rises, blood is shunted to muscles, and the breathing rate increases. You have probably experienced this reaction many times in your life, as when you nearly have an automobile accident or slip and nearly fall. In the case of an amusement park ride, your body adapts and you believe you enjoy the experience.

Gravity is an attractive force, a pull that exists between all objects in the universe. It is a mutual force that, just like all other forces, comes in matched pairs. Since the earth attracts you with a certain force, you must attract the earth with an exact opposite force. The magnitude of this force of mutual attraction depends on several variables. These variables were first described by Newton in *Principia*, his famous book on motion that was printed in 1687. Newton had, however, worked out his ideas much earlier, by the age of twenty-four, along with ideas about his laws of motion and the formula for centripetal acceleration. In a biography written by a friend in 1752, Newton stated that the notion of gravitation came to mind during a time of thinking that "was occasioned by the fall of an apple." He was thinking about why the Moon stays in orbit around Earth rather than moving off in a straight line as would be predicted by the first law of motion. Perhaps the same force that attracts the moon toward the earth, he thought, attracts the apple to the earth. Newton developed a theoretical equation for gravitational force that explained not only the motion of the moon but the motion of the whole solar system. Today, this relationship is known as the **universal law of gravitation:**

> **Every object in the universe is attracted to every other object with a force that is directly proportional to the product of their masses and inversely proportional to the square of the distances between them.**

In symbols, m_1 and m_2 can be used to represent the masses of two objects, d the distance between their centers, and G a constant of proportionality. The equation for the law of universal gravitation is therefore

$$F = G\frac{m_1 m_2}{d^2}$$

equation 3.7

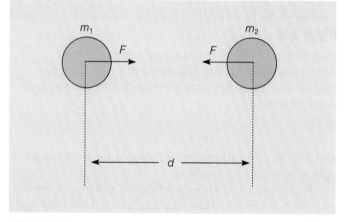

FIGURE 3.12

The variables involved in gravitational attraction. The force of attraction (F) is proportional to the product of the masses (m_1, m_2) and inversely proportional to the square of the distance (d) between the centers of the two masses.

This equation gives the magnitude of the attractive force that each object exerts on the other. The two forces are oppositely directed. The constant G is a universal constant, since the law applies to all objects in the universe. It was first measured experimentally by Henry Cavendish in 1798. The accepted value today is $G = 6.67 \times 10^{-11}$ N·m²/kg². Do not confuse G, the universal constant, with g, the acceleration due to gravity on the surface of the earth.

Thus, the magnitude of the force of gravitational attraction is determined by the mass of the two objects and the distance between them (figure 3.12). The law also states that *every* object

A CLOSER LOOK | Spin Down

Does water rotate one way in the Northern Hemisphere and the other way in the Southern Hemisphere as it drains from a bathtub? This effect is observed with a large wind system, which is moving air—a fluid. People sometimes wonder if the same effect occurs with any fluid that is moving—for example, the direction water moves as it drains from the bathtub when you pull the plug.

The reason that moving air turns different ways in the two hemispheres can be found in the observation that the earth is a rotating sphere. Since it has a spherical shape, all the surface of the earth does not turn at the same rate. As shown in box figure 3.1, the earth has a greater rotational velocity at the equator than at the poles. As air moves north or south, the surface below has a different rotational velocity so it rotates beneath the air as it moves in a straight line. This gives the moving air an apparent deflection to the right of the direction of movement in the Northern Hemisphere and to the left in the Southern Hemisphere. The apparent deflection caused by the earth's rotation is called the *Coriolis effect*. For example, air sinking in the center of a region of atmospheric high pressure produces winds that move outward. In the Northern Hemisphere, the Coriolis effect deflects this wind to the right, producing a clockwise circulation. In the Southern Hemisphere, the wind is deflected to the left, producing a counterclockwise circulation pattern.

Now, back to our much smaller system of water draining from a bathtub. There is a small Coriolis effect on the water that is moving the very short distance to a drain.

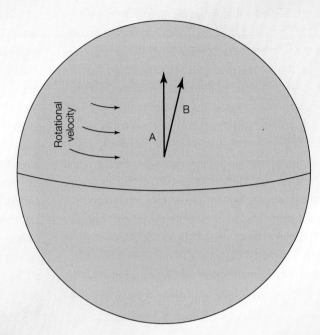

BOX FIGURE 3.1

The earth has a greater rotational velocity at the equator and less toward the poles. As air moves north or south (*A*), it passes over land with a different rotational velocity, which produces a deviation to the right in the Northern Hemisphere (*B*), and to the left in the Southern Hemisphere.

The overall result, however, is way too small to be a factor in determining what happens to the water as it leaves the tub. The perpendicular acceleration of the moving water in the tub is too small to influence which way the water turns. You might want to try a carefully controlled experiment to see for yourself. You will need to eliminate all the possible variables that might influence which way the draining water turns. Consider, for example, absolutely still water, with no noise or other sources of vibrations, and then carefully remove the plug in a way not to disturb the water.

You cannot use the direction water circulates when you flush a toilet for evidence that the Coriolis effect turns the water one way or the other. The water in a flushing toilet rotates in the direction the incoming water pipe is pointing, which has nothing to do with the Coriolis effect in either hemisphere.

CONNECTIONS

Antigravity

Isaac Newton published *Philosophiae Naturalis Principia Mathematica* in 1687, describing the basic ideas and laws of motion that are still widely used today, some 300 years after they were first published. These ideas and laws form the foundation for most of the physics studied in introductory courses, and this foundation is generally called *Newtonian mechanics* or *classical mechanics*. Newtonian mechanics is more than adequate for understanding everyday, ordinary objects that move at ordinary speeds in our everyday world. Today, scientists understand and respect the limits of usefulness of Newtonian mechanics, understanding that different models are needed for things and events that are outside these limits. For example, the model known as *quantum mechanics* is used when dealing with atom-sized or smaller objects. The model known as *general relativity* is used for the solar system or larger objects. Finally, *special relativity* is the model of choice when dealing with objects with very high speeds.

Is it possible to "turn off" or "shield" gravity by use of an "antigravity" machine? Newtonian mechanics considers gravity to be an interaction that takes place instantaneously between two masses. An antigravity machine would violate Newton's law of gravity. It would also provide a perpetual motion machine, thereby violating the principle of conservation of energy.

And what about the other models of how things work? Perhaps there are quantum-gravitational effects that would permit a repulsive gravity-like force, but such speculation requires the existence of a strange material whose existence would violate current understandings about the properties of matter. The general theory of relativity describes gravity as a curvature of space-time caused by the presence of mass (or energy). This curvature cannot be "turned off" by imposing additional forces or materials. So, it appears that to counter gravity we will have to continue doing it the old fashioned way—using energy and the other forces at our disposal to run machines that work against gravity.

---

is attracted to every other object. You are attracted to all the objects around you—chairs, tables, other people, and so forth. Why don't you notice the forces between you and other objects? One or both of the interacting objects must be quite massive before a noticeable force results from the interaction. That is why you do not notice the force of gravitational attraction between you and objects that are not very massive compared to the earth. The attraction between you and the earth overwhelmingly predominates, and that is all you notice.

Newton was able to show that the distance used in the equation is the distance from the center of one object to the center of the second object. This does not mean that the force originates at the center, but that the overall effect is the same as if you considered all the mass to be concentrated at a center point. The weight of an object, for example, can be calculated by using a form of Newton's second law, $F = ma$. This general law shows a relationship between *any* force acting *on* a body, the mass of a body, and the resulting acceleration. When the acceleration is due to gravity, the equation becomes $F = mg$. The law of gravitation deals *specifically with the force of gravity* and how it varies with distance and mass. Since weight is a force, then $F = mg$. You can write the two equations together,

$$mg = G\frac{mm_e}{d^2}$$

where m is the mass of some object on the earth, m_e is the mass of the earth, g is the acceleration due to gravity, and d is the distance between the centers of the masses. Canceling the m's in the equation leaves

$$g = G\frac{m_e}{d^2},$$

which tells you that on the surface of the earth the acceleration due to gravity, 9.8 m/s², is a constant because the other two variables (mass of the earth and the distance to the center of the earth) are constant. Since the m's canceled, you also know that the mass of an object does not affect the rate of free fall; all objects fall at the same rate, with the same acceleration, no matter what their masses are.

The acceleration due to gravity, g, is about 9.8 m/s² and is practically a constant for relatively short distances above the surface. Notice, however, that Newton's law of gravitation is an inverse square law. This means if you double the distance, the force is 1/(2)² or 1/4 as great. If you triple the distance, the force is 1/(3)² or 1/9 as great. In other words, the force of gravitational attraction and g decrease inversely with the square of the distance from the earth's center. The weight of an object and the value of g are shown for several distances in figure 3.13. If you have the time, a good calculator, and the inclination, you could check the values given in figure 3.13 for a 70.0 kg person. In fact, you could even calculate the mass of the earth, since you already have the value of g.

Newton was able to calculate the acceleration of the Moon toward Earth, about 0.0027 m/s². The Moon "falls" toward Earth

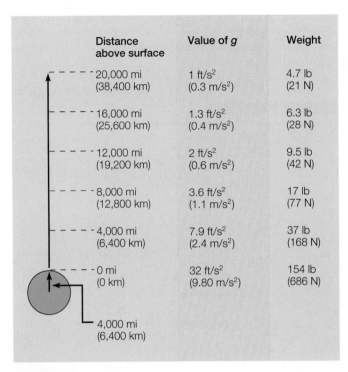

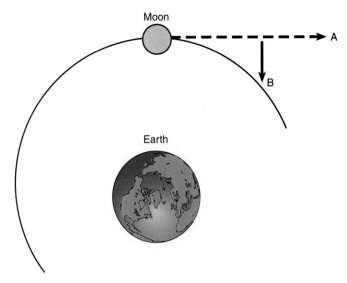

Distance above surface	Value of *g*	Weight
20,000 mi (38,400 km)	1 ft/s² (0.3 m/s²)	4.7 lb (21 N)
16,000 mi (25,600 km)	1.3 ft/s² (0.4 m/s²)	6.3 lb (28 N)
12,000 mi (19,200 km)	2 ft/s² (0.6 m/s²)	9.5 lb (42 N)
8,000 mi (12,800 km)	3.6 ft/s² (1.1 m/s²)	17 lb (77 N)
4,000 mi (6,400 km)	7.9 ft/s² (2.4 m/s²)	37 lb (168 N)
0 mi (0 km)	32 ft/s² (9.80 m/s²)	154 lb (686 N)

4,000 mi (6,400 km)

FIGURE 3.14

Gravitational attraction acts as a centripetal force that keeps the Moon from following the straight-line path shown by the dashed line to position A. It was pulled to position B by gravity (0.0027 m/s²) and thus "fell" toward Earth the distance from the dashed line to B, resulting in a somewhat circular path.

FIGURE 3.13

The force of gravitational attraction decreases inversely with the square of the distance from the earth's center. Note the weight of a 70.0 kg person at various distances above the earth's surface.

C O N N E C T I O N S

Equal Gravity?

As Earth moves in its orbit, the Sun appears to be north or south of the equator. However, there are two times when it is directly over the equator. At these times, daylight and night are of equal length and are called the *equinoxes* after the Latin term meaning "equal nights." The spring equinox (also called the vernal equinox) occurs on about March 21 and identifies the beginning of the spring season. The autumnal equinox occurs on about September 23 and identifies the beginning of the fall season.

Now, sometimes you might hear that eggs balance only on the equinox and fall over at other times. In fact, around the time of an equinox you might see a newscast of kids trying to balance eggs. The voice-over might suggest that the eggs balance on this day because gravity is "balanced" when the Sun is over Earth's equator.

Does this make sense, or does it sound like pseudoscience? Recall that pseudoscience takes on the flavor of science but is not supportable as valid or reliable. The focus in the typical news report about balancing eggs on the equinox seems to be on the eggs that *do* balance rather than the fact that some eggs balance, and some do not. In reality, no more or no less eggs balance on an equinox than they do on any other day.

Some things to think about: Would the eggs balance at the instant the Sun is over the equator, or all day? Would eggs balance the same at all latitudes? How would you know for sure that eggs balanced better on an equinox than any other day of the year?

because it is accelerated by the force of gravitational attraction. This attraction acts as a *centripetal force* that keeps the Moon from following a straight-line path as would be predicted from the first law. Thus, the acceleration of the Moon keeps it in a somewhat circular orbit around Earth. Figure 3.14 shows that the Moon would be in position A if it followed a straight-line path instead of "falling" to position B as it does. The Moon thus "falls" around Earth. Newton was able to analyze the motion of the Moon quantitatively as evidence that it is gravitational force that keeps the Moon in its orbit. The law of gravitation was extended to the Sun, other planets, and eventually the universe. The quantitative predictions of observed relationships among the planets were strong evidence that all objects obey the same law of gravitation. In addition, the law provided a means to calculate the mass of Earth, the Moon, the planets, and the Sun. Newton's law of gravitation, laws of motion, and work with mathematics formed the basis of most physics and technology for the next two centuries, as well as accurately describing the world of everyday experience.

SUMMARY

Isaac Newton developed a complete explanation of motion with three laws of motion. The laws explain the role of a *force* and the *mass* of an object involved in a *change of motion.*

Newton's *first law of motion* is concerned with the motion of an object and the *lack* of an unbalanced force. Also known as the *law of inertia,* the first law states that an object will retain its state of straight-line motion (or state of rest) unless an unbalanced force acts on it. The amount of resistance to a change of motion, *inertia,* describes the *mass* of an object.

The *second law of motion* describes a relationship between *net force, mass,* and *acceleration.* The relationship is $F = ma$. A *newton* of force is the force needed to give a 1.0 kg mass an acceleration of 1.0 m/s². *Weight* is the downward force that results from the earth's gravity acting on the mass of an object. Weight can be calculated from $w = mg$, a special case of $F = ma$. Weight is measured in *newtons* in the metric system and *pounds* in the English system.

Newton's *third law of motion* states that forces are produced by the interaction of *two different* objects and that these forces *always* occur in *matched pairs* that are *equal in size* and *opposite in direction.* These forces are capable of producing an acceleration in accord with the second law of motion.

Momentum is the product of the mass of an object and its velocity. In the absence of external forces, the momentum of a group of interacting objects always remains the same. This relationship is the *law of conservation of momentum.*

An object moving in a circular path must have a force acting on it, since it does not move off in a straight line. The force that pulls an object out of its straight-line path is called a *centripetal force.* The centripetal force needed to keep an object in a circular path depends on the mass (m) of the object, its velocity (v), and the radius of the circle (r), or

$$F = \frac{mv^2}{r}$$

The *universal law of gravitation* is a relationship between the masses of two objects, the distance between the objects, and a proportionality constant. The relationship is

$$F = G\frac{m_1 m_2}{d^2}$$

Newton was able to use this relationship to show that gravitational attraction provides the centripetal force that keeps the Moon in its orbit. This relationship was found to explain the relationship between all parts of the solar system.

Summary of Equations

3.1
$$\text{force} = \text{mass} \times \text{acceleration}$$
$$F = ma$$

3.2
$$\text{weight} = \text{mass} \times \text{acceleration due to gravity}$$
$$w = mg$$

3.3
$$\text{force on object A} = \text{force on object B}$$
$$F_{\text{A due to B}} = F_{\text{B due to A}}$$

3.4
$$\text{momentum} = \text{mass} \times \text{velocity}$$
$$p = mv$$

3.5
$$\text{centripetal acceleration} = \frac{\text{velocity squared}}{\text{radius of circle}}$$
$$a_c = \frac{v^2}{r}$$

3.6
$$\text{centripetal force} = \frac{\text{mass} \times \text{velocity squared}}{\text{radius of circle}}$$
$$F = \frac{mv^2}{r}$$

3.7
$$\text{gravitational force} = \text{constant} \times \frac{\text{one mass} \times \text{another mass}}{\text{distance squared}}$$
$$F = G\frac{m_1 m_2}{d^2}$$

KEY TERMS

centripetal force (p. **57**)

first law of motion (p. **48**)

law of conservation of momentum (p. **56**)

mass (p. **49**)

momentum (p. **56**)

newton (p. **52**)

second law of motion (p. **51**)

third law of motion (p. **54**)

universal law of gravitation (p. **58**)

A CLOSER LOOK | Space Station Weightlessness

When do astronauts experience weightlessness, or "zero gravity"? Theoretically, the gravitational field of Earth extends to the whole universe. You know that it extends to the Moon, and indeed, even to the Sun some 93 million miles away. There is a distance, however, at which the gravitational force must become immeasurably small. But even at an altitude of 20,000 miles above the surface of Earth, gravity is measurable. At 20,000 miles, the value of g is about 1 ft/s^2 (0.3 m/s^2) compared to 32 ft/s^2 (9.8 m/s^2) on the surface. Since gravity does exist at these distances, how can an astronaut experience "zero gravity"?

Gravity does act on astronauts in spacecraft that are in orbit around Earth. The spacecraft stays in orbit, in fact, because of the gravitational attraction and because it has the correct tangential speed. If the tangential speed were less than 5 mi/s, the spacecraft would return to Earth. Astronauts fire their retro-rockets, which slow the tangential speed, causing the spacecraft to fall down to Earth. If the tangential speed were more than 7 mi/s, the spacecraft would fly off into space. The spacecraft stays in orbit because it has the right tangential speed to continuously "fall" around and around Earth. Gravity provides the necessary centripetal force that causes the spacecraft to fall out of its natural straight-line motion.

Since gravity is acting on the astronaut and spacecraft, the term *zero gravity* is not an accurate description of what is happening. The astronaut, spacecraft, and everything in it are experiencing *apparent weightlessness* because they are continuously falling toward Earth (box figure 3.2). Everything seems to float because everything is falling together. But, strictly speaking, everything still has weight, because weight is defined as a gravitational force acting on an object ($w = mg$).

Whether weightlessness is apparent or real, however, the effects on people are the same. Long-term orbital flights have provided evidence that the human body changes from the effect of weightlessness. Bones lose calcium and minerals, the heart shrinks to a much smaller size, and leg muscles shrink so much on prolonged flights that astronauts cannot walk when they return to Earth. These and other problems resulting from prolonged weightlessness must be worked out before long-term weightless flights can take place. One solution to these problems might be a large, uniformly spinning spacecraft. The astronauts tend to move in a straight line, and the side of the turning spacecraft (now the "floor") exerts a force on them to make them go in a curved path. This force would act as an artificial gravity.

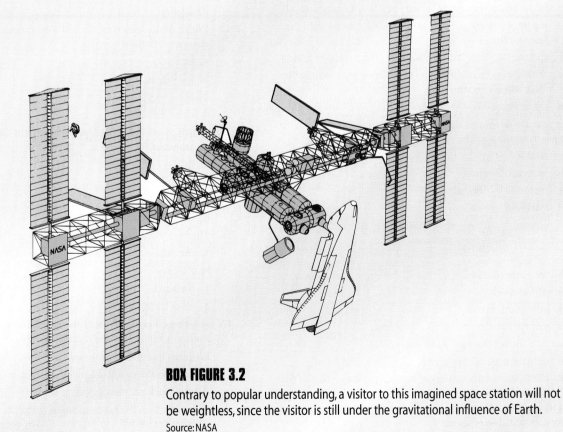

BOX FIGURE 3.2

Contrary to popular understanding, a visitor to this imagined space station will not be weightless, since the visitor is still under the gravitational influence of Earth.

Source: NASA

APPLYING THE CONCEPTS

1. The law of inertia is another name for Newton's
   a. first law of motion.
   b. second law of motion.
   c. third law of motion.
   d. none of the above.

2. The extent of resistance to a change of motion is determined by an object's
   a. weight.
   b. mass.
   c. density.
   d. all of the above.

3. Mass is
   a. a measure of inertia.
   b. a measure of how difficult it is to stop a moving object.
   c. a measure of how difficult it is to change the direction of travel of a moving object.
   d. all of the above.

4. A change in the state of motion is evidence of
   a. a force.
   b. an unbalanced force.
   c. a force that has been worn out after an earlier application.
   d. any of the above.

5. Considering the forces on the system of you and a bicycle as you pedal the bike at a constant velocity in a horizontal straight line,
   a. the force you are exerting on the pedal is greater than the resisting forces.
   b. all forces are in balance, with the net force equal to zero.
   c. the resisting forces of air and tire friction are less than the force you are exerting.
   d. the resisting forces are greater than the force you are exerting.

6. If you double the unbalanced force on an object of a given mass, the acceleration will be
   a. doubled.
   b. increased fourfold.
   c. increased by one-half.
   d. increased by one-fourth.

7. If you double the mass of a cart while it is undergoing a constant unbalanced force, the acceleration will be
   a. doubled.
   b. increased fourfold.
   c. half as much.
   d. one-fourth as much.

8. The acceleration of any object depends on
   a. only the net force acting on the object.
   b. only the mass of the object.
   c. both the net force acting on the object and the mass of the object.
   d. neither the net force acting on the object nor the mass of the object.

9. Which of the following is a measure of mass?
   a. pound
   b. newton
   c. kilogram
   d. all of the above.

10. Which of the following is a correct unit for a measure of weight?
    a. kilogram
    b. newton
    c. kg·m/s
    d. none of the above.

11. From the equation of $F = ma$, you know that m is equal to
    a. a/F
    b. Fa
    c. F/a
    d. $F - a$

12. A newton of force has what combination of units?
    a. kg·m/s
    b. kg/m^2
    c. kg/m^2/s^2
    d. kg·m/s^2

13. Which of the following is *not* a unit or combination of units for a downward force?
    a. lb
    b. kg
    c. N
    d. kg·m/s^2

14. Which of the following is a unit for a measure of resistance to a change of motion?
    a. lb
    b. kg
    c. N
    d. all of the above

15. Which of the following represents *mass* in the English system of measurement?
    a. lb
    b. lb/ft/s^2
    c. kg
    d. none of the above

16. The force that accelerates a car over a road comes from
    a. the engine.
    b. the tires.
    c. the road.
    d. all of the above.

17. Ignoring all external forces, the momentum of the exhaust gases from a flying jet airplane that was initially at rest is
    a. much less than the momentum of the airplane.
    b. equal to the momentum of the airplane.
    c. somewhat greater than the momentum of the airplane.
    d. greater than the momentum of the airplane when the aircraft is accelerating.

18. Doubling the distance between the center of an orbiting satellite and the center of the earth will result in what change in the gravitational attraction of the earth for the satellite?

 a. one-half as much

 b. one-fourth as much

 c. twice as much

 d. four times as much

19. If a ball swinging in a circle on a string is moved twice as fast, the force on the string will be

 a. twice as great.

 b. four times as great.

 c. one-half as much.

 d. one-fourth as much.

20. A ball is swinging in a circle on a string when the string length is doubled. At the same velocity, the force on the string will be

 a. twice as great.

 b. four times as great.

 c. one-half as much.

 d. one-fourth as much.

Answers

1. a **2.** b **3.** d **4.** b **5.** b **6.** a **7.** c **8.** c **9.** c **10.** b **11.** c **12.** d **13.** b **14.** b **15.** b **16.** c **17.** b **18.** b **19.** b **20.** c

QUESTIONS FOR THOUGHT

1. Is it possible for a small car to have the same momentum as a large truck? Explain.
2. What net force is needed to maintain the constant velocity of a car moving in a straight line? Explain.
3. How can there ever be an unbalanced force on an object if every action has an equal and opposite reaction?
4. Compare the force of gravity on an object at rest, the same object in free fall, and the same object moving horizontally to the surface of the earth.
5. Is it possible for your weight to change as your mass remains constant? Explain.
6. What maintains the *speed* of Earth as it moves in its orbit around the Sun?
7. Suppose you are standing on the ice of a frozen lake and there is no friction whatsoever. How can you get off the ice? (*Hint:* Friction is necessary to crawl or walk, so that will not get you off the ice.)
8. A rocket blasts off from a platform on a space station. An identical rocket blasts off from free space. Considering everything else to be equal, will the two rockets have the same acceleration? Explain.
9. An astronaut leaves a spaceship, moving through free space to adjust an antenna. Will the spaceship move off and leave the astronaut behind? Explain.
10. Is a constant force necessary for a constant acceleration? Explain.
11. Is an unbalanced force necessary to maintain a constant speed? Explain.
12. Use Newton's laws of motion to explain why water leaves the wet clothes when a washer is in the spin cycle.

PARALLEL EXERCISES

The exercises in groups A and B cover the same concepts. Solutions to group A exercises are located in appendix D.

Note: Neglect all frictional forces in all exercises.

Group A

1. What force is needed to give a 40.0 kg grocery cart an acceleration of 2.4 m/s^2?
2. What is the resulting acceleration when an unbalanced force of 100 N is applied to a 5 kg object?
3. What force is needed to accelerate a 1,000 kg car from 72 to 108 km/hr over a time period of 5 s.
4. An unbalanced force of 18 N is needed to give an object an acceleration of 3 m/s^2. What force is needed to give this very same object an acceleration of 10 m/s^2?
5. A rocket pack with a thrust of 100 N accelerates a weightless astronaut at 0.5 m/s^2 through free space. What is the mass of the astronaut and equipment?
6. A 1,500 kg car is to be pulled across level ground by a towline from another car. If the pulling car accelerates at 2 m/s^2, what force (tension) must the towline support without breaking?
7. What is the gravitational force Earth exerts on the Moon if Earth has a mass of 5.98×10^{24} kg, the Moon has a mass of 7.36×10^{22} kg, and the Moon is on the average a center-to-center distance of 3.84×10^8 m from Earth?

Group B

1. How much would an 80.0 kg person weigh (a) on Mars, where the acceleration of gravity is 3.93 m/s^2 and (b) on Earth's Moon, where the acceleration of gravity is 1.63 m/s^2?
2. What force would an asphalt road have to give a 6,000 kg truck in order to accelerate it at 2.2 m/s^2 over a level road?
3. Find the resulting acceleration from a 300 N force acting on an object with a mass of 3,000 kg.
4. How much time would be required to stop a 2,000 kg car that is moving at 80.0 km/hr if the braking force is 8,000 N?
5. If a space probe weighs 39,200 N on the surface of Earth, what will be the mass of the probe on the surface of Mars?
6. On Earth, an astronaut and equipment weigh 1,960 N. Weightless in space, the motionless astronaut and equipment are accelerated by a rocket pack with a 100 N thruster that fires for 2 s. What is the resulting final velocity?
7. What force is applied to the seatback of an 80.0 kg passenger on a jet plane that is accelerating at 5 m/s^2? How does this force compare to the weight of the person?

8. What is the acceleration of gravity at an altitude of 500 kilometers above the earth's surface?

9. What would a student who weighs 591 N on the surface of the earth weigh at an altitude of 1,100 kilometers from the surface?

10. What is the momentum of a 100 kg football player who is moving at 6 m/s?

11. A car weighing 13,720 N is speeding down a highway with a velocity of 91 km/hr. What is the momentum of this car?

12. Car A has a mass of 1,200 kg and is driving north at 25 m/s when it collides head-on with car B, which has a 2,200 kg mass and is moving at 15 m/s. If the collision is exactly head-on and the cars lock together, which way will the wreckage move?

13. A 15 g bullet is fired with a velocity of 200 m/s from a 6 kg rifle. What is the recoil velocity of the rifle?

14. An astronaut and equipment weigh 2,156 N on Earth. Weightless in space, the astronaut throws away a 5.0 kg wrench with a velocity of 5.0 m/s. What is the resulting velocity of the astronaut in the opposite direction?

15. A student and her boat have a combined mass of 100.0 kg. Standing in the motionless boat in calm water, she tosses a 5.0 kg rock out the back of the boat with a velocity of 5.0 m/s. What will be the resulting speed of the boat?

16. (a) What is the weight of a 1.25 kg book? (b) What is the acceleration when a net force of 10.0 N is applied to the book?

17. What net force is needed to accelerate a 1.25 kg book 5.00 m/s²?

18. What net force does the road exert on a 70.0 kg bicycle and rider to give them an acceleration of 2.0 m/s²?

19. A 1,500 kg car accelerates uniformly from 44.0 km/hr to 80.0 km/hr in 10.0 s. What was the net force exerted on the car?

20. A net force of 5,000.0 N accelerates a car from rest to 90.0 km/hr in 5.0 s. (a) What is the mass of the car? (b) What is the weight of the car?

21. What is the weight of a 70.0 kg person?

22. A 1,000.0 kg car at rest experiences a net force of 1,000.0 N for 10.0 s. What is the final speed of the car?

23. What is the momentum of a 50 kg person walking at a speed of 2 m/s?

24. How much centripetal force is needed to keep a 0.20 kg ball on a 1.50 m string moving in a circular path with a speed of 3.0 m/s?

25. What is the velocity of a 100.0 g ball on a 50.0 cm string moving in a horizontal circle that requires a centripetal force of 1.0 N?

26. A 1,000.0 kg car moves around a curve with a 20.0 m radius with a velocity of 10.0 m/s. (a) What centripetal force is required? (b) What is the source of this force?

27. On Earth, an astronaut and equipment weigh 1,960.0 N. While weightless in space, the astronaut fires a 100 N rocket backpack for 2.0 s. What is the resulting velocity of the astronaut and equipment?

8. Calculations show that acceleration due to gravity is 1.63 m/s² at the surface of the Moon, and the distance to the center of the Moon is 1.74×10^6 m. Use these figures to calculate the mass of the Moon.

9. An astronaut and space suit weigh 1,960 N on Earth. What weight would they have on the Moon, where gravity has a value of 1.63 m/s²?

10. An astronaut and space suit weigh 1,960 N on Earth. What weight would they have on the planet Mars, where gravity has a value of 3.92 m/s²?

11. To what altitude above Earth's surface would you need to travel to have half your weight?

12. What is the momentum of a 30.0 kg shell fired from a cannon with a velocity of 500 m/s?

13. What is the momentum of a 39.2 N bowling ball with a velocity of 7.00 m/s?

14. A 39.2 N bowling ball moving at 7.00 m/s collides head-on with a 29.4 N bowling ball that is moving at 9.33 m/s. Which way will the balls move after the impact?

15. A 30.0 kg shell is fired from a 2,000 kg cannon with a velocity of 500 m/s. What is the resulting velocity of the cannon?

16. An 80.0 kg man is standing on a frictionless ice surface when he throws a 4.00 kg book at 20.0 m/s. With what velocity does the man move across the ice?

17. A person with a mass of 70.0 kg steps horizontally from the side of a 300.0 kg boat at 2.00 m/s. What happens to the boat?

18. How fast would a 1,000 kg small car have to drive to have the same momentum as a 5,500 kg truck moving at 10.0 m/s?

19. (a) What is the weight of a 5.00 kg backpack? (b) What is the acceleration of the backpack if a net force of 10.0 N is applied?

20. What net force is required to accelerate a 20.0 kg object to 10.0 m/s²?

21. What forward force must the ground apply to the foot of a 60.0 kg person to result in an acceleration of 1.00 m/s²?

22. A 1,000.0 kg car accelerates uniformly to double its speed from 36.0 km/hr in 5.00 s. What net force acted on this car?

23. A net force of 3,000.0 N accelerates a car from rest to 36.0 km/hr in 5.00 s. (a) What is the mass of the car? (b) What is the weight of the car?

24. How much does a 60.0 kg person weigh?

25. A 60.0 g tennis ball is struck by a racket with a force of 425.0 N for 0.01 s. What is the speed of the tennis ball in km/hr as a result?

26. Compare the momentum of (a) a 2,000.0 kg car moving at 25.0 m/s and (b) a 1,250 kg car moving at 40.0 m/s.

27. What tension must a 50.0 cm length of string support in order to whirl an attached 1,000.0 g stone in a circular path at 5.00 m/s?

28. What is the maximum speed at which a 1,000.0 kg car can move around a curve with a radius of 30.0 m if the tires provide a maximum frictional force of 2,700.0 N?

29. How much centripetal force is needed to keep a 60.0 kg person and skateboard moving at 6.0 m/s in a circle with a 10.0 m radius?

30. A 200.0 kg astronaut and equipment move with a velocity of 2.00 m/s toward an orbiting spacecraft. How long will the astronaut need to fire a 100.0 N rocket backpack to stop the motion relative to the spacecraft?

The wind can be used as a source of energy. All you need is a way to capture the energy—such as these wind turbines in California—and to live someplace where the wind blows enough to make it worthwhile.

CHAPTER | Four

Energy

The term *energy* is closely associated with the concepts of force and motion. Naturally moving matter, such as the wind or moving water, exerts forces. You have felt these forces if you have ever tried to walk against a strong wind or stand in one place in a stream of rapidly moving water. The motion and forces of moving air and moving water are used as *energy sources* (figure 4.1). The wind is an energy source as it moves the blades of a windmill performing useful work. Moving water is an energy source as it forces the blades of a water turbine to spin, turning an electric generator. Thus, moving matter exerts a force on objects in its path, and objects moved by the force can also be used as an energy source.

Matter does not have to be moving to supply energy; matter *contains* energy. Food supplied the energy for the muscular exertion of the humans and animals that accomplished most of the work before this century. Today, machines do the work that was formerly accomplished by muscular exertion. Machines also use the energy contained in matter. They use gasoline, for example, as they supply the forces and motion to accomplish work.

Moving matter and matter that contains energy can be used as energy sources to perform work. The concepts of work and energy and the relationship to matter are the topics of this chapter. You will learn how energy flows in and out of your surroundings as well as a broad, conceptual view of energy.

WORK

You learned earlier that the term *force* has a special meaning in science that is different from your everyday concept of force. In everyday use, you use the term in a variety of associations such as police force, economic force, or the force of an argument. Earlier, force was discussed in a general way as a push or pull. Then a more precise scientific definition of force was developed from Newton's laws of motion—a force is a result of an interaction that is capable of changing the state of motion of an object.

The word *work* represents another one of those concepts that has a special meaning in science that is different from your everyday concept. In everyday use, work is associated with a task to be accomplished or the time spent in performing the task. You might work at understanding science, for example, or you might tell someone that science is a lot of work. You also probably associate physical work, such as lifting or moving boxes, with how tired you become from the effort. The scientific definition of work is not concerned with tasks, time, or how tired you become from doing a task. It is concerned with the application of a force to an object and the distance the object moves as a result of the force. The **work** done on the object is defined as *the magnitude of the applied force multiplied by the parallel distance through which the force acts:*

$$\text{work} = \text{force} \times \text{distance}$$
$$W = Fd$$

equation 4.1

Work, in the scientific sense, is the product of a force and the distance an object moves as a result of the force. There are two important considerations to remember about this definition: (1) something *must move* whenever work is done, and (2) the movement must be in the *same direction* as the direction of the force. When you move a book to a higher shelf in a bookcase you are doing work on the book. You apply a vertically upward force equal to the weight of the book as you move it in the same direction as the direction of the applied force. The work done on the book can therefore be calculated by multiplying the weight of the book by the distance it was moved (figure 4.2).

If you simply stand there holding the book, however, you are doing no work on the book. Your arm may become tired from holding the book, since you must apply a vertically upward force equal to the weight of the book. But this force is not acting through a distance, since the book is not moving. According to equation 4.1, a distance of zero results in zero work (figure 4.3). Only a force that results in motion in the same direction results in work.

Suppose you walk across the room while holding the book. You are exerting a vertically upward force equal to the weight of the book as before, but the direction of movement is perpendicular (90°) to the upward force on the book. Since the movement is perpendicular to the direction of the applied force, no work is done on the book. In this case there is no *relationship* between the applied force and the direction of movement. Since the upward force has nothing to do with the horizontal movement, no work is done by this *particular* force acting on the book (figure 4.4). This may seem odd at this point, but it will make sense after you learn some concepts of energy.

The applied force does not have to be exactly parallel to the direction of movement. A force is a vector that can be resolved into the component force that acts in the same direction as the movement (figure 4.5). The force vector, however, cannot be perpendicular (90°) to the direction of movement. This is true because both the force and the displacement are

FIGURE 4.1

This is Glen Canyon Dam on the Colorado River between Utah and Arizona. The dam is among the tallest in America, so you can imagine the tremendous pressure on the water as it moves through penstocks to generators at the bottom of the dam.

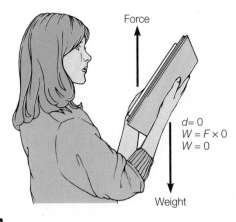

FIGURE 4.3

It is true that a force is exerted simply to hold a book, but the book does not move through a distance. Therefore the distance moved is zero, and the work accomplished is also zero.

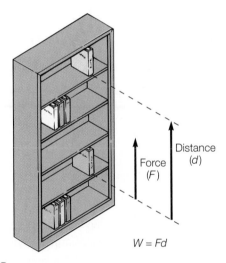

FIGURE 4.2

The force on the book moves it through a vertical distance from the second shelf to the fifth shelf, and work is done, $W = Fd$.

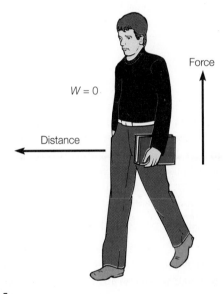

FIGURE 4.4

If the direction of movement (the distance moved) is perpendicular to the direction of the force, no work is done. This person is doing no work by carrying a book across a room.

vector quantities. Additional mathematics, which we will not go into, would show that the component of a vector in a direction perpendicular to its own is zero.

Units of Work

The units of work are defined by the definition of work, $W = Fd$. In the metric system a force is measured in newtons (N), and distance is measured in meters (m), so the unit of work is

$$W = Fd$$

$$W = (\text{newton})(\text{meter})$$

$$W = (\text{N})(\text{m})$$

The newton-meter is therefore the unit of work. This derived unit has a name. The newton-meter is called a **joule** (J) (pronounced "jool").

$$1 \text{ joule} = 1 \text{ newton-meter}$$

The units for a newton are kg·m/s^2, and the unit for a meter is m. It therefore follows that the units for a joule are $\text{kg·m}^2/\text{s}^2$.

In the English system, the force is measured in pounds (lb), and the distance is measured in feet (ft). The unit of work in the English system is therefore the ft·lb. The ft·lb does not have a name of its own as the N·m does (figure 4.6). (Note that although the equation is $W = Fd$, and this means = (pounds)(feet), the unit is called the ft·lb.)

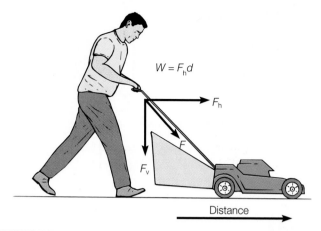

$$W = F_h d$$

F_h

F

F_v

Distance

FIGURE 4.5

A force at some angle (not 90°) to the direction of movement can be resolved into the horizontal component to calculate the work done.

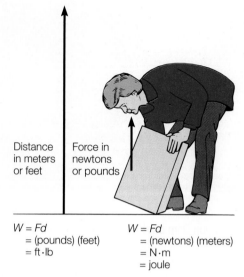

Distance in meters or feet

Force in newtons or pounds

$W = Fd$
= (pounds) (feet)
= ft·lb

$W = Fd$
= (newtons) (meters)
= N·m
= joule

FIGURE 4.6

Work is done against gravity when lifting an object. Work is measured in joules or in foot-pounds.

CONNECTIONS

Energy for Work

When you hear of a person who has a lot of energy, you might think of a person who does things with intensity and vigor, perhaps for an extended period of time. Energy is defined as the ability to do work, and this seems about right for an intense, vigorous person who can do a lot of work. What is the source of all this energy that helps a body do work?

Cells use the molecule adenosine triphosphate (ATP) to transfer chemical-bond energy from energy-releasing to energy-requiring reactions. Each ATP molecule used in the cell is like a rechargeable AAA battery used to power small toys and electronic equipment. Each contains just the right amount of energy to power the job. When the power has been drained, it can be recharged numerous times before it must be recycled.

After the chemical-bond energy has been drained by breaking one of its bonds:

$$ATP \rightarrow ADP + P + energy$$

Plugging it into a high-powered energy source now recharges the discharged molecule (ADP). This source is chemical-bond energy released from cellular use of carbohydrates and fats:

$$energy + ADP + P \rightarrow ATP$$

An ATP molecule is formed from adenine (nitrogenous base), ribose (sugar), and phosphates. These three are chemically bonded to form AMP, adenosine monophosphate. When a second phosphate group is added to the AMP, a molecule of ADP (diphosphate) is formed. The ADP, with the addition of more energy, is able to bond to a third phosphate group and form ATP. The bonds that attach the second and third phosphates to the AMP molecules are easily broken to release energy for energy-requiring cell processes. Because the energy in this bond is so easy for a cell to use, it is called a high-energy phosphate bond. Both ADP and ATP, because they contain high-energy bonds, are very unstable molecules and readily lose their phosphates. When this occurs, the energy held in the high-energy bonds of the phosphate can be transferred to another molecule or released. Within a cell, enzymes speed this release of energy as ATP is broken down to ADP and P.

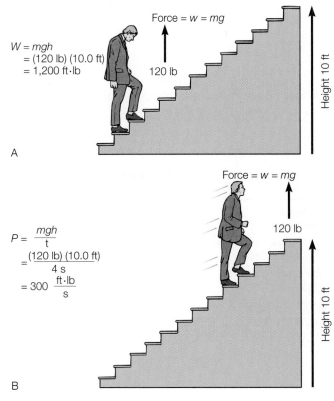

FIGURE 4.7

(A) The work accomplished in climbing a stairway is the person's weight times the vertical distance. (B) The power level is the work accomplished per unit of time.

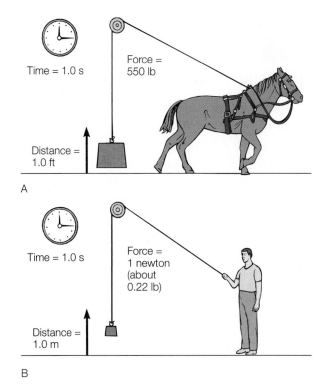

FIGURE 4.8

(A) A horsepower is defined as a power rating of 550 ft·lb/s. (B) A watt is defined as a newton-meter per second, or joule per second.

Power

You are doing work when you walk up a stairway, since you are lifting yourself through a distance. You are lifting your weight (force exerted) the *vertical* height of the stairs (distance through which the force is exerted). Consider a person who weighs 120 lb and climbs a stairway with a vertical distance of 10 ft. This person will do (120 lb)(10 ft) or 1,200 ft·lb of work. Will the amount of work change if the person were to run up the stairs? The answer is no, the same amount of work is accomplished. Running up the stairs, however, is more tiring than walking up the stairs. You use up your energy at a greater *rate* when running. The rate at which energy is transformed or the rate at which work is done is called **power** (figure 4.7). Power is defined as work per unit of time,

$$\text{power} = \frac{\text{work}}{\text{time}}$$

$$P = \frac{W}{t}$$

equation 4.2

The 120 lb person who ran up the 10 ft height of stairs in 4 s would have a power rating of

$$P = \frac{W}{t} = \frac{(120 \text{ lb})(10 \text{ ft})}{4 \text{ s}} = 300 \frac{\text{ft·lb}}{\text{s}}$$

If the person had a time of 3 s on the same stairs, the power rating would be greater, 400 ft·lb/s. This is a greater rate of energy use, or greater power.

When the steam engine was first invented, there was a need to describe the rate at which the engine could do work. Since people at this time were familiar with using horses to do their work, the steam engines were compared to horses. James Watt, who designed a workable steam engine, defined **horsepower** as a power rating of 550 ft·lb/s (figure 4.8A). To convert a power rating in the English units of ft·lb/s to horsepower, divide the power rating by 550 ft·lb/s/hp. For example, the 120 lb person who had a power rating of 400 ft·lb/s had a horsepower of 400 ft·lb/s ÷ 550 ft·lb/s/hp, or 0.7 hp.

In the metric system, power is measured in joules per second. The unit J/s, however, has a name. A J/s is called a **watt** (W). The watt (figure 4.8B) is used with metric prefixes for large numbers: 1,000 W = 1 kilowatt (kW) and 1,000,000 W = 1 megawatt (MW). It takes 746 W to equal 1 horsepower. One

CONNECTIONS

Hibernation and Temperature

What does "hibernation" really mean? What happens during hibernation? *Hibernation* occurs when an animal enters a *long-term, seasonal* condition of hypothermia (lowered body temperature); for example, some squirrels, bats, and badgers hibernate. This condition is different from two other types of hypothermia: *torpor*—a *daily* or *short-term* hypothermia that occurs in response to lowered environmental temperatures (occurs in birds and bears); and *estivation*—the same as torpor but occurs when environmental temperatures rise (occurs in desert animals such as toads and snakes). Hibernation is a mechanism by which animals that have internal temperature-regulating mechanisms—and a constant body temperature—can conserve energy. In comparison to animals that regulate their body temperature by moving to places where they can be most comfortable, ten times the energy is required to keep the animals with a constant body temperature going!

True Squirrel Hibernation	**Bear Torpor**
Body temperature decreases about 30°C	Body temperature decreases about 5°C
Heart rate reduces to 2–3 beats per minute	Heart rate reduces to 12 beats per minute
Metabolic rate reduces by 75%	Metabolic rate reduces by 50%
Body waste elimination occurs	No body waste elimination occurs
Do not eat	Do not eat

kilowatt is equal to about 1 1/3 horsepower. The electric utility company charges you for how much power you have used with your electrical appliances. But they also want to know for how long. Electrical energy is measured by power (kW) times the time of use (hr). Thus, electrical energy is measured in kWhr. The kWhr is a unit of *work*, not power. Since power is

$$p = \frac{W}{t}$$

then it follows that

$$W = Pt$$

So power multiplied by time equals a unit of work, kWhr. We will return to kilowatts and kilowatt-hours later when we discuss electricity.

Activities

1. Find your horsepower rating. Measure the vertical height of a stairway that is more than 3 m (10 ft) high. Have someone time you while you walk up the stairs. Find your walking power in both watts and horsepower.
2. Repeat number 1, but this time go up the stairs as fast as you can. Compare your exerted power rating to your walking power in both watts and horsepower.

MOTION, POSITION, AND ENERGY

Closely related to the concept of work is the concept of **energy.** Energy can be defined as the *ability to do work.* This definition of energy seems consistent with everyday ideas about energy and physical work. After all, it takes a lot of energy to do a lot of work. In fact, one way of measuring the energy of something is to see how much work it can do. Likewise, when work is done *on* something, a change occurs in its energy level. The following examples will help clarify this close relationship between work and energy.

Potential Energy

Consider a book on the floor next to a bookcase. You can do work on the book by vertically raising it to a shelf. You can measure this work by multiplying the vertical upward force applied times the distance that the book is moved. You might find, for example, that you did an amount of work equal to 10 J on the book.

Suppose that the book has a string attached to it, as shown in figure 4.9. The string is threaded over a frictionless pulley and attached to an object on the floor. If the book is caused to fall from the shelf, the object on the floor will be vertically lifted through some distance by the string. The falling book exerts a force on the object through the string, and the object is moved through a distance. In other words, the *book* did work on the object through the string, $W = Fd$.

The book can do more work on the object if it falls from a higher shelf, since it will move the object a greater distance. The

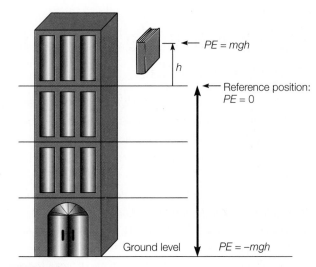

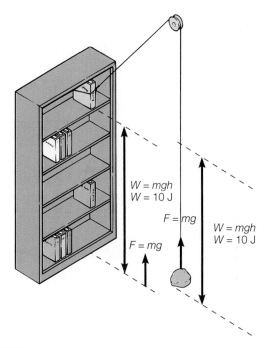

$$W = mgh$$
$$W = 10 \text{ J}$$

$$F = mg$$

$$W = mgh$$
$$W = 10 \text{ J}$$

$$F = mg$$

FIGURE 4.9

If moving a book from the floor to a high shelf requires 10 J of work, then the book will do 10 J of work on an object of the same mass when the book falls from the shelf.

$$PE = mgh$$

h

Reference position:
$PE = 0$

Ground level

$$PE = -mgh$$

FIGURE 4.10

The zero reference level for potential energy is chosen for convenience. Here the reference position chosen is the third floor, so the book will have a negative potential energy at ground level.

higher the shelf, the greater the *potential* for the book to do work. The ability to do work is defined as energy. The energy that an object has because of its position is called **potential energy** (*PE*). Potential energy is defined as *energy due to position.* This type of potential energy is called *gravitational potential energy,* since it is a result of gravitational attraction. There are other types of potential energy, such as that in a compressed or stretched spring.

Note the relationship between work and energy in the example. You did 10 J of work to raise the book to a higher shelf. In so doing, you increased the potential energy of the book by 10 J. The book now has the *potential* of doing 10 J of additional work on something else, therefore,

Work done on an object to change position	=	Increase in potential energy	=	Increase in work the object can do
Work on book	=	Potential energy of book	=	Work by book
(10 J)		(10 J)		(10 J)

As you can see, a joule is a measure of work accomplished on an object. A joule is also a measure of potential energy. And, a joule is a measure of how much work an object can do. Both work and energy are measured in joules (or ft·lbs).

The potential energy of an object can be calculated, as described previously, from the work done *on* the object to change its position. You exert a force equal to its weight as you lift it some height above the floor, and the work you do is the product of the weight and height. Likewise, the amount of work the object *could* do because of its position is the product of its weight and height. For the metric unit of mass, weight is the product of the mass of an object times *g,* the acceleration due to gravity, so

$$\text{potential energy} = \text{weight} \times \text{height}$$
$$PE = mgh$$

equation 4.3

For English units, the pound *is* the gravitational unit of force, or weight, so equation 4.3 becomes $PE = (w)(h)$.

Under what conditions does an object have zero potential energy? Considering the book in the bookcase, you could say that the book has zero potential energy when it is flat on the floor. It can do no work when it is on the floor. But what if that floor happens to be the third floor of a building? You could, after all, drop the book out of a window. The answer is that it makes no difference. The same results would be obtained in either case since it is the *change of position* that is important in potential energy. The zero reference position for potential energy is therefore arbitrary. A zero reference point is chosen as a matter of convenience. Note that if the third floor of a building is chosen as the zero reference position, a book on ground level would have negative potential energy. This means that you would have to do work on the book to bring it back to the zero potential energy position (figure 4.10). You will learn more about negative energy levels later in the chapters on chemistry.

Kinetic Energy

Moving objects have the ability to do work on other objects because of their motion. A rolling bowling ball exerts a force on the bowling pins and moves them through a distance, but

CONNECTIONS

Energy Converter Coaster

No matter how simple or how complex, a roller coaster is basically a potential and kinetic energy converter. The only time an outside energy source is involved is to move the cars to the top of the first hill, where it has the most potential energy it will have for the entire ride. The cars acquire this maximum potential energy from the mechanical energy of the lift-drive mechanism, which is usually chain driven. So, the first hill is always the highest hill above the ground, but the bottom of the first hill is not necessarily where the fastest speed occurs.

When the cars reach the top of the first hill, the lift drive disengages. The cars begin to move down a sloping track with increasing speed as potential energy is converted to kinetic energy. At the bottom of this hill the track then starts up a second hill, and the cars this time convert kinetic energy back to potential energy. Ideally, all of the potential energy is converted to kinetic energy as the cars move down a hill; then all the kinetic energy is converted to potential energy as the cars move up a hill. Of course there is no perfect energy conversion like this, and some energy is lost to air friction, wheel friction, and so on. A modern roller coaster design allows for these losses and still has some residual kinetic energy—enough, in fact, that it is necessary to apply brakes at the end of the ride.

It would be instructive for you to make a sketch of the profile of a roller coaster ride. The profile should show the relative differences in height between the track surface at the crest of each hill and the bottom of each upcoming dip. From such a profile you could find where in a ride the maximum speed would occur, as well as what speed to expect. As you study a profile, keep in mind that the roller coaster is designed to produce many changes of speed—accelerations—rather than high speed alone.

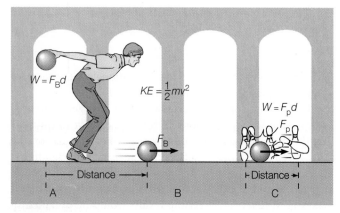

$$W = F_B d$$

$$KE = \frac{1}{2}mv^2$$

$$W = F_p d$$

FIGURE 4.11

(A) Work is done on the bowling ball as a force (F_B) moves it through a distance. (B) This gives the ball a kinetic energy equal in amount to the work done on it. (C) The ball does work on the pins and has enough remaining energy to crash into the wall behind the pins.

the ball loses speed as a result of the interaction (figure 4.11). A moving car has the ability to exert a force on a tree and knock it down, again with a corresponding loss of speed. Objects in motion have the ability to do work, so they have energy. The energy of motion is known as **kinetic energy.** Kinetic energy can be measured (1) in terms of the work done to put the object in motion or (2) in terms of the work the moving object will do in coming to rest. Consider objects that you put into motion by throwing. You exert a force on a football as you accelerate it through a distance before it leaves your hand. The kinetic energy that the ball now has is equal to the work (force times distance) that you did on the ball. You exert a force on a baseball through a distance as the ball increases its speed before it leaves your hand. The kinetic energy that the ball now has is equal to the work that you did on the ball. The ball exerts a force on the hand of the person catching the ball and moves it through a distance. The net work done on the hand is equal to the kinetic energy that the ball had. Therefore,

Work done to put an object in motion	=	Increase in kinetic energy	=	Increase in work the object can do

A baseball and a bowling ball moving with the same velocity do not have the same kinetic energy. You cannot knock down many bowling pins with a slowly rolling baseball. Obviously the more massive bowling ball can do much more work than a less massive baseball with the same velocity. Is it possible for the bowling ball and the baseball to have the same kinetic energy? The answer is yes, if you can give the baseball sufficient velocity. This might require shooting the baseball from a cannon, however. Kinetic energy is proportional to the mass of a moving object, but velocity has a greater influence. Consider two balls of the same mass, but one is moving twice as fast as the other. The ball with twice the velocity will do *four* times as much work as the slower ball. A ball with three times the velocity will do *nine* times as much work as the slower ball. Kinetic energy is proportional

CONNECTIONS

Bungee Jumping Energy

A group of college students in England started the sport of "bungee jumping" by jumping from bridges during the summer of 1979. They jumped with a rubber cord, and a common British word for rubber items is "bungee." Thus jumping from a high place with a long rubber cord became known as "bungee jumping." During the early 1990s towers, cranes, and hot air balloons were soon serving as bungee jumping platforms across the United States.

Today, commercial bungee jumping companies have specially made bungee cords, harnesses, and connectors. Let us consider what happens to a 645 N (mass = 65.8 kg) bungee jumper by first considering some hypothetical measurements. Our jumper is on a crane platform 27 m (90 ft) above the ground, in a body harness attached to a 9 m long (30 ft) bungee cord, which is attached to a 2 m length of cable from the crane. During the jump the bungee cord will stretch to a length 18 m, doubling in length.

Now, let us follow a hypothetical jump. After falling or jumping from the platform our jumper accelerates toward the surface at

9.8 m/s^2 in a free fall that lasts somewhere between one and two seconds. The bungee cord will now straighten to its full length, then stretch to double its length. At this instant of maximum stretch the velocity of the jumper and the cord is zero. For our hypothetical jumper the stretched cord results in a net upward force of three times the jumper's weight as it snaps back. From Newton's second law of motion, we find that the net upward acceleration would be 29.4 m/s^2. This acceleration is three times the normal g, meaning our hypothetical jumper was accelerated upward at about 3 g.

What is an acceleration of 3 g? One g is the normal acceleration due to gravity, which is 9.8 m/s^2. A 3 g acceleration of 29.4 m/s^2 will increase the velocity from zero to about 106 km/h (66 mi/h) in one second. In other words, our hypothetical bungee jumper was whisked back away from the ground at an acceleration equivalent to a car accelerating to a speed of 106 km/h (66 mi/h) in a second.

---

to the square of the velocity ($2^2 = 4$; $3^2 = 9$). The kinetic energy (KE) of an object is

$$\text{kinetic energy} = \frac{1}{2}(\text{mass})(\text{velocity})^2$$

$$KE = \frac{1}{2}mv^2$$

equation 4.4

The unit of mass is the kg, and the unit of velocity is m/s. Therefore the unit of kinetic energy is

$$KE = (\text{kg})\left(\frac{\text{m}}{\text{s}}\right)^2$$

$$= (\text{kg})\left(\frac{\text{m}^2}{\text{s}^2}\right)$$

$$= \frac{\text{kg·m}^2}{\text{s}^2}$$

which is the same thing as

$$\left(\frac{\text{kg·m}}{\text{s}^2}\right)(\text{m})$$

or

$$\text{N·m}$$

or

$$\text{joule (J)}$$

Kinetic energy is measured in joules, as is work ($F \times d$ or N·m) and potential energy (mgh or N·m).

ENERGY FLOW

The key to understanding the individual concepts of work and energy is to understand the close relationship between the two. When you do work on something you give it energy of position (potential energy) or you give it energy of motion (kinetic energy). In turn, objects that have kinetic or potential energy can now do work on something else as the transfer of energy continues. Where does all this energy come from and where does it go? The answer to this question is the subject of this section on energy flow.

Energy Forms

Energy comes in various forms, and different terms are used to distinguish one form from another. Although energy comes in various *forms*, this does not mean that there are different *kinds* of energy. The forms are the result of the more common fundamental forces—gravitational, electromagnetic, and nuclear—and objects that are interacting. There are five forms of energy: (1) *mechanical*, (2) *chemical*, (3) *radiant*, (4) *electrical*, and (5) *nuclear*. The following is a brief discussion of each of the five forms of energy.

A CLOSER LOOK | Environmentally Responsible Choices

Environmental science is an interdisciplinary subject that is concerned with problems of human degradation of the environment and remedies for that damage. Environmental problems can be complex, and it helps to understand at least three things: (1) the physical and biological processes that take place in the natural world, (2) the role of technology in terms of its capacity to alter natural processes as well as solve other problems of human impact, and (3) the complex social processes that characterize human populations, and how some of these processes can cause great amounts of environmental damage.

As an example of environmental damage brought about by humans, consider the thinning of the earth's ozone layer that was discovered over Antarctica in the early 1980s. Scientists found that certain human-made compounds, such as chlorofluorocarbons (CFCs), carbon tetrachloride, methyl chloroform, and halons were interacting with the ozone layer, causing it to become so thin that the press started talking about an "ozone hole" in the atmosphere. A weakened ozone layer allows more ultraviolet (UV) radiation to reach the earth's surface. The consequences of more UV radiation in humans include increased skin damage (such as skin cancer) and increased eye damage (such as cataracts).

Once this threat to the environment and human health and safety was understood, environmental action groups, businesses, and industry began working together to solve the problem. Before 1988 much of the Styrofoam cartons and food packaging were manufactured by using CFCs as blowing agents.

In 1988 the makers of these cartons and packages voluntarily halted the use of CFCs. Other industries followed this example, and new rules concerning the use of these chemicals were added to the Clean Air Act, which is managed by the U.S. Environmental Protection Agency (EPA). Today the EPA is ending production of ozone-depleting substances, and ensuring that refrigerants and halon fire extinguishing agents are recycled properly. As a consequence, ozone depletion is no longer occurring at an accelerated rate. Scientists now predict that CFC levels will continue to become less and less and the ozone layer is now expected to recover completely by the year 2050.

In June 1992, representatives from 178 countries, including 115 heads of state, met in Rio de Janeiro, Brazil, at the Earth Summit. The meeting was titled the United Nations Conference on Environment and Development (UNCED), and it was called to promote better integration of nations' environmental goals with their economic aspirations. The *Rio Declaration on Environment and Development* identified 27 principles to guide the behavior of nations toward more environmentally sustainable patterns of development. A *Framework Convention on Climate Change* asked the signing nations to reduce their emissions of gases believed to contribute to global warming, and the United States agreed to reduce its emissions of greenhouse gases back to their 1990 levels.

But what can an individual do to become more environmentally responsible? First, individuals can become informed about the impacts of their consumption choices. For example, many people continue to avoid use of spray cans or Styrofoam coffee cups decades after the CFCs used in the cans or in the manufacture of Styrofoam were eliminated. Spray cans and Styrofoam cups are no longer responsible for ozone destruction in the earth's upper atmosphere! So spray cans and Styrofoam cups really are not issues today in being environmentally responsible.

What about other environmentally responsible choices? For example, should you use paper or plastic grocery bags? Ride the bus or take your car to the grocery store? Studies have found that the highest amount of environmental damage results from modern consumer activities concerning the use of transportation. Cars, minivans, sport utility vehicles, and pickup trucks (small and large) cause the highest amount of environmental damage when compared to the damage resulting from other consumer decisions. These vehicles contribute about half of the toxic air pollution and more than a quarter of the greenhouse gases. If you really want to be environmentally responsible, you should plan your shopping trips so you drive less or—better yet—use bus or rail transit when possible. Because today's cars and trucks do so much environmental damage, it does not make sense to drive your car, van, or pickup to the grocery store and then worry about using plastic or paper bags. You should be more concerned about significant issues such as the use of fossil fuels, air pollution, and greenhouse gases.

Mechanical energy is the form of energy of familiar objects and machines (figure 4.12). A car moving on a highway has kinetic mechanical energy. Water behind a dam has potential mechanical energy. The spinning blades of a steam turbine have kinetic mechanical energy. The form of mechanical energy is usually associated with the kinetic energy of everyday-sized objects and the potential energy that results from gravity. There are other possibilities (e.g., sound), but this description will serve the need for now.

Chemical energy is the form of energy involved in chemical reactions (figure 4.13). Chemical energy is released in the chemical reaction known as *oxidation*. The fire of burning wood is an example of rapid oxidation. A slower oxidation releases energy from food units in your body. As you will learn in the chemistry unit, chemical energy involves electromagnetic forces between the parts of atoms. Until then, consider the following comparison. Photosynthesis is carried on in green plants. The plants use the energy of sunlight to rearrange carbon dioxide and water

FIGURE 4.12

Mechanical energy is the energy of motion, or the energy of position, of many familiar objects. This boat has energy of motion.

into plant materials and oxygen. This reaction could be represented by the following word equation:

energy + carbon dioxide + water = wood + oxygen

The plant took energy and two substances and made two different substances. This is similar to raising a book to a higher shelf in a bookcase. That is, the new substances have more energy than the original ones did. Consider a word equation for the burning of wood:

wood + oxygen = carbon dioxide + water + energy

Notice that this equation is exactly the reverse of photosynthesis. In other words, the energy used in photosynthesis was released during oxidation. Chemical energy is a kind of potential energy that is stored and later released during a chemical reaction.

Radiant energy is energy that travels through space (figure 4.14). Most people think of light or sunlight when considering this form of energy. Visible light, however, occupies only a small part of the complete electromagnetic spectrum, as shown in figure 4.15. Radiant energy includes light and all other parts of the spectrum. Infrared radiation is sometimes called "heat radiation" because of the association with heating when this type of radiation is absorbed. For example, you feel the interaction of infrared radiation when you hold your hand near a warm range element. However, infrared radiation is another type of radiant energy. In fact, some snakes, such as the sidewinder, can see infrared radiation emitted from warm animals where you see total darkness. Microwaves are another type of radiant energy that is used in cooking. As with other forms of energy, light, infrared, and microwaves will be considered in more detail later. For now, consider all types of radiant energy to be forms of energy that travel through space.

Electrical energy is another form of energy from electromagnetic interactions that will be considered in detail later. You are familiar with electrical energy that travels through wires to your home from a power plant (figure 4.16), electrical energy that is generated by chemical cells in a flashlight, and electrical energy that can be "stored" in a car battery.

Nuclear energy is a form of energy often discussed because of its use as an energy source in power plants. Nuclear energy is

A

B

FIGURE 4.13

Chemical energy is a form of potential energy that is released during a chemical reaction. Both (*A*) wood and (*B*) coal have chemical energy that has been stored through the process of photosynthesis. The pile of wood might provide fuel for a small fireplace for several days. The pile of coal might provide fuel for a power plant for a hundred days.

another form of energy from the atom, but this time the energy involves the nucleus, the innermost part of an atom, and nuclear interactions.

Energy Conversion

Potential energy can be converted to kinetic energy and vice versa. The simple pendulum offers a good example of this conversion. A simple pendulum is an object, called a bob, suspended by a string or wire from a support. If the bob is moved to one side and then released, it will swing back and forth in an arc. At the moment that the bob reaches the top of its swing, it stops for an instant, then begins another swing. At the instant of stopping, the bob has 100 percent potential energy and no kinetic energy. As the bob starts back down through the swing,

A

B

FIGURE 4.14

Radiant energy is energy that travels through space. (*A*) This demonstration solar cell array converts radiant energy from the sun to electrical energy, producing an average of 200,000 watts of electric power (after conversion). (*B*) Solar panels are mounted on the roof of this house.

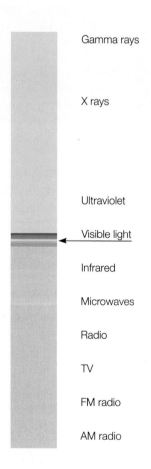

Gamma rays

X rays

Ultraviolet

Visible light ←

Infrared

Microwaves

Radio

TV

FM radio

AM radio

FIGURE 4.15

The electromagnetic spectrum includes many forms of radiant energy. Note that visible light occupies only a tiny part of the entire spectrum.

FIGURE 4.16

The blades of a steam turbine. In a power plant, chemical or nuclear energy is used to heat water to steam, which is directed against the turbine blades. The mechanical energy of the turbine turns an electric generator. Thus, a power plant converts chemical or nuclear energy to mechanical energy, which is then converted to electrical energy.

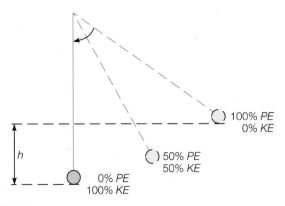

FIGURE 4.17

This pendulum bob loses potential energy (*PE*) and gains an equal amount of kinetic energy (*KE*) as it falls through a distance *h*. The process reverses as the bob moves up the other side of its swing.

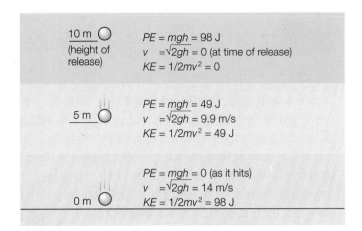

FIGURE 4.18

The ball trades potential energy for kinetic energy as it falls. Notice that the ball had 98 J of potential energy when dropped and has a kinetic energy of 98 J just as it hits the ground.

it is gaining kinetic energy and losing potential energy. At the instant the bob is at the bottom of the swing, it has 100 percent kinetic energy and no potential energy. As the bob now climbs through the other half of the arc, it is gaining potential energy and losing kinetic energy until it again reaches an instantaneous stop at the top, and the process starts over. The kinetic energy of the bob at the bottom of the arc is equal to the potential energy it had at the top of the arc (figure 4.17). Disregarding friction, the sum of the potential energy and the kinetic energy remains constant throughout the swing.

The potential energy lost during a fall equals the kinetic energy gained (figure 4.18). In other words,

$$PE_{\text{lost}} = KE_{\text{gained}}$$

Substituting the values from equations 4.3 and 4.4,

$$mgh = \frac{1}{2} mv^2$$

Canceling the *m* and solving for v_f,

$$v_f = \sqrt{2gh}$$

equation 4.5

Equation 4.5 tells you the final speed of a falling object after its potential energy is converted to kinetic energy. This assumes, however, that the object is in free fall, since the effect of air resistance is ignored. Note that the *m*'s cancel, showing again that the mass of an object has no effect on its final speed.

Any *form* of energy can be converted to another form. In fact, most technological devices that you use are nothing more than *energy-form converters* (figure 4.19). A lightbulb, for example, converts electrical energy to radiant energy. A car converts chemical energy to mechanical energy. A solar cell converts radiant energy to electrical energy, and an electric motor converts electrical energy to mechanical energy. Each technological device converts some form of energy (usually chemical or electrical) to another form that you desire (usually mechanical or radiant).

It is interesting to trace the *flow of energy* that takes place in your surroundings. Suppose, for example, that you are riding a bicycle. The bicycle has kinetic mechanical energy as it moves along. Where did the bicycle get this energy? It came from you, as you use the chemical energy of food units to contract your muscles and move the bicycle along. But where did your chemical energy come from? It came from your food, which consists of plants, animals who eat plants, or both plants and animals. In any case, plants are at the bottom of your food chain. Plants convert radiant energy from the sun into chemical energy. Radiant energy comes to the plants from the sun because of the nuclear reactions that took place in the core of the sun. Your bicycle is therefore powered by nuclear energy that has undergone a number of form conversions!

Energy Conservation

Energy can be transferred from one object to another, and it can be converted from one form to another form. If you make a detailed accounting of all forms of energy before and after a transfer or conversion, the total energy will be *constant*. Consider your bicycle coasting along over level ground when you apply the brakes. What happened to the kinetic mechanical energy of the bicycle? It went into heating the rim and brakes of your bicycle, then eventually radiated to space as infrared radiation. All radiant energy that reaches the earth is eventually radiated back to space (figure 4.20). Thus, throughout all the form conversions and energy transfers that take place, the total sum of energy remains constant.

C O N N E C T I O N S

Roller Coaster Relationships

A roller coaster gains potential energy as it is moved by a motor-driven chain to the top of the first hill. It then rolls down the hill, increasing speed as potential energy is swapped for kinetic. Is the speed of the moving roller coaster at the bottom of the hill directly proportional to the height of the hill? Does the speed depend on how many people are in the coaster; that is, will more people in the cars make the coaster go downhill at a higher speed?

First, let's consider the relationship between hill height and speed. It is commonly thought that the speed of a roller coaster is directly proportional to the height of a hill. This is not true. Ignoring friction, the speed of a coaster at the bottom of a hill is proportional to the square root of the height of the hill. This means that to double the speed of the coaster at the bottom of the hill, you would need to build the hill four times higher. Furthermore, the coaster does not drop straight down (hopefully), and doubling the height increases the theoretical speed only 40 percent. To achieve a speed of 160 km/hr (about 100 mi/hr) would require the hill to be about 122 m (about 400 ft) above ground level.

Second, what is the relationship between the number of people on a roller coaster (meaning the weight) and the speed achieved at the bottom of the hill? The answer is that the number of people does not matter. Ignoring air resistance and friction, a heavy roller coaster full of people and a lighter one with just a few people will both have the same speed at the bottom of the hill.

For most people the fun part of a roller coaster ride comes from unexpected bits of acceleration, not just speed. However, the acceleration is related to the speed as well as the radius of the curve in the track. Acceleration occurs when the coaster changes direction and the magnitude of the acceleration is proportional to the square of the speed of the coaster cars and inversely proportional to the track radius. Thus, doubling the speed around a curve quadruples the force generated unless the radius of the curve is also quadrupled.

The art of building an outstanding roller coaster ride is based on using the science of physical relationships for a ride that is comfortable and safe, but with reasonable forces for fun accelerations.

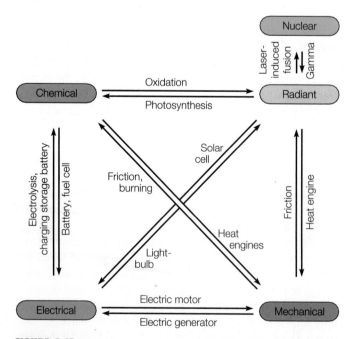

FIGURE 4.19

The energy forms and some conversion pathways.

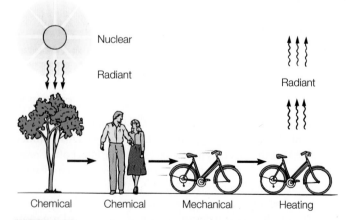

FIGURE 4.20

Energy arrives from the sun, goes through a number of conversions, then radiates back into space. The total sum leaving eventually equals the original amount that arrived.

A CLOSER LOOK | V Energy Saver

Did you ever wonder why geese fly in a V-formation? The answer has something to do with saving energy, which will be explained following a brief introduction to how a bird is able to fly.

There are some similarities and some differences between the flight of a bird and the flight of an airplane. The basic similarity is found in the shape of an airplane wing and the shape of a bird's wing. A bird wing has a rounded leading edge and a sharp trailing edge, in a profile that causes air flowing over the upper surface to move faster than the air underneath. It is built this way to take advantage of the relationship between pressure and velocity, that the pressure of a moving fluid decreases with increased velocity. Thus the faster moving air over the top of a bird wing exerts less air pressure, and the slower moving air on the underside of the wing exerts more pressure.* Just like the airplane wing, wind over the curved top of a bird wing leads to a pressure difference, an upward force. The magnitude of the lift force depends on the velocity of the air moving across the wing and the shape of the wing. To maintain a level flight, the lift force must be equal to the weight of the bird.

There are other factors involved in the mechanics of bird flight, such as thrust and drag. Drag is the air friction force, together with all the other forces that oppose forward motion. In flight, a forward thrust must balance drag. Airplanes provide lift from their wings and forward thrust from a jet engine or propeller. A bird, however, uses its wings to provide lift and the forward thrust at the same time by a flapping action. For many

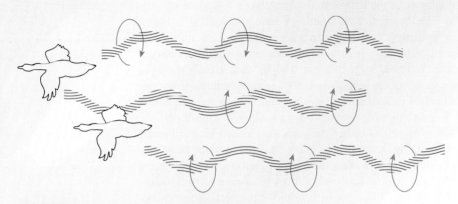

BOX FIGURE 4.1

The downward-flowing air across the top of a bird's wing forms a trailing air vortex behind the wing tips. Migrating birds can conserve energy by taking advantage of the upward side of the airflow of a vortex, forming a V-formation when in the optimal wing-tip-to-wing-tip position. (For clarity, not all vortex wakes are shown.)

birds, it is the downstroke that provides the thrust, but this varies with the bird and with the circumstances.

The downstroke of a wing and the movement of air over the curve of the wing provide thrust and lift, but air must flow downward behind the wing to match the forward momentum of the bird. This forms an intensely twisting "rope" of air, much like a sideways tornado. The twisting rope of air—called a *vortex*—forms in the wake of each wing tip. The vortices are a consequence of the force generated by the wind, and lift cannot exist without them. Different kinds of birds, with different kinds of wing beats, form different types of vortices. Since we are interested in answering the question about why geese fly in a V-formation, we will

consider the vortices formed behind a flying goose, as shown in box figure 4.1.

Geese, and a few other kinds of birds, form a V-formation with others of their flock while migrating. They do this because the upward wing-tip vortices provide lift for another bird flying close enough to be in the slipstream, as you can see in box figure 4.1. According to theoretical calculations, 25 geese flying wing tip to wing tip in a V-formation flock can reduce their energy needs by 12 percent as compared to a solitary bird. Of course, this does not apply to the lead goose, who would expend 12 percent more energy than the others just by being in the lead position. For the other geese, however, the energy savings would be enormous over the total migration distance.

*Hold a sheet of notebook paper by the bottom corners, with the edge you are holding near your lips. Allow the other end to hang freely, forming a curved surface. Blow gently over the curved surface to demonstrate the result of the greater underside pressure.

The total energy is constant in every situation that has been measured. This consistency leads to another one of the conservation laws of science, the **law of conservation of energy:**

> **Energy is never created or destroyed. Energy can be converted from one form to another but the total energy remains constant.**

You may be wondering about the source of nuclear energy. Does a nuclear reaction create energy? Albert Einstein answered this question back in the early 1900s when he formulated his now-famous relationship between mass and energy, $E = mc^2$. This relationship will be discussed in detail in chapter 15. Basically, the relationship states that mass *is* a form of energy, and this has been experimentally verified many times.

Energy Transfer

Earlier it was stated that when you do work on something, you give it energy. The result of work could be increased kinetic mechanical energy, increased gravitational potential energy, or an increase in the temperature of an object. You could summarize this by stating that either *working* or *heating* is always involved any time energy is transformed. This is not unlike your financial situation. In order to increase or decrease your financial status, you need some mode of transfer, such as cash or checks, as a means of conveying assets. Just as with cash flow from one individual to another, energy flow from one object to another requires a mode of transfer. In energy matters the mode of transfer is working or heating. Any time you see working or heating occurring you know that an energy transfer is taking place. The next time you see heating, think about what energy form is being converted to what new energy form. (The final form is usually radiant energy.) Heating is the topic of the next chapter, where you will consider the role of heat in energy matters.

ENERGY SOURCES TODAY

Prometheus, according to ancient Greek mythology, stole fire from heaven and gave it to humankind. Fire has propelled human advancement ever since. All that was needed was something to burn—fuel for Prometheus's fire.

Any substance that burns can be used to fuel a fire, and various fuels have been used over the centuries as humans advanced. First, wood was used as a primary source for heating. Then coal fueled the industrial revolution. Eventually, humankind roared into the twentieth century burning petroleum. Today, petroleum is the most widely used source of energy (figure 4.21). It provides about 40 percent of the total energy used by the United States, but this dependence has been dropping since the 1970s. Natural gas contributes about 23 percent of the total energy used today. The use of coal has been increasing, and today provides about 23 percent of the total. Note that petroleum, coal, and natural gas are all chemical sources of energy, sources that are mostly burned for their energy. These chemical sources supply about 86 percent of the total energy consumed. About a third of this is burned for heating, and the rest is burned to drive engines or generators.

Nuclear energy and hydropower are the nonchemical sources of energy. These sources are used to generate electrical energy. The alternative sources of energy, such as solar and geothermal, provide less than 0.5 percent of the total energy consumed in the United States today.

The energy-source mix has changed from past years, and it will change in the future. Wood supplied 90 percent of the energy until the 1850s, when the use of coal increased. Then, by 1910, coal was supplying about 75 percent of the total energy needs. Then petroleum began making increased contributions to the energy supply. Now increased economic constraints and a decreasing supply of petroleum are producing another supply shift. The present petroleum-based energy era is about to shift to a new energy era.

Over 99 percent of the total energy consumed today is provided by four sources: (1) petroleum (including natural gas),

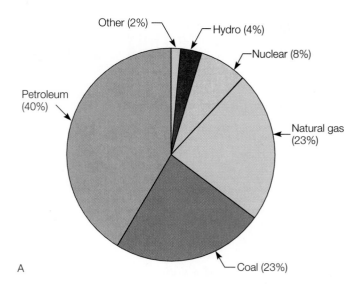

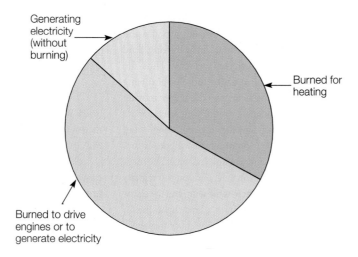

FIGURE 4.21

(*A*) The sources of energy and (*B*) the uses of energy during the 1990s.

(2) coal, (3) hydropower, and (4) nuclear power. The following is a brief introduction to these four sources.

Petroleum

The word *petroleum* is derived from the word "petra," meaning rock, and the word "oleum," meaning oil. Petroleum is oil that comes from oil-bearing rock. Natural gas is universally associated with petroleum and has similar origins. Both petroleum and natural gas form from organic sediments, materials that have settled out of bodies of water. Sometimes a local condition permits the accumulation of sediments that are exceptionally rich in organic material. This could occur under special conditions in a freshwater lake, or it could occur on shallow ocean basins. In either case, most of the organic material is from plankton—tiny free-floating animals and plants such as algae. It is from such accumulations of buried organic material that petroleum and natural gas are formed.

The exact process by which buried organic material becomes petroleum and gas is not understood. It is believed that bacteria, pressure, appropriate temperatures, and time are all important. Natural gas is formed at higher temperatures than is petroleum. Varying temperatures over time may produce a mixture of petroleum and gas or natural gas alone.

Petroleum forms a thin film around the grains of the rock where it formed. Pressure from the overlying rock and water move the petroleum and gas through the rock until it reaches a rock type or structure that stops it. If natural gas is present it occupies space above the accumulating petroleum. Such accumulations of petroleum and natural gas are the sources of supply for these energy sources.

Discussions about the petroleum supply and the cost of petroleum usually refer to a "barrel of oil." The *barrel* is an accounting device of 42 U.S. gallons. Such a 42-gallon barrel does not exist. When or if oil is shipped in barrels, each drum holds 55 U.S. gallons. The various uses of petroleum products are discussed in chapter 14.

The supply of petroleum and natural gas is limited. Most of the continental drilling prospects appear to be exhausted, and the search for new petroleum supplies is now offshore. In general, over 25 percent of the United States' petroleum is estimated to come from offshore wells. The amount of imported petroleum has ranged from 30 to 50 percent over the years, with most imported oil coming from Mexico, Canada, Great Britain, Indonesia, and Saudi Arabia.

Petroleum is used for gasoline (about 45 percent), diesel (about 40 percent), and heating oil (about 15 percent). Petroleum is also used in making medicine, clothing fabrics, plastics, and ink.

Coal

Petroleum and natural gas formed from the remains of tiny plants and animals that lived millions of years ago. Coal, on the other hand, formed from an accumulation of plant materials that collected under special conditions millions of years ago. Thus, petroleum, natural gas, and coal are called **fossil fuels.** Fossil fuels contain the stored radiant energy of plants that lived millions of years ago.

Fossil plants found in coal are similar to plants that grow today in the swamps of Florida, Georgia, and Louisiana. When plants die in such a swamp, they fall into the water, become waterlogged, and sink. There, in the stagnant water, they are protected from consumption by termites, fungi, and bacteria. The plants rot somewhat, and layers of loose plant materials collect at the bottom of swampy lakes. This carbon-rich decayed plant material is called *peat* (not to be confused with peat moss). Peat can be used as a fuel. It is also the earliest stage of coal. Pressure, compaction, and heating brought about by movements of the earth's crust changed the water content and the free carbon in the material as it gradually changed the *rank,* or stage of development of the coal. The lowest rank is lignite (brown coal), then subbituminous, then bituminous (soft coal), and the highest rank is anthracite (hard coal).

Each rank of coal has different burning properties and a different energy content. Coal also contains impurities of clay, silt, iron oxide, and sulfur. The mineral impurities leave an ash when the coal is burned, and the sulfur produces sulfur dioxide, a pollutant.

Most of the coal mined today is burned by utilities to generate electricity (about 80 percent). The coal is ground to a face-powder consistency and blown into furnaces. This greatly increases efficiency but produces *fly ash,* ash that "flies" up the chimney. Industries and utilities are required by the federal Clean Air Act to remove sulfur dioxide and fly ash from plant emissions. About 20 percent of the cost of a new coal-fired power plant goes into air pollution control equipment. Coal is an abundant but dirty energy source.

Water Power

Moving water has been used as a source of energy for thousands of years. It is considered a renewable energy source, inexhaustible as long as the rain falls. Today, hydroelectric plants generate about 4 percent of the nation's *total* energy consumption at about 2,400 power-generating dams across the nation. Hydropower furnished about 40 percent of the United States' electric power in 1940. Today, dams furnish 9 percent of the electric power. It is projected that this will drop even lower, perhaps to 7 percent in the near future. Energy consumption has increased, but hydropower production has not kept pace because geography limits the number of sites that can be built.

Water from a reservoir is conducted through large pipes called penstocks to a powerhouse, where it is directed against turbine blades that turn a shaft on an electric generator. A rough approximation of the power that can be extracted from the falling water can be made by multiplying the depth of the water (in feet) by the amount of water flowing (in cubic feet per second), then dividing by 10. The result is roughly equal to the horsepower.

Nuclear Power

Nuclear power plants use nuclear energy to produce electricity. Energy is released as the nuclei of uranium and plutonium atoms split, or undergo fission. The fissioning takes place in a large steel vessel called a *reactor.* Water is pumped through the reactor to produce steam, which is used to produce electrical energy, just as in the fossil fuel power plants. The nuclear processes are described in detail in chapter 15, and the process of producing electrical energy is described in detail in chapter 7. Nuclear power plants use nuclear energy to produce electricity, but there are opponents to this process. The electric utility companies view nuclear energy as *one* energy source used to produce electricity. They state that they have no allegiance to any one energy source but are seeking to utilize the most reliable and dependable of several energy sources. Petroleum, coal, and hydropower are also presently utilized as energy sources for electric power production. The electric utility companies are concerned that petroleum and natural gas are becoming increasingly expensive, and there are questions about long-term supplies. Hydropower has limited potential for growth, and solar energy is

A CLOSER LOOK | Solar Technologies

An alternative source of energy is one that is different from the typical sources used today. The sources used today are the fossil fuels (coal, petroleum, and natural gas), nuclear, and falling water. Alternative sources could be solar, geothermal, hydrogen gas, fusion, or any other energy source that a new technology could utilize.

The term *solar energy* is used to describe a number of technologies that directly or indirectly utilize sunlight as an alternative energy source (box figure 4.2). There are eight main categories of these solar technologies:

1. *Solar cells.* A solar cell is a thin crystal of silicon, gallium, or some polycrystalline compound that generates electricity when exposed to light. Also called *photovoltaic devices,* solar cells have no moving parts and produce electricity directly, without the need for hot fluids or intermediate conversion states. Solar cells have been used extensively in space vehicles and satellites. Here on the earth, however, use has been limited to demonstration projects, remote site applications, and consumer specialty items such as solar-powered watches and calculators. The problem with solar cells today is that the manufacturing cost is too high (they are essentially handmade). Research is continuing on the development of highly efficient, affordable solar cells that could someday produce electricity for the home.

2. *Power tower.* This is another solar technology designed to generate electricity. One type of planned power tower will have a 171 m (560 ft) tower surrounded by some 9,000 special mirrors called heliostats. The heliostats will focus sunlight on a boiler at the top of the tower where salt (a mixture of sodium nitrate and potassium nitrate) will be heated to about 566°C (about 1,050°F). This molten salt will be pumped to a steam

BOX FIGURE 4.2

Wind is another form of solar energy. This wind turbine generates electrical energy for this sailboat, charging batteries for backup power when the wind is not blowing. In case you are wondering, the turbine cannot be used to make a wind to push the boat. In accord with Newton's laws of motion, this would not produce an unbalanced force on the boat.

generator, and the steam will be used to drive a generator, just like other power plants. Water could be heated directly in the power tower boiler. Molten salt is used because it can be stored in an insulated storage tank for use when the sun is not shining, perhaps for up to twenty hours.

3. *Passive application.* In passive applications energy flows by natural means, without mechanical devices such as motors, pumps, and so forth. A passive solar house would include considerations like orientation of a house to the sun, the size and positioning of windows, and a roof overhang that lets sunlight in during the winter but keeps it out during the summer. There are different design plans to capture, store, and distribute solar energy throughout a house.

4. *Active application.* An active solar application requires a solar collector in which sunlight heats air, water, or some liquid. The liquid or air is pumped through pipes in a house to generate electricity, or it is used directly for hot water. Solar water heating makes more economic sense today than the other applications.

5. *Wind energy.* The wind has been used for centuries to move ships, grind grain into flour, and pump water. The wind blows, however, because radiant energy from the sun heats some parts of the earth's surface more than other parts. This differential heating results in pressure differences and the horizontal movement of air, which is called *wind.* Thus, wind is another

—Continued top of next page

Continued—

form of solar energy. Wind turbines are used to generate electrical energy or mechanical energy. The biggest problem with wind energy is the inconsistency of the wind. Sometimes the wind speed is too great, and other times it is not great enough. Several methods of solving this problem are being researched.

6. *Biomass.* Biomass is any material formed by photosynthesis, including plants, trees, and crops, and any garbage, crop residue, or animal waste. Biomass can be burned directly as a fuel, converted into a gas fuel (methane), or converted into liquid

fuels such as alcohol. The problems with using biomass include the energy expended in gathering the biomass, as well as the energy used to convert it to a gaseous or liquid fuel.

7. *Agriculture and industrial heating.* This is a technology that simply uses sunlight to dry grains, cure paint, or do anything that can be done with sunlight rather than using traditional energy sources.

8. *Ocean thermal energy conversion (OTEC).* This is an electric generating plant that would take advantage of the approximately 22°C (about 40°F) temperature difference between the

surface and the depths of tropical, subtropical, and equatorial ocean waters.

Basically, warm water is drawn into the system to vaporize a fluid, which expands through a turbine generator. Cold water from the depths condenses the vapor back to a liquid form, which is then cycled back to the warm-water side. The concept has been tested several times and was found to be technically successful. The greatest interest seems to be islands that have warm surface waters (and cold depths) such as Hawaii, Puerto Rico, Guam, the Virgin Islands, and others.

Activities

Compare amounts of energy sources needed to produce electric power. Generally, 1 MW (1,000,000 W) will supply the electrical needs of 1,000 people.

1. Use the population of your city to find how many megawatts of electricity are required for your city.

2. Use the following equivalencies to find out how much coal, oil, gas, or uranium would be consumed in one day to supply the electrical needs.

$$\frac{1 \text{ kWhr}}{\text{of electricity}} = \begin{cases} 1 \text{ lb of coal} \\ 0.08 \text{ gal of oil} \\ 9 \text{ cubic ft of gas} \\ 0.00013 \text{ g of uranium} \end{cases}$$

Example

Assume your city has 36,000 people. Then 36 MW of electricity will be needed. How much oil is needed to produce this electricity?

$$36MW \times \frac{1,000 \text{ kW}}{MW} \times \frac{24 \text{ hr}}{\text{day}} \times \frac{0.08 \text{ gal}}{\text{kWhr}} = 69,120 \text{ or about } 70,000 \text{ gal/day}$$

Since there are 42 gallons in a barrel,

$$\frac{70,000 \text{ gal / day}}{42 \text{ gal / barrel}} = \frac{70,000}{42} \times \frac{\text{gal}}{\text{day}} \times \frac{\text{barrel}}{\text{gal}} = 1,666, \text{ or about } 2,000 \text{ barrel/day}$$

---

prohibitively expensive today. Utility companies see two major energy sources that are available for growth: coal and nuclear. There are problems and advantages to each, but the utility companies feel they must use coal and nuclear power until the new technologies, such as solar power, are economically feasible.

SUMMARY

Work is defined as the product of an applied force and the distance through which the force acts. Work is measured in newton-meters, a metric unit called a *joule. Power* is work per unit of time. Power is measured in *watts.* One watt is 1 joule per second. Power is also measured in *horsepower.* One horsepower is 550 ft·lb/s.

Energy is defined as the ability to do work. An object that is elevated against gravity has a potential to do work. The object is said to have *potential energy,* or *energy of position.* Moving objects have the ability to do work on other objects because of their motion. The *energy of motion* is called *kinetic energy.*

Work is usually done *against inertia, fundamental forces, friction, shape,* or *combinations of these.* As a result, there is a gain of *kinetic energy, potential energy, an increased temperature,* or *any combination of these.* Energy comes in the *forms* of *mechanical, chemical, radiant, electrical,* or *nuclear.* Potential energy can be *converted* to kinetic and kinetic can be *converted* to potential. Any form of energy can be *converted* to any other form. Most technological devices are *energy-form converters* that do work for you. Energy flows into and out of the surroundings, but the amount of energy is always constant. The *law of conservation of energy* states that *energy is never created or destroyed.* Energy conversion always takes place through *heating* or *working.*

The basic energy sources today are the chemical *fossil fuels* (petroleum, natural gas, and coal), *nuclear energy,* and *hydropower. Petroleum* and *natural gas* were formed from organic material of plankton, tiny free-floating plants and animals. A barrel of petroleum is 42 U.S. gallons, but such a container does not actually exist. *Coal* formed from plants that were protected from consumption by falling into a swamp. The decayed plant material, *peat,* was changed into the various *ranks* of coal by pressure and heating over some period of time. Coal is a dirty fuel that contains impurities and sulfur. Controlling air pollution from burning coal is costly. Water power and nuclear energy are used for the generation of electricity.

Summary of Equations

4.1

$$\text{work} = \text{force} \times \text{distance}$$
$$W = Fd$$

4.2

$$\text{power} = \frac{\text{work}}{\text{time}}$$
$$P = \frac{W}{t}$$

4.3

$$\text{potential energy} = \text{weight} \times \text{height}$$
$$PE = mgh$$

4.4

$$\text{kinetic energy} = \frac{1}{2}\,(\text{mass})\,(\text{velocity})^2$$
$$KE = \frac{1}{2}\,mv^2$$

4.5

$$\text{final velocity} = \text{square root of }(2 \times \overset{\text{acceleration due}}{\text{to gravity}} \times \text{height of fall})$$
$$v_{\text{f}} = \sqrt{2gh}$$

KEY TERMS

energy (p. **72**)
fossil fuels (p. **83**)
horsepower (p. **71**)
joule (p. **69**)
kinetic energy (p. **74**)
law of conservation of
    energy (p. **81**)

potential energy (p. **73**)
power (p. **71**)
watt (p. **71**)
work (p. **68**)

APPLYING THE CONCEPTS

1. According to the scientific definition of work, pushing on a rock accomplishes no work unless there is
   a. movement.
   b. a net force.
   c. an opposing force.
   d. movement in the same direction as the direction of the force.

2. The metric unit of a joule (J) is a unit of
   a. potential energy.
   b. work.
   c. kinetic energy.
   d. any of the above.

3. Which of the following is the combination of units called a joule?
   a. $kg \cdot m/s^2$
   b. kg/s^2
   c. $kg \cdot m^2/s^2$
   d. $kg \cdot m/s$

4. The newton-meter is called a joule. The similar unit in the English system, the foot-pound, is called a
   a. horsepower.
   b. slug.
   c. gem.
   d. foot-pound, with no other name.

5. Power is
   a. the rate at which work is done.
   b. the rate at which energy is expended.
   c. work per unit time.
   d. any of the above.

6. A power rating of 600 ft·lb/s is
   a. more than 1 horsepower.
   b. exactly 1 horsepower.
   c. less than 1 horsepower.

7. A N·m/s is a unit of
   a. work.
   b. power.
   c. energy.
   d. none of the above.

8. Which of the following is a combination of units called a watt?
   a. N·m/s
   b. $kg \cdot m^2/s^2/s$
   c. J/s
   d. all of the above

9. About how many watts are equivalent to 1 horsepower?
   a. 7.5
   b. 75
   c. 750
   d. 7,500

10. A kilowatt-hour is actually a unit of
    a. power.
    b. work.
    c. time.
    d. electrical charge.

11. In calculating the upward force required to lift an object, it is necessary to use g if the mass is given in kg. The quantity of g is not needed if the weight is given in lb because
    a. the rules of measurement are different in the English system.
    b. the symbol for metric mass has the letter "g" in it, and the symbol for pound does not.
    c. a pound is defined as a measure of force, and a kilogram is not.
    d. a kilogram is a unit of weight.

12. The potential energy of a box on a shelf, relative to the floor, is a measure of
    a. the work that was required to put the box on the shelf from the floor.
    b. the weight of the box times the distance above the floor.
    c. the energy the box has because of its position above the floor.
    d. all of the above.

13. A rock on the ground is considered to have zero potential energy. In the bottom of a well, then, the rock would be considered to have
    a. zero potential energy, as before.
    b. negative potential energy.
    c. positive potential energy.
    d. zero potential energy, but will require work to bring it back to ground level.

14. Which quantity has the greatest influence on the amount of kinetic energy that a large truck has while moving down the highway?
    a. mass
    b. weight
    c. velocity
    d. size

15. Which of the following is a form of energy that is a kind of potential energy?
    a. radiant
    b. electrical
    c. chemical
    d. none of the above

16. Electrical energy can be converted to
    a. chemical energy.
    b. mechanical energy.
    c. radiant energy.
    d. any of the above.

17. Most all energy comes to and leaves the earth in the form of
    a. nuclear energy.
    b. chemical energy.
    c. radiant energy.
    d. kinetic energy.

18. The law of conservation of energy is basically that
    a. energy must not be used up faster than it is created or the supply will run out.
    b. energy should be saved because it is easily destroyed.
    c. energy is never created or destroyed.
    d. you are breaking a law if you needlessly destroy energy.

19. The most widely used source of energy today is
    a. coal.
    b. petroleum.
    c. nuclear.
    d. water power.

20. The accounting device of a "barrel of oil" is defined to hold how many U.S. gallons of petroleum?
    a. 24
    b. 42
    c. 55
    d. 100

Answers

1. d 2. d 3. c 4. d 5. d 6. a 7. b 8. d 9. c 10. b 11. c 12. d 13. b 14. c 15. c 16. d 17. c 18. c 19. b 20. b

QUESTIONS FOR THOUGHT

1. How is work related to energy?
2. What is the relationship between the work done while moving a book to a higher bookshelf and the potential energy that the book has on the higher shelf?
3. Does a person standing motionless in the aisle of a moving bus have kinetic energy? Explain.
4. A lamp bulb is rated at 100 W. Why is a time factor not included in the rating?
5. Is a kWhr a unit of work, energy, power, or more than one of these? Explain.
6. If energy cannot be destroyed, why do some people worry about the energy supplies?
7. A spring clamp exerts a force on a stack of papers it is holding together. Is the spring clamp doing work on the papers? Explain.
8. Why are petroleum, natural gas, and coal called *fossil fuels*?
9. From time to time, people claim to have invented a machine that will run forever without energy input and develops more energy than it uses (perpetual motion). Why would you have reason to question such a machine?
10. Define a joule. What is the difference between a joule of work and a joule of energy?
11. Compare the energy needed to raise a mass 10 m on Earth to the energy needed to raise the same mass 10 m on the Moon. Explain the difference, if any.
12. What happens to the kinetic energy of a falling book when the book hits the floor?

PARALLEL EXERCISES

The exercises in groups A and B cover the same concepts. Solutions to group A exercises are located in appendix D.

Note: Neglect all frictional forces in all exercises.

Group A

1. A force of 200 N is needed to push a table across a level classroom floor for a distance of 3 m. How much work was done on the table?

Group B

1. One thousand two hundred joules of work are done while pushing a crate across a level floor for a distance of 1.5 m. What force was used to move the crate?

2. An 880 N box is pushed across a level floor for a distance of 5.0 m with a force of 440 N. How much work was done on the box?

3. How much work is done in raising a 10.0 kg backpack from the floor to a shelf 1.5 m above the floor?

4. If 5,000 J of work is used to raise a 102 kg crate to a shelf in a warehouse, how high was the crate raised?

5. A 60.0 kg student runs up a 5.00 m high stairway in a time of 3.92 s. (a) How many watts of power did she develop? (b) How many horsepower is this?

6. (a) How many horsepower is a 1,400 W blow dryer? (b) How many watts is a 3.5 hp lawnmower?

7. What is the kinetic energy of a 2,000 kg car moving at 72 km/hr?

8. How much work is needed to stop a 1,000.0 kg car that is moving straight down the highway at 54.0 km/hr?

9. A 1,000 kg car stops on top of a 51.02 m hill. (a) How much energy was used in climbing the hill? (b) How much potential energy does the car have?

10. What is the velocity of water that falls 100.0 m through the penstock of a hydroelectric dam?

11. Consider four identical trucks, all moving at 72 km/hr, but having different total masses of 1,000, 2,000, 3,000, and 4,000 kg because of loads they are carrying. Do some calculations, then make a generalization about how doubling the total mass of a truck changes the kinetic energy of that truck.

12. Consider three identical trucks, all with a mass of 3,000.0 kg, but each moving with a different velocity. One truck moves at 10.0 m/s, another at 20.0 m/s, and a third at 30.0 m/s. Do some calculations, then make a generalization about how doubling the velocity changes the kinetic energy of a given truck.

13. A horizontal force of 10.0 lb is needed to push a bookcase 5 ft across the floor. (a) How much work was done on the bookcase? (b) How much did the gravitational potential energy change as a result?

14. (a) How much work is done in moving a 2.0 kg book to a shelf 2.00 m high? (b) What is the potential energy of the book as a result? (c) How much kinetic energy will the book have as it hits the ground as it falls?

15. A 150 g baseball has a velocity of 30.0 m/s. What is its kinetic energy in J?

16. (a) What is the kinetic energy of a 1,000.0 kg car that is traveling at 90.0 km/hr? (b) How much work was done to give the car this kinetic energy? (c) How much work must be done to now stop the car?

17. A 60.0 kg jogger moving at 2.0 m/s decides to double the jogging speed. How did this change in speed change the kinetic energy?

18. A bicycle and rider have a combined mass of 70.0 kg and are moving at 6.00 m/s. A 70.0 kg person is now given a ride on the bicycle. (Total mass is 140.0 kg.) How did the addition of the new rider change the kinetic energy at the same speed?

19. A 170.0 lb student runs up a stairway to a classroom 25.0 ft above ground level in 10.0 s. (a) How much work did the student do? (b) What was the average power output in hp?

20. (a) How many seconds will it take a 20.0 hp motor to lift a 2,000.0 lb elevator a distance of 20.0 ft? (b) What was the average velocity of the elevator?

21. A ball is dropped from 9.8 ft above the ground. Using energy considerations only, find the velocity of the ball just as it hits the ground.

22. What is the velocity of a 1,000.0 kg car if its kinetic energy is 200 kJ?

23. A Foucault pendulum swings to 3.0 in above the ground at the highest point and is practically touching the ground at the lowest point. What is the maximum velocity of the pendulum?

24. An electric hoist is used to lift a 250.0 kg load to a height of 80.0 m in 39.2 s. (a) What is the power of the hoist motor in kW? (b) in hp?

2. How much work is done by a hammer that exerts a 980.0 N force on a nail, driving it 1.50 cm into a board?

3. A 5.0 kg textbook is raised a distance of 30.0 cm as a student prepares to leave for school. How much work did the student do on the book?

4. A 4.0 hp lawnmower engine moves the mower with a force of 1,492 N over a distance of 100.0 m. How much time did this require?

5. What is the horsepower of a 1,500.0 kg car that can go to the top of a 360.0 m high hill in exactly one minute?

6. (a) What is the kinetic energy of a 30.0 g bullet that is traveling at 200.0 m/s? (b) What velocity would you have to give a 60.0 g bullet to give it the same kinetic energy?

7. How much work will be done by a 30.0 g bullet traveling at 200 m/s?

8. A 10.0 kg box is lifted 15 m above the ground by a construction crane. (a) How much work did the crane do on the box? (b) How much potential energy does the box have relative to the ground? (c) With what velocity would the box strike the ground if it were to fall?

9. A force of 50.0 lb is used to push a box 10.0 ft across a level floor. (a) How much work was done on the box? (b) What is the change of potential energy as a result of this move?

10. (a) How much work is done in raising a 50.0 kg crate a distance of 1.5 m above a storeroom floor? (b) What is the change of potential energy as a result of this move? (c) How much kinetic energy will the crate have as it falls and hits the floor?

11. What is the kinetic energy in J of a 60.0 g tennis ball approaching a tennis racket at 20.0 m/s?

12. (a) What is the kinetic energy of a 1,500.0 kg car with a velocity of 72.0 km/hr? (b) How much work must be done on this car to bring it to a complete stop?

13. The driver of an 800.0 kg car decides to double the speed from 20.0 m/s to 40.0 m/s. What effect would this have on the amount of work required to stop the car, that is, on the kinetic energy of the car?

14. Compare the kinetic energy of an 800.0 kg car moving at 20.0 m/s to the kinetic energy of a 1,600.0 kg car moving at an identical speed.

15. A 175.0 lb hiker is able to ascend a 1,980.0 ft high slope in 1 hour and 45 minutes. (a) How much work did the hiker do? (b) What was the average power output in hp?

16. (a) What distance will a 10.0 hp motor lift a 2,000.0 lb elevator in 30.0 s? (b) What was the average velocity of this elevator during the lift?

17. A ball is dropped from 20.0 ft above the ground. (a) At what height is half of its energy kinetic and half potential? (b) Using energy considerations only, what is the velocity of the ball just as it hits the ground?

18. What is the velocity of a 60.0 kg jogger with a kinetic energy of 1,080.0 J?

19. A small sports car and a pickup truck start coasting down a 10.0 m hill together, side by side. Assuming no friction, what is the velocity of each vehicle at the bottom of the hill?

20. A 70.0 kg student runs up the stairs of a football stadium to a height of 10.0 m above the ground in 10.0 s. (a) What is the power of the student in kW? (b) in hp?

Sparks fly from a plate of steel as it is cut by an infrared laser. Today, lasers are commonly used to cut as well as weld metals, so the cutting and welding is done by light, not by a flame or electric current.

CHAPTER | Five

Heat and Temperature

Heat has been closely associated with the comfort and support of people throughout history. You can imagine the appreciation when your earliest ancestors first discovered fire and learned to keep themselves warm and cook their food. You can also imagine the wonder and excitement about 3000 B.C., when people put certain earthlike substances on the hot, glowing coals of a fire and later found metallic copper, lead, or iron. The use of these metals for simple tools followed soon afterwards. Today, metals are used to produce complicated engines that use heat for transportation and that do the work of moving soil and rock, construction, and agriculture. Devices made of heat-extracted metals are also used to control the temperature of structures, heating or cooling the air as necessary. Thus, the production and control of heat gradually built the basis of civilization today (figure 5.1).

The sources of heat are the energy forms that you learned about in chapter 4. The fossil fuels are *chemical* sources of heat. Heat is released when oxygen is combined with these fuels. Heat also results when *mechanical* energy does work against friction, such as in the brakes of a car coming to a stop. Heat also appears when *radiant* energy is absorbed. This is apparent when solar energy heats water in a solar collector or when sunlight melts snow. The transformation of *electrical* energy to heat is apparent in toasters, heaters, and ranges. *Nuclear* energy provides the heat to make steam in a nuclear power plant. Thus, all the energy forms can be converted to heat.

The relationship between energy forms and heat appears to give an order to nature, revealing patterns that you will want to understand. All that you need is some kind of explanation for the relationships—a model or theory that helps make sense of it all. This chapter is concerned with heat and temperature and their relationship to energy. It begins with a simple theory about the structure of matter, and then uses the theory to explain the concepts of heat, energy, and temperature changes.

THE KINETIC MOLECULAR THEORY

The idea that substances are composed of very small particles can be traced back to certain early Greek philosophers. The earliest record of this idea was written by Democritus during the fifth century B.C. He wrote that matter was empty space filled with tremendous numbers of tiny, indivisible particles called *atoms.* This idea, however, was not acceptable to most of the ancient Greeks, because matter seemed continuous, and empty space was simply not believable. The idea of atoms was rejected by Aristotle as he formalized his belief in continuous matter composed of the earth, air, fire, and water elements. Aristotle's belief about matter, like his beliefs about motion, predominated through the 1600s. Some people, such as Galileo and Newton, believed the ideas about matter being composed of tiny particles, or atoms, since this theory seemed to explain the behavior of matter. Widespread acceptance of the particle model did not occur, however, until strong evidence was developed through chemistry in the late 1700s and early 1800s. The experiments finally led to a collection of assumptions about the small particles of matter and the space around them. Collectively, the assumptions could be called the **kinetic molecular theory.** The following is a general description of some of these assumptions.

Molecules

The basic assumption of the kinetic molecular theory is that all matter is made up of tiny, basic units of structure called *atoms.* Atoms are neither divided, created, nor destroyed during any type of chemical or physical change. There are similar groups of atoms that make up the pure substances known as chemical *elements.* Each element has its own kind of atom, which is different from the atoms of other elements. For example, hydrogen, oxygen, carbon, iron, and gold are chemical elements, and each has its own kind of atom.

In addition to the chemical elements, there are pure substances called *compounds* that have more complex units of structure (figure 5.2). Pure substances, such as water, sugar, and alcohol, are composed of atoms of two or more elements that join together in definite proportions. Water, for example, has structural units that are made up of two atoms of hydrogen tightly bound to one atom of oxygen (H_2O). These units are not easily broken apart and stay together as small physical particles of which water is composed. Each is the smallest particle of water that can exist, a molecule of water. A *molecule* is generally defined as a tightly bound group of atoms in which the atoms maintain their identity. How atoms become bound together to form molecules is discussed in chapters 9–11.

Some elements exist as gases at ordinary temperatures, and all elements are gases at sufficiently high temperatures. At ordinary temperatures, the atoms of oxygen, nitrogen, and other gases are paired in groups of two to form *diatomic molecules.* Other gases, such as helium, exist as single, unpaired atoms at ordinary temperatures. At sufficiently high temperatures, iron, gold, and other metals vaporize to form gaseous, single, unpaired atoms. In the kinetic molecular theory the term *molecule* has the additional meaning of the smallest, ultimate particle of matter that can exist. Thus, the ultimate particle of a gas,

FIGURE 5.1

Heat and modern technology are inseparable. These glowing steel slabs, at over 1,100°C (about 2,000°F), are cut by an automatic flame torch. The slab caster converts 300 tons of molten steel into slabs in about 45 minutes. The slabs are converted to sheet steel for use in the automotive, appliance, and building industries.

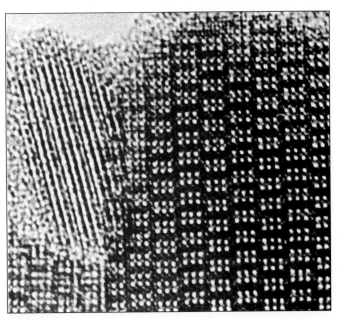

FIGURE 5.2

Metal atoms appear in the micrograph of a crystal of titanium niobium oxide, magnified 7,800,000 times by an electron microscope.

whether it is made up of two or more atoms bound together or of a single atom, is conceived of as a molecule. A single atom of helium, for example, is known as a *monatomic molecule.* For now, a **molecule** is defined as the smallest particle of a compound, or a gaseous element, that can exist and still retain the characteristic properties of that substance.

Molecules Interact

Some molecules of solids and liquids interact, strongly attracting and clinging to each other. When this attractive force is between the same kind of molecules, it is called *cohesion.* It is a stronger cohesion that makes solids and liquids different from gases, and without cohesion all matter would be in the form of gases. Sometimes one kind of molecule attracts and clings to a different kind of molecule. The attractive force between unlike molecules is called *adhesion.* Water wets your skin because the adhesion of water molecules and skin is stronger than the cohesion of water molecules. Some substances, such as glue, have a strong force of adhesion when they harden from a liquid state, and they are called adhesives.

Phases of Matter

Different phases of matter have different molecular arrangements (figure 5.3). You know that matter can occur in the three phases: solid, liquid, and gas. The different characteristics of each phase can be attributed to the molecular arrangements and the strength of attraction between the molecules (table 5.1).

FIGURE 5.3

(*A*) In a solid, molecules vibrate around a fixed equilibrium position and are held in place by strong molecular forces. (*B*) In a liquid, molecules can rotate and roll over each other because the molecular forces are not as strong. (*C*) In a gas, molecules move rapidly in random, free paths.

TABLE 5.1	The shape and volume characteristics of solids, liquids, and gases are reflections of their molecular arrangements*		
	Solids	**Liquids**	**Gases**
Shape	Fixed	Variable	Variable
Volume	Fixed	Fixed	Variable

*These characteristics are what would be expected under ordinary temperature and pressure conditions on the surface of the earth.

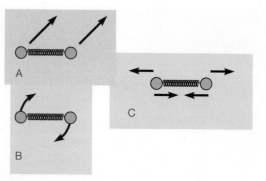

FIGURE 5.4

The basic forms of kinetic energy of molecules. (*A*) Translational motion is the motion of a molecule as a whole moving from place to place. (*B*) Rotational motion is the motion of a turning molecule. (*C*) Vibrational motion is the back-and-forth movement of a vibrating molecule.

Solids have definite shapes and volumes because they have molecules that are fixed distances apart and bound by relatively strong cohesive forces. Each molecule is a nearly fixed distance from the next, but it does vibrate and move around an equilibrium position. The masses of these molecules and the spacing between them determine the density of the solid. The hardness of a solid is the resistance of a solid to forces that tend to push its molecules further apart.

Liquids have molecules that are not confined to an equilibrium position as in a solid. The molecules of a liquid are close together and bound by cohesive forces that are not as strong as in a solid. This permits the molecules to move from place to place within the liquid. The molecular forces are strong enough to give the liquid a definite volume but not strong enough to give it a definite shape. Thus, a pint of milk is always a pint of milk (unless it is under tremendous pressure) and takes the shape of the container holding it. Because the forces between the molecules of a liquid are weaker than the forces between the molecules of a solid, a liquid cannot support the stress of a rock placed on it as a solid does. The liquid molecules *flow*, rolling over each other as the rock pushes its way between the molecules. Yet, the molecular forces are strong enough to hold the liquid together, so it keeps the same volume.

Gases are composed of molecules with weak cohesive forces acting between them. The gas molecules are relatively far apart and move freely in a constant, random motion that is changed often by collisions with other molecules. Gases therefore have neither fixed shapes nor fixed volumes.

There are other distinctions between the phases of matter. The term *vapor* is sometimes used to describe a gas that is usually in the liquid phase. Water vapor, for example, is the gaseous form of liquid water. Liquids and gases are collectively called *fluids* because of their ability to flow, a property that is lacking in most solids. A not-so-ordinary phase of matter on the earth is called *plasma*. Plasma occurs at extremely high temperatures and is made up of charged parts of atoms.

Molecules Move

Suppose you are in an evenly heated room with no air currents. If you open a bottle of ammonia, the odor of ammonia is soon noticeable everywhere in the room. According to the kinetic molecular theory, molecules of ammonia leave the bottle and bounce around among the other molecules making up the air until they are everywhere in the room, slowly becoming more evenly distributed. The ammonia molecules *diffuse*, or spread, throughout the room. The ammonia odor diffuses throughout the room faster if the air temperature is higher and slower if the air temperature is lower. This would imply a relationship between the temperature and the speed at which molecules move about.

The mathematical relationship between the temperature of a gas and the motion of molecules was formulated in 1857 by Rudolf Clausius. He showed that the temperature of a gas is proportional to the average kinetic energy of the gas molecules. This means that ammonia molecules have a greater average velocity at a higher temperature and a slower average velocity at a lower temperature. This explains why gases diffuse at a greater rate at higher temperatures. Recall, however, that kinetic energy involves the mass of the molecules as well as their velocity ($KE = 1/2 \ mv^2$). It is the *average kinetic energy* that is proportional to the temperature, which involves the molecular mass as well as the molecular velocity.

The kinetic energy of molecules can be in three basic forms: vibrational, rotational, or translational (involving the motion of a molecule as a whole) (figure 5.4). The kinetic energy of the molecules of a solid is mostly in the form of vibrational kinetic energy. The molecules of a liquid can have vibrational, rotational, and some translational kinetic energy. The molecules of a liquid have been compared to dancers on a packed dance floor; they are free to move about, but it is difficult to move across the room. Gas molecules, on the other hand, can move about easily since they are far apart. Gas molecules can have all three forms of kinetic energy—translational, rotational, and vibrational. The *total* kinetic energy of the molecules of a substance can be complicated combinations of the three basic forms plus pulsing shape changes, twisting, and other forms of motion. In general,

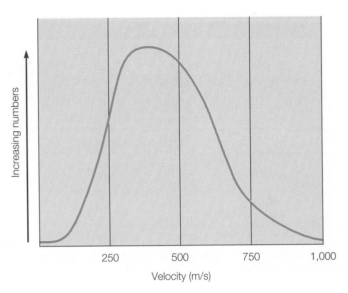

FIGURE 5.5

The number of oxygen molecules with certain velocities that you might find in a sample of air at room temperature. Notice that a few are barely moving and some have velocities over 1,000 m/s at a given time, but the *average* velocity is somewhere around 500 m/s.

the type of motion represented by kinetic energy varies with the state of matter. Whether the kinetic energy is jiggling, vibrating, rotating, or moving from place to place, the **temperature** of a substance is *a measure of the average kinetic energy of the molecules making up the substance* (figure 5.5).

TEMPERATURE

If you ask people about the temperature, they usually respond with a referent ("hotter than the summer of '89") or a number ("68°F or 20°C"). Your response, or feeling, about the referent or number depends on a number of factors, including a *relative* comparison. A temperature of 20°C (68°F), for example, might seem cold during the month of July but warm during the month of January. The 20°C temperature is compared to what is expected at the time, even though 20°C is 20°C, no matter what month it is.

When people ask about the temperature, they are really asking *how hot or how cold something is.* Without a thermometer, however, most people can do no better than *hot* or *cold,* or perhaps *warm* or *cool,* in describing a relative temperature. Even then, there are other factors that confuse people about temperature. Your body judges temperature on the basis of the net *direction* of energy flow. You call energy flowing into your body *warm* and energy flowing out *cool.* Perhaps you have experienced having your hands in snow for some time, then washing your hands in cold water. The cold water feels warm. Your hands are colder than the water, energy flows into your hands, and they communicate "warm."

Animal Temperature Regulation

Unicellular organisms are all *poikilotherms,* organisms whose body temperature varies. The body temperature of a multicellular ant changes as the external temperature changes, which means that at colder temperatures, poikilotherms also have lower metabolic rates. Most animals, including insects, worms, and reptiles, are poikilotherms. Some animals such as deer, however, are *homeotherms,* which means that they maintain a constant body temperature that is generally higher than the environmental temperature.

The body temperature of a homeotherm remains constant regardless of changes in environmental temperature, and the body temperature of a poikilotherm is dependent upon environmental temperature. *Ectothermic* describes animals that regulate their body temperature by moving to places where they can be most comfortable. A good example of this occurs in snakes and other reptiles. *Endothermic* describes animals that have internal temperature-regulating mechanisms and can maintain a relatively constant body temperature in spite of wide variations in the temperature of their environment, as occurs in humans. There is a price to be paid, however, since ten times more energy is required to keep an endotherm going as compared to an ectotherm!

Thermometers

The human body is a poor sensor of temperature, so a device called a *thermometer* is used to measure the hotness or coldness of something. Most thermometers are based on the relationship between some property of matter and changes in temperature. Almost all materials expand with increasing temperatures. A strip of metal is slightly longer when hotter and slightly shorter when cooler, but the change of length is too small to be useful in a thermometer. A more useful, larger change is obtained when two metals that have different expansion rates are bonded together in a strip. The bimetallic strip will bend toward the metal with less expansion when the strip is heated (figure 5.6). Such a bimetallic strip is formed into a coil and used in thermostats and dial thermometers (figure 5.7).

The common glass thermometer is a glass tube with a small bore connected to a relatively large glass bulb. The bulb contains mercury or colored alcohol, which expands up the bore with increases in temperature and contracts back toward the bulb with decreases in temperature. The height of this liquid column is used with a referent scale to measure temperature. Some thermometers, such as a fever thermometer, have a small constriction

Room temperature

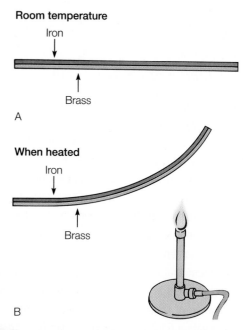

A

When heated

B

FIGURE 5.6

(*A*) A bimetallic strip is two different metals, such as iron and brass, bonded together as a single unit, shown here at room temperature. (*B*) Since one metal expands more than the other, the strip will bend when it is heated. In this example, the brass expands more than the iron, so the bimetallic strip bends away from the brass.

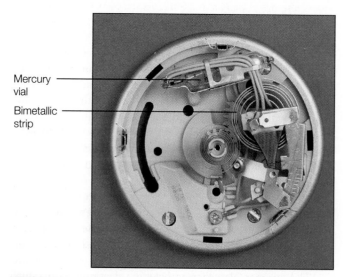

FIGURE 5.7

This thermostat has a coiled bimetallic strip that expands and contracts with changes in the room temperature. The attached vial of mercury is tilted one way or the other, and the mercury completes or breaks an electric circuit that turns the heating or cooling system on or off.

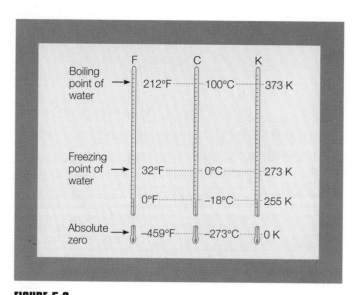

FIGURE 5.8

The Fahrenheit, Celsius, and Kelvin temperature scales.

in the bore so the liquid cannot normally return to the bulb. Thus, the thermometer shows the highest reading, even if the temperature it measures has fluctuated up and down during the reading. The liquid must be forced back into the bulb by a small swinging motion, bulb-end down, then sharply stopping the swing with a snap of the wrist. The inertia of the mercury in the bore forces it past the constriction and into the bulb. The fever thermometer is then ready to use again.

Thermometer Scales

There are several referent scales used to define numerical values for measuring temperatures (figure 5.8). The **Fahrenheit scale** was developed by the German physicist Gabriel D. Fahrenheit in about 1715. Fahrenheit invented a mercury-in-glass thermometer with a scale based on two arbitrarily chosen reference points. The original Fahrenheit scale was based on the temperature of an ice and salt mixture for the lower reference point (0°) and the temperature of the human body as the upper reference point (about 100°). Thus, the original Fahrenheit scale was a centigrade scale with 100 divisions between the high and the low reference points. The distance between the two reference points was then divided into equal intervals called *degrees*. There were problems with identifying a "normal" human body temperature as a reference point, since body temperature naturally changes during a given day and from day to day. The only consistent thing about the human body temperature is constant change. The standards for the Fahrenheit scale were eventually changed to something more consistent, the freezing point and the boiling point of water at normal atmospheric pressure. The original scale was retained with the new reference points, however, so the "odd" numbers of 32°F (freezing point of water) and 212°F (boiling point of water under normal pressure) came to be the

CONNECTIONS

Infrared Fever Thermometers

Today, scientists have developed a new type of thermometer and a way around the problems of using a glass mercury fever thermometer. This new approach measures the internal core temperature by quickly reading infrared radiation from the eardrum. All bodies with a temperature above absolute zero emit infrared radiation, including your body. Since the intensity of the infrared radiation is a sensitive function of body temperature, reading the infrared radiation emitted will tell you about the temperature of that body.

The human eardrum is close to the hypothalamus, the body's thermostat, so a temperature reading taken here must be close to the temperature of the internal core. You cannot use a mercury thermometer in the ear because of the very real danger of puncturing the eardrum, along with obtaining doubtful readings from a mercury bulb. You can use a pyroelectric material to measure the infrared radiation coming from the entrance to the ear canal, however, to quickly obtain a temperature reading. A pyroelectric material is a polarized crystal that generates an electric charge in proportion to a temperature change.

The infrared fever thermometer has a short barrel, which is inserted in the ear canal opening. A button opens a shutter inside the battery-powered device, admitting infrared radiation for about 300 milliseconds. Infrared radiation from the ear canal increases the temperature of a thin pyroelectric crystal, which develops an electric charge. A current spike from the pyroelectric sensor now moves through some filters and converters, and into a microprocessor chip. The chip is programmed with the relationship between the body temperature and the infrared radiation emitted. Using this information, it calculates the temperature by measuring the current spike produced by the infrared radiation falling on the pyroelectric crystal. The microprocessor now sends the temperature reading to an LCD display on the outside of the device, where it can be read almost immediately.

reference points. There are 180 equal intervals, or degrees, between the freezing and boiling points on the Fahrenheit scale.

The **Celsius scale** was invented by Anders C. Celsius, a Swedish astronomer, in about 1735. The Celsius scale uses the freezing point and the boiling point of water at normal atmospheric pressure, but it has different arbitrarily assigned values. The Celsius scale identifies the freezing point of water as 0°C and the boiling point as 100°C. There are 100 equal intervals, or degrees, between these two reference points, so the Celsius scale is sometimes called the *centigrade* scale.

There is nothing special about either the Celsius scale or the Fahrenheit scale. Both have arbitrarily assigned numbers, and one is no more accurate than the other. The Celsius scale is more convenient because it is a decimal scale and because it has a direct relationship with a third scale to be described shortly, the Kelvin scale. Both scales have arbitrarily assigned reference points and an arbitrary number line that indicates *relative* temperature changes. Zero is simply one of the points on each number line and does *not* mean that there is no temperature. Likewise, since the numbers are relative measures of temperature change, 2° is not twice as hot as a temperature of 1° and 10° is not twice as hot as a temperature of 5°. The numbers simply mean some measure of temperature *relative to* the freezing and boiling points of water under normal conditions.

You can convert from one temperature to the other by considering two differences in the scales: (1) the difference in the degree size between the freezing and boiling points on the two scales, and (2) the difference in the values of the lower reference points.

The Fahrenheit scale has 180° between the boiling and freezing points (212°F − 32°F) and the Celsius scale has 100° between the same two points. Therefore each Celsius degree is 180/100 or 9/5 as large as a Fahrenheit degree. Each Fahrenheit degree is 100/180 or 5/9 of a Celsius degree. You know that this is correct since there are more Fahrenheit degrees than Celsius degrees between freezing and boiling. The relationship between the degree sizes is 1°C = 9/5°F and 1°F = 5/9°C. In addition, considering the difference in the values of the lower reference points (0°C and 32°F) gives the equations for temperature conversion.

$$T_F = \frac{9}{5} T_C + 32°$$

equation 5.1

$$T_C = \frac{5}{9} (T_F - 32°)$$

equation 5.2

There is a temperature scale that does not have arbitrarily assigned reference points and zero *does* mean nothing. This is not a relative scale but an absolute temperature scale called the **absolute scale,** or **Kelvin scale.** The zero point on the absolute scale is thought to be the lowest limit of temperature. **Absolute zero** is the *lowest temperature possible,* occurring when all random motion of molecules has ceased. Absolute zero is written as 0 K. A degree symbol is not used, and the K stands for the SI standard scale unit, Kelvin (see table 1.2). The absolute scale uses the same degree size as the Celsius scale and −273°C = 0 K. Note in figure 5.8 that 273 K is the freezing point of water, and

CONNECTIONS

Ice Fishes

Seawater in the Antarctic stays the same temperature of −2.2°C (about 28°F) year-round, just below the freezing point of pure water. During the winter, the air temperature cools to −45° to −75°C (about −50° to −100°F). This freezes the surface water, but the water that does not freeze under the ice stays about −2.2°C (about 28°F). The water also remains about this same temperature when the ice melts or floats away in the summer. Antarctic seawater below the surface is therefore consistently cold, but remains liquid water just below the freezing point.

How can fish live in water below the freezing point? The fish called *notothenioids* indeed do live and flourish in the Antarctic waters that hover just below the freezing point. The strangest of these fish are called ice fishes. Ice fish have a backbone, grow up to 2 feet long, but do not have red blood cells. Without red blood cells their blood is milky white. This blood has much less salt in it than seawater, so you might expect it to freeze in below-freezing water. But the blood of these fish does not freeze. In the blood of these fish are glycoproteins that are resistant to cold temperatures. It is a protein antifreeze that helps to lower the freezing point of their blood to below that of the surrounding seawater.

---

373 K is the boiling point. You could think of the absolute scale as a Celsius scale with the zero point shifted by 273°. Thus, the relationship between the absolute and Celsius scales is

$$T_K = T_C + 273$$

equation 5.3

A temperature of absolute zero has never been reached, but scientists have cooled a sample of sodium to 700 nanokelvins, or 700 billionths of a kelvin above absolute zero.

HEAT

The concept of **heat** is not the same as the concept of temperature, which deals with the hotness or coldness of some object or part of the environment. Temperature is related to the average kinetic energy of the molecules of a body. Heat, on the other hand, involves more than just the average kinetic energy of the molecules. The relationship between mechanical energy and temperature will help explain the difference between heat and temperature, so we will consider it first.

If you rub your hands together a few times they will feel a little warmer. If you rub them together vigorously for a while they will feel a lot warmer, maybe hot. The temperature increase that results from the rubbing friction is proportional to the amount of mechanical energy expended. To understand how this happens, imagine two metal blocks, each a solid system of molecules held in place by strong forces. Suppose you tightly slide the surface of one block against a surface on the other block. The surfaces are not perfectly smooth, so some of the molecules on one surface will catch on the other surface. The solid continues to move and each molecule is gradually pulled away from its home position, stretching the molecular forces that are holding it. A few molecules might be pulled completely

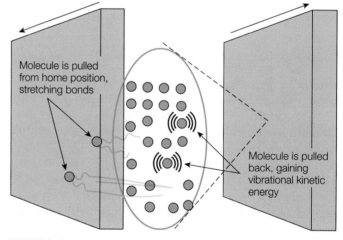

FIGURE 5.9

One theory about how friction results in increased temperatures: Molecules on one moving surface will catch on another surface, stretching the molecular forces that are holding it. They are pulled back to their home position with a snap, resulting in a gain of vibrational kinetic energy.

away, but most are pulled back to their home position with a snap, resulting in a gain of vibrational energy (figure 5.9). With lots of molecular snags and snap-backs, you can see that the average kinetic energy—and the temperature—of the molecules at the surface is going to increase.

A temperature increase takes place anytime mechanical energy causes one surface to rub against another. The two surfaces could be solids, such as the two blocks, but they can also be the surface of a solid and a fluid, such as air. A solid object moving through the air encounters fluid friction, which results in a

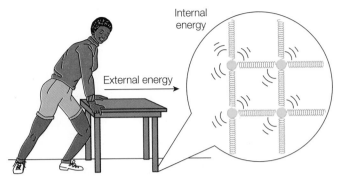

FIGURE 5.10

External energy is the kinetic and potential energy that you can see. *Internal energy* is the total kinetic and potential energy of molecules. When you push a table across the floor, you do work against friction. Some of the external mechanical energy goes into internal kinetic and potential energy, and the bottom surface of the legs becomes warmer.

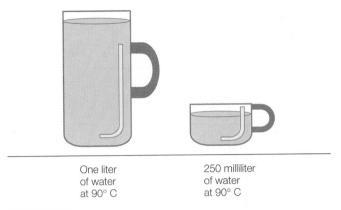

One liter of water at 90° C   250 milliliter of water at 90° C

FIGURE 5.11

Heat and temperature are different concepts, as shown by a liter of water (1,000 mL) and a 250 mL cup of water, both at the same temperature. You know the liter of water contains more heat since it will require more ice cubes to cool it to, say, 25°C than will be required for the cup of water. In fact, you will have to remove 48,750 *additional* calories to cool the liter of water.

higher temperature of the surface. A high-velocity meteor enters the earth's atmosphere and is heated so much from the friction that it begins to glow, resulting in the fireball and smoke trail of a "falling star."

A meteor high in the atmosphere has potential energy because of its position above the earth's surface, and it has kinetic energy by virtue of its motion. The molecules of the meteor, on the other hand, can also have molecular kinetic energy and they also have molecular potential energy. To distinguish between the energy of the object and the energy of its molecules, we use the terms "external" and "internal" energy. **External energy** is the total potential and kinetic energy of an everyday-sized object. All the kinetic and potential energy considerations discussed in previous chapters were about the external energy of an object.

Internal energy is the total kinetic and potential energy of the *molecules* of an object. The kinetic energy of a molecule can be much more complicated than straight-line velocity might suggest, however, as a molecule can have many different types of motion at the same time (pulsing, twisting, turning, etc.). Overall, internal energy is characterized by properties such as temperature, density, heat, volume, pressure of a gas, and so forth.

When you push a table across the floor, the observable *external* kinetic energy of the table is transferred to the *internal* kinetic energy of the molecules between the table legs and the floor, resulting in a temperature increase (figure 5.10). The relationship between external and internal kinetic energy explains why the heating is proportional to the amount of mechanical energy used.

Heat as Energy Transfer

Temperature is a measure of the degree of hotness or coldness of a body, a measure that is based on the average molecular kinetic energy. Heat, on the other hand, is based on the *total internal*

energy of the molecules of a body. You can see one difference in heat and temperature by considering a cup of water and a large tub of water. If both the small and the large amount of water have the same temperature, both must have the same average molecular kinetic energy. Now, suppose you wish to cool both by, say, 20°. The large tub of water would take much longer to cool, so it must be that the large amount of water has more heat (figure 5.11). Heat is a measure based on the *total* internal energy of the molecules of a body, and there is more total energy in a large tub of water than in a cup of water at the same temperature.

How can we measure heat? Since it is difficult to see molecules, internal energy is difficult to measure directly. Thus heat is nearly always measured during the process of a body gaining or losing energy. This measurement procedure will also give us a working definition of heat:

> **Heat is a measure of the internal energy that has been absorbed or transferred from one body to another.**

The *process* of increasing the internal energy is called "heating" and the *process* of decreasing internal energy is called "cooling." The word "process" is italicized to emphasize that heat is energy in transit, not a material thing you can add or take away. Heat is understood to be a measure of internal energy that can be measured as energy flows in or out of an object.

There are two general ways that heating can occur. These are (1) from a temperature difference, with energy moving from the region of higher temperature, and (2) from an object gaining energy by way of an energy-form conversion.

When a *temperature difference* occurs, energy is transferred from a region of higher temperature to a region of lower temperature. Energy flows from a hot range element, for example, to a pot of cold water on a range. It is a natural process for energy to flow from a region of a higher temperature to a region of a lower temperature just as it is natural for a ball to roll downhill. The temperature of an

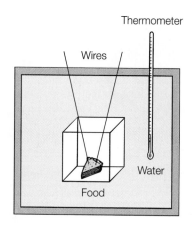

FIGURE 5.12

The Calorie value of food is determined by measuring the heat released from burning the food. If there is 10.0 kg of water and the temperature increased from 10° to 20°C, the food contained 100 Calories (100,000 calories). The food illustrated here would release much more energy than this.

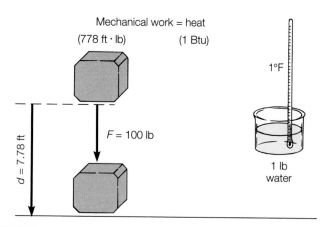

FIGURE 5.13

Joule worked with the English system of measurement used during his time. When a 100 lb object falls 7.78 ft, it can do 778 ft·lb of work. If the work is done against friction, as by stirring 1 lb of water, the heat produced by the work raises the temperature 1°F.

object and the temperature of the surroundings determines if heat will be transferred to or from an object. The terms "heating" and "cooling" describe the direction of energy flow, naturally moving from a region of higher energy to one of lower energy.

The internal energy of an object can be increased during an *energy-form conversion* (mechanical, radiant, electrical, etc.), so we say that heating is taking place. The classical experiments by Joule showed an equivalence between mechanical energy and heating, electrical energy and heating, and other conversions. On a molecular level, the energy forms are doing work on the molecules which can result in an increase of internal energy. Thus heating by energy-form conversion is actually a transfer of energy by *working*. This brings us back to the definition that "energy is the ability to do work." We can mentally note that this includes the ability to do work at the molecular level.

Heating that takes place because of a temperature difference will be considered in more detail after considering how heat is measured.

Measures of Heat

Since heating is a method of energy transfer, a quantity of heat can be measured just like any quantity of energy. The metric unit for measuring work, energy, or heat is the *joule*. However, the separate historical development of the concepts of heat (Black, Rumford, Joule) and the concepts of motion (Galileo, Newton) resulted in separate units, some based on temperature differences.

The metric unit of heat is called the **calorie** (cal), a leftover term from the caloric theory of heat. A calorie is defined as the *amount of energy (or heat) needed to increase the temperature of 1 gram of water 1 degree Celsius*. A more precise definition specifies the degree interval from 14.5°C to 15.5°C because the energy required varies slightly at different temperatures. This precise definition is not needed for a general discussion. A **kilocalorie** (kcal) is the *amount of energy (or heat) needed to increase the tem-*

perature of 1 kilogram of water 1 degree Celsius. The measure of the energy released by the oxidation of food is the kilocalorie, but it is called the Calorie (with a capital C) by nutritionists (figure 5.12). This results in much confusion, which can be avoided by making sure that the scientific calorie is never capitalized (cal) and the dieter's Calorie is always capitalized. The best solution would be to call the Calorie what it is, a kilocalorie (kcal).

The English system's measure of heating is called the **British thermal unit** (Btu). A Btu is *the amount of energy (or heat) needed to increase the temperature of 1 pound of water 1 degree Fahrenheit.* The Btu is commonly used to measure the heating or cooling rates of furnaces, air conditioners, water heaters, and so forth. The rate is usually expressed or understood to be in Btu per hour. A much larger unit is sometimes mentioned in news reports and articles about the national energy consumption. This unit is the *quad,* which is 1 quadrillion Btu (a million billion or 10^{15} Btu).

The relationship between the three units of heat is

$$252 \text{ cal} = 1.0 \text{ Btu} = 0.252 \text{ kcal}$$

The amount of heat generated by mechanical work was first investigated by Count Rumford, then quantified by Joule in 1849. Using English units, Joule determined the *mechanical equivalent of heat* (figure 5.13) to be

$$778 \text{ ft·lb} = 1 \text{ Btu}$$

In metric units, the mechanical equivalent of heat is

$$4.184 \text{ J} = 1 \text{ cal}$$

or

$$4,184 \text{ J} = 1 \text{ kcal}$$

The establishment of this precise proportionality means that, fundamentally, mechanical energy and heat are different forms of the same thing.

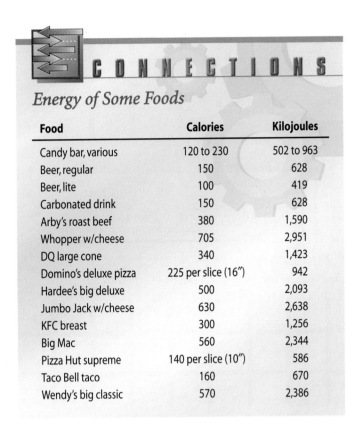

Energy of Some Foods

Food	Calories	Kilojoules
Candy bar, various	120 to 230	502 to 963
Beer, regular	150	628
Beer, lite	100	419
Carbonated drink	150	628
Arby's roast beef	380	1,590
Whopper w/cheese	705	2,951
DQ large cone	340	1,423
Domino's deluxe pizza	225 per slice (16″)	942
Hardee's big deluxe	500	2,093
Jumbo Jack w/cheese	630	2,638
KFC breast	300	1,256
Big Mac	560	2,344
Pizza Hut supreme	140 per slice (10″)	586
Taco Bell taco	160	670
Wendy's big classic	570	2,386

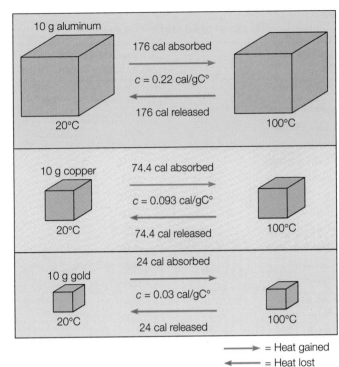

FIGURE 5.14

Of these three metals, aluminum needs the most heat per gram per degree when warmed, and releases the most heat when cooled.

Specific Heat

There are at least three variables that influence the energy transfer that takes place during heating: (1) the *temperature change*, (2) the *mass* of the substance, and (3) the nature of the *material* being heated. In some normal range of temperature (between 0°C and 100°C), you would expect a temperature change that is proportional to the amount of heating or cooling. The quantity of heat (Q) needed to increase the temperature of a pot of water from an initial temperature (T_i) to a final temperature (T_f) is therefore proportional to ($T_f - T_i$), or $Q \propto (T_f - T_i)$. Recalling that the symbol Δ means "a change in," this relationship could also be written as $Q \propto \Delta T$. This simply means that the temperature change of a substance is proportional to the quantity of heat that is added. The quantity ΔT is found from $T_f - T_i$.

The quantity of heat (Q) absorbed or given off during a certain change of temperature is also proportional to the mass (m) of the substance being heated or cooled. A larger mass requires more heat to go through the same temperature change than a smaller mass. In symbols, $Q \propto m$.

Different materials require different quantities of heat (Q) to go through the same temperature range when their masses are equal (figure 5.14). If the units for this relationship are the same units used to measure the quantity of heat (Q), then all the relationships can be written in equation form,

$$Q = mc\Delta T$$

equation 5.4

where c is called the **specific heat.** The specific heat is *the amount of energy (or heat) needed to increase the temperature of 1 gram of a substance 1 degree Celsius.* Specific heat is related to the internal structure of a substance; some of the energy goes into the internal potential energy of the molecules and some goes into the internal kinetic energy of the molecules. The different values for the specific heat of different substances are related to the number of molecules in the 1 gram sample and to the way they form a molecular structure. If you could isolate a single molecule of solid metal, for example, you would find that it would take the same amount of energy for each molecule to produce the same kinetic energy, that is, the same temperature increase. Table 5.2 lists some specific heats of some common substances.

Note that equation 5.4 can be used for problems of heating *or* cooling. A negative result would mean that the energy is leaving the material, or cooling. When two materials of different temperatures are involved in heat transfer and are perfectly insulated from the surroundings, the heat lost by one will equal the heat gained by the other,

heat lost = heat gained

(by warmer material)(by cooler material)

or

$$Q_{lost} = Q_{gained}$$

or

$$(mc\Delta T)_{lost} = (mc\Delta T)_{gained}$$

TABLE 5.2	The specific heat of selected substances
Substance	**Specific Heat (cal/gC° or kcal/kgC°)**
Air	0.17
Aluminum	0.22
Concrete	0.16
Copper	0.093
Glass (average)	0.160
Gold	0.03
Ice	0.500
Iron	0.11
Lead	0.0305
Mercury	0.033
Seawater	0.93
Silver	0.056
Soil (average)	0.200
Steam	0.480
Water	1.00

Note: To convert to specific heat in J/kgC°, multiply each value by 4,184. Also note that 1 cal/gC° = 1 kcal/kgC°.

Heat Flow

In a previous section you learned that heating is a transfer of energy that involves (1) a temperature difference or (2) energy-form conversions. Heat transfer that takes place because of a temperature difference takes place in three different ways: by conduction, convection, or radiation.

Conduction

Anytime there is a temperature difference, there is a natural transfer of heat from the region of higher temperature to the region of lower temperature. In solids, this transfer takes place as heat is *conducted* from a warmer place to a cooler one. Recall that the molecules in a solid vibrate in a fixed equilibrium position and that molecules in a higher temperature region have more kinetic energy, on the average, than those in a lower temperature region. When a solid, such as a metal rod, is held in a flame the molecules in the warmed end vibrate violently. Through molecular interaction, this increased energy of vibration is passed on to the adjacent, slower-moving molecules, which also begin to vibrate more violently. They, in turn, pass on more vibrational energy to the molecules next to them. The increase in activity thus moves from molecule to molecule, causing the region of increased activity to extend along the rod. This is called **conduction,** the transfer of energy from molecule to molecule (figure 5.15).

The *rate* of conduction depends on the temperature difference between the two regions, the area and the thickness of the substance, and the nature of the material. Table 5.3 shows the conduction of heat through various materials when the temperature

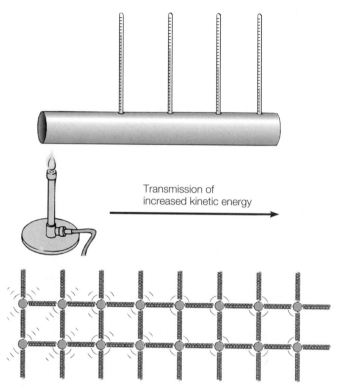

FIGURE 5.15

Thermometers placed in holes drilled in a metal rod will show that heat is conducted from a region of higher temperature to a region of lower temperature. The increased molecular activity is passed from molecule to molecule in the process of conduction.

TABLE 5.3	Rate of conduction of materials*
Silver	0.97
Copper	0.92
Aluminum	0.50
Iron	0.11
Lead	0.08
Concrete	4.0×10^{-3}
Glass	2.5×10^{-3}
Tile	1.6×10^{-3}
Brick	1.5×10^{-3}
Water	1.3×10^{-3}
Wood	3.0×10^{-4}
Cotton	1.8×10^{-4}
Styrofoam	1.0×10^{-4}
Glass wool	9.0×10^{-5}
Air	6.0×10^{-5}
Vacuum	0

*Based on temperature difference of 1°C per cm. Values are cal/s through a square centimeter of the material.

A CLOSER LOOK | Alternative-Fuel Vehicles

There are just a handful of key tasks you really have to do in life to be a good, hard working, and honorable citizen:

1. Never try to drive and read at the same time.
2. Avoid talking on a cell phone while driving.
3. Always leave the seat down.
4. Recycle everything.
5. Drive less, use rail or bus transit, or drive an electric or alternative-fuel vehicle.

The first three tasks listed above do not require any more discussion, and you can get a good start on being eco-conscious by properly disposing of soda cans and newspapers. But what is this about electric and alternative-fuel vehicles?

General Motors, Ford, Chrysler, Mazda, Toyota, Honda, and Nissan manufacture electric vehicles, but they are not yet sold everywhere. There are current problems with the 80- to 125-mile range between charges and problems with use and eventual disposal of heavy lead-acid batteries. However, electric vehicles do improve air quality. The total amount of pollution caused by electric cars is much less than that caused by gasoline-powered cars and trucks. Further-

more, it is easier to control pollution from a stationary power plant—where the electricity is generated—than from many moving individual sources. A hybrid vehicle that runs part of the time on electric power and part of the time on gasoline also has fewer emissions and helps solve the problem of a limited range.

Fuels other than gasoline can reduce the emissions from cars and trucks. In fact, a vehicle powered by hydrogen fuel cells, or by engines that burn the gas, is a renewable resource, zero-emissions technology if the hydrogen is produced from water using solar energy. Research and development of hydrogen vehicles is ongoing, concerned mostly with the problems of the costly and difficult production and storage of hydrogen gas. One promising technology is the combined use of hydrogen-powered fuel cells in an electric vehicle.

Compressed natural gas (CNG) is a reduced-emission fuel that can be used in converted conventional gasoline cars and trucks. It is currently used in thousands of converted fleet vehicles that run 80 percent cleaner than they would using gasoline. CNG is also about half as expensive as gasoline, but requires a heavy, bulky fuel tank and an expensive refueling station. CNG vehicles

also have a lower performance and shorter range than they would if burning gasoline.

Propane is yet another reduced-emission fuel that can be used in converted conventional cars and trucks. Propane is a proven, low-cost fuel that burns about 50 percent cleaner than conventional gasoline. It does require a heavy, bulky fuel tank and seems to make sense for heavy vehicles as a replacement for diesel fuel.

Methanol is yet another reduced-emission fuel that can be used in converted conventional cars and trucks. It is sometimes mixed with gasoline, and an 85 percent methanol mixture burns 30 percent to 50 percent cleaner than gasoline. However, methanol is highly corrosive, has a lower energy content than gasoline, and costs more than pure gasoline.

The use of alternative fuels does lower carbon dioxide emissions and, coupled with other programs, can make a marked change in CO_2 emissions. During the previous decade, for example, environmental groups working with the city of Portland, Oregon, identified the ten things they could do to improve air quality. These efforts resulted in a 3 percent reduction during a time when the rest of the United States averaged an 8 percent increase in per capita CO_2 emissions.

---

difference, area, and thickness are the same for each. The materials with the higher values are good conductors, since heat flows through them quickly. The materials with the smallest values are poor conductors and are called heat *insulators,* since heat flows through them slowly.

Most insulating materials are good insulators because they contain many small air spaces (figure 5.16). The small air spaces are poor conductors because the molecules of air are far apart, compared to a solid, making it more difficult to pass the increased vibrating motion from molecule to molecule. Styrofoam, glass wool, and wool cloth are good insulators because they have many small air spaces, not because of the material they are made of. The best insulator is a vacuum, since there are no molecules to pass on the vibrating motion.

Wooden and metal parts of your desk have the same temperature, but the metal parts will feel cooler if you touch them.

Metal is a better conductor of heat than wood and feels cooler because it conducts heat from your finger faster. This is the same reason that a wood or tile floor feels cold to your bare feet. You use an insulating rug to slow the conduction of heat from your feet.

Convection

Convection is the transfer of heat by a large-scale displacement of groups of molecules with relatively higher kinetic energy. In conduction, increased kinetic energy is passed from molecule to molecule. In convection, molecules with higher kinetic energy are moved from one place to another place. Conduction happens primarily in solids, but convection happens only in liquids and gases, where fluid motion can carry molecules with higher kinetic energy over a distance. When molecules gain energy, they move more rapidly and push more

FIGURE 5.16

Fiberglass insulation is rated in terms of R-value, a ratio of the conductivity of the material to its thickness.

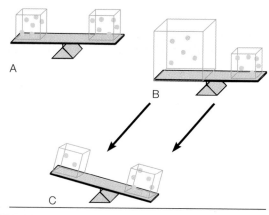

FIGURE 5.17

(*A*) Two identical volumes of air are balanced, since they have the same number of molecules and the same mass. (*B*) Increased temperature causes one volume to expand from the increased kinetic energy of the gas molecules. (*C*) The same volume of the expanded air now contains fewer gas molecules and is less dense, and it is buoyed up by the cooler, more dense air.

Activities

1. Objects that have been in a room with a constant temperature for some time should all have the same temperature. Touch metal, plastic, and wooden parts of a desk or chair to sense their temperature. Explain your findings.

2. Fill one pan with a mixture of ice and water, a second pan with lukewarm water, and a third with the hottest water you can stand. Place one hand in the hot water and the other hand in the cold water for several minutes, then quickly dry your hands and place them both in the lukewarm water. Explain what each hand feels.

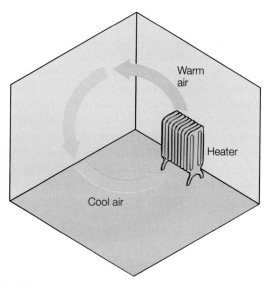

FIGURE 5.18

Convection currents move warm air throughout a room as the air over the heater becomes warmed, expands, and is moved upward by cooler air.

vigorously against their surroundings. The result is an expansion as the region of heated molecules pushes outward and increases the volume. Since the same amount of matter now occupies a larger volume, the overall density has been decreased (figure 5.17).

In fluids, expansion sets the stage for convection. Warm, less dense fluid is pushed upward by the cooler, more dense fluid around it. In general, cooler air is more dense; it sinks and flows downhill. Cold air, being more dense, flows out near the bottom of an open refrigerator. You can feel the cold, dense air pouring from the bottom of a refrigerator to your toes on the floor. On the other hand, you hold your hands *over* a heater because the warm, less dense air is pushed upward. In a room, warm air is pushed upward from a heater. The warm air spreads outward along the ceiling and is slowly displaced as newly warmed air is

pushed upward to the ceiling. As the air cools it sinks over another part of the room, setting up a circulation pattern known as a *convection current* (figure 5.18). Convection currents can also be observed in a large pot of liquid that is heating on a range. You can see the warmer liquid being forced upward over the warmer parts of the range element, then sink over the cooler parts. Overall, convection currents give the pot of liquid an appearance of turning over as it warms.

Convection currents will ventilate a room if the top *and* bottom parts of a window are opened. In the winter, cool and denser air will pour in the bottom opening, forcing the warmer, stale air out the top opening. An open fireplace functions because of convection currents. Smoke (and much of the heat) goes up the chimney as fresh air from the room furnishes oxygen for the fire. One way to start smoldering wood burning in a fireplace is to open a window to let cool, dense air into the room. The resulting convection has the same effect as fanning the smoldering wood, and it will burst into flames.

Radiation

The third way that heat transfer takes place because of a temperature difference is called **radiation.** Radiation involves the form of energy called *radiant energy,* energy that moves through space. As you will learn in chapter 8, radiant energy includes visible light and many other forms as well. All objects with a temperature above absolute zero give off radiant energy. The absolute temperature of the object determines the rate, intensity, and kinds of radiant energy emitted. You know that visible light is emitted if an object is heated to a certain temperature. A heating element on an electric range, for example, will glow with a reddish-orange light when at the highest setting, but it produces no visible light at lower temperatures, although you feel warmth in your hand when you hold it near the element. Your hand absorbs the nonvisible radiant energy being emitted from the element. The radiant energy does work on the molecules of your hand, giving them more kinetic energy. You sense this as an increase in temperature, that is, warmth.

All objects above absolute zero emit radiant energy, but all objects also absorb radiant energy. A hot object, however, emits more radiant energy than a cold object. The hot object will emit more energy than it absorbs from the cold object, and the cold object will absorb more energy from the hot object than it emits. There is, therefore, a net energy transfer that will take place by radiation as long as there is a temperature difference between the two objects.

ENERGY, HEAT, AND MOLECULAR THEORY

The kinetic molecular theory of matter is based on evidence from different fields of science, not just one subject area. Chemists and physicists developed some convincing conclusions about the structure of matter over the past 150 years, using carefully designed experiments and mathematical calculations that explained observable facts about matter. Step by step, the detailed structure of this submicroscopic, invisible world of particles became firmly established. Today, an understanding of this particle structure is basic to physics, chemistry, biology, geology, and practically every other science subject. This understanding has also resulted in present-day technology.

Phase Change

Solids, liquids, and gases are the three common phases of matter, and each phase is characterized by different molecular arrangements. The motion of the molecules in any of the three common phases can be increased by (1) adding heat through a

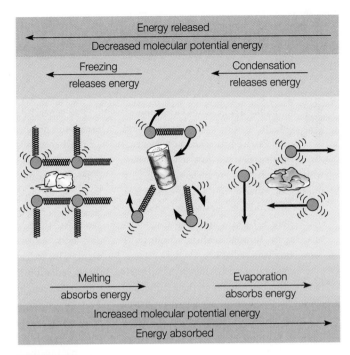

FIGURE 5.19

Each phase change absorbs or releases a quantity of latent heat, which goes into or is released from molecular potential energy.

temperature difference or (2) the absorption of one of the five forms of energy, which results in heating. In either case the temperature of the solid, liquid, or gas increases according to the specific heat of the substance, and more heating generally means higher temperatures.

More heating, however, does not always result in increased temperatures. When a solid, liquid, or gas changes from one phase to another, the transition is called a **phase change.** A phase change always absorbs or releases energy, *a quantity of heat that is not associated with a temperature change.* Since the quantity of heat associated with a phase change is not associated with a temperature change, it is called *latent heat.* Latent heat is a term borrowed from the old caloric theory of heat and means "hidden heat." Today, latent heat refers to the "hidden" energy of phase changes, which is energy (heat) that goes into or comes out of *internal potential energy* (figure 5.19).

There are three kinds of major phase changes that can occur: (1) *solid-liquid,* (2) *liquid-gas,* and (3) *solid-gas.* In each case the phase change can go in either direction. For example, the solid-liquid phase change occurs when a solid melts to a liquid or when a liquid freezes to a solid. Ice melting to water and water freezing to ice are common examples of this phase change and its two directions. Both occur at a temperature called the *freezing point* or the *melting point,* depending on the direction of the phase change. In either case, however, the freezing and melting points are the same temperature.

The liquid-gas phase change also occurs in two different directions. The temperature at which a liquid boils and changes to a gas (or vapor) is called the *boiling point.* The temperature at

A CLOSER LOOK | Passive Solar Design

Passive solar application is an economically justifiable use of solar energy today. Passive solar design uses a structure's construction to heat a living space with solar energy. There are few electric fans, motors, or other energy sources used. The passive solar design takes advantage of free solar energy; it stores and then distributes this energy through natural conduction, convection, and radiation.

Sunlight that reaches the earth's surface is mostly absorbed. Buildings, the ground, and objects become warmer as the radiant energy is absorbed. Nearly all materials, however, reradiate the absorbed energy at longer wavelengths, wavelengths too long to be visible to the human eye. The short wavelengths of sunlight pass readily through ordinary window glass, but the longer, reemitted wavelengths cannot. Therefore, sunlight passes through a window and warms objects inside a house. The reradiated longer wavelengths cannot pass readily back through the glass but are absorbed by certain molecules in the air. The temperature of the air is thus increased. This is called the "greenhouse effect." Perhaps you have experienced the effect when you left your car windows closed on a sunny, summer day.

In general, a passive solar home makes use of the materials from which it is constructed to capture, store, and distribute solar energy to its occupants. Sunlight enters the house through large windows facing south and warms a thick layer of concrete, brick, or stone. This energy "storage mass" then releases energy during the day, and, more important, during the night. This release of energy can be by direct radiation to occupants, by conduction to adjacent air, or by convection of air across the surface of the storage mass. The living space is thus heated without special plumbing or forced air circulation. As you can imagine, the key to a successful passive solar home is to consider every detail of natural energy flow, including the materials of which floors and walls are constructed, convective air circulation patterns, and the size and placement of windows. In addition, a passive solar home requires a different lifestyle and living patterns. Carpets, for example, would defeat the purpose of a storage-mass floor, since it would insulate the storage mass from sunlight. Glass is not a good insulator, so windows must have curtains or movable insulation panels to slow energy loss at night.

This requires the daily activity of closing curtains or moving insulation panels at night and then opening curtains and moving panels in the morning. Passive solar homes, therefore, require a high level of personal involvement by the occupants.

There are three basic categories of passive solar design: (1) direct solar gain, (2) indirect solar gain, and (3) isolated solar gain.

A *direct solar gain* home is one in which solar energy is collected in the actual living space of the home (box figure 5.1). The advantage of this design is the large, open window space with a calculated overhang, which admits maximum solar energy in the winter but prevents solar gain in the summer. The disadvantage is that the occupants are living in the collection and storage components of the design and can place nothing (such as carpets and furniture) that would interfere with warming the storage mass in the floors and walls.

An *indirect solar gain* home uses a massive wall inside a window that serves as a storage mass. Such a wall, called a *Trombe wall*, is shown in box figure 5.2. The Trombe wall collects and stores solar energy, then

—Continued top of next page

---

which a gas or vapor changes back to a liquid is called the *condensation point*. The boiling and condensation points are the same temperature. There are conditions other than boiling under which liquids may undergo liquid-gas phase changes, and these conditions are discussed in the next section.

You probably are not as familiar with solid-gas phase changes, but they are common. A phase change that takes a solid directly to a gas or vapor is called *sublimation*. Mothballs and dry ice (solid CO_2) are common examples of materials that undergo sublimation, but frozen water, meaning common ice, also sublimates under certain conditions. Perhaps you have noticed ice cubes in a freezer become smaller with time as a result of sublimation. The frost that forms in a freezer, on the other hand, is an example of a solid-gas phase change that takes place in the other direction. In this case, water vapor forms the frost without going through the liquid state, a solid-gas phase change that takes place in an opposite direction to sublimation.

For a specific example, consider the changes that occur when ice is subjected to a constant source of heat (figure 5.20). Starting at the left side of the graph, you can see that the temperature of the ice increases from the constant input of heat. The ice warms according to $Q = mc\Delta T$, where c is the specific heat of ice. When the temperature reaches the melting point (0°C), it stops increasing as the ice begins to melt. More and more liquid water appears as the ice melts, but the temperature *remains* at 0°C even though heat is still being added at a constant rate. It takes a certain amount of heat to melt all of the ice. Finally, when all the ice is completely melted, the temperature again increases at a constant rate between the melting and boiling points. Then, at constant temperature the addition of heat produces another phase change, from liquid to gas. The

Continued—

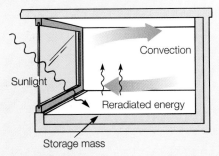

BOX FIGURE 5.1

The direct solar gain design collects and stores solar energy in the living space.

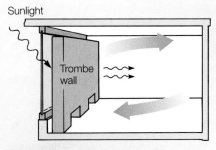

BOX FIGURE 5.2

The indirect solar gain design uses a Trombe wall to collect, store, and distribute solar energy.

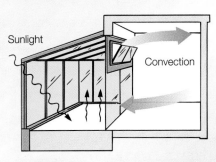

BOX FIGURE 5.3

The isolated solar gain design uses a separate structure to collect and store solar energy.

warms the living space with radiant energy and convection currents. The disadvantage to the indirect solar gain design is that large windows are blocked by the Trombe wall. The advantage is that the occupants are not in direct contact with the solar collection and storage area, so they can place carpets and furniture as they wish. Controls to prevent energy loss at night are still necessary with this design.

An *isolated solar gain* home uses a structure that is separated from the living space to collect and store solar energy. Examples of an isolated gain design are an attached greenhouse or sun porch (box fig-

ure 5.3). Energy flow between the attached structure and the living space can be by conduction, convection, and radiation, which can be controlled by opening or closing off the attached structure. This design provides the best controls, since it can be completely isolated, opened to the living space as needed, or directly used as living space when the conditions are right. Additional insulation is needed for the glass at night, however, and for sunless winter days.

It has been estimated that building a passive solar home would cost about 10 percent more than building a traditional

home of the same size. Considering the possible energy savings, you might believe that most homes would now have a passive solar design. They do not, however, as most new buildings require technology and large amounts of energy to maximize comfort. Yet, it would not require too much effort to consider where to place windows in relation to the directional and seasonal intensity of the sun and where to plant trees. Perhaps in the future you will have an opportunity to consider using the environment to your benefit through the natural processes of conduction, convection, and radiation.

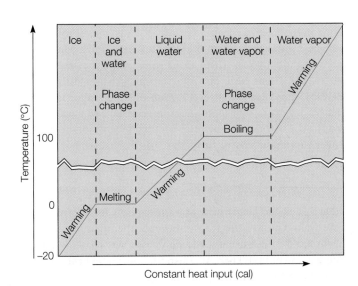

FIGURE 5.20

This graph shows three warming sequences and two phase changes with a constant input of heat. The ice warms to the melting point, then absorbs heat during the phase change as the temperature remains constant. When all the ice has melted, the now-liquid water warms to the boiling point, where the temperature again remains constant as heat is absorbed during this second phase change from liquid to gas. After all the liquid has changed to gas, continued warming increases the temperature of the water vapor.

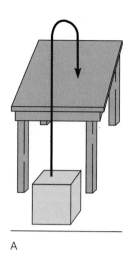

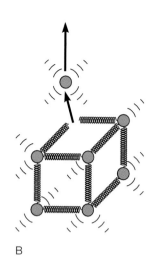

A   B

FIGURE 5.21

(A) Work is done against gravity to lift an object, giving the object more gravitational potential energy. (B) Work is done against intermolecular forces in separating a molecule from a solid, giving the molecule more potential energy.

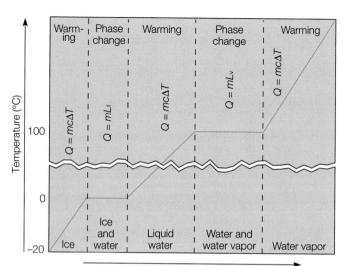

FIGURE 5.22

Compare this graph to the one in figure 5.20. This graph shows the relationships between the quantity of heat absorbed during warming and phase changes as water is warmed from ice at $-20°C$ to water vapor at some temperature above $100°C$. Note that the specific heat for ice, liquid water, and water vapor (steam) have different values.

quantity of heat involved in this phase change is used in doing the work of breaking the molecule-to-molecule bonds in the solid, making a liquid with molecules that are now free to move about and roll over one another. Since the quantity of heat (Q) is absorbed without a temperature change, it is called the **latent heat of fusion** (L_f). The latent heat of fusion is *the heat involved in a solid-liquid phase change in melting or freezing.* You learned in a previous chapter that when you do work on something you give it energy. In this case, the work done in breaking the molecular bonds in the solid gave the molecules more *potential energy* (figure 5.21). This energy is "hidden," or latent, since heat was absorbed but a temperature increase did not take place. This same potential energy is given up when the molecules of the liquid return to the solid state. A melting solid absorbs energy and a freezing liquid releases this *same amount* of energy, warming the surroundings. Thus, you put ice in a cooler because the melting ice absorbs the latent heat of fusion from the beverage cans, cooling them. Citrus orchards are flooded with water when freezing temperatures are expected because freezing water releases the latent heat of fusion, which warms the air around the trees. For water, the latent heat of fusion is 80.0 cal/g (144.0 Btu/lb). This means that every gram of ice that melts in your cooler *absorbs* 80.0 cal of heat. Every gram of water that freezes *releases* 80.0 cal. The total heat involved in a solid-liquid phase change depends on the mass of the substance involved, so

$$Q = mL_f$$

equation 5.5

where L_f is the latent heat of fusion for the substance involved.

Refer again to figure 5.20. After the solid-liquid phase change is complete, the constant supply of heat increases the temperature of the water according to $Q = mc\Delta T$, where c is

now the specific heat of liquid water. When the water reaches the boiling point, the temperature again remains constant even though heat is still being supplied at a constant rate. The quantity of heat involved in the liquid-gas phase change again goes into doing the work of overcoming the attractive molecular forces. This time the molecules escape from the liquid state to become single, independent molecules of gas. The quantity of heat (Q) absorbed or released during this phase change is called the **latent heat of vaporization** (L_v). The latent heat of vaporization is *the heat involved in a liquid-gas phase change where there is evaporation or condensation.* The latent heat of vaporization is the energy gained by the gas molecules as work is done in overcoming molecular forces. Thus, the escaping molecules absorb energy from the surroundings, and a condensing gas (or vapor) releases this *exact same amount of energy.* For water, the latent heat of vaporization is 540.0 cal/g (970.0 Btu/lb). This means that every gram of water vapor that condenses on your bathroom mirror releases 540.0 cal, which warms the bathroom. The total heating depends on how much water vapor condensed, so

$$Q = mL_v$$

equation 5.6

where L_v is the latent heat of vaporization for the substance involved. The relationships between the quantity of heat absorbed during warming and phase changes are shown in figure 5.22. Some physical constants for water and heat are summarized in table 5.4.

C O N N E C T I O N S

Evaporative Coolers

Evaporative cooling is one of the earliest forms of mechanical air conditioning. An evaporative cooler (sometimes called a "swamp cooler") works by cooling outside air by evaporation, then blowing the cooler, but now more humid air through a house. Usually, an evaporative cooler moves a sufficient amount of air through a house to completely change the air every two or three minutes.

An evaporative cooler is a metal or fiberglass box with louvers. Inside the box is a fan and motor, and there might be a small pump to recycle water. Behind the louvers are loose pads made of wood shavings (excelsior) or some other porous material. Water is pumped from the bottom of the cooler and trickles down through the pads, thoroughly wetting them. The fan forces air into the house and dry, warm outside air moves through the wet pads into the house in a steady flow. Windows or a door in the house must be partly open or the air will not be able to flow through the house.

The water in the pads evaporates, robbing heat from the air moving through them. Each liter of water that evaporates could take over 540 kcal of heat from the air. The actual amount removed depends on the temperature of the water, the efficiency of the cooler, and the relative humidity of the outside air.

An electric fan cools you by evaporation, but an evaporative cooler cools the air that blows around you. Relative humidity is the main variable that determines how much an evaporative cooler will cool the air. The following data show the cooling power of a typical evaporative cooler at various relative humidities when the outside air is 38°C (about 100°F).

At This Humidity:	Air Will Be Cooled to:
10%	22°C (72°F)
20%	24°C (76°F)
30%	27°C (80°F)
40%	29°C (84°F)
50%	31°C (88°F)
60%	33°C (91°F)
70%	34°C (93°F)
80%	35°C (95°F)
90%	37°C (98°F)

The advantage to an evaporative cooler is the low operating cost compared to refrigeration. The disadvantages are that it doesn't cool the air that much when the humidity is high, it adds even more humidity to the air, and outside air with its dust, pollen, and pollution is continually forced through the house. Another disadvantage is that mineral deposits left by evaporating hard water could require a frequent maintenance schedule.

TABLE 5.4	Some physical constants for water and heat

Specific Heat (c)

Water	$c = 1.00$ cal/gC°
Ice	$c = 0.500$ cal/gC°
Steam	$c = 0.480$ cal/gC°

Latent Heat of Fusion

L_f (water)	$L_f = 80.0$ cal/g

Latent Heat of Vaporization

L_v (water)	$L_v = 540.0$ cal/g

Mechanical Equivalent of Heat

1 kcal	4,184 J

Evaporation and Condensation

Liquids do not have to be at the boiling point to change to a gas and, in fact, tend to undergo a phase change at any temperature when left in the open. The phase change occurs at any temperature but does occur more rapidly at higher temperatures. The temperature of the water is associated with the *average* kinetic energy of the water molecules. The word *average* implies that some of the molecules have a greater energy and some have less (refer to figure 5.5). If a molecule of water that has an exceptionally high energy is near the surface, and is headed in the right direction, it may overcome the attractive forces of the other water molecules and escape the liquid to become a gas. This is the process of *evaporation*. Evaporation reduces a volume of liquid water as water molecules leave the liquid state to become water vapor in the atmosphere (figure 5.23).

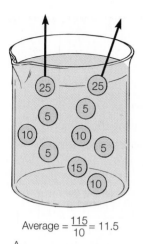

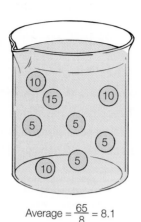

Average = $\frac{115}{10}$ = 11.5

A

Average = $\frac{65}{8}$ = 8.1

B

FIGURE 5.23

Temperature is associated with the average energy of the molecules of a substance. These numbered circles represent arbitrary levels of molecular kinetic energy that, in turn, represent temperature. The two molecules with the higher kinetic energy values [25 in (A)] escape, which lowers the average values from 11.5 to 8.1 (B). Thus evaporation of water molecules with more kinetic energy contributes to the cooling effect of evaporation in addition to the absorption of latent heat.

Evaporation occurs at any temperature, but the energy (heat) required per gram of liquid water changing to water vapor is the same in evaporation as in boiling. A gram of water at 20°C (room temperature) will need 80.0 cal for the equivalent energy of boiling water and an additional 540.0 cal for the latent heat of vaporization. Thus, each gram of room-temperature water requires 620.0 cal to change to a vapor. This supply of energy must be present to maintain the process of evaporation, and the water robs this energy from its surroundings. This explains why water at a higher temperature evaporates more rapidly than water at a cooler temperature. More energy is available at higher temperatures to maintain the process, so the water evaporates more rapidly. It also explains why evaporation is a cooling process. Consider, for example, how perspiring cools your body. Water from sweat glands runs onto your skin and evaporates, removing heat from your body. Since your body temperature is about 37.0°C (98.6°F), each gram of water will require 63.0 cal to give it sufficient energy to evaporate. Each gram of water will then remove 540.0 cal (the latent heat of vaporization) as it changes to water vapor. Each gram of water evaporated therefore removed about 603 cal (540.0 + 63.0) from your body, or about 603 kcal per liter of water.

Water molecules that evaporate move about in all directions, and some will return, striking the liquid surface. The same forces that they escaped from earlier capture the molecules, returning them to the liquid state. This is called the process of condensation. Condensation is the opposite of evaporation. In *evaporation*, more molecules are leaving the liquid state than are returning. In *condensation*, more molecules are returning to the

FIGURE 5.24

The inside of this closed bottle is isolated from the environment, so the space above the liquid becomes saturated. While it is saturated, the evaporation rate equals the condensation rate. When the bottle is cooled, condensation exceeds evaporation and droplets of liquid form on the inside surfaces.

liquid state than are leaving. This is a dynamic, ongoing process with molecules leaving and returning continuously. The net number leaving or returning determines if evaporation or condensation is taking place (figure 5.24).

When the condensation rate *equals* the evaporation rate, the air above the liquid is said to be *saturated*. The air immediately next to a surface may be saturated, but the condensation of water molecules is easily moved away with air movement. There is no net energy flow when the air is saturated, since the heat carried away by evaporation is returned by condensation. This is why you fan your face when you are hot. The moving air from the fanning action pushes away water molecules from the air near your skin, preventing the adjacent air from becoming saturated, thus increasing the rate of evaporation. Think about this process the next time you see someone fanning his or her face.

There are four ways to increase the rate of evaporation. (1) An increase in the temperature of the liquid will increase the average kinetic energy of the molecules and thus increase the number of high-energy molecules able to escape from the liquid state. (2) Increasing the surface area of the liquid will also increase the likelihood of molecular escape to the air. This is why you spread out wet clothing to dry or spread out a puddle you want to evaporate. (3) Removal of water vapor from near the surface of the liquid will prevent the return of the vapor molecules to the liquid state and thus increase the net rate of evaporation. This is why things dry more rapidly on a windy day. (4) *Pressure* is defined as *force per unit area*, which can be measured in lb/in² or N/m². Gases exert a pressure, which is interpreted in terms of the kinetic molecular theory. Atmospheric pressure is discussed in detail in chapter 21. For now, consider that the atmosphere exerts a pressure of about 10.0 N/cm² (14.7 lb/in²) at sea level. The atmospheric pressure, as well as the intermolecular forces, tend to hold water molecules in the liquid state. Thus, reducing the atmospheric pressure will reduce one of the forces holding molecules in a liquid state. Perhaps you have noticed that wet items dry more quickly at higher elevations, where the atmospheric pressure is less.

Relative Humidity

There is a relationship between evaporation-condensation and the air temperature. If the air temperature is decreased, the average kinetic energy of the molecules making up the air is decreased. Water vapor molecules condense from the air when they slow enough that molecular forces can pull them into the liquid state. Fast-moving water vapor molecules are less likely to be captured than slow-moving ones. Thus, as the air temperature increases, there is less tendency for water vapor molecules to return to the liquid state. Warm air can therefore hold more water vapor than cool air. In fact, air at 38°C (100°F) can hold five times as much vapor as air at 10°C (50°F) (figure 5.25).

The ratio of how much water vapor *is* in the air to how much water vapor *could be* in the air at a certain temperature is called **relative humidity.** This ratio is usually expressed as a percentage, and

$$\text{relative humidity} = \frac{\text{water vapor in air}}{\text{capacity at present temperature}} \times 100\%$$

$$R.H. = \frac{\text{g/m}^3 \text{(present)}}{\text{g/m}^3 \text{(max)}} \times 100\%$$

equation 5.7

Figure 5.25 shows the maximum amount of water vapor that can be in the air at various temperatures. Suppose that the air contains 10 g/m³ of water vapor at 10°C (50°F). According to figure 5.25, the maximum amount of water vapor that *can be* in the air when the air temperature is 10°C (50°F) is 10 g/m³. Therefore the relative humidity is (10 g/m³) ÷ (10 g/m³) × 100%, or 100 percent. This air is therefore saturated. If the air held only 5 g/m³ of water vapor at 10°C, the relative humidity would be 50 percent, and 2 g/m³ of water vapor in the air at 10°C is 20 percent relative humidity.

As the air temperature increases, the capacity of the air to hold water vapor also increases. This means that if the *same amount* of water vapor is in the air during a temperature increase, the relative

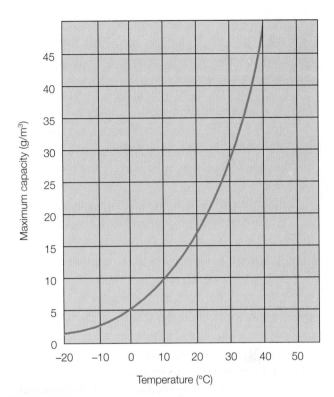

FIGURE 5.25

The curve shows the *maximum* amount of water vapor in g/m³ that can be in the air at various temperatures.

humidity will decrease. Thus, the relative humidity increases every night because the air temperature decreases, not because water vapor has been added to the air. The relative humidity is important because it is one of the things that controls the rate of evaporation, and the evaporation rate is one of the variables involved in how well you can cool yourself in hot weather.

THERMODYNAMICS

The branch of physical science called *thermodynamics* is concerned with the study of heat and its relationship to mechanical energy, including the science of heat pumps, heat engines, and the transformation of energy in all its forms. The *laws of thermodynamics* describe the relationships concerning what happens as energy is transformed to work and the reverse, also serving as useful intellectual tools in meteorology, chemistry, and biology.

Mechanical energy is easily converted to heat through friction, but a special device is needed to convert heat to mechanical energy. A **heat engine** is a device that converts heat into mechanical energy. The operation of a heat engine can be explained by the kinetic molecular theory, as shown in figure 5.26. This illustration shows a cylinder, much like a big can, with a closely fitting piston that traps a sample of air. The piston is like a slightly smaller cylinder and has a weight resting on it, supported by the

CONNECTIONS

Human Responses to Cold and Hot

For an average age and minimal level of activity, many people feel comfortable when the environmental temperature is about 25°C (77°F). Comfort at this temperature probably comes from the fact that the body does not have to make an effort to conserve or get rid of heat.

Conservation of heat occurs when the temperature of the air and clothing directly next to a person becomes less than 20°C, or if the body senses rapid heat loss. First, blood vessels in the skin are constricted. This slows the flow of blood near the surface, which reduces heat loss by conduction. Constriction of skin blood vessels reduces body heat loss, but may also cause the skin and limbs to become three or four degrees or so cooler than the body core temperature (producing cold feet, for example).

Sudden heat loss, or a chill, often initiates another heat-saving action by the body. Skin hair is pulled upright, erected to slow heat loss to cold air moving across the skin. Contraction of a tiny muscle attached to the base of the hair shaft makes a tiny knot, or bump on the skin. These are sometimes called "goose bumps" or "chill bumps."

Further cooling after the blood vessels in the skin have been constricted results in the body taking yet another action. The body now begins to produce *more* heat, making up for heat loss through involuntary muscle contractions called "shivering." The greater the need for more body heat the greater the activity of shivering.

If the environmental temperatures rise above about 25°C (77°F), the body triggers responses that causes it to *lose* heat. One response is to make blood vessels in the skin larger, which increases blood flow in the skin. This brings more heat from the core to be conducted through the skin, then radiated away. It also causes some people to have a red blush from the increased blood flow in the skin. This action increases conduction through the skin, but radiation alone provides insufficient cooling at environmental temperatures above about 29°C (84°F). At about this temperature sweating begins and perspiration pours onto the skin to provide cooling through evaporation. The warmer the environmental temperature, the greater the rate of sweating and cooling through evaporation.

The actual responses to a cool, cold, warm, or hot environment will be influenced by a person's level of activity, age, and gender, and environmental factors such as the relative humidity, air movement, and combinations of these factors. Temperature is the single most important comfort factor. However, when the temperature is high enough to require perspiration for cooling, humidity also becomes an important factor in human comfort.

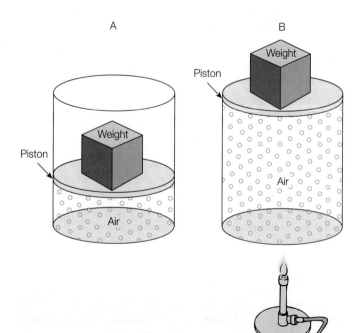

FIGURE 5.26

A very simple heat engine. The air in (*B*) has been heated, increasing the molecular motion and thus the pressure. Some of the heat is transferred to the increased gravitational potential energy of the weight as it is converted to mechanical energy.

A CLOSER LOOK | The Drinking Bird

here is an interesting toy called a "Dippy Bird," "Happy Drinking Bird," (box figure 5.4), or various other names, that "drinks" water from a glass. The bird rocks back and forth, dipping its head into a glass of water, then bobbing up and rocking back and forth again before dipping down for another "drink." The bird will continue rocking and bobbing until the water in the glass is too low to reach.

Why does the Dippy Bird rock back and forth, then dip its head in the water? This question can be answered with some basic concepts of evaporation, heat, and vapor pressure, but first we need some information about the construction of the bird. The head and body of the bird are similar-sized glass bulbs connected with a short glass tube. The tube is attached directly to the upper bulb, but it extends well inside the lower bulb, almost to the bottom of the bulb. The tube and lower body are sealed with a measured amount of a red liquid. The red liquid is probably methylene chloride or some other volatile liquid that vaporizes at room temperature. The volume of liquid is sufficient to cover the tube in the lower bulb when the bird is upright, but not enough to cover both ends of the tube when the bird is nearly horizontal.

A metal pivot bar is clamped on the glass tube, near the middle, with ends resting on a stand sometimes made to look like bird legs. The bar is bent so the bird leans slightly

in a forward direction. The upper bulb is coated with some fuzzy stuff, and has an attached beak, hat, eyes, and feathers. To start the bird on its cycles of dipping, the head and beak are wet with water. The bird is then set upright in front of a glass of water, filled and placed so the bird's beak will dip into the water when it rotates forward.

Before starting, the red liquid inside the two bulbs is at equilibrium with the same temperature, vapor pressure, and volume. Equilibrium is upset when the fuzzy material on the upper bulb is wet. Evaporation of water cools the head, and some of the vapor inside the upper bulb condenses. The greater pressure from the warmer vapor in the lower bulb is now able to push some fluid up the tube and into the upper bulb. The additional fluid in the upper bulb changes the pivot point and the bird dips forward. However, the tube opening is no longer covered by fluid, and some of the warmer vapor is now able to move to the upper bulb. This equalizes the pressure between the two bulbs, and fluid drains back into the lower bulb. This changes the balance, returning the bird to an upright position to rock back and forth, then repeat the events in the cycle.

When it was in the nearly horizontal position, the beak dipped into the glass of water. The material on the beak keeps the head wet, which makes it possible to continue the cycle of cooling and warming the

BOX FIGURE 5.4
A Dippy Bird.

upper bulb. Since the Dippy Bird operation depends on the cooling effects of evaporation, it will work best in dry air and performs poorly in humid air.

Since the operation of the Dippy Bird depends on the relative humidity, perhaps you can figure out some way to use it to indicate the relative humidity.

trapped air. If the air in the large cylinder is now heated, the gas molecules will acquire more kinetic energy. This results in more gas molecule impacts with the enclosing surfaces, which results in an increased pressure. Increased pressure results in a net force, and the piston and weight move upward as shown in figure 5.26B. Thus, some of the heat has now been transformed to the increased gravitational potential energy of the weight.

Thermodynamics is concerned with the *internal energy* (U), the total internal potential and kinetic energies of mole-

cules making up a substance, such as the gases in the simple heat engine. The variables of temperature, gas pressure, volume, heat, and so forth characterize the total internal energy, which is called the *state* of the system. Once the system is identified everything else is called the *surroundings*. A system can exist in a number of states since the variables that characterize a state can have any number of values and combinations of values. Any two systems that have the same values of variables that characterize internal energy are said to be in the same state.

The First Law of Thermodynamics

Any thermodynamic system has a unique set of properties that will identify the internal energy of the system. This state can be changed two ways, by (1) heat flowing into (Q_{in}) or out (Q_{out}) of the system, or (2) by the system doing work (W_{out}) or by work being done on the system (W_{in}). Thus work (W) and heat (Q) can change the internal energy of a thermodynamic system according to

$$Q - W = U_2 - U_1$$

equation 5.8

where ($U_2 - U_1$) is the internal energy difference between two states. This equation represents the **first law of thermodynamics,** which states that the energy supplied to a thermodynamic system in the form of heat, minus the work done by the system, is equal to the change in internal energy. The first law of thermodynamics is an application of the law of conservation of energy, which applies to all energy matters. The first law of thermodynamics is concerned specifically with a thermodynamic system. As an example, consider energy conservation that is observed in the thermodynamic system of a heat engine (figure 5.27). As the engine cycles to the original state of internal energy ($U_2 - U_1 = 0$), all the external work accomplished must be equal to all the heat absorbed in the cycle. The heat supplied to the engine from a high temperature source (Q_H) is partly converted to work (W) and the rest is rejected in the lower temperature exhaust (Q_L). The work accomplished is therefore the difference in the heat input and the heat output ($Q_H - Q_L$), so the work accomplished represents the heat used,

$$W = J(Q_H - Q_L)$$

equation 5.9

where J is the mechanical equivalence of heat ($J = 4.184$ joules/calorie). A schematic diagram of this relationship is shown in figure 5.27. You can increase the internal energy (produce heat) as long as you supply mechanical energy (or do work). The first law of thermodynamics states that the conversion of work to heat is reversible, meaning that heat can be changed to work. There are several ways of converting heat to work, for example, the use of a steam turbine or gasoline automobile engine.

The Second Law of Thermodynamics

A heat pump is the *opposite* of a heat engine as shown schematically in figure 5.28. The heat pump does work (W) in compressing vapors and moving heat from a region of lower temperature (Q_L) to a region of higher temperature (Q_H). That work is required to move heat this way is in accord with the observation that heat naturally flows from a region of higher temperature to a region of lower temperature. Energy is required for the opposite, moving heat from a cooler region to a warmer region. The natural direction of this process is called the **second law of thermodynamics,** which is that heat flows from objects with a higher temperature to objects with a cooler temperature. In other words, if you want heat to flow from a colder region to a warmer one you must *cause* it to do so by using

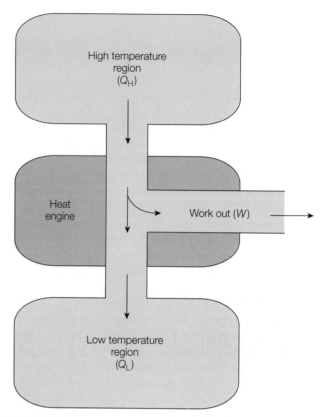

FIGURE 5.27

The heat supplied (Q_H) to a heat engine goes into the mechanical work (W), and the remainder is expelled in the exhaust (Q_L). The work accomplished is therefore the difference in the heat input and output ($Q_H - Q_L$), so the work accomplished represents the heat used, $W = J(Q_H - Q_L)$.

energy. And if you do, such as with the use of a heat pump, you necessarily cause changes elsewhere, particularly in the energy sources used in the generation of electricity. Another statement of the second law is that it is impossible to convert heat completely into mechanical energy. This does not say that you cannot convert mechanical energy completely into heat, for example, in the brakes of a car when the brakes bring it to a stop. The law says that the reverse process is not possible, that you cannot convert 100 percent of a heat source into mechanical energy. Both of the above statements of the second law are concerned with a *direction* of thermodynamic processes, and the implications of this direction will be discussed next.

The Second Law and Natural Processes

Energy can be viewed from two considerations of scale, (1) the observable *external energy* of an object, and (2) the *internal energy* of the molecules, or particles that make up an object. A ball, for example, has kinetic energy after it is thrown through the air and the entire system of particles making up the ball acts like a single massive particle as the ball moves. The motion and energy of the single system can be calculated from the laws of motion and from the equations representing the concepts of

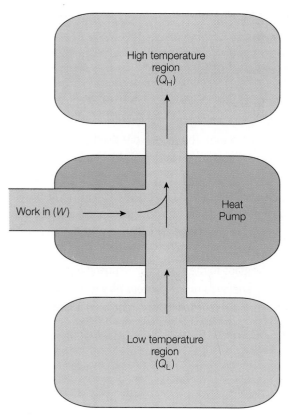

FIGURE 5.28

A heat pump uses work (W) to move heat from a low temperature region (Q_L) to a high temperature region (Q_H). The heat moved (Q_L) requires work (W), so $JQ_L = W$.

work and energy. All of the particles are moving together, in *coherent motion* when the external kinetic energy is considered.

But the particles making up the ball have another kind of kinetic energy, with the movements and vibrations of internal kinetic energy. In this case the particles are not moving uniformly together, but are vibrating with motions in many different directions. Since there is a lack of net motion and a lack of correlation, the particles have a jumbled *incoherent motion*, which is often described as chaotic. This random, chaotic motion is sometimes called *thermal motion*.

Thus there are two kinds of motion that the particles of an object can have, (1) a coherent motion where they move together, in step, and (2) an incoherent, chaotic motion of individual particles. These two types of motion are related to the two modes of energy transfer, working and heating. The relationship is that *work* on an object is associated with its *coherent motion*, while *heating* an object is associated with its internal *incoherent motion*.

The second law of thermodynamics implies a direction to the relationship between work (coherent motion) and heat (incoherent motion) and this direction becomes apparent as you analyze what happens to the motions during energy conversions. Some forms of energy, such as electrical and mechanical, have a greater amount of order since they involve particles moving together in a coherent motion. The term **quality of energy** is used to identify the amount of coherent motion. Energy with

high order and coherence is called a *high-quality energy*. Energy with less order and less coherence, on the other hand, is called *low-quality energy*. In general, high-quality energy can be easily converted to work, but low-quality energy is less able to do work.

High-quality electrical and mechanical energy can be used to do work, but then become dispersed as heat through energy-form conversions and friction. The resulting heat can be converted to do more work only if there is a sufficient temperature difference. The temperature differences do not last long, however, as conduction, convection, and radiation quickly disperse the energy even more. Thus the transformation of high-quality energy into lower-quality energy is a natural process. Energy tends to disperse, both from the conversion of an energy form to heat and from the heat flow processes of conduction, convection, and radiation. Both processes flow in one direction only and cannot be reversed. This is called the *degradation of energy*, which is the transformation of high-quality energy to lower-quality energy. In every known example it is a natural process of energy to degrade, becoming less and less available to do work. The process is *irreversible* even though it is possible to temporarily transform heat to mechanical energy through a heat engine or to upgrade the temperature through the use of a heat pump. Eventually the upgraded mechanical energy will degrade to heat and the increased heat will disperse through the processes of heat flow.

The apparent upgrading of energy by a heat pump or heat engine is always accompanied by a greater degrading of energy someplace else. The electrical energy used to run the heat pump, for example, was produced by the downgrading of chemical or nuclear energy at an electrical power plant. The overall result is that the *total* energy was degraded toward a more disorderly state.

A *thermodynamic measure of disorder* is called **entropy.** Order means patterns and coherent arrangements. Disorder means dispersion, no patterns, and a randomized, or spread-out arrangement. Entropy is therefore a measure of chaos, and this leads to another statement about the second law of thermodynamics and the direction of natural change, that

the total entropy of the universe continually increases.

Note the use of the words *total* and *universe* in this statement of the second law. The entropy of a system can decrease (more order), for example when a heat pump cools and condenses the random, chaotically moving water vapor molecules into the more ordered state of liquid water. When the energy source for the production, transmission, and use of electrical energy is considered, however, the *total* entropy will be seen as increasing. Likewise, the total entropy increases during the growth of a plant or animal. When all the food, waste products, and products of metabolism are considered, there is again an increase in *total* entropy.

Thus the *natural process* is for a state of order to degrade into a state of disorder with a corresponding increase in entropy. This means that all the available energy of the universe is gradually diminishing, and over time, the universe should therefore approach a limit of maximum disorder called the *heat death* of the universe. The heat death of the universe is the theoretical limit of disorder, with all molecules spread far, far apart, vibrating slowly with a uniform low temperature.

CONNECTIONS

Solar-Powered Refrigeration

The solar technology for using hot water to cool buildings is called absorption refrigeration (figure 5.29). Hot water from a solar collector runs through a heat exchanger coil in the *generator tank,* which has a function of separating the refrigerant (ammonia) from the absorber (water). The heat exchanger heats the ammonia-water mixture, causing the lower boiling point ammonia to vaporize.

The ammonia vapor moves through a pipe to the *condenser tank.* Cool water runs through a coil, cooling the ammonia vapors and condensing them into a liquid. The cooling water could be provided from an evaporative cooler or from some natural source.

The liquid ammonia now moves through an *expansion valve.* The ammonia boils and evaporates inside the cooling coil, absorbing heat. A fan blows air across the coil, which cools the air, then moves it to a room where cooling is desired. Inside the coil, the ammonia vapor, carrying the heat it took from the air, now moves into the *absorber tank.* The vapor is dissolved (absorbed) by water being cycled from the generator tank. As the ammonia vapor dissolves, it gives up the heat removed from the cooling coils. Cooling water is circulated through a heat exchanger coil in the tank to remove the released heat. The cooled ammonia-water solution is now pumped to the generator tank to repeat the process in an ongoing cycle.

The advantage to the absorption refrigeration is the low, fixed cost for the energy to run the system. All you need is a source of hot water, a source of cooling water, and a small amount of electrical energy to run the small pumps.

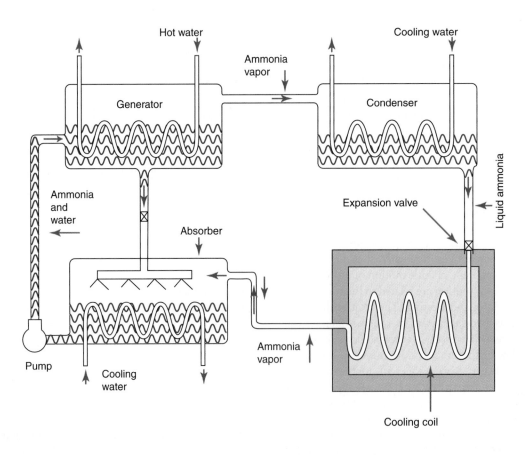

FIGURE 5.29

Absorption refrigeration schematic.

The heat death of the universe seems to be a logical consequence of the second law of thermodynamics, but scientists are not certain if the second law should apply to the whole universe. What do you think? Will the universe with all its complexities of organization end with the simplicity of spread-out and slowly vibrating molecules? As has been said, nature is full of symmetry—so why should the universe begin with a bang and end with a whisper?

SUMMARY

The kinetic theory of matter assumes that all matter is made up of tiny, ultimate particles of matter called *molecules*. A molecule is defined as the smallest particle of a compound, or a gaseous element, that can exist and still retain the characteristic properties of that substance. Molecules interact, attracting each other through a force of *cohesion*. Liquids, solids, and gases are the *phases of matter* that are explained by the molecular arrangements and forces of attraction between their molecules. A *solid* has a definite shape and volume because it has molecules that vibrate in a fixed equilibrium position with strong cohesive forces. A *liquid* has molecules that have cohesive forces strong enough to give it a definite volume but not strong enough to give it a definite shape. The molecules of a liquid can flow, rolling over each other. A *gas* is composed of molecules that are far apart, with weak cohesive forces. Gas molecules move freely in a constant, random motion.

Molecules can have *vibrational, rotational,* or *translational* kinetic energy. The *temperature* of an object is related to the *average kinetic energy* of the molecules making up the object. A measure of temperature tells how hot or cold an object is on two arbitrary scales, the *Fahrenheit scale* and the *Celsius scale*. The *absolute scale*, or *Kelvin scale*, has the coldest temperature possible ($-273°C$) as zero (0 K).

The observable potential and kinetic energy of an object is the *external energy* of that object, while the potential and kinetic energy of the molecules making up the object is the *internal energy* of the object. Heat refers to the total internal energy and is a transfer of energy that takes place (1) because of a *temperature difference* between two objects, or (2) because of an *energy-form conversion*. An energy-form conversion is actually an energy conversion involving work at the molecular level, so all energy transfers involve *heating* and *working*.

A quantity of heat can be measured in *joules* (a unit of work or energy) or *calories* (a unit of heat). A *kilocalorie* is 1,000 calories, another unit of heat. A *Btu*, or *British thermal unit*, is the English system unit of heat. The *mechanical equivalent of heat* is 4,184 J = 1 kcal.

The *specific heat* of a substance is the amount of energy (or heat) needed to increase the temperature of 1 gram of a substance 1 degree Celsius. The specific heat of various substances is not the same because the molecular structure of each substance is different.

Energy transfer that takes place because of a temperature difference does so through conduction, convection, or radiation. *Conduction* is the transfer of increased kinetic energy from molecule to molecule. Substances vary in their ability to conduct heat, and those that are poor conductors are called *insulators*. Gases, such as air, are good insulators. The best insulator is a vacuum. *Convection* is the transfer of heat by the displacement of large groups of molecules with higher kinetic energy. Convection takes place in fluids, and the fluid movement that takes place because of density differences is called a *convection current*. *Radiation* is radiant energy that moves through space. All objects with an absolute temperature above zero give off radiant energy, but all objects absorb it as well. Energy is transferred from a hot object to a cold one through radiation.

The transition from one phase of matter to another is called a *phase change*. A phase change always absorbs or releases a quantity of *latent heat* not associated with a temperature change. Latent heat is energy that goes into or comes out of *internal potential energy*. The *latent heat of fusion* is absorbed or released at a solid-liquid phase change. The latent heat of fusion for water is 80.0 cal/g (144.0 Btu/lb). The *latent heat of vaporization* is absorbed or released at a liquid-gas phase change. The latent heat of vaporization for water is 540.0 cal/g (970.0 Btu/lb).

Molecules of liquids sometimes have a high enough velocity to escape the surface through the process called *evaporation*. Evaporation is a cooling process, since the escaping molecules remove the latent heat of vaporization in addition to their high molecular energy. Vapor molecules return to the liquid state through the process called *condensation*. Condensation is the opposite of evaporation and is a warming process. When the condensation rate equals the evaporation rate, the air is said to be *saturated*. The rate of evaporation can be *increased* by (1) increased temperature, (2) increased surface area, (3) removal of evaporated molecules, and (4) reduced atmospheric pressure.

Warm air can hold more water vapor than cold air, and the ratio of how much water vapor is in the air to how much could be in the air at that temperature (saturation) is called *relative humidity*.

Thermodynamics is the study of heat and its relationship to mechanical energy, and the *laws of thermodynamics* describe these relationships: *The first law of thermodynamics* states that the energy supplied to a thermodynamic system in the form of heat, minus the work done by the system, is equal to the change in internal energy. The *second law of thermodynamics* states that heat flows from objects with a higher temperature to objects with a cooler temperature. The second law implies a *degradation of energy* as *high-quality* (more ordered) energy sources undergo *degradation* to *low-quality* (less ordered) sources. *Entropy* is a thermodynamic measure of disorder, and entropy is seen as continually increasing in the universe and may result in the maximum disorder called the *heat death* of the universe.

Summary of Equations

5.1

$$T_F = \frac{9}{5} T_C + 32°$$

5.2

$$T_C = \frac{5}{9} (T_F - 32°)$$

5.3

$$T_K = T_C + 273$$

5.4

Quantity of heat = (mass)(specific heat)(temperature change)

$$Q = mc\Delta T$$

5.5

Heat absorbed or released = (mass)(latent heat of fusion)

$$Q = mL_f$$

5.6

Heat absorbed or released = (mass)(latent heat of vaporization)

$$Q = mL_v$$

5.7

$$\text{relative humidity} = \frac{\text{water vapor in air}}{\text{capacity at present temperature}} \times 100\%$$

$$R.H. = \frac{\text{g/m}^3(\text{present})}{\text{g/m}^3(\text{max})} \times 100\%$$

5.8

$$(\text{heat}) - (\text{work}) = \text{internal energy difference between two states}$$

$$Q - W = U_2 - U_1$$

5.9

$$\text{work} = (\text{mechanical equivalent of heat})(\text{difference in heat input and heat output})$$

$$W = J(Q_H - Q_L)$$

KEY TERMS

absolute scale (p. **95**)

absolute zero (p. **95**)

British thermal unit (p. **98**)

calorie (p. **98**)

Celsius scale (p. **95**)

conduction (p. **100**)

convection (p. **101**)

entropy (p. **113**)

external energy (p. **97**)

Fahrenheit scale (p. **94**)

first law of thermodynamics (p. **112**)

heat (p. **96**)

heat engine (p. **109**)

internal energy (p. **97**)

Kelvin scale (p. **95**)

kilocalorie (p. **98**)

kinetic molecular theory (p. **90**)

latent heat of fusion (p. **106**)

latent heat of vaporization (p. **106**)

molecule (p. **91**)

phase change (p. **103**)

quality of energy (p. **113**)

radiation (p. **103**)

relative humidity (p. **109**)

second law of thermodynamics (p. **112**)

specific heat (p. **99**)

temperature (p. **93**)

APPLYING THE CONCEPTS

1. The temperature of a gas is proportional to the
   a. average velocity of the gas molecules.
   b. internal potential energy of the gas.
   c. number of gas molecules in a sample.
   d. average kinetic energy of the gas molecules.

2. The kinetic molecular theory explains the expansion of a solid material with increases of temperature as basically the result of
   a. individual molecules expanding.
   b. increased translational kinetic energy.
   c. molecules moving a little farther apart.
   d. heat taking up the spaces between molecules.

3. Two degree intervals on the Celsius temperature scale are
   a. equivalent to 3.6 Fahrenheit degree intervals.
   b. equivalent to 35.6 Fahrenheit degree intervals.
   c. twice as hot as 1 Celsius degree.
   d. none of the above.

4. A temperature reading of 2°C is
   a. equivalent to 3.6°F.
   b. equivalent to 35.6°F.
   c. twice as hot as 1°C.
   d. none of the above.

5. The temperature known as *room temperature* is nearest to
   a. 0°C.
   b. 20°C.
   c. 60°C.
   d. 100°C.

6. Using the absolute temperature scale, the freezing point of water is correctly written as
   a. 0 K.
   b. 0°K.
   c. 273 K.
   d. 273°K.

7. The metric unit of heat called a *calorie* is
   a. the specific heat of water.
   b. the energy needed to increase the temperature of 1 gram of water 1 degree Celsius.
   c. equivalent to a little over 4 joules of mechanical work.
   d. all of the above.

8. Which of the following is a shorthand way of stating that "the temperature change of a substance is directly proportional to the quantity of heat added"?
   a. $Q \propto m$
   b. $m \propto T_f - T_i$
   c. $Q \propto \Delta T$
   d. $Q = T_f - T_i$

9. The quantity known as *specific heat* is
   a. any temperature reported on the more specific absolute temperature scale.
   b. the energy needed to increase the temperature of 1 gram of a substance 1 degree Celsius.
   c. any temperature of a 1 kg sample reported in degrees Celsius.
   d. the heat needed to increase the temperature of 1 pound of water 1 degree Fahrenheit.

10. Table 5.1 lists the specific heat of soil as 0.20 kcal/kgC° and the specific heat of water as 1.00 kcal/kgC°. This means that if 1 kg of soil and 1 kg of water each receive 1 kcal of energy, ideally,
    a. the water will be warmer than the soil by 0.8°C.
    b. the soil will be 5°C warmer than the water.
    c. the water will be 5°C warmer than the soil.
    d. the water will warm by 1°C, and the soil will warm by 0.2°C.

11. The heat transfer that takes place by energy moving directly from molecule to molecule is called
    a. conduction.
    b. convection.
    c. radiation.
    d. none of the above.

12. The heat transfer that does not require matter is
    a. conduction.
    b. convection.
    c. radiation.
    d. impossible, for matter is always required.

13. Styrofoam is a good insulating material because
    a. it is a plastic material that conducts heat poorly.
    b. it contains many tiny pockets of air.
    c. of the structure of the molecules making up the Styrofoam.
    d. it is not very dense.

14. The transfer of heat that takes place because of density difference in fluids is
    a. conduction.
    b. convection.
    c. radiation.
    d. none of the above.

15. When a solid, liquid, or gas changes from one physical state to another, the change is called
    a. melting.
    b. entropy.
    c. a phase change.
    d. sublimation.

16. Latent heat is "hidden" because it
    a. goes into or comes out of internal potential energy.
    b. is a fluid (caloric) that cannot be sensed.
    c. does not actually exist.
    d. is a form of internal kinetic energy.

17. As a solid undergoes a phase change to a liquid state, it
    a. releases heat while remaining at a constant temperature.
    b. absorbs heat while remaining at a constant temperature.
    c. releases heat as the temperature decreases.
    d. absorbs heat as the temperature increases.

18. The condensation of water vapor actually
    a. warms the surroundings.
    b. cools the surroundings.
    c. sometimes warms and sometimes cools the surroundings, depending on the relative humidity at the time.
    d. neither warms nor cools the surroundings.

19. Water molecules move back and forth between the liquid and the gaseous state
    a. only when the air is saturated.
    b. at all times, with evaporation, condensation, and saturation defined by the net movement.
    c. only when the outward movement of vapor molecules produces a pressure equal to the atmospheric pressure.
    d. only at the boiling point.

20. No water vapor is added to or removed from a sample of air that is cooling, so the relative humidity of this sample of air will
    a. remain the same.
    b. be lower.
    c. be higher.
    d. be higher or lower, depending on the extent of change.

21. Compared to cooler air, warm air can hold
    a. more water vapor.
    b. less water vapor.
    c. the same amount of water vapor.
    d. less water vapor, the amount depending on the humidity.

22. A heat engine is designed to
    a. drive heat from a cool source to a warmer location.
    b. drive heat from a warm source to a cooler location.
    c. convert mechanical energy into heat.
    d. convert heat into mechanical energy.

23. The work that a heat engine is able to accomplish is ideally equivalent to the
    a. difference in the heat supplied and the heat rejected.
    b. heat that was produced in the cycle.
    c. heat that appears in the exhaust gases.
    d. sum total of the heat input and the heat output.

Answers

1. d **2.** c **3.** a **4.** b **5.** b **6.** c **7.** d **8.** c **9.** b **10.** b **11.** a **12.** c **13.** b **14.** b **15.** c **16.** a **17.** b **18.** a **19.** b **20.** c **21.** a **22.** d **23.** a

QUESTIONS FOR THOUGHT

1. What is temperature? What is heat?
2. Explain why most materials become less dense as their temperature is increased.
3. Would the tight packing of more insulation, such as glass wool, in an enclosed space increase or decrease the insulation value? Explain.
4. A true vacuum bottle has a double-walled, silvered bottle with the air removed from the space between the walls. Describe how this design keeps food hot or cold by dealing with conduction, convection, and radiation.
5. Why is cooler air found in low valleys on calm nights?
6. Why is air a good insulator?
7. Explain the meaning of the mechanical equivalent of heat.
8. What do people really mean when they say that a certain food "has a lot of Calories"?
9. A piece of metal feels cooler than a piece of wood at the same temperature. Explain why.
10. Explain how latent heat of fusion and latent heat of vaporization are "hidden."
11. What is condensation? Explain, on a molecular level, how the condensation of water vapor on a bathroom mirror warms the bathroom.
12. Which provides more cooling for a Styrofoam cooler, one with 10 lb of ice at 0°C or one with 10 lb of ice water at 0°C? Explain your reasoning.
13. Explain why a glass filled with a cold beverage seems to "sweat." Would you expect more sweating inside a house during the summer or during the winter? Explain.

14. Explain why a burn from 100°C steam is more severe than a burn from water at 100°C.

15. The relative humidity increases almost every evening after sunset. Explain how this is possible if no additional water vapor is added to or removed from the air.

16. Briefly describe, using sketches as needed, how a heat pump is able to move heat from a cooler region to a warmer region.

17. Which has more entropy—ice, liquid water, or water vapor? Explain your reasoning.

18. Suppose you use a heat engine to do the work to drive a heat pump. Could the heat pump be used to provide the temperature difference to run the heat engine? Explain.

PARALLEL EXERCISES

The exercises in groups A and B cover the same concepts. Solutions to group A exercises are located in appendix D.

Note: Neglect all frictional forces in all exercises.

Group A

1. The average human body temperature is 98.6°F. What is the equivalent temperature on the Celsius scale?

2. A bank temperature display indicates 20°C, which is sometimes defined as "room temperature." What is the equivalent temperature on the Fahrenheit scale?

3. A science article refers to a temperature of 300.0 K. (a) What is the equivalent Celsius temperature? (b) What is the equivalent Fahrenheit temperature?

4. An electric current heats a 221 g copper wire from 20.0°C to 38.0°C. How much heat was generated by the current? (c_{copper} = 0.093 kcal/kgC°)

5. A bicycle and rider have a combined mass of 100.0 kg. How many calories of heat are generated in the brakes when the bicycle comes to a stop from a speed of 36.0 km/hr?

6. A 15.53 kg loose bag of soil falls 5.50 m at a construction site. If all the energy is retained by the soil in the bag, how much will its temperature increase? (c_{soil} = 0.200 kcal/kgC°)

7. A 75.0 kg person consumes a small order of french fries (250.0 Cal) and wishes to "work off" the energy by climbing a 10.0 m stairway. How many vertical climbs are needed to use all the energy?

8. A 0.5 kg glass bowl (c_{glass} = 0.2 kcal/kgC°) and a 0.5 kg iron pan (c_{iron} = 0.11 kcal/kgC°) have a temperature of 68°F when placed in a freezer. How much heat will the freezer have to remove from each to cool them to 32°F?

9. A sample of silver at 20.0°C is warmed to 100.0°C when 896 cal is added. What is the mass of the silver? (c_{silver} = 0.056 kcal/kgC°)

10. A 300.0 W immersion heater is used to heat 250.0 g of water from 10.0°C to 70.0°C. About how many minutes did this take?

11. A 100.0 g sample of metal is warmed 20.0°C when 60.0 cal is added. What is the specific heat of this metal?

12. How much heat is needed to change 250.0 g of ice at 0°C to water at 0°C?

13. How much heat is needed to change 250.0 g of water at 80.0°C to steam at 100.0°C?

14. A 100.0 g sample of water at 20.0°C is heated to steam at 125.0°C. How much heat was absorbed?

15. In an electric freezer, 400.0 g of water at 18.0°C is cooled, frozen, and the ice is chilled to −5.00°C. (a) How much total heat was removed from the water? (b) If the latent heat of vaporization of the Freon refrigerant is 40.0 cal/g, how many grams of Freon must be evaporated to absorb this heat?

16. A heat engine is supplied with 300.0 cal and rejects 200.0 cal in the exhaust. How many joules of mechanical work was done?

17. A refrigerator removes 40.0 kcal of heat from the freezer and releases 55.0 kcal through the condenser on the back. How much work was done by the compressor?

Group B

1. The Fahrenheit temperature reading is 98° on a hot summer day. What is this reading on the Kelvin scale?

2. A 0.25 kg length of aluminum wire is warmed 10.0°C by an electric current. How much heat was generated by the current? ($c_{aluminum}$ = 0.22 kcal/kgC°)

3. A 1,000.0 kg car with a speed of 90.0 km/hr brakes to a stop. How many cal of heat are generated by the brakes as a result?

4. A 1.0 kg metal head of a geology hammer strikes a solid rock with a velocity of 5.0 m/s. Assuming all the energy is retained by the hammer head, how much will its temperature increase? (c_{head} = 0.11 kcal/kgC°)

5. A 60.0 kg person will need to climb a 10.0 m stairway how many times to "work off" each excess Cal (kcal) consumed?

6. A 50.0 g silver spoon at 20.0°C is placed in a cup of coffee at 90.0°C. How much heat does the spoon absorb from the coffee to reach a temperature of 89.0°C?

7. If the silver spoon placed in the coffee in problem 6 causes it to cool 0.75°C, what is the mass of the coffee? (Assume c_{coffee} = 1.0 cal/gC°)

8. How many minutes would be required for a 300.0 W immersion heater to heat 250.0 g of water from 20.0°C to 100.0°C?

9. A 200.0 g china serving bowl is warmed 65.0°C when it absorbs 2.6 kcal of heat from a serving of hot food. What is the specific heat of the china dish?

10. A 1.00 kg block of ice at 0°C is added to a picnic cooler. How much heat will the ice remove as it melts to water at 0°C?

11. A 500.0 g pot of water at room temperature (20.0°C) is placed on a stove. How much heat is required to change this water to steam at 100.0°C?

12. Spent steam from an electric generating plant leaves the turbines at 120.0°C and is cooled to 90.0°C liquid water by water from a cooling tower in a heat exchanger. How much heat is removed by the cooling tower water for each kg of spent steam?

13. Lead is a soft, dense metal with a specific heat of 0.028 kcal/kgC°, a melting point of 328.0°C, and a heat of fusion of 5.5 kcal/kg. How much heat must be provided to melt a 250.0 kg sample of lead with a temperature of 20.0°C?

14. A heat engine converts 100.0 cal from a supply of 400.0 cal into work. How much mechanical work was done?

15. A heat pump releases 60.0 kcal as it removes 40.0 kcal at the evaporator coils. How much work does this heat pump ideally accomplish?

Compared to the sounds you hear on a calm day in the woods, the sounds from a waterfall can carry up to a million times more energy.

CHAPTER | Six

Wave Motions and Sound

S ometimes you can feel the floor of a building shake for a moment when something heavy is dropped. You can also feel prolonged vibrations in the ground when a nearby train moves by. The floor of a building and the ground are solids that transmit vibrations from a disturbance. Vibrations are common in most solids because the solids are elastic, having a tendency to rebound, or snap back, after a force or an impact deforms them. Usually you cannot see the vibrations in a floor or the ground, but you sense they are there because you can feel them.

There are many examples of vibrations that you can see. You can see the rapid blur of a vibrating guitar string (figure 6.1). You can see the vibrating up-and-down movement of a bounced-upon diving board. Both the vibrating guitar string and the diving board set up a vibrating motion of air that you identify as a sound. You cannot see the vibrating motion of the air, but you sense it is there because you hear sounds.

There are many kinds of vibrations that you cannot see but can sense. Heat, as you have learned, is associated with molecular vibrations that are too rapid and too tiny for your senses to detect other than as an increase in temperature. Other invisible vibrations include electrons that vibrate, generating spreading electromagnetic radio waves or visible light. Thus vibrations take place as an observable motion of objects but are also involved in sound, heat, electricity, and light. The vibrations involved in all these phenomena are fundamentally alike in many ways and all involve energy. Therefore, many topics of science are concerned with vibrational motion. In this chapter you will learn about the nature of vibrations and how they produce waves in general. These concepts will be applied to sound in this chapter and to electricity, light, and radio waves in later chapters.

FORCES AND ELASTIC MATERIALS

An *elastic* material is one that is capable of recovering its shape and form after some force has deformed it. A spring, for example, can be stretched or compressed, two changes that deform the shape of the spring. As you can imagine, there is a direct relationship between the extent of stretching or compression of a spring and the amount of force applied to it. As long as the applied force does not exceed the elastic limit of the spring, the spring always returns to its original shape when the applied force is removed. There are three important considerations about the applied force and deformation relationship: (1) the greater the applied force, the greater the compression or stretch of the spring from its original shape, (2) the spring appears to have an *internal restoring force,* which returns it to its original shape, and (3) the farther the spring is pushed or pulled, the *stronger* the restoring force that returns the spring to its original shape.

Forces and Vibrations

A **vibration** is a back-and-forth motion that repeats itself. Almost any solid can be made to vibrate if it is elastic. To see how forces are involved in vibrations, consider the spring and mass in figure 6.2. The spring and mass are arranged so that the mass can freely move back and forth on a frictionless surface. When the mass has not been disturbed, it is at rest at an *equilibrium position* (figure 6.2A). At the equilibrium position the spring is not compressed or stretched, so it applies no force on the mass. If, however, the mass is pulled to the right (figure 6.2B) the spring is stretched and applies a restoring force on the mass toward the left. The farther the mass is displaced, the greater the stretch of the spring and thus the greater the restoring force. The restoring force is proportional to the displacement and is in the opposite direction of the applied force.

If the mass is now released, the restoring force is the only force acting (horizontally) on the mass, so it accelerates back toward the equilibrium position. This force will continuously decrease until the moving mass arrives back at the equilibrium position, where the force is zero. The mass will have a maximum velocity when it arrives, however, so it overshoots the equilibrium position and continues moving to the left (figure 6.2C). As it moves to the left of the equilibrium position, it compresses the spring, which exerts an increasing force on the mass. The moving mass comes to a temporary halt, but now the restoring force again starts it moving back toward the equilibrium position. The whole process repeats itself again and again as the mass moves back and forth over the same path.

The periodic vibration, or oscillation, of the mass is similar to many vibrational motions found in nature called *simple harmonic motion*. Simple harmonic motion is defined as the vibratory motion that occurs when there is a restoring force opposite to and proportional to a displacement.

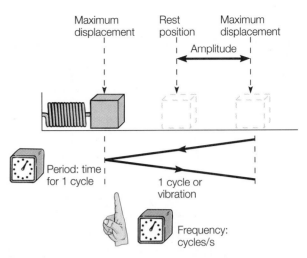

FIGURE 6.3

A vibrating mass attached to a spring is displaced from the rest, or equilibrium, position, and then released. The maximum displacement is called the *amplitude* of the vibration. A cycle is one complete vibration. The period is the time required for one complete cycle. The frequency is a count of how many cycles it completes in 1 s.

FIGURE 6.1

Vibrations are common in many elastic materials, and you can see and hear the results of many in your surroundings. Other vibrations in your surroundings, such as those involved in heat, electricity, and light, are invisible to the senses.

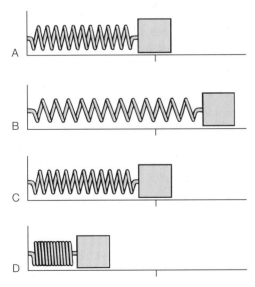

FIGURE 6.2

A mass on a frictionless surface is at rest at an equilibrium position (*A*) when undisturbed. When the spring is stretched (*B*) or compressed (*D*), then released (*C*), the mass vibrates back and forth because restoring forces pull opposite to and proportional to the displacement.

The vibrating mass and spring system will continue to vibrate for a while, slowly decreasing with time until the vibrations stop completely. The slowing and stopping is due to air resistance and internal friction. If these could be eliminated or compensated for with additional energy, the mass would continue to vibrate with a repeating, or *periodic*, motion.

Describing Vibrations

A vibrating mass is described by measuring several variables (figure 6.3). The extent of displacement from the equilibrium position is called the **amplitude.** A vibration that has a mass displaced a greater distance from equilibrium thus has a greater amplitude than a vibration with less displacement.

A complete vibration is called a **cycle.** A cycle is the movement from some point, say the far left, all the way to the far right, and back to the same point again, the far left in this example. The **period** (T) is simply the time required to complete one cycle. For example, suppose 0.1 s is required for an object to move through one complete cycle, to complete the back-and-forth motion from one point, then back to that point. The period of this vibration is 0.1 s.

Sometimes it is useful to know how frequently a vibration completes a cycle every second. The number of cycles per second is called the **frequency** (f). For example, a vibrating object moves through 10 cycles in 1 s. The frequency of this vibration is 10 cycles per second. Frequency is measured in a unit called a **hertz** (Hz). The unit for a hertz is 1/s because a cycle does not have dimensions. Thus a frequency of 10 cycles per second is referred to as 10 hertz or 10 1/s.

The period and frequency are two ways of describing the time involved in a vibration. Since the period (T) is the total

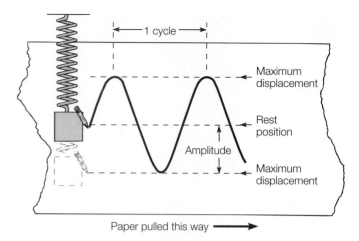

FIGURE 6.4

A graph of simple harmonic motion is described by a sinusoidal curve.

time involved in one cycle, and the frequency (f) is the number of cycles per second, the relationship is

$$T = \frac{1}{f}$$

equation 6.1

or

$$f = \frac{1}{T}$$

equation 6.2

You can obtain a graph of a vibrating object, which makes it easier to measure the amplitude, period, and frequency. If a pen is fixed to a vibrating mass and a paper is moved beneath it at a steady rate, it will draw a curve, as shown in figure 6.4. The greater the amplitude of the vibrating mass, the greater the height of this curve. The greater the frequency, the closer together the peaks and valleys. Note the shape of this curve. This shape is characteristic of simple harmonic motion and is called a *sinusoidal*, or sine, graph. It is so named because it is the same shape as a graph of the sine function in trigonometry.

WAVES

A vibration can be a repeating, or *periodic*, type of motion that disturbs the surroundings. A *pulse* is a disturbance of a single event of short duration. Both pulses and periodic vibrations can create a physical *wave* in the surroundings. A wave is a disturbance that moves through a medium such as a solid or the air. A heavy object dropped on the floor, for example, makes a pulse that sends a mechanical wave that you feel. It might also make a sound wave in the air that you hear. In either case, the medium that transported a wave (solid floor or air) returns to its normal state after the wave has passed. The medium does not travel from place to place, the wave does. Two major considerations about a wave are that (1) a wave is a traveling disturbance, and (2) a wave transports energy.

You can observe waves when you drop a rock into a still pool of water. The rock pushes the water into a circular mound as it enters the water. Since it is forcing the water through a distance, it is doing work to make the mound. The mound starts to move out in all directions, in a circle, leaving a depression behind. Water moves into the depression, and a circular wave—mound and depression—moves from the place of disturbance outward. Any floating object in the path of the wave, such as a leaf, exhibits an up-and-down motion as the mound and depression of the wave pass. But the leaf merely bobs up and down and after the wave has passed, it is much in the same place as before the wave. Thus, it was the disturbance that traveled across the water, not the water itself. If the wave reaches a leaf floating near the edge of the water, it may push the leaf up and out of the water, doing work on the leaf. Thus, the wave is a moving disturbance that transfers energy from one place to another.

Kinds of Waves

If you could see the motion of an individual water molecule near the surface as a water wave passed, you would see it trace out a circular path as it moves up and over, down and back. This

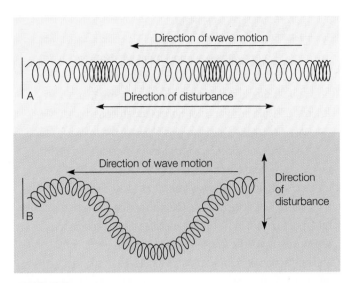

FIGURE 6.5

(A) Longitudinal waves are created in a spring when the free end is moved back and forth parallel to the spring. (B) Transverse waves are created in a spring when the free end is moved up and down.

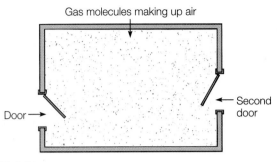

FIGURE 6.6

When you open one door into this room, the other door closes. Why does this happen? The answer is that the first door creates a pulse of compression that moves through the air like a sound wave. The pulse of compression pushes on the second door, closing it.

circular motion is characteristic of the motion of a particle reacting to a water wave disturbance. There are other kinds of waves, and each involves particles in a characteristic motion.

A **longitudinal wave** is a disturbance that causes particles to move closer together or farther apart in the same direction that the wave is moving. If you attach one end of a coiled spring to a wall and pull it tight, you will make longitudinal waves in the spring if you grasp the spring and then move your hand back and forth parallel to the spring. Each time you move your hand toward the length of the spring, a pulse of closer-together coils will move across the spring (figure 6.5A). Each time you pull your hand back, a pulse of farther-apart coils will move across the spring. The coils move back and forth in the same direction that the wave is moving, which is the characteristic movement in reaction to a longitudinal wave.

You will make a different kind of wave in the stretched spring if you now move your hand up and down perpendicular to the length of the spring. This creates a **transverse wave.** A transverse wave is a disturbance that causes motion perpendicular to the direction that the wave is moving. Particles responding to a transverse wave do not move closer together or farther apart in response to the disturbance; rather, they vibrate back and forth or up and down in a direction perpendicular to the direction of the wave motion (see figure 6.5B).

Whether you make mechanical longitudinal or transverse waves depends not only on the nature of the disturbance creating the waves, but also on the nature of the medium. Transverse waves can move through a material only if there is some interaction, or attachment, between the molecules making up the medium. In a gas, for example, the molecules move about freely without attachments to one another. A pulse can cause these molecules to move closer together or farther apart, so a gas can carry a longitudinal wave. But if a gas molecule is caused to move up or down, there is no reason for other molecules to do the same, since they are not attached. Thus, a gas will carry longitudinal waves but not transverse waves. Likewise a liquid will carry longitudinal waves but not transverse waves since the liquid molecules simply slide past one another. The surface of a liquid, however, is another story because of surface tension. A surface water wave is, in fact, a combination of longitudinal and transverse wave patterns that produce the circular motion of a disturbed particle. Solids can and do carry both longitudinal and transverse waves because of the strong attachments between the molecules.

Waves in Air

Waves that move through the air are longitudinal, so sound waves must be longitudinal waves. A familiar situation will be used to describe the nature of a longitudinal wave moving through air before considering sound specifically. The situation involves a small room with no open windows and two doors that open into the room (figure 6.6). When you open one door into the room the other door closes. Why does this happen? According to the kinetic molecular theory, the room contains many tiny, randomly moving gas molecules that make up the air. As you opened the door, it pushed on these gas molecules, creating a jammed-together zone of molecules immediately adjacent to the door. This jammed-together zone of air now has a greater density and pressure, which immediately spreads outward from the door as a pulse. The disturbance is rapidly passed from molecule to molecule, and the pulse of compression spreads through the room. It is not unlike a pulse of movement that can sometimes be seen in a swarm of flying insects. A momentary disturbance will cause nearby individual insects to momentarily move away from the disturbance and toward another flying insect, then back toward their original position. The movement is passed on through the swarm. During the movement of the disturbance, each insect maintains its own random motion. The overall effect, however, is that of a pulse moving through the swarm. In the example of the closing door, the pulse of greater density and pressure of air reached the door at the other side of the room, and the composite effect of the molecules impacting the door, that is, the increased pressure, caused it to close.

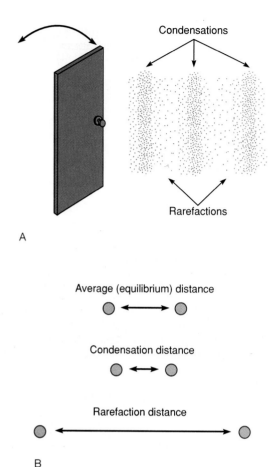

Condensations

Rarefactions

A

Average (equilibrium) distance

Condensation distance

Rarefaction distance

B

FIGURE 6.7

(A) Swinging the door inward produces pulses of increased density and pressure called *condensations*. Pulling the door outward produces pulses of decreased density and pressure called *rarefactions*. (B) In a condensation, the average distance between gas molecules is momentarily decreased as the pulse passes. In a rarefaction, the average distance is momentarily increased.

If the door at the other side of the room does not latch, you can probably cause it to open again by pulling on the first door quickly. By so doing, you send a pulse of thinned-out molecules of lowered density and pressure. The door you pulled quickly pushed some of the molecules out of the room. Other molecules quickly move into the region of less pressure, then back to their normal positions. The overall effect is the movement of a thinned-out pulse that travels through the room. When the pulse of slightly reduced pressure reaches the other door, molecules exerting their normal pressure on the other side of the door cause it to move. After a pulse has passed a particular place, the molecules are very soon homogeneously distributed again due to their rapid, random movement.

If you were to swing a door back and forth it would be a vibrating object. As it vibrates back and forth it would have a certain frequency in terms of the number of vibrations per second. As the vibrating door moves toward the room, it creates a pulse of jammed-together molecules called a **condensation** (or com-

pression) that quickly moves throughout the room. As the vibrating door moves away from the room, a pulse of thinned-out molecules called a **rarefaction** quickly moves throughout the room. The vibrating door sends repeating pulses of condensation (increased density and pressure) and rarefaction (decreased density and pressure) through the room as it moves back and forth (figure 6.7). You know that the pulses transmit energy because they produce movement, or do work on, the other door. Individual molecules execute a harmonic motion about their equilibrium position and can do work on a movable object. Energy is thus transferred by this example of longitudinal waves.

Hearing Waves in Air

You cannot hear a vibrating door because the human ear normally hears sounds originating from vibrating objects with a frequency between 20 and 20,000 Hz. Longitudinal waves with frequencies less than 20 Hz are called **infrasonic.** You usually *feel* sounds below 20 Hz rather than hearing them, particularly if you are listening to a good sound system. Longitudinal waves above 20,000 Hz are called **ultrasonic.** Although 20,000 Hz is usually considered the upper limit of hearing, the actual limit varies from person to person and becomes lower and lower with increasing age. Humans do not hear infrasonic nor ultrasonic sounds, but various animals have different limits. Dogs, cats, rats, and bats can hear higher frequencies than humans. Dogs can hear an ultrasonic whistle when a human hears nothing, for example. Some bats make and hear sounds of frequencies up to 100,000 Hz as they navigate and search for flying insects in total darkness.

A tuning fork that vibrates at 260 Hz makes longitudinal waves much like the swinging door, but these longitudinal waves are called *sound waves* because they are within the frequency range of human hearing. The prongs of a struck tuning fork vibrate, moving back and forth. This is more readily observed if the prongs of the fork are struck, then held against a sheet of paper

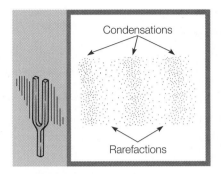

FIGURE 6.8

A vibrating tuning fork produces a series of condensations and rarefactions that move away from the tuning fork. The pulses of increased and decreased pressure reach your ear, vibrating the eardrum. The ear sends nerve signals to the brain about the vibrations, and the brain interprets the signals as sounds.

or plunged into a beaker of water. In air, the vibrating prongs first move toward you, pushing the air molecules into a condensation of increased density and pressure. As the prongs then move back, a rarefaction of decreased density and pressure is produced. The alternation of increased and decreased pressure pulses moves from the vibrating tuning fork and spreads outward in all directions, much like the surface of a rapidly expanding balloon (figure 6.8). When the pulses reach your eardrum, it is forced in and out by the pulses. It now vibrates with the same frequency as the tuning fork. The vibrations of the eardrum are transferred by three tiny bones to a fluid in a coiled chamber. Here, tiny hairs respond to the frequency and size of the disturbance, activating nerves that transmit the information to the brain. The brain interprets a frequency as a sound with a certain **pitch.** High-frequency sounds are interpreted as high-pitched musical notes, for example, and low-frequency sounds are interpreted as low-pitched musical notes. The brain then selects certain sounds from all you hear, and you "tune" to certain ones, enabling you to listen to whatever sounds you want while ignoring the background noise, which is made up of all the other sounds.

WAVE TERMS

A tuning fork vibrates with a certain frequency and amplitude, producing a longitudinal wave of alternating pulses of increased-pressure condensations and reduced-pressure rarefactions. A graph of the frequency and amplitude of the vibrations is shown in figure 6.9A, and a representation of the condensations and rarefactions is shown in figure 6.9B. The wave pattern can also be represented by a graph of the changing air pressure of the traveling sound wave, as shown in figure 6.9C. This graph can be used to define some interesting concepts associated with sound waves. Note the correspondence between the (1) amplitude, or displacement, of the vibrating prong, (2) the pulses of condensations and rarefactions, and (3) the changing air pressure. Note also the correspondence between the frequency of the vibrating prong and the frequency of the wave cycles.

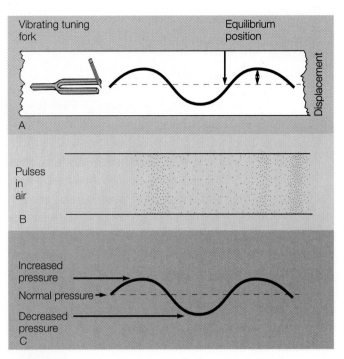

FIGURE 6.9

Compare the (A) back-and-forth vibrations of a tuning fork with (B) the resulting condensations and rarefactions that move through the air and (C) the resulting increases and decreases of air pressure on a surface that intercepts the condensations and rarefactions.

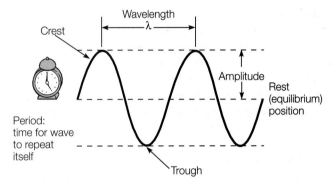

FIGURE 6.10

Here are some terms associated with periodic waves. The *wavelength* is the distance from a part of one wave to the same part in the next wave, such as from one crest to the next. The *amplitude* is the displacement from the rest position. The *period* is the time required for a wave to repeat itself, that is, the time for one complete wavelength to move past a given location.

Figure 6.10 shows the terms commonly associated with waves from a continuously vibrating source. The wave *crest* is the maximum disturbance from the undisturbed (rest) position. For a sound wave, this would represent the maximum increase of air pressure. The wave *trough* is the maximum disturbance in the opposite direction from the rest position. For a sound wave, this would represent the maximum decrease of air pressure. The

CONNECTIONS

Musical Instruments: Membranes

One group of musical instruments includes all those using a vibrating membrane, which can be made from any relatively thin layer of a natural or synthetic material that is elastic. Such materials include animal skins, paper, metal, plastic, or glass. The various kinds of drums and the tambourine, for example, use flat, elastic membranes stretched across a frame or a resonator. Cymbals and gongs use a thin layer of metal that vibrates, so they belong in this group, too, as do bells. A bell is a thin metal layer that has been formed into a different shape, so it becomes its own sound resonator. Bells, gongs, drums, and the other instruments that use a membrane generally produce complicated waveforms made up of nonharmonic partials. Since the sound maker is formed into a curved surface, the waveforms are more complicated than those produced by solid metal bars.

Compared to the solid bars, strings, and reeds of other instruments, the membrane instruments have a much larger area of sound-maker material that does the vibrating. This larger area is able to make sound waves that involve a greater mass of air, making sounds that are much louder. In the past, the loud sounds from drums and bells were used as signals and communication devices since they can be heard at great distances. This is impossible today, however, as you can no longer hear the distant drum or bell because of other noise. The church bell that was once heard across town is today not heard a city block away because of traffic noise.

The sounds from bells are not periodic, yet you know that one can play tunes with bells. This is possible because each bell has a set of frequencies that are nonharmonic partials, yet they are very close to being harmonics. In addition, the timing of nonharmonic partials can sometimes suggest a pitch, so periodic noises can give an impression of musical sounds. This is how someone once "made music" by changing a radio dial to one station, then another, with calculated periodic timing.

---

amplitude of a wave is the displacement from rest to the crest *or* from rest to the trough. The time required for a wave to repeat itself is the *period* (T). To repeat itself means the time required to move through one full wave, such as from the crest of one wave to the crest of the next wave. This length in which the wave repeats itself is called the **wavelength** (the symbol is λ, which is the Greek letter lambda). Wavelength is measured in centimeters or meters just like any other length.

There is a relationship between the wavelength, period, and speed of a wave. Recall that speed is

$$v = \frac{\text{distance}}{\text{time}}$$

Since it takes one period (T) for a wave to move one wavelength (λ), then the speed of a wave can be measured from

$$v = \frac{\text{one wavelength}}{\text{one period}} = \frac{\lambda}{T}$$

The *frequency*, however, is more convenient than the period for dealing with waves that repeat themselves rapidly. Recall the relationship between frequency (f) and the period (T) is

$$f = \frac{1}{T}$$

Substituting f for $1/T$ yields

$$v = \lambda f$$

equation 6.3

which we will call the *wave equation*. This equation tells you that the velocity of a wave can be obtained from the product of the wavelength and the frequency. Note that it also tells you that the wavelength and frequency are inversely proportional at a given velocity.

SOUND WAVES

The transmission of a sound wave requires a medium, that is, a solid, liquid, or gas to carry the disturbance. Therefore, sound does not travel through the vacuum of outer space, since there is nothing to carry the vibrations from a source. The nature of the molecules making up a solid, liquid, or gas determines how well or how rapidly the substance will carry sound waves. The two variables are (1) the inertia of the molecules and (2) the strength of the interaction, if the molecules are attached to one another. Thus, hydrogen gas, with the least massive molecules with no interaction or attachments, will carry a sound wave at 1,284 m/s (4,213 ft/s) when the temperature is 0°C. More massive helium gas molecules have more inertia and carry a sound wave at only 965 m/s (3,166 ft/s) at the same temperature. A solid, however, has molecules that are strongly attached so vibrations are passed rapidly from molecule to molecule. Steel, for example, is highly elastic, and sound will move through a steel rail at 5,940 m/s (19,488 ft/s). Thus there is a reason for the old saying, "Keep your ear to the ground," because sounds move through solids more rapidly than through a gas (table 6.1).

TABLE 6.1 Speed of sound in various materials		
Medium	**m/s**	**ft/s**
Carbon dioxide (0°C)	259	850
Dry air (0°C)	331	1,087
Helium (0°C)	965	3,166
Hydrogen (0°C)	1,284	4,213
Water (25°C)	1,497	4,911
Seawater (25°C)	1,530	5,023
Lead	1,960	6,430
Glass	5,100	16,732
Steel	5,940	19,488

Velocity of Sound in Air

Most people have observed that sound takes some period of time to move through the air. If you watch a person hammering on a roof a block away, the sounds of the hammering are not in sync with what you see. Light travels so rapidly that you can consider what you see to be simultaneous with what is actually happening for all practical purposes. Sound, however, travels much more slowly and the sounds arrive late in comparison to what you are seeing. This is dramatically illustrated by seeing a flash of lightning, then hearing thunder seconds later. Perhaps you know of a way to estimate the distance to a lightning flash by timing the interval between the flash and boom. You will learn a precise way to measure this distance shortly.

The air temperature influences how rapidly sound moves through the air. The gas molecules in warmer air have a greater kinetic energy than those of cooler air. The molecules of warmer air therefore transmit an impulse from molecule to molecule more rapidly. More precisely, the speed of a sound wave increases 0.60 m/s (2.0 ft/s) for *each* Celsius degree increase in temperature. In *dry* air at sea-level density (normal pressure) and 0°C (32°F), the velocity of sound is about 331 m/s (1,087 ft/s). Therefore, the velocity of sound at different temperatures can be calculated from the following relationships:

$$v_{T_p}(\text{m/s}) = v_0 + \left(\frac{0.60 \text{ m/s}}{°C}\right)(T_p)$$

equation 6.4

where v_{T_p} is the velocity of sound at the present temperature, v_0 is the velocity of sound at 0°C, and T_p is the present temperature. This equation tells you that the velocity of a sound wave increases 0.6 m/s for each degree Celsius above 0°C. For units of ft/s,

$$v_{T_p}(\text{ft/s}) = v_0 + \left(\frac{2.0 \text{ ft/s}}{°C}\right)(T_p)$$

equation 6.5

Equation 6.5 tells you that the velocity of a sound wave increases 2.0 ft/s for each degree Celsius above 0°C.

CONNECTIONS

Musical Instruments: Air Columns

This group includes any instrument that uses an air column as a sound maker. This includes all the wind instruments such as the clarinet, flute, trombone, trumpet, pipe organ, and many others. All of these instruments use open or closed air columns, except the pipe organ, which usually has a combination of open and closed air columns. Basically, the length of the air column is what determines the frequency, and the various wind instruments have different ways of doing this, as well as different ways of making the air column vibrate. In the flute, air vibrates as it moves over a sharp edge, while in the clarinet, saxophone, and other reed instruments it vibrates through fluttering thin reeds. The air column in brass instruments, on the other hand, is vibrated by the tightly fluttering lips of the player.

Woodwind instruments have holes in the side of a tube, holes that can be opened or closed to change the effective length of the air column inside the tube. By changing the length of the vibrating air column, the placement of the open end and its antinode is changed. The resulting tone depends on the length of the air column and the resonance overtones. Brass instruments have an air column in a long metal tube, which is often coiled into a form that has a flared, horn-shaped end.

Refraction and Reflection

When you drop a rock into a still pool of water, circular patterns of waves move out from the disturbance. These water waves are on a flat, two-dimensional surface. Sound waves, however, move in three-dimensional space like a rapidly expanding balloon. Sound waves are *spherical waves* that move outward from the source. Spherical waves of sound move as condensations and rarefactions from a continuously vibrating source at the center. If you identify the same part of each wave in the spherical waves, you have identified a *wave front*. For example, the crests of each condensation could be considered as a wave front. From one wave front to the next, therefore, identifies one complete wave or wavelength. At some distance from the source, a small part of a spherical wave front can be considered a *linear wave front* (figure 6.11).

A CLOSER LOOK | Hearing Loss

A hearing test will tell you if you are losing your hearing ability or not. The test is made by identifying the dB threshold needed to first hear a pure tone. The threshold is found for frequencies of 1,000, 2,000, 3,000, 4,000, and 6,000 Hz for each ear. To find if hearing damage has occurred or not, the results are compared to earlier baseline data, and allowance is made for the contribution of aging to the change in hearing level. You might be interested in the tables used to predict the normal loss of hearing because of aging. Box table 6.1 lists the threshold of hearing the different frequencies, in dB, that can be expected for males at different ages. Box table 6.2 shows the same data for females. These tables give the thresholds expected if you have not damaged your ears. If you spend a lot of time in places with very loud noises, for example, close to a loud band or group, you will lose your hearing ability at a much younger age.

How do you read the tables? For an example, suppose you are a 20-year-old female. According to box table 6.2, you will be able to hear a 6 dB pure tone that has a frequency of 6,000 Hz. However, if you are a 20-year-old male, you will not be able to hear a 6,000 Hz pure tone until it is 2 dB louder, at 8 dB. Ten years later, at age 30, the normal loss of hearing that comes with aging finds the female not hearing the 6,000 Hz tone until it is at 9 dB. The 30-year-old male, on the other hand, hears nothing below a 12 dB 6,000 Hz tone. You can follow the table down through the years for a particular frequency, and you can compare the hearing loss expected for the different frequencies.

In general, there are losses at all frequencies, but the ability to hear higher frequencies appears to be greater than the loss of ability to hear lower ones. You cannot stop this natural loss of ability to hear, but you can prevent it from being even worse by protecting your ears from long exposure to very loud noise. Think about what the drivers are doing to their hearing ability next time you hear a "boom car," one with a very loud radio you can hear "booming" a half-block away.

—Continued top of next page

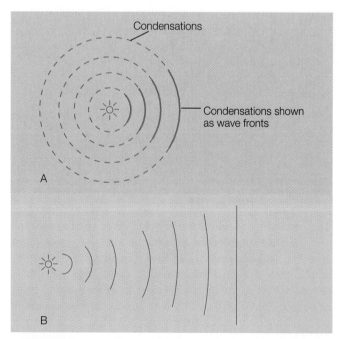

FIGURE 6.11

(*A*) Spherical waves move outward from a sounding source much as a rapidly expanding balloon. This two-dimensional sketch shows the repeating condensations as spherical wave fronts. (*B*) Some distance from the source, a spherical wave front is considered a linear, or plane, wave front.

Waves move within a homogeneous medium such as a gas or a solid at a fairly constant rate but gradually lose energy to friction. When a wave encounters a different condition, however, drastic changes may occur rapidly. The division between two physical conditions is called a *boundary*. Boundaries are usually encountered (1) between different materials or (2) between the same materials with different conditions. An example of a wave moving between different materials is a sound made in the next room that moves through the air to the wall and through the wall to the air in the room where you are. The boundaries are air-wall and wall-air. If you have ever been in a room with "thin walls," it is obvious that sound moved through the wall and air boundaries.

An example of sound waves moving through the same material with different conditions is found when a wave front moves through air of different temperatures. Since sound travels faster in warm air than in cold air, the wave front becomes bent. The bending of a wave front between boundaries is called *refraction*. Refraction changes the direction of travel of a wave front. Consider, for example, that on calm, clear nights the air near the earth's surface is cooler than air farther above the surface. Air at rooftop height above the surface might be four or five degrees warmer under such ideal conditions. Sound will travel faster in the higher, warmer air than it will in the lower, cooler air close to the surface. A wave front will therefore become bent, or refracted, toward the ground on a cool night and you will be able to hear sounds

Continued—

BOX TABLE 6.1

Age correction values in decibels for males

Years	Audiometric Test Frequency (Hz)				
	1,000	2,000	3,000	4,000	6,000
20 or younger	5	3	4	5	8
21	5	3	4	5	8
22	5	3	4	5	8
23	5	3	4	6	9
24	5	3	5	6	9
25	5	3	5	7	10
26	5	4	5	7	10
27	5	4	6	7	11
28	6	4	6	8	11
29	6	4	6	8	12
30	6	4	6	9	12
31	6	4	7	9	13
32	6	5	7	10	14
33	6	5	7	10	14
34	6	5	8	11	15
35	7	5	8	11	15
36	7	5	9	12	16
37	7	6	9	12	17
38	7	6	9	13	17
39	7	6	10	14	18
40	7	6	10	14	19
41	7	6	10	14	20
42	8	7	11	16	20
43	8	7	12	16	21
44	8	7	12	17	22
45	8	7	13	18	23
46	8	8	13	19	24
47	8	8	14	19	24
48	9	8	14	20	25
49	9	9	15	21	26
50	9	9	16	22	27
51	9	9	16	23	28
52	9	10	17	24	29
53	9	10	18	25	30
54	10	10	18	26	31
55	10	11	19	27	32
56	10	11	20	28	34
57	10	11	21	29	35
58	10	12	22	31	36
59	11	12	22	32	37
60 or older	11	13	23	33	38

Source: OSHA 29 CFR 1910.95 Appendix F.

BOX TABLE 6.2

Age correction values in decibels for females

Years	Audiometric Test Frequency (Hz)				
	1,000	2,000	3,000	4,000	6,000
20 or younger ...	7	4	3	3	6
21	7	4	4	3	6
22	7	4	4	4	6
23	7	5	4	4	7
24	7	5	4	4	7
25	8	5	4	4	7
26	8	5	5	4	8
27	8	5	5	5	8
28	8	5	5	5	8
29	8	5	5	5	9
30	8	6	5	5	9
31	8	6	6	5	9
32	9	6	6	6	10
33	9	6	6	6	10
34	9	6	6	6	10
35	9	6	7	7	11
36	9	7	7	7	11
37	9	7	7	7	12
38	10	7	7	7	12
39	10	7	8	8	12
40	10	7	8	8	13
41	10	8	8	8	13
42	10	8	9	9	13
43	11	8	9	9	14
44	11	8	9	9	14
45	11	8	10	10	15
46	11	9	10	10	15
47	11	9	10	11	16
48	12	9	11	11	16
49	12	9	11	12	16
50	12	10	11	12	17
51	12	10	12	12	17
52	12	10	12	13	18
53	12	10	13	13	18
54	13	11	13	14	19
55	13	11	14	14	19
56	13	11	14	15	20
57	13	11	15	15	20
58	14	12	15	16	21
59	14	12	16	16	21
60 or older	14	12	16	17	22

Source: OSHA 29 CFR 1910.95 Appendix F.

Noise Pollution Solution?

Bells, jet planes, sirens, motorcycles, jackhammers, and construction noises all contribute to the constant racket that has become commonplace in many areas. Such noise pollution is everywhere, and it is a rare place where you can escape the ongoing din. Earplugs help reduce the noise level, but they also block sounds that you want to hear, such as music and human voices. In addition, earplugs are more effective against high-frequency sounds than they are against lower ones like aircraft engines or wind noise.

A better solution for noise pollution might be the relatively new "antinoise" technology that cancels sound waves before they reach your ears. A microphone detects background noise and transmits information to microprocessors about the noise waveform. The microprocessors then generate an "antinoise" signal that is 180 degrees out of phase with the background noise. When the noise and antinoise meet, they undergo destructive interference and significantly reduce the final sound loudness.

The noise-canceling technology is today available as a consumer portable electronic device with a set of headphones, look-ing much like a portable tape player. The headphones can be used to combat steady, ongoing background noise such as you might hear from a spinning computer drive, a vacuum cleaner, or while inside the cabin of a jet plane. Some vendors claim their device cancels up to 40 percent of whirring air conditioner noise, 80 percent of ongoing car noise, or up to 95 percent of constant airplane cabin noise. You hear all the other sounds—people talking, warning sounds, and music—as the microprocessors are not yet fast enough to match anything but a constant sound source.

Noise-canceling microphones can also be used to limit background noise that muddles the human voice in telephone booths, in teleconferencing, and during cellular phone use from a number of high-noise places. With improved microprocessor and new digital applications, noise-canceling technology will help bring higher sound quality to voice-driven applications. It may also find extended practical use in helping to turn down the volume on our noisy world.

from farther away than on warm nights (figure 6.12A). The opposite process occurs during the day as the earth's surface becomes warmer from sunlight (figure 6.12B). Wave fronts are refracted upward because part of the wave front travels faster in the warmer air near the surface. Thus, sound does not seem to carry as far in the summer as it does in the winter. What is actually happening is that during the summer the wave fronts are refracted away from the ground before they travel very far.

When a wave front strikes a boundary that is parallel to the front the wave may be absorbed, transmitted, or undergo *reflection*, depending on the nature of the boundary medium, or the wave may be partly absorbed, partly transmitted, partly reflected, or any combination thereof. Some materials, such as hard, smooth surfaces, reflect sound waves more than they absorb them. Other materials, such as soft, ruffly curtains, absorb sound waves more than they reflect them. If you have ever been in a room with smooth, hard walls and with no curtains, carpets, or furniture, you know that sound waves may be reflected several times before they are finally absorbed. Sounds seem to increase in volume because of all the reflections and the apparent increase in volume due to **reverberation.** Reverberation is the mixing of sound with reflections, and is one of the factors that determine the acoustical qualities of a room, lecture hall, or auditorium (figure 6.13).

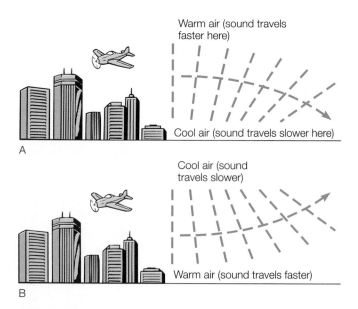

FIGURE 6.12

(*A*) Since sound travels faster in warmer air, a wave front becomes bent, or refracted, toward the earth's surface when the air is cooler near the surface. (*B*) When the air is warmer near the surface, a wave front is refracted upward, away from the surface.

FIGURE 6.13

This closed-circuit TV control room is acoustically treated by covering the walls with sound-absorbing baffles.

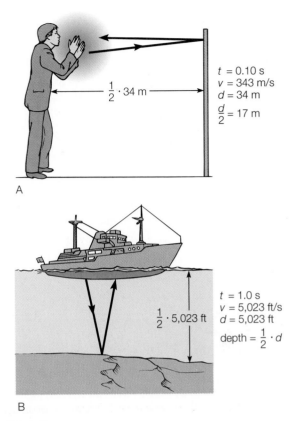

FIGURE 6.14

(*A*) At room temperature, sound travels at 343 m/s. In 0.10 s, sound would travel 34 m. Since the sound must travel to a surface and back in order for you to hear an echo, the distance to the surface is one-half the total distance. (*B*) Sonar measures a depth by measuring the elapsed time between an ultrasonic sound pulse and the echo. The depth is one-half the round trip.

If a reflected sound arrives after 0.10 s, the human ear can distinguish the reflected sound from the original sound. A reflected sound that can be distinguished from the original is called an **echo.** Thus, a reflected sound that arrives before 0.10 s is perceived as an increase in volume and is called a *reverberation,* but a sound that arrives after 0.10 s is perceived as an echo.

Sound wave echoes are measured to determine the depth of water or to locate underwater objects by a *sonar* device. The word *sonar* is taken from *so*und *na*vigation *r*anging. The device generates an underwater ultrasonic sound pulse, then measures the elapsed time for the returning echo. Sound waves travel at about 1,531 m/s (5,023 ft/s) in seawater at 25°C (77°F). A 1 s lapse between the ping of the generated sound and the echo return would mean that the sound traveled 5,023 ft for the round trip. The bottom would be half this distance below the surface (figure 6.14).

Interference

Waves interact with a boundary much as a particle would, reflecting or refracting because of the boundary. A moving ball, for example, will bounce from a surface at the same angle it strikes the surface, just as a wave does. A particle or a ball, however, can be in only one place at a time, but waves can be spread over a distance at the same time. You know this since many different people in different places can hear the same sound at the same time.

Another difference between waves and particles is that two or more waves can exist in the same place at the same time. When two patterns of waves meet, they pass through each other without refracting or reflecting. However, at the place where they meet the waves interfere with each other, producing a *new* disturbance. This new disturbance has a different amplitude, which is the algebraic sum of the amplitudes of the two separate wave patterns. If the wave crests or wave troughs arrive at the same place at the same time, the two waves are said

to be *in phase.* The result of two waves arriving in phase is a new disturbance with a crest and trough that has greater displacement than either of the two separate waves. This is called *constructive interference* (figure 6.15A). If the trough of one wave arrives at the same place and time as the crest of another wave, the waves are completely *out of phase.* When two waves are completely out of phase, the crest of one wave (positive displacement) will cancel the trough of the other wave (negative displacement), and the result is zero total disturbance, or no wave. This is called *destructive interference* (figure 6.15B). If the two sets of wave patterns do not have the exact same amplitudes or wavelengths, they will be neither completely in phase nor completely out of phase. The result will be partly constructive or destructive interference, depending on the exact nature of the two wave patterns.

Suppose that two vibrating sources produce sounds that are in phase, equal in amplitude, and equal in frequency. The resulting sound will be increased in volume because of constructive interference. But suppose the two sources are slightly different in frequency, for example, 350 and 352 Hz. You will

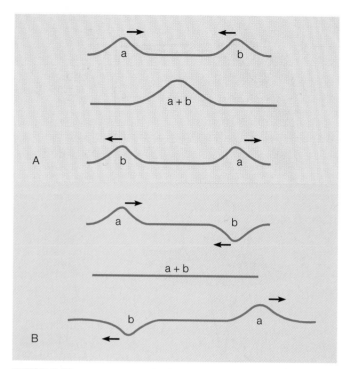

FIGURE 6.15

(*A*) Constructive interference occurs when two equal, in-phase waves meet. (*B*) Destructive interference occurs when two equal, out-of-phase waves meet. In both cases, the wave displacements are superimposed when they meet, but they then pass through one another and return to their original amplitudes.

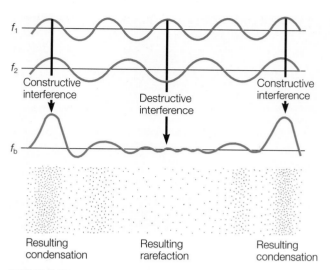

FIGURE 6.16

Two waves of equal amplitude but slightly different frequencies interfere destructively and constructively. The result is an alternation of loudness called a *beat.*

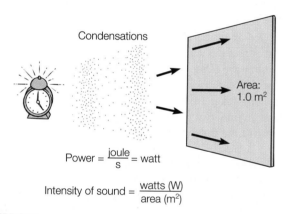

FIGURE 6.17

The intensity, or energy, of a sound wave is the rate of energy transferred to an area perpendicular to the waves. Intensity is measured in watts per square meter, W/m².

hear a regularly spaced increase and decrease of sound known as **beats.** Beats occur because the two sound waves experience alternating constructive and destructive interferences (figure 6.16). The phase relationship changes because of the difference in frequency, as you can see in the illustration. These alternating constructive and destructive interference zones are moving from the source to the receiver, and the receiver hears the results as a rapidly rising and falling sound level. The beat frequency is the difference between the frequencies of the two sources. A 352 Hz source and 350 Hz source sounded together would result in a beat frequency of 2 Hz. Thus, the frequencies are closer and closer together, and fewer beats will be heard per second. You may be familiar with the phenomenon of beats if you have ever flown in an airplane with two engines. If one engine is running slightly faster than the other, you hear a slow beat. The same phenomenon can sometimes be heard from two or more engines of a diesel locomotive and from two snow tires on a car driving down a highway. The beat frequency (f_b) is equal to the absolute difference in frequency of two interfering waves with slightly different frequencies, or

$$f_b = f_2 - f_1$$

equation 6.6

ENERGY AND SOUND

All waves involve the transportation of energy, including sound waves. The vibrating mass and spring in figure 6.2 vibrate with an amplitude that depends on how much work you did on the mass in moving it from its equilibrium position. More work on the mass results in a greater displacement and a greater amplitude of vibration. A vibrating object that is producing sound waves will produce more intense condensations and rarefactions if it has a greater amplitude. The intensity of a sound wave is a measure of the energy the sound wave is carrying (figure 6.17). **Intensity** is defined as the power (in watts) transmitted by a wave to a unit area (in square meters) that is perpendicular to the waves. Intensity is therefore measured in watts per square meter (W/m²) or

$$\text{Intensity} = \frac{\text{Power}}{\text{Area}}$$

$$I = \frac{P}{A}$$

equation 6.7

Loudness

The *loudness* of a sound is a subjective interpretation that varies from person to person. Loudness is also related to (1) the energy of a vibrating object, (2) the condition of the air the sound wave travels through, and (3) the distance between you and the vibrating source. Furthermore, doubling the amplitude of the vibrating source will quadruple the *intensity* of the resulting sound wave, but the sound will not be perceived as four times as loud. The relationship between perceived loudness and the intensity of a sound wave is not a linear relationship. In fact, a sound that is perceived as twice as loud requires ten times the intensity, and quadrupling the loudness requires a one-hundred-fold increase in intensity.

The human ear is very sensitive, capable of hearing sounds with intensities as low as 10^{-12} W/m² and is not made uncomfortable by sound until the intensity reaches about 1 W/m². The second intensity is a million million (10^{12}) times greater than the first. Within this range, the subjective interpretation of intensity seems to vary by powers of ten. This observation led to the development of the **decibel scale** to measure relative intensity. The scale is a ratio of the intensity level of a given sound to the threshold of hearing, which is defined as 10^{-12} W/m² at 1,000 Hz. In keeping with the power-of-ten subjective interpretations of intensity, a logarithmic scale is used rather than a linear scale. Originally the scale was the logarithm of the ratio of the intensity level of a sound to the threshold of hearing. This definition set the zero point at the threshold of human hearing. The unit was named the *bel* in honor of Alexander Graham Bell. This unit was too large to be practical, so it was reduced by one-tenth and called a *decibel*. The intensity level of a sound is therefore measured in decibels (table 6.2). Compare the decibel noise level of familiar sounds listed in table 6.2, and note that each increase of 10 on the decibel scale is matched by a *multiple* of 10 on the intensity level. For example, moving from a decibel level of 10 to a decibel level of 20 requires *ten times* more intensity. Likewise, moving from a decibel level of 20 to 40 requires a one-hundred-fold increase in the intensity level. As you can see, the decibel scale is not a simple linear scale.

Resonance

You know that sound waves transmit energy when you hear a thunderclap rattle the windows. In fact, the sharp sounds from an explosion have been known not only to rattle but also break windows. The source of the energy is obvious when thunderclaps or explosions are involved. But sometimes energy transfer occurs through sound waves when it is not clear what is hap-

TABLE 6.2	Comparison of noise levels in decibels with intensity		
Example	**Response**	**Decibels**	**Intensity W/m²**
Least needed for hearing	Barely perceived	0	1×10^{-12}
Calm day in woods	Very, very quiet	10	1×10^{-11}
Whisper (15 ft)	Very quiet	20	1×10^{-10}
Library	Quiet	40	1×10^{-8}
Talking	Easy to hear	65	3×10^{-6}
Heavy street traffic	Conversation difficult	70	1×10^{-5}
Pneumatic drill (50 ft)	Very loud	95	3×10^{-3}
Jet plane (200 ft)	Discomfort	120	1

pening. A truck drives down the street, for example, and one window rattles but the others do not. A singer shatters a crystal water glass by singing a single note, but other objects remain undisturbed. A closer look at the nature of vibrating objects and the transfer of energy will explain these phenomena.

Almost any elastic object can be made to vibrate and will vibrate freely at a constant frequency after being sufficiently disturbed. Entertainers sometimes discover this fact and appear on late-night talk shows playing saws, wrenches, and other odd objects as musical instruments. All material objects have a **natural frequency** of vibration determined by the materials and shape of the objects. The natural frequencies of different wrenches enable an entertainer to use the suspended tools as if they were the bars of a xylophone.

If you have ever pumped a swing, you know that small forces can be applied at any frequency. If the frequency of the applied forces matches the natural frequency of the moving swing, there is a dramatic increase in amplitude. When the two frequencies match, energy is transferred very efficiently. This condition, when the frequency of an external force matches the natural frequency, is called **resonance.** The natural frequency of an object is thus referred to as the *resonant frequency,* that is, the frequency at which resonance occurs.

A silent tuning fork will resonate if a second tuning fork with the same frequency is struck and vibrates nearby (figure 6.18). You will hear the previously silent tuning fork sounding if you stop the vibrations of the struck fork by touching it. The waves of condensations and rarefactions produced by the struck tuning fork produce a regular series of impulses that match the natural frequency of the silent tuning fork. This illustrates that at resonance, relatively little energy is required to start vibrations.

A truck causing vibrations as it is driven past a building may cause one window to rattle while others do not. Vibrations caused by the truck have matched the natural frequency of this window but not the others. The window is undergoing resonance

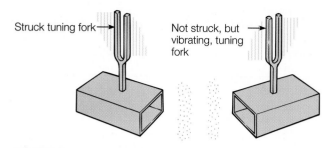

Struck tuning fork

Not struck, but vibrating, tuning fork

FIGURE 6.18

When the frequency of an applied force, including the force of a sound wave, matches the natural frequency of an object, energy is transferred very efficiently. The condition is called *resonance*.

Activities

Set up two identical pendulums as shown in figure 6.19A. The bobs should be identical and suspended from identical strings of the same length attached to a tight horizontal string. Start one pendulum vibrating by pulling it back, then releasing it. Observe the vibrations and energy exchange between the two pendulums for the next several minutes. Now change the frequency of vibrations of *one* of the pendulums by shortening the string (figure 6.19B). Again start one pendulum vibrating and observe for several minutes. Compare what you observe when the frequencies are matched and when they are not. Explain what happens in terms of resonance.

from the sound wave impulses that matched its natural frequency. It is also resonance that enables a singer to break a water glass. If the tone is at the resonant frequency of the glass, the resulting vibrations may be large enough to shatter it.

Resonance considerations are important in engineering. A large water pump, for example, was designed for a nuclear power plant. Vibrations from the electric motor matched the resonant frequency of the impeller blades, and they shattered after a short period of time. The blades were redesigned to have a different natural frequency when the problem was discovered. Resonance vibrations are particularly important in the design of buildings.

SOURCES OF SOUNDS

All sounds have a vibrating object as their source. The vibrations of the object send pulses or waves of condensations and rarefactions through the air. These sound waves have physical properties that can be measured, such as frequency and intensity. Subjectively, your response to frequency is to identify a certain pitch. A high-frequency sound is interpreted as a high-pitched sound, and a low-frequency sound is interpreted as a low-pitched sound. Likewise, a greater intensity is interpreted as increased

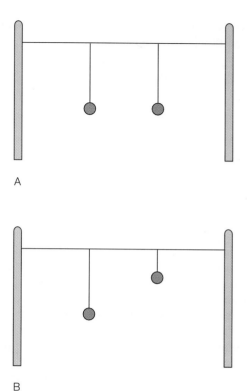

A

B

FIGURE 6.19

The "Activities" section demonstrates that one of these pendulum arrangements will show resonance, and the other will not. Can you predict which one will show resonance?

loudness, but there is not a direct relationship between intensity and loudness as there is between frequency and pitch.

There are other subjective interpretations about sounds. Some sounds are bothersome and irritating to some people but go unnoticed by others. In general, sounds made by brief, irregular vibrations such as those made by a slamming door, dropped book, or sliding chair are called *noise*. Noise is characterized by sound waves with mixed frequencies and jumbled intensities (figure 6.20). On the other hand, there are sounds made by very regular, repeating vibrations such as those made by a tuning fork. A tuning fork produces a *pure tone* with a sinusoidal curved pressure variation and regular frequency. Yet a tuning fork produces a tone that most people interpret as bland. You would not call a tuning fork sound a musical note! Musical sounds from instruments have a certain frequency and loudness, as do noise and pure tones, but you can readily identify the source of the very same musical note made by two different instruments. You recognize it as a musical note, not noise and not a pure tone. You also recognize if the note was produced by a violin or a guitar. The difference is in the wave form of the sounds made by the two instruments, and the difference is called the *sound quality*. How does a musical instrument produce a sound of a characteristic quality? The answer may be found by looking at the two broad categories of instruments, (1) those that make use of vibrating strings and (2) those that make use of vibrating columns of air. These two categories will be considered separately.

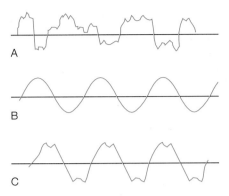

FIGURE 6.20

Different sounds that you hear include (*A*) noise, (*B*) pure tones, and (*C*) musical notes.

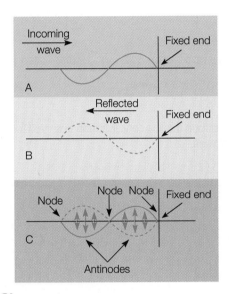

FIGURE 6.21

An incoming wave on a cord with a fixed end (*A*) meets a reflected wave (*B*) with the same amplitude and frequency, producing a standing wave (*C*). Note that a standing wave of one wavelength has three nodes and two antinodes.

Vibrating Strings

A stringed musical instrument, such as a guitar, has strings that are stretched between two fixed ends. When a string is plucked, waves of many different frequencies travel back and forth on the string, reflecting from the fixed ends. Many of these waves quickly fade away but certain frequencies resonate, setting up patterns of waves. Before considering these resonant patterns in detail, keep in mind that (1) two or more waves can be in the same place at the same time, traveling through one another from opposite directions; (2) a confined wave will be reflected at a boundary, and the reflected wave will be inverted (a crest becomes a trough); and (3) reflected waves interfere with incoming waves of the same frequency to produce **standing waves.** Figure 6.21 is a graphic "snapshot" of what happens when reflected wave patterns meet incoming wave pat-

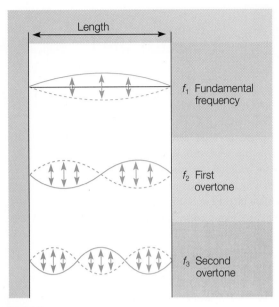

FIGURE 6.22

A stretched string of a given length has a number of possible resonant frequencies. The lowest frequency is the fundamental, f_1; the next higher frequencies, or overtones, shown are f_2 and f_3.

terns. The incoming wave is shown as a solid line, and the reflected wave is shown as a dotted line. The result is (1) places of destructive interference, called *nodes*, which show no disturbance, and (2) loops of constructive interference, called *antinodes*, which take place where the crests and troughs of the two wave patterns produce a disturbance that rapidly alternates upward and downward. This pattern of alternating nodes and antinodes does not move along the string and is thus called a *standing wave*. Note that the standing wave for *one wavelength* will have a node at both ends and in the center and also two antinodes. Standing waves occur at the natural, or resonant, frequencies of the string, which are a consequence of the nature of the string, the string length, and the tension in the string. Since the standing waves are resonant vibrations, they continue as all other waves quickly fade away.

Since the two ends of the string are not free to move, the ends of the string will have nodes. The *longest* wave that can make a standing wave on such a string has a wavelength (λ) that is twice the length (L) of the string. Since frequency (f) is inversely proportional to wavelength ($f = v/\lambda$ from equation 6.3), this longest wavelength has the lowest frequency possible, called the **fundamental frequency.** The fundamental frequency has one antinode, which means that the length of the string has one-half a wavelength. The fundamental frequency (f_1) determines the pitch of the *basic* musical note being sounded and is called the first harmonic. Other resonant frequencies occur at the same time, however, since other standing waves can also fit onto the string. A higher frequency of vibration (f_2) could fit two half-wavelengths between the two fixed nodes. An even higher frequency (f_3) could fit three half-wavelengths between the two fixed nodes (figure 6.22). Any whole number of halves of the wavelength will permit a standing wave to form. The frequencies (f_2, f_3, etc.) of these

A CLOSER LOOK | Resonance and the Greenhouse Effect

There is no doubt that the amount of carbon dioxide (CO_2) in the earth's atmosphere has been increasing (box figure 6.1). By the year 2050, this increase could lead to levels of 600 parts per million or more. Why is CO_2 increasing like this? The levels are increasing because cars, factories, and industries are burning more and more fossil fuels, releasing CO_2 faster than plants and the oceans can absorb it.

No one disputes that the amount of CO_2 is increasing, as you can clearly see in box figure 6.1. Scientists have not been able to fully agree, however, on the coming impact of the increasing CO_2 levels. Some things can affect weather patterns over time—such as a simple volcanic eruption, for example—and this makes forecasting climate change even more difficult than forecasting the weather next month. So, we do not know for sure what the exact climatic consequences might be from increasing CO_2 in the atmosphere.

Forecasting global climate changes is complicated since any prediction must take into account all the complex weather interactions that can take place on a global scale. This is done by using numerical models of atmospheric and oceanic general circulation as well as all the complex factors that influence weather over a long term. Forecasters explore many possible outcomes, and the effects of assumptions and approximations are averaged into a global model. Such models that are used in climate projections are still undergoing development, but they still can serve as valuable tools to help predict outcomes in various situations. According to the Intergovernmental Panel on Climate Change (IPCC), a U.N.-sponsored group of 2,500 scientists, current climate models point to an eventual increase in the temperature—perhaps as much as about 1.7° to 4.5°C (3° to 8°F) by 2100. Higher temperatures will mean melting of more polar ice and the subsequent rising of sea level. Sea level is projected to rise 18 to 99 cm (7 to 39 in), with a "best estimate" of about 50 cm (about 20 in). This is enough water to submerge coastal cities and wide areas of some states.

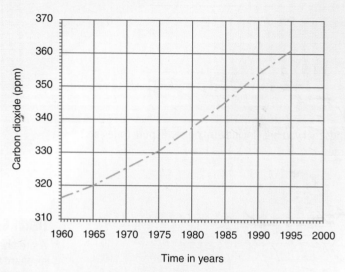

BOX FIGURE 6.1

Averaged yearly data of atmospheric carbon dioxide concentration observed for the last 40 years at Mauna Loa Observatory, Hawaii.

How does CO_2 result in global heating? Here is the scoop: About one-half of incoming sunlight usually reaches the earth's surface, as the rest is reflected back to space or absorbed by the atmosphere. The value of sunlight received by the surface has a global average of about 240 watts per square meter. Wide variations from this average occur with latitude as well as with the season. Rocks, soil, and water absorb sunlight that does reach the earth's surface and the ground becomes warmer as a result. The ground emits the absorbed solar energy as infrared radiation, wavelengths longer than the visible part of the spectrum. This longer wavelength infrared radiation has a frequency that matches some of the natural frequencies of vibration of CO_2. This match means that CO_2 readily absorbs infrared radiation that is emitted from the surface of the earth, even though the shorter wavelength visible light passed by it mostly unaffected. The absorbed infrared energy from the surface shows up as an increased kinetic energy, which is indicated by an increase in temperature. The warmer molecules of CO_2 now emit infrared radiation of their own, in

all directions, as they seek a lower energy state. Some of this reemitted radiation is again absorbed by other nearby CO_2 molecules in the atmosphere, some is emitted to space, and significantly, some is directed back to the surface to be absorbed and start the process all over again. The net result is that less of the energy from the sun escapes immediately to space after being absorbed and emitted as infrared radiation. It is retained for a while through the process of being redirected to the surface, increasing the temperature more than it would have been otherwise. Thus, more carbon dioxide present in the atmosphere means more energy will be bounced around and redirected back toward the surface, increasing the temperature near the surface.

The process of heating the atmosphere in the lower parts by the absorption and reemission of infrared radiation is called the *greenhouse effect*. It is called the greenhouse effect because greenhouse glass allows short wavelengths of solar radiation to pass into the greenhouse, but does not allow the

—Continued top of next page

Continued—

longer infrared radiation to leave. This analogy is somewhat misleading, however, because CO_2 molecules do not "trap" infrared radiation, but are involved in a dynamic absorption and downward reemission process that increases the surface temperature. The analogy breaks down as you realize that more layers of glass on a greenhouse will not increase the temperature inside the greenhouse. However, increased amounts of CO_2 in the atmosphere *will* continue to increase the temperature near the surface.

Other gases also contribute to the greenhouse effect, including methane and nitrogen oxides. Methane is the simplest of the hydrocarbons and is produced by bacteria acting on plant materials (cattle and termites), in landfills, and any time organic matter is decomposed while not in air. Nitrogen oxides are pollutants from cars and power plants. The greenhouse gases come from many sources of human activity, but these are activities that can be altered to slow the greenhouse process.

Another significant step to slow global warming is to move away from fossil fuels—oil, gas, and coal—as the primary energy sources. This will require, however, full use of all the alternative energy sources, such as solar energy technologies, geothermal energy, and hydrogen gas. The greatest danger is waiting until it is too late to develop and use these alternative energy sources.

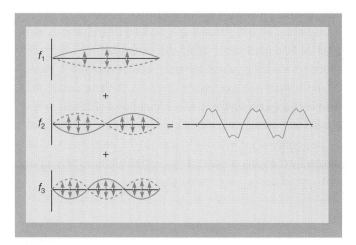

FIGURE 6.23

A combination of the fundamental and overtone frequencies produces a composite waveform with a characteristic sound quality.

wavelengths are called the *overtones*, or harmonics. It is the presence and strength of various overtones that give a musical note from a certain instrument its characteristic quality. The fundamental and the overtones add together to produce the characteristic *sound quality*, which is different for the same-pitched note produced by a violin and by a guitar (figure 6.23).

Since nodes must be located at the ends, only half-wavelengths $(1/2\lambda)$ can fit on a string of a given length (L), so the fundamental frequency of a string is $1/2\lambda = L$, or $\lambda = 2L$. Substituting this value in the wave equation (solved for frequency, f) will give the relationship for finding the fundamental frequency and the overtones when the string length and velocity of waves on the string are known. The relationship is

$$f_n = \frac{nv}{2L}$$

equation 6.8

where $n = 1, 2, 3, 4 \ldots$, and $n = 1$ is the fundamental frequency and $n = 2$, $n = 3$, and so forth are the overtones.

CONNECTIONS

Musical Instruments: Strings

This group of musical instruments includes those that use one or more vibrating strings as sound makers, which is a large and rather diverse number of instruments, such as the piano, guitar, violin, cello, harp, viola, mandolin, and many others. The vibrating string produces a waveform with overtones, so each instrument in this group can be used as a harmonic instrument. A vibrating string has a small area, so it is not able to make very loud sound waves compared to the solid bars or membrane instruments. For this reason, the strings are coupled to a sound resonator, which amplifies the sound by giving it a source of a larger vibrating surface. The violin, for example, has a structure called a "bridge" through which the strings transmit vibrations to the inside of the violin. This vibrates the enclosed air, which acts as a resonator that intensifies and prolongs certain overtones, while reducing the intensities of others. It is the combination of increasing some overtones while reducing the intensity of others that determines the quality of a particular violin, producing notes that can be described as smooth and rich rather than harsh and bland. The sound waves—smooth or harsh—move from the body of the violin to the surroundings with an increased intensity. Thus the body of the violin is its resonator structure, and the body amplifies the sound by moving a larger mass of air than a string alone could accomplish. All the stringed instruments have resonator structures, and these structures generally determine which overtones are amplified and which are suppressed.

CONNECTIONS

The Human Ear

Figure 6.24 shows the anatomy of the ear. The sound that arrives at the ear is first funneled by the external ear to the tympanum, also known as the eardrum. The cone-shaped nature of the external ear focuses sound on the tympanum and causes it to vibrate at the same frequency as the sound waves reaching it. Attached to the tympanum are three tiny bones known as the malleus (hammer), incus (anvil), and stapes (stirrup). The malleus is attached to the tympanum, the incus is attached to the malleus and stapes, and the stapes is attached to a small, membrane-covered opening called the oval window in a snail-shaped structure known as the cochlea. The vibration of the tympanum causes the tiny bones (malleus, incus, and stapes) to vibrate, and they, in turn, cause a corresponding vibration in the membrane of the oval window.

The cochlea of the ear is the structure that detects sound, and it consists of a snail-shaped set of fluid-filled tubes. When the oval window vibrates, the fluid in the cochlea begins to move, causing a membrane in the cochlea, called the basilar membrane, to vibrate. High-pitched, short-wavelength sounds cause the basilar membrane to vibrate at the base of the cochlea near the oval window. Low-pitched, long-wavelength sounds vibrate the basilar mem-

brane far from the oval window. Loud sounds cause the basilar membrane to vibrate more vigorously than do faint sounds. Cells on this membrane depolarize when they are stimulated by its vibrations. Since they synapse with neurons, messages can be sent to the brain.

Because sounds of different wavelengths stimulate different portions of the cochlea, the brain is able to determine the pitch of a sound. Most sounds consist of a mixture of pitches that are heard. Louder sounds stimulate the membrane more forcefully, causing the sensory cells in the cochlea to send more nerve impulses per second. Thus, the brain is able to perceive the loudness of various sounds, as well as the pitch.

Associated with the cochlea are two fluid-filled chambers and a set of fluid-filled tubes called the semicircular canals. These structures are not involved in hearing but are involved in maintaining balance and posture. In the walls of these canals and chambers are cells similar to those found on the basilar membrane. These cells are stimulated by movements of the head and by the position of the head with respect to the force of gravity. The constantly changing position of the head results in sensory input that is important in maintaining balance.

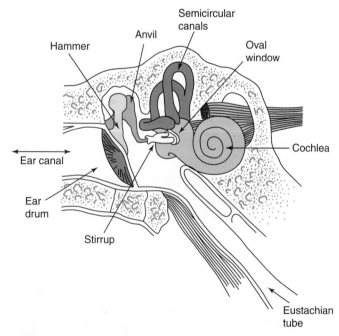

FIGURE 6.24

A schematic sketch of the human ear.

Sounds from Moving Sources

When the source of a sound is stationary, waves of condensation and rarefaction expand from the source like the surface of an expanding balloon. These concentric waves have equal spacing in all directions, and an observer standing in any location will hear the same pitch (frequency) as another observer standing at any other location. However, if the source of the sound moves then the center of each successive wave will be displaced in the direction of movement. The successive crests are therefore crowded closer together in the direction of the motion and spread farther apart in the opposite direction. An observer in front of the moving source will encounter more wave crests per second than are being generated by the source. The frequency of the sound will therefore seem higher than it really is, and the observer will interpret this as a higher pitch. As the moving source passes the observer, the source is moving away from the waves sent backward. The frequency will then seem lower than it really is, and the observer will interpret this as a lower pitch. The overall effect is that of a higher pitch as the source approaches, then a lower pitch as it moves away. This apparent shift of frequency of a sound from a moving source is called the **Doppler effect.** The Doppler effect is

CONNECTIONS

How Animals Make Sounds

There are several ways that animals intentionally make sounds for the purpose of communication. Many animals force air through openings or over surfaces to cause sounds. In the simplest case the air may simply produce a whistling or hissing sound as the air is forced through a small opening. If the opening is larger, the sound may be a snort or booming sound. In many vertebrate animals air is forced past vocal cords that are caused to vibrate in a manner similar to a violin string. The tension on the vocal cords can be changed to vary the pitch of the sound that can then be used to encode different messages.

Many insects such as crickets and katydids rub parts of their bodies (wings) together and generate vibrations that cause sounds. Different species produce different frequencies or durations of sound.

Other insects, such as cicadas, use muscles to deform a portion of their external skeleton. When the portion of the skeleton pops back to its normal shape, a sound is produced. By repeating this process very rapidly, cicadas produce the high-pitched buzzing sound you can hear high in trees in late summer.

Some animals, like woodpeckers, actually select specific trees that have good sound-producing properties and hammer at them to produce sounds that advertise their presence. Some have even been known to select aluminum flagpoles.

Some birds purposely make sounds with their wings. Ruffed grouse rapidly move their wings forward and stop them abruptly. The first few beats are slow and gradually speed up causing a blur of sound at the end of the episode. This action creates a booming sound that can be heard for several hundred meters.

evident if you stand by a street and an approaching car sounds its horn as it drives by you. You will hear a higher-pitched horn as the car approaches, which shifts to a lower-pitched horn as the waves go by you. The driver of the car, however, will hear the continual, true pitch of the horn since the driver is moving with the source (figure 6.25).

A Doppler shift is also noted if the observer is moving and the source of sound is stationary. When the observer moves toward the source, the wave fronts are encountered more frequently than if the observer were standing still. As the observer moves away from the source, the wave fronts are encountered less frequently than if the observer were not moving. An observer on a moving train approaching a crossing with a sounding bell thus hears a high-pitched bell that shifts to a lower-pitched bell as the train whizzes by the crossing.

When an object moves through the air at the speed of sound, it keeps up with its own sound waves. All the successive wave fronts pile up on one another, creating a large wave disturbance called a *shock wave.* The shock wave from a supersonic airplane is a cone-shaped shock wave of intense condensations trailing backward at an angle dependent on the speed of the aircraft. Wherever this cone of superimposed crests passes, a **sonic boom** occurs. The many crests have been added together, each contributing to the pressure increase. The human ear cannot differentiate between such a pressure wave created by a supersonic aircraft and a pressure wave created by an explosion.

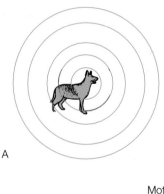

A

Motion of source

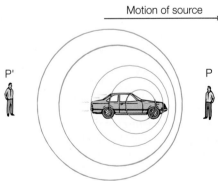

B

FIGURE 6.25

(*A*) Spherical sound waves from a stationary source spread out evenly in all directions. (*B*) If the source is moving, an observer at position P will experience more wave crests per second than an observer at position P′. The observer at P interprets this as a higher pitch, and the phenomenon is called the *Doppler effect.*

A CLOSER LOOK | Ultrasonics

Ultrasonic waves are mechanical waves that have frequencies above the normal limit of hearing of the human ear. The arbitrary upper limit is about 20,000 Hz, so an ultrasonic wave has a frequency of 20,000 Hz or greater. Intense ultrasonic waves are used in many ways in industry and medicine.

Industrial and commercial applications of ultrasound utilize lower-frequency ultrasonic waves in the 20,000 to 60,000 Hz range. Commercial devices that send ultrasound through the air include burglar alarms and rodent repellers. An ultrasonic burglar alarm sends ultrasonic spherical waves through the air of a room. The device is adjusted to ignore echoes from the contents of the room. The presence of a person provides a new source of echoes, which activates the alarm. Rodents emit ultrasonic frequencies up to 150,000 Hz that are used in rodent communication when they are disturbed or during aggressive behavior. The ultrasonic rodent repeller generates ultrasonic waves of similar frequency. Other commercial applications of ultrasound include sonar and depth measurements, cleaning and drilling, welding plastics and metals, and material flaw detection. Ultrasonic cleaning baths are used to remove dirt and foreign matter from solid surfaces, usually within a liquid solvent. The ultrasonic

waves create vapor bubbles in the liquid, which vibrate and emit audible and ultrasonic sound waves. The audible frequencies are often heard as a hissing or frying sound.

Medical applications of ultrasound use frequencies in the 1,000,000 to 20,000,000 Hz range. Ultrasound in this frequency range cannot move through the air because the required displacement amplitudes of the gas molecules in the air are less than the average distance between the molecules. Thus, a gas molecule that is set into motion in this frequency range cannot collide with other gas molecules to transmit the energy of the wave. Intense ultrasound is used for cleaning teeth and disrupting kidney stones. Less intense ultrasound is used for therapy (heating and reduction of pain). The least intense ultrasounds are used for diagnostic imaging. The largest source of exposure of humans to ultrasound is for the purpose of ultrasonic diagnostic imaging, particularly in fertility and pregnancy cases. In the United States, it has been estimated that more than half of the children born in the 1980s were scanned at least once by ultrasound before birth.

Originally developed in the late 1950s, ultrasonic medical machines have been improved and today make images of outstanding detail and clarity. One type of

ultrasonic medical machine uses a transducer probe, which emits an ultrasonic pulse that passes into the body. Echoes from the internal tissues and organs are reflected back to the transducer, which sends the signals to a computer. Another pulse of ultrasound is then sent out after the echoes from the first pulse have returned. The strength of and number of pulses per second vary with the application, ranging from hundreds to thousands of pulses per second. The computer constructs a picture from the returning echoes, showing an internal view without the use of more dangerous X rays.

Ultrasonic scanners have been refined to the point that the surface of the ovaries can now be viewed, showing the number and placement of developing eggs. The ovary scan is typically used in conjunction with fertility-stimulating drugs, where multiple births are possible, and to identify the exact time of ovulation. As early as four weeks after conception, the ultrasonic scanner is used to identify and monitor the fetus. By the thirteenth week, an ultrasonic scan can show the fetal heart movement, bone and skull size, and internal organs. The ultrasonic scanner is often used to show the position of the fetus and placenta for the purpose of amniocentesis, which involves withdrawing a sample of amniotic fluid from the uterus for testing.

SUMMARY

Elastic objects *vibrate,* or move back and forth, in a repeating motion when disturbed by some external force. They are able to do this because they have an *internal restoring force* that returns them to their original positions after being deformed by some external force. If the internal restoring force is opposite to and proportional to the deforming displacement, the vibration is called a *simple harmonic motion.* The extent of displacement is called the *amplitude,* and one complete back-and-forth motion is one *cycle.* The time required for one cycle is a *period.* The *frequency* is the number of cycles per second, and the unit of frequency is the *hertz.* A *graph* of the displacement as a function of time for a simple harmonic motion produces a *sinusoidal* graph.

Periodic, or repeating, vibrations or the *pulse* of a single disturbance can create *waves,* disturbances that carry energy through a medium. A wave that disturbs particles in a back-and-forth motion in

the direction of the wave travel is called a *longitudinal wave.* A wave that disturbs particles in a motion perpendicular to the direction of wave travel is called a *transverse wave.* The nature of the medium and the nature of the disturbance determine the type of wave created.

Waves that move through the air are longitudinal and cause a back-and-forth motion of the molecules making up the air. A zone of molecules forced closer together produces a *condensation,* a pulse of increased density and pressure. A zone of reduced density and pressure is a *rarefaction.* A vibrating object produces condensations and rarefactions that expand outward from the source. If the frequency is between 20 and 20,000 Hz, the human ear perceives the waves as *sound* of a certain *pitch.* High frequency is interpreted as high-pitched sound, and low frequency as low-pitched sound.

A graph of pressure changes produced by condensations and rarefactions can be used to describe sound waves. The condensations produce *crests*, and the rarefactions produce *troughs*. The *amplitude* is the maximum change of pressure from the normal. The *wavelength* is the distance between any two successive places on a wave train, such as the distance from one crest to the next crest. The *period* is the time required for a wave to repeat itself. The *velocity* of a wave is how quickly a wavelength passes. The *frequency* can be calculated from the *wave equation*, $v = \lambda f$.

Sound waves can move through any medium but not a vacuum. The velocity of sound in a medium depends on the molecular inertia and strength of interactions. Sound, therefore, travels most rapidly through a solid, then a liquid, then a gas. In air, sound has a greater velocity in warmer air than in cooler air because the molecules of air are moving about more rapidly, therefore transmitting a pulse more rapidly.

Sound waves are *reflected* or *refracted* from a *boundary*, which means a change in the transmitting medium. Reflected waves that are *in phase* with incoming waves undergo *constructive interference* and waves that are *out of phase* undergo *destructive interference*. Two waves that are otherwise alike but with slightly different frequencies produce an alternating increasing and decreasing of loudness called *beats*.

The *energy* of a sound wave is called the wave *intensity*, which is measured in watts per square meter. The intensity of sound is expressed on the *decibel scale*, which relates it to changes in loudness as perceived by the human ear.

All elastic objects have *natural frequencies* of vibration that are determined by the materials they are made of and their shapes. When energy is transferred at the natural frequencies, there is a dramatic increase of amplitude called *resonance*. The natural frequencies are also called *resonant frequencies*.

Sounds are compared by pitch, loudness, and *quality*. The quality is determined by the instrument sounding the note. Each instrument has its own characteristic quality because of the resonant frequencies that it produces. The basic, or *fundamental*, *frequency* is the longest standing wave that it can make. The fundamental frequency determines the basic note being sounded, and other resonant frequencies, or standing waves called *overtones* or *harmonics*, combine with the fundamental to give the instrument its characteristic quality.

A moving source of sound or a moving observer experiences an apparent shift of frequency called the *Doppler effect*. If the source is moving as fast or faster than the speed of sound, the sound waves pile up into a *shock wave* called a *sonic boom*. A sonic boom sounds very much like the pressure wave from an explosion.

Summary of Equations

6.1

$$period = \frac{1}{frequency}$$

$$T = \frac{1}{f}$$

6.2

$$frequency = \frac{1}{period}$$

$$f = \frac{1}{T}$$

6.3

$$velocity = (wavelength)(frequency)$$

$$v = \lambda f$$

6.4

$$\begin{array}{l} \text{velocity of} \\ \text{sound (m/s)} \\ \text{at present} \\ \text{temperature} \end{array} = \begin{array}{l} \text{velocity} \\ \text{of sound} \\ \text{at 0°C} \end{array} + \begin{array}{l} \text{0.60 m/s} \\ \text{increase per} \\ \text{degree Celsius} \end{array} \times \begin{array}{l} \text{present} \\ \text{temperature} \\ \text{in °C} \end{array}$$

$$v_{Tp}(\text{m/s}) = v_0 + \left(\frac{0.60\,\text{m/s}}{°C}\right)(T_p)$$

6.5

$$\begin{array}{l} \text{velocity of} \\ \text{sound (ft/s)} \\ \text{at present} \\ \text{temperature} \end{array} = \begin{array}{l} \text{velocity} \\ \text{of sound} \\ \text{at 0°C} \end{array} + \begin{array}{l} \text{2.0 ft/s} \\ \text{increase per} \\ \text{degree Celsius} \end{array} \times \begin{array}{l} \text{present} \\ \text{temperature} \\ \text{in °C} \end{array}$$

$$v_{Tp}(\text{ft/s}) = v_0 + \left(\frac{2.0\,\text{ft/s}}{°C}\right)(T_p)$$

6.6

$$\text{beat frequency} = \text{one frequency} - \text{other frequency}$$

$$f_b = f_2 - f_1$$

6.7

$$\text{Intensity} = \frac{\text{Power}}{\text{Area}}$$

$$I = \frac{P}{A}$$

6.8

$$\text{resonant frequency} = \frac{\text{number} \times \text{velocity on string}}{2 \times \text{length of string}}$$

where number 1 = fundamental frequency, and numbers 2, 3, 4, and so on = overtones.

$$f_n = \frac{nv}{2L}$$

KEY TERMS

APPLYING THE CONCEPTS

1. Simple harmonic motion is
   a. any motion that results in a pure tone.
   b. a combination of overtones.
   c. a vibration with a restoring force proportional and opposite to a displacement.
   d. all of the above.

2. The displacement of a vibrating object is measured by
   a. a cycle.
   b. amplitude.
   c. the period.
   d. frequency.

3. The time required for a vibrating object to complete one full cycle is the
   a. frequency.
   b. amplitude.
   c. period.
   d. hertz.

4. The number of cycles that a vibrating object moves through during a time interval of 1 second is the
   a. amplitude.
   b. period.
   c. full cycle.
   d. frequency.

5. The unit of cycles per second is called a
   a. hertz.
   b. lambda.
   c. wave.
   d. watt.

6. The period of a vibrating object is related to the frequency, since they are
   a. directly proportional.
   b. inversely proportional.
   c. frequently proportional.
   d. not proportional.

7. A wave is
   a. the movement of material from one place to another place.
   b. a traveling disturbance that carries energy.
   c. a wavy line that moves through materials.
   d. all of the above.

8. A longitudinal mechanical wave causes particles of a material to move
   a. back and forth in the same direction the wave is moving.
   b. perpendicular to the direction the wave is moving.
   c. in a circular motion in the direction the wave is moving.
   d. in a circular motion opposite the direction the wave is moving.

9. A transverse mechanical wave causes particles of a material to move
   a. back and forth in the same direction the wave is moving.
   b. perpendicular to the direction the wave is moving.
   c. in a circular motion in the direction the wave is moving.
   d. in a circular motion opposite the direction the wave is moving.

10. Transverse mechanical waves will move only through
    a. solids.
    b. liquids.
    c. gases.
    d. all of the above.

11. Longitudinal mechanical waves will move only through
    a. solids.
    b. liquids.
    c. gases.
    d. all of the above.

12. A pulse of jammed-together molecules that quickly moves away from a vibrating object
    a. is called a condensation.
    b. causes an increased air pressure when it reaches an object.
    c. has a greater density than the surrounding air.
    d. includes all of the above.

13. The characteristic of a wave that is responsible for what you interpret as pitch is the wave
    a. amplitude.
    b. shape.
    c. frequency.
    d. height.

14. The extent of displacement of a vibrating tuning fork is related to the resulting sound wave characteristic of
    a. frequency.
    b. amplitude.
    c. wavelength.
    d. period.

15. The number of cycles that a vibrating tuning fork experiences each second is related to the resulting sound wave characteristic of
    a. frequency.
    b. amplitude.
    c. wave height.
    d. quality.

16. From the wave equation of $v = \lambda f$, you know that the wavelength and frequency at a given velocity are
    a. directly proportional.
    b. inversely proportional.
    c. directly or inversely proportional, depending on the pitch.
    d. not related.

17. Since $v = \lambda f$, then f must equal
    a. $v\lambda$.
    b. v/λ.
    c. λ/v.
    d. $v\lambda/\lambda$.

18. Sound waves travel faster in
    a. solids as compared to liquids.
    b. liquids as compared to gases.
    c. warm air as compared to cooler air.
    d. all of the above.

19. The difference between an echo and a reverberation is
    a. an echo is a reflected sound; reverberation is not.
    b. the time interval between the original sound and the reflected sound.
    c. the amplitude of an echo is much greater.
    d. reverberation comes from acoustical speakers; echoes come from cliffs and walls.

20. Sound interference is necessary to produce the phenomenon known as
    a. resonance.
    b. decibels.
    c. beats.
    d. reverberation.

21. The efficient transfer of energy that takes place at a natural frequency is known as
    a. resonance.
    b. beats.
    c. the Doppler effect.
    d. reverberation.

22. The fundamental frequency of a standing wave on a string has
    a. one node and one antinode.
    b. one node and two antinodes.
    c. two nodes and one antinode.
    d. two nodes and two antinodes.

23. An observer on the ground will hear a sonic boom from an airplane traveling faster than the speed of sound
    a. only when the plane breaks the sound barrier.
    b. as the plane is approaching.
    c. when the plane is directly overhead.
    d. after the plane has passed by.

Answers

1. c 2. b 3. c 4. d 5. a 6. b 7. b 8. a 9. b 10. a 11. d 12. d 13. c 14. b 15. a 16. b 17. b 18. d 19. b 20. c 21. a 22. c 23. d

QUESTIONS FOR THOUGHT

1. What is a wave?
2. Is it possible for a transverse wave to move through air? Explain.
3. A piano tuner hears three beats per second when a tuning fork and a note are sounded together and six beats per second after the string is tightened. What should the tuner do next, tighten or loosen the string? Explain.
4. Why do astronauts on the moon have to communicate by radio even when close to one another?
5. What is resonance?
6. Explain why sounds travel faster in warm air than in cool air.
7. Do all frequencies of sound travel with the same velocity? Explain your answer by using the wave equation.
8. What eventually happens to a sound wave traveling through the air?
9. What gives a musical note its characteristic quality?
10. Does a supersonic aircraft make a sonic boom only when it cracks the sound barrier? Explain.
11. What is an echo?
12. Why are fundamental frequencies and overtones also called resonant frequencies?

PARALLEL EXERCISES

The exercises in groups A and B cover the same concepts. Solutions to group A exercises are located in appendix D.

Group A

1. A vibrating object produces periodic waves with a wavelength of 50 cm and a frequency of 10 Hz. How fast do these waves move away from the object?
2. The distance between the center of a condensation and the center of an adjacent rarefaction is 1.50 m. If the frequency is 112.0 Hz, what is the speed of the wave front?
3. Water waves are observed to pass under a bridge at a rate of one complete wave every 4.0 s. (a) What is the period of these waves? (b) What is the frequency?

Group B

1. A tuning fork vibrates 440.0 times a second, producing sound waves with a wavelength of 78.0 cm. What is the velocity of these waves?
2. The distance between the center of a condensation and the center of an adjacent rarefaction is 65.23 cm. If the frequency is 256.0 Hz, how fast are these waves moving?
3. A warning buoy is observed to rise every 5.0 s as crests of waves pass by it. (a) What is the period of these waves? (b) What is the frequency?

4. A sound wave with a frequency of 260 Hz moves with a velocity of 330 m/s. What is the distance from one condensation to the next?

5. The following sound waves have what velocity?
   (a) Middle C, or 256 Hz and 1.34 m λ.
   (b) Note A, or 440.0 Hz and 78.0 cm λ.
   (c) A siren at 750.0 Hz and λ of 45.7 cm.
   (d) Note from a stereo at 2,500.0 Hz and λ of 13.72 cm.

6. What is the speed of sound, in ft/s, if the air temperature is:
   (a) 0.0°C
   (b) 20.0°C
   (c) 40.0°C
   (d) 80.0°C

7. An echo is heard from a cliff 4.80 s after a rifle is fired. How many feet away is the cliff if the air temperature is 43.7°F?

8. The air temperature is 80.0°F during a thunderstorm, and thunder was timed 4.63 s after lightning was seen. How many feet away was the lightning strike?

9. If the velocity of a 440 Hz sound is 1,125 ft/s in the air and 5,020 ft/s in seawater, find the wavelength of this sound (a) in air, (b) in seawater.

4. Sound from the siren of an emergency vehicle has a frequency of 750.0 Hz and moves with a velocity of 343.0 m/s. What is the distance from one condensation to the next?

5. The following sound waves have what velocity?
   (a) 20.0 Hz, λ of 17.15 m
   (b) 200.0 Hz, λ of 1.72 m
   (c) 2,000.0 Hz, λ of 17.15 cm
   (d) 20,000.0 Hz, λ of 1.72 cm

6. How much time is required for a sound to travel 1 mile (5,280.0 ft) if the air temperature is:
   (a) 0.0°C
   (b) 20.0°C
   (c) 40.0°C
   (d) 80.0°C

7. A ship at sea sounds a whistle blast, and an echo returns from the coastal land 10.0 s later. How many km is it to the coastal land if the air temperature is 10.0°C?

8. How many seconds will elapse between seeing lightning and hearing the thunder if the lightning strikes 1 mile (5,280 ft) away and the air temperature is 90.0°F?

9. A 600.0 Hz sound has a velocity of 1,087.0 ft/s in the air and a velocity of 4,920.0 ft/s in water. Find the wavelength of this sound (a) in the air, (b) in the water.

A thunderstorm produces an interesting display of electrical discharge. Each bolt can carry over 150,000 amperes of current with a voltage of 100 million volts.

CHAPTER | Seven

Electricity

The previous chapters have been concerned with *mechanical* concepts, explanations of the motion of objects that exert forces on one another. These concepts were used to explain straight-line motion, the motion of free fall, and the circular motion of objects on the earth as well as the circular motion of planets and satellites. The mechanical concepts were based on Newton's laws of motion and are sometimes referred to as Newtonian physics. The mechanical explanations were then extended into the submicroscopic world of matter through the kinetic molecular theory. The objects of motion were now particles, molecules that exert force on one another, and concepts associated with heat were interpreted as the motion of these particles. In a further extension of Newtonian concepts, mechanical explanations were given for concepts associated with sound, a mechanical disturbance that follows the laws of motion as it moves through the molecules of matter.

You might wonder, as did the scientists of the 1800s, if mechanical interpretations would also explain other natural phenomena such as electricity, chemical reactions, and light. A mechanical model would be very attractive since it already explained so many other facts of nature, and scientists have always looked for basic, unifying theories. Mechanical interpretations were tried, as electricity was considered as a moving fluid, and light was considered as a mechanical wave moving through a material fluid. There were many unsolved puzzles with such a model and gradually it was recognized that electricity, light, and chemical reactions could not be explained by mechanical interpretations. Gradually, the point of view changed from a study of particles to a study of the properties of the *space* around the particles. In this chapter you will learn about electric charge in terms of the space around particles. This model of electric charge, called the *field model,* will be used to develop understandings about electric current, the electric circuit, and electrical work and power. A relationship between electricity and the fascinating topic of magnetism is discussed next, including what magnetism is and how it is produced (figure 7.1). The relationship is then used to explain the mechanical production of electricity, how electricity is measured, and how electricity is used in everyday technological applications.

ELECTRIC CHARGE

You are familiar with the use of electricity in many electrical devices such as lights, toasters, radios, and calculators. You are also aware that electricity is used for transportation and for heating and cooling places where you work and live. Many people accept electrical devices as part of their surroundings, with only a hazy notion of how they work. To many people electricity seems to be magical. Electricity is not magical, and it can be understood, just as we understand any other natural phenomenon. There are theories that explain observations, quantities that can be measured, and relationships between these quantities, or laws, that lead to understanding. All of the observations, measurements, and laws begin with an understanding of *electric charge.*

Electron Theory of Charge

One of the first understandings about the structure of matter was discovered in 1897 by the physicist Joseph J. Thomson. From his experiments with cathode ray tubes, Thomson concluded that negatively charged particles, now called *electrons,* are present in all matter. He thought that matter was made up of electrons

FIGURE 7.1

The importance of electrical power seems obvious in a modern industrial society. What is not so obvious is the role of electricity in magnetism, light, chemical change, and as the very basis for the structure of matter. All matter, in fact, is electrical in nature, as you will see.

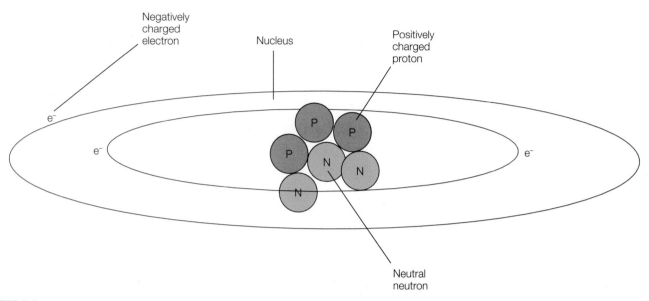

FIGURE 7.2

A very highly simplified model of an atom has most of the mass in a small, dense center called the *nucleus.* The nucleus has positively charged protons and neutral neutrons. Negatively charged electrons move around the nucleus at a much greater distance than is suggested by this simplified model. Ordinary atoms are neutral because there is balance between the number of positively charged protons and negatively charged electrons.

embedded in a positive "fluid." This model of the atom was modified by Ernest Rutherford in 1911. As a result of his work with radioactivity, Rutherford concluded that most of the mass of an atom is concentrated in a small dense center of the atom he called the *nucleus* and that the nucleus contains positively charged particles called *protons.* In 1932 James Chadwick discovered another particle in the nucleus, the *neutron.* Neutrons have no charge and are slightly more massive than protons (figure 7.2).

Today, the atom is considered to be made up of a number of subatomic particles with the electron, proton, and neutron being the most stable. The electron belongs to a family of elementary particles, of which at least six are known. The proton and neutron belong to a family of composite particles, of which hundreds are known. These composite particles are believed to be made up of even more elementary particles. For your information, you can read about all of the families of particles and their members in the reading at the end of chapter 9. For understanding electricity, you need only to consider the positively charged protons in the nucleus and the negatively charged electrons that move around the nucleus. Electrons can be moved from an atom and caused to move through a material, but the forces required to do this vary from one substance to another. Basically, the electrical, light, and chemical phenomena involve the *electrons* and not the more massive nucleus. The massive nuclei remain in a relatively fixed position in a solid, but some of the electrons can move about from atom to atom.

Electric Charge and Electrical Forces

Electrons have a *negative electric charge* and protons have a *positive electric charge.* The negative charge on an electron and the positive charge on a proton describe the way that electrons and protons *are* and how they behave. Charge is as fundamental to these subatomic particles as gravitational attraction is fundamental to masses. This means that you cannot separate gravity from masses, and you cannot separate charge from electrons and protons.

There are only two kinds of electric charge, and the charges interact to produce a force that is called the *electrical force. Like charges produce a repulsive electrical force* as positive repels positive, and negative repels negative. *Unlike charges produce an attractive electrical force* as positive and negative charges attract each other. The electrical *force* is as fundamental to subatomic particles as the force of gravitational attraction is between two masses. The electrical force is billions and billions of times stronger than the gravitational force between the tiny particles with their tiny masses. Thus, when dealing with subatomic particles such as the electron and the proton, the electrical force is the force of consequence, and the gravitational force is ignored for all practical purposes.

Ordinary atoms are usually neutral because there is a balance between the number of positively charged protons and the number of negatively charged electrons. But when there is an imbalance, as occurs from an electron being torn away by friction, the

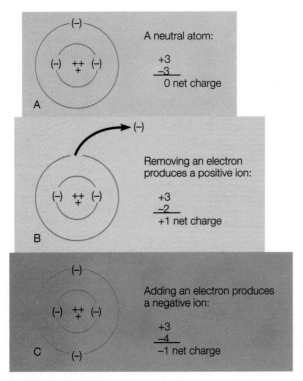

FIGURE 7.3

(*A*) A neutral atom has no net charge because the numbers of electrons and protons are balanced. (*B*) Removing an electron produces a net positive charge; the charged atom is called a positive ion. (*C*) The addition of an electron produces a net negative charge and a negative ion.

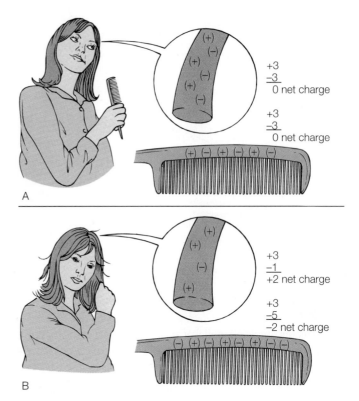

FIGURE 7.4

Arbitrary numbers of protons (+) and electrons (−) on a comb and in hair (*A*) before and (*B*) after combing. Combing transfers electrons from the hair to the comb by friction, resulting in a negative charge on the comb and a positive charge on the hair.

remaining atom has a net positive charge and is now called a *positive ion* (figure 7.3). The removed electron might continue to be a free negative charge, or it might become attached to a neutral atom to make a *negative ion*. Of course, the electron could also become attached to a positive ion to make a neutral atom.

Electrostatic Charge

Electrons can be moved from atom to atom to create ions. They can also be moved from one object to another by friction and by other means that will be discussed soon. Since electrons are negatively charged, an object that acquires an excess of electrons becomes a negatively charged body. The loss of electrons by another body results in a deficiency of electrons, which results in a positively charged object. Thus, *electric charges on objects result from the gain or loss of electrons.* Because the electric charge is confined to an object and is not moving, it is called an **electrostatic charge.** You probably call this charge *static electricity.* Static electricity is an accumulated electric charge at rest, that is, one that is not moving. When you comb your hair with a hard rubber comb, the comb becomes negatively charged because electrons are transferred *from* your hair to the comb. Your hair becomes positively charged with a charge equal in magnitude to the charge gained by the comb (figure 7.4). Both the negative charge on the comb from

an excess of electrons and the positive charge on your hair from a deficiency of electrons are charges that are momentarily at rest, so they are electrostatic charges.

Once charged by friction, objects such as the rubber comb soon return to a neutral, or balanced, state by the movement of electrons. This happens more quickly on a humid day because water vapor assists with the movement of electrons to or from charged objects. Thus, static electricity is more noticeable on dry days than on humid ones.

An object can become electrostatically charged (1) by *friction,* which transfers electrons from one object to another, (2) by *contact* with another charged body, which results in the transfer of electrons, or (3) by *induction.* Induction produces a charge by a redistribution of charges in a material. When you comb your hair, for example, the comb removes electrons from your hair and acquires a negative charge. When the negatively charged comb is held near small pieces of paper, it repels some electrons in the paper to the opposite side of the paper. This leaves the side of the paper closest to the comb with a positive charge, and there is an attraction between the pieces of paper and the comb, since unlike charges attract. Note that no transfer of electrons takes place in induction; the attraction results from a reorientation of the charges in the paper (figure 7.5).

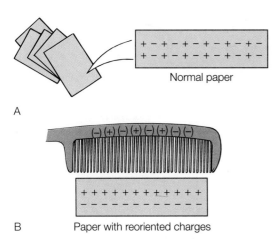

TABLE 7.1	**Electrical conductors and insulators**
Conductors	**Insulators**
Silver	Rubber
Copper	Glass
Gold	Carbon (diamond)
Aluminum	Plastics
Carbon (graphite)	Wood
Tungsten	
Iron	
Lead	
Nichrome	

FIGURE 7.5

Charging by induction. The comb has become charged by friction, acquiring an excess of electrons. The paper (*A*) normally has a random distribution of (+) and (−) charges. (*B*) When the charged comb is held close to the paper, there is a reorientation of charges because of the repulsion of like charges. This leaves a net positive charge on the side close to the comb, and since unlike charges attract, the paper is attracted to the comb.

A c t i v i t i e s

1. This activity works best on a day with low humidity. Tie a string around the lip of a small glass test tube. Have a partner hold one end of the tube with a cloth while rubbing the tube with a silk cloth. You hold a second test tube with a cloth and also rub it with a silk cloth for several minutes. As your partner allows the tube to hang freely from the string, bring your tube close, but not touching, and observe any interactions. (If nothing happens, try rubbing the tubes longer.)
2. Your partner should again rub the test tube with the silk cloth while you rub a comb with fur or flannel for several minutes. Bring the comb near the hanging test tube and observe any interactions.
3. Tie a string on a second comb. Your partner should rub this comb with fur or flannel as you rub your comb again for several minutes. Bring your comb near the hanging comb and observe any interactions.
4. Describe what you observe during this procedure, and give evidence that two kinds of electric charge exist.

Electrical Conductors and Insulators

When you slide across a car seat or scuff your shoes across a carpet, you are rubbing some electrons from the materials and acquiring an excess of negative charges. Because the electric charge is confined to you and is not moving, it is an electrostatic charge. The electrostatic charge is produced by friction between two surfaces and will remain until the electrons can move away because of their mutual repulsion. This usually happens when you reach for a metal doorknob, and you know when it happens because the electron movement makes a spark. Materials like the metal of a doorknob are good **electrical conductors** because they have electrons that are free to move throughout the metal. If you touch plastic or wood, however, you will not feel a shock. Materials like plastic and wood do not have electrons that are free to move throughout the material, and they are called *electrical nonconductors*. Nonconductors are also called *electrical insulators*. Electrons do not move easily through an insulator, but electrons can be added or removed, and the charge tends to remain. In fact, your body is a poor conductor, which is why you become charged by friction in the first place (table 7.1).

Materials vary in their ability to conduct charges, and this ability is determined by how tightly or loosely the electrons are held to the nucleus. Metals have millions of free electrons that can take part in the conduction of an electric charge. Materials such as rubber, glass, and plastics hold tightly to their electrons and are good insulators. Thus, metal wires are used to conduct an electric current from one place to another, and rubber, glass, and plastics are used as insulators to keep the current from going elsewhere.

There is a third class of materials, such as silicon and germanium, that sometimes conduct and sometimes insulate, depending on the conditions and how pure they are. These materials are called *semiconductors,* and their special properties make possible a number of technological devices such as the electrostatic copying machine, solar cells, and so forth.

Measuring Electrical Charges

As you might have experienced, sometimes you receive a slight shock after walking across a carpet, and sometimes you are really zapped. You receive a greater shock when you have accumulated a greater electric charge. Since there is less electric charge at one time and more at another, it should be evident that charge is a measurable quantity. The magnitude of an electric charge is identified with the number of electrons that have been transferred onto or away from an object. The quantity of such a charge (q) is measured in a unit called a **coulomb** (C). A coulomb unit is equivalent to the charge resulting from the transfer of 6.24×10^{18} of the charge carried by particles such as the electron. The coulomb is a fundamental metric unit of measure like the meter, kilogram, and second. There is not, however, a direct way of measuring coulombs as there is for measuring meters, kilograms, or seconds because of the difficulty of measuring tiny charged particles.

The coulomb is a *unit* of electric charge that is used with other metric units such as meters for distance and newtons for force. Thus, a quantity of charge (q) is described in units of coulomb (C). This is just like the process of a quantity of mass (m) being described in units of kilogram (kg). The concepts of charge and coulomb may seem less understandable than the concepts of mass and kilogram, since you cannot *see* charge or how it is measured. But charge does exist and it can be measured, so you can understand both the concept and the unit by working with them. Consider, for example, that an object has a net electric charge (q) because it has an unbalanced number (n) of electrons (e^-) and protons (p^+). The net charge on you after walking across a carpet depends on how many electrons you rubbed from the carpet. The net charge in this case would be the excess of electrons, or

quantity of charge = (number of electrons)(electron charge)

or

$$q = ne$$

equation 7.1

Since 1.00 coulomb is equivalent to the transfer of 6.24×10^{18} particles such as the electron, the charge on one electron must be

$$e = \frac{q}{n}$$

where q is 1.00 C, and n is 6.24×10^{18} electrons,

$$e = \frac{1.00 \text{ coulomb}}{6.24 \times 10^{18} \text{ electron}}$$

$$= 1.60 \times 10^{-19} \frac{\text{coulomb}}{\text{electron}}$$

This charge, 1.60×10^{-19} coulomb, is the *smallest* common charge known (more exactly $1.6021892 \times 10^{-19}$ C). It is the **fundamental charge** of the electron ($e^- = 1.60 \times 10^{-19}$ C) and the proton ($p^+ = 1.60 \times 10^{-19}$ C). All charged objects have multiples of this fundamental charge.

Measuring Electrical Forces

Recall that two objects with like charges, ($-$) and ($-$) or ($+$) and ($+$), produce a repulsive force, and two objects with unlike charges, ($+$) and ($-$), produce an attractive force. These forces were investigated by Charles Coulomb, a French military engineer, in 1785. Coulomb invented a sensitive balance to measure the forces between two pith balls (figure 7.6). Pith is a light material made from the dried inside of the stem of a plant or

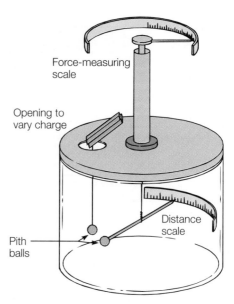

FIGURE 7.6

Coulomb constructed a torsion balance to test the relationships between a quantity of charge, the distance between the charges, and the electrical force produced. He found the inverse square law held accurately for various charges and distances.

from dried potatoes. Coulomb found that the force between two electrically charged pith balls was (1) directly proportional to the product of the electric charge and (2) inversely proportional to the square of the distance between them. In symbols, q_1 represents the quantity of electric charge on object 1, and q_2 represents the quantity of electric charge on object 2, d is the distance between objects 1 and 2, and k is a constant of proportionality. The magnitude of the electrical force between object 1 and object 2, F, is either attractive (unlike charges) or repulsive (like charges) depending on the charges on the two objects. This relationship, known as **Coulomb's law,** is

$$F = k\frac{q_1 q_2}{d^2}$$

equation 7.2

where k has the value of 9.00×10^9 newton-meters^2/coulomb^2 (9.00×10^9 N·m^2/C^2).

Force Fields

Does it seem odd to you that gravitational forces and electrical forces can act on objects that are not touching? How can gravitational forces act through the vast empty space between the earth and the sun? How can electrical forces act through a distance to pull pieces of paper to your charged comb? Such questions have bothered people since the early discovery of small, light objects being attracted to rubbed amber. There was no mental model of how such a force could act through a distance without touching. The idea of "invisible fluids" was an early attempt to develop a mental model that would help people

A CLOSER LOOK | Hydrogen and Fuel Cells

There is more than one way to power an electric vehicle and a rechargeable lead-acid battery is not the best answer. There is a newly developing technology that uses a hydrogen-powered fuel cell to power an electric vehicle. Hydrogen fuel cells can produce electricity without pollution and they do have the potential to eliminate the reliance on petroleum to power our transportation. The new technology has now been tested by Ballard Power Systems, Inc., operating six hydrogen-powered fuel-cell buses in Vancouver, British Columbia, and in Chicago, Illinois. Simply put, the technology succeeded, operating the buses economically without any problems.

Fuel cells are able to generate electricity directly onboard an electric vehicle, so heavy batteries are not needed. A short driving range is not a problem, and time lost charging the batteries is not a problem. Instead, electricity is produced electrochemically, in a device without any moving parts. As you probably know, energy is required to separate water into its component gases of hydrogen and oxygen. Thus, as you might expect, energy is released when hydrogen and oxygen combine to form water. It is this energy that a fuel cell uses to produce an electric current.

One design of a fuel cell has porous membranes separated by an electrolyte bath, with attached anode and cathode terminals. Hydrogen molecules are forced through a platinum-coated membrane that separates them into hydrogen ions and electrons. The electrons move through the circuit toward the cathode plate and the ions move into the electrolyte bath. Oxygen molecules from the air move through the cathode plate, where they are separated into oxygen atoms. They combine with hydrogen ions and electrons from the anode to create water and heat. An electric current is available between the anode and cathode terminals, and this current can be used to run the electric motors of an electric vehicle.

The silent-running, nonpolluting fuel cell with no moving parts sounds too good to be true, but the technology works. The technology has been too expensive for everyday use until recently, but is now more affordable. Fuel-cell-powered vehicles can operate directly on compressed hydrogen gas or liquid hydrogen and when they do, the only emission is water vapor. Direct use of hydrogen is also very efficient, with a 50 to 60 percent efficiency compared to the typical 15 to 20 percent efficiency for automobiles that run on petroleum in an internal combustion engine. Other fuels can also be used by running them through an onboard reformer, which transforms the fuel to hydrogen. Methanol or natural gas, for example, can be used with significantly less CO_2, CO, HC, and NO_x emissions than produced by an internal combustion engine. An added advantage to the use of methanol is that the existing petroleum fuel distribution system (tanks, pumps, etc.) can be used to distribute this liquid fuel. Liquid or compressed hydrogen, on the other hand, require a completely new type of distribution system.

A fuel-cell powered electric vehicle gives the emission benefits of a battery-powered vehicle without the problems of constantly recharging the batteries. A 1999 Daimler-Chrysler test car using hydrogen gas to power the fuel cells in a converted Mercedes-Benz reached top speeds of 145 km/hr (90 mi/hr) and traveled 450 km (280 mi) before refueling. Before long, you may see a fuel-cell vehicle in your neighborhood. It is the car of the future, which is needed now for the environment.

---

visualize how a force could act over a distance without physical contact. Then Newton developed the law of universal gravitation, which correctly predicted the magnitude of gravitational forces acting through space. Coulomb's law of electrical forces had similar success in describing and predicting electrostatic forces acting through space. "Invisible fluids" were no longer needed to explain what was happening, because the two laws seemed to explain the results of such actions. But it was still difficult to visualize what was happening physically when forces acted through a distance, and there were a few problems with the concept of action at a distance. Not all observations were explained by the model.

The work of Michael Faraday and James Maxwell in the early 1800s finally provided a new mental model for interaction at a distance. This new model did *not* consider the force that one object exerts on another one through a distance. Instead, it considered *the condition of space* around an object. The condition of space around an electric charge is considered to be changed by the presence of the charge. The charge produces a **force field** in the space around it. Since this force field is produced by an electrical charge, it is called an **electric field.** Imagine a second electric charge, called a *test charge,* that is far enough away from the electric charge that no forces are experienced. As you move the test charge closer and closer, it will experience an increasing force as it enters the electric field. The test charge is assumed not to change the field that it is entering and can be used to identify the electric field that spreads out and around the space of an electric charge.

All electric charges are considered to be surrounded by an electric field. All *masses* are considered to be surrounded by a *gravitational field.* The earth, for example, is considered to change the condition of space around it because of its mass. A spaceship far, far from the earth does not experience a measurable force. But as it approaches the earth, it enters the earth's gravitational field, and thus it experiences a measurable force. Likewise, a magnet creates a *magnetic field* in the space around

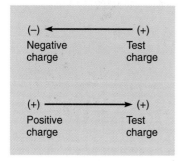

FIGURE 7.7

A *positive test charge* is used by convention to identify the properties of an electric field. The vector arrow points in the direction of the force that the test charge would experience.

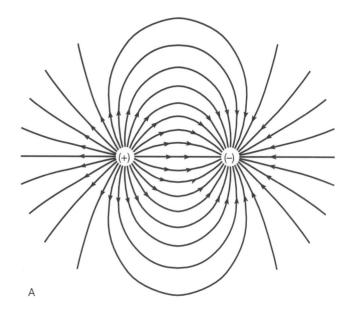

A

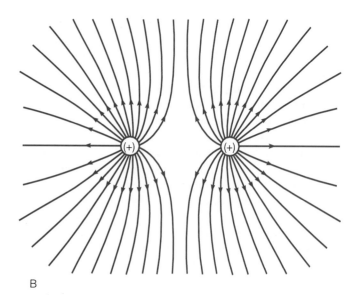

B

FIGURE 7.8

Lines of force diagrams for (*A*) a negative charge and (*B*) a positive charge when the charges have the same magnitude as the test charge.

it. You can visualize a magnetic field by moving a magnetic compass needle around a bar magnet. Far from the bar magnet the compass needle does not respond. Moving it closer to the bar magnet, you can see where the magnetic field begins. Another way to visualize a magnetic field is to place a sheet of paper over a bar magnet, then sprinkle iron filings on the paper. The filings will clearly identify the presence of the magnetic field.

Another way to visualize a field is to make a map of the field. Consider a small positive test charge that is brought into an electric field. A *positive* test charge is always used by convention. As shown in figure 7.7, a positive test charge is brought near a negative charge and a positive charge. The vector arrow points in the direction of the force that the *test charge experiences.* Thus, when brought near a negative charge, the test charge is attracted toward the unlike charge, and the arrow points that way. When brought near a positive charge, the test charge is repelled, so the arrow points away from the positive charge.

An electric field is represented by drawing *lines of force* or *electric field lines* that show the direction of the field. The vector arrows in figure 7.8 show field lines that could extend outward forever from isolated charges, since there is always some force on a distant test charge (review Coulomb's law; the force ideally never reaches zero). The field lines between pairs of charges in figure 7.8 show curved field lines that originate on positive charges and end on negative charges. By convention, the field lines are closer together where the field is stronger and farther apart where the field is weaker.

The field concept explains some observations that were not explained with the Newtonian concept of action at a distance. Suppose, for example, that a charge produces an electric field. This field is not instantaneously created all around the charge, but it is seen to build up and spread into space. If the charge is suddenly neutralized, the field that it created continues to spread outward, and then collapses back at some speed, even though the source of the field no longer exists. Consider an example with the gravitational field of the sun. If the mass of the sun were to instantaneously disappear, would the earth notice this instantaneously? Or would the gravitational field of the sun collapse at some speed, say the speed of light, to be noticed by the earth some eight minutes later? The Newtonian concept of action at a

distance did not consider any properties of space, so according to this concept the gravitational force from the sun would disappear instantly. The field concept, however, explains that the disappearance would be noticed after some period of time, about eight minutes. This time delay agrees with similar observations of objects interacting with fields, so the field concept is more useful than a mysterious action-at-a-distance concept, as you will see.

Actually there are three models for explaining how gravitational, electrical, and magnetic forces operate at a distance. (1) The *action-at-a-distance model* recognizes that masses are attracted gravitationally and that electric charges and magnetic poles attract and repel each other through space, but it gives no further explanation; (2) the *field model* considers a field to be a condition of space around a mass, electric charge, or magnet, and the properties of fields are described by field lines; and (3) the *field-particle model* is a complex and highly mathematical explanation of attractive and repulsive forces as the rapid emission and absorption of subatomic particles. This model explains electrical and magnetic forces as the exchange of *virtual photons*, gravitational forces as the exchange of *gravitons*, and strong nuclear forces as the exchange of *mesons*.

Electric Potential

Recall from chapter 4 that work is accomplished as you move an object to a higher location on the earth, say by moving a book from the first shelf of a bookcase to a higher shelf. By virtue of its position, the book now has gravitational potential energy that can be measured by *mgh* (the force of the book's weight × distance) joules of gravitational potential energy. Using the field model, you could say that this work was accomplished against the gravitational field of the earth. Likewise, an electric charge has an electric field surrounding it, and work must be done to move a second charge into or out of this field. Bringing a like charged particle *into* the field of another charged particle will require work, since like charges repel, and separating two unlike charges will also require work, since unlike charges attract. In either case, the *electric potential energy* is changed, just as the gravitational potential energy is changed by moving a mass in the earth's gravitational field.

One useful way to measure electric potential energy is to consider the *potential difference (PD)* that occurs when a certain amount of work (*W*) is used to move a certain quantity of charge (*q*). For example, suppose there is a firmly anchored and insulated metal sphere that has a positive charge (figure 7.9). The sphere will have a positive electric field in the space around it. Suppose also that you have a second sphere that has exactly 1.00 coulomb of positive charge. You begin moving the coulomb of positive charge toward the anchored sphere. As you enter the electric field you will have to push harder and harder to overcome the increasing repulsion. If you stop moving when you have done exactly 1.00 joule of work, the repulsion will *do* one joule of work if you now release the sphere. The sphere has potential energy in the same way that a compressed spring has potential energy. In electrical matters, the *potential difference (PD) that is created by doing 1.00 joule of work in moving 1.00 coulomb of charge is defined to be 1.00 volt.* The **volt** (V) is a measure of potential difference between two points, or

$$\text{electric potential} = \frac{\text{work to create potential}}{\text{charge moved}}$$

$$PD = \frac{W}{q}$$

equation 7.3

In units,

$$1.00 \text{ volt (V)} = \frac{1.00 \text{ joule (J)}}{1.00 \text{ coulomb (C)}}$$

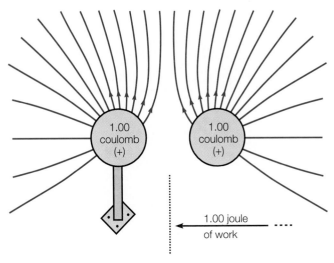

FIGURE 7.9

Electric potential results from moving a positive coulomb of charge into the electric field of a second positive coulomb of charge. When 1.00 joule of work is done in moving 1.00 coulomb of charge, 1.00 volt of potential results. A volt is a joule/coulomb.

The voltage of any electric charge, either static or moving, is the energy transfer per coulomb. The energy transfer can be measured by the *work that is done to move the charge* or by the *work that the charge can do* because of its position in the field. This is perfectly analogous to the work that must be done to give an object gravitational potential energy or to the work that the object can potentially do because of its new position. Thus, when a 12 volt battery is charged, 12.0 joules of work are done to transfer 1.00 coulomb of charge from an outside source against the electric field of the battery terminal. When the 12 volt battery is used, it does 12.0 joules of work for each coulomb of charge transferred from one terminal of the battery through the electrical system and back to the other terminal.

ELECTRIC CURRENT

So far, we have considered electric charges that have been instantaneously moved by friction but then generally stayed in one place. Experiments with static electricity played a major role in the development of the understanding of electricity by identifying charge, the attractive and repulsive forces between charges, and the field concept. The work of Franklin, Coulomb, Faraday, and Maxwell thus increased studies of electrical phenomena and eventually led to insights into connections between electricity and magnetism, light, and chemistry. These connections will be discussed later, but for now, consider the sustained flowing or moving of charge, an *electric current (I)*. **Electric current** means a flow of charge in the same way that "water current" means a flow of water. Since the word "current" *means* flow, you are being redundant if you speak of "flow of current." It is the *charge* that flows, and the current is defined as the flow of charge. Note that the symbol for a quantity of current (*I*) is not an abbreviation for the word "current."

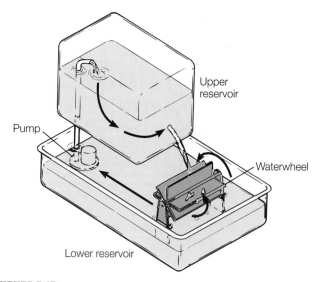

FIGURE 7.10

The falling water can do work in turning the waterwheel only as long as the pump maintains the potential difference between the upper and lower reservoirs.

The Electric Circuit

When you slide across a car seat, you are acquiring electrons on your body by friction. Through friction, you did *work* on the electrons as you removed them from the seat covering. You now have a net negative charge from the imbalance of electrons, which tend to remain on you because you are a poor conductor. But the electrons are now closer than they want to be, within a repulsive electric field, and there is an electrical potential difference (*PD*) between you and some uncharged object, say a metal door handle. When you touch the handle, the electrons will flow, creating a momentary current in the form of a spark, which lasts only until the charge on you is neutralized.

In order to keep an electric current going, you must maintain the separation of charges and therefore maintain the electric field (or potential difference), which can push the charges through a conductor. This might be possible if you could somehow continuously slide across the car seat, but this would be a hit-and-miss way of maintaining a separation of charges and would probably result in a series of sparks rather than a continuous current. This is how electrostatic machines work.

A useful analogy for understanding the requirements for a sustained electric current is the decorative waterwheel device (figure 7.10). Water in the upper reservoir has a greater gravitational potential energy than water in the lower reservoir. As water flows from the upper reservoir, it can do work in turning the waterwheel, but it can continue to do this only as long as the pump does the work to maintain the potential difference between the two reservoirs. This "water circuit" will do work in turning the waterwheel as long as the pump returns the water to a higher potential continuously as the water flows back to the lower potential.

So, by a water circuit analogy, a steady electric current is maintained by pumping charges to a higher potential, and the

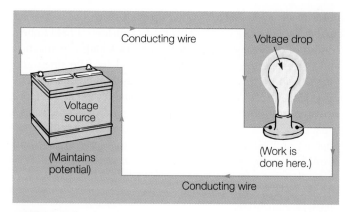

FIGURE 7.11

A simple electric circuit has a voltage source (such as a generator or battery) that maintains the electrical potential, some device (such as a lamp or motor) where work is done by the potential, and continuous pathways for the current to follow.

charges do work as they move back to a lower potential. The higher electric potential energy is analogous to the gravitational potential energy in the waterwheel example (figure 7.10). An **electric circuit** contains some device, such as a battery or electric generator, that acts as a source of energy as it gives charges a higher potential against an electric field. The charges do work in another part of the circuit as they light bulbs, run motors, or provide heat. The charges flow through connecting wires to make a continuous path. An electric switch is a means of interrupting or completing this continuous path.

The electrical potential difference between the two connecting wires shown in figure 7.11 is one factor in the work done *by* the device that creates a higher electrical potential (battery, for example) and the work done *in* some device (lamp, for example). Disregarding any losses, the work done in both places would be the same. Recall that work done per unit of charge is joules/coulomb, or volts (equation 7.3). The source of the electrical potential difference is therefore referred to as a *voltage source*. The device where the charges do their work causes a *voltage drop*. Electrical potential difference is measured in volts, so the term *voltage* is often used for it. Household circuits usually have a difference of potential of 120 or 240 volts. A voltage of 120 means that each coulomb of charge that moves through the circuit can do 120 joules of work in some electrical device.

Voltage describes the potential difference, in joules/coulomb, between two places in an electric circuit. By way of analogy to pressure on water in a circuit of water pipes, this potential difference is sometimes called an "electrical force" or "electromotive force" (emf). Note that in electrical matters, however, the potential difference is the *source* of a force rather than being a force such as water under pressure. Nonetheless, just as you can have a small water pipe and a large water pipe under the same pressure, the two pipes would have a different rate of water flow in gallons per minute. The *rate* at which an electric current (*I*) flows is the quan-

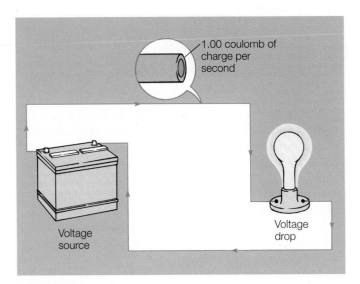

FIGURE 7.12

A simple electric circuit carrying a current of 1.00 coulomb per second through a cross section of a conductor has a current of 1.00 amp.

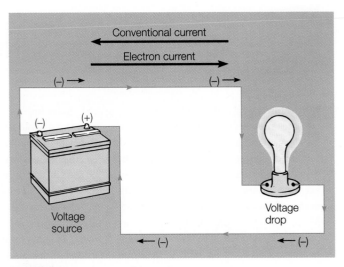

FIGURE 7.13

A conventional current describes positive charges moving from the positive terminal (+) to the negative terminal (−). An electron current describes negative charges (−) moving from the negative terminal (−) to the positive terminal (+).

tity of charge (q) that moves through a cross section of a conductor in a unit of time (t), or

$$\text{electric current} = \frac{\text{quantity of charge}}{\text{time}}$$

$$I = \frac{q}{t}$$

equation 7.4

The units of current are thus coulombs/second. A coulomb/second is called an **ampere** (A), or **amp** for short. In units, current is therefore

$$1.00 \text{ amp (A)} = \frac{1.00 \text{ coulomb (C)}}{1.00 \text{ second (s)}}$$

A 1.00 amp current is 1.00 coulomb of charge moving through a conductor each second, a 2.00 amp current is 2.00 coulombs per second, and so forth (figure 7.12).

Using the water circuit analogy, you would expect a greater rate of water flow (gallons/minute) when the water pressure is produced by a greater gravitational potential difference. The rate of water flow is thus directly proportional to the difference in gravitational potential energy. In an electric circuit, the rate of current (coulombs/second, or amps) is directly proportional to the difference of electrical potential (joules/coulombs, or volts) between two parts of the circuit, $I \propto PD$.

The Nature of Current

There are two ways to describe the current that flows outside the power source in a circuit, (1) a historically based description called *conventional current* and (2) a description based on a flow of charges called *electron current*. The *conventional* current

describes current as positive charges that flow from the positive to the negative terminal of a battery. This description has been used by convention ever since Ben Franklin first misnamed the charge of an object based on an accumulation, or a positive amount, of "electrical fluid." Conventional current is still used in circuit diagrams. The *electron* current description is in an opposite direction to the conventional current. The electron current describes current as the drift of negative charges that flow from the negative to the positive terminal of a battery. Today, scientists understand the role of electrons in a current, something that was unknown to Franklin. But conventional current is still used by tradition. It actually does not make any difference which description is used, since positive charges moving from the positive terminal are mathematically equivalent to negative charges moving from the negative terminal (figure 7.13).

The description of an electron current also retains historical traces of the earlier fluid theories of electricity. Today, people understand that electricity is not a fluid but still speak of current, rate of flow, and resistance to flow. Fluid analogies can be helpful because they describe the overall electrical effects. But they can also lead to incorrect concepts such as the following: (1) an electric current is the movement of electrons through a wire just as water flows through a pipe; (2) electrons are pushed out one end of the wire as more electrons are pushed in the other end; and (3) electrons must move through a wire at the speed of light since a power plant failure hundreds of miles away results in an instantaneous loss of power. Perhaps you have held one or more of these misconceptions from fluid analogies.

What is the exact nature of an electric current? First, consider the nature of a metal conductor without a current. The atoms making up the metal have unattached electrons that are free to move about, much as the molecules of a gas in a container.

A CLOSER LOOK | Cars, Lights, and the Future of Earth

The threat of global warming is probably the most severe environmental challenge that humans have ever faced. Increased concentrations of greenhouse gases are expected to increase global average temperatures by a few degrees over the next century, resulting in a 18 to 99 cm (7 to 39 in) rise in sea levels that could swamp coastal communities. The so-called greenhouse gases do occur naturally in the atmosphere, but the rapid increase of CO_2 and CH_4 is a direct consequence of human activity.

Since 1800, atmospheric concentration of CO_2 has increased 30 percent, and CH_4 has increased 145 percent. The primary source of CO_2 is fossil fuel combustion and the largest source of CH_4 is decomposition of wastes in landfills, manure, and intestinal gas released by cattle being raised to match demands for the consumption of red meat. In other words, the threat is really from activities that burn fossil fuels—transportation, power plants, and industry—and the raising of cattle for the consumption of red meat.

According to the United States Environmental Protection Agency*, transportation activities account for about 31 percent, electric utilities are responsible for about 36 percent, and industries account for about 33 percent of the total U.S. emissions of CO_2.

Based on this information, there are several things you can do to reduce the threat of global warming. Driving less, or using bus or train transit facilities, will cut down on CO_2 emission, and this is one the best things you can do for the environment. The second best thing you can do is to replace the beef in your diet with grains or chicken. Less consumption of red meat means fewer livestock, and fewer livestock means less CH_4 produced by the cattle. You can also improve the lighting efficiency where you live. About 40 percent of the electricity consumed is used for lighting, so improvements in lighting efficiency would significantly reduce the demand for electrical energy. The incandescent light bulb is only about 4 percent efficient in converting

electrical energy to light and the wasted energy ends up as heat. High-intensity lamps, on the other hand, are 32 percent efficient and fluorescent lamps are about 65 percent efficient. Replacement of incandescent light bulbs reduces the amount of waste heat, which reduces the amount of electricity needed to cool buildings. Increased lighting efficiency also reduces the demand for electricity.

Replacing incandescent bulbs can make a big difference. Det Norske Veritas (DNV) is a certification foundation that measures things such as reduction of CO_2 emissions. One of their tests in Mexico (called ILUMEX) found that 171,168 tons of CO_2 emissions were eliminated between 1995 and 1998 when the residents of Monterrey and Guadalajara replaced incandescent with fluorescent light bulbs.

Drive less, do not eat red meat, and replace incandescent light bulbs. The tasks seem simple enough, but they are powerful enough to begin helping the earth's environment!

*See http://www.epa.gov/globalwarming/inventory.

They randomly move at high speed in all directions, often colliding with each other and with stationary positive ions of the metal. This motion is chaotic, and there is no net movement in any one direction, but the motion does increase with increases in the absolute temperature of the conductor.

When a potential difference is applied to the wire in a circuit, an electric field is established everywhere in the circuit. The *electric field* travels through the conductor at nearly the speed of light as it is established. A force is exerted on each electron by the field, which accelerates the free electrons in the direction of the force. The resulting increased velocity of the electrons is superimposed on their existing random, chaotic movement. This added motion is called the *drift velocity* of the electrons. The drift velocity of the electrons is a result of the imposed electric field. The electrons do not drift straight through the conductor, however, because they undergo countless collisions with other electrons and stationary positive ions. This results in a random zigzag motion with a net motion in one direction. *This net motion constitutes a current,* a flow of charge (figure 7.14).

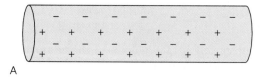

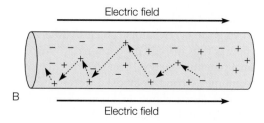

FIGURE 7.14

(*A*) A metal conductor without a current has immovable positive ions surrounded by a swarm of chaotically moving electrons. (*B*) An electric field causes the electrons to shift positions, creating a separation charge as the electrons move with a zigzag motion from collisions with stationary positive ions and other electrons.

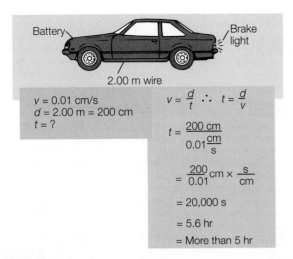

FIGURE 7.15

Electrons move very slowly in a direct current circuit. With a drift velocity of 0.01 cm/s, more than 5 hr would be required for an electron to travel 200 cm from a car battery to the brake light. It is the electric field, not the electrons, that moves at near the speed of light in an electric circuit.

When the voltage across a conductor is zero the drift velocity is zero, and there is no current. The current that occurs when there is a voltage depends on (1) the number of free electrons per unit volume of the conducting material, (2) the charge on each electron (the fundamental charge), (3) the drift velocity, which depends on the electronic structure of the conducting material and the temperature, and (4) the cross-sectional area of the conducting wire.

The relationship between the number of free electrons, charge, drift velocity, area, and current can be used to determine the drift velocity when a certain current flows in a certain size wire made of copper. A 1.0 amp current in copper bell wire (#18), for example, has an average drift velocity on the order of 0.01 cm/s. At that rate, it would take over 5 hr for an electron to travel the 200 cm from your car battery to the brake light of your car (figure 7.15). Thus, it seems clear that it is the *electric field*, not electrons, that causes your brake light to come on almost instantaneously when you apply the brake. The electric field accelerates the electrons already in the filament of the brake light bulb. Collisions between the electrons in the filament cause the bulb to glow.

Conclusions about the nature of an electric current are that (1) an electric potential difference establishes, at near the speed of light, an electric field throughout a circuit, (2) the field causes a net motion that constitutes a flow of charge, or current, and (3) the average velocity of the electrons moving as a current is very slow, even though the electric field that moves them travels with a speed close to the speed of light.

Another aspect of the nature of an electric current is the direction the charge is flowing. A circuit like the one described with your car battery has a current that always moves in one direction, a **direct current** (dc). Chemical batteries, fuel cells, and solar cells produce a direct current, and direct currents are utilized in elec-

tronic devices. Electric utilities and most of the electrical industry, on the other hand, use an **alternating current** (ac). An alternating current, as the name implies, moves the electrons alternately one way, then the other way. Since the electrons are simply moving back and forth, there is no electron drift along a conductor in an alternating current. Since household electric circuits use alternating current, there is no movement of electrons from the electrical outlets through the circuits. The electric field moves back and forth through the circuit near the speed of light, moving electrons back and forth. This movement constitutes a current that flows one way, then the other with the changing field. The current changes like this 120 times a second in a 60 hertz alternating current.

Electrical Resistance

Recall the natural random and chaotic motion of electrons in a conductor and their frequent collisions with each other and with the stationary positive ions. When these collisions occur, electrons lose energy that they gained from the electric field. The stationary positive ions gain this energy, and their increased energy of vibration results in a temperature increase. Thus, there is a resistance to the movement of electrons being accelerated by an electric field and a resulting energy loss. Materials have a property of opposing or reducing a current, and this property is called **electrical resistance** (R).

Recall that the current (I) through a conductor is directly proportional to the potential difference (PD) between two points in a circuit. If a conductor offers a small resistance, less voltage would be required to push an amp of current through the circuit. If a conductor offers more resistance, then more voltage will be required to push the same amp of current through the circuit. Resistance (R) is therefore a *ratio* between the potential difference (PD) between two points and the resulting current (I). This ratio is

$$\text{resistance} = \frac{\text{electrical potential}}{\text{current}}$$

$$R = \frac{PD}{I}$$

In units, this ratio is

$$1.00 \text{ ohm } (\Omega) = \frac{1.00 \text{ volt (V)}}{1.00 \text{ amp (A)}}$$

The ratio of volts/amps is the unit of resistance called an **ohm** (Ω) after the German physicist who discovered the relationship. The resistance of a conductor is therefore 1.00 ohm if 1.00 volt is required to maintain a 1.00 amp current. The ratio of volt/amp is *defined as* an ohm. Therefore,

$$\text{ohm} = \frac{\text{volt}}{\text{amp}}$$

Another way to show the relationship between the voltage, current, and resistance is

$$V = IR$$

equation 7.5

A CLOSER LOOK | Household Circuits and Safety

The example household circuit in box figure 7.1 shows a light and four wall outlets in a parallel circuit. The use of the term "parallel" means that a current can flow through separate branches, but does not imply that the branches are necessarily lined up with each other. Lights and outlets in this circuit all have at least two wires, one that carries the electrical load and one that maintains a potential difference by serving as a system ground. The load-carrying wire is usually black (or red) and the system ground is usually white. A third wire, usually bare or green, might serve as an appliance ground.

Too many appliances running in a circuit—or a short circuit—can result in a very large current, perhaps great enough to cause strong heating and a possibly a fire. A fuse or circuit breaker prevents this by disconnecting the circuit when it reaches a preset value, usually 15 or 20 amps. A fuse contains a short piece of metal that melts by design, creating a gap when the current through the circuit reaches the preset rating. The gap disconnects the circuit, just as cutting the wire or closing a switch. The circuit breaker has the same purpose, but uses the proportional relationship between the magnitude of a current and the strength of the magnetic field that forms around the conductor. When the current reaches a preset level, the magnetic field is strong enough to open a spring-loaded switch. The circuit breaker is reset by flipping the switch back to its original position.

Besides a fuse or circuit breaker, a modern household electric circuit has three-pronged plugs, polarized plugs, and ground fault interrupters (GFI) to help protect people and property from electrical damage. A *three-pronged plug* provides a grounding wire

A

BOX FIGURE 7.1

(*A*) One circuit breaker in this photo of a circuit breaker panel has tripped, indicating an overload or short circuit.
(*B*) If this is the tripped circuit perhaps the toaster was the problem.

B

—*Continued top of next page*

Continued—

through a (usually round) prong on the plug. The grounding wire connects the housing of an appliance directly to the ground. If there is a short circuit, the current will take the path of least resistance—through the grounding wire—rather than through a person.

A *polarized plug* has one of the two flat prongs larger than the other. An alternating current moves back and forth with a frequency of 60 Hz, and polarized in this case has nothing to do with positive or negative. A polarized plug in an ac circuit means that one prong always carries the load. The smaller plug is connected to the load-carrying wire and the larger one is connected to the neutral wire. An ordinary, nonpolarized plug can fit

into an outlet either way, which means there is a 50–50 chance that one of the wires will be the one that carries the load. The polarized plug always has the load-carrying wire on the same side of the circuit, so the switch can be wired in so it is always on the load-carrying side. The switch will function the same on either wire, but when it is on the ground wire the appliance contains a load-carrying wire, just waiting for a short circuit. When the switch is on the load-carrying side the appliance does not have this potential safety hazard.

Yet another safety device called a *ground-fault interrupter* (GFI) offers a different kind of protection. Normally, the current

in the load-carrying and system ground wire is the same. If a short circuit occurs, some of the current might be diverted directly to the ground or to the appliance ground. A GFI device monitors the load-carrying and system ground wires and if any difference is detected, it trips, opening the circuit within a fraction of a second. This is much quicker than the regular fuse or general circuit breaker can react, and the difference might be enough to prevent a fatal shock. The GFI device is usually placed in an outside or bathroom circuit, places where currents might be diverted through people with wet feet. Note the GFI can also be tripped by a line surge that might occur during an electrical thunderstorm.

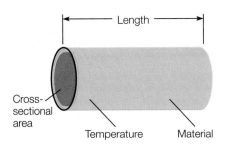

FIGURE 7.16

The four factors that influence the resistance of an electrical conductor are the length of the conductor, the cross-sectional area of the conductor, the material the conductor is made of, and the temperature of the conductor.

CONNECTIONS

Effects of Electric Current on People

Current (A)	Effect (varies with individual)
0.001 to 0.005	Perception threshold
0.005 to 0.01	Mild shock
0.01 to 0.02	Cannot let go of wire
0.02 to 0.05	Breathing difficult
0.05 to 0.1	Breathing stops, heart stops
0.1 and higher	Severe burns, death

which is known as **Ohm's law.** This is one of three ways to show the relationship, but this way (solved for *V*) is convenient for easily solving the equation for other unknowns.

The magnitude of the electrical resistance of a conductor depends on four variables: (1) the length of the conductor, (2) the cross-sectional area of the conductor, (3) the material the conductor is made of, and (4) the temperature of the conductor (figure 7.16). Different materials have different resistances, as shown by the list of conductors in table 7.1 (p. 149). Silver, for example, is at the top of the list because it offers the least resistance, followed by copper, gold, then aluminum. Of the materials listed in table 7.1, nichrome is the conductor with the greatest resistance. By definition, conductors have

less electrical resistance than insulators, which have a very large electrical resistance.

The length of a conductor varies directly with the resistance; that is, a longer wire has more resistance and a shorter wire has less resistance. In addition, the cross-sectional area of a conductor varies inversely with the resistance. A thick wire has a greater area and therefore has less resistance than a thin wire. For most materials, the resistance increases with increases in temperature. As previously discussed, this is a consequence of the increased motion of electrons and ions at higher temperatures, which increases the number of collisions. At very low temperatures, the resistance of some materials approaches zero, and the materials are said to be *superconductors.*

Electrical Power and Electrical Work

All electric circuits have three parts in common: (1) a *voltage source,* such as a battery or electric generator that uses some non-electric source of energy to do work on electrons, moving them *against* an electric field to a higher potential; (2) an *electric device,* such as a lightbulb or electric motor where work is done *by* the electric field; and (3) *conducting wires* that maintain the potential difference across the electrical device. In a direct current circuit, the electric field moves from one terminal of a battery to the electric device through one wire. The second wire from the device carries the now low-potential field back to the other terminal, maintaining the potential difference. In an alternating current circuit, such as a household circuit, one wire supplies the alternating electric field from the electric generator of a utility company. The second wire from the device is connected to a pipe in the ground and is at the same potential as the earth. The observation that a bird can perch on a current-carrying wire without harm is explained by the fact that there is no potential difference across the bird's body. If the bird were to come into contact with the earth through a second, grounded wire, a potential difference would be established and a current would flow through it.

The work done by a voltage source (battery, electric generator) is equal to the work done by the electric field in an electric device (lightbulb, electric motor) *plus* the energy lost to resistance. Resistance is analogous to friction in a mechanical device, so low-resistance conducting wires are used to reduce this loss. Disregarding losses to resistance, electrical work can therefore be measured where the voltage source creates a potential difference by doing work (W) to move charges (q) to a higher potential (PD). From equation 7.3, this relationship is

$$\text{work} = (\text{potential})(\text{charge})$$

or

$$W = (PD)(q)$$

In units, the electrical potential is measured in joules/coulomb, and a quantity of charge is measured in coulombs. Therefore the unit of electrical work is the *joule,*

$$(PD)(q) = W$$

$$\frac{\text{joules}}{\text{coulomb}} \times \text{coulomb} = \text{joules}$$

Recall that a joule is a unit of work in mechanics (a newton-meter). In electricity, a joule is also a unit of work, but it is derived from moving a quantity of charge (coulomb) to higher potential difference (joules/coulomb). In mechanics the work put into a simple machine equals the work output when you disregard *friction.* In electricity the work put into an electric circuit equals the work output when you disregard *resistance.* Thus, the work done by a voltage source is ideally equal to the work done by electrical devices in the circuit.

Recall also that mechanical power (P) was defined as work (W) per unit time (t), or

$$P = \frac{W}{t}$$

equation 7.6

Since electrical work is $W = PDq$, then electrical power must be

$$P = \frac{PDq}{t}$$

equation 7.7

Equation 7.4 defined a quantity of charge (q) per unit time (t) as a current (I), or $I = q/t$. Therefore electrical power is

$$P = \left(\frac{q}{t}\right)(PD)$$

In units, you can see that multiplying the current A = C/s by the potential ($V = $ J/C) yields

$$\frac{\text{coulombs}}{\text{second}} \times \frac{\text{joules}}{\text{coulombs}} = \frac{\text{joules}}{\text{second}}$$

A joule/second is a unit of power called the **watt.** Therefore, electrical power is measured in units of watts, and

$$P = \text{A} \cdot \text{V}$$

$$\text{power (in watts)} = \text{current (in amps)} \times \text{potential (in volts)}$$

$$\text{watts} = \text{volts} \times \text{amps}$$

This relationship is often shown by the following equation, using the unit symbol (V) rather than the quantity symbol for electrical potential (PD):

$$P = IV$$

equation 7.8

Household electrical devices are designed to operate on a particular voltage, usually 120 or 240 volts (figure 7.17). They therefore draw a certain current to produce the designed power. Information about these requirements is usually found somewhere on the device. A lightbulb, for example, is usually stamped with the designed power, such as 100 watts. Other electrical devices may be stamped with amp and volt requirements. You can determine the power produced in these devices by using equation 7.8, that is, amps × volts = watts. Another handy conversion factor to remember is that 746 watts are equivalent to 1.00 horsepower.

An electric utility company measures the electrical work done in a household with a meter located near where the wires enter the house. The meter measures kilowatt-hours (kWhr) of work. Since household electrical devices are designed to operate at a particular potential (V), the utility company wants to know the current (I) you used for what time period. When you multiply volts × amps × time, you are really multiplying power × time and the answer is the work done, $W = P \times t$ (from equation 7.6). Since $P = IV$ (equation 7.8), then an expression of work can be obtained by combining the two equations, or

$$(IV)(t) = W$$

equation 7.9

In units,

$$\frac{\text{coulomb}}{\text{second}} \times \frac{\text{joules}}{\text{coulomb}} \times \text{second} = \text{joules}$$

A

B

C

FIGURE 7.17

What do you suppose it would cost to run each of these appliances for one hour? (*A*) This lightbulb is designed to operate on a potential difference of 120 volts and will do work at the rate of 100 W. (*B*) The finishing sander does work at the rate of 1.6 amp × 120 volts, or 192 W. (*C*) The garden shredder does work at the rate of 8 amps × 120 volts, or 960 W.

Charge It!

The Leyden jar is a glass jar covered inside and out with metal or metal foil (figure 7.18). Electrons are delivered to a metal ball on top of the jar, and these electrons are mutually repelled down to the inner metal liner. This accumulation of electrons on the inner metal liner repels electrons on the outer metal covering. Repelled electrons on the outer covering can move away through a ground wire, so the outer covering now has a positive charge. The overall effect is that electrons are attracted to the positive charge of the outer metal covering, creating a greater-than-usual accumulation of electrons on the inner metal liner. Charges accumulate inside the jar, producing a large spark when a conductor is brought between the metal ball on the top and the outside layer of metal foil.

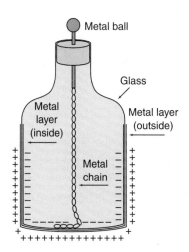

FIGURE 7.18

A schematic sketch of a Leyden jar.

Thus, the utility meter shows the amount of work that was done in a household for a month. The joule is a small unit of work, and its use would result in some very large numbers. The electric utility therefore uses a kilowatt- (1,000 × amp × volt) hour (3,600 s) to measure the electrical work done (figure 7.19). One kilowatt-hour of work is thus equivalent to 3,600,000 joules. A typical monthly electric bill shows the charge for 1,000 kWhr, not 3,600,000,000 joules.

The electric utility charge for the electrical work done is at a rate of cents per kWhr. The rate varies from place to place across the country, depending on the cost of producing the

FIGURE 7.19

This meter measures the amount of electric *work* done in the circuits, usually over a time period of a month. The work is measured in kWhr.

TABLE 7.2	Summary of electrical quantities and units	
Quantity	**Definition***	**Units**
Charge	$q = ne$	1.00 coulomb (C) = charge equivalent to 6.24×10^{18} particles such as the electron
Electric potential difference	$PD = \dfrac{W}{q}$	$1.00 \text{ volt (V)} = \dfrac{1.00 \text{ joule (J)}}{1.00 \text{ coulomb (C)}}$
Electric current	$I = \dfrac{q}{t}$	$1.00 \text{ amp (A)} = \dfrac{1.00 \text{ coulomb (C)}}{1.00 \text{ second (s)}}$
Electrical resistance	$R = \dfrac{PD}{I}$	$1.00 \text{ ohm } (\Omega) = \dfrac{1.00 \text{ volt (V)}}{1.00 \text{ amp (A)}}$
Electrical power	$P = IV$	$1.00 \text{ watt (W)} = \dfrac{C}{s} \times \dfrac{J}{C}$

*See Summary of Equations for more information.

power. You can predict the cost of running a particular electrical appliance by first finding the work done in kWhr from equation 7.9. In units,

$$\text{kWhr} = \frac{(\text{volts})(\text{amps})(\text{time})}{1,000\ \dfrac{W}{kW}}$$

Note that volts $\times$ amps = watts, so a watt power rating can be substituted for amps $\times$ volts. Also note that the time unit is in hours, so if you want to know the cost of running an appliance for a number of minutes, this must be converted to the decimal

Inside a Dry Cell

The common dry cell used in a flashlight produces electrical energy from a chemical reaction between ammonium chloride and the zinc can (figure 7.20). The reaction leaves a negative charge on the zinc and a positive charge on the carbon rod. Manganese dioxide takes care of hydrogen gas, which is a by-product of the reaction. Dry cells always produce 1.5 volts, regardless of their size. Larger voltages are produced by combinations of smaller cells (making a true "battery").

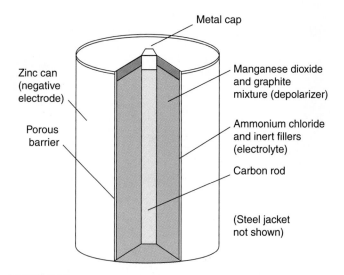

FIGURE 7.20

A schematic sketch of a dry cell.

equivalent of an hour. Once the work in kWhr is found, the cost of running the appliance is determined by multiplying the rate by the kWhr used, or

$$(\text{work})(\text{rate}) = \text{cost}$$

equation 7.10

MAGNETISM

The ability of a certain naturally occurring rock to attract iron has been known since at least 600 B.C. The early Greeks called this rock "Magnesian stone," since it was discovered near the ancient city of Magnesia in western Turkey. Knowledge about the iron-attracting properties of the Magnesian stone grew slowly. About A.D. 100, the Chinese learned to magnetize a piece of iron with a Magnesian stone, and sometime before A.D. 1000, they learned to use the magnetized iron or stone as a direction

FIGURE 7.21

Every magnet has ends, or poles, about which the magnetic properties seem to be concentrated. As this photo shows, more iron filings are attracted to the poles, revealing their location.

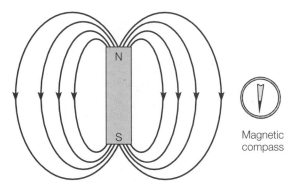

Magnetic compass

FIGURE 7.22

These lines are a map of the magnetic field around a bar magnet. The needle of a magnetic compass will follow the lines, with the north end showing the direction of the field.

finder (compass). Today, the rock that attracts iron is known to be the black iron oxide mineral named *magnetite* after the city of Magnesia. If a sample of magnetite acts as a natural magnet, it is called *lodestone*, after its use as a "leading stone," or compass.

A lodestone is a *natural magnet* and strongly attracts iron and steel but also attracts cobalt and nickel. Such substances that are attracted to magnets are said to have *ferromagnetic properties*, or simply *magnetic* properties. Iron, cobalt, and nickel are considered to have magnetic properties, and most other common materials are considered not to have magnetic properties. Most of these nonmagnetic materials, however, are slightly attracted or slightly repelled by a strong magnet. In addition, certain rare earth elements (see chapter 10), as well as certain metal oxides, exhibit strong magnetic properties.

Magnetic Poles

The ancient Chinese were the first to discover that a lodestone has two **magnetic poles,** or ends, about which the force of attraction seems to be concentrated. Iron filings or other small pieces of iron are attracted to the poles of a lodestone, for example, revealing their location (figure 7.21). The Chinese also discovered that when a lodestone is free to turn, such as a lodestone floating on a piece of wood, one pole moves toward the north as the other pole moves toward the south. Today, the north-seeking pole is simply called the *north pole* and the south-seeking pole is called the *south pole.* Such floating lodestones were used as early magnetic compasses. A modern magnetic compass is a magnetic needle on a pivot, usually with the north-seeking end colored blue or black.

You are probably familiar with the fact that two magnets exert forces on each other. For example, if you move the north pole of one magnet near the north pole of a second magnet resting on a tabletop, each experiences a repelling force. If you move your magnet slowly, you can push the second magnet across the

table without the magnets ever touching. A repelling force also occurs if two south poles are moved close together. But if the north pole of one magnet is brought near the south pole of a second magnet, an attractive force occurs. Moving two like poles together usually causes a magnet resting on a table to repel, rotate, then move toward the magnet you are holding. This occurs because *like magnetic poles repel* and *unlike magnetic poles attract.*

Magnetic Fields

A magnet moved into the space near a second magnet experiences a magnetic force as it enters the **magnetic field** of the second magnet. Recall that the electric field in the space near a charged particle was represented by electric field lines of force. A magnetic field can be represented by *magnetic field lines.* By convention, magnetic field lines are drawn to indicate how the *north pole* of a tiny imaginary magnet would point when in various places in the magnetic field. Arrowheads indicate the direction that the north pole would point, thus defining the direction of the magnetic field. The strength of the magnetic field is greater where the lines are closer together and weaker where they are farther apart. Figure 7.22 shows the magnetic field lines around the familiar bar magnet. Note that magnetic field lines emerge from the magnet at the north pole and enter the magnet at the south pole. Magnetic field lines always form closed loops.

The north end of a magnetic compass needle points north because the earth has a magnetic field (figure 7.23). Since the north, or north-seeking, pole of the needle is attracted to the geographic North Pole, it must be a magnetic *south* pole since unlike poles attract. However, it is conventionally called the *north magnetic pole*, since it is located near the geographic North Pole. This can be confusing. Also, the two poles are not in the same place. The north magnetic pole is presently about 2,000 km (1,200 mi) south of the geographic North Pole. Thus, depending on your location, the north pole of a compass needle does not always point to the exact geographic north, or true north. The angle between the magnetic north and the geographic true north is called the *magnetic declination.* The magnetic declination map

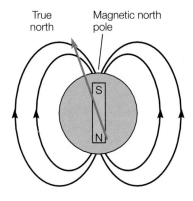

FIGURE 7.23

The earth's magnetic field. Note that the magnetic north pole and the geographic North Pole are not in the same place. Note also that the magnetic north pole acts as if the south pole of a huge bar magnet were inside the earth. You know that it must be a magnetic south pole, since the north end of a magnetic compass is attracted to it, and opposite poles attract.

in figure 7.24 shows how many degrees east or west of true north a compass needle will point in different locations.

The typical compass needle pivots in a horizontal plane, moving to the left or right without up or down motion. Inspection of figure 7.23, however, shows that the earth's magnetic field is horizontal to the surface only at the magnetic equator. A compass needle that is pivoted so that it moves only up and down will be horizontal only at the magnetic equator. Elsewhere, it shows the angle of the field from the horizontal, called the *magnetic dip*. The angle of dip is the vertical component of the earth's magnetic field. As you travel from the equator, the angle of magnetic dip increases from zero to a maximum of 90° at the magnetic poles.

The Source of Magnetic Fields

The observation that like magnetic poles repel and unlike magnetic poles attract might remind you of the forces involved with like and unlike charges. Recall that electric charges exist as single isolated units of positive protons and units of

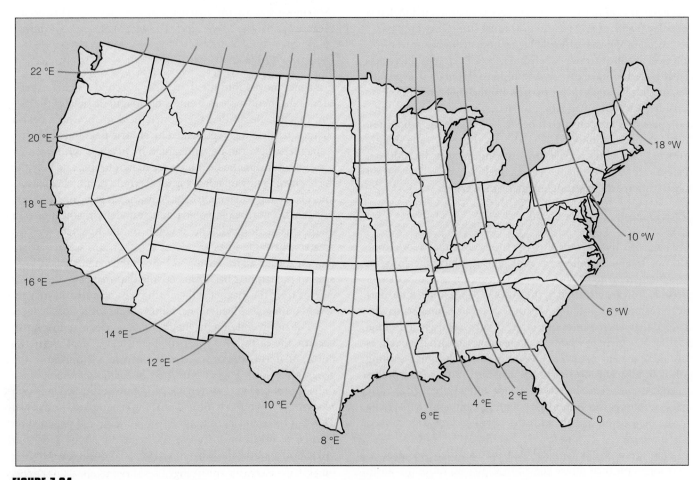

FIGURE 7.24

This magnetic declination map shows the approximate number of degrees east or west of the true geographic north that a magnetic compass will point in various locations.

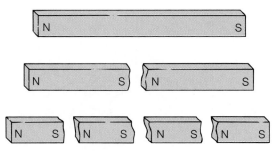

FIGURE 7.25

A bar magnet cut into halves always makes new, complete magnets with both a north and a south pole. The poles always come in pairs, and the separation of a pair into single poles, called monopoles, has never been accomplished.

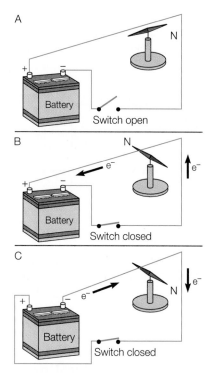

FIGURE 7.26

Oersted discovered that a compass needle below a wire (A) pointed north when there was not a current, (B) moved at right angles when a current flowed one way, and (C) moved at right angles in the opposite direction when the current was reversed.

negative electrons. An object becomes electrostatically charged when charges are separated, and the object acquires an excess or deficiency of negative charges. You might wonder, by analogy, if the poles of a magnet are similarly made up of an excess or deficiency of magnetic poles. The answer is no; magnetic poles are different from electric charges. Positive and negative charges *can* be separated and isolated. But suppose that you try to separate and isolate the poles of a magnet by cutting a magnet into two halves. Cutting a magnet in half will produce two new magnets, each with north and south poles. You could continue cutting each half into new halves, but each time the new half will have its own north and south poles (figure 7.25). It seems that no subdivision will ever separate and isolate a single magnetic pole, called a *monopole.* Magnetic poles always come in matched pairs of north and south and a monopole has never been found. Scientists continue to search for monopoles, seeking a symmetry between magnetism and electricity. The two poles are always found to come together, and as it is understood today, magnetism is thought to be produced by *electric currents,* not an excess of monopoles. The modern concept of magnetism is electric in origin, and magnetism is understood to be a secondary property of electricity.

The key discovery about the source of magnetic fields was reported in 1820 by a Danish physics professor named Hans Christian Oersted. Oersted found that a wire conducting an electric current caused a magnetic compass needle below the wire to move. When the wire was not connected to a battery the needle of the compass was lined up with the wire and pointed north as usual. But when the wire was connected to a battery, the compass needle moved perpendicular to the wire (figure 7.26). When he reversed the current in the wire the magnetic needle swung 180°, again perpendicular to the wire. Oersted had discovered that an electric current produces a magnetic field. An electric current is understood to be the movement of electric charges, so Oersted's discovery suggested that magnetism is a property of charges in motion. Recall that every electric charge is surrounded by an electric field. If the charge is moving, it is surrounded by an electric field *and* a magnetic field. The electric field and magnetic field are different, and the differences all

point to the understanding that magnetism is a secondary property of electricity. The electric field of a charge, for example, is fixed according to the fundamental charge of the particle. The magnetic field, however, changes with the velocity of the moving charge. The magnetic field does not exist at all if the charge is not moving, and the strength of the magnetic field increases with increases in velocity. It seems clear that magnetic fields are produced by the motion of charges, or electric currents. Thus, a magnetic field is a *property* of the space around a moving charge.

Permanent Magnets

The magnetic fields of bar magnets, horseshoe magnets, and other so-called permanent magnets are explained by the relationship between magnetism and moving charges. According to modern atomic theory, all matter is made up of atoms. An extremely simplified view of an atom pictures electrons moving about the nucleus of the atom. Since electrons are charges in motion, they produce magnetic fields. In most materials these magnetic fields cancel one another and neutralize the overall magnetic effect. In other materials, such as iron, cobalt, and nickel, the electrons are arranged and oriented in a complicated way that imparts a magnetic property to the atomic structure. These atoms are grouped in a tiny region called a **magnetic domain.** A magnetic domain is roughly 0.01 to 1 mm in length

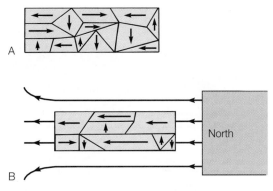

FIGURE 7.27

(*A*) In an unmagnetized piece of iron, the magnetic domains have a random arrangement that cancels any overall magnetic effect. (*B*) When an external magnetic field is applied to the iron, the magnetic domains are realigned, and those parallel to the field grow in size at the expense of the other domains, and the iron is magnetized.

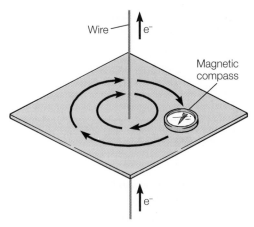

FIGURE 7.28

A magnetic compass shows the presence and direction of the magnetic field around a straight length of current-carrying wire.

or width and does not have a fixed size (figure 7.27). The atoms in each domain are magnetically aligned, contributing to the polarity of the domain. Each domain becomes essentially a tiny magnet with a north and south pole. In an unmagnetized piece of iron the domains are oriented in all possible directions and effectively cancel any overall magnetic effect. The net magnetism is therefore zero or near zero.

When an unmagnetized piece of iron is placed in a magnetic field, the orientation of the domain changes to align with the magnetic field, and the size of aligned domains may grow at the expense of unaligned domains. This explains why a "string" of iron paper clips is picked up by a magnet. Each paper clip has domains that become temporarily and slightly aligned by the magnetic field, and each paper clip thus acts as a temporary magnet while in the field of the magnet. In a strong magnetic field, the size of the aligned domains grows to such an extent that the paper clip becomes a "permanent magnet." The same result can be achieved by repeatedly stroking a paper clip with the pole of a magnet. The magnetic effect of a "permanent magnet" can be reduced or destroyed by striking, dropping, or heating the magnet to a sufficiently high temperature. These actions randomize the direction of the magnetic domains, and the overall magnetic field disappears.

Earth's Magnetic Field

Earth's magnetic field is believed to originate deep within the earth. Like all other magnetic fields, Earth's magnetic field is believed to originate with moving charges. Earthquake waves and other evidence suggest that the earth has a solid inner core with a radius of about 1,200 km (about 750 mi), surrounded by a fluid outer core some 2,200 km (about 1,400 mi) thick. This core is probably composed of iron and nickel, which flows as the earth rotates, creating electric currents that result in Earth's magnetic field. How the electric currents are generated is not yet understood.

Other planets have magnetic fields, and there seems to be a relationship between the rate of rotation and the strength of the planet's magnetic field. Jupiter and Saturn rotate faster than Earth and have stronger magnetic fields than Earth. Venus and Mercury rotate more slowly than Earth and have weaker magnetic fields. This is indirect evidence that the rotation of a planet is associated with internal fluid movements, which somehow generate electric currents and produce a magnetic field.

In addition to questions about how the electric current is generated, there are puzzling questions from geologic evidence. Lava contains magnetic minerals that act like tiny compasses that are oriented to Earth's magnetic field when the lava is fluid but become frozen in place as the lava cools. Studies of these rocks by geologic dating and studies of the frozen magnetic mineral orientation show that Earth's magnetic field has undergone sudden reversals in polarity: the north magnetic pole becomes the south magnetic pole and vice versa. This has happened many times over the distant geologic past. The cause of such magnetic field reversals is unknown, but it must be related to changes in the flow patterns of the earth's fluid outer core of iron and nickel.

ELECTRIC CURRENTS AND MAGNETISM

As Oersted discovered, electric charges in motion produce a magnetic field around the charges. You can map the magnetic field established by the current in a wire by running a straight wire vertically through a sheet of paper. The wire is connected to a battery, and iron filings are sprinkled on the paper. The filings will become aligned as the domains in each tiny piece of iron are forced parallel to the field. Overall, filings near the wire form a pattern of concentric circles with the wire in the center.

The direction of the magnetic field around a current-carrying wire can be determined by using the common device for finding

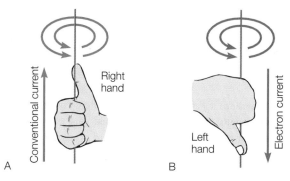

FIGURE 7.29

Use (*A*) a right-hand rule of thumb to determine the direction of a magnetic field around a conventional current and (*B*) a left-hand rule of thumb to determine the direction of a magnetic field around an electron current.

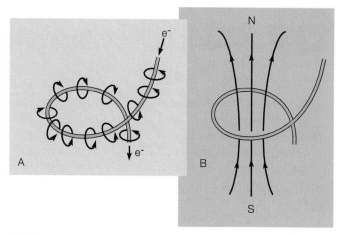

FIGURE 7.30

(*A*) Forming a wire into a loop causes the magnetic field to pass through the loop in the same direction. (*B*) This gives one side of the loop a north pole and the other side a south pole.

the direction of a magnetic field, the magnetic compass. The north-seeking pole of the compass needle will point in the direction of the magnetic field lines (by definition). If you move the compass around the wire the needle will always move to a position that is tangent to a circle around the wire. Evidently the magnetic field lines are closed concentric circles that are at right angles to the length of the wire (figure 7.28).

If you *reverse* the direction of the current in the wire and again move a compass around the wire, the needle will again move to a position that is tangent to a circle around the wire. But this time the north pole direction is reversed. Thus, the magnetic field around a current-carrying wire has closed concentric field lines that are perpendicular to the length of the wire. The direction of the magnetic field is determined by the direction of the current.

You can also determine the direction of a magnetic field around a current-carrying wire by using a "rule of thumb." If you are considering a *conventional* (positive) current, use your *right* hand to grasp a current-carrying wire with your thumb pointing in the direction of the conventional current. Your curled fingers will point in the circular direction of the magnetic field (figure 7.29). If you are considering an *electron* (negative) current, use your *left* hand to grasp the wire with your thumb pointing in the direction of the electron current. This rule is easily understood when you realize that a conventional current runs in the opposite direction of an electron current. When you point upward with your right thumb and downward with your left thumb, the curled fingers of both hands will point in the same direction.

Current Loops

The magnetic field around a current-carrying wire will interact with another magnetic field, one formed around a permanent magnet or one from a second current-carrying wire. The two fields interact, exerting forces just like the forces between the fields of two permanent magnets. The force could be increased by increasing the current, but there is a more efficient way to

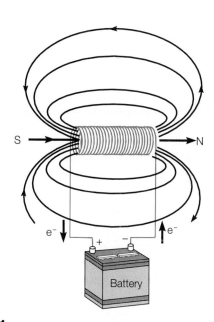

FIGURE 7.31

When a current is run through a cylindrical coil of wire, a solenoid, it produces a magnetic field like the magnetic field of a bar magnet.

obtain a larger force. A current-carrying wire that is formed into a loop has perpendicular, circular field lines that pass through the inside of the loop in the same direction. This has the effect of concentrating the field lines, which increases the magnetic field intensity. Since the field lines all pass through the loop in the same direction, one side of the loop will have a north pole and the other side a south pole (figure 7.30).

Many loops of wire formed into a cylindrical coil are called a *solenoid*. When a current is in a solenoid, each loop contributes field lines along the length of the cylinder (figure 7.31). The overall effect is a magnetic field around the solenoid that acts

A CLOSER LOOK | Electromagnetic Fields and Health

Should people be concerned about ordinary electrical devices because they produce electromagnetic fields and radiation? Ordinary electrical devices mean overhead power lines, household wiring, appliances, computers, and anything that produces, transmits, or uses electricity around people. It is true that all these do in fact create invisible fields and some might give off electromagnetic radiation. Is there any evidence that these fields and radiation could be harmful? This article is about such magnetic fields, electric fields, and electromagnetic radiation and human health. It begins by describing what the fields and radiation are before looking at the evidence about their effect on humans.

Electromagnetic fields are formed around all current-carrying wires, and the greater the current the greater the electromagnetic field strength. An electromagnetic field is a condition of the space around the source, and the field does not move away from the source. The electromagnetic field associated with power lines and household wiring is similar to other forms of electromagnetic energy such as radio waves, infrared radiation, visible light, ultraviolet, and X rays. Each of these can be identified and characterized by frequency, the rate at which the field changes direction. All commercial electric current in the United States is 60 Hz alternating current, so the frequency of electromagnetic fields around wires carrying this current is also 60 Hz. Electromagnetic fields associated with 60 Hz power lines are often referred to as extremely low frequencies, as compared to frequencies of radio waves of about one million Hz, ultraviolet radiation at millions of Hz, and X rays with frequencies of billions of Hz.

An electromagnetic field stays in the space around the source, but *electromagnetic radiation* is transmitted, moving from the source. It moves away and through space until it is absorbed by interacting with matter. A source of electromagnetic radiation will also have an electromagnetic field surrounding it, usually measurable out to a distance of about a wavelength of the radiation. In general, at a distance beyond a wavelength the interaction of matter is with the transmitted energy. At a distance of less than a wavelength the field effect dominates as matter interacts with the field.

The effect of electromagnetic radiation on biological material depends on the frequency of the source. High-frequency radiation (ultraviolet, X ray, gamma ray) has high-energy photons, with enough energy to break chemical bonds. Radiation from this part of the electromagnetic spectrum acts more like particles than waves and is generally referred to as "ionizing radiation" because it can form ions by breaking chemical bonds. The biological damage caused by ultraviolet, X ray, and gamma radiation is from the ionizing effect on biological molecules. If the biological molecule happens to be DNA, the ionizing radiation can cause changes in the structure that can lead to cancer. This has been observed, as the ultraviolet radiation part of sunlight is known to cause skin cancer.

Electromagnetic radiation from the other end of the spectrum (radio waves, infrared radiation, visible light) does not have enough energy to break chemical bonds. However, the shorter radio waves (microwaves and radar) can induce heating in biological tissue through the torquing of water molecules. The longer electromagnetic waves (long radio waves and fields around current-carrying wires) do not interact with water molecules, and do not easily produce heating in biological tissue.

Thus there are three kinds of effects produced by electromagnetic radiation: (1) the high-energy, high-frequency gamma, X ray, and ultraviolet radiation that breaks chemical bonds and ionizes molecules; (2) the middle-energy, mid-frequency infrared, visible light, and microwave radiation that can produce heating, and (3) the lower-energy, low-frequency radio and power frequencies that seldom interact. What is the effect of electromagnetic radiation associated with electric-carrying wires? The energy of electromagnetic fields associated with electric-carrying wires is too low to cause ionization. In fact, the radiation associated with current-carrying wires cannot pass through walls or other materials, so they can produce no biological effects at all.

Magnetic fields can move through walls and other materials, but they typically have too low of an energy to cause heating in any nearby biological material from induced currents. A magnetic field of more than 5 gauss is required to produce an electric current that is close to the magnitude of natural currents already occurring in the human body. Compare this figure to the typical measurements of magnetic fields near a home appliance of 0.2 milligauss in the center of a room, to one gauss when a few centimeters from an appliance. The magnetic field beneath a high-voltage transmission line might be about 100 milligauss, but this varies somewhat with the actual voltage and distance from the line.

The concern about electromagnetic fields appears to have originated from studies suggesting a relationship between numbers of children living near high-voltage transmission lines and higher than average rates of leukemia and other forms of cancer. There was no correlation between measurements of the field intensities and rate of cancer and there was no correlation between adults with cancer living near high-voltage transmission lines. In addition, laboratory studies have shown little evidence of a correlation between power-frequency fields and cancer. Finally, scientists have compared historical data on the generation and consumption of electrical power since 1900 to corresponding data on cancer death and incidence rates. They found the United States' total electric power generation per capita had increased by 350 times since 1902, but trends in cancer incidence show only a 0.9 percent increase per year. Furthermore, most of the increase was attributable to forms

—Continued top of next page

Continued—

of cancer for which there are known causes other than electromagnetic fields.

On the other hand, are magnetic fields beneficial? Some people believe they are, strapping or taping magnets to their body to kill pain or make themselves feel better. They sleep on magnetic mattress pads, walk on magnetic shoe inserts, wear magnetic wrist wraps, and more, proclaiming that magnetic therapy works. Sports medicine physicians, however, can find no evidence that magnets are beneficial, and there have been no authoritative studies on the positive effects of magnetic therapy. Without such scientific studies, magnetic therapy seems to belong in the same category of folk medicine as the wearing of copper bracelets to treat arthritis or the wearing of amber beads for protection against colds. Magnets, copper, and amber lack intrinsic remedial value but serve instead as placebos.

just like the magnetic field of a bar magnet. This magnet, called an **electromagnet,** can be turned on or off by turning the current on or off. In addition, the strength of the electromagnet depends on the magnitude of the current and the number of loops (ampere-turns). The strength of the electromagnet can also be increased by placing a piece of soft iron in the coil. The domains of the iron become aligned by the influence of the magnetic field. This induced magnetism increases the overall magnetic field strength of the solenoid as the magnetic field lines are gathered into a smaller volume within the core.

Applications of Electromagnets

The discovery of the relationship between an electric current, magnetism, and the resulting forces created much excitement in the 1820s and 1830s. This excitement was generated because it was now possible to explain some seemingly separate phenomena in terms of an interrelationship and because people began to see practical applications almost immediately. Within a year of Oersted's discovery, André Ampère had fully explored the magnetic effects of currents, combining experiments and theory to find the laws describing these effects. Soon after Ampère's work, the possibility of doing mechanical work by sending currents through wires was explored. The electric motor, similar to motors in use today, was invented in 1834, only fourteen years after Oersted's momentous discovery.

The magnetic field produced by an electric current is used in many practical applications, including electrical meters, electromagnetic switches that make possible the remote or programmed control of moving mechanical parts, and electric motors. In each of these applications, an electric current is applied to an electromagnet.

Electric Meters

Since you cannot measure electricity directly, it must be measured indirectly through one of the effects that it produces. The strength of the magnetic field produced by an electromagnet is proportional to the electric current in the electromagnet. Thus, one way to measure a current is to measure the magnetic field that it produces. A device that measures currents from their magnetic fields is called a *galvanometer* (figure 7.32). A galvanometer has a coil of wire that can rotate on pivots in the magnetic field of a permanent

Activities

1. You can make a simple compass galvanometer that will detect a small electric current (figure 7.33). All you need is a magnetic compass and some thin insulated wire (the thinner the better).
2. Wrap the thin insulated wire in parallel windings around the compass. Make as many parallel windings as you can, but leave enough room to see both ends of the compass needle. Leave the wire ends free for connections.
3. To use the galvanometer, first turn the compass so the needle is parallel to the wire windings. When a current passes through the coil of wire the magnetic field produced will cause the needle to move from its north-south position, showing the presence of a current. The needle will deflect one way or the other depending on the direction of the current.
4. Test your galvanometer with a "lemon battery." Roll a soft lemon on a table while pressing on it with the palm of your hand. Cut two slits in the lemon about 1 cm apart. Insert a 1 cm by 8 cm (approximate) copper strip in one slit and a same-sized strip of zinc in the other slit, making sure the strips do not touch inside the lemon. Connect the galvanometer to the two metal strips. Try the two metal strips in other fruits, vegetables, and liquids. Can you find a pattern?

magnet. The coil has an attached pointer that moves across a scale and control springs that limit its motion and return the pointer to zero when there is no current. When there is a current in the coil the electromagnetic field is attracted and repelled by the field of the permanent magnet. The larger the current the greater the force and the more the coil will rotate until it reaches an equilibrium position with the control springs. The amount of movement of the coil (and thus the pointer) is proportional to the current in the coil. With certain modifications and applications, the galvanometer can be used to measure current (ammeter), potential difference (voltmeter), and resistance (ohmmeter).

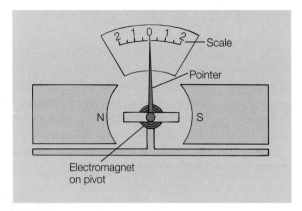

FIGURE 7.32

A galvanometer measures the direction and relative strength of an electric current from the magnetic field it produces. A coil of wire wrapped around an iron core becomes an electromagnet that rotates in the field of a permanent magnet. The rotation moves a pointer on a scale.

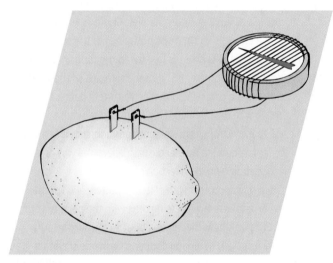

FIGURE 7.33

You can use the materials shown here to create and detect an electric current. See the Activities section on page 169.

Electromagnetic Switches

A *relay* is an electromagnetic switch device that makes possible the use of a low-voltage control current to switch a larger, high-voltage circuit on and off (figure 7.34). A thermostat, for example, utilizes two thin, low-voltage wires in a glass tube of mercury. The glass tube of mercury is attached to a metal coil that expands and contracts with changes in temperature, tipping the attached glass tube. When the temperature changes enough to tip the glass tube, the mercury flows to the bottom end, which makes or breaks contact with the two wires, closing or opening the circuit. When contact is made, a weak current activates an electromagnetic switch, which closes the circuit on the large-current furnace or heat pump motor.

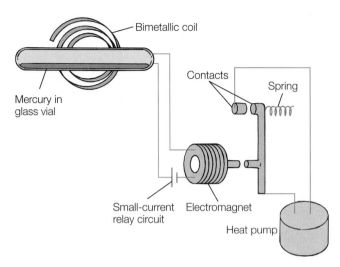

FIGURE 7.34

A schematic of a relay circuit. The mercury vial turns as changes in temperature expand or contract the coil, moving the mercury and making or breaking contact with the relay circuit. When the mercury moves to close the relay circuit, a small current activates the electromagnet, which closes the contacts on the large-current circuit.

A solenoid is a coil of wire with a current. Some solenoids have a spring-loaded movable piece of iron inside. When a current flows in such a coil the iron is pulled into the coil by the magnetic field, and the spring returns the iron when the current is turned off. This device could be utilized to open a water valve, turning the hot or cold water on in a washing machine or dishwasher, for example. Solenoids are also used as mechanical switches on VCRs, automobile starters, and signaling devices such as bells and buzzers. The dot matrix computer printer works with a group of small solenoids that are activated by electric currents from the computer. Seven or more of these small solenoids work together to strike a print ribbon, forming the letters or images as they rapidly move in and out.

Telephones and Loudspeakers

The mouthpiece of a typical telephone contains a cylinder of carbon granules with a thin metal diaphragm facing the front. When someone speaks into the telephone, the diaphragm moves in and out with the condensations and rarefactions of the sound wave (figure 7.35). This movement alternately compacts and loosens the carbon granules, increasing and decreasing the electric current that increases and decreases with the condensations and rarefactions of the sound waves.

The moving electric current is fed to the earphone part of a telephone at another location. The current runs through a coil of wire that attracts and repels a permanent magnet attached to a speaker cone. When repelled forward, the speaker cone makes a condensation, and when attracted back the cone makes a rarefaction. The overall result is a series of condensations and rarefactions that, through the changing electric current, accurately match the sounds made by the other person.

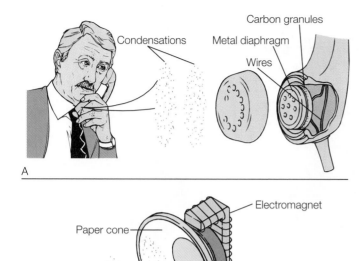

FIGURE 7.35

(*A*) Sound waves are converted into a changing electrical current in a telephone. (*B*) Changing electrical current can be changed to sound waves in a speaker by the action of an electromagnet pushing and pulling on a permanent magnet. The permanent magnet is attached to a stiff paper cone or some other material that makes sound waves as it moves in and out.

The loudspeaker in a radio or stereo system works from changes in an electric current in a similar way, attracting and repelling a permanent magnet attached to the speaker cone. You can see the speaker cone in a large speaker moving back and forth as it creates condensations and rarefactions.

Electric Motors

An electric motor is an electromagnetic device that converts electrical energy to mechanical energy. Basically, a motor has two working parts, a stationary electromagnet called a *field magnet* and a cylindrical, movable electromagnet called an *armature.* The armature is on an axle and rotates in the magnetic field of the field magnet. The axle turns fan blades, compressors, drills, pulleys, or other devices that do mechanical work.

Different designs of electric motors are used for various applications, but the simple demonstration motor shown in figure 7.36 can be used as an example of the basic operating principle. Both the field coil and the armature are connected to an electric current. The armature turns, and it receives the current through a *commutator* and *brushes.* The brushes are contacts that brush against the commutator as it rotates, maintaining contact. When the current is turned on, the field coil and the armature become electromagnets, and the unlike poles attract, rotating the armature. If the current is dc, the armature would turn no farther, stopping as it does in a galvanometer. But the commutator has insulated segments so when it turns halfway, the commutator segments switch brushes

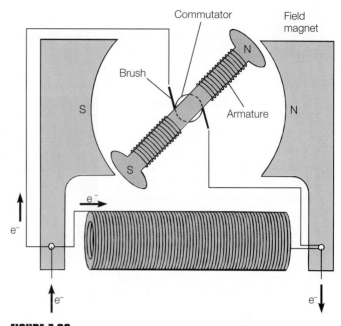

FIGURE 7.36

A schematic of a simple electric motor.

and the current flows through the armature in the *opposite* direction. This switches the armature poles, which are now repelled for another half-turn. The commutator again reverses the polarity, and the motion continues in one direction. An actual motor has many coils (called "windings") in the armature to obtain a useful force, and many commutator segments. This gives the motor a smoother operation with a greater turning force.

ELECTROMAGNETIC INDUCTION

So far, you have learned that (1) a moving charge and a current-carrying wire produce a magnetic field; (2) a second magnetic field exerts a force on a moving charge and exerts a force on a current-carrying wire as their magnetic fields interact; and (3) the direction of the maximum force produced on a moving charge or moving charges is at right angles to their velocity and to the interacting magnetic field lines.

Soon after the discovery of these relationships by Oersted and Ampère, people began to wonder if the opposite effect was possible; that is, would a magnetic field produce an electric current? The discovery was made independently in 1831 by Joseph Henry in the United States and by Michael Faraday in England. They found that *if a loop of wire is moved in a magnetic field, or if the magnetic field is changed, a voltage is induced in the wire.* The voltage is called an *induced voltage,* and the resulting current in the wire is called an *induced current.* The overall interaction is called **electromagnetic induction.**

One way to produce electromagnetic induction is to move a bar magnet into or out of a coil of wire (figure 7.37). A galvanometer shows that the induced current flows one way when the bar magnet is moved toward the coil and flows the other way

A CLOSER LOOK | The Home Computer and Data Storage

A

B

I f you magnify a newspaper photograph, you will see that it is actually made up of many, many dots that vary in density from place to place on the photograph. Your brain combines what your eye sees to form the impression of a large, continuous image. A computer can construct such a photographic image by *digitizing*, or converting all the dots into a string of numbers. The numbers will identify the location and brightness of every tiny piece of the image, which is called a *pixel*. Once digitized, an image can be manipulated, reproduced by a printer, and stored for future use. In fact, many personal (home) computers can now produce, manipulate, and store photographs, other color graphics, and sounds as well as the results of word processing. This article describes how such digitized information is stored by magnetic media such as the floppy disk (box figure 7.2).

A computer processes information by recognizing if a particular solid-state circuit carries an electric current. There are only two possibilities: either a circuit is on, or it is off. Logical calculations are achieved by using combinations of individual circuit elements called *gates*, but the computer still processes information by counting in twos. Counting in twos is called the *binary system*, which is different from our ordinary decimal system way of counting in tens. The binary system has only two symbols, with 1 representing on (current) and 0 representing off (no current). Each of the binary numbers (1 or 0) is a *binary digit*, so it is called a *bit*. A *byte* is a string of binary numbers, usually eight in number (that is, eight bits), used to represent ordinary numbers and letters. For example, the letter *a* could be represented by the byte 01000001. However, the computer must be told what to do with the byte 01000001 when it receives it.

A computer processes data according to instructions from a type of internal memory called *read-only memory* or *ROM*. The ROM memory is stored in unchangeable miniaturized circuits called *microchips*. These microchips are part of the *hardware* that makes up the computer, and they contain the commands and programs that enable the computer to start up

BOX FIGURE 7.2

(*A*) The home computer can produce, manipulate, and store photographs, color graphics, sounds, and text. Storing the information and data from such activities is similar to the process of magnetically recording music or the spoken word. (*B*) Here are four so-called floppy disks, which can be used to magnetically store data from a home computer. The external shell of the smaller disk is not so floppy, but the thin plastic disk inside is flexible.

(called *boot up*) and carry out basic operations. The *software* is a program that works with the hardware to tell the computer specifically how to perform a particular task.

While the computer is operating, it temporarily stores data and the results of operations on microchips in its *random-access memory* or *RAM*. The size of the random-access memory of a home computer can usually be increased by adding more RAM chips, which usually come mounted on cards that plug into the computer circuit board. The RAM chips contain millions of circuit gates and capacitors that respond to changes in electric currents. Each capacitor can hold one bit of data, 1 for charged and 0 for not charged. This is also called "read-and-write" memory because it can be changed many times before it is saved, meaning stored either in the computer's main memory (such as an internal magnetic disk) or stored in an auxiliary unit (such as a magnetic floppy disk). Unless it is saved, RAM information is lost forever when the computer is turned off or if power is lost.

The process of reading and writing computer information to an internal magnetic or floppy disk is similar to the process of magnetically recording music, sounds, or the spoken word. Binary electrical impulses are sent to the

recorder's writing head, which may be as close as 15 millionths of an inch above a spinning disk. The disk is usually a thin plastic circle coated with microscopic pieces of any one of several magnetizable powders. Patterns of electromagnetism in the recording head (corresponding to on and off patterns in bytes of information) produce corresponding magnetic patterns in the particle coating on the disk. This process is reversed when the disk is read. The disk bearing magnetic patterns is spun very close to the head, which is also an electromagnet. The magnetic patterns stored on the disk produce corresponding electromagnetic patterns in the head that are converted to a series of electric impulses. These electric impulses are sent to the computer, where they are converted to information identical in pattern to the originally written data. The distance of 15 millionths of an inch between the reading/writing head and spinning disk does not leave much room for error. A head crash sometimes occurs, resulting in distortion of the stored binary magnetic fields (called file corruption), resulting in the loss of data and other information. A head crash is not a common problem with laser optical disc systems (CD, or compact discs), which will be discussed in a reading in chapter 8.

Activities

1. Make a coil of wire from insulated bell wire (#18 copper wire) by wrapping fifty windings around a narrow jar. Tape the coil at several places so it does not come apart.
2. Connect the coil to a compass galvanometer (see previous activity).
3. Move a strong bar magnet into and out of the stationary coil of wire and observe the galvanometer. Note the magnetic pole, direction of movement, and direction of current for both in and out movements.
4. Move the coil of wire back and forth over the stationary bar magnet.

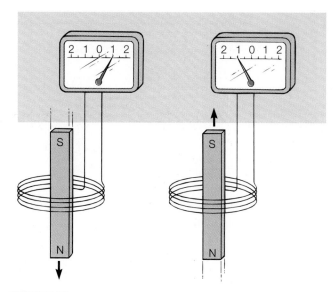

FIGURE 7.37

A current is induced in a coil of wire moved through a magnetic field. The direction of the current depends on the direction of motion.

when the bar magnet is moved away from the coil. The same effect occurs if you move the coil back and forth over a stationary magnet. Furthermore, no current is detected when the magnetic field and the coil of wire are not moving. Thus, electromagnetic induction depends on the relative motion of the magnetic field and the coil of wire. It does not matter which moves or changes, but one must move or change relative to the other for electromagnetic induction to occur.

Electromagnetic induction occurs when the loop of wire cuts across magnetic field lines or when magnetic field lines cut across the loop. The magnitude of the induced voltage is proportional to (1) the number of wire loops cutting the magnetic field lines, (2) the strength of the magnetic field, and (3) the rate at which magnetic field lines are cut by the wire.

Generators

Soon after the discovery of electromagnetic induction the **electric generator** was developed. This development began the tremendous advance of technology that followed soon after. The generator is essentially an axle with many wire loops that rotates in a magnetic field. The axle is turned by some form of mechanical energy, such as a water turbine or a steam turbine, which uses steam generated from fossil fuels or nuclear energy. The use of the electric generator began in the 1890s, when George Westinghouse designed and built a waterwheel-powered 75 kilowatt ac generator near Ouray, Colorado (figure 7.38). Thomas Edison built a dc power plant in New York City about the same time (figure 7.39). In 1896 a 3.7 megawatt power plant at Niagara Falls sent electrical power to factories in Buffalo, New York. The factories no longer had to be near a source of energy such as a waterfall, since electric energy could now be transmitted by wires.

Transformers

In the 1890s the production of electrical power from electromagnetic induction began in the United States with generators built by George Westinghouse and Thomas Edison. A controversy arose, however, because Edison built dc generators, and Westing-

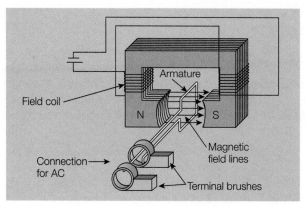

A

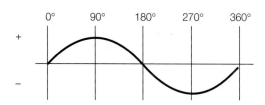

B

FIGURE 7.38

(*A*) Schematic of a simple alternator (ac generator) with one output loop. (*B*) Output of the single loop turning in a constant magnetic field, which alternates the induced current each half-cycle.

house built ac generators. Edison believed that alternating current was dangerous and argued for the use of direct current only. Alternating current eventually won because (1) the voltage of ac could be easily changed to meet different applications, but the voltage of

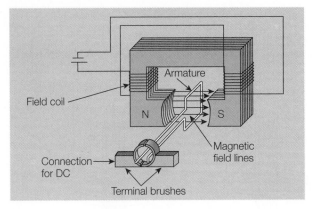

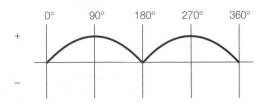

A

B

FIGURE 7.39

(A) Schematic of a simple dc generator with one output loop. (B) Output of the single loop turning in a constant magnetic field. The split ring (commutator) reverses the sign of the output when the voltage starts to reverse, so the induced current has half-cycle voltages of a constant sign, which is the definition of direct current.

dc could *not* be easily changed, and (2) dc transmission suffered from excessive power losses in transmission while ac power was made possible by a **transformer,** a device that uses electromagnetic induction to increase or decrease ac voltage.

A transformer has two basic parts: (1) a *primary coil,* which is connected to a source of alternating current, and (2) a *secondary coil,* which is close by. Both coils are often wound on a single iron core but are always fully insulated from each other. When an alternating current flows through the primary coil, a magnetic field grows around the coil to a maximum size, collapses to zero, then grows to a maximum size with an opposite polarity. This happens 120 times a second as the alternating current oscillates at 60 hertz. The magnetic field is strengthened and directed by the iron core. The growing and collapsing magnetic field moves across the wires in the secondary coil, inducing a voltage in the secondary coil. The growing and collapsing magnetic field from the primary coil thus induces a voltage in the secondary coil, just as an induced voltage occurs in the wire loops of a generator.

The transformer increases or decreases the voltage in an alternating current because the magnetic field grows and collapses past the secondary coil, inducing a voltage. If a direct current is applied to the primary coil the magnetic field grows around the primary coil as the current is established but then becomes stationary. Recall that electromagnetic induction occurs when there is relative motion between the magnetic field lines and a wire loop. Thus, an induced voltage occurs from a direct current (1) only for an instant when the current is established and the growing field moves across the secondary coil and (2) only for an instant when the current is turned off and the field collapses back across the secondary coil. In order to use dc in a transformer, the current must be continually interrupted to produce a changing magnetic field.

When an alternating current or a continually interrupted direct current is applied to the primary coil, the magnitude of the induced voltage in the secondary coil is proportional to the ratio of wire loops in the two coils. If they have the same number of loops, the primary coil produces just as many magnetic field lines as are intercepted by the secondary coil. In this case, the induced voltage in the secondary coil will be the same as the voltage in the primary coil. Suppose, however, that the secondary coil has one-tenth as many loops as the primary coil. This means that the secondary loops will cut one-tenth as many field lines as the primary coil produces. As a result, the induced voltage in the secondary coil will be one-tenth the voltage in the primary coil. This is called a *step-down transformer* since the voltage was stepped down in the secondary coil. On the other hand, more wire loops in the secondary coil will intercept more magnetic field lines. If the secondary coil has ten times *more* loops than the primary coil, then the voltage will be *increased* by a factor of 10. This is a *step-up transformer.* How much the voltage is stepped up or stepped down depends on the ratio of wire loops in the primary and secondary coils (figure 7.40). Note that the *volts per wire loop* are the same in each coil. The relationship is

$$\frac{\text{volts}_{\text{primary}}}{(\text{number of loops})_{\text{primary}}} = \frac{\text{volts}_{\text{secondary}}}{(\text{number of loops})_{\text{secondary}}}$$

or

$$\frac{V_p}{N_p} = \frac{V_s}{N_s}$$

equation 7.11

A step-up or step-down transformer steps up or steps down the *voltage* of an alternating current according to the ratio of wire loops in the primary and secondary coils. Assuming no losses in the transformer, the *power input* on the primary coil equals the *power output* on the secondary coil. Since $P = IV$, you can see that when the voltage is stepped up the current is correspondingly decreased, as

$$\text{power input} = \text{power output}$$

$$\text{watts input} = \text{watts output}$$

$$(\text{amps} \times \text{volts})_{\text{in}} = (\text{amps} \times \text{volts})_{\text{out}}$$

or

$$V_p I_p = V_s I_s$$

equation 7.12

Energy losses in transmission are reduced by stepping up the voltage. Recall that electrical resistance results in an energy loss and a corresponding absolute temperature increase in the conducting wire. If the current is large, there are many collisions between the moving electrons and positive ions of the wire, resulting in a large energy loss. Each collision takes energy from the

CONNECTIONS

Magnetic Fields and Instinctive Behavior

Do animals use the earth's magnetic fields to navigate? Since animals move from place to place to meet their needs it is useful to be able to return to a nest, water hole, den, or favorite feeding spot. This requires some sort of memory of their surroundings (a mental map) and a way of determining direction. Often it is valuable to have information about distance as well. Direction can be determined by such things as magnetic fields, identified landmarks, scent trails, or reference to the sun or stars. If the sun or stars are used for navigation, then some sort of time sense is also needed since these bodies move in the sky.

Instinctive behaviors are automatic, preprogrammed, and genetically determined. Such behaviors are found in a wide range of organisms from simple one-celled protozoans to complex vertebrates. These behaviors are performed correctly the first time without previous experience when the proper stimulus is given. A stimulus is some change in the internal or external environment of the organism that causes it to react. The reaction of the organism to the stimulus is called a response.

An organism can respond only to stimuli it can recognize. For example, it is difficult for us as humans to appreciate what the

world seems like to a bloodhound. The bloodhound is able to identify individuals by smell, whereas we have great difficulty detecting, let alone distinguishing, many odors. Some animals, like dogs, deer, and mice, are color-blind and are able to see only shades of gray. Others, such as honeybees, can see ultraviolet light, which is invisible to us. Some birds and other animals are able to detect the magnetic field of the earth.

There is evidence that some birds navigate by compass direction, that is, they fly as if they had a compass in their heads. They seem to be able to sense magnetic north. Their ability to sense magnetic fields has been proven at the U.S. Navy's test facility in Wisconsin. The weak magnetism radiated from this test site has changed the flight pattern of migrating birds, but it is yet to be proven that birds use the magnetism of the earth to guide their migration. Homing pigeons are famous for their ability to find their way home. They make use of a wide variety of clues, but it has been shown that one of the clues they use involves magnetism. Birds with tiny magnets glued to the sides of their heads were very poor navigators, while others with nonmagnetic objects attached to the sides of their heads did not lose their ability to navigate.

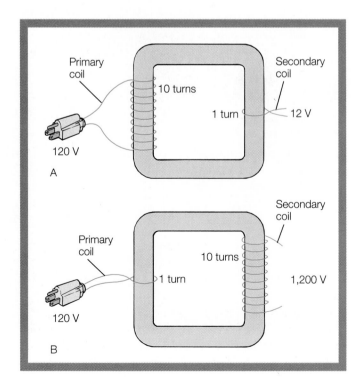

electric field, diverting it into increased kinetic energy of the positive ions and thus increased temperature of the conductor. The energy lost to resistance is therefore reduced by *lowering* the current, which is what a transformer does by increasing the voltage. Hence electric power companies step up the voltage of generated power for economical transmission. A step-up transformer at a power plant, for example, might step up the voltage from 22,000 volts to 500,000 volts for transmission across the country to a city. This step up in voltage correspondingly reduces the current, lowering the resistance losses to a more acceptable 4 or 5 percent over long distances. A step-down transformer at a substation near the city reduces the voltage to several thousand volts for transmission around the city. Step-down transformers reduce this voltage to 120 volts for transmission to three or four houses (figure 7.41).

FIGURE 7.40

(*A*) This step-down transformer has ten turns on the primary for each turn on the secondary and reduces the voltage from 120 V to 12 V. (*B*) This step-up transformer increases the voltage from 120 V to 1,200 V, since there are ten turns on the secondary to each turn on the primary.

A

FIGURE 7.41

Energy losses in transmission are reduced by increasing the voltage, so the voltage of generated power is stepped up at the power plant. (*A*) These transformers, for example, might step up the voltage from tens to hundreds of thousands of volts. After a stepdown transformer reduces the voltage at a substation, still another transformer (*B*) reduces the voltage to 120 volts for transmission to three or four houses.

B

SUMMARY

The first electrical phenomenon recognized was the charge produced by friction, which today is called *static electricity*. By the early 1900s, the *electron theory of charge* was developed from studies of the *atomic nature of matter*. These studies lead to the understanding that matter is made of *atoms*, which are composed of *negatively charged electrons* moving about a central *nucleus*, which contains *positively charged protons*. The two kinds of charges interact as *like charges produce a repellant force*, and *unlike charges produce an attractive force*. An object acquires an *electric charge* when it has an excess or deficiency of electrons, which is called an *electrostatic charge*.

A *quantity of charge* (q) is measured in units of *coulombs* (C), the charge equivalent to the transfer of 6.24×10^{18} charged particles such as the electron. The *fundamental charge* of an electron or proton is 1.60×10^{-19} coulomb. The *electrical forces* between two charged objects can be calculated from the relationship between the quantity of charge and the distance between two charged objects. The relationship is known as *Coulomb's law*.

A charged object in an electric field has *electric potential energy* that is related to the charge on the object and the work done to move it into a field of like charge. The resulting *electric potential difference* (PD) is a ratio of the work done (W) to move a quantity of charge (q). In units, a joule of work done to move a coulomb of charge is called a *volt*.

A flow of electric charge is called an *electric current* (I). A current requires some device, such as a generator or battery, to maintain a potential difference. The device is called a *voltage source*. An *electric circuit* contains (1) a voltage source, (2) a *continuous path* along which the current flows, and (3) a device such as a lamp or motor where work is done, called a *voltage drop*. Current (I) is measured as the *rate* of flow of charge, the quantity of charge (q) through a conductor in a period of time (t). The unit of current in coulomb/second is called an *ampere* or *amp* for short (A).

Current occurs in a conductor when a *potential difference* is applied and an *electric field* travels through the conductor at near the speed of light. The electrons drift *very slowly*, accelerated by the electric field. The field moves the electrons in one direction in a *direct current* (dc) and moves them back and forth in an *alternating current* (ac).

Materials have a property of opposing or reducing an electric current called *electrical resistance* (R). Resistance is a ratio between the potential difference (V) between two points and the resulting current (I), or $R = V/I$. The unit is called the *ohm* (Ω), and $1.00\ \Omega = 1.00$ volt/1.00 amp. The relationship between voltage, current, and resistance is called *Ohm's law*.

Disregarding the energy lost to resistance, the *work* done by a voltage source is equal to the work accomplished in electrical devices in a circuit. The *rate* of doing work is *power*, or *work per unit time*, $P = W/t$. *Electrical power* can be calculated from the relationship of $P = IV$, which gives the power unit of *watts*.

Magnets have two poles about which their attraction is concentrated. When free to turn, one pole moves to the north and the other to the south. The north-seeking pole is called the *north pole* and the south-seeking pole is called the *south pole*. *Like poles repel* one another and *unlike poles attract*.

The property of magnetism is *electric in origin*, produced by charges in motion. *Permanent magnets* have tiny regions called *magnetic domains*, each with its own north and south pole. An *unmagnetized* piece of iron has randomly arranged domains. When *magnetized*, the domains become aligned and contribute to the overall magnetic effect.

A *current-carrying wire* has magnetic field lines of closed, *concentric circles* that are at right angles to the length of wire. The *direction* of the magnetic field depends on the direction of the current. A coil of many loops is called a *solenoid* or *electromagnet*. The electromagnet is the working part in electrical meters, electromagnetic switches, and the electric motor.

When a loop of wire is moved in a magnetic field, or if a magnetic field is moved past a wire loop, a voltage is induced in the wire loop. The interaction is called *electromagnetic induction*. An electric generator is a rotating coil of wire in a magnetic field. The coil is rotated by mechanical energy, and electromagnetic induction induces a voltage, thus converting mechanical energy to electrical energy. A *transformer* steps up or steps down the voltage of an alternating current. The ratio of input and output voltage is determined by the number of loops in the primary and secondary coils. Increasing the voltage decreases the current, which makes long-distance transmission of electrical energy economically feasible.

Summary of Equations

7.1

$$\text{Quantity of charge} = (\text{number of electrons})\,(\text{electron charge})$$

$$q = ne$$

7.2

$$\text{Electrical force} = (\text{constant}) \times \frac{\text{charge on one object} \times \text{charge on second object}}{\text{distance between objects squared}}$$

$$F = k\frac{q_1 q_2}{d^2}$$

where $k = 9.00 \times 10^9$ newton·meters^2/coulomb^2

7.3

$$\text{Electric potential} = \frac{\text{work to create potential}}{\text{charge moved}}$$

$$PD = \frac{W}{q}$$

7.4

$$\text{Electric current} = \frac{\text{quantity of charge}}{\text{time}}$$

$$I = \frac{q}{t}$$

7.5

$$\text{Volts} = \text{current} \times \text{resistance}$$

$$V = IR$$

7.6

$$\text{Power} = \frac{\text{work}}{\text{time}}$$

$$P = \frac{W}{t}$$

A CLOSER LOOK | Solar Cells

You may be familiar with many solid-state devices such as calculators, computers, word processors, digital watches, VCRs, digital stereos, and camcorders. All of these are called solid-state devices because they use a solid material, such as the semiconductor silicon, in an electric circuit in place of vacuum tubes. Solid-state technology developed from breakthroughs in the use of semiconductors during the 1950s, and the use of thin pieces of silicon crystal is common in many electric circuits today.

A related technology also uses thin pieces of a semiconductor such as silicon but not as a replacement for a vacuum tube. This technology is concerned with photovoltaic devices, also called *solar cells,* that generate electricity when exposed to light (box figure 7.3). A solar cell is unique in generating electricity since it produces electricity directly, without moving parts or chemical reactions, and potentially has a very long lifetime. This reading is concerned with how a solar cell generates electricity.

Light is an electromagnetic wave in a certain wavelength range. This wave is a pulse of electric and magnetic fields that are locked together, moving through space as they exchange energy back and forth, regenerating each other in an endless cycle until the wave is absorbed, giving up its energy. Light also has the ability to give its energy to an electron, knocking it out of a piece of metal. This phenomenon, called the *photoelectric effect,* is described in the next chapter.

Solid-state technology is mostly concerned with crystals and electrons in crystals. A *crystal* is a solid with an ordered, three-dimensional arrangement of atoms. A crystal of common table salt, sodium chloride, has an ordered arrangement that produces a cubic structure. You can see this cubic structure of table salt if you look closely at a few grains. The regular geometric arrangement of atoms (or ions) in a crystal is called the *lattice,* or framework, of the crystal structure.

A highly simplified picture of a crystal of silicon has each silicon atom bonded with four other silicon atoms, with each pair of

A   B

BOX FIGURE 7.3

Solar cells are economical in remote uses such as (*A*) navigational aids and (*B*) communications. The solar panels in both of these examples are oriented toward the south.

atoms sharing two electrons. Normally, this ties up all the electrons and none are free to move and produce an electric current. This happens in the dark in silicon crystals, and the silicon crystal is an insulator. Light, however, can break electrons free in a silicon crystal, so that they can move in a current. Silicon is therefore a semiconductor.

The conducting properties of silicon can be changed by *doping,* that is, artificially forcing atoms of other elements into the crystal lattice. Phosphorus, for example, has five electrons in its outermost shell compared to the four in a silicon atom. When phosphorus atoms replace silicon atoms in the crystal lattice, there are extra electrons not tied up in the two electron bonds. The extra electrons move easily through the crystal lattice, carrying a charge. Since the phosphorus-doped silicon carries a negative charge it is called an *n-type* semiconductor. The n means negative charge carrier.

A silicon crystal doped with boron will have atoms in the lattice with only three electrons in the outermost shell. This results in a deficiency, that is, electron "holes" that act as positive charges. A hole can move as an electron is attracted to it, but it leaves another hole elsewhere, from where it moved. Thus, a flow of electrons in one direction is equivalent to a flow of holes in the opposite direction. A hole, therefore, behaves as a positive charge. Since the boron-doped silicon carries a positive charge it is called a *p-type* semiconductor. The p means positive charge carrier.

The basic operating part of a silicon solar cell is typically an 8 cm wide and 3×10^{-1} mm (about one-hundredth of an inch) thick wafer cut from a silicon crystal. One side of the wafer is doped with boron to make p-silicon, and the other side is doped with phosphorus to make n-silicon. The place of contact

—*Continued top of next page*

Continued—

between the two is called the p-n junction, which creates a *cell barrier*. The cell barrier forms as electrons are attracted from the n-silicon to the holes in the p-silicon. This creates a very thin zone of negatively charged p-silicon and positively charged n-silicon (box figure 7.4). Thus, an internal electric field is established at the p-n junction, and the field is the cell barrier. The barrier is about 3×10^{-5} mm (about one-millionth of an inch) thick.

A metal base plate is attached to the p-silicon side of the wafer, and a grid of metal contacts to the n-silicon side for electrical contacts. The grid is necessary to allow light into the cell. The entire cell is then coated with a transparent plastic covering.

The cell is thin, and light can penetrate through the p-n junction. Light impacts the p-silicon, freeing electrons. Low-energy free electrons might combine with a hole, but high-energy electrons cross the cell barrier into the n-silicon. The electron loses some of its energy, and the barrier prevents it from returning, creating an excess negative charge in the n-silicon and a positive charge in the p-silicon. This establishes a potential of about 0.5 volt, which will drive a 2 amp current (8 cm cell) through a circuit connected to the electrical contacts.

Solar cells are connected different ways for specific electrical energy requirements in about a 1 meter arrangement called a *module.* Several modules are used to make a solar *panel,* and panels are the units used to design a solar cell *array.*

Today, solar cells are essentially handmade and are economical only in remote

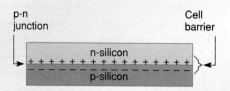

BOX FIGURE 7.4

The cell barrier forms at the p-n junction between the n-silicon and the p-silicon. The barrier creates a "one-way" door that accumulates negative charges in the n-silicon.

power uses (navigational aids, communications, or irrigation pumps) and in consumer specialty items (solar-powered watches and calculators). Research continues on finding methods of producing highly efficient, highly reliable solar cells that are affordably priced.

7.7

$$\text{Electrical power} = \frac{(\text{electric potential})(\text{charge})}{\text{time}}$$

$$P = \frac{PDq}{t}$$

7.8

$$\text{Electrical power} = (\text{amps})(\text{volts})$$

$$P = IV$$

7.9

$$\text{Electric current} \times \text{electric potential} \times \text{time} = \text{electrical work}$$

$$(IV)(t) = W$$

7.10

$$(\text{work})(\text{rate}) = \text{cost}$$

7.11

$$\frac{\text{volts}_{\text{primary}}}{(\text{number of loops})_{\text{primary}}} = \frac{\text{volts}_{\text{secondary}}}{(\text{number of loops})_{\text{secondary}}}$$

$$\frac{V_p}{N_p} = \frac{V_s}{N_s}$$

7.12

$$(\text{volts}_{\text{primary}})(\text{current}_{\text{primary}}) = (\text{volts}_{\text{secondary}})(\text{current}_{\text{secondary}})$$

$$V_P I_P = V_s I_s$$

KEY TERMS

alternating current (p. 157)
amp (p. 155)
ampere (p. 155)
coulomb (p. 150)
Coulomb's law (p. 150)
direct current (p. 157)
electric circuit (p. 154)
electric current (p. 153)
electric field (p. 151)
electric generator (p. 173)
electrical conductors (p. 149)
electrical resistance (p. 157)
electromagnet (p. 169)

electromagnetic induction (p. 171)
electrostatic charge (p. 148)
force field (p. 151)
fundamental charge (p. 150)
magnetic domain (p. 165)
magnetic field (p. 163)
magnetic poles (p. 163)
ohm (p. 157)
Ohm's law (p. 159)
transformer (p. 174)
volt (p. 153)
watt (p. 160)

APPLYING THE CONCEPTS

1. How does an electron acquire a negative charge?
   a. from an imbalance of subatomic particles
   b. by induction or contact with charged objects
   c. from the friction of certain objects rubbing together
   d. Charge is a fundamental property of an electron.

2. An object that acquires an excess of electrons becomes a (an)
   a. ion.
   b. negatively charged object.
   c. positively charged object.
   d. electrical conductor.

3. Which of the following is most likely to acquire an electrostatic charge?
   a. electrical conductor
   b. electrical nonconductor
   c. Both are equally likely.
   d. None of the above is correct.

4. A quantity of electric charge is measured in a unit called a (an)
   a. coulomb.
   b. volt.
   c. amp.
   d. watt.

5. The unit that describes the potential difference that occurs when a certain amount of work is used to move a certain quantity of charge is called the
   a. ohm.
   b. volt.
   c. amp.
   d. watt.

6. Which of the following units are measures of *rates?*
   a. amp and volt
   b. coulomb and joule
   c. volt and watt
   d. amp and watt

7. An electric current is measured in units of
   a. coulomb.
   b. volt.
   c. amp.
   d. watt.

8. If you consider a current to be positive charges that flow from the positive to the negative terminal of a battery, you are using which description of a current?
   a. electron current
   b. conventional current
   c. proton current
   d. alternating current

9. In an electric current the electrons are moving
   a. at a very slow rate.
   b. at the speed of light.
   c. faster than the speed of light.
   d. at a speed described as "Warp 8."

10. In which of the following currents is there no electron movement from one end of a conducting wire to the other end?
    a. electron current
    b. direct current
    c. alternating current
    d. None of the above is correct.

11. If you multiply amps × volts, the answer will be in units of
    a. resistance.
    b. work.
    c. current.
    d. power.

12. The unit of resistance is the
    a. watt.
    b. ohm.
    c. amp.
    d. volt.

13. Which of the following is a measure of electrical work?
    a. kilowatt
    b. C
    c. kWhr
    d. C/s

14. Compared to a thick wire, a thin wire of the same length, material, and temperature has
    a. less electrical resistance.
    b. more electrical resistance.
    c. the same electrical resistance.
    d. None of the above is correct.

15. If an electric charge is somehow suddenly neutralized, the electric field that surrounds it will
    a. immediately cease to exist.
    b. collapse inward at some speed.
    c. continue to exist until neutralized.
    d. move off into space until it finds another charge.

16. The earth's north magnetic pole
    a. is located at the geographic North Pole.
    b. is a magnetic south pole.
    c. has always had the same orientation.
    d. None of the above is correct.

17. A permanent magnet has magnetic properties because
    a. the magnetic fields of its electrons are balanced.
    b. of an accumulation of monopoles in the ends.
    c. the magnetic domains are aligned.
    d. All of the above are correct.

18. A current-carrying wire has a magnetic field around it because
    a. a moving charge produces a magnetic field of its own.
    b. the current aligns the magnetic domains in the metal of the wire.
    c. the metal was magnetic before the current was established and the current enhanced the magnetic effect.
    d. None of the above is correct.

19. If you reverse the direction that a current is running in a wire, the magnetic field around the wire
    a. is oriented as it was before.
    b. is oriented with an opposite north direction.
    c. flips to become aligned parallel to the length of the wire.
    d. ceases to exist.
20. A step-up transformer steps up (the)
    a. power.
    b. current.
    c. voltage.
    d. All of the above are correct.

Answers

1. d 2. b 3. b 4. a 5. b 6. d 7. c 8. b 9. a 10. c 11. d 12. b 13. c 14. b 15. b
16. b 17. c 18. a 19. b 20. c

QUESTIONS FOR THOUGHT

1. Explain why a balloon that has been rubbed sticks to a wall for a while.
2. Explain what is happening when you walk across a carpet and receive a shock when you touch a metal object.
3. Why does a positively or negatively charged object have multiples of the fundamental charge?
4. Explain how you know that it is an electric field, not electrons, that moves rapidly through a circuit.
5. Is a kWhr a unit of power or a unit of work? Explain.
6. What is the difference between ac and dc?
7. What is a magnetic pole? How are magnetic poles named?
8. How is an unmagnetized piece of iron different from the same piece of iron when it is magnetized?
9. Explain why the electric utility company increases the voltage of electricity for long-distance transmission.
10. Describe how an electric generator is able to generate an electric current.
11. Why does the north pole of a magnet point to the geographic North Pole if like poles repel?
12. Explain what causes an electron to move toward one end of a wire when the wire is moved across a magnetic field.

PARALLEL EXERCISES

The exercises in groups A and B cover the same concepts. Solutions to group A exercises are located in appendix D.

Group A

1. A rubber balloon has become negatively charged from being rubbed with a wool cloth, and the charge is measured as 1.00×10^{-14} C. According to this charge, the balloon contains an excess of how many electrons?
2. One rubber balloon with a negative charge of 3.00×10^{-14} C is suspended by a string and hangs 2.00 cm from a second rubber balloon with a negative charge of 2.00×10^{-12} C. (a) What is the direction of the force between the balloons? (b) What is the magnitude of the force?
3. A dry cell does 7.50 J of work through chemical energy to transfer 5.00 C between the terminals of the cell. What is the electric potential between the two terminals?
4. An electric current through a wire is 6.00 C every 2.00 s. What is the magnitude of this current?
5. A 1.00 A electric current corresponds to the charge of how many electrons flowing through a wire per second?
6. A current of 4.00 A flows through a toaster connected to a 120.0 V circuit. What is the resistance of the toaster?
7. What is the current in a 60.0 Ω resistor when the potential difference across it is 120.0 V?
8. A lightbulb with a resistance of 10.0 Ω allows a 1.20 A current to flow when connected to a battery. (a) What is the voltage of the battery? (b) What is the power of the lightbulb?
9. A small radio operates on 3.00 V and has a resistance of 15.0 Ω. At what rate does the radio use electric energy?

Group B

1. An inflated rubber balloon is rubbed with a wool cloth until an excess of a billion electrons is on the balloon. What is the magnitude of the charge on the balloon?
2. What is the force between two balloons with a negative charge of 1.6×10^{-10} C if the balloons are 5.0 cm apart?
3. How much energy is available from a 12 V storage battery that can transfer a total charge equivalent to 100,000 C?
4. A wire carries a current of 2.0 A. At what rate is the charge flowing?
5. What is the magnitude of the least possible current that could theoretically exist?
6. A current of 0.83 A flows through a lightbulb in a 120 V circuit. What is the resistance of this lightbulb?
7. What is the voltage across a 60.0 Ω resistor with a current of 3⅓ amp?
8. A 10.0 Ω lightbulb is connected to a 12.0 V battery. (a) What current flows through the bulb? (b) What is the power of the bulb?
9. A lightbulb designed to operate in a 120.0 V circuit has a resistance of 192 Ω. At what rate does the bulb use electric energy?
10. What is the monthly energy cost of leaving a 60 W bulb on continuously if electricity costs 10¢ per kWhr?
11. An electric motor draws a current of 11.5 A in a 240 V circuit. (a) What is the power of this motor in W? (b) How many hp is this?

10. A 1,200 W hair dryer is operated on a 120 V circuit for 15 min. If electricity costs $0.10/kWhr, what was the cost of using the blow dryer?

11. An automobile starter rated at 2.00 hp draws how many amps from a 12.0 V battery?

12. An average-sized home refrigeration unit has a 1/3 hp fan motor for blowing air over the inside cooling coils, a 1/3 hp fan motor for blowing air over the outside condenser coils, and a 3.70 hp compressor motor. (a) All three motors use electric energy at what rate? (b) If electricity costs $0.10/kWhr, what is the cost of running the unit per hour? (c) What is the cost for running the unit 12 hours a day for a 30 day month?

13. A 15 ohm toaster is turned on in a circuit that already has a 0.20 hp motor, three 100 W lightbulbs, and a 600 W electric iron that are on. Will this trip a 15 A circuit breaker? Explain.

14. A power plant generator produces a 1,200 V, 40 A alternating current that is fed to a step-up transformer before transmission over the high lines. The transformer has a ratio of 200 to 1 wire loops. (a) What is the voltage of the transmitted power? (b) What is the current?

15. A step-down transformer has an output of 12 V and 0.5 A when connected to a 120 V line. Assuming no losses: (a) What is the ratio of primary to secondary loops? (b) What current does the transformer draw from the line? (c) What is the power output of the transformer?

16. A step-up transformer on a 120 V line has 50 loops on the primary and 150 loops on the secondary, and draws a 5.0 A current. Assuming no losses: (a) What is the voltage from the secondary? (b) What is the current from the secondary? (c) What is the power output?

12. A swimming pool requiring a 2.0 hp motor to filter and circulate the water runs for 18 hours a day. What is the monthly electrical cost for running this pool pump if electricity costs 10¢ per kWhr?

13. Is it possible for two people to simultaneously operate 1,300 W hair dryers on the same 120 V circuit without tripping a 15 A circuit breaker? Explain.

14. A step-up transformer has a primary coil with 100 loops and a secondary coil with 1,500 loops. If the primary coil is supplied with a household current of 120 V and 15 A, (a) what voltage is produced in the secondary circuit? (b) What current flows in the secondary circuit?

15. The step-down transformer in a local neighborhood reduces the voltage from a 7,200 V line to 120 V. (a) If there are 125 loops on the secondary, how many are on the primary coil? (b) What current does the transformer draw from the line if the current in the secondary is 36 A? (c) What are the power input and output?

16. A step-down transformer connected to a 120 V electric generator has 30 loops on the primary for each loop in the secondary. (a) What is the voltage of the secondary? (b) If the transformer has a 90 A current in the primary, what is the current in the secondary? (c) What are the power input and output?

This fiber optics bundle carries pulses of light from an infrared laser to carry much more information than could be carried by electrons moving through wires. This is part of a dramatic change underway, a change that will first find a hybrid "optoelectronics" replacing the more familiar "electronics" of electrons and wires.

CHAPTER | Eight

Light

You use light and your eyes more than any other sense to learn about your surroundings. All of your other senses—touch, taste, sound, and smell—involve matter, but the most information is provided by light. Yet, light seems more mysterious than matter. You can study matter directly, measuring its dimensions, taking it apart, and putting it together to learn about it. Light, on the other hand, can only be studied indirectly in terms of how it behaves (figure 8.1). Once you understand its behavior, you know everything there is to know about light. Anything else is thinking about what the behavior means.

The behavior of light has stimulated thinking, scientific investigations, and debate for hundreds of years. The investigations and debate have occurred because light cannot be directly observed, which makes the exact nature of light very difficult to pin down. For example, you know that light moves energy from one place to another place. You can feel energy from the sun as sunlight warms you, and you know that light has carried this energy across millions of miles of empty space. The ability of light to move energy like this could be explained (1) as energy transported by waves, just as sound waves carry energy from a source, or (2) as the kinetic energy of a stream of moving particles, which give up their energy when they strike a surface. The movement of energy from place to place could be explained equally well by a wave model of light or by a particle model of light. When two possibilities exist like this in science, experiments are designed and measurements are made to support one model and reject the other. Light, however, presents a baffling dilemma. Some experiments provide evidence that light consists of waves and not a stream of moving particles. Yet other experiments provide evidence of just the opposite, that light is a stream of particles and not a wave. Evidence for accepting a wave or particle model seems to depend on which experiments are considered.

The purpose of using a model is to make new things understandable in terms of what is already known. When these new things concern light, three models are useful in visualizing separate behaviors. Thus, the electromagnetic wave model will be used to describe how light is created at a source. Another model, a model of light as a ray, a small beam of light, will be used to discuss some common properties of light such as reflection and the refraction, or bending, of light. Finally, properties of light that provide evidence for a particle model will be discussed before ending with a discussion of the present understanding of light.

SOURCES OF LIGHT

The sun, stars, lightbulbs, and burning materials all give off light. When something produces light it is said to be **luminous.** The sun is a luminous object that provides almost all of the *natural* light on the earth. A small amount of light does reach the earth from the stars but not really enough to see by on a moonless night. The moon and planets shine by reflected light and do not produce their own light, so they are not luminous.

Burning has been used as a source of *artificial* light for thousands of years. A wood fire and a candle flame are luminous because of their high temperatures. When visible light is given off as a result of high temperatures, the light source is said to be **incandescent.** A flame from any burning source, an ordinary lightbulb, and the sun are all incandescent sources because of high temperatures.

How do incandescent objects produce light? One explanation is given by the electromagnetic wave model. This model describes a relationship between electricity, magnetism, and light. The model pictures an electromagnetic wave as forming whenever an electric charge is *accelerated* by some external force. The acceleration produces a wave consisting of electrical and magnetic fields that become isolated from the accelerated

charge, moving off into space. As the wave moves through space, the two fields exchange energy back and forth, continuing on until they are absorbed by matter and give up their energy.

The frequency of an electromagnetic wave depends on the acceleration of the charge; the greater the acceleration, the higher the frequency of the wave that is produced. The complete range of frequencies is called the *electromagnetic spectrum.* The spectrum ranges from radio waves at the low-frequency end of the spectrum to gamma rays at the high-frequency end. Visible light occupies only a small part of the middle portion of the complete spectrum.

Visible light is emitted from incandescent sources at high temperatures, but actually electromagnetic radiation is given off from matter at *any* temperature. This radiation is called **blackbody radiation,** which refers to an idealized material (the *blackbody*) that perfectly absorbs and perfectly emits electromagnetic radiation. From the electromagnetic wave model, the radiation originates from the acceleration of charged particles near the surface of an object. The frequency of the blackbody radiation is determined by the energy available for accelerating charged particles, that is, the temperature of the object. Near absolute zero, there is little energy available and no radiation is given off. As the temperature of an object is increased, more energy is available,

FIGURE 8.1

Light, sounds, and odors can identify the pleasing environment of this garden, but light provides the most information. Sounds and odors can be identified and studied directly, but light can only be studied indirectly, that is, in terms of how it behaves. As a result, the behavior of light has stimulated thinking, scientific investigations, and debate for hundreds of years. Perhaps you have wondered about light and its behaviors. What is light?

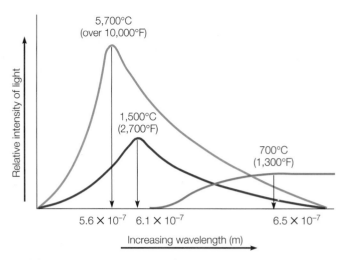

FIGURE 8.2

Three different objects emitting blackbody radiation at three different temperatures. The intensity of blackbody radiation increases with increasing temperature, and the peak wavelength emitted shifts toward shorter wavelengths.

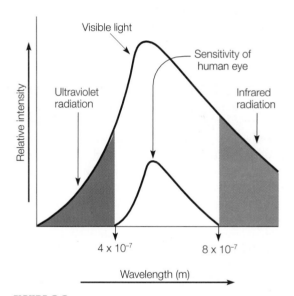

FIGURE 8.3

Sunlight is about 9 percent ultraviolet radiation, 40 percent visible light, and 51 percent infrared radiation before it travels through the earth's atmosphere.

and this energy is distributed over a range of values, so more than one frequency of radiation is emitted. A graph of the frequencies emitted from the range of available energy is thus somewhat bell-shaped. The steepness of the curve and the position of the peak depend on the temperature (figure 8.2). As the temperature of an object increases, there is an increase in the *amount* of radiation given off, and the peak radiation emitted progressively *shifts* toward higher and higher frequencies.

At room temperature the radiation given off from an object is in the infrared region, invisible to the human eye. When the temperature of the object reaches about 700°C (about 1,300°F), the peak radiation is still in the infrared region, but the peak has shifted enough toward the higher frequencies that a little visible light is emitted as a dull red glow. As the temperature of the object continues to increase, the amount of radiation increases, and the peak continues to shift toward shorter wavelengths. Thus, the object begins to glow brighter, and the color changes from red, to orange, to yellow, and eventually to white. The association of this color change with temperature is noted in the referent description of an object being "red hot," "white hot," and so forth.

The incandescent flame of a candle or fire results from the blackbody radiation of carbon particles in the flame. At a blackbody temperature of 1,500°C (about 2,700°F), the carbon particles emit visible light in the red to yellow frequency range. The tungsten filament of an incandescent lightbulb is heated to about 2,200°C (about 4,000°F) by an electric current. At this temperature, the visible light emitted is in the reddish, yellow-white range.

The radiation from the sun, or sunlight, comes from the sun's surface, which has a temperature of about 5,700°C (about 10,000°F). As shown in figure 8.3, the sun's radiation has a broad spectrum centered near the yellow-green wavelength. Your eye is most sensitive to this wavelength of sunlight. The spectrum of sunlight before it travels through the earth's atmosphere is infrared (about 51 percent), visible light (about 40 percent), and ultraviolet (about 9 percent). Sunlight originated as energy released from nuclear reactions in the sun's core. This energy requires about a million years to work its way up to the surface. At the surface, the energy from the core accelerates charged

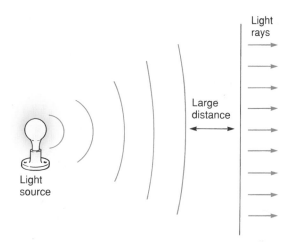

Light rays

Large distance

Light source

FIGURE 8.4

Light rays are perpendicular to a wave front. A wave front that has traveled a long distance is a plane wave front, and its rays are parallel to each other. The rays show the direction of the wave motion.

particles, which then emit light like tiny antennas. The sunlight requires about eight minutes to travel the distance from the sun's surface to the earth.

PROPERTIES OF LIGHT

You can see luminous objects from the light they emit, and you can see nonluminous objects from the light they reflect, but you cannot see the path of the light itself. For example, you cannot see a flashlight beam unless you fill the air with chalk dust or smoke. The dust or smoke particles reflect light, revealing the path of the beam. This simple observation must be unknown to the makers of science fiction movies, since they always show visible laser beams zapping through the vacuum of space.

Some way to represent the invisible travels of light is needed in order to discuss some of its properties. Throughout history a **light ray model** has been used to describe the travels of light. The meaning of this model has changed over time, but it has always been used to suggest that "something" travels in *straight-line paths*. The light ray is a line that is drawn to represent the straight-line travel of light. It is often used with a discussion of waves, since many properties of light can be explained in terms of the behavior of waves. The light ray is, nonetheless, a line that represents an imaginary thin beam of light (figure 8.4). A line is drawn to represent this imaginary beam to illustrate the law of reflection (as from a mirror) and the law of refraction (as through a lens). There are limits to using a light ray for explaining some properties of light, but it works very well in explaining mirrors, prisms, and lenses.

Light Interacts with Matter

A ray of light travels in a straight line from a source until it encounters some object or particles of matter (figure 8.5). What happens next depends on several factors, including (1) the smoothness of the surface, (2) the nature of the material, and (3) the angle at which the light ray strikes the surface.

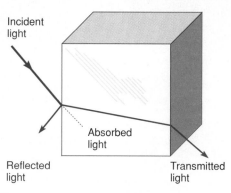

Incident light

Absorbed light

Reflected light

Transmitted light

FIGURE 8.5

Light that interacts with matter can be reflected, absorbed, or transmitted through transparent materials. Any combination of these interactions can take place, but a particular substance is usually characterized by what it mostly does to light.

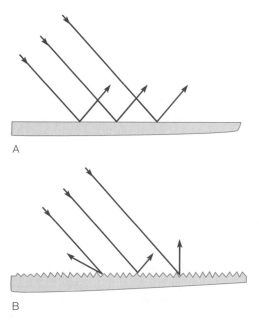

A

B

FIGURE 8.6

(*A*) Rays reflected from a perfectly smooth surface are parallel to each other. (*B*) Diffuse reflection from a rough surface causes rays to travel in many random directions.

The *smoothness* of the surface of an object can range from perfectly smooth to extremely rough. If the surface is perfectly smooth, rays of light undergo *reflection*, leaving the surface parallel to each other. A mirror is a good example of a very smooth surface that reflects light this way (figure 8.6A). If a surface is not smooth, the light rays are reflected in many random directions as *diffuse reflection* takes place (figure 8.6B). Rough and irregular surfaces and dust in the air make diffuse reflections. It is diffuse reflection that provides light in places not in direct lighting, such as under a table or under a tree. Such shaded areas would be very dark without diffuse reflection of light.

Some *materials* allow much of the light that falls on them to move through the material without being reflected. Materials

FIGURE 8.7

Light travels in a straight line, and the color of an object depends on which wavelengths of light the object reflects. Each of these flowers absorbs the colors of white light and reflects the color that you see.

A

B

FIGURE 8.8

(A) A one-way mirror reflects most of the light that strikes. (B) It also transmits some light to a person behind the mirror in a darkened room.

that allow transmission of light through them are called *transparent.* Glass and clear water are examples of transparent materials. Many materials do not allow transmission of any light and are called *opaque.* Opaque materials reflect light, absorb light, or some combination of partly absorbing and partly reflecting light (figure 8.7). The light that is reflected varies with wavelength and gives rise to the perception of color, which will be discussed shortly. Absorbed light gives up its energy to the material and may be reemitted at a different wavelength, or it may simply show up as a temperature increase.

The *angle* of the light ray to the surface and the nature of the material determine if the light is absorbed, transmitted through a transparent material, or reflected. Vertical rays of light, for example, are mostly transmitted through a transparent material with some reflection and some absorption. If the rays strike the surface at some angle, however, much more of the light is reflected, bouncing off the surface. Thus, the glare of reflected sunlight is much greater around a body of water in the late afternoon than when the sun is directly overhead.

Light that interacts with matter is reflected, transmitted, or absorbed, and all combinations of these interactions are possible. Materials are usually characterized by which of these interactions they *mostly* do, but this does not mean that other interactions are not occurring too. For example, a window glass is usually characterized as a transmitter of light. Yet, the glass always *reflects* about 4 percent of the light that strikes it. The reflected light usually goes unnoticed during the day because of the bright light that is transmitted from the outside. When it is dark outside you notice the reflected light as the window glass now appears to act much like a mirror. A one-way mirror is another example of both reflection and transmission occurring (figure 8.8). A mirror is usually characterized as a reflector of light. A one-way mirror, however, has a very thin silvering that reflects most of the light but still transmits a little. In a lighted room, a one-way mirror appears to reflect light just as any other mirror does. But a person behind the mirror in a dark room can see into the lighted room by means of the transmitted light. Thus, you know that this mirror transmits as well as reflects light. One-way mirrors are used to unobtrusively watch for shoplifters in many businesses.

Reflection

Most of the objects that you see are visible from diffuse reflection. For example, consider some object such as a tree that you see during a bright day. Each *point* on the tree must reflect light in all directions, since you can see any part of the tree from any angle (figure 8.9). As a model, think of bundles of light rays entering your eye, which enable you to see the tree. This means that you can see any part of the tree from any angle because different bundles of reflected rays will enter your eye from different parts of the tree.

Groups of rays move from each point on the tree, traveling in all directions. Figure 8.10 shows a two-dimensional ray representation of this diffuse reflection. If you consider just two of the light rays, you will notice that they are spreading farther and farther apart as they travel. You are not aware of it, but at a relatively close distance the rate of spreading carries information that your eyes and brain interpret in terms of distance. At some

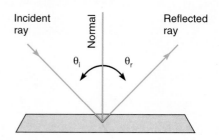

FIGURE 8.11
The law of reflection states that the angle of incidence (θ_i) is equal to the angle of reflection (θ_r). Both angles are measured from the *normal,* a reference line drawn perpendicular to the surface at the point of reflection.

FIGURE 8.9
Bundles of light rays are reflected diffusely in all directions from every point on an object. Only a few light rays are shown from only one point on the tree in this illustration. The light rays that move to your eyes enable you to see the particular point from which they were reflected.

FIGURE 8.10
Adjacent light rays spread farther and farther apart after reflecting from a point. Close to the point, the rate of spreading is great. At a great distance, the rays are almost parallel. The rate of ray spreading carries information about distance.

the reflecting surface and is located at the point where the incident ray struck the surface. This line is called the *normal.* The angle between the incident ray and the normal is called the *angle of incidence,* θ_i, and the angle between the reflected ray and the normal is called the *angle of reflection,* θ_r (figure 8.11). The *law of reflection,* which was known to the ancient Greeks, is that the *angle of incidence equals the angle of reflection,* or

$$\theta_i = \theta_r$$

equation 8.1

Figure 8.12 shows how the law of reflection works when you look at a flat mirror. Light is reflected from all points on the box, and of course only the rays that reach your eyes are detected. These rays are reflected according to the law of reflection, with the angle of reflection equaling the angle of incidence. If you move your head slightly, then a different bundle of rays reaches your eyes. Of all the bundles of rays that reach your eyes, only two rays from a point are shown in the illustration. After these two rays are reflected, they continue to spread apart at the same rate that they were spreading before reflection. Your eyes and brain do not know that the rays have been reflected, and the diverging rays appear to come from behind the mirror, as the dashed lines show. The image, therefore, appears to be the same distance *behind* the mirror as the box is from the front of the mirror. Thus, a mirror image is formed where the rays of light *appear* to originate. This is called a **virtual image.** A virtual image is the result of your eyes' and brain's interpretations of light rays, not actual light rays originating from an image. Light rays that do originate from the other kind of image are called a **real image.** A real image is like the one displayed on a movie screen, with light originating from the image. A virtual image cannot be displayed on a screen, since it results from an interpretation.

Curved mirrors are either *concave,* with the center part curved inward, or *convex,* with the center part bulging outward. A concave mirror can be used to form an enlarged virtual image, such as a shaving or makeup mirror, or it can be used to form a real image, as in a reflecting telescope. Convex mirrors, for example, the mirrors on the sides of trucks and vans, are often used to increase the field of vision. Convex mirrors are also used above an aisle in a store to show a wide area.

great distance from a source of light, only rays that are almost parallel will reach your eye. Light rays from great distances, such as the distance to the sun or more, are almost parallel. Overall, the rate at which groups of rays are spreading or their essentially parallel orientation carries information about distances.

Light rays that are diffusely reflected move in all possible directions, but rays that are reflected from a smooth surface, such as a mirror, leave the mirror in a definite direction. Suppose you look at a tree in a mirror. There is only one place on the mirror where you look to see any one part of the tree. Light is reflecting off the mirror from all parts of the tree, but the only rays that reach your eye are the rays that are reflected at a certain angle from the place where you look. The relationship between the light rays moving from the tree and the direction in which they are reflected from the mirror to reach your eyes can be understood by drawing three lines: (1) a line representing an original ray from the tree, called the *incident ray,* (2) a line representing a reflected ray, called the *reflected ray,* and (3) a reference line that is perpendicular to

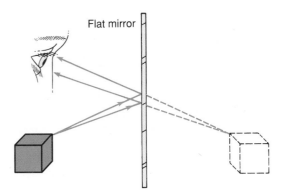

FIGURE 8.12

Light rays leaving a point on the block are reflected according to the law of reflection, and those reaching your eye are seen. After reflecting, the rays continue to spread apart at the same rate. You interpret this to be a block the same distance behind the mirror. You see a virtual image of the block, because light rays do not actually move from the image.

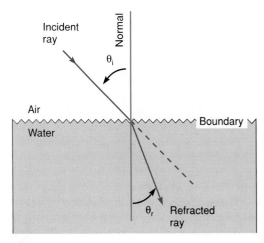

FIGURE 8.13

A ray diagram shows refraction at the boundary as a ray moves from air through water. Note that θ_i does not equal θ_r in refraction.

Refraction

You may have observed that an object that is partly in the air and partly in water appears to be broken, or bent, where the air and water meet. When a light ray moves from one transparent material to another, such as from water through air, the ray undergoes a change in the direction of travel at the boundary between the two materials. This change of direction of a light ray at the boundary is called **refraction.** The amount of change can be measured as an angle from the normal, just as it was for the angle of reflection. The incoming ray is called the *incident ray* as before, and the new direction of travel is called the *refracted ray.* The angles of both rays are measured from the normal (figure 8.13).

Refraction results from a *change in speed* when light passes from one transparent material into another. The speed of light in a vacuum is 3.00×10^8 m/s, but it is slower when moving through a transparent material. In water, for example, the speed of light is reduced to about 2.30×10^8 m/s. The speed of light has a magnitude that is specific for various transparent materials.

When light moves from one transparent material to another transparent material with a *slower* speed of light, the ray is refracted *toward* the normal (figure 8.14A). For example, light travels through air faster than through water. Light traveling from air into water is therefore refracted toward the normal as it enters the water. On the other hand, if light has a *faster* speed in the new material, it is refracted *away* from the normal. Thus, light traveling from water into the air is refracted away from the normal as it enters the air (figure 8.14B).

The magnitude of refraction depends on (1) the angle at which light strikes the surface and (2) the ratio of the speed of light in the two transparent materials. An incident ray that is perpendicular (90°) to the surface is not refracted at all. As the angle of incidence is increased, the angle of refraction is also increased. There is a limit, however, that occurs when the angle of refraction reaches 90°, or along the water surface.

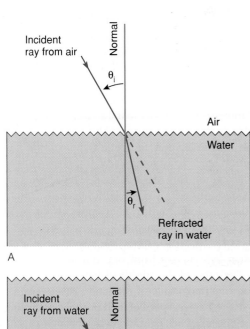

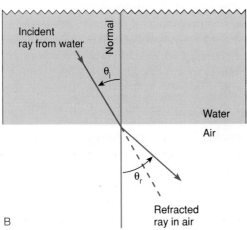

FIGURE 8.14

(*A*) A light ray moving to a new material with a slower speed of light is refracted toward the normal ($\theta_i > \theta_r$). (*B*) A light ray moving to a new material with a faster speed is refracted away from the normal ($\theta_i < \theta_r$).

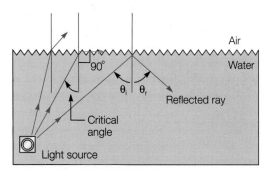

FIGURE 8.15

When the angle of incidence results in an angle of refraction of 90°, the refracted light ray is refracted along the water surface. The angle of incidence for a material that results in an angle of refraction of 90° is called the *critical angle*. When the incident ray is at this critical angle or greater, the ray is reflected internally. The critical angle for water is about 49°, and for a diamond it is about 25°.

Figure 8.15 shows rays of light traveling from water to air at various angles. When the incident ray is about 49°, the angle of refraction that results is 90°, along the water surface. This limit to the angle of incidence that results in an angle of refraction of 90° is called the *critical angle* for a water-to-air surface (figure 8.15). At any incident angle greater than the critical angle, the light ray does not move from the water to the air but is *reflected* back from the surface as if it were a mirror. This is called **total internal reflection** and implies that the light is trapped inside if it arrived at the critical angle or beyond. Faceted transparent gemstones such as the diamond are brilliant because they have a small critical angle and thus reflect much light internally. Total internal reflection is also important in fiber optics, as discussed in the reading at the end of this chapter.

As was stated earlier, refraction results from a change in speed when light passes from one transparent material into another. The ratio of the speeds of light in the two materials determines the magnitude of refraction at any given angle of incidence. The greatest speed of light possible, according to current theory, occurs when light is moving through a vacuum. The speed of light in a vacuum is accurately known to nine decimals but is usually rounded to 3.00×10^8 m/s for general discussion. The speed of light in a vacuum is a very important constant in physical science, so it is given a symbol of its own, c. The ratio of c to the speed of light in some transparent material, v, is called the **index of refraction,** n, of that material or

$$n = \frac{c}{v}$$

equation 8.2

The indexes of refraction for some substances are listed in table 8.1. The values listed are constant physical properties and can be used to identify a specific substance. Note that a larger value means a greater refraction at a given angle. Of the materials listed, diamond refracts light the most and air the least. The index for air is nearly 1, which means that light is slowed only slightly in air.

TABLE 8.1	Index of refraction
Substance	**$n = c/v$**
Glass	1.50
Diamond	2.42
Ice	1.31
Water	1.33
Benzene	1.50
Carbon tetrachloride	1.46
Ethyl alcohol	1.36
Air (0°C)	1.00029
Air (30°C)	1.00026

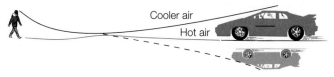

FIGURE 8.16

Mirages are caused by hot air near the ground refracting, or bending light rays upward into the eyes of a distant observer. The observer believes he is seeing an upside down image reflected from water on the highway.

Note that table 8.1 shows that colder air at 0°C (32°F) has a higher index of refraction than warmer air at 30°C (86°F), which means that light travels faster in warmer air. This difference explains the "wet" highway that you sometimes see at a distance in the summer. The air near the road is hotter on a clear, calm day. Light rays traveling toward you in this hotter air are refracted upward as they enter the cooler air. Your brain interprets this refracted light as *reflected* light. Light traveling downward from other cars is also refracted upward toward you, and you think you are seeing cars "reflected" from the wet highway (figure 8.16). When you reach the place where the "water" seemed to be, it disappears, only to appear again farther down the road.

Sometimes convection currents produce a mixing of warmer air near the road with the cooler air just above. This mixing refracts light one way, then the other, as the warmer and cooler air mix. This produces a shimmering or quivering that some people call "seeing heat." They are actually seeing changing refraction, which is a *result* of heating and convection. In addition to causing distant objects to quiver, the same effect causes the point source of light from stars to appear to twinkle. The light from closer planets does not twinkle because the many light rays from the disklike sources are not refracted together as easily as the fewer rays from the point sources of stars. The light from planets will appear to quiver, however, if the atmospheric turbulence is great.

Dispersion and Color

Electromagnetic waves travel with the speed of light with a whole spectrum of waves of various frequencies and wavelengths. The

FIGURE 8.17
The flowers appear to be red because they reflect light in the 7.9×10^{-7} m to 6.2×10^{-7} m range of wavelengths.

CONNECTIONS

Diamond Quality

The beauty of a diamond depends largely on optical properties, including degree of refraction, but how well the cutter shapes the diamond determines the brilliance and reflection of light from a given stone. The "finish" describes the precision of facet placement on a cut stone, but overall quality involves much more. There are four factors that are generally used to determine the quality of a cut diamond. These are sometimes called the four C's of Carat, Color, Clarity, and Cut.

Carat is a measure of the weight of a diamond. When measuring diamonds (as opposed to gold), one metric carat is equivalent to 0.2 g, so a 5 carat diamond would have a weight of 1 g. Diamonds are weighed in to the nearest hundredth of a carat, and 100 points equal 1 carat. Thus a 3 point diamond weighs 3/100 of a carat, or about 0.006 g.

Color is a measure of how much (or how little) color a diamond has. The color grade is based on the color of the body of a diamond, not the color of light dispersion from the stone. The color grades range from colorless, near colorless, faint yellow, very light yellow, and down to light yellow. It is measured by comparing a diamond to a standard of colors, but the color of a diamond weighing one-half carat or less is often difficult to determine.

Clarity is a measure of inclusions (grains or crystals), cleavages, or other blemishes. Such inclusions and flaws may affect transparency, brilliance, and even the durability of a diamond. The clarity grade ranges from flawless down to industrial grade stones, which may appear milky from too many tiny inclusions.

Cut involves the relationships between the sizes of a stone's major physical features and their various angles. They determine, above all else, the limit to which a diamond will accomplish its optical potential. There are various types of faceted cuts, including the round brilliant cut, with a flat top, a flat bottom, and about fifty-eight facets. Facet shapes of square, triangular, diamond-shaped, and trapezoidal cuts might be used.

The details of a given diamond's four C's determine how expensive the diamond is going to be, and a top grade in all factors would be very expensive. No two stones are alike, however, and it is possible to find good value in a diamond with a fewer lower grades that is still brilliant.

speed of electromagnetic waves (c) is related to the wavelength (λ) and the frequency (f) by a form of the wave equation, or

$$c = \lambda f$$

equation 8.3

Visible light is the part of the electromagnetic spectrum that your eyes can detect, a narrow range of wavelength from about 7.90×10^{-7} m to 3.90×10^{-7} m. In general, this range of visible light can be subdivided into ranges of wavelengths that you perceive as colors (figure 8.17). These are the colors of the rainbow, and there are six distinct colors that blend one into another. These colors are *red, orange, yellow, green, blue,* and *violet*. The corresponding ranges of wavelengths and frequencies of these colors are given in table 8.2.

TABLE 8.2	Range of wavelengths and frequencies of the colors of visible light	
Color	**Wavelength (in meters)**	**Frequency (in hertz)**
Red	7.9×10^{-7} to 6.2×10^{-7}	3.8×10^{14} to 4.8×10^{14}
Orange	6.2×10^{-7} to 6.0×10^{-7}	4.8×10^{14} to 5.0×10^{14}
Yellow	6.0×10^{-7} to 5.8×10^{-7}	5.0×10^{14} to 5.2×10^{14}
Green	5.8×10^{-7} to 4.9×10^{-7}	5.2×10^{14} to 6.1×10^{14}
Blue	4.9×10^{-7} to 4.6×10^{-7}	6.1×10^{14} to 6.6×10^{14}
Violet	4.6×10^{-7} to 3.9×10^{-7}	6.6×10^{14} to 7.7×10^{14}

A CLOSER LOOK | Optics

istorians tell us there are many early stories and legends about the development of ancient optical devices. The first glass vessels were made about 1500 B.C., so it is possible that samples of clear, transparent glass were available soon after. One legend claimed that the ancient Chinese invented eyeglasses as early as 500 B.C. A burning glass (lens) was mentioned in an ancient Greek play written about 424 B.C. Several writers described how Archimedes saved his hometown of Syracuse with a burning glass in about 214 B.C. Syracuse was besieged by Roman ships when Archimedes supposedly used the burning glass to focus sunlight on the ships, setting them on fire. It is not known if this story is true or not, but it is known that the Romans indeed did have burning glasses. Glass spheres, which were probably used to start fires, have been found in Roman ruins, including a convex lens recovered from the ashes of Pompeii.

Today, lenses are no longer needed to start fires, but they are common in cameras, scanners, optical microscopes, eyeglasses, lasers, binoculars, and many other optical devices. Lenses are no longer just made from glass, and today many are made from a transparent, hard plastic that is shaped into a lens.

The primary function of a lens is to form an image of a real object by refracting incoming parallel light rays. Lenses have two basic shapes, with the center of a surface either bulging in or bulging out. The outward bulging shape is thicker at the center than around the outside edge and this is called a *convex lens* (box figure 8.1A). The other basic lens shape is just the opposite, thicker around the outside edge than at the center and is called a *concave lens* (box figure 8.1B).

Convex lenses are used to form images in magnifiers, cameras, eyeglasses, projectors, telescopes, and microscopes (box figure 8.2). Concave lenses are used in some eyeglasses and in combinations with the convex lens to correct for defects. The convex lens is the most commonly used lens shape.

Your eyes are optical devices with convex lenses. Box figure 8.3 shows the basic

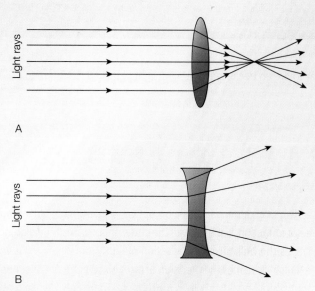

BOX FIGURE 8.1

(*A*) Convex lenses are called converging lenses since they bring together, or converge, parallel rays of light. (*B*) Concave lenses are called diverging lenses since they spread apart, or diverge, parallel rays of light.

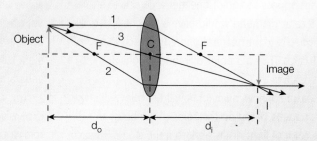

BOX FIGURE 8.2

A convex lens forms an inverted image from refracted light rays of an object outside the focal point. Convex lenses are mostly used to form images in cameras, file or overhead projectors, magnifying glasses, and eyeglasses.

structure. First, a transparent hole called the *pupil* allows light to enter the eye. The size of the pupil is controlled by the *iris,* the colored part that is a muscular diaphragm. The *lens* focuses a sharp image on the back surface of the eye, the *retina*. The retina is made up of millions of light-sensitive structures, and nerves carry signals (impulses) from the retina through the optic nerve to the brain.

The lens is a convex, pliable material held in place and changed in shape by the attached *ciliary muscle*. When the eye is focused on a distant object the ciliary muscle is completely relaxed. Looking at a closer object requires the contraction of the ciliary muscles to change the curvature of the lens. This adjustment of focus by action of the cil-

—*Continued top of next page*

Continued—

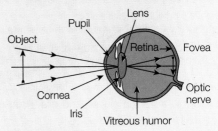

BOX FIGURE 8.3

Light rays from a distant object are focused by the lens onto the retina, a small area on the back of the eye.

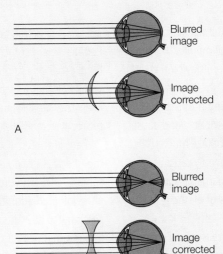

BOX FIGURE 8.4

Eyeglasses can help (*A*) a person with presbyopia (farsighted) or (*B*) a person with myopia (nearsighted).

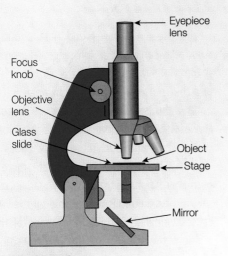

BOX FIGURE 8.5

A simple microscope uses a system of two lenses, which are an objective lens that makes an enlarged image of the specimen, and an eyepiece lens that makes an enlarged image of that image.

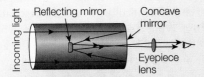

BOX FIGURE 8.6

This illustrates how the path of light moves through a simple reflecting astronomical telescope. Several different designs and placement of mirrors are possible.

iary muscle is called *accommodation*. The closest distance an object can be seen without a blurred image is called the *near point*, and this is the limit to accommodation.

The near point moves outward with age as the lens becomes less pliable. By middle age the near point may be twice this distance or greater, creating the condition known as farsightedness. The condition of farsightedness, or *hyperopia*, is a problem associated with aging (called presbyopia). Hyperopia can be caused at an early age by an eye that is too short or by problems with the cornea or lens that focuses the image behind the retina. Farsightedness can be corrected with a convex lens as shown in box figure 8.4A.

Nearsightedness, or *myopia*, is a problem caused by an eye that is too long or problems with the cornea or lens that focus the image in front of the retina. Nearsightedness can be corrected with a concave lens as shown in box figure 8.4B.

The microscope is an optical device used to make things look larger. It is essentially a system of two lenses, one to produce an image of the object being studied, and the other to act as a magnifying glass and enlarge that image. The power of the microscope is basically determined by the *objective lens*, which is placed close to the specimen on the stage of the microscope. Light is projected up through the specimen, and the objective lens makes an enlarged image of the specimen inside the tube between the two lenses. The *eyepiece lens* is positioned so that it makes a

sharp enlarged image of the image produced by the objective lens (box figure 8.5).

Telescopes are optical instruments used to provide enlarged images of near and distant objects. There are two major types of telescopes, *refracting* telescopes that use two lenses, and *reflecting* telescopes that use combinations of mirrors, or a mirror and a lens. The refracting telescope has two lenses, with the objective lens forming a reduced image, which is viewed with an eyepiece lens to enlarge that image. In reflecting telescopes, mirrors are used instead of lenses to collect the light (box figure 8.6).

Finally, the *digital camera* is one of the more recently developed light-gathering and photograph-taking optical instruments. This camera has a group of small photocells, with perhaps thousands lined up on the focal plane behind a converging lens. An image falls on the array, and each photocell stores a charge that is proportional to the amount of light falling on the cell. A microprocessor measures the amount of charge registered by each photocell and considers it

as a pixel, a small bit of the overall image. A shade of gray or a color is assigned to each pixel and the image is ready to be enhanced, transmitted to a screen, printed, or magnetically stored for later use. Modern digital cameras have an advantage since the picture is available instantly, without the use of darkrooms, chemicals, or light-sensitive paper. A photo can be immediately downloaded from the camera into a computer, and then perhaps loaded onto your website, ready for the whole world to view.

CONNECTIONS

Made in the Shade?

Health researchers have found that too much ultraviolet ray exposure today can result in skin cancer later on in your life. Two types of ultraviolet radiation reach the earth's surface: ultraviolet A (UVA) and the shorter-wavelength ultraviolet B (UVB). UVA radiation is about a thousand times less effective in causing burns than UVB. However, UVA radiation penetrates skin more deeply and causes you to tan or burn much more slowly. It does not take much UVB to cause skin damage, but UVA penetrates more deeply, causing more permanent damage than UVB. UVA radiation damages blood vessels in the skin and is responsible for the loss of elasticity. It also causes wrinkling, sagging, and many of the effects known as "aging" of the skin. UVA in high dosages also increases your chances of developing skin cancer, in addition to premature skin aging.

It is important to remember that UVB radiation is most predominant in sunlight between the hours of 9 A.M. and 3 P.M. However, UVA radiation penetrates equally any time the sun is shining. People have learned to avoid ultraviolet exposure of direct sunlight, especially during the peak exposure times. They wear protective clothing, such as a broad-brimmed hat and long sleeves, keep the car window next to a person closed, and use a sun-protective cream or lotion when exposure is necessary.

People should also know that they might be exposed when in the shade. Ultraviolet radiation bounces around in the atmosphere, reaching you even in the shade of a tree. The amount of ultraviolet radiation that reaches you in the shade of a tree is proportional to how much sky you can see. The more sky you can see, the more ultraviolet radiation is reaching you. Thus you will receive less ultraviolet radiation in the woods, under many trees, than you would in the shade of a single tree.

In general, light is interpreted to be white if it has the same mixture of colors as the solar spectrum. That sunlight is made up of component colors was first investigated in detail by Isaac Newton. While a college student, Newton became interested in grinding lenses, light, and color. At the age of twenty-three, Newton visited a local fair and bought several triangular glass prisms and proceeded to conduct a series of experiments with a beam of sunlight in his room. From the beginning, Newton intended to learn from direct observation and avoid speculation about the nature of light. In 1672, he reported the results of his experiments with prisms and color, concluding that white light is a mixture of all the independent colors. Newton found that a beam of sunlight falling on a glass prism in a darkened room produced a band of colors he called a *spectrum*. Further, he found that a second glass prism would not subdivide each separate color but would combine all the colors back into white sunlight. Newton concluded that sunlight consists of a mixture of the six colors. Today, light that is composed, or made up, of several colors of light is called *polychromatic* light. Light that is one wavelength only is called *monochromatic* light, but each color of the spectrum is a range of wavelengths.

A glass prism separates sunlight into a spectrum of colors because the index of refraction is different for different wavelengths of light. The same processes that slow the speed of light in a transparent substance have a greater effect on short wavelengths than they do on longer wavelengths. As a result, violet light is refracted most, red light is refracted least, and the other colors are refracted between these extremes. This results in a beam of white light being separated, or dispersed, into a spectrum when it is refracted. Any transparent material in which the index of refraction varies with wavelength has the property of *dispersion*. The dispersion of light by ice crystals sometimes produces a colored halo around the sun and the moon.

EVIDENCE FOR WAVES

The nature of light became a topic of debate toward the end of the 1600s as Isaac Newton published his *particle theory* of light. He believed that the straight-line travel of light could be better explained as small particles of matter that traveled at great speed from a source of light. Particles, reasoned Newton, should follow a straight line according to the laws of motion. Waves, on the other hand, should bend as they move, much as water waves on a pond bend into circular shapes as they move away from a disturbance.

About the same time that Newton developed his particle theory of light, Christian Huygens (pronounced "hi-ganz") was concluding that light is not a stream of particles but rather a longitudinal wave that moves through the "ether." Huygens' *wave theory* proposed that waves were created by a light source. The wave front, according to this theory, did not simply move off through the ether. Each point on the wave front was pictured as impacting an "ether molecule." Each "ether molecule" produced a small *wavelet* as a result. Each wavelet was a small hemisphere that spread outward in the direction of the movement. A line connecting all the leading surfaces of the wavelets

A CLOSER LOOK | The Rainbow

A rainbow is a spectacular, natural display of color that is supposed to have a pot of gold under one end. Understanding the why and how of a rainbow requires information about water droplets and knowledge of how light is reflected and refracted. This information will also explain why the rainbow seems to move when you move—making it impossible to reach the end to obtain that mythical pot of gold.

First, note the pattern of conditions that occurs when you see a rainbow. It usually appears when the sun is shining low in one part of the sky and rain is falling in the opposite part. With your back to the sun, you are looking at a zone of raindrops that are all showing red light, another zone that are all showing violet light, with zones of the other colors between (ROYGBV). For a rainbow to form like this requires a surface that refracts and reflects the sunlight, a condition met by spherical raindrops.

Water molecules are put together in such a way that they have a positive side and a negative side, and this results in strong molecular attractions. It is the strong attraction of water molecules for one another that results in the phenomenon of surface tension. Surface tension is the name given to the surface of water acting as if it is covered by an ultra-thin elastic membrane that is contracting. It is surface tension that pulls raindrops into a spherical shape as they fall through the air.

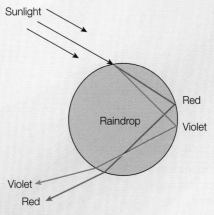

BOX FIGURE 8.7

Light is refracted when it enters a raindrop and when it leaves. The part that leaves the front surface of the raindrop is the source of the light in thousands upon thousands of raindrops from which you see zones of color—a rainbow.

Box Figure 8.7 shows one thing that can happen when a ray of sunlight strikes a single spherical raindrop near the top of the drop. At this point some of the sunlight is reflected, and some is refracted into the raindrop. The refraction disperses the light into its spectrum colors, with the violet light being refracted most and red the least. The refracted light travels through the drop to the opposite side, where some of it might be

reflected back into the drop. The reflected part travels back through the drop again, leaving the front surface of the raindrop. As it leaves, the light is refracted for a second time. The combined refraction, reflection, and second refraction is the source of the zones of colors you see in a rainbow. This also explains why you see a rainbow in the part of the sky opposite from the sun.

The light from any one raindrop is one color, and that color comes from all drops on the arc of a circle that is a certain angle between the incoming sunlight and the refracted light. Thus the raindrops in the red region refract red light toward your eyes at an angle of 42°, and all other colors are refracted over your head by these drops. Raindrops in the violet region refract violet light toward your eyes at an angle of 40° and the red and other colors toward your feet. Thus the light from any one drop is seen as one color, and all drops showing this color are on the arc of a circle. An arc is formed because the angle between the sunlight and the refracted light of a color is the same for each of the spherical drops.

There is sometimes a fainter secondary rainbow, with colors reversed, that forms from sunlight entering the bottom of the drop, reflecting twice, and then refracting out the top. The double reflection reverses the colors, and the angles are 50° for the red and 54° for the violet.

described a new wave front, which now impacted other "ether molecules," which made new wavelets, which formed a new wave front, and so on. The forming and reforming of the wave front resulted in a plane wave that moved through space in a straight line. Thus, Huygens' wave theory explained the straight-line travel of light as the continual formation of new waves (figure 8.18).

Both theories had advocates during the 1700s, but the majority favored Newton's particle theory. By the beginning of the 1800s, new evidence was found that favored the wave theory, evidence that could not be explained in terms of anything but waves.

Diffraction

One of Newton's arguments for a particle nature of light concerned the observation that light moved in straight lines through openings such as a window. He felt that if light were waves, they would "bend around" obstacles such as the edge of the window. By analogy, the water waves in figure 8.19 strike the edge of a wall, an obstacle, and noticeably bend around the corner, some bending enough to travel parallel to the wall. Light does not bend around an obstacle like this and appears to make a sharp shadow behind an obstacle. So Newton argued for a particle nature of light, since particles would move straight past an obstacle and make a sharp shadow.

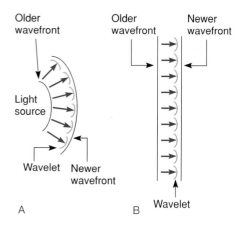

FIGURE 8.18

(*A*) Huygens' wave theory described wavelets that formed from an older wave front from a light source. The wavelets then moved out to form a new wave front. (*B*) As the wavelets spread, the wave front eventually becomes a plane wave, or straight wave, at some distance. This continuous process explained how waves could travel in a straight line.

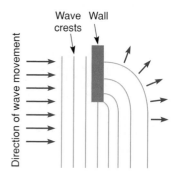

FIGURE 8.19

Water waves are observed to bend around an obstacle such as a wall in the water. Light appears to move straight past an obstacle, forming a shadow. This is one reason that Newton thought light must be particles, not waves.

In 1665, Francesco Grimaldi reported that there was not a sharp edge to a shadow moving through a small pinhole, and the light formed a larger spot than would be expected from light traveling in a straight line. Grimaldi had made the first description of **diffraction,** the bending of light around the edge of an opaque object. Any small opening or a sharp-edged object was found to produce diffraction. Newton could not imagine the extremely small wavelengths of light that would be required to produce this result, so he dismissed the observation as an interaction between particles of light and the edges of the pinhole. As shown in figure 8.20, the determining factor is the size of an opening (or obstacle) as *compared to the wavelength.* Ordinary-sized openings and obstacles are much larger than the wavelengths of light, so little diffraction occurs, and the light travels in straight lines.

Diffraction can be explained by using Huygens' wave theory. Recall that Huygens described a light wave as a disturbance in the

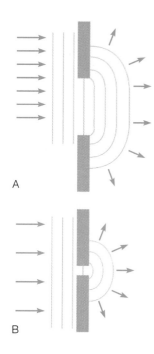

FIGURE 8.20

(*A*) When an opening is large as compared to the wavelength, waves appear to move mostly straight through the opening. (*B*) When the opening is about the same size as the wavelengths, the waves diffract, spreading into an arc.

form of waves moving away from a light source. Each point on a crest is considered to be a source of *small wavelets* that spread in the direction of movement. The wavelets are small hemispheres, and a line connecting all the surfaces at the leading edge forms a new wave front that is made up of all the little wavelets. Arrows perpendicular to the wave front are called *rays,* and the rays show the direction of movement. From a distant source such as the sun, the wave front is considered to be in a plane, which means that the perpendicular rays are parallel to each other.

According to the Huygens model, a wave front passing through a large opening will continue to generate wavelets, so the original shape of the wave front moves straight through the opening. If the opening is very small, about the same size as the wavelength, a *single* wavelet will move out in all directions as an expanding arc from the opening. This explains diffraction from an opening (figure 8.21).

When a wave front strikes an obstacle the same size as the wavelength, a wavelet will move off in all directions from the corners, including into the shadow behind the obstacle. Thus, the use of light rays is restricted to describing the behavior of light when obstacles and openings are large compared to the wavelength of light, where little diffraction occurs.

Interference

In 1801, Thomas Young published evidence of a behavior of light that could only be explained in terms of a wave model of light. Young's experiment is illustrated in figure 8.22. Light from a single source is used to produce two beams of light that are in

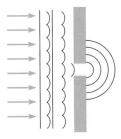

FIGURE 8.21

Huygens' wave theory explains diffraction of light. When the opening is large as compared to the wavelength, the wavelets form a new wave front as usual, and the front continues to move in a straight line. When the opening is the size of the wavelength, a single wavelet moves through, expanding in an arc.

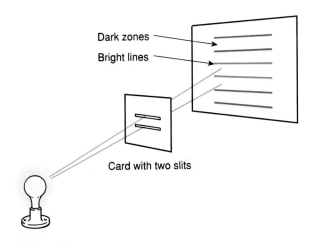

FIGURE 8.22

Young's double-slit experiment produced a pattern of bright lines and dark zones when light from a single source passed through the slits.

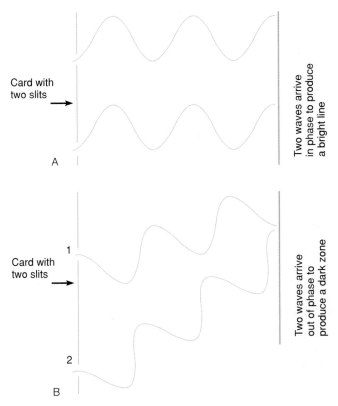

FIGURE 8.23

An interference pattern of bright lines and dark zones is produced because of the different distances that waves must travel from the two slits. (*A*) When they arrive together, a bright line is produced. (*B*) Light from slit 2 must travel farther than light from slit 1, so they arrive out of phase.

phase, that is, having their crests and troughs together as they move away from the source. This light falls on a card with two slits, each less than a millimeter in width. The light is diffracted from each slit, moving out from each as an expanding arc. Beyond the card the light from one slit crosses over the light from the other slit to produce a series of bright lines on a screen. Young had produced a phenomenon of light called **interference,** and interference can only be explained by waves.

The pattern of bright lines and dark zones is called an *interference pattern* (figure 8.23). The light moved from each slit in phase, crest to crest and trough to trough. Light from both slits traveled the same distance directly across to the screen, so they arrived in phase. The crests from the two slits are superimposed here, and constructive interference produces a bright line in the center of the pattern. But for positions above and below the center, the light from the two slits must travel different distances to the screen. At a certain distance above and below the bright center line, light from one slit had to travel a greater distance and arrives one-half wavelength after light from the other slit. Destructive interference produces a zone of darkness at these

positions. Continuing up and down the screen, a bright line of constructive interference will occur at each position where the distance traveled by light from the two slits differs by any whole number of wavelengths. A dark zone of destructive interference will occur at each position where the distance traveled by light from the two slits differs by any half-wavelength. Thus, bright lines occur above and below the center bright line at positions representing differences in paths of 1, 2, 3, 4, and so on wavelengths. Similarly, zones of darkness occur above and below the center bright line at positions representing differences in paths of ½, 1½, 2½, 3½, and so on wavelengths. Young found all of the experimental data such as this in full agreement with predictions from a wave theory of light. About fifteen years later, A. J. Fresnel (pronounced "fray-nel") demonstrated mathematically that diffraction as well as other behaviors of light could be fully explained with the wave theory. In 1821 Fresnel determined that the wavelength of red light was about 8×10^{-7} m and of violet light about 4×10^{-7} m, with other colors in between these two extremes. The work of Young and Fresnel seemed to resolve the issue of considering light to be a stream of particles or a wave, and it was generally agreed that light must be waves. The waves, however, were considered to be mechanical waves in the "ether" until after the work of Maxwell, Lorentz, and Einstein.

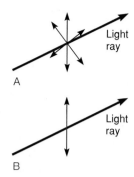

FIGURE 8.24

(*A*) Unpolarized light has transverse waves vibrating in all possible directions perpendicular to the direction of travel. (*B*) Polarized light vibrates only in one plane. In this illustration, the wave is vibrating in a vertical direction only.

Polarization

Huygens' wave theory and Newton's particle theory could explain some behaviors of light satisfactorily, but there were some behaviors that neither (original) theory could explain. Both theories failed to explain some behaviors of light, such as light moving through certain transparent crystals. For example, a slice of the mineral tourmaline transmits what appears to be a low-intensity greenish light. But if a second slice of tourmaline is placed on the first and rotated, the transmitted light passing through both slices begins to dim. The transmitted light is practically zero when the second slice is rotated 90°. Newton suggested that this behavior had something to do with "sides" or "poles" and introduced the concept of what is now called the *polarization* of light.

The waves of Huygens' wave theory were longitudinal, moving like sound waves, with wave fronts moving in the direction of travel. A longitudinal wave could not explain the polarization behavior of light. In 1817, Young modified Huygens' theory by describing the waves as *transverse*, vibrating at right angles to the direction of travel. This modification helped explain the polarization behavior of light transmitted through the two crystals and provided firm evidence that light is a transverse wave. As shown in figure 8.24A, **unpolarized light** is assumed to consist of transverse waves vibrating in all conceivable random directions. Polarizing materials, such as the tourmaline crystal, transmit light that is vibrating in one direction only, such as the vertical direction in figure 8.24B. Such a wave is said to be **polarized**, or *plane-polarized*, since it vibrates only in one plane. The single crystal polarized light by transmitting only waves that vibrate parallel to a certain direction while selectively absorbing waves that vibrate in all other directions. Your eyes cannot tell the difference between unpolarized and polarized light, so the light transmitted through a single crystal looks just like any other light. When a second crystal is placed on the first, the amount of light transmitted depends on the alignment of the two crystals (figure 8.25). When the two crystals are *aligned*, the polarized light from the first crystal passes through the second with little absorption. When the crystals are *crossed*

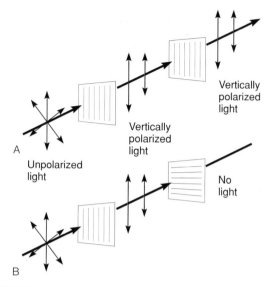

FIGURE 8.25

(*A*) Two crystals that are aligned both transmit vertically polarized light that looks like any other light. (*B*) When the crystals are crossed, no light is transmitted.

at 90°, the light transmitted by the first is vibrating in a plane that is absorbed by the second crystal, and practically all the light is absorbed. At some other angle, only a fraction of the polarized light from the first crystal is transmitted by the second.

You can verify whether or not a pair of sunglasses is made of polarizing material by rotating a lens of one pair over a lens of a second pair. Light is transmitted when the lenses are aligned but mostly absorbed at 90° when the lenses are crossed.

Light is completely polarized when all the waves are removed except those vibrating in a single direction. Light is partially polarized when some of the waves are in a particular orientation, and any amount of polarization is possible. There are several means of producing partially or completely polarized light, including (1) selective absorption, (2) reflection, and (3) scattering.

Selective absorption is the process that takes place in certain crystals, such as tourmaline, where light in one plane is transmitted and all the other planes are absorbed. A method of manufacturing a polarizing film was developed in the 1930s by Edwin H. Land. The film is called *Polaroid*. Today Polaroid is made of long chains of hydrocarbon molecules that are aligned in a film. The long-chain molecules ideally absorb all light waves that are parallel to their lengths and transmit light that is perpendicular to their lengths. The direction that is *perpendicular* to the oriented molecular chains is thus called the polarization direction or the *transmission axis*.

Reflected light with an angle of incidence between 1° and 89° is partially polarized as the waves parallel to the reflecting surface are reflected more than other waves. Complete polarization, with all waves parallel to the surface, occurs at a particular angle of incidence. This angle depends on a number of variables, including the nature of the reflecting material. Figure 8.26 illustrates polarization by reflection. Polarizing sunglasses reduce

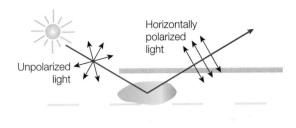

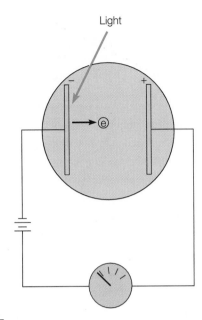

FIGURE 8.26

Light that is reflected becomes partially or fully polarized in a horizontal direction, depending on the incident angle and other variables.

the glare of reflected light because they have vertically oriented transmission axes. This absorbs the horizontally oriented reflected light. If you turn your head from side to side so as to rotate your sunglasses while looking at a reflected glare, you will see the intensity of the reflected light change. This means that the reflected light is partially polarized.

The phenomenon called *scattering* occurs when light is absorbed and reradiated by particles about the size of gas molecules that make up the air. Sunlight is initially unpolarized. When it strikes a molecule, electrons are accelerated and vibrate horizontally and vertically. The vibrating charges reradiate polarized light. Thus, if you look at the blue sky with a pair of polarizing sunglasses and rotate them, you will observe that light from the sky is polarized. Bees are believed to be able to detect polarized skylight and use it to orient the direction of their flights. Violet and blue light have the shortest wavelengths of visible light, and red and orange light have the longest. The violet and blue rays of sunlight are scattered the most. However, sunlight has more blue than violet light (see figure 8.3), and this is why the sky appears blue when the sun is high in the sky. At sunset the path of sunlight through the atmosphere is much longer than when the sun is more directly overhead. Much of the blue and violet have been scattered away as a result of the longer path through the atmosphere at sunset. The remaining light that comes through is mostly red and orange, so these are the colors you see at sunset.

EVIDENCE FOR PARTICLES

The evidence from diffraction, interference, and polarization of light was very important in the acceptance of the wave theory because there was simply no way to explain these behaviors with a particle theory. Then, in 1850, J. L. Foucault was able to prove that light travels much slower in transparent materials than it does in air. This was in complete agreement with the wave theory and completely opposed to the particle theory. By the end of the 1800s, Maxwell's theoretical concept of electric and magnetic fields changed the concept of light from mechanical waves to waves of changing electric and magnetic fields. Further evidence removed the necessity for ether, the material supposedly needed for waves to move through. Light was now seen as

FIGURE 8.27

A setup for observing the photoelectric effect. Light strikes the negatively charged plate, and electrons are ejected. The ejected electrons move to the positively charged plate and can be measured as a current in the circuit.

electromagnetic waves that could move through empty space. By this time it was possible to explain all behaviors of light moving through empty space or through matter with a wave theory. Yet, there were nagging problems that the wave theory could not explain. In general, these problems concerned light that is absorbed by or emitted from matter.

Photoelectric Effect

Light is a form of energy, and it gives its energy to matter when it is absorbed. Usually the energy of absorbed light results in a temperature increase, such as the warmth you feel from absorbed sunlight. Sometimes, however, the energy from absorbed light results in other effects. In some materials, the energy is acquired by electrons, and some of the electrons acquire sufficient energy to jump out of the material. The movement of electrons as a result of energy acquired from light is known as the **photoelectric effect.** The photoelectric effect is put to a practical use in a solar cell, which transforms the energy of light into an electric current (figure 8.27).

The energy of light can be measured with great accuracy. The kinetic energy of electrons after they absorb light can also be measured with great accuracy. When these measurements were made of the light and electrons involved in the photoelectric effect, some unexpected results were observed. Monochromatic light, that is, light of a single, fixed frequency, was used to produce the photoelectric effect. First, a low-intensity, or dim, light was used, and the numbers and energy of the ejected electrons were measured. Then a high-intensity light was used, and the numbers and energy of the ejected electrons were again measured. Measurement showed that

CONNECTIONS

Bioluminescence and Photosynthesis

Many kinds of organisms have the ability to produce light, including bacteria, protozoa, fungi, fish, squid, jellyfish, and insects. There are several different mechanisms by which light is produced, but they all involve a chemical reaction that releases light. What makes bioluminescence different from most chemical reactions that produce light is that there is very little heat produced with the light. The chemical reactions are encouraged by specific enzymes and generally require the use of a source of chemical energy (ATP) to produce the light. The chemical that is broken down to produce the light is often referred to as *lucifrin* and the enzyme that causes its breakdown is called *lucifrinase*. The genes for bioluminescence have been transferred from one organism to another by genetic engineering techniques. One interesting case involves transferring bioluminescence genes into tobacco plants.

Some fish and other animals produce light but lack the ability to do so themselves. They have special organs that harbor bacteria that are bioluminescent, however.

Organisms such as green plants, algae, and certain bacteria are capable of trapping light energy and holding it in the chemical bonds of molecules such as carbohydrates. The process of convert-

ing sunlight energy to chemical-bond energy, called *photosynthesis*, is a major biochemical pathway. Photosynthetic organisms produce food molecules, such as carbohydrates, for themselves as well as for all the other organisms that feed upon them.

The following equation summarizes the chemical reactions green plants and many other photosynthetic organisms use to make organic molecules for growth and reproduction:

$$\text{light} + 6\,CO_2 + 6\,H_2O \rightarrow C_6H_{12}O_6 + 6\,O_2$$

There are two major activities involved in photosynthesis:

1. The chlorophyll is used to trap light energy. The chlorophyll has excited electrons that can be used to cause certain chemical reactions. One of the primary activities of the excited chlorophyll molecules is to separate a water molecule (H_2O) into its components of hydrogen (H) and oxygen (O). The oxygen is released into the atmosphere as O_2.
2. The second activity involves combining carbon dioxide (CO_2) from the atmosphere and hydrogen from the previous step to form a carbohydrate as sugar ($C_6H_{12}O_6$).

---

(1) low-intensity light caused fewer electrons to be ejected, and high-intensity light caused many to be ejected, and (2) all electrons ejected from low- or high-intensity light ideally had the *same* kinetic energy. Surprisingly, the kinetic energy of the ejected electrons was found to be *independent* of the light intensity. This was contrary to what the wave theory of light would predict, since a stronger light should mean that waves with more energy have more energy to give to the electrons. Here is a behavior involving light that the wave theory could not explain.

Quantization of Energy

In addition to the problem of the photoelectric effect, there were problems with blackbody radiation, light emitted from hot objects. The experimental measurements of light emitted through blackbody radiation did not match predictions made from theory. In 1900, Max Planck (pronounced "plonk"), a German physicist, found that he could fit the experimental measurements and theory together by assuming that the vibrating molecules that emitted the light could only have a *fixed amount* of energy. Instead of energy existing through a continuous range of amounts, Planck found that the vibrating molecules could only have energy in multiples of energy in certain amounts, or **quanta** (meaning "fixed amounts"; *quantum* is singular, and *quanta* plural). Planck thus developed the concept of quantization of energy; that is,

vibrating molecules involved in blackbody radiation vibrate with quantized energy E according to the relationship

$$E = nhf$$

equation 8.4

where n is a positive whole number, f is the frequency of vibration of the molecules, and h is a proportionality constant known today as *Planck's constant*. The value of Planck's constant is 6.63×10^{-34} J·s.

Planck's discovery of quantized energy states was a radical, revolutionary development, and most scientists, including Planck, did not believe it at the time. Planck, in fact, spent considerable time and effort trying to disprove his own discovery. It was, however, the beginning of the quantum theory, which was eventually to revolutionize physics.

Five years later, in 1905, Albert Einstein applied Planck's quantum concept to the problem of the photoelectric effect. Einstein described the energy in a light wave as quanta of energy called **photons.** Each photon has an energy E that is related to the frequency f of the light through Planck's constant h, or

$$E = hf$$

equation 8.5

This relationship says that higher-frequency light (e.g., blue light at 6.50×10^{14} Hz) has more energy than lower-frequency light (e.g., red light at 4.00×10^{14} Hz). The energy of such high- and low-frequency light can be verified by experiment.

The photon theory also explained the photoelectric effect. According to this theory, light is a stream of moving photons. It is the number of photons in this stream that determines if the light is dim or intense. A high-intensity light has many, many photons, and a low-intensity light has only a few photons. At any particular fixed frequency, all the photons would have the same energy, the product of the frequency and Planck's constant (hf). When a photon interacts with matter, it is absorbed and gives up all of its energy. In the photoelectric effect, this interaction takes place between photons and electrons. When an intense light is used, there are more photons to interact with the electrons, so more electrons are ejected. The energy given up by each photon is a function of the frequency of the light, so at a fixed frequency, the energy of each photon, hf, is the same, and the acquired kinetic energy of each ejected electron is the same. Thus, the photon theory explains the measured experimental results of the photoelectric effect.

The photoelectric effect is explained by considering light to be photons with quanta of energy, not a wave of continuous energy. This is not the only evidence about the quantum nature of light, and more will be presented in the next chapter. But, as you can see, there is a dilemma. The electromagnetic wave theory and the photon theory seem incompatible. Some experiments cannot be explained by the wave theory and seem to support the photon theory. Other experiments are contradictions, providing seemingly equal evidence to reject the photon theory in support of the wave theory.

THE PRESENT THEORY

Today, light is considered to have a dual nature, sometimes acting like a wave and sometimes acting like a particle. A wave model is useful in explaining how light travels through space and how it exhibits such behaviors as refraction, interference, and diffraction. A particle model is useful in explaining how light is emitted from and absorbed by matter, exhibiting such behaviors as blackbody radiation and the photoelectric effect. Together, both of these models are part of a single theory of light, a theory that pictures light as having both particle and wave properties. Some properties are more useful when explaining some observed behaviors, and other properties are more useful when explaining other behaviors.

Frequency is a property of a wave, and the energy of a photon is a property of a particle. Both frequency and the energy of a photon are related in equation 8.5, $E = hf$. It is thus possible to describe light in terms of a frequency (or wavelength) or in terms of a quantity of energy. Any part of the electromagnetic spectrum can thus be described by units of frequency, wavelength, or energy, which are alternative means of describing light. The radio radiation parts of the spectrum are low-frequency, low-energy, and long-wavelength radiations. Radio radiations have more wave properties and practically no particle properties, since the energy levels are low. Gamma radiation, on the other hand, is high-frequency, high-energy, and short-wavelength radiation. Gamma

FIGURE 8.28

It would seem very strange if there were not a sharp distinction between objects and waves in our everyday world. Yet this appears to be the nature of light.

radiation has more particle properties, since the extremely short wavelengths have very high energy levels. The more familiar part of the spectrum, visible light, is between these two extremes and exhibits both wave and particle properties, but it never exhibits both properties at the same time in the same experiment.

Part of the problem in forming a concept or mental image of the exact nature of light is understanding this nature in terms of what is already known. The things you already know about are observed to be particles, or objects, or they are observed to be waves. You can see objects that move through the air, such as baseballs or footballs, and you can see waves on water or in a field of grass. There is nothing that acts like a moving object in some situations but acts like a wave in other situations. Objects are objects, and waves are waves, but objects do not become waves, and waves do not become objects. If this dual nature did exist it would seem very strange. Imagine, for example, holding a book at a certain height above a lake (figure 8.28). You can make measurements and calculate the kinetic energy the book will have when dropped into the lake. When it hits the water, the book disappears, and water waves move away from the point of impact in a circular pattern that moves across the water. When the waves reach another person across the lake, a book identical to the one you dropped pops up out of the water as the waves disappear. As it leaves the water across the lake, the book has the same kinetic energy that your book had when it hit the water in front of you. You and the other person could measure things about either book, and you could measure things about the waves, but you could not measure both at the same time. You might say that this behavior is not only strange but impossible. Yet, it is an analogy to the observed behavior of light.

As stated, light has a dual nature, sometimes exhibiting the properties of a wave and sometimes exhibiting the properties of moving particles but never exhibiting both properties at the same time. Both the wave and the particle nature are accepted as being part of one model today, with the understanding that the exact nature of light is not describable in terms of anything that is known to exist in the everyday-sized world. Light is an extremely small-scale phenomenon that must be different, without a sharp

A CLOSER LOOK | The Compact Disc (CD)

A compact disc (CD) is a laser-read (also called *optically read*) data storage device on which music, video, or any type of computer data can be stored. Two types in popular use today are the CD audio and video discs used for recording music and television and the CD-ROM used to store any type of computer data. CD-ROM stands for Compact Disc Read-Only Memory. A CD-ROM can be read by a computer, and a CD audio disc can be played by a compact audio disc player. Both the CD-ROM and the CD audio discs are made the same way and are identical in structure. There are slight differences in the way CD audio drives and CD-ROM drives retrieve information, since music is continuous and computer data is stored in small, discrete chunks. Both drives, however, have an optical sensor head with a tiny diode laser, detection optics, and a means of focusing. Both rotate between 200 to 500 revolutions per minute, compared to the constant rotation of 33⅓ revolutions per minute for the old vinyl records. The CD drive changes speed to move the head at a constant linear velocity over the recording track, faster near the inner hub and slower near the outer edge of the disc. Furthermore, the CD drive reads from the inside out, so the disc will slow as it is played.

The CD itself is a 12 cm diameter, 1.3 mm thick sandwich of a hard plastic core, a mirrorlike layer of metallic aluminum, and a tough, clear plastic overcoating that protects the thin layer of aluminum (box figure 8.8). Technically, the disc does not contain any operational memory, and the "Read-Only Memory" is not actually memory either. It is digitized data; music, video, or computer data that have been converted into a string of binary numbers. (See the reading on data storage in chapter 7.) First, a master disc is made. The binary numbers are translated into a series of pulses that are fed to a laser. The laser is focused onto a photosensitive material on a spinning master disc. Whenever there is a pulse in the signal, the laser burns a small oval pit into the surface, making a pattern of pits and bumps on the track of the master disc. The laser beam is incredibly small, making marks about a micron or so in diameter. A micron is one-millionth of

BOX FIGURE 8.8

This plastic disc with a 12 cm diameter can store about 600 megabytes of data or other information, the equivalent of about 300,000 typewritten pages.

—Continued top of next page

distinction between a particle and a wave. Evidence about this strange nature of an extremely small-scale phenomenon will be considered again in the next chapter as a basis for introducing the quantum theory of matter.

SUMMARY

Electromagnetic radiation is emitted from all matter with a temperature above absolute zero, and as the temperature increases, more radiation and shorter wavelengths are emitted. Visible light is emitted from matter hotter than about 700°C, and this matter is said to be *incandescent*. The sun, a fire, and the ordinary lightbulb are incandescent sources of light.

The behavior of light is shown by a light ray model that uses straight lines to show the straight-line path of light. Light that interacts with matter is *reflected* with parallel rays, moves in random directions by *diffuse reflection* from points, or is *absorbed,* resulting in a temperature increase. Matter is *opaque,* reflecting light, or *transparent,* transmitting light.

In reflection, the incoming light, or *incident ray,* has the same angle as the *reflected ray* when measured from a perpendicular from the point of reflection, called the *normal.* That the two angles are equal is

called the *law of reflection.* The law of reflection explains how a flat mirror forms a *virtual image,* one from which light rays do not originate. Light rays do originate from the other kind of image, a *real image.*

Light rays are bent, or *refracted,* at the boundary when passing from one transparent media to another. The amount of refraction depends on the *incident angle* and the *index of refraction,* a ratio of the speed of light in a vacuum to the speed of light in the media. When the refracted angle is 90°, *total internal reflection* takes place. This limit to the angle of incidence is called the *critical angle,* and all light rays with an incident angle at or beyond this angle are reflected internally.

Each color of light has a range of wavelengths that forms the *spectrum* from red to violet. A glass prism has the property of *dispersion,* separating a beam of white light into a spectrum. Dispersion occurs because the index of refraction is different for each range of colors, with short wavelengths refracted more than larger ones.

A wave model of light can be used to explain diffraction, interference, and polarization, all of which provide strong evidence for the wavelike nature of light. *Diffraction* is the bending of light around the edge of an object or the spreading of light in an arc after passing through a tiny opening. *Interference* occurs when light passes through two small slits or holes and produces an *interference pattern* of bright lines and dark zones. *Polarized light* vibrates in one direction only, in a plane. Light

Continued—

a meter, so you can fit a tremendous number of data tracks onto the disc, which has each track spaced 1.6 microns apart. Next, the CD audio or CD-ROM discs are made by using the master disc as a mold. Soft plastic is pressed against the master disc in a vacuum-forming machine so the small physical marks—the pits and bumps made by the laser—are pressed into the plastic. This makes a record of the strings of binary numbers that were etched into the master disc by the strong but tiny laser beam. During playback, a low-powered laser beam is reflected off the track to read the binary marks on it. The optical sensor head contains a tiny diode laser, a lens, mirrors, and tracking devices that can move the head in three directions. The head moves side to side to keep the head over a single track (within 1.6 micron), it moves up and down to keep the laser beam in focus, and it moves forward and backward as a fine adjustment to maintain a constant linear velocity.

The advantages of the CD audio and video discs over conventional records or tapes include more uniform and accurate frequency response, a complete absence of background noise, and absence of wear—since

nothing mechanical touches the surface of the disc when it is played. The advantages of the CD-ROM over the traditional magnetic floppy disks or magnetic hard disks include storage capacity, long-term reliability, and the impossibility of a head crash with a resulting loss of data. Each CD-ROM, because of the incredibly tiny size of the recording track, can store about 600 megabytes (millions of bytes) of data. This means that one 12 cm CD-ROM disc will hold the same amount of data as 429 high-density, 1.4 megabyte, 3.5-inch floppy disks (or 750 of the common double-sided, 800 kilobyte floppy disks).

The disadvantage of the CD audio and CD-ROM discs is the lack of ability to do writing or rewriting. Rewritable optical media are available, for example, using the magneto-optical method (M-O), which combines magnetic recording with optical techniques. M-O uses a plastic disc that has a layer of magnetic particles embedded in the plastic. To record digitized data a laser heats a section of the track. At the same time an electromagnet magnetizes the metallic particle layer: north end up for a binary 1, and south end up for a binary 0. The plastic cools and "freezes" the binary information

in the magnetic particles of the track. The disc is read as a linearly polarized laser beam is rotated clockwise or counterclockwise according to the magnetic orientation of the layer it is focused upon. The light is reflected to a photodetector, where changes in light intensity are interpreted as binary data. This process can be used repeatedly as necessary.

Newer optical data storage technologies include a Compact Disc-Write-Once method (CD-WO). This method uses a dye-based optical medium that absorbs heat from the writing laser, changing color and reflecting light differently for the reading laser. Another common new technology is the Write-Once Read-Many (WORM) system that writes data to a disc only once; the data is then permanently stored. Currently, WORM systems are used for records management and office functions that require a large storage capacity and relatively fast access. In the future, you may see systems that store tremendous amounts of data on a simple paper card the size of a credit card. Could you imagine all of your textbooks stored on cards that you could carry around in one pocket? Perhaps you will read them on a small, pocket-sized TV.

can be polarized by certain materials, by reflection, or by scattering. Polarization can only be explained by a transverse wave model.

A wave model fails to explain observations of light behaviors in the *photoelectric effect* and *blackbody radiation*. Max Planck found that he could modify the wave theory to explain blackbody radiation by assuming that vibrating molecules could only have fixed amounts, or *quanta*, of energy and found that the quantized energy is related to the frequency and a constant known today as *Planck's constant*. Albert Einstein applied Planck's quantum concept to the photoelectric effect and described a light wave in terms of quanta of energy called *photons*. Each photon has an energy that is related to the frequency and Planck's constant.

Today, the properties of light are explained by a model that incorporates both the wave and the particle nature of light. Light is considered to have both wave and particle properties and is not describable in terms of anything known in the everyday-sized world.

Summary of Equations

8.1

$$\text{angle of incidence} = \text{angle of reflection}$$

$$\theta_i = \theta_r$$

8.2

$$\text{index of refraction} = \frac{\text{speed of light in vacuum}}{\text{speed of light in material}}$$

$$n = \frac{c}{v}$$

8.3

$$\text{speed of light in vacuum} = (\text{wavelength})(\text{frequency})$$

$$c = \lambda f$$

8.4

$$\begin{pmatrix}\text{quantized} \\ \text{energy}\end{pmatrix} = \begin{pmatrix}\text{positive} \\ \text{whole} \\ \text{number}\end{pmatrix}\begin{pmatrix}\text{Planck's} \\ \text{constant}\end{pmatrix}(\text{frequency})$$

$$E = nhf$$

8.5

$$\begin{pmatrix}\text{energy of} \\ \text{photon}\end{pmatrix} = \begin{pmatrix}\text{Planck's} \\ \text{constant}\end{pmatrix}(\text{frequency})$$

$$E = hf$$

A CLOSER LOOK | Fiber Optics

The communication capacities of fiber optics are enormous. When fully developed and exploited, a single fiber the size of a human hair could in principle carry all the telephone conversations and all the radio and television broadcasts in the United States at one time without interfering with one another. This enormous capacity is possible because of a tremendous range of frequencies in a beam of light. This range determines the information-carrying capacity, which is called *bandwidth*. The traditional transmission of telephone communication depends on the movement of electrons in a metallic wire, and a single radio station usually broadcasts on a single frequency. Light has a great number of different frequencies, as you can imagine when looking at table 8.2. An optic fiber has a bandwidth about a million times greater than the frequency used by a radio station.

Physically, an optical fiber has two main parts. The center part, called the *core*, is practically pure silica glass with an index of refraction of about 1.6. Larger fibers, about the size of a human hair or larger, transmit light by total internal reflection. Light rays internally reflect back and forth about a thousand times or so in each 15 cm length of such an optical fiber core. Smaller fibers, about the size of a light wavelength, act as waveguides, with rays moving straight through the core. A layer of transparent material called the *cladding* is used to cover the core. The cladding has a lower index of refraction, perhaps 1.50, and helps keep the light inside the core.

Presently fiber optics systems are used to link telephone offices across the United States and other nations, and in a submarine cable that links many nations. The electrical signals of a telephone conversation are converted to light signals, which travel over optical fibers. The light signals are regenerated by devices called *repeaters* until the signal reaches a second telephone office. Here, the light signals are transformed to electrical signals, which are then transmitted by more traditional means to homes and businesses. Today the laying of fiber optics lines to homes and businesses is a goal of nearly every major telephone company in the United States.

One way of transmitting information over a fiber optics system is to turn the transmitting light on and off rapidly. Turning the light on and off is not a way of sending ordinary codes or signals, but is a rapid and effective way of sending *digital information*. Digital information is information that has been converted into digits, that is, into a string of numbers. Fax machines, CD-ROM players, computers, and many other modern technologies use digitized information. Closely examine a black-and-white newspaper photograph, and you will see that it is made up of many tiny dots that vary in density from place to place. The image can be digitized by identifying the location of every tiny piece of the image, which is called a *pixel*, with a string of numbers. Once digitized, an image can be manipulated by a computer, reproduced by a printer, sent over a phone line by a fax machine, or stored for future use. Most fax machines transmit images with a resolution of about 240 pixels per inch, which is also called 240 dots per inch (dpi).

How is a telephone conversation digitized? The mouthpiece part of a telephone converts the pressure fluctuations of a voice sound wave into electric current fluctuations. This fluctuating electric current is digitized by a logic circuit that samples the amplitude of the wave at regular intervals, perhaps thousands of times per second. The measured amplitude of a sample (i.e., how far the sample is above or below the baseline) is now described by numbers. The succession of sample heights becomes a succession of numbers, and the frequency and amplitude pattern of a voice is now digitized.

Digitized information is transmitted and received by using the same numbering system that is used by a computer. A computer processes information by recognizing if a particular circuit carries an electric current or not, that is, if a circuit is on or off. The computer thus processes information by counting in twos, using what is called the *binary system*. Unlike our ordinary decimal system, in which we count by tens, the binary system has only two symbols, the digits 0 and 1. For example, the decimal number 4 has a binary notation of 0100; the decimal number 5 is 0101, 6 is 0110, 7 is 0111, 8 is 1000, and so on. In good sound-recording systems, a string of 16 binary digits is used to record the magnitude of each sample.

Digitized information is transmitted by translating the binary numbers, the 0s and 1s, into a series of pulses that are fed to a laser (with 0 meaning off, and 1 meaning on). Millions of bits per second can be transmitted over the fiber optics system to a receiver at the other end. The receiver recognizes the very rapid on-and-off patterns as binary numbers, which are converted into the original waveform. The waveform is now transformed into electric current fluctuations, just like the electric current fluctuations that occurred at the other end of the circuit. These fluctuations are amplified and converted back into the pressure fluctuations of the original sound waveform by the earpiece of the receiving telephone.

Currently, new optical fiber systems have increased almost tenfold the number of conversations that can be carried through the much smaller, lighter-weight fiber optics. Extension of the optical fiber to the next generation of computers will be the next logical step, with hundreds of available television channels, two-way videophone conversations, and unlimited data availability. The use of light for communications will have come a long way since the first use of signaling mirrors and flashing lamps from mountaintops and church steeples. Some people now call the new data availability the *information superhighway*.

KEY TERMS

blackbody radiation (p. **184**)
diffraction (p. **196**)
incandescent (p. **184**)
index of refraction (p. **190**)
interference (p. **197**)
light ray model (p. **186**)
luminous (p. **184**)
photoelectric effect (p. **199**)
photons (p. **200**)

polarized (p. **198**)
quanta (p. **200**)
real image (p. **188**)
refraction (p. **189**)
total internal
  reflection (p. **190**)
unpolarized light (p. **198**)
virtual image (p. **188**)

APPLYING THE CONCEPTS

1. A luminous object is an object that
   a. gives off a dim blue-green light in the dark.
   b. produces light of its own by any method.
   c. shines by reflected light only, such as the moon.
   d. glows only in the absence of light.

2. According to the electromagnetic wave model, visible light is produced when
   a. an electric charge is accelerated with a magnitude within a given range.
   b. an electric charge is moved at a constant velocity.
   c. a blackbody is heated to any temperature above absolute zero.
   d. an object absorbs electromagnetic radiation.

3. An object is hot enough to emit a dull red glow. When this object is heated even more, it will
   a. emit shorter-wavelength, higher-frequency radiation.
   b. emit longer-wavelength, lower-frequency radiation.
   c. emit the same wavelengths as before, but with more energy.
   d. emit more of the same wavelengths with more energy.

4. The difference in the light emitted from a candle, an incandescent lightbulb, and the sun is basically from differences in
   a. energy sources.
   b. materials.
   c. temperatures.
   d. phases of matter.

5. Before it travels through the earth's atmosphere, sunlight is mostly
   a. infrared radiation.
   b. visible light.
   c. ultraviolet radiation.
   d. blue light.

6. You are able to see in shaded areas, such as under a tree, because light has undergone
   a. refraction.
   b. incident bending.
   c. a change in speed.
   d. diffuse reflection.

7. An image that is not produced by light rays coming from the image but is the result of your brain's interpretations of light rays is called a(n)
   a. real image.
   b. imagined image.
   c. virtual image.
   d. phony image.

8. Light traveling at some angle as it moves from water into the air is refracted away from the normal as it enters the air, so the fish you see under water is actually (draw a sketch if needed)
   a. above the refracted image.
   b. below the refracted image.
   c. beside the refracted image.
   d. in the same place as the refracted image.

9. When viewed straight down (90° to the surface), a fish under water is
   a. above the image (away from you).
   b. below the image (closer to you).
   c. beside the image.
   d. in the same place as the image.

10. The ratio of the speed of light in a vacuum to the speed of light in some transparent materials is called
    a. the critical angle.
    b. total internal reflection.
    c. the law of reflection.
    d. the index of refraction.

11. Any part of the electromagnetic spectrum, including the colors of visible light, can be measured in units of
    a. wavelength.
    b. frequency.
    c. energy.
    d. any of the above.

12. A prism separates the colors of sunlight into a spectrum because
    a. each wavelength of light has its own index of refraction.
    b. longer wavelengths are refracted more than shorter wavelengths.
    c. red light is refracted the most, violet the least.
    d. all of the above.

13. Light moving through a small pinhole does not make a shadow with a distinct, sharp edge because of
    a. refraction.
    b. diffraction.
    c. polarization.
    d. interference.

14. Which of the following can only be explained by a wave model of light?
    a. reflection
    b. refraction
    c. interference
    d. photoelectric effect

15. The polarization behavior of light is best explained by considering light to be
    a. longitudinal waves.
    b. transverse waves.
    c. particles.
    d. particles with ends, or poles.

16. The sky appears to be blue when the sun is high in the sky because
    a. blue is the color of air, water, and other fluids in large amounts.
    b. red light is scattered more than blue.
    c. blue light is scattered more than the other colors.
    d. none of the above.

17. The photoelectric effect proved to be a problem for a wave model of light because
    a. the number of electrons ejected varied directly with the intensity of the light.
    b. the light intensity had no effect on the energy of the ejected electrons.
    c. the energy of the ejected electrons varied inversely with the intensity of the light.
    d. the energy of the ejected electrons varied directly with the intensity of the light.

18. Max Planck made the revolutionary discovery that the energy of vibrating molecules involved in blackbody radiation existed only in
    a. multiples of certain fixed amounts.
    b. amounts that smoothly graded one into the next.
    c. the same, constant amount of energy in all situations.
    d. amounts that were never consistent from one experiment to the next.

19. Einstein applied Planck's quantum discovery to light and found
    a. a direct relationship between the energy and frequency of light.
    b. that the energy of a photon divided by the frequency of the photon always equaled a constant known as Planck's constant.
    c. that the energy of a photon divided by Planck's constant always equaled the frequency.
    d. all of the above.

20. Today, light is considered to be
    a. tiny particles of matter that move through space, having no wave properties.
    b. electromagnetic waves only, with no properties of particles.
    c. a small-scale phenomenon without a sharp distinction between particle and wave properties.
    d. something that is completely unknown.

Answers

1. b 2. a 3. a 4. c 5. a 6. d 7. c 8. b 9. d 10. d 11. d 12. a 13. b 14. c 15. b 16. c 17. b 18. a 19. d 20. c

QUESTIONS FOR THOUGHT

1. What determines if an electromagnetic wave emitted from an object is a visible light wave or a wave of infrared radiation?
2. What model of light does the polarization of light support? Explain.
3. Which carries more energy, red light or blue light? Should this mean anything about the preferred color of warning and stop lights? Explain.
4. What model of light is supported by the photoelectric effect? Explain.
5. What happens to light that is absorbed by matter?
6. One star is reddish, and another is bluish. Do you know anything about the relative temperatures of the two stars? Explain.
7. When does total internal reflection occur? Why does this occur in the diamond more than other gemstones?
8. Why does a highway sometimes appear wet on a hot summer day when it is not wet?
9. How can you tell if a pair of sunglasses is polarizing or not?
10. What conditions are necessary for two light waves to form an interference pattern of bright lines and dark areas?
11. Explain why the intensity of reflected light appears to change if you tilt your head from side to side while wearing polarizing sunglasses.
12. Why do astronauts in orbit around Earth see a black sky with stars that do not twinkle but see a blue Earth?
13. What was so unusual about Planck's findings about blackbody radiation? Why was this considered revolutionary?
14. Why are both the photon model and the electromagnetic wave model accepted today as a single theory? Why was this so difficult for people to accept at first?

PARALLEL EXERCISES

The exercises in groups A and B cover the same concepts. Solutions to group A exercises are located in appendix D.

Group A

1. What is the speed of light while traveling through (a) water and (b) ice?
2. How many minutes are required for sunlight to reach the earth if the sun is 1.50×10^8 km from the earth?
3. How many hours are required before a radio signal from a space probe near the planet Pluto reaches Earth, 6.00×10^9 km away?
4. A light ray is reflected from a mirror with an angle $10°$ to the normal. What was the angle of incidence?
5. Light travels through a transparent substance at 2.20×10^8 m/s. What is the substance?
6. The wavelength of a monochromatic light source is measured to be 6.00×10^{-7} m in a diffraction experiment. (a) What is the frequency? (b) What is the energy of a photon of this light?
7. At a particular location and time, sunlight is measured on a 1 square meter solar collector with an intensity of 1,000.0 W. If the peak intensity of this sunlight has a wavelength of 5.60×10^{-7} m, how many photons are arriving each second?
8. A light wave has a frequency of 4.90×10^{14} cycles per second. (a) What is the wavelength? (b) What color would you observe (see table 8.2)?
9. What is the energy of a gamma photon of frequency 5.00×10^{20} Hz?
10. What is the energy of a microwave photon of wavelength 1.00 mm?

Group B

1. (a) What is the speed of light while traveling through a vacuum? (b) While traveling through air at 30°C? (c) While traveling through air at 0°C?
2. How much time is required for reflected sunlight to travel from the Moon to Earth if the distance between Earth and the Moon is 3.85×10^5 km?
3. How many minutes are required for a radio signal to travel from Earth to a space station on Mars if the planet Mars is 7.83×10^7 km from Earth?
4. An incident light ray strikes a mirror with an angle of $30°$ to the surface of the mirror. What is the angle of the reflected ray?
5. The speed of light through a transparent substance is 2.00×10^8 m/s. What is the substance?
6. A monochromatic light source used in a diffraction experiment has a wavelength of 4.60×10^{-7} m. What is the energy of a photon of this light?
7. In black-and-white photography, a photon energy of about 4.00×10^{-19} J is needed to bring about the changes in the silver compounds used in the film. Explain why a red light used in a darkroom does not affect the film during developing.
8. The wavelength of light from a monochromatic source is measured to be 6.80×10^{-7} m. (a) What is the frequency of this light? (b) What color would you observe?
9. How much greater is the energy of a photon of ultraviolet radiation ($\lambda = 3.00 \times 10^{-7}$ m) than the energy of an average photon of sunlight ($\lambda = 5.60 \times 10^{-7}$ m)?
10. At what rate must electrons in a wire vibrate to emit microwaves with a wavelength of 1.00 mm?

You can see the tracks of hundreds of atomic particles in this bubble chamber. Current research indicates that protons and neutrons are made of even smaller particles called *quarks*.

CHAPTER | Nine

Atomic Structure

Many materials used today are relatively new, created in the last few decades. These new materials are the result of modern chemical research, produced and manufactured through controlled chemical reactions. The new materials include synthetic fibers, from nylon to polyesters, and plastics, from polyethylene to Teflon. They also include water-based paints and superadhesives used in construction. The manufactured materials are lighter, stronger, and have special properties not found in natural materials. Today, such synthetic materials are used extensively in buildings, clothing, automobiles, and airplanes. The packaging, preserving, and marketing of many convenience foods are also made possible by the products of chemical research, as are manufactured vitamins and drugs that help keep you healthy. From synthetic fibers to synthetic drugs, there are millions of products today that are the direct result of chemical research.

The countless numbers of new products resulting from chemical research demonstrate understandings about matter and how it is put together. These understandings start with the most basic unit of matter, the atom. Perhaps you have wondered how incredibly tiny atoms were discovered and how they can be studied. Atoms are so tiny that they are invisible to any optical device. Even more incredible is the study of the innermost parts of these invisible atoms and the development of knowledge of how they are put together. You will soon know the answer to questions about how atoms were discovered and studied (figure 9.1). This chapter contains the essence of the fascinating story of how the atomic concept was discovered and developed.

The development of the modern atomic model illustrates how modern scientific understanding comes from many different fields of study. For example, you will learn how studies of electricity led to the discovery that atoms have subatomic parts called *electrons*. The discovery of radioactivity led to the discovery of more parts, a central nucleus that contains protons and neutrons. Information from the absorption and emission of light was used to construct a model of how these parts are put together, a model resembling a miniature solar system with electrons circling the nucleus. The solar system model had initial, but limited, success and was inconsistent with other understandings about matter and energy. Modifications of this model were attempted, but none solved the problems. Then the discovery of wave properties of matter led to an entirely new model of the atom.

The atomic model will be put to use in later chapters to explain the countless varieties of matter and the changes that matter undergoes. In addition, you will learn how these changes can be manipulated to make new materials, from drugs to ceramics. In short, you will learn how understanding the atom and all the changes it undergoes not only touches your life directly but shapes and affects all parts of civilization.

FIRST DEFINITION OF THE ATOM

The atomic concept is very old, dating back to ancient Greek philosophers some 2,500 years ago. The ancient Greeks thought, reasoned, and speculated about the basis for their surroundings. Consider, as an example, that you have observed water in puddles, ponds, lakes, rivers, and perhaps an ocean. You recognize rain, snow, sleet, dew, and ice to be water. You have also observed that plants and animals contain water, and that water can be produced by heating wood and other substances. If you think about all the forms, places, and things that water is a part of, you might begin to get the idea that water is a basic, fundamental substance that makes up other materials. Over time, this kind of reasoning led the ancient Greeks to the concept that everything is made up of four basic, fundamentally different substances, or *elements:* earth, air, fire, and water. From logical considerations of what they observed, the ancient Greeks considered all substances to be composed of these four elements in varying proportions.

The ancient Greeks also reasoned about the way that pure substances are put together. A glass of water, for example, appears to be the same throughout. Is it the same? Two plausible, but conflicting, ideas were possible as an intellectual exercise. The water could have a continuous structure, that is, it could be completely homogeneous throughout. The other idea was that the water only appears to be continuous but is actually *discontinuous.* This means that if you continue to divide the water into smaller and smaller volumes, you would eventually reach a limit to this dividing, a particle that could not be further subdivided. This model was developed by Leucippus and Democritus in the fourth century B.C. Democritus called the indivisible particle an *atom,* from a Greek word meaning "uncuttable."

Democritus speculated that the atoms of a substance were separated by *empty space* and that the atoms of the four elements had different shapes. Since this is an intellectual exercise, you can imagine anything you wish—as long as it is logically consistent with what is observed. For example, you might imagine that

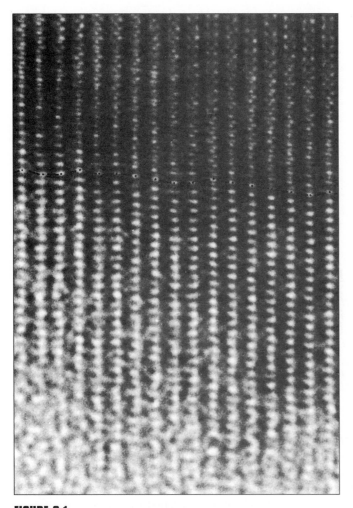

FIGURE 9.1

This electron-microscope high-resolution image shows magnification of the thin edge of a piece of mica. The white dots are "empty tunnels" between layers of silicon-oxygen tetrahedrons, and the black dots are potassium atoms that bond the tetrahedrons together. Note the 10 Angstrom width, which is 0.000001 mm.

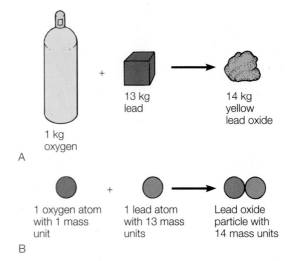

FIGURE 9.2

(*A*) Oxygen and lead combine to form yellow lead oxide in a ratio of 1:13. (*B*) If 1 atom of oxygen combines with 1 atom of lead, the fixed ratio in which oxygen and lead combine must mean that 1 atom of lead is 13 times more massive than 1 atom of oxygen.

ATOMIC STRUCTURE DISCOVERED

In the 1600s, Robert Boyle's work with gases provided evidence to reject Aristotle's idea that a vacuum could not exist. In 1661, Boyle published *The Sceptical Chymist,* in which he rejected Aristotle's four-element theory, too. Boyle defined an element as a "simple substance" that could not be broken down into anything simpler. Today, an **element** is defined as a pure substance that cannot be broken down to anything simpler by chemical or physical means. Water is a pure substance, but it can be broken down into oxygen and hydrogen, so water is not an element. Oxygen and hydrogen are pure substances that cannot be broken down into anything simpler, so they are elements. Oxygen, silicon, iron, gold, and aluminum are common elements, and over one hundred elements are known today. Elements will be considered in the next chapter. The development of a model of the atom is the topic of interest in this chapter, a model that will explain how elements are different.

It was information about how elements combine that led John Dalton in the early 1800s to bring back the ancient Greek idea of hard, indivisible atoms. In general, he had noted, as others had, that certain elements always combined with other elements in fixed ratios. For example, 1 gram of oxygen always combined with 13 grams of lead to produce 14 grams of a yellow compound called lead oxide (figure 9.2). Dalton reasoned that this must mean that the oxygen and lead were made up of individual particles called *atoms,* not a form of matter that was completely homogeneous in structure. If elements were continuous, there would be no reason for the oxygen and lead to combine in fixed ratios. If elements were composed of atoms, however, then whole atoms would combine to make the compound. By way of analogy, on a macroscopic scale you could consider

atoms of water might be round and smooth since water pours and flows. Atoms of earth might have cubic shapes that would prevent them from pouring and flowing. Democritus speculated that one substance changed to another by separation or combination of atoms.

When Aristotle organized and recorded all the ancient Greek knowledge that was available during this time, it was about one hundred years after Democritus had proposed the existence of atoms. Aristotle adopted the continuous, four-element model because of perceived logical problems with the empty space between the atoms—he did not believe a vacuum could exist in nature. The idea of four elements—earth, air, fire, and water—with a continuous structure became the accepted model for the next thousand years. If anyone during this period did support the idea of atoms, and a few did, they were definitely in the minority.

FIGURE 9.3

Reasoning the existence of atoms from the way elements combine in fixed-weight ratios. (*A*) If matter were a continuous, infinitely divisible material, there would be no reason for one amount to go with another amount. (*B*) If matter is made up of discontinuous, discrete units (atoms), then the units would combine in a fixed-weight ratio. Since discrete units combine in a fixed-piece ratio, they must also combine in a fixed-weight-based ratio. See text for a discussion of this analogy.

both peanut butter and jelly to be homogeneous, continuous substances. If you combine peanut butter and jelly there is no reason for a particular amount of peanut butter to go with a particular amount of jelly. Crackers and uniform slices of cheese, however, could both be considered on a macroscopic scale to be individual units. If you combine a slice of cheese and a cracker, the combination would have a total weight with part contributed by a cracker and part contributed by the cheese slice. A fixed ratio of the weight of a slice of cheese to the weight of an individual cracker would result in the same total weight each time. Using this kind of reasoning, Dalton theorized that elements must be composed of atoms (figure 9.3).

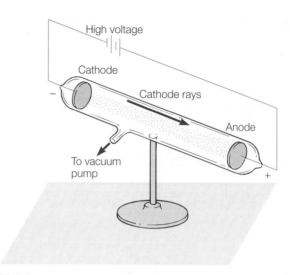

FIGURE 9.4

A vacuum tube with metal plates attached to a high-voltage source produces a greenish beam called *cathode rays*. These rays move from the cathode (negative charge) to the anode (positive charge).

During the 1800s, Dalton's concept of hard, indivisible atoms was familiar to most scientists. Yet, the existence of atoms was not generally accepted by all scientists. There was skepticism about something that could not be observed directly. Strangely, full acceptance of the atom came in the early 1900s with the discovery that the atom was not indivisible after all. The atom has parts that give it an internal structure. The first part to be discovered was the *electron*, a part that was discovered through studies of electricity.

Discovery of the Electron

Scientists of the 1800s were interested in understanding the nature of electricity, but it was impossible to see anything but the effects of a current as it ran through a wire. To observe a current directly, they tried to produce a current by itself, away from matter, by pumping the air from a tube and then running a current through the empty space. By 1885, a good air pump was invented that could evacuate the air from a glass tube until the pressure was 1/10,000 of normal air pressure. When metal plates inside such a tube were connected to the negative and positive terminals of a high-voltage source (figure 9.4), a greenish beam was observed that seemed to move from the cathode (negative terminal) through the empty tube and collect at the anode (positive terminal). Since this mysterious beam seemed to come out of the cathode it was said to be made of *cathode rays.*

Just what the cathode rays were became a source of controversy. Some scientists suggested that cathode rays were atoms of the cathode material. But the properties of cathode rays were found to be the same when the cathode was made of any material. Perhaps the rays were a form of light (figure 9.5)? But the rays were observed to be deflected by a magnet, not a behavior observed with light. What *were* the cathode rays?

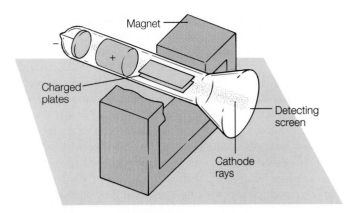

FIGURE 9.6

A cathode ray passed between two charged plates is deflected toward the positively charged plate. The ray is also deflected by a magnetic field. By measuring the deflection by both, J. J. Thomson was able to calculate the ratio of charge to mass. He was able to measure the deflection because the detecting screen was coated with zinc sulfide, a substance that produces a visible light when struck by a charged particle.

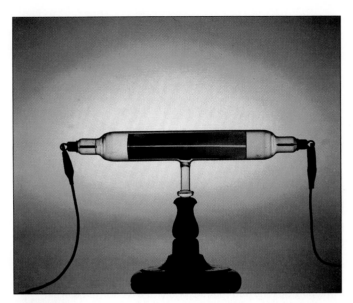

FIGURE 9.5

What appears to be visible light coming through the slit in this vacuum tube is produced by cathode ray particles striking a detecting screen. You know it is not light, however, since the beam can be pulled or pushed away by a magnet and since it is attracted to a positively charged metal plate. These are not the properties of light, so cathode rays must be something other than light.

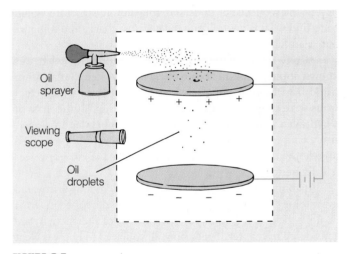

FIGURE 9.7

Millikan measured the charge of an electron by balancing the pull of gravity on oil droplets with an upward electrical force. Knowing the charge-to-mass ratio that Thomson had calculated, Millikan was able to calculate the charge on each droplet. He found that all the droplets had a charge of 1.60×10^{-19} coulomb or multiples of that charge. The conclusion was that this had to be the charge of an electron.

The mystery of the cathode rays was finally solved by the English physicist J. J. Thomson in 1897. Thomson had placed a positively charged metal plate on one side of the beam and a negatively charged metal plate on the other side (figure 9.6). The beam was deflected toward the positive plate and away from the negative plate. Since it was known that unlike charges attract and like charges repel, this meant that the beam must be composed of *negatively charged particles.*

The cathode ray was also deflected when caused to pass between the poles of a magnet. Thomson knew that moving charges are deflected by a magnetic field. He found more information by adjusting the electric charge on the plates above and below the beam, then measuring the deflection. The same procedure was used with a measured magnetic field. A greater charge on a particle would result in a greater deflection by an electric field, and a greater moving mass would be more difficult to deflect than a lesser moving mass. By balancing the deflections made by the magnet with the deflections made by the electric field, Thomson could thus determine the *ratio of the charge to mass* for an individual particle. Today, the charge-to-mass ratio is considered to be 1.7584×10^{11} coulomb/kilogram. The significant part of Thomson's experiments was finding that the charge-to-mass ratio was the *same* no matter what gas was in the tube and of what materials the electrodes were made. Thomson was convinced that he had discovered a fundamental particle, the stuff of which atoms are made.

Thomson did not propose any special name for the particle. Some time earlier, the term **electron** had been proposed as a name for the unit of charge gained or lost when atoms became ions. Not long after Thomson reported his findings in 1897, the existence of the fundamental particle was generally accepted, and everyone started calling the particles *electrons.*

A method for measuring the charge and mass of the electron was worked out by an American physicist, Robert A. Millikan, around 1906. Millikan used an apparatus like the one illustrated in figure 9.7 to measure the charge indirectly. Small droplets of mineral oil sprayed into the apparatus could be observed with a magnifier, and measurements could be made as the droplets drifted downwards. With a vertical electric field turned on, the droplets would drift upwards at a rate depending on the electric charge on each droplet. Measuring the rise and fall of the droplets as the electric field was turned on and off enabled Millikan to deduce the charge from the speed of fall and rise.

Millikan found that none of the droplets had a charge less than one particular value (1.60×10^{-19} coulomb) and that larger charges on various droplets were always multiples of this unit of charge. Since all of the droplets carried the single unit of charge or multiples of the single unit, the unit of charge was assumed to be the charge of a single electron.

Knowing the charge of a single electron and knowing the charge-to-mass ratio that Thomson had measured now made it possible to calculate the mass of a single electron. The mass of an electron was thus determined to be about 9.11×10^{-31} kg, or about 1/1,840 of the mass of the lightest atom, hydrogen.

Thomson had discovered the negatively charged electron, and Millikan had measured the charge and mass of the electron. But atoms themselves are electrically neutral. If an electron is part of an atom, there must be something else that is positively charged, canceling the negative charge of the electron. The next step in the sequence of understanding atomic structure would be to find what is neutralizing the negative charge and to figure out how all the parts are put together.

About 1900, Thomson proposed a model for what was known about the atom at the time. He suggested that an atom could be a blob of positively charged matter in which electrons were stuck like "raisins in plum pudding." If the mass of a hydrogen atom is due to the electrons embedded in a positively charged matrix, then 1,840 electrons would be needed together with sufficient positive matter to make the atom electrically neutral.

At about this same time, 1896, radioactivity was discovered by Antoine-Henri Becquerel, and it was soon described in terms of alpha, beta, and gamma rays. The details of radioactivity are considered in chapter 15. For now, all you need to know is what was known in 1907, that alpha particles are very fast, massive, and positively charged particles that are spontaneously given off from radioactive elements. The experimental work with alpha particles would lead to the discovery of the nucleus and then other parts of the atom.

The Nucleus

The nature of radioactivity and matter were the research interests of a British physicist, Ernest Rutherford. In 1907, Rutherford was studying the scattering of alpha particles directed toward a thin sheet of metal. As shown in figure 9.8, alpha particles from a radioactive source were allowed to move through a small opening in a lead container, so only a narrow beam of the massive, fast-moving particles would penetrate a very thin sheet of gold. The alpha particles were then detected by plates covered with zinc sulfide, which produced a small flash of light when struck by the positively charged alpha particle.

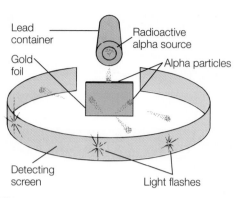

FIGURE 9.8

Rutherford and his co-workers studied alpha particle scattering from a thin metal foil. The alpha particles struck the detecting screen, producing a flash of visible light. Measurements of the angles between the flashes, the metal foil, and the source of the alpha particles showed that the particles were scattered in all directions, including straight back toward the source.

Activities

The luminous dial of a watch or clock contains a mixture of zinc sulfide and a radioactive substance. Obtain a watch or clock with a luminous dial and a magnifying glass. Wait in a completely dark room about ten minutes until your eyes adjust to the darkness. Observe the glowing parts of the dial with the magnifying glass. Is the glow continuous or is it made up of tiny flashes of light? What would flashes of light mean?

Rutherford and his co-workers found that most of the alpha particles went straight through the foil, and just as expected, none were scattered by more than a few degrees. Then he suggested that one of his young co-workers check to see if any alpha particles were scattered through very large angles. He really did not expect any great amount of scattering since alpha particles were very fast, massive particles with a great deal of energy and electrons were not very massive in comparison. When a young co-worker reported that alpha particles were deflected at very large angles and some were even reflected backwards, Rutherford was astounded. He could account for the large deflections and backward scattering only by assuming that the massive, positively charged particles were repelled by a massive positive charge concentrated in a small region of the atom (figure 9.9). He concluded that an atom must have a tiny, massive, and positively charged **nucleus** surrounded by electrons. Since the charge of the electrons was opposite that of the nucleus, the electrons must be moving, or else they would be attracted to it.

From measurements of the scattering, Rutherford estimated the radius of the nucleus to be approximately 10^{-13} cm. Other researchers had estimated the radius of the atom to be on

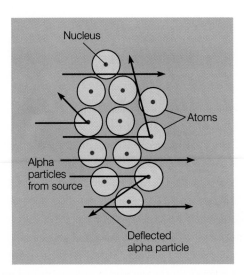

FIGURE 9.9

Rutherford's nuclear model of the atom explained the alpha-scattering results as positive alpha particles experiencing a repulsive force from the positive nucleus. Measurements of the percent of alpha particles passing straight through and of the various angles of scattering of those coming close to the nuclei gave Rutherford a means of estimating the size of the nucleus.

the order of 10^{-8} cm, so the electrons must be moving around the nucleus at a distance 100,000 times the radius of the nucleus. To visualize this spatial relationship, think of the thickness of a dime, which is about 1 mm thick. A distance 100,000 times the thickness of a dime is 100,000 mm, or 100 m. Thus, if the radius of a nucleus were the thickness of a dime, the electrons would be moving around the nucleus at a distance of about 100 m away, or about the length of a football field. If the radius of a nucleus were about the same size as the thickness of a dime, the atom would be about two football fields wide (figure 9.10). As you can see, the volume of an atom is mostly empty space.

Rutherford announced his conclusions and evidence for the existence of the atomic nucleus in 1911. This was a revolutionary development that would soon lead to more developments and more evidence about atomic structure. In 1917, Rutherford was able to break up the nucleus of a nitrogen atom by using alpha particles and to identify the discrete unit of positive charge which we now call a **proton.** Rutherford also speculated about the existence of a neutral particle in the nucleus, a **neutron.** The neutron was eventually identified in 1932 by James Chadwick.

Today, the number of *protons* in the nucleus of an atom is called the **atomic number.** An element is made up of atoms that all have the same number of protons in their nucleus, so all atoms of an element have the same atomic number. Hydrogen has an atomic number of 1, so any atom that has one proton in its nucleus is an atom of the element hydrogen. Today, there are 110 different kinds of elements, each with a different number of protons. The *neutrons* of the nucleus, along with the protons, contribute to the mass of an atom. Atomic mass (and atomic weight) will be considered in the next chapter.

A

B

FIGURE 9.10

From measurements of alpha particle scattering, Rutherford estimated the radius of an atom to be 100,000 times greater than the radius of the nucleus. This ratio is comparable to that of the (A) thickness of a dime to the (B) length of a football field.

Thus, the atom has a tiny, massive nucleus containing positively charged protons and neutral neutrons. Negatively charged electrons, equal in number to the protons, are moving around at a distance of about 100,000 times the radius of the nucleus. How are these electrons moving? It might occur to you, as it did to Rutherford and others, that an atom might be similar to a miniature solar system. In this analogy, the nucleus is in the role of the sun, electrons in the role of moving planets in their orbits, and electrical attractions between the nucleus and electrons in the role of gravitational attraction. There are, however, significant problems with this idea. If electrons were moving in circular orbits, they would continually change their direction of travel and would therefore be accelerating. According to the Maxwell model of electromagnetic radiation, an accelerating electric charge emits electromagnetic radiation such as light. If an electron gave off light, it would lose energy. The energy loss would mean that the electron could not maintain its orbit, and it would

Night Vision

How is it possible to see when photons do not have the energy of visible light? The charge-coupled device (CCD) is used in applications with visible light, including the home video camera, commercial-studio and news-bureau photography, and astronomy. The light-gathering part of a CCD device is basically a group of small photocells, with perhaps thousands in an array behind a lens. An image falls on the array, and each photocell stores a charge that is proportional to the amount of light that the cell absorbs. A microprocessor measures the amount of charge in each photocell and considers it as a pixel, a small bit of the overall image. A shade of gray or a color is assigned to each pixel and the image is transmitted to a screen, printed, or magnetically stored for future use as a digitized movie or photograph. Modern digital cameras have the ability to take high-resolution, 36-bit color photographs at 1/250 second, and record as many as five pictures in about 3 seconds onto a magnetic card.

Modern, still-life, digital CCD cameras bring electronic photography very close to the image quality and control that is needed by professional photographers. A CCD-array digital camera has an advantage since the picture is available instantly, without the use of darkrooms, chemicals, or light-sensitive paper. Another advantage is found in the way that a picture can be manipulated electronically, easily undergoing image processing. This means a computer can be used to manipulate a picture, and the process is called "computer enhancement." Computer enhancement is usu-

ally done to enhance details, visibility of a feature, or to correct for a defect, such as a hazy atmosphere in an outdoor photograph.

The CCD requires visible light, so how could the device be used to see in the dark? All objects with a temperature above absolute zero emit infrared radiation, which is the part of the electromagnetic spectrum sometimes called "heat radiation." Infrared radiation lies just outside the visible light region of the spectrum, but it behaves much like visible light. A lens can focus infrared radiation, for example, just as visible light. Portable night-vision devices use a lens to focus infrared radiation onto an array of charged-coupled devices, much like the digital camera that uses visible light. But in this case the electronic pixels measure temperature differences of a fraction of a degree, then electronically create an image of the heat differences. This image of heat differences is electronically converted into a black-and-white video image, which can be viewed as if a video camera could see things in the dark.

Night-vision devices are made possible by the detector material, a pyroelectric ceramic formed of barium strontium titanate. This ceramic has pyroelectric properties that change sharply as a function of temperature, so each pixel generates an electric signal that matches the temperature differences in fractions of a degree. The focused infrared radiation from a scene is thus converted into an electric signal of the scene that appears almost as realistic as an image made from reflected visible light.

---

be pulled into the oppositely charged nucleus. The atom would collapse as electrons spiraled into the nucleus. Since atoms do not collapse like this, there is a significant problem with the solar system model of the atom.

THE BOHR MODEL

Niels Bohr was a young Danish physicist who visited Rutherford's laboratory in 1912 and became very interested in questions about the solar system model of the atom. He wondered what determined the size of the electron orbits and the energies of the electrons. He wanted to know why orbiting electrons did not give off electromagnetic radiation. Seeking answers to questions such as these led Bohr to incorporate the *quantum concept* of Planck and Einstein with Rutherford's model to describe the electrons in the outer part of the atom. This quan-

tum concept will be briefly reviewed before proceeding with the development of Bohr's model of the hydrogen atom.

The Quantum Concept

In the year 1900, Max Planck introduced the idea that matter emits and absorbs energy in discrete units that he called **quanta.** Planck had been trying to match data from spectroscopy experiments with data that could be predicted from the theory of electromagnetic radiation. In order to match the experimental findings with the theory, he had to assume that specific, discrete amounts of energy were associated with different frequencies of radiation. In 1905, Albert Einstein extended the quantum concept to light, stating that light consists of discrete units of energy that are now called **photons.** The energy of a photon is directly proportional to the frequency of vibration, and the higher the frequency of light, the greater the energy of the individual photons. In addition, the inter-

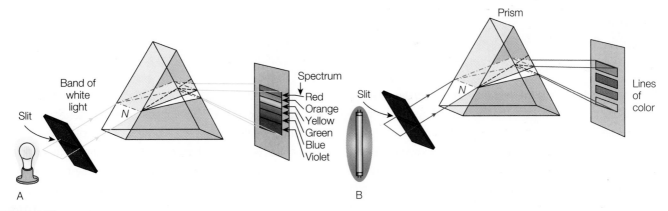

FIGURE 9.11

(A) Light from incandescent solids, liquids, or dense gases produces a continuous spectrum as atoms interact to emit all frequencies of visible light. (B) Light from an incandescent gas produces a line spectrum as atoms emit certain frequencies that are characteristic of each element.

action of a photon with matter is an "all-or-none" affair, that is, matter absorbs an entire photon or none of it. The relationship between frequency (f) and energy (E) is

$$E = hf$$

equation 9.1

where h is the proportionality constant known as *Planck's constant* (6.63×10^{-34} J·s). This relationship means that higher-frequency light, such as ultraviolet, has more energy than lower-frequency light, such as red light.

Atomic Spectra

Planck was concerned with hot solids that emit electromagnetic radiation. The nature of this radiation, called *blackbody radiation,* depends on the temperature of the source. When this light is passed through a prism, it is dispersed into a *continuous spectrum,* with one color gradually blending into the next as in a rainbow. Today, it is understood that a continuous spectrum comes from solids, liquids, and dense gases because the atoms interact, and all frequencies within a temperature-determined range are emitted. Light from an incandescent gas, on the other hand, is dispersed into a **line spectrum,** narrow lines of colors with no light between the lines (figure 9.11). The atoms in the incandescent gas are able to emit certain characteristic frequencies, and each frequency is a line of color that represents a definite value of energy. The line spectra are specific for a substance, and increased or decreased temperature changes only the intensity of the lines of colors. Thus hydrogen always produces the same colors of lines in the same position. Helium has its own specific set of lines, as do other substances. Line spectra are a kind of fingerprint that can be used to identify a gas. A line spectrum might also extend beyond visible light into ultraviolet, infrared, and other electromagnetic regions.

In 1885, a Swiss mathematics teacher named J. J. Balmer was studying the regularity of spacing of the hydrogen line spectra. Balmer was able to develop an equation that fit all the visible

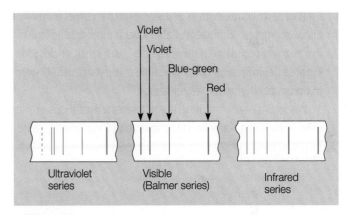

FIGURE 9.12

Atomic hydrogen produces a series of characteristic line spectra in the ultraviolet, visible, and infrared parts of the total spectrum. The visible light spectra always consist of two violet lines, a blue-green line, and a bright red line.

lines. By assigning values (n) of 3, 4, 5, and 6 to the four lines, he found the wavelengths fit the equation

$$\frac{1}{\lambda} = R\left(\frac{1}{2^2} - \frac{1}{n^2}\right)$$

equation 9.2

when R is a constant of 1.097×10^7 1/m.

Balmer's findings were:

Violet line	($n = 6$)	$\lambda = 4.1 \times 10^{-7}$ m
Violet line	($n = 5$)	$\lambda = 4.3 \times 10^{-7}$ m
Blue-green line	($n = 4$)	$\lambda = 4.8 \times 10^{-7}$ m
Red line	($n = 3$)	$\lambda = 6.6 \times 10^{-7}$ m

These four lines became known as the *Balmer series.* Other series were found later, outside the visible part of the spectrum (figure 9.12). The equations of the other series were different only in the value of n and the number in the other denominator.

C O N N E C T I O N S

Depression and Lack of Light

There is a definite order in the light from atoms and there are evidently expectations of order in the light sensed by people. Some people have special problems during the autumn and winter seasons, experiencing a lack of energy, wanting to sleep more, and having increased eating habits with an accompanying weight gain. These people might be suffering from winter depression, a condition also known as *seasonal affective disorder* (SAD). The problem is "seasonal" since it occurs only during the autumn and winter, and "affective" is a psychiatric term meaning something that results from emotions or feelings rather than from thought.

Note that SAD is different from *cabin fever,* which is the term used to describe the distress or uneasiness that comes from a lack of environmental stimulation. Cabin fever is more related to living in a sparsely populated region, especially in a small, enclosed living area. Cabin fever can occur any time of the year, but SAD seems to occur only during the winter season.

When SAD was first investigated, researchers thought they were on to something when they found that it was more common

in northern latitudes and less common in the "Sun Belt" states. This seemed to indicate that SAD might be a result of shorter days and less light in winter. Perhaps this is a problem of photoperiodism, a result of less exposure to light that somehow affects the clock in the brain that regulates sleeping, eating, and perhaps mood. Another theory is that chemical neurotransmitters such as serotonin and dopamine might become imbalanced, but corrected by light therapy. Yet another theory concerns a reduced physical sensitivity to light during the winter season, a change that can be corrected by light therapy.

The first step in testing these theories would be for people diagnosed as SAD patients to spend more time outdoors and to expose them to bright artificial light for at least 30 minutes a day. Such "light therapy" seems to have helped about half the patients. Ten years of research on "light therapy" has been inconclusive, however, as researchers have not yet been able to confirm the cause of the winter-related depression. Further research is needed to determine the causes of SAD and how light therapy helps.

Such regularity of observable spectral lines must reflect some unseen regularity in the atom. At this time it was known that hydrogen had only one electron. How could one electron produce series of spectral lines with such regularity?

Bohr's Theory

An acceptable model of the hydrogen atom would have to explain the characteristic line spectra and their regularity as described by Balmer. In fact, a successful model should be able to *predict* the occurrence of each color line as well as account for its origin. By 1913, Bohr was able to do this by applying the quantum concept to a solar system model of the atom. He began by considering the single hydrogen electron to be a single "planet" revolving in a circular orbit around the nucleus. He assumed that the electron could occupy more than one circular orbit and that it would have different states of energy, which depended on the radius of the orbit it was in at any particular time. Each orbit was assigned a number *n*, which could be any whole number 1, 2, 3, and so on out from the nucleus. These were known as the orbit *quantum numbers.* The following describes how orbit quantum numbers were used with definitions of (1) allowed orbits, (2) radiationless orbits, and (3) quantum jumps to describe the **Bohr model** of the atom.

Allowed Orbits

An electron can revolve around an atom only in specific allowed orbits. Bohr considered the electron to be a particle with a known mass in motion around the nucleus. Rotational motion is measured by *angular momentum,* a product of the mass (*m*), velocity (*v*), and radius of the orbit (*r*), or *mvr.* Conservation of angular momentum requires a greater velocity for a smaller orbit and less velocity for a larger orbit. It was Bohr's assumption that the allowed orbits are those for which the angular momentum of the electron (*mvr*) equaled the orbit quantum number (*n* = 1, 2, 3, ...) times Planck's constant (*h*) divided by 2π, or

$$mvr = n\frac{h}{2\pi}$$

equation 9.3

(Note that $2\pi r$ describes a circumference, and that *mv* describes momentum.) Bohr used this relationship to determine the allowed orbits because differences between the energy levels it describes fit exactly with the differences between the frequencies (and thus energies) of the line spectra described by Balmer. Bohr did not have a reason for this assumption; he used it simply because it worked.

Bohr also assumed that the force of electrical attraction between the electron (q_1) and proton (q_2) must be equal to the centripetal force if the electron moves in a circular orbit. This means that the force according to Coulomb's law of electrical attraction must be equal to the centripetal force described by Newton's second law of motion, or

$$\frac{kq_1q_2}{r^2} = \frac{mv^2}{r}$$

equation 9.4

This describes a relationship that is very similar to the relationship between the gravitational forces between the earth and moon and the centripetal force that keeps the moon in its orbit around the earth (see chapter 3). The earth and moon are attracted by the force of gravity, however, not electrical forces.

Equations 9.3 and 9.4 can be combined to eliminate v, and the known values of k, q, h, m, and π, yielding $2 = n^2 (0.529 \times 10^{-10}$ m). The value of r represents the distances, or radii, of the allowed orbits. When r is obtained for the first quantum number ($n = 1$), the radius of the closest orbit is found to be 0.529×10^{-10} m, which is known as the calculated *Bohr radius*. The next allowed orbit has a radius of 2.12×10^{-10} m, the next a radius of 4.76×10^{-10} m, and so on. According to the Bohr model, electrons can exist *only* in one of these allowed orbits and nowhere else.

Radiationless Orbits

An electron in an allowed orbit does not emit radiant energy as long as it remains in the orbit. It had been understood since the development of Maxwell's theory of electromagnetic radiation that an accelerating electron should emit an electromagnetic wave, such as light, which would move off into space from the electron. Bohr recognized that electrons moving in a circular orbit are accelerating, since they are changing direction continuously. Yet, light was not observed to be emitted from hydrogen atoms in their normal state. Bohr decided that the situation must be different for orbiting electrons, and that electrons could stay in their allowed orbits and *not* give off light. He postulated this rule as a way to make his theory consistent with other scientific theories.

Quantum Jumps

An electron gains or loses energy only by moving from one allowed orbit to another (figure 9.13). The reference level for the potential energy of an electron is considered to be zero when the electron is *removed* from an atom. The electron, therefore, has a lower and lower potential energy at closer and closer distances to the nucleus and has a negative value when it is in some allowed orbit. By way of analogy, you could consider ground level as a reference level where the potential energy of some object equals zero. But suppose there are two basement levels below the ground. An object on either basement level would have a gravitational potential energy less than zero, and work would have to be done on each object to bring it back to the zero level. Thus, each object would have a negative potential energy. The object on the lowest level would have the largest negative value of energy, since more work would have to be done on it to bring it back to the zero level. Therefore, the object on the lowest level would have the *least* potential energy, and this would be expressed as the *largest negative value*.

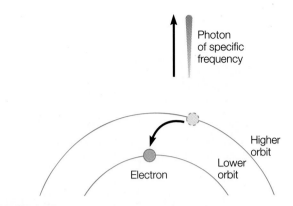

FIGURE 9.13

Each time an electron makes a "quantum leap," moving from a higher-energy orbit to a lower-energy orbit, it emits a photon of a specific frequency and energy value.

Just as the objects on different basement levels have negative potential energy, the electron has a definite negative potential energy in each of the allowed orbits. Bohr calculated the energy of an electron in the orbit closest to the nucleus to be -2.18×10^{-18} J, which is called the energy of the lowest state. The energy of electrons is expressed in units of the **electron volt** (eV). An electron volt is defined as the energy of an electron moving through a potential of one volt. Since this energy is charge times voltage (from $V = W/q$), 1.00 eV is equivalent to 1.60×10^{-19} J. Therefore the energy of an electron in the innermost orbit is its energy in joules divided by 1.60×10^{-19} J/eV, or -13.6 eV.

Bohr found that the energy of each of the allowed orbits could be found from the simple relationship of

$$E_n = \frac{E_L}{n^2}$$

equation 9.5

where E_L is the energy of the innermost orbit (-13.6 eV), and n is the quantum number for an orbit, or 1, 2, 3, and so on. Thus, the energy for the second orbit ($n = 2$) is $E_2 = -13.6$ eV/4 = -3.40 eV. The energy for the third orbit out ($n = 3$) is $E_3 = -13.6$ eV/9 = -1.51 eV, and so forth (figure 9.14). Thus, the energy of each orbit is *quantized*, occurring only as a definite value.

In the Bohr model, the energy of the electron is determined by which allowable orbit it occupies. The only way that an electron can change its energy is to jump from one allowed orbit to another in quantum "jumps." An electron must *acquire* energy to jump from a lower orbit to a higher one. Likewise an electron *gives up* energy when jumping from a higher orbit to a lower one. Such jumps must be all at once, not part way and not gradual. By way of analogy, this is very much like the gravitational potential energy that you have on the steps of a staircase. You have the lowest potential on the bottom step and the greatest amount on the top step. Your potential energy is quantized because you can increase or decrease it by going up or down a number of steps, but you cannot stop between the steps.

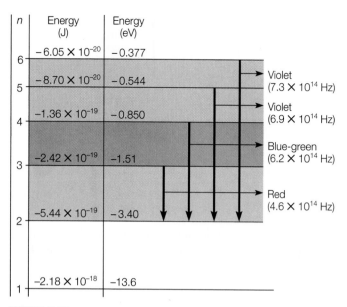

FIGURE 9.14

An energy level diagram for a hydrogen atom, not drawn to scale. The energy levels (*n*) are listed on the left side, followed by the energies of each level in J and eV. The color and frequency of the visible light photons emitted are listed on the right side, with the arrow showing the orbit moved from and to.

FIGURE 9.15

These fluorescent lights emit light as electrons of mercury atoms inside the tubes gain energy from the electric current. As soon as they can, the electrons drop back to a lower-energy orbit, emitting photons with ultraviolet frequencies. Ultraviolet radiation strikes the fluorescent chemical coating inside the tube, stimulating the emission of visible light.

An electron acquires energy from high temperatures or from electrical discharges to jump to a higher orbit. An electron jumping from a higher to a lower orbit gives up energy in the form of light. A single photon is emitted when a downward jump occurs, and the *energy of the photon is exactly equal to the difference in the energy level* of the two orbits. If E_L represents the lower-energy level (closest to the nucleus) and E_H represents a higher-energy level (farthest from the nucleus), the energy of the emitted photon is

$$hf = E_H - E_L$$

equation 9.6

where h is Planck's constant, and f is the frequency of the emitted light (figure 9.15).

The energy level diagram in figure 9.14 shows the energy states for the orbits of a hydrogen atom. The lowest energy state, $n = 1$, is known as the **ground state** (or normal state). The higher states, $n = 2$, $n = 3$, and so on, are known as the **excited states.** The electron in a hydrogen atom would, under normal conditions, be located in the ground state ($n = 1$), but high temperatures or electric discharge can give the electron sufficient energy to jump to one of the excited states. Once in an excited state, the electron immediately jumps back to a lower state, as shown by the arrows in the figure. The length of the arrow represents the frequency of the photon that the electron emits in the process. A hydrogen atom can give off only one photon at a time, and the many lines of a hydrogen line spectrum come from many atoms giving off many photons at the same time.

As you can see, the energy level diagram in figure 9.14 shows how the change of known energy levels from known orbits results in the exact energies of the color lines in the Balmer series. Bohr's theory did offer an explanation for the lines in the hydrogen spectrum with a remarkable degree of accuracy. However, the model did not have much success with larger atoms. Larger atoms had spectra lines that could not be explained by the Bohr model with its single quantum number. A German physicist, A. Sommerfield, tried to modify Bohr's model by adding elliptical orbits in addition to Bohr's circular orbits. It soon became apparent that the "patched up" model, too, was not adequate. Bohr had made the rule that there were radiationless orbits without an explanation, and he did not have an explanation for the quantized orbits. There was something fundamentally incomplete about the model.

QUANTUM MECHANICS

The Bohr model of the atom successfully accounted for the line spectrum of hydrogen and provided an understandable mechanism for the emission of photons by atoms. However, the model did not predict the spectra of any atom larger than hydrogen, and there were other limitations. A new, better theory was needed. The roots of a new theory would again come from experiments with light. Experiments with light had established that sometimes light behaves like a stream of particles, and at other times it behaves like a wave (see chapter 8). Eventually scientists began

CONNECTIONS

Emitted Light and What You See

All visible light consists of photons that have been emitted by electrons moving from an excited state to a ground state. But this does not explain why the same frequency light would give different results at different times. For example, why does the same sunlight sometimes make a blue sky, other times make a red sky, while clouds appear to be white? The answer is related to the frequency of the emitted photons and the nature of matter doing the interacting with the photons.

The sky is blue because sunlight is scattered by nitrogen and oxygen molecules in the atmosphere. These molecules are much smaller than the wavelength of light, and the relationship is that the amount of scattering is proportional to the frequency of light to the 4th power. Thus the higher frequency blue light from the sun is scattered in all directions, much more than the frequency of other colors, and you see blue in the sky wherever you look. This also explains why a sunrise or a sunset is often red. Light travels through more air when the sun is rising or setting and the blue and other higher frequency colors of light are scattered away, leaving the red.

Clouds are white because the particles making up the cloud (water droplets) are larger than all the ROYGBV light wavelengths. Thus all wavelengths are scattered the same way and white light is reflected.

---

to accept that light has both wave properties and particle properties, which is now referred to as the *wave-particle duality of light*. This dual nature of light was recognized in 1905 when Einstein applied Planck's quantum concept to the energy of a photon with the relationship found in equation 9.1, $E = hf$, where E is the energy of a photon particle, f is the frequency of the associated wave, and h is Planck's constant.

Matter Waves

In 1923 Louis de Broglie, a French physicist, reasoned that symmetry is usually found in nature, so if a particle of light has a dual nature, then particles such as electrons should too. De Broglie reasoned further that if this is true, an electron in its circular path around the nucleus would have to have a particular wavelength that would fit into the circumference of the orbit (figure 9.16). The circumference of the orbit must be a whole number of wavelengths long, or

$$\text{circumference} = (\text{number})(\text{wavelength})$$

or

$$2\pi r = n\lambda$$

where $n = 1, 2, 3, \ldots$

equation 9.7

De Broglie derived a relationship from equations concerning light and energy, which was

$$\lambda = \frac{h}{mv}$$

equation 9.8

where λ is the wavelength, m is mass, v is velocity, and h is again Planck's constant. This equation means that any moving particle has a wavelength that is associated with its mass and velocity.

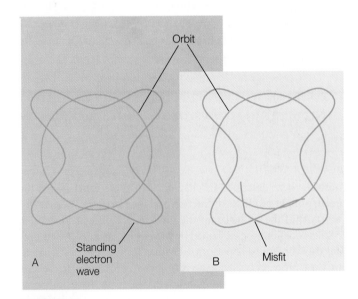

FIGURE 9.16

(*A*) Schematic of de Broglie wave, where the standing wave pattern will just fit in the circumference of an orbit. This is an allowed orbit. (*B*) This orbit does not have a circumference that will match a whole number of wavelengths; it is not an allowed orbit.

In other words, de Broglie was proposing a wave-particle duality of matter, the existence of *matter waves*. According to equation 9.8, *any* moving object should exhibit wave properties. However, an ordinary-sized object would have wavelengths so small that they could not be observed. This is different for electrons because they have such a tiny mass.

The idea of matter waves was soon tested after de Broglie published his theory. Experiments with a beam of light passing through a very small opening or by the edge of a sharp-edged

Life and Light

There is a close connection between atoms, experiments with the nature of light, and the life sciences. Physics typically deals with relationships between the nature of light and how it is produced. Chronobiology is a field of life science that deals with the nature of light and life, concerned with how light changes atoms and molecules. The overall result of this light and molecular involvement results in biological rhythms.

You may have heard of jet lag as a disruption of bodily rhythms caused by high-speed jet travel across several time zones. Many of our internal cycles such as temperature, sleep, cravings for sweets, and so on, are programmed by habit or by nature and many such cycles are linked to environmental cues. Travelling across time zones can change the cues and disrupt the rhythms, resulting in jet lag.

Humans are not the only organisms that experience biological rhythms. All organisms seem to have some variation of biological rhythms, from bacteria to complex plants, fungi, and animals. This is how organisms respond to a changing environment that is a result of living on a rotating world. The relationship between autumn and trees dropping their leaves, for example, is a biological rhythm based on light. Experiments in a greenhouse have shown that shortening the length of days will cause trees to drop their leaves even though there is no change in temperature. So, leaf dropping is a rhythm that involves changes in the amount of daylight each day. Many biological rhythms are based on changes in the amount of daylight or moonlight.

Early research in chronobiology was concerned with cataloging the existence of biological rhythms; then the focus changed to finding the parts of the nervous system that could serve as a clock, the regulator of the rhythms. Today, research in chronobiology has moved to the molecular level of analysis, identifying genes whose protein products create a biological clock. Biological clocks function to create an awareness of the passage of time, but also to anticipate change in the environment. At the molecular level, the clock might be a system of different proteins that change according to external cues, reacting by turning certain genes on or off.

Why is research in chronobiology important? One reason is that many human applications exist, such as problems of jet lag, shift-worker problems, and sleeping disorders. Understanding the biologic basis of 24-hour periodicity rhythms is necessary to find the causes or origins of these and other disorders.

---

obstacle were described in chapter 8. In these experiments the beam of light produces diffraction and interference patterns. This was part of the evidence for the wave nature of light, since such results could only be explained by waves, not particles. When similar experiments were performed with a beam of electrons, *identical* wave property behaviors were observed. This and many related experiments showed without doubt that electrons have both wave properties and particle properties. And, as was the case with light waves, measurements of the electron interference patterns provided a means to measure the wavelength of electron waves.

Recall that waves confined on a fixed string establish resonant modes of vibration called *standing waves* (see chapter 6). Only certain fundamental frequencies and harmonics can exist on a string, and the combination of the fundamental and overtones gives the stringed instrument its particular quality. The same result of resonant modes of vibrations is observed in *any* situation where waves are confined to a fixed space. Characteristic standing wave patterns depend on the wavelength and wave velocity for waves formed on strings, in enclosed columns of air, or for any kind of wave in a confined space. Electrons are confined to the space near a nucleus, and electrons have wave properties, so an electron in an atom must be a confined wave. Does an electron form a characteristic wave pattern? This was the question being asked in about 1925 when Heisenberg, Schrödinger, Dirac, and others applied the wave

nature of the electron to develop a new model of the atom based on the mechanics of electron waves. The new theory is now called *wave mechanics*, or **quantum mechanics.**

Wave Mechanics

Erwin Schrödinger, an Austrian physicist, treated the atom as a three-dimensional system of waves to derive what is now called the *Schrödinger equation*. Instead of the simple circular planetary orbits of the Bohr model, solving the Schrödinger equation results in a description of three-dimensional shapes of the patterns that develop when electron waves are confined by a nucleus. Schrödinger first considered the hydrogen atom, calculating the states of vibration that would be possible for an electron wave confined by a nucleus. He found that the frequency of these vibrations, when multiplied by Planck's constant, matched exactly, to the last decimal point, the observed energies of the quantum states of the hydrogen atom ($E = hf$). The conclusion is that the wave nature of the electron is the important property to consider for a successful model of the atom.

The quantum mechanics theory of the atom proved to be very successful; it confirmed all the known experimental facts and predicted new discoveries. The theory does have some of the same quantum ideas as the Bohr model; for example, an electron emits a photon when jumping from a higher state to a lower one. The Bohr model, however, considered the particle nature of an electron moving in a circular orbit with a definitely assigned position at a given time. Quantum mechanics considers the wave nature, with the electron as a confined wave with well-defined shapes and frequencies. A wave is not localized like a particle and is spread out in space. The quantum mechanics model is, therefore, a series of orbitlike smears, or fuzzy statistical representations, of where the electron might be found.

The Quantum Mechanics Model

The quantum mechanics model is a highly mathematical treatment of the mechanics of matter waves. In addition, the wave properties are considered as three-dimensional problems, and three quantum numbers are needed to describe the fuzzy electron cloud. The mathematical detail will not be presented here. The following is a qualitative description of the main ideas in the quantum mechanics model. It will describe the results of the mathematics and will provide a mental visualization of what it all means.

First, understand that the quantum mechanical theory is not an extension or refinement of the Bohr model. The Bohr model considered electrons as particles in circular orbits that could be only certain distances from the nucleus. The quantum mechanical model, on the other hand, considers the electron as a wave and considers the energy of its harmonics, or modes, of standing waves. In the Bohr model, the location of an electron was certain—in an orbit. In the quantum mechanical model, the electron is a spread-out wave.

Quantum mechanics describes the energy state of an electron wave with four *quantum numbers,* in terms of its (1) distance from the nucleus, (2) energy sublevel, (3) orientation in space, and (4) direction of spin.

The *principal quantum number,* called *n,* describes the *main energy level* of an electron in terms of its most probable distance from the nucleus. The lowest energy state possible is closest to the nucleus and is assigned the principal quantum number of 1 ($n = 1$). Higher states are assigned progressively higher positive whole numbers of $n = 2$, $n = 3$, $n = 4$, and so on. Electrons with higher principal quantum numbers have higher energies and are located farther from the nucleus.

The *angular momentum quantum number* defines energy sublevels within the main energy levels. Each sublevel is identified with a letter. The first four of these letters, in order of

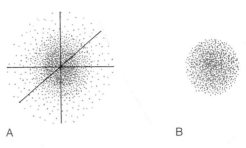

FIGURE 9.17

(*A*) An electron distribution sketch representing probability regions where an electron is most likely to be found. (*B*) A boundary surface, or contour, that encloses about 90 percent of the electron distribution shown in (*A*). This three-dimensional space around the nucleus, where there is the greatest probability of finding an electron, is called an *orbital.*

increasing energy, are s, p, d, and f. The choice of these letters goes back to spectral studies when the spectral lines were described as *s*harp, *p*rincipal, *d*iffuse, and *f*ine. The letter s represents the lowest sublevel, and the letter f represents the highest sublevel. A principal quantum number and a letter indicating the angular momentum quantum number are combined to identify the main energy state and energy sublevel of an electron. For an electron in the lowest main energy level, $n = 1$, and in the lowest sublevel, s, the number and letter are 1s (read as "one-s"). Thus 1s indicates an electron that is as close to the nucleus as possible in the lowest energy sublevel possible.

As stated in the Bohr model, the location of an electron was certain, that is, in an orbit. In the quantum mechanical model, the electron is spread out, and knowledge of its location is very uncertain. The **Heisenberg uncertainty principle** states that you cannot measure both the momentum and the exact position of an electron at the same time. The location of the electron can only be described in terms of *probabilities* of where it might be at a given instant. The probability of location is described by a fuzzy region of space called an **orbital.** An orbital defines the space where an electron is likely to be found. Orbitals have characteristic three-dimensional shapes and sizes and are identified with electrons of characteristic energy levels (figure 9.17).

An orbital shape represents where an electron could probably be located at any particular instant. This "probability cloud" could likewise have any particular orientation in space, and the direction of this orientation is uncertain. On the other hand, an external magnetic field applied to an atom produces different energy levels that are related to the orientation of the orbital to the magnetic field. The orientation of an orbital in space is described by the *magnetic quantum number.* This number is related to the energies of orbitals as they are oriented in space relative to an external magnetic field, a kind of energy subsublevel. In general, the lowest-energy sublevel (s) has only one orbital orientation. The next higher-energy sublevel (p) can have three orbital orientations (figure 9.18). The d sublevel can have five orbital orientations, and the highest sublevel, f, can have a total of seven different orientations (table 9.1).

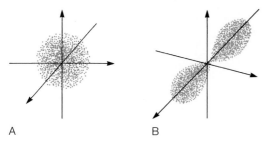

FIGURE 9.18

(*A*) A contour representation of an s orbital. (*B*) A contour representation of a p orbital.

TABLE 9.1	Quantum numbers and electron distribution to $n = 4$		
Main Energy Level	**Energy Sublevels**	**Maximum Number of Electrons**	**Maximum Number of Electrons per Main Energy Level**
$n = 1$	s	2	2
$n = 2$	s	2	
	p	6	8
$n = 3$	s	2	
	p	6	
	d	10	18
$n = 4$	s	2	
	p	6	
	d	10	
	f	14	32

High-resolution studies of the hydrogen line spectra revealed details that were not known when the Bohr model of the atom was first developed. These studies showed, for example, that what was previously believed to be a single red line was actually two lines that were very close together. The only way to explain the splitting was to consider the electron to be spinning on its axis like a top. Such a spinning movement would cause the electron to produce a magnetic field. The energy of the electron would depend on which way the electron magnetic field was aligned with an external magnetic field. Thus, an electron spinning one way (say clockwise) would have a different energy than one spinning the other way (say counterclockwise). These two spin orientations are described by the *spin quantum number* (figure 9.19).

Electron spin is an important property of electrons that helps determine the electronic structure of an atom. As it turns out, two electrons spinning in opposite directions produce unlike magnetic fields that are attractive, balancing some of the normal repulsion from two like charges. Two electrons of opposite spin, called an **electron pair,** can thus occupy the same orbital. This was summarized in 1924 by Wolfgang Pauli, a German physicist. His summary, now known as the **Pauli exclusion principle,** states that *no two electrons in an atom can have the same four quantum numbers.* This provides the key for understanding the electron structure of atoms.

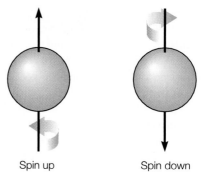

Spin up   Spin down

FIGURE 9.19

Experimental evidence supports the concept that electrons can be considered to spin one way or the other as they move about an orbital under an external magnetic field.

Electron Configuration

Recall that the energy of electrons is measured by considering an unattached electron to have an energy value of zero. It takes energy to remove an electron from an atom, so a bound electron must have *less* energy than a free one. The arrangement of electrons in orbitals is called the *electron configuration*. When in the ground state, electrons always adopt the lowest possible energies consistent with the Pauli exclusion principle, arranging themselves in the lowest energy orbitals as close to the nucleus as possible. The lowest possible energy level is $n = 1$, and the lowest sublevel is s. *The number of electrons that can occupy this orbital is limited.* According to the Pauli exclusion principle, no two electrons in an atom can have all four of their quantum numbers exactly the same. If one electron is in the first energy level ($n = 1$) in the orbital of the lowest energy (s), a second electron can occupy this same $n = 1$, s orbital only if it has a different spin orientation. Thus, the exclusion principle states that there can only be two electrons in the $n = 1$, s orbital and, as it works out, there can only be *a maximum of two electrons in any given orbital*. An atom of helium has two electrons and both can occupy the $n = 1$, s orbital because they have opposite spins. In general, electrons occupy the orbitals in an order starting with the lowest energy. Before you can describe the electron arrangement, you need to know how many electrons are present in an atom.

An atom is electrically neutral, so the number of protons (positive charge) must equal the number of electrons (negative charge). The atomic number therefore identifies the number of electrons as well as the number of protons:

atomic number = number of protons = number of electrons

Now that you have a means of finding the number of electrons, consider the various energy levels to see how the electron configuration is determined. There are four things to consider: (1) the main energy level, (2) the energy sublevel, (3) the number of orbital orientations, and (4) the electron spin. Recall that the lowest-energy level is $n = 1$, and successive numbers identify progressively higher-energy levels. Recall also that the energy sublevels, in order of increasing energy, are s, p, d, and f. This electron configuration is written in shorthand, with 1s standing for the lowest-energy sub-

CONNECTIONS

NSLS: The National Synchrotron Light Source

The National Synchrotron Light Source (NSLS) is a research facility that shows close connections between the concepts presented in this chapter and the growing nature of interdisciplinary research. Scientists from all over the country do research at NSLS, but all projects are reviewed and approved based on scientific importance. In return for using the facility for free, the researchers agree to publish their results in open literature where others can read about their experiments and use the results in their own work.

A synchrotron is a machine that uses powerful magnets to guide electrons through a circular tube, a ring with almost all the air pumped out of it. An electron gun sends bunches of electrons through the vacuum and, as they round each bend in the ring, they give off energy in the form of light. This is called "synchrotron light."

More than 2,000 researchers use the NSLS each year. Many of these researchers are working in the field of materials sciences, a field interested in the development of transistors and chip technology for computers and processors. A fast-growing field is life sciences, particularly in the area of structural biology. Many major pharmaceutical companies are using synchrotron light to solve the structures of certain biological molecules, and then to design drugs to inhibit the molecules' actions. Medical centers and laboratories are using the NSLS in their research on antidepressants, joint diseases such as arthritis, and HIV. The Protein Data Bank is an archive of experimentally determined three-dimensional structures of biological macromolecules, serving a global community of researchers, educators, and students.*

*Information from the Internet: http://www.nsls.bnl.gov/

level of the first energy level. A superscript gives the number of electrons present in a sublevel. Thus, the electron configuration for a helium atom, which has two electrons, is written as

$$1s^2$$

This combination of symbols has the following meaning:

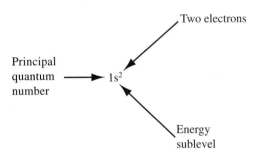

The symbols mean an atom with two electrons in the s sublevel of the first main energy level.

Table 9.2 gives the electron configurations for the first twenty elements. The configurations of the p energy sublevel have been condensed in this table. There are three possible orientations of the p orbital, each with two electrons (figure 9.20). This is shown as p^6, which is a condensation of the three possible p orientations. Note that the sum of the electrons in all the orbitals equals the atomic number. Note also that as you proceed from a lower atomic number to a higher one, the higher element has the same configuration as the element before it with the addition of one more electron. In general, it is then possible to begin with the simplest atom, hydrogen, and add one electron at a time to the order

TABLE 9.2	Electron configuration for the first twenty elements	
Atomic Number	**Element**	**Electron Configuration**
1	Hydrogen	$1s^1$
2	Helium	$1s^2$
3	Lithium	$1s^2 2s^1$
4	Beryllium	$1s^2 2s^2$
5	Boron	$1s^2 2s^2 2p^1$
6	Carbon	$1s^2 2s^2 2p^2$
7	Nitrogen	$1s^2 2s^2 2p^3$
8	Oxygen	$1s^2 2s^2 2p^4$
9	Fluorine	$1s^2 2s^2 2p^5$
10	Neon	$1s^2 2s^2 2p^6$
11	Sodium	$1s^2 2s^2 2p^6 3s^1$
12	Magnesium	$1s^2 2s^2 2p^6 3s^2$
13	Aluminum	$1s^2 2s^2 2p^6 3s^2 3p^1$
14	Silicon	$1s^2 2s^2 2p^6 3s^2 3p^2$
15	Phosphorus	$1s^2 2s^2 2p^6 3s^2 3p^3$
16	Sulfur	$1s^2 2s^2 2p^6 3s^2 3p^4$
17	Chlorine	$1s^2 2s^2 2p^6 3s^2 3p^5$
18	Argon	$1s^2 2s^2 2p^6 3s^2 3p^6$
19	Potassium	$1s^2 2s^2 2p^6 3s^2 3p^6 4s^1$
20	Calcium	$1s^2 2s^2 2p^6 3s^2 3p^6 4s^2$

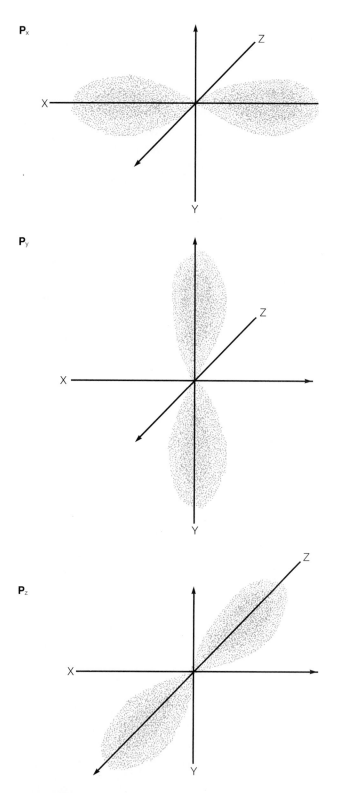

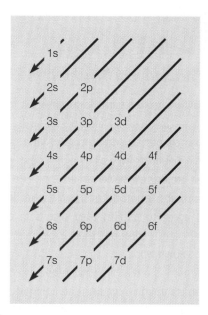

FIGURE 9.21

A matrix showing the order in which the orbitals are filled. Start at the top left, then move from the head of each arrow to the tail of the one immediately below it. This sequence moves from the lowest-energy level to the next higher level for each orbital.

of energy sublevels and obtain the electron configuration for all the elements. The exclusion principle limits the number of electrons in any orbital, and allowances will need to be made for the more complex behavior of atoms with many electrons.

The energies of the orbital are not fixed as you progress through the atomic numbers, and there are several factors that influence their energies. The first orbitals are filled in a straightforward 1s, 2s, 2p, 3s, then 3p order. Then the order becomes contrary to what you might expect. One useful way of figuring out the order in which orbitals are filled is illustrated in figure 9.21. Each row of this matrix represents a principal energy level with possible energy sublevels increasing from left to right. The order of filling is indicated by the diagonal arrows. There are exceptions to the order of filling shown by the matrix, but it works for most of the elements.

SUMMARY

Attempts at understanding matter date back to ancient Greek philosophers, who viewed matter as being composed of *elements*, or simpler substances. Two models were developed that considered matter to be (1) *continuous*, or infinitely divisible, or (2) *discontinuous*, made up of particles called *atoms*.

During the 1600s, Robert Boyle provided experimental evidence to reject the ancient Greek ideas of continuous matter made up of earth, air, fire, and water. Boyle reasoned that there were *elements*, which could not be broken down to anything simpler, and compounds, which were made up of combinations of elements.

In the early 1800s, Dalton published an *atomic theory*, reasoning that matter was composed of hard, indivisible atoms that were joined together or dissociated during chemical change.

FIGURE 9.20

There are three possible orientations of the p orbital, and these are called p_x, p_y, and p_z. Each orbital can hold two electrons, so a total of six electrons are possible in the three orientations; thus the notation p^6.

A CLOSER LOOK | The Quark

Some understanding about how matter is put together came with the discovery of the electron, proton, and neutron—three elementary particles that make up an atom. In the early 1900s, a particle outside the atom, the *photon,* was verified by experimental evidence. Two other particles were verified in the 1930s, the *neutrino* ("little neutral one") and the *positron* (a positively charged electron). By the mid-1930s a total of six elementary particles were known. Since that time, high-energy accelerator experiments have made it possible to collide particles with great violence, probing the inner parts of atoms and how they are put together. A multitude of elementary particles are now known to exist.

There are now thought to be twelve elementary particles that make up matter. They can be divided into three main groups: (1) *leptons,* which exist independently, (2) *quarks,* which exist together, making up a third group, and (3) *hadrons.*

Leptons are a group of fundamental particles that include the familiar electron, the muon (an overweight relative of the electron), and three types of neutrinos. In radioactive decay, an electron (beta particle) is emitted with an electron neutrino. For each lepton there is a corresponding antiparticle, or antilepton, with the same mass but opposite electric charge.

Hadrons are a group of composite particles with an internal structure, so they are not elementary particles. There are two subgroups of hadrons: (1) the *mesons* (meaning "intermediate mass," between electrons and protons) and (2) the *baryons* (meaning "greater mass"). Hundreds of short-lived hadrons have been identified that exist briefly after high-energy collisions. Among the more stable are the baryons named *protons* and *neutrons.*

Hadrons are composed of different combinations of fundamental particles called *quarks.* Five kinds of quarks, fancifully called *flavors,* have been identified, and a sixth is believed to exist. The existing quarks are called *up, down, sideways* (or *strange*), *charm,* and *bottom* (or *beauty*). The sixth flavor is named *top* (or *truth*). Each flavor carries a fractional charge that is either $-1/3$ or $+2/3$. Antiquarks have equal but opposite charges. In order to explain how identical quarks could combine as observed, each flavor was assigned three quantum states that are called *color.* Each flavor can carry a charge of red, green, or blue. Antiquarks carry a corresponding anticolor,

for example, a red quark has an antiquark of the color cyan, a green quark has an antiquark of the color magenta, and a blue quark has an antiquark of the color yellow. The idea of quark color was designed to follow the allowable combinations of quarks and antiquarks according to the exclusion principle. Hadrons do not have a color charge, so the sum of the quark colors making up the hadron must result in a white hadron. Baryons are made up of three quarks, so a combination of a red, a green, and a blue quark would be acceptable, since this would result in a white baryon. Mesons are made up of a quark and an antiquark so a combination of a blue quark and a yellow antiquark, would be acceptable since this would result in a white meson.

The quark model holds that all matter is made from combinations of six quarks and the six independent leptons. The story of subnuclear elementary particles is by no means complete. There is no explanation for why quarks and leptons exist. Are there more fundamental particles? Answers to this and more questions await further research.

When a good air pump to provide a vacuum was invented in 1885, *cathode rays* were observed to move from the negative terminal in an evacuated glass tube. The nature of cathode rays was a mystery. The mystery was solved in 1887 when Thomson discovered they were negatively charged particles now known as *electrons.* Thomson had discovered the first elementary particle of which atoms are made and measured their charge-to-mass ratio.

Rutherford developed a solar system model based on experiments with alpha particles scattered from a thin sheet of metal. This model had a small, massive, and positively charged *nucleus* surrounded by moving electrons. These electrons were calculated to be at a distance from the nucleus of 100,000 times the radius of the nucleus, so the volume of an atom is mostly empty space. Later, Rutherford proposed that the nucleus contained two elementary particles: *protons* with a positive charge and *neutrons* with no charge. The *atomic number* is the number of protons in an atom.

Bohr developed a model of the hydrogen atom to explain the characteristic *line spectra* emitted by hydrogen. His model specified that (1) electrons can move only in allowed orbits, (2) electrons do not emit radiant energy when they remain in an orbit, and (3) electrons move from one allowed orbit to another when they gain or lose energy. When an electron jumps from a higher orbit to a lower one, it gives up energy in the form of a single photon. The energy of the photon corresponds to the difference in energy between the two levels. The Bohr model worked well for hydrogen but not for other atoms.

De Broglie proposed that moving particles of matter (electrons) should have wave properties like moving particles of light (photons). His derived equation, $\lambda = h/mv$, showed that these *matter waves* were only measurable for very small particles such as electrons. De Broglie's proposal was tested experimentally, and the experiments confirmed that electrons do have wave properties.

Schrödinger and others used the wave nature of the electron to develop a new model of the atom called *wave mechanics,* or *quantum mechanics.* This model was found to confirm exactly all the experimental data as well as predict new data. The quantum mechanical model describes the energy state of the electron in terms of quantum numbers based on the wave nature of the electron. The quantum numbers defined the *probability* of the location of an electron in terms of fuzzy regions of space called *orbitals.*

Summary of Equations

9.1

$$\text{energy} = (\text{Planck's constant})(\text{frequency})$$

$$E = hf$$

where $h = 6.63 \times 10^{-34}\text{J·s}$

9.2

$$\frac{1}{\text{wavelength}} = \text{constant}\left(\frac{1}{2^2} - \frac{1}{\text{number}^2}\right)$$

$$\frac{1}{\lambda} = R\left(\frac{1}{2^2} - \frac{1}{n^2}\right)$$

where $R = 1.097 \times 10^7 \text{ 1/m}$

9.3

$$\text{angular momentum} = \left(\begin{array}{c}\text{orbit}\\\text{quantum}\\\text{number}\end{array}\right)\left(\frac{\text{Planck's constant}}{2\pi}\right)$$

$$mvr = n\frac{h}{2\pi}$$

where $h = 6.63 \times 10^{-34}$ J·s, and $n = 1, 2, 3, \ldots$ for an orbit

9.4

$$\text{electrical force} = \text{centripetal force}$$

$$\text{Coulomb's law} = \text{Newton's second law for circular motion}$$

$$\frac{kq_1 q_2}{r^2} = \frac{mv^2}{r}$$

9.5

$$\text{energy state of orbit number} = \frac{\text{energy state of innermost orbit}}{\text{number squared}}$$

$$E_n = \frac{E_L}{n^2}$$

where $E_L = -13.6$ eV, and $n = 1, 2, 3, \ldots$

9.6

$$\begin{array}{c}\text{energy}\\\text{of}\\\text{photon}\end{array} = \left(\begin{array}{c}\text{energy state}\\\text{of}\\\text{higher orbit}\end{array}\right) - \left(\begin{array}{c}\text{energy state}\\\text{of}\\\text{lower orbit}\end{array}\right)$$

$$hf = E_H - E_L$$

where $h = 6.63 \times 10^{-34}$ J·s; E_H and E_L must be in joules

9.7

$$\text{circumference of orbit} = (\text{whole number})(\text{wavelength})$$

$$2\pi r = n\lambda$$

where $n = 1, 2, 3, \ldots$

9.8

$$\text{wavelength} = \frac{\text{Planck's constant}}{(\text{mass})(\text{velocity})}$$

$$\lambda = \frac{h}{mv}$$

where $h = 6.63 \times 10^{-34}$ J·s

KEY TERMS

atomic number (p. **215**)

Bohr model (p. **218**)

electron (p. **213**)

electron pair (p. **224**)

electron volt (p. **219**)

element (p. **211**)

excited states (p. **220**)

ground state (p. **220**)

Heisenberg uncertainty principle (p. **223**)

line spectrum (p. **217**)

neutron (p. **215**)

nucleus (p. **214**)

orbital (p. **223**)

Pauli exclusion principle (p. **224**)

photons (p. **216**)

proton (p. **215**)

quanta (p. **216**)

quantum mechanics (p. **222**)

APPLYING THE CONCEPTS

1. According to the modern definition, which of the following is an element?
   a. water
   b. iron
   c. air
   d. All of the above.

2. John Dalton reasoned that atoms exist from the evidence that
   a. elements could not be broken down into anything simpler.
   b. water pours and flows when in the liquid state.
   c. elements always combined in certain fixed ratios.
   d. peanut butter and jelly could be combined in any ratio.

3. The electron was discovered through experiments with
   a. radioactivity.
   b. light.
   c. matter waves.
   d. electricity.

4. Thomson was convinced that he had discovered a subatomic particle, the electron, from the evidence that
   a. the charge-to-mass ratio was the same for all materials.
   b. cathode rays could move through a vacuum.
   c. electrons were attracted toward a negatively charged plate.
   d. the charge was always 1.60×10^{-19} coulomb.

5. The existence of a tiny, massive, and positively charged nucleus was deduced from the observation that
   a. fast, massive, and positively charged alpha particles all move straight through metal foil.
   b. alpha particles were deflected by a magnetic field.
   c. some alpha particles were deflected by metal foil.
   d. None of the above are correct.

6. According to Rutherford's calculations, the volume of an atom is mostly
   a. occupied by protons and neutrons.
   b. filled with electrons.
   c. occupied by tightly bound protons, electrons, and neutrons.
   d. empty space.

7. The atomic number is the number of
   a. protons.
   b. protons plus neutrons.
   c. protons plus electrons.
   d. protons, neutrons, and electrons in an atom.

8. All neutral atoms of an element have the same
   a. atomic number.
   b. number of electrons.
   c. number of protons.
   d. All of the above are correct.

9. The main problem with a solar system model of the atom is that
   a. electrons move in circular, not elliptical orbits.
   b. the electrons should lose energy since they are accelerating.
   c. opposite charges should attract one another.
   d. the mass ratio of the nucleus to the electrons is wrong.

10. The energy of a photon
    a. varies inversely with the frequency.
    b. is directly proportional to the frequency.
    c. varies directly with the velocity, not the frequency.
    d. is inversely proportional to the velocity.

11. The frequency of a particular color of light is equal to
    a. *Eh.*
    b. *h/E.*
    c. *E/h.*
    d. *Eh/2.*

12. A photon of which of the following has the most energy?
    a. red light
    b. orange light
    c. green light
    d. blue light

13. The lines of color in a line spectrum from a given element
    a. change colors with changes in the temperature.
    b. are always the same, with a regular spacing pattern.
    c. are randomly spaced, having no particular pattern.
    d. have the same colors, with a spacing pattern that varies with the temperature.

14. Hydrogen, with its one electron, produces a line spectrum in the visible light range with
    a. one color line.
    b. two color lines.
    c. three color lines.
    d. four color lines.

15. Using the laws of motion for moving particles and the laws of electrical attraction, Bohr calculated that electrons could
    a. move only in orbits of certain allowed radii.
    b. move, as do the planets, in orbits at any distance from the nucleus.
    c. move in orbits at distances from the nucleus that matched the distances between colors in the line spectrum.
    d. move in orbits at variable distances from the nucleus that are directly proportional to the velocity of the electrons.

16. According to the Bohr model, an electron gains or loses energy only by
    a. moving faster or slower in an allowed orbit.
    b. jumping from one allowed orbit to another.
    c. being completely removed from an atom.
    d. jumping from one atom to another atom.

17. When an electron in a hydrogen atom jumps from an orbit farther from the nucleus to an orbit closer to the nucleus, it
    a. emits a single photon with an energy equal to the energy difference of the two orbits.
    b. emits four photons, one for each of the color lines observed in the line spectrum of hydrogen.
    c. emits a number of photons dependent on the number of orbit levels jumped over.
    d. None of the above are correct.

18. The Bohr model of the atom
    a. explained the color lines in the hydrogen spectrum.
    b. could not explain the line spectrum of atoms larger than hydrogen.
    c. had some made-up rules without explanations.
    d. All of the above are correct.

19. The proposal that matter, like light, has wave properties in addition to particle properties was
    a. verified by diffraction experiments with a beam of electrons.
    b. never tested, since it was known to be impossible.
    c. tested mathematically, but not by actual experiments.
    d. verified by physical measurement of a moving baseball.

20. The quantum mechanics model of the atom is based on
    a. the quanta, or measured amounts of energy of a moving particle.
    b. the energy of a standing electron wave that can fit into an orbit.
    c. calculations of the energy of the three-dimensional shape of a circular orbit of an electron particle.
    d. Newton's laws of motion, but scaled down to the size of electron particles.

21. The Bohr model of the atom described the energy state of electrons with one quantum number. The quantum mechanics model uses how many quantum numbers to describe the energy state of an electron?
    a. one
    b. two
    c. four
    d. ten

22. An electron in the second main energy level and the second sublevel is described by the symbols
    a. 1s.
    b. 2s.
    c. 1p.
    d. 2p.

23. The space in which it is probable that an electron will be found is described by a(an)
    a. circular orbit.
    b. elliptical orbit.
    c. orbital.
    d. geocentric orbit.

24. Two electrons can occupy the same orbital because they have different
    a. principal quantum numbers.
    b. angular momentum quantum numbers.
    c. magnetic quantum numbers.
    d. spin quantum numbers.

Answers

1. b **2.** c **3.** d **4.** a **5.** c **6.** d **7.** a **8.** d **9.** b **10.** b **11.** c **12.** d **13.** b **14.** d **15.** a **16.** b **17.** a **18.** d **19.** a **20.** b **21.** c **22.** d **23.** c **24.** d

QUESTIONS FOR THOUGHT

1. What reason did Dalton have for bringing back the ancient Greek idea of matter being composed of hard, indivisible atoms?
2. What was the experimental evidence that Thomson had discovered the existence of a subatomic particle when working with cathode rays?
3. Describe the experimental evidence that led Rutherford to the concept of a nucleus in an atom.
4. What is the main problem with a solar system model of the atom?
5. Compare the size of an atom to the size of its nucleus.
6. What does *atomic number* mean? How does the atomic number identify the atoms of a particular element? How is the atomic number related to the number of electrons in an atom?
7. An atom has 11 protons in the nucleus. What is the atomic number? What is the name of this element? What is the electron configuration of this atom?
8. Describe the three main points in the Bohr model of the atom.
9. Why do the energies of electrons in an atom have negative values? (*Hint:* It is *not* because of the charge of the electron.)
10. Which has the lowest energy, an electron in the first energy level ($n = 1$) or an electron in the third energy level ($n = 3$)? Explain.
11. What is similar about the Bohr model of the atom and the quantum mechanical model? What are the fundamental differences?
12. What is the difference between a hydrogen atom in the ground state and one in the excited state?

PARALLEL EXERCISES

The exercises in groups A and b cover the same concepts. Solutions to group A exercises are located in appendix D.

Group A

1. A neutron with a mass of 1.68×10^{-27} kg moves from a nuclear reactor with a velocity of 3.22×10^3 m/s. What is the de Broglie wavelength of the neutron?
2. Calculate the energy (a) in eV and (b) in joules for the sixth energy level ($n = 6$) of a hydrogen atom.
3. How much energy is needed to move an electron in a hydrogen atom from $n = 2$ to $n = 6$? Give the answer (a) in joules and (b) in eV. (See figure 9.14 for needed values.)
4. What frequency of light is emitted when an electron in a hydrogen atom jumps from $n = 6$ to $n = 2$? What color would you see?
5. How much energy is needed to completely remove the electron from a hydrogen atom in the ground state?
6. Thomson determined the charge-to-mass ratio of the electron to be -1.76×10^{11} coulomb/kilogram. Millikan determined the charge on the electron to be -1.60×10^{-19} coulomb. According to these findings, what is the mass of an electron?
7. Assume that an electron wave making a standing wave in a hydrogen atom has a wavelength of 1.67×10^{-10} m. Considering the mass of an electron to be 9.11×10^{-31} kg, use the de Broglie equation to calculate the velocity of an electron in this orbit.
8. Using any reference you wish, write the complete electron configurations for (a) boron, (b) aluminum, and (c) potassium.
9. Explain how you know that you have the correct *total* number of electrons in your answers for 8a, 8b, and 8c.
10. Refer to figure 9.21 *only,* and write the complete electron configurations for (a) argon, (b) zinc, and (c) bromine.

Group B

1. An electron with a mass of 9.11×10^{-31} kg has a velocity of 4.3×10^6 m/s in the innermost orbit of a hydrogen atom. What is the de Broglie wavelength of the electron?
2. Calculate the energy (a) in eV and (b) in joules of the third energy level ($n = 3$) of a hydrogen atom.
3. How much energy is needed to move an electron in a hydrogen atom from the ground state ($n = 1$) to $n = 3$? Give the answer (a) in joules and (b) in eV.
4. What frequency of light is emitted when an electron in a hydrogen atom jumps from $n = 2$ to the ground state ($n = 1$)?
5. How much energy is needed to completely remove an electron from $n = 2$ in a hydrogen atom?
6. If the charge-to-mass ratio of a proton is 9.58×10^7 coulomb/kilogram and the charge is 1.60×10^{-19} coulomb, what is the mass of the proton?
7. An electron wave making a standing wave in a hydrogen atom has a wavelength of 8.33×10^{-11} m. If the mass of the electron is 9.11×10^{-31} kg, what is the velocity of the electron according to the de Broglie equation?
8. Using any reference you wish, write the complete electron configurations for (a) nitrogen, (b) phosphorus, and (c) chlorine.
9. Explain how you know that you have the correct *total* number of electrons in your answers for 8a, 8b, and 8c.
10. Referring to figure 9.21 *only,* write the complete electron configuration for (a) neon, (b) sulfur, and (c) calcium.

This is a picture of pure zinc, one of the 89 naturally occurring elements found on the earth.

CHAPTER | Ten

Elements and the Periodic Table

I n chapter 9 you learned how the concept of the atom was developed, evolving into the modern model of a fuzzy cloud of matter with an internal structure. We considered this internal structure and how the electrons make up the fuzzy cloud in atoms of different elements. But there was no discussion about the meaning or implications of the different structures or how they could be used to understand the properties of matter. That is the goal of this chapter, to understand matter based on the electron structure of atoms.

The behavior of matter seems bewildering when you consider all of its different kinds, forms, and shapes and all the changes it undergoes. Things seem bewildering and confusing because you cannot make connections between behaviors; that is, there are no apparent patterns. Often in such situations, the act of grouping or classifying things by similar properties is helpful. Classifying helps you to find patterns of similarities and identities of groups. Once you have a pattern, you will want to find a reason for its existence. You are now on your way to an understanding that not only accounts for the patterns but also provides a means of organizing all the information as well.

This chapter presents an example of gaining understanding through grouping, which is also known as *classifying* (figure 10.1). We will consider several different ways to classify matter, for example, into classes of mixtures and pure substances. Pure substances can be further subdivided into groups of elements and compounds. Some of the more interesting elements will be described, along with how they were named and their symbols. Matter can be classified other ways, including changes in matter. All changes are either physical or chemical, and chemical changes are the key to grouping according to the electron structure of atoms. After an introduction of a few new properties of atoms, the periodic table will be discussed in terms of its systematic classification of elements. This periodic classification can be used to predict how elements react with one another, making the study of matter much easier.

CLASSIFYING MATTER

The universe is made up of just two basics, matter and energy. **Matter** is usually defined as anything that occupies space and has mass. It is the substance of any particle or object, from the parts that make up atoms to the bulk of a giant star. Between these two extremes of size, there is an overwhelming and complex variety of observable sizes, shapes, and kinds of matter on the earth. One way to make thinking about such a variety a little less complex is to *classify*. Classifying is the act of mentally making groups based on similar properties. Classifying helps to organize your thinking and reveals patterns that might otherwise go unnoticed. The following discussion considers several different ways of classifying matter by focusing on different properties.

Metals and Nonmetals

Humans learned thousands of years ago that certain earthlike materials placed on a bed of glowing coals produced a new substance, a metal. The smelting of copper and tin dates back to about 3500 B.C., the beginning of the Bronze Age. Iron smelting dates back to about 1500 B.C., and steel was first made about 1200 B.C. These metals all have very different properties than the earthlike materials they were extracted from. Today a **metal** is recognized as a kind of matter having the following physical properties:

1. metallic *luster*, the way a shiny metal reflects light (figure 10.2);
2. high heat and electrical *conductivity*;
3. *malleability,* which means that you can roll it or pound it into a thin sheet; and
4. *ductility,* which means that you can pull it into a wire.

All metals are solids at room temperature except mercury, which is a liquid.

A **nonmetal** is a kind of matter that does not have a metallic luster, is a poor conductor of heat and electricity, and when solid, is a brittle material that cannot be pounded or pulled into new shapes. Nonmetals occur as solids, liquids, or gases at room temperature.

Most unprotected metals do not last long when exposed to air. Iron, for example, rusts to a reddish, earthlike material with nonmetallic properties. In time, most metals seem to return to the same kind of nonmetallic, earthlike materials from which they were extracted. Eventually, people began to wonder about the transformation of nonmetallic, earthlike materials to metals and why they returned to nonmetallic forms. Today, such questions are considered in **chemistry,** the science concerned with the study of the composition, structure, and properties of substances and the transformations they undergo.

Solids, Liquids, and Gases

The Greek philosophers of some 2,500 years ago thought of all matter as being made up of four elements that were identified as earth, air, water, and fire. It is interesting that, as we know now, the ancient Greeks were not describing elements as we understand the term today. They were describing the general

FIGURE 10.1

Classification is arranging items into groups or categories according to some criteria. The act of classifying creates a pattern that helps you to recognize and understand the behavior of fish, chemicals, or any matter in your surroundings. These fish, for example, are classified as salmon because they live in the northern Pacific Ocean, have a pinkish colored flesh, and characteristically swim from saltwater to freshwater to spawn.

FIGURE 10.2

Most matter can be classified as metals or nonmetals according to physical properties. Aluminum, for example, is a lightweight kind of matter that can be melted and rolled into a thin sheet or pulled into a wire. Here you see aluminum pop cans that have been compressed into 1,600 lb bales for recycling, destined to be formed again into new pop cans, aluminum foil, or perhaps aluminum wire.

forms in which all matter exists and one form of energy. Earth, air, and water are the most common examples of the solid, gaseous, and liquid phases of matter. Fire is the most common example of heat, the energy involved in changes of matter such as nonmetallic ores to metals. Thus, the ancient Greeks were actually describing the basics that make up the universe, matter and energy.

Solid, liquid, and gas are called the *phases of matter,* and these phases represent another way to classify matter. On the earth, all matter generally belongs to one of the three groups, or phases, which are defined by two general properties. These properties are how well a sample of matter maintains (1) its shape and (2) its volume (figure 10.3). For example, the *gaseous* phase of matter is not able to maintain a definite shape or a definite volume. A sample of gas released into a completely evacuated, rigid container will disperse throughout the entire space inside the container, taking the shape and volume of the con-

tainer. Since density is mass per unit volume, the density of the gas will depend on how much gas is placed in the container as well as the volume of the container.

Liquids, like gases, will flow and can be poured from one container to another. Both liquids and gases are fluids. Both have an indefinite shape and take the shape of their container. A liquid, however, has a definite volume and does not fill any container as a gas does. A 500 cm^3 sample of gas will disperse to fill a 1,000 cm^3 container completely. A 500 cm^3 sample of liquid will still have that volume when placed into a 1,000 cm^3 container. Liquids have a much greater density than gases and are not very compressible in comparison. Most common liquids have a density not much different from water at 1 g/cm^3 (1 kg/L) at 4°C.

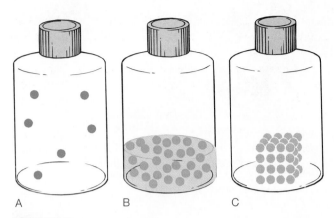

FIGURE 10.3

(*A*) A gas disperses throughout a container, taking the shape and volume of the container. (*B*) A liquid takes the shape of the container but retains its own volume. (*C*) A solid retains its own shape and volume.

Solids, like liquids, have a definite volume. Unlike gases and liquids, a solid has a definite shape. A solid with a cubic shape placed in a container of some other shape maintains both its volume and its original shape. Most solids are more dense when in the solid state than in the liquid state. Ice is the common exception to this generalization; ice floats on water, which is more dense. This exception will be explained when water is discussed in a later chapter.

Mixtures and Pure Substances

Matter that you see around you seems to come in a wide variety of sizes, shapes, forms, and kinds. Are there any patterns in the apparent randomness that will help us comprehend matter? Yes, there are many patterns and one of the more obvious is that all matter occurs as either a mixture or as a pure substance (figure 10.4). A *mixture* has unlike parts and a composition that varies from sample to sample. For example, sand from a beach is a variable mixture of things such as bits of rocks, minerals, and sea shells.

There are two distinct ways that mixtures can have unlike parts with a variable composition. A *heterogeneous mixture* has physically distinct parts with different properties. Beach sand, for example, is usually a heterogeneous mixture of tiny pieces of rocks and tiny pieces of shells that you can see. It is said to be heterogeneous because any two given samples will have a different composition with different kinds of particles.

A solution of salt dissolved in water also meets the definition of a mixture since it has unlike parts and can have a variable composition. A solution, however, is different from a sand mixture since it is a *homogeneous mixture*, meaning it is the same throughout a given sample. A homogeneous mixture, or solution, is the same throughout. The key to understanding that a solution is a mixture is found in its variable composition, that is, a given solution might be homogeneous, but one solution can vary from the next. Thus you can have a salt solution with a 1 percent concentration, another with a 7 percent concentration, yet another with

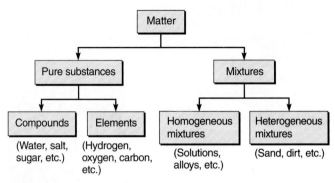

FIGURE 10.4

A classification scheme for matter.

a 10 percent concentration, and so on. Solutions, mixed gases, and metal alloys are all homogeneous mixtures since they are made of unlike parts and do not have a fixed, definite composition.

Mixtures can be separated into their component parts by physical means. For example, you can physically separate the parts making up a sand mixture by using a magnifying glass and tweezers to move and isolate each part. A solution of salt in water is a mixture since the amount of salt dissolved in water can have a variable composition. But how do you separate the parts of a solution? One way is to evaporate the water, leaving the salt behind. There are many methods for separating mixtures, but all involve a **physical change.** A physical change does not alter the *identity* of matter. When water is evaporated from the salt solution, for example, it changes to a different state of matter (water vapor) but is still recognized as water. Physical changes involve physical properties only; no new substances are formed. Examples of physical changes include evaporation, condensation, melting, freezing, and dissolving, as well as reshaping processes such as crushing or bending.

Mixtures can be physically separated into *pure substances,* materials that are the same throughout and have a fixed, definite composition. If you closely examine a sample of table salt, you will see that it is made up of hundreds of tiny cubes. Any one of these cubes will have the same properties as any other cube, including a salty taste. Sugar, like table salt, has all of its parts alike. Unlike salt, sugar grains have no special shape or form, but each grain has the same sweet taste and other properties as any other sugar grain.

If you heat salt and sugar in separate containers, you will find very different results. Salt, like some other pure substances, undergoes a physical change and melts, changing back to a solid upon cooling with the same properties that it had originally. Sugar, however, changes to a black material upon heating, while it gives off water vapor. The black material does not change back to sugar upon cooling. The sugar has *decomposed* to a new substance, while the salt did not. The new substance is carbon, and it has properties completely different from sugar. The original substance (sugar) and its properties have changed to a new substance (carbon) with new properties. The sugar has gone through a **chemical change.** A chemical change alters the iden-

A   B

FIGURE 10.5

Sugar (A) is a compound that can be easily decomposed to simpler substances by heating. (B) One of the simpler substances is the black element carbon, which cannot be further decomposed by chemical or physical means.

tity of matter, producing new substances with different properties. In this case, the chemical change was one of decomposition. Heat produced a chemical change by decomposing the sugar into carbon and water vapor.

The decomposition of sugar always produces the same mass ratio of carbon to water, so sugar has a fixed, definite composition. A **compound** is a pure substance that can be decomposed by a chemical change into simpler substances with a fixed mass ratio. This means that sugar is a compound (figure 10.5).

A pure substance that cannot be broken down into anything simpler by chemical or physical means is an **element.** Sugar is decomposed by heating into carbon and water vapor. Carbon cannot be broken down further, so carbon is an element. It has been known since about 1800 that water is a compound that can be broken down by electrolysis into hydrogen and oxygen, two gases that cannot be broken down to anything simpler. So sugar is a compound made from the elements of carbon, hydrogen, and oxygen.

But what about the table salt? Is table salt a compound? Table salt is a stable compound that is not decomposed by heating. It melts at a temperature of about 800°C, and then returns to the solid form with the same salty properties upon cooling. Electrolysis was used in the early 1800s to decompose table salt into the elements sodium and chlorine, positively proving that it is a compound. Heat brings about a chemical change and decomposes some compounds, such as sugar. Heat will not bring about a chemical change in table salt, so other means are needed.

Pure substances are either compounds or elements. Decomposition through heating and decomposition through electrolysis are two means of distinguishing between compounds and elements. If a substance can be decomposed into something simpler, you know for sure that it is a compound. If the substance cannot be decomposed, it might be an element, or it might be a stable compound that resists decomposition. More testing would be necessary before you can be confident that you have identified an element. Most pure substances are compounds. There are millions of different compounds but only 112 known elements at the present time. These elements are the fundamental materials of which all matter is made.

ELEMENTS

Modern science is usually identified as beginning just after the time of Galileo and Newton, about three hundred years ago. Understandings about matter at this time were still based on the writings of Aristotle of some two thousand years earlier. There were alternative ideas about the number of elements, but earth, air, fire, and water were generally accepted as the basic elements.

Reconsidering the Fire Element

During the early 1700s, the use of fuels such as coal, the process of burning, and the production of steam became topics of general interest. The first working steam engine was invented by Thomas Newcomen in 1711, and the time was ripe for new ideas, new theories, and answers to questions about burning and what was going on during the burning process.

About 1700 a new theory about burning was introduced by a German physician. This theory considered all burnable materials to contain a substance called *phlogiston,* a word from the Greek meaning "fire." Burning was considered to be the escape of phlogiston from fuels into the air. Materials that did not burn were considered not to contain phlogiston, either because they never had it or because they had already lost it (such as ashes). So far, phlogiston might remind you of Aristotle's fire element with a different name. But this theory continued on in detail to explain other observations. For example, a candle in a closed container would burn for a few minutes, then go out. The explanation for this observation was that the air could hold only so much phlogiston, which was released during burning. When the air was saturated, it could accept no more phlogiston. The escape of more phlogiston was thus prevented, and the fire would go out.

Other observations explained by the theory involved (1) the conversion of nonmetallic, earthlike materials (ores) into metals by fire and (2) the rusting of metals, returning them to nonmetallic, earthlike matter over a period of time. The ores were considered to be phlogiston-poor. Placing the materials in a fire permitted them to absorb phlogiston, becoming metals. The metals could not hold on to the phlogiston, and over time it leaked away, returning the metals to their nonmetallic form (figure 10.6). The phlogiston theory, with its convincing explanations, became the accepted understanding of burning, metal smelting, rusting, and the role of air in these processes during the 1700s.

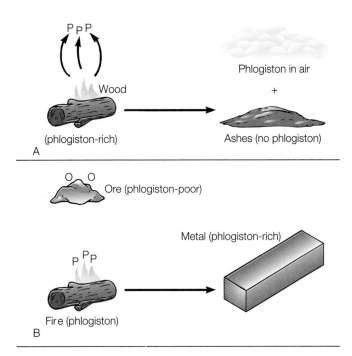

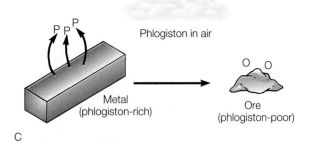

FIGURE 10.6

The phlogiston theory. (*A*) In this theory, burning was considered to be the escape of phlogiston into the air. (*B*) Smelting combined phlogiston-poor ore with phlogiston from a fire to make a metal. (*C*) Metal rusting was considered to be the slow escape of phlogiston from a metal into the air.

Discovery of Modern Elements

A series of events led to the downfall of the phlogiston theory and the discovery of modern elements in the 1770s. The first of these events was an experiment in gas chemistry conducted by an English minister named Joseph Priestley. Priestley was an amateur chemist with a natural flair for experimental research. One of his first discoveries was a method for producing carbon dioxide by reacting chalk and sulfuric acid, then forcing the gas into a container of water. Priestley had invented soda water, which soon would become modern-day cola. During Priestley's time, a navy officer won approval to supply ships with soda water, believing it might help with scurvy. The officer, coincidentally, was named Lord Sandwich.

Priestley is recognized today for the discovery of oxygen. In one of his experiments, Priestley heated a nonmetallic red pow-

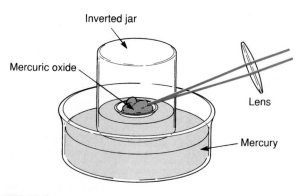

FIGURE 10.7

Priestley produced a gas (oxygen) by using sunlight to heat mercuric oxide kept in a closed container. The oxygen forced some of the mercury out of the jar as it was produced, increasing the volume about five times.

der that was then called *mercurius calcinatus* (mercuric oxide) in a jar inverted in a bowl of mercury (figure 10.7). The powder gave off a gas, forcing some of the mercury out of the inverted jar. From this, Priestley could tell that the red powder gave off about five times its own volume of gas. He found that this gas caused a candle to burn vigorously, a glowing piece of wood to sparkle, and a mouse to jump about with vigor.

Priestley had discovered oxygen, but he interpreted his findings in terms of the then-current phlogiston theory. He believed that the red powder combined with phlogiston from the air. That was why things burned vigorously in the air. The air was depleted of phlogiston, Priestley thought, which rapidly "pulled" it from flaming candles and glowing pieces of wood. The phlogiston theory provided an explanation for what was observed, and Priestley was not much concerned with measurements.

Antoine Lavoisier was already a recognized French chemist in 1772 when, at the age of twenty-nine, he began to experiment with the role of phlogiston in burning and the rusting of metals. He had observed that sulfur and phosphorus *gained* weight when they burned; they did not lose weight as the phlogiston theory predicted. He supposed that the sulfur and phosphorus combined with air to make the additional weight. When he read about Priestley's "dephlogisticated air," Lavoisier repeated Priestley's work with a carefully measured analysis. Following through with his earlier observations about burning and weight gain, Lavoisier proved that Priestley's "dephlogisticated air" was actually a substance he called *oxygen* (figure 10.8). He replaced the phlogiston theory with a theory that burning is a chemical combination of some material with oxygen.

In the meantime, Henry Cavendish had isolated a very light, highly flammable gas by reacting metals with acids. Cavendish found that when this gas burned, pure water was produced. Lavoisier repeated the Cavendish experiment and concluded that the light gas combined with oxygen to produce water. Lavoisier named the gas *hydrogen* from Greek words meaning "water

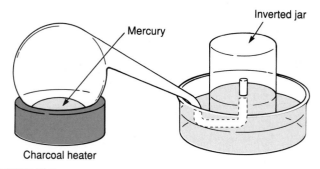

FIGURE 10.8

Lavoisier heated a measured amount of mercury to form the red oxide of mercury. He measured the amount of oxygen removed from the jar and the amount of red oxide formed. When the reaction was reversed, he found the original amounts of mercury and oxygen.

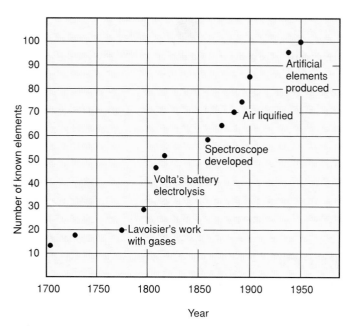

FIGURE 10.9

The number of known elements increased as new chemical and analytical techniques were developed.

former." This proved to be the final blow to the ancient Greek concepts of water and air as elements as well as the end of the phlogiston theory. Lavoisier recognized the need for a whole new concept of elements, compounds, and chemical change. The concept of elements was reconsidered, and by working with other leading chemists of the time, a whole new list of elements was developed.

Lavoisier published the results of his experiments and a list of the known elements in 1789. Lavoisier's list of thirty-three elements included twenty-three that are recognized as elements today and eight that are recognized as compounds, along with light and heat. This was the beginning of modern chemistry and of understanding elements as they are known today.

Figure 10.9 shows the number of known elements since the time of Lavoisier. New elements were discovered as new chemical and analytical techniques were developed. The increase just before 1800 resulted from an interest in rocks, minerals, and the materials of the earth. Just after the 1800s, the invention of the electric battery was followed by experiments in which compounds were decomposed into elements by electrolysis. These experiments resulted in the discovery of more elements. By 1809, about half of the natural elements had been discovered. The development of the spectroscope in the late 1850s led to the identification of more elements by their spectral lines. Mendeleev found patterns in elements in 1869, which resulted in even more discoveries. By the 1890s, technology was developed that made it possible to produce very cold, liquified gases, and this resulted in even more discoveries. All the natural elements were discovered by 1940. The naturally-occurring elements generally range from hydrogen (atomic number 1) to uranium (atomic number 92). The exceptions follow:

- Technetium (43) and promethium (61), which are not found anywhere on earth unless they are artificially produced.
- Francium (87) and astatine (85), which exist only a short time before undergoing radioactive decay, so they do not exist in any significant quantities. Half a sample of francium decays in 21 minutes. The most stable form of astatine decays in 7 1/2 hours.
- Plutonium (94), which has been found occurring naturally in small amounts.

TABLE 10.1	The known heavier elements, atomic numbers 104–118, and their current IUPAC and temporary IUPAC systematic names that are suggested for use until the appropriate names are agreed upon

Names of Elements 104–118		
Element	**IUPAC Names**	**Symbol**
104	Rutherfordium	Rf
105	Dubnium	Db
106	Seaborgium	Sg
107	Bohrium	Bh
108	Hassium	Hs
109	Meitnerium	Mt
110	Ununnilium	Uun
111	Unununium	Uuu
112	Ununbium	Uub
114	Ununquadium	Uuq
116	Ununhexium	Uuh
118	Ununoctium	Uuo

The discovers of elements 110–118 have not yet suggested names for these elements.

The other elements that do not occur naturally were artificially made during nuclear physics experiments and are not found on the earth. Thus, there are 89 naturally-occurring elements that occur in significant quantities and 26 short-lived and artificially prepared elements, for a total of 115 that are known at the present time (table 10.1).

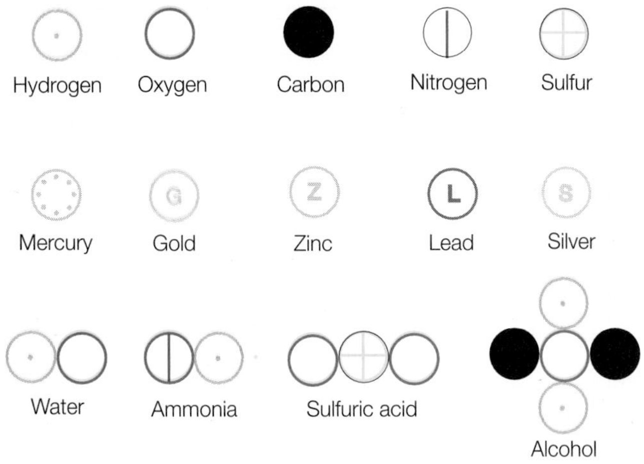

FIGURE 10.10

Here are some of the symbols Dalton used for atoms of elements and molecules of compounds. He probably used a circle for each because, like the ancient Greeks, he thought of atoms as tiny, round, hard spheres.

Names of the Elements

Elements such as sulfur, zinc, tin, and iron have been known since ancient times but were not recognized as elements. The original meanings of their names are ancient and have become obscured with time. More recently, the right to name an element belonged to the discoverer, who could call it anything as long as the name of a metal ended with "-um." The first 103 elements have internationally accepted names. Sources of these names have included the following:

1. the compound or substance in which the element was discovered;
2. an unusual or identifying property of the element;
3. places, cities, and countries;
4. famous scientists;
5. Greek mythology or some other mythology;
6. astronomical objects.

Chemical Symbols

When John Dalton introduced his atomic theory in the early 1800s, he introduced a system of symbols to represent atoms and how they formed compounds. Each element had its own special symbol, which stood for one atom of that element. Some of the symbols Dalton used to explain his atomic theory are shown in figure 10.10. Each atom was a circle, probably because Dalton thought of atoms as tiny, indivisible spheres. He indicated the different elements by symbols or letters within each circle, for example, a dot for hydrogen and a G for gold. Dalton represented compounds with combinations of element symbols as shown in the figure.

Letter symbols came into use about 1913 after an earlier recommendation by Jöns Berzelius, a Swedish chemist. Berzelius recommended that the letter symbol should be the capitalized first letter of the name of an element, for example, the symbol H

TABLE 10.2 Elements with symbols from Latin or German names

Atomic Number	Name	Source of Symbol	Symbol
11	Sodium	Latin: Natrium	Na
19	Potassium	Latin: Kalium	K
26	Iron	Latin: Ferrum	Fe
29	Copper	Latin: Cuprum	Cu
47	Silver	Latin: Argentum	Ag
50	Tin	Latin: Stannum	Sn
51	Antimony	Latin: Stibium	Sb
74	Tungsten	German: Wolfram	W
79	Gold	Latin: Aurum	Au
80	Mercury	Latin: Hydrargyrum	Hg
82	Lead	Latin: Plumbum	Pb

for hydrogen. Today, there are about a dozen common elements that have single capitalized first letters as their symbols. All the rest have two letters with the first letter *always* capitalized and the second letter *never* capitalized. The second, lowercase letter is either (1) the second letter in the name of the element or (2) a letter representing a strong consonant heard when the name of the element is spoken. Examples of using the first two letters of the name are Ca for calcium and Si for silicon. Examples of using the first letter and the letter that is heard when the name is spoken are Cl for chlorine and Cr for chromium.

Some elements have symbols from the earlier use of Latin names for the elements. For example, the symbol Au is used for gold because the metal was earlier known by its Latin name of *aurum,* meaning "shining dawn." There are ten elements with symbols from Latin names and one with a symbol from a German name. These eleven elements are listed in table 10.2, together with the sources of their names and their symbols.

Chemical symbols are like the vocabulary of a new language and are used to describe chemical changes with clarity and precision. The first key to understanding this language is to understand that a chemical symbol both identifies a specific element and represents an atom of that element. Thus, the symbol He can refer either to the element helium or to one atom of helium. The symbol Hg can mean either mercury in general, one atom of the element mercury, and so on.

You usually learn the symbols by using them and looking up a symbol as needed from a table such as the one located on the inside back cover of this text. This is really not as big a task as it might seem at first, because the elements are not equally abundant and only a few are common. In table 10.3, for example, you can see that only eight elements make up about 99 percent of the solid surface of the earth. Oxygen is most abundant, making up about 50 percent of the weight of the earth's crust. Silicon makes up more than 25 percent, so these two nonmetals alone make up about 75 percent of the earth's solid surface. Almost all the rest is made up of just six metals, as shown in the table.

TABLE 10.3	Elements making up 99 percent of Earth's crust
Element (Symbol)	**Percent by Weight**
Oxygen (O)	46.6
Silicon (Si)	27.7
Aluminum (Al)	8.1
Iron (Fe)	5.0
Calcium (Ca)	3.6
Sodium (Na)	2.8
Potassium (K)	2.6
Magnesium (Mg)	2.1

FIGURE 10.11

The elements of aluminum, iron, oxygen, and silicon make up about 88 percent of the earth's solid surface. Water on the surface and in the air as clouds and fog is made up of hydrogen and oxygen. The air is 99 percent nitrogen and oxygen. Hydrogen, oxygen, and carbon make up 97 percent of a person. Thus, almost everything you see in this picture is made up of just six elements.

		Weight				Weight
	Hydrogen	1		Z	Zinc	56
	Nitrogen	5		L	Lead	90
	Carbon	5		S	Silver	190
	Oxygen	7		G	Gold	190
	Sulfur	13			Mercury	167

FIGURE 10.12

Using information from the fixed mass ratios of combining elements, Dalton was able to calculate the relative atomic masses of some of the elements. Many of his findings were wrong, as you can see from this sample of his table.

The number of common elements is limited elsewhere, too. Only two elements make up about 99 percent of the atmospheric air around the earth. Air is mostly nitrogen (about 78 percent) and oxygen (about 21 percent), with traces of five other elements and compounds. Water on the earth is hydrogen and oxygen, of course, but seawater also contains elements in solution. These elements are chlorine (55 percent), sodium (31 percent), sulfur (8 percent), and magnesium (4 percent). Only three elements make up about 97 percent of your body. These elements are hydrogen (60 percent), oxygen (26 percent), and carbon (11 percent). Generally, all of this means that the elements are not equally distributed or equally abundant in nature (figure 10.11).

Symbols and Atomic Structures

When Dalton developed his atomic theory, he pictured atoms as tiny, hard spheres. He considered these spheres to differ only in their masses, with all the atoms of one element having the same mass, which was different from the atomic masses of the other elements. An important contribution of Dalton's theory was the attempt to determine a comparison of the atomic masses. He used the fixed mass ratios of combining elements to determine the relative atomic masses. For example, hydrogen always combines with oxygen in a mass ratio of 1:8 to make water. This means that 1 gram of hydrogen combines with 8 grams of oxygen to form 9 grams of water. Dalton assumed that one atom of hydrogen combined with one atom of oxygen to form a particle of water. Today, the particle of water is called a **molecule**. For now, consider a molecule to be a particle composed of two or more atoms held together by an attractive force called a *chemical bond*.

From the assumption that one atom of hydrogen joined with one atom of oxygen to form a molecule of water, Dalton reasoned that the mass ratio of 1:8 could be explained if one atom of oxygen is eight times more massive than one atom of hydrogen. Measuring mass ratios of hydrogen, oxygen, and other elements, Dalton was able to construct the table of relative atomic masses (figure 10.12).

Dalton's relative atomic mass table was a step in the right direction, but many of his findings were wrong. They were wrong because of measurement errors and also because of the assumption that one atom of an element always combines with one atom of another element. Today, for example, it is known that two atoms of hydrogen combine with one atom of oxygen to form water. Chemists eventually worked out techniques for determining the number of atoms of elements that combine to form compounds, and these techniques will be discussed in later chapters. It took about one hundred years of carefully measured chemical experimentation before accurate values of relative atomic masses were established.

Radioactivity was discovered in the late 1890s, and before long new but puzzling data were observed. Recall that Dalton

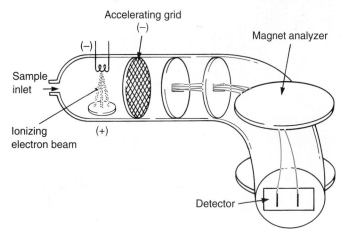

FIGURE 10.13

A schematic of a mass spectrometer. The atoms of a sample of gas become positive ions after being bombarded by a beam of electrons. The ions are deflected into a curved path by a magnetic field, which separates them according to their charge-to-mass ratio. Less massive ions are deflected the most, so the device identifies different groups of particles with different masses.

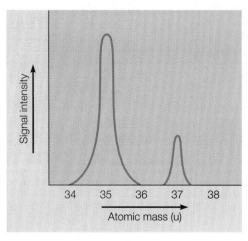

FIGURE 10.14

A mass spectrum of chlorine from a mass spectrometer. Note that two separate masses of chlorine atoms are present, and their abundance can be measured from the signal intensity. The greater the signal intensity, the more abundant the isotope.

considered all atoms of an element to have the same mass and all to act the same when combining with other elements. That is, all the atoms of an element were considered to have the same physical properties and the same chemical properties. Analysis of radioactive elements found that the masses of atoms were *not* always the same. One sample of the metal lead, for example, was radioactive, but another sample was not radioactive. Further analysis found that the two samples had different masses, even though the two samples were the same element with identical chemical behaviors. In 1910 Frederic Soddy called such varieties of an element **isotopes** because they were in the same place chemically (from the Greek *iso* meaning "same" and *tope* meaning "place"). So isotopes are atoms of an element with identical chemical properties but with different masses.

The masses and abundance of isotopes were measured by Francis William Aston in the 1900s. Aston had been an assistant to J. J. Thomson and was familiar with his method of measuring the charge-to-mass ratio of the electron. But Aston now had a new device called a *mass spectrometer* (figure 10.13). This instrument uses a beam of electrons to ionize, for example, vaporized atoms of an element. Positive ions are formed, then accelerated by high voltage through a tube. In the tube they are deflected by a magnetic field, and the amount of deflection depends on their masses. A wide range of mass values is scanned by varying the magnetic field and accelerating voltage. A detector is thus not only able to measure the presence of various masses but also the abundance of each mass from the intensity. A plot of the intensity measured by the detector of the mass is called a *mass spectrum.*

Aston was able to provide the first clear, undisputed evidence for the existence of isotopes with the mass spectrograph.

He confirmed that suspected isotopes existed for some elements and discovered new ones as well. Figure 10.14, for example, shows that a sample of chlorine gas has two chlorine isotopes in different proportions, and this was unknown before Aston analyzed chlorine in the mass spectrograph. In 1919, Aston reported that the atomic mass of each substance was very close to, but not exactly, a whole number. Chlorine, for example, has one isotope with an atomic mass very close to 35 (34.969 units). Thus, the atomic mass of an element with more than one isotope must be the *average* of the atomic masses of the isotopes. The contribution of a particular isotope to this average value depends on its abundance. For example, if the chlorine isotope with an atomic mass of 34.969 units makes up 75.53 percent of a sample of chlorine gas, and the chlorine isotope with an atomic mass of 36.966 units makes up the other 24.47 percent, then

$$\text{average atomic mass} = (75.53\% \times 34.939 \text{ units})$$
$$+ (24.47\% \times 36.966 \text{ units})$$
$$= 26.41 \text{ units} + 9.05 \text{ units}$$
$$= 35.46 \text{ units}$$

The average atomic mass of chlorine in this example is 35.46 units, which is a *weighted average* of the masses of the chlorine isotopes as they occur. **Atomic weight** is the name given to the weighted average of the masses of stable isotopes of an element as they occur in nature.

Dalton had originally assigned hydrogen a mass of 1 unit, and then compared the mass of other atoms to this standard. Today, the mass of any isotope is compared to the mass of a particular isotope of carbon. This isotope is *assigned* a mass of exactly 12.00 units called **atomic mass units** (u). This isotope, called carbon-12, provides the standard to which the masses of

TABLE 10.4	Selected atomic weights calculated from mass and abundance of isotopes		
Stable Isotopes	Mass of Isotope Compared to C-12	Abundance	Atomic Weight
$^{1}_{1}H$	1.007	99.985%	
$^{2}_{1}H$	2.0141	0.015%	1.0079
$^{9}_{4}Be$	9.01218	100.%	9.01218
$^{14}_{7}N$	14.00307	99.63%	
$^{15}_{7}N$	15.00011	0.37%	14.0067
$^{16}_{8}O$	15.99491	99.759%	
$^{17}_{8}O$	16.99914	0.037%	
$^{18}_{8}O$	17.00016	0.204%	15.9994
$^{19}_{9}F$	18.9984	100.%	18.9984
$^{20}_{10}Ne$	19.99244	90.92%	
$^{21}_{10}Ne$	20.99395	0.257%	
$^{22}_{10}Ne$	21.99138	8.82%	20.179
$^{27}_{13}Al$	26.9815	100.%	26.9815

TABLE 10.5	Summary of chlorine's numbers	
Facts		**Numbers**
Atomic number		17
Symbol and mass numbers		$^{36}_{17}Cl$
Mass of isotopes		
chlorine-35		34.969 u
chlorine-37		36.966 u
Abundance of isotopes		
chlorine-35		75.53%
chlorine-37		24.47%
Atomic weight		35.46 u

all other isotopes are compared. The mass of any isotope is based on the mass of a carbon-12 isotope. A table of atomic weights, such as the one on the inside back cover of this text, lists the atomic weight of carbon as slightly more than 12 u. Carbon occurs naturally as two stable isotopes, one with an atomic mass of exactly 12.00 u (98.9 percent) and one with an atomic mass of about 13 u (1.11 percent). Thus, the weighted average, or *atomic weight,* of carbon is slightly greater than about 12.01 u. Other examples are illustrated in table 10.4.

The abundance of each isotope making up a naturally occurring element is always the same, no matter where the sample is measured. While isotopes have different masses, they have the *same chemical behavior.* Isotopes occur because of varying numbers of neutrons in the nuclei of atoms with the same atomic number. The atomic number is the number of protons in the nucleus and the number of electrons in the atom. What element an atom belongs to is identified by the number of protons, and the chemical nature is identified by the number of electrons. Since the isotopes of an element all have the same number of protons and electrons with the same configuration, they all have the same chemical properties. Since they have different numbers of neutrons, they have different atomic masses.

A neutron is slightly more massive than a proton, but an electron has a comparatively trivial mass (about 1/1,840 the mass of a proton). Thus, neutrons and protons make up almost all the mass of an atom. The sum of the number of protons and neutrons in a nucleus is called the **mass number** of an atom. Mass numbers are used to identify isotopes. A chlorine atom

with 17 protons and 20 neutrons has a mass number of 17 + 20, or 37, and is referred to as chlorine-37. A chlorine atom with 17 protons and 18 neutrons has a mass number of 17 + 18, or 35, and is referred to as chlorine-35. Using symbols, chlorine-37 is written as

$$^{37}_{17}Cl$$

where Cl is the chemical symbol for chlorine, the subscript to the bottom left is the atomic number, and the superscript to the top left is the mass number. Consider the following rules when you use a chemical symbol to identify isotopes with a mass number:

1. The mass number is the closest whole number to the atomic mass of an isotope. Only carbon-12 has an atomic mass with a whole number (by definition).
2. The number of protons in the nucleus of an atom equals the atomic number.
3. The number of neutrons in the nucleus of an atom equals the mass number minus the atomic number.

Atomic numbers are always whole numbers because they represent the number of protons in the nucleus of an atom. *Mass numbers* are always whole numbers because they represent the number of protons and neutrons in the nucleus of an isotope. The *atomic mass of an isotope* is *not* a whole number, with the exception of carbon-12, which was assigned a mass of exactly 12 units to set up a relative scale from which the masses of all other atoms are compared. Why are the atomic masses of other isotopes not whole numbers? The answer is found in a mass contribution from the internal energy of the nucleus. There is a relationship between energy and mass, which will be explained in a future chapter on nuclear energy. *Atomic weights* are also *not* whole numbers, but you might expect this because atomic weight is a weighted average of the masses of the isotopes. Compare these values in table 10.4. The numbers for chlorine are summarized in table 10.5.

CONNECTIONS

Nutrient Elements

All molecules required to support living things are called nutrients. Some nutrients are inorganic molecules such as calcium, iron, or potassium, and others are organic molecules such as carbohydrates, proteins, fats, and vitamins. All minerals are elements and cannot be synthesized by the body. Because they are elements, they cannot be broken down or destroyed by metabolism or cooking. They commonly occur in many foods and in water. Minerals retain their characteristics whether they are in foods or in the body, and each plays a different role in metabolism.

Minerals can function as regulators, activators, transmitters, and controllers of various enzymatic reactions. For example, sodium ions (Na^+) and potassium ions (K^+) are important in the transmission of nerve impulses, while magnesium ions (Mg^{++}) facilitate energy release during reactions involving ATP. Without iron, not enough hemoglobin would be formed to transport oxygen, a con-

dition called anemia, and a lack of calcium may result in osteoporosis. Osteoporosis is a condition that results from calcium loss leading to painful, weakened bones. There are many minerals that are important in your diet. In addition to those just mentioned, you need chlorine, cobalt, copper, iodine, phosphorus, potassium, sulfur, and zinc to remain healthy. With few exceptions, adequate amounts of minerals are obtained in a normal diet. Calcium and iron supplements may be necessary, particularly in women.

Osteoporosis is a nutritional deficiency disease that results in a change in the density of the bones as a result of the loss of bone mass. Bones that have undergone this change look lacy or like Swiss cheese, with larger than normal holes. A few risk factors found to be associated with this disease are being female and fair skinned; having a sedentary lifestyle; using alcohol, caffeine, and tobacco; being anorexic; and having reached menopause.

THE PERIODIC LAW

Many new elements were discovered in a short period of time in the early 1800s, nearly doubling the number of known elements to sixty by the 1840s. As the list grew, information about the behavior of elements and their compounds created a large body of apparently unrelated facts. Seeking to organize and make sense out of this mass of information, chemists tried to classify the elements according to their properties. They were looking for some underlying order, a pattern that would not only organize the large body of facts but also perhaps form the basis of a theory that would account for the facts. The process began with attempts to relate the behaviors of elements to their atomic weights.

As information about the behavior of elements increased, chemists began to recognize certain patterns of similarities. Lithium, sodium, and potassium were found to be shiny, soft metals that reacted vigorously with water to form an alkaline solution. Calcium, strontium, and barium were found to be another group of soft metals with similar properties, but their properties were different from those of the sodium group. Iron, cobalt, and nickel were similar hard metals. Chlorine, bromine, and iodine were similar nonmetals. By 1829, a German chemist named Johann Dobereiner described the existence of *triads*, groups of three elements with similar chemical properties. Dobereiner found that when the elements of a triad were listed in order of atomic weights, the atomic weight of the middle element was almost equal to the average atomic weight of the other two elements. In most cases, the values for density, melting point, and other properties of the middle element were also midway between values for the other two elements. Three of

TABLE 10.6	Examples of element triads		
Element	**Atomic Weight (u)**	**Average Atomic Weight (First and Third)**	**Density (g/cm³)**
Lithium (Li)	6.9		0.53
Sodium (Na)	23.0	23.0	0.97
Potassium (K)	39.1		0.86
Calcium (Ca)	40.1		1.55
Strontium (Sr)	87.6	88.7	2.54
Barium (Ba)	137.3		3.50
Chlorine (Cl)	35.5		1.56
Bromine (Br)	79.9	81.2	3.12
Iodine (I)	126.9		4.93

these triads are identified in table 10.6. There was no explanation that would account for the occurrence of triads at this time, and they were considered an interesting curiosity.

A workable classification scheme of the elements was developed independently by two scientists. The Russian chemist Dmitri Mendeleev and the German physicist Lothar Meyer published similar classification schemes in 1869. Mendeleev's scheme was based mostly on the chemical properties of elements and their atomic weight, and Meyer's was based on physical properties and atomic weight. But both arranged the elements in order of increasing atomic weight and observed that the properties recur

A CLOSER LOOK | Elements 116 and 118 Discovered at Berkeley Lab*

Berkeley, CA—Discovery of two new "superheavy" elements has been announced by scientists at the U.S. Department of Energy's Lawrence Berkeley National Laboratory. Element 118 and its immediate decay product, element 116, were discovered at Berkeley Lab's 88-Inch Cyclotron by bombarding targets of lead with an intense beam of high-energy krypton ions. Although both new elements almost instantly decay into other elements, the sequence of decay events is consistent with theories that have long predicted an "island of stability" for nuclei with approximately 114 protons and 184 neutrons.

The isotope of element 118 with mass number 293 identified at Berkeley Lab contains 118 protons and 175 neutrons in its nucleus. By comparison, the heaviest element found in nature in sizeable quantities is uranium, which in its most common form contains 92 protons and 146 neutrons. Transuranic elements in the periodic table can only be synthesized in nuclear reactors or particle accelerators. Though often short-lived, these artificial elements provide scientists with valuable insights into the structure of atomic nuclei and offer opportunities to study the chemical properties of the heaviest elements beyond uranium.

Within less than a millisecond after its creation, the element 118 nucleus decays by emitting an alpha particle, leaving behind an isotope of element 116 with mass number 289, containing 116 protons and 173 neutrons. This daughter, element 116, is also radioactive, alpha-decaying to an isotope of element 114. The chain of successive alpha decays continues until at least element 106.

Elements 118 and 116 were discovered by accelerating a beam of krypton-86 ions to an energy of 449 million electron volts and directing the beam into targets of lead-208. This yielded heavy compound nuclei at low excitation energies.

During the last several years, low excitation energy reactions failed to take scientists beyond element 112, and it was assumed that production rates for heavier elements were too small to extend the periodic table further using this approach. However, recent calculations indicated increased production rates for the Kr-86 + Pb-208 reaction prompted the experimental search for element 118 at Berkeley Lab.

The key to the success of this experiment was the newly constructed Berkeley Gas-filled Separator (BGS). The BGS design resulted in a separator with unsurpassed efficiency and background suppression, allowing scientists to investigate nuclear reactions with production rates smaller than one atom per week. For these experiments, the strong magnetic fields in the BGS focused the element 118 ions and separated them from all of the interfering reaction products, which were produced in much larger quantities.

Another important factor for the experiment's success was the unique ability of the 88-Inch Cyclotron to accelerate neutron-rich isotopes such as krypton-86 to high-energy and high-intensity beams with an average current of approximately 2 trillion ions per second.

In operation since 1961, the 88-inch Cyclotron has been upgraded with the addition of a high-performance ion source and can now accelerate beams of ions as light as hydrogen or as heavy as uranium. The 88-Inch Cyclotron is a national user facility serving researchers from around the world for basic and applied studies.

Berkeley Lab is an U.S. Department of Energy national laboratory located in Berkeley, California. It conducts unclassified scientific research and is managed by the University of California.

*From the Internet. http://www.lbl.gov/Science-Articles/Archive/elements-116-118.html

periodically. Both had devised schemes that would systematize the study of chemistry and lead to the modern periodic table.

Mendeleev and Meyer both arranged the elements in rows of increasing atomic weights, arranged so that elements with similar properties made vertical columns. These vertical columns contained Dobereiner's triads of elements with similar properties. Mendeleev and Meyer were not restricted by trying to fit their individual schemes to some perceived "law" as others before them had been. Both left blank spaces if the element with the next highest atomic weight did not fit with a vertical family. For example, the next known element following zinc (Zn) in the sequence of increasing atomic weight was arsenic (As). But placing As after Zn in the table would place it in the same vertical column with aluminum (Al). The properties of As suggested that it belonged in the column with phosphorus (P), not Al. So two blank spaces were left after Zn, suggesting undiscovered elements.

Mendeleev demonstrated his understanding and daring by predicting that elements would be discovered to fill the gaps in his table and predicting the physical and chemical properties of the yet to be discovered elements. The gaps were directly below boron, aluminum, and silicon in the 1871 version of his table. The gaps below aluminum and silicon are illustrated in figure 10.15. Mendeleev named the elements eka-boron, eka-aluminum, and eka-silicon (*eka* is Sanskrit for "one" and in this context means "next one"). Eka-aluminum was discovered in 1875 and is now called gallium (Ga), eka-boron was discovered in 1879 and is now called scandium (Sc), and eka-silicon was discovered in 1886 and is now called germanium (Ge). There was an impressive correspondence between Mendeleev's predicted properties of these elements and the properties that were observed after their discovery. Mendeleev is usually given credit for developing the periodic

CONNECTIONS

Bismuth, Bowels, and Bugs

In its pure state, *bismuth* is a solid, lustrous, reddish white element that has been known since ancient times. Today, it is commonly found in over-the-counter medications such as Pepto-Bismol, Bismatral, and Pink Bismuth, in the form of the ingredient bismuth subsalicylate. It is used to control traveler's diarrhea (TD). This is the diarrhea that affects people who visit countries or regions with poor public sanitation and hygiene. TD may be caused by a number of microbes, including the bacterium *E. coli,* protozoans such as *Crytosporidium* and *Cyclospora,* and some viruses.

These microbes may be found in contaminated water and food. Taking eight pills per day has been shown to reduce the symptoms of TD by as much as 60 percent. However, there can be some unde-

sirable side effects. Nausea and vomiting, along with the temporary blackening of the stool, teeth, and tongue, may occur.

How does bismuth do its job? The hypothesis is that it promotes the formation of a protective layer of fluids on the surface of the intestinal tract which prevents the microbes from doing their dirty work. However, children and teens have been advised by the FDA (Food and Drug Administration) to avoid bismuth subsalicylate-containing medications when they have the flu or chicken pox since they are at an increased risk of developing Reye syndrome. More recently, bismuth subsalicylate is being used to treat patients with peptic ulcers since it also inhibits the growth of *Helicobacter pylori,* the bacterium that causes many peptic ulcers.

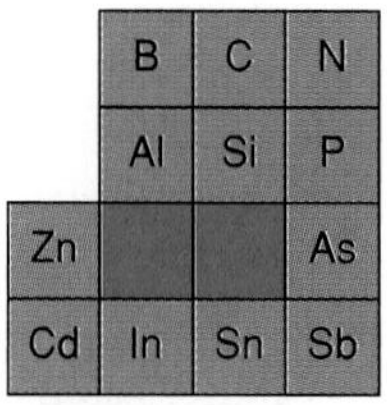

	B	C	N
	Al	Si	P
Zn			As
Cd	In	Sn	Sb

FIGURE 10.15

Mendeleev left blank spaces in his table when the properties of the elements above and below did not seem to match. The existence of unknown elements was predicted by Mendeleev on the basis of the blank spaces. When the unknown elements were discovered, it was found that Mendeleev had closely predicted the properties of the elements as well as their discovery.

table, probably because of his dramatic and highly publicized predictions about the unknown elements.

There were some problems with Mendeleev's original periodic table: not all atomic weights were quite right for the proper placement, and there was no theoretical accounting for similar chemical and physical properties in terms of atomic weight. After the work of Rutherford and Moseley on the atomic nucleus, it became clear that the *atomic number,* not the atomic weight, was the fundamental factor. The atomic number, that is, the number of protons in the nucleus and the number of electrons around the nucleus, is the significant, essential basis for the modern periodic table. Thus, the modern **periodic law** follows:

> **Similar physical and chemical properties recur periodically when the elements are listed in order of increasing atomic number.**

THE MODERN PERIODIC TABLE

The periodic table is made up of rows and columns of cells, with each element having its own cell in a specific location. The cells are not arranged symmetrically. The arrangement has a meaning, both about atomic structure and about chemical behaviors. The key to meaningful, satisfying use of the table is to understand the code of this structure. The following explains some of what the code means. It will facilitate your understanding of the code if you refer frequently to a periodic table during the following discussion.

An element is identified in each cell with its chemical symbol. The number above the symbol is the atomic number of the element, and the number below the symbol is the rounded atomic weight of the element. Horizontal *rows* of elements run from left to right with increasing atomic numbers. Each row is called a **period.** The periods are numbered from 1 to 7 on the left side. Period 1, for example, has only two elements, H (hydrogen) and He (helium). Period 2 starts with Li (lithium) and ends with Ne (neon). The two rows at the bottom of the table are actually part of periods 6 and 7 (between atomic numbers 57 and 72, 89 and 104). They are moved so that the table is not so wide.

A vertical *column* of elements is called a **family** (or *group*) of elements. Elements in families have similar properties, but this is more true of some families than others. Note that the families are identified with Roman numerals and letters at the top of each column. Group IIA, for example, begins with Be (beryllium) at the top and has Ra (radium) at the bottom. The A families are in sequence from left to right. The B families are not in sequence, and one group contains more elements than the others (figure 10.16).

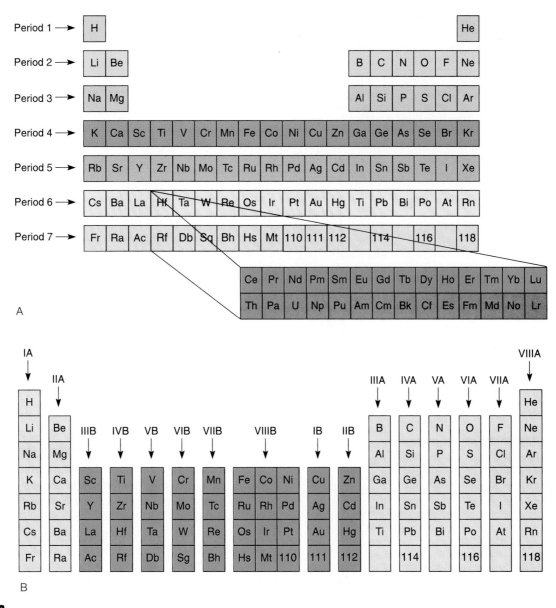

FIGURE 10.16

(*A*) Periods of the periodic table, and (*B*) families of the periodic table.

Periodic Patterns

Hydrogen is a colorless, odorless, light gas that burns explosively, combining with oxygen to form water. Helium is a colorless, odorless, light gas that does not burn and, in fact, will not react with other elements at all. Why are hydrogen and helium different? Why does one have atoms that are very reactive while the other has atoms that do not react at all? The answer to these questions is found in the atomic structures of the two gases. Recall that the number of protons determines the identity of an atom. Any atom that has only one proton is an atom of hydrogen. Any atom that has two protons is an atom of helium. The

chemical behavior of the two gases is determined by their electron configuration, which, in turn, is determined by the number of electrons they have.

Since the energy levels of electrons are quantized, they roughly correspond to a series of distances that resemble the spherical layers of an onion. These spherical layers are sometimes referred to as **shells.** The term *shell* is used to represent all the electrons with the same value of *n* (see table 9.1). The shell concept first came from studies of X-ray spectra, and X-ray vocabulary is used to identify the shells. The shortest X rays are called K X rays, and they were experimentally found to be produced by electrons closest to the nucleus. This energy level closest to the

TABLE 10.7 Electron configurations for periods 2 and 3

Period 2 from the End of Period 1 Where He: $1s^2$

Element (Atomic Number and Symbol)	Electron Configuration	Number of Electrons in K (First) Shell	Number of Electrons in L (Second) Shell
Lithium ($_3$Li)	[He] $2s^1$	2	1
Beryllium ($_4$Be)	[He] $2s^2$	2	2
Boron ($_5$B)	[He] $2s^2 2p^1$	2	3
Carbon ($_6$C)	[He] $2s^2 2p^2$	2	4
Nitrogen ($_7$N)	[He] $2s^2 2p^3$	2	5
Oxygen ($_8$O)	[He] $2s^2 2p^4$	2	6
Fluorine ($_9$F)	[He] $2s^2 2p^5$	2	7
Neon ($_{10}$Ne)	[He] $2s^2 2p^6$	2	8

Period 3 from the End of Period 2 Where Ne: $1s^2 2s^2 2p^6$

Element (Atomic Number and Symbol)	Electron Configuration	Electrons in K Shell	Electrons in L Shell	Electrons in M Shell
Sodium ($_{11}$Na)	[Ne] $3s^1$	2	8	1
Magnesium ($_{12}$Mg)	[Ne] $3s^2$	2	8	2
Aluminum ($_{13}$Al)	[Ne] $3s^2 3p^1$	2	8	3
Silicon ($_{14}$Si)	[Ne] $3s^2 3p^2$	2	8	4
Phosphorus ($_{15}$P)	[Ne] $3s^2 3p^3$	2	8	5
Sulfur ($_{16}$S)	[Ne] $3s^2 3p^4$	2	8	6
Chlorine ($_{17}$Cl)	[Ne] $3s^2 3p^5$	2	8	7
Argon ($_{18}$Ar)	[Ne] $3s^2 3p^6$	2	8	8

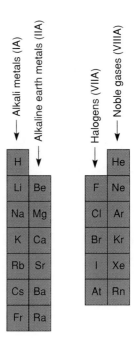

FIGURE 10.17

Four chemical families of the periodic table: the alkali metals (IA), alkaline earth metals (IIA), halogens (VIIA), and the noble gases (VIIIA).

nucleus, $n = 1$, came to be called the K shell, and the shells farther out were called the L shell, M shell, N shell, and so on. As noted in chapter 9, the number of electrons that can occupy a given orbital is limited, so *there is a limit to how many electrons can occupy a given shell.* We will return to this important key concept shortly.

As already noted, the first period contains just hydrogen (H) and helium (He). Hydrogen has an atomic number of 1, so it has one proton and one electron. This electron is at the lowest possible energy level, in the K shell. Helium has two protons in the nucleus, so it has two electrons. Both electrons can occupy the K shell, which means that the helium atom has a filled outside shell since a maximum of two electrons is allowed in the K shell. Note that period 1 ends with the filling of the orbital in the K shell.

After helium, lithium (Li) is the next element, and it begins the second period. Li has three electrons, since it has an atomic number of 3. Two of these electrons are accommodated in the K shell, but the third must go to the next higher level, in the L shell. Lithium is followed by beryllium, boron, carbon, nitrogen, oxygen, fluorine, and neon in order. Each element adds one electron to the outermost shell, in this case, the L shell. Note that period 2 ends with the filling of the orbitals in the L shell (table 10.7).

The third period also contains eight elements, from sodium (Na) to argon (Ar). The outer shell is filled just as for the elements in the second period, but this time at the third energy level. The third period ends with the filling of the orbitals in the M shell.

By now a couple of patterns are becoming apparent. First, note that the first three periods contain just A families. Each period *begins* with a single electron in a new outer shell. Second, each period *ends* with the filling of an orbital in an outer shell, completing the maximum number of electrons that can occupy that shell. Since the first A family is identified as IA, this means that all the atoms of elements in this family have one electron in their outer shells. All the atoms of elements in family IIA have two electrons in their outer shells. This pattern continues on to family VIIIA, in which all the atoms of elements have eight electrons in their outer shells except helium. Thus, the number identifying the A families *also identifies the number of electrons in the outer shells,* with the exception of helium. Helium is nonetheless similar to the other elements in this family, since all have filled orbitals in their outer shells. The electron theory of chemical bonding, which is discussed in the next chapter, states that only the electrons in the outermost shells of an atom are involved in chemical reactions. Thus, *the outer shell electrons are mostly responsible for the chemical properties of an element.* Since the members of a family all have similar outer configurations, you would expect them to have similar chemical behaviors, and they do.

The members of the A-group families are called the *main-group* or **representative elements.** The members of the B-group families are called the **transition elements** (or metals). How the members of the families (figure 10.17) in the representative elements resemble one another will be discussed next.

Q & A

How many outer shell electrons are found in an atom of (a) oxygen, (b) calcium, and (c) aluminum?

Answer

(a) According to the list of elements on the inside back cover of this text, oxygen has the symbol O and an atomic number of 8. The square with the symbol O and the atomic number 8 is located in the column identified as VIA. Since the A family number is the same as the number of electrons in the outer shell, oxygen has six outer shell electrons. (b) Calcium has the symbol Ca (atomic number 20) and is located in column IIA, so a calcium atom has two outer shell electrons. (c) Aluminum has the symbol Al (atomic number 13) and is located in column IIIA, so an aluminum atom has three outer shell electrons.

Chemical Families

As shown in table 10.8, all of the elements in group IA have an outside electron configuration of one electron. With the exception of hydrogen, the IA elements are shiny, low-density metals that are so soft you can cut them easily with a knife. These IA metals are called the **alkali metals** because they react violently with water to form an alkaline solution. The alkali metals do not occur in nature as free elements because they are so reactive. Hydrogen is a unique element in the periodic table. It is not an alkali metal and is placed in the IA group because it seems to fit there with one outer shell electron.

The elements in group IIA all have an outside configuration of two electrons and are called the **alkaline earth metals.** The alkaline earth metals are soft, reactive metals but not as reactive or soft as the alkali metals. Calcium and magnesium, in the form of numerous compounds, are familiar examples of this group.

The elements in group VIIA all have an outside configuration of seven electrons, needing only one more electron to completely fill the outer shell. These elements are called the **halogens.** The halogens are very reactive nonmetals. The halogens fluorine and chlorine are greenish colored gases. Bromine is a reddish brown liquid and iodine is a dark purple solid. Halogens are used as disinfectants, bleaches, and combined with a metal as a source of light in halogen headlights. Halogens react with metals to form a group of chemicals called *salts,* such as sodium chloride. In fact, the word *halogen* is Greek, meaning "salt former."

As shown in table 10.9, the elements in group VIIIA have orbitals that are filled to capacity in the outside shells. These elements are colorless, odorless gases that almost never react with other elements to form compounds. Sometimes they are called the **noble gases** because they are chemically inert, perhaps indicating they are above the other elements. They have also been called the *rare gases* because of their scarcity and *inert gases* because they are mostly chemically inert, not forming compounds. The noble gases are inert because they have filled outer electron configurations, a particularly stable condition.

TABLE 10.8	Electron structures of the alkali metal family							
Element	**Electron Configuration**	**Number of Electrons in Shell**						
		1st	**2nd**	**3rd**	**4th**	**5th**	**6th**	**7th**
Lithium (Li)	[He] $2s^1$	2	1	—	—	—	—	—
Sodium (Na)	[Ne] $3s^1$	2	8	1	—	—	—	—
Potassium (K)	[Ar] $4s^1$	2	8	8	1	—	—	—
Rubidium (Rb)	[Kr] $5s^1$	2	8	18	8	1	—	—
Cesium (Cs)	[Xe] $6s^1$	2	8	18	18	8	1	—
Francium (Fr)	[Rn] $7s^1$	2	8	18	32	18	8	1

TABLE 10.9	Electron structures of the noble gas family							
Element	**Electron Configuration**	**Number of Electrons in Shell**						
		1st	**2nd**	**3rd**	**4th**	**5th**	**6th**	**7th**
Helium (He)	$1s^2$	2	—	—	—	—	—	—
Neon (Ne)	[He] $2s^2 2p^6$	2	8	—	—	—	—	—
Argon (Ar)	[Ne] $3s^2 3p^6$	2	8	8	—	—	—	—
Krypton (Kr)	[Ar] $4s^2 3d^{10} 4p^6$	2	8	18	8	—	—	—
Xenon (Xe)	[Kr] $5s^2 4d^{10} 5p^6$	2	8	18	18	8	—	—
Radon (Rn)	[Xe] $6s^2 4f^{14} 5d^{10} 6p^6$	2	8	18	32	18	8	—

Q & A

(a) To what chemical family does chlorine belong? (b) How many electrons does an atom of chlorine have in its outer shell?

Answer

(a) According to the list of elements on the inside back cover of this text, chlorine has the symbol Cl and an atomic number of 17. The square with the symbol Cl and the atomic number 17 is located in the third period and in the column identified as VIIA. Column VIIA is the chemical family known as the halogens. (b) Each A family number is the same as the number of electrons in the outer shell, so an atom of chlorine has seven electrons in its outer shell.

Metals, Nonmetals, and Semiconductors

As indicated earlier, chemical behavior is mostly concerned with the outer shell electrons. The outer shell electrons, that is, the highest energy level electrons, are conveniently represented with an **electron dot notation,** made by writing the chemical symbol with dots around it indicating the number of outer shell electrons. Electron dot notations are shown for the representative elements in figure 10.18. Again, note the pattern in figure 10.18—all the noble gases are in group VIIIA, and all

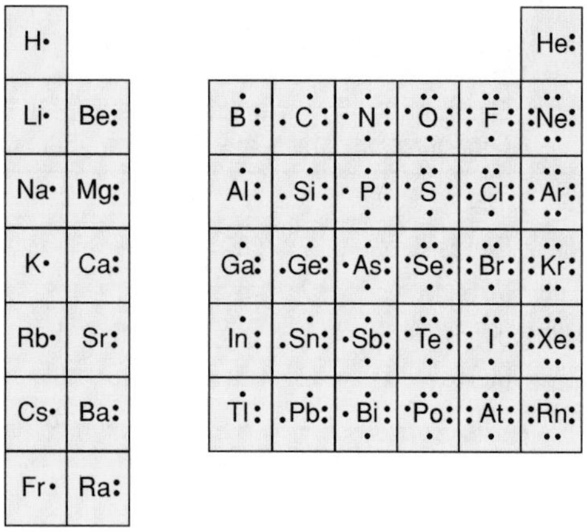

H·							He:
Li·	Be:		Ḃ:	·Ċ:	·N̈:	·Ö:	·F̈: :Ne:
Na·	Mg:		Äl:	·Si:	·P̈:	·S̈:	:Cl: :Ar:
K·	Ca:		Ga:	·Ge:	·As:	·Se:	:Br: :Kr:
Rb·	Sr:		In:	·Sn:	·Sb:	·Te:	: I : :Xe:
Cs·	Ba:		Tl:	·Pb:	·Bi:	·Po:	:Ät: :Rn:
Fr·	Ra:						

FIGURE 10.18

Electron dot notation for the representative elements.

(except helium) have eight outer electrons. All the group IA elements (alkali metals) have one dot, all the IIA elements have two dots, and so on. This pattern will explain the difference in metals, nonmetals, and a third group of in-between elements called semiconductors.

This chapter began with a discussion of several ways to classify matter according to properties. One example given was to group substances according to the physical properties of metals and nonmetals—luster, conductivity, malleability, and ductility. Metals and nonmetals also have certain chemical properties that are related to their positions in the periodic table. Figure 10.19 shows where the *metals, nonmetals,* and *semiconductors* are located. Note that about 80 percent of all the elements are metals.

The noble gases have completely filled outer orbitals in their highest energy levels, and this is a particularly stable arrangement. Other elements react chemically, either *gaining or losing electrons to attain a filled outermost energy level like the noble gases.*

When an atom loses or gains electrons, it acquires an unbalanced electron charge and is called an **ion.** An atom of lithium, for example, has three protons (plus charges) and three electrons (negative charges). If it loses the outermost electron, it now has an outer filled orbital structure like helium, a noble gas. It is also now an ion, since it has three protons (3+) and two electrons (2−), for a net charge of 1+. A lithium ion thus has a 1+ charge.

(a) Is strontium a metal, nonmetal, or semiconductor? (b) What is the charge on a strontium ion?

Answer

(a) The list of elements inside the back cover identifies the symbol for strontium as Sr (atomic number 38). In the periodic table, Sr is located in family IIA, which means that an atom of strontium has two electrons in its outer shell. For several reasons, you know that strontium is a metal: (1) An atom of strontium has two electrons in its outer shell and atoms with one, two, or three outer electrons are identified as metals; (2) strontium is located in the IIA family, the alkaline earth metals; and (3) strontium is located on the left side of the periodic table and, in general, elements located in the left two-thirds of the table are metals. (b) Elements with one, two, or three outer electrons tend to lose electrons to form positive ions. Since strontium has an atomic number of 38, you know that it has thirty-eight protons (38+) and thirty-eight electrons (38−). When it loses its two outer shell electrons, it has 38+ and 36− for a charge of 2+.

(a) Is iodine a metal, nonmetal, or semiconductor? (b) What is the charge on an iodine ion? (Answer: Nonmetal with a charge of 1−)

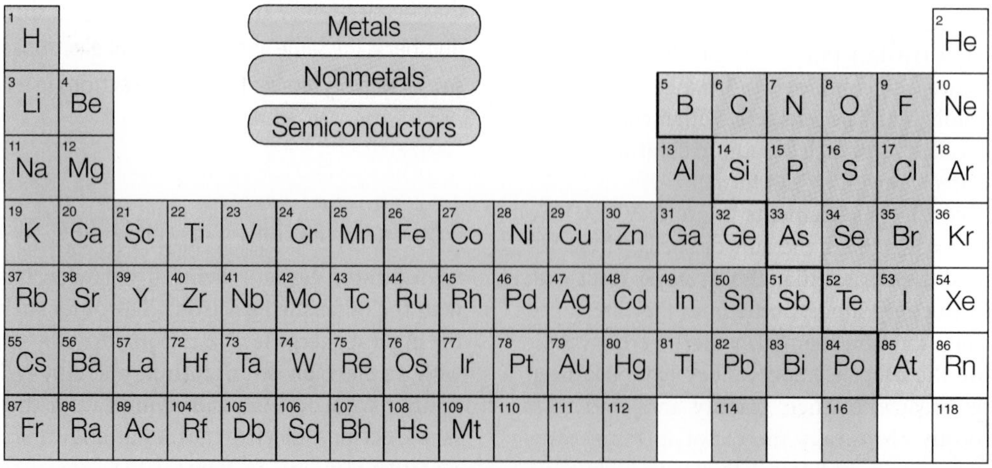

1 H																	2 He
3 Li	4 Be			Metals								5 B	6 C	7 N	8 O	9 F	10 Ne
11 Na	12 Mg			Nonmetals								13 Al	14 Si	15 P	16 S	17 Cl	18 Ar
19 K	20 Ca	21 Sc	22 Ti	23 V	24 Cr	25 Mn	26 Fe	27 Co	28 Ni	29 Cu	30 Zn	31 Ga	32 Ge	33 As	34 Se	35 Br	36 Kr
37 Rb	38 Sr	39 Y	40 Zr	41 Nb	42 Mo	43 Tc	44 Ru	45 Rh	46 Pd	47 Ag	48 Cd	49 In	50 Sn	51 Sb	52 Te	53 I	54 Xe
55 Cs	56 Ba	57 La	72 Hf	73 Ta	74 W	75 Re	76 Os	77 Ir	78 Pt	79 Au	80 Hg	81 Tl	82 Pb	83 Bi	84 Po	85 At	86 Rn
87 Fr	88 Ra	89 Ac	104 Rf	105 Db	106 Sq	107 Bh	108 Hs	109 Mt	110	111	112		114		116		118

FIGURE 10.19

The location of metals, nonmetals, and semiconductors in the periodic table.

Elements with one, two, or three outer electrons tend to lose electrons to form positive ions. The metals lose electrons like this, and the *metals are elements that lose electrons to form positive ions* (figure 10.20). Nonmetals, on the other hand, are elements with five to seven outer electrons that tend to acquire electrons to fill their outer orbitals. *Nonmetals are elements that gain electrons to form negative ions.* In general, elements located in the left two-thirds or so of the periodic table are metals. The nonmetals are on the right side of the table (figure 10.21).

The dividing line between the metals and nonmetals is a steplike line from the left top of group IIIA down to the bottom left of group VIIA. This is not a line of sharp separation between the metals and nonmetals, and elements *along* this line sometimes act like metals, sometimes like nonmetals, and sometimes like both. These hard-to-classify elements are called **semiconductors** (or *metalloids*). Silicon, germanium, and arsenic have physical properties of nonmetals; for example, they are brittle materials that cannot be hammered into a new shape. Yet these elements conduct electric currents under certain conditions. The ability to conduct an electric current is a property of a metal, and nonmalleability is a property of nonmetals, so as you can see, these semiconductors have the properties of both metals and nonmetals.

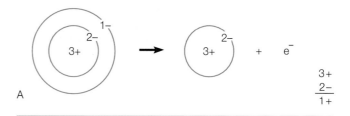

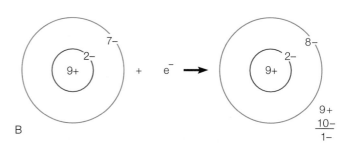

FIGURE 10.20

(*A*) Metals lose their outer electrons to acquire a noble gas structure and become positive ions. Lithium becomes a 1+ ion as it loses its one outer electron. (*B*) Nonmetals gain electrons to acquire an outer noble gas structure and become negative ions. Fluorine gains a single electron to become a 1− ion.

IA																	VIIIA
1 H 1.01	IIA											IIIA	IVA	VA	VIA	VIIA	2 He 4.0
3 Li 6.94	4 Be 9.01											5 B 10.8	6 C 12.0	7 N 14.0	8 O 16.0	9 F 19.0	10 Ne 20.2
11 Na 23.0	12 Mg 24.3	IIIB	IVB	VB	VIB	VIIB	VIIIB			IB	IIB	13 Al 27.0	14 Si 28.1	15 P 31.0	16 S 32.1	17 Cl 35.5	18 Ar 39.9
19 K 39.1	20 Ca 40.1	21 Sc 45.0	22 Ti 47.9	23 V 50.9	24 Cr 52.0	25 Mn 54.9	26 Fe 55.8	27 Co 58.9	28 Ni 58.7	29 Cu 63.5	30 Zn 65.4	31 Ga 69.7	32 Ge 72.6	33 As 74.9	34 Se 79.0	35 Br 79.9	36 Kr 83.8
37 Rb 85.5	38 Sr 87.6	39 Y 88.9	40 Zr 91.2	41 Nb 92.9	42 Mo 95.9	43 Tc 98.9	44 Ru 101.1	45 Rh 102.9	46 Pd 106.4	47 Ag 107.9	48 Cd 112.4	49 In 114.8	50 Sn 118.7	51 Sb 121.8	52 Te 127.6	53 I 126.9	54 Xe 131.3
55 Cs 132.9	56 Ba 137.3	57 La 138.9	72 Hf 168.5	73 Ta 180.9	74 W 183.9	75 Re 186.2	76 Os 190.2	77 Ir 192.2	78 Pt 195.1	79 Au 197.0	80 Hg 200.6	81 Tl 204.4	82 Pb 207.2	83 Bi 209.0	84 Po (210)	85 At (210)	86 Rn (222)
87 Fr (223)	88 Ra 226.0	89 Ac (227)	104 Rf (261)	105 Db (262)	106 Sg (266)	107 Bh (264)	108 Hs (269)	109 Mt (268)	110 (269)	111 (272)	112 (277)		114 (285)		116 (289)		118 (293)

Atomic number — 1
Symbol — H
Atomic weight (rounded value) — 1.01

58 Ce 140.1	59 Pr 140.9	60 Nd 144.2	61 Pm (145)	62 Sm 150.4	63 Eu 152.0	64 Gd 157.3	65 Tb 158.9	66 Dy 162.5	67 Ho 164.9	68 Er 167.3	69 Tm 168.9	70 Yb 173.0	71 Lu 175.0
90 Th 232.0	91 Pa (231)	92 U 238.0	93 Np (237)	94 Pu (244)	95 Am (243)	96 Cm (247)	97 Bk (247)	98 Cf (251)	99 Es (252)	100 Fm (257)	101 Md (258)	102 No (259)	103 Lr (260)

( ) represents an isotope

FIGURE 10.21

The periodic table of the elements.

A CLOSER LOOK | The Rare Earths

Compounds of the rare earths were first identified when they were isolated from uncommon minerals in the late 1700s. The elements are very reactive and have similar chemical properties, so they were not recognized as elements until some fifty years later. Thus, they were first recognized as earths, that is, nonmetal substances, when in fact they are metallic elements. They were also considered to be rare since, at that time, they were known to occur only in uncommon minerals. Today, these metallic elements are known to be more abundant in the earth than gold, silver, mercury, or tungsten. The rarest of the rare earths, thulium, is twice as abundant as silver. The rare earth elements are neither rare nor earths, and they are important materials in glass, electronic, and metallurgical industries.

You can identify the rare earths in the two lowermost rows of the periodic table.

These rows contain two series of elements that actually belong in periods 6 and 7, but they are moved below so that the entire table is not so wide. Together, the two series are called the inner transition elements. The top series is fourteen elements wide from elements 58 through 71. Since this series belongs next to element 57, lanthanum, it is sometimes called the *lanthanide series*. This series is also known as the rare earths. The second series of fourteen elements is called the *actinide series*. These are mostly the artificially prepared elements that do not occur naturally.

You may never have heard of the rare earth elements, but they are key materials in many advanced or high-technology products. Lanthanum, for example, gives glass special refractive properties and is used in optic fibers and expensive camera lenses. Samarium, neodymium, and dysprosium are used to manufacture crystals used in lasers. Samar-

ium, ytterbium, and terbium have special magnetic properties that have made possible new electric motor designs, magnetic-optical devices in computers, and the creation of a ceramic superconductor. Other rare earth metals are also being researched for use in possible high-temperature superconductivity materials. Many rare earths are also used in metal alloys; for example, an alloy of cerium is used to make heat-resistant jet-engine parts. Erbium is also used in high-performance metal alloys. Dysprosium and holmium have neutron-absorbing properties and are used in control rods to control nuclear fission. Europium should be mentioned because of its role in making the red color of color television screens. The rare earths are relatively abundant metallic elements that play a key role in many common and high-technology applications. They may also play a key role in superconductivity research.

Some of the elements near the semiconductors exhibit **allotropic forms,** which means the same element can have several different structures with very different physical properties. Carbon atoms, for example, are joined differently in diamond, graphite, and charcoal. All three are made of the element carbon, but all three have different physical properties. They are allotropic forms of carbon. Phosphorus, sulfur, and tin are other elements near the dividing line that also have several allotropic forms.

The transition elements, which are all metals, are located in the B-group families. Unlike the representative elements, which form vertical families of similar properties, the transition elements tend to form horizontal groups of elements with similar properties. Iron (Fe), cobalt (Co), and nickel (Ni) in group VIIIB, for example, are three horizontally arranged metallic elements that show magnetic properties.

A family of representative elements all form ions with the same charge. Alkali metals, for example, all lose an electron to form a 1+ ion. The transition elements have *variable charges*. Some transition elements, for example, lose their one outer electron to form 1+ ions (copper, silver). Copper, because of its special configuration, can also lose an additional electron to form a 2+ ion. Thus, copper can form either a 1+ ion or a 2+ ion. Most

Activities

Make a periodic table. List the elements in boxes on a roll of adding tape. Spiral the tape so that the noble gases appear one below the other. Use cellophane tape to hold the noble gas family together, then cut the tape just to the right of this group. When this cut spiral is spread flat, a long form of the periodic table will be the result. Moving the inner transition elements as shown in figure 10.21 will produce the familiar short form of the periodic table.

transition elements have two outer s orbital electrons and lose them both to form 2+ ions (iron, cobalt, nickel), but some of these elements also have special configurations that permit them to lose more of their electrons. Thus, iron and cobalt, for example, can form either a 2+ ion or a 3+ ion. Much more can be interpreted from the periodic table, and more generalizations will be made as the table is used in the following chapters.

SUMMARY

Matter can be *classified,* that is, mentally grouped into sets with common properties to make thinking about the wide variety of matter a little less complex. One classification scheme identifies matter as *metals* or *nonmetals* according to the *metallic properties* of *luster, conductivity, malleability,* and *ductility.* A second classification scheme identifies matter as *solids, liquids,* and *gases,* according to how well it maintains its shape and volume. A third scheme identifies groups of matter according to how it is put together. *Mixtures* are made up of *unlike parts* with a *variable composition. Pure substances* are the *same throughout* and have a *definite composition.* Mixtures can be separated into their components by *physical changes,* changes that do not alter the identity of matter. Some pure substances can be broken down into simpler substances by a *chemical change,* a change that *alters the identity of matter as it produces new substances with different properties.* A pure substance that can be decomposed by chemical change into simpler substances with a definite composition is a *compound.* A pure substance that cannot be broken down into anything simpler is an *element.* There are 89 naturally occurring elements, 22 that have been made artificially, and millions of known compounds.

Elements are identified by *letter symbols,* and each symbol for an element identifies the element and *represents one atom* of that element. Studies of radioactive elements found that some atoms of the same element have identical chemical behaviors with different masses. The name *isotope* was given to atoms of the same element with different masses. The masses of isotopes were accurately measured with a *mass spectroscope,* and new isotopes were discovered for many elements. These isotopes were found to occur in different proportions with a mass very near a whole number.

The mass of each isotope was compared to the mass of the carbon-12 isotope, which was assigned a mass of exactly 12.00 *atomic mass units* (u). The mass contribution of isotopes according to their abundance is used to determine a *weighted average* of all the isotopes of an element. The weighted average is called the *atomic weight* of that element. Isotopes are identified by their *mass number,* defined as the sum of the number of protons and neutrons in the nucleus. The mass number is the closest whole number to the actual atomic mass of an isotope. Isotopes are identified by a chemical symbol with the atomic number (the number of protons) shown as a subscript, and the mass number (the number of protons and neutrons) shown as a superscript.

The *periodic law* states that the properties of elements recur periodically when the elements are listed in order of increasing atomic number. The periodic table has horizontal rows of elements called *periods* and vertical columns of elements called *families.* Families have the same outer shell electron configurations, and it is the electron configuration that is mostly responsible for the chemical properties of an element. The *chemical families* of *alkali metals, alkaline earth metals, halogens,* and *noble gases* all have elements with similar properties and identical outer electron arrangements. The outer electrons are represented by *electron dot notations,* which use dots around a chemical symbol to represent the outer electrons.

Elements react chemically to *gain or lose electrons to attain a filled outer orbital structure like the noble gases.* When an atom gains or loses electrons, it acquires an imbalanced charge and is called an *ion.* Metals *lose electrons to form positive ions.* Nonmetals *gain electrons to form negative ions.* Elements on the dividing line between metals and nonmetals are *semiconductors.* Elements near the dividing line exhibit *allotropic forms* with different physical properties.

KEY TERMS

alkali metals (p. **247**)	mass number (p. **241**)
alkaline earth metals (p. **247**)	matter (p. **232**)
allotropic forms (p. **250**)	metal (p. **232**)
atomic mass units (p. **240**)	molecule (p. **239**)
atomic weight (p. **240**)	noble gases (p. **247**)
chemical change (p. **234**)	nonmetal (p. **232**)
chemistry (p. **232**)	period (p. **244**)
compound (p. **235**)	periodic law (p. **244**)
electron dot notation (p. **247**)	physical change (p. **234**)
element (p. **235**)	representative elements (p. **246**)
family (p. **244**)	semiconductors (p. **249**)
halogens (p. **247**)	shells (p. **245**)
ion (p. **248**)	transition elements (p. **246**)
isotopes (p. **240**)	

APPLYING THE CONCEPTS

1. Matter is usually defined as anything that
   a. has a consistent weight unless subdivided.
   b. has a specific volume, but in any shape or size.
   c. occupies space and emits energy in the form of visible light.
   d. occupies space and has mass.

2. One of the physical properties of a metal, compared to a nonmetal, is that the metal is malleable. This means you can
   a. see your reflection in it.
   b. use it to conduct electricity.
   c. pound it into a thin sheet.
   d. pull it into a wire that will conduct electricity.

3. Chemistry is the science concerned with
   a. how the world around you moves and changes.
   b. the study of matter and the changes it undergoes.
   c. atomic theory and the structure of an atom.
   d. the making of synthetic materials.

4. A sample of matter has an indefinite shape and takes the shape of its container. This sample must be in which phase?
   a. gas
   b. liquid
   c. either gaseous or liquid
   d. None of the above.

5. A sample of a salt and water solution is homogeneous throughout. Is this sample a mixture or a pure substance?
   a. a pure substance, since it is the same throughout
   b. a mixture, because it can have a variable composition
   c. a pure substance, because it has a definite composition
   d. a mixture, because it has a fixed, definite composition

6. Which of the following represents a chemical change?
   a. heating a sample of ice until it melts
   b. tearing a sheet of paper into tiny pieces
   c. burning a sheet of paper
   d. All of the above are correct.

7. Which of the following represents a physical change?
   a. the fusion of hydrogen in the sun
   b. making an iron bar into a magnet
   c. burning a sample of carbon
   d. All of the above are correct.

8. A pure substance that can be decomposed by a chemical change into simpler substances with a fixed mass ratio is called a (an)
   a. element.
   b. compound.
   c. mixture.
   d. isotope.

9. A pure substance that cannot be decomposed into anything simpler by chemical or physical means is called a (an)
   a. element.
   b. compound.
   c. mixture.
   d. isotope.

10. If you cannot decompose a pure substance into anything simpler, you know for sure that the substance is a (an)
    a. element.
    b. compound.
    c. element or stable compound.
    d. element, compound, or mixture.

11. Priestley discovered oxygen when he heated a red powder, but interpreted his findings in terms of the phlogiston theory. He would not have made this interpretation if, in his experiment, he had
    a. made measurements.
    b. smelled the air.
    c. forced the gas into a container of water.
    d. poured the gas onto a flaming candle.

12. How many elements are found naturally on the earth in significant quantities, not including the ones that were artificially produced in nuclear physics experiments?
    a. 81
    b. 89
    c. 103
    d. 109

13. You see the symbols "CO" in a newspaper article. According to the list of elements inside the back cover of this book, these symbols mean
    a. one atom of cobalt.
    b. one atom of copper.
    c. one atom of carbon and one atom of oxygen.
    d. one atom of copper and one atom of oxygen.

14. The chemical symbol for sodium is
    a. S.
    b. So.
    c. Na.
    d. Nd.

15. The chemical symbol "Fe" can mean
    a. the element fluorine.
    b. one atom of the element fermium.
    c. one atom of the element iron.
    d. the fifth isotope of the element fluorine.

16. About 75 percent of the earth's solid crust is made up of how many different kinds of atoms, not including isotopes?
    a. 2
    b. 8
    c. 91
    d. 109

17. Dalton was able to construct a table of relative atomic masses by
    a. using an instrument called a mass spectrometer.
    b. comparing the fixed mass ratios of combining elements.
    c. weighing the molecules of a compound before and after decomposition.
    d. finding the masses of combining elements, then dividing by the number of atoms.

18. Two isotopes of the same element have
    a. the same number of protons, neutrons, and electrons.
    b. the same number of protons and neutrons, but different numbers of electrons.
    c. the same number of protons and electrons, but different numbers of neutrons.
    d. the same number of neutrons and electrons, but different numbers of protons.

19. Atomic weight is
    a. the weight of an atom in grams.
    b. the average atomic mass of the isotopes as they occur in nature.
    c. the number of protons and neutrons in the nucleus.
    d. All of the above are correct.

20. The mass of any isotope is based on the mass of
    a. hydrogen, which is assigned the number 1 since it is the lightest element.
    b. oxygen, which is assigned a mass of 16.
    c. an isotope of carbon, which is assigned a mass of 12.
    d. its most abundant isotope as found in nature.

21. The isotopes of a given element always have
    a. the same mass and the same chemical behavior.
    b. the same mass and a different chemical behavior.
    c. different masses and different chemical behaviors.
    d. different masses and the same chemical behavior.

22. If you want to know the number of protons in an atom of a given element, you would look up the
    a. mass number.
    b. atomic number.
    c. atomic weight.
    d. abundance of isotopes compared to the mass number.

23. If you want to know the number of neutrons in an atom of a given element, you would
    a. round the atomic weight to the nearest whole number.
    b. add the mass number and the atomic number.
    c. subtract the atomic number from the mass number.
    d. add the mass number and the atomic number, then divide by two.

24. Which of the following is always a whole number?

   a. atomic mass of an isotope

   b. mass number of an isotope

   c. atomic weight of an element

   d. None of the above are correct.

25. The chemical family of elements called the noble gases are found in what column of the periodic table?

   a. IA

   b. IIA

   c. VIIA

   d. VIIIA

26. A particular element is located in column IVA of the periodic table. How many dots would be placed around the symbol of this element in its electron dot notation?

   a. 1

   b. 3

   c. 4

   d. 8

Answers

1. d 2. c 3. b 4. c 5. b 6. c 7. b 8. b 9. a 10. c 11. a 12. b 13. c 14. c 15. c
16. a 17. b 18. c 19. b 20. c 21. d 22. b 23. c 24. b 25. d 26. c

QUESTIONS FOR THOUGHT

1. How was Mendeleev able to predict the chemical and physical properties of elements that were not yet discovered?
2. What is an isotope? Are *all* atoms isotopes? Explain.
3. Which of the following are whole numbers, and which are not whole numbers? Explain why for each.

   (a) atomic number

   (b) isotope mass

   (c) mass number

   (d) atomic weight
4. Why does the carbon-12 isotope have a whole-number mass but not the other isotopes?
5. What two things does a chemical symbol represent?
6. What do the members of the noble gas family have in common? What are their differences?
7. How are the isotopes of an element similar? How are they different?
8. What is the difference between a chemical change and a physical change? Give three examples of each.
9. What is an ion? How are ions formed?
10. What patterns are noted in the electron structures of elements found in a period and in a family in the periodic table?
11. What is a semiconductor?
12. Why do chemical families exist?

PARALLEL EXERCISES

The exercises in groups A and b cover the same concepts. Solutions to group A exercises are located in appendix D.

Group A

1. Write the chemical symbols for the following chemical elements:

   (a) Silicon

   (b) Silver

   (c) Helium

   (d) Potassium

   (e) Magnesium

   (f) Iron

2. Lithium has two naturally occurring isotopes, lithium-6 and lithium-7. Lithium-6 has a mass of 6.01512 relative to carbon-12 and makes up 7.42 percent of all naturally occurring lithium. Lithium-7 has a mass of 7.016 compared to carbon-12 and makes up the remaining 92.58 percent. According to this information, what is the atomic weight of lithium?

3. Identify the number of protons, neutrons, and electrons in the following isotopes:

   (a) $^{12}_{6}C$

   (b) $^{1}_{1}H$

   (c) $^{40}_{18}Au$

   (d) $^{2}_{1}H$

   (e) $^{197}_{79}Au$

   (f) $^{235}_{92}U$

Group B

1. Write the chemical symbols for the following chemical elements:

   (a) Argon

   (b) Gold

   (c) Neon

   (d) Sodium

   (e) Calcium

   (f) Tin

2. Boron has two naturally occurring isotopes, boron-10 and boron-11. Boron-10 has a mass of 10.0129 relative to carbon-12 and makes up 19.78 percent of all naturally occurring boron. Boron-11 has a mass of 11.00931 compared to carbon-12 and makes up the remaining 80.22 percent. What is the atomic weight of boron?

3. Identify the number of protons, neutrons, and electrons in the following isotopes:

   (a) $^{14}_{7}N$

   (b) $^{7}_{3}Li$

   (c) $^{35}_{17}Cl$

   (d) $^{48}_{20}Ca$

   (e) $^{63}_{29}Cu$

   (f) $^{230}_{92}U$

4. Identify the period and the family in the periodic table for the following elements:
   (a) Radon
   (b) Sodium
   (c) Copper
   (d) Neon
   (e) Iodine
   (f) Lead

5. How many outer shell electrons are found in an atom of:
   (a) Li
   (b) N
   (c) F
   (d) Cl
   (e) Ra
   (f) Be

6. Write electron dot notations for the following elements:
   (a) Boron
   (b) Bromine
   (c) Calcium
   (d) Potassium
   (e) Oxygen
   (f) Sulfur

7. Identify the charge on the following ions:
   (a) Boron
   (b) Bromine
   (c) Calcium
   (d) Potassium
   (e) Oxygen
   (f) Nitrogen

8. Use the periodic table to identify if the following are metals, nonmetals, or semiconductors:
   (a) Krypton
   (b) Cesium
   (c) Silicon
   (d) Sulfur
   (e) Molybdenum
   (f) Plutonium

9. From their charges, predict the periodic table family number for the following ions:
   (a) Br^{-1}
   (b) K^{+1}
   (c) Al^{+3}
   (d) S^{-2}
   (e) Ba^{+2}
   (f) O^{-2}

10. Use chemical symbols and numbers to identify the following isotopes:
    (a) Oxygen-16
    (b) Sodium-23
    (c) Hydrogen-3
    (d) Chlorine-35

4. Identify the period and the family in the periodic table for the following elements:
   (a) Xenon
   (b) Potassium
   (c) Chromium
   (d) Argon
   (e) Bromine
   (f) Barium

5. How many outer shell electrons are found in an atom of:
   (a) Na
   (b) P
   (c) Br
   (d) I
   (e) Te
   (f) Sr

6. Write electron dot notations for the following elements:
   (a) Aluminum
   (b) Fluorine
   (c) Magnesium
   (d) Sodium
   (e) Carbon
   (f) Chlorine

7. Identify the charge on the following ions:
   (a) Aluminum
   (b) Chlorine
   (c) Magnesium
   (d) Sodium
   (e) Sulfur
   (f) Hydrogen

8. Use the periodic table to identify if the following are metals, nonmetals, or semiconductors:
   (a) Radon
   (b) Francium
   (c) Arsenic
   (d) Phosphorus
   (e) Hafnium
   (f) Uranium

9. From their charges, predict the periodic table family number for the following ions:
   (a) F^{-1}
   (b) Li^{+1}
   (c) B^{+3}
   (d) O^{-2}
   (e) Be^{+2}
   (f) Si^{+4}

10. Use chemical symbols and numbers to identify the following isotopes:
    (a) Potassium-39
    (b) Neon-22
    (c) Tungsten-184
    (d) Iodine-127

A chemical change occurs when iron rusts, and rust is a different substance with physical and chemical properties different from iron. This rusted anchor makes a colorful display on the bow of a grain ship.

CHAPTER | Eleven

Compounds and Chemical Change

In the previous two chapters, you learned how the modern atomic theory is used to describe the structures of atoms of different elements. The electron structures of different atoms successfully account for the position of elements in the periodic table as well as for groups of elements with similar properties. On a large scale, all metals were found to have a similarity in electron structure, as were nonmetals. On a smaller scale, chemical families such as the alkali metals were found to have the same outer electron configurations. Thus, the modern atomic theory accounts for observed similarities between elements in terms of atomic structure.

So far, only individual, isolated atoms have been discussed; we have not considered how atoms of elements join together to produce compounds. There is a relationship between the electron structure of atoms and the reactions they undergo to produce specific compounds. Understanding this relationship will explain the changes that matter itself undergoes. For example, hydrogen is a highly flammable, gaseous element that burns with an explosive reaction. Oxygen, on the other hand, is a gaseous element that supports burning. As you know, hydrogen and oxygen combine to form water. Water is a liquid that neither burns nor supports burning. What happens when atoms of elements such as hydrogen and oxygen join to form molecules such as water? Why do such atoms join and why do they stay together? Why does water have properties different from the elements that combine to produce it? And finally, why is water H_2O and not H_3O or H_4O?

Answers to questions about why and how atoms join together in certain numbers are provided by considering the electronic structures of the atoms. Chemical substances are formed from the interactions of electrons as their structures merge, forming new patterns that result in molecules with new properties (figure 11.1). It is the new electron pattern of the water molecule that gives water properties that are different from the oxygen or hydrogen from which it formed. Understanding how electron structures of atoms merge to form new patterns is understanding the changes that matter itself undergoes, the topic of this chapter.

COMPOUNDS AND CHEMICAL CHANGE

The air you breathe, the liquids you drink, and all the things around you are elements, compounds, or mixtures. Most are compounds, however, and very few are pure elements. Water, sugar, gasoline, and chalk are examples of compounds. Each can be broken down into the elements that make it up. Recall that elements are basic substances that cannot be broken down into simpler substances. Examples of elements are hydrogen, carbon, and calcium. Why and how these elements join together in different ways to form different compounds is the subject of this chapter.

You have already learned that elements are made up of atoms that can be described by the modern atomic theory. You can also consider an **atom** to be *the smallest unit of an element that can exist alone or in combination with other elements.* Compounds are formed when atoms are held together by an attractive force called a *chemical bond.* The chemical bond binds individual atoms together in a compound. A molecule is generally thought of as a tightly bound group of atoms that maintains its identity. More specifically, a **molecule** is defined as *the smallest particle of a compound, or a gaseous element, that can exist and still retain the characteristic chemical properties of a substance.* Compounds with one type of chemical bond, as you will see, have molecules that are electrically neutral groups of atoms held together strongly enough to be considered independent units. For example, water is a compound. The smallest unit of water that can exist alone is an electrically neutral unit made up of two hydrogen atoms and one oxygen atom held together by chemical bonds. The concept of a molecule will be expanded as chemical bonds are discussed.

Compounds occur naturally as gases, liquids, and solids. Many common gases occur naturally as molecules made up of two or more atoms. For example, at ordinary temperatures hydrogen gas occurs as molecules of two hydrogen atoms bound together. Oxygen gas also usually occurs as molecules of two oxygen atoms bound together. Both hydrogen and oxygen occur naturally as *diatomic molecules* ("di-" means "two"). Oxygen sometimes occurs as molecules of three oxygen atoms bound together. These *triatomic molecules* ("tri-" means "three") are called *ozone*. The noble gases are unique, occurring as single atoms called *monatomic* ("mon-" or "mono-" means "one") (figure 11.2). These monatomic particles are sometimes called *monatomic molecules* since they are the smallest units of the noble gases that can exist alone. Helium and neon are examples of the monatomic noble gases.

When multiatomic molecules of any size are formed or broken down into simpler substances, new materials with new properties are produced. This kind of a change in matter is called a chemical change, and the process is called a chemical reaction. A **chemical reaction** is defined as

a change in matter in which different chemical substances are created by forming or breaking chemical bonds.

In general, chemical bonds are formed when atoms of elements are bound together to form compounds. Chemical bonds are broken when a compound is decomposed into simpler substances.

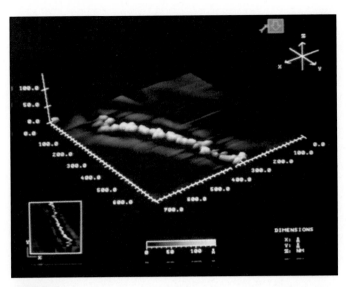

FIGURE 11.1

This is a scanning tunneling electron microscope image of DNA, showing the actual grooves of the double helix. The scale is in Angstroms (1 Angstrom = 10^{-10} m), and each twist of the helix is about 35 Angstroms. A molecule like this one determined the characteristics that other people recognize as you.

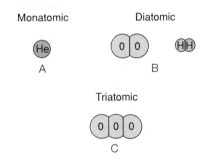

FIGURE 11.2

(*A*) The nobel gases are monatomic, occurring as single atoms. (*B*) Many gases, such as hydrogen and oxygen, are diatomic, with two atoms per molecule. (*C*) Ozone is a form of oxygen that is triatomic, occurring with three atoms per molecule.

Chemical bonds are electrical in nature, formed by electrical attractions, as discussed in chapter 7.

Chemical reactions happen all the time, all around you. A growing plant, burning fuels, and your body's utilization of food all involve chemical reactions. These reactions produce different chemical substances with greater or smaller amounts of internal potential energy (see chapter 5 for a discussion of internal potential energy). Energy is *absorbed* to produce new chemical substances with more internal potential energy. Energy is *released* when new chemical substances are produced with less internal potential energy. In general, changes in internal potential energy are called **chemical energy.** For example, new chemical substances are produced in green plants through the process

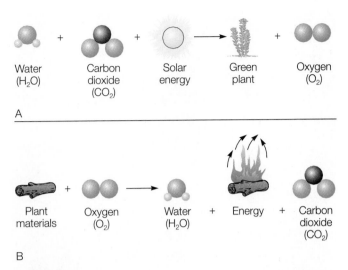

FIGURE 11.3

(*A*) New chemical bonds are formed as a green plant makes new materials and stores solar energy through the photosynthesis process. (*B*) The chemical bonds are later broken, and the same amount of energy and the same original materials are released. The same energy and the same materials are released rapidly when the plant materials burn, and they are released slowly when the plant decomposes.

called *photosynthesis*. A green plant uses radiant energy (sunlight), carbon dioxide, and water to produce new chemical materials and oxygen. These new chemical materials, the stuff that leaves, roots, and wood are made of, contain more chemical energy than the carbon dioxide and water they were made from.

A **chemical equation** is a way of describing what happens in a chemical reaction. Later, you will learn how to use formulas in a chemical reaction. For now, the chemical reaction of photosynthesis will be described by using words in an equation:

$$\begin{array}{ccccc} \text{energy} & \text{carbon} & \text{water} & \text{plant} & \text{oxygen} \\ \text{(sunlight)} + \text{dioxide} + \text{molecules} \rightarrow \text{material} + \text{molecules} \\ & \text{molecules} & & \text{molecules} \end{array}$$

The substances that are changed are on the left side of the word equation and are called *reactants*. The reactants are carbon dioxide molecules and water molecules. The equation also indicates that energy is absorbed, since the term *energy* appears on the left side. The arrow means *yields*. The new chemical substances are on the right side of the word equation and are called *products*. Reading the photosynthesis reaction as a sentence you would say, "Carbon dioxide and water use energy to react, yielding plant materials and oxygen."

The plant materials produced by the reaction have more internal potential energy, also known as *chemical energy,* than the reactants. You know this from the equation because the term *energy* appears on the left side but not the right. This means that the energy on the left went into internal potential energy on the right. You also know this because the reaction can be reversed to release the stored energy (figure 11.3). When plant materials (such as wood) are burned, the materials react with oxygen, and chemical

FIGURE 11.4

Magnesium is an alkaline earth metal that burns brightly in air, releasing heat and light. As chemical energy is released, a new chemical substance is formed. The new chemical material is magnesium oxide, a soft powdery material that forms an alkaline solution in water (called *milk of magnesia*).

energy is released in the form of radiant energy (light) and high kinetic energy of the newly formed gases and vapors. In words,

plant
material + oxygen → carbon + water + energy
molecules   molecules   dioxide   molecules
                          molecules

If you compare the two equations, you will see that burning is the opposite of the process of photosynthesis! The energy released in burning is exactly the same amount of solar energy that was stored as internal potential energy by the plant. Such chemical changes, in which chemical energy is stored in one reaction and released by another reaction, are the result of the making, then the breaking, of chemical bonds. Chemical bonds were formed by utilizing energy to produce new chemical substances. Energy was released when these bonds were broken to produce the original substances (figure 11.4). In this example, chemical reactions and energy flow can be explained by the making and breaking of chemical bonds. Chemical bonds can be explained in terms of changes in the electron structures of atoms. Thus, the place to start in seeking understanding about chemical reactions is the electron structure of the atoms themselves.

VALENCE ELECTRONS AND IONS

As discussed in chapter 10, it is the number of electrons in the outermost shell that usually determines the chemical properties of an atom. These outer electrons are called **valence electrons,** and it is the valence electrons that participate in chemical bonding. The inner electrons are in stable, fully occupied orbitals and do not participate in chemical bonds. The representative elements (the A-group families) have valence electrons in the outermost orbitals, which contain from one to eight valence electrons. Recall that you can easily find the number of valence electrons by referring to a periodic table. The number at the top of each representative family is the same as the number of outer shell electrons (with the exception of helium).

The noble gases have filled outer orbitals and do not normally form compounds. Apparently, half-filled and filled orbitals are particularly stable arrangements. Atoms have a tendency to seek such a stable, filled outer orbital arrangement such as the one found in the noble gases. For the representative elements, this tendency is called the **octet rule.** The octet rule states that *atoms attempt to acquire an outer orbital with eight electrons* through chemical reactions. This rule is a generalization, and a few elements do not meet the requirement of eight electrons but do seek the same general trend of stability. There are a few other exceptions, and the octet rule should be considered a generalization that helps keep track of the valence electrons in most representative elements.

The family number of the representative element in the periodic table tells you the number of valence electrons and what the atom must do to reach the stability suggested by the octet rule. For example, consider sodium (Na). Sodium is in family IA, so it has one valence electron. If the sodium atom can get rid of this outer valence electron through a chemical reaction, it will have the same outer electron configuration as an atom of the noble gas neon (Ne) (compare figure 11.5B and 11.5C).

When a sodium atom (Na) loses an electron to form a sodium ion (Na^+), it has the same, stable outer electron configuration as a neon atom (Ne). The sodium ion (Na^+) is still a form of sodium since it still has eleven protons. But it is now a sodium *ion,* not a sodium *atom,* since it has eleven protons (eleven positive charges) and now has ten electrons (ten negative charges) for a total of

$$11 + \text{(protons)}$$
$$\underline{10 - \text{(electrons)}}$$
$$1 + \text{(net charge on sodium ion)}$$

This charge is shown on the chemical symbol of Na^+ for the *sodium ion.* Note that the sodium nucleus and the inner orbitals do not change when the sodium atom is ionized. The sodium ion is formed when a sodium atom loses its valence electron, and the process can be described by

$$\text{energy} + \text{Na} \cdot \longrightarrow Na^+ + e^-$$

equation 11.1

where Na · is the electron dot symbol for sodium, and the e^- is the electron that has been pulled off the sodium atom.

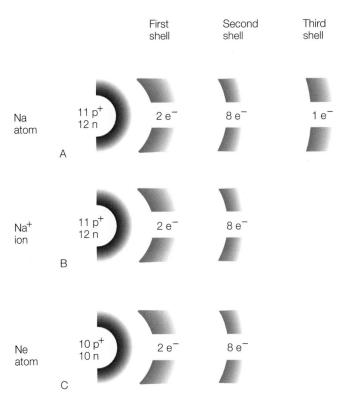

First shell	Second shell	Third shell

Na atom — $11\ p^+$ $12\ n$ — $2\ e^-$ — $8\ e^-$ — $1\ e^-$

A

Na$^+$ ion — $11\ p^+$ $12\ n$ — $2\ e^-$ — $8\ e^-$

B

Ne atom — $10\ p^+$ $10\ n$ — $2\ e^-$ — $8\ e^-$

C

FIGURE 11.5

(*A*) A sodium atom has two electrons in the first energy level, eight in the second energy level, and one in the third level. (*B*) When it loses its one outer, or valence, electron, it becomes a sodium ion with the same electron structure as an atom of neon (*C*).

What is the symbol and charge for a calcium ion?

Answer

From the list of elements on the inside back cover, the symbol for calcium is Ca, and the atomic number is 20. The periodic table tells you that Ca is in family IIA, which means that calcium has 2 valence electrons. According to the octet rule, the calcium ion must lose 2 electrons to acquire the stable outer arrangement of the noble gases. Since the atomic number is 20, a calcium atom has 20 protons (20 +) and 20 electrons (20 −). When it is ionized, the calcium ion will lose 2 electrons for a total charge of (20 +) + (18 −), or 2 +. The calcium ion is represented by the chemical symbol for calcium and the charge shown as a superscript: Ca^{2+}. What is the symbol and charge for an aluminum ion? (Answer: Al^{3+})

CHEMICAL BONDS

Atoms gain or lose electrons through a chemical reaction to achieve a state of lower energy, the stable electron arrangement of the noble gas atoms. Such a reaction results in a **chemical bond,** an *attractive force that holds atoms together in a compound.* There are three general classes of chemical bonds: (1) ionic bonds, (2) covalent bonds, and (3) metallic bonds.

Ionic bonds are formed when atoms *transfer* electrons to achieve the noble gas electron arrangement. Electrons are given up or acquired in the transfer, forming positive and negative ions. The electrostatic attraction between oppositely charged ions forms ionic bonds, and ionic compounds are the result. In general, ionic compounds are formed when a metal from the left side of the periodic table reacts with a nonmetal from the right side.

Covalent bonds result when atoms achieve the noble gas electron structure by *sharing* electrons. Covalent bonds are generally formed between the nonmetallic elements on the right side of the periodic table.

Metallic bonds are formed in solid metals such as iron, copper, and the other metallic elements that make up about 80 percent of all the elements. The atoms of metals are closely packed and share many electrons in a "sea" that is free to move throughout the metal, from one metal atom to the next. Metallic bonding accounts for metallic properties such as high electrical conductivity.

Ionic, covalent, and metallic bonds are attractive forces that hold atoms or ions together in molecules and crystals. There are two ways to describe what happens to the electrons when one of these bonds is formed, by considering (1) the new patterns formed when atomic orbitals overlap to form a combined orbital called a *molecular orbital* or (2) the atoms in a molecule as *isolated atoms* with changes in their outer shell arrangements. The molecular orbital description considers that the electrons belong to the whole molecule and form a molecular orbital with its own shape, orientation, and energy levels. The isolated atom description considers the electron energy levels as if the atoms in the molecule were alone, isolated from the molecule. The isolated atom description is less accurate than the molecular orbital description, but it is less complex and more easily understood. Thus, the following details about chemical bonding will mostly consider individual atoms and ions in compounds.

Ionic Bonds

An **ionic bond** is defined as the *chemical bond of electrostatic attraction* between negative and positive ions. Ionic bonding occurs when an atom of a metal reacts with an atom of a nonmetal. The reaction results in a transfer of one or more valence electrons from the metal atom to the valence shell of the nonmetal atom. The atom that loses electrons becomes a positive ion, and the atom that gains electrons becomes a negative ion. Oppositely charged ions attract one another, and when pulled together, they form an ionic solid with the ions arranged in an orderly geometric structure (figure 11.6). This

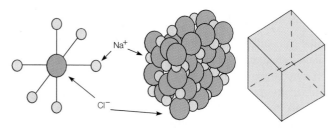

FIGURE 11.6

Sodium chloride crystals are composed of sodium and chlorine ions held together by electrostatic attraction. Each sodium ion is surrounded by six chlorine ions, and each chlorine ion is surrounded by six sodium ions. A crystal builds up like this, giving the sodium chloride crystal a cubic structure.

FIGURE 11.7

You can clearly see the cubic structure of these ordinary table salt crystals because they have been magnified about ten times.

results in a crystalline solid that is typical of salts such as sodium chloride (figure 11.7).

As an example of ionic bonding, consider the reaction of sodium (a soft reactive metal) with chlorine (a pale yellow-green gas). When an atom of sodium and an atom of chlorine collide, they react violently as the valence electron is transferred from the sodium to the chlorine atom. This produces a sodium ion and a chlorine ion. The reaction can be illustrated with electron dot symbols as follows:

$$\text{Na} \cdot \ + \ \cdot \ddot{\underset{..}{\text{Cl}}} : \longrightarrow \text{Na}^+ \ \left(: \ddot{\underset{..}{\text{Cl}}} : \right)^-$$

equation 11.2

As you can see, the sodium ion transferred its valence electron, and the resulting ion now has a stable electron configuration. The chlorine atom accepted the electron in its outer orbital to acquire a stable electron configuration. Thus, a stable positive ion and a stable negative ion are formed. Because of opposite electrical charges, the ions attract each other to produce an ionic bond. When many ions are involved, each Na^+ ion is surrounded by six Cl^- ions, and each Cl^- ion is surrounded by six Na^+ ions. This gives the resulting solid NaCl its crystalline cubic structure, as shown in figure 11.7. In the solid state, all the sodium ions and all the chlorine ions are bound together in one giant unit. Thus, the term *molecule* is not really appropriate for ionic solids such as sodium chloride. But the term is sometimes used anyway, since any given sample will have the same number of Na^+ ions as Cl^- ions.

Energy and Electrons in Ionic Bonding

The sodium-chlorine reaction can be represented with electron dot notation as occurring in three steps:

1. $\text{energy} \ + \ \text{Na} \cdot \longrightarrow \text{Na}^+ \ + \ e^-$

2. $\cdot \ddot{\underset{..}{\text{Cl}}} : \ + \ e^- \longrightarrow \left(: \ddot{\underset{..}{\text{Cl}}} : \right)^- \ + \ \text{energy}$

3. $\text{Na}^+ \ + \ \left(: \ddot{\underset{..}{\text{Cl}}} : \right)^- \longrightarrow \text{Na}^+ \left(: \ddot{\underset{..}{\text{Cl}}} : \right)^- + \ \text{energy}$

The energy released in steps 2 and 3 is greater than the energy absorbed in step 1, and an ionic bond is formed. The energy released is called the **heat of formation.** It is also the amount of energy required to decompose the compound (sodium chloride) into its elements. The reaction does not take place in steps as described, however, but occurs all at once. Note again, as in the photosynthesis-burning reactions described earlier, that the total amount of chemical energy is conserved. The energy released by the formation of the sodium chloride compound is the *same* amount of energy needed to decompose the compound.

Ionic bonds are formed by electron transfer, and electrons are conserved in the process. This means that electrons are not created or destroyed in a chemical reaction. The same total number of electrons exists after a reaction that existed before the reaction. There are two rules you can use for keeping track of electrons in ionic bonding reactions:

1. Ions are formed as atoms gain or lose valence electrons to achieve the stable noble gas structure.
2. There must be a balance between the number of electrons lost and the number of electrons gained by atoms in the reaction.

The sodium-chlorine reaction follows these two rules. The loss of one valence electron from a sodium atom formed a stable sodium ion. The gain of one valence electron by the chlorine atom formed a stable chlorine ion. Thus, both ions have noble gas configurations (rule 1), and one electron was lost and one was gained, so there is a balance in the number of electrons lost and the number gained (rule 2).

Ionic Compounds and Formulas

The **formula** of a compound *describes what elements are in the compound and in what proportions.* Sodium chloride contains one positive sodium ion for each negative chlorine ion. The formula of the compound sodium chloride is NaCl. If there are no subscripts at the lower right part of each symbol, it is understood that the symbol has a number "1." Thus, NaCl indicates a compound made up of the elements sodium and chlorine, and there is one sodium atom for each chlorine atom.

Calcium (Ca) is an alkaline metal in family IIA, and fluorine (F) is a halogen in family VIIA. Since calcium is a metal and fluorine is a nonmetal, you would expect calcium and fluorine atoms to react, forming a compound with ionic bonds. Calcium must lose two valence electrons to acquire a noble gas configuration. Fluorine needs one valence electron to acquire a noble gas configuration. So calcium needs to lose two electrons and fluorine needs to gain one electron to achieve a stable configuration (rule 1). Two fluorine atoms, each acquiring one electron, are needed to balance the number of electrons lost and the number of electrons gained. The compound formed from the reaction, calcium fluoride, will therefore have a calcium ion with a charge of plus two for every fluorine ion with a charge of minus one. Recalling that electron dot symbols show only the outer valence electrons, you can see that the reaction is

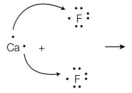

which shows that a calcium atom transfers two electrons, one each to two fluorine atoms. Now showing the results of the reaction, a calcium ion is formed from the loss of two electrons (charge 2+) and two fluorine ions are formed by gaining one electron each (charge 1−):

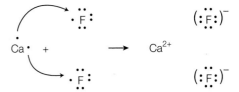

equation 11.3

The formula of the compound is therefore CaF_2, with the subscript 2 for fluorine and the understood subscript 1 for calcium. This means that there are two fluorine atoms for each calcium atom in the compound.

Sodium chloride (NaCl) and magnesium fluoride (MgF_2) are examples of compounds held together by ionic bonds. Such compounds are called **ionic compounds.** Ionic compounds of the representative elements are generally white, crystalline solids that form colorless solutions. Sodium chloride, the most common example, is common table salt. Many of the transition elements form colored compounds that make colored solutions. Ionic compounds dissolve in water, producing a solution of ions that can conduct an electric current.

In general, the elements in families IA and IIA of the periodic table tend to form positive ions by losing electrons. The ion charge for these elements equals the family number of these elements. The elements in families VIA and VIIA tend to form negative ions by gaining electrons. The ion charge for these elements equals their family number minus 8. The elements in

TABLE 11.1	Common ions of some representative elements	
Element	**Symbol**	**Ion**
Lithium	Li	1+
Sodium	Na	1+
Potassium	K	1+
Magnesium	Mg	2+
Calcium	Ca	2+
Barium	Ba	2+
Aluminum	Al	3+
Oxygen	O	2−
Sulfur	S	2−
Hydrogen	H	1+,1−
Fluorine	F	1−
Chlorine	Cl	1−
Bromine	Br	1−
Iodine	I	1−

TABLE 11.2	Common ions of some transition elements	
Single-Charge Ions		
Element	*Symbol*	*Charge*
Zinc	Zn	2+
Tungsten	W	6+
Silver	Ag	1+
Cadmium	Cd	2+
Variable-Charge Ions		
Element	*Symbol*	*Charge*
Chromium	Cr	2+,3+,6+
Manganese	Mn	2+,4+,7+
Iron	Fe	2+,3+
Cobalt	Co	2+,3+
Nickel	Ni	2+,3+
Copper	Cu	1+,2+
Tin	Sn	2+,4+
Gold	Au	1+,3+
Mercury	Hg	1+,2+
Lead	Pb	2+,4+

families IIIA and VA have less of a tendency to form ionic compounds, except for those in higher periods. Common ions of representative elements are given in table 11.1. The transition elements form positive ions of several different charges. Some common ions of the transition elements are listed in table 11.2.

The single-charge representative elements and the variable-charge transition elements form single, monatomic negative ions. There are also many polyatomic (*poly* means "many") negative ions, charged groups of atoms that act like a single unit in ionic compounds. Polyatomic ions are held together by covalent bonds, which are discussed in the next section.

Use electron dot notation to predict the formula of a compound formed when aluminum (Al) combines with fluorine (F).

Answer

Aluminum, atomic number 13, is in family IIIA so it has three valence electrons and an electron dot notation of

$$\overset{\displaystyle \cdot}{\underset{\displaystyle \cdot}{Al}} \cdot$$

According to the octet rule, the aluminum atom would need to lose three electrons to acquire the stable noble gas configuration. Fluorine, atomic number 9, is in family VIIA so it has seven valence electrons and an electron dot notation of

$$\cdot \overset{\displaystyle ..}{\underset{\displaystyle ..}{F}} :$$

Fluorine would acquire a noble gas configuration by accepting one electron. Three fluorine atoms, each acquiring one electron, are needed to balance the three electrons lost by aluminum. The reaction can be represented as

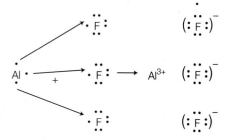

The ratio of aluminum atoms to fluorine atoms in the compound is 1:3. The formula for aluminum fluoride is therefore AlF_3.

Predict the formula of the compound formed between aluminum and oxygen using electron dot notation. (Answer: Al_2O_3).

Covalent Bonds

Most substances do not have the properties of ionic compounds since they are not composed of ions. Most substances are molecular, composed of electrically neutral groups of atoms that are tightly bound together. As noted earlier, many gases are diatomic, occurring naturally as two atoms bound together as an electrically neutral molecule. Hydrogen, for example, occurs as molecules of H_2 and no ions are involved. The hydrogen atoms are held together by a covalent bond. A **covalent bond** is a *chemical bond formed by the sharing of a pair of electrons*. In the diatomic hydro-

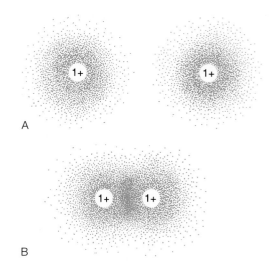

FIGURE 11.8

(*A*) Two hydrogen atoms, each with its own probability distribution of electrons about the nucleus. (*B*) When the hydrogen atoms bond, a new electron distribution pattern forms around the entire molecule, and both electrons occupy the molecular orbital.

gen molecule each hydrogen atom contributes a single electron to the shared pair. Both hydrogen atoms count the shared pair of electrons in achieving their noble gas configuration. Hydrogen atoms both share one pair of electrons, but other elements might share more than one pair to achieve a noble gas structure.

Consider how the covalent bond forms between two hydrogen atoms by imagining two hydrogen atoms moving toward one another. Each atom has a single electron. As the atoms move closer and closer together, their orbitals begin to overlap. Each electron is attracted to the oppositely charged nucleus of the other atom and the overlap tightens. Then the repulsive forces from the like-charged nuclei will halt the merger. A state of stability is reached between the two nuclei and two electrons, and an H_2 molecule has been formed. The two electrons are now shared by both atoms, and the attraction of one nucleus for the other electron and vice versa holds the atoms together (figure 11.8).

Covalent Compounds and Formulas

Electron dot notation can be used to represent the formation of covalent bonds. For example, the joining of two hydrogen atoms to form an H_2 molecule can be represented as

$$H\cdot \ + \ H\cdot \ \longrightarrow \ H:H$$

Since an electron pair is *shared* in a covalent bond, the two electrons move throughout the entire molecular orbital. Since each hydrogen atom now has both electrons on an equal basis, each can be considered now to have the noble gas configuration of helium. A dashed circle around each symbol shows that both atoms have two electrons:

$$H\cdot \ + \ H\cdot \ \longrightarrow \ (H:H)$$

equation 11.4

Hydrogen and fluorine react to form a covalent molecule (how this is known will be discussed shortly), and this bond can be represented with electron dots. Fluorine is in the VIIA family, so you know an atom of fluorine has seven valence electrons in the outermost energy level. The reaction is

$$H\cdot \; + \; \cdot \overset{\displaystyle ..}{\underset{\displaystyle ..}{F}}: \; \longrightarrow \; \left(H : \overset{\displaystyle ..}{\underset{\displaystyle ..}{F}} : \right)$$

equation 11.5

Each atom shares a pair of electrons to achieve a noble gas configuration. Hydrogen achieves the helium configuration, and fluorine achieves the neon configuration. All the halogens have seven valence electrons and all need to gain one electron (ionic bond) or share an electron pair (covalent bond) to achieve a noble gas configuration. This also explains why the halogen gases occur as diatomic molecules. Two fluorine atoms can achieve a noble gas configuration by sharing a pair of electrons:

$$\cdot \overset{\displaystyle ..}{\underset{\displaystyle ..}{F}}: \; + \; \cdot \overset{\displaystyle ..}{\underset{\displaystyle ..}{F}}: \; \longrightarrow \; \left(: \overset{\displaystyle ..}{\underset{\displaystyle ..}{F}} : \overset{\displaystyle ..}{\underset{\displaystyle ..}{F}} : \right)$$

equation 11.6

Each fluorine atom thus achieves the neon configuration by bonding together. Note that there are two types of electron pairs: (1) orbital pairs and (2) bonding pairs. Orbital pairs are not shared, since they are the two electrons in an orbital, each with a separate spin. Orbital pairs are also called *lone pairs,* since they are not shared. *Bonding pairs,* as the name implies, are the electron pairs shared between two atoms. Considering again the F_2 molecule,

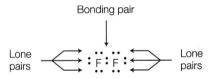

Often, the number of bonding pairs that are formed by an atom is the same as the number of single, *unpaired* electrons in the atomic electron dot notation. For example, hydrogen has one unpaired electron, and oxygen has two unpaired electrons. Hydrogen and oxygen combine to form an H_2O molecule, as shown below

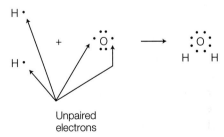

The diatomic hydrogen (H_2) and fluorine (F_2), hydrogen fluoride (HF), and water (H_2O) are examples of compounds held together by covalent bonds. A compound held together by covalent bonds is called a **covalent compound.** In general,

TABLE 11.3	Structures and compounds of nonmetallic elements combined with hydrogen	
Nonmetallic Elements	**Element (E Represents Any Element of Family)**	**Compound**
Family IVA: C, Si, Ge	$\cdot \dot{E} \cdot$	H H:E:H H
Family VA: N, P, As, Sb	$\cdot \dot{E} \cdot$	H:E:H H
Family VIA: O, S, Se, Te	$\cdot \ddot{E} \cdot$	H:E:H
Family VIIA: F, Cl, Br, I	$\cdot \ddot{E} :$	H:E:

covalent compounds form from nonmetallic elements on the right side of the periodic table. For elements in families IVA through VIIA, the number of unpaired electrons (and thus the number of covalent bonds formed) is eight minus the family number. You can get a lot of information from the periodic table from generalizations like this one. For another generalization, compare table 11.3 with the periodic table. The table gives the structures of nonmetals combined with hydrogen and the resulting compounds.

Multiple Bonds

Two dots can represent a lone pair of valence electrons, or they can represent a bonding pair, a single pair of electrons being shared by two atoms. Bonding pairs of electrons are often represented by a simple line between two atoms. For example,

H : H is shown as H — H

and

$: \overset{\displaystyle ..}{\underset{\displaystyle }{O}} :$ is shown as (bent structure)
  H H   H   H

Note that the line between the two hydrogen atoms represents an electron pair, so each hydrogen atom has two electrons in the outer shell, as does helium. In the water molecule, each hydrogen atom has two electrons as before. The oxygen atom has two lone pairs (a total of four electrons) and two bonding pairs (a total of four electrons) for a total of eight electrons. Thus, oxygen has acquired a stable octet of electrons.

A covalent bond in which a single pair of electrons is shared by two atoms is called a *single covalent bond* or simply a **single bond.** Some atoms have two unpaired electrons and can share more than one electron pair. A **double bond** is a covalent bond formed when *two pairs* of electrons are shared by two atoms. This happens mostly in compounds involving atoms of the elements C, N, O, and S.

FIGURE 11.9

Acetylene is a hydrocarbon consisting of two carbon atoms and two hydrogen atoms held together by a triple covalent bond between the two carbon atoms. When mixed with oxygen gas (the tank to the right), the resulting flame is hot enough to cut through most metals.

Ethylene, for example, is a gas given off from ripening fruit. The electron dot formula for ethylene is

The ethylene molecule has a double bond between two carbon atoms. Since each line represents two electrons, you can simply count the lines around each symbol to see if the octet rule has been satisfied. Each H has one line, so each H atom is sharing two electrons. Each C has four lines so each C atom has eight electrons, satisfying the octet rule.

A **triple bond** is a covalent bond formed when *three pairs* of electrons are shared by two atoms. Triple bonds occur mostly in compounds with atoms of the elements C and N. Acetylene, for example, is a gas often used in welding torches (figure 11.9). The electron dot formula for acetylene is

The acetylene molecule has a triple bond between two carbon atoms. Again, note that each line represents two electrons. Each C atom has four lines, so the octet rule is satisfied.

Coordinate Covalent Bonds

The single, double, and triple covalent bonds are formed when an atom shares one, two, or three pairs of electrons. For each pair, each atom contributes one electron, and another atom shares one of its electrons in return. There is another type of covalent bond that is a "hole-and-plug" kind of sharing. A **coordinate covalent bond** is formed when one atom contributes both electrons of a shared

TABLE 11.4	Some common polyatomic ions
Ion Name	**Formula**
Acetate	$(C_2H_3O_2)^-$
Ammonium	$(NH_4)^+$
Borate	$(BO_3)^{3-}$
Carbonate	$(CO_3)^{2-}$
Chlorate	$(ClO_3)^-$
Chromate	$(CrO_4)^{2-}$
Cyanide	$(CN)^-$
Dichromate	$(Cr_2O_7)^{2-}$
Hydrogen carbonate (or bicarbonate)	$(HCO_3)^-$
Hydrogen sulfate (or bisulfate)	$(HSO_4)^-$
Hydroxide	$(OH)^-$
Hypochlorite	$(ClO)^-$
Nitrate	$(NO_3)^-$
Nitrite	$(NO_2)^-$
Perchlorate	$(ClO_4)^-$
Permanganate	$(MnO_4)^-$
Phosphate	$(PO_4)^{3-}$
Phosphite	$(PO_3)^{3-}$
Sulfate	$(SO_4)^{2-}$
Sulfite	$(SO_3)^{2-}$

pair. The coordinate covalent bond is like the other covalent bonds, since a pair of electrons is shared between two atoms. The difference is that both of these electrons come from one atom, not one each from two atoms. Ammonia, for example, has an electron dot structure of

which appears to be a stable structure, since the octet rule is satisfied. However, notice the lone pair of electrons on the ammonia molecule. Ammonia can contribute its pair of nonbonding electrons to, say, a hydrogen ion, which shares the pair with ammonia as shown below,

equation 11.7

forming an ammonium ion. All of the covalent bonds with hydrogen are now identical, with each hydrogen atom sharing a pair of electrons. The molecule is now an ion, however, since it has a net charge. An ion made up of many atoms is called a **polyatomic ion.** Coordinate covalent bonding is common in many polyatomic ions. These ions are important in many common chemicals, which will be discussed later. Some of the common polyatomic ions are listed in table 11.4.

IA																	
1 **H** 2.1	IIA												IIIA	IVA	VA	VIA	VIIA
3 **Li** 1.0	4 **Be** 1.5												5 **B** 2.0	6 **C** 2.5	7 **N** 3.0	8 **O** 3.5	9 **F** 4.0
11 **Na** 0.9	12 **Mg** 1.2	IIIB	IVB	VB	VIB	VIIB		VIIIB		IB	IIB		13 **Al** 1.5	14 **Si** 1.8	15 **P** 2.1	16 **S** 2.5	17 **Cl** 3.0
19 **K** 0.8	20 **Ca** 1.0	21 **Sc** 1.3	22 **Ti** 1.5	23 **V** 1.6	24 **Cr** 1.6	25 **Mn** 1.5	26 **Fe** 1.8	27 **Co** 1.8	28 **Ni** 1.8	29 **Cu** 1.9	30 **Zn** 1.6	31 **Ga** 1.6	32 **Ge** 1.8	33 **As** 2.0	34 **Se** 2.4	35 **Br** 2.8
37 **Rb** 0.8	38 **Sr** 1.0	39 **Y** 1.2	40 **Zr** 1.4	41 **Nb** 1.6	42 **Mo** 1.8	43 **Tc** 1.9	44 **Ru** 2.2	45 **Rh** 2.2	46 **Pd** 2.2	47 **Ag** 1.9	48 **Cd** 1.7	49 **In** 1.7	50 **Sn** 1.8	51 **Sb** 1.9	52 **Te** 2.1	53 **I** 2.5
55 **Cs** 0.7	56 **Ba** 0.9	57 **La** 1.1	72 **Hf** 1.3	73 **Ta** 1.5	74 **W** 1.7	75 **Re** 1.9	76 **Os** 2.2	77 **Ir** 2.2	78 **Pt** 2.2	79 **Au** 2.4	80 **Hg** 1.9	81 **Tl** 1.8	82 **Pb** 1.8	83 **Bi** 1.9	84 **Po** 2.0	85 **At** 2.2
87 **Fr** 0.7	88 **Ra** 0.9	89 **Ac** 1.1														

FIGURE 11.10

Electronegativities of the elements. These values are comparative only, assigned an arbitrary scale to indicate the relative tendency of atoms to attract shared electrons.

Note that (1) all have negative charges except the ammonium ion, (2) all are formed exclusively of nonmetals except three that contain metals, and (3) some are similar with different *-ite* and *-ate* endings. The *-ate* ion always has one more oxygen than the *-ite* ion.

BOND POLARITY

How do you know if a bond between two atoms will be ionic or covalent? In general, ionic bonds form between metal atoms and nonmetal atoms, especially those from the opposite sides of the periodic table. Also in general, covalent bonds form between the atoms of nonmetals. If an atom has a much greater electron-pulling ability than another atom, the electron is pulled completely away from the atom with lesser pulling ability, and an ionic bond is the result. If the electron-pulling ability is more even between the two atoms, the electron is shared, and a covalent bond results. As you can imagine, all kinds of reactions are possible between atoms with different combinations of electron-pulling abilities. The result is that it is possible to form many gradations of bonding between completely ionic and completely covalent bonding. Which type of bonding will result can be found by comparing the electronegativity of the elements involved. **Electronegativity** is the *comparative ability of atoms of an element to attract bonding electrons.* The assigned numerical values for electronegativities are given in figure 11.10. Elements with higher values have the great-

TABLE 11.5	The meaning of absolute differences in electronegativity		
Absolute Difference	→		**Type of Bond Expected**
1.7 or greater	means		ionic bond
between 0.5 and 1.7	means		polar covalent bond
0.5 or less	means		covalent bond

est attraction for bonding electrons, and elements with the lowest values have the least attraction for bonding electrons.

The absolute ("absolute" means without plus or minus signs) difference in the electronegativity of two bonded atoms can be used to predict if a bond is ionic or covalent (table 11.5). A large difference means that one element has a much greater attraction for bonding electrons than the other element. *If the absolute difference in electronegativity is 1.7 or more,* one atom pulls the bonding electron completely away and *an ionic bond results.* For example, sodium (Na) has an electronegativity of 0.9. Chlorine (Cl) has an electronegativity of 3.0. The difference is 2.1, so you can expect sodium and chloride to form ionic bonds. *If the absolute difference in electronegativity is 0.5 or less,* both atoms have about the same ability to attract bonding electrons. The result is that the electron is shared, and *a covalent bond results.* A given hydrogen atom (H) has an electronegativity of another hydrogen atom, so the difference is 0. Zero is less

Electron distribution and kinds of bonding

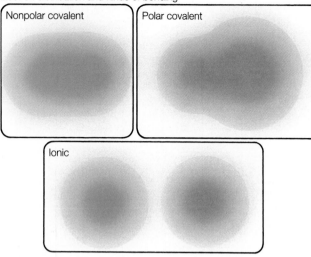

FIGURE 11.11

The absolute difference in electronegativities determines the kind of bond formed.

than 0.5 so you can expect a molecule of hydrogen gas to have a nonpolar covalent bond.

An ionic bond can be expected when the difference in electronegativity is 1.7 or more, and a covalent bond can be expected when the difference is less than 0.5. What happens when the difference is between 0.5 and 1.7? A covalent bond is formed, but there is an inequality since one atom has a greater bonding electron attraction than the other atom. Thus, the bonding electrons are shared unequally. A **polar covalent bond** is *a covalent bond in which there is an unequal sharing of bonding electrons.* Thus, the bonding electrons spend more time around one atom than the other. The term "polar" means "poles," and that is what forms in a polar molecule. Since the bonding electrons spend more time around one atom than the other, one end of the molecule will have a negative pole, and the other end will have a positive pole. Since there are two poles, the molecule is sometimes called a *dipole.* Note that the molecule as a whole still contains an equal number of electrons and protons, so it is overall electrically neutral. The poles are created by an uneven charge distribution, not an imbalance of electrons and protons. Figure 11.11 shows this uneven charge distribution for a polar covalent compound. The bonding electrons spend more time near the atom on the right, giving this side of the molecule a negative pole.

Figure 11.11 also shows a molecule that has an even charge distribution. The electron distribution around one atom is just like the charge distribution around the other. This molecule is thus a *nonpolar molecule* with a *nonpolar bond.* Thus, a polar bond can be viewed as an intermediate type of bond between a nonpolar covalent bond and an ionic bond. Many gradations are possible between the transition from a purely nonpolar covalent bond and a purely ionic bond.

Q & A

Predict if the following bonds are nonpolar covalent, polar covalent, or ionic: (a) H-O; (b) C-Br; and (c) K-Cl

Answer

From the electronegativity values in figure 11.10, the absolute differences are

(a) H-O, 1.4
(b) C-Br, 0.3
(c) K-Cl, 2.2

Since an absolute difference of less than 0.5 means nonpolar covalent, between 0.5 and 1.7 means polar covalent, and greater than 1.7 means ionic, then

(a) H-O, polar covalent
(b) C-Br, nonpolar covalent
(c) K-Cl, ionic

Predict if the following bonds are nonpolar covalent, polar covalent, or ionic: (a) Ca-O; (b) H-Cl; and (c) C-O. (Answer: (a) ionic; (b) polar covalent; (c) polar covalent)

COMPOSITION OF COMPOUNDS

As you can imagine, there are literally millions of different chemical compounds from all the possible combinations of over ninety natural elements held together by ionic or covalent bonds. Each of these compounds has its own name, so there are millions of names and formulas for all the compounds. In the early days, compounds were given *common names* according to how they were used, where they came from, or some other means of identifying them. Thus, sodium carbonate was called soda, and closely associated compounds were called baking soda (sodium bicarbonate), washing soda (sodium carbonate), caustic soda (sodium hydroxide), and the bubbly drink made by reacting soda with acid was called soda water, later called soda pop (figure 11.12). Potassium carbonate was extracted from charcoal by soaking in water and came to be called potash. Such common names are colorful, and some are descriptive, but it was impossible to keep up with the names as the number of known compounds grew. So a systematic set of rules was developed to determine the name and formula of each compound. Once you know the rules, you can write the formula when you hear the name. Conversely, seeing the formula will tell you the systematic name of the compound. This can be an interesting intellectual activity and can also be important when reading the list of ingredients to understand the composition of a product.

There is a different set of systematic rules to be used with ionic compounds and covalent compounds, but there are a few

FIGURE 11.12

These substances are made up of sodium and some form of a carbonate ion. All have common names with the term "soda" for this reason. Soda water (or "soda pop") was first made by reacting soda (sodium carbonate) with an acid, so it was called "soda water."

TABLE 11.6	Modern names of some variable-charge ions
Ion	**Name of Ion**
Fe^{2+}	Iron(II) ion
Fe^{3+}	Iron(III) ion
Cu^+	Copper(I) ion
Cu^{2+}	Copper(II) ion
Pb^{2+}	Lead(II) ion
Pb^{4+}	Lead(IV) ion
Sn^{2+}	Tin(II) ion
Sn^{4+}	Tin(IV) ion
Cr^{2+}	Chromium(II) ion
Cr^{3+}	Chromium(III) ion
Cr^{6+}	Chromium(VI) ion

rules in common. For example, a compound made of only two different elements always ends with the suffix "-ide." So when you hear the name of a compound ending with "-ide" you automatically know that the compound is made up of only two elements. Sodium chlor*ide* is an ionic compound made up of sodium and chlorine ions. Carbon diox*ide* is a covalent compound with carbon and oxygen atoms. Thus, the systematic name tells you what elements are present in a compound with an "-ide" ending.

Ionic Compound Names

Ionic compounds formed by representative metal ions are named by stating the name of the metal (positive ion) first, then the name of the nonmetal (negative ion). Ionic compounds formed by variable-charge ions of the transition elements have an additional rule to identify which variable-charge ion is involved. There was an old way of identifying the charge on the ion by adding either "-ic" or "-ous" to the name of the metal. The suffix "-ic" meant the higher of two possible charges, and the suffix "-ous" meant the lower of two possible charges. For example, iron has two possible charges, 2 + or 3 +. The old system used the Latin name for the root. The Latin name for iron is ferrum, so a higher charged iron ion (3+) was named a ferric ion. The lower charged iron ion (2+) was called a ferrous ion.

You still hear the old names sometimes, but chemists now have a better way to identify the variable-charge ion. The newer system uses the English name of the metal with Roman numerals in parentheses to indicate the charge number. Thus, an iron ion with a charge of 2 + is called an iron(II) ion and an iron ion with a charge of 3 + is an iron(III) ion. Table 11.6 gives some of the modern names for variable-charge ions. These names are used with the name of a nonmetal ending in "-ide," just like the single-charge ions in ionic compounds made up of two different elements.

Some ionic compounds contain three or more elements, and so are more complex than a combination of a metal ion and a nonmetal ion. This is possible because they have *polyatomic ions*,

groups of two or more atoms that are bound together tightly and behave very much like a single monatomic ion. For example, the OH^- ion is an oxygen atom bound to a hydrogen atom with a net charge of $1-$. This polyatomic ion is called a *hydroxide ion*. The hydroxide compounds make up one of the main groups of ionic compounds, the *metal hydroxides*. A metal hydroxide is an ionic compound consisting of a metal with the hydroxide ion. Another main group consists of the salts with polyatomic ions.

The metal hydroxides are named by identifying the metal first and the term *hydroxide* second. Thus, NaOH is named sodium hydroxide and KOH is potassium hydroxide. The salts are similarly named, with the metal (or ammonium ion) identified first, then the name of the polyatomic ion. Thus, $NaNO_3$ is named sodium nitrate and $NaNO_2$ is sodium nitrite. Note that the suffix "-ate" means the polyatomic ion with one more oxygen atom than the "-ite" ion. For example, the chlor*ate* ion is $(ClO_3)^-$, and the chlor*ite* ion is $(ClO_2)^-$. Sometimes more than two possibilities exist, and more oxygen atoms are identified with the prefix "per-", and less with the prefix "hypo-".

Activities

Dissolving an ionic compound in water results in ions being pulled from the crystal lattice to form free ions. A solution that contains ions will conduct an electric current, so electrical conductivity is one way to test a dissolved compound to see if it is an ionic compound or not.

Make a conductivity tester like the one illustrated in figure 11.13. This one is a 9 V radio battery and a miniature Christmas tree bulb with two terminal wires. Sand these wires to make a good electrical contact. Try testing dry (1) baking soda, (2) table salt, and (3) sugar. Test solutions of these substances dissolved in distilled water. Test vinegar and rubbing alcohol. Explain which contain ions and why.

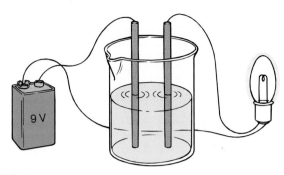

FIGURE 11.13

A battery and bulb will tell you if a solution contains ions. See "Activities" section on page 267.

Thus, the *perchlorate* ion is $(ClO_4)^-$ and the *hypochlorite* ion is $(ClO)^-$.

Ionic Compound Formulas

The formulas for ionic compounds are easy to write. There are two rules: (1) the symbol for the positive element is written first, followed by the symbol for the negative element, just as in the order in the name, and (2) subscripts are used to indicate the numbers of ions needed to produce an electrically neutral compound. For example, the name "calcium chloride" tells you that this compound consists of positive calcium ions and negative chlorine ions. Again, "-ide" means only two elements are present. The calcium ion is Ca^{2+}, and the chlorine ion is Cl^- (you know this by applying the atomic theory, knowing their positions in the periodic table, or by using a table of ions and their charges). To be electrically neutral, the compound must have an equal number of pluses and minuses. Thus, two negative chlorine ions are needed for every calcium ion with its 2+ charge. The formula is $CaCl_2$. The total charge of two chlorines is thus 2−, which balances the 2+ charge on the calcium ion.

One easy way to write a formula showing that a compound is electrically neutral is to cross over the absolute charge numbers (without plus or minus signs) and use them as subscripts. For example, the symbols for the calcium ion and the chlorine ion are

$$Ca^{2+}Cl^{1-}$$

Crossing the absolute numbers as subscripts, as follows

$$Ca_1^{2+} \quad Cl_2^{1-}$$

and then dropping the charge numbers gives

$$Ca_1 \ Cl_2$$

No subscript is written for 1; it is understood. The formula for calcium chloride is thus

$$CaCl_2$$

The crossover technique works because ionic bonding results from a transfer of electrons, and the net charge is conserved. A calcium ion has a 2+ charge because the atom lost two electrons and two chlorine atoms gain one electron each, for a total of two electrons gained. Two electrons lost equals two electrons gained, and the net charge on calcium chloride is zero, as it has to be.

When using the crossover technique it is sometimes necessary to reduce the ratio to the lowest common multiple. Thus, Mg_2O_2 means an equal ratio of magnesium and oxygen ions, so the correct formula is MgO.

The formulas for variable-charge ions are easy to write, since the Roman numeral tells you the charge number. The formula for tin(II) fluoride is written by crossing over the charge numbers (Sn^{2+}, F^{1-}), and the formula is SnF_2.

Q & A

Name the following compounds: (a) LiF and (b) PbF_2. Write the formulas for the following compounds: (c) potassium bromide and (d) copper(I) sulfide.

Answer

(a) The formula LiF means that the positive metal ions are lithium, the negative nonmetal ions are fluorine, and there are only two elements in the compound. Lithium ions are Li^{1+} (family IA), and fluorine ions are F^{1-} (family VIIA). The name is lithium fluoride.

(b) Lead is a variable-charge transition element (table 11.6), and fluorine ions are F^{1-}. The lead ion must be Pb^{2+} because the compound PbF_2 is electrically neutral. Therefore, the name is lead(II) fluoride.

(c) The ions are K^{1+} and Br^{1-}. Crossing over the charge numbers and dropping the signs gives the formula KBr.

(d) The Roman numeral tells you the charge on the copper ion, so the ions are Cu^{1+} and S^{2-}. The formula is Cu_2S.

The formulas for ionic compounds with polyatomic ions are written from combinations of positive metal ions or the ammonium ion with the polyatomic ions, as listed in table 11.4. Since the polyatomic ion is a group of atoms that has a charge and stays together in a unit, it is sometimes necessary to indicate this with parentheses. For example, magnesium hydroxide is composed of Mg^{2+} ions and $(OH)^{1-}$ ions. Using the crossover technique to write the formula, you get

$$Mg^{2+} \quad OH_2^{1-} \quad \text{or} \quad Mg(OH)_2$$

CONNECTIONS

Household Chemicals

A survey of typical household products will show that many different kinds of chemicals are in use. In the kitchen you might find sodium chloride, sodium bicarbonate, and acetic acid. In cleaning supplies you might find ammonia, sodium hypochlorite, and sodium hydroxide. In yard and garden supplies you might find ammonium nitrate, phosphate compounds, potassium oxide, and pesticides.

Some of these household chemicals can be dangerous if the chemicals are not used in accordance with directions. Ammonia, sodium hypochlorite (bleach), and sodium hydroxide (drain cleaner), for example, can have a caustic action on plant and animal tissue, converting the tissue into soluble materials. This is, in fact, how these chemicals are able to clean greasy spots from a surface or drain.

Disposal of unwanted paints, solvents, and household chemicals can be a problem. Many of the chemicals can be dangerous if mixed, such as bleach and fertilizers, and should not be dumped into the sewage system. You can make a difference in water quality by not dumping household chemicals into the sewage system. Instead, use completely the contents of oven, toilet bowl, and drain cleaners, bleaches, paints, rust removers, and solvents. Also, choose water-based latex paint instead of oil-based paint if you have that choice where you live, and use it up instead of storing or dumping it. There is more you can do; begin by reading up on environmental issues for your local area.

The parentheses are used and the subscript is written *outside* the parenthesis to show that the entire hydroxide unit is taken twice. The formula $Mg(OH)_2$ means

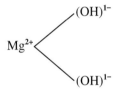

which shows that the pluses equal the minuses. Parentheses are not used, however, when only one polyatomic ion is present. Sodium hydroxide is $NaOH$, not $Na(OH)_1$.

Covalent Compound Names

Covalent compounds are molecular, and the molecules are composed of two *nonmetals,* as opposed to the metal and nonmetal elements that make up ionic compounds. The combinations of nonmetals alone do not present simple names as the ionic compounds did, so a different set of rules for naming and formula writing is needed.

Ionic compounds were named by stating the name of the positive metal ion, then the name of the negative nonmetal ion with an "-ide" ending. This system is not adequate for naming the covalent compounds. To begin, covalent compounds are composed of two or more nonmetal atoms that form a molecule. It is possible for some atoms to form single, double, or even triple bonds with other atoms, including atoms of the same element, and coordinate covalent bonding is also possible in some compounds. The net result is that the same two elements can form more than one kind of covalent compound. Carbon and oxygen, for example, can combine to form the gas released from burning and respiration, carbon dioxide (CO_2). Under certain conditions

TABLE 11.7	Prefixes and element stem names		
Prefixes		**Stem Names**	
Prefix	*Meaning*	*Element*	*Stem*
Mono-	1	Hydrogen	Hydr-
Di-	2	Carbon	Carb-
Tri-	3	Nitrogen	Nitr-
Tetra-	4	Oxygen	Ox-
Penta-	5	Fluorine	Fluor-
Hexa-	6	Phosphorus	Phosph-
Hepta-	7	Sulfur	Sulf-
Octa-	8	Chlorine	Chlor-
Nona-	9	Bromine	Brom-
Deca-	10	Iodine	Iod-

Note: the *a* or *o* ending on the prefix is often dropped if the stem name begins with a vowel, e.g., "tetroxide," not "tetraoxide."

the very same elements combine to produce a different gas, the poisonous carbon monoxide (CO). Similarly, sulfur and oxygen can combine differently to produce two different covalent compounds. A successful system for naming covalent compounds must therefore provide a means of identifying different compounds made of the same elements. This is accomplished by using a system of Greek prefixes (see table 11.7). The rules are as follows:

1. The first element in the formula is named first with a prefix indicating the number of atoms if the number is greater than one.
2. The stem name of the second element in the formula is next. A prefix is used with the stem if two elements form more than one compound. The suffix "-ide" is again used to indicate a compound of only two elements.

A CLOSER LOOK | Microwave Ovens and Molecular Bonds

A microwave oven rapidly cooks foods that contain water, but paper, glass, and plastic products remain cool in the oven. If they are warmed at all, it is from the heat conducted from the food. The explanation of how the microwave oven heats water, but not most other substances, begins with the nature of the chemical bond.

A chemical bond acts much like a stiff spring, resisting both compression and stretching as it maintains an equilibrium distance between the atoms. As a result, a molecule tends to vibrate when energized or buffeted by other molecules. The rate of vibration depends on the "stiffness" of the spring, which is determined by the bond strength, and the mass of the atoms making up the molecule. Each kind of molecule therefore has its own set of characteristic vibrations, a characteristic natural frequency.

Disturbances with a wide range of frequencies can impact a vibrating system. When the frequency of a disturbance matches the natural frequency, energy is transferred very efficiently, and the system undergoes a large increase in amplitude. Such a frequency match is called resonance. When the disturbance is visible light or some other form of radiant energy, a resonant match results in absorption of the radiant energy and an increase in the molecular kinetic energy of vibration. Thus, a resonant match results in a temperature increase.

The natural frequency of a water molecule matches the frequency of infrared radiation, so resonant heating occurs when infrared radiation strikes water molecules. It is the water molecules in your skin that absorb infrared radiation from the sun, a fire, or some hot object, resulting in the warmth that you feel. Because of this match between the frequency of infrared radiation and the natural frequency of a water molecule, infrared is often called "heat radiation." Since infrared radiation is absorbed by water molecules, it is mostly absorbed on the surface of an object, penetrating only a short distance.

The frequency ranges of visible light, infrared radiation, and microwave radiation are given in box table 11.1. Most microwave ovens operate at the lower end of the microwave frequency range, between 1×10^9 Hz to 3×10^9 Hz, or 1 to 3 gigahertz. This range is too low for a resonant match with water molecules, so something else must transfer energy from the microwaves to heat the water. This something else is a result of another characteristic of the water mole-cule, the type of covalent bond holding the molecule together.

The difference in electronegativity between a hydrogen and oxygen atom is 1.4, meaning the water molecule is held together by a polar covalent bond. The electrons are strongly shifted toward the oxygen end of the molecule, creating a negative pole at the oxygen end and a positive pole at the hydrogen ends. The water molecule is thus a dipole, as shown in box figure 11.1A.

—Continued top of next page

BOX TABLE 11.1

Approximate ranges of visible light, infrared radiation, and microwave radiation

Radiation	Frequency Range (Hz)
Visible light	4×10^{14} to 8×10^{14}
Infrared radiation	3×10^{11} to 4×10^{14}
Microwave radiation	1×10^9 to 3×10^{11}

For example, CO is carbon monoxide and CO_2 is carbon dioxide. The compound BF_3 is boron trifluoride and N_2O_4 is dinitrogen tetroxide. Knowing the formula and the prefix and stem information in table 11.7, you can write the name of any covalent compound made up of two elements by ending it with "-ide." Conversely, the name will tell you the formula. However, there are a few polyatomic ions with "-ide" endings that are compounds made up of more than just two elements (hydroxide and cyanide). Compounds formed with the ammonium will also have an "-ide" ending, and these are also made up of more than two elements.

Covalent Compound Formulas

The systematic name tells you the formula for a covalent compound. The gas that dentists use as an anesthetic, for example, is dinitrogen monoxide. This tells you there are two nitrogen atoms and one oxygen atom in the molecule, so the formula is N_2O. A different molecule composed of the very same elements is nitrogen dioxide. Nitrogen dioxide is the pollutant responsible for the brownish haze of smog. The formula for nitrogen dioxide is NO_2. Other examples of formulas from systematic names are carbon dioxide (CO_2) and carbon tetrachloride (CCl_4) (figure 11.14).

Formulas of covalent compounds indicate a pattern of how many atoms of one element combine with atoms of another. Carbon, for example, combines with no more than two oxygen atoms to form carbon dioxide. Carbon combines with no more than four chlorine atoms to form carbon tetrachloride. Electron dot formulas show these two molecules as

Continued—

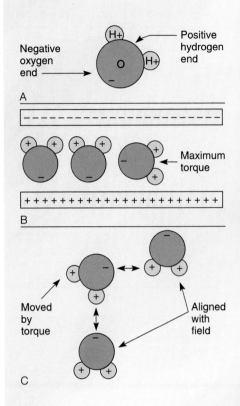

A

Negative oxygen end

Positive hydrogen end

B

Maximum torque

C

Moved by torque

Aligned with field

BOX FIGURE 11.1

(*A*) A water molecule is polar, with a negative pole on the oxygen end and positive poles on the hydrogen ends. (*B*) An electric field aligns the water dipoles, applying a maximum torque at right angles to the dipole vector. (*C*) Electrostatic attraction between the dipoles holds groups of water molecules together.

The dipole of water molecules has two effects: (1) the molecule can be rotated by the electric field of a microwave (see box figure 11.1B) and (2) groups of individual molecules are held together by an electrostatic attraction between the positive hydro-gen ends of a water molecule and the nega-tive oxygen end of another molecule (see box figure 11.1C).

One model to explain how microwaves heat water involves a particular group of three molecules, arranged so that the end molecules of the group are aligned with the microwave electric field, with the center mol-ecule not aligned. The microwave torques the center molecule, breaking its hydrogen bond. The energy of the microwave goes into doing the work of breaking the hydrogen bond, and the molecule now has increased potential energy as a consequence. The detached water molecule reestablishes its hydrogen bond, giving up its potential energy, which goes into the vibration of the group of molecules. Thus, the energy of the microwaves is con-verted into a temperature increase of the water. The temperature increase is high enough to heat and cook most foods.

Microwave cooking is different from conventional cooking because the heating results from energy transfer in polar water molecules, not conduction and convection. The surface of the food never reaches a tem-perature over the boiling point of water, so a microwave oven does not brown food (a conventional oven may reach temperatures almost twice as high). Large food items con-tinue to cook for a period of time after being in a microwave oven as the energy is con-ducted from the water molecules to the food. Most recipes allow for this continued cooking by specifying a waiting period after removing the food from the oven.

Microwave ovens are able to defrost frozen foods because ice always has a thin layer of liquid water (which is what makes it slippery). To avoid "spot cooking" of small pockets of liquid water, many microwave ovens cycle on and off in the defrost cycle. The electrons in metals, like the dipole water molecules, are affected by the electric field of a microwave. A piece of metal near the wall of a microwave oven can result in sparking, which can ignite paper. Metals also reflect microwaves, which can damage the radio tube that produces the microwaves.

Using a dash to represent bonding pairs, we have

$$O = C = O$$

$$Cl - \underset{\underset{Cl}{|}}{\overset{\overset{Cl}{|}}{C}} - Cl$$

In both of these compounds, the carbon atom forms four covalent bonds with another atom. The number of **covalent bonds** that an atom can form is called its **valence.** Carbon has a valence of four and can form single, double, or triple bonds. Here are the pos-sibilities for a single carbon atom (combining elements not shown):

$$-\overset{|}{\underset{|}{C}}- \qquad -\overset{|}{C}= \qquad =C= \qquad -C\equiv$$

Hydrogen has only one unshared electron, so the hydrogen atom has a valence of one. Oxygen has a valence of two and nitrogen has a valence of three. Here are the possibilities for hydrogen, oxygen, and nitrogen:

$$H- \qquad -\overset{..}{\underset{..}{O}}- \qquad \overset{..}{:}\overset{}{O}=$$

$$-\overset{..}{\underset{|}{N}}- \qquad -\overset{..}{N}= \qquad :N\equiv$$

Note the lone pairs shown on the oxygen and nitrogen atoms. Such lone pairs create the possibility of forming a coordinate covalent bond with another atom. The number of bonds and the number of lone pairs also determine the *shape* of a molecule. Molecules are not flat like the formulas on paper. Molecules have three-dimensional shapes, as illus-trated in figure 11.15. More will be said about molecular shapes later.

A

B

FIGURE 11.14

Once you understand chemical names and formulas, you can figure out what chemical compounds are contained in different household products. For example, (A) washing soda is sodium carbonate (Na_2CO_3), and (B) oven cleaner is sodium hydroxide (NaOH), which is also known as *lye*.

Electron pairs

Bonding	Lone	Shape		Example	
2	0	Straight	○—○—○	$BaCl_2$	Cl—Ba—Cl
3	0	Trigonal		BF_3	
2	1	Bent		SnF_2	
4	0	Tetrahedral		CH_4	
3	1	Pyramidal		NH_3	
2	2	Bent		H_2O	

Other shapes

Electron pairs		Shape	Electron pairs		Shape
Bonding	Lone		Bonding	Lone	
5	0	Bipyramidal	6	0	Octahedral
4	1	Seesaw	5	1	Square pyramidal
3	2	T-shaped			
2	3	Linear	4	2	Square planar

FIGURE 11.15

As you can see by studying these two charts, there is a relationship between the number of bonding electron pairs and the number of lone electron pairs and the shape of a molecule.

SUMMARY

Elements are basic substances that cannot be broken down into anything simpler, and an *atom* is the smallest unit of an element. *Compounds* are combinations of two or more elements and can be broken down into simpler substances. Compounds are formed when atoms are held together by an attractive force called a *chemical bond*. A *molecule* is the smallest unit of a compound, or a gaseous element, that can exist and still retain the characteristic properties of a substance.

A *chemical change* produces new substances with new properties, and the new materials are created by making or breaking chemical bonds. The process of chemical change in which different chemical substances are created by forming or breaking chemical bonds is called a *chemical reaction*. During a chemical reaction, different chemical substances with greater or lesser amounts of internal potential energy are produced. *Chemical energy* is the change of internal potential energy during a chemical reaction, and other reactions absorb energy. A *chemical equation* is a shorthand way of describing a chemical reaction. An equation shows the substances that are changed, the *reactants,* on the left side, and the new substances produced, the *products,* on the right side.

Chemical reactions involve *valence electrons,* the electrons in the outermost shell of an atom. Atoms tend to lose or acquire electrons to achieve the configuration of the noble gases with stable, filled outer orbitals. This tendency is generalized as the *octet rule,* that atoms lose or gain electrons to acquire the noble gas structure of eight electrons in the outer orbital. Atoms form negative or positive *ions* in the process.

A chemical bond is an attractive force that holds atoms together in a compound. Chemical bonds that are formed when atoms transfer electrons to become ions are *ionic bonds.* An ionic bond is an electrostatic attraction between oppositely charged ions. Chemical bonds formed when ions share electrons are *covalent bonds.*

Ionic bonds result in *ionic compounds* with a crystalline structure. The energy released when an ionic compound is formed is called the *heat of formation.* It is the same amount of energy that is required to decompose the compound into its elements. A *formula* of a compound uses symbols to tell what elements are in a compound and in what proportions. Ions of representative elements have a single, fixed charge, but many transition elements have variable charges. Electrons are conserved when ionic compounds are formed, and the ionic compound is electrically neutral. The formula shows this overall balance of charges.

Covalent compounds are molecular, composed of electrically neutral groups of atoms bound together by *covalent bonds.* A *single covalent bond* is formed by the sharing of a pair of electrons, with each atom contributing a single electron to the shared pair. Covalent bonds formed when two pairs of electrons are shared are called *double bonds.* A *triple bond* is the sharing of three pairs of electrons. If a shared electron pair comes from a single atom, the bond is called a *coordinate covalent bond.* Coordinate covalent bonding sometimes results in a group of atoms with a charge that acts together as a unit. The charged unit is called a *polyatomic ion.*

The electron-pulling ability of an atom in a bond is compared with arbitrary values of *electronegativity.* A high electronegative value means a greater attraction for bonding electrons. If the absolute difference in electronegativity of two bonded atoms is 1.7 or more, one atom pulls the bonding electron away, and an ionic bond results. If the difference is less than 0.5, the electrons are equally shared in a covalent bond. Between 0.5 and 1.7, the electrons are shared unequally in a *polar covalent bond.* A polar covalent bond results in electrons spending more time around the atom or atoms with the greater pulling ability, creating a negative pole at one end and a positive pole at the other. Such a molecule is called a *dipole,* since it has two poles, or centers, of charge.

Compounds are named with systematic rules for ionic and covalent compounds. Both ionic and covalent compounds that are made up of only two different elements always end with an *-ide* suffix, but there are a few *-ide* names for compounds that have more than just two elements.

The modern systematic system for naming variable-charge ions states the English name and gives the charge with Roman numerals in parentheses. Ionic compounds are electrically neutral, and formulas must show a balance of charge. The *crossover technique* is an easy way to write formulas that show a balance of charge.

Covalent compounds are molecules of two or more nonmetal atoms held together by a covalent bond. The system for naming covalent compounds uses Greek prefixes to identify the numbers of atoms, since more than one compound can form from the same two elements (CO and CO_2, for example).

Summary of Equations

11.1

Ionization of a metal atom:

$$\text{energy} + \text{Na}\cdot \longrightarrow \text{Na}^+ + e^-$$

11.2

Ionic bonding reaction (single charges):

$$\text{Na}\cdot + \cdot\ddot{\underset{\cdot\cdot}{\text{Cl}}}: \longrightarrow \text{Na}^+(:\ddot{\underset{\cdot\cdot}{\text{Cl}}}:)^-$$

11.3

Ionic bonding reaction (single and double charges):

$$\text{Ca}\cdot + \begin{matrix}\cdot\ddot{\underset{\cdot\cdot}{\text{F}}}: \\ \\ \cdot\ddot{\underset{\cdot\cdot}{\text{F}}}:\end{matrix} \longrightarrow \text{Ca}^{2+} \begin{matrix}(:\ddot{\underset{\cdot\cdot}{\text{F}}}:)^- \\ \\ (:\ddot{\underset{\cdot\cdot}{\text{F}}}:)^-\end{matrix}$$

11.4

Covalent bonding (hydrogen-hydrogen):

$$\text{H}\cdot + \text{H}\cdot \longrightarrow (\text{H}\!:\!\text{H})$$

11.5

Covalent bonding (hydrogen-fluorine):

$$\text{H}\cdot + \cdot\ddot{\underset{\cdot\cdot}{\text{F}}}: \longrightarrow (\text{H}\!:\!\ddot{\underset{\cdot\cdot}{\text{F}}}:)$$

11.6

Covalent bonding (fluorine-fluorine):

$$\cdot\ddot{\underset{\cdot\cdot}{\text{F}}}: + \cdot\ddot{\underset{\cdot\cdot}{\text{F}}}: \longrightarrow (:\ddot{\underset{\cdot\cdot}{\text{F}}}\!:\!\ddot{\underset{\cdot\cdot}{\text{F}}}:)$$

11.7

Coordinate covalent bonding (ammonium ion):

$$\text{H}^+ + \begin{matrix}\text{H} \\ :\text{N}\!:\!\text{H} \\ \text{H}\end{matrix} \longrightarrow \begin{bmatrix}\text{H} \\ \text{H}\!:\!\text{N}\!:\!\text{H} \\ \text{H}\end{bmatrix}^-$$

KEY TERMS

atom (p. **256**)

chemical bond (p. **259**)

chemical energy (p. **257**)

chemical equation (p. **257**)

chemical reaction (p. **256**)

coordinate covalent
  bond (p. **264**)

covalent bond (p. **262**)

covalent compound (p. **263**)

double bond (p. **263**)

electronegativity (p. **265**)

formula (p. **260**)

heat of formation (p. **260**)

ionic bond (p. **259**)

ionic compounds (p. **261**)

molecule (p. **256**)

octet rule (p. **258**)

polar covalent bond (p. **266**)

polyatomic ion (p. **264**)

single bond (p. **263**)

triple bond (p. **264**)

valence (p. **271**)

valence electrons (p. **258**)

APPLYING THE CONCEPTS

1. The smallest unit of an element that can exist alone or in combination with other elements is the
   a. electron.
   b. atom.
   c. molecule.
   d. chemical bond.

2. The smallest unit of a covalent compound that can exist while retaining the chemical properties of the compound is the
   a. electron.
   b. atom.
   c. molecule.
   d. ionic bond.

3. You know that a chemical reaction is taking place if
   a. the temperature of a substance increases.
   b. electrons move in a steady current.
   c. chemical bonds are formed or broken.
   d. All of the above are correct.

4. Chemical reactions that involve changes in the internal potential energy of molecules always involve changes of
   a. the mass of the reactants as compared to the products.
   b. chemical energy.
   c. radiant energy.
   d. the weight of the reactants.

5. The energy released in burning materials produced by photosynthesis has what relationship to the solar energy that was absorbed in making the materials? It is
   a. less than the solar energy absorbed.
   b. the same as the solar energy absorbed.
   c. more than the solar energy absorbed.
   d. variable, having no relationship to the energy absorbed.

6. The electrons that participate in chemical bonding are (the)
   a. valence electrons.
   b. electrons in fully occupied orbitals.
   c. stable inner electrons.
   d. all of the above.

7. Atoms of the representative elements have a tendency to seek stability through
   a. acquiring the noble gas structure.
   b. filling or emptying their outer orbitals.
   c. any situation that will satisfy the octet rule.
   d. all of the above.

8. An ion is formed when an atom of a representative element
   a. gains or loses protons.
   b. shares electrons to achieve stability.
   c. loses or gains electrons to satisfy the octet rule.
   d. All of the above are correct.

9. An atom of an element that is in family VIA will have what charge when it is ionized?
   a. 2+
   b. 6+
   c. 6−
   d. 2−

10. Which type of chemical bond is formed by a transfer of electrons?
    a. ionic
    b. covalent
    c. metallic
    d. coordinate covalent

11. Which type of chemical bond is formed between two atoms by the sharing of two electrons, with one electron from each atom?
    a. ionic
    b. covalent
    c. metallic
    d. coordinate covalent

12. Salts, such as sodium chloride, are what type of compounds?
    a. ionic compounds
    b. covalent compounds
    c. polar compounds
    d. Any of the above are correct.

13. If there are two bromide ions for each barium ion in a compound, the chemical formula is
    a. $_2Br_1Ba$.
    b. Ba_2Br.
    c. $BaBr_2$.
    d. none of the above.

14. Which combination of elements forms crystalline solids that will dissolve in water, producing a solution of ions that can conduct an electric current?
    a. metal and metal
    b. metal and nonmetal
    c. nonmetal and nonmetal
    d. All of the above are correct.

15. In a single covalent bond between two atoms,
    a. a single electron from one of the atoms is shared.
    b. a pair of electrons from one of the atoms is shared.
    c. a pair of electrons, one from each atom, is shared.
    d. a single electron is transferred from one atom.

16. The number of pairs of shared electrons in a covalent compound is often the same as the number of
    a. unpaired electrons in the electron dot notation.
    b. valence electrons.
    c. orbital pairs.
    d. protons in the nucleus.

17. Sulfur and oxygen are both in the VIA family of the periodic table. If element X combines with oxygen to form the compound X_2O, element X will combine with sulfur to form the compound
    a. XS_2.
    b. X_2S.
    c. X_2S_2.
    d. It is impossible to say without more information.

18. One element is in the IA family of the periodic table, and a second element is in the VIIA family. What type of compound will the two elements form?
    a. ionic
    b. covalent
    c. They will not form a compound.
    d. More information is needed to answer this question.

19. One element is in the VA family of the periodic table, and a second is in the VIA family. What type of compound will these two elements form?
    a. ionic
    b. covalent
    c. They will not form a compound.
    d. More information is needed to answer this question.

20. A covalent bond in which there is an unequal sharing of bonding electrons is a
    a. single covalent bond.
    b. double covalent bond.
    c. triple covalent bond.
    d. polar covalent bond.

21. An inorganic compound made of only two different elements has a systematic name that always ends with the suffix
    a. -ite.
    b. -ate.
    c. -ide.
    d. -ous.

22. Dihydrogen monoxide is the systematic name for a compound that has the common name of
    a. laughing gas.
    b. water.
    c. smog.
    d. rocket fuel.

Answers

1. b **2.** c **3.** c **4.** b **5.** b **6.** a **7.** d **8.** c **9.** d **10.** a **11.** b **12.** a **13.** c **14.** b **15.** c **16.** a **17.** b **18.** a **19.** b **20.** d **21.** c **22.** b

QUESTIONS FOR THOUGHT

1. Describe how the following are alike and how they are different: (a) a sodium atom and a sodium ion, and (b) a sodium ion and a neon atom.
2. What is the difference between a polar covalent bond and a nonpolar covalent bond?
3. What is the difference between an ionic and covalent bond? What do atoms forming the two bond types have in common?
4. What is the octet rule?
5. Is there a relationship between the number of valence electrons and how many covalent bonds an atom can form? Explain.
6. Write electron dot formulas for molecules formed when hydrogen combines with (a) chlorine, (b) oxygen, and (c) carbon.
7. Sodium fluoride is often added to water supplies to strengthen teeth. Is sodium fluoride ionic, nonpolar covalent, or polar covalent? Explain the basis of your answer.
8. What is the modern systematic name of a compound with the formula (a) SnF_2? (b) PbS?
9. What kinds of elements are found in (a) ionic compounds with a name ending with an "-ide" suffix? (b) covalent compounds with a name ending with an "-ide" suffix?
10. Why is it necessary to use a system of Greek prefixes to name binary covalent compounds?
11. What are variable-charge ions? Explain how variable-charge ions are identified in the modern system of naming compounds.
12. What is a polyatomic ion? Give the names and formulas for several common polyatomic ions.
13. Write the formula for magnesium hydroxide. Explain what the parentheses mean.
14. What is a double bond? A triple bond?

PARALLEL EXERCISES

The exercises in groups A and B cover the same concepts. Solutions to group A exercises are located in appendix D.

Group A

1. Use electron dot symbols in equations to predict the formula of the ionic compound formed from the following:
   (a) K and I
   (b) Sr and S
   (c) Na and O
   (d) Al and O

Group B

1. Use electron dot symbols in equations to predict the formula of the ionic compound formed between the following:
   (a) Li and F
   (b) Be and S
   (c) Li and O
   (d) Al and S

2. Name the following ionic compounds formed from variable-charge transition elements:
   (a) CuS
   (b) Fe_2O_3
   (c) CrO
   (d) PbS

3. Name the following polyatomic ions:
   (a) $(OH)^-$
   (b) $(SO_3)^{2-}$
   (c) $(ClO)^-$
   (d) $(NO_3)^-$
   (e) $(CO_3)^{2-}$
   (f) $(ClO_4)^-$

4. Use the crossover technique to write formulas for the following compounds:
   (a) Iron(III) hydroxide
   (b) Lead(II) phosphate
   (c) Zinc carbonate
   (d) Ammonium nitrate
   (e) Potassium hydrogen carbonate
   (f) Potassium sulfite

5. Write formulas for the following covalent compounds:
   (a) Carbon tetrachloride
   (b) Dihydrogen monoxide
   (c) Manganese dioxide
   (d) Sulfur trioxide
   (e) Dinitrogen pentoxide
   (f) Diarsenic pentasulfide

6. Name the following covalent compounds:
   (a) CO
   (b) CO_2
   (c) CS_2
   (d) N_2O
   (e) P_4S_3
   (f) N_2O_3

7. Predict if the bonds formed between the following pairs of elements will be ionic, polar covalent, or nonpolar covalent:
   (a) Si and O
   (b) O and O
   (c) H and Te
   (d) C and H
   (e) Li and F
   (f) Ba and S

2. Name the following ionic compounds formed from variable-charge transition elements:
   (a) $PbCl_2$
   (b) FeO
   (c) Cr_2O_3
   (d) PbO

3. Name the following polyatomic ions:
   (a) $(C_2H_3O_2)^-$
   (b) $(HCO_3)^-$
   (c) $(SO_4)^{2-}$
   (d) $(NO_2)^-$
   (e) $(MnO_4)^-$
   (f) $(CO_3)^{2-}$

4. Use the crossover technique to write formulas for the following compounds:
   (a) Aluminum hydroxide
   (b) Sodium phosphate
   (c) Copper(II) chloride
   (d) Ammonium sulfate
   (e) Sodium hydrogen carbonate
   (f) Cobalt(II) chloride

5. Write formulas for the following covalent compounds:
   (a) Silicon dioxide
   (b) Dihydrogen sulfide
   (c) Boron trifluoride
   (d) Dihydrogen dioxide
   (e) Carbon tetrafluoride
   (f) Nitrogen trihydride

6. Name the following covalent compounds:
   (a) N_2O
   (b) SO_2
   (c) SiC
   (d) PF_5
   (e) $SeCl_6$
   (f) N_2O_4

7. Predict if the bonds formed between the following pairs of elements will be ionic, polar covalent, or nonpolar covalent:
   (a) Si and C
   (b) Cl and Cl
   (c) S and O
   (d) Sr and F
   (e) O and H
   (f) K and F

This is a computer-generated model of a beryllium atom, showing the nucleus and
1s, 2s electron orbitals. This configuration can also be predicted from information
on a periodic table.

CHAPTER | Twelve

Chemical Formulas and Equations

We live in a chemical world that has been partly manufactured through controlled chemical change. Consider all of the synthetic fibers and plastics that are used in clothing, housing, and cars. Consider all the synthetic flavors and additives in foods, how these foods are packaged, and how they are preserved. Consider also the synthetic drugs and vitamins that keep you healthy. There are millions of such familiar products that are the direct result of chemical research. Most of these products simply did not exist sixty years ago.

Many of the products of chemical research have remarkably improved the human condition. For example, synthetic fertilizers have made it possible to supply food in quantities that would not otherwise be possible. Chemists learned how to take nitrogen from the air and convert it into fertilizers on an enormous scale. Other chemical research resulted in products such as weed killers, insecticides, and mold and fungus inhibitors. The fertilizers and these products have made it possible to supply food for millions of people who would otherwise have starved (figure 12.1).

Yet, we also live in a world with concerns about chemical pollutants, the greenhouse effect, acid rain, and a disappearing ozone shield. The very nitrogen fertilizers that have increased food supplies also wash into rivers, polluting the waterways and bays. Such dilemmas require an understanding of chemical products and the benefits and hazards of possible alternatives. Understanding requires a knowledge of chemistry, since the benefits, and risks, are chemical in nature.

The previous chapters were about the modern atomic theory and how it explains elements and how compounds are formed in chemical change. This chapter is concerned with describing chemical changes and the different kinds of chemical reactions that occur. These reactions are explained with balanced chemical equations, which are concise descriptions of reactions that produce the products used in our chemical world.

CHEMICAL FORMULAS

In chapter 11 you learned how to name and write formulas for ionic and covalent compounds, including the ionic compound of table salt and the covalent compound of ordinary water. Recall that a formula is a shorthand way of describing the elements or ions that make up a compound. There are basically three kinds of formulas that describe compounds: (1) *empirical* formulas, (2) *molecular* formulas, and (3) *structural* formulas. Empirical and molecular formulas, and their uses, will be considered in this chapter. Structural formulas will be considered in chapter 14.

An **empirical formula** identifies the elements present in a compound and describes the *simplest whole number ratio* of atoms of these elements with subscripts. For example, the empirical formula for ordinary table salt is NaCl. This tells you that the elements sodium and chlorine make up this compound, and there is one atom of sodium for each chlorine atom. The empirical formula for water is H_2O, meaning there are two atoms of hydrogen for each atom of oxygen.

Covalent compounds exist as molecules. A chemical formula that identifies the *actual numbers* of atoms in a molecule is known as a **molecular formula.** Figure 12.2 shows the structure of some common molecules and their molecular formulas. Note that each formula identifies the elements and numbers of atoms in each molecule. The figure also indicates how molecular formulas can be written to show how the atoms are arranged in the molecule. Formulas that show the relative arrangements are called *structural formulas.* Compare the structural formulas in the illustration with the three-dimensional representations and the molecular formulas.

How do you know if a formula is empirical or molecular? First, you need to know if the compound is ionic or covalent. You know that ionic compounds are usually composed of metal and nonmetal atoms with an electronegativity difference greater than 1.7. Formulas for ionic compounds are *always* empirical formulas. Ionic compounds are composed of many positive and negative ions arranged in an electrically neutral array. There is no discrete unit, or molecule, in an ionic compound, so it is only possible to identify ratios of atoms with an empirical formula.

Covalent compounds are generally nonmetal atoms bonded to nonmetal atoms in a molecule. You could therefore assume that a formula for a covalent compound is a molecular formula unless it is specified otherwise. You can be certain it is a molecular formula if it is not the simplest whole-number ratio. Glucose, for example, is a simple sugar (also known as dextrose) with the formula $C_6H_{12}O_6$. This formula is divisible by six, yielding a formula with the simplest whole-number ratio of CH_2O. Therefore, CH_2O is the empirical formula for glucose, and $C_6H_{12}O_6$ is the molecular formula.

FIGURE 12.1

The products of chemical research have substantially increased food supplies but have also increased the possibilities of pollution. Balancing the benefits and hazards of the use of chemicals requires a knowledge of chemistry and a knowledge of the alternatives.

Name	Molecular formula	Sketch	Structural formula
Water	H_2O		
Ammonia	NH_3		
Hydrogen peroxide	H_2O_2		
Carbon dioxide	CO_2		$O=C=O$

FIGURE 12.2

The name, molecular formula, sketch, and structural formula of some common molecules. Compare the kinds and numbers of atoms making up each molecule in the sketch to the molecular formula.

Molecular and Formula Weights

The **formula weight** of a compound is the sum of the atomic weights of all the atoms in a chemical formula. For example, the formula for water is H_2O. Hydrogen and oxygen are both nonmetals, so the formula means that one atom of oxygen is bound to two hydrogen atoms in a molecule. From the periodic table, you know that the approximate (rounded) atomic weight of

hydrogen is 1.0 u and oxygen is 16.0 u. Adding the atomic weights for *all* the atoms,

Atoms	Atomic Weight		Totals
2 of H	2×1.0 u	=	2.0 u
1 of O	1×16.0 u	=	16.0 u
	Formula weight	=	18.0 u

Thus, the formula weight of a water molecule is 18.0 u.

The formula weight of an ionic compound is found in the same way, by adding the rounded atomic weights of atoms (or ions) making up the compound. Sodium chloride is NaCl, so the formula weight is 23.0 u plus 35.5 u, or 58.5 u. The *formula weight* can be calculated for an ionic or molecular substance. The **molecular weight** is the formula weight of a molecular substance. The term *molecular weight* is sometimes used for all substances, whether or not they have molecules. Since ionic substances such as NaCl do not occur as molecules, this is not strictly correct. Both molecular and formula weights are calculated in the same way, but formula weight is a more general term.

Q & A

What is the formula weight of table sugar (sucrose), which has the formula $C_{12}H_{22}O_{11}$?

Answer

The formula identifies the numbers of each atom, and the atomic weights are from a periodic table:

Atoms	Atomic Weight		Totals
12 of C	12×12.0 u	=	144.0 u
22 of H	22×1.0 u	=	22.0 u
11 of O	11×16.0 u	=	176.0 u
	Formula weight	=	342.0 u

What is the molecular weight of ethyl alcohol, C_2H_5OH? (Answer: 46.0 u)

Percent Composition of Compounds

The formula weight of a compound can provide useful information about the elements making up a compound (figure 12.3). For example, suppose you want to know how much calcium is provided by a dietary supplement. The label lists the main ingredient as calcium carbonate, $CaCO_3$. To find how much calcium is supplied by a pill with a certain mass, you need to find the *mass percentage* of calcium in the compound.

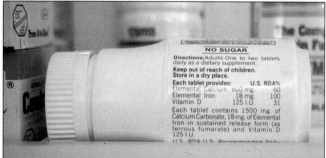

A

B

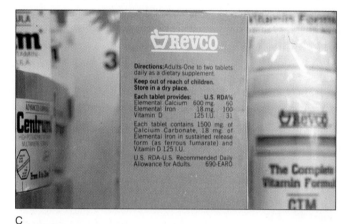

C

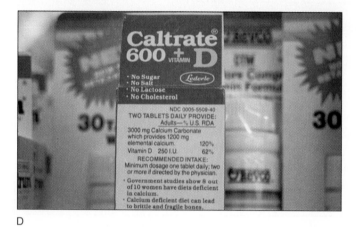

D

FIGURE 12.3

If you know the name of an ingredient, you can write a chemical formula, and the percent composition of a particular substance can be calculated from the formula. This can be useful information for consumer decisions.

Percent is simply the fractional part of the whole times 100 percent (meaning "per 100"), or

$$\left(\frac{\text{part}}{\text{whole}}\right)(100\% \text{ of whole}) = \% \text{ of part}$$

equation 12.1

For example, if 13 students in a class of 50 are freshmen, the percentage of freshmen in the class is

$$\left(\frac{13 \text{ freshmen}}{50 \text{ classmates}}\right)(100\% \text{ of classmates})$$

$$= \left(.026 \frac{\text{freshmen}}{\text{classmates}}\right)(100\% \text{ of classmates})$$

$$= 26\% \text{ freshmen}$$

Note the classmate units cancel, giving the answer in percent freshmen.

Since the formula weight of a compound represents all of its composition, the formula weight is the "whole" in equation 12.1, with all the atoms contributing a part of the whole weight. The "part" in equation 12.1 is the atomic weight times the number of atoms of the element in which you are interested. Thus, the mass percentage of an element in a compound can be found from

$$\frac{\left(\begin{array}{c}\text{atomic weight}\\\text{of element}\end{array}\right)\left(\begin{array}{c}\text{number of atoms}\\\text{of element}\end{array}\right)}{\text{formula weight of compound}} \times \begin{array}{c}100\% \text{ of}\\\text{compound}\end{array} = \begin{array}{c}\% \text{ of}\\\text{element}\end{array}$$

equation 12.2

The mass percentage of calcium in $CaCO_3$ can be found in two steps:

Step 1:   Determine formula weight:

Atoms	Atomic Weight		Totals
1 of Ca	1×40.1 u	=	40.1 u
1 of C	1×12.0 u	=	12.0 u
3 of O	3×16.0 u	=	48.0 u
	Formula weight	=	100.1 u

Step 2:   Determine percentage of Ca:

$$\frac{(40.1 \text{ u Ca})(1)}{100.1 \text{ u } CaCO_3} \times 100\% \ CaCO_3 = 40.1\% \text{ Ca}$$

Knowing the percentage of the total mass contributed by the calcium, you can multiply this fractional part (as a decimal) by the mass of the supplement pill to find the calcium supplied. The mass percentage of the other elements can also be determined with equation 12.2.

Q & A

Sodium fluoride is added to water supplies and to some toothpastes for fluoridation. What is the percentage composition of the elements in sodium fluoride?

Answer

Step 1: Write the formula for sodium fluoride, NaF.

Step 2: Determine the formula weight.

Atoms	Atomic Weight		Totals
1 of Na	1×23.0 u	=	23.0 u
1 of F	1×19.0 u	=	19.0 u
	Formula weight	=	42.0 u

Step 3: Determine the percentage of Na and F.

For Na:

$$\frac{(23.0 \text{ u Na})(1)}{42.0 \text{ u NaF}} \times 100\% \text{ NaF} = \boxed{54.7\% \text{ Na}}$$

For F:

$$\frac{(19.0 \text{ u F})(1)}{42.0 \text{ u NaF}} \times 100\% \text{ NaF} = \boxed{45.2\% \text{ F}}$$

The percentage often does not total to exactly 100 percent because of rounding.

Calculate the percentage composition of carbon in table sugar, sucrose, which has a formula of $C_{12}H_{22}O_{11}$. (Answer: 42.1% C)

Activities

Chemical fertilizers are added to the soil when it does not contain sufficient elements essential for plant growth. The three critical elements are nitrogen, phosphorus, and potassium, and these are the basic ingredients in most chemical fertilizers. In general, lawns require fertilizers high in nitrogen and gardens require fertilizers high in phosphorus.

Read the labels on commercial packages of chemical fertilizers sold in a garden shop. Find the name of the chemical that supplies each of these critical elements; for example, nitrogen is sometimes supplied by ammonium sulfate $(NH_4)_2SO_4$. Calculate the mass percentage of each critical element supplied according to the label information. Compare these percentages to the grade number of the fertilizer, for example, 10–20–10. Determine which fertilizer brand gives you the most nutrients for the money.

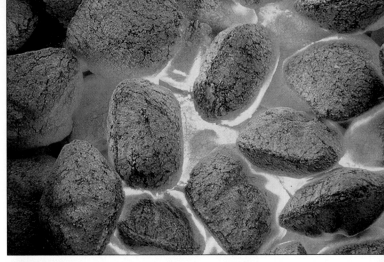

FIGURE 12.4

The charcoal used in a grill is basically carbon. The carbon reacts with oxygen to yield carbon dioxide. The chemical equation for this reaction, $C + O_2 \rightarrow CO_2$, contains the same information as the English sentence but has quantitative meaning as well.

CHEMICAL EQUATIONS

Chemical reactions occur when bonds between the outermost parts of atoms are formed or broken. Bonds are formed, for example, when a green plant uses sunlight—a form of energy—to create molecules of sugar, starch, and plant fibers. Bonds are broken and energy is released when you digest the sugars and starches or when plant fibers are burned. Chemical reactions thus involve changes in matter, the creation of new materials with new properties, and energy exchanges. So far you have considered chemical symbols as a concise way to represent elements, and formulas as a concise way to describe what a compound is made of. There is also a concise way to describe a chemical reaction, the **chemical equation.**

Balancing Equations

Word equations are useful in identifying what has happened before and after a chemical reaction. The substances that existed before a reaction are called *reactants,* and the substances that exist after the reaction are called the *products.* The equation has a general form of

$$\text{reactants} \rightarrow \text{products}$$

where the arrow signifies a separation in time; that is, it identifies what existed before the reaction and what exists after the reaction. For example, the charcoal used in a barbecue grill is carbon (figure 12.4). The carbon reacts with oxygen while burning, and the reaction (1) releases energy and (2) forms carbon dioxide. The reactants and products for this reaction can be described as

$$\text{carbon} + \text{oxygen} \rightarrow \text{carbon dioxide}$$

The arrow means *yields,* and the word equation is read as, "Carbon reacts with oxygen to yield carbon dioxide." This word equation describes what happens in the reaction but says nothing about the quantities of reactants or products.

Chemical symbols and formulas can be used in the place of words in an equation and the equation will have a whole new meaning. For example, the equation describing carbon reacting with oxygen to yield carbon dioxide becomes

$$C + O_2 \rightarrow CO_2$$

(balanced)

The new, added meaning is that one atom of carbon (C) reacts with one molecule of oxygen (O_2) to yield one molecule of carbon dioxide (CO_2). Note that the equation also shows one atom of carbon and two atoms of oxygen (recall that oxygen occurs as a diatomic molecule) as reactants on the left side and one atom of carbon and two atoms of oxygen as products on the right side. Since the same number of each kind of atom appears on both sides of the equation, the equation is said to be *balanced.*

You would not want to use a charcoal grill in a closed room because there might not be enough oxygen. An insufficient supply of oxygen produces a completely different product, the poisonous gas, carbon monoxide (CO). An equation for this reaction is

$$C + O_2 \rightarrow CO$$

(not balanced)

As it stands, this equation describes a reaction that violates the **law of conservation of mass,** that matter is neither created nor destroyed in a chemical reaction. From the point of view of an equation, this law states that

$$\text{mass of reactants} = \text{mass of products}$$

Mass of reactants here means all that you start with, including some that might not react. Thus elements are neither created nor destroyed, and this means the elements present and their mass. In any chemical reaction the kind and mass of the reactive elements are identical to the kind and mass of the product elements.

From the point of view of atoms, the law of conservation of mass means that *atoms are neither created nor destroyed in the chemical reaction.* A chemical reaction is the making or breaking of chemical bonds between atoms or groups of atoms. Atoms are not lost or destroyed in the process, nor are they changed to a different kind. The equation for the formation of carbon monoxide has two oxygen atoms in the reactants (O_2) but only one in the product (CO). An atom of oxygen has disappeared somewhere, and that violates the law of conservation of mass. You cannot fix the equation by changing the CO to a CO_2, because this would change the identity of the compounds. Carbon monoxide is a poisonous gas that is different from carbon dioxide, a relatively harmless product of burning and respiration. *You cannot change the subscript in a formula* because that would change the formula. A different formula means a different composition and thus a different compound.

You cannot change the subscripts of a formula, but you can place a number called a *coefficient* in *front* of the formula. Changing a coefficient changes the *amount* of a substance, not the identity. Thus 2 CO means two molecules of carbon monoxide and 3 CO means three molecules of carbon monoxide. If there is no coefficient, 1 is understood as with subscripts. The meaning of coefficients and subscripts is illustrated in figure 12.5.

C	means		One atom of carbon
O	means		One atom of oxygen
O_2	means	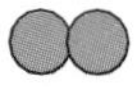	One molecule of oxygen consisting of two atoms of oxygen
CO	means	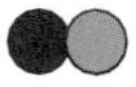	One molecule of carbon monoxide consisting of one atom of carbon attached to one atom of oxygen
CO_2	means	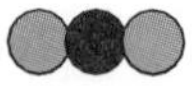	One molecule of carbon dioxide consisting of one atom of carbon attached to two atoms of oxygen
3 CO_2	means	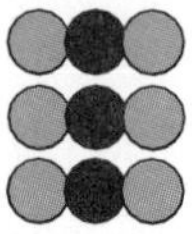	Three molecules of carbon dioxide, each consisting of one atom of carbon attached to two atoms of oxygen

FIGURE 12.5

The meaning of subscripts and coefficients used with a chemical formula. The subscripts tell you how many atoms of a particular element are in a compound. The coefficient tells you about the quantity, or number, of molecules of the compound.

Placing a coefficient of 2 in front of the C and a coefficient of 2 in front of the CO in the equation will result in the same numbers of each kind of atom on both sides:

$$2\,C + O_2 \rightarrow 2\,CO$$

Reactants:	2 C	Products:	2 C
	2 O		2 O

The equation is now balanced.

Suppose your barbecue grill burns natural gas, not charcoal (figure 12.6). Natural gas is mostly methane, CH_4. Methane burns by reacting with oxygen (O_2) to produce carbon dioxide (CO_2) and water vapor (H_2O). A balanced chemical equation for this reaction can be written by following a procedure of four steps.

Step 1: Write the correct formulas for the reactants and products in an unbalanced equation. The reactants and products could have been identified by chemical experiments, or they could have been predicted from what is known about chemical properties. This will be discussed in more detail later. For now, assume that the reactants and products are known and are given in words. For the burning of methane, the unbalanced, but otherwise correct, formula equation would be

$$CH_4 + O_2 \rightarrow CO_2 + H_2O$$

(not balanced)

FIGURE 12.6

Tanks like these grow larger as they are filled with natural gas, then collapse back to the ground as the gas is removed. Why do you suppose the tanks are designed to inflate and collapse? One reason is to keep the gas under a constant pressure. The height of each tank varies with the amount of gas inside, so more gas means a greater volume rather than a greater pressure. A rigid gas tank with a constant volume would be under very high pressure when full and very low pressure when nearly empty, which would make it difficult to pump gas into or out of the tank.

Step 2:   Inventory the number of each kind of atom on both sides of the unbalanced equation. In the example there are

Reactants:	1 C	Products:	1 C
	4 H		2 H
	2 O		3 O

This shows that the H and O are unbalanced.

Step 3:   Determine where to place coefficients in front of formulas to balance the equation. It is often best to focus on the simplest thing you can do with whole number ratios. The H and the O are unbalanced, for example, and there are 4 H atoms on the left and 2 H atoms on the right. Placing a coefficient 2 in front of H_2O will balance the H atoms:

$$CH_4 + O_2 \rightarrow CO_2 + 2\,H_2O$$

(not balanced)

Now take a second inventory:

Reactants:	1 C	Products:	1 C
	4 H		4 H
	2 O		4 O (O_2 + 2 O)

This shows the O atoms are still unbalanced with 2 on the left and 4 on the right. Placing a coefficient of 2 in front of O_2 will balance the O atoms.

$$CH_4 + 2\,O_2 \rightarrow CO_2 + 2\,H_2O$$

(balanced)

Step 4:   Take another inventory to determine (a) if the numbers of atoms on both sides are now equal and, if so, (b) if the coefficients are in the lowest possible whole-number ratio. The inventory is now

Reactants:	1 C	Products:	1 C
	4 H		4 H
	4 O		4 O

(a) The number of each kind of atom on each side of the equation is the same, and (b) the ratio of $1:2 \rightarrow 1:2$ is the lowest possible whole-number ratio. The equation is balanced, which is illustrated with sketches of molecules in figure 12.7.

Balancing chemical equations is mostly a trial-and-error procedure. But with practice, you will find there are a few generalized "role models" that can be useful in balancing equations for many simple reactions. The key to success at balancing equations is to think it out step-by-step while remembering the following:

1. Atoms are neither lost nor gained nor do they change their identity in a chemical reaction. The same kind and number of atoms in the reactants must appear in the products, meaning atoms are conserved.
2. A correct formula of a compound cannot be changed by altering the number or placement of subscripts. Changing subscripts changes the identity of a compound and the meaning of the entire equation.
3. A coefficient in front of a formula multiplies everything in the formula by that number.

There are also a few generalizations that can be helpful for success in balancing equations:

1. Look first to formulas of compounds with the most atoms and try to balance the atoms or compounds they were formed from or decomposed to.
2. Polyatomic ions that appear on both sides of the equation should be treated as independent units with a charge. That is, consider the polyatomic ion as a unit while taking an inventory rather than the individual atoms making up the polyatomic ion. This will save time and simplify the procedure.
3. Both the "crossover technique" and the use of "fractional coefficients" can be useful in finding the least common multiple to balance an equation.

The physical state of reactants and products in a reaction is often identified by the symbols (g) for gas, (l) for liquid, (s) for solid, and (aq) for an aqueous solution ("aqueous" means water). If a gas escapes, this is identified with an arrow pointing up ($\uparrow$). A solid formed from a solution is identified with an arrow pointing down ($\downarrow$). The Greek symbol delta (Δ) is often used under or over the yield sign to indicate a change of temperature or other physical values.

FIGURE 12.7

Compare the numbers of each kind of atom in the balanced equation with the numbers of each kind of atom in the sketched representation. Both the equation and the sketch have the same number of atoms in the reactants and in the products.

Reaction:	Methane reacts with oxygen to yield carbon dioxide and water
Balanced equation:	$CH_4 + 2\ O_2 \longrightarrow CO_2 + 2\ H_2O$

Sketches representing molecules:

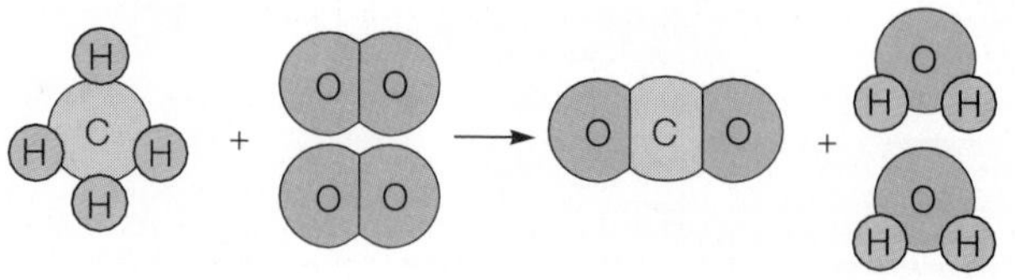

| Meaning: | 1 molecule of methane | + | 2 molecules of oxygen | → | 1 molecule of carbon dioxide | + | 2 molecules of water |

Q & A

Propane is a liquified petroleum gas (LPG) that is often used as a bottled substitute for natural gas (figure 12.8). Propane (C_3H_8) reacts with oxygen (O_2) to yield carbon dioxide (CO_2) and water vapor (H_2O). What is the balanced equation for this reaction?

Answer

Step 1:   Write the correct formulas of the reactants and products in an unbalanced equation.

$$C_3H_8(g) + O_2(g) \rightarrow CO_2(g) + H_2O(g)$$

(unbalanced)

Step 2:   Inventory the numbers of each kind of atom.

Reactants:	3 C	Products:	1 C
	8 H		2 H
	2 O		3 O

Step 3:   Determine where to place coefficients to balance the equation. Looking at the compound with the most atoms (generalization 1), you can see that a propane molecule has 3 C and 8 H. Placing a coefficient of 3 in front of CO_2 and a 4 in front of H_2O will balance these atoms (3 of C and $4 \times 2 = 8$ H atoms on the right has the same number of atoms as C_3H_8 on the left),

$$C_3H_8(g) + O_2(g) \rightarrow 3\ CO_2(g) + 4\ H_2O(g)$$

(not balanced)

A second inventory shows

Reactants:	3 C	Products:	3 C
	8 H		8 H ($4 \times 2 = 8$)
	2 O		10 O [$(3 \times 2) + (4 \times 1) = 10$]

The O atoms are still unbalanced. Place a 5 in front of O_2, and the equation is balanced ($5 \times 2 = 10$). Remember that you cannot change the subscripts and that oxygen occurs as a diatomic molecule of O_2.

$$C_3H_8(g) + 5\ O_2(g) \rightarrow 3\ CO_2(g) + 4\ H_2O(g)$$

(balanced)

Step 4:   Another inventory shows (a) the number of atoms on both sides are now equal, and (b) the coefficients are 1:5 → 3:4, the lowest possible whole-number ratio. The equation is balanced.

Q & A

One type of water hardness is caused by the presence of calcium bicarbonate in solution, $Ca(HCO_3)_2$. One way to remove the troublesome calcium ions from wash water is to add washing soda, which is sodium carbonate, Na_2CO_3. The reaction yields sodium bicarbonate ($NaHCO_3$) and calcium carbonate ($CaCO_3$), which is insoluble. Since $CaCO_3$ is insoluble, the reaction removes the calcium ions from solution. Write a balanced equation for the reaction.

Answer

Step 1:   Write the unbalanced equation

$$Ca(HCO_3)_2(aq) + Na_2CO_3(aq) \rightarrow NaHCO_3(aq) + CaCO_3\downarrow$$

(not balanced)

Step 2:   Inventory the numbers of each kind of atom. This reaction has polyatomic ions that appear on both sides, so they should be

FIGURE 12.8

One of two burners is operating at the moment as this hot air balloon ascends. The burners are fueled by propane (C_3H_8), a liquified petroleum gas (LPG). Like other forms of petroleum, propane releases large amounts of heat during the chemical reaction of burning.

treated as independent units with a charge (generalization 2). The inventory is

Reactants:	1 Ca	Products:	1 Ca
	2 $(HCO_3)^{1-}$		1 $(HCO_3)^{1-}$
	2 Na		1 Na
	1 $(CO_3)^{2-}$		1 $(CO_3)^{2-}$

Step 3: Placing a coefficient of 2 in front of $NaHCO_3$ will balance the equation,

$$Ca(HCO_3)_2(aq) + Na_2CO_3(aq) \rightarrow 2\,NaHCO_3(aq) + CaCO_3\downarrow$$

(balanced)

Step 4: An inventory shows

Reactants:	1 Ca	Products:	1 Ca
	2 $(HCO_3)^{1-}$		2 $(HCO_3)^{1-}$
	2 Na		2 Na
	1 $(CO_3)^{2-}$		1 $(CO_3)^{2-}$

The coefficient ratio of 1:1 → 2:1 is the lowest whole-number ratio. The equation is balanced.

Q & A

Gasoline is a mixture of hydrocarbons, including octane (C_8H_{18}). Combustion of octane produces CO_2 and H_2O, with the release of energy. Write a balanced equation for this reaction.

Answer

Step 1: Write the correct formulas in an unbalanced equation,

$$C_8H_{18}(g) + O_2(g) \rightarrow CO_2(g) + H_2O(g)$$

(not balanced)

Step 2: Take an inventory,

Reactants:	8 C	Products:	1 C
	18 H		2 H
	2 O		3 O

(not balanced)

Step 3: Start with the compound with the most atoms (generalization 1) and place coefficients to balance these atoms,

$$C_8H_{18}(g) + O_2(g) \rightarrow 8\,CO_2(g) + 9\,H_2O(g)$$

(not balanced)

Redo the inventory,

Reactants:	8 C	Products:	8 C
	18 H		18 H
	2 O		25 O

The O atoms are still unbalanced. There are 2 O atoms in the reactants but 25 O atoms in the products. Since the subscript cannot be changed, it will take 12.5 O_2 to produce 25 oxygen atoms (generalization 3).

$$C_8H_{18}(g) + 12.5\,O_2(g) \rightarrow 8\,CO_2(g) + 9\,H_2O(g)$$

(balanced)

Step 4: (a) An inventory will show that the atoms balance,

Reactants:	8 C	Products:	8 C
	18 H		18 H
	25 O		25 O

(b) The coefficients are not in the lowest whole-number ratio (one-half an O_2 does not exist). To make the lowest possible whole-number ratio, all coefficients are multiplied by 2. This results in a correct balanced equation of

$$2\,C_8H_{18}(g) + 25\,O_2(g) \rightarrow 16\,CO_2(g) + 18\,H_2O(g)$$

(balanced)

FIGURE 12.9

Hydrocarbons are composed of the elements hydrogen and carbon. Propane (C_3H_8) and gasoline, which contain octane (C_8H_{18}) are examples of hydrocarbons. *Carbohydrates* are composed of the elements of hydrogen, carbon, and oxygen. Table sugar, for example, is the carbohydrate $C_{12}H_{22}O_{11}$. Generalizing, all hydrocarbons and carbohydrates react completely with oxygen to yield CO_2 and H_2O.

Generalizing Equations

In the previous chapters you learned that the act of classifying, or grouping, something according to some property makes the study of a large body of information less difficult. Generalizing from groups of chemical reactions also makes it possible to predict what will happen in similar reactions. For example, you have studied equations in the previous section describing the combustion of methane (CH_4), propane (C_3H_8), and octane (C_8H_{18}). Each of these reactions involves a *hydrocarbon*, a compound of the elements hydrogen and carbon. Each hydrocarbon reacted with O_2, yielding CO_2 and releasing the energy of combustion. Generalizing from these reactions, you could predict that the combustion of any hydrocarbon would involve the combination of atoms of the hydrocarbon molecule with O_2 to produce CO_2 and H_2O with the release of energy. Such reactions could be analyzed by chemical experiments, and the products could be identified by their physical and chemical properties. You would find your predictions based on similar reactions would be correct, thus justifying predictions from such generalizations. Butane, for example, is a hydrocarbon with the formula C_4H_{10}. The balanced equation for the combustion of butane is

$$2\,C_4H_{10}(g) + 13\,O_2(g) \rightarrow 8\,CO_2(g) + 10\,H_2O(g)$$

You could extend the generalization further, noting that the combustion of compounds containing oxygen as well as carbon and hydrogen also produces CO_2 and H_2O (figure 12.9). These

compounds are *carbohydrates,* composed of carbon and water. Glucose, for example, was identified earlier as a compound with the formula $C_6H_{12}O_6$. Glucose combines with oxygen to produce CO_2 and H_2O, and the balanced equation is

$$C_6H_{12}O_6(s) + 6\,O_2(g) \rightarrow 6\,CO_2(g) + 6\,H_2O(g)$$

Note that three molecules of oxygen were not needed from the O_2 reactant since the other reactant, glucose, contains six oxygen atoms per molecule. An inventory of atoms will show that the equation is thus balanced.

Combustion is a rapid reaction with O_2 that releases energy, usually with a flame. A very similar, although much slower reaction takes place in plant and animal respiration. In respiration, carbohydrates combine with O_2 and release energy used for biological activities. This reaction is slow compared to combustion and requires enzymes to proceed at body temperature. Nonetheless, CO_2 and H_2O are the products.

Oxidation-Reduction Reactions

The reactions involving hydrocarbons and carbohydrates with oxygen are examples of an important group of chemical reactions called *oxidation-reduction* reactions. When the term "oxidation" was first used, it specifically meant reactions involving the combination of oxygen with other atoms. But fluorine, chlorine, and other nonmetals were soon understood to have similar reactions as oxygen, so the definition was changed to one concerning the shifts of electrons in the reaction.

An **oxidation-reduction reaction** (or **redox reaction**) is broadly defined as a reaction in which electrons are transferred from one atom to another. As is implied by the name, such a reaction has two parts and each part tells you what happens to the electrons. *Oxidation* is the part of a redox reaction in which there is a loss of electrons by an atom. *Reduction* is the part of a redox reaction in which there is a gain of electrons by an atom. The name also implies that in any reaction in which oxidation occurs reduction must take place, too. One cannot take place without the other.

Substances that take electrons from other substances are called **oxidizing agents.** Oxidizing agents take electrons from the substances being oxidized. Oxygen is the most common oxidizing agent, and several examples have already been given about how it oxidizes foods and fuels. Chlorine is another commonly used oxidizing agent, often for the purposes of bleaching or killing bacteria (figure 12.10).

A **reducing agent** supplies electrons to the substance being reduced. Hydrogen and carbon are commonly used reducing agents. Carbon is commonly used as a reducing agent to extract metals from their ores. For example, carbon (from coke, which is coal that has been baked) reduces Fe_2O_3, an iron ore, in the reaction

$$2\,Fe_2O_3(s) + 3\,C(s) \rightarrow 4\,Fe(s) + 3\,CO_2 \uparrow$$

The Fe in the ore gained electrons from the carbon, the reducing agent in this reaction.

FIGURE 12.10

Oxidizing agents take electrons from other substances that are being oxidized. Oxygen and chlorine are commonly used, strong oxidizing agents.

Types of Chemical Reactions

Many chemical reactions can be classified as redox or nonredox reactions. Another way to classify chemical reactions is to consider what is happening to the reactants and products. This type of classification scheme leads to four basic categories of chemical reactions, which are (1) *combination,* (2) *decomposition,* (3) *replacement,* and (4) *ion exchange* reactions. The first three categories are subclasses of redox reactions. It is in the ion exchange reactions that you will find the first example of a reaction that is not a redox reaction.

FIGURE 12.11

Rusting iron is a common example of a combination reaction, where two or more substances combine to form a new compound. Rust is iron(III) oxide formed on these screws from the combination of iron and oxygen under moist conditions.

Combination Reactions

A **combination reaction** is a synthesis reaction in which two or more substances combine to form a single compound. The combining substances can be (1) elements, (2) compounds, or (3) combinations of elements and compounds. In generalized form, a combination reaction is

$$X + Y \rightarrow XY$$

equation 12.3

Many redox reactions are combination reactions. For example, metals are oxidized when they burn in air, forming a metal oxide. Consider magnesium, which gives off a bright white light as it burns:

$$2\,Mg(s) + O_2(g) \rightarrow 2\,MgO(s)$$

Note how the magnesium-oxygen reaction follows the generalized form of equation 12.3.

The rusting of metals is oxidation that takes place at a slower pace than burning, but metals are nonetheless oxidized in the process (figure 12.11). Again noting the generalized form of a combination reaction, consider the rusting of iron:

$$4\,Fe(s) + 3\,O_2(g) \rightarrow 2\,Fe_2O_3(s)$$

287

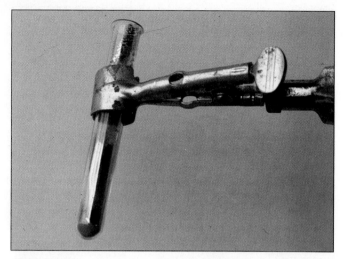

FIGURE 12.12

Mercury(II) oxide is decomposed by heat, leaving the silver-colored element mercury behind as oxygen is driven off. This is an example of a decomposition reaction, $2\,HgO \rightarrow 2\,Hg + O_2\uparrow$. Compare this equation to the general form of a decomposition reaction.

Nonmetals are also oxidized by burning in air, for example, when carbon burns with a sufficient supply of O_2:

$$C(s) + O_2(g) \rightarrow CO_2(g)$$

Note that all the combination reactions follow the generalized form of $X + Y \rightarrow XY$.

Decomposition Reactions

A **decomposition reaction,** as the term implies, is the opposite of a combination reaction. In decomposition reactions a compound is broken down (1) into the elements that make up the compound, (2) into simpler compounds, or (3) into elements and simpler compounds. Decomposition reactions have a generalized form of

$$XY \rightarrow X + Y$$

equation 12.4

Decomposition reactions generally require some sort of energy, which is usually supplied in the form of heat or electrical energy. An electric current, for example, decomposes water into hydrogen and oxygen:

$$2\,H_2O(l) \xrightarrow{\text{electricity}} 2\,H_2(g) + O_2(g)$$

Mercury(II) oxide is decomposed by heat, an observation that led to the discovery of oxygen (figure 12.12):

$$2HgO(s) \xrightarrow{\Delta} 2\,Hg(s) + O_2\uparrow$$

Plaster is a building material made from a mixture of calcium hydroxide, $Ca(OH)_2$, and plaster of Paris, $CaSO_4$. The

calcium hydroxide is prepared by adding water to calcium oxide (CaO), which is commonly called quicklime. Calcium oxide is made by heating limestone or chalk ($CaCO_3$), and

$$CaCO_3(s) \xrightarrow{\Delta} CaO(s) + CO_2\uparrow$$

Note that all the decomposition reactions follow the generalized form of $XY \rightarrow X + Y$.

Replacement Reactions

In a **replacement reaction,** an atom or polyatomic ion is replaced in a compound by a different atom or polyatomic ion. The replaced part can be either the negative or positive part of the compound. In generalized form, a replacement reaction is

$$XY + Z \rightarrow XZ + Y$$
(negative part replaced)

equation 12.5

or

$$XY + A \rightarrow AY + X$$
(positive part replaced)

equation 12.6

Replacement reactions occur because some elements have a stronger electron-holding ability than other elements. Elements that have the least ability to hold on to their electrons are the most chemically active. Figure 12.13 shows a list of chemical activity of some metals, with the most chemically active at the top. Hydrogen is included because of its role in acids (see chapter 13). Take a few minutes to look over the generalizations listed in figure 12.13. The generalizations apply to combination, decomposition, and replacement reactions.

Replacement reactions take place as more active metals give up electrons to elements lower on the list with a greater electron-holding ability. For example, aluminum is higher on the activity series than copper. When aluminum foil is placed in a solution of copper(II) chloride, aluminum is oxidized, losing electrons to the copper. The loss of electrons from metallic aluminum forms aluminum ions in solution, and the copper comes out of solution as a solid metal (figure 12.14).

$$2\,Al(s) + 3\,CuCl_2(aq) \rightarrow 2\,AlCl_3(aq) + 3\,Cu(s)$$

A metal will replace any metal ion in solution that it is above in the activity series. If the metal is listed below the metal ion in solution, no reaction occurs. For example, $Ag(s) + CuCl_2(aq) \rightarrow$ no reaction.

The very active metals (lithium, potassium, calcium, and sodium) react with water to yield metal hydroxides and hydrogen. For example,

$$2\,Na(s) + 2\,H_2O(l) \rightarrow 2\,NaOH(aq) + H_2\uparrow$$

Acids yield hydrogen ions in solution, and metals above hydrogen in the activity series will replace hydrogen to form a metal salt. For example,

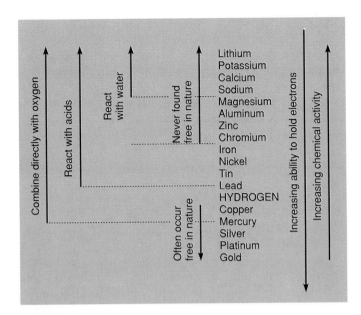

FIGURE 12.13

The activity series for common metals, together with some generalizations about the chemical activities of the metals. The series is used to predict which replacement reactions will take place and which reactions will not occur. (Note that hydrogen is not a metal and is placed in the series for reference to acid reactions.)

FIGURE 12.14

This shows a reaction between metallic aluminum and the blue solution of copper(II) chloride. Aluminum is above copper in the activity series, and aluminum replaces the copper ions from the solution as copper is deposited as a metal. The aluminum loses electrons to the copper and forms aluminum ions in solution.

$$Zn(s) + H_2SO_4(aq) \rightarrow ZnSO_4(aq) + H_2\uparrow$$

In general, the energy involved in replacement reactions is less than the energy involved in combination or decomposition reactions.

Ion Exchange Reactions

An **ion exchange reaction** is a reaction that takes place when the ions of one compound interact with the ions of another compound, forming (1) a solid that comes out of solution (a precipitate), (2) a gas, or (3) water.

A water solution of dissolved ionic compounds is a solution of ions. For example, solid sodium chloride dissolves in water to become ions in solution,

$$NaCl(s) \rightarrow Na^+(aq) + Cl^-(aq)$$

If a second ionic compound is dissolved with a solution of another, a mixture of ions results. The formation of a precipitate, a gas, or water, however, removes ions from the solution, and this must occur before you can say that an ionic exchange reaction has taken place. For example, water being treated for domestic use sometimes carries suspended matter that is removed by adding aluminum sulfate and calcium hydroxide to the water. The reaction is

$$3\,Ca(OH)_2(aq) + Al_2(SO_4)_3(aq) \rightarrow 3\,CaSO_4(aq) + 2\,Al(OH)_3\downarrow$$

The aluminum hydroxide is a jellylike solid, which traps the suspended matter for sand filtration. The formation of the insoluble aluminum hydroxide removed the aluminum and hydroxide ions from the solution, so an ion exchange reaction took place.

In general, an ion exchange reaction has the form

$$AX + BY \rightarrow AY + BX$$

equation 12.7

where one of the products removes ions from the solution. The calcium hydroxide and aluminum sulfate reaction took place as the aluminum and calcium ions traded places. A solubility table such as the one in appendix B will tell you if an ionic exchange reaction has taken place. Aluminum hydroxide is insoluble, according to the table, so the reaction did take place. No ionic exchange reaction occurred if the new products are both soluble.

Another way for an ion exchange reaction to occur is if a gas or water molecule forms to remove ions from the solution. When an acid reacts with a base (an alkaline compound), a salt and water are formed

$$HCl(aq) + NaOH(aq) \rightarrow NaCl(aq) + H_2O(l)$$

The reactions of acids and bases are discussed in chapter 13.

Q & A

Write complete balanced equations for the following, and identify if each reaction is combination, decomposition, replacement, or ion exchange:

(a) silver(s) + sulfur(g) → silver sulfide(s)

(b) aluminum(s) + iron(III) oxide(s) → aluminum oxide(s) + iron

(c) sodium chloride(aq) + silver nitrate(aq) → ?

(d) potassium chlorate(s) $\overset{\Delta}{\rightarrow}$ potassium chloride(s) + oxygen(g)

Answer

(a) The reactants are two elements, and the product is a compound, following the general form $X + Y \rightarrow XY$ of a combination reaction. Table 11.2 gives the charge on silver as Ag^{1+}, and sulfur (as the other nonmetals in family VIA) is S^{2-}. The balanced equation is

$$2 \, Ag(s) + S(g) \rightarrow Ag_2S(s)$$

Silver sulfide is the tarnish that appears on silverware.

(b) The reactants are an element and a compound that react to form a new compound and an element. The general form is $XY + Z \rightarrow XZ + Y$, which describes a replacement reaction. The balanced equation is

$$2 \, Al(s) + Fe_2O_3(s) \rightarrow Al_2O_3(s) + 2 \, Fe(s)$$

This is known as a "thermite reaction," and in the reaction aluminum reduces the iron oxide to metallic iron with the release of sufficient energy to melt the iron. The thermite reaction is sometimes used to weld large steel pieces, such as railroad rails.

(c) The reactants are water solutions of two compounds with the general form of $AX + BY \rightarrow$, so this must be the reactant part of an ion exchange reaction. Completing the products part of the equation by exchanging parts as shown in the general form and balancing,

$$NaCl(aq) + AgNO_3(aq) \rightarrow NaNO_3(?) + AgCl(?)$$

Now consult the solubility chart in appendix B to find out if either of the products is insoluble. $NaNO_3$ is soluble and $AgCl$ is insoluble. Since at least one of the products is insoluble, the reaction did take place, and the equation is rewritten as

$$NaCl(aq) + AgNO_3(aq) \rightarrow NaNO_3(aq) + AgCl\downarrow$$

(d) The reactant is a compound, and the products are a simpler compound and an element, following the generalized form of a decomposition reaction, $XY \rightarrow X + Y$. The delta sign (Δ) also means that heat was added, which provides another clue that this is a decomposition reaction. The formula for the chlorate ion is in table 11.4. The balanced equation is

$$2 \, KClO_3(s) \overset{\Delta}{\rightarrow} 2 \, KCl(s) + 3 \, O_2\uparrow$$

INFORMATION FROM CHEMICAL EQUATIONS

A balanced chemical equation describes what happens in a chemical reaction in a concise, compact way. The balanced equation also carries information about (1) atoms, (2) molecules, and (3) atomic weights. The balanced equation for the combustion of hydrogen, for example, is

$$2 \, H_2(g) + O_2(g) \rightarrow 2 \, H_2O(l)$$

An inventory of each kind of atom in the reactants and products shows

	Reactants:	4 hydrogen		Products:	4 hydrogen
		2 oxygen			2 oxygen
	Total:	6 atoms		Total:	6 atoms

There are six atoms before the reaction and there are six atoms after the reaction, which is in accord with the law of conservation of mass.

In terms of molecules, the equation says that two diatomic molecules of hydrogen react with one (understood) diatomic molecule of oxygen to yield two molecules of water. The number of coefficients in the equation is the number of molecules involved in the reaction. If you are concerned how two molecules plus one molecule could yield two molecules, remember that *atoms* are conserved in a chemical reaction, not molecules.

Since atoms are conserved in a chemical reaction, their atomic weights should be conserved, too. One hydrogen atom has an atomic weight of 1.0 u, so the formula weight of a diatomic hydrogen molecule must be 2×1.0 u, or 2.0 u. The formula weight of O_2 is 2×16.0 u, or 32 u. If you consider the equation in terms of atomic weights, then

Equation

$$2 \, H_2 + O_2 \rightarrow 2 \, H_2O$$

Formula weights

$$2 \, (1.0 \, u + 1.0 \, u) + (16.0 \, u + 16.0 \, u) \rightarrow 2 \, (2 \times 1.0 \, u + 16.0 \, u)$$

$$4 \, u + 32 \, u \rightarrow 36 \, u$$

$$36 \, u \rightarrow 36 \, u$$

The formula weight for H_2O is $(1.0 \, u \times 2) + 16 \, u$, or 18 u. The coefficient of 2 in front of H_2O means there are two molecules of H_2O, so the mass of the products is 2×18 u, or 36 u. Thus, the reactants had a total mass of 4 u + 32 u, or 36 u, and the products had a total mass of 36 u. Again, this is in accord with the law of conservation of mass.

The equation says that 4 u of hydrogen will combine with 32 u of oxygen. Thus hydrogen and oxygen combine in a mass ratio of 4:32, which reduces to 1:8. So 1 gram of hydrogen will combine with 8 grams of oxygen, and, in fact, they will combine in this ratio no matter what the measurement units are (gram, kilogram, pound, etc.). They always combine in this mass ratio because this is the mass of the individual reactants.

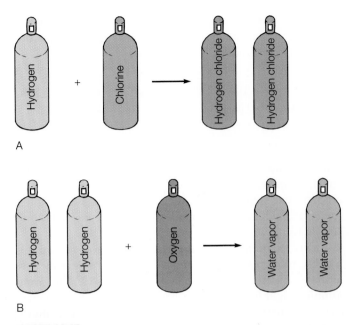

A

B

FIGURE 12.15

Reacting gases combine in ratios of small, whole-number volumes when the temperature and pressure are the same for each volume. (*A*) One volume of hydrogen gas combines with one volume of chlorine gas to yield two volumes of hydrogen chloride gas. (*B*) Two volumes of hydrogen gas combine with one volume of oxygen gas to yield two volumes of water vapor.

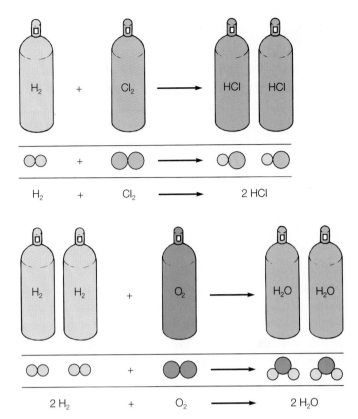

FIGURE 12.16

Avogadro's hypothesis of equal volumes of gas having equal numbers of molecules offered an explanation for the law of combining volumes.

Back in the early 1800s John Dalton attempted to work out a table of atomic weights as he developed his atomic theory (see chapter 10). Dalton made two major errors in determining the atomic weights, including (1) measurement errors about mass ratios of combining elements and (2) incorrect assumptions about the formula of the resulting compound. For water, for example, Dalton incorrectly measured that 5.5 g of oxygen combined with 1.0 g of hydrogen. He assumed that one atom of hydrogen combined with one atom of oxygen, resulting in a formula of HO. Thus, Dalton concluded that the atomic mass of oxygen was 5.5 u, and the atomic mass of hydrogen was 1.0 u. Incorrect atomic weights for hydrogen and oxygen led to conflicting formulas for other substances, and no one could show that the atomic theory worked.

The problem was solved during the first decade of the 1800s through the separate work of a French chemistry professor, Joseph Gay-Lussac, and an Italian physics professor, Amedeo Avogadro. In 1808, Gay-Lussac reported that reacting gases combined in small, whole number *volumes* when the temperature and pressure were constant. Two volumes of hydrogen, for example, combined with one volume of oxygen to form two volumes of water vapor. The term "volume" means any measurement unit, for example, a liter. Other reactions between gases were also observed to combine in small, whole number ratios, and the pattern became known as the *law of combining volumes* (figure 12.15).

Avogadro proposed an explanation for the law of combining volumes in 1811. He proposed that equal volumes of all gases at the same temperature and pressure *contain the same number of molecules*. Avogadro's hypothesis had two important implications for the example of water. First, since two volumes of hydrogen combine with one volume of oxygen, it means that a molecule of water contains twice as many hydrogen atoms as oxygen atoms. The formula for water must be H_2O, not HO. Second, since *two* volumes of water vapor were produced, each molecule of hydrogen and each molecule of oxygen must be diatomic. Diatomic molecules of hydrogen and oxygen would double the number of hydrogen and oxygen atoms, thus producing twice as much water vapor. These two implications are illustrated in figure 12.16, along with a balanced equation for the reaction. Note that the coefficients in the equation now have two meanings, (1) the number of molecules of each substance involved in the reaction and (2) the ratios of combining volumes. The coefficient of 2 in front of the H_2, for example, means two molecules of H_2. It also means two volumes of H_2 gas when all volumes are measured at the same temperature and pressure. Recall that equal volumes of any two gases at the same temperature and pressure contain the same number of molecules. Thus, the ratio of coefficients in a balanced equation means a ratio of *any number* of molecules, from 2 of H_2 and 1 of O_2, 20 of H_2 and 10 of O_2, 2,000 of H_2 and 1,000 of O_2, or however many are found in 2 L of H_2 and 1 L of O_2.

Q & A

Propane is a hydrocarbon with the formula C_3H_8 that is used as a bottled gas. (a) How many liters of oxygen are needed to burn 1 L of propane gas? (b) How many liters of carbon dioxide are produced by the reaction? Assume all volumes to be measured at the same temperature and pressure.

Answer

The balanced equation is

$$C_3H_8(g) + 5\,O_2(g) \rightarrow 3\,CO_2(g) + 4\,H_2O(g)$$

The coefficients tell you the relative number of molecules involved in the reaction, that 1 molecule of propane reacts with 5 molecules of oxygen to produce 3 molecules of carbon dioxide and 4 molecules of water. Since equal volumes of gases at the same temperature and pressure contain equal numbers of molecules, the coefficients also tell you the relative volumes of gases. Thus 1 L of propane (a) requires 5 L of oxygen and (b) yields 3 L of carbon dioxide (and 4 L of water vapor) when reacted completely.

Units of Measurement Used with Equations

The coefficients in a balanced equation represent a ratio of any *number* of molecules involved in a chemical reaction. The equation has meaning about the atomic *weights* and formula *weights* of reactants and products. The counting of numbers and the use of atomic weights are brought together in a very important measurement unit called a *mole* (from the Latin meaning "a mass"). Here are the important ideas in the mole concept:

1. Recall that the atomic weights of elements are average relative masses of the isotopes of an element. The weights are based on a comparison to carbon-12, with an assigned mass of exactly 12.00 (see chapter 10).
2. The *number* of C-12 atoms in exactly 12.00 g of C-12 has been measured experimentally to be 6.02×10^{23}. This number is called **Avogadro's number,** named after the scientist who reasoned that equal volumes of gases contain equal numbers of molecules.
3. An amount of a substance that contains Avogadro's number of atoms, ions, molecules, or any other chemical unit is defined as a **mole** of the substance. Thus a mole is 6.02×10^{23} atoms, ions, etc., just as a dozen is 12 eggs, apples, etc. The mole is the chemist's measure of atoms, molecules, or other chemical units. A mole of Na^+ ions is 6.02×10^{23} Na^+ ions.
4. A mole of C-12 atoms is defined as having a mass of exactly 12.00 g, a mass that is numerically equal to its atomic mass. So the mass of a mole of C-12 atoms is 12.00 g, or

mass of one atom	$\times$	one mole	=	mass of a mole of C-12
(12.00 u)		(6.02×10^{23}) =		12.00 g

The masses of all the other isotopes are *based* on a comparison to the C-12 atom. Thus a He-4 atom has one-third the mass of a C-12 atom. An atom of Mg-24 is twice as massive as a C-12 atom. Thus

	1 Atom	$\times$	1 Mole	=	Mass of Mole
C-12:	12.00 u	$\times$	6.02×10^{23}	=	12.00 g
He-4:	4.00 u	$\times$	6.02×10^{23}	=	4.00 g
Mg-24:	24.00 u	$\times$	6.02×10^{23}	=	24.00 g

Therefore, the mass of a mole of any element is numerically equal to its atomic mass. Samples of elements with masses that are the same numerically as their atomic masses are 1 mole measures, and each will contain the same number of atoms (figure 12.17).

This reasoning can be used to generalize about formula weights, molecular weights, and atomic weights since they are all based on atomic mass units relative to C-12. The **gram-atomic weight** is the mass in grams of one mole of an element that is numerically equal to its atomic weight. The atomic weight of carbon is 12.01 u; the gram-atomic weight of carbon is 12.01 g. The atomic weight of magnesium is 24.3 u; the gram-atomic weight of magnesium is 24.3 g. Any gram-atomic weight contains Avogadro's number of atoms. Therefore the gram-atomic weights of the elements all contain the same number of atoms.

Similarly, the **gram-formula weight** of a compound is the mass in grams of one mole of the compound that is numerically equal to its formula weight. The **gram-molecular weight** is the gram-formula weight of a molecular compound. Note that one mole of Ne atoms (6.02×10^{23} neon atoms) has a gram-atomic weight of 20.2 g, but one mole of O_2 molecules (6.02×10^{23} oxygen molecules) has a gram-molecular weight of 32.0 g. Stated the other way around, 32.0 g of O_2 and 20.2 g of Ne both contain the same Avogadro's number of particles.

Q & A

(a) A 100 percent silver chain has a mass of 107.9 g. How many silver atoms are in the chain? (b) What is the mass of one mole of sodium chloride, NaCl?

Answer

The mole concept and Avogadro's number provide a relationship between numbers and masses. (a) The atomic weight of silver is 107.9 u, so the gram-atomic weight of silver is 107.9 g. A gram-atomic weight is one mole of an element, so the silver chain contains 6.02×10^{23} silver atoms. (b) The formula weight of NaCl is 58.5 u, so the gram-formula weight is 58.5 g. One mole of NaCl has a mass of 58.5 g.

A. Each of the following represents one mole of an element:

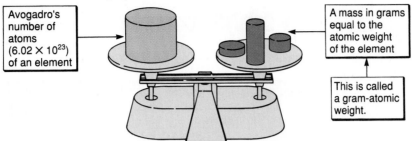

Avogadro's number of atoms (6.02×10^{23}) of an element

A mass in grams equal to the atomic weight of the element

This is called a gram-atomic weight.

B. Each of the following represents one mole of a compound:

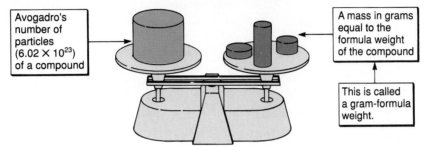

Avogadro's number of particles (6.02×10^{23}) of a compound

A mass in grams equal to the formula weight of the compound

This is called a gram-formula weight.

C. Each of the following represents one mole of a molecular substance:

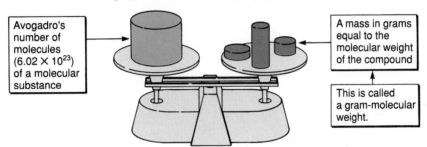

Avogadro's number of molecules (6.02×10^{23}) of a molecular substance

A mass in grams equal to the molecular weight of the compound

This is called a gram-molecular weight.

FIGURE 12.17

The mole concept for (A) elements, (B) compounds, and (C) molecular substances. A mole contains 6.02×10^{23} particles. Since every mole contains the same number of particles, the ratio of the mass of any two moles is the same as the ratio of the masses of individual particles making up the two moles.

Quantitative Uses of Equations

A balanced chemical equation can be interpreted in terms of (1) *a molecular ratio* of the reactants and products, (2) *a mole ratio* of the reactants and products, or (3) *a mass ratio* of the reactants and products. Consider, for example, the balanced equation for reacting hydrogen with nitrogen to produce ammonia,

$$3 \, H_2(g) + N_2(g) \rightarrow 2 \, NH_3(g)$$

From a *molecular* point of view, the equation says that three molecules of hydrogen combine with one molecule of N_2 to form two molecules of NH_3. The coefficients of $3:1 \rightarrow 2$ thus express a molecular ratio of the reactants and the products.

The molecular ratio leads to the concept of a *mole ratio* since any number of molecules can react as long as they are in the ratio of $3:1 \rightarrow 2$. The number could be Avogadro's number, so $(3) \times (6.02 \times 10^{23})$ molecules of H_2 will combine with $(1) \times (6.02 \times 10^{23})$ molecules of N_2 to form $(2) \times (6.02 \times 10^{23})$ mol-

ecules of ammonia. Since 6.02×10^{23} molecules is the number of particles in a mole, the coefficients therefore represent the *numbers of moles* involved in the reaction. Thus, three moles of H_2 react with one mole of N_2 to produce two moles of NH_3.

The mole ratio of a balanced chemical equation leads to the concept of a *mass ratio* interpretation of a chemical equation. The gram-formula weight of a compound is the mass in grams of *one mole* that is numerically equal to its formula weight. Therefore, the equation also describes the mass ratios of the reactants and the products. The mass ratio can be calculated from the mole relationship described in the equation. The three interpretations are summarized in table 12.1.

Thus, the coefficients in a balanced equation can be interpreted in terms of molecules, which leads to an interpretation of moles, mass, or any formula unit. The mole concept thus provides the basis for calculations about the quantities of reactants and products in a chemical reaction.

A CLOSER LOOK | The Catalytic Converter

The modern automobile produces two troublesome products in the form of (1) nitrogen monoxide (NO) and (2) hydrocarbons from the incomplete combustion of gasoline. These products from the exhaust enter the air to react in sunlight, eventually producing an irritating haze known as *photochemical smog*. To reduce photochemical smog, modern automobiles are fitted with a catalytic converter as part of the automobile exhaust system (box figure 12.1). This reading is about how the catalytic converter combats smog-forming pollutants.

Chemical reactions proceed at a rate that is affected by (1) the concentration of the reactants, (2) the temperature at which a reaction occurs, and (3) the surface area of the reaction. In general, a higher concentration, higher temperatures, and greater surface area mean a faster reaction. The problem with nitrogen monoxide is that it is easily oxidized to nitrogen dioxide (NO_2), a reddish brown, damaging gas that also plays a key role in the formation of photochemical smog. Nitrogen dioxide and the hydrocarbons are oxidized slowly in the air when left to themselves. What is needed is a means to decompose nitrogen monoxide and uncombusted hydrocarbons rapidly before they are released into the air.

The rate at which a chemical reaction proceeds is affected by a *catalyst*, a material that speeds up a chemical reaction without being permanently changed by the reaction. Apparently, molecules require a certain amount of energy to change the chemical bonds that tend to keep them as they are, unreacted. This certain amount of energy is called the *activation energy*, and it represents an energy barrier that must be overcome before a chemical reaction can take place. This explains why chemical reactions proceed at a faster rate at higher temperatures. At higher temperatures, molecules have greater average kinetic energies; thus they already have part of the minimum energy needed for a reaction to take place.

A catalyst appears to speed a chemical reaction by lowering the activation energy. Molecules become temporarily attached to the surface of the catalyst, which weakens the chemical bonds holding the molecule together. Thus, the weakened molecule is easier to break apart and the activation energy is lowered. Some catalysts do this better with some specific compounds than others, and extensive chemical research programs are devoted to finding new and more effective catalysts.

Automobile catalytic converters use unreactive metals, such as platinum, and transition metal oxides such as copper(II) oxide and chromium(III) oxide. Catalytic reactions that occur in the converter include the following:

$$H_2O + CO \xrightarrow{catalyst} H_2 + CO_2$$

$$2\,NO + 2\,CO \xrightarrow{catalyst} N_2 + 2\,CO_2$$

$$2\,NO + 2\,H_2 \xrightarrow{catalyst} N_2 + 2\,H_2O$$

$$O_2 + CO^* \xrightarrow{catalyst} CO_2 + H_2O$$
*or a hydrocarbon

Thus nitrogen monoxide is reduced to nitrogen gas, and hydrocarbons are oxidized to CO_2 and H_2O.

A catalytic converter can reduce or oxidize about 90 percent of the hydrocarbons, 85 percent of the carbon monoxide, and 40 percent of the nitrogen monoxide from exhaust gases. Other controls, such as exhaust gas recirculation (EGR), are used to further reduce NO formation.

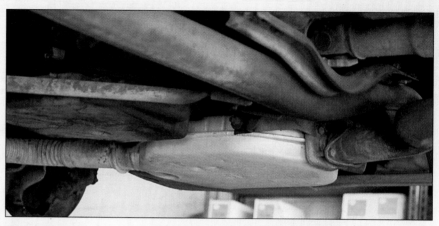

BOX FIGURE 12.1

This silver-colored canister is the catalytic converter. The catalytic converter is located between the engine and the muffler, which is farther back toward the rear of the car.

TABLE 12.1	Three interpretations of a chemical equation

Equation: $3 H_2 + N_2 \rightarrow 2 NH_3$

Molecular Ratio:

$\qquad$ 3 molecules H_2 + 1 molecule $N_2 \rightarrow$ 2 molecules NH_3

Mole Ratio:

$\qquad$ 3 moles H_2 + 1 mole $N_2 \rightarrow$ 2 moles NH_3

Mass Ratio:

$\qquad$ 6.0 g H_2 + 28.0 g $N_2 \rightarrow$ 34.0 g NH_3

Activities

Pick a household product that has a list of ingredients with names of covalent compounds or of ions you have met in this chapter. Write the brand name of the product and the type of product (Example: Sani-Flush; toilet-bowl cleaner), then list the ingredients as given on the label, writing them one under the other (column 1). Beside each name put the formula, if you can figure out what it should be (column 2). Also, in a third column, put whatever you know or can guess about the function of that substance in the product. (Example: This is an acid; helps dissolve mineral deposits.)

SUMMARY

A chemical formula is a shorthand way of describing the composition of a compound. An *empirical formula* identifies the simplest whole number ratio of element. A *molecular formula* identifies the actual number of atoms in a molecule.

The sum of the atomic weights of all the atoms in any formula is called the *formula weight.* The *molecular weight* is the formula weight of a molecular substance. The formula weight of a compound can be used to determine the *mass percentage* of elements making up a compound.

A concise way to describe a chemical reaction is to use formulas in a *chemical equation.* A chemical equation with the same number of each kind of atom on both sides is called a *balanced equation.* A balanced equation is in accord with the *law of conservation of mass,* which states that atoms are neither created nor destroyed in a chemical reaction. To balance a chemical equation, *coefficients* are placed in front of chemical formulas. Subscripts of formulas may not be changed since this would change the formula, meaning a different compound.

One important group of chemical reactions is called *oxidation-reduction reactions,* or *redox* reactions for short. Redox reactions are reactions where shifts of electrons occur. The process of losing electrons is called *oxidation,* and the substance doing the losing is said to be *oxidized.* The process of gaining electrons is called *reduction,* and the substance doing the gaining is said to be *reduced.* Substances that take electrons from other substances are called *oxidizing agents.* Substances that supply electrons are called *reducing agents.*

Chemical reactions can also be classified as (1) *combination,* (2) *decomposition,* (3) *replacement,* or (4) *ion exchange.* The first three of these are redox reactions, but ion exchange is not.

A balanced chemical equation describes chemical reactions and has quantitative meaning about numbers of atoms, numbers of molecules, and conservation of atomic weights. The coefficients also describe the *volumes* of combining gases. At a constant temperature and pressure gases combine in small, whole number ratios that are given by the coefficients. Each volume at the same temperature and pressure contains the *same number of molecules.*

The number of atoms in exactly 12.00 g of C-12 is called *Avogadro's number,* which has a value of 6.02×10^{23}. Any substance that contains Avogadro's number of atoms, ions, molecules, or any chemical unit is called a *mole* of that substance. The mole is a measure of a number of atoms, molecules, or other chemical units. The mass of a mole of any substance is equal to the atomic mass of that substance.

The mass, number of atoms, and mole concepts are generalized to other units. The *gram-atomic weight* of an element is the mass in grams that is numerically equal to its atomic weight. The *gram-formula weight* of a compound is the mass in grams that is numerically equal to the formula weight of the compound. The *gram-molecular weight* is the gram-formula weight of a molecular compound. The relationships between the mole concept and the mass ratios can be used with a chemical equation for calculations about the quantities of reactants and products in a chemical reaction.

Summary of Equations

12.1

$$\left(\frac{part}{whole}\right)(100\% \text{ of whole}) = \% \text{ of part}$$

12.2

$$\frac{\left(\begin{array}{c}\text{atomic weight}\\\text{of element}\end{array}\right)\left(\begin{array}{c}\text{number of atoms}\\\text{of element}\end{array}\right)}{\text{formula weight of compound}} \times \begin{array}{c}100\% \text{ of}\\\text{compound}\end{array} = \begin{array}{c}\% \text{ of}\\\text{element}\end{array}$$

12.3

Combination reaction, general form

$$X + Y \rightarrow XY$$

12.4

Decomposition reaction, general form

$$XY \rightarrow X + Y$$

12.5

Replacement reaction, general form for negative part replaced

$$XY + Z \rightarrow XZ + Y$$

12.6

Replacement reaction, general form for positive part replaced

$$XY + A \rightarrow AY + X$$

12.7

Ion exchange reaction, general form

$$AX + BY \rightarrow AY + BX$$

(Note: One of the products must remove ions by forming an insoluble product, water, or a gas.)

KEY TERMS

Avogadro's number (p. **292**)

chemical equation (p. **281**)

combination
   reaction (p. **287**)

decomposition
   reaction (p. **288**)

empirical formula (p. **278**)

formula weight (p. **279**)

gram-atomic weight (p. **292**)

gram-formula weight (p. **292**)

gram-molecular
   weight (p. **292**)

ion exchange reaction (p. **289**)

law of conservation of
   mass (p. **282**)

mole (p. **292**)

molecular formula (p. **278**)

molecular weight (p. **279**)

oxidation-reduction
   reaction (p. **286**)

oxidizing agents (p. **286**)

redox reaction (p. **286**)

reducing agent (p. **286**)

replacement reaction (p. **288**)

APPLYING THE CONCEPTS

1. A formula for a compound is given as KC1. This is a (an)
   a. empirical formula.
   b. molecular formula.
   c. structural formula.
   d. formula, but type unknown without further information.

2. A formula for a compound is given as C_8H_{18}. This is a (an)
   a. empirical formula.
   b. molecular formula.
   c. structural formula.
   d. formula, but type unknown without further information.

3. The formula weight of sulfuric acid, H_2SO_4 is
   a. 49 u.
   b. 50 u.
   c. 98 u.
   d. 194 u.

4. A balanced chemical equation has
   a. the same number of molecules on both sides of the equation.
   b. the same kinds of molecules on both sides of the equation.
   c. the same number of each kind of atom on both sides of the equation.
   d. all of the above.

5. The law of conservation of mass means that
   a. atoms are not lost or destroyed in a chemical reaction.
   b. the mass of a newly formed compound cannot be changed.
   c. in burning, part of the mass must be converted into fire in order for mass to be conserved.
   d. molecules cannot be broken apart because this would result in less mass.

6. A chemical equation is balanced by changing (the)
   a. subscripts.
   b. superscripts.
   c. coefficients.
   d. any of the above as necessary to achieve a balance.

7. Since wood is composed of carbohydrates, you should expect what gases to exhaust from a fireplace when complete combustion takes place?
   a. carbon dioxide, carbon monoxide, and pollutants
   b. carbon dioxide and water vapor
   c. carbon monoxide and smoke
   d. It depends on the type of wood being burned.

8. When carbon burns with an insufficient supply of oxygen, carbon monoxide is formed according to the following equation: $2 C + O_2 \rightarrow 2 CO$. What category of chemical reaction is this?
   a. combination
   b. ion exchange
   c. replacement
   d. None of the above, because the reaction is incomplete.

9. According to the activity series for metals, adding metallic iron to a solution of aluminum chloride should result in
   a. a solution of iron chloride and metallic aluminum.
   b. a mixed solution of iron and aluminum chloride.
   c. the formation of iron hydroxide with hydrogen given off.
   d. no metal replacement reaction.

10. In a replacement reaction, elements that have the most ability to hold onto their electrons are
    a. the most chemically active.
    b. the least chemically active.
    c. not generally involved in replacement reactions.
    d. none of the above.

11. Of the elements listed below, the one with the greatest electron-holding ability is
    a. sodium.
    b. zinc.
    c. copper.
    d. platinum.

12. Of the elements listed below, the one with the greatest chemical activity is
    a. aluminum.
    b. zinc.
    c. iron.
    d. mercury.

13. You know that an expected ion exchange reaction has taken place if the products include
    a. a precipitate.
    b. a gas.
    c. water.
    d. any of the above.

14. The incomplete equation of $2 KClO_3(s) \xrightarrow{\Delta}$ probably represents which type of chemical reaction?
    a. combination
    b. decomposition
    c. replacement
    d. ion exchange

15. In the equation of $2 H_2(g) + O_2(g) \rightarrow 2 H_2O(g)$,

    a. the total mass of the gaseous reactants is less than the total mass of the liquid product.

    b. the total number of molecules in the reactants is equal to the total number of molecules in the products.

    c. one volume of oxygen combines with two volumes of hydrogen to produce 2 volumes of water.

    d. All of the above are correct.

16. An amount of a substance that contains Avogadro's number of atoms, ions, or molecules is (a)

    a. mole.

    b. gram-atomic weight.

    c. gram-formula weight.

    d. any of the above.

17. If you have 6.02×10^{23} atoms of metallic iron, you will have how many grams of iron?

    a. 26

    b. 55.8

    c. 334.8

    d. 3.4×10^{25}

Answers

1. a **2.** b **3.** c **4.** c **5.** a **6.** c **7.** b **8.** a **9.** d **10.** b **11.** d **12.** a **13.** d **14.** b **15.** c **16.** d **17.** b

QUESTIONS FOR THOUGHT

1. How is an empirical formula like and unlike a molecular formula?

2. Describe the basic parts of a chemical equation. Identify how the physical state of elements and compounds is identified in an equation.

3. What is the law of conservation of mass? How do you know if a chemical equation is in accord with this law?

4. Describe in your own words how a chemical equation is balanced.

5. What is a hydrocarbon? What is a carbohydrate? In general, what are the products of complete combustion of hydrocarbons and carbohydrates?

6. Define and give an example in the form of a balanced equation of (a) a combination reaction, (b) a decomposition reaction, (c) a replacement reaction, and (d) an ion exchange reaction.

7. What must occur in order for an ion exchange reaction to take place? What is the result if this does not happen?

8. Predict the products for the following reactions: (a) The combustion of ethyl alcohol, C_2H_5OH, (b) the rusting of aluminum, and (c) the reaction between iron and sodium chloride.

9. The formula for butane is C_4H_{10}. Is this an empirical formula or a molecular formula? Explain the reason(s) for your answer.

10. How is the activity series for metals used to predict if a replacement reaction will occur or not?

11. What is a gram-formula weight? How is it calculated?

12. What is the meaning and the value of Avogadro's number? What is a mole?

PARALLEL EXERCISES

The exercises in groups A and B cover the same concepts. Solutions to group A exercises are located in appendix D.

Group A

1. Identify the following as empirical formulas or molecular formulas and indicate any uncertainty with (?):

    (a) $MgCl_2$

    (b) C_2H_2

    (c) BaF_2

    (d) C_8H_{18}

    (e) CH_4

    (f) S_8

2. What is the formula weight for each of the following compounds?

    (a) Copper(II) sulfate

    (b) Carbon disulfide

    (c) Calcium sulfate

    (d) Sodium carbonate

3. What is the mass percentage composition of the elements in the following compounds?

    (a) Fool's gold, FeS_2

    (b) Boric acid, H_3BO_3

    (c) Baking soda, $NaHCO_3$

    (d) Aspirin, $C_9H_8O_4$

Group B

1. Identify the following as empirical formulas or molecular formulas and indicate any uncertainty with (?):

    (a) CH_2O

    (b) $C_6H_{12}O_6$

    (c) NaCl

    (d) CH_4

    (e) F_6

    (f) CaF_2

2. Calculate the formula weight for each of the following compounds:

    (a) Dinitrogen monoxide

    (b) Lead(II) sulfide

    (c) Magnesium sulfate

    (d) Mercury(II) chloride

3. What is the mass percentage composition of the elements in the following compounds?

    (a) Potash, K_2CO_3

    (b) Gypsum, $CaSO_4$

    (c) Saltpeter, KNO_3

    (d) Caffeine, $C_8H_{10}N_4O_2$

4. Write balanced chemical equations for each of the following un-balanced reactions:
   (a) $SO_2 + O_2 \rightarrow SO_3$
   (b) $P + O_2 \rightarrow P_2O_5$
   (c) $Al + HCl \rightarrow AlCl_3 + H_2$
   (d) $NaOH + H_2SO_4 \rightarrow Na_2SO_4 + H_2O$
   (e) $Fe_2O_3 + CO \rightarrow Fe + CO_2$
   (f) $Mg(OH)_2 + H_3PO_4 \rightarrow Mg_3(PO_4)_2 + H_2O$

5. Identify the following as combination, decomposition, replacement, or ion exchange reactions:
   (a) $NaCl_{(aq)} + AgNO_{3(aq)} \rightarrow NaNO_{3(aq)} + AgCl\downarrow$
   (b) $H_2O_{(l)} + CO_{2(g)} \rightarrow H_2CO_{3(l)}$
   (c) $2\,NaHCO_{3(s)} \rightarrow Na_2CO_{3(s)} + H_2O_{(g)} + CO_{2(g)}$
   (d) $2\,Na_{(s)} + Cl_{2(g)} \rightarrow 2\,NaCl_{(s)}$
   (e) $Cu_{(s)} + 2\,AgNO_{3(aq)} \rightarrow Cu(NO_3)_{2(aq)} + 2\,Ag_{(s)}$
   (f) $CaO_{(s)} + H_2O_{(l)} \rightarrow Ca(OH)_{2(aq)}$

6. Write complete, balanced equations for each of the following reactions:
   (a) $C_5H_{12(g)} + O_{2(g)} \rightarrow$
   (b) $HCl_{(aq)} + NaOH_{(aq)} \rightarrow$
   (c) $Al_{(s)} + Fe_2O_{3(s)} \rightarrow$
   (d) $Fe_{(s)} + CuSO_{4(aq)} \rightarrow$
   (e) $MgCl_{(aq)} + Fe(NO_3)_{2(aq)} \rightarrow$
   (f) $C_6H_{10}O_{5(s)} + O_{2(g)} \rightarrow$

7. Write complete, balanced equations for each of the following decomposition reactions. Include symbols for physical states, heating, and others as needed:
   (a) Solid potassium chloride and oxygen gas are formed when solid potassium chlorate is heated.
   (b) Upon electrolysis, molten bauxite (aluminum oxide) yields solid aluminum metal and oxygen gas.
   (c) Upon heating, solid calcium carbonate yields solid calcium oxide and carbon dioxide gas.

8. Write complete, balanced equations for each of the following replacement reactions. If no reaction is predicted, write "no reaction" as the product:
   (a) $Na_{(s)} + H_2O_{(l)} \rightarrow$
   (b) $Au_{(s)} + HCl_{(aq)} \rightarrow$
   (c) $Al_{(s)} + FeCl_{3(aq)} \rightarrow$
   (d) $Zn_{(s)} + CuCl_{2(aq)} \rightarrow$

9. Write complete, balanced equations for each of the following ion exchange reactions. If no reaction is predicted, write "no reaction" as the product:
   (a) $NaOH_{(aq)} + HNO_{3(aq)} \rightarrow$
   (b) $CaCl_{2(aq)} + KNO_{3(aq)} \rightarrow$
   (c) $Ba(NO_3)_{2(aq)} + Na_3PO_{4(aq)} \rightarrow$
   (d) $KOH_{(aq)} + ZnSO_{4(aq)} \rightarrow$

10. The gas welding torch is fueled by two tanks, one containing acetylene (C_2H_2) and the other pure oxygen (O_2). The very hot flame of the torch is produced as acetylene burns,

$$2\,C_2H_{2(g)} + O_{2(g)} \rightarrow 4\,CO_{2(g)} + H_2O_{(g)}$$

According to this equation, how many liters of oxygen are required to burn 1 liter of acetylene?

4. Write balanced chemical equations for each of the following un-balanced reactions:
   (a) $NO + O_2 \rightarrow NO_2$
   (b) $KClO_3 \rightarrow KCl + O_2$
   (c) $NH_4Cl + Ca(OH)_2 \rightarrow CaCl_2 + NH_3 + H_2O$
   (d) $NaNO_3 + H_2SO_4 \rightarrow Na_2SO_4 + HNO_3$
   (e) $PbS + H_2O_2 \rightarrow PbSO_4 + H_2O$
   (f) $Al_2(SO_4)_3 + BaCl_2 \rightarrow AlCl_3 + BaSO_4$

5. Identify the following as combination, decomposition, replacement, or ion exchange reactions:
   (a) $ZnCO_{3(s)} \rightarrow ZnO_{(s)} + CO_2\uparrow$
   (b) $2\,NaBr_{(aq)} + Cl_{2(g)} \rightarrow 2\,NaCl_{(aq)} + Br_{2(g)}$
   (c) $2\,Al_{(s)} + 3\,Cl_{2(g)} \rightarrow 2\,AlCl_{3(s)}$
   (d) $Ca(OH)_{2(aq)} + H_2SO_{4(aq)} \rightarrow CaSO_{4(aq)} + 2\,H_2O_{(l)}$
   (e) $Pb(NO_3)_{2(aq)} + H_2S_{(g)} \rightarrow 2\,HNO_{3(aq)} + PbS\downarrow$
   (f) $C_{(s)} + ZnO_{(s)} \rightarrow Zn_{(s)} + CO\uparrow$

6. Write complete, balanced equations for each of the following reactions:
   (a) $C_3H_{6(g)} + O_{2(g)} \rightarrow$
   (b) $H_2SO_{4(aq)} + KOH_{(aq)} \rightarrow$
   (c) $C_6H_{12}O_{6(s)} + O_{2(g)} \rightarrow$
   (d) $Na_3PO_{4(aq)} + AgNO_{3(aq)} \rightarrow$
   (e) $NaOH_{(aq)} + Al(NO_3)_{3(aq)} \rightarrow$
   (f) $Mg(OH)_{2(aq)} + H_3PO_{4(aq)} \rightarrow$

7. Write complete, balanced equations for each of the following decomposition reactions. Include symbols for physical states, heating, and others as needed:
   (a) When solid zinc carbonate is heated, solid zinc oxide and carbon dioxide gas are formed.
   (b) Liquid hydrogen peroxide decomposes to liquid water and oxygen gas.
   (c) Solid ammonium nitrite decomposes to liquid water and nitrogen gas.

8. Write complete, balanced equations for each of the following replacement reactions. If no reaction is predicted, write "no reaction" as the product:
   (a) $Zn_{(s)} + FeCl_{2(aq)} \rightarrow$
   (b) $Zn_{(s)} + AlCl_{3(aq)} \rightarrow$
   (c) $Cu_{(s)} + HgCl_{2(aq)} \rightarrow$
   (d) $Al_{(s)} + HCl_{(aq)} \rightarrow$

9. Write complete, balanced equations for each of the following ion exchange reactions. If no reaction is predicted, write "no reaction" as the product:
   (a) $Ca(OH)_{2(aq)} + H_2SO_{4(aq)} \rightarrow$
   (b) $NaCl_{(aq)} + AgNO_{3(aq)} \rightarrow$
   (c) $NH_4NO_{3(aq)} + Mg_3(PO_4)_{2(aq)} \rightarrow$
   (d) $Na_3PO_{4(aq)} + AgNO_{3(aq)} \rightarrow$

10. Iron(III) oxide, or hematite, is one mineral used as an iron ore. Other iron ores are magnetite (Fe_3O_4) and siderite ($FeCO_3$). Assume that you have pure samples of all three ores that will be reduced by reaction with carbon monoxide. Which of the three ores will have the highest yield of metallic iron?

Water is often referred to as the *universal solvent* because it makes so many different kinds of solutions. Eventually, moving water can dissolve solid rock, carrying it away in solution.

CHAPTER | Thirteen

Water and Solutions

The previous three chapters were concerned with elements, compounds, and their chemical reactions. Elements and compounds are pure substances, materials with a definite, fixed composition that is the same throughout. Mixtures, on the other hand, have a composition that may vary from one sample to the next. A mixture contains the particles of one substance physically dispersed throughout the particles of another substance. These particles can be any size, from the size of rock particles in a gravel mixture down to the smallest size possible—particles the size of ions or molecules. The size of the particles gives a mixture its appearance. When the particles are relatively large and visible to the eye, a mixture appears to be heterogeneous. Thus, a pile of gravel appears to be a heterogeneous mixture. At the other end of the size scale, a uniform mixture of ion- or molecule-sized particles appears to be homogeneous, or the same throughout. Such homogeneous mixtures are called *solutions*. This chapter is concerned with solutions, those involving water in particular.

Many common liquids are solutions. Solutions are commonly used in everyday household activities, as well as in the chemistry laboratory. The household activities of cooking, cleaning, and painting all involve solutions that will be considered in this chapter. Detergents, cleaners, and drain openers all function because a solution is a medium for rapid chemical change. Solids react slowly, if at all, because they have limited contact only at their immediate surfaces. Thus, solutions are commonly used to speed reactions. The water you drink is also a solution (figure 13.1). Hard water results in certain reactions that occur between soap and the water solution. This chapter considers hard water and how it is softened, in addition to acids, bases, the pH scale, and many common solutions used in everyday activities.

SOLUTIONS

A **solution** is defined as a homogeneous mixture of ions or molecules of two or more substances. The process of making a solution is called *dissolving*, and during dissolving the different components that make up the solution become mixed. The components could be sugar and water, for example, and when sugar dissolves in water, molecules of sugar are uniformly dispersed throughout the molecules of water. The uniform taste of sweetness of any part of the sugar solution is a result of this uniform mixing. In a salt and water solution, however, the salt dissolves into sodium and chlorine ions. The components of a salt and water solution are sodium and chlorine ions dissolved in water.

Solutions are not limited to solids, such as sugar or salt, dissolved in liquids, such as water. There are three states of matter, and a solution can involve any combination of the ions or molecules of gases, liquids, or solids (figure 13.2). Thus, it is possible to have nine kinds of solutions. Table 13.1 gives examples of some kinds of solutions.

The amounts of the components of a solution are identified by the general terms of *solvent* and *solute*. The solvent is the component present in the larger amount. The solute is the component that dissolves in the solvent. Atmospheric air, for example, is about 78 percent nitrogen, so nitrogen is considered the solvent. Oxygen (about 21 percent), argon (about 0.9 percent), and other gases make up the solutes.

If one of the components of a solution is a liquid, it is usually identified as the solvent. An *aqueous solution* is a solution of a solid, a liquid, or a gas in water. Water is the solvent in an aqueous solution. A *tincture* is a solution of something dissolved in alcohol. Tincture of iodine, for example, is iodine dissolved in the solvent alcohol.

Concentration of Solutions

The relative amounts of solute and solvent are described by the concentration of a solution. In general, a solution with a large amount of solute is *concentrated*, and a solution with much less solute is *dilute*. The terms "dilute" and "concentrated" are somewhat arbitrary, and it is sometimes difficult to know the difference between a solution that is "weakly concentrated" and one that is "not very diluted." More meaningful information is provided by measurement of the *amount of solute in a solution*. There are different ways to express concentration measurements, each lending itself to a particular kind of solution or to how the information will be used. For example, you read about concentrations of parts per million in an article about pollution, but most of the concentration of solutions sold in stores are reported in percent by volume or percent by weight (figure 13.3). Each of these concentrations is concerned with the amount of *solute* in the *solution*.

Concentration ratios that describe small concentrations of solute are sometimes reported as a ratio of *parts per million* (ppm) or *parts per billion* (ppb). This ratio could mean ppm by volume or ppm by weight, depending if the solution is a gas or a liquid. For example, a drinking water sample with 1 ppm Na^+ by weight has 1 weight measure of solute, sodium ions, *in* every 1,000,000 weight measures of the total solution. By way of analogy, 1 ppm expressed in money means 1 cent in every $10,000 (which is one million

FIGURE 13.1

Water is the most abundant liquid on the earth and is necessary for all life. Because of water's great dissolving properties, any sample is a solution containing solids, other liquids, and gases from the environment. This stream also carries suspended, ground-up rocks, called *rock flour,* from a nearby glacier.

TABLE 13.1	Examples of solutions
Kind of Solution	**Example**
Gas in gas	*Air* is O_2, CO_2, and other gases dissolved in nitrogen gas.
Liquid in gas	*Humid air* is water vapor dissolved in air.
Solid in gas	*Smoke* can be solid particles dissolved in air (many smoke particles are larger than molecules, so they are not part of the solution).
Gas in liquid	*Soda water* is CO_2 dissolved in water.
Liquid in liquid	*Alcohol* is alcohol molecules dissolved in water (unless 200 proof, which is pure alcohol).
Solid in liquid	*Seawater* is the ions of salts dissolved in water (mostly sodium chloride).
Liquid in solid	*Dental fillings* are prepared from a solution of mercury in silver.
Solid in solid	*Brass* is zinc dissolved in copper.

FIGURE 13.2

Above this city there are at least three kinds of solutions. These are (1) gas in gas—oxygen dissolved in nitrogen, (2) liquid in gas—water vapor dissolved in air, and (3) solid in gas—tiny particles of smoke dissolved in the air.

FIGURE 13.3

There are different ways to express the concentration of a solution. How many different ways can you identify in this photograph?

cents). A concentration of 1 ppb means 1 cent in $10,000,000. Thus, the concentrations of very dilute solutions, such as certain salts in seawater, minerals in drinking water, and pollutants in water or in the atmosphere are often reported in ppm or ppb.

Sometimes it is useful to know the conversion factors between ppm or ppb and the more familiar percent concentration by weight. These factors are ppm $\div (1 \times 10^4) =$ percent concentration and ppb $\div (1 \times 10^7) =$ percent concentration. For example, very hard water (water containing Ca^{2+} or Mg^{2+} ions), by definition, contains more than 300 ppm of the ions. This is a percent concentration of $300 \div 1 \times 10^4$, or 0.03 percent. To be suitable for agricultural purposes, irrigation water must not contain more than 700 ppm of total dissolved salts, which means a concentration no greater than 0.07 percent salts.

The concentration term of *percent by volume* is defined as the *volume of solute in 100 volumes of solution.* This concentration term is just like any other percentage ratio, that is, "part" divided by the "whole" times 100 percent. The distinction is that the part and the whole are concerned with a volume of solute and a volume of solution. Knowing the meaning of percent by

A Solution strength by parts

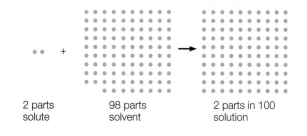

| 2 parts solute | 98 parts solvent | 2 parts in 100 solution |

B Solution strength by percent (volume)

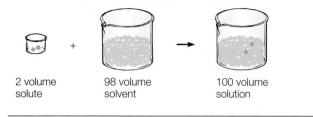

| 2 volume solute | 98 volume solvent | 100 volume solution |

C Solution strength by percent (weight)

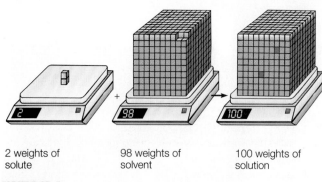

| 2 weights of solute | 98 weights of solvent | 100 weights of solution |

FIGURE 13.4

Three ways to express the amount of solute in a solution: (A) as parts (e.g., parts per million), this is 2 parts per 100; (B) as a percent by volume, this is 2 percent by volume; (C) as percent by weight, this is 2 percent by weight.

volume can be useful in consumer decisions. Rubbing alcohol, for example, can be purchased at a wide range of prices. The various brands range from a concentration, according to the labels, of "12% by volume" to "70% by volume." If the volume unit is mL, a "12% by volume" concentration contains 12 mL of pure isopropyl (rubbing) alcohol in every 100 mL of solution. The "70% by volume" contains 70 mL of isopropyl alcohol in every 100 mL of solution. The relationship for percent by volume is

$$\frac{\text{volume solute}}{\text{volume solution}} \times 100\% \text{ solution} = \% \text{ solute}$$

or

$$\frac{V_{\text{solute}}}{V_{\text{solution}}} \times 100\% \text{ solution} = \% \text{ solute}$$

equation 13.1

The concentration term of *percent by weight* is defined as the *weight of solute in 100 weight units of solution*. This concentration term is just like any other percentage composition, the difference being that it is concerned with the weight of solute (the part) in a weight of solution (the whole) (figure 13.4). Hydrogen peroxide, for example, is usually sold in a concentration of "3% by weight." This means that 3 oz (or other weight units) of pure hydrogen peroxide are in 100 oz of solution. Since weight is proportional to mass in a given location, mass units such as grams are sometimes used to calculate a percent by weight. The relationship for percent by weight (using mass units) is

$$\frac{\text{mass of solute}}{\text{mass of solution}} \times 100\% \text{ solution} = \% \text{ solute}$$

or

$$\frac{m_{\text{solute}}}{m_{\text{solution}}} \times 100\% \text{ solution} = \% \text{ solute}$$

equation 13.2

Q & A

Vinegar that is prepared for table use is a mixture of acetic acid in water, usually 5.00% by weight. How many grams of pure acetic acid are in 25.0 g of vinegar?

Answer

The percent by weight is given (5.00%), the mass of the solution is given (25.0 g), and the mass of the solute ($H_2C_2H_3O_2$) is the unknown. The relationship between these quantities is found in equation 13.2, which can be solved for the mass of the solute:

% solute = 5.00%

m_{solution} = 25.0 g

m_{solute} = ?

$$\frac{m_{\text{solute}}}{m_{\text{solution}}} \times 100\% \text{ solution} = \% \text{ solute}$$

$$\therefore$$

$$m_{\text{solute}} = \frac{(m_{\text{solution}})(\% \text{ solute})}{100\% \text{ solution}}$$

$$= \frac{(m_{\text{solution}})(\% \text{ solute})}{100\% \text{ solution}}$$

$$= \frac{(25.0 \text{ g})(5.00)}{100} \text{ solute}$$

$$= \boxed{1.25 \text{ g solute}}$$

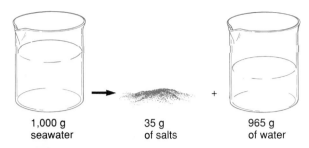

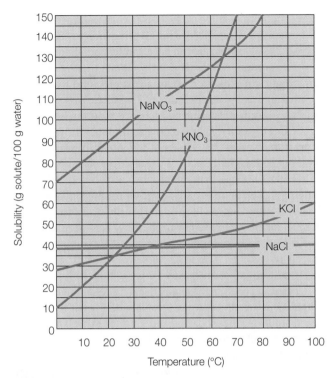

FIGURE 13.5

Salinity is a measure of the amount of salts dissolved in 1 kg of solution. If 1,000 g of seawater were evaporated, 35.0 g of salts would remain as 965.0 g of water leave.

A solution used to clean contact lenses contains 0.002% by volume of thimerosal as a preservative. How many mL of this preservative are needed to make 100,000 L of the cleaning solution? (Answer: 2.0 mL)

FIGURE 13.6

Approximate solubility curves for sodium nitrate, potassium nitrate, potassium chloride, and sodium chloride.

Both percent by volume and percent by weight are defined as the volume or weight per 100 units of solution because percent *means* parts per hundred. The measure of dissolved salts in seawater is called *salinity*. **Salinity** is defined as the mass of salts dissolved in 1,000 g of solution. As illustrated in figure 13.5, evaporation of 965 g of water from 1,000 g of seawater will leave an average of 35 g salts. Thus, the average salinity of the seawater is 35‰. Note the ‰, which means parts per thousand just as % means parts per hundred. Thus, the average salinity of seawater is 35‰, which means there are 35 g of salts dissolved in every 1,000 g of seawater. The equivalent percent measure for salinity is 3.5%, which equals 35‰.

Chemists use a measure of concentration that is convenient for considering chemical reactions of solutions. The measure is based on moles of solute since a mole is a known number of particles (atoms, molecules, or ions). The concentration term of **molarity (M)** is defined as the number of moles of solute dissolved in one liter of solution. Thus,

$$\text{Molarity (M)} = \frac{\text{moles of solute}}{\text{liters of solution}}$$

equation 13.3

An aqueous solution that is 1.0 M NaCl contains 1.0 M NaCl per liter of solution. To make such a solution you would place 58.5 g (1.0 mole) NaCl in a beaker, then add water to make 1 liter of solution.

Solubility

Gases and liquids appear to be soluble in all proportions, but there is an obvious limit to how much solid can be dissolved in a liquid. You may have noticed that a cup of hot tea will dissolve several teaspoons of sugar, but the limit of solubility is reached quickly in a glass of iced tea. The limit of how much sugar will dissolve seems to depend on the temperature of the tea. More sugar added to the cold tea after the limit is reached will not dissolve, and solid sugar granules begin to accumulate at the bottom of the glass. At this limit the sugar and tea solution is said to be *saturated*. Dissolving does not actually stop when a solution becomes saturated and undissolved sugar continues to enter the solution. However, dissolved sugar is now returning to the undissolved state at the same rate. The overall equilibrium condition of sugar dissolving and sugar coming out of solution is called a **saturated solution.** A saturated solution is a *state of equilibrium that exists between dissolving solute and solute coming out of solution.* You actually cannot see the dissolving and coming out of solution that occurs in a saturated solution because the exchanges are taking place with particles the size of molecules or ions.

Not all compounds dissolve as sugar does, and more or less of a given compound may be required to produce a saturated solution at a particular temperature. In general, the difficulty of dissolving a given compound is referred to as *solubility*. More specifically, the **solubility** of a solute is defined as the *concentration that is reached in a saturated solution at a particular temperature.* Solubility varies with the temperature as the sodium and potassium salt examples show in figure 13.6. These solubility curves describe the amount of solute required to reach the saturation equilibrium at a particular temperature. In general, the solubilities of most ionic solids increase with temperature, but there are exceptions. In addition, some salts release heat when dissolved in water, and other salts absorb heat when dissolved. The "instant cold pack" used for first aid is a bag of water containing a second bag of ammonium nitrate (NH_4NO_3). When the bag of ammonium nitrate is broken, the compound dissolves and absorbs heat.

You can usually dissolve more of a solid, such as salt or sugar, as the temperature of the water is increased. Contrary to what you might expect, gases usually become *less* soluble in water as the temperature increases. As a glass of water warms, small bubbles collect on the sides of the glass as dissolved air comes out of solution. The first bubbles that appear when warming a pot of water to boiling are also bubbles of dissolved air coming out of solution. This is why water that has been boiled usually tastes "flat." The dissolved air has been removed by the heating. The "normal" taste of water can be restored by pouring the boiled water back and forth between two glasses. The water dissolves more air during this process, restoring the usual taste.

Changes in pressure have no effect on the solubility of solids in liquids but greatly affect the solubility of gases. The fizz of an opened bottle or can of soda occurs because pressure is reduced on the beverage and dissolved carbon dioxide comes out of solution. In general, *gas solubility decreases with temperature and increases with pressure.* As usual, there are exceptions to this generalization.

WATER SOLUTIONS

Water is the one chemical that is absolutely essential for all organisms. Organisms use water to transport food and essential elements and to carry biological molecules in a solution. In fact, living organisms are cells filled with water solutions. Your foods are mostly water, with fruits and vegetables consisting of up to 95 percent water and meat consisting of 50 percent water. Your body consists of over 70 percent water by weight. It is the specific properties of water that make it so important for life, in particular the unusual ability of water to act as a solvent. The properties of a water molecule must be considered in order to account for the solvent abilities of water.

Properties of Water Molecules

A water molecule is composed of two atoms of hydrogen and one atom of oxygen joined by a polar covalent bond. The electron dot formula for a water molecule is

Notice the four pairs of electrons around the oxygen atom, consisting of two lone pairs and two bonding pairs. These four electron pairs are arranged in the direction of a tetrahedral arrangement (figure 13.7C). Since there are only two bonding pairs, however, a water molecule has a bent molecular arrangement, represented with a structural formula as follows:

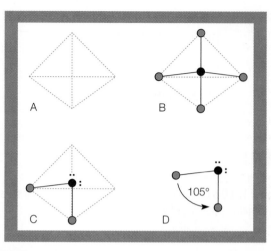

FIGURE 13.7

(*A*) A tetrahedron is a four-sided pyramid. (*B*) Molecules with four bonding electron pairs have a tetrahedral shape like that of the CCl_4 molecule illustrated here. Note that the carbon atom is in the center of the pyramid, with chlorine atoms at each of the four corners. (*C*) A water molecule has two bonding pairs and two lone pairs in a tetrahedral arrangement. (*D*) The water molecule has an angular arrangement called *bent*. If something were attached to the two lone pairs, it would be a tetrahedral arrangement (see also figure 13.9).

The angle between the two bonds is not 90° but has been experimentally measured to be about 105°, which is very close to the tetrahedral angle (figure 13.7D).

Oxygen has a stronger electronegativity (3.5) than hydrogen (2.1), and thus it has a greater ability to attract shared electrons. This results in the bonding electrons spending more time near the oxygen atom than the hydrogen atoms, essentially leaving the hydrogen as exposed protons on one end of the molecule. Thus, the water molecule has polar covalent bonds and is a *polar molecule* with centers of negative or positive charges.

The polar water molecule, with its negative oxygen end and positive hydrogen end, sets the stage for *intermolecular forces*, forces of interaction between molecules. The positive end of the water molecule can attract the negative end of another molecule, including the hydrogen end of another water molecule. The general term for weak attractive intermolecular forces is a *van der Waals force*, named after the Dutch physicist who first proposed the concept. Specifically, the intermolecular force of attraction between a hydrogen atom in a polar molecule and electrons around an electronegative atom such as oxygen is called a **hydrogen bond.** A hydrogen bond is a weak to moderate bond between the hydrogen end (+) of a polar molecule and the negative end (−) of a second polar molecule (figure 13.8). In general, hydrogen bonding occurs between the hydrogen atom of one molecule and the oxygen, fluorine, or nitrogen atom of another molecule.

Hydrogen bonding accounts for the physical properties of water, including its unusually high heat of fusion and heat of vaporization as well as its unusual density changes. Figure 13.9 shows the hydrogen bonded structure of ice. Each oxygen atom

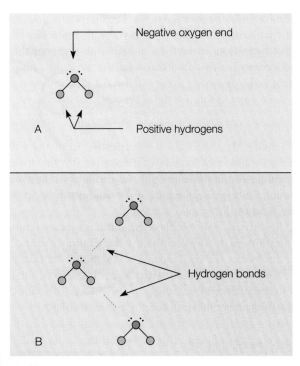

FIGURE 13.8

(*A*) The water molecule is polar, with centers of positive and negative charges. (*B*) Attractions between these positive and negative centers establish hydrogen bonds between adjacent molecules.

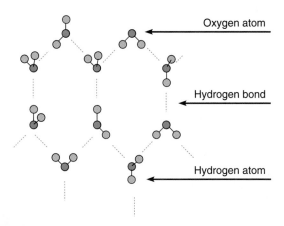

FIGURE 13.9

The hexagonal structure of ice. Hydrogen bonding between the oxygen atom and two hydrogen atoms of other water molecules results in a tetrahedral arrangement, which forms the open, hexagonal structure of ice.

is associated with four hydrogen atoms, two in the H_2O molecule and two from other water molecules, held by hydrogen bonds. The arrangement is tetrahedral, forming a six-sided hexagonal structure. The open space of the hexagonal channel in this structure results in ice being less dense than water. The shape of the channel also suggests why snowflakes always have six sides.

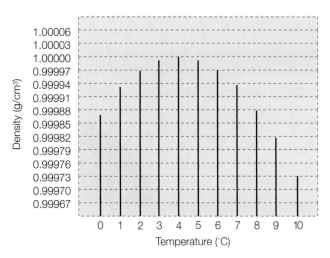

FIGURE 13.10

The density of water from 0°C to 10°C. The density of water is at a maximum at 4°C, becoming less dense as it is cooled or warmed from this temperature. Hydrogen bonding explains this unusual behavior.

When ice is warmed, the increased vibrations of the molecules begin to expand and stretch the hydrogen bond structure. When ice melts, about 15 percent of the hydrogen bonds break and the open structure collapses into the more compact arrangement of liquid water. As the liquid water is warmed from 0°C still more hydrogen bonds break down, and the density of the water steadily increases. At 4°C the expansion of water from the increased molecular vibrations begins to predominate and the density decreases steadily with further warming (figure 13.10). Thus, water has its greatest density at a temperature of 4°C.

The heat of fusion, specific heat, and heat of vaporization of water are unusually high when compared to other, but chemically similar, substances. These high values are accounted for by the additional energy needed to break hydrogen bonds.

The Dissolving Process

A solution is formed when the molecules or ions of two or more substances become homogeneously mixed. But the process of dissolving must be more complicated than the simple mixing together of particles because (1) solutions become saturated, meaning there is a limit on solubility, and (2) some substances are *insoluble*, not dissolving at all or at least not noticeably. In general, the forces of attraction between molecules or ions of the solvent and solute determine if something will dissolve and any limits on the solubility. These forces of attraction and their role in the dissolving process will be considered in the following examples.

First, consider the dissolving process in gaseous and liquid solutions. In a gas, the intermolecular forces are small, so gases can mix in any proportion. Fluids that can mix in any proportion like this are called **miscible fluids.** Fluids that do not mix are called *immiscible fluids.* Air is a mixture of gases and vapors, so gases and vapors are miscible.

Liquid solutions can have a gas, another liquid, or a solid as a solute. Gases are miscible in liquids, and a carbonated beverage

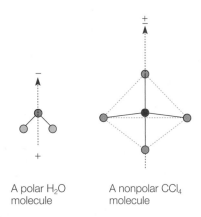

A polar H_2O molecule   A nonpolar CCl_4 molecule

FIGURE 13.11

Water is polar, and carbon tetrachloride is nonpolar. Since like dissolves like, water and carbon tetrachloride are immiscible.

(your favorite cola) is the common example, consisting of carbon dioxide dissolved in water. Whether or not two given liquids form solutions depends on some similarities in their molecular structures. The water molecule, for example, is a polar molecule with a negative end and a positive end. On the other hand, carbon tetrachloride (CCl_4) is a molecule with polar bonds that are symmetrically arranged. Because of the symmetry, CCl_4 has no negative or positive ends, so it is nonpolar. Thus some liquids have polar molecules, and some have nonpolar molecules. The general rule for forming solutions is *like dissolves like* (figure 13.11). A nonpolar compound, such as carbon tetrachloride, will dissolve oils and greases because they are nonpolar compounds. Water, a polar compound, will not dissolve the nonpolar oils and greases. Carbon tetrachloride was at one time used as a cleaning solvent because of its oil and grease dissolving abilities. Its use is no longer recommended because it is also a possible health hazard (liver damage).

Some molecules, such as ethyl alcohol and soap, have a part of the molecule that is polar and a part that is nonpolar. Washing with water alone will not dissolve oils because water and oil are immiscible. When soap is added to the water, however, the polar end of the soap molecule is attracted to the polar water molecules, and the nonpolar end is absorbed into the oil. A particle (larger than a molecule) is formed, and the oil is washed away with the water.

The "like dissolves like" rule applies to solids and liquid solvents as well as liquids and liquid solvents. Polar solids, such as salt, will readily dissolve in water, which has polar molecules, but do not dissolve readily in oil, grease, or other nonpolar solvents. Polar water readily dissolves salt because the charged polar water molecules are able to exert an attraction on the ions, pulling them away from the crystal structure. Thus, ionic compounds dissolve in water.

As noted in figure 13.6, ionic compounds vary in their solubilities in water. This difference is explained by the existence of two different forces involved in an ongoing "tug of war." One force is the attraction between an ion on the surface of the crystal and a water molecule, an *ion-polar molecule force*. When solid sodium chloride and water are mixed together, the negative ends of the water molecules (the oxygen ends) become oriented toward the positive sodium ions on the crystal. Likewise, the positive ends of

water molecules (the hydrogen ends) become oriented toward the negative chlorine ions. The attraction of water molecules for ions is called **hydration.** If the force of hydration is greater than the attraction between the ions in the solid, they are pulled away from the solid, and dissolving occurs (figure 13.12). Not considering the role of water in this dissolving process, the equation is

$$Na^+Cl^-(s) \rightarrow Na^+(aq) + Cl^-(aq)$$

which shows that the ions were separated from the solid to become a solution of ions. In other compounds the attraction between the ions in the solid might be greater than the energy of hydration. In this case, the ions of the solid would win the "tug of war," and the ionic solid is insoluble.

The saturation of soluble compounds is explained in terms of hydration eventually occupying a large number of the polar water molecules. Fewer available water molecules means less attraction on the ionic solid, with more solute ions being pulled back to the surface of the solid. The tug of war continues back and forth as an equilibrium condition is established.

Properties of Solutions

Pure solvents have characteristic physical and chemical properties that are changed by the presence of the solute. Following are some of the more interesting changes.

Electrolytes

Water solutions of ionic substances will conduct an electric current, so they are called **electrolytes.** Ions must be present and free to move in a solution to carry the charge, so electrolytes are solutions containing ions. Pure water will not conduct an electric current as it is a covalent compound, which ionizes only very slightly. Water solutions of sugar, alcohol, and most other covalent compounds are nonconductors, so they are called *nonelectrolytes.* Nonelectrolytes are covalent compounds that form molecular solutions, so they cannot conduct an electric current (figure 13.13).

Some covalent compounds are nonelectrolytes as pure liquids but become electrolytes when dissolved in water. Pure hydrogen chloride (HCl), for example, does not conduct an electric current, so you can assume that it is a molecular substance. When dissolved in water, hydrogen chloride does conduct a current, so it must now contain ions. Evidently, the hydrogen chloride has become *ionized* by the water. The process of forming ions from molecules is called *ionization.* Hydrogen chloride, just as water, has polar molecules. The positive hydrogen atom on the HCl molecule is attracted to the negative oxygen end of a water molecule, and the force of attraction is strong enough to break the hydrogen-chlorine bond, forming charged particles (figure 13.14). The reaction is

$$HCl(l) + H_2O(l) \rightarrow H_3O^+(aq) + Cl^-(aq)$$

The H_3O^+ ion is called a **hydronium ion.** A hydronium ion is basically a molecule of water with an attached hydrogen ion. The presence of the hydronium ion gives the solution new chemical properties; the solution is no longer hydrogen chloride but is *hydrochloric acid.* Hydrochloric acid, and other acids, will be discussed shortly.

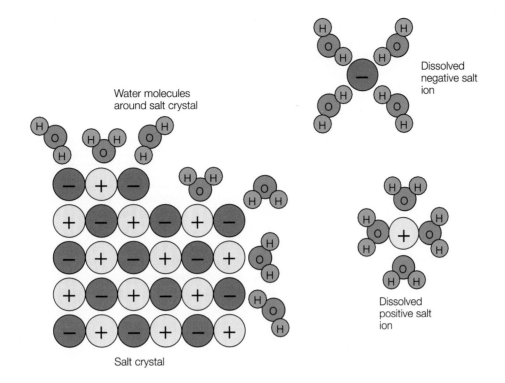

FIGURE 13.12

An ionic solid dissolves in water because the number of water molecules around the surface is greater than the number of other ions of the solid. The attraction between polar water molecules and a charged ion enables the water molecules to pull ions away from the crystal, a process called *dissolving*.

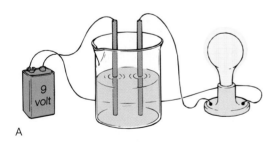

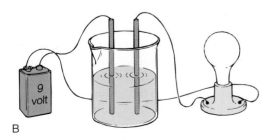

FIGURE 13.13

(*A*) Water solutions that conduct an electric current are called *electrolytes*. (*B*) Water solutions that do not conduct electricity are called *nonelectrolytes*.

FIGURE 13.14

Three representations of water and hydrogen chloride in an ionizing reaction. (*A*) Sketches of molecules involved in the reaction. (*B*) Electron dot equation of the reaction. (*C*) The chemical equation for the reaction. Each of these representations shows the hydrogen being pulled away from the chlorine atom to form H_3O^+, the hydronium ion.

A CLOSER LOOK | Domestic Water Use

Water is an essential resource, not only because it is required for life processes but also because of its role in a modern society. Water is used in the home for drinking and cooking (2%), cleaning dishes (6%), laundry (11%), bathing (23%), toilets (29%), and for maintaining lawns and gardens (29%).

The water supply is obtained from streams, lakes, and reservoirs on the surface, or from groundwater pumped from below the surface. Surface water contains more sediments, bacteria, and possible pollutants than water from a well because it is exposed to the atmosphere and it is also more active. Surface water requires filtering to remove suspended particles, treatment to kill bacteria, and sometimes processing to remove pollution. Well water, on the other hand, might require treatment to kill bacteria and processing to remove pollution that has seeped through the ground from waste dumps or from industrial sites.

Most pollutants are usually too dilute to be considered a significant health hazard, but there are exceptions. There are five types of contamination found in U.S. drinking water that are responsible for the most widespread danger, and these are listed box table 13.1. In spite of these widespread concerns and other occasional local problems, the U.S. water supply is considered to be among the cleanest in the world.

BOX TABLE 13.1

Possible pollution problems in the U.S. water supply

Pollutant	Source	Risk
Lead	Lead pipes in older homes; solder in copper pipes, brass fixtures	Nerve damage, miscarriage, birth defects, high blood pressure, hearing problems
Chlorinated solvents	Industrial pollution	Cancer
Trihalomethanes	Chlorine disinfectant reacting with other pollutants	Liver damage, kidney damage, possibly cancer
PCBs	Industrial waste, older transformers	Liver damage, possibly cancer
Bacteria and viruses	Septic tanks, outhouses, overflowing sewer lines	Gastrointestinal problems, serious diseases

Demand for domestic water sometimes exceeds the immediate supply in some growing metropolitan areas. Demand also reaches the limits of available water in many areas during the summer, when water demand is high and rainfall is often low. Communities in these areas often have public education campaigns designed to help reduce the demand for water. For example, did you know that taking a tub bath can use up to 135 liters (about 36 gal) of water compared to only 95 liters (about 25 gal) for a regular shower? Even more water is saved by a shower that does not run continuously—wetting down, soaping up, and rinsing off uses only 15 liters (about 4 gal) of water. You can also save about 35 liters (about 9 gal) of water by not letting the water run continuously while brushing your teeth.

It is often difficult to convince people to conserve water when it seems to be an inexpensive, limitless supply. However, history has shown that efforts to conserve water increase dramatically as the cost of maintaining an adequate domestic supply of water increases.

Boiling Point

Boiling occurs when the pressure of the vapor escaping from a liquid is equal to the atmospheric pressure on the liquid. The *normal* boiling point is defined as the temperature at which the vapor pressure is equal to the average atmospheric pressure at sea level. For pure water, this temperature is 100°C (212°F). It is important to remember that boiling is a purely physical process. No bonds within water molecules are broken during boiling.

During the later 1880s, a French chemist named Francois Raoult observed that the vapor pressure over a solution is *less* than the vapor pressure over the pure solvent at the same temperature. Molecules of a liquid can escape into the air only at the surface of the liquid, and the presence of molecules of a solute means that fewer solvent molecules can be at the surface to escape. Thus, the vapor pressure over a solution is less than the vapor pressure over a pure solvent (figure 13.15).

Because the vapor pressure over a solution is less than that over the pure solvent, the solution boils at a higher temperature. A higher temperature is required to increase the vapor pressure to that of the atmospheric pressure. Some cooks have been observed to add a "pinch" of salt to a pot of water before boiling. Is this to increase the boiling point, and therefore cook the food more quickly? How much does a pinch of salt increase the boiling temperature? The answers are found in the relationship between the concentration of a solute and the boiling point of the solution.

It is the number of solute particles (ions or molecules) at the surface of a solution that increases the boiling point. Recall from chapter 12 that a mole is a measure that can be defined as a number of particles called Avogadro's number. Since the number of particles at the surface is proportional to the ratio of particles in the solution, the concentration of the solute will directly influence the increase in the boiling point. In other words, the

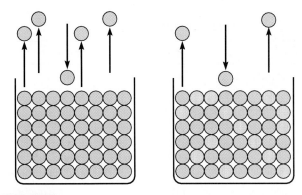

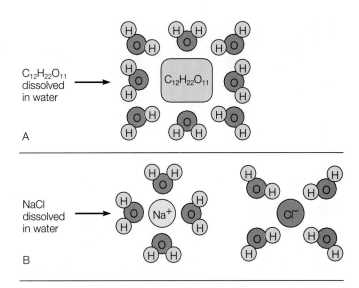

FIGURE 13.15

The rate of evaporation, and thus the vapor pressure, is less for a solution than for a solvent in the pure state. The greater the solute concentration, the less the vapor pressure.

boiling point of any dilute solution is increased proportional to the concentration of the solute. For water, the boiling point is increased 0.521°C for every mole of solute dissolved in 1,000 g of water. Thus, any water solution will boil at a higher temperature than pure water. Since it boils at a higher temperature, it also takes a longer time to reach the boiling point.

It makes no difference what substance is dissolved in the water; one mole of solute in 1,000 g of water will elevate the boiling point by 0.521°C. A mole contains Avogadro's number of particles, so a mole of any solute will lower the vapor pressure by the same amount. Sucrose, or table sugar, for example, is $C_{12}H_{22}O_{11}$ and has a gram-formula weight of 342 g. Thus 342 g of sugar in 1,000 g of water (about a liter) will increase the boiling point by 0.521°C. Therefore, if you measure the boiling point of a sugar solution you can determine the concentration of sugar in the solution. For example, pancake syrup that boils at 100.261°C (sea-level pressure) must contain 171 g of sugar dissolved in 1,000 g of water. You know this because the increase of 0.261°C over 100°C is one-half of 0.521°C. If the boiling point were increased by 0.521°C over 100°C, the syrup would have the full gram-formula weight (342 g) dissolved in a kg of water.

Since it is the number of particles of solute in a specific sample of water that elevates the boiling point, different effects are observed in dissolved covalent and dissolved ionic compounds (figure 13.16). Sugar is a covalent compound, and the solute is molecules of sugar moving between the water molecules. Sodium chloride, on the other hand, is an ionic compound and dissolves by the separation of ions, or

$$Na^+Cl^-(s) \rightarrow Na^+(aq) + Cl^-(aq)$$

This equation tells you that one mole of NaCl separates into one mole of sodium ions and one mole of chlorine ions for a total of *two* moles of solute. The boiling point elevation of a solution made from one mole of NaCl (58.5 g) is therefore multiplied by two, or $2 \times 0.521°C = 1.04°C$. The boiling point of a solution made by adding 58.5 g of NaCl to 1,000 g of water is therefore 101.04°C at normal sea-level pressure.

Now back to the question of how much a pinch of salt increases the boiling point of a pot of water. Assuming the pot

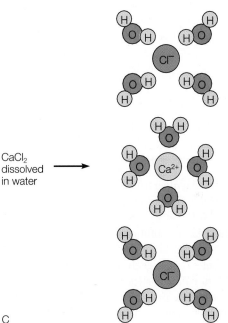

FIGURE 13.16

Since ionic compounds dissolve by separation of ions, they provide more particles in solution than molecular compounds. (A) A mole of sugar provides Avogadro's number of particles. (B) A mole of NaCl provides two times Avogadro's number of particles. (C) A mole of $CaCl_2$ provides three times Avogadro's number of particles.

contains about a liter of water (about a quart), and assuming that a "pinch" of salt has a mass of about 0.2 gram, the boiling point will be increased by 0.0037°C. Thus, there must be some reason other than increasing the boiling point that a cook adds a pinch of salt to a pot of boiling water.

Freezing Point

Freezing occurs when the kinetic energy of molecules has been reduced sufficiently so the molecules can come together, forming the crystal structure of the solid. Reduced kinetic energy of the

TABLE 13.2	Some common acids	
Name	**Formula**	**Comment**
Acetic acid	CH_3COOH	A weak acid found in vinegar
Boric acid	H_3BO_3	A weak acid used in eyedrops
Carbonic acid	H_2CO_3	The weak acid of carbonated beverages
Formic acid	$HCOOH$	Makes the sting of insects and certain plants
Hydrochloric acid	HCl	Also called muriatic acid; used in swimming pools, soil acidifiers, and stain removers
Lactic acid	$CH_3CHOHCOOH$	Found in sour milk, sauerkraut, and pickles; gives tart taste to yogurt
Nitric acid	HNO_3	A strong acid
Phosphoric acid	H_3PO_4	Used in cleaning solutions; added to carbonated beverages for tartness
Sulfuric acid	H_2SO_4	Also called oil of vitriol; used as battery acid and in swimming pools

TABLE 13.3	Some common bases	
Name	**Formula**	**Comment**
Sodium hydroxide	$NaOH$	Also called lye or caustic soda; a strong base used in oven cleaners and drain cleaners
Potassium hydroxide	KOH	Also called caustic potash; a strong base used in drain cleaners
Ammonia	NH_3	A weak base used in household cleaning solutions
Calcium hydroxide	$Ca(OH)_2$	Also called slaked lime; used to make brick mortar
Magnesium hydroxide	$Mg(OH)_2$	Solution is called milk of magnesia; used as antacid and laxative

molecules, that is, reduced temperature, results in a specific freezing point for each pure liquid. The *normal* freezing point for pure water, for example, is 0°C (32°F) under normal pressure. The presence of solute particles in a solution interferes with the water molecules as they attempt to form the six-sided hexagonal structure. The water molecules cannot get by the solute particles until the kinetic energy of the solute particles is reduced, that is, until the temperature is below the normal freezing point. Thus, the presence of solute particles lowers the freezing point, and solutions freeze at a lower temperature than the pure solvent.

The freezing-point depression of a solution has a number of interesting implications for solutions such as seawater. When seawater freezes, the water molecules must work their way around the salt particles as was described earlier. Thus, the solute particles are *not* normally included in the hexagonal structure of ice. Ice formed in seawater is practically pure water. Since the solute was *excluded* when the ice formed, the freezing of seawater increases the salinity. Increased salinity means increased concentration, so the freezing point of seawater is further depressed and more ice forms only at a lower temperature. When this additional ice forms, more pure water is removed and the process goes on. Thus, seawater does not have a fixed freezing point but has a lower and lower freezing point as more and more ice freezes.

The depression of the freezing point by a solute has a number of interesting applications in colder climates. Salt, for example, is spread on icy roads to lower the freezing point (and thus the melting point) of the ice. Calcium chloride, $CaCl_2$, is a salt that is often used for this purpose. Water in a car radiator would

also freeze in colder climates if a solute, called antifreeze, were not added to the radiator water. Methyl alcohol has been used as an antifreeze because it is soluble in water and does not damage the cooling system. Methyl alcohol, however, has a low boiling point and tends to boil away. Ethylene glycol has a higher boiling point, so it is called a "permanent" antifreeze. Like other solutes, ethylene glycol also raises the boiling point, which is an added benefit for summer driving.

ACIDS, BASES, AND SALTS

The electrolytes known as *acids, bases,* and *salts* are evident in environmental quality, foods, and everyday living. Environmental quality includes the hardness of water, which is determined by the presence of certain salts, the acidity of soils, which determines how well plants grow, and acid rain, which is a by-product of industry and automobiles. Many concerns about air and water pollution are often related to the chemistry concepts of acids, bases, and salts. These concepts, and uses of acids, bases, and salts, will be considered in this section.

Properties of Acids and Bases

Acids and bases are classes of chemical compounds that have certain characteristic properties. These properties can be used to identify if a substance is an acid or a base (tables 13.2 and 13.3). The following are the properties of *acids* dissolved in water:

1. Acids have a sour taste such as the taste of citrus fruits.
2. Acids change the color of certain substances; for example, litmus changes from blue to red when placed in an acid solution (figure 13.17A).
3. Acids react with active metals, such as magnesium or zinc, releasing hydrogen gas.
4. Acids *neutralize* bases, forming water and salts from the reaction.

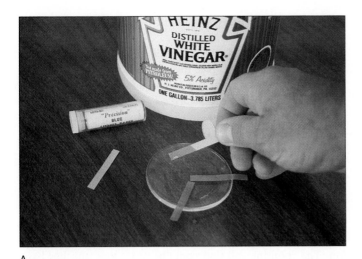

A

B

FIGURE 13.17

(*A*) Acid solutions will change the color of blue litmus to red. (*B*) Solutions of bases will change the color of red litmus to blue.

Likewise, *bases* have their own characteristic properties. Bases are also called alkaline substances, and the following are the properties of bases dissolved in water:

1. Bases have a bitter taste, for example, the taste of caffeine.
2. Bases reverse the color changes that were caused by acids. Red litmus is changed back to blue when placed in a solution containing a base (figure 13.17B).
3. Basic solutions feel slippery on the skin. They have a *caustic* action on plant and animal tissue, converting tissue into soluble materials. A strong base, for example, reacts with fat to make soap and glycerine. This accounts for the slippery feeling on the skin.
4. Bases *neutralize* acids, forming water and salts from the reaction.

Tasting an acid or base to see if it is sour or bitter can be hazardous, since some are highly corrosive or caustic. Many organic acids are not as corrosive and occur naturally in foods. Citrus fruit, for example, contains citric acid, vinegar is a solution of acetic acid, and sour milk contains lactic acid. The stings or bites of some insects (bees, wasps, and ants) and some plants (stinging nettles) are painful because an organic acid, formic acid, is injected by the insect or plant. Your stomach contains a solution of hydrochloric acid. In terms of relative strength, the hydrochloric acid in your stomach is about ten times stronger than the carbonic acid (H_2CO_3) of carbonated beverages.

Examples of bases include solutions of sodium hydroxide (NaOH), which has a common name of lye or caustic soda, and potassium hydroxide (KOH), which has a common name of caustic potash. These two bases are used in products known as drain cleaners. They open plugged drains because of their caustic action, turning grease, hair, and other organic "plugs" into soap and other soluble substances that are washed away. A weaker base is a solution of ammonia (NH_3), which is often used as a household cleaner. A solution of magnesium hydroxide,

$Mg(OH)_2$, has a common name of milk of magnesia and is sold as an antacid and laxative.

Many natural substances change color when mixed with acids or bases. You may have noticed that tea changes color slightly, becoming lighter, when lemon juice (which contains citric acid) is added. Some plants have flowers of one color when grown in acidic soil and flowers of another color when grown in basic soil. A vegetable dye that changes color in the presence of acids or bases can be used as an *acid-base indicator.* An indicator is simply a vegetable dye that is used to distinguish between acid and base solutions by a color change. Litmus, for example, is an acid-base indicator made from a dye extracted from certain species of lichens. The dye is applied to paper strips, which turn red in acidic solutions and blue in basic solutions.

Activities

To see how acids and bases change the color of certain vegetable dyes, consider the dye that gives red cabbage its color. Shred several leaves of red cabbage and boil them in a pan of water to extract the dye. After you have a purple solution, squeeze the juice from the cabbage into the pan and allow the solution to cool. Add vinegar in small amounts as you stir the solution, continuing until the color changes. Add ammonia in small amounts, stirring until the color changes again. Reverse the color change again by adding vinegar in small amounts. Will this purple cabbage acid-base indicator tell you if other substances are acids or bases?

Explaining Acid-Base Properties

Comparing the lists in tables 13.2 and 13.3, you can see that acids and bases appear to be chemical opposites. Notice in table 13.2 that the acids all have an H, or hydrogen atom, in their formulas. In table 13.3, most of the bases have a hydroxide ion, OH^-, in their formulas. Could this be the key to acid-base properties?

Lavoisier was one of the first to attempt an explanation for the properties of acids. Lavoisier thought that all acids contained oxygen, so in 1777 he proposed the name *oxygen*, which means "acid former" in Greek. The formulas of acids listed in table 13.2 show that most acids do contain oxygen, with the exception of hydrochloric acid, which is HCl. During Lavoisier's time chlorine was believed to be an oxygen compound. Not until 1810 was chlorine discovered as an element. The modern understanding of acids and bases originated with the introduction of a *theory of ionization,* developed by Svante Arrhenius in 1883. Arrhenius proposed that electrolytes such as acids, bases, and salts produced equal numbers of positive and negative ions when dissolved in dilute solutions. In concentrated solutions, these ions were in equilibrium with solute molecules that were not ionized.

Chemical equilibrium occurs when two opposing reactions happen at the same time and at the same rate. Pure water, for example, ionizes very slightly to produce a hydronium ion, H_3O^+, and a hydroxide ion, OH^-, from the dissociation of a water molecule:

$$H_2O(l) + H_2O(l) \rightarrow H_3O^+(aq) + OH^-(aq)$$

This is termed the *forward reaction,* which occurs to about 1 molecule in 500 million molecules of pure water. The H_3O^+ and OH^- now react in a *reverse reaction* to produce a molecule of water. Chemical equilibrium occurs when the forward reaction occurs at the same rate as the reverse reaction. This is shown by a double arrow:

$$H_2O(l) + H_2O(l) \rightleftharpoons H_3O^+(aq) + OH^-(aq)$$

In a condition of chemical equilibrium the concentration of molecules and ions remains constant, even though both reactions are always occurring.

The modern concept of an acid considers the properties of acids in terms of the hydronium ion, H_3O^+. As was mentioned earlier, the hydronium ion is a water molecule to which an H+ ion is attached. Since a hydrogen ion is a hydrogen atom without its single electron, it could be considered as an ion consisting of a single proton. Thus the H^+ ion can be called a *proton.* An **acid** is defined as any substance that is a *proton donor* when dissolved in water, increasing the hydronium ion concentration. For example, hydrogen chloride dissolved in water has the following reaction:

The dotted circle and arrow were added to show that the hydrogen chloride donated a proton to a water molecule. The result-

ing solution contains H_3O^+ ions and has acid properties, so the solution is called hydrochloric acid.

The bases listed in table 13.3 all appear to have a hydroxide ion, OH^-. Water solutions of these bases do contain OH^- ions, but the definition of a base is much broader. A **base** is defined as any substance that is a *proton acceptor* when dissolved in water, increasing the hydroxide ion concentration. For example, ammonia dissolved in water has the following reaction:

The dotted circle and arrow show that the ammonia molecule accepted a proton from a water molecule, providing a hydroxide ion. The resulting solution contains OH^- ions and has basic properties, so a solution of ammonium hydroxide is a base.

Carbonates, such as sodium carbonate (Na_2CO_3), form basic solutions because the carbonate ion reacts with water to produce hydroxide ions.

$$(CO_3)^{2-}(aq) + H_2O(l) \rightarrow (HCO_3)^-(aq) + OH^-(aq)$$

Thus, sodium carbonate produces a basic solution.

Acids could be thought of as simply solutions of hydronium ions in water, and bases could be considered solutions of hydroxide ions in water. The proton donor and proton acceptor definition is much broader, and it does include the definition of acids and bases as hydronium and hydroxide compounds. The broader, more general definition covers a wider variety of reactions and is therefore more useful.

The modern concept of acids and bases explains why the properties of acids and bases are **neutralized,** or lost, when acids and bases are mixed together. For example, consider the hydronium ion produced in the hydrochloric acid solution and the hydroxide ion produced in the ammonia solution. When these solutions are mixed together, the hydronium ion reacts with the hydroxide ion,

$$H_3O^+(aq) + OH^+(aq) \rightarrow H_2O(l) + H_2O(l)$$

Thus, a proton is transferred from the hydronium ion (an acid), and the proton is accepted by the hydroxide ion (a base). A molecule of water is produced, and both the acid and base properties disappear or are neutralized.

Strong and Weak Acids and Bases

Acids and bases are classified according to their degree of ionization when placed in water. *Strong acids* ionize completely in water, with all molecules dissociating into ions. Nitric acid, for example, reacts completely in the following equation:

$$HNO_3(aq) + H_2O(l) \rightarrow H_3O^+(aq) + (NO_3)^-(aq)$$

Nitric acid, hydrochloric acid (figure 13.18), and sulfuric acid are common strong acids.

Acids that react only partially produce fewer hydronium ions, so they are weaker acids. *Weak acids* are only partially ionized because of an equilibrium reaction with water. Vinegar, for

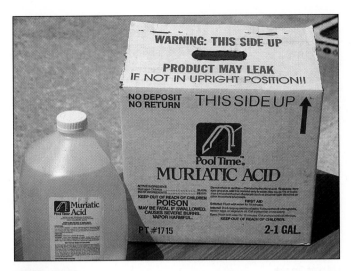

FIGURE 13.18

Hydrochloric acid (HCl) has the common name of *muriatic* acid. Hydrochloric acid is a strong acid used in swimming pools, soil acidifiers, and stain removers.

TABLE 13.4	The pH and hydronium ion concentration (moles/L)	
Hydronium Ion Concentration (moles/L)	Reciprocal of Hydronium Ion Concentration	pH
1×10^{0}	1×10^{0}	0
1×10^{-1}	1×10^{1}	1
1×10^{-2}	1×10^{2}	2
1×10^{-3}	1×10^{3}	3
1×10^{-4}	1×10^{4}	4
1×10^{-5}	1×10^{5}	5
1×10^{-6}	1×10^{6}	6
1×10^{-7}	1×10^{7}	7
1×10^{-8}	1×10^{8}	8
1×10^{-9}	1×10^{9}	9
1×10^{-10}	1×10^{10}	10
1×10^{-11}	1×10^{11}	11
1×10^{-12}	1×10^{12}	12
1×10^{-13}	1×10^{13}	13
1×10^{-14}	1×10^{14}	14

example, contains acetic acid that reacts with water in the following reaction:

$$HC_2H_3O_2 + H_2O \rightleftharpoons H_3O^+ + (C_2H_3O_2)^-$$

The double yield arrows indicate an equilibrium reaction, which means that at any given time, not many of the hydronium ions are in solution. In fact, only about 1 percent or less of the acetic acid molecules ionize, depending on the concentration.

Bases are also classified as strong or weak. A *strong base* is completely ionic in solution and has hydroxide ions. Sodium hydroxide, or lye, is the most common example of a strong base. It dissolves in water to form sodium and hydroxide ions:

$$Na^+OH^-(s) \rightarrow Na^+(aq) + OH^-(aq)$$

A *weak base* is only partially ionized because of an equilibrium reaction with water. Ammonia, magnesium hydroxide, and calcium hydroxide are examples of weak bases. Magnesium and calcium hydroxide are only slightly soluble in water, and this reduces the *concentration* of hydroxide ions in a solution. It would appear that $Ca(OH)_2$ would produce two moles of hydroxide ions. It would, if it were completely soluble and reacted completely. It is the concentration of hydroxide ions in solution that determines if a base is weak or strong, not the number of ions per mole.

The pH Scale

The strength of an acid or a base is usually expressed in terms of a range of values called a **pH scale.** The pH scale is based on the concentration of the hydronium ion (in moles/L) in an acidic or a basic solution. To understand how the scale is able to express both acid and base strength in terms of the hydronium ion, first recall that pure water is very slightly ionized in the equilibrium reaction:

$$H_2O(l) + H_2O(l) \rightleftharpoons H_3O^+(aq) \ 1 \ OH^-(aq)$$

The amount of self-ionization by water has been determined through measurements. In pure water at 25°C or any neutral water solution at that temperature, the H_3O^+ concentration is 1×10^{-7} moles/L, and the OH^- concentration is also 1×10^{-7} moles/L. Since both ions are produced in equal numbers, then the H_3O^+ concentration equals the OH^- concentration, and pure water is neutral, neither acidic nor basic.

In general, adding an acid substance to pure water increases the H_3O^+ concentration. Adding a base substance to pure water increases the OH^- concentration. Adding a base also *reduces* the H_3O^+ concentration as the additional OH^- ions are able to combine with more of the hydronium ions to produce un-ionized water. Thus, at a given temperature, an increase in OH^- concentration is matched by a *decrease* in H_3O^+ concentration. The concentration of the hydronium ion can be used as a measure of acidic, neutral, and basic solutions. In general, (1) acidic solutions have H_3O^+ concentrations above 1×10^{-7} moles/L, (2) neutral solutions have H_3O^+ concentrations equal to 1×10^{-7} moles/L, and (3) basic solutions have H_3O^+ concentrations less than 1×10^{-7} moles/L. These three statements lead directly to the pH scale, which is named from the French *pouvoir hydrogene,* meaning "hydrogen power." Power refers to the exponent of the hydronium ion concentration, and the pH is a *power of ten notation that expresses the H_3O^+ concentration* (table 13.4).

A neutral solution has a pH of 7.0. Acidic solutions have pH values below 7, and smaller numbers mean greater acidic properties. Increasing the OH^- concentration decreases the H_3O^+ concentration, so the strength of a base is indicated on the same

H$_3$O$^+$ Concentration (moles/liters)	pH	Meaning
1 X 10^{-0} (=1)	0	↑ Increasing acidity
1 X 10^{-1}	1	
1 X 10^{-2}	2	
1 X 10^{-3}	3	
1 X 10^{-4}	4	
1 X 10^{-5}	5	
1 X 10^{-6}	6	
1 X 10^{-7}	7	Neutral
1 X 10^{-8}	8	
1 X 10^{-9}	9	
1 X 10^{-10}	10	↓ Increasing basicity
1 X 10^{-11}	11	
1 X 10^{-12}	12	
1 X 10^{-13}	13	
1 X 10^{-14}	14	

FIGURE 13.19
The pH scale.

TABLE 13.5	The approximate pH of some common substances
Substance	**pH (or pH Range)**
Hydrochloric acid (4%)	0
Gastric (stomach) solution	1.6–1.8
Lemon juice	2.2–2.4
Vinegar	2.4–3.4
Carbonated soft drinks	2.0–4.0
Grapefruit	3.0–3.2
Oranges	3.2–3.6
Acid rain	4.0–5.5
Tomatoes	4.2–4.4
Potatoes	5.7–5.8
Natural rainwater	5.6–6.2
Milk	6.3–6.7
Pure water	7.0
Seawater	7.0–8.3
Blood	7.4
Sodium bicarbonate solution	8.4
Milk of magnesia	10.5
Ammonia cleaning solution	11.9
Sodium hydroxide solution	13.0

scale with values greater than 7. Note that the pH scale is logarithmic, so a pH of 2 is ten times as acidic than a pH of 3. Likewise, a pH of 10 is one hundred times as basic than a pH of 8. Figure 13.19 is a diagram of the pH scale, and table 13.5 compares the pH of some common substances (figure 13.20).

FIGURE 13.20
The pH increases as the acidic strength of these substances decreases from left to right. Did you know that lemon juice is more acidic than vinegar? That a soft drink is more acidic than orange juice or grapefruit juice?

Activities

Pick some household product that probably has an acid or base character (Example: pH increaser for aquariums). On a separate paper write the listed ingredients and identify any you believe would be distinctly acidic or basic in a water solution. Tell whether you expect the product to be an acid or a base. Describe your findings of a litmus paper test.

Properties of Salts

Salt is produced by a neutralization reaction between an acid and a base. A **salt** is defined as any ionic compound except those with hydroxide or oxide ions. Table salt, NaCl, is but one example of this large group of ionic compounds. As an example of a salt produced by a neutralization reaction, consider the reaction of HCl (an acid in solution) with Ca(OH)$_2$ (a base in solution). The reaction is

$$2\ HCl(aq) + Ca(OH)_2(aq) \rightarrow CaCl_2(aq) + 2\ H_2O(l)$$

This is an ionic exchange reaction that forms molecular water, leaving Ca^{2+} and Cl$^-$ in solution. As the water is evaporated, these ions begin forming ionic crystal structures as the solution concentration increases. When the water is all evaporated, the white crystalline salt of CaCl$_2$ remains.

If sodium hydroxide had been used as the base instead of calcium hydroxide, a different salt would have been produced:

$$HCl(aq) + NaOH(aq) \rightarrow NaCl(aq) + H_2O(l)$$

Salts are also produced when elements combine directly, when an acid reacts with a metal, and by other reactions. Salts

TABLE 13.6	Some common salts and their uses	
Common Name	Formula	Use
Alum	$KAl(SO_4)_2$	Medicine, canning, baking powder
Baking soda	$NaHCO_3$	Fire extinguisher, antacid, deodorizer, baking powder
Bleaching powder (chlorine tablets)	$CaOCl_2$	Bleaching, deodorizer, disinfectant in swimming pools
Borax	$Na_2B_4O_7$	Water softener
Chalk	$CaCO_3$	Antacid tablets, scouring powder
Chile saltpeter	$NaNO_3$	Fertilizer
Cobalt chloride	$CoCl_2$	Hygrometer (pink in damp weather, blue in dry weather)
Epsom salt	$MgSO_4 \cdot 7\,H_2O$	Laxative
Fluorspar	CaF_2	Metallurgy flux
Gypsum	$CaSO_4 \cdot 2\,H_2O$	Plaster of Paris, soil conditioner
Lunar caustic	$AgNO_3$	Germicide and cauterizing agent
Niter (or saltpeter)	KNO_3	Meat preservative, makes black gunpowder (75 parts KNO_3, 15 of carbon, 10 of sulfur)
Potash	K_2CO_3	Makes soap, glass
Rochelle salt	$KNaC_4H_4O_6$	Baking powder ingredient
TSP	Na_3PO_4	Water softener, fertilizer

TABLE 13.7	Generalizations about salt solubilities	
Salts	Solubility	Exceptions
Sodium Potassium Ammonium	Soluble	None
Nitrate Acetate Chlorate	Soluble	None
Chlorides	Soluble	Ag and Hg (I) are insoluble
Sulfates	Soluble	Ba, Sr, and Pb are insoluble
Carbonates Phosphates Silicates	Insoluble	Na, K, and NH_4 are soluble
Sulfides	Insoluble	Na, K, and NH_4 are soluble: Mg, Ca, Sr, and Ba decompose

are usually prepared commercially by a neutralization reaction between an acid and a base that furnishes the desired ions.

Salts are essential in the diet both as electrolytes and as a source of certain elements, usually called *minerals* in this context. Plants must have certain elements that are derived from water-soluble salts. Potassium, nitrates, and phosphate salts are often used to supply the needed elements. There is no scientific evidence that plants prefer to obtain these elements from natural sources, as compost, or from chemical fertilizers. After all, a nitrate ion is a nitrate ion, no matter what its source. Table 13.6 lists some common salts and their uses.

Hard and Soft Water

Salts vary in their solubility in water, and a solubility chart appears in appendix B. Table 13.7 lists some generalizations concerning the various common salts. Some of the salts are dissolved by water that will eventually be used for domestic supply. When the salts are soluble calcium or magnesium compounds, the water will contain calcium or magnesium ions in solution. A solution of Ca^{2+} or Mg^{2+} ions is said to be *hard water* because it is hard to make soap lather in the water. "Soft" water, on the other hand, makes a soap lather easily. The difficulty occurs because soap is a sodium or potassium compound that is solu-

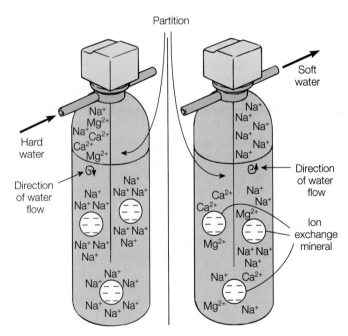

FIGURE 13.21

A water softener exchanges sodium ions for the calcium and magnesium ions of hard water. Thus, the water is now soft, but it contains the same number of ions as before.

ble in water. The calcium or magnesium ions, when present, replace the sodium or potassium ions in the soap compound, forming an insoluble compound. It is this insoluble compound that forms a "bathtub ring" and also collects on clothes being washed, preventing cleansing.

The key to "softening" hard water is to remove the troublesome calcium and magnesium ions (figure 13.21). If the hardness is caused by magnesium or calcium bicarbonates, the removal is accomplished by simply heating the water. Upon

A CLOSER LOOK | Acid Rain

Acid rain is a general term used to describe any acidic substances, wet or dry, that fall from the atmosphere. Wet acidic deposition could be in the form of rain, but snow, sleet, and fog could also be involved. Dry acidic deposition could include gases, dust, or any solid particles that settle out of the atmosphere to produce an acid condition.

Pure, unpolluted rain is naturally acidic. Carbon dioxide in the atmosphere is absorbed by rainfall, forming carbonic acid (H_2CO_3). Carbonic acid lowers the pH of pure rainfall to a range of 5.6 to 6.2. Decaying vegetation in local areas can provide more CO_2, making the pH even lower. A pH range of 4.5 to 5.0, for example, has been measured in remote areas of the Amazon jungle. Human-produced exhaust emissions of sulfur and nitrogen oxides can lower the pH of rainfall even more, to a 4.0 to 5.5 range. This is the pH range of acid rain.

The sulfur and nitrogen oxides that produce acid rain come from exhaust emissions of industries and electric utilities that burn coal and from the exhaust of cars, trucks, and buses (box figure 13.1). The emissions are sometimes called "SO_x" and "NO_x," which is read "socks" and "knox." The x subscript implies the variable presence of any or all of the oxides, for example, nitrogen monoxide (NO), nitrogen dioxide (NO_2), and dinitrogen tetroxide (N_2O_4) for NO_x.

SO_x and NO_x are the raw materials of acid rain and are not themselves acidic. They react with other atmospheric chemicals to form sulfates and nitrates, which combine with water vapor to form sulfuric acid (H_2SO_4) and nitric acid (HNO_3). These are the chemicals of concern in acid rain.

Many variables influence how much and how far SO_x and NO_x are carried in the atmosphere and if they are converted to acid rain or simply return to the surface as a dry gas or particles. During the 1960s and 1970s, concerns about local levels of pollution led to the replacement of short smokestacks of

BOX FIGURE 13.1

Natural rainwater has a pH of 5.6 to 6.2. Exhaust emissions of sulfur and nitrogen oxides can lower the pH of rainfall to a range of 4.0 to 5.5. The exhaust emissions come from industries, electric utilities, and automobiles. Not all emissions are as visible as those pictured in this illustration.

about 60 m (about 200 ft) with taller smokestacks of about 200 m (about 650 ft). This did reduce the local levels of pollution by dumping the exhaust higher in the atmosphere where winds could carry it away. It also set the stage for longer-range transport of SO_x and NO_x and their eventual conversion into acids.

There are two main reaction pathways by which SO_x and NO_x are converted to acids: (1) reactions in the gas phase and (2) reactions in the liquid phase, such as in water droplets in clouds and fog. In the gas phase, SO_x and NO_x are oxidized to acids, mainly by hydroxyl ions and ozone, and the acid is absorbed by cloud droplets and precipitated as rain or snow. Most of the nitric

acid in acid rain and about one-fourth of the sulfuric acid is formed in gas-phase reactions. Most of the liquid-phase reactions that produce sulfuric acid involve the absorbed SO_x and hydrogen peroxide (H_2O_2), ozone, oxygen, and particles of carbon, iron oxide, and manganese oxide particles. These particles also come from the exhaust of fossil fuel combustion.

Acid rain falls on the land, bodies of water, forests, crops, buildings, and people. The concerns about acid rain center on its environmental impact on lakes, forests, crops, materials, and human health. Lakes in different parts of the world, for example, have been increasing in acidity over the past fifty years. Lakes in northern New England, the Adirondacks, and parts of Canada now have a pH of less than 5.0, and correlations have been established between lake acidity and decreased fish populations. Trees, mostly conifers, are dying at unusually rapid rates in the northeastern United States. Red spruce in Vermont's Green Mountains and the mountains of New York and New Hampshire have been affected by acid rain as have pines in New Jersey's Pine Barrens. It is believed that acid rain leaches essential nutrients, such as calcium, from the soil and also mobilizes aluminum ions. The aluminum ions disrupt the water equilibrium of fine root hairs, and when the root hairs die, so do the trees.

Human-produced emissions of sulfur and nitrogen oxides from burning fossil fuels are the cause of acid rain. The heavily industrialized northeastern part of the United States, from the Midwest through New England, releases sulfur and nitrogen emissions that result in a precipitation pH of 4.0 to 4.5. This region is the geographic center of the nation's acid rain problem. The solution to the problem is found in (1) using fuels other than fossil fuels and (2) reducing the thousands of tons of SO_x and NO_x that are dumped into the atmosphere per day when fossil fuels are used.

heating, they decompose, forming an insoluble compound that effectively removes the ions from solution. The decomposition reaction for calcium bicarbonate is

$$Ca^{2+}(HCO_3)_2(aq) \rightarrow CaCO_3(s) + H_2O(l) + CO_2\uparrow$$

The reaction is the same for magnesium bicarbonate. As the solubility chart in appendix B shows, magnesium and calcium carbonates are insoluble, so the ions are removed from solution in the solid that is formed. Perhaps you have noticed such a white compound forming around faucets if you live where bicarbonates are a problem. Commercial products to remove such deposits usually contain an acid, which reacts with the carbonate to make a new, soluble salt that can be washed away.

Water hardness is also caused by magnesium or calcium sulfate, which requires a different removal method. Certain chemicals such as sodium carbonate (washing soda), trisodium phosphate (TSP), and borax will react with the troublesome ions, forming an insoluble solid that removes them from solution. For example, washing soda and calcium sulfate react as follows:

$$Na_2CO_3(aq) + CaSO_4(aq) \rightarrow Na_2SO_4(aq) + CaCO_3\downarrow$$

Calcium carbonate is insoluble; thus the calcium ions are removed from solution before they can react with the soap. Many laundry detergents have Na_2CO_3, TSP, or borax ($Na_2B_4O_7$) added to soften the water. TSP causes other problems, however, as the additional phosphates in the waste water can act as a fertilizer, stimulating the growth of algae to such an extent that other organisms in the water die.

A water softener unit is an ion exchanger. The unit contains a mineral that exchanges sodium ions for calcium and magnesium ions as water is run through it. The softener is regenerated periodically by flushing with a concentrated sodium chloride solution. The sodium ions replace the calcium and magnesium ions, which are carried away in the rinse water. The softener is then ready for use again. The frequency of renewal cycles depends on the water hardness, and each cycle can consume from four to twenty pounds of sodium chloride per renewal cycle. In general, water with less than 75 ppm calcium and magnesium ions is called soft water; with greater concentrations, it is called hard water. The greater the concentration above 75 ppm, the harder the water.

Buffers

A **buffer solution** consists of a weak acid together with a salt and has the same negative ion as the acid. A buffer has the ability to resist changes in the pH when small amounts of an acid or a base are added. Acetic acid, for example, is a weak acid that forms hydronium ions and acetate ions in equilibrium:

$$HC_2H_3O_2(aq) + H_2O(l) \rightarrow H_3O^+(aq) + (C_2H_3O_2)^-(aq)$$

When sodium acetate, $NaC_2H_3O_2$, is added to the solution, it becomes a buffer solution. If a small amount of an acid is added, the hydronium ions are neutralized by reacting with the acetate ions in solution:

$$(C_2H_3O_2)^-(aq) + H_3O^+(aq) \rightarrow HC_2H_3O_2(aq) + H_2O(l)$$

If a small amount of a base is added, the hydroxide ions are neutralized by reacting with acetic acid:

$$HC_2H_3O_2(aq) + OH^-(aq) \rightarrow (C_2H_3O_2)^-(aq) + H_2O(l)$$

Thus, the addition of an acid or a base does not change the pH, but it changes the ratio of $HC_2H_3O_2$ and $(C_2H_3O_2)^-$ instead. The solution will continue its buffering action as long as the number of H_3O^+ or OH^- added does not exceed the number of $HC_2H_3O_2$ molecules or acetate ions in the solution. Your blood contains buffer solutions that maintain the pH at about 7.4. Seawater is a buffer solution that maintains a pH of about 8.2. Buffers are also added to medicines and to foods. Many lemon-lime carbonated beverages, for example, contain citric acid and sodium citrate (check the label), which forms a buffer in the acid range. Sometimes the label says that these chemicals are to impart and regulate "tartness." Any acid will produce a tart taste. In this case, the tart taste comes from the citric acid and the addition of sodium citrate makes it a buffered solution.

SUMMARY

A *solution* is a homogeneous mixture of ions or molecules of two or more substances. The substance present in the large amount is the *solvent,* and the *solute* is dissolved in the solvent. If one of the components is a liquid, however, it is called the solvent. The relative amount of solute in a solvent is called the *concentration* of a solution. Concentrations are measured (1) in *parts per million* (ppm) or *parts per billion* (ppb), (2) *percent by volume,* the volume of a solute per 100 volumes of solution, (3) *percent by weight,* the weight of solute per 100 weight units of solution, and (4) *salinity,* the mass of salts in 1 kg of solution.

A limit to dissolving solids in a liquid occurs when the solution is *saturated.* A *saturated solution* is one with equilibrium between solute dissolving and solute coming out of solution. The *solubility* of a solid is the concentration of a saturated solution at a particular temperature.

A water molecule consists of two hydrogen atoms and an oxygen atom with bonding and electron pairs in a tetrahedral arrangement. This results in a *bent molecular arrangement,* with 105° between the hydrogen atoms. Oxygen is more electronegative than hydrogen, so electrons spend more time around the oxygen, producing a *polar molecule,* with centers of negative and positive charge. Polar water molecules interact with an *intermolecular force,* or *van der Waals force,* between the negative center of one molecule and the positive center of another. The force of attraction is called a *hydrogen bond.* The hydrogen bond accounts for the decreased density of ice, the high heat of fusion, and the high heat of vaporization of water. The hydrogen bond is also involved in the *dissolving* process.

Fluids that mix in any proportion are called *miscible fluids,* and *immiscible fluids* do not mix. Polar substances dissolve in polar solvents, but not nonpolar solvents, and the general rule is *like dissolves like.* Thus oil, a nonpolar substance, is immiscible in water, a polar substance. When a polar substance is added to a polar solvent, the substance dissolves if the *ion-polar molecular force* is greater than the *ion-ion force.* If the ion-ion force is greater, the substance is *insoluble.*

Water solutions that carry an electric current are called *electrolytes,* and nonconductors are called *nonelectrolytes.* In general, ionic substances make electrolyte solutions, and molecular substances make nonelectrolyte solutions. Polar molecular substances may be *ionized* by polar water molecules, however, making an electrolyte from a molecular solution.

The *boiling point of a solution* is greater than the boiling point of the pure solvent, and the increase depends only on the concentration of the solute (at a constant pressure). For water, the boiling point is increased 0.521°C for each mole of solute in each kg of water. The *freezing point of a solution* is lower than the freezing point of the pure solvent, and the depression also depends on the concentration of the solute.

Acids, bases, and salts are chemicals that form ionic solutions in water, and each can be identified by simple properties. These properties are accounted for by the modern concepts of each. *Acids* are *proton donors* that form *hydronium ions* (H_3O^+) in water solutions. *Bases* are *proton acceptors* that form *hydroxide ions* (OH^-) in water solutions. *Strong acids* and *strong bases* ionize completely in water, and *weak acids* and *weak bases* are only partially ionized because of an *equilibrium reaction* with the solvent. The strength of an acid or base is measured on the *pH scale,* a power of ten notation of the hydronium ion concentration. On the scale, numbers from 0 up to 7 are acids, 7 is neutral, and numbers above 7 and up to 14 are bases. Each unit represents a tenfold increase or decrease in acid or base properties.

A *salt* is any ionic compound except those with hydroxide or oxide ions. Salts provide plants and animals with essential elements. The solubility of salts varies with the ions that make up the compound. Solutions of magnesium or calcium produce *hard water,* water in which it is hard to make soap lather. Hard water is softened by removing the magnesium and calcium ions. A *buffer* solution is a solution of a weak acid and one of its salts. The solution resists changes in pH by reacting with acids or bases that are added.

Summary of Equations

13.1

Percent by volume

$$\frac{V_{solute}}{V_{solution}} \times 100\% \text{ solution} = \% \text{ solute}$$

13.2

Percent by weight (mass)

$$\frac{m_{solute}}{m_{solution}} \times 100\% \text{ solution} = \% \text{ solute}$$

13.3

$$\text{Molarity (M)} = \frac{\text{moles of solute}}{\text{liters of solution}}$$

KEY TERMS

acid (p. 312)

base (p. 312)

buffer solution (p. 317)

chemical equilibrium (p. 312)

electrolytes (p. 306)

hydration (p. 306)

hydrogen bond (p. 304)

hydronium ion (p. 306)

miscible fluids (p. 305)

molarity (p. 303)

neutralized (p. 312)

pH scale (p. 313)

salinity (p. 303)

salt (p. 314)

saturated solution (p. 303)

solubility (p. 303)

solution (p. 300)

APPLYING THE CONCEPTS

1. Which of the following is *not* a solution?
   a. seawater
   b. carbonated water
   c. sand
   d. brass

2. Atmospheric air is a homogeneous mixture of gases that is mostly nitrogen gas. The nitrogen is therefore (the)
   a. solvent.
   b. solution.
   c. solute.
   d. none of the above.

3. A homogeneous mixture is made up of 95 percent alcohol and 5 percent water. In this case the water is (the)
   a. solvent.
   b. solution.
   c. solute.
   d. none of the above.

4. The solution concentration terms of parts per million, percent by volume, and percent by weight are concerned with the amount of
   a. solvent in the solution.
   b. solute in the solution.
   c. solute compared to solvent.
   d. solvent compared to solute.

5. A concentration of 500 ppm is reported in a news article. This is the same concentration as
   a. 0.005%.
   b. 0.05%.
   c. 5%.
   d. 50%.

6. According to the label, a bottle of vodka has a 40% by volume concentration. This means the vodka contains 40 mL of pure alcohol
   a. in each 140 mL of vodka.
   b. to every 100 mL of water.
   c. to every 60 mL of vodka.
   d. mixed with water to make 100 mL vodka.

7. A bottle of vinegar is 4% by weight, so you know that the solution contains 4 weight units of pure vinegar with
   a. 96 weight units of water.
   b. 100 weight units of water.
   c. 104 weight units of water.

8. If a salt solution has a salinity of 40‰, what is the equivalent percentage measure?
   a. 400%
   b. 40%
   c. 4%
   d. 0.4%

9. A salt solution has solid salt on the bottom of the container and salt is dissolving at the same rate that it is coming out of solution. You know the solution is
    a. an electrolyte.
    b. a nonelectrolyte.
    c. a buffered solution.
    d. a saturated solution.

10. As the temperature of water *decreases,* the solubility of carbon dioxide gas in the water
    a. increases.
    b. decreases.
    c. remains the same.
    d. increases or decreases, depending on the specific temperature.

11. Water has the greatest density at what temperature?
    a. 100°C
    b. 20°C
    c. 4°C
    d. 0°C

12. An example of a hydrogen bond is a weak-to-moderate bond between
    a. any two hydrogen atoms.
    b. a hydrogen of one polar molecule and an oxygen of another polar molecule.
    c. two hydrogen atoms on two nonpolar molecules.
    d. a hydrogen atom and any nonmetal atom.

13. Whether two given liquids form solutions or not depends on some similarities in their
    a. electronegativities.
    b. polarities.
    c. molecular structures.
    d. hydrogen bonds.

14. A solid salt is insoluble in water so the strongest force must be the
    a. ion-water molecule force.
    b. ion-ion force.
    c. force of hydration.
    d. polar molecule force.

15. Which of the following will conduct an electric current?
    a. pure water
    b. a water solution of a covalent compound
    c. a water solution of an ionic compound
    d. All of the above are correct.

16. Ionization occurs upon solution of
    a. ionic compounds.
    b. some polar molecules.
    c. nonpolar molecules.
    d. none of the above.

17. Adding sodium chloride to water raises the boiling point of water because
    a. sodium chloride has a higher boiling point.
    b. sodium chloride ions occupy space at the water surface.
    c. sodium chloride ions have stronger ion-ion bonds than water.
    d. the energy of hydration is higher.

18. The ice that forms in freezing seawater is
    a. pure water.
    b. the same salinity as liquid seawater.
    c. more salty than liquid seawater.
    d. more dense than liquid seawater.

19. Salt solutions freeze at a lower temperature than pure water because
    a. more ionic bonds are present.
    b. salt solutions have a higher vapor pressure.
    c. ions get in the way of water molecules trying to form ice.
    d. salt naturally has a lower freezing point than water.

20. Which of the following would have a pH of *less* than 7?
    a. a solution of ammonia
    b. a solution of sodium chloride
    c. pure water
    d. carbonic acid

21. Which of the following would have a pH of *more* than 7?
    a. a solution of ammonia
    b. a solution of sodium chloride
    c. pure water
    d. carbonic acid

22. The condition of two opposing reactions happening at the same time and at the same rate is called
    a. neutralization.
    b. chemical equilibrium.
    c. a buffering reaction.
    d. cancellation.

23. Solutions of acids, bases, and salts have what in common? All have
    a. proton acceptors.
    b. proton donors.
    c. ions.
    d. polar molecules.

24. When a solution of an acid and a base are mixed together,
    a. a salt and water are formed.
    b. they lose their acid and base properties.
    c. both are neutralized.
    d. All of the above are correct.

25. A substance that ionizes completely into hydronium ions is known as a
    a. strong acid.
    b. weak acid.
    c. strong base.
    d. weak base.

26. A scale of values that expresses the hydronium ion concentration of a solution is known as
    a. an acid-base indicator.
    b. the pH scale.
    c. the solubility scale.
    d. the electrolyte scale.

27. Substance A has a pH of 2 and substance B has a pH of 3. This means that

    a. substance A has more basic properties than substance B.

    b. substance B has more acidic properties than substance A.

    c. substance A is ten times more acidic than substance B.

    d. substance B is ten times more acidic than substance A.

28. A solution that is able to resist changes in the pH when small amounts of an acid or base are added is called a

    a. neutral solution.

    b. saturated solution.

    c. balanced solution.

    d. buffer solution.

Answers

1. c 2. a 3. c 4. b 5. b 6. d 7. a 8. c 9. d 10. a 11. c 12. b 13. c 14. b 15. c 16. b 17. b 18. a 19. c 20. d 21. a 22. b 23. c 24. d 25. a 26. b 27. c 28. d

QUESTIONS FOR THOUGHT

1. How is a solution different from other mixtures?
2. Explain why some ionic compounds are soluble while others are insoluble in water.
3. Explain why adding salt to water increases the boiling point.
4. A deep lake in Minnesota is covered with ice. What is the water temperature at the bottom of the lake? Explain your reasoning.
5. Explain why water has a greater density at 4°C than at 0°C.
6. What is hard water? How is it softened?
7. According to the definition of an acid and the definition of a base, would the pH increase, decrease, or remain the same when NaCl is added to pure water? Explain.
8. What is a hydrogen bond? Explain how a hydrogen bond forms.
9. What feature of a soap molecule gives it cleaning ability?
10. What ion is responsible for (a) acidic properties? (b) for basic properties?
11. Explain why a pH of 7 indicates a neutral solution—why not some other number?
12. What is a buffer solution?

PARALLEL EXERCISES

The exercises in groups A and B cover the same concepts. Solutions to group A exercises are located in appendix D.

Group A

1. A 50.0 g sample of a saline solution contains 1.75 g NaCl. What is the percentage by weight concentration?
2. A student attempts to prepare a 3.50 percent by weight saline solution by dissolving 3.50 g NaCl in 100 g of water. Since equation 13.2 calls for 100 g of solution, the correct amount of solvent should have been 96.5 g water ($100 - 3.5 = 96.5$). What percent by weight solution did the student actually prepare?
3. Seawater contains 30,113 ppm by weight dissolved sodium and chlorine ions. What is the percent by weight concentration of sodium chloride in seawater?
4. What is the mass of hydrogen peroxide, H_2O_2, in 250 grams of a 3.0% by weight solution?
5. How many mL of pure alcohol are in a 200 mL glass of wine that is 12 percent alcohol by volume?
6. How many mL of pure alcohol are in a single cocktail made with 50 mL of 40% vodka? (Note: "Proof" is twice the percent, so 80 proof is 40%.)
7. If fish in a certain lake are reported to contain 5 ppm by weight DDT, (a) what percentage of the fish meat is DDT? (b) How much of this fish would have to be consumed to reach a poisoning accumulation of 17.0 grams of DDT?
8. For each of the following reactants, draw a circle around the proton donor and a box around the proton acceptor. Label which acts as an acid and which acts as a base.

    (a) $HC_2H_3O_{2(aq)} + H_2O_{(l)} \rightleftharpoons H_3O^+_{(aq)} + C_2H_3O_2^-_{(aq)}$

    (b) $C_6H_5NH_{2(l)} + H_2O_{(l)} \rightleftharpoons C_6H_5NH_3^+_{(aq)} + OH^-_{(aq)}$

    (c) $HClO_{4(aq)} + HC_2H_3O_{2(aq)} \rightleftharpoons H_2C_2H_3O_2^+_{(aq)} + ClO_4^-_{(aq)}$

    (d) $H_2O_{(l)} + H_2O_{(l)} \rightleftharpoons H_3O^+_{(aq)} + OH^-_{(aq)}$

Group B

1. What is the percent by weight of a solution containing 2.19 g NaCl in 75 g of the solution?
2. What is the percent by weight of a solution prepared by dissolving 10 g of NaCl in 100 g of H_2O?
3. A concentration of 0.5 ppm by volume SO_2 in air is harmful to plant life. What is the percent by volume of this concentration?
4. What is the volume of water in a 500 mL bottle of rubbing alcohol that has a concentration of 70% by volume?
5. If a definition of intoxication is an alcohol concentration of 0.05 percent by volume in blood, how much alcohol would be present in the average (155 lb) person's 6,300 mL of blood if that person was intoxicated?
6. How much pure alcohol is in a 355 mL bottle of a "wine cooler" that is 5.0% alcohol by volume?
7. In the 1970s, when lead was widely used in "ethyl" gasoline, the blood level of the average American contained 0.25 ppm lead. The danger level of lead poisoning is 0.80 ppm. (a) What percent of the average person was lead? (b) How much lead would be in an average 80 kg person? (c) How much more lead would the average person need to accumulate to reach the danger level?
8. Draw a circle around the proton donor and a box around the proton acceptor for each of the reactants and label which acts as an acid and which acts as a base.

    (a) $H_3PO_{4(aq)} + H_2O_{(l)} \rightleftharpoons H_3O^+_{(aq)} + H_2PO_4^-_{(aq)}$

    (b) $N_2H_{4(l)} + H_2O_{(l)} \rightleftharpoons N_2H_5^+_{(aq)} + OH^-_{(aq)}$

    (c) $HNO_{3(aq)} + HC_2H_3O_{2(aq)} \rightleftharpoons H_2C_2H_3O_2^+_{(aq)} + NO_3^-_{(aq)}$

    (d) $2\ NH_4^+_{(aq)} + Mg_{(s)} \rightleftharpoons Mg^{2+}_{(aq)} + 2\ NH_3^+_{(aq)} + H_{2(g)}$

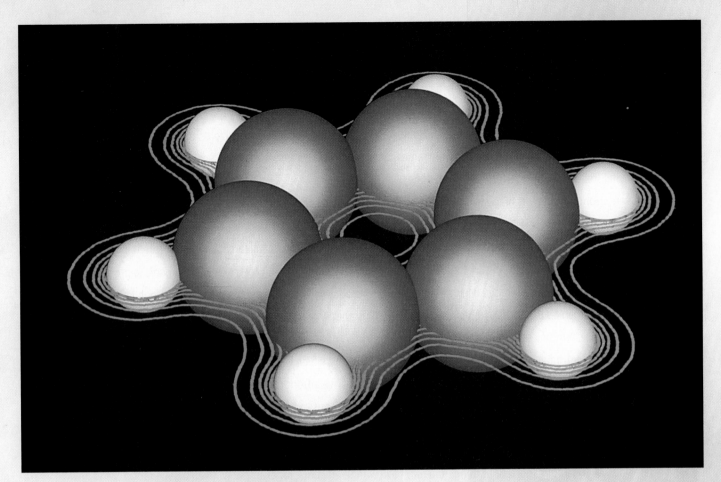

This is a computer-generated model of a benzene molecule, showing six carbon atoms (gold) and six hydrogen atoms (white). Benzene is a hydrocarbon, an organic compound made up of the elements carbon and hydrogen.

CHAPTER | Fourteen

Organic Chemistry

The impact of ancient Aristotelian ideas on the development of understandings of motion, elements, and matter was discussed in earlier chapters. Historians also trace the "vitalist theory" back to Aristotle. According to Aristotle's idea, all living organisms are composed of the four elements (earth, air, fire, and water) and have in addition an *actuating force,* the life or soul that makes the organism different from nonliving things made of the same four elements. Plants, as well as animals, were considered to have this actuating, or vital, force in the Aristotelian scheme of things.

There were strong proponents of the vitalist theory as recent as the early 1800s. Their basic argument was that organic matter, the materials and chemical compounds recognized as being associated with life, could not be produced in the laboratory. Organic matter could only be produced in a living organism, they argued, because the organism had a vital force that is not present in laboratory chemicals. Then, in 1828, a German chemist named Fredrich Wohler decomposed a chemical that was *not organic* to produce urea (N_2H_4CO), a known *organic* compound that occurs in urine. Wohler's production of an organic compound was soon followed by the production of other organic substances by other chemists. The vitalist theory gradually disappeared with each new reaction, and a new field of study, organic chemistry, emerged.

This chapter is an introductory survey of the field of organic chemistry, which is concerned with compounds and reactions of compounds that contain carbon. You will find this an interesting, informative introduction, particularly if you have ever wondered about synthetic materials, natural foods and food products, or any of the thousands of carbon-based chemicals you use every day. The survey begins with the simplest of organic compounds, those consisting of only carbon and hydrogen atoms, compounds known as hydrocarbons. Hydrocarbons are the compounds of crude oil, which is the source of hundreds of petroleum products (figure 14.1). In this section you will find information about things you may have wondered about, for example, what an octane rating is and how petroleum products differ.

Most common organic compounds can be considered derivatives of the hydrocarbons, such as alcohols, ethers, fatty acids, and esters. Some of these are the organic compounds that give flavors to foods, and others are used to make hundreds of commercial products, from face cream to oleo. The main groups, or classes, of derivatives will be briefly introduced, along with some interesting examples of each group. Some of the important organic compounds of life, including proteins, carbohydrates, and fats, are discussed next. The chapter concludes with an introduction to synthetic polymers, what they are, and how they are related to the fossil fuel supply.

ORGANIC COMPOUNDS

Organic compounds are sensitive to increases in temperature, decomposing or burning when heated to 400°C (about 750°F) or greater. When sugar is decomposed by heating, for example, it often leaves a black residue of carbon. When burned completely, sugar and other organic materials produce carbon dioxide (and other products). Carbon is the essential element of organic matter, and today, **organic chemistry** is defined as the study of compounds in which carbon is the principal element, whether the compound was formed by living things or not. The study of all the other elements and compounds is called **inorganic chemistry.** An *organic compound* is thus a compound in which carbon is the principal element, and an *inorganic compound* is any other compound.

Organic compounds, by definition, must contain carbon while all the other compounds can contain all the other elements. Yet, the majority of known compounds are organic. Several million organic compounds are known and thousands of new ones are discovered every year. You use organic compounds every day, including gasoline, plastics, grain alcohol, foods, flavorings, and many others. How can the carbon atom make so many different and diverse compounds compared to all the other elements?

It is the unique properties of carbon that allow it to form so many different compounds. A carbon atom has a simple $1s^2 2s^2 2p^2$ electron structure, and there is room for four more electrons in the outer shell. Carbon has a valence of four and can form four electron pairs, with no lone pairs. The molecular shape of a carbon compound such as CH_4 is therefore tetrahedral. The carbon atom has a valence of four, and can combine with one, two, three, or four *other carbon atoms* in addition to a wide range of other kinds of atoms (figure 14.2). The number of possible molecular combinations is almost limitless, which explains why there are so many organic compounds. Fortunately, there are patterns of groups of carbon atoms and groups of other atoms that lead to similar chemical characteristics,

FIGURE 14.1

Refinery and tank storage facilities, like this one in Texas, are needed to change the hydrocarbons of crude oil to many different petroleum products. The classes and properties of hydrocarbons form one topic of study in organic chemistry.

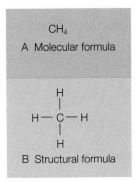

FIGURE 14.3

Recall that a molecular formula (*A*) describes the numbers of different kinds of atoms in a molecule, and a structural formula (*B*) represents a two-dimensional model of how the atoms are bonded to each other. Each dash represents a bonding pair of electrons.

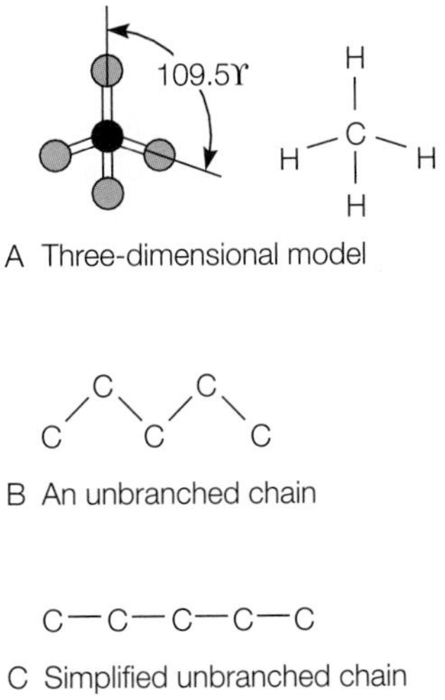

A Three-dimensional model

B An unbranched chain

C — C — C — C — C

C Simplified unbranched chain

FIGURE 14.2

(*A*) The carbon atom forms bonds in a tetrahedral structure with a bond angle of 109.5°. (*B*) Carbon-to-carbon bond angles are 109.5°, so a chain of carbon atoms makes a zigzag pattern. (*C*) The unbranched chain of carbon atoms is usually simplified in a way that looks like a straight chain, but it is actually a zigzag, as shown in (*B*).

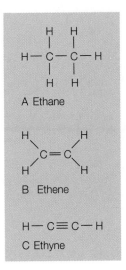

FIGURE 14.4

Carbon-to-carbon bonds can be single (*A*), double (*B*), or triple (*C*). Note that in each example, each carbon atom has four dashes, which represent four bonding pairs of electrons, satisfying the octet rule.

making the study of organic chemistry less difficult. The key to success in studying organic chemistry is to recognize patterns and to understand the code and meaning of organic chemical names. The first patterns to be discussed will be those of the simplest organic compounds, consisting of only two elements.

HYDROCARBONS

A **hydrocarbon** is an organic compound consisting of only two elements. As the name implies, these elements are hydrogen and carbon. The simplest hydrocarbon has one carbon atom and

four hydrogen atoms (figure 14.3), but since carbon atoms can combine with one another, there are thousands of possible structures and arrangements. The carbon-to-carbon bonds are nonpolar covalent and can be single, double, or triple (figure 14.4). Recall that the dash in a structural formula means one shared electron pair. To satisfy the octet rule, this means that each carbon atom must have a total of four dashes around it, no more and no less. Note that when the carbon atom has double or triple bonds, fewer hydrogen atoms can be attached as the octet rule is satisfied. There are four groups of hydrocarbons that are classified according to how the carbon atoms are put

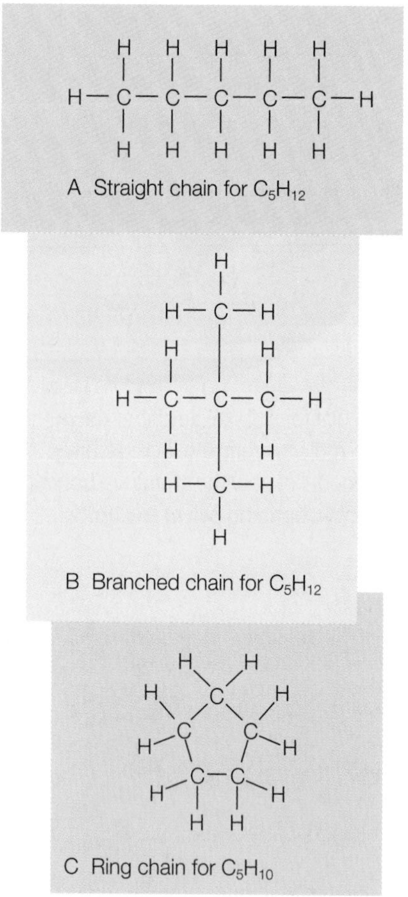

A Straight chain for C_5H_{12}

B Branched chain for C_5H_{12}

C Ring chain for C_5H_{10}

FIGURE 14.5

Carbon-to-carbon chains can be (A) straight, (B) branched, or (C) in a closed ring. (Some carbon bonds are drawn longer, but are actually the same length.)

together, the (1) *alkanes,* (2) *alkenes,* (3) *alkynes,* and (4) *aromatic hydrocarbons.*

The **alkanes** are *hydrocarbons with single covalent bonds* between the carbon atoms. Alkanes that are large enough to form chains of carbon atoms occur with a straight structure, a branched structure, or a ring structure as shown in figure 14.5. (The "straight" structure is actually a zigzag as shown in figure 14.2.) You are familiar with many alkanes, for they make up the bulk of petroleum and petroleum products, which will be discussed shortly. The clues and codes in the names of the alkanes will be considered first.

The alkanes are also called the *paraffin series.* The alkanes are not as chemically reactive as the other hydrocarbons, and the term *paraffin* means "little affinity." They are called a series because *each higher molecular weight alkane has an additional* CH_2. The simplest alkane is methane, CH_4, and the next highest molecular weight alkane is ethane, C_2H_6. As you can see, C_2H_6 is CH_4 with an additional CH_2. If you compare the first ten alkanes in table 14.1, you will find that each successive compound in the series always has an additional CH_2.

Note the names of the alkanes listed in table 14.1. After pentane the names have a consistent prefix and suffix pattern. The prefix and suffix pattern is a code that provides a clue about the compound. The Greek prefix tells you the *number of carbon atoms* in the molecule, for example, "oct-" means eight, so *oct*ane has eight carbon atoms. The suffix "-ane" tells you this hydrocarbon is a member of the alk*ane* series, so it has single bonds only. With the general alkane formula of C_nH_{2n+2}, you can now write the formula when you hear the name. Octane has eight carbon atoms with single bonds and $n = 8$. Two times 8 plus 2 ($2n + 2$) is 18, so the formula for octane is C_8H_{18}. Most organic chemical names provide clues like this, as you will see.

The alkanes in table 14.1 all have straight chains. A straight, continuous chain is identified with the term *normal,* which is abbreviated *n.* Figure 14.6A shows *n*-butane with a straight chain and a molecular formula of C_4H_{10}. Figure 14.6B shows a different branched structural formula that has the same C_4H_{10} molecular formula. Compounds with the same molecular formulas with different structures are called **isomers.** Since the straight-chained isomer is called *n*-butane, the branched isomer is called *isobutane.* The isomers of a particular alkane, such as butane, have different physical and chemical properties because they have different structures. Isobutane, for example, has a boiling point of $-10°C$. The boiling point of *n*-butane, on the other hand, is $-0.5°C$. In the next section you will learn that the various isomers of the octane hydrocarbon perform differently in automobile engines, requiring the "reforming" of *n*-octane to *iso-octane* before it can be used.

Methane, ethane, and propane can have only one structure each, and butane has two isomers. The number of possible isomers for a particular molecular formula increases rapidly as the number of carbon atoms increase. After butane, hexane has five isomers, octane eighteen isomers, and decane seventy-five isomers. Because they have different structures, each isomer has different physical properties. A different naming system is needed because there are just too many isomers to keep track of. The system of naming the branched-chain alkanes is described by rules agreed upon by the International Union of Pure and Applied Chemistry, or IUPAC. Here are the steps in naming the alkane isomers.

Step 1: The longest continuous chain of carbon atoms determines the *base name* of the molecule. The longest continuous chain is not necessarily straight and can take any number of right-angle turns as long as the continuity is not broken. The base name corresponds to the number of carbon atoms in this chain as in table 14.1. For

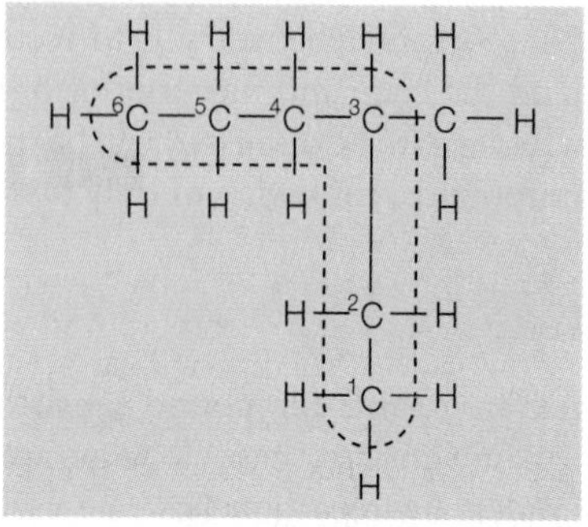

TABLE 14.1		The first ten straight-chain alkanes				

Name	Molecular Formula	Structural Formula	Name	Molecular Formula	Structural Formula
Methane	CH_4		Hexane	C_6H_{14}	
Ethane	C_2H_6		Heptane	C_7H_{16}	
Propane	C_3H_8		Octane	C_8H_{18}	
Butane	C_4H_{10}		Nonane	C_9H_{20}	
Pentane	C_5H_{12}		Decane	$C_{10}H_{22}$	

example, the structure has six carbon atoms in the longest chain, so the base name is *hexane.*

Step 2: The locations of other groups of atoms attached to the base chain are identified by counting carbon atoms from either the left or from the right. The direction selected is the one that results in the *smallest* numbers for attachment locations. For example, the hexane chain has a CH_3 attached to the third or the fourth carbon atom, depending on which way you count. The third atom direction is chosen since it results in a smaller number.

Step 3: The hydrocarbon groups attached to the base chain are named from the number of carbons in the group by changing the alkane suffix "-ane" to "-yl." Thus, a hydrocarbon group attached to a base chain that has one carbon atom is called meth*yl*. Note that the "-yl" hydrocarbon groups have one less hydrogen than the corresponding alkane. Therefore, methane is CH_4, and a *methyl group* is CH_3. The first ten alkanes and their corresponding hydrocarbon group names are listed in table 14.2. In the example, a methyl group is attached to the third carbon atom of the base hexane chain. The name and address of this hydrocarbon group is 3-methyl. The compound is named 3-methylhexane.

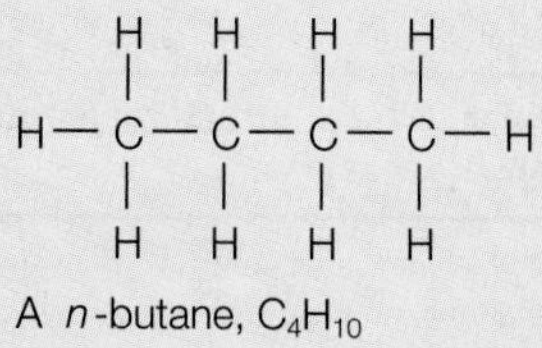

A *n*-butane, C_4H_{10}

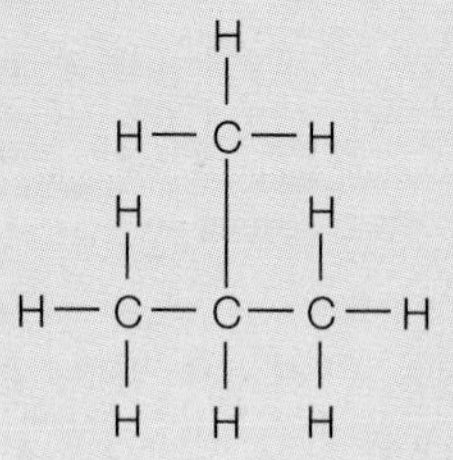

B Isobutane (2-methylpropane), C_4H_{10}

FIGURE 14.6

(*A*) A straight-chain alkane is identified by the prefix *n*- for "normal" in the common naming system. (*B*) A branched-chain alkane isomer is identified by the prefix *iso*- for "isomer" in the common naming system. In the IUPAC name, isobutane is 2-methylpropane. (Carbon bonds are actually the same length.)

TABLE 14.2	Alkane hydrocarbons and corresponding hydrocarbon groups		
Alkane Name	**Molecular Formula**	**Hydrocarbon Group**	**Molecular Formula**
Methane	CH_4	Methyl	$-CH_3$
Ethane	C_2H_6	Ethyl	$-C_2H_5$
Propane	C_3H_8	Propyl	$-C_3H_7$
Butane	C_4H_{10}	Butyl	$-C_4H_9$
Pentane	C_5H_{12}	Amyl	$-C_5H_{11}$
Hexane	C_6H_{14}	Hexyl	$-C_6H_{13}$
Heptane	C_7H_{16}	Heptyl	$-C_7H_{15}$
Octane	C_8H_{18}	Octyl	$-C_8H_{17}$
Nonane	C_9H_{20}	Nonyl	$-C_9H_{19}$
Decane	$C_{10}H_{22}$	Decyl	$-C_{10}H_{21}$

Note: $-CH_3$ means * $-$ C $-$ H where * denotes unattached. The attachment takes place on a base chain or functional group.

Step 4: The prefixes "di-," "tri-," and so on are used to indicate if a particular hydrocarbon group appears on the main chain more than once. For example,

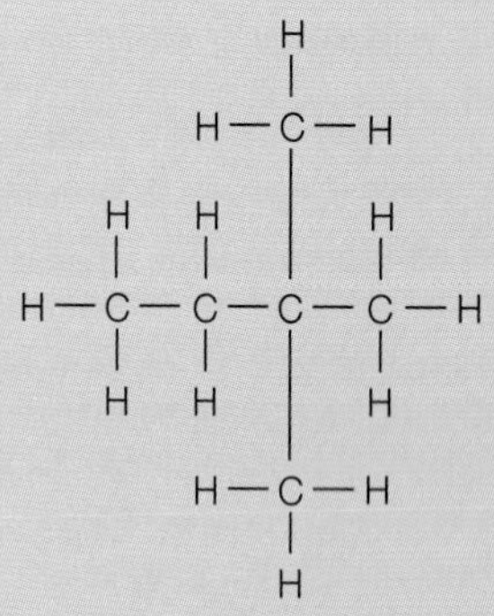

(or)

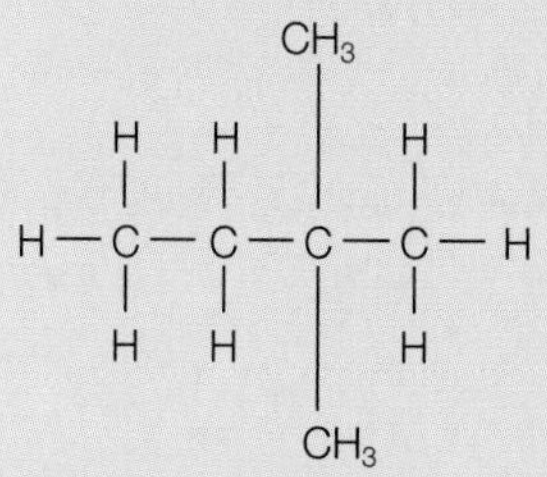

is 2,2-dimethylbutane and

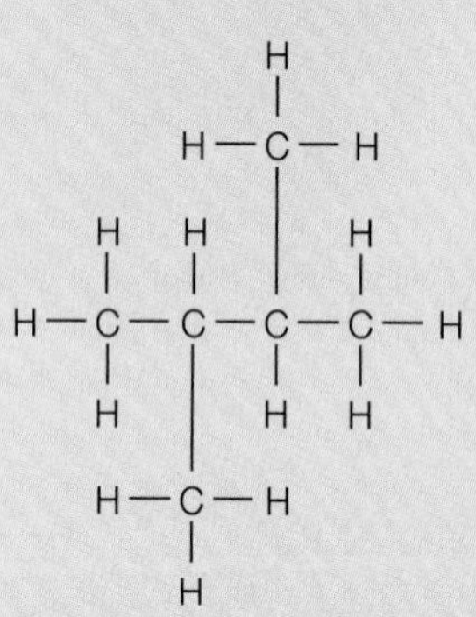

(or)

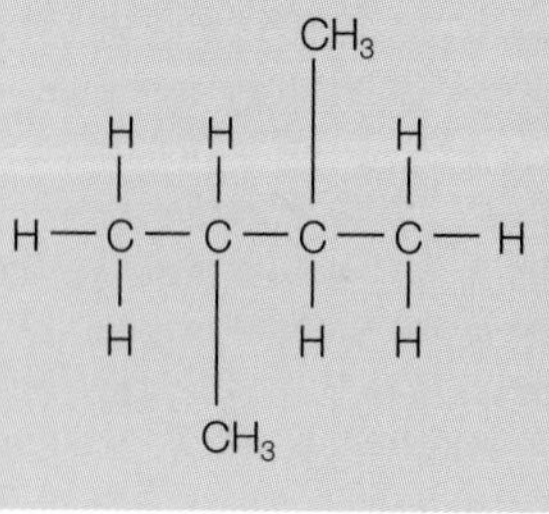

is 2,3-dimethylbutane.

If hydrocarbon groups with different numbers of carbon atoms are on a main chain, they are listed in alphabetic order. For example,

(or)

is named 3-ethyl-2-methylpentane. Note how numbers are separated from names by hyphens.

Q & A

What is the name of an alkane with the following formula?

Answer

The longest continuous chain has seven carbon atoms, so the base name is heptane. The smallest numbers are obtained by counting from right to left and counting the carbons on this chain; there is a methyl group in carbon atom 2, a second methyl group on atom 4,

and an ethyl group on atom 5. There are two methyl groups, so the prefix "di-" is needed, and the "e" of the ethyl group comes first in the alphabet so ethyl is listed first. The name of the compound is 5-ethyl-2,4-dimethylheptane.

Q & A

Write the structural formula for 2,2-dichloro-3-methyloctane. Answer:

TABLE 14.3	The general molecular formulas and molecular structures of the alkanes, alkenes, and alkynes		
Group	**General Molecular Formula**	**Example Compound**	**Molecular Structure**
Alkanes	C_nH_{2n+2}	Ethane	
Alkenes	C_nH_{2n}	Ethene	
Alkynes	C_nH_{2n-2}	Ethyne	

Alkenes and Alkynes

The alkanes are hydrocarbons with single carbon-to-carbon bonds. The **alkenes** are *hydrocarbons with a double covalent carbon-to-carbon bond*. To denote the presence of a double bond the "-ane" suffix of the alkanes is changed to "-ene" as in alk*ene* (table 14.3). Figure 14.4 shows the structural formula for (A) ethane, C_2H_6, and (B) ethene, C_2H_4. Alkenes have room for

FIGURE 14.7

Ethylene is the gas that ripens fruit, and a ripe fruit emits the gas, which will act on unripe fruit. Thus, a ripe tomato placed in a sealed bag with green tomatoes will help ripen them.

two fewer hydrogen atoms because of the double bond, so the general alkene formula is C_nH_{2n}. Note the simplest alkene is called ethene, but is commonly known as ethylene.

Ethylene is an important raw material in the chemical industry. Obtained from the processing of petroleum, about half of the commercial ethylene is used to produce the familiar polyethylene plastic. It is also produced by plants to ripen fruit, which explains why unripe fruit enclosed in a sealed plastic bag with ripe fruit will ripen more quickly (figure 14.7). The ethylene produced by the ripe fruit acts on the unripe fruit. Commercial fruit packers sometimes use small quantities of ethylene gas to ripen fruit quickly that was picked while green.

Perhaps you have heard the terms "saturated" and "unsaturated" in advertisements for cooking oil and margarine. The meaning of these terms with reference to foods will be discussed shortly. First, you need to understand the meaning of the terms. An organic molecule, such as a hydrocarbon, that does not contain the maximum number of hydrogen atoms is an **unsaturated** hydrocarbon. For example, ethylene can add more hydrogen atoms by reacting with hydrogen gas to form ethane:

The ethane molecule has all the hydrogen atoms possible, so ethane is a **saturated** hydrocarbon. Unsaturated molecules are less stable, which means that they are more chemically reactive than saturated molecules. Again, the role of saturated and unsaturated fats in foods will be discussed later.

Alkenes are named as the alkanes are, except (1) the longest chain of carbon atoms must contain the double bond, (2) the base name now ends in "-ene," (3) the carbon atoms are numbered from the end nearest the double bond, and (4) the base name is given a number of its own, which identifies the address of the double bond. For example,

is named 4-methyl-1-pentene. The 1-pentene tells you there is a double bond (-ene), and the 1 tells you the double bond is after the first carbon atom in the longest chain containing the double bond. The methyl group is on the fourth carbon atom in this chain.

An **alkyne** is a *hydrocarbon with a carbon-to-carbon triple bond* and the general formula of C_nH_{2n-2}. The alkynes are highly reactive, and the simplest one, ethyne, has a common name of acetylene. Acetylene is commonly burned with oxygen gas in a welding torch because the flame reaches a temperature of about 3,000°C. Acetylene is also an important raw material in the production of plastics. The alkynes are named as the alkenes are, except the longest chain must contain the triple bond, and the base name suffix is changed to "-yne."

Cycloalkanes and Aromatic Hydrocarbons

The hydrocarbons discussed up until now have been straight or branched open-ended chains of carbon atoms. Carbon atoms can also bond to each other to form a ring, or cyclic, structure. Figure 14.8 shows the structural formulas for some of these cyclic structures. Note that the cycloalkanes have the same molecular formulas as the alkenes, and thus they are isomers of the alkenes. They are, of course, very different compounds, with different physical and chemical properties. This shows the importance of structural, rather than simply molecular, formulas in referring to organic compounds.

The six-carbon ring structure shown in figure 14.9A has three double bonds that do not behave like the double bonds in the alkenes. In this six-carbon ring the double bonds are not localized in one place but are spread over the whole molecule. Instead of alternating single and double bonds, all the bonds are something in between. This gives the C_6H_6 molecule increased stability. As a result, the molecule does not act unsaturated, that is, it does not readily react in order to add hydrogen to the ring. The C_6H_6 molecule is the organic compound named *benzene*. Organic compounds that are based on the benzene ring structure are called *aromatic hydrocarbons*. To denote the six-carbon ring with delocalized electrons, benzene is represented by the symbol shown in figure 14.9B.

The circle in the six-sided benzene symbol represents the delocalized electrons. Figure 14.9B illustrates how this benzene ring symbol is used to show the structural formula of some aromatic hydrocarbons. You may have noticed some of the names

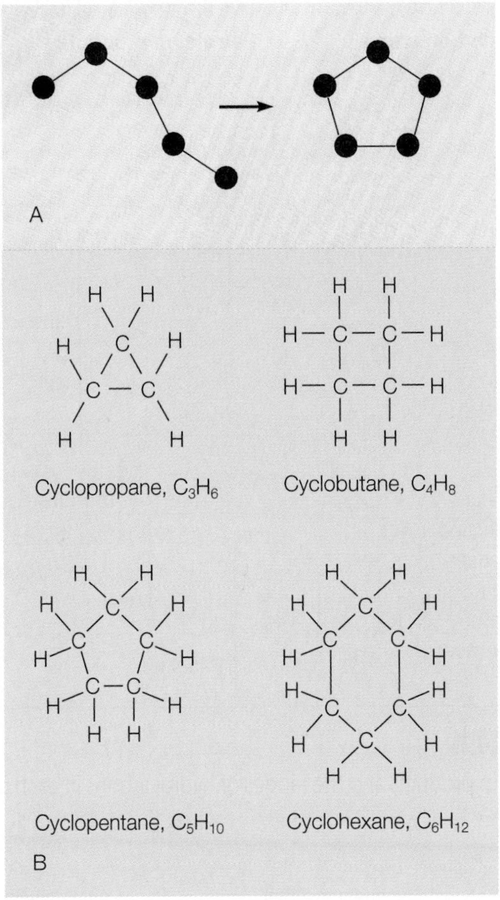

FIGURE 14.8

(A) The "straight" chain has carbon atoms that are able to rotate freely around their single bonds, sometimes linking up in a closed ring. (B) Ring compounds of the first four cycloalkanes.

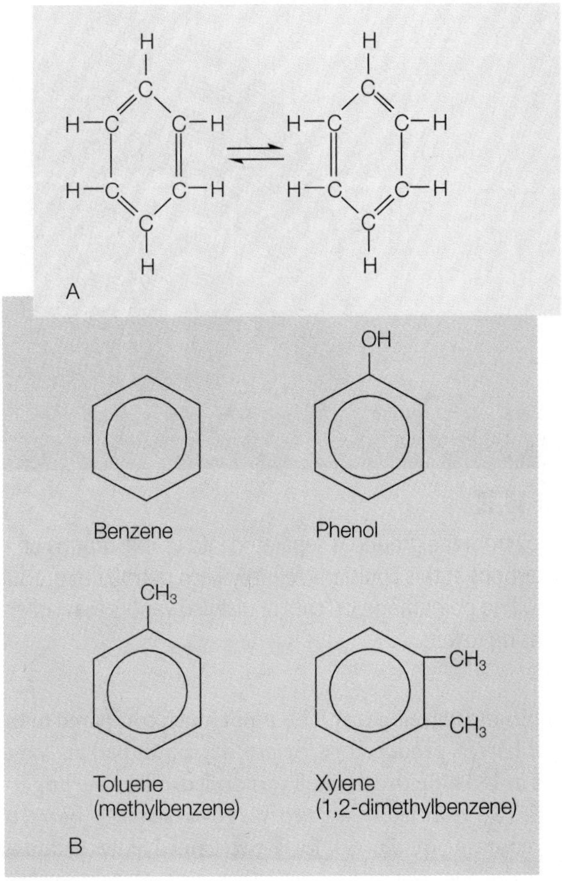

FIGURE 14.9

(A) The bonds in C_6H_6 are something between single and double, which gives it different chemical properties than double-bonded hydrocarbons. (B) The six-sided symbol with a circle represents the benzene ring. Organic compounds based on the benzene ring are called *aromatic hydrocarbons* because of their aromatic character.

on labels of paints, paint thinners, and lacquers. Toluene and the xylenes are commonly used in these products as solvents. A benzene ring attached to another molecule or functional group is given the name *phenyl*.

PETROLEUM

Petroleum is a mixture of alkanes, cycloalkanes, and some aromatic hydrocarbons. The origin of petroleum is uncertain, but it is believed to have formed from the slow anaerobic decomposition of buried marine life, primarily plankton and algae. Time, temperature, pressure, and perhaps bacteria are considered important in the formation of petroleum. As the petroleum formed, it was forced through porous rock until it reached a rock type or rock structure that stopped it. Here, it accumulated to saturate the porous rock, forming an accumulation called an **oil field.** The composition of petroleum varies from one oil field to the next. The oil from a given field might be dark or light in color, and it might have an asphalt base or a paraffin base. Some oil fields contain oil with a high quantity of sulfur, referred to as

"sour crude." Because of such variations, some fields have oil with more desirable qualities than oil from other fields.

In some locations an oil field occurs close to the surface, and petroleum seeps to the surface, often floating on water from a spring. Such seepage is the source of petroleum that has been collected and used since about 3000 B.C. Ancient Babylonians, Egyptians, and Roman civilizations used this oil for medicinal purposes, for paving roads, and when thickened by drying, as a caulking compound in early wooden ships.

Early settlers found oil seeps in the eastern United States and collected the oil for medicinal purposes. One enterprising oil peddler tried to improve the taste by running the petroleum through a whiskey still. He obtained a clear liquid by distilling the petroleum and, by accident, found that the liquid made an excellent lamp oil. This was fortunate timing, for the lamp oil used at that time was whale oil, and whale oil production was declining. This clear liquid obtained by distilling petroleum is today known as kerosene.

The first oil well was drilled in Titusville, Pennsylvania, in 1859. The well struck oil at a depth of seventy feet and produced

FIGURE 14.10

Crude oil from the ground is separated into usable groups of hydrocarbons at this Louisiana refinery. Each petroleum product has a boiling point range, or "cut," of distilled vapors that collect in condensing towers.

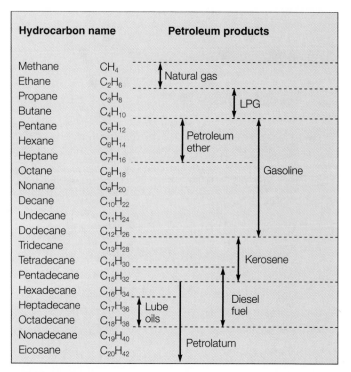

FIGURE 14.11

Petroleum products and the ranges of hydrocarbons in each product.

TABLE 14.4	Petroleum products	
Name	**Boiling Range (°C)**	**Carbon Atoms per Molecule**
Natural gas	Less than 0	C_1 to C_4
Petroleum ether	35–100	C_5 to C_7
Gasoline	35–215	C_5 to C_{12}
Kerosene	35–300	C_{12} to C_{15}
Diesel fuel	300–400	C_{15} to C_{18}
Motor oil, grease	350–400	C_{16} to C_{18}
Paraffin	Solid, melts at about 55	C_{20}
Asphalt	Boiler residue	C_{40} or more

two thousand barrels a year. This is not much compared to the billions of barrels produced per year today, but it had an economic impact in 1859. Before the well was drilled, oil was selling for $40 a barrel. Two years later the price was 10¢ a barrel. A "barrel of oil" is an accounting measure of forty-two United States gallons. Such a barrel size does not really exist. When or if oil is shipped in barrels, each drum holds fifty-five United States gallons.

Wells were drilled, and crude oil refineries were built to produce the newly discovered lamp oil. Gasoline was a by-product of the distillation process and was used primarily as a spot remover. With Henry Ford's automobile production and Edison's electric light invention, the demand for gasoline increased, and the demand for kerosene decreased. The refineries were converted to produce gasoline, and the petroleum industry grew to become one of the world's largest industries.

Crude oil is petroleum that is pumped from the ground, a complex and variable mixture of hydrocarbons with one or more carbon atoms, with an upper limit of about fifty atoms. This thick, smelly black mixture is not usable until it is refined, that is, separated into usable groups of hydrocarbons called petroleum products. The petroleum products are separated by distillation, and any particular product has a boiling point range, or "cut" of the distilled vapors (figure 14.10). Thus, each product, such as gasoline, heating oil, and so forth, is made up of hydrocarbons within a range of carbon atoms per molecule (figure 14.11). The products, their boiling ranges, and ranges of carbon atoms per molecule are listed in table 14.4.

The hydrocarbons that have one to four carbon atoms (CH_4 to C_4H_{10}) are gases at room temperature. They can be pumped from certain wells as a gas, but they also occur dissolved in crude oil. *Natural gas* is a mixture of hydrocarbon gases, but it is about 95 percent methane (CH_4). Propane (C_3H_8) and butane (C_4H_{10}) are liquified by compression and cooling and are sold as liquified petroleum gas, or *LPG*. LPG is used where natural gas is not

available for cooking or heating and is widely used as a fuel in barbecue grills and camp stoves.

Hydrocarbons with five to seven carbon atoms per molecule are volatile liquids at room temperature. Different groups of these closely related volatile hydrocarbons are used for various commercial purposes under the general heading of *petroleum ether,* also called "petroleum distillates," and naphtha. Petroleum ether is used as a cleaning fluid. It is also used as a solvent. Naphtha is also present in gasoline.

Gasoline is a mixture of over 500 hydrocarbons that may have five to twelve carbon atoms per molecule. Gasoline distilled from crude oil consists mostly of straight-chain molecules not suitable for use as an automotive fuel. Straight-chain molecules

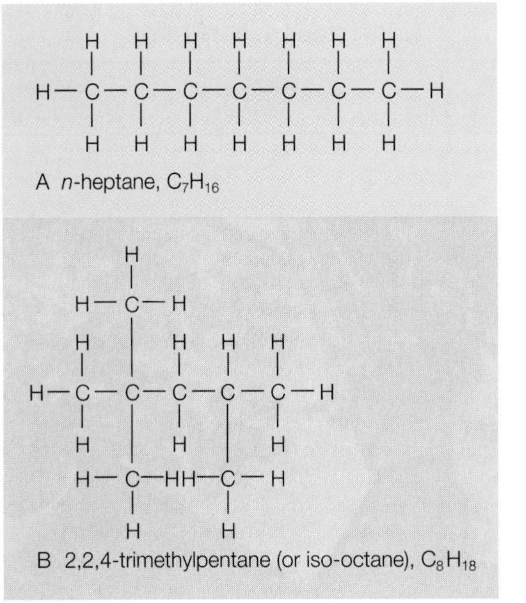

A *n*-heptane, C₇H₁₆

B 2,2,4-trimethylpentane (or iso-octane), C₈H₁₈

FIGURE 14.12

The octane rating scale is a description of how rapidly gasoline burns. It is based on (A) *n*-heptane, with an assigned octane number of 0, and (B) 2,2,4-trimethylpentane, with an assigned number of 100.

burn too rapidly in an automobile engine, producing more of an explosion than a smooth burn. You hear these explosions as a knocking or pinging in the engine, and they mean poor efficiency and could damage the engine. On the other hand, branched-chain molecules burn comparatively slower, without the pinging or knocking explosions. The burning rate of gasoline is described by the *octane number* scale. The scale is based on pure *n*-heptane, straight-chain molecules that are assigned an octane number of 0, and a multiple branched isomer of octane, 2,2,4-trimethylpentane, which is assigned an octane number of 100 (figure 14.12). Most unleaded gasolines have an octane rating of 87, which could be obtained with a mixture that is 87 percent 2,2,4-trimethylpentane and 13 percent *n*-heptane. Gasoline, however, is a much more complex mixture.

The octane rating of gasoline can be improved in one of two ways: (1) by adding a substance that slows the burning rate, such as tetraethyl lead, $(C_2H_5)_4Pb$ ("ethyl"), or (2) by converting some of the straight-chain hydrocarbons into branched-chain ones. The use of tetraethyl lead is less expensive but has been phased out because of increased concerns over lead pollution. It is more expensive to produce unleaded gasoline because some of the straight-chain hydrocarbon molecules must be converted into branched molecules. The process is one of "cracking and reforming" some of the straight-chain molecules. First, the gasoline is passed through metal tubes heated to 500°C to 800°C (932°F to 1,470°F). At this high temperature, and in the absence of oxygen, the hydrocarbon molecules decompose by breaking into smaller carbon-chain units. These smaller hydrocarbons are then passed through tubes containing a catalyst, which causes them to reform

into branched-chain molecules. Unleaded gasoline is produced by the process. Without the reforming that produces unleaded gasoline, low-numbered hydrocarbons (such as ethylene) can be produced. Ethylene is used as a raw material for many plastic materials, antifreeze, and other products. Cracking is also used to convert higher-numbered hydrocarbons, such as heating oil, into gasoline.

Kerosene is a mixture of hydrocarbons that have from twelve to fifteen carbon atoms. The petroleum product called kerosene is also known by other names, depending on its use. Some of these names are lamp oil (with coloring and odorants added), jet fuel (with a flash flame retardant added), heating oil, #1 fuel oil, and in some parts of the country, "coal oil."

Diesel fuel is a mixture of a group of hydrocarbons that have from fifteen to eighteen carbon atoms per molecule. Diesel fuel also goes by other names, again depending on its use. This group of hydrocarbons is called diesel fuel, distillate fuel oil, heating oil, or #2 fuel oil. During the summer season there is a greater demand for gasoline than for heating oil, so some of the supply is converted to gasoline by the cracking process.

Motor oil and *lubricating oils* have sixteen to eighteen carbon atoms per molecule. Lubricating grease is heavy oil that is thickened with soap. *Petroleum jelly,* also called petrolatum (or Vaseline), is a mixture of hydrocarbons with sixteen to thirty-two carbon atoms per molecule. *Mineral oil* is a light lubricating oil that has been decolorized and purified.

Depending on the source of the crude oil, varying amounts of *paraffin* wax (C_{20} or greater) or *asphalt* (C_{36} or more) may be present. Paraffin is used for candles, waxed paper, and home canning. Asphalt is mixed with gravel and used to surface roads. It is also mixed with refinery residues and lighter oils to make a fuel called #6 fuel oil or residual fuel oil. Industries and utilities often use this semisolid material that must be heated before it will pour. Number 6 fuel oil is used as a boiler fuel, costing about half as much as #2 fuel oil.

HYDROCARBON DERIVATIVES

The hydrocarbons account for only about 5 percent of the known organic compounds, but the other 95 percent can be considered as hydrocarbon derivatives. **Hydrocarbon derivatives** are formed when *one or more hydrogen atoms on a hydrocarbon have been replaced by some element or group of elements other than hydrogen.* For example, the halogens (F_2, Cl_2, Br_2) react with an alkane in sunlight or when heated, replacing a hydrogen:

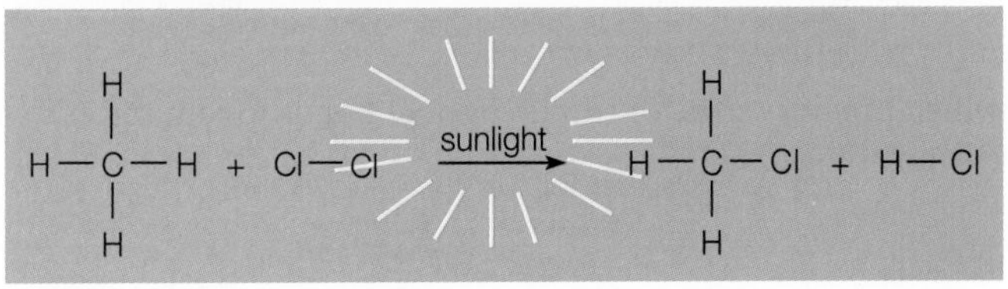

In this particular *substitution reaction* a hydrogen atom on methane is replaced by a chlorine atom to form methyl chloride. Replacement of any number of hydrogen atoms is possible, and a few *organic halides* are illustrated in figure 14.13.

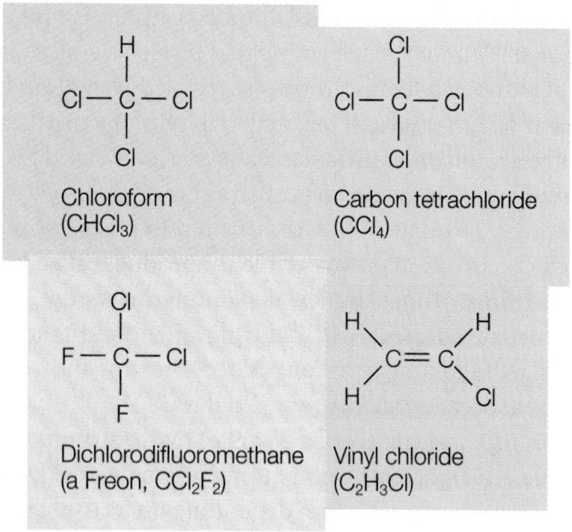

FIGURE 14.13

Common examples of organic halides.

If a hydrocarbon molecule is unsaturated (has a multiple bond), a hydrocarbon derivative can be formed by an *addition reaction:*

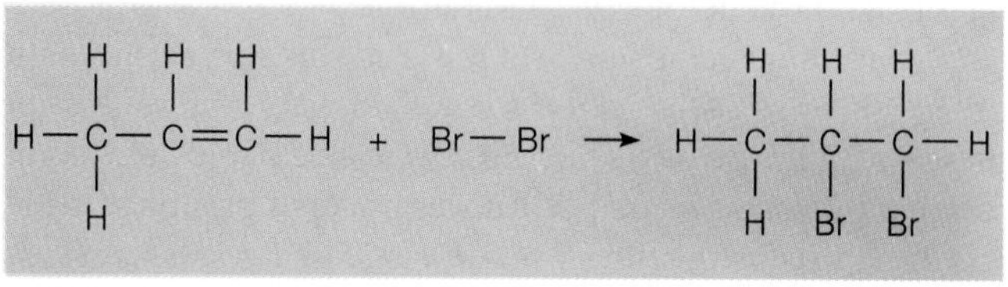

The bromine atoms add to the double bond on propene, forming 1,2-dibromopropane.

Alkene molecules can also add to each other in an addition reaction to form a very long chain consisting of hundreds of molecules. A long chain of repeating units is called a **polymer,** and the reaction is called *addition polymerization.* Ethylene, for example, is heated under pressure with a catalyst to form *polyethylene.* Heating breaks the double bond,

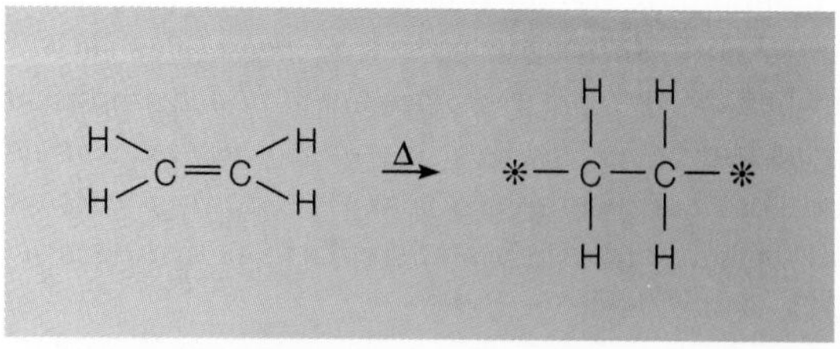

which provides sites for single covalent bonds to join the ethylene units together,

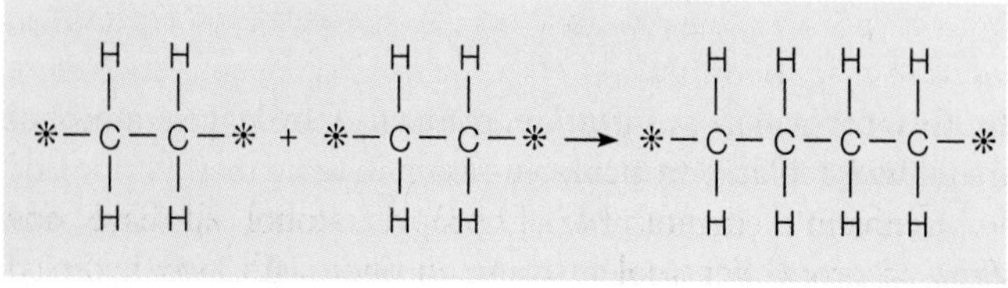

TABLE 14.5 **Selected organic functional groups**

Name of Functional Group	General Formula	General Structure
Organic Halide	RCl	$R-\overset{..}{\underset{..}{C}}l:$
Alcohol	ROH	$R-\overset{..}{\underset{..}{O}}-H$
Ether	ROR'	$R-\overset{..}{\underset{..}{O}}-R'$
Aldehyde	RCHO	$R-\overset{\parallel}{\underset{:O:}{C}}-H$
Ketone	RCOR'	$R-\overset{\parallel}{\underset{:O:}{C}}-R'$
Organic Acid	RCOOH	$R-\overset{\parallel}{\underset{:O:}{C}}-\overset{..}{O}-H$
Ester	RCOOR'	$R-\overset{\parallel}{\underset{:O:}{C}}-\overset{..}{O}-R'$
Amine	RNH$_2$	$R-\overset{..}{N}-H$ with H below

which continues the addition polymerization until the chain is hundreds of units long. Synthetic polymers such as polyethylene are discussed in a later section.

The addition reaction and the addition polymerization reaction can take place because of the double bond of the alkenes, and, in fact, the double bond is the site of most alkene reactions. The atom or group of atoms in an organic molecule that is the site of a chemical reaction is identified as a **functional group.** *It is the functional group that is responsible for the chemical properties of an organic compound.* Functional groups usually have (1) multiple bonds or (2) lone pairs of electrons that cause them to be sites of reactions. Table 14.5 lists some of the common hydrocarbon functional groups. Look over this list, comparing the structure of the functional group with the group name. Some of the more interesting examples from a few of these groups will be considered next. Note that the R and R' stand for one or more of the hydrocarbon groups from table 14.2. For example, in the reaction between methane and chlorine, the product is methyl chloride. In this case the R in RCl stands for methyl, but it could represent any hydrocarbon group.

Methanol

$$
\begin{array}{c}
\text{H} \\
| \\
\text{H} - \text{C} - \text{OH} \\
| \\
\text{H}
\end{array}
$$

(methyl alcohol)

Ethanol

$$
\begin{array}{c}
\text{H}\quad\text{H} \\
|\qquad| \\
\text{H} - \text{C} - \text{C} - \text{OH} \\
|\qquad| \\
\text{H}\quad\text{H}
\end{array}
$$

(ethyl alcohol)

1-propanol

$$
\begin{array}{c}
\text{H}\quad\text{H}\quad\text{H} \\
|\qquad|\qquad| \\
\text{H} - \text{C} - \text{C} - \text{C} - \text{OH} \\
|\qquad|\qquad| \\
\text{H}\quad\text{H}\quad\text{H}
\end{array}
$$

(n-propyl alcohol)

2-propanol

$$
\begin{array}{c}
\text{H}\quad\text{OH}\quad\text{H} \\
|\qquad|\qquad| \\
\text{H} - \text{C} - \text{C} - \text{C} - \text{H} \\
|\qquad|\qquad| \\
\text{H}\quad\text{H}\quad\text{H}
\end{array}
$$

(isopropyl alcohol)

FIGURE 14.14

Four different alcohols. The IUPAC name is given above each structural formula, and the common name is given below.

Alcohols

An *alcohol* is an organic compound formed by replacing one or more hydrogens on an alkane with a hydroxyl functional group (−OH). The hydroxyl group should not be confused with the hydroxide ion, OH⁻. The hydroxyl group is attached to an organic compound and does not form ions in solution as the hydroxide ion does. It remains attached to a hydrocarbon group (R), giving the compound its set of properties that are associated with alcohols.

The name of the hydrocarbon group (table 14.2) determines the name of the alcohol. If the hydrocarbon group in ROH is methyl, for example, the alcohol is called *methyl alcohol*. Using the IUPAC naming rules, the name of an alcohol has the suffix "-ol." Thus, the IUPAC name of methyl alcohol is *methanol*. If the molecule has a sufficient number of carbon atoms to require further definition, the base name is determined from the longest continuous chain of carbon atoms that has the −OH. The location of the hydroxyl group is identified with a number (figure 14.14).

All alcohols have the hydroxyl functional group, and all are chemically similar. Alcohols are toxic to humans, for example,

FIGURE 14.15

Gasoline is a mixture of hydrocarbons (C_8H_{18} for example) that contain no atoms of oxygen. Gasohol contains ethyl alcohol, C_2H_5OH, which does contain oxygen. The addition of alcohol to gasoline, therefore, adds oxygen to the fuel. Since carbon monoxide forms when there is an insufficient supply of oxygen, the addition of alcohol to gasoline helps cut down on carbon monoxide emissions. An atmospheric inversion, with increased air pollution, is likely during the dates shown on the pump, so that is when the ethanol is added.

except that ethanol can be consumed in limited quantities. Consumption of other alcohols such as 2-propanol (isopropyl alcohol) can result in serious gastric distress. Consumption of methanol can result in blindness and death. Ethanol, C_2H_5OH, is produced by the action of yeast or by a chemical reaction of ethylene derived from petroleum refining. Yeast acts on sugars to produce ethanol and CO_2. When beer, wine, and other such beverages are the desired products, the CO_2 escapes during fermentation, and the alcohol remains in solution. In baking, the same reaction utilizes the CO_2 to make the dough rise, and the alcohol is evaporated during baking. Most alcoholic beverages are produced by the yeast fermentation reaction, but some are made from ethanol derived from petroleum refining.

The hydroxyl group is strongly polar, and alcohols with six or fewer carbon atoms per molecule are soluble in both alkanes and water. A solution of ethanol and gasoline is called *gasohol* (figure 14.15). Alcoholic beverages are a solution of ethanol and water. The *proof* of such a beverage is double the ethanol concentration by volume. Therefore, a solution of 40 percent ethanol by volume in water is 80 proof, and wine that is 12 percent alcohol by volume is 24 proof. Distillation alone will produce a 190 proof concentration, but other techniques are necessary to obtain 200 proof absolute alcohol. *Denatured alcohol* is ethanol with acetone, formaldehyde, and other chemicals in solution that are difficult to separate by distillation. Since these denaturants make consumption impossible, denatured alcohol is sold without the consumption tax.

Methanol, ethanol, and isopropyl alcohol all have one hydroxyl group per molecule. An alcohol with two hydroxyl

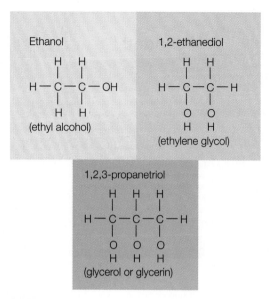

FIGURE 14.16

Common examples of alcohols with one, two, and three hydroxyl groups per molecule. The IUPAC name is given above each structural formula, and the common name is given below.

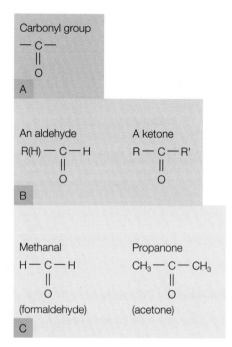

FIGURE 14.17

The carbonyl group (*A*) is present in both aldehydes and ketones, as shown in (*B*). (*C*) The simplest example of each, with the IUPAC name above and the common name below each formula.

groups per molecule is called a *glycol*. Ethylene glycol is perhaps the best-known glycol since it is used as an antifreeze. An alcohol with three hydroxyl groups per molecule is called *glycerol* (or *glycerin*). Glycerol is a by-product in the making of soap. It is added to toothpastes, lotions, and some candies to retain moisture and softness. Ethanol, ethylene glycol, and glycerol are compared in figure 14.16.

Glycerol reacts with nitric acid in the presence of sulfuric acid to produce glyceryl trinitrate, commonly known as *nitroglycerine*. Nitroglycerine is a clear oil that is violently explosive, and when warmed, it is extremely unstable. In 1867, Alfred Nobel discovered that a mixture of nitroglycerine and siliceous earth was more stable than pure nitroglycerine but was nonetheless explosive. The mixture is packed in a tube and is called *dynamite*. Old dynamite tubes, however, leak pure nitroglycerine that is again sensitive to a slight shock.

Ethers, Aldehydes, and Ketones

An *ether* has a general formula of ROR′, and the best-known ether is diethylether. In a molecule of diethylether, both the R and the R′ are ethyl groups. Diethylether is a volatile, highly flammable liquid that was used as an anesthetic in the past. Today, it is used as an industrial and laboratory solvent.

Aldehydes and *ketones* both have a functional group of a carbon atom doubly bonded to an oxygen atom called a *carbonyl group*. The *aldehyde* has a hydrocarbon group, R (or a hydrogen in one case), and a hydrogen attached to the carbonyl group. A *ketone* has a carbonyl group with two hydrocarbon groups attached (figure 14.17).

The simplest aldehyde is *formaldehyde*. Formaldehyde is soluble in water, and a 40 percent concentration called *formalin* has been used as an embalming agent and to preserve biological

specimens. Formaldehyde is also a raw material used to make plastics such as Bakelite. All the aldehydes have odors, and the odors of some aromatic hydrocarbons include the odors of almonds, cinnamon, and vanilla. The simplest ketone is *acetone*. Acetone has a fragrant odor and is used as a solvent in paint removers and nail polish removers. By sketching the structural formulas you can see that ethers are isomers of alcohols while aldehydes and ketones are isomers of each other. Again, the physical and chemical properties are quite different.

Organic Acids and Esters

Mineral acids, such as hydrochloric and sulfuric acid, are made of inorganic materials. Acids that were derived from organisms are called **organic acids.** Because many of these organic acids can be formed from fats, they are sometimes called *fatty acids.* Chemically, they are known as the *carboxylic acids* because they contain the carboxyl functional group, −COOH, and have a general formula of RCOOH.

The simplest carboxylic acid has been known since the Middle Ages, when it was isolated by the distillation of ants. The Latin word *formica* means "ant," so this acid was given the name *formic acid* (figure 14.18). Formic acid is

$$H - C - OH$$
$$\parallel$$
$$O$$

It is formic acid, along with other irritating materials, that causes the sting of bees, ants, and certain plants.

FIGURE 14.18

These red ants, like other ants, make the simplest of the organic acids, formic acid. The sting of bees, ants, and some plants contains formic acid, along with some other irritating materials. Formic acid is HCOOH.

TABLE 14.6	Flavors and esters	
Ester Name	**Formula**	**Flavor**
Amyl Acetate	$CH_3 - C - O - C_5H_{11}$ $\parallel$ O	Banana
Octyl Acetate	$CH_3 - C - O - C_8H_{17}$ $\parallel$ O	Orange
Ethyl Butyrate	$C_3H_7 - C - O - C_2H_5$ $\parallel$ O	Pineapple
Amyl Butyrate	$C_3H_7 - C - O - C_5H_{11}$ $\parallel$ O	Apricot
Ethyl Formate	$H - C - O - C_2H_5$ $\parallel$ O	Rum

Acetic acid, the acid of vinegar, has been known since antiquity. Acetic acid forms from the oxidation of ethanol. An oxidized bottle of wine contains acetic acid in place of the alcohol, which gives the wine a vinegar taste. Before wine is served in a restaurant, the person ordering is customarily handed the bottle cork and a glass with a small amount of wine. You first break the cork in half to make sure it is dry, which tells you that the wine has been sealed from oxygen. The small sip is to taste for vinegar before the wine is served. If the wine has been oxidized, the reaction is

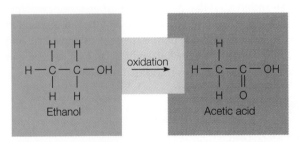

Ethanol   oxidation →   Acetic acid

Organic acids are common in many foods. The juice of citrus fruit, for example, contains citric acid, which relieves a thirsty feeling by stimulating the flow of saliva. Lactic acid is found in sour milk, buttermilk, sauerkraut, and pickles. Lactic acid also forms in your muscles as a product of carbohydrate metabolism, causing a feeling of fatigue. Citric and lactic acids are small molecules compared to some of the carboxylic acids that are formed from fats. Palmitic acid, for example, is $C_{16}H_{32}O_2$ and comes from palm oil. The structure of palmitic acid is a chain of fourteen CH_2 groups with CH_3- at one end and $-COOH$ at the other. Again, it is the functional carboxyl group, $-COOH$, that gives the molecule its acid properties. Organic acids are also raw materials used in the making of polymers of fabric, film, and paint.

Esters are common in both plants and animals, giving fruits and flowers their characteristic odor and taste. Esters are also used in perfumes and artificial flavorings. A few of the flavors that particular esters are responsible for are listed in table 14.6. These liquid esters can be obtained from natural sources or they can be chemically synthesized. Whatever the source, amyl acetate, for example, is the chemical responsible for what you identify as the flavor of banana. Natural flavors, however, are complex mixtures of these esters along with other organic compounds. Lower molecular weight esters are fragrant-smelling liquids, but higher molecular weight esters are odorless oils and fats. These are discussed in the next section along with carbohydrates and proteins.

Activities

Pick a household product that has ingredients sounding like they could be organic compounds. On a separate sheet of paper write the brand name of the product and the type of product (Example: Oil of Olay; skin moisturizer), then list the ingredients one under the other (column 1). In a second column beside each name put the type of compound if you can figure it out from its name or find it in any reference (Example: cetyl palmitate—an ester of cetyl alcohol and palmitic acid.) In a third column, put the structural formula if you can figure it out or find it in any reference such as a CRC handbook or the Merck Index. Finally, in a fourth column put whatever you know or can find out about the function of that substance in the product.

ORGANIC COMPOUNDS OF LIFE

Aristotle and the later proponents of the vitalist theory were *partly* correct in their concept that living organisms are different from inorganic substances made of the same elements. Living organisms, for example, have the ability to (1) exchange matter and energy with their surroundings and (2) transform matter and energy into different forms as they (3) respond to changes in their surroundings. In addition, living organisms can use the transformed matter and energy to (4) grow and (5) reproduce. Living organisms are able to do these things through a great variety of organic reactions that are catalyzed by enzymes, however, and not through some mysterious "vital force." These enzyme-regulated organic reactions take place because living organisms are highly organized and have an incredible number of relationships between many different chemical processes.

The chemical processes regulated by living organisms begin with relatively small organic molecules and water. The organism uses energy and matter from the surroundings to build large *macromolecules.* A **macromolecule** is a very large molecule that is a combination of many smaller, similar molecules joined together in a chainlike structure. Macromolecules have molecular weights of thousands or millions of atomic mass units. There are three main types of macromolecules: (1) proteins, (2) carbohydrates, and (3) nucleic acids, in addition to fats. A living organism, even a single-cell organism such as a bacterium, contains six thousand or so different kinds of macromolecules. The basic unit of an organism is called a *cell.* Cells are made of macromolecules that are formed inside the cell. The cell decomposes organic molecules taken in as food and uses energy from the food molecules to build more macromolecules. The process of breaking down organic molecules and building up macromolecules is called *metabolism.* Through metabolism, the cell grows, then divides into two cells. Each cell is a genetic duplicate of the other, even down to the number and kinds of macromolecules it contains. Each new cell continues the process of growth, then reproduces again, making more cells. This is the basic process of life. The complete process is complicated and very involved, easily filling a textbook in itself, so the details will not be presented here. The following discussion will be limited to three groups of organic molecules involved in the metabolic processes: proteins, carbohydrates, and fats and oils.

Proteins

Proteins are macromolecular polymers made up of smaller molecules called amino acids. These very large macromolecules have molecular weights that vary from about six thousand to fifty million. Some proteins are simple straight-chain polymers of amino acids, but others contain metal ions such as Fe^{2+} or parts of organic molecules derived from vitamins. Proteins serve as major structural and functional materials in living things. *Structurally,* proteins are major components of muscles, connective tissue, and the skin, hair, and nails. *Functionally,* some proteins are enzymes, which catalyze metabolic reactions; hormones, which regulate body activities; hemoglobin, which carries oxygen to cells; and antibodies, which protect the body.

Proteins are formed from 20 **amino acids,** which are organic acid molecules with acid and amino functional groups with the general formula of

Note the carbon atom labeled "alpha" in the general formula. The amino functional group (NH_2) is attached to this carbon atom, which is next to the carboxylic group (COOH). This arrangement is called an *alpha-amino acid,* and the building blocks of proteins are all alpha-amino acids. The 20 amino acids differ in the nature of the R group, also called the *side chain.* It is the side chain that determines the properties of a protein. Figure 14.19 gives the structural formula for the 20 amino acids found in most proteins and the three-letter abbreviations of the name of each amino acid.

Amino acids are linked to form a protein by a peptide bond between the amino group of one amino acid and the carboxyl group of a second amino acid. A polypeptide is a polymer formed from linking many amino acid molecules. If the polypeptide is involved in a biological structure or function, it is called a *protein.* A protein chain can consist of different combinations of the 20 amino acids with hundreds or even thousands of amino acid molecules held together with peptide bonds (figure 14.20). The arrangement or sequence of these amino acid molecules determines the structure that gives the protein its unique set of biochemical properties. Insulin, for example, is a protein hormone that biochemically regulates the blood sugar level. Insulin contains 86 amino acids that begin (at the amino group) with phenylalanine, valine, asparagine, and then 83 other amino acid molecules in the chain. Hemoglobin is the protein that carries oxygen in the bloodstream, and its biochemical characteristics are determined by its chain of 146 amino acid molecules.

Activities

Pick a food product and write the number of grams of fats, proteins, and carbohydrates per serving according to the information on the label. Multiply the number of grams of proteins and carbohydrates each by 4 Cal/g and the number of grams of fat by 9 Cal/g. Add the total Calories per serving. Does your total agree with the number of Calories per serving given on the label? Also examine the given serving size. Is this a reasonable amount to be consumed at one time or would you probably eat two or three times this amount? Write the rest of the nutrition information (vitamins, minerals, sodium content, etc.) and then write the list of ingredients. Tell what ingredient you think is providing which nutrient. (Example: vegetable oil—source of fat; milk—provides calcium and vitamin A; MSG—source of sodium.)

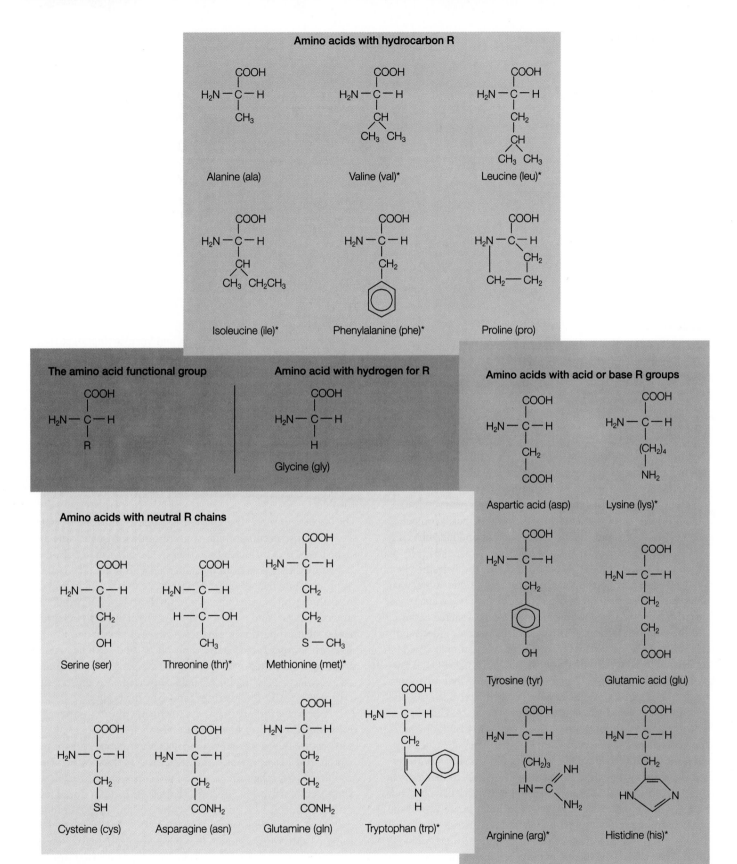

FIGURE 14.19

The twenty amino acids that make up proteins, with three-letter abbreviations. The carboxyl group of one amino acid bonds with the amino group of a second acid to yield a dipeptide and water. Proteins are polypeptides.

*Essential amino acids, which are amino acids that cannot be made by the human body and, therefore, must be obtained in the diet.

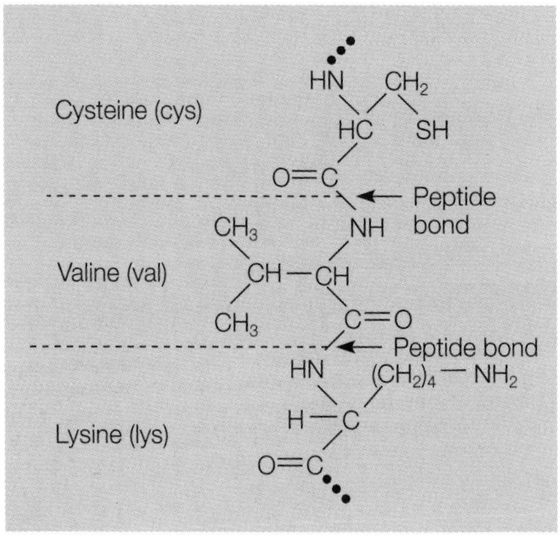

FIGURE 14.20

Part of a protein polypeptide made up of the amino acids cysteine (cys), valine (val), and lysine (lys). A protein can have from fifty to one thousand of these amino acid units; each protein has its own unique sequence.

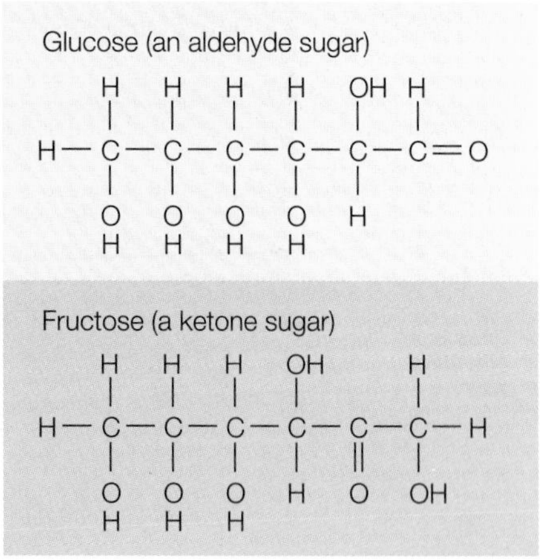

Carbohydrates

Carbohydrates are an important group of organic compounds that includes sugars, starches, and cellulose, and they are important in plants and animals for structure, protection, and food. Cellulose is the skeletal substance of plants and plant materials, and chitin is a similar material that forms the hard, protective covering of insects and shellfish such as crabs and lobsters. *Glucose,* $C_6H_{12}O_6$, is the most abundant carbohydrate and serves as a food and a basic building block for other carbohydrates.

Carbohydrates were named when early studies found that water vapor was given off and carbon remained when sugar was heated. The name *carbohydrate* literally means "watered carbon," and the empirical formulas for most carbohydrates indeed indicate carbon (C) and water (H_2O). Glucose, for example, could be considered to be six carbons with six waters, or $C_6(H_2O)_6$. However, carbohydrate molecules are more complex than just water attached to a carbon atom. They are polyhydroxyl aldehydes and ketones, two of which are illustrated in figure 14.21. The two carbohydrates in this illustration belong to a group of carbohydrates known as *monosaccharides,* or *simple sugars.* They are called simple sugars because they are the smallest units that have characteristics of carbohydrates and they can be combined to make larger complex carbohydrates. There are many kinds of simple sugars but 6-carbon molecules such as glucose and fructose are two of the most common simple sugars found in nature. Glucose (also called dextrose) is found in the sap of plants, and in the human bloodstream it is called *blood sugar.* Corn syrup, which is often used as a sweetener, is mostly glucose. Fructose, as its name implies, is the sugar that occurs in fruits, and it is sometimes called *fruit sugar.* Both glucose and fructose have the same molecular formula, but glucose is an aldehyde sugar and fructose is a ketone sugar (figure 14.21). A mixture of glucose and fructose is found in honey. This mixture also is formed when table sugar (sucrose) is reacted with water in the presence of an acid, a reaction that takes place in the

FIGURE 14.21

Glucose (blood sugar) is an aldehyde, and fructose (fruit sugar) is a ketone. Both have a molecular formula of $C_6H_{12}O_6$.

preparation of canned fruit and candies. The mixture of glucose and fructose is called *invert sugar.* Thanks to fructose, invert sugar is about twice as sweet to the taste as the same amount of sucrose.

Two monosaccharides are joined together to form *disaccharides* with the loss of a water molecule, for example,

$$C_6H_{12}O_6 + C_6H_{12}O_6 \rightarrow C_{12}H_{22}O_{11} + H_2O$$
$$\text{glucose} \quad \text{fructose} \quad \text{sucrose}$$

The most common disaccharide is *sucrose,* or ordinary table sugar. Sucrose occurs in high concentrations in sugar cane and sugar beets. It is extracted by crushing the plant materials, then dissolving the sucrose from the materials with water. The water is evaporated and the crystallized sugar is decolorized with charcoal to produce white sugar. Other common disaccharides include *lactose* (milk sugar) and *maltose* (malt sugar). All three disaccharides have similar properties, but maltose tastes only about one-third as sweet as sucrose. Lactose tastes only about one-sixth as sweet as sucrose. No matter which disaccharide sugar is consumed (sucrose, lactose, or maltose), it is converted into glucose and transported by the bloodstream for use by the body.

Polysaccharides are polymers consisting of monosaccharide units joined together in straight or branched chains. Polysaccharides are the energy-storage molecules of plants and animals (starch and glycogen) and the structural molecules of plants (cellulose). **Starches** are a group of complex carbohydrates composed of many glucose units that plants use as a stored food source. Potatoes, rice, corn, and wheat contain starch granules and serve as an important source of food for humans. The human body breaks down the starch molecules to glucose, which is transported by the bloodstream and utilized just like any other glucose. This digestive process begins with enzymes secreted with saliva in the mouth. You may have noticed a result of this enzyme-catalyzed reaction as you eat bread. If you chew the bread for awhile it begins to taste sweet.

Plants store sugars in the form of starch polysaccharides, and animals store sugars in the form of the polysaccharide *glycogen.* Glycogen is a starchlike polysaccharide that is synthesized by the

FIGURE 14.22

These plants and their flowers are made up of a mixture of carbohydrates that were manufactured from carbon dioxide and water, with the energy of sunlight. The simplest of the carbohydrates are the monosaccharides, simple sugars (fruit sugar) that the plant synthesizes. Food is stored as starches, which are polysaccharides made from the simpler monosaccharide, glucose. The plant structure is held upright by fibers of cellulose, another form of a polysaccharide.

human body and stored in the muscles and liver. Glycogen, like starch, is a very high molecular weight polysaccharide but it is more highly branched. These highly branched polysaccharides serve as a direct reserve source of energy in the muscles. In the liver, they serve as a reserve source to maintain the blood sugar level.

Cellulose is a polysaccharide that is abundant in plants, forming the fibers in cell walls that preserve the structure of plant materials (figure 14.22). Cellulose molecules are straight chains, consisting of large numbers of glucose units. These glucose units are arranged in a way that is very similar to the arrangement of the glucose units of starch but with differences in the bonding arrangement that holds the glucose units together (figure 14.23). This difference turns out to be an important one where humans are concerned, because enzymes that break down starches do not affect cellulose. Humans do not have the necessary enzymes to break down the cellulose chain (digest it), so humans receive no food value from cellulose. Cattle and termites that do utilize cellulose as a source of food have protozoa and bacteria (with the necessary enzymes) in their digestive systems. Cellulose is still needed in the human diet, however, for fiber and bulk.

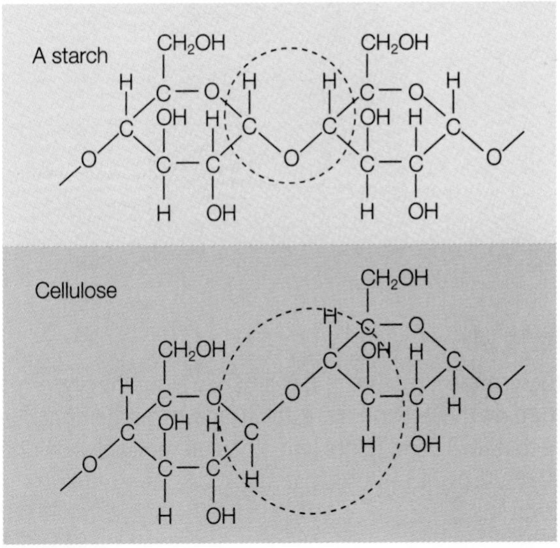

FIGURE 14.23

Starch and cellulose are both polymers of glucose, but humans cannot digest cellulose. The difference in the bonding arrangement might seem minor, but enzymes must fit a molecule very precisely. Thus, enzymes that break down starch do nothing to cellulose.

Fats and Oils

The human body can normally synthesize all of the amino acids needed to build proteins except for ten called the *essential amino acids* (see figure 14.19). An adequate diet must contain the ten essential amino acids, or health problems result. Meat and dairy products usually provide the essential amino acids, but they can also be acquired by combining cereal grains (corn, wheat, rice, etc.) with a legume (beans, peanuts, etc.). Interestingly, many ethnic foods have such a combination, for example, corn and beans (Mexican), rice and soybeans (tofu) (Japanese), and rice and red beans (Cajun).

Cereal grains and legumes also provide carbohydrates, the human body's preferred food for energy. When a sufficient amount of carbohydrates is consumed, the body begins to store some of its energy source in the form of glycogen in the muscles and liver. Beyond this storage for short-term needs, the body begins to store energy in a different chemical form for longer-term storage. This chemical form is called **fats** in animals and **oils** in plants. Fats and oils are esters formed from glycerol (1,2,3-trihydroxypropane) and three long-chain carboxylic acids (fatty acids). This ester is called a **triglyceride,** and its structural formula is shown in figure 14.24. Fats are solids and oils are liquids at room temperature, but they both have this same general structure.

Fats and oils usually have two or three different fatty acids, and several are listed in table 14.7. Animal fats can be either saturated or unsaturated, but most are saturated. Oils are liquids at room temperature because they contain a higher number of unsaturated units. These unsaturated oils (called "polyunsaturated" in news and advertisements), such as safflower and corn oils, are used as liquid cooking oils because unsaturated oils are believed to lead to lower cholesterol levels in the bloodstream.

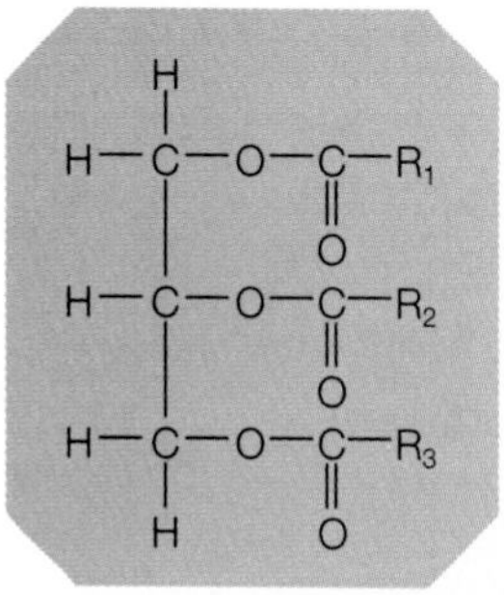

FIGURE 14.24

The triglyceride structure of fats and oils. Note the glycerol structure on the left and the ester structure on the right. Also notice that R_1, R_2, and R_3 are long-chained molecules of 12, 14, 16, 18, 20, 22, or 24 carbons that might be saturated or unsaturated.

TABLE 14.7 Some fatty acids occurring in fats

Common Name	Condensed Structure	Source
Lauric Acid	$CH_3(CH_2)_{10}COOH$	Coconuts
Palmitic Acid	$CH_3(CH_2)_{14}COOH$	Palm oil
Stearic Acid	$CH_3(CH_2)_{16}COOH$	Animal fats
Oleic Acid	$CH_3(CH_2)_7CH=CH(CH_2)_7COOH$	Corn oil
Linoleic Acid	$CH_3(CH_2)_4CH=CHCH_2=CH(CH_2)_7COOH$	Soybean oil
Linolenic Acid	$CH_3CH_2(CH=CHCH_2)_3(CH_2)_6COOH$	Fish oils

Saturated fats, along with cholesterol, are believed to contribute to hardening of the arteries over time.

Cooking oils from plants, such as corn and soybean oil, are hydrogenated to convert the double bonds of the unsaturated oil to the single bonds of a saturated one. As a result, the liquid oils are converted to solids at room temperature. For example, one brand of oleomargarine lists ingredients as "liquid soybean oil (nonhydrogenated) and partially hydrogenated cottonseed oil with water, salt, preservatives, and coloring." Complete hydrogenation would result in a hard solid, so the cottonseed oil is partially hydrogenated and then mixed with liquid soybean oil. Coloring is added because oleo is white, not the color of butter. Vegetable shortening is the very same product without added coloring. Reaction of a triglyceride with a strong base such as KOH or NaOH yields a fatty acid of salt and glycerol. A sodium or potassium fatty acid is commonly known as *soap.*

Excess food from carbohydrate, protein, or fat and oil sources is converted to fat for long-term energy storage in *adipose tissue,* which also serves to insulate and form a protective padding. In terms of energy storage, fats yield more than twice the energy per gram oxidized than carbohydrates or proteins.

Activities

Pick two competing brands of a product you use (example: Tylenol and a store-brand acetaminophen) and write the following information: All ingredients, amount of each ingredient (if this is not given, remember that by law ingredients have to be listed in order of their percent by weight), and the cost per serving or dose. Comment on whether each of the listed ingredients is a single substance, and thus something with the same properties wherever it is found (example: salt, as a label ingredient, means sodium chloride no matter in what product it appears) or a mixture and thus possibly different in different products (example: tomatoes, as a ketchup ingredient, might be of better quality in one brand of ketchup than another). Then draw a reasonably informed conclusion as to whether there is any significant difference between the two brands or whether the more expensive one is worth the difference in price. Finally, do your own consumer test to check your prediction.

SYNTHETIC POLYMERS

Polymers are huge, chainlike molecules made of hundreds or thousands of smaller, repeating molecular units called *monomers.* Polymers occur naturally in plants and animals. Cellulose, for example, is a natural plant polymer made of glucose monomers. Wool and hair are natural animal polymers made of protein monomers. Synthetic polymers are now manufactured from a wide variety of substances, and you are familiar with these polymers as synthetic fibers such as nylon and the inexpensive light plastic used for wrappings and containers (figure 14.25).

The first synthetic polymer was a modification of the naturally existing cellulose polymer. Cellulose was chemically modified in 1862 to produce celluloid, the first *plastic.* The term "plastic" means that celluloid could be molded to any desired shape. Celluloid was produced by first reacting cotton with a mixture of nitric and sulfuric acids, which produced an ester of cellulose nitrate. This ester is an explosive compound known as "guncotton," or smokeless gunpowder. When made with ethanol and camphor, the product is less explosive and can be formed and molded into useful articles. This first plastic, celluloid, was used to make dentures, combs, eyeglass frames, and photographic film. Before the discovery of celluloid, many of these articles, including dentures, were made from wood. Today, only Ping-Pong balls are made from cellulose nitrate.

Cotton reacted with acetic acid and sulfuric acid produces a cellulose acetate ester. This polymer, through a series of chemical reactions, produces viscose rayon filaments when forced through small holes. The filaments are twisted together to form viscose rayon thread. When forced through a thin slit, a sheet is formed rather than filaments, and the transparent sheet is called

Name	Chemical unit	Uses	Name	Chemical unit	Uses
Polyethylene	$-CH_2-CH_2-$ (repeat, n)	Squeeze bottles, containers, laundry and trash bags, packaging	Polyvinyl acetate	$[-CH_2-CH(O-CO-CH_3)-]_n$	Mixed with vinyl chloride to make vinylite; used as an adhesive and resin in paint
Polypropylene	$[-CH_2-CH(CH_3)-]_n$	Indoor-outdoor carpet, pipe valves, bottles	Styrene-butadiene rubber	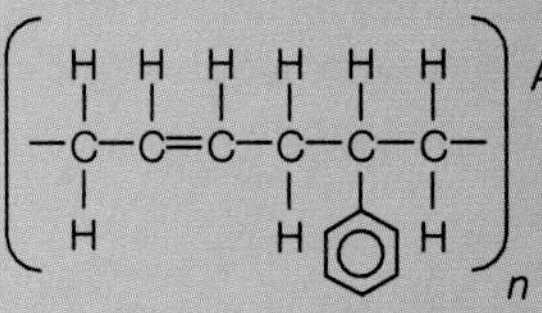 $[-CH_2-CH=CH-CH_2-CH_2-CH(C_6H_5)-]_n$	Automobile tires
Polyvinyl chloride (PVC)	$[-CH_2-CHCl-]_n$	Plumbing pipes, synthetic leather, plastic tablecloths, phonograph records, vinyl tile	Polychloroprene (Neoprene)	$[-CH_2-CCl=CH-CH_2-]_n$	Shoe soles, heels
Polyvinylidene chloride (Saran)	$[-CH_2-CCl_2-]_n$	Flexible food wrap	Polymethyl methacrylate (Plexiglas, Lucite)	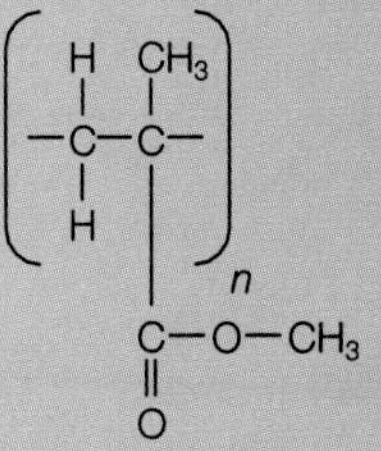 $[-CH_2-C(CH_3)(CO-O-CH_3)-]_n$	Moldings, transparent surfaces on furniture, lenses, jewelry, transparent plastic "glass"
Polystyrene (Styrofoam)	$[-CH_2-CH(C_6H_5)-]_n$	Coolers, cups, insulating foam, shock-resistant packing material, simulated wood furniture	Polycarbonate (Lexan)	$[-C_6H_4-C(CH_3)_2-C_6H_4-O-CO-O-]_n$	Tough, molded articles such as motorcycle helmets
Polytetrafluoroethylene (Teflon)	$[-CF_2-CF_2-]_n$	Gears, bearings, coating for nonstick surface of cooking utensils	Polyacrylonitrile (Orlon, Acrilan, Creslan)	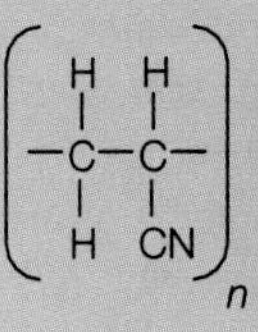 $[-CH_2-CH(CN)-]_n$	Textile fibers

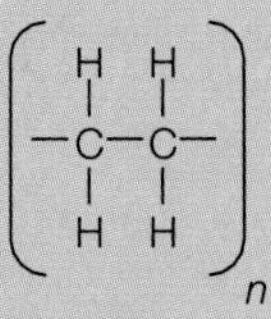

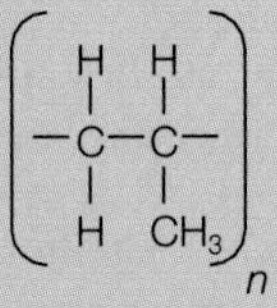

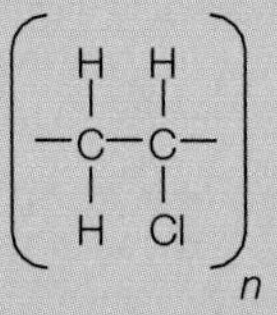

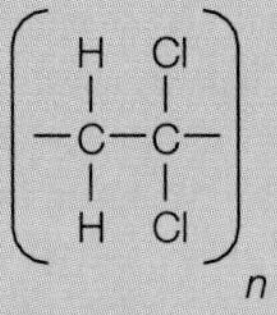

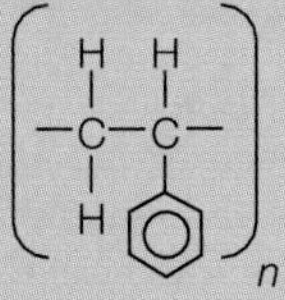

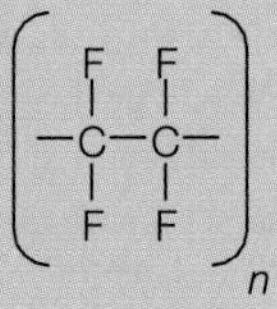

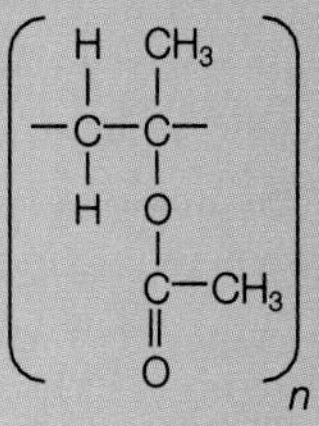

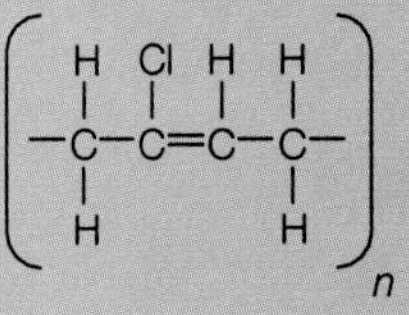

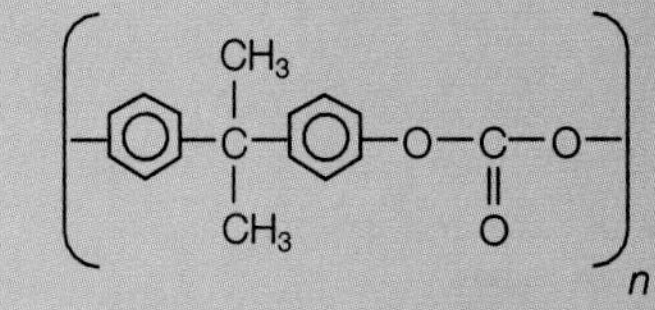

FIGURE 14.25

Synthetic polymers, the polymer unit, and some uses of each polymer.

cellophane. Both rayon and cellophane, as celluloid, are manufactured by modifying the natural polymer of cellulose.

The first truly synthetic polymer was produced in the early 1900s by reacting two chemicals with relatively small molecules rather than modification of a natural polymer. Phenol, an aromatic hydrocarbon, was reacted with formaldehyde, the simplest aldehyde, to produce the polymer named *Bakelite.* Bakelite is a *thermosetting* material that forms cross-links between the polymer chains. Once the links are formed during production, the plastic becomes permanently hardened and cannot be softened or made to flow. Some plastics are *thermoplastic* polymers

and soften during heating and harden during cooling because they do not have cross-links.

Polyethylene is a familiar thermoplastic polymer used for vegetable bags, dry cleaning bags, grocery bags, and plastic squeeze bottles. Polyethylene is a polymer produced by a polymerization reaction of ethylene, which is derived from petroleum. Polyethylene was invented just before World War II and was used as an electrical insulating material during the war. Today, there are many variations of polyethylene that are produced by different reaction conditions or by substitution of one or more hydrogen atoms in the ethylene molecule. When soft polyethylene near the melting point is rolled in alternating perpendicular directions or expanded and compressed as it is cooled, the polyethylene molecules become ordered in a way that improves the rigidity and tensile strength. This change in the microstructure produces *high-density polyethylene* with a superior rigidity and tensile strength compared to *low-density polyethylene*. High-density polyethylene is used in milk jugs, as liners in screw-on jar tops and bottle caps, and as a material for toys.

The properties of polyethylene are changed by replacing one of the hydrogen atoms in a molecule of ethylene. If the hydrogen is replaced by a chlorine atom the compound is called vinyl chloride, and the polymer formed from vinyl chloride is

$$
\begin{array}{ccc}
\text{H} & & \text{H} \\
 \diagdown & & \diagup \\
 & \text{C} = \text{C} & \\
 \diagup & & \diagdown \\
\text{H} & & \text{Cl}
\end{array}
$$

polyvinyl chloride (PVC). Polyvinyl chloride is used to make plastic water pipes, synthetic leather, and other vinyl products. It differs from the waxy plastic of polyethylene because of the chlorine atom that replaces hydrogen on each monomer.

Replacement of a hydrogen atom with a benzene ring makes a monomer called *styrene*. Styrene is

$$
\begin{array}{ccc}
\text{H} & & \text{H} \\
 \diagdown & & \diagup \\
 & \text{C} = \text{C} & \\
 \diagup & & \diagdown \\
\text{H} & & \bigcirc
\end{array}
$$

and polymerization of styrene produces *polystyrene.* Polystyrene is puffed full of air bubbles to produce the familiar Styrofoam coolers, cups, and insulating materials.

If all hydrogens of an ethylene molecule are replaced with atoms of fluorine, the product is polytetrafluoroethylene, a tough plastic that resists high temperatures and acts more like a metal than a plastic. Since it has a low friction it is used for bearings, gears, and as a nonsticking coating on frying pans. You probably know of this plastic by its trade name of *Teflon.*

There are many different polymers in addition to PVC, Styrofoam, and Teflon, and the monomers of some of these are shown in figure 14.25. There are also polymers of isoprene, or synthetic rubber, in wide use. Fibers and fabrics may be polyamides (such as nylon), polyesters (such as Dacron), or polyacrylonitriles (Orlon, Acrilan, Creslan), which have a CN in place of a hydrogen atom on an ethylene molecule and are called acrylic materials. All of these synthetic polymers have added much to practically every part of your life. It would be impossible to list all of their uses here; however, they present problems since (1) they are manufactured from raw materials obtained from coal and a dwindling petroleum supply (figure 14.26), and (2) they do not readily decompose when dumped into rivers, oceans, or other parts of the environment. However, research in the polymer sciences is beginning to reflect new understandings learned from research on biological tissues. This could lead to whole new molecular designs for synthetic polymers that will be more compatible with the ecosystems.

SUMMARY

Organic chemistry is the study of compounds that have carbon as the principal element. Such compounds are called *organic compounds,* and all the rest are *inorganic compounds.* There are millions of organic compounds because a carbon atom can link with other carbon atoms as well as atoms of other elements.

A *hydrocarbon* is an organic compound consisting of hydrogen and carbon atoms. The simplest hydrocarbon is one carbon atom and four hydrogen atoms, or CH_4. All hydrocarbons larger than CH_4 have one or more carbon atoms bonded to another carbon atom. The bond can be single, double, or triple, and this forms a basis for classifying hydrocarbons. A second basis is whether the carbons are in a ring or not. The *alkanes* are hydrocarbons with single carbon-to-carbon bonds, the *alkenes* have a double carbon-to-carbon bond, and the *alkynes* have a triple carbon-to-carbon bond. The alkanes, alkenes, and alkynes can have straight- or branched-chain molecules. When the number of carbon atoms is greater than three, there are different arrangements that can occur for a particular number of carbon atoms. The different arrangements with the same molecular formula are called isomers. *Isomers* have different physical properties, so each isomer is given its own name. The name is determined by (1) identifying the longest continuous carbon chain as the base name, (2) locating the attachment of other atoms or hydrocarbon groups by counting from the direction that results in the smallest numbers, (3) identifying attached hydrocarbon groups by changing the "-ane" suffix of alkanes to "-yl," (4) identifying the number of these hydrocarbon groups with prefixes, and (5) identifying the location of the groups with the carbon atom number.

The alkanes have all the hydrogen atoms possible, so they are *saturated* hydrocarbons. The alkenes and the alkynes can add more hydrogens to the molecule, so they are *unsaturated* hydrocarbons. Unsaturated hydrocarbons are more chemically reactive than saturated molecules.

Hydrocarbons that occur in a ring or cycle structure are cyclohydrocarbons. A six-carbon cyclohydrocarbon with three double bonds has different properties than the other cyclohydrocarbons because the double bonds are not localized. This six-carbon molecule is *benzene,* the basic unit of the *aromatic hydrocarbons.*

Petroleum is a mixture of alkanes, cycloalkanes, and a few aromatic hydrocarbons that formed from the slow decomposition of buried marine plankton and algae. Petroleum from the ground, or *crude oil,* is distilled into petroleum products of *natural gas, LPG, petroleum ether, gasoline, kerosene, diesel fuel,* and *motor oils.* Each group contains a range of hydrocarbons and is processed according to use.

In addition to oxidation, hydrocarbons react by *substitution, addition,* and *polymerization* reactions. Reactions take place at sites of multiple bonds or lone pairs of electrons on the *functional groups.* The functional group determines the chemical properties of organic compounds. Functional group results in the *hydrocarbon derivatives* of *alcohols, ethers, aldehydes, ketones, organic acids, esters,* and *amines.*

Living organisms have an incredible number of highly organized chemical reactions that are catalyzed by *enzymes,* using food and energy to grow and reproduce. The process involves building large *macromolecules* from smaller molecules and units. The organic molecules involved in the process are proteins, carbohydrates, and fats and oils.

Proteins are macromolecular polymers of *amino acids* held together by *peptide bonds.* There are 20 amino acids that are used in various polymer combinations to build structural and functional pro-

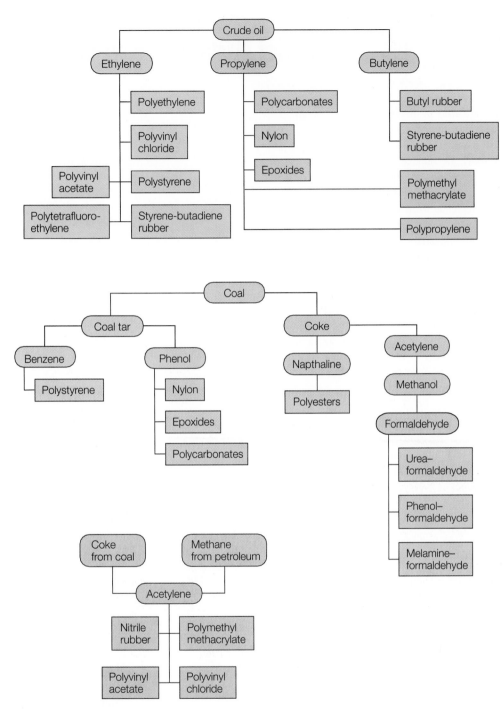

FIGURE 14.26

Petroleum and coal as sources of raw materials for manufacturing synthetic polymers.

teins. *Structural proteins* are muscles, connective tissue, and the skin, hair, and nails of animals. *Functional proteins* are enzymes, hormones, and antibodies.

Carbohydrates are polyhydroxyl aldehydes and ketones that form three groups, the monosaccharides, disaccharides, and polysaccharides. The *monosaccharides* are simple sugars such as *glucose* and *fructose.* Glucose is *blood sugar,* a source of energy. The disaccharides are *sucrose* (table sugar), *lactose* (milk sugar), and *maltose* (malt

sugar). The disaccharides are broken down (digested) to glucose for use by the body. The polysaccharides are polymers or glucose in straight or branched chains used as a near-term source of stored energy. Plants store the energy in the form of *starch,* and animals store it in the form of *glycogen. Cellulose* is a polymer similar to starch that humans cannot digest.

Fats and oils are esters formed from three fatty acids and glycerol into a *triglyceride. Fats* are usually solid triglycerides associated with

A CLOSER LOOK | How to Sort Plastic Bottles for Recycling

Plastic containers are made of different types of plastic resins; some are suitable for recycling and some are not. How do you know which are suitable and how to sort them? Most plastic containers have a code stamped on the bottom. The code is a number in the recycling arrow logo, sometimes appearing with some letters. Here is what the numbers and letters mean in terms of (a) the plastic, (b) how it is used, and (c) if it is usually recycled or not.

a. Polyethylene terephthalate (PET)
b. Large soft-drink bottles, salad dressing bottles
c. Frequently recycled

a. High-density polyethylene (HDPE)
b. Milk jugs, detergent and bleach bottles, others
c. Frequently recycled

a. Polyvinyl chloride (PVC or PV)
b. Shampoos, hair conditioners, others
c. Rarely recycled

a. Low-density polyethylene (LDPE)
b. Plastic wrap, laundry and trash bags
c. Rarely recycled

a. Polypropylene (PP)
b. Food containers
c. Rarely recycled

a. Polystyrene (PS)
b. Styrofoam cups, burger boxes, plates
c. Occasionally recycled

a. Mixed resins
b. Catsup squeeze bottles, other squeeze bottles
c. Rarely recycled

animals, and *oils* are liquid triglycerides associated with plant life, but both represent a high-energy storage material.

Polymers are huge, chain-like molecules of hundreds or thousands of smaller, repeating molecular units called *monomers*. Polymers occur naturally in plants and animals, and many *synthetic polymers* are made today from variations of the ethylene-derived monomers.

Among the more widely used synthetic polymers derived from ethylene are polyethylene, polyvinyl chloride, polystyrene, and Teflon. Problems with the synthetic polymers include that (1) they are manufactured from fossil fuels that are also used as the primary energy supply, and (2) they do not readily decompose and tend to accumulate in the environment.

A CLOSER LOOK | Organic Molecules and Heredity

D id you ever wonder how hereditary characteristics are passed on from parents to their children and how one protein becomes a muscle but another forms hair? Hereditary information and the synthesis of proteins are controlled and carried out by certain organic molecules. This reading is a brief introduction to the complicated and involved chemical processes that determine everything biological, from what kinds of proteins are formed as you grow, to why you are you.

Nucleic acids are organic polymers that give the self-replicating ability to living cells (box figure 14.1). There are two types of nucleic acid polymers: deoxyribonucleic acids (DNA) and ribonucleic acids (RNA). Both of these polymers consist of repeating units called *nucleotides*. The nucleotides of DNA are made up of (1) the organic base amines of thymine, adenine, guanine, and cytosine; (2) a phosphoric acid molecule; and (3) the simple sugar deoxyribose. RNA is different in structure in that thymine is replaced by a different organic base, uracil, and the deoxyribose is replaced by ribose.

In general, DNA is found in the nucleus of a cell. The linear sequences of nucleotides are the genetic codes of *chromosomes,* the cell structures that contain the DNA. A DNA molecule is very large, with molecular weights up to sixteen million, and is configured in a double helix. The phosphate and sugar groups are on the outside of this spiral arrangement, and the amines are inside, bonded in a specific way. When a cell divides, the DNA strands are separated, and each single strand takes other organic molecules to build two new, identical double helixes. The sequences of amines determine the hereditary information for the synthesis of proteins that work together to produce the particular organism being replicated.

Protein synthesis is not carried out by the DNA. The DNA only carries the genetic information. Information coded in the DNA is transmitted to structures called *ribosomes,* where the protein synthesis takes place, by RNA. Actually, there are three kinds of RNA: (1) messenger RNA, (2) transfer RNA, and

(3) ribosomal RNA. The transfer RNA is a complementary strand of the DNA, which produces a smaller messenger RNA molecule. Messenger RNA carries the pattern of protein synthesis in a sequence of three amines. The messenger RNA molecules move about the cell, attaching to the ribosomes, where they serve as templates for protein synthesis. The sequence of this template determines the kind of protein formed and thus the function and purpose of that protein. There are sixty-one

different arrangements, or codes, that the RNA template can carry that determine if muscle proteins or the proteins of hormones are produced, for example. It is a complicated and involved process and any small error, from the replication of DNA to the production of proteins, can lead to mutations or genetic diseases that result from faulty protein synthesis. Research in *genetic engineering*, the manipulation of DNA molecules, may someday enable scientists to correct such genetic errors.

A Ribose

B Deoxyribose

C Adenine

D Guanine

E Cytosine

F Uracil

G Thymine

H Phosphoric acid

BOX FIGURE 14.1

The building blocks of nucleotides are two simple sugars, (*A*) and (*B*), five amines, (*C*) through (*G*), and phosphoric acid molecules, (*H*).

KEY TERMS

alkanes (p. **324**)

alkenes (p. **327**)

alkyne (p. **328**)

amino acids (p. **336**)

carbohydrates (p. **338**)

cellulose (p. **339**)

fats (p. **339**)

functional group (p. **332**)

hydrocarbon (p. **323**)

hydrocarbon
    derivatives (p. **331**)

inorganic chemistry (p. **322**)

isomers (p. **324**)

macromolecule (p. **336**)

oil field (p. **329**)

oils (p. **339**)

organic acids (p. **334**)

organic chemistry (p. **322**)

petroleum (p. **329**)

polymer (p. **332**)

proteins (p. **336**)

saturated (p. **328**)

starches (p. **338**)

triglyceride (p. **339**)

unsaturated (p. **328**)

APPLYING THE CONCEPTS

1. An organic compound is a compound that
    a. contains carbon and was formed only by a living organism.
    b. is a natural compound that has not been synthesized.
    c. contains carbon, no matter if it was formed by a living thing or not.
    d. was formed by a plant.

2. There are millions of organic compounds but only thousands of inorganic compounds because
    a. organic compounds were formed by living things.
    b. there is more carbon on the earth's surface than any other element.
    c. atoms of elements other than carbon never combine with themselves.
    d. carbon atoms can combine with up to four other atoms, including other carbon atoms.

3. You know for sure that the compound named decane has
    a. more than 10 isomers.
    b. 10 carbon atoms in each molecule.
    c. only single bonds.
    d. all of the above.

4. An alkane with 4 carbon atoms would have how many hydrogen atoms in each molecule?
    a. 4
    b. 8
    c. 10
    d. 16

5. Isomers are compounds with the same
    a. molecular formula with different structures.
    b. molecular formula with different atomic masses.
    c. atoms, but different molecular formulas.
    d. structures, but different formulas.

6. Isomers have
    a. the same chemical and physical properties.
    b. the same chemical, but different physical properties.
    c. the same physical, but different chemical properties.
    d. different physical and chemical properties.

7. The organic compound 2,2,4-trimethylpentane is an isomer of
    a. propane.
    b. pentane.
    c. heptane.
    d. octane.

8. The hydrocarbons with a double covalent carbon-carbon bond are called
    a. alkanes.
    b. alkenes.
    c. alkynes.
    d. none of the above.

9. According to their definitions, which of the following would not occur as unsaturated hydrocarbons?
    a. alkanes
    b. alkenes
    c. alkynes
    d. None of the above are correct.

10. Petroleum is believed to have formed mostly from the anaerobic decomposition of buried
    a. dinosaurs.
    b. fish.
    c. pine trees.
    d. plankton and algae.

11. The label on a container states that the product contains "petroleum distillates." Which of the following hydrocarbons is probably present?
    a. CH_4
    b. C_5H_{12}
    c. $C_{16}H_{34}$
    d. $C_{40}H_{82}$

12. Tetraethyl lead ("ethyl") was added to gasoline to increase the octane rating by
    a. increasing the power by increasing the heat of combustion.
    b. increasing the burning rate of the gasoline.
    c. absorbing the pings and knocks.
    d. decreasing the burning rate.

13. The reaction of $C_2H_2 + Br_2 \rightarrow C_2H_2Br_2$ is a
    a. substitution reaction.
    b. addition reaction.
    c. addition polymerization reaction.
    d. substitution polymerization reaction.

14. Ethylene molecules can add to each other in a reaction to form a long chain called a
    a. monomer.
    b. dimer.
    c. trimer.
    d. polymer.

15. Chemical reactions usually take place on an organic compound at the site of (a)

    a. double bond.

    b. lone pair of electrons.

    c. functional group.

    d. any of the above.

16. The R in ROH represents

    a. a functional group.

    b. a hydrocarbon group with a name ending in "-yl."

    c. an atom of an inorganic element.

    d. a polyatomic ion that does not contain carbon.

17. The OH in ROH represents

    a. a functional group.

    b. a hydrocarbon group with a name ending in "-yl."

    c. the hydroxide ion, which ionizes to form a base.

    d. the site of chemical activity in a strong base.

18. What is the proof of a "wine cooler" that is 5 percent alcohol by volume?

    a. 2.5 proof

    b. 5 proof

    c. 10 proof

    d. 50 proof

19. An alcohol with two hydroxyl groups per molecule is called

    a. ethanol.

    b. glycerol.

    c. glycerin.

    d. glycol.

20. A bottle of wine that has "gone bad" now contains

    a. CH_3OH.

    b. CH_3OCH_3.

    c. CH_3COOH.

    d. CH_3COOCH_3.

21. A protein is a polymer formed from the linking of many

    a. glucose units.

    b. DNA molecules.

    c. amino acid molecules.

    d. monosaccharides.

22. Which of the following is *not* converted to blood sugar by the human body?

    a. lactose

    b. dextrose

    c. cellulose

    d. glycogen

23. Fats from animals and oils from plants have the general structure of a (an)

    a. aldehyde.

    b. ester.

    c. amine.

    d. ketone.

24. Liquid oils from plants can be converted to solids by adding what to the molecule?

    a. metal ions

    b. carbon

    c. polyatomic ions

    d. hydrogen

25. The basic difference between a monomer of polyethylene and a monomer of polyvinyl chloride is

    a. the replacement of a hydrogen by a chlorine.

    b. the addition of four fluorines.

    c. the elimination of double bonds.

    d. the removal of all hydrogens.

26. Many synthetic polymers become a problem in the environment because they

    a. decompose to nutrients, which accelerates plant growth.

    b. do not readily decompose and tend to accumulate.

    c. do not contain vitamins as natural materials do.

    d. become a source of food for fish, but ruin the flavor of fish meat.

Answers

1. c 2. d 3. d 4. c 5. a 6. d 7. d 8. b 9. a 10. d 11. b 12. d 13. b 14. d
15. d 16. b 17. a 18. c 19. d 20. c 21. c 22. c 23. b 24. d 25. a 26. b

QUESTIONS FOR THOUGHT

1. What is an organic compound?

2. There are millions of organic compounds but only thousands of inorganic compounds. Explain why this is the case.

3. What is cracking and reforming? For what purposes are either or both used by the petroleum industry?

4. Is it possible to have an isomer of ethane? Explain.

5. Suggest a reason that ethylene is an important raw material used in the production of plastics but ethane is not.

6. What is the size of the "barrel of oil" that is described in news reports?

7. What are (a) natural gas, (b) LPG, and (c) petroleum ether?

8. What does the octane number of gasoline describe? On what is the number based?

9. Why is unleaded gasoline more expensive than gasoline with lead additives?

10. What is a functional group? What is it about the nature of a functional group that makes it the site of chemical reactions?

11. Draw a structural formula for alcohol. Describe how alcohols are named.

12. A soft drink is advertised to "contain no sugar." The label lists ingredients of carbonated water, dextrose, corn syrup, fructose, and flavorings. Evaluate the advertising and the list of ingredients.

13. What are fats and oils? What are saturated and unsaturated fats and oils?

14. What is a polymer? Give an example of a naturally occurring plant polymer. Give an example of a synthetic polymer.

15. Explain why a small portion of wine is customarily poured before a bottle of wine is served. Sometimes the cork is handed to the person doing the ordering with the small portion of wine. What is the person supposed to do with the cork and why?

PARALLEL EXERCISES

The exercises in groups A and B cover the same concepts. Solutions to group A exercises are located in appendix D.

Group A

1. Draw the structural formulas for (a) *n*-pentane, and (b) an isomer of pentane with the maximum possible branching. (c) Give the IUPAC name of this isomer.
2. Write structural formulas for all the hexane isomers you can identify. Write the IUPAC name for each isomer.
3. Write structural formulas for
   a. 3,3,4-trimethyloctane.
   b. 2-methyl-1-pentene.
   c. 5,5-dimethyl-3-heptyne.
4. Write the IUPAC name for each of the following.

Group A (Continued)

5. Which would have the higher octane rating, 2,2,3-trimethylbutane or 2,2-dimethylpentane? Explain with an illustration.

6. Use the information in table 14.5 to classify each of the following as an alcohol, ether, organic acid, ester, or amide.

A

$$H-\underset{\underset{H}{|}}{\overset{\overset{H}{|}}{C}}-\underset{\underset{H}{|}}{\overset{\overset{H}{|}}{C}}-\underset{\underset{H}{|}}{\overset{\overset{H}{|}}{C}}-OH$$

D

$$H-\underset{\underset{H}{|}}{\overset{\overset{H}{|}}{C}}-\underset{\underset{H}{|}}{\overset{\overset{H}{|}}{C}}-O-\underset{\underset{O}{\|}}{C}-\underset{\underset{H}{|}}{\overset{\overset{H}{|}}{C}}-\underset{\underset{H}{|}}{\overset{\overset{H}{|}}{C}}-\underset{\underset{H}{|}}{\overset{\overset{H}{|}}{C}}-H$$

B

$$H-\underset{\underset{H}{|}}{\overset{\overset{H}{|}}{C}}-\underset{\underset{H}{|}}{\overset{\overset{H}{|}}{C}}-\underset{\underset{H}{|}}{\overset{\overset{H}{|}}{C}}-NH_2$$

E

$$H-\underset{\underset{H}{|}}{\overset{\overset{H}{|}}{C}}-\underset{\underset{H}{|}}{\overset{\overset{H}{|}}{C}}-\underset{\underset{H}{|}}{\overset{\overset{H}{|}}{C}}-\underset{\underset{H}{|}}{\overset{\overset{H}{|}}{C}}-\underset{\underset{O}{\|}}{C}-OH$$

C

$$H-\underset{\underset{H}{|}}{\overset{\overset{H}{|}}{C}}-\underset{\underset{H}{|}}{\overset{\overset{H}{|}}{C}}-\underset{\underset{H}{|}}{\overset{\overset{H}{|}}{C}}-O-\underset{\underset{H}{|}}{\overset{\overset{H}{|}}{C}}-\underset{\underset{H}{|}}{\overset{\overset{H}{|}}{C}}-\underset{\underset{H}{|}}{\overset{\overset{H}{|}}{C}}-H$$

Group B

1. Write structural formulas for (a) *n*-octane, and (b) an isomer of octane with the maximum possible branching. (c) Give the IUPAC name of this isomer.

2. Write the structural formulas for all the heptane isomers you can identify. Write the IUPAC name for each isomer.

3. Write structural formulas for
   a. 2,3-dimethylpentane.
   b. 1-butene.
   c. 3-ethyl-2-methyl-3-hexene.

4. Write the IUPAC name for each of the following.

A

$$H-\underset{\underset{H}{|}}{\overset{\overset{H}{|}}{C}}-\underset{\underset{H}{|}}{\overset{\overset{\overset{\overset{H}{|}}{C}}{|}}{C}}-\underset{\underset{Br}{|}}{\overset{\overset{H}{|}}{C}}-\underset{\underset{H}{|}}{\overset{\overset{H}{|}}{C}}-H$$

C

$$H-\underset{\underset{H}{|}}{\overset{\overset{H}{|}}{C}}-\underset{\underset{Cl}{|}}{\overset{\overset{H}{|}}{C}}=C-\underset{\underset{H}{|}}{\overset{\overset{H}{|}}{C}}=\underset{\underset{H}{|}}{C}-\underset{\underset{H}{|}}{\overset{\overset{H}{|}}{C}}-H$$

B

$$H-\underset{\underset{H}{|}}{\overset{\overset{H}{|}}{C}}-\underset{}{\overset{\overset{H}{|}}{C}}=\underset{}{C}-\underset{\underset{|}{}}{\overset{\overset{H}{|}}{C}}-\underset{\underset{|}{}}{\overset{\overset{H}{|}}{C}}-\underset{\underset{H}{|}}{\overset{\overset{H}{|}}{C}}-H$$

D

$$Br-\underset{\underset{H}{|}}{\overset{\overset{H}{|}}{C}}-\underset{}{\overset{\overset{H}{|}}{C}}=\underset{}{C}-\underset{\underset{H}{|}}{\overset{\overset{H}{|}}{C}}-H$$

Group B (Continued)

5. Which would have the higher octane rating, 2-methyl-butane or dimethylpropane? Explain with an illustration.

6. Classify each of the following as an alcohol, ether, organic acid, ester, or amide.

With the top half of the steel vessel and control rods removed, fuel rod bundles can be replaced in the water-flooded nuclear reactor.

CHAPTER | Fifteen

Nuclear Reactions

T he ancient alchemist dreamed of changing one element into another, such as lead into gold. The alchemist was never successful, however, because such changes were attempted with chemical reactions. Chemical reactions are reactions that involve only the electrons of atoms. Electrons are shared or transferred in chemical reactions, and the internal nucleus of the atom is unchanged. Elements thus retain their identity during the sharing or transferring of electrons. This chapter is concerned with a different kind of reaction, one that involves the *nucleus* of the atom. In nuclear reactions, the nucleus of the atom is often altered, changing the identity of the elements involved. The ancient alchemist's dream of changing one element into another was actually a dream of achieving a nuclear change, that is, a nuclear reaction.

Understanding nuclear reactions is important because although fossil fuels are the major source of energy today, there are growing concerns about (1) air pollution from fossil fuel combustion, (2) increasing levels of CO_2 from fossil fuel combustion, which may be warming the earth (the greenhouse effect), and (3) the dwindling fossil fuel supply itself, which cannot last forever. Energy experts see nuclear energy as a means of meeting rising energy demands in an environmentally acceptable way. However, the topic of nuclear energy is controversial, and discussions of it often result in strong emotional responses. Decisions about the use of nuclear energy require some understandings about nuclear reactions and some facts about radioactivity and radioactive materials (figure 15.1). These understandings and facts are the topics of this chapter.

NATURAL RADIOACTIVITY

Radioactivity was discovered in 1896 by Henri Becquerel, a French scientist who was very interested in the recent discovery of X rays. Becquerel was experimenting with fluorescent minerals, minerals that give off visible light after being exposed to sunlight. He wondered if fluorescent minerals emitted X rays in addition to visible light. From previous work with X rays, Becquerel knew that they would penetrate a wrapped, light-tight photographic plate, exposing it as visible light exposes an unprotected plate. Thus, Becquerel decided to place a fluorescent uranium mineral on a protected photographic plate while the mineral was exposed to sunlight. Sure enough, he found a silhouette of the mineral on the plate when it was developed. Believing the uranium mineral emitted X rays, he continued his studies until the weather turned cloudy. Storing a wrapped, protected photographic plate and the uranium mineral together during the cloudy weather, Becquerel returned to the materials later and developed the photographic plate to again find an image of the mineral (figure 15.2). He concluded that the mineral was emitting an "invisible radiation" that was not induced by sunlight. Becquerel named the emission of invisible radiation *radioactivity*. Materials that have the property of radioactivity are called *radioactive* materials.

Becquerel's discovery led to the beginnings of the modern atomic theory and to the discovery of new elements. Ernest Rutherford studied the nature of radioactivity and found that there are three kinds, which are today known by the first three letters of the Greek alphabet—alpha (α), beta (β), and gamma (γ). These Greek letters were used at first before the nature of the radiation was known. Today, an **alpha particle** (sometimes called an alpha ray) is known to be the nucleus of a helium atom,

that is, two protons and two neutrons. A **beta particle** (or beta ray) is a high-energy electron. A **gamma ray** is electromagnetic radiation, as is light, but of very short wavelength (figure 15.3).

It was Rutherford's work with alpha particles that resulted in the discovery of the nucleus and the proton (see chapter 9). At Becquerel's suggestion, Madame Marie Curie searched for other radioactive materials and, in the process, discovered two new elements, polonium and radium. More radioactive elements have been discovered since that time, and, in fact, all the isotopes of all the elements with an atomic number greater than 83 (bismuth) are radioactive. Today, **radioactivity** is defined as the *spontaneous emission of particles or energy from an atomic nucleus* as it disintegrates. As a result of the disintegration, the nucleus of an atom often undergoes a change of identity, becoming a simpler nucleus. The spontaneous disintegration of a given nucleus is a purely natural process and cannot be controlled or influenced. The natural spontaneous disintegration or decomposition of a nucleus is also called **radioactive decay.** Although it is impossible to know *when* a given nucleus will undergo radioactive decay, as you will see later, it is possible to deal with the *rate* of decay for a given radioactive material with precision.

Nuclear Equations

There are two main subatomic particles in the nucleus, the proton and the neutron. The proton and neutron are called **nucleons.** Recall that the number of protons, the *atomic number*, determines what element an atom is, and that all atoms of a given element have the same number of protons. The number of neutrons varies in *isotopes*, which are atoms with the same atomic number but different numbers of neutrons (figure 15.4).

A

FIGURE 15.1

Decisions about nuclear energy require some understanding of nuclear reactions and the nature of radioactivity. This is one of the three units of the Palo Verde Nuclear Generating Station in Arizona. With all three units running, enough power is generated to meet the electrical needs of nearly 4 million people.

The number of protons and neutrons together determines the *mass number*, so different isotopes of the same element are identified with their mass numbers. Thus, the two most common, naturally occurring isotopes of uranium are referred to as uranium-238 and uranium-235, and the 238 and 235 are the mass numbers of these isotopes. Isotopes are also represented by the following symbol:

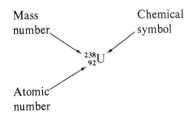

Subatomic particles involved in nuclear reactions are represented by symbols with the following form:

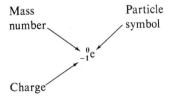

Symbols for these particles are illustrated in table 15.1.

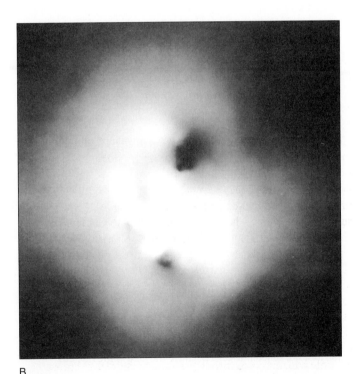

B

FIGURE 15.2

Radioactivity was discovered by Henri Becquerel when he exposed a light-tight photographic plate to a radioactive mineral, then developed the plate. (*A*) A photographic film is exposed to an uranite ore sample. (*B*) The film, developed normally after a four-day exposure to uranite. Becquerel found an image like this one and deduced that the mineral gave off invisible radiation that he called radioactivity.

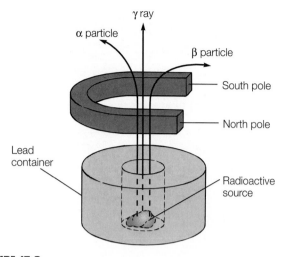

FIGURE 15.3

Radiation passing through a magnetic field shows that massive, positively charged alpha particles are deflected one way, and less massive beta particles with their negative charge are greatly deflected in the opposite direction. Gamma rays, like light, are not deflected.

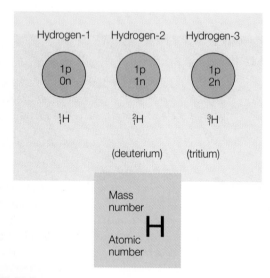

FIGURE 15.4

The three isotopes of hydrogen have the same number of protons but different numbers of neutrons. Hydrogen-1 is the most common isotope. Hydrogen-2, with an additional neutron, is named *deuterium,* and hydrogen-3 is called *tritium.* Neutrons and protons are called *nucleons* because they are in the nucleus.

Symbols are used in an equation for a nuclear reaction that is written much like a chemical reaction with reactants and products. When a uranium-238 nucleus emits an alpha particle ($_2^4\text{He}$), for example, it loses two protons and two neutrons. The nuclear reaction is written in equation form as

$$_{92}^{238}\text{U} \rightarrow {}_{90}^{234}\text{Th} + {}_2^4\text{He}$$

The *products* of this nuclear reaction from the decay of a uranium-238 nucleus are (1) the alpha particle ($_2^4\text{He}$) given off and (2) the

TABLE 15.1 Names, symbols, and properties of particles in nuclear equations

Name	Symbol	Mass Number	Charge
Proton	$_1^1\text{H}$ (or $_1^1\text{p}$)	1	1+
Electron	$_{-1}^0\text{e}$ (or $_{-1}^0\beta$)	0	1−
Neutron	$_0^1\text{n}$	1	0
Gamma photon	$_0^0\gamma$	0	0

nucleus, which remains after the alpha particle leaves the original nucleus. What remains is easily determined since all nuclear equations must show conservation of charge and conservation of the total number of nucleons. Therefore, (1) the number of protons (positive charge) remains the same, and the sum of the subscripts (atomic number, or numbers of protons) in the reactants must equal the sum of the subscripts in the products; and (2) the total number of nucleons remains the same, and the sum of the superscripts (atomic mass, or number of protons plus neutrons) in the reactants must equal the sum of the superscripts in the products. The new nucleus remaining after the emission of an alpha particle, therefore, has an atomic number of 90 (92 − 2 = 90). According to the table of atomic numbers on the inside back cover of this text, this new nucleus is thorium (Th). The mass of the thorium isotope is 238 minus 4, or 234. The emission of an alpha particle thus decreases the number of protons by 2 and the mass number by 4. From the subscripts, you can see that the total charge is conserved (92 = 90 + 2). From the superscripts, you can see that the total number of nucleons is also conserved (238 = 234 + 4). The mass numbers (superscripts) and the atomic numbers (subscripts) are *balanced* in a correctly written nuclear equation. Such nuclear equations are considered to be independent of any chemical form or chemical reaction. Nuclear reactions are independent and separate from chemical reactions, whether or not the atom is in the pure element or in a compound. Each particle that is involved in nuclear reactions has its own symbol with a superscript indicating mass number and a subscript indicating the charge. These symbols, names, and numbers are given in table 15.1.

Q & A

A plutonium-242 nucleus undergoes radioactive decay, emitting an alpha particle. Write the nuclear equation for this nuclear reaction.

Answer

Step 1: The table of atomic weights on the inside back cover gives the atomic number of plutonium as 94. Plutonium-242 therefore has a symbol of $_{94}^{242}\text{Pu}$. The symbol for an alpha particle is ($_2^4\text{He}$), so the nuclear equation so far is

$$_{94}^{242}\text{Pu} \rightarrow {}_2^4\text{He} + \text{?}$$

Step 2: From the subscripts, you can see that $94 = 2 + 92$, so the new nucleus has an atomic number of 92. The table of atomic weights identifies element 92 as uranium with a symbol of U.

Step 3: From the superscripts, you can see that the mass number of the uranium isotope formed is $242 - 4 = 238$, so the product nucleus is $^{238}_{92}U$ and the complete nuclear equation is

$$^{242}_{94}Pu \rightarrow \, ^{4}_{2}He \, + \, ^{238}_{92}U$$

Step 4: Checking the subscripts $(94 = 2 + 92)$ and the superscripts $(242 = 4 + 238)$, you can see that the nuclear equation is balanced.

What is the product nucleus formed when radium emits an alpha particle? (Answer: Radon-222, a chemically inert, radioactive gas)

The Nature of the Nucleus

The modern atomic theory does not picture the nucleus as a group of stationary protons and neutrons clumped together by some "nuclear glue." The protons and neutrons are understood to be held together by a *nuclear force*, a strong fundamental force of attraction that is functional only at very short distances, on the order of 10^{-15} m or less. At distances greater than about 10^{-15} m the nuclear force is negligible, and the weaker *electromagnetic force*, the force of repulsion between like charges, is the operational force. Thus, like-charged protons experience a repulsive force when they are farther apart than about 10^{-15} m. When closer together than 10^{-15} m, the short-range, stronger nuclear force predominates, and the protons experience a strong attractive force. This explains why the like-charged protons of the nucleus are not repelled by their like electric charges.

Observations of radioactive decay reactants and products and experiments with nuclear stability have led to a *shell model of the nucleus*. This model considers the protons and neutrons moving in energy levels, or shells, in the nucleus analogous to the shell structure of electrons in the outermost part of the atom. As in the electron shells, there are certain configurations of nuclear shells that have a greater stability than others. Considering electrons, filled and half-filled shells are more stable than other arrangements, and maximum stability occurs with the noble gases and their 2, 10, 18, 36, 54, and 86 electrons. Considering the nucleus, atoms with 2, 8, 20, 28, 50, 82, or 126 protons or neutrons have a maximum nuclear stability. The stable numbers are not the same for electrons and nucleons because of differences in nuclear and electromagnetic forces.

Isotopes of uranium, radium, and plutonium, as well as other isotopes, emit an alpha particle during radioactive decay to a simpler nucleus. The alpha particle is a helium

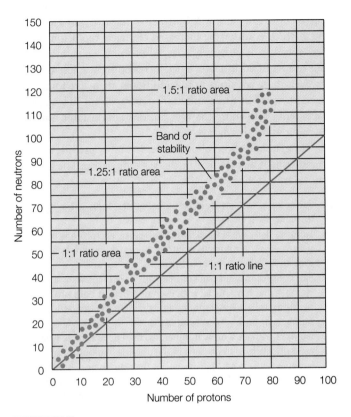

FIGURE 15.5

The dots indicate stable nuclei, which group in a band of stability according to their neutron-to-proton ratio. As the size of nuclei increases, so does the neutron-to-proton ratio that represents stability. Nuclei outside this band of stability are radioactive.

nucleus, $^{4}_{2}He$. The alpha particle contains two protons as well as two neutrons, which is one of the nucleon numbers of stability, so you would expect the helium nucleus (or alpha particle) to have a stable nucleus, and it does. *Stable* means it does not undergo radioactive decay. Pairs of protons and pairs of neutrons have increased stability, just as pairs of electrons in a molecule do. As a result, nuclei with an *even number* of both protons and neutrons are, in general, more stable than nuclei with odd numbers of protons and neutrons. There are a little more than 150 stable isotopes with an even number of protons and an even number of neutrons, but there are only 5 stable isotopes with odd numbers of each. Just as in the case of electrons, there are other factors that come into play as the nucleus becomes larger and larger with increased numbers of nucleons.

The results of some of these factors are shown in figure 15.5, which is a graph of the number of neutrons versus the number of protons in nuclei. As the number of protons increases, the neutron-to-proton ratio of the *stable nuclei* also increases in a **band of stability**. Within the band the neutron-to-proton ratio increases from about 1:1 at the bottom left to about 1½:1 at the top right. The increased ratio of neutrons is needed to produce a stable nucleus as the number of protons increases.

Neutrons provide additional attractive *nuclear* (not electrical) forces, which counter the increased electrical repulsion from a larger number of positively charged protons. Thus, more neutrons are required in larger nuclei to produce a stable nucleus. However, there is a limit to the additional attractive forces that can be provided by more and more neutrons, and all isotopes of all elements with more than 83 protons are unstable and thus undergo radioactive decay.

The generalizations about nuclear stability provide a means of predicting if a particular nucleus is radioactive. The generalizations are as follows:

1. All isotopes with an atomic number greater than 83 have an unstable nucleus.
2. Isotopes that contain 2, 8, 20, 28, 50, 82, or 126 protons or neutrons in their nucleus occur in more stable isotopes than those with other numbers of protons or neutrons.
3. Pairs of protons and pairs of neutrons have increased stability, so isotopes that have nuclei with even numbers of both protons and neutrons are generally more stable than nuclei with odd numbers of both protons and neutrons.
4. Isotopes with an atomic number less than 83 are stable when the ratio of neutrons to protons in the nucleus is about 1:1 in isotopes with up to 20 protons, but the ratio increases in larger nuclei in a band of stability (see figure 15.5). Isotopes with a ratio to the left or right of this band are unstable and thus will undergo radioactive decay.

Q & A

Would you predict the following isotopes to be radioactive or stable?

(a) $^{60}_{27}$Co

(b) $^{222}_{86}$Rn

(c) $^{3}_{1}$H

(d) $^{40}_{20}$Ca

Answer

(a) Cobalt-60 has 27 protons and 33 neutrons, both odd numbers, so you might expect $^{60}_{27}$Co to be radioactive.

(b) Radon has an atomic number of 86, and all isotopes of all elements beyond atomic number 83 are radioactive. Radon-222 is therefore radioactive.

(c) Hydrogen-3 has an odd number of protons and an even number of neutrons, but its 2:1 neutron-to-proton ratio places it outside the band of stability. Hydrogen-3 is radioactive.

(d) Calcium-40 has an even number of protons and an even number of neutrons, containing 20 of each. The number 20 is a particularly stable number of protons or neutrons, and calcium-40 has 20 of each. In addition, the neutron-to-proton ratio is 1:1, placing it within the band of stability. All indications are that calcium-40 is stable, not radioactive.

Which of the following would you predict to be radioactive?

(a) $^{127}_{53}$I

(b) $^{131}_{53}$I

(c) $^{206}_{82}$Pb

(d) $^{214}_{82}$Pb

(Answer: (b) and (d))

Types of Radioactive Decay

Through the process of radioactive decay, an unstable nucleus becomes a more stable one with less energy. The three more familiar types of radiation emitted—alpha, beta, and gamma—were introduced earlier. There are five common types of radioactive decay, and three of these involve alpha, beta, and gamma radiation.

1. *Alpha emission.* Alpha (α) emission is the expulsion of an alpha particle ($^{4}_{2}$He) from an unstable, disintegrating nucleus. The alpha particle, a helium nucleus, travels from 2 to 12 cm through the air, depending on the energy of emission from the source. An alpha particle is easily stopped by a sheet of paper close to the nucleus. As an example of alpha emission, consider the decay of a radon-222 nucleus,

$$^{222}_{86}\text{Rn} \rightarrow ^{218}_{84}\text{Po} + ^{4}_{2}\text{He}$$

The spent alpha particle eventually acquires two electrons and becomes an ordinary helium atom.

2. *Beta emission.* Beta (β^-) emission is the expulsion of a different particle, a beta particle, from an unstable disintegrating nucleus. A beta particle is simply an electron ($^{0}_{-1}$e) ejected from the nucleus at a high speed. The emission of a beta particle *increases the number of protons* in a nucleus. It is as if a neutron changed to a proton by emitting an electron, or

$$^{1}_{0}\text{n} \rightarrow ^{1}_{1}\text{p} + ^{0}_{-1}\text{e}$$

Carbon-14 is a carbon isotope that decays by beta emission:

$$^{14}_{6}\text{C} \rightarrow ^{14}_{7}\text{N} + ^{0}_{-1}\text{e}$$

Note that the number of protons increased from six to seven, but the mass number remained the same. The mass number is unchanged because the mass of the expelled electron (beta particle) is negligible.

Beta particles are more penetrating than alpha particles and may travel several hundred centimeters through the air. They can be stopped by a thin layer of metal close to the emitting nucleus, such as a 1 cm thick piece of aluminum. A spent beta particle may eventually join an ion to become part of an atom, or it may remain a free electron.

3. *Gamma emission.* Gamma (γ) emission is a high-energy burst of electromagnetic radiation from an excited nucleus. It is a burst of light (photon) of a wavelength much too short to be detected by the eye. Other types of radioactive decay, such as alpha or beta emission, sometimes leave the nucleus with an excess of energy, a

TABLE 15.2	Radioactive decay		
Unstable Condition	Type of Decay	Emitted	Product Nucleus
More than 83 protons	Alpha emission	^4_2He	Lost 2 protons and 2 neutrons
Neutron-to-proton ratio too large	Beta emission	$^0_{-1}e$	Gained 1 proton, no mass change
Excited nucleus	Gamma emission	$^0_0\gamma$	No change
Neutron-to-proton ratio too small	Other emission	^0_1e	Lost 1 proton, no mass change

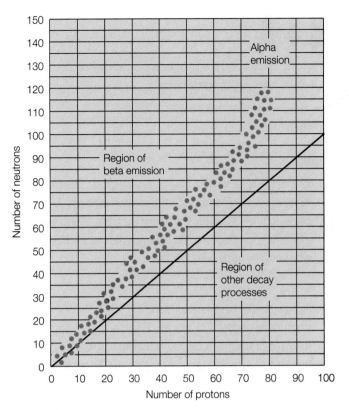

FIGURE 15.6

Unstable nuclei undergo different types of radioactive decay to obtain a more stable nucleus. The type of decay depends, in general, on the neutron-to-proton ratio, as shown.

condition called an *excited state.* As in the case of excited electrons, the nucleus returns to a lower energy state by emitting electromagnetic radiation. From a nucleus, this radiation is in the high-energy portion of the electromagnetic spectrum. Gamma is the most penetrating of the three common types of nuclear radiation. Like X rays, gamma rays can pass completely through a person, but all gamma radiation can be stopped by a 5 cm thick piece of lead close to the source. As with other types of electromagnetic radiation, gamma radiation is absorbed by and gives its energy to materials. Since the product nucleus changed from an excited state to a lower energy state, there is no change in the number of nucleons. For example, radon-222 is an isotope that emits gamma radiation:

$$^{222}_{86}Rn^* \rightarrow {}^{222}_{86}Rn + {}^0_0\gamma$$

(* denotes excited state)

Radioactive decay by alpha, beta, and gamma emission is summarized in table 15.2, which also lists the unstable nuclear conditions that lead to the particular type of emission. Just as electrons seek a state of greater stability, a nucleus undergoes radioactive decay to achieve a balance between nuclear attractions, electromagnetic repulsions, and a low quantum of nuclear shell energy. The key to understanding the types of reactions that occur is found in the band of stable nuclei illustrated in figure 15.5. The isotopes within this band have achieved the state of stability, and other isotopes above, below, or beyond the band are unstable and thus radioactive.

Nuclei that have a neutron-to-proton ratio beyond the upper right part of the band are unstable because of an imbalance between the proton-proton electromagnetic repulsions and all the combined proton and neutron nuclear attractions. Recall that the neutron-to-proton ratio increases from about 1:1 to about 1½:1 in the larger nuclei. The additional neutron provided additional nuclear attractions to hold the nucleus together, but atomic number 83 appears to be the upper limit to this additional stabilizing contribution. Thus, all nuclei with an atomic number greater than 83 are outside the upper right

limit of the band of stability. Emission of an alpha particle reduces the number of protons by two and the number of neutrons by two, moving the nucleus more toward the band of stability. Thus, you can expect a nucleus that lies beyond the upper right part of the band of stability to be an alpha emitter (figure 15.6).

A nucleus with a neutron-to-proton ratio that is too large will be on the left side of the band of stability. Emission of a beta particle decreases the number of neutrons and increases the number of protons, so a beta emission will lower the neutron-to-proton ratio. Thus, you can expect a nucleus with a large neutron-to-proton ratio, that is, one to the left of the band of stability, to be a beta emitter.

A nucleus that has a neutron-to-proton ratio that is too small will be on the right side of the band of stability. These nuclei can increase the number of neutrons and reduce the number of protons in the nucleus by other types of radioactive decay. As usual when dealing with broad generalizations and trends, there are exceptions to the summarized relationships between neutron-to-proton ratios and radioactive decay.

Q & A

Refer to figure 15.6 and predict the type of radioactive decay for each of the following unstable nuclei:

(a) $^{131}_{53}I$

(b) $^{241}_{94}Pu$

Answer

(a) Iodine-131 has a nucleus with 53 protons and 131 minus 53, or 78 neutrons, so it has a neutron-to-proton ratio of 1.47:1. This places iodine-131 on the left side of the band of stability, with a high neutron-to-proton ratio that can be reduced by beta emission. The nuclear equation is

$$^{131}_{53}I \rightarrow {^{131}_{54}Xe} + {^{0}_{-1}e}$$

(b) Plutonium-241 has 94 protons and 241 minus 94, or 147 neutrons, in the nucleus. This nucleus is to the upper right, beyond the band of stability. It can move back toward stability by emitting an alpha particle, losing 2 protons and 2 neutrons from the nucleus. The nuclear equation is

$$^{241}_{94}Pu \rightarrow {^{238}_{92}U} + {^{4}_{2}He}$$

Radioactive Decay Series

A radioactive decay reaction produces a simpler, and eventually more stable nucleus than the reactant nucleus. As discussed in the previous section, large nuclei with an atomic number greater than 83 decay by alpha emission, giving up two protons and two neutrons with each alpha particle. A nucleus with an atomic number greater than 86, however, will emit an alpha particle and *still* have an atomic number greater than 83, which means the product nucleus will also be radioactive. This nucleus will also undergo radioactive decay, and the process will continue through a series of decay reactions until a stable nucleus is achieved. Such a series of decay reactions that (1) begins with one radioactive nucleus, which (2) decays to a second nucleus, which (3) then decays to a third nucleus, and so on until (4) a stable nucleus is reached is called a *radioactive decay series*. There are three naturally occurring radioactive decay series. One begins with thorium-232 and ends with lead-208, another begins with uranium-235 and ends with lead-207, and the third series begins with uranium-238 and ends with lead-206. Figure 15.7 shows the uranium-238 radioactive decay series.

As figure 15.7 illustrates, the uranium-238 begins with uranium-238 decaying to thorium-234 by alpha emission. Thorium has a new position on the graph because it now has a new atomic number and a new mass number. Thorium-234 is unstable and decays to protactinium-234 by beta emission, which is

also unstable and decays by beta emission to uranium-234. The process continues with five sequential alpha emissions, then two beta-beta-alpha decay steps before the series terminates with the stable lead-206 nucleus.

Uranium-238 is radioactive and decays to thorium-234 by emitting an alpha particle. Yet not all uranium-238 has decayed to lead-206, and, in fact, a sample of uranium-238 will continue to give off alpha particles for millions of years. Uranium-238, like uranium-235 and thorium-232, undergoes radioactive decay very slowly, and any given nucleus may disintegrate today or it may disintegrate millions of years later. It is not possible to predict when a nucleus will decay because it is a random process. It is possible, however, to deal with nuclear disintegration statistically since the rate of decay is not changed by any external conditions of temperature, pressure, or any chemical state. When dealing with a large number of nuclei, the ratio of the rate of nuclear disintegration per unit of time to the total number of radioactive nuclei will be a constant, or

$$\text{radioactive decay constant} = \frac{\text{decay rate}}{\text{number of nuclei}}$$

or, in symbols,

$$k = \frac{\text{rate}}{n}$$

equation 15.1

The *radioactive decay constant, k,* is a specific constant for a particular isotope, and each isotope has its own decay constant. For example, a 238 g sample of uranium-238 (1 mole) that has 2.93×10^6 disintegrations per second would have a decay constant of

$$k = \frac{\text{rate}}{n} = \frac{2.93 \times 10^6 \text{ nuclei/s}}{6.02 \times 10^{23} \text{ nuclei}}$$

$$= 4.87 \times 10^{-18}/s$$

The rate of radioactive decay is usually described in terms of its *half-life.* The **half-life** is the time required for one-half of the unstable nuclei to decay. Since each isotope has a characteristic decay constant, each isotope has its own characteristic half-life. Half-lives of some highly unstable isotopes are measured in fractions of seconds, and other isotopes have half-lives measured in seconds, minutes, hours, days, months, years, or billions of years. Table 15.3 lists half-lives of some of the isotopes, and the process is illustrated in figure 15.8.

As an example of the half-life measure, consider a hypothetical isotope that has a half-life of one day. The half-life is independent of the amount of the isotope being considered, but suppose you start with a 1.0 kg sample of this element with a half-life of one day. One day later, you will have half of the original sample, or 500 g. The other half did not disappear, but it is now the decay product, that is, some new element. During the next day half of the remaining nuclei will disintegrate, and only 250 g of the initial sample is still the original element. One-half of the remaining sample will disintegrate each day until the original sample no longer exists.

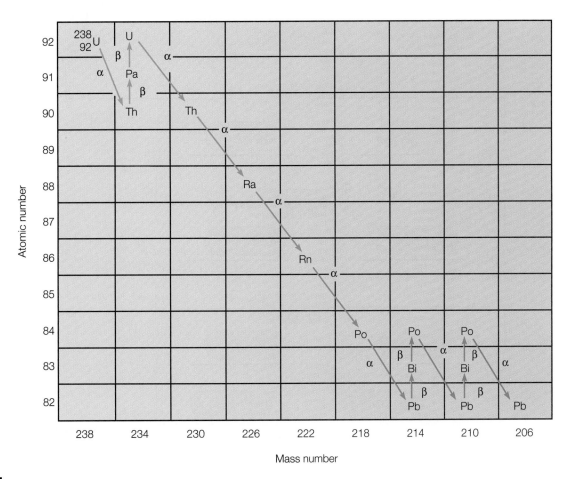

FIGURE 15.7

The radioactive decay series for uranium-238. This is one of three naturally occurring series.

TABLE 15.3	Half-lives of some radioactive isotopes	
Isotope	**Half-Life**	**Mode of Decay**
$^{3}_{1}H$ (tritium)	12.26 years	Beta
$^{14}_{6}C$	5,730 years	Beta
$^{90}_{38}Sr$	28 years	Beta
$^{131}_{53}I$	8 days	Beta
$^{133}_{54}Xe$	5.27 days	Beta
$^{238}_{92}U$	4.51×10^9 years	Alpha
$^{242}_{94}Pu$	3.79×10^5 years	Alpha
$^{240}_{94}Pu$	6,760 years	Alpha
$^{239}_{94}Pu$	24,360 years	Alpha
$^{40}_{19}K$	1.3×10^9 years	Alpha

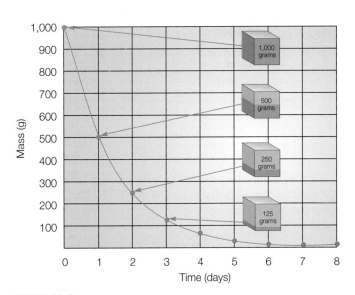

FIGURE 15.8

Radioactive decay of a hypothetical isotope with a half-life of one day. The sample decays each day by one-half to some other element. Actual half-lives may be in seconds, minutes, or any time unit up to billions of years.

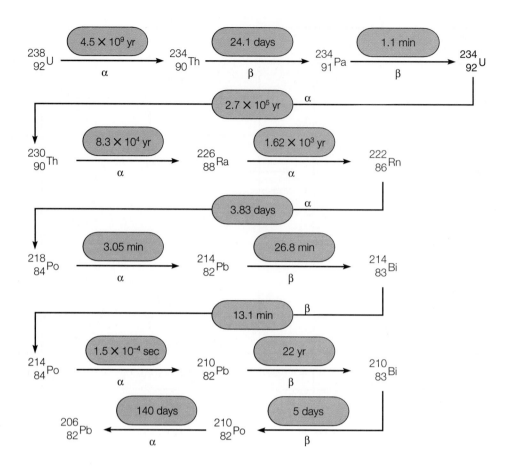

FIGURE 15.9

The half-life of each step in the uranium-238 radioactive decay series.

The half-life of a radioactive nucleus is related to its radioactive decay constant by

$$\text{half-life} = \frac{\text{a mathematical constant}}{\text{decay constant}}$$

or

$$t_{1/2} = \frac{0.693}{k}$$

equation 15.2

For example, the radioactive decay constant for uranium-238 was determined earlier to be $4.87 \times 10^{-18}/s$. The half-life of uranium-238 is therefore

$$t_{1/2} = \frac{0.693}{4.87 \times 10^{-18}/s} = 1.42 \times 10^{17}s$$

This is the half-life of uranium-238 in seconds. There are $60 \times 60 \times 24 \times 365$, or 3.15×10^7 s in a year, so

$$\frac{1.42 \times 10^{17} \text{ s}}{3.15 \times 10^7 s/yr} = 4.5 \times 10^9 \text{ yr}$$

The half-life of uranium-238 is thus 4.5 billion years. Figure 15.9 gives the half-life for each step in the uranium-238 decay series.

As you can see from equations 15.1 and 15.2, the half-life of a radioactive isotope is directly proportional to its rate of disintegration. Thus, isotopes with a shorter half-life are more active and are disintegrating at a faster rate. On the other hand, longer half-lives mean less activity and lower rates of radiation.

MEASUREMENT OF RADIATION

The measurement of radiation is important in determining the half-life of radioactive isotopes, as you learned in the previous section. Radiation measurement is also important in considering biological effects, which will be discussed in the next section. As is the case with electricity, it is not possible to make direct measurements on things as small as electrons and other parts of atoms. Indirect measurement methods are possible, however, by considering the effects of the radiation.

Measurement Methods

As Becquerel discovered, radiation affects photographic film, exposing it as visible light does. Since the amount of film exposure is proportional to the amount of radiation, photographic film can be used as an indirect measure of radiation. Today,

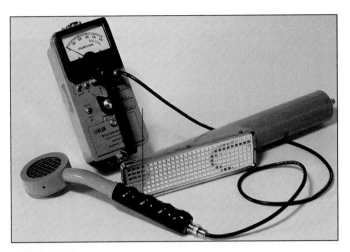

FIGURE 15.10

This is a beta-gamma probe, which can measure beta and gamma radiation in millirems per unit of time.

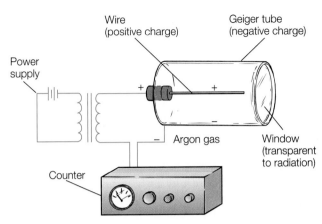

FIGURE 15.11

The working parts of a Geiger counter.

people who work around radioactive materials or X rays carry light-tight film badges. The film is replaced periodically and developed. The optical density of the developed film provides a record of the worker's exposure to radiation.

There are also devices that indirectly measure radiation by measuring an effect of the radiation. An *ionization counter* is one type of device that measures ions produced by radiation. A second type of device is called a *scintillation counter*. *Scintillate* is a word meaning "sparks or flashes," and a scintillation counter measures the flashes of light produced when radiation strikes a phosphor.

The most common example of an ionization counter is known as a *Geiger counter* (figure 15.10). The working components of a Geiger counter are illustrated in figure 15.11. Radiation is received in a metal tube filled with an inert gas, such as argon, through a thin plastic window that is transparent to alpha, beta, and gamma radiation. An insulated wire inside the tube is connected to the positive terminal of a direct current source. The metal cylinder around the insulated wire is connected to the negative terminal. There is not a current between the center wire and the metal cylinder because the gas acts as an insulator. When radiation passes through the window, however, it ionizes some of the gas atoms, releasing free electrons. These electrons are accelerated by the field between the wire and cylinder, and the accelerated electrons ionize more gas molecules, which results in an *avalanche* of free electrons. The avalanche creates a pulse of current that is amplified and then measured. More radiation means more avalanches, so the pulses are an indirect means of measuring radiation. When connected to a speaker or earphone, each avalanche produces a "pop" or "click."

Some materials are *phosphors*, substances that emit a flash of light when excited by radiation. Zinc sulfide, for example, is used in television screens and luminous watches, and it was used by Rutherford to detect alpha particles. A luminous watch dial has a mixture of zinc sulfide and a small amount of radium sulfate. A zinc sulfide atom gives off a tiny flash of light when struck by radiation from a disintegrating radium nucleus. A scintilla-

tion counter measures the flashes of light through the photoelectric effect, producing free electrons that are accelerated to produce a pulse of current. Again, the pulses of current are used as an indirect means to measure radiation.

Radiation Units

You have learned that *radioactivity* is a property of isotopes with unstable, disintegrating nuclei and *radiation* is emitted particles (alpha or beta) or energy traveling in the form of photons (gamma). Radiation can be measured (1) at the source of radioactivity or (2) at a place of reception, where the radiation is absorbed.

The *activity* of a radioactive source is a measure of the number of nuclear disintegrations per unit of time. The unit of activity at the source is called a **curie** (Ci), which is defined as 3.70×10^{10} nuclear disintegrations per second. The radioactivity can be measured by a radiation counter, or it can be calculated from the radioactive decay rate. For example, a 238 g sample of uranium-238 has a decay rate of 2.93×10^6 disintegrations per second, so the activity in curies is

$$\frac{2.93 \times 10^6}{3.7 \times 10^{10}} = 7.92 \times 10^{-5} \text{ Ci}$$

A unit frequently mentioned is a *picocurie*, which is a millionth of a millionth of a curie.

As radiation from a source moves out and strikes a material, it gives the material energy. The amount of energy released by radiation striking living tissue is usually very small, but it can cause biological damage nonetheless. Chemical bonds are broken and free polyatomic ions are produced by radiation, and the broken bonds and free polyatomic ions are the damaging results.

One measure of radiation received by a material is called the **rad.** The term *rad* is from *r*adiation *a*bsorbed *d*ose, and one rad releases 1×10^{-2} J/kg. Another measure of radiation received considers the biological effect from a rad. This unit is called a **rem,** which takes into account the possible biological damage produced by different types of radiation. The term *rem* is from *r*oentgen *e*quivalent *m*an (a roentgen is another measure of radiation). The equivalent measure is needed because alpha radiation, for example, has a greater ionizing power than beta or gamma radiation, so fewer rads of alpha are required to produce a rem. Beta and gamma radiation are the most penetrating, however, and alpha radiation barely penetrates the skin. Alpha radiation can be very damaging if the source gets inside the body, which is the reason so many people are concerned about exposure to radon gas. Radon is chemically inert and cannot be filtered or absorbed by a gas mask or any other means. Most common isotopes of radon are alpha emitters.

Overall, there are many factors and variables that affect the possible damage from radiation, including the distance from the source and what shielding materials are between a person and a source. A *millirem* is 1/1,000 of a rem and is the unit of choice when low levels of radiation are discussed.

Radiation Exposure

Natural radioactivity is a part of your environment, and you receive between 100 and 500 millirems each year from natural sources. This radiation from natural sources is called **background radiation.** Background radiation comes from outer space in the form of cosmic rays and from unstable isotopes in the ground, building materials, and foods. Many activities and situations will increase your yearly exposure to radiation. For example, the atmosphere absorbs some of the cosmic rays from space, so the less atmosphere above you, the more radiation you will receive. You are exposed to one additional millirem per year for each 100 feet you live above sea level. You receive approximately 0.3 millirem for each hour spent on a jet flight. Airline crews receive an additional 300 to 400 millirems per year because they spend so much time high in the atmosphere. Additional radiation exposure comes from medical X rays, television sets, and luminous objects such as watch and clock dials. In general, the background radiation exposure for the average person is about 130 millirems per year.

What are the consequences of radiation exposure? Radiation can be a hazard to living organisms because it produces ionization along its path of travel. This ionization can (1) disrupt chemical bonds in essential macromolecules such as DNA and (2) produce molecular fragments, which are free polyatomic ions that can interfere with enzyme action and other essential cell functions. Tissues with highly active cells are more vulnerable to radiation damage than others, such as blood-forming tissue. Thus, one of the symptoms of an excessive radiation exposure is an altered blood count. Table 15.4 compares the estimated results of various levels of acute radiation exposure.

Radiation is not a mysterious, unique health hazard. It is a hazard that should be understood and neither ignored nor exaggerated. Excessive radiation exposure should be avoided, just as you avoid excessive exposure to other hazards such as certain chemicals, electricity, or even sunlight. Everyone agrees that *excessive* radiation exposure should be avoided, but there is some controversy about long-term, low-level exposure and its possible role in cancer. Some claim that tolerable low-level exposure does not exist because that is not possible. Others point to many studies comparing high and low background radioactivity with cancer mortality data. For example, no cancer mortality

TABLE 15.4	Approximate single dose, whole body effects of radiation exposure
Level	**Comment**
0.130 rem	Average annual exposure to natural background radiation
0.500 rem	Upper limit of annual exposure to general public
25.0 rem	Threshold for observable effects such as blood count changes
100.0 rem	Fatigue and other symptoms of radiation sickness
200.0 rem	Definite radiation sickness, bone marrow damage, possibility of developing leukemia
500.0 rem	Lethal dose for 50 percent of individuals
1,000.0 rem	Lethal dose for all

differences could be found between people receiving 500 or more millirems a year and those receiving less than 100 millirems a year. The controversy continues, however, because of lack of knowledge about long-term exposure. Two models of long-term, low-level radiation exposure have been proposed: (1) a linear model and (2) a threshold model. The *linear model* proposes that any radiation exposure above zero is damaging and can produce cancer and genetic damage. The *threshold model* proposes that the human body can repair damage and get rid of damaging free polyatomic ions up to a certain exposure level called the threshold (figure 15.12). The controversy over long-term, low-level radiation exposure will probably continue until there is clear evidence about which model is correct. Whichever is correct will not lessen the need for rational risks versus cost-benefit analyses of all energy alternatives.

NUCLEAR ENERGY

As discussed, some nuclei are unstable because they are too large or because they have an unstable neutron-to-proton ratio. These unstable nuclei undergo radioactive decay, eventually forming products of greater stability. An example of this radioactive decay is the alpha emission reaction of uranium-238 to thorium-234,

$$^{238}_{92}U \rightarrow {}^{234}_{90}Th + {}^{4}_{2}He$$

$$238.0003 \text{ u} \rightarrow 233.9942 \text{ u} + 4.00150 \text{ u}$$

The numbers below the nuclear equation are the *nuclear* masses (u) of the reactant and products. As you can see, there seems to be a loss of mass in the reaction,

$$233.9942 + 4.00150 - 238.0003 = -0.0046 \text{ u}$$

This change in mass is related to the energy change according to the relationship that was formulated by Albert Einstein in 1905. The relationship is

$$E = mc^2$$

equation 15.3

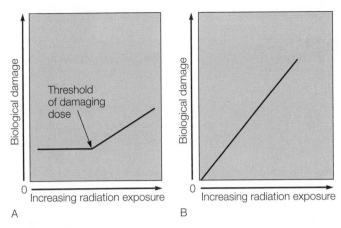

FIGURE 15.12

Graphic representation of the (A) threshold model and (B) linear model of low-level radiation exposure. The threshold model proposes that the human body can repair damage up to a threshold. The linear model proposes that any radiation exposure is damaging.

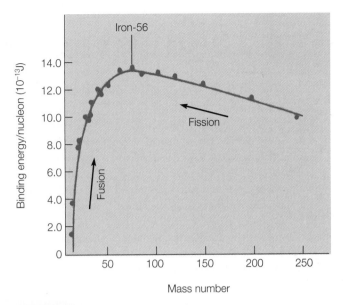

FIGURE 15.13

The maximum binding energy per nucleon occurs around mass number 56, then decreases in both directions. As one result, fission of massive nuclei and fusion of less massive nuclei both release energy.

where E is a quantity of energy, m is a quantity of mass, and c is a constant equal to the speed of light in a vacuum, 3.00×10^8 m/s. According to this relationship, matter and energy are the same thing, and energy can be changed to matter and vice versa. Since the mass of a mole in grams is numerically equal to the atomic mass unit (u) of a nucleus, the mass change for a mole of decaying uranium-238 is -0.0046 g, or -4.6×10^{-6} kg. Using this mass loss (Δm) in equation 15.3, you can calculate the energy change (ΔE),

$$\Delta E = \Delta mc^2$$

$$= (-4.6 \times 10^{-6} \text{kg})\left(3.00 \times 10^8 \frac{\text{m}}{\text{s}}\right)^2$$

$$= (-4.6 \times 10^{-6} \text{kg})\left(9.00 \times 10^{16} \frac{\text{m}^2}{\text{s}^2}\right)$$

$$= (-4.6 \times 9.00) \times 10^{(-6+16)} \frac{\text{kg}\cdot\text{m}^2}{\text{s}^2}$$

$$= -4.14 \times 10^{11} \text{ J}$$

Thus, the products of a mole of uranium-238 decaying to more stable products (1) have a lower energy of 4.14×10^{11} J and (2) lost a mass of 4.6×10^{-6} kg. As you can see, a very small amount of matter was converted into a large amount of energy in the process, forming products of lower energy.

The relationship between mass and energy explains why the mass of a nucleus is always *less* than the sum of the masses of the individual particles of which it is made. For example, the masses of the particles making up a helium-4 nucleus are

$$2 \text{ protons} = 2(1.00728 \text{ u}) = 2.01456 \text{ u}$$
$$2 \text{ neutrons} = 2(1.00867 \text{ u}) = \underline{2.01734 \text{ u}}$$
$$4.03190 \text{ u}$$

But the mass of a helium-4 nucleus is 4.00150 u, a difference of 0.03040. The difference between (1) the mass of the individual nucleons making up a nucleus and (2) the actual mass of the

nucleus is called the **mass defect** of the nucleus. The explanation for the mass defect is again found in $E = mc^2$. When nucleons join to make a nucleus, energy is released as the more stable nucleus is formed. A mole of helium-4 nuclei would release a very large amount of energy,

$$\Delta E = \Delta mc^2$$

$$= (-3.04 \times 10^{-5} \text{kg})\left(3.00 \times 10^8 \frac{\text{m}}{\text{s}}\right)^2$$

$$= (-3.04 \times 10^{-5} \text{kg})\left(9.00 \times 10^{16} \frac{\text{m}^2}{\text{s}^2}\right)$$

$$= (-3.04 \times 9.00) \times 10^{(-5+16)} \frac{\text{kg}\cdot\text{m}^2}{\text{s}^2}$$

$$= -2.74 \times 10^{12} \text{ J}$$

By comparison, hydrogen atoms coming together to form one mole of H_2 molecules release about 4.3×10^5 J of chemical energy, or about 10,000,000 times less energy per mole.

The energy equivalent released when a nucleus is formed is the same as the **binding energy,** the energy required to break the nucleus into individual protons and neutrons. The binding energy of the nucleus of any isotope can be calculated from the mass defect of the nucleus.

The ratio of binding energy to nucleon number is a reflection of the stability of a nucleus (figure 15.13). The greatest binding energy per nucleon occurs near mass number 56, with about 1.4×10^{-12} J per nucleon, then decreases for both more massive and less massive nuclei. This means that more massive nuclei can gain stability by splitting into smaller nuclei with the release of energy. It also means that less massive nuclei can gain stability by joining together with the release of energy. The slope

A CLOSER LOOK | Nuclear Medicine

Nuclear medicine had its beginnings in 1946 when radioactive iodine was first successfully used to treat thyroid cancer patients. Then physicians learned that radioactive iodine could also be used as a diagnostic tool, providing a way to measure the function of the thyroid and to diagnose thyroid disease. More and more physicians began to use nuclear medicine to diagnose thyroid disease as well as to treat hyperthyroidism and other thyroid problems. Nuclear medicine is a branch of medicine using radiation or radioactive materials to diagnose as well as treat diseases.

The development of new nuclear medicine technologies, such as new cameras, detection instruments, and computers, has led to a remarkable increase in the use of nuclear medicine as a diagnostic tool. Today, there are nearly 100 different nuclear medicine imaging procedures. These provide unique, detailed information about virtually every major organ system within the body, information that was unknown just years ago. Treatment of disease with radioactive materials continues to be a valuable part of nuclear medicine, too. The material that follows will consider some techniques of using nuclear medicine as a diagnostic tool, followed by a short discussion of the use of radioactive materials in the treatment of disease.

Nuclear medicine provides diagnostic information about organ function, compared to conventional radiology, which provides images about the structure. For example, a conventional X-ray image will show if a bone is broken or not, while a bone imaging nuclear scan will show changes caused by tumors, hair-line fractures, or arthritis. There are procedures for making detailed structural X-ray pictures of internal organs such as the liver, kidney, or heart, but these images often cannot provide diagnostic information, showing only the structure. Nuclear medicine scans, on the other hand, can provide information about how much heart tissue is still alive after a heart attack or if a kidney is working, even when there are no detectable changes in organ appearance.

An X-ray image is produced when X rays pass through the body and expose photographic film on the other side. Some X-ray exams improve photographic contrast by introducing certain substances. A barium sulfate "milk shake," for example, can be swallowed to highlight the esophagus, stomach, and intestine. More information is provided if X rays are used in a CAT scan (CAT stands for "Computed Axial Tomography"). The CAT scan is a diagnostic test that combines the use of X rays with computer technology. The CAT scan shows organs of interest by making X-ray images from many different angles as the source of the X rays moves around the patient. Contrast-improving substances, such as barium sulfate, might also be used with a CAT scan. In any case, CAT scan images are assembled by a computer into a three-dimensional picture that can show organs, bones, and tissues in great detail.

The gamma camera is a key diagnostic imaging tool used in nuclear medicine. Its use requires a radioactive material, called a radiopharmaceutical, to be injected into or swallowed by the patient. A given radiopharmaceutical tends to go to a specific organ of the body, for example, radioactive iodine tends to go to the thyroid gland, and others go to other organs. Gamma emitting radiopharmaceuticals are used with the gamma camera, and the gamma camera collects and processes these gamma rays to produce images. These images provide a way of studying the structure as well as measuring the function of the selected organ. Together, the structure and function provide a way of identifying tumors, areas of infection, or other problems. The patient experiences little or no discomfort and the radiation dose is small.

A SPECT scan (Single Photon Emission Computerized Tomography) is an imaging technique employing a gamma camera that rotates around the patient, measuring gamma rays and computing their point of origin. Cross-sectional images of a three-dimensional internal organ can be obtained from such data, resulting in images that have higher resolution and thus more diagnostic

information than a simple gamma camera image. A gallium radiopharmaceutical is often used in a scan to diagnose and follow the progression of tumors or infections. Gallium scans also can be used to evaluate the heart, lungs, or any other organ that may be involved with inflammatory disease.

Use of MRI (Magnetic Resonance Imaging) also produces images as an infinite number of projections through the body. Unlike CAT, gamma, or SPECT scans, MRI does not use any form of ionizing radiation. MRI uses magnetic fields, radio waves, and a computer to produce detailed images. As the patient enters an MRI scanner, his body is surrounded by a large magnet. The technique requires a very strong magnetic field, a field so strong that it aligns the nuclei of the person's atoms. The scanner sends a strong radio signal, temporarily knocking the nuclei out of alignment. When the radio signal stops, the nuclei return to the aligned position, releasing their own faint radio frequencies. These radio signals are read by the scanner, which uses them in a computer program to produce very detailed images of the human anatomy.

The PET scan (Positron Emission Tomography) is the most recent technological development in nuclear medicine imaging, producing 3D images superior to gamma camera images. This technique is built around a radiopharmaceutical that emits positrons (like an electron with a positive charge). Positrons collide with electrons, releasing a burst of energy in the form of photons. Detectors track the emissions and feed the information into a computer. The computer has a program to plot the source of radiation and translates the data into an image. Positron-emitting radiopharmaceuticals used in a PET scan can be low atomic weight elements like carbon, nitrogen, and oxygen. This is important for certain purposes since these are the same elements found in many biological substances like sugar, urea, or carbon dioxide. Thus a PET scan can be used to study processes in organs such as the brain and

—Continued top of next page

Continued—

heart where glucose is being broken down or oxygen is being consumed. This diagnostic method can be used to detect epilepsy or brain tumors, among other problems.

Radiopharmaceuticals used for diagnostic examinations are selected for their affinity for certain organs, if they emit sufficient radiation to be easily detectable in the body, and if they have a rather short half-life, preferably no longer than a few hours. Useful radioisotopes that meet these criteria for

diagnostic purposes are technetium-99, gallium-67, indium-111, iodine-123, iodine-131, thallium-201, and krypton-81.

The goal of therapy in nuclear medicine is to use radiation to destroy diseased or cancerous tissue while sparing adjacent healthy tissue. Few radioactive therapeutic agents are injected or swallowed, with the exception of radioactive iodine—mentioned earlier as a treatment for cancer of the thyroid. Useful radioisotopes for therapeutical

purposes are iodine-131, phosphorus-32, iridium-192, and gold-198. The radioactive source placed in the body for local irradiation of a tumor is normally iridium-192. A nuclear pharmaceutical is a physiologically active carrier to which a radioisotope is attached. Today, it is possible to manufacture chemical or biological carriers that migrate to a particular part of the human body, and this is the subject of much on-going medical research.

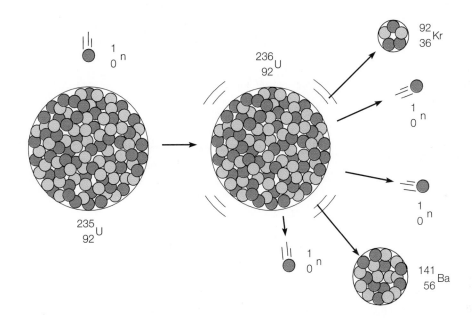

FIGURE 15.14

The fission reaction occurring when a neutron is absorbed by a uranium-235 nucleus. The deformed nucleus splits any number of ways into lighter nuclei, releasing neutrons in the process.

also shows that more energy is released in the coming-together process than in the splitting process.

The nuclear reaction of splitting a massive nucleus into more stable, less massive nuclei with the release of energy is **nuclear fission** (figure 15.14). Nuclear fission occurs rapidly in an atomic bomb explosion and occurs relatively slowly in a nuclear reactor. The nuclear reaction of less massive nuclei, coming together to form more stable, and more massive, nuclei with the release of energy is **nuclear fusion.** Nuclear fusion occurs rapidly in a hydrogen bomb explosion and occurs continually in the sun, releasing the energy essential for the continuation of life on the earth. Nuclear fission and nuclear fusion are the topics of the next sections.

Nuclear Fission

Nuclear fission was first accomplished in the late 1930s when researchers were attempting to produce isotopes by bombarding massive nuclei with neutrons. In 1938 two German scientists, Otto Hahn and Fritz Strassman, identified the element barium in a uranium sample that had been bombarded with neutrons. Where the barium came from was a puzzle at the time, but soon afterward Lise Meitner, an associate who had moved to Sweden, deduced that the uranium nuclei had split, producing barium. The reaction might have been

$$\,_0^1 \text{n} + \,_{92}^{235}\text{U} \rightarrow \,_{56}^{141}\text{Ba} + \,_{36}^{92}\text{Kr} + 3\,_0^1\text{n}$$

TABLE 15.5 | Fragments and products from nuclear reactors using fission of uranium-235

Isotope	Major Mode of Decay	Half-Life	Isotope	Major Mode of Decay	Half-Life
Tritium	Beta	12.26 years	Cerium-144	Beta, gamma	285 days
Carbon-14	Beta	5,930 years	Promethium-147	Beta	2.6 years
Argon-41	Beta, gamma	1.83 hours	Samarium-151	Beta	90 years
Iron-55	Electron capture	2.7 years	Europium-154	Beta, gamma	16 years
Cobalt-58	Beta, gamma	71 days	Lead-210	Beta	22 years
Cobalt-60	Beta, gamma	5.26 years	Radon-222	Alpha	3.8 days
Nickel-63	Beta	92 years	Radium-226	Alpha, gamma	1,620 years
Krypton-85	Beta, gamma	10.76 years	Thorium-229	Alpha	7,300 years
Strontium-89	Beta	5.4 days	Thorium-230	Alpha	26,000 years
Strontium-90	Beta	28 years	Uranium-234	Alpha	2.48×10^5 years
Yttrium-91	Beta	59 days	Uranium-235	Alpha, gamma	7.13×10^8 years
Zirconium-93	Beta	9.5×10^5 years	Uranium-238	Alpha	4.51×10^9 years
Zirconium-95	Beta, gamma	65 days	Neptunium-237	Alpha	2.14×10^6 years
Niobium-95	Beta, gamma	35 days	Plutonium-238	Alpha	89 years
Technetium-99	Beta	2.1×10^5 years	Plutonium-239	Alpha	24,360 years
Ruthenium-106	Beta	1 year	Plutonium-240	Alpha	6,760 years
Iodine-129	Beta	1.6×10^7 years	Plutonium-241	Beta	13 years
Iodine-131	Beta, gamma	8 days	Plutonium-242	Alpha	3.79×10^5 years
Xenon-133	Beta, gamma	5.27 days	Americium-241	Alpha	458 years
Cesium-134	Beta, gamma	2.1 years	Americium-243	Alpha	7,650 years
Cesium-135	Beta	2×10^6 years	Curium-242	Alpha	163 days
Cesium-137	Beta	30 years	Curium-244	Alpha	18 years
Cerium-141	Beta	32.5 days			

The phrase "might have been" is used because a massive nucleus can split in many different ways, producing different products. About thirty-five different, less massive elements have been identified among the fission products of uranium-235. Some of these products are fission fragments, and some are produced by unstable fragments that undergo radioactive decay. Selected fission fragments are listed in table 15.5, together with their major modes of radioactive decay and half-lives. Some of the isotopes are the focus of concern about nuclear wastes, the topic of the reading at the end of this chapter.

The fission of a uranium-235 nucleus produces two or three neutrons along with other products. These neutrons can each move to other uranium-235 nuclei where they are absorbed, causing fission with the release of more neutrons, which move to other uranium-235 nuclei to continue the process. A reaction where the products are able to produce more reactions in a self-sustaining series is called a **chain reaction.** A chain reaction is self-sustaining until all the uranium-235 nuclei have fissioned or until the neutrons fail to strike a uranium-235 nucleus (figure 15.15).

You might wonder why all the uranium in the universe does not fission in a chain reaction. Natural uranium is mostly uranium-238, an isotope that does not fission easily. Only about 0.7 percent of natural uranium is the highly fissionable uranium-235. This low ratio of readily fissionable uranium-235 nuclei makes it unlikely that a stray neutron would be able to achieve a chain reaction.

In order to achieve a chain reaction, there must be (1) a sufficient mass with (2) a sufficient concentration of fissionable nuclei. When the mass and concentration are sufficient to sustain a chain reaction the amount is called a **critical mass.** Likewise, a mass too small to sustain a chain reaction is called a *subcritical mass.* A mass of sufficiently pure uranium-235 (or plutonium-239) that is large enough to produce a rapidly accelerating chain reaction is called a *supercritical mass.* An atomic bomb is simply a device that uses a small, conventional explosive to push subcritical masses of fissionable material into a supercritical mass. Fission occurs almost instantaneously in the supercritical mass, and tremendous energy is released in a violent explosion.

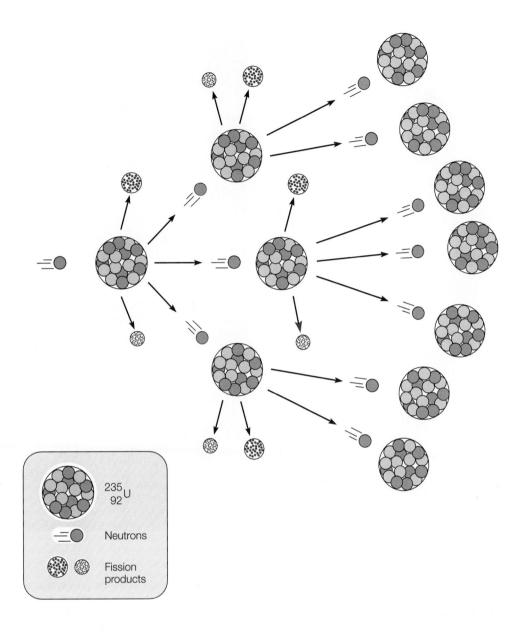

FIGURE 15.15

A schematic representation of a chain reaction. Each fissioned nucleus releases neutrons, which move out to fission other nuclei. The number of neutrons can increase quickly with each series.

Nuclear Power Plants

The nuclear part of a nuclear power plant is the *nuclear reactor,* a steel vessel in which a controlled chain reaction of fissionable material releases energy (figure 15.16). In the most popular design, called a pressurized light-water reactor, the fissionable material is enriched 3 percent uranium-235 and 97 percent uranium-238 that has been fabricated in the form of small ceramic pellets (figure 15.17A). The pellets are encased in a long zirconium alloy tube called a *fuel rod.* The fuel rods are locked into a *fuel rod assembly* by locking collars, arranged to permit pressurized water to flow around each fuel rod (figure 15.17B)

and to allow the insertion of *control rods* between the fuel rods. *Control rods* are constructed of materials, such as cadmium, that absorb neutrons. The lowering or raising of control rods within the fuel rod assemblies slows or increases the chain reaction by varying the amount of neutrons absorbed. When they are lowered completely into the assembly, enough neutrons are absorbed to stop the chain reaction.

It is physically impossible for the low-concentration fuel pellets to form a supercritical mass. A nuclear reactor in a power plant can only release energy at a comparatively slow rate, and it is impossible for a nuclear power plant to produce a nuclear explosion. In a pressurized water reactor the energy released is

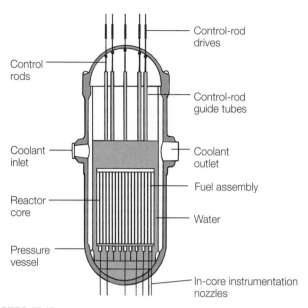

A

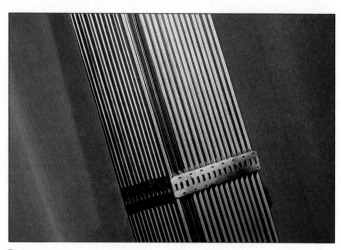

B

FIGURE 15.16

A schematic representation of the basic parts of a nuclear reactor. The largest commercial nuclear power plant reactors are nine- to eleven-inch-thick steel vessels with a stainless steel liner, standing about 40 feet high with a diameter of 16 feet. Such a reactor has four pumps, which move 440,000 gallons of water per minute through the primary loop.

FIGURE 15.17

(A) These are uranium oxide fuel pellets that are stacked inside fuel rods, which are then locked together in a fuel rod assembly. (B) A fuel rod assembly. See also figure 15.20, which shows a fuel rod assembly being loaded into a reactor.

carried away from the reactor by pressurized water in a closed pipe called the *primary loop* (figure 15.18). The water is pressurized at about 150 atmospheres (about 2,200 lb/in²) to keep it from boiling, since its temperature may be 350°C (about 660°F).

In the pressurized light-water (ordinary water) reactor the circulating pressurized water acts as a coolant, carrying heat away from the reactor. The water also acts as a *moderator*, a substance that slows neutrons so they are more readily absorbed by uranium-235 nuclei. Other reactor designs use heavy water (deuterium dioxide) or graphite as a moderator.

Water from the closed primary loop is circulated through a heat exchanger called a *steam generator* (figure 15.18). The pressurized high-temperature water from the reactor moves through hundreds of small tubes inside the generator as *feedwater* from the *secondary loop* flows over the tubes. The water in the primary loop heats feedwater in the steam generator and then returns to the nuclear reactor to become heated again. The feedwater is heated to steam at about 235°C (455°F) with a pressure of about 68 atmospheres (1,000 lb/in²). This steam is piped to the turbines, which turn an electric generator (figure 15.19).

After leaving the turbines, the spent steam is condensed back to liquid water in a second heat exchanger receiving water from the cooling towers. Again, the cooling water does not mix with the closed secondary loop water. The cooling-tower water enters the condensing heat exchanger at about 32°C (90°F) and leaves at about 50°C (about 120°F) before returning to a cooling tower, where it is cooled by evaporation. The feedwater is preheated, then recirculated to the steam generator to start the cycle

over again. The steam is condensed back to liquid water because of the difficulty of pumping and reheating steam.

After a period of time the production of fission products in the fuel rods begins to interfere with effective neutron transmission, so the reactor is shut down annually for refueling. During refueling about one-third of the fuel that had the longest exposure in the reactor is removed as "spent" fuel. New fuel rod assemblies are inserted to make up for the part removed (figure 15.20). However, only about 4 percent of the "spent" fuel is unusable waste, about 94 percent is uranium-238, 0.8 percent is uranium-235, and about 0.9 percent is plutonium (figure 15.21). Thus, "spent" fuel rods contain an appreciable amount of usable uranium and plutonium. For now, spent reactor fuel rods are mostly stored in cooling pools at the nuclear plant sites. In the future, a decision will be made either to reprocess the spent fuel, recovering the uranium and plutonium through chemical

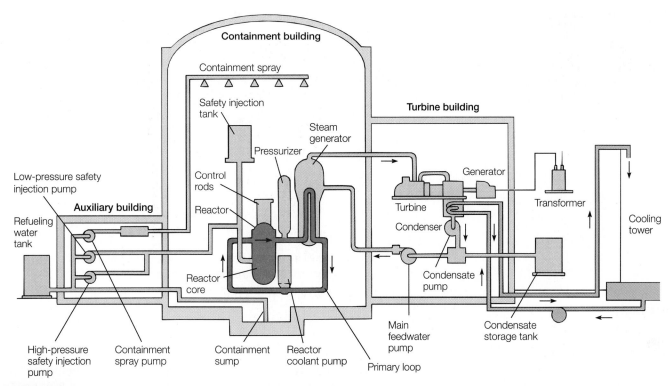

FIGURE 15.18

A schematic general system diagram of a pressurized water nuclear power plant, not to scale. The containment building is designed to withstand an internal temperature of 300°F at a pressure of 60 lbs/in² and still maintain its leak-tight integrity.

FIGURE 15.19

The turbine deck of a nuclear generating station. There is one large generator in line with four steam turbines in this non-nuclear part of the plant. The large silver tanks are separators that remove water from the steam after it has left the high-pressure turbine and before it is recycled back into the low-pressure turbines.

FIGURE 15.20

Spent fuel rod assemblies are removed and new ones are added to a reactor head during refueling. This shows an initial fuel load to a reactor, which has the upper part removed and set aside for the loading.

reprocessing, or put the fuel in terminal storage. Concerns about reprocessing are based on the fact that plutonium-239 and uranium-235 are fissionable and could possibly be used by terrorist groups to construct nuclear explosive devices. Six other countries do have reprocessing plants, however, and the spent fuel rods represent an energy source that will accumulate by the year 2000 to an amount equivalent to more than 25 billion barrels of petroleum. Some energy experts say that it would be inappropriate to dispose of such an energy source.

The technology to dispose of fuel rods exists if the decision is made to do so. The longer half-life waste products are mostly alpha emitters. These metals could be converted to

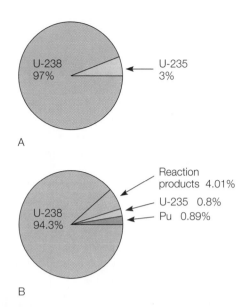

A

B

FIGURE 15.21

The composition of the nuclear fuel in a fuel rod (*A*) before and (*B*) after use over a three-year period in a nuclear reactor.

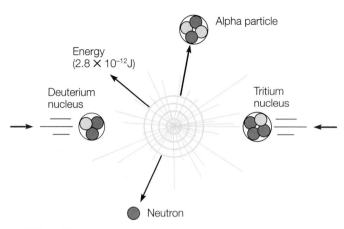

FIGURE 15.22

A fusion reaction between a tritium nucleus and a deuterium nucleus requires a certain temperature, density, and time of containment to take place.

oxides, mixed with powdered glass (or a ceramic), melted, and then poured into stainless steel containers. The solidified canisters would then be buried in a stable geologic depository. The glass technology is used in France for disposal of high-level wastes. Buried at two-thousand- to three-thousand-foot depths in solid granite, the only significant means of the radioactive wastes reaching the surface would be through groundwater dissolving the stainless steel, glass, and waste products and then transporting them back to the surface. Many experts believe that if such groundwater dissolving were to take place it would require thousands of years. The radioactive isotopes would thus undergo natural radioactive decay by the time they could reach the surface. Nonetheless, research is continuing on nuclear waste and its disposal. In the meantime, the question of whether it is best to reprocess fuel rods or place them in permanent storage remains unanswered.

What is the volume of nuclear waste under question? If all the spent fuel rods from all the commercial nuclear plants accumulated up to the year 2000 were reprocessed, then mixed with glass, the total amount of glassified waste would make a pile on one football field an estimated 4 m (about 13 ft) high.

Nuclear Fusion

As the graph of nuclear binding energy versus mass numbers shows (see figure 15.13), nuclear energy is released when (1) massive nuclei such as uranium-235 undergo fission and (2) when less massive nuclei come together to form more massive nuclei through nuclear fusion. Nuclear fusion is responsible for the energy released by the sun and other stars. At the present halfway point in the sun's life—with about 5 billion years to

go—the core is now 35% hydrogen and 65% helium. Through fusion, the sun converts about 650 million tons of hydrogen to 645 million tons of helium every second. The other roughly 5 million tons of matter are converted into energy. Even at this rate, the sun has enough hydrogen to continue the process for an estimated 5 billion years. There are several fusion reactions that take place between hydrogen and helium isotopes, including the following:

$$_1^1\text{H} + _1^1\text{H} \rightarrow _1^2\text{H} + _1^0\text{e}$$

$$_1^2\text{H} + _1^2\text{H} \rightarrow _2^3\text{H} + _0^1\text{n}$$

$$_2^3\text{He} + _2^3\text{He} \rightarrow _2^4\text{He} + 2_1^1\text{H}$$

The fusion process would seem to be a desirable energy source on earth because (1) two isotopes of hydrogen, deuterium ($_1^2\text{H}$) and tritium ($_1^3\text{H}$), undergo fusion at a relatively low temperature; (2) the supply of deuterium is practically unlimited, with each gallon of seawater containing about a teaspoonful of deuterium dioxide; and (3) enormous amounts of energy are released with no radioactive by-products.

The oceans contain enough deuterium to generate electricity for the entire world for millions of years, and tritium can be constantly produced by a fusion device. Researchers know what needs to be done to tap this tremendous energy source. The problem is *how* to do it in an economical, continuous energy-producing fusion reactor. The problem, one of the most difficult engineering tasks ever attempted, is meeting three basic fusion reaction requirements of (1) temperature, (2) density, and (3) time (figure 15.22):

1. *Temperature*. Nuclei contain protons and are positively charged, so they experience the electromagnetic repulsion of like charges. This force of repulsion can be overcome, moving the nuclei close enough to fuse together, by giving the nuclei sufficient kinetic energy. The fusion reaction of deuterium and tritium, which has the lowest temperature requirements of any fusion reaction known at the

A CLOSER LOOK | Nuclear Waste

There are two general categories of nuclear wastes: (1) low-level wastes and (2) high-level wastes. The *low-level wastes* are produced by the normal operation of a nuclear reactor. Radioactive isotopes sometimes escape from fuel rods in the reactor and in the spent fuel storage pools. These isotopes are removed from the water by ion-exchange resins and from the air by filters. The used resins and filters will contain the radioactive isotopes and will become low-level wastes. In addition, any contaminated protective clothing, tools, and discarded equipment also become low-level wastes.

Low-level liquid wastes are evaporated, mixed with cement, then poured into fifty-five-gallon steel drums. Solid wastes are compressed and placed in similar drums. The drums are currently disposed of by burial in government-licensed facilities. In general, low-level waste has an activity of less than 1.0 curie per cubic foot. Contact with the low-level waste could expose a person to up to 20 millirems per hour of contact.

High-level wastes from nuclear power plants are spent nuclear fuel rods. At the present time most of the commercial nuclear power plants have these rods in temporary storage at the plant site. These rods are "hot" in

BOX FIGURE 15.1

This is a standard warning sign for a possible radioactive hazard. Such warning signs would have to be maintained around a nuclear waste depository for thousands of years.

the radioactive sense, producing about 100,000 curies per cubic foot. They are also hot in the thermal sense, continuing to generate heat for months after removal from the reactor. The rods are cooled by heat exchangers connected to storage pools; they could otherwise achieve

an internal temperature as high as 800°C for several decades. In the future, these spent fuel rods will be reprocessed or disposed of through terminal storage.

Agencies of the United States federal government have also accumulated millions of gallons of high-level wastes from the manufacture of nuclear weapons and nuclear research programs. These liquid wastes are stored in million-gallon stainless steel containers that are surrounded by concrete. The containers are located in the states of Washington, Idaho, and South Carolina. The future of this large amount of high-level wastes may be evaporation to a solid form or mixture with a glass or ceramic matrix, which is melted and poured into stainless steel containers. These containers would be buried in solid granite rock in a stable geologic depository. Such high-level wastes must be contained for thousands of years as they undergo natural radioactive decay (box figure 15.1). Burial at a depth of two thousand to three thousand feet in solid granite would provide protection from exposure by explosives, meteorite impact, or erosion. One major concern about this plan is that a hundred generations later, people might lose track of what is buried in the nuclear garbage dump.

---

present time, requires temperatures on the order of 100 million°C.

2. *Density.* There must be a sufficiently dense concentration of heavy hydrogen nuclei, on the order of $10^{14}/cm^3$, so many reactions occur in a short time.

3. *Time.* The nuclei must be confined at the appropriate density up to a second or longer at pressures of at least 10 atmospheres to permit a sufficient number of reactions to take place.

The temperature, density, and time requirements of a fusion reaction are interrelated. A short time of confinement, for example, requires an increased density, and a longer confinement time requires less density. The primary problems of fusion research are the high-temperature requirements and confinement. No material in the world can stand up to a temperature of 100 million degrees Celsius, and any material container would be instantly vaporized.

Thus, research has centered on meeting the fusion reaction requirements without a material container. Two approaches are being tested, *magnetic confinement* and *inertial confinement.*

Magnetic confinement utilizes a *plasma,* a very hot gas consisting of atoms that have been stripped of their electrons because of the high kinetic energies. The resulting positively and negatively charged particles respond to electrical and magnetic forces, enabling researchers to develop a "magnetic bottle," that is, magnetic fields that confine the plasma and avoid the problems of material containers that would vaporize. A magnetically confined plasma is very unstable, however, and researchers have compared the problem to trying to carry a block of jello on a pair of rubber bands. Different magnetic field geometries and magnetic "mirrors" are the topics of research in attempts to stabilize the hot, wobbly plasma. Electric currents, injection of fast ions, and radio frequency (microwave) heating methods are also being studied.

Inertial confinement is an attempt to heat and compress small frozen pellets of deuterium and tritium with energetic laser beams or particle beams, producing fusion. The focus of this research is new and powerful lasers, light ion and heavy ion beams. If successful, magnetic or inertial confinement will provide a long-term solution for future energy requirements.

The Source of Nuclear Energy

When elements undergo the natural radioactive decay process, energy is released, and the decay products have less energy than the original reactant nucleus. When massive nuclei undergo fission, much energy is rapidly released along with fission products that continue to release energy through radioactive decay. What is the source of all this nuclear energy? The answer to this question is found in current theories about how the universe started and in theories about the life cycle of the stars. Theories about the life cycle of stars are discussed in chapters 16 and 17. For now, consider just a brief introduction to the life cycle of a star in order to understand the ultimate source of nuclear energy.

The current universe is believed to have started with a "big bang" of energy, which created a plasma of protons and neutrons. This primordial plasma cooled rapidly and, after several minutes, began to form hydrogen nuclei. Throughout the newly formed universe massive numbers of hydrogen atoms—on the order of 10^{57} nuclei—were gradually pulled together by gravity into masses that would become the stars. As the hydrogen atoms fell toward the center of each mass of gas, they accelerated, just like any other falling object. As they accelerated, the contracting mass began to heat up because the average kinetic energy of the atoms increased from acceleration. Eventually, after say, ten million years or so of collapsing and heating, the mass of hydrogen condensed to a sphere with a diameter of 1.5 million miles or so, or about twice the size of the sun today. At the same time the interior temperature increased to millions of degrees, reaching the critical points of density, temperature, and containment for a fusion reaction to begin. Thus, a star was born as hydrogen nuclei fused into helium nuclei, releasing enough energy that the star began to shine.

Hydrogen nuclei in the newborn star had a higher energy per nucleon than helium nuclei, and helium nuclei had more energy per nucleon than other nuclei up to around iron. The fusion process continued for billions of years, releasing energy as heavier and heavier nuclei were formed. Eventually, the star materials were fused into nuclei around iron, the element with the lowest amount of energy per nucleon, and the star used up its energy source. Larger, more massive dying stars explode into supernovas (discussed in chapter 16). Such an explosion releases a flood of neutrons, which bombard medium-weight nuclei and build them up to more massive nuclei, all the way from iron up to uranium. Thus, the more massive elements were born from an exploding supernova, then spread into space as dust. In a process to be dis-

cussed later, this dust became the materials of which planets were made, including Earth. The point for the present discussion, however, is that the energy of naturally radioactive elements, and the energy released during fission, can be traced back to the force of gravitational attraction, which provided the initial energy for the whole process.

SUMMARY

Radioactivity is the spontaneous emission of particles or energy from an unstable atomic nucleus. The modern atomic theory pictures the nucleus as protons and neutrons held together by a short-range *nuclear force* that has moving *nucleons* (protons and neutrons) in *energy shells* analogous to the shell structure of electrons. A graph of the number of neutrons to the number of protons in a nucleus reveals that stable nuclei have a certain neutron-to-proton ratio in a *band of stability*. Nuclei that are above or below the band of stability, and nuclei that are beyond atomic number 83, are radioactive and undergo *radioactive decay*.

Three common examples of radioactive decay involve the emission of an *alpha particle*, a *beta particle*, and a *gamma ray*. An alpha particle is a helium nucleus, consisting of two protons and two neutrons. A beta particle is a high-speed electron that is ejected from the nucleus. A gamma ray is a short-wavelength electromagnetic radiation from an excited nucleus. In general, nuclei with an atomic number of 83 or larger become more stable by alpha emission. Nuclei with a neutron-to-proton ratio that is too large become more stable by beta emission. Gamma ray emission occurs from a nucleus that was left in a high-energy state by the emission of an alpha or beta particle.

Each radioactive isotope has its own specific *radioactive decay constant* (k), a ratio of the rate of nuclear disintegration to the total number of nuclei (n), or $k = \text{rate}/n$. The rate is usually described in terms of *half-life*, the time required for one-half the unstable nuclei to decay. Half-life is related to the decay constant by half-life $= 0.693/k$, where 0.693 is a mathematical constant for exponential decay (the natural log of 2).

Radiation is measured by (1) its effects on photographic film, (2) the number of ions it produces, or (3) the flashes of light produced on a phosphor. It is measured at a source in units of a *curie*, defined as 3.70×10^{10} nuclear disintegrations per second. It is measured where received in units of a *rad*, defined as 1×10^{-5} J. A *rem* is a measure of radiation that takes into account the biological effectiveness of different types of radiation damage. In general, the natural environment exposes everyone to 100 to 500 millirems per year, an exposure called *background radiation*. Life-style and location influence the background radiation received, but the average is 130 millirems per year.

Energy and mass are related by Einstein's famous equation of $E = mc^2$, which means that *matter can be converted to energy and energy to matter*. The mass of a nucleus is always less than the sum of the masses of the individual particles of which it is made. This *mass defect* of a nucleus is equivalent to the energy released when the nucleus was formed according to $E = mc^2$. It is also the *binding energy*, the energy required to break the nucleus apart into nucleons.

When the binding energy is plotted against the mass number, the greatest binding energy per nucleon is seen to occur for an atomic number near that of iron. More massive nuclei therefore release energy by fission, or splitting to more stable nuclei. Less massive nuclei release energy by fusion, the joining of less massive nuclei to produce a more stable, more massive nucleus. Nuclear fission provides the energy for atomic explosions and nuclear power plants. Nuclear fusion is the energy source of the sun and other stars and also holds promise as a future energy source for humans. The source of the energy of a nucleus can be traced back to the gravitational attraction that formed a star.

Summary of Equations

15.1

$$\text{radioactive decay constant} = \frac{\text{decay rate}}{\text{number of nuclei}}$$

$$k = \frac{\text{rate}}{n}$$

15.2

$$\text{half-life} = \frac{\text{a mathematical constant}}{\text{decay constant}}$$

$$t_{1/2} = \frac{0.693}{k}$$

15.3

$$\text{energy} = \text{mass} \times \text{the speed of light squared}$$

$$E = mc^2$$

KEY TERMS

alpha particle (p. **352**)   half-life (p. **358**)
background radiation (p. **362**)   mass defect (p. **363**)
band of stability (p. **355**)   nuclear fission (p. **365**)
beta particle (p. **352**)   nuclear fusion (p. **365**)
binding energy (p. **363**)   nucleons (p. **352**)
chain reaction (p. **366**)   rad (p. **361**)
critical mass (p. **366**)   radioactive decay (p. **352**)
curie (p. **361**)   radioactivity (p. **352**)
gamma ray (p. **352**)   rem (p. **361**)

APPLYING THE CONCEPTS

1. A high-speed electron ejected from a nucleus during radioactive decay is called a (an)
   a. alpha particle.
   b. beta particle.
   c. gamma ray.
   d. none of the above.

2. The ejection of an alpha particle from a nucleus results in
   a. an increase in the atomic number by one.
   b. an increase in the atomic mass by four.
   c. a decrease in the atomic number by two.
   d. none of the above.

3. The emission of a gamma ray from a nucleus results in
   a. an increase in the atomic number by one.
   b. an increase in the atomic mass by four.
   c. a decrease in the atomic number by two.
   d. none of the above.

4. An atom of radon-222 loses an alpha particle to become a more stable atom of
   a. radium.
   b. bismuth.
   c. polonium.
   d. radon.

5. The nuclear force is
   a. attractive when nucleons are closer than 10^{-15} m.
   b. repulsive when nucleons are closer than 10^{-15} m.
   c. attractive when nucleons are farther than 10^{-15} m.
   d. repulsive when nucleons are farther than 10^{-15} m.

6. Which of the following is more likely to be radioactive?
   a. nuclei with an even number of protons and neutrons
   b. nuclei with an odd number of protons and neutrons
   c. nuclei with the same number of protons and neutrons
   d. Number of protons and neutrons have nothing to do with radioactivity.

7. Which of the following isotopes is more likely to be radioactive?
   a. magnesium-24
   b. calcium-40
   c. astatine-210
   d. ruthenium-101

8. Hydrogen-3 is a radioactive isotope of hydrogen. Which type of radiation would you expect an atom of this isotope to emit?
   a. an alpha particle
   b. a beta particle
   c. either of the above
   d. neither of the above

9. A sheet of paper will stop a (an)
   a. alpha particle.
   b. beta particle.
   c. gamma ray.
   d. none of the above.

10. The most penetrating of the three common types of nuclear radiation is the
    a. alpha particle.
    b. beta particle.
    c. gamma ray.
    d. All have equal penetrating ability.

11. An atom of an isotope with an atomic number greater than 83 will probably emit a (an)

    a. alpha particle.

    b. beta particle.

    c. gamma ray.

    d. none of the above.

12. An atom of an isotope with a large neutron-to-proton ratio will probably emit a (an)

    a. alpha particle.

    b. beta particle.

    c. gamma ray.

    d. none of the above.

13. All of the naturally occurring radioactive decay series end when the radioactive elements have decayed to

    a. lead.

    b. bismuth.

    c. uranium.

    d. hydrogen.

14. The rate of radioactive decay can be increased by increasing the

    a. temperature.

    b. pressure.

    c. size of the sample.

    d. None of the above are correct.

15. The radioactive decay constant is a specific constant only for (a)

    a. particular isotope.

    b. certain temperature.

    c. certain sample size.

    d. all of the above.

16. Isotope A has a half-life of seconds, and isotope B has a half-life of millions of years. Which isotope is more radioactive?

    a. It depends on the sample size.

    b. isotope A

    c. isotope B

    d. Unknown, from the information given.

17. A Geiger counter indirectly measures radiation by measuring

    a. ions produced.

    b. flashes of light.

    c. speaker static.

    d. curies.

18. A measure of radioactivity at the *source* is (the)

    a. curie.

    b. rad.

    c. rem.

    d. any of the above.

19. A measure of radiation received that considers the biological effect resulting from the radiation is (the)

    a. curie.

    b. rad.

    c. rem.

    d. any of the above.

20. The mass of a nucleus is always _____ the sum of the masses of the individual particles of which it is made.

    a. equal to

    b. less than

    c. more than

    d. Unable to say without more information.

21. When protons and neutrons join together to make a nucleus, energy is

    a. released.

    b. absorbed.

    c. neither released nor absorbed.

    d. unpredictably absorbed or released.

22. Used fuel rods from a nuclear reactor contain about

    a. 96% usable uranium and plutonium.

    b. 33% usable uranium and plutonium.

    c. 4% usable uranium and plutonium.

    d. 0% usable uranium and plutonium.

23. The source of energy from the sun is

    a. chemical (burning).

    b. fission.

    c. fusion.

    d. radioactive decay.

24. The energy released by radioactive decay and the energy released by nuclear reactions can be traced back to the energy that isotopes acquired from

    a. fusion.

    b. the sun.

    c. gravitational attraction.

    d. the big bang.

Answers

1. b **2.** c **3.** d **4.** c **5.** a **6.** b **7.** c **8.** b **9.** a **10.** c **11.** a **12.** b **13.** a **14.** c **15.** a **16.** b **17.** a **18.** a **19.** c **20.** b **21.** a **22.** a **23.** c **24.** c

QUESTIONS FOR THOUGHT

1. How is a radioactive material different from a material that is not radioactive?

2. What is radioactive decay? Describe how the radioactive decay rate can be changed if this is possible.

3. Describe three kinds of radiation emitted by radioactive materials. Describe what eventually happens to each kind of radiation after it is emitted.

4. How are positively charged protons able to stay together in a nucleus since like charges repel?

5. What is half-life? Give an example of the half-life of an isotope, describing the amount remaining and the time elapsed after five half-life periods.

6. Would you expect an isotope with a long half-life to be more, the same, or less radioactive than an isotope with a short half-life? Explain.

7. What is (a) a curie? (b) a rad? (c) a rem?

8. What is meant by background radiation? What is the normal radiation dose for the average person from background radiation?

9. Why is there controversy about the effects of long-term, low levels of radiation exposure?

10. What is a mass defect? How is it related to the binding energy of a nucleus? How can both be calculated?

11. Compare and contrast nuclear fission and nuclear fusion.

PARALLEL EXERCISES

The exercises in groups A and B cover the same concepts. Solutions to group A exercises are located in appendix D.

Group A

Note: You will need the table of atomic weights inside the back cover of this text.

1. Give the number of protons and the number of neutrons in the nucleus of each of the following isotopes:
   (a) cobalt-60
   (b) potassium-40
   (c) neon-24
   (d) lead-208

2. Write the nuclear symbols for each of the nuclei in exercise 1.

3. Predict if the nuclei in exercise 1 are radioactive or stable, giving your reasoning behind each prediction.

4. Write a nuclear equation for the decay of the following nuclei as they give off a beta particle:
   (a) $^{56}_{26}Fe$
   (b) $^{7}_{4}Be$
   (c) $^{64}_{29}Cu$
   (d) $^{24}_{11}Na$
   (e) $^{214}_{82}Pb$
   (f) $^{32}_{15}P$

5. Write a nuclear equation for the decay of the following nuclei as they undergo alpha emission:
   (a) $^{235}_{92}U$
   (b) $^{226}_{88}Ra$
   (c) $^{239}_{94}Pu$
   (d) $^{214}_{83}Bi$
   (e) $^{230}_{90}Th$
   (f) $^{210}_{84}Po$

6. The half-life of iodine-131 is 8 days. How much of a 1.0 oz sample of iodine-131 will remain after 32 days?

7. If the half-life of strontium-90 is 27.6 years, what is the decay constant for strontium-90?

8. Using the decay constant for strontium-90 obtained in exercise 7, find the number of nuclear disintegrations over a period of time for a molar mass of strontium-90.

9. What is the activity in curies of the molar mass of strontium-90 described in exercise 8?

10. How much energy must be supplied to break a single iron-56 nucleus into separate protons and neutrons? (The mass of an iron-56 nucleus is 55.9206 u, one proton is 1.00728 u, and one neutron is 1.00867 u.)

Group B

Note: You will need the table of atomic weights inside the back cover of this text.

1. Give the number of protons and the number of neutrons in the nucleus of each of the following isotopes:
   (a) aluminum-25
   (b) technetium-95
   (c) tin-120
   (d) mercury-200

2. Write the nuclear symbols for each of the nuclei in exercise 1.

3. Predict if the nuclei in exercise 1 are radioactive or stable, giving your reasoning behind each prediction.

4. Write a nuclear equation for the beta emission decay of each of the following:
   (a) $^{14}_{6}C$
   (b) $^{60}_{27}Co$
   (c) $^{64}_{11}Na$
   (d) $^{241}_{94}Pu$
   (e) $^{131}_{53}I$
   (f) $^{210}_{82}Pb$

5. Write a nuclear equation for each of the following alpha emission decay reactions:
   (a) $^{241}_{95}Am$
   (b) $^{232}_{90}Th$
   (c) $^{223}_{88}Ra$
   (d) $^{234}_{92}U$
   (e) $^{242}_{96}Cm$
   (f) $^{237}_{93}Np$

6. If the half-life of cesium-137 is 30 years, how much time will be required to reduce a 1.0 kg sample to 1.0 g?

7. The half-life of tritium ($^{3}_{1}H$) is 12.26 years. What is the radioactive decay constant for tritium?

8. What is the number of disintegrations per unit of time for a molar mass of tritium? The decay constant is obtained from exercise 7.

9. Calculate the activity in curies of the molar mass of tritium described in exercise 8.

10. How much energy is needed to separate the nucleons in a single lithium-7 nucleus? (The mass of a lithium-7 nucleus is 7.01435 u, one proton is 1.00728 u, and one neutron is 1.00867 u.)

This is a planetary nebula in the constellation Aquarius. Planetary nebulae are clouds of ionized gases with no relationship to any planet. They were named long ago, when they appeared similar to the planets Neptune and Uranus when viewed through early telescopes.

CHAPTER | Sixteen

The Universe

Astronomy is an exciting and mind-expanding field of science that has fascinated and intrigued people since the beginnings of recorded history. Ancient civilizations searched the heavens in wonder, some recording on clay tablets what they observed. Many religious and philosophical beliefs were originally based on interpretations of these ancient observations. Today, we are still awed by the heavens and space, but now we are fascinated with ideas of space travel, black holes, and the search for extraterrestrial life. Throughout history, people have speculated about the universe and their place in it and watched the sky and wondered (figure 16.1). What is out there and what does it all mean? Are there other people such as ourselves on other planets, looking at the star in their sky that is our sun, wondering if we exist?

Until about thirty years ago progress in astronomy was limited to what could be observed and photographed. Developments in astronomy, in technology, and in other branches of science then began to provide the details of what is happening in the larger expanses of space away from Earth. These developments included understandings about nuclear reactions and implications about what must be going on inside a star and the nature of the light emitted; new data made available from the development of infrared telescopes, radio telescopes, and X-ray telescopes; and detailed spectral analysis of light from the stars. All of these developments and the discovery and theoretical meaning of pulsars, neutron stars, and black holes began to fit together like the pieces of a puzzle. Theoretical models emerged about how stars evolve, about what galaxies are and how they evolve, and eventually, about the explosive beginnings of the universe and the chain of events that led to the formation of the Sun and Earth. This chapter is concerned with these topics, beginning with historical attempts to understand how the stars are arranged in space. The chapter concludes with theoretical models of how the universe began and what may happen to it in the future.

ANCIENT IDEAS

Early civilizations had a much better view of the night sky before city lights, dust, and pollution obscured much of the sky. Today, you must travel far from the cities, perhaps to a remote mountaintop, to see a clear night sky as early people observed it. Back then, people could clearly see the motion of the moon and stars night after night, observing recurring cycles of motion. These cycles became important as they became associated with the timing of certain events. Thus, watching the sun, moon, and star movements became a way to identify when to plant crops, when to harvest, and when it was time to plan for other events. Observing the sky was an important activity, and many early civilizations built observatories with sighting devices to track and record astronomical events. Stonehenge, for example, was an ancient observatory built in England around 2600 B.C. by Neolithic people (figure 16.2).

There were many different centers of ancient civilizations, but historians identify those of the Euphrates River valleys (Babylon and Chaldea), the Nile River valleys (Egypt), and the Greek and Roman countries as civilizations that contributed the first known astronomical knowledge. As early as 2000 B.C., for example, the Babylonians kept track of long periods of time by dividing the year into 12 months, with 7 days to a week and 360 days to a year. Later, the Babylonians maintained observatories, compiled star catalogs, and were able to predict certain eclipses. They did not, however, attempt to explain any of the observed motions or cycles. The Babylonian concept of nature did not include the notion of physical cause and effect. To them, individual gods created and controlled the different parts of nature, so any explanation beyond a religious myth was not possible. They conceived of the universe as being the valley where they lived, with other celestial objects doing the moving. This concept of a **geocentric,** or earth-centered, universe was typical of ancient concepts. To the Babylonians, the stars were attached to a shell, or dome, that rested on mountaintops. The sun and the moon entered and exited the dome through locked gates maintained by a god. The Egyptian concept of the universe was also geocentric, but their universe was surrounded by water. They conceived of stars as lamps hung from the dome and the sun as a disk of fire carried in a boat by the sun god Ra. The idea of stars being attached to an inverted bowl or dome was expanded upon by the ancient Greeks, who believed a *celestial sphere* surrounded the earth. The stars do seem fixed on a celestial sphere, and Earth's rotation provides the illusion that the entire sphere turns while the earth stands still. The Greek concept of a celestial sphere turning around a fixed Earth, like the Babylonian and Egyptian concepts, was geocentric.

The early Babylonian observers of the night sky noted that the Sun, the Moon, and the five planets known at the time (Mercury, Venus, Mars, Jupiter, and Saturn) moved across the dome of the sky only along a certain path. The stars followed the motions of these seven celestial bodies along the path but kept the same position relative to each other as if they were fixed on the dome. The seven celestial bodies moved independently of the stars and only within a narrow band across the celestial sphere, with the sun's movement in the center of this band (figure 16.3). This path

A

FIGURE 16.2

The stone pillars of Stonehenge were positioned so that the movement of the sun and moon could be followed with the seasons of the year.

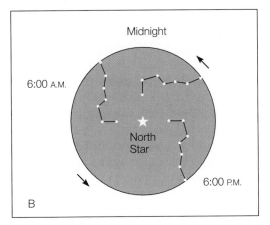

B

FIGURE 16.1

Ancient civilizations used celestial cycles of motion as clocks and calendars. (*A*) This photograph shows the path of stars around the North Star. (*B*) A "snapshot" of the position of the Big Dipper over a period of 24 hours as it turns around the North Star one night. This shows how the Big Dipper can be used to help you keep track of time.

along which the sun appears to move among the stars is today called the *ecliptic*. As viewed from the earth, the sun appears to move completely around the ecliptic each year. To keep track of the sun in its travels, the Babylonians imagined the arrangements of certain stars to be the shapes of some mythical god, object, or animal. These imagined patterns of stars, or **constellations,** were used to identify twelve equal divisions of the ecliptic through which the sun passed in monthly succession. Thus, there are twelve constellations around the ecliptic, and all twelve are called the *zodiac*, which means "circle of animals." The sun moves across a constellation, or "sign," each month, so there are twelve *signs of the zodiac* (figure 16.4). Table 16.1 gives the names and original meanings of these zodiac signs.

The early Babylonians used the zodiac as a basis for keeping time. The twelve signs of the zodiac identified the twelve months, with each new month beginning with the new moon.

Seven days to a week originated from the motion of the seven celestial bodies according to how long it took each to move across the sky. Thus "Saturn's day" became Saturday, "Sun's day" became Sunday, and "Moon's day" became Monday. The days of Tuesday, Wednesday, Thursday, and Friday are still named after the other four planets in the French, Italian, and Spanish languages, but other names are used today in English and German. By using a vertical rod as a sundial, the Babylonians also divided a day into hours, minutes, and seconds.

The later Babylonians, or Chaldeans, maintained astronomical observatories, but from their beliefs in the gods of nature they also developed astrology, the belief that the stars influence humans. Fully developed by 540 B.C., astrology became the art of studying the stars to guide human affairs. Some people today still have confidence in astrology, even though it is a fanciful extension of beliefs in mythical gods and an animistic conception of nature.

The zodiac signs of three thousand years ago are not the zodiac signs of today. Earth's precession (to be discussed in the next chapter) has shifted the constellations of the zodiac to the west. Thus, three thousand years ago the sun entered the "house" (constellation) of Virgo in August. The astrological "forecasts" of today are still based on the sun being in the constellation of Virgo in August. But today, the sun is in the "house of Leo" in August. Because of Earth's precession, the sun will circle through the entire zodiac in 25,780 years (figure 16.5).

Activities

Drive a rod vertically into the ground and measure the length of the shadow three times a day for a month. Graph the results and see how many models you can think of that would explain your findings.

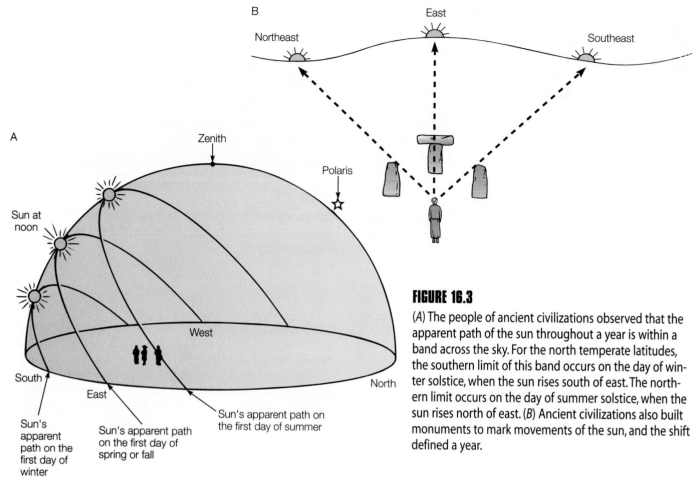

FIGURE 16.3

(*A*) The people of ancient civilizations observed that the apparent path of the sun throughout a year is within a band across the sky. For the north temperate latitudes, the southern limit of this band occurs on the day of winter solstice, when the sun rises south of east. The northern limit occurs on the day of summer solstice, when the sun rises north of east. (*B*) Ancient civilizations also built monuments to mark movements of the sun, and the shift defined a year.

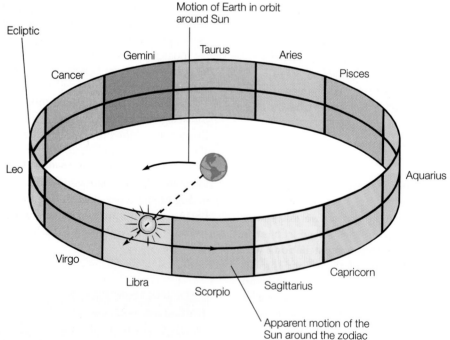

FIGURE 16.4

The sun, moon, and planets move across the constellations of the zodiac, with the sun moving around all twelve constellations during a year. From the earth, the sun will appear to be "in" Libra at sunrise in this sketch. As the earth revolves around the sun, the sun will seem to move from Libra into Scorpio, then through each constellation in turn.

TABLE 16.1 The zodiac: The object, person, or animal (sign) and its original meaning

Ram (Aries)

The Babylonians sacrificed rams during the first month of their year, so the position of the sun in this constellation identified the first month.

Bull (Taurus)

Ancient association of the sun with a bull. The sun rose in this constellation on the first day of spring.

Twins (Gemini)

After an ancient legend of a wolf raising twins who built Rome.

Crab (Cancer)

The sun retreated this month and a crab moves backward.

Lion (Leo)

The lion was an ancient symbol for hot, which was the weather for the month that the sun was in this constellation.

Virgin (Virgo)

After Babylonian myth of Ishtar.

Balance (Libra)

Day and night were the same length this month, so they were balanced when the sun was in this constellation.

Scorpion (Scorpio)

A scorpion comes out at night, or with darkness, and the sun in this constellation marked the approach of winter with longer nights.

Archer (Sagittarius)

The Babylonian god of war.

Goat (Capricorn)

After the ancient legend of the god of the sun who was nursed by a goat when young.

Water Bearer (Aquarius)

This was the month when floods began on the Nile, so the sun in this constellation marked the time of floods.

Fishes (Pisces)

This was the spring month when people returned to work after a winter of darkness; the fish was the ancient symbol for life after death, and the sun in this constellation marked the time when everything returned to life.

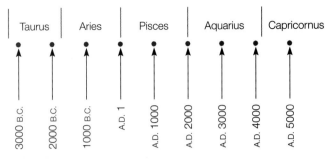

FIGURE 16.5

During the time of the Babylonians, the sun rose with the constellation Taurus on the first day of spring. Today, however, the sun rises with the constellation Pisces on the first day of spring. The earth's precession will continue to change the position of the sun during a particular month, and 25,780 years after the time of the Babylonians, the sun will again rise with the constellation Taurus on the first day of spring.

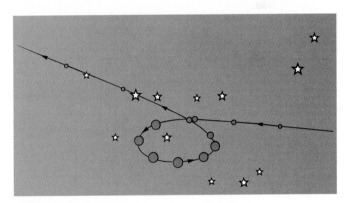

FIGURE 16.6

The apparent position of Mars against the background of stars as it goes through retrograde motion. Each position is observed approximately two weeks after the previous position.

The ancient Greek civilization, beginning about 600 B.C., added observations, reasoned theories, and generally removed the contrived influence of gods from the study of astronomy. They observed, for example, that over time the seven celestial bodies moving on the ecliptic did not hold the same relative positions and sometimes moved backward, making a loop over a period of several months. They called the five bodies that did this the Greek word for wanderer, "planetes," from which the word *planet* is derived. The Greek theory of the universe (which was discussed in chapter 2) had the earth fixed and motionless at the center. In general, the sun, the moon, planets, and stars were thought to move around the earth attached to spherical shells. Attempting to account for the movements of the sun, the moon, and planets required a model that eventually had fifty-five concentric shells, all centered on the earth and all moving at different rates. The model could not, how-

ever, explain why a planet such as Mars would move eastward across the ecliptic one month, slow down during the next month, stop, then move backward (westward) for the next two months, then resume its normal eastward motion. Today, this looping change of motion is called *retrograde motion* (figure 16.6). The time required for retrograde motion ranges from 34 days for Mercury to 139 days for Saturn.

Later Greek astronomers attempted to explain retrograde motion as the result of a planet being fixed on the outer edge of a wheel as it rolled across the sky. Thus, they established a means of ranking the distances of the planets from Earth. Since Saturn took the longest time period to go through retrograde motion, the Greeks reasoned that it must be farthest away. Mercury must be the closest planet according to this theory. The moon was reasoned to be the closest of all because it occasionally moved in front of the planets. Ptolemy recorded this ancient Greek geocentric model of

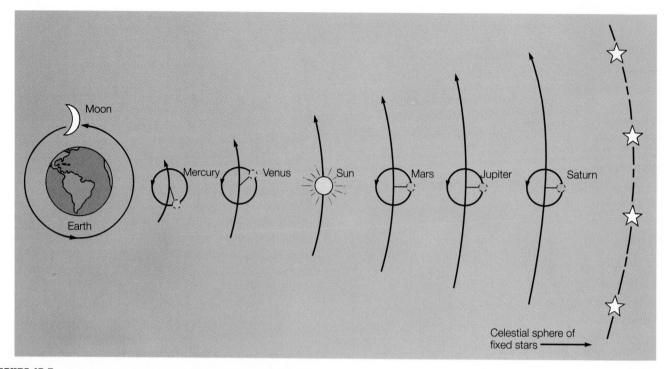

FIGURE 16.7

A schematic representation of the geocentric Ptolemaic system.

the universe in a seven-volume work. The earth-centered Ptole-maic system (figure 16.7) became the accepted model of the universe when, along with Aristotle's theories, it became a part of Church doctrine. How the sun-centered model became established will be discussed with the solar system in the next chapter.

THE NIGHT SKY

Away from city lights, dust, and pollution you can make observations of the stars and planets as the ancient astronomers did without any special equipment. Light from the stars and planets must pass through the earth's atmosphere to reach you, however, and the atmosphere affects the light. Stars appear as point *sources* of light, and each star generates its own light. The stars seem to twinkle because density differences in the atmosphere refract the point of starlight one way, then the other, as the air moves. The result is the slight dancing about and change in intensity called twinkling. The points of starlight are much steadier when viewed on a calm night or when viewed from high in the mountains where there is less atmosphere for the starlight to pass through. Astronauts outside the atmosphere see no twinkling, and the stars appear as steady point sources of light.

Back at ground level, within the atmosphere, the *reflected* light from a planet does not seem to twinkle. A planet appears as a disk of light rather than a point source, so refraction from moving air of different densities does not affect the image as much. Sufficient air movement can cause planets to appear to shimmer, however, just as a road appears to shimmer on a hot summer day.

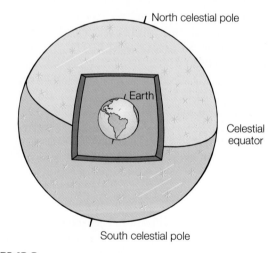

FIGURE 16.8

The celestial sphere with the celestial equator directly above Earth's equator, and the celestial poles directly above Earth's poles.

Celestial Location

To locate the ecliptic, planets, or anything else in the sky, you need something to refer to, a referent system. A referent system is easily established by first imagining the sky to be a *celestial sphere* just as the ancient Greeks did (figure 16.8). A coordinate system of lines can be visualized on this celestial sphere just as you think of the coordinate system of latitude and longitude lines on the

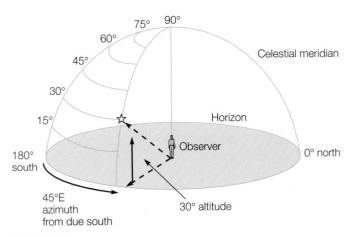

FIGURE 16.9

Once you have established the celestial equator, the celestial poles, and the celestial meridian, you can use a two-coordinate horizon system to locate positions in the sky. One popular method of using this system identifies the altitude angle (in degrees) from the horizon up to an object on the celestial sphere and the azimuth angle (again in degrees) the object on the celestial sphere is east or west of due south, where the celestial meridian meets the horizon. The illustration shows an altitude of 30° and an azimuth of 45° east of due south.

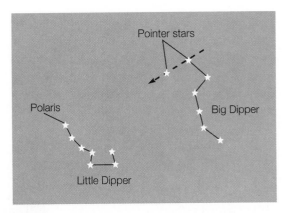

FIGURE 16.10

The North Star, or Polaris, is located by using the pointer stars of the Big Dipper.

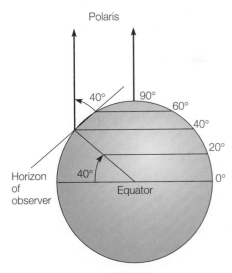

FIGURE 16.11

The altitude of Polaris above the horizon is approximately the same as the observer's latitude in the Northern Hemisphere.

earth's surface. Imagine that you could inflate the earth until its surface touched the celestial sphere. If you now transfer the latitude and longitude lines to the celestial sphere, you will have a system of sky coordinates. The line of the equator of the earth on the celestial sphere is called the *celestial equator*. The North Pole of the earth touches the celestial sphere at a point called the *north celestial pole*. From the surface of the earth, you can see that the celestial equator is a line on the celestial sphere directly above the earth's equator, and the north celestial pole is a point directly above the North Pole of the earth. Likewise, the *south celestial pole* is a point directly above the South Pole of the earth.

You can only see half of the overall celestial sphere from any one place on the surface of the earth. Imagine a point on the celestial sphere directly above where you are located. An imaginary line that passes through this point, then passes north through the north celestial pole, continuing all the way around through the south celestial pole and back to the point directly above you makes a big circle called the *celestial meridian* (figure 16.9). Note that the celestial meridian location is determined by where *you* are on the earth. The celestial equator and the celestial poles, on the other hand, are always in the same place no matter where you are.

Overall, the celestial sphere appears to spin, turning on an axis through the celestial poles. A photograph made by pointing a camera at the north celestial pole and leaving the shutter open for several hours will show the apparent motion of the celestial sphere with star trails (see figure 16.1A). The moderately bright star near the center is the North Star, *Polaris*. Polaris is almost, but not exactly, at the north celestial pole. You can locate Polaris by finding the *Big Dipper*. The two stars on the end of the dipper opposite the handle are called the pointers. Imagine a line

moving from the bottom of the dipper upward through the two pointers. The first bright star that this line meets is Polaris (figure 16.10). The angle that you see Polaris above the horizon is your approximate latitude in the Northern Hemisphere. Figure 16.11 shows the geometric relationships between your latitude and the angle of Polaris above the horizon.

If you observe the constellations night after night, you will see that the stars maintain their positions relative to one another as they turn counterclockwise around Polaris. Those near Polaris pivot around it and are called "circumpolar." Those farther out rise in the east, move in an arc, then set in the west.

Celestial Distance

When you look at Polaris, or any other star, it seems impossible to know anything about the distance to such a point of light. The distance *between* stars on the celestial sphere is measured by a

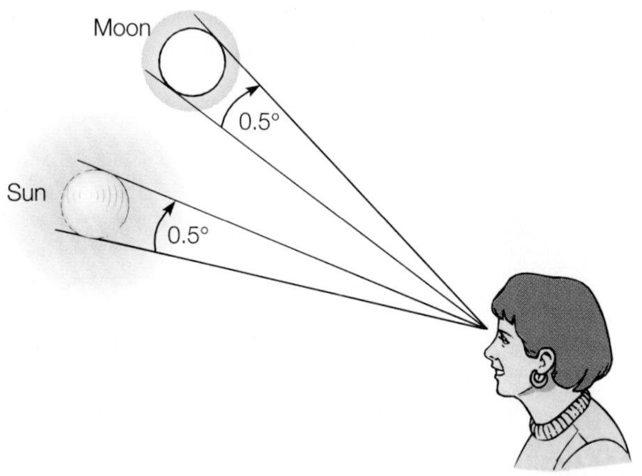

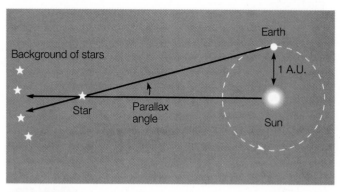

FIGURE 16.12

The moon and the sun both have an angular size of 0.5°, but the sun is much farther away. The observed angular size depends on distance and the true size of an object. Thus during a solar eclipse the moon appears to nearly cover the sun (images are separated here for clarity).

technique that was first developed by the Babylonians. The Babylonians considered a circle, such as the celestial meridian, to have 360 divisions, or degrees. Thus 1° is 1/360 of the circle, which is referred to as 1° *of arc* so it is not confused with temperature. An arc is a segment of the circumference of a circle, and the length of a segment can be used to indicate how far apart things are on the celestial sphere. The entire celestial meridian, for example, contains 360° of arc, but only 180° of arc are visible from one horizon to the other. A star directly overhead is 90° of arc above the horizon. The pointer stars in the Big Dipper are about 5° of arc apart. Polaris is 1° of arc away from the north celestial pole. For smaller measurements of arc, each degree is divided into 60 minutes (60′), and each minute is divided into 60 seconds (60″). These subdivisions are called "arc minutes" and "arc seconds" so they are not confused with time measurements.

The apparent size of an object is measured by the angle an object makes, which is called the angular size or *angular diameter*. This is an apparent size because the angular diameter of an object decreases as its distance increases. Both the sun and the moon, for example, have an angular diameter of about 0.5° of arc when viewed from the earth, but the sun is much farther away. Thus, the angular size of an object depends both on its distance and its true size (figure 16.12).

The distance to a star can be measured by *parallax,* the apparent shift of position by closer objects against a background when viewed from different observation points. As shown in figure 16.13, a relatively nearby star viewed from the earth in two different parts of the earth's orbit will appear in different locations against the background of stars. The radius of the earth's orbit is called an **astronomical unit** (A.U.). One A.U. is about 1.5×10^8 km (about 9.3×10^7 mi). The baseline of 1 A.U. defines the astronomical unit of distance called a **parsec.** A parsec is the distance at which the angle made from a 1 A.U. baseline is 1 arc second (figure 16.14).

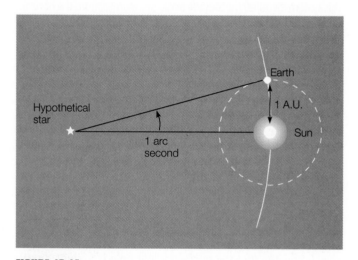

FIGURE 16.13

The angle through which a star seems to move against the background of stars between two observations that are 1 A.U. apart defines the parallax angle. The distance to relatively nearby stars can be determined from the parallax angle.

FIGURE 16.14

A *parsec* is the distance at which the angle made from a 1 A.U. baseline is 1 arc second.

The parsec is the distance unit of choice of astronomers since it historically provided a means of measuring distances with a telescope. The distance to a star in parsecs is simply the reciprocal of the parallax angle in seconds of arc, distance = 1/seconds of arc. The other unit of astronomical distance is the **light-year** (ly), which is the distance that light travels in one year, about 9.5×10^{12} km (about 6×10^{12} mi). One parsec is approximately equal to 3.26 light-years (figure 16.15). There is an obvious limit to using the parallax measurement since the angle becomes smaller and smaller with greater distances. At the present time this limit is about 500 light-years from the sun. Distances beyond this require other measurement techniques. Standard referent units of length such as kilometers or miles have little meaning in astronomy since there are no referent points of comparison. Thus, the light-year measures distance in terms of time, and the parsec measures distance in terms of angles.

CONNECTIONS

Navigation and Migration

Since animals move from place to place to meet their needs it is useful to be able to return to a nest, water hole, den, or favorite feeding spot. This requires some sort of memory of their surroundings (a mental map) and a way of determining direction. Often it is valuable to have information about distance as well. Direction can be determined by such things as magnetic fields, identifying landmarks, scent trails, or reference to the sun or stars. If the sun or stars are used for navigation, then some sort of time sense is also needed since these bodies move in the sky.

Like honeybees, some daytime-migrating birds use the sun to guide them. We need two instruments to navigate by the sun—an accurate clock and a sextant for measuring the angle between the sun and the horizon. Can a bird perform such measurements without instruments when we, with our much bigger brains, need these

instruments to help us? It is unquestionably true! For nighttime migration, some birds use the stars to help them find their way. In one interesting experiment, warblers, which migrate at night, were placed in a planetarium. The pattern of stars as they appear at any season could be projected onto a large domed ceiling. During autumn, when these birds would normally migrate southward, the stars of the autumn sky were shown on the ceiling. The birds responded with much fluttering activity at the south side of the cage, as if they were trying to migrate southward. Then the experimenters tried projecting the stars of the spring sky, even though it was autumn. Now the birds tended to try to fly northward, although there was less unity in their efforts to head north; the birds seemed somewhat confused. Nevertheless, the experiment showed that the birds recognized star patterns and were influenced by them.

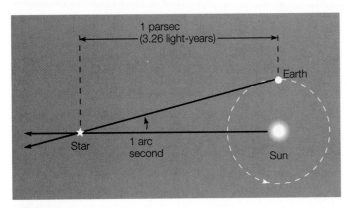

FIGURE 16.15

A parsec is approximately equal to 3.26 light-years.

STARS

If you could travel by spaceship a few hundred light-years from the earth, you would observe the sun shrink to a bright point of light among the billions and billions of other stars. The sun is just an ordinary star with an average brightness. Like the other stars, the sun is a massive, dense ball of gases with a surface heated to incandescence by energy released from fusion reactions deep within. Since the sun is an average star, it can be used as a reference for understanding all the other stars.

Origin of Stars

Theoretically, a star with the mass of the sun was born from swirling clouds of hydrogen gas in the deep space between other stars. Such interstellar clouds are called **nebulae.** Most nebulae cannot be seen without a telescope, but one in the constellation Orion can be seen on a clear winter night in the Northern Hemisphere. Such clouds have a random, swirling motion, with gas atoms passing by one another mostly unaffected because they are not massive enough for gravity to exert much of a force. Complex motion of stars, however, can produce a shock wave that causes particles to collide, making local compressions. Their mutual gravitational attraction then begins to pull them together in a cluster. The cluster grows as more atoms are pulled in, which increases the mass and thus the gravitational attraction, and still more atoms are pulled in from farther away. Staggeringly huge numbers of atoms must be present to create the mutual attractions necessary to hold the atoms together in a cluster. Theoretical calculations indicate that on the order of 1×10^{57} atoms are necessary, all within a distance of several trillion miles. When these conditions do occur, the cloud of gas atoms begins to condense by gravitational attraction to a **protostar,** an accumulation of gases that will become a star.

Gravitational attraction pulls the average protostar from a cloud with a diameter of trillions of miles down to a dense sphere with a diameter of 1.5 million miles or so. As gravitational attraction accelerates the atoms toward the center, they gain kinetic energy, and the interior temperature increases. Over a period of some 10 million years of contracting and heating, the temperature

CONNECTIONS

The Biology behind Fall Colors

Do plants receive information from the stars and the sun as birds and insects do? After all, many trees change the color of their leaves every fall. How do the plants know when it is fall? How do they change color?

The autumn color change you see in leaves is the result of green chlorophyll breaking down. Other pigments (red, yellow, orange, and brown) are present in leaves all summer but are masked by the green chlorophyll pigments. In the fall, a layer of waterproof tissue called an *abscission layer* forms at the base of each leaf and cuts the flow of water and other nutrients to the leaves. The leaf cells die and their chlorophyll disintegrates, revealing the reds, oranges, yellows, and browns that make a trip through the countryside a colorful experience.

Perhaps you can think of an experiment you could do to find out how plants "know" when it is fall. Could the plants recognize that the days are becoming shorter and shorter? Or perhaps they recognize that the daily temperature is decreasing? If plants are exposed to stars or not, the number of hours of light, and the air temperature are all variables you can control in an experimental situation. How will you test these variables?

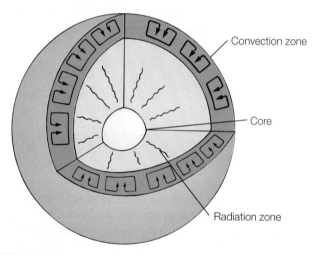

Convection zone

Core

Radiation zone

FIGURE 16.16

The structure of an average, mature star such as the sun. Hydrogen fusion reactions occur in the core, releasing gamma and X-ray radiation. This radiation moves through the radiation zone from particle to particle, eventually heating gases at the bottom of the convection zone. Convection cells carry energy to the surface, where it is emitted to space as visible light, ultraviolet radiation, and infrared radiation.

and density conditions at the center of the protostar are sufficient to start nuclear fusion reactions. Pressure from hot gases and energy from increasing fusion reactions begin to balance the gravitational attraction over the next 17 million years, and the newborn, average star begins its stable life, which will continue for the next 10 billion years.

The interior of an average star, such as the sun, is modeled after the theoretical pressure, temperature, and density conditions that would be necessary to produce the observed energy and light from the surface. This model describes the interior as a set of three shells: (1) the core, (2) a radiation zone, and (3) the convection zone (figure 16.16).

The **core** is the dense, very hot region where nuclear fusion reactions release gamma and X-ray radiation. The pressure at the core is about 3×10^{11} atmospheres with a central temperature of $1.5 \times 10^{7}°$C. At this pressure, the density of the core is about 150 g/cm^3, or about twelve times that of solid lead. Because of the plasma conditions, however, the core remains in a gaseous state even at this density. The core contains about a third of the total mass of an average star but reaches only one-fourth of the distance to the surface.

The **radiation zone** is less dense than the core, having a density of about 1 g/cm^3 (which is the density of water) halfway from the center to the surface. Energy in the form of gamma and X rays from the core is absorbed and reemitted by collisions with atoms in this zone. The radiation slowly diffuses outward because of the countless collisions over a distance comparable to the distance between Earth and the Moon. Each collision red-shifts the radiation toward ultraviolet radiation, and it could take millions of years before this radiation finally escapes the radiation zone.

The **convection zone** begins about seven-tenths of the way to the surface, and the density of the gases here is much less, about 1 percent the density of water. Gases at the bottom of this zone are heated by radiation from the zone below, expand from the heating, and are buoyed to the surface by convection. At the surface, the gases emit energy in the form of visible light, ultraviolet radiation, and infrared radiation, which moves out into space. The now cooler gases contract in volume and sink back to the radiation zone to become heated again, continuously carrying energy from the radiation zone to the surface in convection cells. The surface is continuously heated by the convection cells as it gives off energy to space, maintaining a temperature of about 5,800 K (about 5,500°C).

As an average star, the sun converts about 1.4×10^{17} kg of matter to energy every year as hydrogen nuclei are fused to produce helium. The sun was born about five billion years ago and has sufficient hydrogen in the core to continue shining for another four or five billion years. Other stars, however, have masses that are much greater or much less than the mass of the sun so they have different life spans. More massive stars generate higher temperatures in the core because they have a greater gravitational contraction from their greater masses. Higher temperatures mean increased kinetic energy, which results in increased numbers of collisions between hydrogen nuclei with the end result an increased number of fusion reactions. Thus, a more massive star uses up its hydrogen more rapidly than a less massive star. On the other hand, stars that are less massive than the sun use their hydrogen at a slower rate so they have longer life spans. The life spans of the stars range from a few million years for large, massive stars, to ten billion years for average stars like the sun, to trillions of years for small, less massive stars.

Brightness of Stars

Stars generate their own energy and light, but some stars appear brighter than others in the night sky. As you can imagine, this difference in brightness could be related to (1) the amount of energy and light produced by the stars, (2) the size of each star, or (3) the distance to a particular star. A combination of these factors is responsible for the brightness of a star as it appears to you in the night sky. A classification scheme for different levels of brightness that you see is called the **apparent magnitude** scale (table 16.2). The apparent magnitude scale is based on a system established by a Greek astronomer over two thousand years ago. Hipparchus made a catalog of the stars he could see and assigned a numerical value to each to identify its relative brightness. The brightness values ranged from 1 to 6, with the number 1 assigned to the brightest star and the number 6 assigned to the faintest star that could be seen. Stars assigned the number 1 came to be known as first-magnitude stars, those a little dimmer as second-magnitude stars, and so on to the faintest stars visible, the sixth-magnitude stars.

When technological developments in the nineteenth century made it possible to measure the brightness of a star, Hipparchus's system of brightness values acquired a precise, quantitative meaning. Today, a first-magnitude star is defined as one that is 100 times brighter than a sixth-magnitude star, with five uniform multiples of decreasing brightness on a scale from the first magnitude to the sixth magnitude. Each magnitude is about 2.51 times fainter than the next highest magnitude number. Thus a first-magnitude star is 2.51 times brighter than a second-magnitude star, $(2.51)^2$ or 6.31 times brighter than a third-magnitude star, and so on. Table 16.3 shows the brightness ratio that can be used to compare the brightness of two stars based on apparent magnitude differences. When using this table, recall that the *brighter* star has the *lower* magnitude number. This seems backward, but this convention has been followed ever since the first scale was devised by Hipparchus. Also note that some stars were found, by measurement, to be brighter than the apparent magnitude of +1, which extends the scale into negative numbers. The brightest star in the night sky is Sirius, for example, with an apparent magnitude of −1.42.

The apparent magnitude of a star depends on how far away stars are in addition to differences in the stars themselves. Stars at

a farther distance will appear fainter, and those closer will appear brighter, just as any other source of light. To compensate for distance differences, astronomers calculate the brightness that stars would appear to have if they were all at a defined, standard distance. The standard distance is defined as 10 parsecs (32.6 light-years), and the brightness of a star at this distance is called the **absolute magnitude.** The sun, for example, is the closest star and has an apparent magnitude of −26.7 at an average distance from the earth. When viewed from the standard distance of 10 parsecs, the sun would have an absolute magnitude of +4.8, which is about the brightness of a faint star (figure 16.17).

The absolute magnitude is an expression of **luminosity,** the total amount of energy radiated into space each second from the surface of a star. The sun, for example, radiates 4×10^{26} joules per second from the surface. The luminosity of stars is often

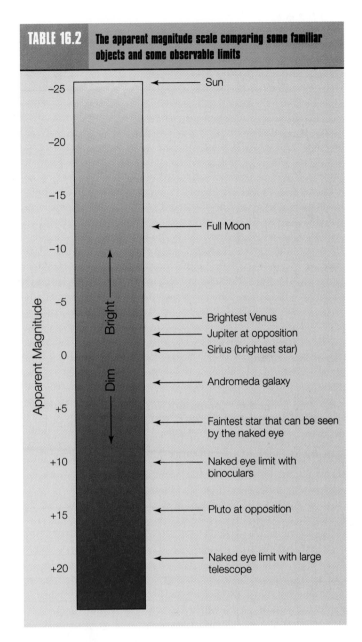

TABLE 16.2 The apparent magnitude scale comparing some familiar objects and some observable limits

TABLE 16.3	Brightness comparisons for apparent magnitude differences
Difference of Two Apparent Magnitudes	**Ratio of Brightness between the Two Stars**
0.0	1.0
0.5	1.6
1.0	2.5
1.5	4.0
2.0	6.3
2.5	10
3.0	16
3.5	25
4.0	40
4.5	63
5.0	100
5.5	160
6.0	250
6.5	400
7.0	630

Star Temperature

If you observe the stars on a clear night you will notice that some are brighter than others, but you will also notice some color differences. Some stars have a reddish color, some have a bluish white color, and others have a yellowish color. This color difference is understood to be a result of the relationship that exists between the color and the temperature of an incandescent object. The colors of the various stars are a result of the temperatures of the stars. You see a cooler star as reddish in color and comparatively hotter stars as bluish white. Stars with in-between temperatures, such as the Sun, appear to have a yellowish color (figure 16.18).

Astronomers analyze starlight to measure the temperature and luminosity as well as the chemical composition of a star. Light from a star is focused through a telescope on a solid-state photocell device, and the amount of radiation measured is used to calculate the luminosity of the star. By using light filters, the intensity of blue light from a star can be compared to the intensity of red light. Comparison of the two intensities indicates the temperature, since more intense blue light means higher temperatures and more intense red light means cooler temperatures. When the starlight is analyzed in a spectroscope, specific elements can be identified from the unique set of spectral lines that each element emits.

Astronomers use information about the star temperature and spectra as the basis for a star classification scheme. Originally, the classification scheme was based on sixteen categories according to the strength of the hydrogen line spectra. The groups were identified alphabetically with A for the group with the strongest hydrogen line spectrum, B for slightly weaker lines,

compared to the sun's luminosity, with the sun considered to have a luminosity of 1 unit. When this is done the luminosity of the stars ranges from a low of 10^{-6} sun units for the dimmest stars up to a high of 10^5 sun units. Thus, the sun is somewhere in the middle of the range of star luminosity.

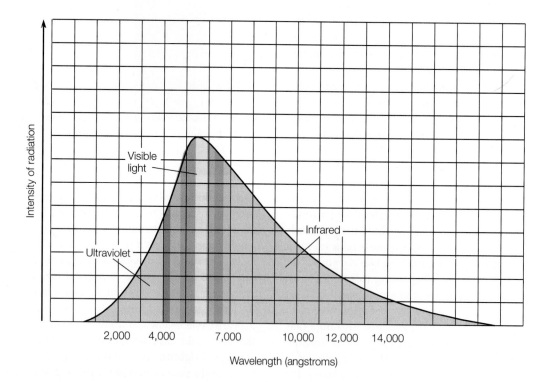

FIGURE 16.17

Not all energy from a star goes into visible light. The graph shows the distribution of radiant energy emitted from the sun, which has an absolute magnitude of +4.8.

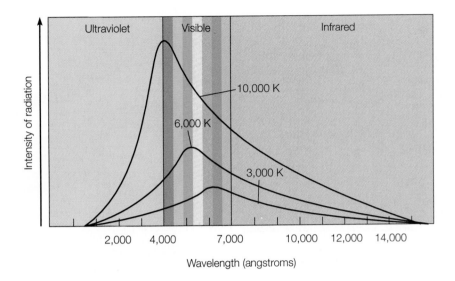

FIGURE 16.18

The distribution of radiant energy emitted is different for stars with different surface temperatures. Note that the peak radiation of a cooler star is more toward the red part of the spectrum, and the peak radiation of a hotter star is more toward the blue part of the spectrum.

TABLE 16.4	Major stellar spectral types and temperatures		
Type	Color	Temperature (K)	Comment
O	Bluish	30,000–80,000	Spectrum with ionized helium and hydrogen but little else; short-lived and rare stars
B	Bluish	10,000–30,000	Spectrum with neutral helium, none ionized
A	Bluish	7,500–10,000	Spectrum with no helium, strongest hydrogen, some magnesium and calcium
F	White	6,000–7,500	Spectrum with ionized calcium, magnesium, neutral atoms of iron
G	Yellow	5,000–6,000	The spectral type of the sun. Spectrum shows sixty-seven elements in the sun
K	Orange-red	3,500–5,000	Spectrum packed with lines from neutral metals
M	Reddish	2,000–3,500	Band spectrum of molecules, e.g., titanium oxide; other related spectral types (R, N, and S) are based on other molecules present in each spectral type

and on to the last group with the faintest lines. Later, astronomers realized that the star temperature was the important variable, so they rearranged the categories according to decreasing temperatures. The original letter categories were retained, however, resulting in classes of stars with the hottest temperature first and the coolest last with the sequence O B A F G K M. Table 16.4 compares the color, temperature ranges, and other features of the stellar spectra classification scheme.

Star Types

In 1910, Henry Russell in the United States and Ejnar Hertzsprung in Denmark independently developed a scheme to classify stars with a temperature-luminosity graph. Today, the graph is called the **Hertzsprung-Russell diagram,** or the *H-R diagram* for short. The diagram is a plot with temperature indicated by spectral types, and the true brightness indicated by absolute magnitude. The diagram, as shown in figure 16.19, plots temperature by spectral types sequenced O through M types, so the temperature decreases from left to right. The hottest, brightest stars are thus located at the top left of the diagram, and the coolest, faintest stars are located at the bottom right.

Each dot is a data point representing the surface temperature and brightness of a particular star. The sun, for example, is a type G star with an absolute magnitude of about +5, which places the data point for the sun almost in the center of the diagram. This means that the sun is an ordinary, average star with respect to both surface temperature and true brightness.

Most of the stars plotted on an H-R diagram fall in or close to a narrow band that runs from the top left to the lower right. This band is made up of **main sequence stars.** Stars along the main sequence band are normal, mature stars that are using their nuclear fuel at a steady rate. Those stars on the upper left

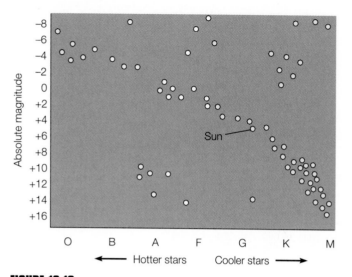

FIGURE 16.19

The Hertzsprung-Russell diagram.

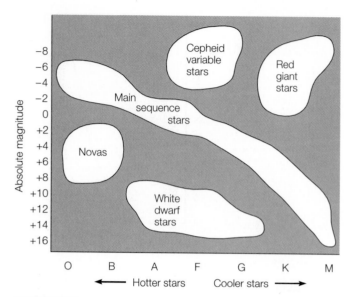

FIGURE 16.20

The region of the Cepheid variable, red giant, main sequence, and white dwarf stars and novas on the H-R diagram.

of the main sequence are the brightest, bluest, and most massive stars on the sequence. Those at the lower right are the faintest, reddest, and least massive of the stars on the main sequence. In general, most of the main sequence stars have masses that fall between a range from ten times greater than the mass of the sun (upper left) to one-tenth the mass of the sun (lower right). The extremes, or ends, of the main sequence range from about sixty times more massive than the sun to one-twenty-fifth of the sun's mass. It is the *mass* of a main sequence star that determines its brightness, its temperature, and its location on the H-R diagram. High-mass stars on the main sequence are brighter, hotter, and have shorter lives than low-mass stars. These relationships do not apply to the other types of stars in the H-R diagram.

As shown in figure 16.20, there are two groups of stars that have a different set of properties than the main sequence stars. The **red giant stars** are bright, but low-temperature, giants. These reddish stars are enormously bright for their temperature because they are very large, with an enormous surface area giving off light. A red giant might be one hundred times larger but have the same mass as the sun. These low-density red giants are located in the upper right part of the H-R diagram. The **white dwarf stars,** on the other hand, are located at the lower left because they are faint, white-hot stars. A white dwarf is faint because it is small, perhaps twice the size of Earth. It is also very dense, with a mass approximately equal to the sun's. During its lifetime a star will be found in different places on the H-R diagram as it undergoes changes. Red giants and white dwarfs are believed to be evolutionary stages that aging stars pass through, and the path a star takes across the diagram is called an evolutionary track. During the lifetime of the sun, it will be a main sequence star, a red giant, and then a white dwarf.

Stars such as the sun emit a steady light because the force of gravitational contraction is balanced by the outward flow of energy. *Variable stars,* on the other hand, are stars that change in brightness over a period of time. The changes in brightness can range from hundredths of a magnitude up to 20 magnitudes, and with a period that can range from seconds to years. The brightness changes and the time interval depends on the kind of variable star, for example, those that physically pulse, pulses from eclipsing binaries, or pulses that come from rotating stars. There are almost 30,000 stars that have been identified as being variable, and many are important to astronomers because they provide information about the properties of stars and distances as well.

A **Cepheid variable** is a bright variable star that is used to measure distances. In 1912, the American astronomer Henrietta Leavitt discovered an interesting relationship in Cepheids, one that holds true for stars of this type, but in two distinct populations. The general relationship is the longer the time needed for one pulse, the greater the apparent brightness of that star. Another American astronomer, Harlow Shapley, calibrated this period-brightness relationship to distance by comparing the apparent brightness with the absolute magnitude (true brightness) of a Cepheid at a known distance with a known period. This calibration allowed astronomers to calculate the distance to a star by using the inverse-square law of brightness.

In the 1920s, Edwin Hubble used the Cepheid period-brightness relationship to find the distances to other galaxies and discovered yet another relationship, that the greater the distance to a galaxy, the greater a shift in spectral lines toward the red end of the spectrum (red shift). This relationship is called Hubble's law, and it forms the foundation for understandings about our expanding universe. Measuring red shift provides another means of establishing distances to other, far-out galaxies. Such distance measurements are used to calculate the rate at which the universe is expanding, called the *Hubble constant*. The Hubble constant is used to estimate the size and age of the universe. The Hubble

Space Telescope is being used to observe Cepheid variables in distant galaxies about 56 million light-years from Earth, the farthest intergalactic distances that have been determined precisely.

The Life of a Star

A star is born in a gigantic cloud of gas and dust in interstellar space, then settles to a period of millions or billions of years of calmly shining while it fuses hydrogen nuclei in the core. How long a star shines and what happens to it when it uses up the hydrogen in the core depends on the mass of the star. Some stars, such as the sun, will slowly expand to a bloated red giant, then violently blow off their outer shells to become a white dwarf star. Other, more massive stars become a collapsed core of nuclear matter called a *neutron star*. The collapsed core of still more massive stars might collapse into a *black hole* in space. Of course no one has observed a life cycle of over millions or billions of years. The life cycle of a star is a theoretical outcome of computations concerning relationships between the measured values of the mass of a star, its surface temperature, and luminosity with models based on what is known about nuclear reactions and the theoretical changes that would occur as the fuel for the nuclear reactions is used up. The resulting model of the life cycle of a star and the predicted outcomes seem to agree with observations of stars today, with different groups of stars that can be plotted on the H-R diagram. Thus, the groups of stars on the diagram—main sequence, red giants, and white dwarfs, for example—are understood to be stars in various stages of their lives.

The first stage in the theoretical model of the life cycle of a star is the formation of the protostar. As gravity pulls the gas of a protostar together, the density, pressure, and temperature increase from the surface down to the center. Eventually, the conditions are right for nuclear fusion reactions to begin in the core, which requires a temperature of 10 million kelvins. The initial fusion reaction essentially combines four hydrogen nuclei to form a helium nucleus with the release of much energy. This energy heats the core beyond the temperature reached by gravitational contraction, eventually to 16 million kelvins. Since the star is a gas, the increased temperature expands the volume of the star. The outward pressure of expansion balances the inward pressure from gravitational collapse, and the star settles down to a balanced condition of calmly converting hydrogen to helium in the core, radiating the energy released into space (figure 16.21). The theoretical time elapsed from the initial formation and collapse of the protostar to the main sequence is about 50 million years for a star of a solar mass.

Where the star is located on the main sequence and what happens to it next depend only on how massive it is. The more massive stars have higher core temperatures and use up their hydrogen more rapidly as they shine at higher surface temperatures (O type stars). Less massive stars shine at lower surface temperatures (M type stars) as they use their fuel at a slower rate. The overall life span on the main sequence ranges from millions of years for O type stars to trillions of years for M type stars. An average one-solar-mass star will last about 10 billion years.

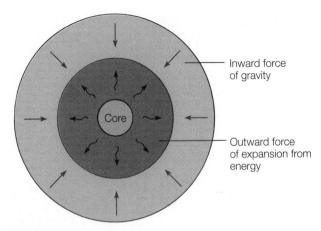

FIGURE 16.21

A star becomes stable when the outward forces of expansion from the energy released in nuclear fusion reactions balance the inward forces of gravity.

The next stage in the theoretical life of a star begins as much of the hydrogen in the core has been fused into helium. With fewer hydrogen fusion reactions, less energy is released and less outward balancing pressure is produced, so the star begins to gravitationally collapse. The collapse heats the now-helium core and the surrounding shell where hydrogen still exists. The hydrogen in the shell begins to undergo fusion from the increased temperatures, and the increased energy released now expands the outer layers of the star. With an increased surface area, the amount of radiation emitted per unit area is less, and the star acquires the properties of a brilliant red giant. Its data point position on the H-R diagram changes since it now has different luminosity and temperature properties. (The star has not physically *moved*, as is often said. The changing properties move its temperature-luminosity data point, not the star, to a new position.)

After about five hundred million years as a red giant, the star now has a surface temperature of about 4,000 kelvins compared to its main sequence surface temperature of 6,000 kelvins. The radius of the red giant is now a thousand times greater, a distance that will engulf the earth when the sun reaches this stage. Even though the surface temperature has decreased from the expansion, the helium core is continually heating and eventually reaches a temperature of 100 million kelvins, the critical temperature necessary for the helium nuclei to undergo fusion to produce carbon nuclei. The red giant now has helium fusion reactions in the core and hydrogen fusion reactions in a shell around the core. This changes the radius, the surface temperature, and the luminosity, with the overall result depending on the composition of the star. In general, the radius and luminosity decrease when this stage is reached, moving the star back toward the main sequence (figure 16.22).

After millions of years of helium fusion reactions, the core is gradually converted to a carbon core and helium fusion begins in the shell surrounding the core. The core reactions decrease as the star now has a helium-fusing shell surrounded by a second, hydrogen-fusing shell. This releases additional energy, and the

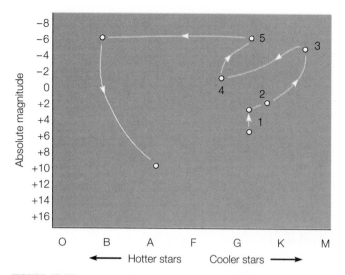

FIGURE 16.22

The evolution of a star of solar mass as it depletes hydrogen in the core (1), fuses hydrogen in the shell to become a red giant (2 to 3), becomes hot enough to produce helium fusion in the core (3 to 4), then expands to a red giant again as helium and hydrogen fusion reactions move out into the shells (4 to 5). It eventually becomes unstable and blows off the outer shells to become a white dwarf star.

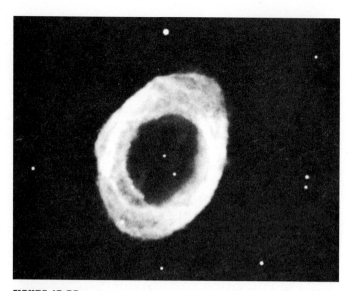

FIGURE 16.23

The blown-off outer layers of stars form ringlike structures called *planetary nebulae.*

star again expands to a red giant for the second time. A star the size of the sun or less massive may cool enough at this point that nuclei at the surface become neutral atoms rather than a plasma. As neutral atoms they can absorb radiant energy coming from within the star, heating the outer layers. Changes in temperature produce changes in pressure, which change the balance between the temperature, pressure, and the internal energy generation rate. The star begins to expand outward from heating. The expanded gases are cooled by the expansion process, however, and are pulled back to the star by gravity, only to be heated and expand outward again. In other words, the outer layers of the star begin to pulsate in and out. Finally, a violent expansion blows off the outer layers of the star, leaving the hot core. Such blown-off outer layers of a star form circular nebulae called *planetary nebulae* (figure 16.23). The nebulae continue moving away from the core, eventually adding to the dust and gases between the stars. The remaining carbon core and helium-fusing shell begin gravitationally to contract to a small, dense *white dwarf* star. A star with the original mass of the sun, or less, slowly cools from white to red, then to a black lump of carbon in space (figure 16.24).

A more massive star will have a different theoretical ending than the slow cooling of a white dwarf. A massive star will contract, just as the less massive stars, after blowing off its outer shells. In a more massive star, however, heat from the contraction may reach the critical temperature of 600 million kelvins to begin carbon fusion reactions. Thus, a more massive star may go through a carbon fusing stage and other fusion reaction stages that will continue to produce new elements until the element iron is reached. After iron, energy is no longer released by the fusion process (see chapter 15), and the star has used up all of its

energy sources. Lacking an energy source, the star is no longer able to maintain its internal temperature. The star loses the outward pressure of expansion from the high temperature, which had previously balanced the inward pressure from gravitational attraction. The star thus collapses, then rebounds like a compressed spring into a catastrophic explosion called a **supernova.** A supernova produces a brilliant light in the sky that may last for months before it begins to dim as the new elements that were created during the life of the star diffuse into space. These include all the elements up to iron that were produced by fusion reactions during the life of the star and heavier elements that were created during the instant of the explosion. All the elements heavier than iron were created as some less massive nuclei disintegrated in the explosion, joining with each other and with lighter nuclei to produce the nuclei of the elements from iron to uranium. These fusion-produced and supernova-produced elements are scattered through space by the supernova explosion. As you will see in the next chapter, these newly produced, scattered elements will later become the building blocks for new stars and planets such as the sun and the earth.

The remains of the compressed core after the supernova have still yet another fate if they have (1) a mass greater than 1.4 solar masses or (2) a mass greater than about 3 solar masses or more. Both of these solar masses represent a crucial point in what happens to the remaining core. Matter is fantastically compressed in a white dwarf star, so tightly that there is practically no space between the electrons. The mutual repulsion of electrons forced so close together at such a great density balances the contracting forces of gravity, and the white dwarf has a stable size as it cools, with a density equivalent to 10^3 kg/cm^3. If the core of a supernova has a remaining mass greater than 1.4 solar masses, the gravitational forces on the remaining matter, together with the compressional forces of the supernova explosion, are great

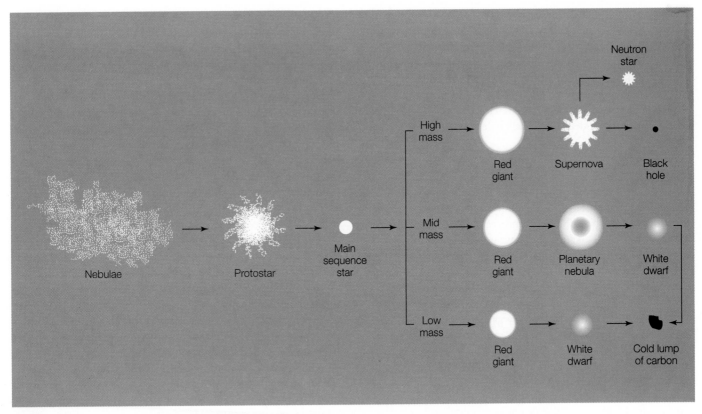

FIGURE 16.24

This flowchart shows some of the possible stages in the birth and aging of a star. The differences are determined by the mass of the star.

enough to collapse nuclei, forcing protons and electrons together into neutrons, forming the core of a **neutron star.** A neutron star is the very small (10 to 20 km diameter), superdense (10^{11} kg/cm^3 or greater) remains of a supernova with a center core of pure neutrons.

Because it is a superconcentrated form of matter, the neutron star also has an extremely powerful magnetic field, capable of becoming a pulsar. A **pulsar** is a very strongly magnetized neutron star that emits a uniform series of equally spaced electromagnetic pulses. Evidently, the magnetic field of a rotating neutron star makes it a powerful electric generator, capable of accelerating charged particles to very high energies. These accelerated charges are responsible for emitting a beam of electromagnetic radiation, which sweeps through space with amazing regularity. The pulsating radio signals from a pulsar were a big mystery when first discovered. For a time extraterrestrial life was considered as the source of the signals, so they were jokingly identified as LGM (for Little Green Men). Over 300 pulsars have been identified, and most emit radiation in the form of radio waves. Two, however, emit visible light, two emit beams of gamma radiation, and one emits X-ray pulses.

Another theoretical limit occurs if the remaining core has a mass of about 3 solar masses or more. At this limit the force of gravity overwhelms *all* nucleon forces, including the repulsive forces between like charged particles. If this theoretical limit is

reached, nothing can stop the collapse, and the collapsed star will become so dense that even light cannot escape. The star is now a **black hole** in space. Since nothing can stop the collapsing star, theoretically a black hole would continue to collapse to a pinpoint and then to a zero radius called a *singularity*. This event seems contrary to anything that can be directly observed in the physical universe, but it does agree with the general theory of relativity and concepts about the curvature of space produced by such massively dense objects. Black holes are theoretical and none has been seen, of course, because a black hole theoretically pulls in radiation of all wavelengths and emits nothing. Evidence for the existence of a black hole is sought by studying binary systems and X rays that would be given off by matter as it is accelerated into a black hole.

The existence of a black hole has been used as a possible explanation for many unresolved space phenomena, from quasars to the creation of positrons at the center of the Milky Way galaxy. Quasars (quasi-stars) are point sources of light that are strongly redshifted, with spectral properties similar to cores of some galaxies. Photographs of the cores of some galaxies show a pileup of starlight, suggesting that a massive black hole might be present, but some astronomers want stronger evidence. They insist on seeing the actual movement of matter orbiting the presumed black hole. Such evidence has now been established observationally about as well as is possible by photographs from the Hubble Space

Telescope. Hubble pictured a disk of gas only about 60 light-years out from the center of a galaxy (M87), moving at more than 1.6 million km/hr (about 1 million mi/hr). The only known possible explanation for such a massive disk of gas moving with this velocity at the distance observed would require the presence of a 1–2 billion solar-mass black hole. This gas disk could only be resolved by the Hubble Space Telescope, so this telescope has provided the first observational evidence of a black hole.

GALAXIES

Stars are associated with other stars on many different levels, from double stars that orbit a common center of mass, to groups of tens or hundreds of stars that have gravitational links and a common origin, to the billions and billions of stars that form the basic unit of the universe, a **galaxy.** The sun is but one of an estimated one hundred billion stars that are held together by gravitational attraction in the Milky Way galaxy. The numbers of stars and vastness of the Milky Way galaxy alone seem almost beyond comprehension, but there is more to come. The Milky Way is but one of *billions* of galaxies that are associated with other galaxies in clusters, and these clusters are associated with one another in superclusters. Through a large telescope you can see more galaxies than individual stars in any direction, each galaxy with its own structure of billions of stars. Yet, there are similarities that point to a common origin. Some of the similarities and associations of stars will be introduced in this section along with the Milky Way galaxy, the vast, flat, spiraling arms of stars, gas, and dust where the sun is located (figure 16.25).

The Milky Way Galaxy

Away from city lights, you can clearly see the faint, luminous band of the Milky Way galaxy on a moonless night. Through a telescope or a good pair of binoculars, you can see that the luminous band is made up of countless numbers of stars. You may also be able to see the faint glow of nebulae, concentrations of gas and dust. There are dark regions in the Milky Way that also give an impression of something blocking starlight, such as dust. You can also see small groups of stars called **galactic clusters.** Galactic clusters are gravitationally bound subgroups of as many as one thousand stars that move together within the Milky Way. Other clusters are more symmetrical and tightly packed, containing as many as a million stars, and are known as **globular clusters.**

Viewed from a distance in space, the Milky Way would appear to be a huge, flattened cloud of spiral arms radiating out from the center. There are three distinct parts: (1) the spherical concentration of stars at the center of the disk called the *galactic nucleus;* (2) the rotating *galactic disk,* which contains most of the bright, blue stars along with much dust and gas; and (3) a spherical *galactic halo,* which contains some 150 globular clusters located outside the galactic disk (figure 16.26). The sun is located in one of the arms of the galactic disk, some 25,000 to 30,000 light-years from the center. The galactic disk rotates, and the sun completes one full rotation every 200 million years.

The diameter of the *galactic disk* is about 100,000 light-years, or the distance that light traveling at 186,000 miles per second would cover in 100,000 years (5.86×10^{12} miles/year $\times$ 100,000, or 5.86×10^{17} miles). Yet, in spite of the one hundred billion stars in the Milky Way, it is mostly full of emptiness. By way of analogy, imagine reducing the size of the Milky Way disk until stars like the sun were reduced to the size of tennis balls. The distance between two of these tennis-ball-sized stars would now compare to the distance across the state of Texas. The space between the stars is not actually empty since it contains a thin concentration of gas, dust, and molecules of chemical compounds. The gas particles outnumber the dust particles about 10^{12} to 1. The gas is mostly hydrogen, and the dust is mostly solid iron, carbon, and silicon compounds. Over forty different chemical molecules have been discovered in the space between the stars, including many organic molecules. Some nebulae consist of clouds of molecules with a maximum density of about 10^6 molecules/cm^3. The gas, dust, and chemical compounds make up part of the mass of the galactic disk, and the stars make up the remainder. The gas plays an important role in the formation of new stars, and the dust and chemical compounds play an important role in the formation of planets.

The spherical *galactic nucleus* is hidden from Earth by the clouds of dust in the central plane of the galactic disk. Studies of infrared and radio waves from the nucleus and studies of the nuclei of galaxies similar to the Milky Way raise many questions about the nucleus. Studies of other galaxies indicate that the nucleus contains old, red stars with little interstellar dust or gas, with the stars clustered close together within a radius of some 5,000 light-years. Studies of infrared and radio waves indicate that the nucleus has a central part with a 1 or 2 light-year radius that is emitting enormous amounts of energy, with matter both streaming out and falling in. The source of this energy and matter is not known, but it is believed to be a massive black hole.

The *halo* around the Milky Way consists mostly of groups of massive, old, red stars in globular clusters, a few individual stars, and not much gas or dust. The stars in these globular clusters are much closer together than those in the disk, and they are believed to be the oldest stars in the Milky Way.

The stars in the *galactic disk* occur in clusters of tens to hundreds of stars gravitationally linked in galactic clusters or binary stars that revolve around a common center of mass, or they occur as individual stars. A galactic cluster (also called an open cluster) of the galactic disk contains a much lower concentration of stars than the globular clusters of the halo. Like the globular clusters, the stars of a galactic cluster are linked by gravity and have a common origin, age, and composition. The stars of a galactic cluster generally fall on the main sequence on an H-R diagram, meaning that they are young-to-mature stars of various masses. Since the galactic clusters and the bulk of the dust and gas are found in the galactic disk, this is the active part of the Milky Way where young O and B type stars are forming from the gas and dust. These stars shine relatively briefly, then explode to replenish the dust and gas of the spiral arms. Periodically, a longer-lived phenomenon, an average size star such as the sun and its planets, forms from the dust and gas.

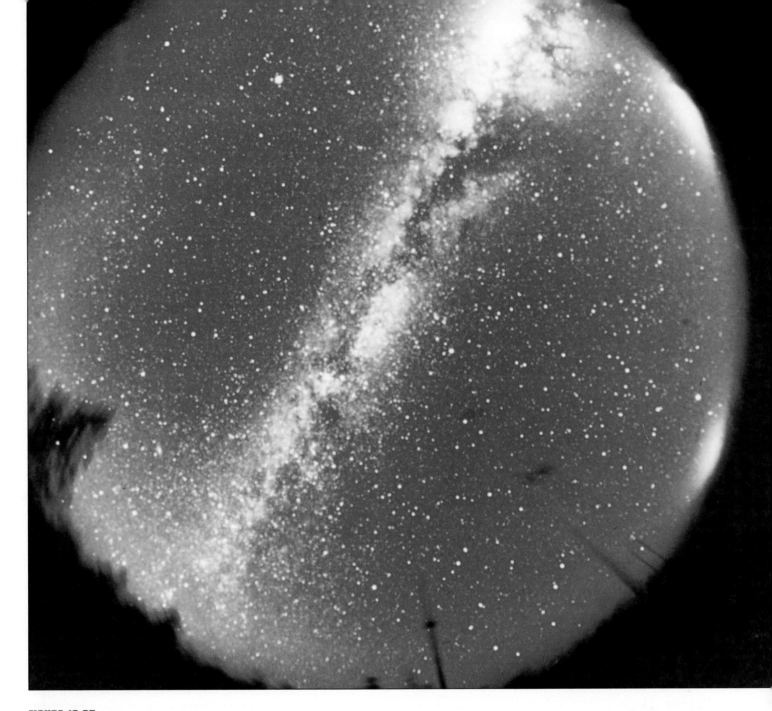

FIGURE 16.25

A wide-angle view toward the center of the Milky Way galaxy. Parts of the white, milky band are obscured from sight by gas and dust clouds in the galaxy.

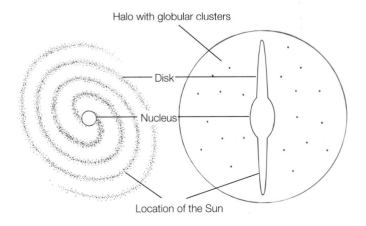

FIGURE 16.26

The structure of the Milky Way galaxy.

FIGURE 16.27

The Andromeda galaxy, which is believed to be similar in size, shape, and structure to the Milky Way galaxy.

Other Galaxies

Outside the Milky Way is a vast expanse of emptiness, lacking even the few molecules of gas and dust spread thinly through the galactic nucleus. There is only the light from faraway galaxies and the time that it takes for this light to travel across the vast vacuum of intergalactic space. How far away is the nearest galaxy? Recall that the Milky Way is so large that it takes light 100,000 years to travel the length of its diameter.

The nearest galactic neighbor to the Milky Way is a dwarf spherical galaxy only 80,000 light-years from the solar system, in the constellation Sagittarius. This nearest neighbor is also the newest known neighbor since it was only recently discovered in its location on the other side of the center of the Milky Way. The nearby galaxy is called a dwarf because it has a diameter of only about 1,000 light-years. It is apparently in the process of being pulled apart by the gravitational pull of the Milky Way, which now is known to have eleven satellite galaxies.

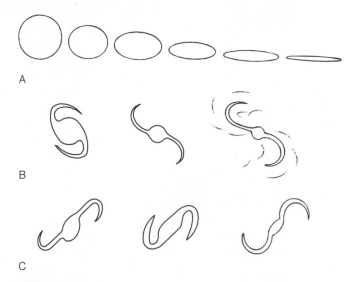

FIGURE 16.28

Different subgroups in the Hubble classification scheme: (*A*) elliptical galaxies, (*B*) spiral galaxies, and (*C*) barred galaxies.

The nearest galactic neighbor similar to the Milky Way is Andromeda, about 2 million light-years away. Andromeda is similar to the Milky Way in size and shape, with about one hundred billion stars, gas, and dust turning in a giant spiral pinwheel (figure 16.27). Other galaxies have other shapes and other characteristics. The American astronomer Edwin Hubble developed a classification scheme for the structure of galaxies based on his 1926 study of some six hundred different galaxies. The basic galactic structures were identified as elliptical, spiral, barred, and irregular (figure 16.28).

Elliptical galaxies appear to be spherical to flattened elliptical shapes with rotational symmetry. Hubble identified eight subclasses of elliptical galactic shapes based on the degree of flatness. This shape made up about 20 percent of Hubble's sample, and all galaxies with this shape have properties in common. Elliptical galaxies contain only old stars with no O or B types and very little gas or dust. Elliptical galaxies have a great range of sizes, and all of the evidence indicates that they are old.

Spiral galaxies have a small spherical nucleus with usually two but sometimes more spiral arms radiating out from the nucleus in a flattened disk. Spiral galaxies made up about half of Hubble's sample. They contain stars of all ages, but the young stars are found in the spirals and the older stars occur in the globular clusters of the halo. Spiral galaxies are slightly less massive than elliptical galaxies, with about 10^{11} solar masses.

Barred galaxies have the shape of a bar across the nucleus with spiral arms radiating from the ends of the bars. About 30 percent of Hubble's sample were barred galaxies. As you would expect, barred galaxies have properties very similar to spiral galaxies.

Irregular galaxies are those that Hubble could not fit into the other categories because they lacked symmetry. The irregular galaxies are relatively few in number, the least massive with only 10^9 solar masses, and contain mostly young stars with the greatest accumulation of dust and gas of all the galaxies.

The Life of a Galaxy

Hubble's classification of galaxies into distinctly different categories of shape was an exciting accomplishment because it suggested that some relationship or hidden underlying order may exist in the shapes. Finding underlying order is important because it leads to the discovery of the physical laws that govern the universe. Soon after Hubble published his classification results in 1926, two models of galactic evolution were proposed. One model, which was suggested by Hubble, had extremely slowly spinning spherical galaxies forming first, which gradually flattened out as their rate of spin increased while they condensed. This is a model of spherical galaxies flattening out to increasingly elliptical shapes, eventually spinning off spirals until they finally broke up into irregular shapes over a long period of time.

Another model of galactic evolution had the galaxies beginning as irregular shapes that were condensing into the spherical shape. In this model the center part would not rotate as rapidly as the outer parts, so gravity would pull the rapidly moving outer regions into a spherical shape. This is a model of irregular galaxies collapsing to spiral galaxies, which eventually condensed to spherical shapes over a long period of time.

When these two opposing models of galactic evolution were formulated, astronomers did not know about the life of a star and technology had not yet developed to the stage where they could know the spectra types of stars in the classes of galaxies. When this knowledge and information did become available, it was clear that both models were unacceptable. The flattening-out elliptical-to-spiral model is unacceptable because it is inconsistent with the ages of stars observed in these galaxies. Elliptical galaxies contain only old, red stars with no young O or B types and very little gas or dust. Thus, they could not evolve into spiral galaxies with half their masses in hydrogen gas and dust and half in stars, many of which are new young stars. Furthermore, the winding-up spiral-to-elliptical model is also unacceptable because spiral galaxies contain old stars, with little gas or dust in the globular clusters of their halos. Since these old stars are just as old as the old stars in the elliptical galaxies, both types of galaxies must be of about the same age.

Globular clusters are usually extremely old structures, but recent Hubble Space Telescope pictures of colliding spiral galaxies suggest that *new* globular clusters may be formed in the collision process. If this observation proves to be more than a suggestion, this would support the view that elliptical galaxies are created out of collisions of spiral galaxies.

Today, the current model of galactic evolution is based on the **big bang theory** of the creation of the universe. Evidence for the big bang theory comes from (1) present-day microwave radiation from outer space, (2) current data on the expansion of the universe, (3) the relative abundance of elements that were altered in the core of older stars is in agreement with predictions based on analysis of the big bang, and (4) the Cosmic Background Explorer (COBE) spacecraft studied diffuse cosmic background radiation to help answer such questions as how matter is distributed in the universe, if the universe is uniformly expanding, and how and when galaxies first formed. Findings announced in 1991 confirmed several important predictions from the big bang theory.

One problem with understanding the big bang theory is that the theory is based on Einstein's general theory of relativity, a model that does not match our everyday ideas of time and space. For example, the idea of the big bang as an explosion in space, like the detonation of a bomb, is incorrect. The big bang did not occur somewhere in space—it created space, so it is not a small universe expanding like a sphere in space. Furthermore, galaxies are not expanding, nor are they speeding through space away from one another. Instead, galaxies are understood to be moving relatively slowly relative to one another. It is *space* that is expanding, continuing something that began with the big bang.

Newtonian physics considers the force of gravitation to depend on two things, the mass of the interacting objects and the distance between them. Einstein's general theory of relativity has a different interpretation, that space affects matter, but matter also affects space. The general theory of relativity considers matter to change the local geometry of space and a gravitational field is a curved space. The earth, for example, is attracted to the sun because the mass of the sun distorts, or curves, the geometry of space. Thus the force of gravity between the earth and the sun is a consequence of the motion of the earth following the geometry of the underlying curved space.

Another mathematical model of the primordial universe identifies a beginning with nine-dimensional space locked up in the tiny space of a point. At the big bang three of the dimensions expanded, leaving the other six dimensions curled up in a little ball too small to be observed today. The physical object in this space is not a point, but is more like a smoke ring called a *superstring*. When the features of superstrings are enlarged a billion billion times they mathematically begin to have the properties of elementary subatomic particles such as the electron. The superstring theory seems to unify space, matter, and the properties of matter. However, the existence of invisible matter and tiny smoke rings of matter will be very difficult to verify.

Another difficult concept to grasp is a universe with no "edge" or "outside" because the universe contains all the matter, all the energy, and all the presently existing dimensions. Before the big bang there was no space, no matter, and no time—a difficult-to-imagine void. This brings us back to the use of two-dimensional analogies that are not fully accurate, but perhaps will provide some understanding. All of the different astronomical and physical "clocks" indicate that the universe was created in a "big bang" some 18 billion years ago (give or take a billion), expanding as an intense and brilliant explosion from a primeval fireball with a temperature of some 10^{12} kelvins. An often-used analogy for the movement of galaxies after the big bang is to consider galaxies as spots on the surface of a balloon that is being inflated. As the balloon expands, the distances between all the spots increase, with the result that all the spots find themselves farther and farther apart. In other words, the galaxies are not expanding into space that is already there. It is space itself that is expanding, and the galaxies move with the expanding space (figure 16.29). This concept agrees with the theory of general relativity, that space is generated by the presence of matter and that the contour of space is determined by the distribution of matter.

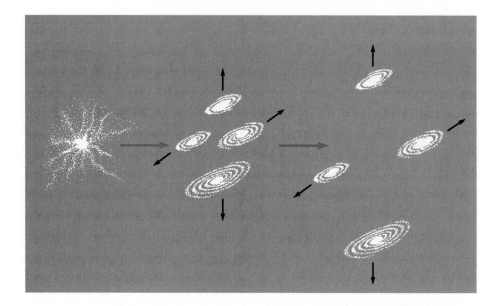

FIGURE 16.29

Will the universe continue expanding as the dust and gas in galaxies become locked up in white dwarf stars, neutron stars, and black holes?

The temperature and the intensity of the initial fireball of radiation, according to the theory, diminished as the universe expanded, and some of the remaining radiation should be measurable today. The presence of the remaining radiation was a theoretical topic of discussion among astronomers, but it was actually discovered in 1965 when scientists with Bell Telephone Laboratories were puzzled by radiation detected by a new radio antenna they were testing. The radiation was a consistent, uniform signal that seemed to come from every direction in space, and the signal intensity was equivalent to a blackbody radiation of about 3 kelvins. Later, it was determined that they had found the remnant of the original primeval fireball. This measured temperature allowed scientists to trace back the evolution of the universe as its temperature cooled. To cool to a temperature of about 3 kelvins requires the universe to have formed about 18 billion years ago.

The second set of evidence for the big bang theory came from Edwin Hubble and his earlier work with galaxies. Hubble had determined the distances to some of the galaxies that had red-shifted spectra. From the Doppler effect (see chapter 6), it was known that these galaxies were moving away from the Milky Way. Hubble found a relationship between the distance to a galaxy and the velocity with which it was moving away. He found the velocity to be directly proportional to the distance; that is, the greater the distance to a galaxy the greater the velocity. This means that a galaxy twice as far from the Milky Way as a second galaxy is moving away from the Milky Way at twice the speed as the second galaxy. Since this relationship was seen in all directions, it meant that the universe is expanding uniformly. The same effect would be viewed from any particular galaxy; that is, all the other galaxies are moving away with a velocity proportional to the distance to the other galaxies. This points to a common beginning, a time when all matter in the universe was together. Assuming that the universe has expanded uniformly, Hubble's law of expansion can be used to estimate how far back the expansion began. The result of this calculation gives the answer that the universe began expanding about 18 billion years ago.

Current theoretical ideas and models about the universe can lead to interesting speculative ideas about how galaxies formed and what may eventually happen to them. For example, consider a universe that initially consisted of a very hot gas of photons that cooled as the universe expanded. As it cooled, the contents—protons, electrons, and other particles and antiparticles—changed over time, forming other particles. After about a million years of cooling, the temperature reached about 3,000 kelvins and clouds of hydrogen gas began to condense, pulling together in a swirling supercloud. As more hydrogen atoms condensed, their mutual gravitational attraction pulled in more atoms and distinct swirls of hydrogen gas were separated. Such clouds of hydrogen gas are called **protogalaxies.** Smaller pockets of hydrogen gas continually form in eddies of the protogalaxy and eventually form protostars, then stars, as was discussed in the previous section. Different rates of swirling gas clouds probably resulted in the shapes of galaxies observed today.

The first stars formed were probably very massive and short-lived, ending as supernova. The supernova blasted out heavy elements, which condensed in the disk as the dust and gas that would form the next generation of stars. With each generation of stars, more and more heavy elements accumulate and more material is tied up in neutron stars. Thus, less material is available for the next generation, and the next generation are stars of lower mass, which, in turn, do not experience a supernova ending. Eventually, only less massive, slowly reacting red stars will remain with no gas or dust to create another generation of young hot stars. Only cooling white dwarfs, neutron stars, and black holes will remain. Eventually, everything cools to the same temperature with all matter now locked in cold, black dwarfs and black holes.

A CLOSER LOOK | Dark Matter

Will the universe continue to expand forever, or will it be pulled back together into a really big crunch? Whether the universe will continue expanding or gravity will pull it back together again depends on the *critical density* of the universe. If there is enough matter the actual density of the universe will be above the critical density, and gravity will stop the current expansion and pull everything back together again. On the other hand, if the actual density of the universe is less than or equal to the critical density, the universe will be able to escape its own gravity and continue expanding. Is the actual density small enough for the universe to escape itself?

All the detailed calculations of astrophysicists point to an actual density that is *less* than the critical density, meaning the universe will continue to expand forever. However, these calculations also show that there is more matter in the universe than can be accounted for in the stars and galaxies. This means there must be matter in the universe that is not visible, or not shining at least, so we cannot see it. Three examples follow that show why astrophysicists believe more matter exists than you can see.

1. There is a relationship between the light emitted from a star and the mass of that star. Thus you can indirectly measure the amount of matter in a galaxy by measuring the light output of that galaxy. Clusters of galaxies have been observed moving, and the motion of galaxies within a cluster does not agree with matter information from the light. The motion suggests that galaxies are attracted by a gravitational force from about ten times more matter than can be accounted for by the light coming from the galaxies.

2. There are mysterious variations in the movement of stars within an individual galaxy. The rate of rotation of stars about the center of rotation of a galaxy is related to the distribution of matter in that galaxy. The outer stars are observed to rotate too fast for the amount of matter that can be seen in the galaxy. Again, measurements indicate there is about ten times more matter in each galaxy than can be accounted for by the light coming from the galaxies.

3. Finally, there are estimates of the total matter in the universe that can be made by measuring ratios such as deuterium to ordinary hydrogen, lithium-7 to helium-3, and helium to hydrogen. In general, theoretical calculations all seem to account for only about 10 percent of all the matter in the universe.

Calculations from a variety of sources all seem to agree that 90 percent or more of the matter that makes up the universe is missing. The missing matter and the mysterious variations in the orbits of galaxies and stars can be accounted for by the presence of **dark matter,** which is invisible and unseen. What is the nature of this dark matter? You could speculate that dark matter is simply normal matter that has been overlooked, such as dark galaxies, brown dwarfs, or planetary material such as rock, dust, and the like. You could also speculate that dark matter is dark because it is in the form of subatomic particles, too small to be seen. As you can see, scientists have reasons to believe that dark matter exists, but they can only speculate about its nature.

The nature of dark matter represents one of the major unsolved problems in astronomy today. There are really two dark matter problems, (1) the nature of the dark matter and (2) how much dark matter contributes to the actual density of the universe. Not everyone believes that dark matter is simply normal matter that has been overlooked. There are at least two schools of thought on the nature of dark matter and what should be the focus of research. One school of thought focuses on *particle physics* and contends that dark matter consists primarily of exotic undiscovered particles or known particles such as the neutrino. Neutrinos are electrically neutral, stable subatomic particles in the lepton family (see the reading on quarks in chapter 9). This school of thought is concerned with forms of dark matter called **WIMPs** (after **W**eakly **I**nteracting **M**assive **P**articles). In spite of the name, particles under study are not always massive. It is entirely possible that some low-mass species of neutrino—or some as of yet unidentified WIMP—will be experimentally discovered and found to have the mass needed to account for the dark matter and thus close the universe.

The other school of thought about dark matter focuses on *astrophysics* and contends that dark matter is ordinary matter in the form of brown dwarfs, unseen planets outside the solar system, and galactic halos. This school of thought is concerned with forms of dark matter called **MACHOs** (after **M**assive **A**strophysical **C**ompact **H**alo **O**bjects). Dark matter considerations in this line of reasoning consider massive objects, but ordinary matter (protons and neutrons) is also considered as the material making up galactic halos. Protons and neutrons belong to a group of subatomic particles called the *baryons,* so they are sometimes referred to as *baryonic dark matter* (see "The Quark" in chapter 9). Some astronomers feel that baryonic dark matter is the most likely candidate for halo dark matter.

In general, astronomers can calculate the probable cosmological abundance of WIMPs, but there is no proof of their existence. By contrast, MACHOs dark matter candidates are known to exist, but there is no way to calculate their abundance. MACHOs astronomers assert that halo dark matter and probably all dark matter may be baryonic. Do you believe WIMPs or MACHOs will provide an answer about the future of the universe? What is the nature of dark matter, how much dark matter exists, and what is the fate of the universe? Answers to these and more questions await further research.

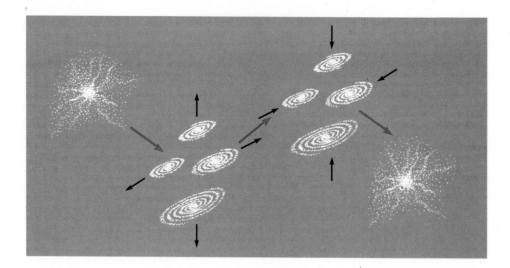

FIGURE 16.30

The oscillating theory of the universe assumes that the space between the galaxies is expanding as does the big bang theory, but in the oscillating theory, the galaxies gradually come back together to begin all over in another big bang.

The ultimate fate of the universe will strongly depend on the mass of the universe and if the expansion is slowing or continuing. All the evidence tells us it began by expanding with a big bang. Will it continue to expand, becoming increasingly cold and diffuse? Will it slow to a halt, then start to contract to a fiery finale of a big crunch (figure 16.30)? Researchers continue searching for answers with experimental data and theoretical models, looking for clues in matches between the data and models.

SUMMARY

The early conception of the universe was *geocentric,* and the ancient Greeks pictured the stars and planets as being attached to a *celestial sphere* that rotated around the earth. Today, the idea of a celestial sphere is still used to locate stars and plot how they move.

The distance between the stars as they appear on the celestial sphere is measured by a segment of the circumference on the sphere. There are 360° in a circle and 1/360 of the circumference is called *1 degree (1°) of arc,* with subdivisions called *arc minutes* and *arc seconds.* The apparent size of an object is measured by the angle between the sides of the object, which is called the *angular size* or the *angular diameter.*

The distance to a relatively close star can be measured by *parallax,* the apparent shift of position of an object against a background. The radius of the earth's orbit is called one *astronomical unit,* or A.U. When a baseline of 1 A.U. is used and the apparent shift of position is 1 arc second, the distance is defined as the astronomical unit of distance called a *parsec.* Another unit of astronomical distance is the *light-year,* the distance that light travels in one year.

Stars are theoretically born in clouds of hydrogen gas and dust in the space between other stars. Gravity pulls huge masses of hydrogen gas together into a *protostar,* a mass of gases that will become a star. The protostar contracts, becoming increasingly hotter at the center, eventually reaching a temperature high enough to start *nuclear fusion* reactions between hydrogen atoms. Pressure from hot gases balances the gravitational contraction, and the average newborn star will shine quietly for billions of years. The average star has a dense, hot *core* where nuclear fusion releases radiation, a less dense *radiation zone* where radi-

ation moves outward, and a thin *convection zone* that is heated by the radiation at the bottom, then moves to the surface to emit light to space.

The brightness of a star is related to the amount of energy and light it is producing, the size of the star, and the distance to the star. The *apparent magnitude* is the brightness of a star as it appears to you. To compensate for differences in brightness due to distance, astronomers calculate the brightness that stars would have at 10 parsecs (32.6 light-years). This standard-distance brightness is called the *absolute magnitude.* Absolute magnitude is an expression of *luminosity,* the total amount of energy radiated into space each second from the surface of a star.

Stars appear to have different colors because they have different surface temperatures. A graph of temperature by spectral types and brightness by absolute magnitude is called the *Hertzsprung-Russell diagram,* or H-R diagram for short. Such a graph shows that normal, mature stars fall on a narrow band called the *main sequence* of stars. Where a star falls on the main sequence is determined by its brightness and temperature, which in turn are determined by the mass of the star. Other groups of stars on the H-R diagram have different sets of properties that are determined by where they are in their evolution.

The life of a star consists of several stages, the longest of which is the *main sequence* stage after a relatively short time as a *protostar.* After using up the hydrogen in the core, a star with an average mass expands to a *red giant,* then blows off the outer shell to become a *white dwarf star,* which slowly cools to a black dwarf. The blown-off outer shell forms a *planetary nebula,* which disperses over time to become the gas and dust of interstellar space. More massive stars collapse into *neutron stars* or *black holes* after a violent *supernova* explosion.

Galaxies are the basic units of the universe. The Milky Way galaxy has three distinct parts: (1) the *galactic nucleus,* (2) a rotating *galactic disk,* and (3) a *galactic halo.* The galactic disk contains subgroups of stars that move together as *galactic clusters.* The halo contains symmetrical and tightly packed clusters of millions of stars called *globular clusters.*

All the billions of galaxies can be classified into groups of four structures: *elliptical, spiral, barred,* and *irregular.* Evidence from four different astronomical and physical "clocks" indicates that the galaxies formed some 18 billion years ago, expanding ever since from a common origin in a *big bang.* The *big bang theory* describes how the universe began by expanding.

A CLOSER LOOK | Extraterrestrials?

Extraterrestrial is a descriptive term, meaning a thing or event outside the earth or its atmosphere. The term is also used to describe a being, a life-form that originated away from the earth. This reading is concerned with the search for extraterrestrials, intelligent life that might exist beyond the earth and outside the solar system.

Why do people believe that extraterrestrials might exist? The affirmative answer comes from a mixture of current theories about the origin and development of stars, statistical odds, and faith. Considering the statistical odds, note that our sun is one of the some 300 billion stars that make up our Milky Way galaxy. The Milky Way galaxy is one of some 10 billion galaxies in the observable universe. Assuming an average of about 300 billion stars per galaxy, this means there are some 300 billion times 10 billion stars, or 3×10^{21} stars in the observable universe. There is nothing special or unusual about our sun, and astronomers believe it to be quite ordinary among all the other stars (all 3,000,000,000,000,000,000,000 or so).

So the sun is an ordinary star, but what about our planet? Not too long ago most people, including astronomers, thought our solar system with its life-supporting planet (Earth) to be unique. Evidence collected over the past decade or so, however, has strongly suggested that this is not so. Planets are now believed to be formed as a natural part of the star-forming process. Evidence of planets around other stars has also been detected by astronomers. One of the stars with planets is "only" 53 light-years from Earth.

Even with a very low probability of planetary systems forming with the development of stars, a population of 3×10^{21} stars means there are plenty of planetary systems in existence, some with the conditions necessary to

BOX FIGURE 16.1
The 300 m (984 ft) diameter radio telescope at Arecibo, Puerto Rico, is the largest fixed-dish radio telescope in the world.

support life (Note: If 1 percent have planetary systems, this means 3×10^{19} stars have planets). Thus, it is a statistical observation that suitable planets for life are very likely to exist. In addition, radio astronomers have found that many organic molecules exist, even in the space between the stars. Based on statistics alone, there should be life on other planets, life that may have achieved intelligence and developed into a technological civilization.

If extraterrestrial life exists, why have we not detected them or why have they not contacted us? The answer to this question is found in the unbelievable distances involved in interstellar space. For example, a logical way to search for extraterrestrial intelligence is to send radio signals to space and analyze those coming from space. Modern radio telescopes can send powerful radio beams and present-day computers and data processing

techniques can now search through incoming radio signals for the patterns of artificially generated radio signals (box figure 16.1). Radio signals, however, travel through space at the speed of light. The diameter of our Milky Way galaxy is about 100,000 light-years, which means 100,000 years would be required for a radio transmission to travel across the galaxy. If we were to transmit a super, super strong radio beam from the earth, it would travel at the speed of light and cross the distance of our galaxy in 100,000 years. If some extraterrestrials on the other side of the Milky Way galaxy did detect the message and send a reply, it could not arrive at Earth until 200,000 years after the message was sent. Now consider the fact that of all the 10 billion other galaxies in the observable universe, our nearest galactic neighbor similar to the Milky Way is Andromeda. Andromeda is 2 million light-years from the Milky Way galaxy.

In addition to problems with distance/time, there are questions about in which part of the sky you should send and look for radio messages, questions about which radio frequency to use, and problems with the power of present-day radio transmitters and detectors. Realistically, the hope for any exchange of radio-transmitted messages would be restricted to within several hundred light-years of Earth.

Considering all the limitations, what sort of signals should we expect to receive from extraterrestrials? Probably a series of pulses that somehow indicate counting, such as 1, 2, 3, 4, and so on, repeated at regular intervals. This is the most abstract, while at the same time the simplest concept that an intelligent being anywhere would have. It could provide the foundation for communications between the stars.

KEY TERMS

absolute magnitude (p. **387**)

apparent magnitude (p. **387**)

astronomical unit (p. **384**)

big bang theory (p. **397**)

black hole (p. **393**)

Cepheid variable (p. **390**)

constellations (p. **379**)

convection zone (p. **386**)

core (p. **386**)

dark matter (p. **399**)

galactic clusters (p. **394**)

galaxy (p. **394**)

geocentric (p. **378**)

globular clusters (p. **394**)

Hertzsprung-Russell
diagram (p. **389**)

light-year (p. **384**)

luminosity (p. **387**)

main sequence stars
(p. **389**)

nebulae (p. **385**)

neutron star (p. **393**)

parsec (p. **384**)

protogalaxies (p. **398**)

protostar (p. **385**)

pulsar (p. **393**)

radiation zone (p. **386**)

red giant stars (p. **390**)

supernova (p. **392**)

white dwarf stars (p. **390**)

APPLYING THE CONCEPTS

1. A planet that is in the middle of its period of retrograde motion
   a. is catching up with the other planet in its orbit.
   b. has drawn even with the other planet in its orbit.
   c. has passed the other planet in its orbit.
   d. None of the above are correct.

2. Stars twinkle and planets do not twinkle because
   a. planets shine by reflected light, and stars produce their own light.
   b. all stars are pulsing light sources.
   c. stars appear as point sources of light, and planets are disk sources.
   d. All of the above are correct.

3. How much of the celestial meridian can you see from any given point on the surface of the earth?
   a. one-fourth
   b. one-half
   c. three-fourths
   d. all of it

4. Which of the following of the coordinate system of lines depends on where you are on the surface of the earth?
   a. celestial meridian
   b. celestial equator
   c. north celestial pole
   d. None of the above are correct.

5. The angle that you see Polaris, the North Star, above the horizon is about the same as your approximate location on
   a. the celestial meridian.
   b. the celestial equator.
   c. a northern longitude.
   d. a northern latitude.

6. Polaris is almost at the north celestial pole and nearby stars appear to move _?_ relative to Polaris.
   a. straight by
   b. with a looping motion
   c. counterclockwise
   d. clockwise

7. If you were at the north celestial pole looking down on the earth, how would it appear to be moving? (Use a globe if you wish.)
   a. clockwise
   b. counterclockwise
   c. one way, then the other as a pendulum
   d. It would not appear to move from this location.

8. Your answer to question 7 means that the earth turns
   a. from the west toward the east.
   b. from the east toward the west.
   c. at the same rate it is moving in its orbit.
   d. not at all.

9. Your answer to question 8 means that the moon, sun, and stars that are not circumpolar appear to rise in the (You may go back and change answers to previous questions if you wish.)
   a. west, move in an arc, then set in the east.
   b. north, move in an arc, then set in the south.
   c. east, move in an arc, then set in the west.
   d. south, move in an arc, then set in the north.

10. A star half-way between directly over your head and the horizon is how many degrees of arc above the horizon?
    a. 45°
    b. 90°
    c. 135°
    d. 180°

11. The angular size of an object depends on
    a. only its distance from you.
    b. only its true size.
    c. both its true size and its distance from you.
    d. its true size, distance, and its mass to volume ratio.

12. An astronomical unit is a referent unit of length based on
    a. the distance from the earth to the sun.
    b. the distance across the earth's orbit.
    c. the distance across the solar system.
    d. the radius of the solar system.

13. Against the background of stars, a relatively nearby star viewed from the earth in two different parts of the earth's orbit will appear
    a. in the same location.
    b. in different locations.
    c. to undergo retrograde motion.
    d. to move in an arc.

14. A parsec is the astronomer's measure of
    a. time.
    b. distance.
    c. an arc.
    d. a couple of seconds.

15. Which of the following represents the greatest distance?
    a. 10,000,000,000 km
    b. 10,000,000 miles
    c. 2.0 light-years
    d. 1.0 parsec

16. In which part of a newborn star does the nuclear fusion take place?
    a. convection zone
    b. radiation zone
    c. core
    d. All of the above are correct.

17. Which of the following stars would have the longer life spans?
    a. the less massive
    b. between the more massive and the less massive
    c. the more massive
    d. All have the same life span.

18. A bright blue star on the main sequence is probably
    a. very massive.
    b. less massive.
    c. between the more massive and the less massive.
    d. None of the above are correct.

19. The brightest of the stars listed are the
    a. first magnitude.
    b. second magnitude.
    c. fifth magnitude.
    d. sixth magnitude.

20. The basic property of a main sequence star that determines most of its other properties, including its location on the H-R diagram, is
    a. brightness.
    b. color.
    c. temperature.
    d. mass.

21. All the elements that are more massive than the element iron were formed in a
    a. nova.
    b. white dwarf.
    c. supernova.
    d. black hole.

22. If the core remaining after a supernova has a mass between 1.5 and 3 solar masses, it collapses to form a
    a. white dwarf.
    b. neutron star.
    c. red giant.
    d. black hole.

23. The basic unit of the universe is a
    a. star.
    b. solar system.
    c. galactic cluster.
    d. galaxy.

24. The relationship between the different shapes of galaxies is
    a. spherical galaxies form first, which flatten out to elliptical galaxies, then spin off spirals until they break up in irregular shapes.
    b. irregular shapes form first, which collapse to spiral galaxies, then condense to spherical shapes.
    c. There is no relationship as the different shapes probably resulted from different rates of swirling gas clouds.
    d. None of the above are correct.

25. Microwave radiation from space, measurements of the expansion of the universe, the age of the oldest stars in the Milky Way galaxy, and ratios of radioactive decay products all indicate that the universe is about how old?
    a. 6,000 years
    b. 4.5 billion years
    c. 20 billion years
    d. 100,000 billion years

26. Whether the universe will continue to expand or will collapse back into another big bang seems to depend on what property of the universe?
    a. the density of matter in the universe
    b. the age of galaxies compared to the age of their stars
    c. the availability of gases and dust between the galaxies
    d. the number of black holes

Answers

1. b 2. c 3. b 4. a 5. d 6. c 7. b 8. a 9. c 10. a 11. c 12. a 13. b 14. b 15. d 16. c 17. a 18. a 19. a 20. d 21. c 22. b 23. d 24. c 25. c 26. a

QUESTIONS FOR THOUGHT

1. Describe briefly how the people of ancient civilizations viewed the universe.
2. What is a constellation? For what purpose are constellations used?
3. What was the origin of the zodiac? What was the origin, in general, of the meaning of the signs of the zodiac?
4. How did the present system of twelve months to a year and seven days to a week originate? Suggest a reason why the number seven is considered by some people today to be a mystical, lucky number.
5. Explain how the ancient Greeks reasoned the distances to the seven celestial bodies to create their model of the celestial bodies moving on concentric shells that turned around the earth.
6. Explain why a geocentric model of the universe was so common among the ancient civilizations.
7. Would you ever observe the sun to move along the celestial meridian? Explain.
8. What is the meaning of a distance between stars that is expressed in degrees of arc? What are arc minutes and arc seconds?
9. What is a parsec and how is it defined? What is a light-year and how is it defined?
10. Why are astronomical distances not measured with standard referent units of distance such as kilometers or miles?
11. Explain why a protostar heats up internally as it gravitationally contracts.

12. About how much time is required for the energy released from nuclear fusion reactions in the center part of an average star such as the sun to reach the surface and be emitted as light?

13. Describe in general the structure and interior density, pressure, and temperature conditions of an average star such as the sun.

14. Which size of star has the longest life span, a star sixty times more massive than the sun, one just as massive as the sun, or a star that has a mass of one-twenty-fifth that of the sun? Explain.

15. What is the difference between apparent magnitude and absolute magnitude?

16. What does the color of a star indicate about the surface temperature of the star? What is the relationship between the temperature of a star and the spectrum of the star? Describe in general the spectral classification scheme based on temperature and stellar spectra.

17. What is the Hertzsprung-Russell diagram? What is the significance of the diagram?

18. What is meant by the main sequence of the H-R diagram? What one thing determines where a star is plotted on the main sequence?

19. Describe in general the life history of a star with an average mass like the sun.

20. What, if anything, is the meaning of the Hubble classification scheme of the galaxies?

21. What is a nova? What is a supernova?

22. Describe the theoretical physical circumstances that lead to the creation of (a) a white dwarf star, (b) a red giant, (c) a neutron star, (d) a black hole, and (e) a supernova.

23. Describe the two forces that keep a star in a balanced, stable condition while it is on the main sequence. Explain how these forces are able to stay balanced for a period of billions of years or longer.

24. What is the source of all the elements in the universe that are more massive than helium but less massive than iron? What is the source of all the elements in the universe that are more massive than iron?

25. Why must the internal temperature of a star be hotter for helium fusion reactions than for hydrogen fusion reactions?

26. When does a protostar become a star? Explain.

27. What is a red giant star? Explain the conditions that lead to the formation of a red giant. How can a red giant become brighter than it was as a main sequence star if it now has a lower surface temperature?

28. Why is an average star like the sun unable to have carbon fusion reactions in its core?

29. Describe the structure of the Milky Way galaxy. Where are new stars being formed in the Milky Way? Explain why they are formed in this part of the structure and not elsewhere.

30. If the universe is expanding, are the galaxies becoming larger? Explain.

31. What is the evidence that supports a big bang theory of the universe?

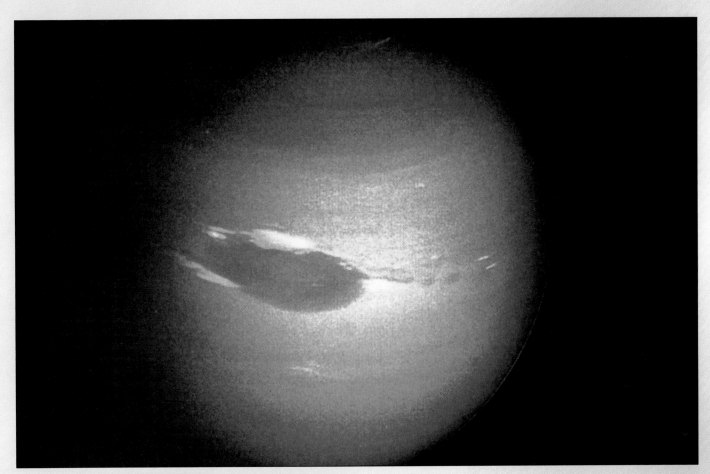

Neptune, the most distant and smallest of the gas giant planets, is a cold and interesting place. It has a Great Dark Spot, as you can see in this photograph, made by *Voyager*. This spot is about the size of Earth, and is similar to the Great Red Spot on Jupiter. Neptune has the strongest winds of any planet of the solar system—up to 2,000 km/hr (1,200 mi/hr). Clouds were observed by *Voyager* to be "scooting" around Neptune every 16 hours or so. *Voyager* scientists called these fast-moving clouds "scooters."

CHAPTER | Seventeen

The Solar System

For generations people have observed the sky in awe, wondering about the bright planets moving across the background of stars, but they could do no more than wonder. You are among the first generations on Earth to see close-up photographs of the planets and to see Earth as it appears from space. Spacecrafts have now made thousands of photographs of the other planets and their moons, measured properties of the planets, and in some cases, studied their surfaces with landers. Astronauts have left Earth and visited the Moon, bringing back rock samples, data, and photographs of Earth as seen from the Moon (figure 17.1). All of these photographs and findings have given your generation a unique new perspective of Earth, the planets, and the moons, comets, and asteroids that make up the solar system.

Viewed from the Moon, Earth is a spectacular blue globe with expanses of land and water covered by huge changing patterns of white clouds. Viewed from a spacecraft, the planets present a very different picture, each unique in its own way. Mercury has a surface that is covered with craters, looking very much like the surface of Earth's Moon. Venus is covered with clouds of sulfuric acid over an atmosphere of mostly carbon dioxide, which is under great pressures with surface temperatures hot enough to melt lead. The surface of Mars has great systems of canyons, inactive volcanoes, and surprisingly, dry riverbeds and tributaries. The giant planets Jupiter and Saturn have orange, red, and white bands of organic and sulfur compounds and storms with gigantic lightning discharges compared to anything ever seen on Earth. One moon of Jupiter has active volcanoes spewing out molten sulfur and sulfur dioxide. The outer giant planets Uranus and Neptune have moons and particles in rings that appear to be covered with powdery, black carbon.

These and many more findings, some fascinating surprises and some expected, have stimulated the imagination as well as added to our comprehension of the frontiers of space. The new information about the Sun's impressive system of planets, moons, comets, and asteroids has also added to speculations and theories about the planets and how they evolved over time in space. This information, along with the theories and speculations, will be presented in this chapter to give you a picture of Earth's immediate neighborhood, the solar system.

IDEAS ABOUT THE SOLAR SYSTEM

If you observe the daily motion of the Sun and Moon and the nightly motion of the Moon and stars over a period of time, you can easily convince yourself that all the heavenly bodies revolve around a fixed, motionless Earth. The Sun, the Moon, and the planets do appear to move from east to west across the sky, and the stars seem to be fixed on a turning sphere, maintaining the same positions relative to one another as they move as units on the turning sphere. To consider that Earth moves rather than the Sun, the Moon, and the planets seems contrary to all these observations.

There is a problem with the observed motion of the planets, however, one troublesome observation that spoils this model of a motionless Earth with everything moving around it. When the planets are observed over a period of a year or more, they are observed not to hold the same relative positions as would be expected of this model. The stars and the planets both rise in the east and set in the west, but the planets do not maintain the same positions relative to the background of stars. Over time, the planets appear to move across the background of stars, sometimes slowing, then reversing their direction in a loop before resuming their normal motion. This *retrograde,* or reverse, motion for the planet Mars is shown in figure 16.6.

The Geocentric Model

Early Greek astronomers and philosophers had attempted to explain the observed motions of the Sun, Moon, and stars with a geometric model, a model of perfect geometrical spheres with attached celestial bodies rotating around a fixed Earth in perfect circles. To explain the occasional retrograde motion of the planets with this model, later Greek astronomers had to modify it by assuming a secondary motion of the planets. The modification required each planet to move in a secondary circular orbit as it moved along with the turning sphere. The small circular orbit, or **epicycle,** was centered on the surface of the sphere as it turned around Earth. The planet was understood to move around the epicycle once as the sphere turned around Earth. The combined motion of the movement of the planet around the epicycle as the sphere turned resulted in a loop with retrograde motion (figure 17.2). Thus, the earlier model of the solar system, modified with epicycles, was able to explain all that was known about the movement of the stars and the seven heavenly bodies and all that was observed during the days of the later ancient Greek civilization.

A version of the explanation of retrograde motion—perfectly circular epicycle motion on perfectly spherical turning spheres—was published by Ptolemy in the second century A.D. and came to be known as the **Ptolemaic system.** This system did account for the facts that were known about the solar system at

FIGURE 17.1
This view of the rising Earth was seen by the *Apollo 11* astronauts after they entered the orbit around the Moon. Earth is just above the lunar horizon in this photograph.

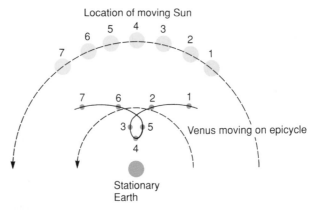

FIGURE 17.2
The paths of Venus and the Sun at equal time intervals according to the Ptolemaic system. The combination of epicycle and Sun movement explains retrograde motion with a stationary Earth.

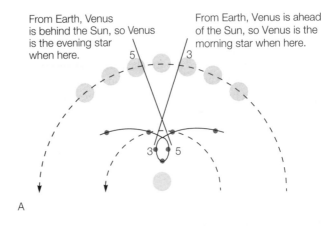

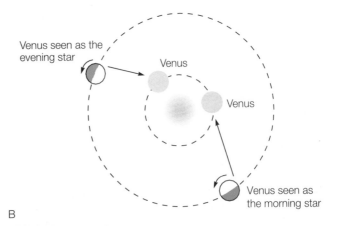

FIGURE 17.3
(*A*) The Ptolemaic system explanation of Venus as a morning star and evening star. (*B*) The heliocentric system explanation of Venus as a morning star and evening star.

that time. Not only did the system describe retrograde motion with complex paths as shown in figure 17.2, but it also explained other observations such as why Venus is sometimes observed for a short time near the Sun at sunrise (the morning "star") and other times observed near the Sun for a short time at sunset (the evening "star") (figure 17.3).

Over the years, newly discovered inconsistencies of observed and predicted positions of the planets were discovered, and epicycles were added to epicycles in an attempt to fit the system with observations. The system became increasingly complicated and unmanageable, but it did seem to agree with other ideas of that time. The agreements were that (1) humans were at the center of a universe that was created for them and (2) heaven was a perfect place, so it would naturally be a place of perfectly circular epicycles moving on perfect spheres, which were in turn moving in perfect circles. Thus, the Ptolemaic system of a geometric, geocentric solar system came to be supported by the Church, and this was the accepted understanding of the solar system for the next fourteen centuries.

The Heliocentric Model

The idea that Earth revolves around the Sun rather than the Sun moving around Earth was proposed by a Polish astronomer, Nicolas Copernicus, in a book published in 1543. In his book, Copernicus pointed out that the observed motions of the planets

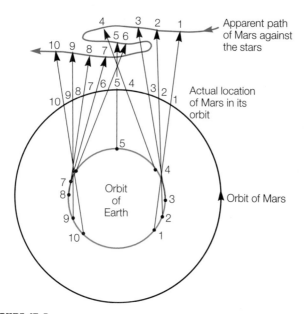

FIGURE 17.4

The heliocentric system explanation of retrograde motion.

could be explained by a model of Earth and the other planets revolving around the Sun as well as by the Ptolemaic system. In this model each planet moved around the Sun in perfect circles at different distances, moving at faster speeds in orbits closer to the Sun. When viewed from a moving Earth, the other planets would appear to undergo retrograde motion because of the combined motions of Earth and the planets. Earth, for example, moves along its inner orbit with about twice the angular speed as Mars in its outer orbit, which is about one and one-half times farther from the Sun. Since Earth is moving faster in the inside orbit it will move even with, then pass, Mars in the outer orbit (figure 17.4). As this happens Mars will appear to slow to a stop, move backward, stop again, then move on. This combined motion is similar to what you observe when you pass a slower moving car, which appears to move backward against the background of the landscape as you pass it. In an outer circular path the car would appear to slow, move backward, then forward again as you pass it.

The **Copernican system** of a heliocentric, or Sun-centered, solar system provided a simpler explanation for retrograde motion than the Ptolemaic system, but it was only an alternative way to consider the solar system. The Copernican system offered no compelling reasons why the alternative Ptolemaic system should be rejected. Furthermore, Copernicus had retained the old Greek idea of planets moving in perfect circles, so there were inconsistencies in predicted and observed motions with this model, too. Clear-cut evidence for rejecting the Ptolemaic system would have to await the detailed measurements of planetary motions made by Tycho Brahe and analysis of those measurements by Johannes Kepler.

Tycho Brahe was a Danish nobleman who constructed highly accurate observatories for his time, which was before the time of the telescope. From his observatory on a little island about twenty miles from Copenhagen, Brahe spent about twenty years (1576–1597) making systematic, uninterrupted

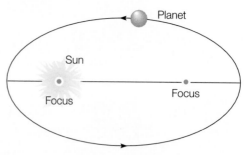

FIGURE 17.5

Kepler's first law describes the shape of a planetary orbit as an ellipse, which is exaggerated in this figure. The Sun is located at one focus of the ellipse.

measurements of the Sun, Moon, planets, and stars. His skilled observations resulted in the first precise, continuous record of planetary position. In 1600, Brahe hired a young German, Johannes Kepler, as an assistant. When Brahe died in 1601, Kepler was promoted to Brahe's position and was given access to the vast collection of observation records. Kepler spent the next twenty-five years analyzing the data to find if planets followed circular paths or if they followed the paths of epicycles. Using the careful observations of Tycho Brahe, Kepler found that the planets did not move in epicycles nor did they move in perfect circles. Planets move in the path of an *ellipse* of certain dimensions. He published his findings in 1609 and 1619, establishing the actual paths of planetary movement. Kepler had found the first evidence that the Ptolemaic system of complicated epicycles was unnecessary and unacceptable as a model of the solar system. His findings also required adjustments of the heliocentric Copernican system, which are described by his three laws of planetary motion. Today, his findings are called **Kepler's laws of planetary motion.**

Kepler's first law states that each planet moves in an orbit that has the shape of an ellipse, with the Sun located at one focus (figure 17.5). *Kepler's second law* states that an imaginary line between the Sun and a planet moves over equal areas of the ellipse during equal time intervals (figure 17.6). This means that the orbital velocity of a planet varies with where the planet is in the orbit, since the distance from the focus to a given position varies around the ellipse. The point at which an orbit comes closest to the Sun is called the *perihelion,* and the point at which an orbit is farthest from the Sun is called the *aphelion.* The shortest line from a planet to the Sun at perihelion means that the planet moves most rapidly when here. The short line and rapidly moving planet would sweep out a certain area in a certain time period, for example, one day. The longest line from a planet to the Sun at aphelion means that the planet moves most slowly at aphelion. The long line and slowly moving planet would sweep out the same area in one day as was swept out at perihelion. Earth travels fastest in its orbit at perihelion on about January 3 and slowest at aphelion on about July 1.

Kepler's third law states that the square of the period of a planet's orbit is proportional to the cube of that planet's semi-

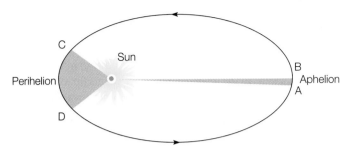

FIGURE 17.6

Kepler's second law. A line from the Sun to a planet at point A sweeps over a certain area as the planet moves to point B in a given time interval. A line from the Sun to a planet at point C will sweep over the same area as the planet moves to point D during the same time interval. The time required to move from point A to point B is the same as the time required to move from point C to point D, so the planet moves faster in its orbit at perihelion.

major axis, or $t^2 \propto d^3$. When the time is expressed in the earth units of one year for a revolution and a radius of astronomical unit, the distance to a planet can be determined by observing the period of revolution and comparing the orbit of the planet with that of Earth. For example, suppose a planet is observed to require eight Earth years to complete one orbit. Then,

$$\frac{t(\text{planet})^2}{t(\text{Earth})^2} = \frac{d(\text{planet})^3}{d(\text{Earth})^3}$$

$$\frac{(8)^2}{1} = \frac{(\text{distance})^3}{1}$$

$$64 = (\text{distance})^3$$

$$\text{distance} = \sqrt[3]{64}\ \text{A.U.}$$

$$\text{distance} = 4\ \text{A.U.}$$

Thus, a planet that takes eight times as long to complete an orbit is four times as far from the Sun as Earth. In general, Kepler's third law means that the more distant a planet is from the Sun, the longer the time required to complete one orbit. Figure 17.7 shows this relationship for the planets of the solar system. Kepler's third law applies to moons, satellites, and comets in addition to the planets. In the case of moons and satellites, the distance is to the planet the moon or satellite is orbiting, not to the Sun.

Kepler's laws were empirically derived from the data collected by Tycho Brahe, and the reason why planets followed these relationships would not be known or understood until Isaac Newton published the law of gravitation some sixty years later (see chapter 3). In the meantime, Galileo constructed his telescope and added observational support to the heliocentric theory. By the time Newton derived the law of gravitation and then improved the accuracy of Kepler's third law, the solar system was understood to consist of planets that move around the Sun in elliptical orbits, paths that could be predicted by applying the law of gravitation. The heliocentric model of the solar system had evolved to a conceptual model that both explained and predicted what was observed.

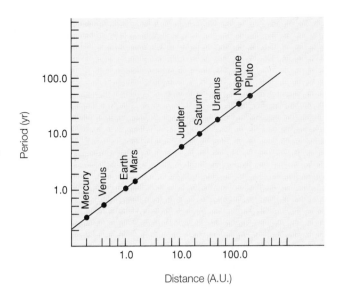

FIGURE 17.7

Kepler's third law describes a relationship between the time required for a planet to move around the Sun and its average distance from the Sun. The relationship is that the time squared is proportional to the distance cubed.

ORIGIN OF THE SOLAR SYSTEM

Kepler used Brahe's continuous record of measurements to figure out how the planets moved, which he generalized in the laws of planetary motion. Later, this was done mathematically with Newton's law of gravity as well. Thus, the heliocentric model of the solar system was built from (1) observations and measurements of the movement of the planets over a period of time, (2) analysis of these measurements, and (3) verification of the model from independent calculations that fit the observed measurements. The heliocentric model is accepted because the model can be independently tested and verified by observation.

The theoretical model of the life of a star (see chapter 16), such as the Sun, presents a different kind of problem in testing and verification. The life of a star takes place over billions of years, so it is not possible to make observations and measurements to test the model. However, there are billions of stars in various stages of their life cycles that can be observed and measured to test the model of the life cycle. Thus, the model of the life of the Sun is acceptable because the model can be verified by observing other stars in different stages of their life cycles.

A model of how the solar system originated presents still another kind of problem in testing and verification. This problem is that the solar system originated a long time ago, some five billion years ago according to a number of different independent sources of evidence, and that there are no other planetary systems that can be directly observed, either in existence or in the process of being formed. At the distance they occur, even the Hubble Space Telescope would not be able to observe planets around their suns. However, there is strong evidence supporting a conclusion that indeed there are other planets orbiting other

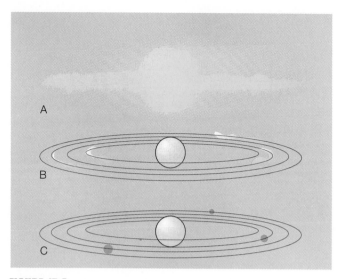

FIGURE 17.8

Formation of the solar system according to the protoplanet nebular model, not drawn to scale. (*A*) The process starts with a nebula of gas, dust, and chemical elements from previously existing stars. (*B*) The nebula is pulled together by gravity, collapsing into the protosun and protoplanets. (*C*) As the planets form, they revolve around the Sun in orbits that are described by Kepler's laws of planetary motion.

stars beyond our solar system. The Hubble Space Telescope has observed the star Beta Pictoris, for example, and found a flat disk of gas and dust that appears to be involved in the early stages of planetary formation. The disk is slightly warped, which might be from the gravitational influence of large bodies—planets in the making—within the disk.

Using infrared radiation data and a doppler shift technique, astronomers have also found that a large number of stars provide evidence of the existence of planets. Evidence of a planet or planets around a star named 50 Pegasi has been found, and then confirmed by other, independent observers. In addition, there is evidence of planets orbiting the stars named 70 Virginis, 47 Ursae majoris, and 55 Cancri. The planet around 70 Virginis is especially interesting since the calculated heat flow would mean a surface temperature of about 85°C (about 185°F), which is cool enough for water and organic molecules to exist.

The Sun-centered model of the solar system enjoys general acceptance because data can be collected to verify and support this model. The model of the life of a star, on the other hand, is generally accepted because data and observations support the model even though questions still exist because some of the details are difficult to observe. The most widely accepted theory of the origin of the solar system is called the **protoplanet nebular model.** A *protoplanet* is the earliest stage in the formation of a planet. The model can be considered in stages, which are not really a part of the model but are simply a convenient way to organize the total picture (figure 17.8).

The first important event in the formation of the solar system involves stars that disappeared billions of years ago, long before

the Sun was born. Earth, the other planets, and all the members of the solar system are composed of elements that were manufactured by these former stars. In a sequence of nuclear reactions, which was described in chapter 16, hydrogen fusion in the core of large stars results in the formation of the elements up to iron. Elements heavier than iron are formed in rare supernova explosions of dying massive stars. Thus *stage A* of the formation of the solar system consisted of the formation of elements heavier than hydrogen in many, many previously existing stars, including the supernovas of more massive stars. Many stars had to live out their life cycles to provide the raw materials of the solar system. The death of each star, including supernovas, added newly formed elements to the accumulating gases and dust in interstellar space. Over a long period of time these elements began to concentrate in one region of space as dust, gases, and chemical compounds, but hydrogen was still the most abundant element in the nebula that was destined to become the solar system.

During *stage B,* the hydrogen gas, dust, elements, and chemical compounds from former stars began to form a large, slowly rotating nebula that was much, much larger than the present solar system. Under the influence of gravity, the large but diffuse, slowly rotating nebula began to contract, increasing its rate of spin. The largest mass pulled together in the center, contracting to the protostar, which eventually would become the Sun. The remaining gases, elements, and dusts formed an enormous, fat, bulging disk called an *accretion disk,* which would eventually form the planets and smaller bodies. The fragments of dust and other solid matter in the disk began to stick together in larger and larger accumulations from numerous collisions over the first million years or so. All of the present-day elements of the planets must have been present in the nebula along with the most abundant elements of hydrogen and helium. The elements and the familiar chemical compounds accumulated into basketball-sized or larger chunks of matter.

You could speculate about how and when the chemical compounds were formed and also about how the chunks of matter could come together to form what would become the planets and smaller bodies of the solar system. Logical arguments could be made for completely different speculations. You could speculate, for example, that if the chunks of matter were cold and icy, they would be more likely to stick when pulled together by gravity. Thus, over a period of time, perhaps 100 million years or so, huge accumulations of frozen water, frozen ammonia, and frozen crystals of methane began to accumulate, together with silicon, aluminum, and iron oxide plus other metals in the form of rock and mineral grains. Such a slushy mixture would no doubt have been surrounded by an atmosphere of hydrogen, helium, and other vapors thinly interspersed with smaller rocky grains of dust. Local concentrations of certain minerals might have occurred throughout the whole accretion disk, with a greater concentration of iron, for example, in the disk where the protoplanet Mars was forming compared to where the protoplanet Earth was forming.

All of the protoplanets might have started out somewhat similarly as huge accumulations of a slushy mixture with an atmosphere of hydrogen and helium gases. Gravitational attraction must have compressed the protoplanets as well as the protosun. During this period of contraction and heating, gravitational adjustments continue, and about a fifth of the disk nearest to the protosun must have been pulled into the central body of the protosun, leaving a larger accumulation of matter in the outer part of the accretion disk.

During *stage C,* the warming protosun became established as a star, perhaps undergoing an initial flare-up that has been observed today in other newly forming stars. Such a flare-up might have been of such a magnitude that it blasted away the hydrogen and helium atmospheres of the interior planets out past Mars, but it did not reach far enough out to disturb the hydrogen and helium atmospheres of the outer planets. The innermost of the outer planets, Jupiter and Saturn, might have acquired some of the matter blasted away from the inner planets, becoming the giants of the solar system by comparison. This is just speculation, however, and the two giants may have simply formed from greater concentrations of matter in that part of the accretion disk.

The evidence shows that the protoplanets, some more than others, underwent heating early in their formation. Much of the heating may have been provided by gravitational contraction, the same process that gave the protosun sufficient heat to begin its internal nuclear fusion reactions. Heat was also provided from radioactive decay processes inside the protoplanets, and the initial greater heating from the Sun may have played a role in the protoplanet heating process. Larger bodies were able to retain this heat better than smaller ones, which radiated it to space more readily. Thus, the larger bodies underwent a more thorough heating and melting, perhaps becoming completely molten early in their history. In the larger bodies the heavier elements, such as iron, were pulled to the center of the now molten mass, leaving the lighter elements near the surface. The overall heating and cooling process took millions of years as the planets and smaller bodies were formed. Gases from the hot interiors formed secondary atmospheres of water vapor, carbon dioxide, and nitrogen on the larger interior planets. How these initial secondary atmospheres evolved into the present-day atmospheres of Mars, Venus, and Earth will be discussed in the next section.

Between the orbits of Mars and Jupiter is a belt of thousands of small rocky bodies called **asteroids.** Interestingly, the belt of asteroids was discovered from a prediction made by the German astronomer Bode at the end of the eighteenth century. Bode had noticed a pattern of regularity in the spacing of the planets that were known at the time. He found that by expressing the distances of the planets from the Sun in astronomical units, these distances could be approximated by the relationship $(n + 4)/10$, where n is a number in the sequence 0, 3, 6, 12, and so on where each number (except the first) is doubled in succession. When these calculations were done the distances turned out to be very close to the distances of all the planets known at that time, but the numbers also predicted a planet between Mars and Jupiter where there was none. Later a belt of asteroids was found where the Bode numbers predicted there should be a planet. This suggested to some people that a planet had existed between Mars and Jupiter in the past, and somehow this planet was broken into pieces, perhaps by a collision with another large body (table 17.1).

TABLE 17.1	Distances from the Sun to planets known in the 1790s		
Planet	**n**	**Distance Predicted by $(n + 4)/10$ (A.U.)**	**Actual Distance (A.U.)**
Mercury	0	0.4	0.39
Venus	3	0.7	0.72
Earth	6	1.0	1.0
Mars	12	1.6	1.5
(Asteroid belt)	24	2.8	—
Jupiter	48	5.2	5.2
Saturn	96	10.0	9.5
Uranus	192	19.6	19.2

Such patterns of apparent regularity in the spacing of planetary orbits were of great interest because, if a true pattern existed, it could hold meaning about the mechanism that determined the location of planets at various distances from the Sun. Many attempts have been made to explain the mechanism of planetary spacing and why a belt of asteroids exists where the Bode numbers predict there should be a planet. The most successful explanations concern Jupiter and the influence of its gigantic gravitational field on the formation of clumps of matter at certain distances from the Sun. In other words, a planet does not exist today between Mars and Jupiter because there never was a planet there. The gravitational influence of Jupiter prevented the clumps of matter from joining together to form a planet, and a belt of asteroids formed instead.

The protoplanet nebular model seems to account for the origin and nature of the planets observed today, and the existence of the asteroid belt. Yet, there are gaps in the model, with missing explanations of how, for example, Earth's Moon formed. On the other hand, relatively recent information from spacecraft missions to the planets has revealed much about the planets that seems to agree with the model. This information is included with the survey of the planets other than Earth, which follows in the next section.

THE PLANETS

Today, the solar system consists of a middle-aged main sequence G type star called the Sun with nine planets, nearly fifty moons, thousands of asteroids, and many comets all revolving around it (figure 17.9). The Sun has an average mass compared to the other main sequence stars, about halfway between the most massive and the least massive. It is, however, 333,400 times more massive than Earth and 1,000 times more massive than Jupiter, the largest planet. The Sun, in fact, has 700 times the mass of all the planets, moons, and minor members of the solar system together. It is the force of gravitational attraction between the comparatively massive Sun and the rest of the solar system that holds it all together. This force of gravitational attraction varies with the product of the mass of each planet and the Sun and the inverse square of the distance between them.

A CLOSER LOOK | Planets and Astrology

Do you read the astrology forecasts in daily newspapers or from Internet portals? Below is a brief background of astrology as developed by the Babylonians, followed by some questions intended for class or small group discussions.

As early as 2000 B.C. the Babylonians began keeping track of time by dividing the year into 12 months, with 7 days to a week and 360 days to a year. They maintained observatories, noting for example that the Sun, the Moon, and five planets known at the time (Mercury, Venus, Mars, Jupiter, and Saturn) moved across the sky only along a certain path, which is today called the ecliptic. This movement was independent of the stars, which followed the motions of the seven celestial bodies but kept the same position relative to each other. The Sun appeared to move completely around the ecliptic each year.

In order to keep track of the Sun and the time of year, the Babylonians imagined the arrangements of certain stars to be the shapes of gods, objects, or animals. These patterns, today called constellations, were used to identify twelve equal divisions of the ecliptic through which the Sun passed in monthly succession. The twelve constellations are called the zodiac, so there are twelve signs of the zodiac in a year (see figure 16.4 and table 16.1). By 540 B.C. the Babylonians had fully developed the art of studying the zodiac, the Sun, and planets as a guide to human affairs and this activity today is known as astrology.

First, consider astrology forecasts as they are made today. A daily horoscope might include a forecast for those with a birthday on this day and forecasts for people born during each of the twelve signs of the zodiac. In the horoscope you can find such predictions as "your computer could be out of order today," "focus on your ability to tear down before you rebuild," and "take a chance on romance with a Virgo." Discuss such forecasts with your group. Consider, for example, the population of a nation and how many people are forecasted to have their computer out of order. How many computers would be out of order without the forecast?

Next, discuss why your passage through the birth canal (your birthday) is so important. Perhaps the time when your embryo formed might be more important if you were going to be "marked" by the planets for certain things to happen to you.

Finally, discuss the topic of how Earth's axis has a slow wobble, called precession, which causes it to swing in a slow circle like the wobble of a spinning top. The axis takes about 26,000 years to complete one turn, or wobble. The moving pole changes over time in which particular signs of the zodiac appear, for example, with the spring equinox. Because of precession, the occurrence of the spring equinox has been moving backward through the zodiac constellations at about 1 degree every 72 years. So, 3,000 years ago the Sun entered the constellation of Virgo in August, which is still the basis for horoscopes today. However, the Sun is now in the Leo constellation in August.

Asteroid belt

FIGURE 17.9

The order of the planets out from the Sun. The orbits and the planet sizes are not drawn to scale.

TABLE 17.2	Properties of the planets								
	Mercury	**Venus**	**Earth**	**Mars**	**Jupiter**	**Saturn**	**Uranus**	**Neptune**	**Pluto**
Average Distance from the Sun:									
in 10^6 km	58	108	150	228	778	1,400	3,000	4,497	5,914
in A.U.	0.38	0.72	1.0	1.5	5.2	9.5	19.2	30.1	39.5
Inclination to Ecliptic	7°	3.4°	0°	1.9°	1.3°	2.5°	0.8°	1.8°	17.2°
Revolution Period (Earth Years)	0.24	0.62	1.00	1.88	11.86	29.46	84.01	164.8	247.7
Rotation Period (Earth Days, hr, min, and s)	59 days	−243 days	23 hr 56 min 4 s	24 hr 37 min 23 s	9 hr 50 min 30 s	10 hr 39 min	−17 hr 14 min	16 hr 6.7 min	6.38 days
Mass (Earth = 1)	0.05	0.82	1.00	0.11	317.9	95.2	14.6	17.2	0.002
Equatorial Dimensions:									
diameter in km	4,880	12,104	12,756	6,787	142,984	120,536	57,118	49,528	2,274
in Earth radius = 1	0.38	0.95	1.00	0.53	11	9	4	4	0.18
Density (g/cm³)	5.43	5.25	5.52	3.95	1.33	0.69	1.29	1.64	2.0
Atmosphere (Major Compounds)	None	CO_2	N_2,O_2	CO_2	H_2,He	H_2,He	H_2,He,CH_4	H_2,He,CH_4	N,CO,CH_4
Solar Energy Received (cal/cm²/s)	13.4	3.8	2.0	0.86	0.08	0.02	0.006	0.002	0.001

Table 17.2 compares the basic properties of the nine planets with properties of the orbital plane, period of revolution, and mass of each compared to Earth. From this table you can see that the planets can be classified into two major groups based on size, density, and atmospheric chemical composition. The interior planets of Mercury, Venus, and Mars have densities and compositions similar to those of Earth, so these planets, along with Earth, are known as the **terrestrial planets.**

Outside the orbit of Mars are four **giant planets,** which are similar in density and chemical composition. The terrestrial planets are mostly composed of rocky materials and iron with a density range of about 4 to 5.5 g/cm³. The giant planets of Jupiter, Saturn, Uranus, and Neptune, on the other hand, are massive giants mostly composed of hydrogen, helium, and methane with a density of less than 2 g/cm³. The density of the giant planets suggests the presence of rocky materials and iron as the terrestrial planets have, but they occur as a core surrounded by a deep layer of compressed gases beneath a deep atmosphere of vapors and gases.

The terrestrial planets are separated from the giant planets by the asteroid belt. The outermost planet, Pluto, has properties that do not fit with either the terrestrial planets or the giant planets and has very strange orbital properties. This has led to the suggestion that perhaps Pluto is not a true planet at all, but may be a captured body that orbits the Sun.

As you found in the previous section, the protoplanet nebular model leads to speculative ideas about how the terrestrial planets came to be so different from the giant planets. In general, these ideas concern (1) the loss of matter from the interior planets to the Sun early in the nebular stage, (2) the loss of the primary, or first, atmospheres of the interior planets as they were "blasted away" by an early flare-up of the newly forming Sun,

and (3) different initial amounts of melting in the larger bodies compared to the smaller bodies because of different amounts of heating from compression and radioactive decay compared to the loss of heat radiated to space. When rock materials are heated to melting, gases and water vapor are expelled from the molten mixture by a process called *degassing*. This process is still going on today on Earth as volcanoes emit much degassed water and carbon dioxide along with molten lava and ash. The idea here is that this same degassing process occurred in the past. The amount of melting was thus related to the size of a planet, or body, and greater amounts of melting would have resulted in greater amounts of degassing. When this idea is coupled with the observation that more massive bodies have greater gravity, and thus a greater ability to hold on to degassed vapors and gases, you have all the information you need to speculate how the secondary atmospheres formed on the terrestrial planets and why these atmospheres are so different today.

Smaller bodies, such as the planet Mercury and Earth's Moon, did not produce much water through degassing since they did not undergo long-term, thorough melting. The water and gases that were released through the process were lost to space because of their low gravities, so not much of an atmosphere, if any, is found on the planet Mercury and Earth's Moon today. The planet Mars, being more massive, did produce more water and carbon dioxide gas and was able to hold on to some of both as it formed a secondary atmosphere.

Earth and Venus are similar in size, the two largest of the terrestrial planets, so both produced appreciable amounts of carbon dioxide and water vapor through the degassing processes. There are complications involving water and carbon dioxide on these two planets, however, because (1) both planets receive intense, energetic ultraviolet solar radiation that dissociates water molecules and

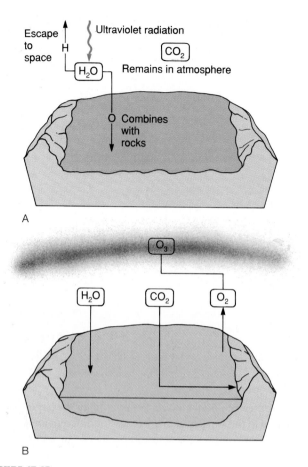

A

B

FIGURE 17.10

(A) The early secondary atmosphere of Venus lost hydrogen to space and oxygen became combined with rocks as ultraviolet radiation decomposed water molecules. (B) On Earth, water formed the oceans, carbon dioxide was removed, and plants released oxygen, which in turn formed an ozone layer in the atmosphere, protecting the water and life below.

(2) carbon dioxide in the atmosphere absorbs infrared radiation emitted from the surface to increase the temperature through the greenhouse effect (see chapter 21). On Venus, the ultraviolet radiation dissociated water molecules to hydrogen and oxygen gas. The less massive hydrogen gas eventually escaped to space from Venus, and the oxygen combined with other elements, becoming locked up in the surface rocks. Thus, Venus was left with a secondary atmosphere consisting mostly of carbon dioxide. Meanwhile, Earth received less intense ultraviolet radiation than Venus since it was farther from the Sun. Much of the water vapor on Earth was able to condense, forming the oceans, which began recycling through the hydrologic cycle. Much of the carbon dioxide in Earth's atmosphere dissolved in its surface waters, reacting with minerals to become locked up as calcium carbonate (limestone). Most of the original carbon dioxide was thus removed from the early secondary atmosphere of Earth. Early forms of plant life, in the meantime, were able to grow profusely in the oceans since the waters protected them from the deadly ultraviolet radiation. Growth slowly increased the oxygen content of Earth's atmosphere. Gradually an ozone layer

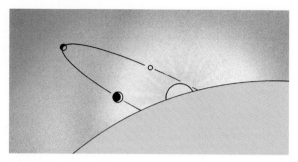

FIGURE 17.11

Mercury is close to the Sun and is visible only briefly before or after sunrise or sunset, showing phases. Mercury actually appears much smaller in an orbit that is not tilted as much as shown in this figure.

accumulated in the upper atmosphere, which now absorbed the deadly ultraviolet radiation from the Sun. New life-forms on the land as well as the water supply were now protected from the ultraviolet radiation, and the atmosphere continued to evolve into its present-day form, consisting mostly of nitrogen and oxygen with traces of carbon dioxide and water vapor (figure 17.10).

Today, carbon dioxide is found on Earth (1) in the atmosphere in trace amounts, (2) dissolved in the oceans as a gas, (3) locked up in carbonate rocks of the surface, and (4) locked up in calcium carbonate seashells and other plant and animal structures. When all this carbon dioxide is added together, the total amount is about equal to the amount now found in the atmosphere of Venus. The formation of Earth's secondary atmosphere and its abundance of water are thus results of Earth (1) being sufficiently massive to generate water and carbon dioxide through degassing; (2) being sufficiently massive to keep the water vapor from escaping to space; and (3) being at a distance from the Sun so the water was not immediately dissociated and could form oceans that removed carbon dioxide from the atmosphere through the hydrologic cycle, which permitted it to become locked up in rock compounds. Thus, with the carbon dioxide removed, the ultra-high temperatures that occur on Venus from the greenhouse effect did not happen on Earth. The overall outcome is that Earth has cooler temperatures than Venus, it is watery, and life occurs on it. There is no evidence of life on the other planets, and why this is so should become apparent as the other planets are considered in more detail.

Mercury

Mercury is the innermost planet, moving rapidly in a highly elliptical orbit that averages about 58 million km (about 36 million mi) from the Sun. This average distance is about 0.4 astronomical unit, or about 0.4 of the average distance of Earth from the Sun. Mercury is the eighth largest planet, with a diameter of less than 5,000 km (about 3,000 mi), which means that it is slightly larger than Earth's Moon. Mercury is very bright because it is so close to the Sun, but it is difficult to observe because it only appears briefly for a few hours immediately after sunset or before sunrise. This appearance, low on the horizon, means that Mercury must be viewed through more of Earth's atmosphere, making the study of such a small object difficult at best (figure 17.11).

FIGURE 17.12

A photomosaic of Mercury made from pictures taken by the *Mariner 10* spacecraft. The surface of Mercury is heavily cratered, looking much like the surface of Earth's Moon. All the interior planets and the Moon were bombarded early in the life of the solar system.

In accord with Kepler's third law, as the innermost planet Mercury has the shortest period of revolution around the Sun. Mercury's period of revolution is about three Earth months at eighty-eight days, so Mercury has the shortest "year" of all the planets. With the highest orbital velocity of all the planets, Mercury was appropriately named after the mythical Roman messenger of speed. Oddly, however, this speedy planet has a rather long day in spite of its very short year. With respect to the stars, Mercury rotates once every fifty-nine days. This means that Mercury rotates three times every two orbits.

The long Mercury day with a nearby large, hot Sun in the sky means high temperatures. High temperatures mean higher gas kinetic energies, and with a low gravity, gases easily escape from Mercury so it has only trace gases for an atmosphere. The lack of an atmosphere to even the heat gains from the long days and heat losses from the long nights results in some very large temperature differences. The temperature of the surface of Mercury ranges from a high of about 427°C (about 800°F) in the sunlight to a low of about −180°C (about −350°F) on the dark side. These extreme temperatures are above the melting point of lead (328°C) and below the boiling point of liquid oxygen (−183°C).

Mercury has been visited by only one spacecraft, *Mariner 10,* which flew by three times in 1973 and 1974. Only 45 percent of the surface was mapped. Mercury is much too close to the Sun to be safely viewed by the Hubble Space Telescope. *Mariner 10* was launched in late 1973 on a trajectory that took it past Venus, then into orbit around the Sun after passing by Mercury. The spacecraft passed by Mercury three times in 1974 and 1975, transmitting data and thousands of photographs back to Earth. The photographs transmitted by *Mariner 10* revealed that the surface of Mercury is covered with craters, and the planet has a surface that very much resembles the surface of Earth's Moon. There are large craters, small craters, superimposed craters, and craters with lighter colored rays coming from them just like the craters on the Moon. Also as on Earth's Moon, there are hills and smooth areas with light and dark colors that were covered by lava in the past, some time after most of the impact craters were formed (figure 17.12).

The cratering on Mercury is believed to be related to the cratering observed on Earth's Moon and, in fact, to the cratering that appears on all the interior planets and their satellites. Almost all of this cratering on the planets and their satellites took place about the same time in the past. Studies of the Moon's craters by *Apollo* astronauts indicate that the bombardment took place at the beginning of the solar system and continued up to about four billion years ago. The evidence from radioactive dating of moon rocks and from rates of erosion of the craters by micrometeorites fits with other evidence that the craters are impact craters formed from collisions with leftover bits and pieces of rocks moving throughout the newly formed solar system. The cratering was produced on all the planets early in the history of the solar system. The craters on Earth were long ago obliterated by the action of water and other erosional agents that are lacking on the Moon and other planets. During this same time period the rocky crust of Mercury has been weathered by solar wind and tiny micrometeorites. The surface of Mercury, like the surface of Earth's Moon, is thus covered by a thin layer of fine dust.

Mercury has no natural satellites or moons, it has a weak magnetic field, and it has an average density more similar to Venus or Earth than to the Moon. The presence of the magnetic field and the relatively high density for such a small body must mean that Mercury probably has a relatively large core of iron with at least part of the core molten. This could mean that Mercury lost much of its less dense, outer layer of rock materials sometime during its formation.

Venus

Venus is the brilliant evening and morning "star" that appears near sunrise or sunset, sometimes shining so brightly that you can see it while it is still daylight. Venus orbits the Sun at an average distance of about 108 million km (about 67 million mi), or 0.7 A.U. in a 225-day orbit. This means that Venus is sometimes to the left of the Sun, appearing as the evening star, and sometimes to the right of the Sun, appearing as the morning star. Venus also has phases just as the Moon does. When Venus is in the full phase, however, it is small and farthest away from Earth. A crescent Venus appears much larger and thus the brightest when it is closest to Earth. You can see the phases of Venus with a good pair of binoculars.

Venus shines brightly because it is covered with clouds that reflect about 80 percent of the sunlight, making it the brightest object in the sky after the Sun and the Moon. These same clouds prevented any observations of the surface of Venus until the early 1960s, when radar astronomers were able to penetrate the clouds and measure the planet's rotation rate. Venus was found to rotate in 243 days with respect to the stars. Since Venus has a 225-day rate of revolution, this means that each day on Venus is longer than a Venus year! Also a surprise, Venus was found to rotate in the *opposite* direction of its direction of revolution. On Venus, you would observe the Sun to rise in the west and set in the east, if you could see it, that is, through all the clouds. The backward rotation is called *retrograde rotation*, since it is opposite the rotation experienced by most of the other planets.

In addition to early studies by radio astronomers, Venus has been the target of many American and Soviet spacecraft probes (table 17.3). The Soviet *Venera* spacecraft series has landed on the surface of Venus, sending back measurements of temperature, pressure, radioactivity, and chemical data, in addition to photographs of the nearby landscape. The American *Mariner* spacecraft series has collected data on the environment of Venus and sent back photographs of the cloud layers. The American *Pioneer* orbiter spacecraft sent probes through the atmosphere to the surface and mapped much of the surface by radar.

TABLE 17.3	Spacecraft missions to Venus		
Date	**Name**	**Owner**	**Remark**
Feb 12, 1961	Venera 1	U.S.S.R.	Flyby
Aug 27, 1962	Mariner 2	U.S.	Flyby
Apr 2, 1964	Zond 1	U.S.S.R.	Flyby
Nov 12, 1965	Venera 2	U.S.S.R.	Flyby
Nov 16, 1965	Venera 3	U.S.S.R.	Crashed on Venus
June 12, 1967	Venera 4	U.S.S.R.	Impacted Venus
June 14, 1967	Mariner 5	U.S.	Flyby
Jan 5, 1969	Venera 5	U.S.S.R.	Impacted Venus
Jan 10, 1969	Venera 6	U.S.S.R.	Impacted Venus
Aug 17, 1970	Venera 7	U.S.S.R.	Venus landing
Mar 27, 1972	Venera 8	U.S.S.R.	Venus landing
Nov 3, 1973	Mariner 10	U.S.	Venus, Mercury flyby photos
June 8, 1975	Venera 9	U.S.S.R.	Lander/orbiter
June 14, 1975	Venera 10	U.S.S.R.	Lander/orbiter
May 20, 1978	Pioneer 12 (also called Pioneer Venus 1 or Pioneer Venus)	U.S.	Orbital studies of Venus
Aug 8, 1978	Pioneer 13 (also called Pioneer Venus 2 or Pioneer Venus)	U.S.	Orbital studies of Venus
Sept 9, 1978	Venera 11	U.S.S.R.	Lander; sent photos
Sept 14, 1978	Venera 12	U.S.S.R.	Lander; sent photos
Oct 30, 1981	Venera 13	U.S.S.R.	Lander; sent photos
Nov 4, 1981	Venera 14	U.S.S.R.	Lander; sent photos
June 2, 1983	Venera 15	U.S.S.R.	Radar mapper
June 7, 1983	Venera 16	U.S.S.R.	Radar mapper
Dec 15, 1984	Vega 1	U.S.S.R.	Venus/Comet Halley probe
Dec 21, 1984	Vega 2	U.S.S.R.	Venus/Comet Halley probe
May 4, 1989	Magellan	U.S.	Orbital Radar Mapper
Oct 18, 1989	Galileo	U.S.	Flyby measurements and photos

The National Aeronautics and Space Administration *Magellan* spacecraft mapped Venus from September, 1990, until September, 1992. The *Magellan* mapping instrument was a sophisticated radar device that saw through the clouds to reveal details on the surface of Venus (figure 17.13).

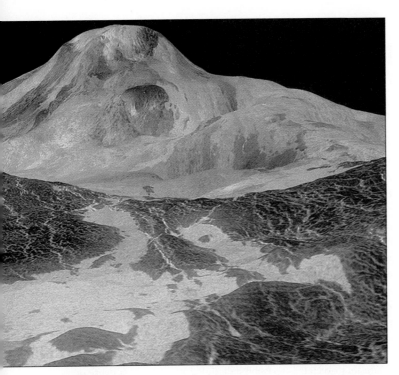

FIGURE 17.13

This is an image of an 8 km (5 mile) high volcano on the surface of Venus. The image was created by a computer using *Magellan* radar data, simulating a viewpoint elevation of 1.7 km (1 mile) above the surface. The lava flows extend for hundreds of km across the fractured plains shown in the foreground. The simulated colors are based on color images recorded by the Soviet *Venera 13* and *14* spacecraft.

Venus has long been called Earth's sister planet since its mass, size, and density are very similar. Venus has an equatorial diameter of about 12,104 km (about 7,517 mi) compared to Earth's equatorial diameter of about 12,756 km (about 7,921 mi). Size, mass, and density, however, are where the similarities with Earth end. Spacecraft probes found a hot, dry surface under tremendous atmospheric pressure. The atmosphere consists mostly of carbon dioxide, a few percent of nitrogen, and traces of water vapor and other gases. The atmospheric pressure at the surface of Venus is almost one hundred times the pressure at the surface of Earth, a pressure many times what a human could tolerate. The average surface temperature of about 480°C (about 900°F) is comparable to the surface temperature on Mercury, which is hot enough to melt lead. The comparably hot temperature on Venus, which is nearly twice the distance from the Sun as Mercury, is a result of the greenhouse effect. Sunlight filters through the atmosphere of Venus, warming the surface. The surface reemits the energy in the form of infrared radiation, which is absorbed by the almost pure carbon dioxide atmosphere. Carbon dioxide molecules absorb the infrared radiation, increasing their kinetic energy and the temperature.

The tremendously hot surface and crushing atmospheric pressure have taken their toll on space probes sent to the surface of Venus. None of the probes has operated for longer than about an hour before it ceased to function. The probes did verify the extreme temperature and pressure measurements before they gave in to the pressure, temperature, or perhaps the clouds and rain consisting of sulfuric acid. Soviet *Venera* landers 9, 10, 13, and 14 were able to land on Venus and transmit images of the surface back to Earth for up to two hours before failing in the tremendous heat and pressure of the atmosphere.

The surface of Venus is mostly a flat, rolling plain but there are several raised areas, or "continents," on about 5 percent of the surface, a mountain larger than Mount Everest, a great valley deeper and wider than the Grand Canyon, and many large, old impact craters. In general, the surface of Venus appears to have evolved much as did the surface of Earth, but without the erosion by ice, rain, and running water.

Venus, like Mercury, has no satellites. Venus also does not have a magnetic field, as expected. Two conditions seem to be necessary in order for a planet to generate a magnetic field: a molten center part and a relatively rapid rate of rotation. With a 243-day rate of rotation, Venus rotates the slowest of all the planets and thus does not have a magnetic field even if some of the interior of Venus is still liquid as in Earth.

Mars

Mars has always attracted attention because of its unique, bright reddish color and its swift retrograde motion against the background of stars. The properties and surface characteristics have also attracted attention, particularly since Mars seems to have similarities to Earth. It orbits the Sun at an average distance of about 228 million km (about 142 million mile), or about 1.5 A.U. It makes a complete orbit every 687 days, about twice the time that Earth takes. Mars rotates in twenty-four hours, thirty-seven minutes, so the length of a day on Mars is about the same as the length of a day on Earth. The observations that Mars has an atmosphere, light and dark regions that appear to be greenish and change colors with the seasons, and white polar caps that grow and shrink with the seasons led to early speculations (and many fantasies) about the possibilities of life on Mars. These speculations increased dramatically in 1877 when Schiaparelli, an Italian astronomer, reported seeing "channels" on the Martian surface. Other astronomers began interpreting the dark greenish regions as vegetation and the white polar caps as ice caps as Earth has. In the early part of the twentieth century, the American astronomer who founded the Lowell Observatory in Arizona, Percival Lowell, published a series of popular books showing a network of hundreds of canals on Mars. Lowell, and other respectable astronomers, interpreted what they believed to be canals as evidence of intelligent life on Mars. Other astronomers, however, interpreted the greenish colors and the canals to be illusions, imagined features of astronomers working with the limited telescopes of that time. Since canals never appeared in photographs, said the skeptics, the canals were the result of the human tendency to see patterns in random markings where no patterns actually exist.

Speculation about Martian canals, vegetation, polar caps of ice, and intelligent life on Mars ended in the late 1960s and early

A CLOSER LOOK | A Recent Magellan Expedition

In September, 1519, Ferdinand Magellan led an expedition of five sailing ships from Seville hoping to find a passage to a new ocean that Balboa had reported finding on the far side of the New World. Fourteen months later they left the Atlantic Ocean, rounded what would later be called the Straits of Magellan, and sailed on to the new ocean. Magellan named it the *Pacific Ocean* because it seemed so peaceful and serene compared to where they had been. The expedition sailed to Guam, then the Philippines where Magellan was killed by natives. The remaining crew crossed the Indian Ocean, the Atlantic Ocean, and finally returned to Seville three years after departing. Magellan had found, named, and crossed the vast Pacific Ocean and his expedition completed the first circumnavigation of Earth.

The spacecraft *Magellan* was named for the leader of Earth's first circumnavigation expedition. This spacecraft was launched in May, 1989, from the space shuttle *Atlantis* and arrived 15 months later at Venus. The spacecraft *Magellan's* mission was to circumnavigate Venus, mapping and exploring the planet in the process. The spacecraft was a bit faster than its namesake, completing one orbit of Venus every 3 hours and 15 minutes compared to the three years required for the first circumnavigation of planet Earth.

The *Magellan* spacecraft was placed in a highly elliptical orbit around Venus, which ranged from 300 km (185 mi) to 8,340 km (5,170 mi) above the surface. *Magellan* completed its original primary mission from this elliptical orbit between August, 1990, and May, 1991, by returning radar images covering about 99 percent of the planet. *Magellan* ended the radar mapping portion of its operation in September, 1992.

Starting in May, 1993, thruster firings were used to lower *Magellan's* orbit to a more circular one of about 600 km (about 375 mi) above the surface, one better suited for measurement of gravity in the polar regions. In this new orbit, *Magellan* was positioned to make gravity measurements at the planet's mid- and higher latitudes as well as at equatorial regions.

It then began to gather gravity data to help scientists develop a model of the planet's interior.

Magellan completed its high-resolution Venus gravity mapping activities in October, 1994. Then the spacecraft conducted a "windmill" experiment that resulted in aerodynamic and atmospheric data that will be useful for future mission designs. This experiment was one of applying a torque to the spacecraft in the upper atmosphere of Venus, and the collection of aerodynamic data as it spun slowly on its axis. The windmill experiment was repeated at three lower altitudes, which finally terminated the *Magellan* mission as it crashed to the surface.

What did *Magellan* find about Venus? Radar mapping from the *Magellan* spacecraft revealed the presence of a wide diversity of geologic features on the surface of Venus. The planet is still geologically active in places, even though radar images indicate that little has changed in the past half-billion years. Many geologic characteristics of different landforms suggest igneous or volcanic activity that accompanied faulting, with multiple episodes of faulting and volcanism. There are also large irregular impact craters, which are interpreted to have formed by multiple impacts. These may have formed from the breakup of a very large object, or perhaps from a series of objects such as a comet. Bright circular regions were also found, and these were interpreted as possible sites of sedimentary deposits.

There are visible streaks on the surface of Venus, line-like features that are interpreted to have formed from the scouring away of volcanic deposits by the wind, revealing fractured bedrock below. The deposits may have formed by fallout from volcanic explosion plumes, characteristic features that are typical of explosive volcanic deposits on Earth.

Another geologic region with line-like features was found to have two sets of parallel lines which intersect almost at right angles. The brighter lines appear in places to begin and end where they intersect a fainter line. It is not yet clear whether the two sets of lines represent faults or fractures. The bright lines are also associated with pit-craters and other volcanic fea-

tures. This type of terrain has not been seen previously, either on Venus or the other planets.

Volcanic domes and flows are seen throughout some regions of the surface. The domes are hills averaging 25 km (about 15.5 mi) in diameter with maximum heights of 750 m (about 2,500 ft). These features are interpreted to result from eruptions of thick lava on relatively level ground, which allowed the lava to flow in an even pattern. Fracture patterns on the surface of the domes suggest that they were extruded, and then a solid outer layer formed. Further intrusion in the interior stretched the cooled surface, forming the fracture patterns. These domes may be analogous to volcanic domes on Earth. Bright margins possibly indicate the presence of rock debris or talus at the slopes of the domes. Fractures on the surrounding plains are both older and younger than the dome-shaped hills.

Some regions have a variety of different terrain types that indicate something about the interior of Venus. For example, there are dark lava plains that also have small volcanic structures on them. In one place a long ridge belt was observed to rise about 1 km (about 0.6 mi) above the surrounding plains. This is interpreted to be a zone of compression and crustal thickening. In places ridges are seen to be grouped into "bands" that are about 20 km (about 12.5 mi) wide. The size of these bands gives a measure of the scale of deformation of the brittle upper part of the planet's crust, which in turn indicates a thickness of about 4 km (about 2.5 mi).

Planetary geologists continue to carefully study the *Magellan* image and gravity data to discover what tectonic and volcanic processes formed the surface of Venus. Understanding the relationship of topography to the stresses and motions in the outer layers of the planet helps geologists and geophysicists to formulate and test models for the formation of the surface of Venus. The results of these studies will add to our understanding of the interior forces that shape the surface of Venus, which adds to our understanding of the forces that shaped the other planets of our solar system, including planet Earth.

TABLE 17.4	Completed spacecraft missions to Mars		
Date	**Name**	**Owner**	**Remark**
Nov 5, 1964	Mariner 3	U.S.	Flyby
Nov 28, 1964	Mariner 4	U.S.	First photos
Feb 24, 1969	Mariner 6	U.S.	Flyby
Mar 27, 1969	Mariner 7	U.S.	Flyby
May 19, 1971	Mars 2	U.S.S.R.	Lander
May 28, 1971	Mars 3	U.S.S.R.	Orbiter/lander
May 30, 1971	Mariner 9	U.S.	Orbiter
Jul 21, 1973	Mars 4	U.S.S.R.	Probe
Jul 25, 1973	Mars 5	U.S.S.R.	Orbiter
Aug 5, 1973	Mars 6	U.S.S.R.	Lander
Aug 9, 1973	Mars 7	U.S.S.R.	Flyby/lander
Aug 20, 1975	Viking 1	U.S.	Lander/orbiter
Sept 9, 1975	Viking 2	U.S.	Lander/orbiter
July 7, 1988	Phobos 1	U.S.S.R.	Orbiter/Phobos lander
July 12, 1988	Phobos 2	U.S.S.R.	Orbiter/Phobos lander
Nov 7, 1996	Global Surveyor	U.S.	Orbiter
Dec 4, 1997	Pathfinder	U.S.	Lander/surface rover

FIGURE 17.14

Surface picture taken by the *Viking 1* lander found reddish, fine-grained material, rocks coated with a reddish stain, and groups of blue-black volcanic rocks.

1970s with extensive studies and probes by spacecraft (table 17.4). Limited photographs by *Mariner* flybys in 1965 and 1969 had provided some evidence that the surface of Mars was much like the Moon, with no canals, vegetation, or much of anything else. Then in 1971, *Mariner 9* became the first spacecraft to orbit Mars, photographing the entire surface as well as making extensive measurements of the Martian atmosphere, temperature ranges, and chemistry. For about a year *Mariner 9* sent a flood of new and surprising information about Mars back to Earth (see figure 17.14 for a photo of the surface taken by *Viking 1*).

Mariner 9 found the surface of Mars not to be a crater-pitted surface as is found on the Moon. Mars has had a geologically active past and has four provinces, or regions, of related surface features. There are (1) volcanic regions with inactive volcanoes, one larger than any found on Earth, (2) regions with systems of canyons, some larger than any found on Earth, (3) regions of terraced plateaus near the poles, and (4) flat regions pitted with impact craters (figure 17.15). There are also regions of sand dunes and powdery, salmon-colored dust that is whipped into monstrous dust storms by seasonal winds. It was the bright dust periodically covering dark rock surfaces that produced the surface color changes visible from Earth and thought to be seasonal vegetation changes. The white polar caps were found to be mostly dry ice, solid carbon dioxide that covers some underlying water ice. Surprisingly, dry channels suggesting former water erosion were discovered near the cratered regions. These are not the straight, long channels that were imagined by earlier astronomers but are sinuous, dry riverbed features with dry tributaries. At one time Mars must have had an abundance of liquid water.

Liquid water may have been present on Mars in the past, but none is to be found today. The atmosphere of Mars is very thin, exerting an average pressure at the surface that is only 0.6 percent of the average atmospheric pressure on Earth's surface. Moreover, this thin Martian atmosphere is about 95 percent carbon dioxide, and 20 percent of this freezes as dry ice at the Martian South Pole every winter. The average Martian surface temperature is −53°C (about −63°F), and any water present is frozen in permafrost or layered sheets, perhaps forming the terraced plateaus near the poles. Today, Mars is much too cold and the air pressure is too low for liquid water to exist.

If liquid water cannot exist on Mars, what formed the dry riverbed features that appear to have been formed by water? In theory, Mars at one time had an atmosphere of much denser concentrations of carbon dioxide and water vapor. A denser Martian atmosphere of carbon dioxide and water vapor could have heated the atmosphere of Mars to much warmer temperatures through the greenhouse effect, even though it is half again as far from the Sun as Earth. According to this theory, both Earth and Mars had similar secondary atmospheres of carbon dioxide and water that were originally degassed from molten rock. Mars, however, is about half the size of Earth. It is the size of a planet— or a body such as the Moon—that determines the extent of internal melting and how rapidly the internal molten rock cools back to the solid state. Thus, a small body such as the Moon has less internal melting and cools more rapidly since it is smaller.

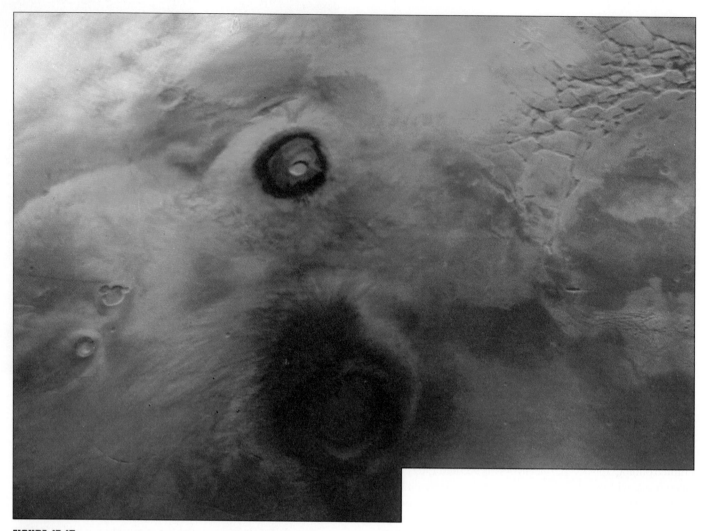

FIGURE 17.15

A view of the surface of Mars taken by the *Viking Orbiter 1* cameras. The scene shows three volcanoes that rise an average of 17 km (about 11 mi) above the top of a 10 km (about 6 mi) high ridge. Clouds can be seen in the upper portion of the photograph, and haze is present in the valleys at the lower right.

Evidence from the Moon, for example, points to molten rock and volcanic activity for only about the first 700 million years. Evidence from Mars, which is twice as large as the Moon, points to molten rock and volcanic activity during the first billion years. Earth, on the other hand, is twice as large as Mars and has had molten rocks and volcanic activity for 4.6 billion years.

The carbon dioxide concentration on Mars was reduced, perhaps by the formation of carbonate rocks as on Earth and perhaps by some other mechanism. As the volcanic activity on Mars abated, the planet lost its source of water vapor and the carbon dioxide in the atmosphere, so the greenhouse effect also declined. The planet then cooled to its present harsh, cold conditions. The water-carved channels observed today were carved when Mars was much warmer with an active greenhouse effect.

Does life exist on Mars? Two *Viking* spacecraft were sent to Mars in 1975 to search for signs of life. The two *Viking* spacecraft were identical, each consisting of an orbiter and a lander. After 11 months of travel time *Viking 1* entered an orbit around Mars in June, 1976, and spent a month sending high-resolution images of the surface back to Earth. From these images a landing site was selected for the *Viking 1* lander. Using retrorockets, parachutes, and descent rockets, the *Viking 1* lander arrived on a dusty, rocky slope in the southern hemisphere on July 20, 1976. The *Viking 2* lander arrived 45 days later, but farther to the north. The *Viking* landers contained a mechanical soil-retrieving arm and a miniature computerized lab to analyze the soil for evidence of metabolism, respiration, and other life processes. Neither lander detected any evidence of life processes or any organic compounds that would indicate life now or in the past.

The *Viking* spacecraft continued sending images and weather data back to Earth until 1982. During their six-year life the orbiters sent about 52,000 images and mapped about 97 percent of the Martian surface. The landers sent an additional 4,500 images, recorded a major "Marsquake," and recorded data about regular dust storms that occur on Mars with seasonal changes.

Some answers about the geology and history of Mars were provided by Mars *Pathfinder*. On July 4, 1997, a *Pathfinder* lander and rover started sending images and data from the Ares Vallis area of Mars. The lander served as a base communications station, but also had cameras and measurement instruments of its own. The first vehicle to roam the surface of another planet, a skateboard-sized rover named *Sojourner*, was designed to move about and analyze the chemical makeup of the surface of Mars. It was programmed to move from rock to rock by instructions relayed to it through the lander, then send analysis information back to Earth—again through the lander. This provided data of the geochemistry and petrology of soils and rocks, which in turn provided data about the early environments and conditions that have existed on Mars. After the *Pathfinder* lander completed its primary 30-day mission and fell silent, it was renamed the Carl Sagan Memorial Station.

Important results of the *Pathfinder* mission were:

1. Scientists were given a true representation of the makeup and conditions of the Martian surface.
2. Scientists found they could accurately predict from orbit what a Mars landing site would be like.
3. On the surface, the *Sojourner* rover found unexpected Earth-like rocks containing high amounts of silica.
4. Temperatures on Mars were found to fluctuate greatly, sometimes as much as 20 degrees in a matter of seconds.
5. Overall, *Sojourner* returned 1.2 gigabits of information, conducted 114 movements on command, traveled 52 meters, and returned 384 images of the Martian surface. A camera from the landing site returned 9,669 images.

Pathfinder also set some records back on Earth, where NASA's Internet site had some 565,902,373 visitors. Nearly 47,000,000 computer users visited the *Pathfinder* site in one day, establishing a new Internet record.

Surveyor, launched November 7, 1996, was designed to provide data for the creation of a detailed map of the surface as well as data about the climate and weather patterns of Mars. The detailed map will make possible the creation of a topographic map of the entire planet, a tool scientists intend to use in choosing landing sites or research prospects for future missions. More unmanned spacecraft missions to Mars are planned, with soil samples returned to Earth by 2010.

Mars was named for a mythical god of war, so the two satellites that circle Mars were named Deimos and Phobos after the two companions of the Roman god. Both satellites are small, irregularly shaped, and highly cratered. Phobos is the larger of the two, about 22 km (about 14 mi) across the longest dimension, and Deimos is about 13 km (about 8 mi) across. Both satellites reflect light poorly and have a much lower density than Mars. They are assumed to be captured asteroids rather than naturally occurring moons.

Jupiter

Jupiter is the largest of all the planets, with a mass equivalent to some 318 Earths and, in fact, is more than twice as massive as all the other planets combined. This massive planet is

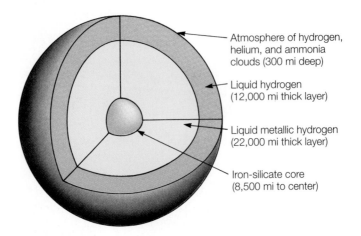

FIGURE 17.16

The interior structure of Jupiter.

located an average 778 million km (about 483 million mi), or 5 A.U., from the Sun in an orbit that takes about twelve years for one revolution. The internal heating from gravitational contraction was tremendous when this giant formed, and today it still radiates twice the energy that it receives from the Sun. The source of this heat is the slow gravitational compression of the planet, not from nuclear reactions as in the Sun. Jupiter would have to be about 80 times as massive to create the internal temperatures needed to start nuclear fusion reactions, or in other words, to become a star itself. Nonetheless, the giant Jupiter and its system of sixteen satellites seem almost like a smaller version of a planetary system within the solar system.

Jupiter has been observed, photographed, and probed by two *Pioneer* spacecraft in 1973 and 1974 and by two *Voyager* spacecraft in 1979. Information from these four spacecraft, together with Earth-based observations, make it possible to derive a model of the interior structure of Jupiter. Measurement of the motion of its satellites, for example, provided the needed data to calculate the mass, and thus the density, of the planet. Jupiter has an average density of 1.3 g/cm^3, which is about a quarter of the density of Earth. This low density indicates that Jupiter is mostly made of light elements, such as hydrogen and helium, but does contain a percentage of heavier rocky substances. The model of Jupiter's interior (figure 17.16) is derived from this and other information from spectral studies, studies of rotation rates, and measurements of heat flow. The model indicates a solid, rocky core with a radius of about 14,000 km (about 8,500 mi). This rocky core is thus more than twice the size of Earth. Above this core is an approximate 35,000 km (about 22,000 mi) thick layer of liquid hydrogen, compressed so tightly by millions of atmospheres of pressure that it is able to conduct electric currents. Liquid hydrogen with this property is called *metallic hydrogen* because it has the conductive ability of metals. Above the layer of metallic hydrogen is a 20,000 km (about 12,000 mi) thick layer of ordinary liquid hydrogen, which is under less pressure. The outer layer, or atmosphere, of Jupiter is

A

B

FIGURE 17.17

Photos of Jupiter taken by *Voyager 1*. (*A*) From a distance of about 36 million km (about 22 million mi). (*B*) A closer view, from the Great Red Spot to the South Pole, showing organized cloud patterns. In general, dark features are warmer, and light features are colder. The Great Red Spot soars about 25 km (about 15 mi) above the surrounding clouds and is the coldest place on the planet.

a 500 km or so (about 300 mi) zone with hydrogen, helium, ammonia gas, crystalline compounds, and a mixture of ice and water. The *Galileo* atmospheric probe found much less water than expected. It is the uppermost ammonia clouds, perhaps mixed with sulphur and organic compounds, that form the bright orange, white, and yellow bands around the planet. The banding is believed to be produced by atmospheric convection, in which bright, hot gases are forced to the top where they cool, darken in color, and sink back to the surface.

Jupiter's famous Great Red Spot is located near the equator. This permanent, deep, red oval feature was first observed by Robert Hooke in the 1600s and has generated much speculation over the years. The red oval, some 40,000 km (about 25,000 mi) long has been identified by infrared observations to be a high-pressure region, with higher and colder clouds, that has lasted for some three hundred years or longer. The energy source for such a huge, long-lasting feature is unknown (figure 17.17).

Jupiter has sixteen satellites, and the four brightest and largest can be seen from Earth with a good pair of binoculars. These four are called the *Galilean moons* because they were discovered by Galileo in 1610. The Galilean moons are named Io, Europa, Ganymede, and Callisto (figure 17.18). Chains of craters on Callisto and Ganymede are now explained as being mostly due to split comets like comet Shoemaker-Levy, which had broken into a chain of twenty-two fragments and impacted Jupiter in July, 1994.

Observations by the *Pioneer* and *Voyager* spacecrafts revealed some fascinating and intriguing information about the moons of Jupiter. Io, for example, was discovered to have active volcanoes that eject enormous plumes of molten sulfur and sulfur dioxide. Europa is covered with smooth water ice, which has a network of long, straight, dark cracks. Ganymede has valleys, ridges, folded mountains, and other evidence of an active geologic history. Callisto, the most distant of the Galilean moons, was found to be the most heavily cratered object in the solar system. In addition to all the new information about Jupiter's satellites, *Voyager* discovered three new satellites and a system of rings between the satellites and the planet. These rings are much smaller than the well-known rings of Saturn.

Comet Shoemaker-Levy lined up in a "string of pearls," (figure 17.19A) and then produced a once-in-a-lifetime spectacle as it proceeded to leave its imprint on Jupiter as well as people of Earth watching from the sidelines. The string of twenty-two comet fragments fell onto Jupiter during July, 1994, creating a show eagerly photographed by telescopes around the world (figure 17.19B). The fragments impacted the upper atmosphere of Jupiter, producing visible, energetic fireballs. The aftereffects of these fireballs were visible for about a year. There are chains of craters on two of the Galilean moons that may have been formed by similar events.

Galileo made the most recent mission to Jupiter, a spacecraft consisting of an orbiter and an atmospheric probe. Launched in October, 1989, *Galileo* observed two asteroids and photographed the impact of comet Shoemaker-Levy with

FIGURE 17.18

The four Galilean moons pictured by *Voyager 1*. Clockwise from upper left, Io, Europa, Ganymede, and Callisto. Io and Europa are about the size of Earth's Moon; Ganymede and Callisto are larger than Mercury.

Jupiter on its way to arriving in orbit in July, 1995. The high-gain antenna would not deploy, so only 70 percent of the planned science objectives could be achieved through the lower speed low-gain antenna. In orbit, *Galileo* released the atmospheric probe, then began a flyby study of the Galilean moons. Comprehensive analysis of the data will take years.

Saturn

Saturn is slightly smaller and substantially less massive than Jupiter, and it has similar surface features, but it is readily identified by its unique, beautiful system of rings. Saturn is about 9.5 A.U. from the Sun, but its system of rings is readily identified with a good pair of binoculars. Saturn also has the lowest average density of any of the planets, 0.7 g/cm^3, which is less than the density of water.

The surface of Saturn, like Jupiter's surface, has bright and dark bands that circle the planet parallel to the equator. Saturn also has a smaller version of Jupiter's Great Red Spot, but in general the bands and spot are not as highly contrasted or brightly colored as they are on Jupiter.

Saturn's rings were found by *Voyager 1* and *Voyager 2* to consist of thousands of narrow rings made up of particles, some rings with particles large enough to be measured in meters and some rings with particles that are dust-sized. Rings with small particles were observed to have waves, and the center ring had radial streaks or "spokes." The waves are thought to be a result of collisions between particles that are caused by the gravity of a nearby moon, or perhaps they are electrically charged particles interacting with Saturn's magnetic field. The changing pattern of "spokes," which rotate with the ring, are thought to be related to Saturn's magnetic field and electrical

A

B

FIGURE 17.19

(A) This image, made by the Hubble Space Telescope, clearly shows the large impact site made by fragment G of former comet Shoemaker-Levy 9 when it collided with Jupiter. (B) This is a picture of comet Shoemaker-Levy 9 after it broke into twenty-two pieces, lined up in this row, then proceeded to plummet into Jupiter during July, 1994. The picture was made by the Hubble Space Telescope.

discharges that are up to 100,000 times more energetic than lightning on Earth. The wealth of information collected on the rings of Saturn by the *Voyager* spacecraft is still being analyzed and interpreted (figure 17.20).

Before the *Voyager* spacecraft mission, Saturn was known to have ten satellites: Janus, Mimas, Enceladus, Tethys, Dione, Rhea, Titan, Hyperion, Iapetus, and Phoebe. The *Voyager* observations added eight more smaller satellites to the list. All the ten known moons were observed to be icy and heavily cratered except Titan, which is covered with clouds and impossible to observe. Titan is the only moon in the solar system that has a substantial atmosphere. Titan is larger than the planet Mercury and is covered with a 240 km (about 150 mi) deep layer of reddish clouds. Titan's atmosphere is mostly nitrogen, with some hydrocarbons. This could be similar to Earth's atmosphere before life began adding oxygen to the atmosphere. The pressure on the surface is 1.5 atmospheres, but with a surface temperature of about −180°C (about −290°F) it is doubtful that life has developed on Titan.

The *Cassini* spacecraft is the next U.S. mission to Saturn. It was launched on October 15, 1997, and will take seven years to reach Saturn. The *Cassini* spacecraft is similar to the *Galileo* craft, with an atmospheric probe and an orbiter, and will orbit Saturn and release a probe to land on Titan. The orbiter will study Saturn, its rings, and its satellites for an extended period of time. The atmospheric probe is scheduled to visit Titan because there is a possibility that lakes of liquid hydrocarbons exist on its surface that may have resulted from photochemical processes in its upper atmosphere. The *Cassini* orbiter will also use its radar to peek through Titan's cloud cover and determine if liquid is present on the surface.

Uranus, Neptune, and Pluto

Uranus and Neptune are two more giant planets that are far, far away from Earth. Uranus revolves around the Sun at an average distance of over 19 A.U., taking about 84 years to circle the Sun once. Neptune is an average 30 A.U. from the Sun and takes about 165 years for one complete orbit. Thus, Uranus is about twice as far away from the Sun as Saturn, and Neptune is three times as far away. To give you an idea of these tremendous distances, consider the time required for radio signals to travel from the *Voyager 2* spacecraft to Earth. *Voyager* photographed and made observations of Uranus in 1986, sending this information back to Earth in the form of radio signals. If you consider Uranus to be 2,720 million km from Earth, a radio signal traveling through space at the speed of light would require 2.72×10^{12} m divided by 3.00×10^8 m/s, or 9,067 s, which is more than 2.5 hours! It would be most difficult to carry on a conversation by radio with someone such a distance away. Even farther away, a radio signal from a transmitter near Neptune would require over 4 hours to reach

FIGURE 17.20

A part of Saturn's system of rings, pictured by *Voyager 2* from a distance of about 3 million km (about 2 million mi). More than sixty bright and dark ringlets are seen here; different colors indicate different surface compositions.

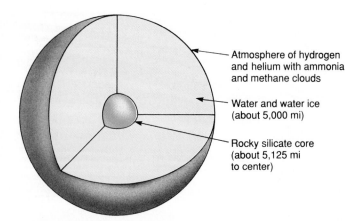

FIGURE 17.22

The interior structure of Uranus and Neptune.

FIGURE 17.21

This is a photo image of Neptune taken by *Voyager*. Neptune has a turbulent atmosphere over a very cold surface of frozen hydrogen and helium.

Earth, which means 8 hours would be required for two people just to say "Hello" to each other!

Uranus and Neptune are more similar to each other than Saturn is to Jupiter (figure 17.21). Both are the smallest of the giant planets, with a diameter of about 50,000 km (about 30,000 mi) and about a third the mass of Jupiter. Both planets are thought to have similar interior structures (figure 17.22), which consist of water and water ice surrounding a rocky core with an atmosphere of hydrogen and helium. Because of their great distances from the Sun, both have very low average surface temperatures, −210°C (about −350°F) for Uranus and −235°C (about −391°F) for Neptune.

In addition to retrograde rotation, Uranus has an odd orientation of its axis to the plane of its orbit. Most planets have their axes of rotation tilted less than 30° from a vertical line to the plane of the orbit (Earth's tilt is 23.5°). Uranus's tilt, however, is 82°, meaning that it is practically on its side. Moving in its orbit around the Sun on its side, one pole receives direct sunlight for twenty-one years, then the equator receives direct sunlight for the next twenty-one years, then the other pole receives the direct sunlight for the next twenty-one years, and so on. This would produce some interesting climatic patterns on Uranus, that is, if the weak sunlight is able to penetrate the atmosphere.

Voyager 2 discovered ten new satellites, in addition to the five that were previously known, and new rings in the system of rings around Uranus for a total of ten narrow rings and a number of dusty bands. Interestingly, the rock particles making up the rings and the newly discovered moons are dark colored,

reflecting less than 5 percent of the sunlight falling on them. By comparison, Earth's Moon has a reflectivity of about 10 percent. One explanation of the dark colors is that the surfaces were originally covered with frozen methane, which has been decomposed by radiation, leaving a layer of black carbon.

In 1989, *Voyager 2* visited Neptune and discovered six new satellites, making a total of eight, and a system of rings. *Voyager* found that Neptune has a churning, turbulent atmosphere and that the largest moon, Triton, is about 2,705 km (about 1,680 mi) in diameter. Triton orbits in the opposite direction from the rotation of its planet, unlike other large moons in the solar system.

The planet Neptune was named for the Roman god of the sea, known to the ancient Greeks as Poseidon. It seems logical, then, that the six Neptunian moons discovered by the *Voyager 2* spacecraft in 1989 would be named after other mythological characters associated with Poseidon. The six Neptunian moons were named after two water nymphs named *Naiad* and *Galatea*; after two lovers of Poseidon named *Thalassa* and *Larissa*; and after two children of Poseidon, a son named *Proteus* and a daughter named *Despina*.

Pluto, the outermost planet, is so small that seven of the moons of the solar system are larger—the Earth's Moon, Io, Europa, Ganymede, Callisto, Titan, and Triton. It is so far away at its approximate 40 A.U. average distance from the Sun that little is well known about the tiny planet.

Based on a calculated density of about 2 g/cm³, Pluto might be composed of about 70 percent rock and 30 percent water ice. The tenuous, thin atmosphere is probably mostly nitrogen, with some methane and carbon dioxide. With an estimated surface temperature of about −233°C (−287°F), you can understand why the surface is believed to be covered with frozen nitrogen, with some ices of methane and carbon dioxide. The plane of its orbit is tilted about 17° from the ecliptic, the plane of Earth's orbit, while all the other planets have a much smaller tilt. In addition, the orbit of Pluto is the most eccentric, that is, the least circular, of all the planetary orbits (figure 17.23). Pluto's orbit is so eccentric that it is sometimes nearer the Sun than Neptune and sometimes farther out than Neptune. The two planets do

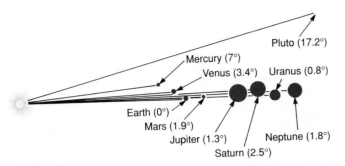

FIGURE 17.23

The orbital planes of the planets. Notice the exceptional tilt of the orbit of Pluto.

TABLE 17.5	Completed spacecraft missions to study comets		
Date	**Name**	**Owner**	**Remark**
Sep 11, 1985	*ISSE 3* (or *ICE*)	U.S.	Studies of electric and magnetic fields around Giacobini-Zinner comet from 7,860 km (4,880 mi)
Mar 6, 1986	*Vega 1*	U.S.S.R.	Photos and studies of nucleus of Halley's comet from 8,892 km (5,525 mi)
Mar 8, 1986	*Suisei*	Japan	Studied hydrogen halo of Halley's comet from 151,000 km (93,800 mi)
Mar 9, 1986	*Vega 2*	U.S.S.R.	Photos and studies of nucleus of Halley's comet from 8,034 km (4,992 mi)
Mar 11, 1986	*Sakigake*	Japan	Studied solar wind in front of Halley's comet 7.1 million km (4.4 million mi)
Mar 28, 1986	*Giotto*	ESA	Photos and studies of Halley's comet from 541 km (336 mi)
Mar 28, 1986	*ISSE 3* (or *ICE*)	U.S.	Studies of electric and magnetic fields around Halley's comet from 32 million km (20 million mi)

not come close enough together to collide, but because their orbits cross, it might be an escaped moon of Neptune rather than a true planet. No other planets have such strange, crossed orbits, and Pluto remains a puzzle.

Pluto's moon, Charon, was discovered in 1978. It is about half the size of the planet, with the largest ratio of the size of a moon to its planet in the solar system. Since it has the same period of revolution as the period of rotation for Pluto, Charon would appear stationary from the surface of the planet.

SMALL BODIES OF THE SOLAR SYSTEM

In the earlier description of how the planets formed from the accretion disk, they were described as forming from protoplanets, icy collections of slushy materials, gases, and rocky particles. Not all of the materials in the accretion disk ended up in the planets, however; some ended up in smaller bodies of the solar system. Bodies of some of the original icy collections that were too small to form planets themselves and were not incorporated into the planets or the Sun still exist today, sometimes orbiting close enough to the Sun to become illuminated. These illuminated masses of icy materials may form a tail and are called *comets*. Other leftover materials are more solid, with a rocky or iron composition. These *asteroids* or *minor planets* are thought to be the stuff of a planet that never quite made it to the final stage before being pulled apart. Rare collisions between asteroids break off small pieces, some of which fall to Earth as *meteorites*. Comets, asteroids, and meteorites are the leftovers from the formation of the Sun and planets. Presently, the total mass of all these leftovers in and around the solar system may account for a significant fraction of the mass of the solar system, perhaps as much as two-thirds of the total mass. It must have been much greater in the past, however, as evidenced by the intense bombardment that took place on the Moon and other planets up to some four billion years ago.

Comets

A **comet** is known to be a relatively small, solid body of frozen water, carbon dioxide, ammonia, and methane, along with dusty and rocky bits of materials mixed in. Until the 1950s most astronomers believed that comet bodies were mixtures of sand

and gravel. Fred Whipple proposed what became known as the *dirty-snowball cometary model*, which was recently verified when spacecraft probes observed Halley's comet in 1986 (table 17.5).

Based on calculation of their observed paths, comets are estimated to originate some 30 A.U. to a light-year or more from the Sun. Here, according to other calculations and estimates, is a region of space containing billions and billions of objects. The objects are porous aggregates of water ice, frozen methane, frozen ammonia, dry ice, and the dust of mineral grains. There is a spherical "cloud" of the objects beyond the orbit of Pluto from about 30,000 A.U. out to a light-year or more from the Sun, called the **Oort cloud.** The icy aggregates of the Oort cloud are understood to be the source of long-period comets.

There is also a disk-shaped region of small icy bodies, which ranges from about 30 to 100 A.U. from the Sun, called the **Kuiper Belt** (figure 17.24). The small icy bodies in the Kuiper Belt are understood to be the source of short-period comets. There are thousands of Kuiper Belt objects that are larger than 100 km in diameter, and six are known to be orbiting between Jupiter and Neptune. Called *Centaurs,* these objects are believed to have escaped the Kuiper Belt. Centaurs might be small, icy bodies similar to Pluto. Indeed, some speculate that Pluto and Charon are large Kuiper Belt objects.

The current theory of the origin of comets was developed by the Dutch astronomer Jan Oort in 1950. According to the theory, the huge cloud and belt of icy, dusty aggregates are leftovers from the formation of the solar system and have been slowly

FIGURE 17.24

Comets originate from the *Oort cloud,* a large spherical cloud of icy aggregates, or from the *Kuiper Belt,* a disk-shaped region of icy bodies. The Kuiper Belt is illustrated here, extending from 30 to 100 A.U. from the Sun. The Oort cloud is a spherical "cloud" that ranges from about 30,000 A.U. to a light-year or more from the Sun.

circling the solar system ever since it formed. Something, perhaps a gravitational nudge from a passing star, moves one of the icy bodies enough that it is pulled toward the Sun in what will become an extremely elongated elliptical orbit. The icy, dusty body forms the only substantial part of a comet, and the body is called the comet *nucleus.*

Observations by the *Vega* and *Giotto* spacecrafts found the nucleus of Halley's comet to be an elongated mass of about 8 by 11 km (about 5 by 7 mi) with an overall density less than one-fourth that of solid water ice. As the comet nucleus moves toward the Sun, it warms from the increasing intense solar radiation. Somewhere between Jupiter and Mars the ice and frozen gases begin to vaporize, releasing both grains of dust and evaporated ices. These materials form a large hazy head around the comet called a *coma.* The coma grows larger with increased vaporization, perhaps several hundred or thousands of kilometers across. The coma reflects sunlight as well as producing its own light, making it visible from Earth. The coma generally appears when a comet is within about 3 astronomical units of the Sun. It reaches its maximum diameter about 1.5 astronomical units from the Sun. The nucleus and coma together are called the *head* of the comet. In addition, a large cloud of invisible hydrogen gas surrounds the head and this hydrogen *halo* may be hundreds of thousands of kilometers across.

As the comet nears the Sun, the solar wind and solar radiation ionize gases and push particles from the coma, pushing both into the familiar visible *tail* of the comet. Comets may have two types of tails, (1) ionized gases, and (2) dust. The dust is pushed from the coma by the pressure from sunlight. It is visible because of reflected sunlight. The ionized gases are pushed into the tail by magnetic fields carried by the solar wind. The spacecraft intercept of comet Giacobini-Zinner in September, 1985, measured the complex magnetic field-plasma interactions that set the form of the tail. Electron densities in the center of the

FIGURE 17.25

As a comet nears the Sun it grows brighter, with the tail always pointing away from the Sun.

tail were found to be greater than 10^9 electrons per cubic meter. The ionized gases of the tail are fluorescent, emitting visible light because they are excited by ultraviolet radiation from the Sun. The tail generally points away from the Sun, so it follows the comet as it approaches the Sun, but leads the comet as it moves away from the Sun (figure 17.25).

Comets are not very massive or solid, and the porous, snow-like mass has a composition more similar to the giant planets than to the terrestrial planets in comparison. Each time a comet passes near the Sun, it loses some of its mass through evaporation of gases and loss of dust to the solar wind. After passing the Sun the surface forms a thin, fragile crust covered with carbon and other dust particles. Each pass by the Sun means a loss of matter, and the coma and tail are dimmer with each succeeding pass. About 20 percent of the approximately six hundred comets that are known have orbits that return them to the Sun within a two-hundred-year period, some of which return as often as

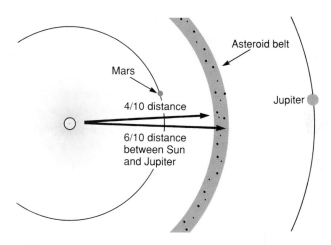

Mars

Asteroid belt

Jupiter

4/10 distance

6/10 distance
between Sun
and Jupiter

FIGURE 17.26

Most of the asteroids in the asteroid belt are about halfway between the Sun and Jupiter.

TABLE 17.6	Some annual meteor showers	
Name	**Date of Maximum**	**Hour Rate**
Quadrantid	January 3	30
Aqaurid	May 4	5
Perseid	August 12	40
Orionid	October 22	15
Taurids	November 1, 16	5
Leonid	November 17	5
Geminid	December 12	55

every five or ten years. The other 80 percent have long elliptical orbits that return them at intervals exceeding two hundred years. The famous Halley's comet has a smaller elliptical orbit and returns about every seventy-six years. Halley's comet, like all other comets, may eventually break up into a trail of gas and dust particles that orbit the Sun.

Hale-Bopp was probably the most-observed comet in history. It was bright—very bright—with a visual magnitude of −1 and appeared in the early evening sky for two months through May 1997. It was visible during this time in the Northern Hemisphere to the naked eye alone, even in an urban setting. Calculations indicate an orbital period of several thousand years, extending out some 300 A.U. from the Sun. Hale-Bopp visited the inner solar system about 4,200 years ago and will return again in about 2,300 years. The period has grown shorter because of the gravitational influence of the major planets.

Asteroids

The asteroid belt between Mars and Jupiter was introduced earlier (figure 17.26). This belt contains thousands of *asteroids* that range in size from 1 km or less up to the largest asteroid, named Ceres, which has a diameter of about 1,000 km (over 600 mi). The asteroids are thinly distributed in the belt, 1 million km or so apart (about 600,000 mi), but there is evidence of collisions occurring in the past. Most asteroids larger than 50 km (about 30 mi) have been studied by analyzing the sunlight reflected from their surfaces. These spectra provide information about the composition of the asteroids. Asteroids on the inside of the belt are made of stony materials, and those on the outside of the belt are dark with carbon minerals. Still other asteroids are metallic, containing iron and nickel. These spectral composition studies, analyses of the orbits of asteroids, and studies of meteorites that have fallen to Earth all indicate that the asteroids are not the remains of a planet or planets that were broken up. The asteroids are now believed to have formed some 4.6 billion years ago from the original solar nebula. During their formation, or shortly thereafter, their interi-

ors were partly melted, perhaps from the heat of short-lived radioactive decay reactions. Their location close to Jupiter, with its gigantic gravitational field, prevented the slow gravitational clumping-together process that would have formed a planet.

Jupiter's gigantic gravitational field also captured some of the asteroids, pulling them into its orbit. Today there are two groups of asteroids, called the *Trojan asteroids,* which lead and follow Jupiter in its orbit. They lead and follow at a distance where the gravitational forces of Jupiter and the Sun balance to keep them in the orbit. A third group of asteroids, called the *Apollo asteroids,* has orbits that cross the orbit of Earth. It is possible that one of the Apollo asteroids could collide with Earth. One theory about what happened to the dinosaurs is based on evidence that such a collision indeed did occur some sixty-five million years ago. This theory will be discussed in a later chapter. The chemical and physical properties of the two satellites of Mars, Phobos and Deimos, are more similar to the asteroids than to Mars. It is probable that the Martian satellites are captured asteroids.

Meteors and Meteorites

Comets leave trails of dust and rock particles after encountering the heat of the Sun, and collisions between asteroids in the past have ejected fragments of rock particles into space. In space, the remnants of comets and asteroids are called **meteoroids.** When a meteoroid encounters Earth moving through space, it accelerates toward the surface with a speed that depends on its direction of travel and the relative direction that Earth is moving. It soon begins to heat from air friction in the upper atmosphere, melting into a visible trail of light and smoke. The streak of light and smoke in the sky is called a **meteor.** The "falling star" or "shooting star" is a meteor. Most meteors burn up or evaporate completely within seconds after reaching an altitude of about 100 km (about 60 mi) because they are nothing more than a speck of dust. A **meteor shower** occurs when Earth passes through a stream of particles left by a comet in its orbit. Earth might meet the stream of particles concentrated in such an orbit on a regular basis as it travels around the Sun, resulting in predictable meteor showers (table 17.6). In the third week of October, for example, Earth crosses the orbital path of Halley's comet, resulting in a shower of some ten to fifteen meteors per hour. Meteor showers are named for the constellation in which

A CLOSER LOOK | Strange Meteorites

Meteorites are a valuable means of learning about the solar system since they provide direct evidence about objects away from Earth in space. Over two thousand meteorites have been collected, classified, and analyzed for this purpose. Most appear to have remained unchanged since they were formed in the asteroid belt between Jupiter and Mars, along with the rest of the solar system, some five billion years ago. Yet a few meteorites are very unusual or strange, not fitting in with the rest. The mystery of two types of these strange meteorites is the topic of this reading.

The first example of a meteorite mystery is the case of the carbonaceous chondrites, which are soft, black stony meteorites with a high carbon content. Fewer than twenty-five of these soft and sooty black fragments have been recovered since the first one was identified in France in 1806. The carbonaceous chondrite is similar to the ordinary chondrite except that it contains carbon compounds, including amino acids. Amino acids are organic compounds basic to life. The mystery of these meteorites is in the meaning of the presence of amino acids. Are they the remains of some former life? Or are they precursors to life, organic compounds produced naturally when the solar system formed? If this is true, what is the relationship between the amino acids and life on

Earth, and what are the implications for elsewhere? The answers to these questions will probably not be available until more of these strange meteorites are available for study.

The second example of a meteorite mystery is the case of the shergottites, which are meteorites of volcanic origin with strange properties. The shergottites are named after a town in India where the first one identified was observed to fall in 1865. A second shergottite fell in Africa in 1962, and a third was recently found on an ice sheet in Antarctica. The shergottites are known to be meteorites because the first two were actually observed to fall and all three have the typical glassy crust that forms from partial melting during the fall.

The strangeness of the shergottites comes from their age, chemical composition, and physical structure as compared to all the other meteorites. Radioactive dating shows that the shergottites are the youngest meteorites known, cooling from a molten state somewhere between 650 million and 1 billion years ago. All of the other meteorites were originally formed some 4.6 billion years ago in the asteroid belt. In addition, the shergottites do not have the same composition as the asteroids according to spectral analysis. The conclusion so far is that they are not from the asteroid belt. Their chemical composition is more similar to that of a planet such as Earth or Mars than it is to an asteroid.

One interesting feature of the shergottites is the special kind of glass that comprises about half of their masses. This special glass forms only when certain minerals are subjected to tremendous pressures by impact shock. Evidently these meteorites were subjected to a tremendous shock sometime after cooling from the molten state.

The last piece of the puzzle to the mystery comes from the quantitative analysis of the soil on Mars that was performed by the *Viking* landers. The analysis showed that the chemical composition of the shergottites is more similar to Mars than it is to Earth. Large volcanoes were observed on Mars by the *Viking* spacecraft, and calculations indicate that Mars was volcanically active when the shergottites were formed.

Meteorites are pieces of the solar system that have fallen to Earth. Most come from asteroids, but 15 have been identified as originating on the Moon and 16 from Mars. One of the Martian meteorites, known as ALH84001, has indications of early life on Mars. This particular meteorite crystallized some 4.5 billion years ago, so it is much older than the other known Martian meteorites. The Martian meteorites, and other meteorites as well, are being carefully studied for information about how they formed, their history, and evidence of alien life.

---

they appear to originate. The October meteor shower resulting from an encounter with the orbit of Halley's comet, for example, is called the Orionid shower because it appears to come from the constellation Orion.

Did you know that atom-bomb-sized meteoroid explosions often occur high in Earth's atmosphere? Most smaller meteors melt into the familiar trail of light and smoke. Larger meteors may fragment upon entering the atmosphere, and the smaller fragments will melt into multiple light trails. Still larger meteors may actually explode at altitudes of about 32 km (about 20 mi) or so. Military satellites that watch Earth for signs of rockets blasting off or nuclear explosions record an average of eight meteor explosions a year. These are big explosions, with an energy equivalent estimated to be similar to a small nuclear bomb. Actual explo-

sions, however, may be 10 times larger than the estimation. Based on statistical data, scientists have estimated that every 10 million years Earth should be hit by a very, very large meteor. The catastrophic explosion and aftermath would devastate life over much of the planet, much like the theoretical dinosaur-killing impact of 65 million years ago.

If a meteoroid survives its fiery trip through the atmosphere to strike the surface of Earth, it is called a **meteorite.** Most meteors are from fragments of comets, but most meteorites generally come from particles that resulted from collisions between asteroids that occurred long ago. Meteorites are classified into three basic groups according to their composition: (1) *iron meteorites,* (2) *stony meteorites,* and (3) *stony-iron meteorites* (figure 17.27). The most common meteorites are

A

B

FIGURE 17.27

(*A*) A stony meteorite. The smooth, black surface was melted by friction with the atmosphere. (*B*) An iron meteorite that has been cut, polished, and etched with acid. The pattern indicates that the original material cooled from a molten material over millions of years.

stony, composed of the same minerals that make up rocks on Earth. The stony meteorites are further subdivided into two groups according to their structure, the *chondrites* and the *achondrites.* Chondrites have a structure of small spherical lumps of silicate minerals or glass, called *chondrules,* held together by a fine-grained cement. The achondrites do not have the chondrules, as their name implies, but have a homogeneous texture more like volcanic rocks such as basalt that cooled from molten rock.

The iron meteorites are about half as abundant as the stony meteorites. They consist of variable amounts of iron and nickel, with traces of other elements. In general, there is proportionally much more nickel than is found in the rocks of Earth. When cut, polished, and etched, beautiful crystal patterns are observed on the surface of the iron meteorite. The patterns mean that the iron was originally molten, then cooled very slowly over millions of years as the crystal patterns formed.

A meteorite is not, as is commonly believed, a ball of fire that burns up the landscape where it lands. The iron or rock has been in the deep freeze of space for some time, and it travels rapidly through the earth's atmosphere. The outer layers become hot enough to melt, but there is insufficient time for this heat to be conducted to the inside. Thus, a newly fallen iron meteorite will be hot since metals are good heat conductors, but it will not be hot enough to start a fire. A stone meteorite is a poor conductor of heat so it will be merely warm.

According to the radioactive "clock," which is read by determining radioactive decay ratios, meteorites formed some 4.6 billion years ago, which fits with other evidence about the age of the solar system. Since cosmic rays in outer space cause isotopic changes in materials, physicists are able to measure how long a meteoroid has existed as a small body; that is, when it was broken off from a larger asteroid. This data indicates that the meteoroids were broken off in a few major collisions as opposed to a series of many small collisions.

SUMMARY

The ancient Greeks explained observations of the sky with a *geocentric model* of a motionless Earth surrounded by turning geometrical spheres with attached celestial bodies. To explain the observed *retrograde motion* of the planets, later Greek astronomers added small circular orbits, or *epicycles,* to the model, which is called the *Ptolemaic system.* A *heliocentric,* or Sun-centered, model of the solar system was proposed by Nicolas Copernicus. The *Copernican system* considered the planets to move around the Sun in circular orbits. Tycho Brahe made the first precise, continuous record of planetary motions. His assistant, Johannes Kepler, analyzed the data to find if planets followed circular paths or if they followed the paths of epicycles. Today, his findings are called *Kepler's laws of planetary motion. Kepler's first law* describes the orbit of a planet as having the shape of an *ellipse* with the Sun at one focus. *Kepler's second law* describes the motion of the planet along this ellipse, that a line from the planet to the Sun will move over equal areas of the orbital plane during equal periods of time. The point of the orbit closest to the Sun is called the *perihelion,* and the point of the orbit farthest from the Sun is called the *aphelion. Kepler's third law* states a relationship between the time required for a planet to complete one orbit and its distance from the Sun. In general, the farther away from the Sun, the longer the period of time to complete an orbit.

The *protoplanet nebular model* is the most widely accepted theory of the origin of the solar system, and this theory can be considered as a series of events, or stages. *Stage A* is the creation of all the elements heavier than hydrogen in previously existing stars. *Stage B* is the formation of a nebula from the raw materials created in stage A. The nebula contracts from gravitational attraction, forming the *protosun* in the center with a fat, bulging *accretion disk* around it. The Sun will form from the protosun, and the planets will form in the accretion disk. *Stage C* begins as the protosun becomes established as a star. The icy remains of the original nebula are the birthplace of *comets*. *Asteroids* are other remains that did undergo some melting.

The plane of Earth's orbit is called the *ecliptic*. The planets can be classified into two major groups: (1) the *terrestrial planets* of Mercury, Venus, Mars, and Earth and (2) the *giant planets* of Jupiter, Saturn, Uranus, and Neptune.

The present-day atmospheres of the terrestrial planets are believed to be secondary atmospheres derived from rocks by a *degassing* process, the expulsion of carbon dioxide, other gases, and water vapor from molten rock. The amount released is related to how much melting occurred on a particular planet or body. The size of the planet or body is also related to its ability to hold on to a degassed atmosphere. The present-day atmospheres of Mars, Venus, and Earth all evolved from the same kind of degassed carbon dioxide and water vapor atmospheres. The oxygen in Earth's atmosphere is the result of photosynthetic organisms releasing oxygen.

Comets are porous aggregates of water ice, frozen methane, frozen ammonia, dry ice, and dust. The solar system is surrounded by the *Kuiper Belt* and the *Oort cloud* of these objects. Something nudges one of the icy bodies and it falls into a long elliptical orbit around the Sun. As it approaches the Sun increased radiation evaporates ices and pushes ions and dust into a long visible tail. *Asteroids* are rocky or metallic bodies that are mostly located in a belt between Mars and Jupiter. The remnants of comets, fragments of asteroids, and dust are called *meteoroids*. A meteoroid that falls through Earth's atmosphere and melts to a visible trail of light and smoke is called a *meteor*. A meteoroid that survives the trip through the atmosphere to strike the surface of Earth is called a *meteorite*. Most meteors are fragments and pieces of dust from comets. Most meteorites are fragments that resulted from collisions between asteroids.

KEY TERMS

asteroids (p. **411**)

comet (p. **427**)

Copernican system (p. **408**)

epicycle (p. **406**)

giant planets (p.**413**)

Kepler's laws of planetary
motion (p. **408**)

Kuiper Belt (p. **427**)

meteor (p. **429**)

meteorite (p. **430**)

meteoroids (p. **429**)

meteor shower (p. **429**)

Oort cloud (p. **427**)

protoplanet nebular model
(p. **410**)

Ptolemaic system (p. **406**)

terrestrial planets (p. **413**)

APPLYING THE CONCEPTS

1. A model of a motionless Earth with the Sun, the Moon, and planets moving around it will explain all observations except
   a. stars maintaining the same relative positions.
   b. the Sun, the Moon, and planets rise in the east and set in the west.
   c. the motion of the planets over a year.
   d. None of the above are correct.

2. The Ptolemaic system explained retrograde motion by assuming that planets
   a. actually had two different orbital motions.
   b. moved around an epicycle.
   c. moved in a perfect circle while turning on a moving sphere.
   d. All of the above are correct.

3. Based on careful measurements, Kepler found that planets move in the path of a (an)
   a. perfect circle.
   b. epicycle.
   c. ellipse.
   d. oval with a loop at both ends.

4. Earth is closest to the Sun at what part of its orbit?
   a. perihelion
   b. aphelion
   c. This varies from year to year.
   d. Earth is always the same distance from the Sun.

5. Earth moves most slowly in its orbit during the month of
   a. January.
   b. March.
   c. July.
   d. September.

6. Earth is closest to the Sun during the month of
   a. January.
   b. March.
   c. July.
   d. September.

7. Using Kepler's laws, the distance from the Sun to a planet can be calculated from
   a. the period of rotation of the planet.
   b. the period of revolution of the planet.
   c. the square of its speed when at perihelion.
   d. the cube root of its period of rotation.

8. Planets that are progressively farther from the Sun have years that are progressively
   a. longer periods of time.
   b. shorter periods of time.
   c. longer or shorter, depending on the mass of the planet.
   d. longer or shorter, depending on the rate of revolution.

9. Earth, other planets, and all the members of the solar system
   a. have always existed.
   b. formed thousands of years ago from elements that have always existed.
   c. formed millions of years ago, when the elements and each body were created at the same time.
   d. formed billions of years ago from elements that were created in many previously existing stars.

10. The atmosphere that is found on Earth today
   a. formed when Earth formed.
   b. is the secondary atmosphere that formed from degassing.
   c. is a secondary atmosphere that has been modified over time.
   d. is now as it always has been in the past.

11. The belt of asteroids between Mars and Jupiter is probably
    a. the remains of a planet that exploded.
    b. clumps of matter that condensed from the accretion disk, but never got together as a planet.
    c. the remains of two planets that collided.
    d. the remains of a planet that collided with an asteroid or comet.

12. Comparing the amount of interior heating and melting that has taken place over time, the smaller planets underwent
    a. more heating and melting because they are smaller.
    b. the same heating and melting as larger planets.
    c. less heating and melting for a shorter period of time.
    d. There is no evidence to answer this question.

13. Which of the following planets would be mostly composed of hydrogen, helium, and methane and have a density of less than 2 g/cm^3?
    a. Uranus
    b. Mercury
    c. Mars
    d. Venus

14. Which of the following planets probably still has its original atmosphere?
    a. Mercury
    b. Venus
    c. Mars
    d. Jupiter

15. Of the following planets, which has odd orbital properties compared to the other planets?
    a. Jupiter
    b. Saturn
    c. Neptune
    d. Pluto

16. Venus appears the brightest when it is in the
    a. full phase.
    b. half phase.
    c. quarter phase.
    d. crescent phase.

17. Which of the following planets has a rotation period (day) longer than its revolution period (year)?
    a. Venus
    b. Jupiter
    c. Mars
    d. None of the above are correct.

18. The largest planet is
    a. Saturn.
    b. Jupiter.
    c. Uranus.
    d. Neptune.

19. The small body with a composition and structure closest to the materials that condensed from the accretion disk is a (an)
    a. asteroid.
    b. meteorite.
    c. comet.
    d. None of the above are correct.

20. A small body from space that falls on the surface of the earth is a
    a. meteoroid.
    b. meteor.
    c. meteor shower.
    d. meteorite.

Answers

1. c **2.** d **3.** c **4.** a **5.** c **6.** a **7.** b **8.** a **9.** d **10.** c **11.** b **12.** c **13.** a **14.** d **15.** d **16.** d **17.** a **18.** b **19.** c **20.** d

QUESTIONS FOR THOUGHT

1. Describe the one observation that is difficult to explain with a simple geocentric model of the solar system. Explain how the Ptolemaic system took care of this difficulty.

2. Describe the Copernican system and how it accounted for the motions of the planets. Briefly discuss at least three reasons why the Copernican system was not immediately accepted over the Ptolemaic system.

3. Discuss the contributions each of the following made to the eventual acceptance of a heliocentric model of the solar system: (a) Nicolas Copernicus, (b) Tycho Brahe, (c) Johannes Kepler, (d) Galileo, and (e) Isaac Newton.

4. Describe Kepler's three laws of planetary motion, using diagrams as needed.

5. What are the perihelion and the aphelion? Compare the orbital velocity of a planet at perihelion and at aphelion.

6. Evaluate the comparative difficulties in creating, testing, and verifying models of (a) the structure of the solar system, (b) the life of a star, and (c) the origin of the solar system.

7. Describe the protoplanet nebular model of the origin of the solar system. Which part or parts of this model seem least credible to you? Explain. What information could you look for today that would cause you to accept or modify this least credible part of the model?

8. What are the basic differences between the terrestrial planets and the giant planets? Describe how the protoplanet nebular model accounts for these differences.

9. Identify at least three properties of the terrestrial planets that theoretically determined what kind of atmosphere is present on the terrestrial planets today. Explain the role of each of these planetary properties in determining the atmosphere of a planet.

10. Compare the atmospheres and surface conditions on the two hot planets of Mercury and Venus. Provide reasons why these differences exist.

11. Explain (a) why Venus and Earth are believed to have had similar atmospheres at two different times during their history and (b) why the atmospheres are so different today.

12. Describe the surface and atmospheric conditions on Mars.

13. What evidence exists that Mars at one time had abundant liquid water? If Mars did have liquid water at one time, what happened to it and why?

14. Describe the internal structure of Jupiter and Saturn.

15. What are the rings of Saturn? Name other planets that have ring structures.

16. Describe some of the unusual features found on the moons of Jupiter, Saturn, and Neptune.

17. What are the similarities and the differences between the Sun and Jupiter?

18. Give one idea about why the Great Red Spot exists on Jupiter. Does the existence of a similar spot on Saturn support or not support this idea? Explain.

19. What is so unusual about the motions and orbits of Venus, Uranus, and Pluto?

20. What evidence exists today that the number of rocks and rock particles floating around in the solar system was much greater in the past soon after the planets formed?

21. What was the source of the water found today in Earth's oceans? Explain the reasoning behind your answer.

22. Explain why carbon dioxide is a major component of the terrestrial planets of Mars and Venus but not of (a) Mercury, (b) Earth, and (c) the giant planets.

23. Explain why oxygen is a major component of Earth's atmosphere but not the atmospheres of Venus or Mars.

24. Using the properties of the planets other than Earth, discuss the possibilities of life on each.

25. What are "shooting stars"? Where do they come from? Where do they go?

26. What is an asteroid? What evidence indicates that asteroids are parts of a broken-up planet? What evidence indicates that asteroids are not parts of a broken-up planet?

27. Where do comets come from? Why are astronomers so interested in studying the physical and chemical structure of a comet?

28. What is a meteor? What is the most likely source of meteors?

29. What is a meteorite? What is the most likely source of meteorites?

30. Technically speaking, what is wrong with calling a rock that strikes the surface of the Moon a meteorite? Again speaking technically, what should you call a rock that strikes the surface of the Moon (or any other planet)?

31. Describe the physical structure and composition of the two main kinds of meteorites, including any subdivisions. What is the meaning of the structure, composition, and texture of a meteorite?

32. If a comet is an icy, dusty body, explain why it appears brightly in the night sky.

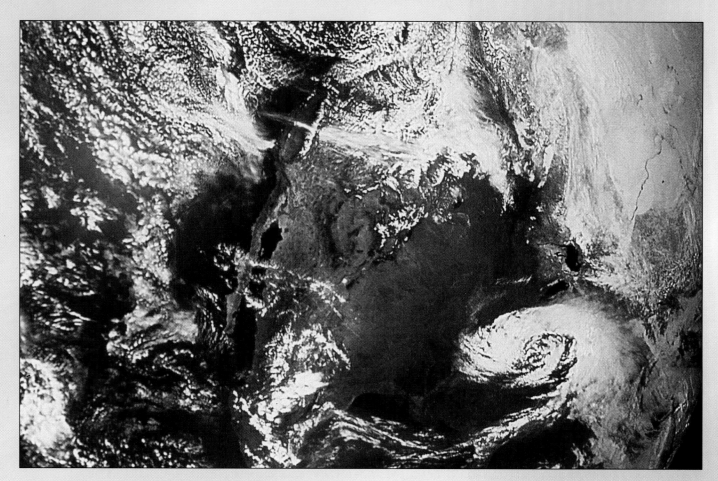

Earth as seen from space. Do you see the United States, with a storm over the East Coast? Do you see Denver and Los Angeles? One topic of this chapter is identifying places on Earth, which should help you find places in the United States.

CHAPTER | Eighteen

Earth in Space

E arth is a common object in the solar system, one of nine planets that goes around the Sun once a year in an almost circular orbit. Earth is the third planet out from the Sun, it is fifth in mass and diameter, and it has the greatest density of all the planets (figure 18.1). Earth is unique because of its combination of an abundant supply of liquid water, a strong magnetic field, and a particular atmospheric composition. In addition to these physical properties, Earth has a unique set of natural motions that humans have used for thousands of years as a frame of reference to mark time and to identify the events of their lives. These references to Earth's motions are called the day, the month, and the year.

Eventually, about three hundred years ago, people began to understand that their references for time came from an Earth that spins like a top as it circles the Sun. It was still difficult, however, for them to understand Earth's place in the universe. The problem was not unlike that of a person trying to comprehend the motion of a distant object while riding a moving merry-go-round being pulled by a cart. Actually, the combined motions of Earth are much more complex than a simple moving merry-go-round being pulled by a cart. Imagine trying to comprehend the motion of a distant object while undergoing a combination of Earth's more conspicuous motions, which are as follows:

1. A daily rotation of 1,670 km/hr (about 1,040 mi/hr) at the equator and less at higher latitudes.
2. A monthly revolution of Earth around the Earth-Moon center of gravity at about 50 km/hr (about 30 mi/hr).
3. A yearly revolution around the Sun at about an average 106,000 km/hr (about 66,000 mi/hr).
4. A motion of the solar system around the core of the Milky Way at about 370,000 km/hr (about 230,000 mi/hr).
5. A motion of the local star group that contains the Sun as compared to other star clusters of about 1,000,000 km/hr (about 700,000 mi/hr).
6. Movement of the entire Milky Way galaxy relative to other, remote galaxies at about 580,000 km/hr (about 360,000 mi/hr).
7. Minor motions such as cycles of change in (a) the size and shape of Earth's orbit, (b) the tilt of Earth's axis, and (c) the time of perihelion. In addition to these slow changes, there is a gradual slowing of the rate of Earth's daily rotation.

Basically Earth is moving through space at fantastic speeds, following the Sun in a spiral path of a giant helix as it spins like a top (figure 18.2). This ceaseless and complex motion in space is relative to various frames of reference, however, and the limited perspective from Earth's surface can result in some very different ideas about Earth and its motions. This chapter is about the more basic, or fundamental, motions of Earth and the Moon. In addition to conceptual understandings and evidences for the motions, some practical human uses of the motions of the planet Earth will be discussed.

SHAPE AND SIZE OF EARTH

The most widely accepted theory about how the solar system formed pictures the planets forming in a disk-shaped nebula with a turning, swirling motion. The planets formed from separate accumulations of materials within this disk-shaped, turning nebula, so the orbit of each planet was established along with its rate of rotation as it formed. Thus, all the planets move around the Sun in the same direction in elliptical orbits that are nearly circular. The flatness of the solar system results in the observable planets moving in, or near, the plane of Earth's orbit, which is called the plane of the ecliptic.

Today, almost everyone has seen pictures of Earth from space, and it is difficult to deny that it has a rounded shape. During the fifth and sixth century B.C., the ancient Greeks decided that Earth must be round because (1) philosophically, they considered the sphere to be the perfect shape and they considered Earth to be perfect, so therefore Earth must be a sphere, (2) Earth was observed to cast a circular shadow on the Moon during a lunar eclipse, and (3) ships were observed to slowly disappear below the horizon as they sailed off into the distance. More abstract evidence of a round Earth was found in the observation that the altitude of the North Star above the horizon appeared to increase as a person traveled northward. This established that Earth's surface was curved, at least, which seemed to fit with other evidence and philosophical reasonings.

One of the first known attempts to calculate the size of the earth from actual data was made by the Greek astronomer Eratosthenes in about 250 B.C. Eratosthenes lived in Alexandria, Egypt. He learned, perhaps from travelers, that an interesting thing happened every first day of summer (June 21) in the town of Syene (now Aswan). On that day when the Sun reached its highest point in the sky (noon), sunlight would move straight down a deep well without making any shadows on the sides of the well. Eratosthenes decided to check this observation at noon on the first day of summer and found the Sun over Alexandria was slightly over 7° from the vertical. Eratosthenes

FIGURE 18.1

Artist's concept of the solar system. Shown are the orbits of the planets, Earth being the third planet from the Sun, and the other planets and their relative sizes and distances from each other and to the Sun. Also shown is the solar system as seen looking toward Earth from the Moon.

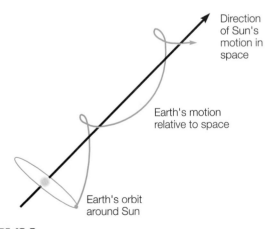

FIGURE 18.2

Earth undergoes many different motions as it moves through space. There are seven more conspicuous motions, three of which are more obvious on the surface. Earth follows the path of a gigantic helix, moving at fantastic speeds as it follows the Sun and the galaxy through space.

considered the earth to be a sphere and, as the Babylonians did before him, divided the circumference of a sphere (a circle) into 360°. Eratosthenes reasoned that since the Sun was directly over Syene *at the same instant* that it was a little over 7° from the vertical at Alexandria, the difference must be a consequence of Earth's curved surface. Further, 7° is about one-fiftieth of 360°, a complete circle, so the distance between Alexandria and Syene must be one-fiftieth of Earth's circumference. Thus, Earth's circumference could be calculated by measuring the distance between Alexandria and Syene. Eratosthenes' calculation was made in the length units used at that time, the stadia. The distance between Syene and Alexandria was 5,000 stadia, so Earth's circumference was calculated to be 5,000 stadia times 50, or 250,000 stadia. The exact length of the ancient Greek stadium unit is unknown today. The unit was based on the length of the track in the local stadium, and this was a time long before standard units were established. The shortest of these tracks was about 185 m (about 607 ft), so Eratosthenes' measurement was probably at least 15 to 20 percent too large. Today, Earth's equatorial circumference is measured at 40,075 km (about

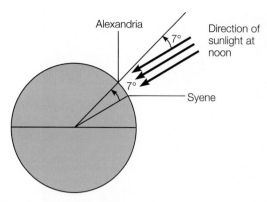

FIGURE 18.3

Eratosthenes calculated the size of Earth's circumference after learning that the Sun's rays were vertical at Syene at noon on the same day they made an angle of a little over 7° at Alexandria. He reasoned that the difference was due to Earth's curved surface. Since 7° is about 1/50 of 360°, then the size of Earth's circumference had to be fifty times the distance between the two towns. (The angle is exaggerated in the diagram for clarity.)

FIGURE 18.4

Earth as seen from space.

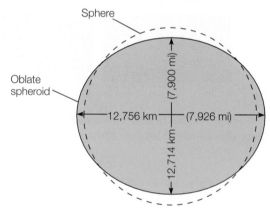

FIGURE 18.5

Earth has an irregular, slightly lopsided, slightly pear-shaped form. In general, it is considered to have the shape of an oblate spheroid, departing from a perfect sphere as shown here.

24,886 mi). Nonetheless, Eratosthenes' calculation showed that the ancient Greeks had a pretty good idea about Earth's size as well as its shape (figure 18.3).

Today, the shape and size of the earth have been precisely measured by artificial satellites circling the earth. These measurements have found that Earth is not a perfectly round sphere as believed by the ancient Greeks. It is flattened at the poles and has an equatorial bulge, as do many other planets. In fact, you can observe through a telescope that both Jupiter and Saturn are considerably flattened at the poles. A shape that is flattened at the poles has a greater distance through the equator than through the poles, which is described as an *oblate* shape. Earth, like a water-filled, round balloon resting on a table, has an oblate shape. It is not perfectly symmetrically oblate, however, since the North Pole is slightly higher and the South Pole is slightly lower than the average surface. In addition, it is not perfectly circular around the equator, with a lump in the Pacific Ocean and a depression in the Indian Ocean. Earth is a slightly pear-shaped, slightly lopsided *oblate spheroid*. All the elevations and depressions are less than 85 m (about 280 ft), however, which is practically negligible compared to the size of the earth (figure 18.4). Thus, the earth is very close to, but not exactly, an oblate spheroid. The significance of this shape will become apparent when the earth's motions are discussed next (figure 18.5).

MOTIONS OF EARTH

Ancient civilizations had a fairly accurate understanding of the size and shape of Earth but had difficulty accepting the idea that Earth moves. The geocentric theory of a motionless Earth with the Sun, the Moon, planets, and stars circling it was discussed in the previous chapter. Ancient people had difficulty with anything but a motionless Earth for at least two reasons: (1) they

could not sense any motion of Earth and (2) they had ideas about being at the center of a universe that was created for them. Thus, it was not until the 1700s that the concept of an Earth in motion became generally accepted. Today, Earth is understood to move a number of different ways, seven of which were identified in the introduction to this chapter. Three of these motions are independent of motions of the Sun and the galaxy. These are (1) a yearly revolution around the Sun, (2) a daily rotation on its axis, and (3) a slow clockwise wobble of its axis.

Revolution

Earth moves constantly around the Sun in a slightly elliptical orbit that requires an average of one year for one complete circuit. Recall that the movement around the Sun is called a *revolution*

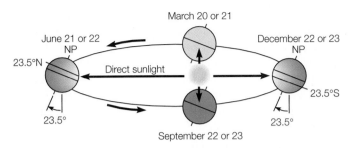

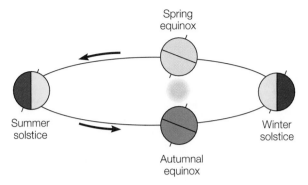

FIGURE 18.6

The consistent tilt and orientation of Earth's axis as it moves around its orbit is the cause of the seasons. The North Pole is pointing toward the Sun during the summer solstice and away from the Sun during the winter solstice.

FIGURE 18.7

The length of daylight during each season is determined by the relationship of Earth's shadow to the tilt of the axis. At the equinoxes, the shadow is perpendicular to the latitudes, and day and night are of equal length everywhere. At the summer solstice, the North Pole points toward the Sun and is completely out of the shadow for a twenty-four-hour day. At the winter solstice, the North Pole is in the shadow for a twenty-four-hour night. The situation is reversed for the South Pole.

and that all points of Earth's orbit lie in a plane called the *plane of the ecliptic.* The average distance between Earth and the Sun is about 150 million km (about 93 million mi).

Earth's orbit is slightly elliptical, so the planet moves with a speed that varies according to Kepler's laws of planetary motion (see chapter 17). It moves fastest when it is closer to the Sun in January, at perihelion, and moves slowest when it is farthest away from the Sun in early July, at aphelion. Earth is about 2.5 million km (about 1.5 million mi) closer to the Sun in January and about the same distance farther away in July than it would be if the orbit were a circle. This total difference of about 5 million km (about 3 million mi) results in a January Sun with an apparent diameter that is 3 percent larger than the July Sun, and Earth as a whole receives about 6 percent more solar energy in January. The effect of being closer to the Sun is much less than the effect of some directional relationships, and winter occurs in the Northern Hemisphere when Earth is closest to the Sun. Likewise, summer occurs in the Northern Hemisphere when the Sun is at its greatest distance from Earth (figure 18.6).

The important directional relationships that override the effect of Earth's distance from the Sun involve the daily *rotation,* or spinning, of Earth around an imaginary line through the geographic poles called Earth's *axis.* The important directional relationships are a *constant inclination* of Earth's axis to the plane of the ecliptic and a *constant orientation* of the axis to the stars. The inclination of Earth's axis to the plane of the ecliptic is about 66.5° (or 23.5° from a line perpendicular to the plane). This relationship between the plane of Earth's orbit and the tilt of its axis is considered to be the same day after day throughout the year, even though small changes do occur in the inclination over time. Likewise, the orientation of Earth's axis to the stars is considered to be the same throughout the year as Earth moves through its orbit. Again, small changes do occur in the orientation over time. Thus, in general, the axis points in the same direction, remaining essentially parallel to its position during any day of the year. The essentially constant orientation and inclination of the axis results in the axis pointing toward the Sun as Earth moves in one part of its orbit, then pointing away from the Sun six months later. The generally constant inclination and orientation of the axis, together with Earth's rotation and revolution, combine to pro-

duce three related effects: (1) days and nights that vary in length, (2) changing seasons, and (3) climates that vary with latitude.

Figure 18.6 shows how the North Pole points toward the Sun on June 21 or 22, then away from the Sun on December 22 or 23 as it maintains its orientation to the stars. When the North Pole is pointed toward the Sun it receives sunlight for a full twenty-four hours, and the South Pole is in Earth's shadow for a full twenty-four hours. This is summer in the Northern Hemisphere with the longest daylight periods and the Sun at its maximum noon height in the sky. Six months later, on December 22 or 23, the orientation is reversed with winter in the Northern Hemisphere, the shortest daylight periods, and the Sun at its lowest noon height in the sky.

The beginning of a season can be recognized from any one of the three related observations: (1) the length of the daylight period, (2) the altitude of the Sun in the sky at noon, or (3) the length of a shadow from a vertical stick at noon. All of these observations vary with changes in the direction of Earth's axis of rotation relative to the Sun. On about June 22 and December 22 the Sun reaches its highest and lowest noon altitudes as Earth moves to point the North Pole directly toward the Sun (June 21 or 22) and directly away from the Sun (December 22 or 23). Thus, the Sun appears to stop increasing or decreasing its altitude in the sky, stop, then reverse its movement twice a year. These times are known as **solstices** after the Latin meaning "Sun stand still." The Northern Hemisphere's *summer solstice* occurs on about June 22 and identifies the beginning of the summer season. At the summer solstice the Sun at noon has the highest altitude, and the shadow from a vertical stick is shorter than any other day of the year. The Northern Hemisphere's *winter solstice* occurs on about December 22 and identifies the beginning of the winter season. At the winter solstice the Sun at noon has the lowest altitude, and the shadow from a vertical stick is longer than any other day of the year (figure 18.7).

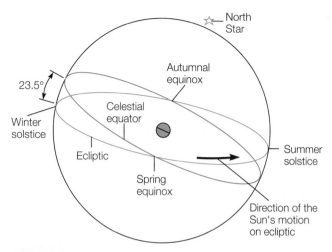

FIGURE 18.8

The position of the Sun on the celestial sphere at the solstices and the equinoxes.

As Earth moves in its orbit between pointing its North Pole toward the Sun on about June 22 and pointing it away on about December 22, there are two times when it is halfway between. At these times Earth's axis is perpendicular to a line between the center of the Sun and Earth, and daylight and night are of equal length. These are called the **equinoxes** after the Latin meaning "equal nights." The *spring equinox* (also called the *vernal equinox*) occurs on about March 21 and identifies the beginning of the spring season. The *autumnal equinox* occurs on about September 23 and identifies the beginning of the fall season.

The relationship between the apparent path of the Sun on the celestial sphere and the seasons is shown in figure 18.8. Recall that the celestial equator is a line on the celestial sphere directly above Earth's equator. The equinoxes are the points on the celestial sphere where the ecliptic, the path of the Sun, crosses the celestial equator. Note also that the summer solstice occurs when the ecliptic is 23.5° north of the celestial equator, and the winter solstice occurs when it is 23.5° south of the celestial equator.

Activities

Make a chart to show the time of sunrise and sunset for a month. Calculate the amount of daylight and darkness. Does the sunrise change in step with the sunset, or do they change differently? What models can you think of that would explain all your findings?

Rotation

Observing the apparent turning of the celestial sphere once a day and seeing the east to west movement of the Sun, the Moon, and stars, it certainly seems as if it is the heavenly bodies and not Earth doing the moving. You cannot sense any movement, and there is little apparent evidence that Earth indeed moves. Evidence of a moving Earth comes from at least three different observations: (1) the observation that the other planets and the Sun rotate, (2) the observation of the changing plane of a long, heavy pendulum at different latitudes on Earth, and (3) the observation of the direction of travel of something moving across, but above, Earth's surface, such as a rocket.

Other planets, such as Jupiter, and the Sun can be observed to rotate by keeping track of features on the surface such as the Great Red Spot on Jupiter and sunspots on the Sun. While such observations are not direct evidence that Earth also rotates, they do show that other members of the solar system spin on their axes. As described earlier, Jupiter is also observed to be oblate, flattened at its poles with an equatorial bulge. Since Earth is also oblate, this is again indirect evidence that it rotates too.

The most easily obtained and convincing evidence about the earth's rotation comes from a *Foucault pendulum,* a heavy mass swinging from a long wire. This pendulum is named after the French physicist Jean Foucault, who first used a long pendulum in 1851 to prove that the earth rotates. Foucault started a long, heavy pendulum moving just above the floor, marking the plane of its back-and-forth movement. Over some period of time, the pendulum appeared to slowly change its position, smoothly shifting its plane of rotation. Science museums often show this shifting plane of movement by setting up small objects for the pendulum to knock down. Foucault demonstrated that the pendulum actually maintains its plane of movement in space (inertia) while the earth rotates eastward (counterclockwise) under the pendulum. It is the earth that turns under the pendulum, causing the pendulum to appear to change its plane of rotation. It is difficult to imagine the pendulum continuing to move in a fixed direction in space while the earth, and everyone on it, turns under the swinging pendulum (figure 18.9).

Figure 18.10 illustrates the concept of the Foucault pendulum. A pendulum is attached to a support on a stool that is free to rotate. If the stool is slowly turned while the pendulum is swinging, you will observe that the pendulum maintains its plane of rotation while the stool turns under it. If you were much smaller and now on the stool below the pendulum, it would appear to turn as you rotate with the turning stool. This is what happens on the earth. Such a pendulum at the North Pole would make a complete turn in about twenty-four hours. Moving south from the North Pole, the change decreases with latitude until, at the equator, the pendulum would not appear to turn at all. At higher latitudes, the plane of the pendulum appears to move clockwise in the Northern Hemisphere and counterclockwise in the Southern Hemisphere.

More evidence that the earth rotates is provided by objects that move above and across the earth's surface. As shown in figure 18.11, the earth has a greater rotational velocity at the equator than at the poles. As an object leaves the surface and moves north or south, the surface has a different rotational velocity, so it rotates beneath the object as it proceeds in a straight line. This gives the moving object an apparent deflection to the right of the direction of movement in the Northern Hemisphere and to the

FIGURE 18.9

As is being demonstrated in this old woodcut, Foucault's insight helped people understand that the earth turns. The pendulum moves back and forth without changing its direction of movement, and we know this is true because no forces are involved. We turn with the earth and this makes the pendulum appear to change its plane of rotation. Thus we know the earth rotates.

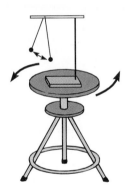

FIGURE 18.10

The Foucault pendulum swings back and forth in the same plane while a stool is turned beneath it. Likewise, a Foucault pendulum on the earth's surface swings back and forth in the same plane while the earth turns beneath it. The amount of turning observed depends on the latitude of the pendulum.

left in the Southern Hemisphere. The apparent deflection caused by the earth's rotation is called the **Coriolis effect.** The Coriolis effect will explain the earth's prevailing wind systems as well as the characteristic direction of wind in areas of high pressure and areas of low pressure (see chapter 21).

Precession

If Earth were a perfect spherically shaped ball, its axis would always point to the same reference point among the stars. The reaction of Earth to the gravitational pull of the Moon and the

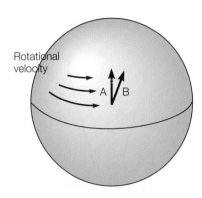

FIGURE 18.11

The earth has a greater rotational velocity at the equator and less toward the poles. As an object moves north or south (A), it passes over land with a different rotational velocity, which produces a deviation to the right in the Northern Hemisphere (B) and to the left in the Southern Hemisphere.

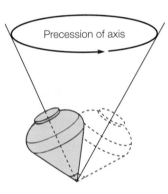

FIGURE 18.12

A spinning top wobbles as it spins, and the axis of the top traces out a small circle. The wobbling of the axis is called *precession*.

Sun on its equatorial bulge, however, results in a slow wobbling of the earth as it turns on its axis. This slow wobble of Earth's axis, called **precession,** causes it to swing in a slow circle like the wobble of a spinning top (figure 18.12). It takes Earth's axis about 26,000 years to complete one turn, or wobble. Today, the axis points very close to the North Star, Polaris, but is slowly moving away to point to another star. In about 12,000 years the star Vega will appear to be in the position above the North Pole, and Vega will be the new North Star. The moving pole also causes changes over time in which particular signs of the zodiac appear with the spring equinox (see chapter 16). Because of precession the occurrence of the spring equinox has been moving backward (westward) through the zodiac constellations at about 1 degree every 72 years. Thus, after about 26,000 years the spring equinox will have moved through all the constellations and will again approach the constellation of Aquarius for the next "age of Aquarius" (figure 18.13).

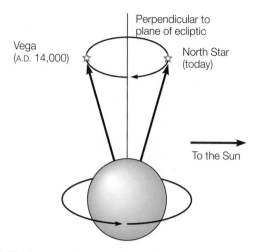

FIGURE 18.13

The slow, continuous precession of Earth's axis results in the North Pole pointing around a small circle over a period of about 26,000 years.

PLACE AND TIME

The continuous rotation and revolution of the earth establishes an objective way to determine direction, location, and time on the earth. If Earth were an unmoving sphere there would be no side, end, or point to provide a referent for direction and location. Earth's rotation, however, defines an axis of rotation, which serves as a reference point for determination of direction and location on the entire surface. Earth's rotation and revolution together define cycles, which define standards of time. The following describes how Earth's movements are used to identify both place and time.

Identifying Place

A system of two straight lines can be used to identify a point, or position, on a flat, two-dimensional surface. The position of the letter *X* on this page, for example, can be identified by making a line a certain number of measurement units from the top of the page and a second line a certain number of measurement units from the left side of the page. Where the two lines intersect will identify the position of the letter *X*, which can be recorded or communicated to another person (figure 18.14).

A system of two straight lines can also be used to identify a point, or position, on a sphere except this time the lines are circles. The reference point for a sphere is not as simple as in the flat, two-dimensional case, however, since a sphere does not have a top or side edge. Earth's axis provides the north-south reference point. The equator is a big circle around the earth that is exactly halfway between the two ends, or poles, of the rotational axis. An infinite number of circles are imagined to run around the earth parallel to the equator as shown in figure 18.15. The east- and west-running parallel circles are called **parallels.** Each parallel is the same distance between the equator and one of the poles all the way around the earth. The distance from the equa-

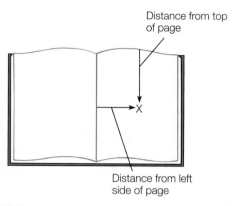

FIGURE 18.14

Any location on a flat, two-dimensional surface is easily identified with two references from two edges. This technique does not work on a motionless sphere because there are no reference points.

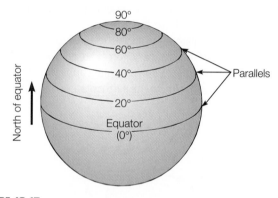

FIGURE 18.15

A circle that is parallel to the equator is used to specify a position north or south of the equator. A few of the possibilities are illustrated here.

tor to a point on a parallel is called the **latitude** of that point. Latitude tells you how far north or south a point is from the equator by telling you the parallel the point is located on. The distance is measured northward from the equator (which is 0°) to the North Pole (90° north) or southward from the equator (0°) to the South Pole (90° south) (figure 18.16). If you are somewhere at a latitude of 35° north, you are somewhere on the earth on the 35° latitude line north of the equator.

Since a parallel is a circle, a location of 40°N latitude could be anyplace on that circle around the earth. To identify a location you need another line, this time one that runs pole to pole and perpendicular to the parallels. These north-south running arcs that intersect at both poles are called **meridians** (figure 18.17). There is no naturally occurring, identifiable meridian that can be used as a point of reference such as the equator serves for parallels, so one is identified as the referent by international agreement. The referent meridian is the one that passes through the Greenwich Observatory near London, England, and this meridian is called the **prime meridian.** The distance from the

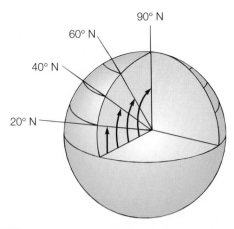

FIGURE 18.16

If you could see to the earth's center, you would see that latitudes run from 0° at the equator north to 90° at the North Pole (or to 90° south at the South Pole).

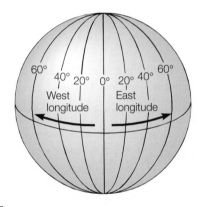

FIGURE 18.17

Meridians run pole to pole perpendicular to the parallels and provide a reference for specifying east and west directions.

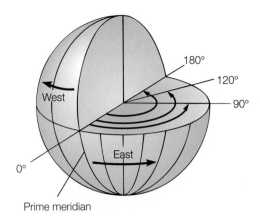

FIGURE 18.18

If you could see inside the earth, you would see 360° around the equator and 180° of longitude east and west of the prime meridian.

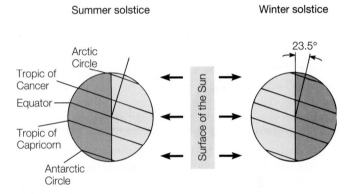

FIGURE 18.19

At the summer solstice, the noon Sun appears directly overhead at the tropic of Cancer (23.5°N) and twenty-four hours of daylight occurs north of the Arctic Circle (66.5°N). At the winter solstice, the noon Sun appears overhead at the tropic of Capricorn (23.5°S) and twenty-four hours of daylight occurs south of the Antarctic Circle (66.5°S).

prime meridian east or west is called the **longitude.** The degrees of longitude of a point on a parallel are measured to the east or to the west from the prime meridian up to 180° (figure 18.18). New Orleans, Louisiana, for example, has a latitude of about 30°N of the equator and a longitude of about 90°W of the prime meridian. The location of New Orleans is therefore described as 30°N, 90°W.

Locations identified with degrees of latitude north or south of the equator and degrees of longitude east or west of the prime meridian are more precisely identified by dividing each degree of latitude into subdivisions of 60 minutes (60′) per degree, and each minute into 60 seconds (60″). On the other hand, latitudes near the equator are sometimes referred to in general as the *low latitudes,* and those near the poles are sometimes called the *high latitudes.*

In addition to the equator (0°) and the poles (90°), the parallels of 23.5°N and 23.5°S from the equator are important references for climatic consideration. The parallel of 23.5°N is called the **tropic of Cancer,** and the parallel of 23.5°S is called the **tropic of Capricorn.** These two parallels identify the limits toward the

poles of where the Sun appears directly overhead during the course of a year. The parallel of 66.5°N is called the **Arctic Circle,** and the parallel of 66.5°S is called the **Antarctic Circle.** These two parallels identify the limits toward the equator of where the Sun appears above the horizon all day during the summer (figure 18.19). This starts with six months of daylight everyday at the pole, then decreases as you get fewer days of full light until reaching the limit of one day of 24-hour daylight at the 66.5° limit.

Measuring Time

Standards of time are determined by intervals between two successive events that repeat themselves in a regular way. Since ancient civilizations, many of the repeating events used to mark time have been recurring cycles associated with the rotation of Earth on its axis and its revolution around the Sun. Thus, the day, month, season, and year are all measures of time based on recurring natural motions of Earth. All other measures of time are based on other events or definitions of events. There are, however, several different ways to describe the day, month, and year, and each depends on a different set of events. These events are described in the following section.

Daily Time

The technique of using astronomical motions for keeping time originated some four thousand years ago with the Babylonian culture. The Babylonians marked the yearly journey of the Sun against the background of the stars, which was divided into twelve periods, or months, after the signs of the zodiac. Based on this system, the Babylonian year was divided into twelve months with a total of 360 days. In addition, the Babylonians invented the week and divided the day into hours, minutes, and seconds. The week was identified as a group of seven days, each based on one of the seven heavenly bodies that were known at the time. The hours, minutes, and seconds of a day were determined from the movement of the shadow around a straight, vertical rod.

As seen from a place in space above the North Pole, Earth rotates counterclockwise turning toward the east. On Earth, this motion causes the Sun to appear to rise in the east, travel across the sky, and set in the west. The changing angle between the tilt of Earth's axis and the Sun produces an apparent shift of the Sun's path across the sky, northward in the summer season and southward in the winter season. The apparent movement of the Sun across the sky was the basis for the ancient as well as the modern standard of time known as the day.

Today, everyone knows that Earth turns as it moves around the Sun, but it is often convenient to regard space and astronomical motions as the ancient Greeks did, as a celestial sphere that turns around a motionless Earth. Recall that the celestial meridian is a great circle on the celestial sphere that passes directly overhead where you are and continues around the earth through both celestial poles. The movement of the Sun across the celestial meridian identifies an event of time called **noon.** As the Sun appears to travel west it crosses meridians that are farther and farther west, so the instant identified as noon moves west with the Sun. The instant of noon at any particular longitude is called the *apparent local noon* for that longitude because it identifies noon from the apparent position of the Sun in the sky. The morning hours before the Sun crosses the meridian are identified as *ante meridiem* (A.M.) hours, which is Latin for "before meridian." Afternoon hours are identified as *post meridiem* (P.M.) hours, which is Latin for "after the meridian."

There are several ways to measure the movement of the Sun across the sky. The ancient Babylonians, for example, used a vertical rod called a *gnomon* to make and measure a shadow that moved as a result of the apparent changes of the Sun's position. The gnomon eventually evolved into a *sundial,* a vertical or slanted gnomon with divisions of time marked on a horizontal plate beneath the gnomon. The shadow from the gnomon indicates the *apparent local solar time* at a given place and a given instant from the apparent position of the Sun in the sky. If you have ever read the time from a sundial, you know that it usually does not show the same time as a clock or a watch (figure 18.20). In addition, sundial time is nonuniform, fluctuating throughout

FIGURE 18.20

A sundial indicates the apparent local solar time at a given instant in a given location. The time read from a sundial, which is usually different from the time read from a clock, is based on an average solar time.

the course of a year, sometimes running ahead of clock time and sometimes running behind clock time.

A sundial shows the apparent local solar time, but clocks are set to measure a uniform standard time based on *mean solar time.* Mean solar time is a uniform time averaged from the apparent solar time. The apparent solar time is nonuniform, fluctuating because (1) Earth moves sometimes faster and sometimes slower in its elliptical orbit around the Sun and (2) the equator of the earth is inclined to the ecliptic. The combined consequence of these two effects is a variable, nonuniform sundial time as compared to the uniform mean solar time, otherwise known as clock time.

A day is defined as the length of time required for Earth to rotate once on its axis. There are different ways to measure this rotation, however, which result in different definitions of the day. A *sidereal day* is the interval between two consecutive crossings of the celestial meridian by a particular star ("sidereal" is a word meaning "star"). This interval of time depends only on the time Earth takes to rotate 360° on its axis. One sidereal day is practically the same length as any other sidereal day since Earth's rate of rotation is constant for all practical purposes.

An *apparent solar day* is the interval between two consecutive crossings of the celestial meridian by the Sun, for example, from one local solar noon to the next solar noon. Since Earth is moving in orbit around the Sun, it must turn a little bit farther to compensate for its orbital movement, bringing the Sun back to local solar noon (figure 18.21). As a consequence, the apparent solar day is about four minutes longer than the sidereal day. This additional time accounts for the observation that the stars and constellations of the zodiac rise about four minutes earlier every night, appearing higher in the sky at the same clock time until they complete a yearly cycle. A sidereal day is twenty-three hours, fifty-six minutes, and four seconds long. A *mean solar day* is

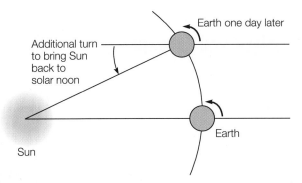

FIGURE 18.21

Because Earth is moving in orbit around the Sun, it must rotate an additional distance each day, requiring about four minutes to bring the Sun back across the celestial meridian (local solar noon). This explains why the stars and constellations rise about four minutes earlier every night.

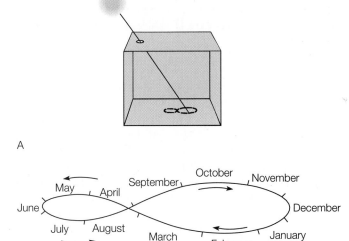

FIGURE 18.22

(A) During a year, a beam of sunlight traces out a lopsided figure eight on the floor if the position of the light is marked at noon every day. (B) The location of the point of light on the figure eight during each month.

twenty-four hours long, averaged from the mean solar time to keep clocks in closer step with the Sun than would be possible using the variable apparent solar day. Just how out of synchronization the apparent solar day can become with a clock can be illustrated with another ancient way of keeping track of the Sun's motions in the sky, the "hole in the wall" sun calendar and clock.

Variations of the "hole in the wall" sun calendar were used all over the world by many different ancient civilizations, including the early Native Americans of the American Southwest. More than one ancient Native American ruin has small holes in the western wall aligned in such a way to permit sunlight to enter a chamber only on the longest and shortest days of the year. This established a basis for identifying the turning points in the yearly cycle of seasons.

A hole in the roof can be used as a sun clock, but it will require a whole year to establish the meaning of a beam of sunlight shining on the floor. Imagine a beam of sunlight passing through a small hole to make a small spot of light on the floor. For a year you mark the position of the spot of light on the floor *each day* when your clock tells you the *mean solar time is noon.* You trace out an elongated, lopsided figure eight with the small end pointing south and the larger end pointing north (figure 18.22A). Note by following the monthly markings shown in figure 18.22B that the figure-eight shape is actually traced out by the spot of sunlight making two S shapes as the Sun changes its apparent position in the sky. Together, the two S shapes make the shape of the figure eight.

Why did the sunbeam trace out a figure eight over a year? The two extreme north-south positions of the figure are easy to understand because by December, Earth is in its orbit with the North Pole tilted away from the Sun. At this time the direct rays of the Sun fall on the tropic of Capricorn (23.5° south of the equator), and the Sun appears low in the sky as seen from the Northern Hemisphere. Thus, on this date, the winter solstice, a beam of sunlight strikes the floor at its northernmost position beneath the hole. By June, Earth has moved halfway around its orbit, and the North Pole is now tilted toward the Sun. The direct rays of the Sun now fall on the tropic of Cancer (23.5°

north of the equator), and the Sun appears high in the sky as seen from the Northern Hemisphere. Thus, on this date, the summer solstice, a beam of sunlight strikes the floor at its southernmost position beneath the hole.

If everything else were constant, the path of the spot would trace out a straight line between the northernmost and southernmost positions beneath the hole. The east and west movements of the point of light as it makes an **S** shape on the floor must mean, however, that the Sun crosses the celestial meridian (noon) earlier one part of the year and later the other part. This early and late arrival is explained in part by Earth moving at different speeds in its orbit. Recall that Earth moves faster when at perihelion, when it is closest to the Sun during the winter season. Starting from perihelion, the faster moving Earth travels farther in its orbit as it completes one rotation. This means that Earth must rotate farther to bring the Sun back across the celestial meridian. As a result, the Sun appears to move more slowly across the sky. By April the apparent local noon occurs almost eight minutes after the mean solar time as shown by a clock. Six months later Earth is moving from aphelion, so it appears to move across the sky more rapidly. In October the apparent local noon occurs almost eight minutes before the mean solar time as shown by a clock. Figure 18.23 is a graph of these differences in time, showing how many minutes the apparent local solar time (sundial time) is ahead or behind the mean solar time (clock time) for each month. Note that the two clocks are together at the times of perihelion and aphelion of Earth in its orbit.

If changes in orbital speed were the only reason that the Sun does not cross the sky at the same rate during the year, the spot of sunlight on the floor would trace out an oval rather than a figure eight. The plane of the ecliptic, however, does not coincide

A CLOSER LOOK | The World's Population and Forests

This chapter is about the shape, orientation, and motions of Earth and how they can be used to identify place and time. Over time the earth changes slowly, but what is happening to life on the surface of the earth? From a global perspective, both human populations and forests are slowly changing, too. What changes are happening to the world's population? The answer is that they seem to be moving to the cities. The United Nations estimates that currently about 45 percent of the world's population lives in cities. This is expected to increase to more than 50 percent by 2005 and reach 60 percent by 2025. The economically developed world is currently about 75 percent urban, and the less-developed world is about 37 percent urban. Therefore, the increase in the urban population will occur primarily in the less-developed world where resources to deal with urban problems are least available.

Urbanization is not necessarily bad. The economic activity of the developed world takes place primarily in cities. Cities can be planned and managed to be healthy, interesting places to live. Cities offer jobs, health care, schools, and other services that are usually lacking in rural areas. Often the rural economy of a country is considered to be of low status, and young people wish to move to cities where higher-status jobs and desirable cultural amenities are available.

Dense populations of people have a concentrated impact on local resources. Often water must be transported long distances, wastes are difficult to get rid of, and air quality drops as each additional person places a burden on the local environment. Unfortunately, in the less-developed world it is impossible to plan for growth and provide basic resources such as drinking water, sewers, transportation, and housing as fast as the population is growing.

About one-third of the world's land area is covered by forests, and interestingly, just over one-half of these forests are in developing countries. New studies indicate the world is losing about 17 million hectares (an area about four times the size of Switzerland) of tropical forests each year. This lost forest is the result of conversion to agricultural land, logging for lumber, increased demand for fuelwood, fodder, and other forest products, in addition to losses from grazing, fire, and drought. The rate of deforestation is highest in Asia (1.2 percent per year from 1981 to 1990), followed by Latin America (0.9 percent) and Africa (0.8 percent). In the same period, more land was deforested annually in Latin America (8.3 million hectares) than in Africa (5 million hectares) or Asia (3.6 million hectares). Deforestation is second only to burning of fossil fuels as a human source of carbon dioxide, one of the major greenhouse-effect gases that has been implicated in global warming. In addition, deforestation leads to soil degradation and destroys habitat, leading to the extinction of plants, mammals, birds, insects, and other animals.

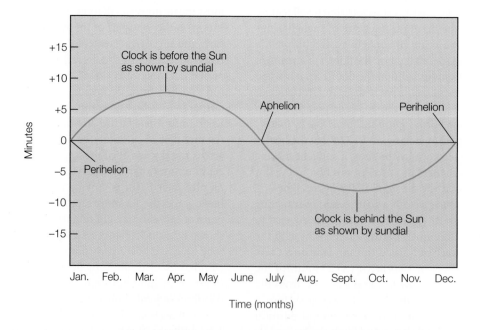

FIGURE 18.23

The difference in sundial time and clock time throughout a year as a consequence of the shape of Earth's orbit. This is not the only factor that causes a difference in the two clocks (see figures 18.24 and 18.25).

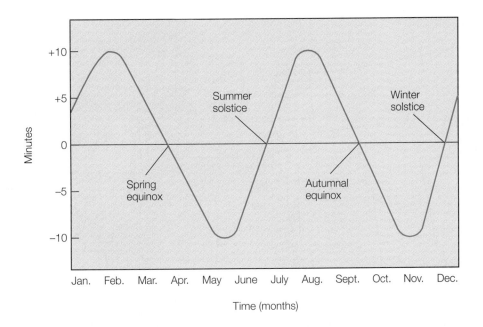

FIGURE 18.24

The difference in sundial time and clock time throughout a year as a consequence of the angle between the plane of the ecliptic and the plane of the equator (see also figures 18.23 and 18.25).

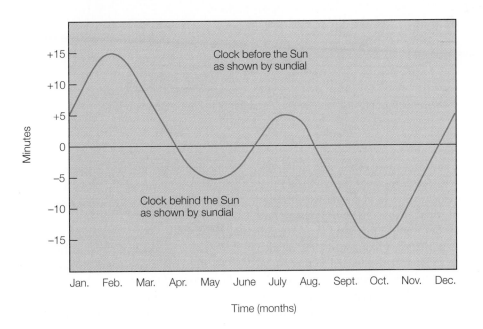

FIGURE 18.25

The equation of time, which shows how many minutes sundial time is faster or slower than clock time during different months of the year. The correction comes from a combination of the two factors shown in figures 18.23 and 18.24.

with the plane of Earth's equator, so the Sun appears at different angles in the sky, and this makes it appear to change its speed during different times of the year. As shown in figure 18.24, this effect changes the length of the apparent solar day by making the Sun up to ten minutes later or earlier than the mean solar time four times a year between the solstices and equinoxes.

The two effects add up to a cumulative variation between the apparent local solar time (sundial time) and the mean solar time (clock time) (figure 18.25). This cumulative variation is known as the *equation of time*, which shows how many minutes sundial time is faster or slower than clock time during different days of the year. The equation of time is often shown on globes in the figure-eight shape called an *analemma*, which also can be used to determine the latitude of direct solar radiation for any day of the year.

Since the local mean time varies with longitude, every place on an east-west line around the earth could possibly have clocks

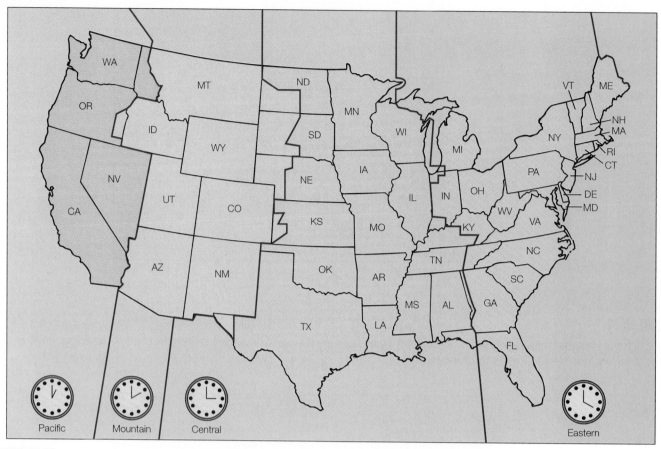

FIGURE 18.26

The standard time zones. Hawaii and most of Alaska are two hours earlier than Pacific Standard Time.

that were a few minutes ahead of those to the west and a few minutes behind those to the east. To avoid the confusion that would result from many clocks set to local mean solar time, the earth's surface is arbitrarily divided into one-hour *standard time zones* (figure 18.26). Since there are 360° around the earth and 24 hours in a day, this means that each time zone is 360° divided by 24, or 15° wide. These 15° zones are adjusted so that whole states are in the same time zone, or the zones are adjusted for other political reasons. The time for each zone is defined as the mean solar time at the middle of each zone. When you cross a boundary between two zones, the clock is set ahead one hour if you are traveling east and back one hour if you are traveling west. Most states adopt *daylight saving time* during the summer, setting clocks ahead one hour in the spring and back one hour in the fall ("spring ahead and fall back"). Daylight saving time results in an extra hour of daylight during summer evenings.

The 180° meridian is arbitrarily called the **international date line,** an imaginary line established to compensate for cumulative time zone changes (figure 18.27). A traveler crossing the date line gains or loses a day just as crossing a time zone boundary results in the gain or loss of an hour. A person moving across the line while traveling westward gains a day; for example, the day after June 2 would be June 4. A person crossing

the line while traveling eastward repeats a day; for example, the day after June 6 would be June 6. Note the date line is curved around land masses to avoid local confusion.

Yearly Time

A *year* is generally defined as the interval of time required for Earth to make one complete revolution in its orbit. As was the case for definitions of a day, there are different definitions of what is meant by a year. The most common definition of a year is the interval between two consecutive spring equinoxes, which is known as the *tropical year* (*trope* is Greek for "turning"). The tropical year is 365 days, 5 hours, 48 minutes, and 46 seconds, or 365.24220 mean solar days.

A *sidereal year* is defined as the interval of time required for Earth to move around its orbit so the Sun is again in the same position relative to the stars. The sidereal year is slightly longer than the tropical year because Earth rotates more than 365.25 times during one revolution. Thus, the sidereal year is 365.25636 mean solar days, which is about 20 minutes longer than the tropical year.

The tropical and sidereal years would be the same interval of time if Earth's axis pointed in a consistent direction. The precession of the axis, however, results in the axis pointing in a slightly different direction with time. This shift of direction over

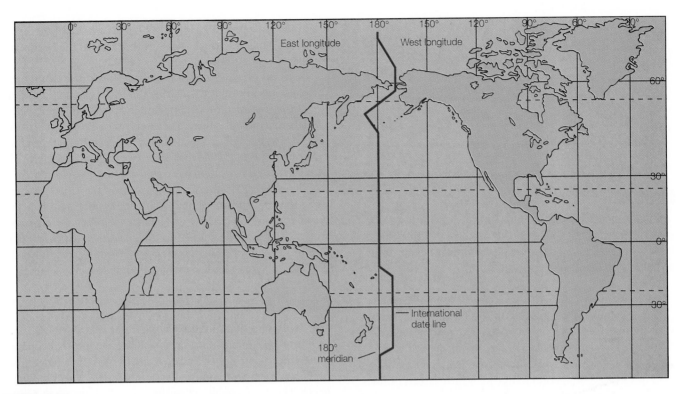

FIGURE 18.27
The international date line follows the 180° meridian but is arranged in a way that land areas and island chains have the same date.

the course of a year moves the position of the spring equinox westward, and the equinox is observed 20 minutes before the orbit has been completely circled. The position of the spring equinox against the background of the stars thus moves westward by some 50 seconds of arc per year.

It is the *tropical year* that is used as a standard time interval to determine the calendar year. Earth does not complete an exact number of turns on its axis while completing one trip around the Sun, so it becomes necessary to periodically adjust the calendar so it stays in step with the seasons. The calendar system that was first designed to stay in step with the seasons was devised by the ancient Romans. Julius Caesar reformed the calendar, beginning in 46 B.C., to have a 365 day year with a 366 day year (leap year) every fourth year. Since the tropical year of 365.24220 mean solar days is very close to 365¼ days, the system, called a *Julian calendar*, accounted for the ¼ day by adding a full day to the calendar every fourth year. The Julian calendar was very similar to the one presently used, except the year began in March, the month of the spring equinox. The month of July was named in honor of Julius Caesar, and the following month was later named after his successor, Augustus.

There was a slight problem with the Julian calendar since it was longer than the tropical year by 365.25 minus 365.24220, or 0.0078 day per year. This small interval (which is 11 minutes, 14 seconds) does not seem significant when compared to the time in a whole year. But over the years the error of minutes and seconds grew to an error of days. By 1582, when Pope Gregory XIII revised

the calendar, the error had grown to 13 days but was corrected for 10 days of error. This revision resulted in the *Gregorian calendar*, which is the system used today. Since the accumulated error of 0.0078 day per year is almost 0.75 day per century, it follows that four centuries will have 0.75 times 4, or 3 days of error. The Gregorian system corrects for the accumulated error by dropping the additional leap year day three centuries out of every four. Thus, the century year of 2000 will be a leap year with 366 days, but the century years of 2100, 2200, and 2300 will not be leap years. You will note that this approximation still leaves an error of 0.0003 day per century, so another calendar revision will be necessary in a few thousand years to keep the calendar in step with the seasons.

Monthly Time

In ancient times, people often used the Moon to measure time intervals that were longer than a day but shorter than a year. The word "month," in fact, has its origins in the word "moon" and its period of revolution. The Moon revolves around Earth in an orbit that is inclined to the plane of Earth's orbit, the plane of the ecliptic, by about 5°. The Moon is thus never more than about ten apparent diameters from the ecliptic. It revolves in this orbit in about 27⅓ days as measured by two consecutive crossings of any star. This period is called a *sidereal month*. The Moon rotates in the same period as the time of revolution, so the sidereal month is also the time required for one rotation. Because the rotation and revolution rates are the same, you always see the same side of the Moon from Earth.

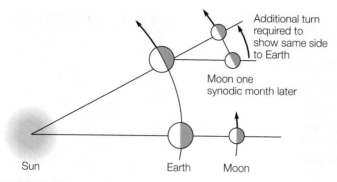

FIGURE 18.28

As the Moon moves in its orbit around Earth, it must revolve a greater distance to bring the same part to face Earth. The additional turning requires about 2.2 days, making the synodic month longer than the sidereal month.

The ancient concept of a month was based on the *synodic month*, the interval of time from new moon to new moon (or any two consecutive identical phases). The synodic month is longer than a sidereal month at a little more than 29½ days. The Moon's phases (see next section) are determined by the relative positions of Earth, the Moon, and the Sun. As shown in figure 18.28, the Moon moves with Earth in its orbit around the Sun. During one sidereal month the Moon has to revolve a greater distance before the same phase is observed on Earth, and this greater distance requires 2.2 days. This makes the synodic month about 29½ days long, only a little less than ½₂ of a year, or the period of time the present calendar identifies as a "month."

THE MOON

Next to the Sun, the Moon is the largest, brightest object in the sky. The Moon is Earth's nearest neighbor at an average distance of 380,000 km (about 238,000 mi), and surface features can be observed with the naked eye. With the aid of a telescope or a good pair of binoculars, you can see light-colored mountainous regions called the *lunar highlands*, smooth dark areas called *maria*, and many sizes of craters, some with bright streaks extending from them (figure 18.29). The smooth dark areas are called maria after a Latin word meaning "sea." They acquired this name from early observers who thought the dark areas were oceans and the light areas were continents. Today, the maria are understood to have formed from ancient floods of molten lava that poured across the surface and solidified to form the "seas" of today. There is no water and no atmosphere on the Moon.

Many facts known about the Moon were established during the *Apollo* missions, the first human exploration of a place away from Earth. A total of twelve *Apollo* astronauts walked on the Moon, taking thousands of photographs, conducting hundreds of experiments, and returning to Earth with over 380 kg (about 840 lb) of moon rocks (table 18.1). In addition, instruments were left on the Moon that continued to radio data back to Earth

after the *Apollo* program ended in 1972. As a result of the *Apollo* missions, many questions were answered about the Moon, but unanswered questions still remain.

A recent unmanned mission to the Moon took place during three months in early 1994. During this short span the spacecraft *Clementine* recorded over 1.5 million images of the Moon with a special multispectral camera. The camera shot the images at eleven different wavelengths, designed to provide information about mineral compositions and rock ages as well as the topography. The excellent spatial resolution has resulted in the best global map of the Moon to date. The vertical topography of the Moon was measured to within 100 m (about 330 ft), an improvement by a factor of 10 over previous maps. The new map includes an extensive study of the little-studied South Pole Aitken basin. At 12 km (7.5 mi) deep and 2,500 km (about 1,500 mi) wide, this is the deepest and largest known impact basin in the solar system.

Composition and Features

The *Apollo* astronauts found that the surface of the Moon is covered by a 3 m (about 10 ft) layer of fine gray dust that contains microscopic glass beads. The dust and beads were formed from millions of years of continuous bombardment of micrometeorites. These very small meteorites generally burn up in Earth's atmosphere. The Moon does not have an atmosphere that protects its surface, so they have continually fragmented and pulverized the surface in a slow, steady rain. The glass beads are believed to have formed when larger meteorite impacts melted part of the surface, which was immediately forced into a fine spray that cooled rapidly while above the surface.

The rocks on the surface of the Moon were found to be mostly *basalts*, a type of rock formed on Earth from the cooling and solidification of molten lava. The dark-colored rocks from the maria are similar to Earth's basalts but contain greater amounts of titanium and iron oxides. The light-colored rocks from the highlands are mostly *breccias*, a kind of rock made up of rock fragments that have been compacted together. On the Moon, the compacting was done by meteorite impacts. The rocks from the highlands contain more aluminum and less iron than the maria basalts and thus have a lower density (2.9 g/cm³) than the darker mare rocks (3.3 g/cm³).

All the moon rocks contained a substantial amount of radioactive elements, which made it possible to precisely measure their age. The light-colored rocks from the highlands were formed some 4 billion years ago. The dark-colored rocks from the maria were much younger, with ages ranging from 3.1 to 3.8 billion years. This indicates a period of repeated volcanic eruptions and lava flooding over a 700-million-year period that ended about 3 billion years ago.

Seismometers left on the Moon by *Apollo* astronauts detected only very weak moonquakes, so weak that they would not be felt by a person. These moonquakes are thought to be produced by the nearby impact of larger meteoroids or by a slight cracking of the crust from gravitational interactions with Earth and the Sun. The movement of these seismic waves through the Moon's interior suggests that the Moon has an internal structure. The outer

FIGURE 18.29

You can easily see the light-colored lunar highlands, smooth and dark maria, and many craters on the surface of Earth's nearest neighbor in space.

TABLE 18.1	The Apollo missions		
Mission	**Date**	**Crew**	**Comments**
Apollo One	Jan 27, 1967	Gus Grissom, Ed White, Roger Chaffee	The crew died in their spacecraft during a test three weeks before they would have flown in space.
Apollo Seven	Oct 11–22, 1968	Wally Schirra, Donn F. Eisele, Walter Cunningham	This was the first *Apollo* Mission in space following the *Apollo One* launchpad fire. It was an eleven-day mission to validate *Apollo* hardware in low Earth orbit.
Apollo Eight	Dec 21–27, 1968	Frank Borman, Jim Lovell, Bill Anders	First space mission to orbit the Moon. The first picture of Earth taken from deep space.
Apollo Nine	Mar 3–13, 1969	James A. McDivitt, David R. Scott, Russel L. Schweickhart	First test of the lunar module in space. *Apollo Nine* was an Earth orbital mission. *Apollo* type rendezvous and docking was tested after a 6 hour, 113 mile separation.
Apollo Ten	May 18–26, 1969	Tom Stafford, John Young, Gene Cernan	Trial rehearsal of Moon landing. The LM was taken to the Moon and separated from the CM, but it did not land on the Moon. The LM was tested in lunar orbit.
Apollo Eleven	Jul 16–24, 1969	Neil Armstrong, Mike Collins, Buzz Aldrin	The lunar module, *Eagle,* landed the first man, Neil Armstrong, on the Moon on July 20, 1969, and established the first manned moon base.
Apollo Twelve	Nov 14–24, 1969	Peter Conrad, Dick Gordon, Al Bean	Landing was very accurate (only 535 ft from *Surveyor* III). The crew conducted two moon walks and put up both a geophysical station and a nuclear power station.
Apollo Thirteen	Apr 11–17, 1970	Jim Lovell, Jack Swigert, Fred Haise	This was the first abort in deep space (200,000 mi from Earth). The lunar module was used as a lifeboat to return the crew safely.
Apollo Fourteen	Jan 31–Feb 9, 1971	Alan Shepard, Stuart A. Roosa, Edgar D. Mitchell	This was the third mission to land on the Moon, landing in the Fra Mauro region. Al Shepard hit two golf balls on the Moon. Lunar specimens (95 lb) were collected.
Apollo Fifteen	Jul 26–Aug 7, 1971	Dave Scott, Alfred M. Worden, James B. Irwin	During their record time on the Moon (66 hr 54 min), the crew placed a subsatellite in lunar orbit and were the first to use the lunar rover.
Apollo Sixteen	Apr 16–27, 1972	John Young, Thomas K. Mattingly II, Charles M. Duke	Highest landing on the Moon (elevation 25,688 ft); lunar rover land speed record of 11.2 mi/hr and distance record of 22.4 miles covered. The crew returned 213 lb of lunar samples.
Apollo Seventeen	Dec 7–19, 1972	Gene Cernan, Ronald E. Evans, Harrison H. Schmitt	This, the last of the *Apollo* flights, was the first time an *Apollo* flight was launched at night. A record of 75 hours was set for time spent on the Moon, and 250 lb of lunar samples were returned.
Apollo-Soyuz Mission	Jul 15–24, 1975	Tom Stafford, Deke Slayton, Vance Brand	First international space rendezvous. This was the first coordinated launch of two spacecraft from different countries.

Source: Data from NASA.

layer of solid rock, or crust, is about 65 km (about 40 mi) thick on the side that always faces Earth and is about twice as thick on the far side. The data also suggest a small, partly molten iron core at a depth of about 900 km (about 600 mi) beneath the surface. A small core would account for the Moon's low density (3.34 g/cm³) as compared to Earth's average (5.5 g/cm³) and would account for the observation that the Moon has no general magnetic field.

History of the Moon

The moon rocks brought back to Earth, the results of the lunar seismographs, and all the other data gathered through the *Apollo* missions have increased our knowledge about the Moon, leading to new understandings of how it formed. This model pictures the present Moon developing through four distinct stages of events.

The *origin stage*, the first stage in the history of the Moon, describes how it originally formed. There was no agreement about how the Moon formed before the *Apollo* missions and a complete study of the returned rock samples. There were three kinds of theories concerning its origins:

1. The *fission* theory—the Moon formed from a part of Earth that broke away early in Earth's formation.
2. The *condensation* theory—the Moon and Earth formed at the same time in neighboring parts of the original solar nebula.
3. The *capture* theory—the Moon formed elsewhere, independently of Earth, and was later captured by Earth's gravitational field when it moved by Earth.

All three theories about how the Moon formed did not work very well and scientists for awhile tried to rule out one or more theories based on a study of the moon rocks. Detailed study of the moon rocks led to yet another concept called the *impact* theory. The impact theory states that the Moon formed from the impact of Earth with a very large object, perhaps as large as Mars or larger. The Moon formed from ejected material produced by this collision. The impact theory is supported by the evidence and is widely accepted, but not all parts of this theory are fully developed.

The *molten surface stage*, the second stage, occurred during the first 200 million years. Heating from a number of sources could have been involved in the melting of the entire lunar surface 100 km (about 60 mi) or so deep. The heating needed to melt the surface is believed to have generally accumulated from the impacts of rock fragments, which were leftover debris from the formation of the solar system that intensely bombarded the Moon. As the bombardment subsided, the molten outer layer cooled and solidified to solid rock. Between 3.9 to 4.2 billion years ago the subsiding meteorite bombardment on the cooling crust formed most of the craters that are seen on the Moon today.

The *molten interior stage*, the third stage in the development of the Moon, involved the melting of the interior of the Moon. Radioactive decay had been slowly heating the interior, and 3.8 billion years ago, or about a billion years after the Moon formed, sufficient heat accumulated to melt the interior. The light and heavier rock materials separated during this period, perhaps producing a small iron core. Molten lava flowed into basins on the surface during this period, forming the smooth, darker maria seen today. The lava flooding continued for about 700 million years, ending about 3.1 billion years ago.

The *cold and quiet stage*, the fourth and last stage in the development of the Moon, began 3.1 billion years ago as the last lava flow cooled and solidified. Since that time the surface of the Moon has been continually bombarded by micrometeorites and a few larger meteorites. With the exception of a few new craters, the surface of the Moon has changed little in the last 3 billion years.

THE EARTH-MOON SYSTEM

Earth and its Moon are unique in the solar system because of the size of the Moon. It is not the largest satellite, but the ratio of its mass to Earth's mass is greater than the mass ratio of any other moon to its planet. This comparison excludes the Pluto-Charon

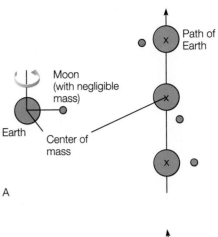

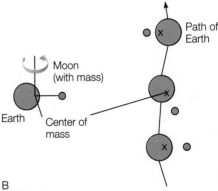

FIGURE 18.30

(*A*) If the Moon had a negligible mass, the center of gravity between the Moon and Earth would be Earth's center, and Earth would follow a smooth orbit around the Sun. (*B*) The actual location of the center of mass between Earth and Moon results in a slightly in and out, or wavy, path around the Sun.

pair, which really appears to be more like a binary planet than a planet and its satellite. The Moon has a diameter of 3,476 km (about 2,159 mi), which is about one-fourth the diameter of Earth, and a mass of about 1/81 of Earth's mass. This is a small fraction of Earth's mass, but it is enough to affect Earth's motion as it revolves around the Sun.

If the Moon had a negligible mass it would circle Earth with a center of rotation (center of mass) located at the center of the earth. In this situation the center of Earth would follow a smooth path around the Sun (figure 18.30A). The mass of the Moon, however, is great enough to move the center of rotation away from the earth's center toward the Moon. As a result, both bodies act as a system, moving around a center of mass. The center of mass between Earth and the Moon follows a smooth orbit around the Sun. Earth follows a slightly wavy path around the Sun as it slowly revolves around the common center of mass (figure 18.30B).

Phases of the Moon

The phases of the Moon are a result of the changing relative positions of Earth, the Moon, and the Sun as the Earth-Moon system moves around the Sun. Sunlight always illuminates half of the

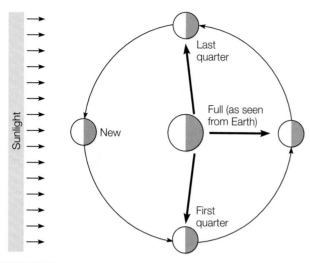

FIGURE 18.31

Half of the Moon is always lighted by the Sun, and half is always in the shadow. The Moon phases result from the view of the lighted and dark parts as the Moon revolves around Earth.

Moon, and half is always in shadow. As the Moon's path takes it between Earth and the Sun, then to the dark side of Earth, you see different parts of the illuminated half called *phases* (figure 18.31). When the Moon is on the dark side of Earth you see the entire illuminated half of the Moon called the *full moon* (or the full phase). Halfway around the orbit, the lighted side of the Moon now faces away from Earth, and the unlighted side now faces Earth. This dark appearance is called the *new moon* (or the new phase). In the new phase the Moon is not *directly* between Earth and the Sun, so it does not produce an eclipse (see the next section).

As the Moon moves from the new phase in its orbit around Earth, you will eventually see half the lighted surface, which is known as the *first quarter*. Often the unlighted part of the Moon shines with a dim light of reflected sunlight from Earth called *earthshine*. Note that the division between the lighted and unlighted part of the Moon's surface is curved in an arc. A straight line connecting the ends of the arc is perpendicular to the direction of the Sun (figure 18.32). After the first quarter the Moon moves to its full phase, then to the *last quarter* (see figure 18.31). The period of time between two consecutive phases, such as new moon to new moon, is the synodic month, or about 29.5 days.

Eclipses of the Sun and the Moon

Sunlight is not visible in the emptiness of space because there is nothing to reflect the light, so the long conical shadow behind each spherical body is not visible either. One side of Earth and one side of the Moon are always visible because they reflect sunlight. The shadow from Earth or from the Moon becomes noticeable only when it falls on the illuminated surface of the other body. This event of Earth's or the Moon's shadow falling on the other body is called an **eclipse.** Most of the time eclipses do not occur because the plane of the Moon's orbit is inclined to Earth's orbit about 5° (figure 18.33). As a result, the shadow

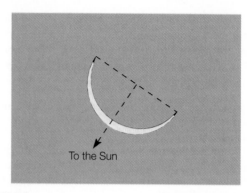

FIGURE 18.32

The cusps, or horns, of the Moon always point away from the Sun. A line drawn from the tip of one cusp to the other is perpendicular to a straight line between the Moon and the Sun.

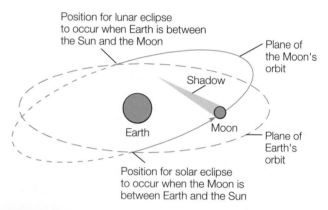

FIGURE 18.33

The plane of the Moon's orbit is inclined to the plane of Earth's orbit by about 5°. An eclipse occurs only where the two planes intersect, and Earth, the Moon, and the Sun are in a line.

from the Moon or the shadow from Earth usually falls above or below the other body, too high or too low to produce an eclipse. An eclipse occurs only when the Sun, Moon, and Earth are in a line with each other.

The shadow from Earth and the shadow from the Moon are long cones that point away from the Sun. Both cones have two parts, an inner cone of a complete shadow called the *umbra* and an outer cone of partial shadow called the *penumbra*. When and where the umbra of the Moon's shadow falls on Earth, people see a **total solar eclipse.** During a total solar eclipse the new Moon completely covers the disk of the Sun. The total solar eclipse is preceded and followed by a partial eclipse, which is seen when the observer is in the penumbra. If the observer is in a location where only the penumbra passes, then only a partial eclipse will be observed (figure 18.34). More people see partial than full solar eclipses because the penumbra covers a larger area. The occurrence of a total solar eclipse is a rare event in a given location, occurring once every several hundred years and then lasting for less than seven minutes.

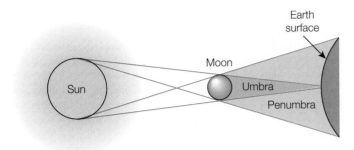

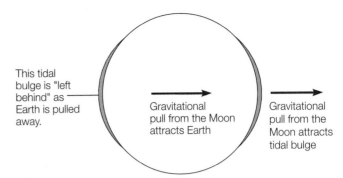

FIGURE 18.34

People in a location where the tip of the umbra falls on the surface of Earth see a total solar eclipse. People in locations where the penumbra falls on Earth's surface see a partial solar eclipse.

FIGURE 18.35

Gravitational attraction pulls on Earth's waters on the side of Earth facing the Moon, producing a tidal bulge. A second tidal bulge on the side of Earth opposite the Moon is produced when Earth, which is closer to the Moon, is pulled away from the waters.

The Moon's cone-shaped shadow averages a length of 375,000 km (about 233,000 mi), which is less than the average distance between Earth and the Moon. The Moon's elliptical orbit brings it sometimes closer to and sometimes farther from Earth. A total solar eclipse occurs only when the Moon is close enough so at least the tip of its umbra reaches the surface of Earth. If the Moon's umbra fails to reach Earth, an **annular eclipse** occurs. Annular means "ring-shaped," and during this eclipse the edge of the Sun is seen to form a bright ring around the Moon. As before, people located in the area where the penumbra falls will see a partial eclipse. The annular eclipse occurs more frequently than the total solar eclipse.

When the Moon is full and the Sun, Moon, and Earth are lined up so Earth's shadow falls on the Moon a **lunar eclipse** occurs. Earth's shadow is much larger than the Moon's diameter, so a lunar eclipse is visible to everyone on the night side of Earth. This larger shadow also means a longer eclipse that may last for hours. As the umbra moves over the Moon, the darkened part takes on a reddish, somewhat copper-colored glow from light refracted and scattered into the umbra by Earth's atmosphere. This light passes through the thickness of Earth's atmosphere on its way to the eclipsed Moon, and it acquires the reddish color for the same reason that a sunset is red: much of the blue light has been removed by scattering in Earth's atmosphere.

Tides

If you live near or have ever visited a coastal area of the ocean, you are familiar with the periodic rise and fall of sea level known as **tides.** The relationship between the motions of the Moon and the magnitude and timing of tides has been known and studied since very early times. These relationships are that (1) the greatest range of tides occurs at full and new moon phases, (2) the least range of tides occurs at quarter moon phases, and (3) in most oceans the time between two high tides or between two low tides is an average of twelve hours and twenty-five minutes. The period of twelve hours and twenty-five minutes is half the average time interval between consecutive passages of the Moon across the celestial meridian. A location on the surface of Earth is directly under the Moon when it crosses the meridian, and

directly opposite it on the far side of Earth an average twelve hours and twenty-five minutes later. There are two *tidal bulges* that follow the Moon as it moves around Earth, one on the side facing the Moon and one on the opposite side. In general, tides are a result of these bulges moving westward around Earth.

A simplified explanation of the two tidal bulges involves two basic factors, the gravitational attraction of the Moon and the motion of the Earth-Moon system (figure 18.35). Water on Earth's surface is free to move, and the Moon's gravitational attraction pulls the water to the tidal bulge on the side of Earth facing the Moon. This tide-raising force directed toward the Moon bulges the water in mid-ocean some .75 m (about 2.5 ft), but it also bulges the land, producing a land tide. Since Earth is much more rigid than water, the land tide is much smaller at about 12 cm (about 4.5 in). Since all parts of the land bulge together, this movement is not evident without measurement by sensitive instruments.

The tidal bulge on the side of Earth opposite the Moon occurs as Earth is pulled away from the ocean by the Earth-Moon gravitational interaction. Between the tidal bulges facing the Moon and the tidal bulge on the opposite side, sea level is depressed across the broad surface. The depression is called a *tidal trough,* even though it does not actually have the shape of a trough. The two tidal bulges, with the trough between, move slowly eastward following the Moon. Earth turns more rapidly on its axis, however, which forces the tidal bulge to stay in front of the Moon moving through its orbit. Thus, the tidal axis is not aligned with the Earth-Moon gravitational axis.

The tides do not actually appear as alternating bulges that move around Earth. There are a number of factors that influence the making and moving of the bulges in complex interactions that determine the timing and size of a tide at a given time in a given location. Some of these factors include (1) the relative positions of Earth, the Moon, and the Sun, (2) the elliptical orbit of the Moon, which sometimes brings it closer to Earth, and (3) the size, shape, and depth of the basin holding the water.

The relative positions of Earth, the Moon, and the Sun determine the size of a given tide because the Sun, as well as the Moon, produces a tide-raising force. The Sun is much more massive than the Moon, but it is so far away that its tide-raising force is about half that of the closer Moon. Thus, the Sun basically modifies lunar tides rather than producing distinct tides of its own. For example, Earth, the Moon, and the Sun are nearly in line during the full and new moon phases. At these times the lunar and solar tide-producing forces act together, producing tides that are unusually high and corresponding low tides that are unusually low. The period of these unusually high and low tides are called periods of *spring tides.* Spring tides occur every two weeks and have nothing to do with the spring season. When the Moon is in its quarter phases, the Sun and Moon are at right angles to one another and the solar tides occur between the lunar tides, causing unusually less pronounced high and low tides called *neap tides.* The period of neap tides also occurs every two weeks.

The size of the lunar-produced tidal bulge varies as the Moon's distance from Earth changes. The Moon's elliptical orbit brings it closest to Earth at a point called **perigee** and farthest from Earth at a point called **apogee.** At perigee the Moon is about 44,800 km (about 28,000 mi) closer to Earth than at apogee, so its gravitational attraction is much greater. When perigee coincides with a new or full moon, especially high spring tides result.

The open basins of oceans, gulfs, and bays are all connected but have different shapes and sizes and have bordering landmasses in all possible orientations to the westward-moving tidal bulges. Water in each basin responds differently to the tidal forces, responding as periodic resonant oscillations that move back and forth much like the water in a bowl shifts when carried. Thus, coastal regions on open seas may experience tides that range between about 1 and 3 m (about 3 to 10 ft), but mostly enclosed basins such as the Gulf of Mexico have tides less than about 1/3 m (about 1 ft). The Gulf of Mexico, because of its size, depth, and limited connections with the open ocean, responds only to the stronger tidal attractions and has only one high and one low tide per day. Even lakes and ponds respond to tidal attractions, but the result is too small to be noticed. Other basins, such as the Bay of Fundy in Nova Scotia, are funnel-shaped and undergo an unusually high tidal range. The Bay of Fundy has experienced as much as a 15 m (about 50 ft) tidal range.

As the tidal bulges are pulled against a rotating Earth, friction between the moving water and the ocean basin tends to slow Earth's rotation over time. This is a very small slowing effect that is increasing the length of each day by about 1.5 seconds per 100,000 years. Evidence for this slowing comes from a number of sources including records of ancient solar eclipses. The solar eclipses of 2,000 years ago occurred 3 hours earlier than would be expected by using today's time but were on the mark if a lengthening day is considered. Fossils of a certain species of coral still living today provide further evidence of a lengthening day. This particular coral adds daily growth rings, and 500-million-year-old fossils show that the day was about 21 hours long at that time. Finally, the Moon is moving away from Earth at a rate of about 4 cm (about 1.5 in) per year. This movement out to a larger orbit is a necessary condition to conserve angular momentum as Earth slows. As the Moon moves away from Earth the length of the month increases. Some time in the distant future both the day and the month will be equal, about fifty of the present days long.

SUMMARY

The earth is an *oblate spheroid* that undergoes three basic motions: (1) a yearly *revolution* around the Sun, (2) a daily *rotation* on its axis, and (3) a slow wobble of its axis called *precession.*

As Earth makes its yearly *revolution* around the Sun, it maintains a generally *constant inclination of its axis* to the *plane of the ecliptic* of 66.5°, or 23.5° from a line perpendicular to the plane. In addition, Earth maintains a generally constant *orientation of its axis* to the stars, which always points in the same direction. The constant inclination and orientation of the axis, together with Earth's rotation and revolution, produce three effects: (1) days and nights that vary in length, (2) seasons that change during the course of a year, and (3) climates that vary with latitude. When Earth is at a place in its orbit so the axis points toward the Sun, the Northern Hemisphere experiences the longest days and the summer season. This begins on June 21 or 22, which is called the *summer solstice.* Six months later, the axis points away from the Sun and the Northern Hemisphere experiences the shortest days and the winter season. This begins on December 22 or 23 and is called the *winter solstice.* On March 20 or 21 Earth is halfway between the solstices and has days and nights of equal length, which is called the *spring* (or *vernal*) *equinox.* On September 22 or 23 the *autumnal equinox,* another period of equal nights and days, identifies the beginning of the fall season.

Precession is a slow wobbling of the axis as Earth spins. Precession is produced by the gravitational tugs of the Sun and Moon on Earth's equatorial bulge.

Lines around the earth that are parallel to the equator are circles called *parallels.* The distance from the equator to a point on a parallel is called the *latitude* of that point. North and south arcs that intersect at the poles are called *meridians.* The meridian that runs through the Greenwich Observatory is a reference line called the *prime meridian.* The distance of a point east or west of the prime meridian is called the *longitude* of that point.

The event of time called *noon* is the instant the Sun appears to move across the celestial meridian. The instant of noon at a particular location is called the *apparent local noon.* The time at a given place that is determined by a sundial is called the *apparent local solar time.* It is the basis for an averaged, uniform standard time called the *mean solar time.* Mean solar time is the time used to set clocks.

A *sidereal day* is the interval between two consecutive crossings of the celestial meridian by a star. An *apparent solar day* is the interval between two consecutive crossings of the celestial meridian by the Sun, from one apparent solar noon to the next. A *mean solar day* is twenty-four hours as determined from mean solar time. The *equation of time* shows how the local solar time is faster or slower than the clock time during different days of the year.

Earth's surface is divided into one-hour *standard time zones* that are about 15° of meridian wide. The *international date line* is the 180° meridian; you gain a day if you cross this line while traveling westward and repeat a day if you are traveling eastward.

A *tropical year* is the interval between two consecutive spring equinoxes. A *sidereal year* is the interval of time between two consecutive crossings of a star by the Sun. It is the tropical year that is used as a standard time interval for the calendar year. A *sidereal month* is the interval of time between two consecutive crossings of a star by the Moon. The *synodic month* is the interval of time from a new moon to the next new moon. The *synodic month* is about 29½ days long, which is about ½ of a year.

The surface of the Moon has light-colored mountainous regions called *highlands,* smooth dark areas called *maria,* and many sizes of

A CLOSER LOOK | Earth's Motions and Ice Ages

The evidence is abundant and widespread that Earth has undergone repeated cycles of glaciation in the past, periods when glaciers and ice sheets covered much of the land. There have been at least ten of these periods of glaciation, or *ice ages* as they are called, in the past million years. The most recent ice age maximum occurred about 18,000 years ago when about one-third of Earth's land was covered with ice. A great sheet of ice, perhaps several kilometers (about 1.25 mi) thick in places, covered all of Canada into the United States as far south as Long Island, along the Ohio and Missouri Rivers, and along a line to Oregon. Half the states in the United States were partly or completely covered by ice during this ice age. At that time, average temperatures were roughly 8°C (about 14°F) lower than the average today in temperate climates. About 10,000 years ago the average temperature began to increase and the North American ice sheet melted back, retreating from its maximum advance. The average temperature then increased gradually until some 5,000 years ago, cooling slightly since. Other temperature fluctuations have followed, most recently causing the rapid retreat of glaciers in Alaska and Canada.

Is a new ice age coming in the future or is the climate becoming warmer? To make such a long-range climatic forecast requires some explanation of why ice ages occur. There are different theories about the cause of ice ages, ranging from variations in solar energy output to the movement of continents, but recent evidence seems to support a theory concerning Earth's motions around the Sun. The theory is called the Milankovitch theory after the geologist who originally proposed it in 1920. This theory proposes that long-term variations in Earth's climate are a result of a combination of three factors: (1) slow changes in the size and shape of Earth's orbit

around the Sun, (2) slow changes in the inclination of Earth's axis, and (3) slow changes in the direction of Earth's axis with respect to the stars. These three factors are seen to result in changes in the amount of solar energy received at different latitudes during different seasons of the year.

The changes in the basic Earth motions of (1) orbital size and shape, (2) tilt, or inclination of axis, and (3) precession are produced by small gravitational attractions of the Moon and the giant planets of Jupiter and Saturn. They produce nearly periodic changes that can be calculated with great accuracy for periods of thousands of years.

The size and shape of Earth's orbit varies with a period of about 100,000 years and changes the solar energy received by Earth by a factor of about 0.3 percent over a million years. The tilt of Earth's axis, on the other hand, varies between 22.1° and 24.5° with a period of about 40,000 years. This period of changes of tilt does not alter the total amount of solar energy received by Earth, but it does vary the amount of sunlight received during the summers and winters of both hemispheres by as much as 20 percent. This would be important because in order for ice to accumulate, growing into an ice sheet, snow must survive the warmer summer months. Less sunlight than normal in the Northern Hemisphere could result in a growing ice sheet, moving the climate toward an ice age.

Earth's precession, with its 26,000-year period, couples with the rotation of Earth's elliptical orbit around the Sun to determine the time of perihelion. The time of perihelion has a period of 20,000 years, presently occurring during the summer of the Southern Hemisphere. As a whole, Earth receives about 6 percent more solar energy at perihelion than six months later. In about another 12,000 years, precession will have moved the

time of perihelion so it occurs during the summer of the Northern Hemisphere.

The combination of changes in Earth's orbit do correlate with the times of the ice ages according to geologic evidence as well as computer analysis. Computer programs analyzed the amount of solar energy received at summertime high-latitude locations in the Northern Hemisphere according to variations in Earth's orbit, axial tilt, and time of perihelion. The analysis predicted major ice ages occurring in a 100,000 year cycle with minor fluctuations at periods of 20,000 and 40,000 years. This prediction correlates roughly with the evidence of past ice ages according to the geologic evidence. The computer model also indicates that Earth will reach another cold period in 60,000 years after some periods of lesser variations.

There are apparently connections between variations in Earth's orbit and the time of ice ages, but it is nonetheless difficult to use this information alone to predict the next ice age. Other major factors are important as well as changes in Earth's orbit. Volcanic eruptions or even a collision with a large meteorite could inject major amounts of dust into the atmosphere. On a small scale, a single volcano in recent times was observed to inject a sufficient amount of dust into the atmosphere to scatter and reflect sunlight, lowering the average summer temperature by several degrees. On the other hand, the continued burning of fossil fuels and destruction of rain forests are increasing the present-day carbon dioxide concentration of the atmosphere. There is a possibility of an increasing greenhouse effect, or warming trend, from such human activities. All of the variations, fluctuations, and possibilities emphasize the difficulty of climatic forecasting, even with understandings about the role of changes in Earth's orbit and the ice ages.

craters and is covered by a layer of fine *dust.* Samples of rocks returned to Earth by *Apollo* astronauts revealed that the highlands are composed of basalt breccias that were formed some 4 billion years ago. The maria are basalts that formed from solidified lava some 3.1 to 3.8 billion years ago. This and other data indicate that the Moon developed through four stages.

Earth and the Moon act as a system, with both bodies revolving around a common center of mass located under Earth's surface. This combined motion around the Sun produces three phenomena: (1) as the Earth-Moon system revolves around the Sun different parts of the illuminated lunar surface, called *phases,* are visible from Earth; (2) a *solar eclipse* is observed where the Moon's shadow falls on Earth, and a *lunar eclipse* is observed where Earth's shadow falls on the Moon; and (3) the *tides,* a periodic rising and falling of sea level, are produced by gravitational attractions of the Moon and the Sun and by the movement of the Earth-Moon system.

KEY TERMS

annular eclipse (p. **455**)
Antarctic Circle (p. **443**)
apogee (p. **456**)
Arctic Circle (p. **443**)
Coriolis effect (p. **441**)
eclipse (p. **454**)
equinoxes (p. **440**)
international date line (p. **448**)
latitude (p. **442**)
longitude (p. **443**)
lunar eclipse (p. **455**)

meridians (p. **442**)
noon (p. **444**)
parallels (p. **442**)
perigee (p. **456**)
precession (p. **441**)
prime meridian (p. **442**)
solstices (p. **439**)
tides (p. **455**)
total solar eclipse (p. **454**)
tropic of Cancer (p. **443**)
tropic of Capricorn (p. **443**)

APPLYING THE CONCEPTS

1. Earth is undergoing a combination of how many different motions?
   a. zero
   b. one
   c. three
   d. seven

2. In the Northern Hemisphere, city A is located a number of miles north of city B. At 12:00 noon in city B, the Sun appears directly overhead. At this very same time, the Sun over city A will appear
   a. to the north of overhead.
   b. directly overhead.
   c. to the south of overhead.

3. Earth as a whole receives more solar energy during what month?
   a. January
   b. March
   c. July
   d. September

4. During the course of a year and relative to the Sun, Earth's axis points
   a. always toward the Sun.
   b. toward the Sun half the year and away the other half.
   c. always away from the Sun.
   d. toward the Sun for half a day and away from the Sun the other half.

5. If you are located at 20°N latitude, when will the Sun appear directly overhead?
   a. never
   b. once a year
   c. twice a year
   d. four times a year

6. If you are located on the equator (0° latitude) when will the Sun appear directly overhead?
   a. never
   b. once a year
   c. twice a year
   d. four times a year

7. If you are located at 40°N latitude, when will the Sun appear directly overhead?
   a. never
   b. once a year
   c. twice a year
   d. four times a year

8. During the equinoxes
   a. a vertical stick in the equator will not cast a shadow at noon.
   b. at noon the Sun is directly overhead at 0° latitude.
   c. daylight and night are of equal length.
   d. All of the above are correct.

9. Evidence that Earth is rotating is provided by
   a. varying lengths of night and day during a year.
   b. seasonal climatic changes.
   c. stellar parallax.
   d. a pendulum.

10. In about 12,000 years, the star Vega will be the North Star, not Polaris, because of Earth's
    a. uneven equinox.
    b. tilted axis.
    c. precession.
    d. recession.

11. The significance of the tropic of Cancer (23.5°N latitude) is that
    a. the Sun appears directly overhead north of this latitude some time during a year.
    b. the Sun appears directly overhead south of this latitude some time during a year.
    c. the Sun appears above the horizon all day for 6 months during the summer north of this latitude.
    d. the Sun appears above the horizon all day for 6 months during the summer south of this latitude.

12. The significance of the Arctic Circle (66.5°N latitude) is that
    a. the Sun appears directly overhead north of this latitude some time during a year.
    b. the Sun appears directly overhead south of this latitude some time during a year.
    c. the Sun appears above the horizon all day at least one day during the summer.
    d. the Sun appears above the horizon all day for 6 months during the summer.

13. In the time 1:00 P.M. the P.M. means
    a. "past morning."
    b. "past midnight."
    c. "before the meridian."
    d. "after the meridian."

14. Clock time is based on
    a. sundial time.
    b. an averaged apparent solar time.
    c. the apparent local solar time.
    d. the apparent local noon.

15. An apparent solar day is
    a. the interval between two consecutive local solar noons.
    b. about four minutes longer than the sidereal day.
    c. of variable length throughout the year.
    d. All of the above are correct.

16. The time as read from a sundial is the same as the time read from a clock
    a. all the time.
    b. only once a year.
    c. twice a year.
    d. four times a year.

17. You are traveling west by jet and cross three time zone boundaries. If your watch reads 3:00 P.M. when you arrive, you should reset it to
    a. 12:00 noon.
    b. 6:00 P.M.
    c. 12:00 midnight.
    d. 6:00 A.M.

18. If it is Sunday when you cross the international date line while traveling westward, the next day is
    a. Wednesday.
    b. Sunday.
    c. Tuesday.
    d. Saturday.

19. What has happened to the surface of the Moon during the last 3 billion years?
    a. heavy meteorite bombardment, producing craters
    b. widespread lava flooding from the interior
    c. both widespread lava flooding and meteorite bombardment
    d. not much

20. If you see a full moon, an astronaut on the Moon looking back at Earth would see a
    a. full Earth.
    b. new Earth.
    c. first quarter Earth.
    d. last quarter Earth.

21. A lunar eclipse can occur only during the moon phase of
    a. full moon.
    b. new moon.
    c. first quarter.
    d. last quarter.

22. A total solar eclipse can occur only during the moon phase of
    a. full moon.
    b. new moon.
    c. first quarter.
    d. last quarter.

23. A lunar eclipse does not occur every month because
    a. the plane of the Moon's orbit is inclined to the ecliptic.
    b. of precession.
    c. Earth moves faster in its orbit when closest to the Sun.
    d. Earth's axis is tilted with respect to the Sun.

24. The least range between high and low tides occurs during
    a. full moon.
    b. new moon.
    c. quarter moon phases.
    d. an eclipse.

Answers

1. d 2. c 3. a 4. b 5. c 6. c 7. a 8. d 9. d 10. c 11. b 12. c 13. d 14. b 15. d 16. d 17. a 18. c 19. d 20. b 21. a 22. b 23. a 24. c

QUESTIONS FOR THOUGHT

1. Briefly describe the more conspicuous of Earth's motions. Identify which of these motions are independent of the Sun and the galaxy.
2. Describe some evidences that (a) Earth is shaped like a sphere and (b) that Earth moves.
3. Use sketches with brief explanations to describe how the constant inclination and constant orientation of Earth's axis produces (a) a variation in the number of daylight hours and (b) a variation in seasons throughout a year.
4. Where on Earth are you if you observe the following at the instant of apparent local noon on September 23? (a) The shadow from a vertical stick points northward. (b) There is no shadow on a clear day. (c) The shadow from a vertical stick points southward.
5. What is the meaning of the word "solstice"? What causes solstices? On about what dates do solstices occur?
6. What is the meaning of "equinox"? What causes equinoxes? On about what dates do equinoxes occur?
7. What is precession?
8. Briefly describe how Earth's axis is used as a reference for a system that identifies locations on Earth's surface.
9. Use a map or a globe to identify the latitude and longitude of your present location.
10. The tropic of Cancer, tropic of Capricorn, Arctic Circle, and Antarctic Circle are parallels that are identified with specific names. What parallels do the names represent? What is the significance of each?
11. What is the meaning of (a) noon, (b) A.M., and (c) P.M.?

12. Explain why the time shown by a sundial does not usually agree with the time shown by a clock. Describe how the sundial time can be corrected to clock time.

13. Explain why standard time zones were established. In terms of longitude, how wide is a standard time zone? Why was this width chosen?

14. When it is 12:00 noon in Texas, what time is it (a) in Jacksonville, Florida? (b) in Bakersfield, California? (c) at the North Pole?

15. On what date is Earth closest to the Sun? What season is occurring in the Northern Hemisphere at this time? Explain this apparent contradiction.

16. Explain why a lunar eclipse is not observed once a month.

17. Use a sketch and briefly describe the conditions necessary for a total eclipse of the Sun.

18. Using sketches, briefly describe the positions of Earth, the Moon, and the Sun during each of the major moon phases.

19. If you were on the Moon as people on Earth observed a full moon, in what phase would you observe Earth?

20. Briefly describe three theories about the origin of the Moon.

21. What are the smooth, dark areas that can be observed on the face of the Moon? When did they form?

22. What made all the craters that can be observed on the Moon? When did this happen?

23. What phase is the Moon in if it rises at sunset? Explain your reasoning.

24. Why doesn't an eclipse of the Sun occur at each new moon when the Moon is between Earth and the Sun?

25. Is the length of time required for the Moon to make one complete revolution around Earth the same length of time required for a complete cycle of moon phases? Explain.

26. What is an annular eclipse? Which is more common, an annular eclipse or a total solar eclipse? Why?

27. Does an eclipse of the Sun occur during any particular moon phase? Explain.

28. Identify the moon phases that occur with (a) a spring tide and (b) a neap tide.

29. What was the basic problem with the Julian calendar? How does the Gregorian calendar correct this problem?

30. What is the source of the dust found on the Moon?

31. Describe the four stages in the Moon's history.

32. Explain why everyone on the dark side of Earth can see a lunar eclipse, but only a limited few ever see a solar eclipse on the lighted side.

33. Explain why there are two tidal bulges on opposite sides of Earth.

34. Describe how the Foucault pendulum provides evidence that Earth turns on its axis.

35. Why are consecutive high tides commonly twelve hours and twenty-five minutes apart?

These rose-red rhodochrosite crystals are a naturally occurring form of manganese carbonate. Rhodochrosite is but one of about 2,500 minerals that are known to exist, making up the solid materials of the earth's crust.

CHAPTER | Nineteen

The Earth

Understandings about the earth come from many branches of science, each with specialized scientists who consider different scales of space and time. *Astronomers* study the motions of Earth in space, its place in the universe, and how the earth formed and evolved over time. *Oceanographers* are primarily concerned with the composition and motions of the earth's oceans. *Meteorologists* are concerned with the composition of the atmosphere and how it changes over time. *Geologists* are concerned primarily with the rocks, landscapes, and the history of the earth through time. Each of these branches of science has its own subdisciplines. For example, there are geologists who study rocks (petrology), geologists who study earthquakes and the earth's interior (seismology), and geologists who study the early history of life on the earth (paleontology). Together, all the scientists who study the earth are known as *earth scientists*.

The scope of the earth sciences has been expanded in recent years to include environmental sciences. Environmental sciences are concerned with the environmental conditions on the earth that affect living organisms. Earth science is a natural home for environmental sciences since studies of environmental conditions include the atmosphere and the oceans of the earth, as well as the land surface.

The separation of the earth sciences into independent branches and subdisciplines was traditionally done for convenience. This made it easier to study a large and complex Earth. In the past, scientists in each branch studied their field without considering the earth as an interacting whole. Today, most earth scientists consider changes in the earth as taking place in an overall dynamic system. The parts of the earth's interior, the rocks on the surface, the oceans, the atmosphere, and the environmental conditions are today understood to be parts of a complex, interacting system with a cyclic movement of materials from one part to another.

How can materials cycle through changes from the interior to the surface, and to the atmosphere and back? As you will see in this and the following chapters, the answer to this question is found in the unique combination of fluids of the earth. No other known planet has Earth's combination of (1) an atmosphere consisting mostly of nitrogen and oxygen, (2) a surface that is mostly covered with liquid water, and (3) an interior that is partly fluid, partly semifluid, and partly solid. Earth's atmosphere is unique both in terms of its composition and in terms of interactions with the liquid water surface (figure 19.1). These interactions have cycled materials, such as carbon dioxide, from the atmosphere to the land and oceans of the earth. The internal flow of rock materials, on the other hand, produces the large-scale motion of the earth's continents and the associated phenomena of earthquakes and volcanoes. Volcanoes cycle carbon dioxide back into the atmosphere and the movement of land cycles rocks from the earth's interior to the surface and back to the interior again. Altogether, the earth's atmosphere, liquid water, and motion of its landmasses make up a dynamic cycling system that is found only on the planet Earth.

Earth also seems to be unique because there is life on Earth, but apparently not on the other planets. The cycling of atmospheric gases and vapors, waters of the surface, and flowing interior rock materials sustain a wide diversity of life on Earth. There are some two million different species of plants and animals. Yet, there is no evidence of even one species of life existing outside the earth. The existence of life on Earth must be directly related to its unique, dynamic system of interacting fluids.

This and the chapters that follow are about the dynamic nature of planet Earth. This chapter is concerned with earth materials, the internal structure of the earth, and the movement of land across the surface of the earth.

EARTH MATERIALS

The earth, like all other solid matter in the universe, is made up of atoms of the chemical elements. The different elements are not distributed equally throughout the mass of the earth, however, nor are they equally abundant. The earth was molten during an early stage in its development, and this has influenced distribution of the elements. During the molten stage the heavier abundant elements, such as iron and nickel, apparently sank to the deep interior of the earth, leaving a thin layer of lighter elements on the surface. This relatively thin layer is called the *crust*. The rocks and rock materials that you see on the surface and the materials sampled in the deepest mines and well holes are all materials of the earth's crust. The bulk of the earth's mass lies below the crust and has not been directly sampled.

Chemical analysis of virtually thousands of rocks from the earth's surface found that only eight elements make up about 98.6 percent of the crust. All the other elements make up the remaining 1.4 percent of the crust. Oxygen is the most abundant element, making up about 50 percent of the weight of the crust.

FIGURE 19.1

No other planet in the solar system has the unique combination of fluids of Earth. Earth has a surface that is mostly covered with liquid water, water vapor in the atmosphere, and both frozen and liquid water on the land.

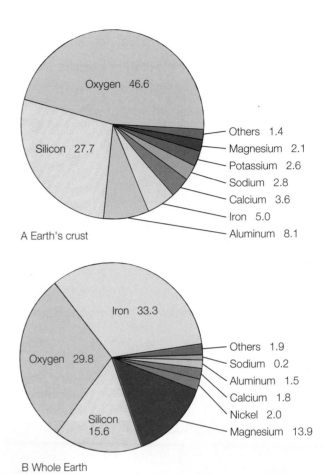

A Earth's crust

B Whole Earth

FIGURE 19.2

(*A*) The percentage by weight of the elements that make up the earth's crust. (*B*) The percentage by weight of the elements that make up the whole earth.

Silicon makes up over 25 percent, so oxygen and silicon alone make up about 75 percent of the earth's solid surface.

Figure 19.2 shows the eight most abundant elements that occur as elements or combine to form the chemical compounds of the earth's crust. They make up the solid materials of the earth's crust that are known as *minerals* and *rocks*. For now, consider a mineral to be a solid material of the earth that has both a known chemical composition and a crystalline structure that is unique to that mineral. About 2,500 minerals are known to exist, but only about 20 are common in the crust. Examples of these common minerals are quartz, calcite, and gypsum.

Minerals are the fundamental building blocks of the rocks making up the earth's crust. A rock is a solid aggregation of one or more minerals that have been cohesively brought together by a rock-forming process. There are many possibilities of different kinds of rocks that could exist from many different variations of mineral mixtures. Within defined ranges of composition, however, there are only about 20 common rocks making up the crust of the earth. Examples of common rocks are sandstone, limestone, and granite.

How the atoms of elements combine depends on the number and arrangement of electrons around the atoms. How they form a mineral, with specific properties and a specific crystalline structure, depends on the electrons and the size of the ions as well. A discussion of the chemical principles that determine the structure and properties of chemical compounds is found in chapters 10 and 11 of this book. You may wish to review these principles before continuing with the discussion of common minerals and rocks.

Minerals

In everyday usage the word "mineral" can have several different meanings. It can mean something your body should have (vitamins and minerals), something a fertilizer furnishes for a plant

(nitrogen, potassium, and phosphorus), or sand, rock, and coal taken from the earth for human use (mineral resources). In the earth sciences, a **mineral** is defined as a naturally occurring, inorganic solid element or compound with a crystalline structure (figure 19.3). This definition means that the element or compound cannot be synthetic (must be naturally occurring), must not be directly produced by a living organism (must be inorganic), and must have atoms arranged in a regular, repeating pattern (a crystal structure). Note that the crystal structure of a mineral can be present on the microscopic scale and it is not necessarily obvious to the unaided eye. Even crystals that could be observed with the unaided eye are sometimes not noticed (figure 19.4).

The crystal structure of a mineral can be made up of atoms of one or more kinds of elements. Diamond, for example, is a mineral with only carbon atoms in a strong crystal structure. Quartz, on the other hand, is a mineral with atoms of silicon and oxygen in a different crystal structure (figure 19.5). No matter how many kinds of atoms are present, each mineral has its own defined chemical composition or range of chemical compositions. A range of chemical composition is possible because the composition of some minerals can vary

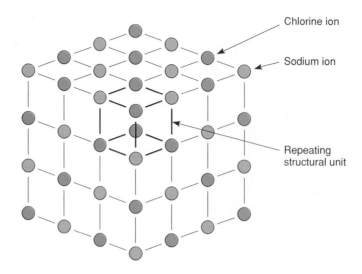

Chlorine ion

Sodium ion

Repeating structural unit

FIGURE 19.3

A crystal is composed of a structural unit that is repeated in three dimensions. This is the basic structural unit of a crystal of sodium chloride, the mineral halite.

FIGURE 19.4

The structural unit for a crystal of table salt, sodium chloride, is cubic, as you can see in the individual grains.

FIGURE 19.5

These quartz crystals are hexagonal prisms.

A CLOSER LOOK | Mineral Resources and the Environment

Ancient humans exploited mineral resources as they mined chert and copper minerals for the making of tools and used salt, clay, and other mineral materials for nutrients and pot making. These early people were few in number and their simple tools made little impact on the environment as they mined what they needed. As the numbers of people grew and technology advanced, more and more mineral resources were exploited to build bigger and better machines and provide increasing amounts of energy. With these advances in population size and technology came increasing impacts on the environment in both size and scope. In addition to chert, copper minerals, and clay, the metal ores of iron, chromium, aluminum, nickel, tin, uranium, manganese, platinum, cobalt, zinc, and many others were now in high demand.

Today, there are three categories of costs recognized with the exploitation of any mineral resource. First, the *economic costs* are the moneys needed to lease or buy land, acquire equipment, and pay for labor to run the equipment. A second category is the *resource costs* of exploiting the resource. It takes energy to concentrate the ore and transport it to smelters or refineries. Sometimes other resources are needed, such as large quantities of water for the extraction or concentration of a mineral resource. If the energy and water are not readily available the resource costs might be converted to economic costs, which could ultimately determine if the operation would be profitable or not. Finally, the *environmental costs* of exploiting the resource must be considered. The environmental costs might include increased air and water pollution, loss of scenic quality, and change or destruction of an ecosystem. Environmental costs are converted to economic costs as controls on pollution are enforced. It is expensive to clean pollution from the land and to restore the ecosystem that was changed by mining operations. Consideration of the conversion of environmental costs to economic costs can also help determine if a mining operation would be profitable or not.

All mining operations start by making a mineral resource accessible so it can be removed. This might take place by strip mining, which begins with the removal of the top layers of soil and rock overlying a resource deposit. This overburden is placed somewhere else, to the side, so the mineral deposit can be easily removed. Access to a smaller, deeper mineral deposit might be gained by building a tunnel to the resource. The debris from building such a tunnel is usually piled outside the entrance. The rock debris from both strip and tunnel mining is an eyesore and it is difficult for vegetation to grow on the barren rock. Since plants are not present, water may wash away small rock particles causing erosion of the land and silting of the streams. The debris might also contain arsenic, lead, and other minerals that can pollute the water supply.

Environmental costs of mining mineral resources are becoming economic costs as regulations on the mining industry require less environmental damage than has been previously tolerated. The cost of finding and processing the minerals is also increasing as the easiest to use, less expensive resources were exploited first. As current mineral resource deposits become exhausted there will be increasing pressure to explore and exploit protected areas. The environmental costs for exploitation of these areas—if permitted—will indeed be large.

---

with the substitution of chemically similar elements. For example, some atoms of magnesium might be substituted for some chemically similar atoms of calcium. Such substitutions might slightly alter some properties, but not enough to make a different mineral.

Crystals can be classified and identified on the basis of the symmetry of their surfaces. This symmetry is an outward expression of the internal symmetry in the arrangement of the atoms making up the crystal. Thus, on the basis of symmetry, crystalline substances are classified into six major systems, which, in turn, are subdivided into smaller groups.

Silicon and oxygen are the most abundant elements in the earth's crust and, as you would expect, the most common minerals contain these two elements. All minerals are classified on the basis of whether the mineral structure contains these two elements or not. The two main groups are thus called the *silicates* and the *nonsilicates* (table 19.1). Note, however, that the silicates can contain some other elements in addition to silicon and oxygen. The silicate minerals are by far the most abundant, making up about 92 percent of the earth's crust. When an atom of silicon (Si^{+4}) combines with four oxygen atoms (O^{-2}), a tetrahedral structure of $(SiO_4)^{-4}$ forms (see figure 19.6). All **silicates** have a basic silicon-oxygen tetrahedral unit either isolated or joined together in the crystal structure. The structure has four unattached electrons on the oxygen atoms that can combine with metallic ions such as iron or magnesium. They can also combine with the silicon atoms of *other* tetrahedral units. Some silicate minerals are thus made up of single tetrahedral units combined with metallic ions. Other silicate minerals are combinations of

TABLE 19.1 **Classification scheme of some common minerals**

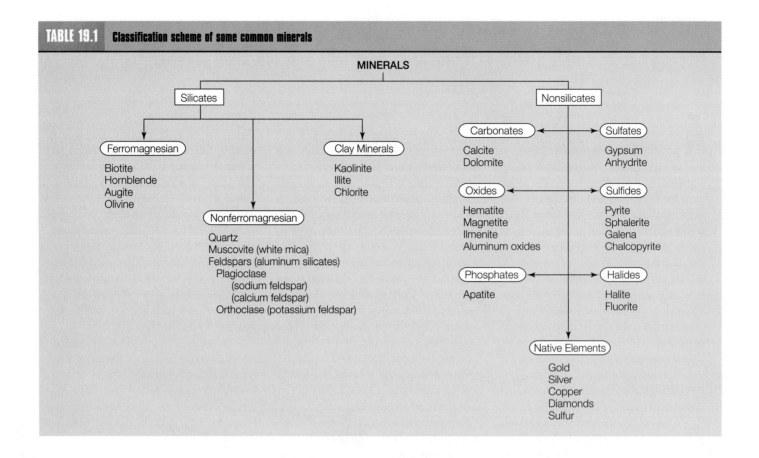

MINERALS

Silicates

Ferromagnesian
Biotite
Hornblende
Augite
Olivine

Nonferromagnesian
Quartz
Muscovite (white mica)
Feldspars (aluminum silicates)
Plagioclase
(sodium feldspar)
(calcium feldspar)
Orthoclase (potassium feldspar)

Clay Minerals
Kaolinite
Illite
Chlorite

Nonsilicates

Carbonates
Calcite
Dolomite

Sulfates
Gypsum
Anhydrite

Oxides
Hematite
Magnetite
Ilmenite
Aluminum oxides

Sulfides
Pyrite
Sphalerite
Galena
Chalcopyrite

Phosphates
Apatite

Halides
Halite
Fluorite

Native Elements
Gold
Silver
Copper
Diamonds
Sulfur

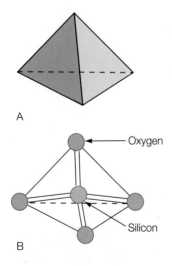

FIGURE 19.6

(A) The geometric shape of a tetrahedron with four equal sides. (B) A silicon and four oxygen atoms are arranged in the shape of a tetrahedron with the silicon in the center. This is the basic building block of all silicate minerals.

tetrahedral units combined in chains, sheets, or an interlocking framework (figure 19.7).

The silicate minerals that form rocks can be conveniently subdivided into two groups based on the presence of iron and magnesium. The basic tetrahedral structure joins with ions of iron, magnesium, calcium, and other elements in the *ferromagnesian silicates*. Examples of ferromagnesian silicates are *olivine, augite, hornblende,* and *biotite* (figure 19.8). They have a greater density and a darker color than the other silicates because of the presence of the metal ions. Augite, hornblende, and biotite are very dark in color, practically black, and olivine is light green.

The *nonferromagnesian silicates* do not contain iron or magnesium ions. These minerals have a light color and a low density compared to the ferromagnesians. This group includes the minerals *muscovite* (*white mica*), the *feldspars,* and *quartz* (figure 19.9).

The silicate minerals can also be classified according to the structural differences, or arrangements of the basic tetrahedral structures. There are four major arrangements of the units: (1) isolated tetrahedrons, (2) chain silicates, (3) sheet

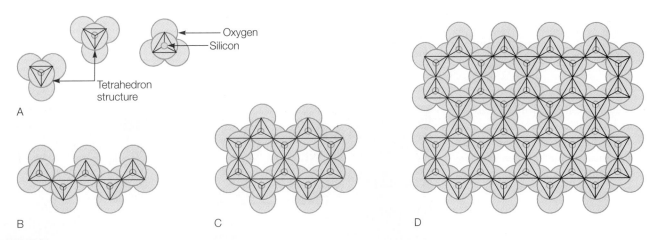

FIGURE 19.7

(A) Isolated silicon-oxygen tetrahedra do not share oxygens. This structure occurs in the mineral olivine. (B) Single chains of tetrahedra are formed by each silicon ion having two oxygens all to itself and sharing two with other silicons at the same time. This structure occurs in augite. (C) Double chains of tetrahedra are formed by silicon ions sharing either two or three oxygens. This structure occurs in hornblende. (D) The sheet structure in which each silicon shares three oxygens occurs in the micas, resulting in layers that pull off easily because of cleavage between the sheets.

FIGURE 19.8

Compare the dark colors of the ferromagnesian silicates augite (right), hornblende (left), and biotite to the light-colored nonferromagnesian silicates in figure 19.9.

FIGURE 19.9

Compare the light colors of the nonferromagnesian silicates mica (front center), white and pink orthoclase (top and center), and quartz to the dark-colored ferromagnesian silicates in figure 19.8.

A CLOSER LOOK | Asbestos

sbestos is a common name for any of several minerals that can be separated into fireproof fibers, fibers that will not melt or ignite. The fibers can be woven into a fireproof cloth, used directly as fireproof insulation material, or mixed with plaster or other materials. People now have a fear of all asbestos because it is presumed to be a health hazard (box figure 19.1). However, there are about six commercial varieties of asbestos. Five of these varieties are made from an amphibole mineral and are commercially called "brown" and "blue" asbestos. The other variety is made from *chrysotile*, a *serpentine* family of minerals and is commercially called "white" asbestos. White asbestos is the asbestos mined and most commonly used in North America. It is only the amphibole asbestos (brown and blue asbestos) that has been linked to cancer, even for a short exposure time. There is, however, no evidence that exposure to white asbestos results in an increased health hazard. It makes sense to ban the use of and remove all the existing amphibole asbestos from public buildings. It does not make sense to ban or remove the serpentine asbestos since it is not a proven health hazard.

BOX FIGURE 19.1
Did you know there are different kinds of asbestos? Are all kinds of asbestos a health hazard?

silicates, and (4) framework silicates. Other structures are possible, but are not as common. The impact of meteorites, for example, creates sufficiently high temperatures and pressures to form a silicate structure with six oxygens around each silicon atom rather than the typical four. A similar structure is believed to exist deep within the earth. Another interesting, less common structure is found in the asbestos minerals. These minerals have a sheet silicate structure that is rolled into fiber-like strands, resulting in a wide variety of silicates that are fibrous.

The remaining 8 percent of minerals making up the earth's crust that do not have silicon-oxygen tetrahedrons in their crystal structure are called *nonsilicates*. There are eight subgroups of nonsilicates: (1) carbonates, (2) sulfates, (3) oxides, (4) sulfides, (5) halides, (6) phosphates, (7) hydroxides, and (8) native elements. Some common nonsilicates are identified in table 19.1. The carbonates are the most abundant of the nonsilicates, but others are important as fertilizers, sources of metals, and sources of industrial chemicals.

Activities

Experiment with growing crystals from solutions of alum, copper sulfate, salt, or potassium permanganate. Write a procedure that will tell others what the important variables are for growing large, well-formed crystals.

Rocks

Elements are *chemically* combined to make minerals. Minerals are *physically* combined to make rocks. A **rock** is defined as an aggregation of one or more minerals and perhaps mineral materials that have been brought together into a cohesive solid. Mineral materials include volcanic glass, a silicate that is not considered a mineral because it lacks a crystalline structure. Thus, a rock can consist of one or more kinds of minerals that are somewhat "glued" together by other materials such as glass. Most

FIGURE 19.10

Granite is a coarse-grained igneous rock composed mostly of light-colored, light-density, nonferromagnesian minerals. The earth's continental areas are dominated by granite and by rocks with the same mineral composition of granite.

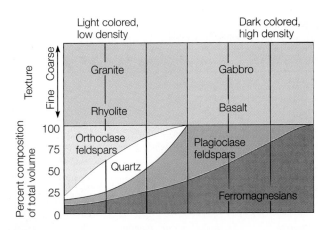

FIGURE 19.11

Igneous rock classification scheme based on mineral composition and texture. There are other blends of minerals with various textures, many of which have specific names.

rocks are composed of silicate minerals, as you might expect since most minerals are silicates. Granite, for example, is a rock that is primarily three silicate minerals: quartz, mica, and feldspar. You can see the grains of these three minerals in a freshly broken surface of most samples of granite (figure 19.10).

Rocks can be described and classified by many different characteristics, such as mineral composition, color, density, or texture. There is a broader classification scheme that is first used, however, and this scheme is based on the way the rocks were formed. There are three main groups in this scheme: (1) *igneous rocks* formed as a hot, molten mass of rock materials cooled and solidified; (2) *sedimentary rocks* formed from particles or dissolved materials from previously existing rocks; and (3) *metamorphic rocks* formed from igneous or sedimentary rocks that were subjected to high temperatures and pressures that deformed or recrystallized the rock without complete melting.

Igneous Rocks

The word *igneous* comes from the Latin *ignis,* which means "fire." This is an appropriate name for **igneous rocks** that are defined as rocks that formed from a hot, molten mass of melted rock materials. The first step in forming igneous rocks is the creation of some very high temperature, hot enough to melt rocks. A mass of melted rock materials is called a *magma.* A magma may cool and crystallize to solid igneous rock either below or on the surface of the earth. The earth has had a history of molten materials, and all rocks of the earth were at one time igneous rocks. Today, about two-thirds of the outer layer, or crust, is made up of igneous rocks. This is not apparent in many locations because the surface is covered by other kinds of rocks and rock materials (sand, soil, etc.).

A general classification scheme for igneous rocks is given in figure 19.11. Igneous rocks are various mixtures of minerals, and this scheme names the rocks according to (1) their mineral

composition and (2) their texture. Note that the mineral composition changes continuously from one side of the chart to the other. There are many intermediate types of igneous rocks possible, but this chart identifies only the most important ones.

Sedimentary Rocks

Sedimentary rocks are rocks that formed from particles or dissolved materials from previously existing rocks. Igneous rocks at the earth's surface, for example, are exposed to conditions very different from the environment in which they formed. Chemical reactions with air and water tend to break down and dissolve some minerals, freeing more chemically stable particles and grains in the process. The more stable particles are mechanically broken down by the action of moving water, wind, and ice. Many such processes are at work altering rocks through a process known as *weathering.* Weathering will be discussed in detail in a future chapter. For now, weathering is of interest because of the role it plays in providing the rocks and mineral grains and dissolved materials that will become sedimentary rocks. The weathered materials are transported by moving water, wind, and ice and deposited as sediments. *Sediments* are accumulations of silt, sand, or gravel that settle out of the atmosphere or out of water. There are actually two sources of sediments, (1) weathered rock fragments and (2) dissolved rock materials (table 19.2).

Most sediments are deposited as many separate particles that accumulate in certain environments as loose sediments. Such accumulations of rock fragments, chemical deposits, or

TABLE 19.2 | **A classification scheme for sedimentary rocks**

Sediment Type	Particle or Composition	Rock
Clastic	Larger than sand	Conglomerate or breccia
Clastic	Sand	Sandstone
Clastic	Silt and clay	Siltstone, claystone, or shale
Chemical	Calcite	Limestone
Chemical	Dolomite	Dolomite
Chemical	Gypsum	Gypsum
Chemical	Halite (sodium chloride)	Salt

TABLE 19.3 | **A simplified classification scheme for clastic sediments and rocks**

Sediment Name	Size Range	Rock
Boulder	Over 256 mm (10 in)	
Gravel	2 to 256 mm (0.08–10 in)	Conglomerate or breccia*
Sand	1/16 to 2 mm (0.025–0.08 in)	Sandstone
Silt (or dust)	1/256 to 1/16 mm (0.00015–0.025 in)	Siltstone**
Clay (or dust)	Less than 1/256 mm (less than 0.00015 in)	Claystone**

*Conglomerate has a rounded fragment; breccia has an angular fragment.
**Both also known as mudstone; called shale if it splits along parallel planes.

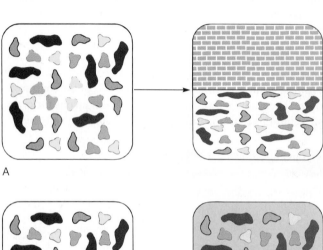

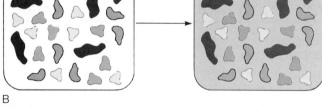

FIGURE 19.12

(*A*) In compaction, the sediment grains are packed more tightly together, often by overlying sediments, as represented by the bricks. (*B*) In cementation, fluids contain dissolved minerals that are precipitated in the space between the grains, cementing them together into a rigid, solid mass.

animal shells must become consolidated into a solid, coherent mass to become sedimentary rock. There are two main parts to this *lithification*, or rock-forming process, (1) compaction and (2) cementation (figure 19.12).

The weight of an increasing depth of overlying sediments causes an increasing pressure on the sediments below. This pressure squeezes the deeper sediments together, gradually reducing the pore space between the individual grains. This *compaction* of the grains reduces the thickness of a sediment deposit, squeezing out water as the grains are packed more tightly together.

Compaction alone is usually not enough to make an unconsolidated deposit into solid rock. Cementation is needed to hold the compacted grains together.

In *cementation* the spaces between the sediment particles are filled with a chemical deposit. The chemical deposit binds the particles together into the rigid, cohesive mass of a sedimentary rock. Compaction and cementation may occur at the same time, but the cementing agent must have been introduced before compaction restricts the movement of the fluid through the open spaces. Many soluble minerals can serve as cementing agents, and calcite (calcium carbonate) and silica (silicon dioxide) are common.

Sediments accumulate from rocks that are in various stages of being broken down, so there is a wide range of sizes of sediments. The size and shape of the sediments are used as criteria to name the sedimentary rocks (table 19.3).

Activities

Collect dry sand from several different locations. Use a magnifying glass to determine the minerals found in each sample.

Metamorphic Rocks

The third group of rocks is called metamorphic. **Metamorphic rocks** are previously existing rocks that have been changed by heat, pressure, or hot solutions into a distinctly different rock. The heat, pressure, or hot solutions that produced the changes are associated with geologic events of (1) movement of the crust, which will be discussed in the next chapter, and with (2) heating and hot solutions from the intrusion of a magma. Pressures from movement of the crust can change the rock texture by flattening, deforming, or realigning mineral grains. Temperatures from an intruded magma must be just right to produce a metamorphic rock. They must be high enough to disrupt the crystal structures to cause them to recrystallize, but not high enough to melt the rocks and form igneous rocks (figure 19.13).

A CLOSER LOOK | Recycling

Nature has the ultimate recycling program, moving materials from the earth's interior to the surface, and to the atmosphere and back. Part of nature's recycling scheme can be found in the rock cycle, as geologic processes act continuously to produce new rocks from old ones. This recycling concept can also be applied to many of the mineral resources that are extracted from ores, and in some cases this makes economic sense. Recycling always makes environmental sense, but it is not always the economic thing to do.

Recycling empty aluminum beverage cans is a good example of recycling that makes economic as well as environmental sense. Recycling aluminum cans requires that the cans are collected (see figure 10.2), then melted and processed into new cans or other aluminum products. When cans are recycled the mining and refining of new aluminum metal from ore is not necessary, and this is good environmentally for several reasons. First, not using aluminum ore conserves this mineral resource. Next, the extraction of aluminum from ore is very energy intensive, so not processing the ore saves energy. Not processing the

ore also means less air pollution, both directly and indirectly from the energy saved.

The motivation for recycling is not always as clear as with the case for aluminum beverage cans. It is less expensive, for example, to mine the raw materials and manufacture new building bricks than it is to recycle used bricks. This is primarily because the labor costs for taking apart a brick wall are greater than the energy costs of making new bricks. As the cost of energy increases, more attention will be directed toward recycling rather than mining and manufacturing.

Recycling is also one of the solutions available for reducing the growing problem of municipal solid waste. The combination of increasing volumes of solid waste and the reliance on landfills for disposal is leading to the problem of many cities running out of acceptable places to put the waste. Comprehensive recycling programs have been started in many U.S. and Canadian cities to help reduce the volume of wastes. In spite of this recycling effort only a small percent of the municipal solid wastes is being recycled. This is mostly aluminum cans (25% recy-

cled), paper (23%), glass (8%), other metals (4%), and plastics (1%).

Not many plastics are recycled because there are seven different kinds that must be separated first. For more information on this problem see "How to Sort Plastic Bottles for Recycling" in chapter 14. Recycling of plastics will increase dramatically when the plastics industry finds a recycling technology that will allow the mixing of different plastics.

In addition to the potential reduction of municipal solid wastes, recycling helps the environment by conserving resources and reducing pollution. For example, it has been estimated that just one Sunday edition of the *New York Times* consumes 62,000 trees. Recycling an equivalent amount of paper saves that many trees. Recycling glass instead of making new glass reduces the temperature requirements of the glass-making process, conserving about half the energy, and thus reduces air pollution. In a similar fashion over 500 million barrels of oil were saved each year during the 1990s by recycling paper, metals, and rubber. In many cases recycling makes good economic sense. In all cases recycling makes good environmental sense.

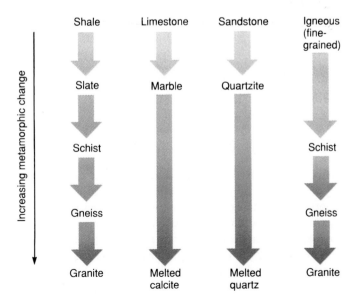

FIGURE 19.13

Increasing metamorphic change occurs with increasing temperatures and pressures. If the melting point is reached, the change is no longer metamorphic, and igneous rocks are formed.

A CLOSER LOOK | The Next Crisis

Most people understand the fact that our mineral resources are limited, finite parts of the earth that we remove from the natural environment. It should be obvious that we cannot go on forever using up limited parts of the environment. Of course, some mineral resources can be recycled, reducing the need to mine ever more minerals. For example, aluminum can be remelted and used repeatedly. Glass, copper, iron, and other metals can similarly be recycled repeatedly. There are other critical resources, however, that cannot be recycled and unfortunately, cannot be replaced. Crude oil, for example, is a dwindling resource that will become depleted sometime in the next decade. Oil is not recyclable once it is burned, and no new supplies are being created, at least not at a rate that would make them available in the immediate future. Even if the earth were a hollow vessel completely filled with oil, it would eventually become depleted, perhaps sooner than you might think.

There is also one of our mineral resources that is critically needed for our survival, but one that will probably soon be depleted. That resource is phosphate fertilizer derived from phosphate rock. Phosphorus is an essential nutrient required for plant growth and if its concentration in soils is too low, plants grow poorly if at all. Most agricultural soils are artificially fertilized with phosphate. Without this amendment, their productivity would decline and, in some cases, cease altogether.

Phosphate occurs naturally as the mineral *apatite*. Deposits of apatite were formed where ocean currents carried cold, deep ocean water rich in dissolved phosphate ions to the upper continental slope and outer continental shelf. Here, phosphate ions replaced the carbonate ions in limestone, forming the mineral apatite. Apatite also occurs as a minor accessory mineral in most igneous, sedimentary, and metamorphic rock types. Some igneous rocks serve as a source of phosphate fertilizer, but most is mined from former submerged coastal areas of limestone such as those found in Florida.

Trends in phosphate production and use over the past 40 years suggest that the proved world reserves of phosphate rock will be exhausted within a few decades. New sources might be discovered, but eventually, some time soon phosphate rock will be mined out and no longer available. When this happens, the food supply will have to be grown on lands that are naturally endowed with an adequate phosphate supply. Estimates are that this worldwide existing land area with adequate phosphate supply will supply food for only 2 billion people on the entire earth. From current trends, demographers predict a worldwide population of about 8 billion by 2010. This is assuming the 8 billion people would be fed, of course, by food grown on lands made agriculturally productive with the application of phosphate fertilizer. But with no phosphate fertilizer, the lands will support only 2 billion people. What will we do, reduce the world population to 2 billion by starvation?

Would concerted conservation efforts or recycling solve the upcoming problem? Unfortunately, phosphate ions are very reactive, forming insoluble salts and compounds. Out of every 2.7 kg of phosphate applied to agricultural land only 1 kg is actually captured by plant roots before being immobilized in the soil. The insoluble, bound phosphate is then flushed into rivers, lakes, and the ocean and then dispersed. Once dispersed, it is too expensive and energy-intensive to reconcentrate.

There are no substitutes for phosphate. It is a fundamental ingredient in cell membranes, in some proteins, in the vital energy transfer molecule adenosine triphosphate (ATP), and in genetic materials (DNA and RNA). Phosphorus is an essential element for all life on earth, and no other element can function in its place.

The Rock Cycle

Earth is a dynamic planet with a constantly changing surface and interior. As you will see in the next chapters, internal changes alter the earth's surface by moving the continents and, for example, building mountains that are soon worn away by weathering and erosion. Seas advance and retreat over the continents as materials are cycled from the atmosphere to the land and from the surface to the interior of the earth and then back again. Rocks are transformed from one type to another through this continual change. There is not a single rock on the earth's surface today that has remained unchanged through the earth's long history. The concept of continually changing rocks through time is called the *rock cycle.* The rock cycle concept views an igneous, a sedimentary, or a metamorphic rock as the present, but temporary stage in the ongoing transformation of rocks to new types. Any particular rock sample today has gone through countless transformations in the 4.6-billion-year history of the earth and will continue to do so in the future.

THE EARTH'S INTERIOR

Many of the properties and characteristics of the earth, including the structure of its interior, can be explained from current theories of how it formed and evolved. Theories and ideas about how the earth and the rest of the solar system formed were discussed in detail in chapter 17. Here is the theoretical summary of how the earth's interior was formed, discussed as if it were a fact. Keep in mind, however, that the following is all theoretical conjecture, even if it is conjecture based on facts.

In brief, the earth is considered to have formed about 4.6 billion years ago in a rotating disk of particles and grains that had condensed around a central protosun. The condensed rock, iron, and mineral grains were pulled together by gravity, growing eventually to a planet-sized mass. Not all the bits and pieces of matter in the original solar nebula were incorporated into the newly formed planets. They were soon being pulled by gravity to the newly born planets and their satellites. All sizes of these left-over bits and pieces of matter thus began bombarding the planets and their moons. Evidently the bombardment was so intense that the heat generated by impact after impact increased the surface temperature to the melting point. Evidence visible on the moon and other planets today indicates that the bombardment was substantial as well as lengthy, continuing for several hundred million years. Calculations of the heating resulting from this tremendous bombardment indicate that sufficient heat was liberated to melt the entire surface of the earth to a layer of glowing, molten lava. Thus, the early earth had a surface of molten lava that eventually cooled and crystallized to solid igneous rocks as the bombardment gradually subsided, then stopped.

Then the earth began to undergo a second melting, this time from the inside. The interior slowly accumulated heat from the radioactive decay of uranium, thorium, and other radioactive isotopes (see chapter 15). Heat conducts slowly through great thicknesses of rock and rock materials. After about a billion years of accumulating heat, parts of the interior became hot enough to melt to pockets of magma. Iron and other metals were pulled from the magma toward the center of the earth, leaving less dense rocks toward the surface. The melting probably did not occur all at one time throughout the interior, but rather in local pockets of magma. Each magma pocket became molten, cooled to a solid, and perhaps repeated the cycle numerous times. With each cyclic melting, the heavier abundant elements were pulled by gravity toward the center of the earth, and additional heat was generated by the release of gravitational energy. Today, the earth's interior still contains an outer core of molten material that is predominantly iron. The environment of the center of the earth today is extreme, with estimates of pressures up to 3.5 million atmospheres (3.5 million times the pressure of the atmosphere at the surface, 14.7 lb/in^2). Recent estimates of the temperatures at the earth's core are about the same as the temperature of the surface of the sun, about 6,000°C (11,000°F).

Seismic waves radiate outward from an earthquake focus, spreading in all directions through the solid earth's interior like sound waves from an explosion. There are basically three kinds of waves: (1) a longitudinal (compressional) wave called a *P-wave,* (2) a transverse (shear) wave called an *S-wave,* and (3) an up-and-down (crest and trough) wave that travels across the surface called a *surface wave* that is much like a water wave that moves across the solid surface of the earth. S- and P-waves provide information about the location and magnitude of an earthquake as well as provide information about the earth's interior.

The melting and flowing of iron to the earth's center were the beginnings of *differentiation,* the separation of materials that gave the earth its present-day stratified or layered interior. The theoretical formation of the earth and the layered structure of its interior

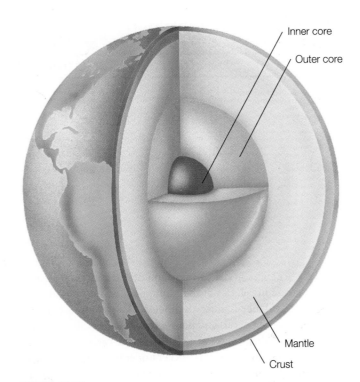

FIGURE 19.14
The structure of the earth's interior.

are supported by indirect evidence from measurements of vibrations in the earth, the earth's magnetic field, gravity, and heat flow.

Using data from vibrations that move through the earth, scientists were able to plot three main parts of the interior (figure 19.14). The *crust* is the outer layer of rock that forms a thin shell around the earth. Below the crust is the *mantle,* a much thicker shell than the crust. The mantle separates the crust from the center part, which is called the *core.* The following section starts on the earth's surface, at the crust, and then digs deeper and deeper into the earth's interior.

The Crust

The earth's **crust** is a thin layer of rock that covers the entire earth, existing below the oceans as well as making up the continents. According to seismic waves, there are differences in the crust making up the continents and the crust beneath the oceans (table 19.4). These differences are (1) the oceanic crust is much thinner than the continental crust and (2) seismic waves move through the oceanic crust faster than they do through the continental crust. The two types of crust vary because they are made up of different kinds of rock.

The Yugoslavian scientist Mohorovicic discovered the boundary between the crust and the mantle in 1909. The boundary is marked by a sharp increase in the velocity of seismic waves as they pass from the crust to the mantle. Today, this boundary is called the **Mohorovicic discontinuity,** or the "Moho" for short. The boundary is actually a zone one or two km thick (about one or two thousand yards) where seismic P-waves increase in velocity

TABLE 19.4	Comparison of oceanic crust and continental crust	
	Oceanic Crust	**Continental Crust**
Age	Less than 200 million years	Up to 3.8 billion years old
Thickness	5 to 8 km (3 to 5 mi)	10 to 70 km (6 to 45 mi)
Density	3.0 g/cm³	2.7 g/cm³
Composition	Basalt	Granite, schist, gneiss

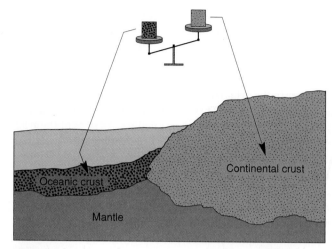

FIGURE 19.15

Continental crust is less dense, granite-type rock, while the oceanic crust is more dense, basaltic rock. Both types of crust behave as if they were floating on the mantle, which is more dense than either type of crust.

because of changes in the composition of the materials. Seismic waves increase in velocity at the Moho because the composition on both sides of the boundary is different. The mantle is richer in ferromagnesian minerals and poorer in silicon than the crust.

Studies of the Moho show that the crust varies in thickness around the earth's surface. It is thicker under the continents, varying from 10 km (about 6 mi) to more than 70 km (about 40 mi) beneath mountain chains. The crust under the oceans is much thinner, ranging from 5 to 8 km (about 3 to 5 mi) thick.

The age of rock samples from the continent has been compared with the age samples of rocks taken from the seafloor by oceanographic ships. This sampling has found the continental crust to be much older, with parts up to 3.8 billion years old. By comparison, the oldest oceanic crust is less than 200 million years old.

Comparative sampling also found that continental crust is a less dense, granite-type rock with a density of about 2.7 g/cm³. Oceanic crust, on the other hand, is made up of basaltic rock with a density of about 3.0 g/cm³. The less dense crust behaves as if it were floating on the mantle, much as less dense ice floats on water. There are explainable exceptions, but in general the thicker, less dense continental crust "floats" in the mantle above sea level and the thin, dense oceanic crust "floats" in the mantle far below sea level (figure 19.15).

The Mantle

The middle part of the earth's interior is called the **mantle** (see figure 19.14). The mantle is a 2,870 km (about 1,780 miles) thick shell between the core and the crust. This shell takes up about 80 percent of the total volume of the earth and accounts for about two-thirds of the earth's total mass. Information about the composition and nature of the mantle comes from (1) studies of seismological data, (2) studies of the nature of meteorites, and (3) studies of materials from the mantle that have been ejected to the earth's surface. The evidence from these separate sources all indicates that the mantle is composed of silicates, predominantly the ferromagnesian silicate *olivine*. Meteorites, as mentioned earlier, are basically either iron meteorites or stony meteorites. Most of the stony meteorites are the carbonaceous chondrites. They are silicates with a composition that would

produce the chemical composition of olivine if they were melted and the heavier elements removed by differentiation. This chemical composition also agrees closely with the composition of basalt, the most common volcanic rock found on the surface of the earth.

The Core

Information about the nature of the **core,** the center part of the earth, comes from studies of three sources of information, (1) seismological data, (2) the nature of meteorites, and (3) geological data at the surface of the earth.

Seismological data provide the primary evidence for the structure of the core of the earth. Seismic P-waves spread throughout the earth from a large earthquake. Figure 19.16 shows how the P-waves spread out, moving to seismic measuring stations all around the world, except between 103° and 142° of arc from the earthquake. This region is called the *P-wave shadow zone,* since no P-waves are received here. The P-wave shadow zone is explained by P-waves being refracted by the core, leaving a shadow. The paths of P-waves can be accurately calculated, so the size and shape of the earth's core can also be accurately calculated.

Seismic S-waves leave a different pattern at seismic receiving stations around the earth. S- (sideways or transverse) waves can travel only through solid materials, ceasing to exist in liquids. An *S-wave shadow zone* also exists and is larger than the P-wave shadow zone (figure 19.17). S-waves are not recorded in the entire region more than 103° away from the epicenter. The S-wave shadow zone seems to indicate that S-waves do not travel through the core at all. If this is true, it implies that the core of the earth is a liquid, or at least acts like a liquid.

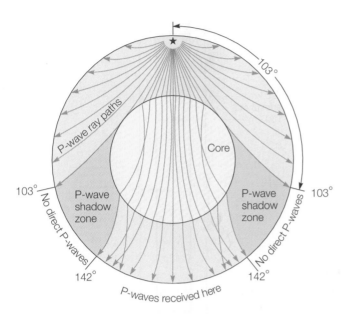

FIGURE 19.16

The P-wave shadow zone, caused by refraction of P-waves within the earth's core.

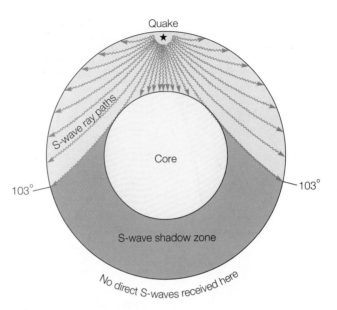

FIGURE 19.17

The S-wave shadow zone. Since S-waves cannot pass through a liquid, at least part of the core is either a liquid or has some of the same physical properties as a liquid.

Analysis of P-wave data suggests that the core has two parts, a *liquid outer core* and a *solid inner core*. Both the P-wave and S-wave data support this conclusion. The data also indicate that the earth's core begins at a depth of 2,900 km (about 1,800 mi) below the surface. Since the earth's center is 6,370 km (about 3,960 mi) below the surface, the core has a radius of 3,470 km (about 2,160 mi). The solid inner core has a radius of 1,200 km (about 750 mi). Overall, the core makes up about 15 percent of the earth's total volume and about one-third of its mass.

Evidence from the nature of meteorites indicates that the earth's core is mostly iron. The earth has a strong magnetic field that has its sources in the turbulent flow of the liquid part of the earth's core. To produce such a field, the material of the core would have to be an electrical conductor, that is, a metal such as iron. There are two general kinds of meteorites that fall to the earth, (1) stony meteorites that are made of silicate minerals and (2) iron meteorites that are made of iron or of a nickel-iron alloy. Since the earth has a silicate-rich crust and mantle, by analogy the earth's core must consist of iron or a nickel and iron alloy.

A Different Structure

There is strong evidence that the earth has a layered structure with a core, mantle, and crust. This description of the structure is important for historical reasons and for understanding how the earth evolved over time. There is another structure that can be described. This structure is far more important in understanding the history and present appearance of the earth's surface, including the phenomena of earthquakes and volcanoes.

The important part of this different structural description of the earth's interior was first identified from seismic data. There is a thin zone in the mantle, ranging in thickness from a depth of 130 km (81 mi) to 160 km (100 mi), where seismic waves undergo a sharp decrease in velocity (figure 19.18). This low-velocity zone is evidently a hot, elastic semiliquid layer that extends around the entire earth. It is called the **asthenosphere** after the Greek word for "weak shell." The asthenosphere is weak because it is plastic, mobile, and yields to stresses. In some regions the asthenosphere is completely liquid, containing pockets of magma.

The rocks above and below the asthenosphere are rigid, solid, and brittle. The solid layer above the asthenosphere is called the **lithosphere** after the Greek word for "stone shell." The lithosphere is also known as the "strong layer" in contrast to the "weak layer" of the asthenosphere. The lithosphere includes the entire crust, the Moho, and the upper part of the mantle. As you will see in the next section, the asthenosphere is one important source of magma that reaches the earth's surface. It is also a necessary part of the mechanism involved in the movement of the crust. The lithosphere is made up of comparatively rigid plates that are moving, floating in the upper mantle like giant tabular ice sheets floating in the ocean.

PLATE TECTONICS

If you observe the shape of the continents on a world map or a globe, you will notice that some of the shapes look as if they would fit together like the pieces of a puzzle. The most obvious is the eastern edge of North and South America, which seem to fit the

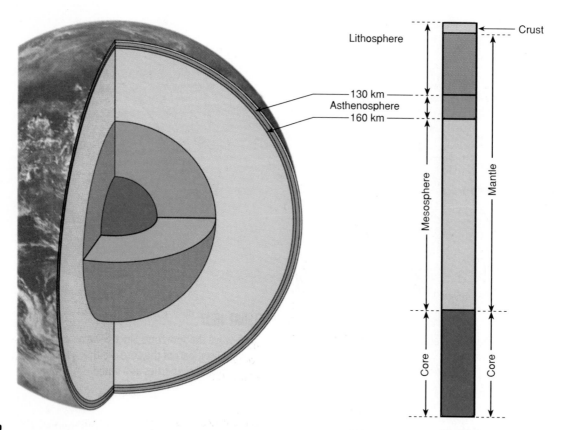

Lithosphere

Crust

130 km

Asthenosphere

160 km

Mesosphere

Mantle

Core

Core

FIGURE 19.18

The earth's interior, showing the weak, plastic layer called the *asthenosphere*. The rigid, solid layer above the asthenosphere is called the *lithosphere*. The lithosphere is broken into plates that move on the upper mantle like giant tabular ice sheets floating on water. This arrangement is the foundation for plate tectonics, which explains many changes that occur on the earth's surface such as earthquakes, volcanoes, and mountain building.

western edge of Europe and Africa in a slight S-shaped curve. Such patterns between continental shapes seem to suggest that the continents were at one time together, breaking apart and moving to their present positions some time in the past. Impressed by the patterns, Antonio Snider published a sketch in 1855 showing Africa and South America fitting together as one landmass. This sketch led to the bold speculation that the continents had at one time been part of a single landmass (figure 19.19).

In the early 1900s a German geologist named Alfred Wegener became enamored with the idea that the continents had shifted positions, and published papers on the subject for nearly two decades. Wegener supposed that at one time there was a single large landmass that he called "Pangaea," which is from the Greek word meaning "all lands." He pointed out that similar fossils found in landmasses on both sides of the Atlantic Ocean today must be from animals and plants that lived in Pangaea, which later broke up and split into smaller continents. Wegener's concept came to be known as *continental drift*, the idea that individual continents could shift positions on the earth's surface. Some people found the idea of continental drift plausible, but most had difficulty imagining how a huge and

massive continent could "drift" around on a solid earth. Since Wegener had provided no explanations of why or how continents might do this, most scientists found the concept unacceptable. The concept of continental drift was dismissed as an interesting, but odd idea. Then new evidence discovered in the 1950s and 1960s began to indicate that the continents have indeed moved. The first of this evidence would come from the bottom of the ocean and would lead to a new, broader theory about movement of the earth's crust.

Evidence from Earth's Magnetic Field

The earth's magnetic field, and its north and south magnetic poles, are discussed in detail in chapter 7. The probable source of the earth's magnetic field is also discussed, that the field is created by electric currents within the slowly circulating liquid part of the iron core. The north magnetic pole is not in the same place as the geographic North Pole, and the north magnetic pole must actually be a south magnetic pole (figure 19.20). This is most confusing, but it must be true since the *north* end of a compass is attracted to the north magnetic pole. Since unlike poles

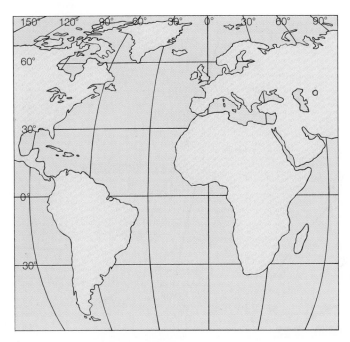

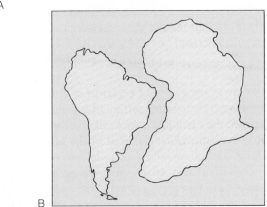

FIGURE 19.19

(A) Normal position of the continents on a world map. (B) A sketch of South America and Africa, suggesting that they once might have been joined together and subsequently separated by a continental drift.

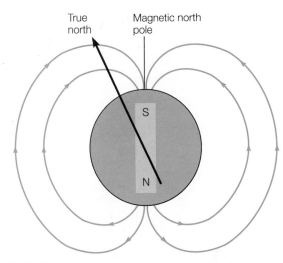

FIGURE 19.20

The earth's magnetic field. Note that the magnetic north pole and the geographic North Pole are not in the same place. Note also that the magnetic north pole acts as if the south pole of a huge bar magnet were inside the earth. You know that it must be a magnetic south pole, since the north end of a magnetic compass is attracted to it, and opposite poles attract.

attract, this must mean that the "north" magnetic pole is actually the south magnetic pole that happens to be located near the geographic North Pole.

There is nothing static about the earth's magnetic poles. Geophysical studies have found that the magnetic poles are moving slowly around the geographic poles. Studies have also found that the earth's magnetic field occasionally undergoes magnetic reversal. Magnetic reversal is the flipping of polarity of the earth's magnetic field. During a magnetic reversal, the north magnetic pole and the south magnetic pole exchange positions. The present magnetic field orientation has persisted for the past 700,000 years, and according to the evidence, is now preparing for another reversal. The evidence, such as the magnetized iron particles found in certain Roman ceramic artifacts, shows that the magnetic field was 40 percent stronger 2,000 years ago than it is today. If the

present decay rate were to continue, the earth's magnetic field would be near zero by the end of the next 2,000 years—if it decays that far before reversing orientation and then increases to its usual value. Decreasing magnetic field strength will present yet another environmental concern for the future since it usually deflects cosmic radiation, solar wind, and other charged particles. Without the magnetic field at its present strength, radiation will reach the earth's surface in sufficient doses to affect organisms.

Many igneous rocks contain a record of the strength and direction of the earth's magnetic field at the time the rocks formed. Iron minerals, such as magnetite (Fe_3O_4), crystallize as particles in a cooling magma and become magnetized and oriented to the earth's magnetic field at the time, like tiny compass needles. When the rock crystallizes to a solid, these tiny compass needles become frozen in the orientations they had at the time. Such rocks thus provide evidence of the direction and distance to the earth's ancient magnetic poles. The study of ancient magnetism, called *paleomagnetics*, provides the information that the earth's magnetic field has undergone twenty-two magnetic reversals during the past 4.5 million years (figure 19.21).

The record shows the time between pole flips is not consistent, sometimes reversing in as little as 10,000 years and sometimes taking as long as 25 million years. Once a reversal starts, however, it takes about 5,000 years to complete the process.

Paleomagnetic studies, along with the determination of the ages of rocks containing the magnetic minerals, indicate that the earth's magnetic field was first formed about 3.5 billion years ago or about a billion years after the earth formed. This fits nicely with other understandings about how the earth formed and the calculated time required for radioactive heating to melt the earth's interior.

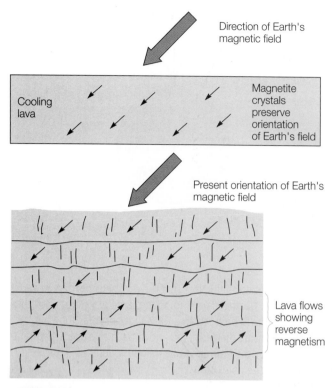

FIGURE 19.21

Magnetite mineral grains align with the earth's magnetic field and are frozen into position as the magma solidifies. This magnetic record shows the earth's magnetic field has reversed itself in the past.

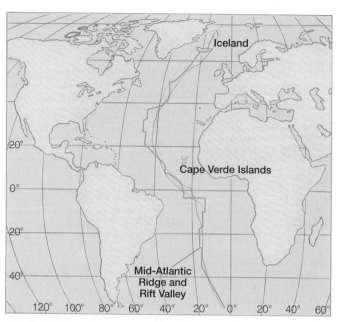

FIGURE 19.22

The Mid-Atlantic Ridge divides the Atlantic Ocean into two nearly equal parts. Where the ridge reaches above sea level, it makes oceanic islands, such as Iceland.

Evidence from the Ocean

The first important studies concerning the movement of continents came from studies of the ocean basin, the bottom of the ocean floor. The basins are covered by 4 to 6 km (about 3 to 4 mi) of water and were not easily observed during Wegener's time. It was not until the development and refinement of sonar and other new technologies during the 1940s and 1950s that scientists began to learn about the nature of the ocean basin. They found that it was not the flat, featureless plain that many had imagined. There are valleys, hills, mountains, and mountain ranges. Long, high, and continuous chains of mountains that seem to run clear around the earth were discovered, and these chains are called **oceanic ridges.** The *Mid-Atlantic Ridge* is one such oceanic ridge, and this one is located in the center of the Atlantic Ocean basin. The Mid-Atlantic Ridge divides the Atlantic Ocean into two nearly equal parts. Where it is high enough to reach sea level, it makes oceanic islands such as Iceland (figure 19.22). The basins also contain **oceanic trenches.** These trenches are long, narrow, and deep troughs with steep sides. Oceanic trenches always run parallel to the edge of continents, a preferred orientation that seems to invite some kind of explanation.

Studies of the Mid-Atlantic Ridge during the late 1950s found at least three related groups of data and observations that

also seemed to invite some kind of explanation. Among these related groups were: (1) submarine earthquakes were discovered and measured, but the earthquakes were all observed to occur mostly in a narrow band under the crest of the Mid-Atlantic Ridge; (2) a long, continuous, and deep valley was observed to run along the crest of the Mid-Atlantic Ridge for its length. This continuous, crest-running valley is called a **rift;** and (3) a large amount of heat was found to be escaping from the rift. One explanation of the related groups of findings is that the rift might be a crack in the earth's crust, a fracture through which basaltic lava flowed to build up the ridge. The evidence of excessive heat flow, earthquakes along the crest of the ridge, and the very presence of the ridge all led to a **seafloor spreading** hypothesis. This hypothesis explained that hot, molten rock moved up from the interior of the earth to emerge along the rift, flowing out in both directions to create new rocks along the ridge. The creation of new rock like this would tend to spread the seafloor in both directions, so thus the name. The test of this hypothesis would come from further studies, this time on the ages and magnetic properties of the seafloor along the ridge (figure 19.23).

Evidence of the age of sections of the seafloor was obtained by drilling into the ocean floor from a research ship. From these drillings, scientists were able to obtain samples of fossils and sediments at progressive distances outward from the Mid-Atlantic Ridge. They found thin layers of sediments near the ridge that became progressively thicker toward the continents. This is a pattern you would expect if the seafloor were spreading, because older layers would have more time to accumulate greater depths of sediments. The fossils and sediments in the bottom of the layer were also progressively older at increasing

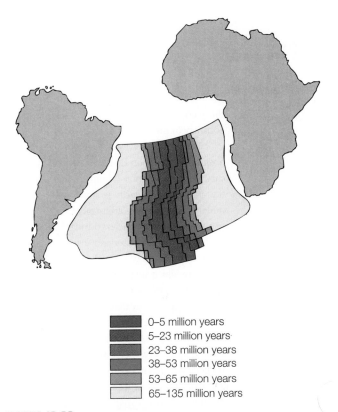

■	0–5 million years
■	5–23 million years
■	23–38 million years
■	38–53 million years
■	53–65 million years
□	65–135 million years

FIGURE 19.23

The pattern of seafloor ages on both sides of the Mid-Atlantic Ridge reflects seafloor spreading activity. Younger rocks are found closer to the ridge.

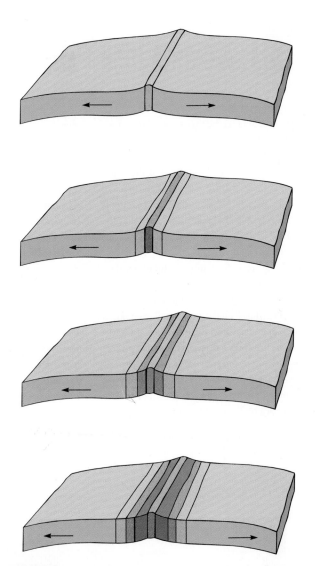

FIGURE 19.24

Formation of magnetic strips on the seafloor. As each new section of seafloor forms at the ridge, iron minerals become magnetized in a direction that depends on the orientation of the earth's field at that time. This makes a permanent record of reversals of the earth's magnetic field.

distances from the ridge. The oldest, which were about 150 million years old, were near the continents. This would seem to indicate that the Atlantic Ocean did not exist until 150 million years ago. At that time a fissure formed between Africa and South America and new materials have been continuously flowing, adding new crust to the edges of the fissure.

More convincing evidence for the support of seafloor spreading came from the paleomagnetic discovery of patterns of magnetic strips in the basaltic rocks of the ocean floor. The earth's magnetic field has been reversed many times in the last 150 million years. The periods of time between each reversal were not equal, ranging from thousands to millions of years. Since iron minerals in molten basalt formed, became magnetized, then frozen in the orientation they had when the rock cooled, they made a record of reversals in the earth's ancient magnetic field (figure 19.24). Analysis of the magnetic pattern in the rocks along the Mid-Atlantic Ridge found identical patterns of magnetic bands on both sides of the ridge. This is just what you would expect if molten rock flowed out of the rift, cooled to solid basalt, then moved away from the rift on both sides. The pattern of magnetic bands also matched patterns of reversals measured elsewhere, providing a means of determining the age of the basalt. This showed that the oceanic crust is like a giant conveyer belt that is moving away from the Mid-Atlantic Ridge in both directions. It is moving at an average 5 cm (about 2 inches) a year,

which is about how fast your fingernails grow. This means that in 50 years the seafloor will have moved 5 cm/yr × 50 yr, or 2.5 meters (about 8 feet). This slow rate is why most people do not recognize that the seafloor—and the continents—move.

Lithosphere Plates and Boundaries

The strong evidence for seafloor spreading soon led to the development of a new theory called **plate tectonics**. According to plate tectonics, the lithosphere is broken into a number of fairly rigid plates that move on the asthenosphere. Some plates, as shown in figure 19.25, contain continents and part of an ocean basin, while other plates contain only ocean basins. The plates move, and the movement is helping to explain why mountains

A CLOSER LOOK | Measuring Plate Movement

According to the theory of plate tectonics, the earth's outer shell is made up of moving plates. The plates making up the continents are about 100 km thick and are gradually drifting at a rate of about 0.5 to 10 cm (0.2 to 4 in) per year. This reading is about one way that scientists know that the earth's plates are moving and how this movement is measured.

The very first human lunar landing mission took place in July, 1969. Astronaut Neil Armstrong stepped onto the lunar surface, stating, "That's one small step for a man, one giant leap for mankind." In addition to fulfilling a dream, the *Apollo* project carried out a program of scientific experiments. The *Apollo 11* astronauts placed a number of experiments on the lunar surface in the Sea of Tranquility. Among the experiments was the first Laser Ranging Retro-reflector Experiment, which was designed to reflect pulses of laser light from the earth. Three more reflectors were later placed on the moon, including two by other *Apollo* astronauts and one by an unmanned Soviet *Lunakhod 2* lander.

The McDonald Observatory in Texas, the Lure Observatory on the island of Maui, Hawaii, and a third observatory in southern France have regularly sent laser beams through optical telescopes to the reflectors. The return signals, which are too weak to be seen with the unaided eye, are detected and measured by sensitive detection equipment at the observatories. The accuracy of these measurements, according to NASA reports, is equivalent to determining the distance between a point on the east and a point on the west coast of the United States to an accuracy of one fiftieth of an inch.

Reflected laser light experiments have found that the moon is pulling away from the earth at about 4 cm/yr (about 1.6 in/yr), that the shape of the earth is slowly changing, undergoing isostatic adjustment from the compression by the glaciers during the last ice age, and that the observatory in Hawaii is slowly moving away from the one in Texas. This provides a direct measurement of the relative drift of two of the earth's tectonic plates. Thus, one way that changes on the surface of the earth are measured is through lunar ranging experiments. Results from lunar ranging, together with laser ranging to artificial satellites in Earth orbit, have revealed the small but constant drift rate of the plates making up Earth's dynamic surface.

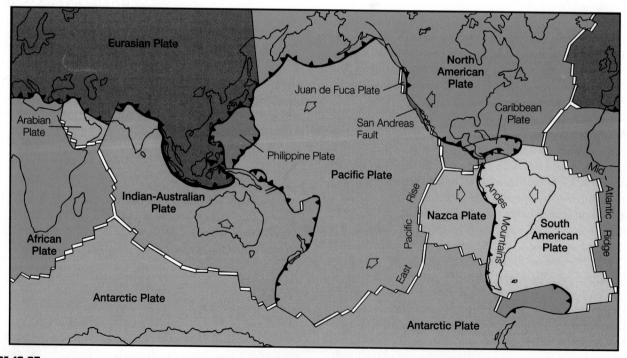

FIGURE 19.25

The major plates of the lithosphere that move on the asthenosphere.
Source: After W. Hamilton, U.S. Geological Survey.

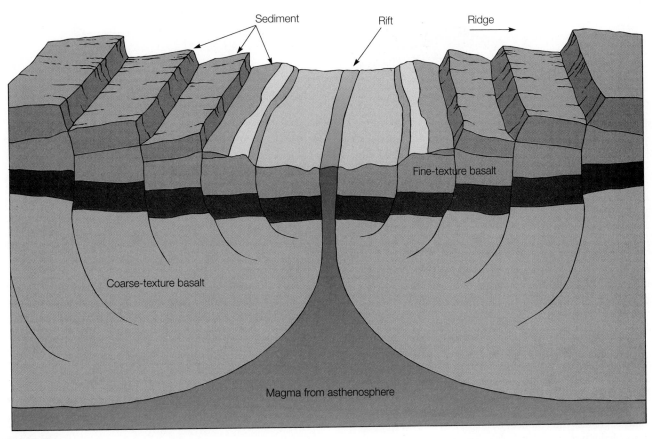

Sediment   Rift   Ridge

Fine-texture basalt

Coarse-texture basalt

Magma from asthenosphere

FIGURE 19.26

A divergent boundary is a new crust zone where molten magma from the asthenosphere rises, cools, and adds new crust to the edges of the separating plates. Magma that cools at deeper depths forms a coarse-grained basalt, while surface lava cools to a fine-grained basalt. Note that deposited sediment is deeper farther from the spreading rift.

form where they do, the occurrence of earthquakes and volcanoes, and in general, the entire changing surface of the earth.

Earthquakes, volcanoes, and most rapid changes in the earth's crust occur at the edge of a plate, which is called a *plate boundary*. There are three general kinds of plate boundaries that describe how one plate moves relative to another:

1. **Divergent boundaries** occur between two plates moving away from each other. Magma forms as the plates separate, decreasing pressure on the mantle below. This molten material from the asthenosphere rises, cools, and adds new crust to the edges of the separating plates. The new crust tends to move horizontally from both sides of the divergent boundary, usually known as an oceanic ridge. A divergent boundary is thus a new crust zone. Most new crust zones are presently on the seafloor, producing seafloor spreading (figure 19.26).

   The Mid-Atlantic Ridge is a divergent boundary between the South American and African Plates, extending north between the North American and Eurasian Plates (see figure 19.25). This ridge is one segment of the global mid-ocean ridge system that encircles the earth. The results of divergent plate movement can be seen in Iceland, where the Mid-Atlantic Ridge runs as it separates the North American and Eurasian Plates. In the northeastern part of Iceland ground cracks are widening, often

accompanied by volcanic activity. The movement was measured extensively between 1975 and 1984, when displacements caused a total separation of about 7 meters (about 23 feet).

The average rate of spreading along the Mid-Atlantic Ridge is about 2.5 centimeters per year (about an inch). This may seem slow, but the process has been going on for millions of years and has caused a tiny inlet of water between the continents of Europe, Africa, and the Americas to grow into the vast Atlantic Ocean that exists today.

Another major ocean may be in the making in East Africa, where a divergent boundary has already moved Saudi Arabia away from the African continent, forming the Red Sea. If this spreading between the African Plate and the Arabian Plate continues, the Indian Ocean will flood the area and the easternmost corner of Africa will become a large island.

2. **Convergent boundaries** occur between two plates moving toward each other. The creation of new crust at a divergent boundary means that old crust must be destroyed somewhere else at the same rate, or else the earth would have a continuously expanding diameter. Old crust is destroyed by returning to the asthenosphere at convergent boundaries. The collision produces an elongated belt of down-bending called a **subduction zone.** The lithosphere of one plate, which contains the crust, is subducted beneath the second plate and partially melts, then

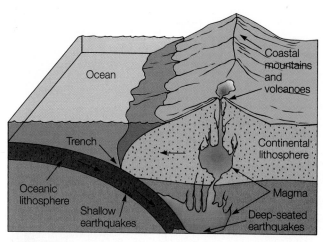

FIGURE 19.27

Ocean-continent plate convergence. This type of plate boundary accounts for shallow and deep-seated earthquakes, an oceanic trench, volcanic activity, and mountains along the coast.

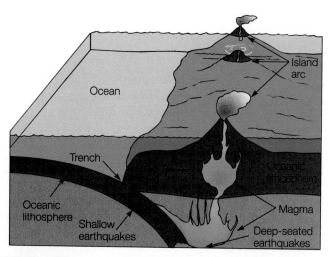

FIGURE 19.28

Ocean-ocean plate convergence. This type of plate convergence accounts for shallow and deep-focused earthquakes, an oceanic trench, and a volcanic arc above the subducted plate.

becoming part of the mantle. The more dense components of this may become igneous materials that remain in the mantle. Some of it may eventually migrate to a spreading ridge to make new crust again. The less dense components may return to the surface as a silicon, potassium, and sodium-rich lava, forming volcanoes on the upper plate, or it may cool below the surface to form a body of granite. Thus, the oceanic lithosphere is being recycled through this process, which explains why ancient seafloor rocks do not exist. Convergent boundaries produce related characteristic geologic features depending on the nature of the materials in the plates, and there are three general possibilities: (1) converging continental and oceanic plates, (2) converging oceanic plates, and (3) converging continental plates.

As an example of *ocean-continent plate convergence,* consider the plate containing the South American continent (the South American Plate) and its convergent boundary with an oceanic plate (the Nazca Plate) along its western edge. Continent-oceanic plate convergence produces a characteristic set of geologic features as the oceanic plate of denser basaltic material is subducted beneath the less dense granite-type continental plate (figure 19.27). The subduction zone is marked by an oceanic trench (the Peru-Chile Trench), deep-seated earthquakes, and volcanic mountains on the continent (the Andes Mountains). The trench is formed from the down-bending associated with subduction and the volcanic mountains from subducted and melted crust that rise up through the overlying plate to the surface. The earthquakes are associated with the movement of the subducted crust under the overlying crust.

Ocean-ocean plate convergence produces another set of characteristics and related geologic features (figure 19.28). The northern boundary of the oceanic Pacific Plate, for example, converges with the oceanic part of the North American Plate near the Bering Sea. The Pacific Plate is subducted, forming the Aleutian oceanic trench with a zone of earthquakes that are shallow near the trench and progressively more deep-seated toward the continent. The deeper earthquakes are associated with the movement of more deeply subducted crust into the mantle. The Aleutian Islands are typical **island arcs,** curving chains of volcanic islands that occur over the belt of deep-seated earthquakes. These islands form where the melted subducted material rises up through the overriding plate

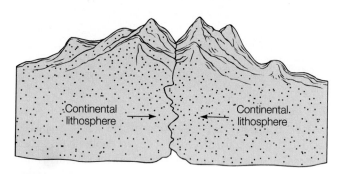

FIGURE 19.29

Continent-continent plate convergence. Rocks are deformed, and some lithosphere thickening occurs, but neither plate is subducted to any great extent.

above sea level. The Japanese, Marianas, and Indonesians are similar groups of arc islands associated with converging oceanic-oceanic plate boundaries.

During *continent-continent plate convergence,* subduction does not occur as the less dense, granite-type materials tend to resist subduction (figure 19.29). Instead, the colliding plates pile up into a deformed and thicker crust of the lighter material. Such a collision produced the thick, elevated crust known as the Tibetan Plateau and the Himalayan Mountains.

3. **Transform boundaries** occur between two plates sliding by each other. Crust is neither created nor destroyed at transform boundaries as one plate slides horizontally past another along a long, vertical fault. The movement is neither smooth nor equal along the length of the fault, however, as short segments move independently with sudden jerks that are separated by periods without motion. The Pacific Plate, for example, is moving slowly to the northwest, sliding past the North American Plate. The San Andreas Fault is one boundary along the California coastline. Vibrations from plate movements along this boundary are the famous California earthquakes.

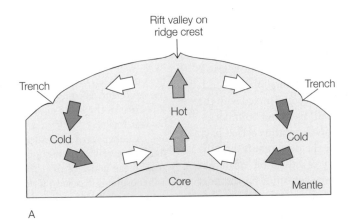

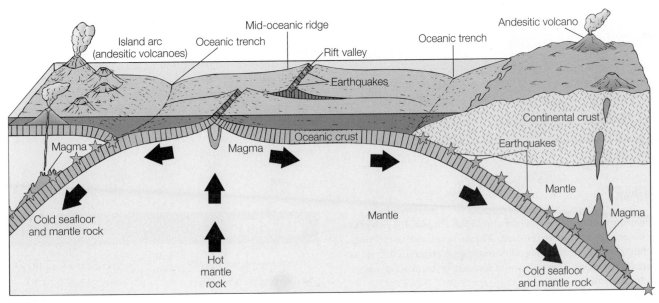

FIGURE 19.30

Not to scale. One idea about convection in the mantle has a convection cell circulating from the core to the lithosphere, dragging the overlying lithosphere laterally away from the oceanic ridge.

Present-Day Understandings

The theory of plate tectonics, developed during the late 1960s and early 1970s, is new compared to most major scientific theories. Measurements are still being made, evidence is being gathered and evaluated, and the exact number of plates and their boundaries are yet to be determined with certainty. The major question that remains to be answered is what drives the plates, moving them apart, together, and by each other? The most widely favored explanation is that slowly turning *convective cells* in the plastic asthenosphere drive the plates (figure 19.30). According to this hypothesis, hot fluid materials rise at the diverging boundaries. Some of the material escapes to form new crust, but most of it spreads out beneath the lithosphere. As it moves beneath the lithosphere, it drags the overlying plate with it. Eventually, it cools and sinks back inward under a subduction zone.

There is uncertainty about the existence of convective cells in the asthenosphere and their possible role because of a lack of clear evidence. Seismic data is not refined enough to show convective cell movement beneath the lithosphere. In addition, deep-seated earthquakes occur to depths of about 700 km, which means that descending materials—parts of a subducted plate—must extend to that depth. This could mean that a convective cell might operate all the way down to the core-mantle boundary some 2,900 km below the surface. This presents another kind of problem because little is known about the lower mantle and how it interacts with the upper mantle. Theorizing without information is called speculation, and that is the best that can be done with existing data. The full answer may include the role of heat and the role of gravity. However, understanding the mechanism of what drives the plates will probably have to await a better understanding of how the lower mantle behaves.

What is generally accepted about the plate tectonic theory is the understanding that the solid materials of the earth are engaged in a continual cycle of change. Oceanic crust is subducted, melted, then partly returned to the crust as volcanic igneous rocks in island arcs and along continental plate boundaries. Other parts of the subducted crust become mixed with the upper mantle, returning as new crust at diverging boundaries. The materials of the crust and the mantle are thus cycled back and forth in a mixing that may include the deep mantle and the core as well. There is more to this story of a dynamic earth that undergoes a constant change. The story continues in the next chapters with different cycles to consider.

Activities

Locate and label the major plates of the lithosphere on an outline map of the world according to the most recent findings in plate tectonics. Show all types of boundaries and associated areas of volcanoes and earthquakes.

SUMMARY

The elements silicon and oxygen make up 75 percent of all the elements in the outer layer, or *crust,* of the earth. The elements combine to make crystalline chemical compounds called minerals. A *mineral* is defined as a naturally occurring, inorganic solid element or compound with a crystalline structure.

About 92 percent of the minerals of the earth's crust are composed of silicon and oxygen, the *silicate minerals.* The basic unit of the silicates is a *tetrahedral structure* that combines with positive metallic ions or with other tetrahedral units to form chains, sheets, or an interlocking framework. The *ferromagnesian silicates* are tetrahedral structures combined with ions of iron, magnesium, calcium, and other elements. The ferromagnesian silicates are darker in color and more dense than other silicates. The *nonferromagnesian silicates* do not have irons or magnesium ions and they are lighter in color and less dense than the ferromagnesians. The *nonsilicate minerals* do not contain silicon and are carbonates, sulfates, oxides, halides, sulfides, and native elements.

A *rock* is defined as an aggregation of one or more minerals that have been brought together into a cohesive solid. *Igneous rocks* formed as hot, molten *magma* cooled and crystallized to firm, hard rocks. *Sedimentary rocks* are formed from *sediments,* accumulations of weathered rock materials that settle out of the atmosphere or out of water. Sediments become sedimentary rocks through a rock-forming process that involves both the *compaction* and *cementation* of the sediments. *Metamorphic rocks* are previously existing rocks that have been changed by heat, pressure, or hot solution into a different kind of rock without melting. The *rock cycle* is a concept that an igneous, a sedimentary, or a metamorphic rock is a temporary stage in the ongoing transformation of rocks to new types.

The earth has a layered interior that formed as the earth's materials underwent *differentiation,* the separation of materials while in the molten state. The center part, or *core,* is predominantly iron with a solid inner part and a liquid outer part. The core makes up about 15 percent of the earth's total volume and about a third of its total mass. The *mantle* is the middle part of the earth's interior that accounts for about two-thirds of the earth's total mass and about 80 percent of its total volume. The *Mohorovicic discontinuity* separates the outer layer, or *crust,* of the earth from the mantle. The crust of the *continents* is composed mostly of less dense granite-type rock. The crust of the *ocean basins* is composed mostly of the more dense basaltic rocks.

Another way to consider the earth's interior structure is to consider the weak layer in the upper mantle, the *asthenosphere* that extends around the entire earth. The rigid, solid, and brittle layer above the asthenosphere is called the *lithosphere.* The lithosphere includes the entire crust, the Moho, and the upper part of the mantle.

Evidence from the ocean floor that was gathered in the 1950s and 1960s revived interest in the idea that continents could move. The evidence for *seafloor spreading* came from related observations concerning oceanic ridge systems, sediment and fossil dating of materials outward from the ridge, and from magnetic patterns of seafloor rocks. Confirmation of seafloor spreading led to the *plate tectonic theory.* According to plate tectonics, new basaltic crust is added at *diverging boundaries* of plates, and old crust is *subducted* at *converging boundaries.* Mountain building, volcanoes, and earthquakes are seen as *related geologic features* that are caused by plate movements. The force behind the movement of plates is uncertain, but it may involve *convection* in the deep mantle.

KEY TERMS

asthenosphere (p. **475**)

convergent boundaries (p. **481**)

core (p. **474**)

crust (p. **473**)

divergent boundaries (p. **481**)

igneous rocks (p. **469**)

island arcs (p. **482**)

lithosphere (p. **475**)

mantle (p. **474**)

metamorphic rocks (p. **470**)

mineral (p. **463**)

Mohorovicic discontinuity (p. **473**)

new crust zone (p. **481**)

oceanic ridges (p. **478**)

oceanic trenches (p. **478**)

plate tectonics (p. **479**)

rift (p. **478**)

rock (p. **468**)

seafloor spreading (p. **478**)

sedimentary rocks (p. **469**)

silicates (p. **465**)

subduction zone (p. **481**)

transform boundaries (p. **482**)

APPLYING THE CONCEPTS

1. Based on its abundance in the earth's crust, most rocks will contain a mineral composed of oxygen and the element
    a. sulfur.
    b. carbon.
    c. silicon.
    d. iron.

2. The most common rock in the earth's crust is
   a. igneous.
   b. sedimentary.
   c. metamorphic.
   d. none of the above.

3. Which of the following formed from previously existing rocks?
   a. sedimentary rocks
   b. igneous rocks
   c. metamorphic rocks
   d. All of the above are correct.

4. Rocks making up the ocean basins and much of the earth's interior have the same chemical composition as
   a. granite.
   b. basalt.
   c. halite.
   d. water.

5. Sedimentary rocks are formed by the processes of compaction and
   a. pressurization.
   b. melting.
   c. cementation.
   d. heating, but not melting.

6. Which type of rock probably existed first, starting the rock cycle?
   a. metamorphic
   b. igneous
   c. sedimentary
   d. All of the above are correct.

7. From seismological data, the earth's shadow zone indicates that part of the earth's interior must be
   a. liquid.
   b. solid throughout.
   c. plastic.
   d. hollow.

8. The Mohorovicic discontinuity is a change in seismic wave velocity that is believed to take place because of
   a. structural changes in minerals of the same composition.
   b. changes in the composition on both sides of the boundary.
   c. a shift in the density of minerals of the same composition.
   d. changes in the temperature with depth.

9. The oldest rocks are found in (the)
   a. continental crust.
   b. oceanic crust.
   c. neither, since both are the same age.

10. The least dense rocks are found in (the)
    a. continental crust.
    b. oceanic crust.
    c. neither, since both are the same density.

11. The idea of seafloor spreading along the Mid-Atlantic Ridge was supported by evidence from
    a. changes in magnetic patterns and ages of rocks moving away from the ridge.
    b. faulting and volcanoes on the continents.
    c. the observation that there was no relationship between one continent and another.
    d. all of the above.

12. According to the plate tectonics theory, seafloor spreading takes place at a
    a. convergent boundary.
    b. subduction zone.
    c. divergent boundary.
    d. transform boundary.

13. The presence of an oceanic trench, a chain of volcanic mountains along the continental edge, and deep-seated earthquakes is characteristic of a (an)
    a. ocean-ocean plate convergence.
    b. ocean-continent plate convergence.
    c. continent-continent plate convergence.
    d. None of the above are correct.

14. The presence of an oceanic trench with shallow earthquakes and island arcs with deep-seated earthquakes is characteristic of a (an)
    a. ocean-ocean plate convergence.
    b. ocean-continent plate convergence.
    c. continent-continent plate convergence.
    d. None of the above are correct.

15. The ongoing occurrence of earthquakes without seafloor spreading, oceanic trenches, or volcanoes is most characteristic of a
    a. convergent boundary between plates.
    b. subduction zone.
    c. divergent boundary between plates.
    d. transform boundary between plates.

16. The evidence that the earth's core is part liquid or acts like a liquid comes from (the)
    a. P-wave shadow zone.
    b. S-wave shadow zone.
    c. meteorites.
    d. all of the above.

Answers

1. c 2. a 3. d 4. b 5. c 6. b 7. a 8. b 9. a 10. a 11. a 12. c 13. b 14. a 15. d 16. b

QUESTIONS FOR THOUGHT

1. Is ice a mineral? Explain.
2. What is a rock?
3. Describe the concept of the rock cycle.
4. Which major kind of rock, based on the way they formed, would you expect to find most of in the earth's crust? Explain.
5. Briefly describe the rock-forming process that changes sediments into solid rock.
6. What are metamorphic rocks? What limits the maximum temperatures possible in metamorphism? Explain.
7. Is the rock cycle unique to the planet Earth? Why is this?
8. What evidence provides information about the nature of the earth's core?
9. What is the asthenosphere? Why is it important in modern understandings of the earth?

10. Rocks, sediments, and fossils around an oceanic ridge have a pattern concerning their ages. What is the pattern? Explain what the pattern means.

11. Describe the origin of the magnetic strip patterns found in the rocks along an oceanic ridge.

12. Explain why ancient rocks are not found on the seafloor.

13. Describe the three major types of plate boundaries and what happens at each.

14. Briefly describe the theory of plate tectonics and how it accounts for the existence of certain geologic features.

15. What is an oceanic trench? What is its relationship to major plate boundaries? Explain this relationship.

16. Describe the probable source of all the earthquakes that occur in southern California.

17. The northwestern coast of the United States has a string of volcanoes running along the coast. According to plate tectonics, what does this mean about this part of the North American Plate? What geologic feature would you expect to find on the seafloor off the northwestern coast? Explain.

18. Explain how the crust of the earth is involved in a dynamic, ongoing recycling process.

Folding, faulting, and lava flows, such as the one you see here, tend to build up, or elevate, the earth's surface.

CHAPTER | Twenty

The Earth's Surface

The central idea of plate tectonics, which was discussed in the previous chapter, is that the earth's surface is made up of rigid plates that are moving slowly across the surface. Since the plates and the continents riding on them are in constant motion, any given map of the world is only a snapshot that shows the relative positions of the continents at a given time. The continents occupied different positions in the distant past. They will occupy different positions in the distant future. The surface of the earth, which seems so solid and stationary, is in fact mobile.

Plate tectonics has changed the accepted way of thinking about the solid, stationary nature of the earth's surface and ideas about the permanence of the surface as well. The surface of the earth is no longer viewed as having a permanent nature, but is understood to be involved in an ongoing cycle of destruction and renewal. Old crust is destroyed as it is plowed back into the mantle through subduction, becoming mixed with the mantle. New crust is created as molten materials move from the mantle through seafloor spreading and volcanoes. Over time, much of the crust must cycle into and out of the mantle.

The mantle-crust cycle is but one of many cycles involving earth materials. Crust exposed to the atmosphere is involved in another kind of cycle that was introduced in the previous chapter as the rock cycle. The atmosphere and rocks of the crust interact, and rocks are weathered to sediments, which are buried and formed into sedimentary rock. Metamorphosis or melting may take place as forces from the movement of plates deform and elevate rocks into mountains. Here, they are weathered to sediments and the rock cycling process begins again.

The movement of plates, the crust-mantle cycle, and the igneous-sedimentary-metamorphic rock cycle all combine to produce a constantly changing surface. There are basically two types of surface changes: (1) changes that originate within the earth, resulting in a building up of the surface (figure 20.1) and (2) changes that result from rocks being exposed to the atmosphere and water, resulting in a sculpturing and tearing down of the surface. This chapter is about the building up of the land. The concepts of this chapter will provide you with something far more interesting about the earth's surface than the scenic aspect. The existence of different features (such as mountains, folded hills, islands) and the occurrence of certain events (such as earthquakes, volcanoes, faulting) are all related. The related features and events also have a story to tell about the earth's past, a story about the here and now, and yet another story about the future.

INTERPRETING EARTH'S SURFACE

Not too long ago the features on the earth's surface—the plains, mountains, and canyons, for example—were not interpreted correctly because the vastness of geologic time was not appreciated. Geologic time was not appreciated because the earth's features change so slowly that a person is able to observe only a little change, if any, in a single lifetime. Statements such as "unchanging as the hills" or "old as the hills" illustrate this lack of appreciation of change over geologic time.

Given a mental framework based on a lack of appreciation of change over geologic time, how do you suppose people interpreted the existence of features such as mountains and canyons? Some believed, as they had observed in their lifetimes, that the mountains and canyons had "always" been there. Others did not believe the features had always been there, but believed they were formed by a sudden, single event (figure 20.2). The Grand Canyon, for example, was not interpreted as the result of incomprehensibly slow river erosion, but as the result of a giant crack or rip that appeared in the surface. The canyon that you see today was interpreted as forming when the earth split open and the Colorado River fell into the split. Early ideas about the earth's past were based on a concept of catastrophes like this. A catastrophe created a fea-

ture of the earth's surface all at once, with little or no change occurring since that time. This interpretation seemed to fit with the observation of no changes in a person's lifetime.

About 200 years ago the idea of unchanging, catastrophically formed landscapes was challenged by James Hutton, a Scottish physician. Hutton, who is known today as the founder of modern geology, traveled widely throughout the British Isles. Hutton was a keen observer of rocks, rock structures, and other features of the landscape. He noted that sandstone, for example, was made up of rock fragments that appeared to be (1) similar to the sand being carried by rivers and (2) similar to the sand making up the beaches next to the sea. He also noted fossil shells of sea animals in sandstone on the land, while the living relatives of these animals were found in the shallow waters of the sea. This and other evidence led Hutton to realize that rocks were being ground into fragments, then carried by rivers to the sea. He surmised that these particles would be reformed into rocks later, then lifted and shaped into the hills and mountains of the land. He saw all this as quiet, orderly change that required only *time* and the ongoing work of the water and some forces to make the sediments back into rocks. With Hutton's logical conclusion came the understanding that the earth's history could be interpreted by tracing it backwards, from the present to the past. This

FIGURE 20.1

An aerial view from the south of the eruption of Mount St. Helens volcano on May 18, 1980.

FIGURE 20.2

Would you believe that this rock island has "always" existed where it is? Would you believe it was formed by a sudden, single event? What evidence would it take to convince you that the rock island formed ever so slowly, starting as a part of southern California and moving very slowly, at a rate of cm/yr, to its present location near the coast of Alaska?

tracing required a frame of reference of slow, uniform change, not the catastrophic frame of reference of previous thinkers. The frame of reference of uniform changes is today called the **principle of uniformity** (also called *uniformitarianism*). The principle of uniformity is often represented by a statement that "the present is the key to the past." This statement means that the same geologic processes you see changing rocks today are the very same processes that changed them in the ancient past, although not necessarily at the same rate. The principle of uniformity does not *exclude* the happening of sudden or catastrophic events on the surface of the earth. A violent volcanic explosion, for example, is a catastrophic event that most certainly modifies the surface of the earth. What the principle of uniformity does state is that the physical and chemical laws we observe today operated exactly the same in the past. The rates of operation may or may not have been the same in the past, but the events you see occurring today are the same events that occurred in the past. Given enough time, you can explain the formation of the structures of the earth's surface with known events and concepts. It is not necessary to imagine magic or supernatural powers to explain why the earth's surface is the way it is. The basic concept in understanding the principle of uniformity is the concept of immense spans of geologic time. Immense spans of time with slow, incomprehensible change taking place is difficult to comprehend since it cannot be observed or experienced in a lifetime. Thus, understanding the principle of uniformity requires a mental model. This model is based on the observable events that build up the surface and wear it down and on an understanding of geologic time.

The principle of uniformity as a mental model for thinking about the earth has been used by geologists since the time of Hutton. The concept of how the constant changes occur has evolved with the development of plate tectonics, but the basic frame of reference is the same. You will see how the principle of uniformity is applied as you make a fascinating trip into the history of a landscape. The trip begins by first considering what can happen to rocks and rock layers that are deeply buried.

PROCESSES THAT BUILD UP THE SURFACE

We now turn to interpreting observable features of the earth's surface in terms of plate tectonics—movement that occurs within the crust. All the possible movements of the earth's plates, including drift toward or away from other plates, isostatic adjustment, and any process that deforms or changes the earth's surface are included in the term *diastrophism*. Diastrophism is the process of deformation that changes the earth's surface. It produces many of the basic structures you see on the earth's surface, such as plateaus, mountains, and folds in the crust. The movement of magma is called *vulcanism* or *volcanism*. Diastrophism, volcanism, and earthquakes are closely related, and their occurrence can most of the time be explained by events involving plate tectonics (chapter 19). The results of diastrophism are discussed in the following section, which is followed by a discussion of earthquakes, volcanoes, and mountain chains. Again, remember that diastrophism, volcanism, earthquakes, and movement of the earth's plates are very closely related. All are involved with the shapes, arrangements, and interrelationships of different parts of the earth's crust and the forces that change it. We will begin with a discussion of some of these forces before discussing what the forces can do.

Stress and Strain

Any solid material responds to a force in a way that depends on the extent of coverage (force per unit area, or pressure), the nature of the material, and other variables such as the temperature. Consider, for example, what happens if you apply an increasing pressure to the outside metal of a car door. (Do not try this on any car without a pressure gauge and information about the limits.) With increasing pressure you can observe at least four different and separate responses:

1. At first, the metal successfully resists a slight pressure and *nothing happens.*
2. At a somewhat greater pressure you will be able to deform, or bend the metal into a concave surface. The metal will return to its original shape, however, when the pressure is removed. This is called an *elastic deformation* since the metal was able to spring back into its original shape.
3. At a still greater pressure the metal is deformed to a concave surface, but this time the metal does not return to its original shape. This means the *elastic limit* of the metal has been exceeded, and it has now undergone a *plastic deformation.* Plastic deformation permanently alters the shape of a material.
4. Finally, at some great pressure the metal will rupture, resulting in a *break* in the material.

Many materials, including rocks, respond to increasing pressures in this way, showing (1) no change, (2) an elastic change with recovery, (3) a plastic change with no recovery, and (4) finally breaking from the pressure.

A **stress** is a force that tends to compress, pull apart, or deform a rock. Rocks in the earth's solid outer crust are subjected to forces as the earth's plates move into, away from, or alongside each other. Thus, there are three types of forces that cause rock stress:

1. *Compressive stress* is caused by two plates moving together or by one plate pushing against another plate that is not moving.
2. *Tensional stress* is the opposite of compressional stress. It occurs when one part of a plate moves away, for example, and another part does not move.
3. *Shear stress* is produced when two plates slide past one another or by one plate sliding past another plate that is not moving.

Just like the metal in your car door, a rock is able to withstand stress up to a limit. Then it might undergo elastic deformation, plastic deformation, or breaking with progressively greater pressures. The adjustment to stress is called **strain.** A rock unit might respond to stress by changes in volume, changes in shape, or by breaking. Thus, there are three types of strain: elastic, plastic, and fracture.

1. In *elastic strain,* rock units recover their original shape after the stress is released.
2. In *plastic strain,* rock units are molded or bent under stress and do not return to their original shape after the stress is released.
3. In *fracture strain,* rock units crack or break as the name suggests.

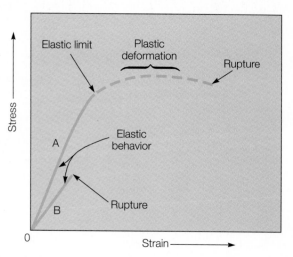

FIGURE 20.3

Stress and deformation relationships for deeply buried, warm rocks under high pressure (*A*) and cooler rocks near the surface (*B*). Breaking occurs when stress exceeds rupture strength.

The relationship between stress and strain, that is, exactly how the rock responds, depends on at least four variables. They are (1) the nature of the rock, (2) the temperature of the rock, (3) how slowly or quickly the stress is applied over time, and (4) the confining pressure on the rock. The temperature and confining pressure are generally a function of how deeply the rock is buried. In general, rocks are better able to withstand compressional rather than pulling-apart stresses. Cold rocks are more likely to break than warm rocks, which tend to undergo plastic deformation. In addition, a stress that is applied quickly tends to break the rock, where stress applied more slowly over time, perhaps thousands of years, tends to result in plastic strain.

In general, rocks at great depths are under great pressure at higher temperatures. These rocks tend to undergo plastic deformation, then plastic flow, so rocks at great depths are bent and deformed extensively. Rocks at less depth can also bend, but they have a lower elastic limit and break more readily (figure 20.3). Rock deformation often results in recognizable surface features called folds and faults, the topics of the next sections.

Folding

Sediments that form most sedimentary rocks are deposited in nearly flat, horizontal layers at the bottom of a body of water. Conditions on the land change over time, and different mixtures of sediments are deposited in distinct layers of varying thickness. Thus, most sedimentary rocks occur naturally as structures of horizontal layers, or beds (figure 20.4).

A sedimentary rock layer that is not horizontal may have been subjected to some kind of compressive stress. The source of such a stress could be from colliding plates, from the intrusion of magma, or from a plate moving over a hot spot. Stress on buried layers of horizontal rocks can result in plastic strain, resulting in a wrinkling of the layers into *folds.* **Folds** are bends in layered bedrock (figure 20.5). They are analogous to layers of rugs or

A

B

FIGURE 20.4

(A) Rock bedding on a grand scale in the Grand Canyon. (B) A closer example of rock bedding can be seen in this roadcut.

FIGURE 20.5

These folded rock layers are in the Calico Hills, California. Can you figure out what might have happened here to fold flat rock layers like this?

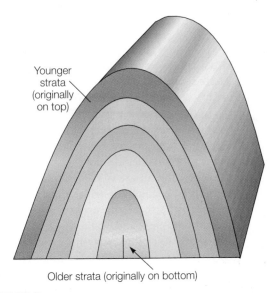

Younger strata (originally on top)

Older strata (originally on bottom)

FIGURE 20.6

An anticline, or arching fold, in layered sediments. Note that the oldest strata are at the center.

blankets that were stacked horizontally, then pushed into a series of arches and troughs. Folds in layered bedrock of all shapes and sizes can occur from plastic strain, depending generally on the regional or local nature of the stress and other factors. Of course, when the folding occurred, the rock layers were in a ductile condition, probably under considerable confining pressure from deep burial. When you see the results of the folding, the rocks are now under very different conditions at the surface.

The most common regional structures from deep plastic deformation are arch-shaped and trough-shaped folds. In general, an arch-shaped fold is called an **anticline** (figure 20.6). The corresponding trough-shaped fold is called a **syncline** (figure 20.7). Anticlines and synclines sometimes alternate across the

land like waves on water. You can imagine that a great compressional stress must have been involved over a wide region to wrinkle the land like this.

Anticlines, synclines, and other types of folds are not always visible as such on the earth's surface. The ridges of anticlines are constantly being weathered into sediments. The sediments, in turn, tend to collect in the troughs of synclines, filling them in. The Appalachian Mountains have ridges of rocks that are more resistant to weathering, forming hills and mountains. The San Joaquin Valley, on the other hand, is a very large syncline in California.

Note that any kind of rock can be folded. Sedimentary rocks are usually the best example of folding, however, since the fold structures of rock layers are easy to see and describe. Folding is

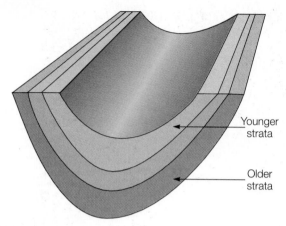

FIGURE 20.7

A syncline, showing the reverse age pattern.

much harder to see in igneous or metamorphic rocks that are blends of minerals without a layered structure.

Faulting

Rock layers do not always respond to stress by folding. Rocks near the surface are cooler and under less pressure, so they tend to be more brittle. A sudden stress on these rocks may reach the rupture point, resulting in a cracking and breaking of the rock structure. When there is relative movement between the rocks on either side of a fracture, the crack is called a **fault.** When faulting occurs, the rocks on one side move relative to the rocks on the other side along the surface of the fault, which is called the *fault plane.* Faults are generally described in terms of (1) the steepness of the fault plane, that is, the angle between the plane and imaginary horizontal plane and (2) the direction of relative movement. There are basically three ways that rocks on one side of a fault can move relative to the rocks on the other side: (1) up and down (called "dip"), (2) horizontally, or sideways (called "strike"), and (3) with elements of both directions of movement (called "oblique").

One classification scheme for faults is based on an orientation referent borrowed from mining (many ore veins are associated with fault planes). Imagine a mine with a fault plane running across a horizontal shaft. Unless the plane is perfectly vertical, a miner would stand on the mass of rock below the fault plane and look up at the mass of rock above. Therefore, the mass of rock below is called the *footwall* and the mass of rock above is called the *hanging wall* (figure 20.8). How the footwall and hanging wall have moved relative to one another describes three basic classes of faults, the (1) normal, (2) reverse, and (3) thrust faults. A **normal fault** is one in which the hanging wall has moved downward relative to the footwall. This seems "normal" in the sense that you would expect an upper block to slide *down* a lower block along a slope (figure 20.9A). Sometimes a huge block of rock bounded by normal faults will drop down, creating a *graben* (figure 20.9B). The opposite of a graben is a *horst,* which is a block bounded by normal faults that is uplifted (fig-

ure 20.9C). A very large block lifted sufficiently becomes a fault-block mountain. Many parts of the western United States are characterized by numerous fault-block mountains separated by adjoining valleys.

In a **reverse fault,** the hanging wall block has moved upward relative to the footwall block. As illustrated in figure 20.10A, a reverse fault is probably the result of horizontal compressive stress.

A reverse fault with a low-angle fault plane is also called a **thrust fault** (figure 20.10B). In some thrust faults the hanging wall block has completely overridden the lower footwall for 10 to 20 km (6 to 12 mi). This is sometimes referred to as an "overthrust."

As shown in figures 20.9 and 20.10, the relative movement of blocks of rocks along a fault plane provides information about the stresses that produced the movement. Reverse and thrust faulting result from compressional stress in the direction of the movement. Normal faulting, on the other hand, results from a pulling-apart stress that might be associated with diverging plates. It might also be associated with the stretching and bulging up of the crust over a hot spot.

EARTHQUAKES

This section is concerned with the nature and origin of earthquakes. The use of seismic waves to determine the earth's internal structure was discussed in the previous chapter. In this section, seismic waves of earthquakes will be considered in more detail, along with how the quakes are measured and located by measuring the waves. The effects of earthquakes, such as ground motion, ground displacement, damage, and tsunamis ("tidal waves") will also be described.

There is a worldwide pattern to the distribution of earthquakes, as most occur in long narrow belts, although they do occasionally occur elsewhere. Of all the intermediate-depth earthquakes in the world, 9 out of 10 occur in a narrow zone, or belt, which encircles the rim of the Pacific Ocean. Essentially all the earth's deep-depth earthquakes also occur within this particular belt. This pattern was once a big puzzle, leading to wild explanations. One, for example, explained that the moon was once part of the earth. Something (?) pulled the moon away from the earth, leaving the Pacific Ocean basin. The earthquake belt around the Pacific was supposedly a zone of weakness that remained from the removal of crust. Today, geologists have a more plausible explanation, understanding the earthquake belt around the rim of the Pacific Ocean through the concept of plate tectonics. This concept explains why earthquakes—and other related phenomena—occur around the rim of the Pacific Ocean.

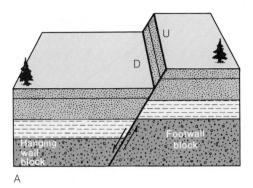

FIGURE 20.8

(*A*) This sketch shows the relationship between the hanging wall, the footwall, and a fault. (*B*) A photo of a fault near Kingman, Arizona, showing how the footwall has moved relative to the hanging wall.

B

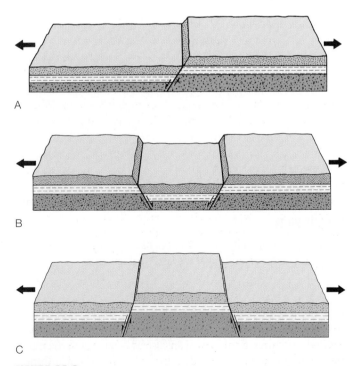

FIGURE 20.9

How tensional stress could produce (*A*) a normal fault, (*B*) a graben, and (*C*) a horst.

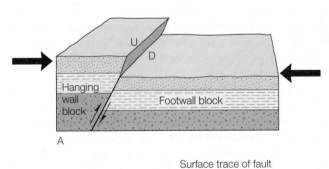

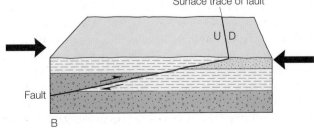

FIGURE 20.10

How compressive stress could produce (*A*) a reverse fault, and (*B*) a thrust fault.

What is an earthquake? An **earthquake** is a quaking, shaking, vibrating, or upheaval of the ground. Earthquakes are the result of the sudden release of energy that comes from *stress* on rock beneath the earth's surface. In the previous section you learned that rock units can bend and become deformed in response to stress, but there are limits as to how much stress rock can take before it fractures. When it does fracture, the sudden movement of blocks of rock produces vibrations that move out as waves throughout the earth. These vibrations are called **seismic waves.** It is strong seismic waves that people feel as a shaking, quaking, or vibrating during an earthquake.

Seismic waves are generated when a huge mass of rock breaks and slides into a different position. As you learned in the previous section, the plane between two rock masses that have moved into new relative positions is called a *fault.* Major earthquakes occur along existing fault planes or when a new fault is formed by the fracturing of rock. In either case, most earthquakes occur along a fault plane when there is displacement of one side relative to the other.

Most earthquakes occur along a fault plane and they occur near the earth's surface. You might expect this to happen since the rocks near the surface are brittle, and those deeper are more ductile from increased temperature and pressure. Shallow-focus earthquakes are typical of those that occur at the boundary of the North American Plate, which is moving against the Pacific Plate. In California, the boundary between these two plates is known as the *San Andreas fault* (figure 20.11). The San Andreas fault runs north-south for some 1,300 km (800 mi) through California, with the Pacific Plate moving on one side and the North American Plate moving on the other. The two plates are tightly pressed against each other, and friction between the rocks along the fault prevents them from moving easily. Stress continues to build along the entire fault as one plate attempts to move along the other. Some elastic deformation does occur from the stress, but eventually the rupture strength of the rock (or the friction) is overcome. The stressed rock, now released of the strain, snaps suddenly into new positions in the phenomenon known as **elastic rebound** (figure 20.12). The rocks are displaced to new positions on either side of the fault, and the vibrations from the sudden movement are felt as an earthquake. The elastic rebound and movement tend to occur along short segments of the fault at different times rather than along long lengths. Thus, the resulting earthquake tends to be a localized phenomenon rather than a regional one.

Most earthquakes are explained by movement of rock blocks along faults, but there are also other causes. Earthquakes have occurred in the eastern United States and without any apparent relationship to known faults at or near the surface. One of the largest earthquakes on record in the United States occurred not in California, but in the region of New Madrid, Missouri in 1811. This quake toppled chimneys 400 miles away in Ohio and was felt from the Rocky Mountains to the Atlantic Coast and from Canada to the Gulf of Mexico. The New Madrid fault zone is theorized to represent a failed rift that is being reactivated by compressional stress from the west and east.

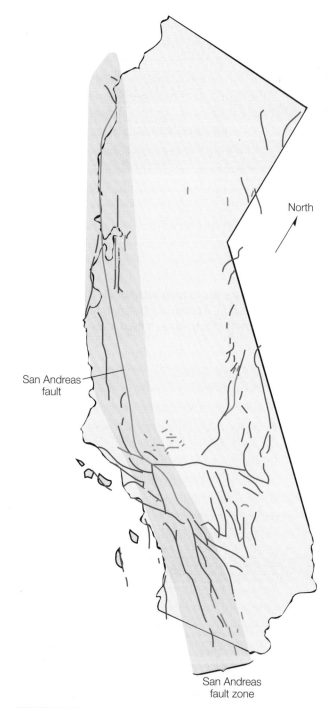

North

San Andreas fault

San Andreas fault zone

FIGURE 20.11

These lines show fault movement that has occurred along the San Andreas and other faults over the last 2 million years.

Some of the few earthquakes that are difficult to explain seem to be associated with deeply buried anticlines or other deeply buried structures. Earthquakes are also associated with the movement of magma that occurs beneath a volcano before an eruption. Earthquakes also occur during an explosive volcanic eruption. Earthquakes associated with volcanic activity, however, are always relatively feeble compared to those associated with faulting.

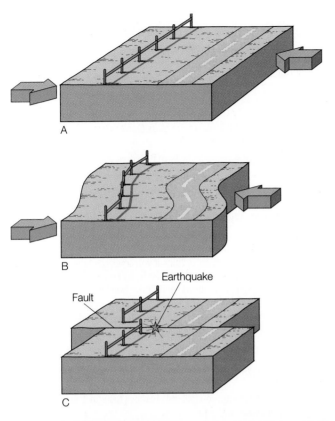

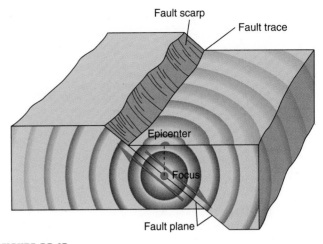

FIGURE 20.13

Simplified diagram of a fault, illustrating component parts and associated earthquake terminology.

FIGURE 20.12

The elastic rebound theory of the cause of earthquakes. (*A*) Rock with stress acting on it. (*B*) Stress has caused strain in the rock. Strain builds up over a long period of time. (*C*) Rock breaks suddenly, releasing energy, with rock movement along a fault. Horizontal motion is shown; rocks can also move vertically. (*D*) Horizontal offset of rows in a lettuce field, 1979, El Centro, California.

Most earthquakes occur near plate boundaries. Occurrences do happen elsewhere, but they are rare. The actual place where seismic waves originate beneath the surface is called the *focus* of the earthquake. The focus is considered to be the center of the earthquake and the place of initial rock movement on a fault. The point on the earth's surface directly above the focus is called the earthquake *epicenter* (figure 20.13).

Seismic waves radiate outward from an earthquake focus, spreading in all directions through the solid earth's interior like the sound waves from an explosion. As introduced in the previous chapter, there are three kinds of waves: *P-waves, S-waves,* and *surface waves.* S- and P-waves provide information about the location and magnitude of an earthquake as well as information about the earth's interior. P-waves are the fastest, moving through surface rocks at speeds of 4 to 7 km/s (9,000 to 16,000 mi/hr) and travel through solid and liquid materials. S-waves are next in speed, moving at 2 to 5 km/s (4,500 to 11,000 mi/hr). S-waves do not travel through liquids since liquids do not have the cohesion necessary to transmit a shear, or side-to-side, motion. Surface waves are the slowest and occur where S- or P-waves reach the surface.

Seismic S- and P-waves leave the focus of an earthquake at essentially the same time. As they travel away from the focus, they gradually separate because the P-waves travel faster than the S-waves. To locate an epicenter, at least three recording stations measure the time lag between the arrival of the P-waves and the slower S-waves. The difference in the speed between the two waves is a constant. Therefore, the farther they travel, the greater the time lag between the arrival of the faster P-waves and the slower S-waves (figure 20.14A). By measuring the time lag and knowing the speed of the two waves, it is possible to calculate the distance to their source. However, the calculated distance provides no information about the direction or location of the source of the waves. The location is found by first using the calculated distance as the radius of a circle drawn on a map. The place where the circles from the three recording stations intersect is the location of the source of the waves (figure 20.14B).

Earthquakes occur in a wide range of intensities, from the many that are barely detectable to the few that cause widespread destruction. Destruction is caused by the seismic waves, which cause the land and buildings to vibrate. Vibrations during small quakes can crack windows and walls, but vibrations in strong quakes can topple bridges and buildings. Injuries and death are

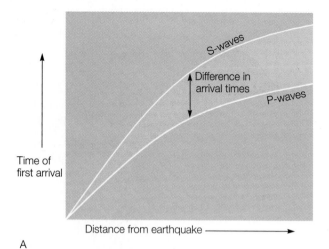

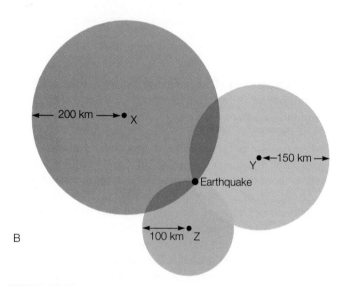

FIGURE 20.14

Use of seismic waves in locating earthquakes. (A) Difference in times of first arrivals of P-waves and S-waves is a function of the distance from the focus. (B) Triangulation using data from several seismograph stations allows location of the earthquake.

usually the result of falling debris, crumbling buildings, or falling bridges. Fire from broken gas pipes was a problem in the 1906 and 1989 earthquakes in San Francisco and the 1994 earthquake in Los Angeles. Broken water mains made it difficult to fight the 1906 San Francisco fires, but in 1989 fireboats and fire hoses using water pumped from the bay were able to extinguish the fires. Table 20.1 lists some recommendations of things to do during and after an earthquake.

Other effects of earthquakes include landslides, displacement of the land surface, and tsunamis. Vertical, horizontal, or both vertical and horizontal displacement of the land can occur during a quake. People sometimes confuse cause and effect when they see a land displacement saying things like, "Look what the earthquake did!" The fact is that the movement of the land probably produced the seismic waves (the earthquake). The seismic waves did not produce the land displacement. Such dis-

TABLE 20.1	Earthquake safety rules from the National Oceanic and Atmospheric Administration

During the Shaking

1. *Don't panic.*
2. If you are indoors, stay there. Seek protection under a table or desk or in a doorway. Stay away from glass. Don't use matches, candles, or any open flame; douse all fires.
3. If you are outside, move away from buildings and power lines and stay in the open. Don't run through or near buildings.
4. If you are in a moving car, bring it to a stop as quickly as possible but stay in it. The car's springs will absorb some of the shaking and it will offer you protection.

After the Shaking

1. Check, but do *not* turn on, utilities. If you smell gas, open windows, shut off the main valve, and leave the building. Report the leak to the utility company and don't reenter the building until it has been checked out. If water mains are damaged, shut off the main valve. If electrical wiring is shorting, close the switch at the main meter box.
2. Turn on radio or television (if possible) for emergency bulletins.
3. Stay off the telephone except to report an emergency.
4. Stay out of severely damaged buildings that could collapse in aftershocks.
5. Don't go sightseeing; you will only hamper the efforts of emergency personnel and repair crews.

Source: National Oceanic and Atmospheric Administration.

placements from a single earthquake can be up to 10 to 15 m (about 30 to 50 ft).

Tsunami is a Japanese term now used to describe the very large ocean waves that can be generated by an earthquake, landslide, or volcanic explosion. Such large waves were formerly called "tidal waves." Since the large, fast waves were not associated with tides or tidal forces in any way, the term *tsunami* or *seismic sea wave* is preferred.

A tsunami, like other ocean waves, is produced by a disturbance. Most common ocean waves are produced by winds, travel at speeds of 90 km/hr (55 mi/hr), and produce a wave height of 0.6 to 3 m (2 to 10 ft) when they break on the shore. A tsunami, on the other hand, is produced by some strong disturbance in the seafloor, travels at speeds of 725 km/hr (450 mi/hr), and produces a wave height of 15 to 30 m (50 to 100 ft) when it breaks on the shore. A tsunami may have a very long wavelength of 160 km (100 mi) compared to the wavelength of ordinary wind-generated waves of 400 m (1,300 ft). Because of its great wavelength, a tsunami does not just break on the shore, then withdraw. Depending on the seafloor topography, the water from a tsunami may continue to rise for 5 to 10 minutes, flooding the coastal region before the wave withdraws. A gently sloping seafloor and a funnel-shaped bay can force tsunamis to great heights as they break on the shore. The current record height was established in 1971 when a tsunami broke at 85 m (278 ft) on an island south of Japan. Waves of such great height and long wavelength can be very destructive, sweeping away trees, lighthouses,

TABLE 20.2	Effects of earthquakes of various magnitudes
Richter Magnitudes	**Description**
0–2	Smallest detectable earthquake.
2–3	Detected and measured but not generally felt.
3–4	Felt as small earthquake but no damage occurs.
4–5	Minor earthquake with local damage.
5–6	Moderate earthquake with structural damage.
6–7	Strong earthquake with destruction.
7–8	Major earthquake with extensive damage and destruction.
8–9	Great earthquake with total destruction.

and buildings up to 30 m (100 ft) above sea level. The size of a particular tsunami depends on the nature of how the seafloor was disturbed. Generally, a "big" earthquake that causes the seafloor suddenly to rise or fall favors the generation of large tsunami.

The size of an earthquake (if it was a "big one" or a "little one") can be measured in terms of vibrations, in terms of displacement, or in terms of the amount of energy released at the site of the earthquake. The energy of the vibrations or motion of the land associated with an earthquake is called its *magnitude*. Earthquake magnitude is often reported by the media using the *Richter scale* (table 20.2). This scale assigns a number that increases with the magnitude of an earthquake. The numbers have meaning about (1) the severity of the ground-shaking vibrations and (2) the energy released by the earthquake. Each higher number indicates about 10 times more ground movement and about 30 times more energy released than the preceding number. An earthquake measuring below 3 on the scale is usually not felt by people near the epicenter. The largest earthquake measured so far had a magnitude near 9, but there is actually no upper limit to the scale.

The Richter scale was developed by Charles Richter, who was a seismologist at the California Institute of Technology in the early 1930s. The scale was based on the widest swing in the back-and-forth line traces of seismograph recording. Today, professional seismologists rate the size of earthquakes in different ways, depending on what they are comparing and why, but each way results in logarithmic scales similar to the Richter scale.

The *moment magnitude* is a logarithmic scale based on the size of a fault and the amount of movement along the fault. The 1965 Alaska quake had a moment magnitude of 9.2, the 1989 San Francisco earthquake had a moment magnitude of 7.0, and the 1994 Los Angeles earthquake had a moment magnitude of 6.7. The January 1995 earthquake that struck Japan had an estimated magnitude of 6.8 in the Kobe-Osaka region. The *surface-wave magnitude,* on the other hand, is based on the displacement of the earth's surface from the earthquake, resulting in a different logarithmic scale.

FIGURE 20.15

The folded structure of the Appalachian Mountains, revealed by weathering and erosion, is obvious in this *Skylab* photograph of the Virginia-Tennessee-Kentucky boundary area. The clouds are over the Blue Ridge Mountains.

ORIGIN OF MOUNTAINS

Folding and faulting have created most of the interesting features of the earth's surface, and the most prominent of these features are **mountains.** Mountains are elevated parts of the earth's crust that rise abruptly above the surrounding surface. Most mountains do not occur in isolation, but occur in groups that are associated in chains or belts. These long, thin belts are generally found along the edges of continents rather than in the continental interior. There are a number of complex processes involved in the origin of mountains and mountain chains, and no two mountains are exactly alike. For convenience, however, mountains can be classified according to three basic origins: (1) folding, (2) faulting, and (3) volcanic activity.

The major mountain ranges of the earth—the Appalachian, Rocky, and Himalayan Mountains, for example—have a great vertical relief that involves complex folding on a very large scale. The crust was thickened in these places as compressional forces produced tight, almost vertical folds. Thus, folding is a major feature of the major mountain ranges, but faulting and igneous intrusions are invariably also present. Differential weathering of different rock types produced the parallel features of the Appalachian Mountains that are so prominent in satellite photographs (figure 20.15). The folded sedimentary rocks of the Rockies are evident in the almost upright beds along the flanks of the front range.

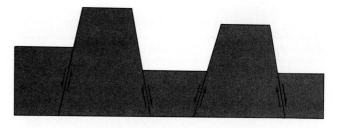

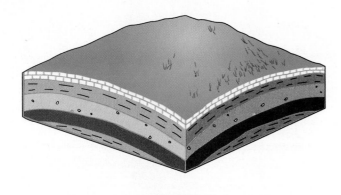

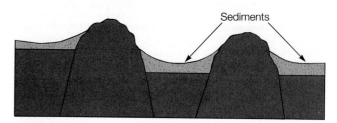

Original sharply bounded
fault blocks softened by
erosion and sedimentation

Sediments

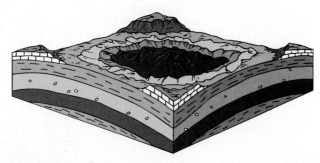

FIGURE 20.17

Fault block mountains are weathered and eroded as they are elevated, resulting in a rounded shape and sedimentation rather than sharply edged fault blocks.

FIGURE 20.16

A domed mountain begins as a broad, upwarped fold, or dome. The overlying rock of the dome is eroded away, leaving more resistant underlying rock as hills and mountains in a somewhat circular shape. These mountains are surrounded by layers of the rock that formerly covered the dome.

A broad arching fold, which is called a dome, produced the Black Hills of South Dakota. The sedimentary rocks from the top of the dome have been weathered away, leaving a somewhat circular area of more resistant granite hills surrounded by upward-tilting sedimentary beds (figure 20.16). The Adirondack Mountains of New York are another example of this type of mountain formed from folding, called domed mountains.

Compression and relaxation of compressional forces on a regional scale can produce large-scale faults, shifting large crustal blocks up or down relative to one another. Huge blocks of rocks can be thrust to mountainous heights, creating a series of fault block mountains. Fault block mountains rise sharply from the surrounding land along the steeply inclined fault plane. The mountains are not in the shape of blocks, however, as weathering has carved them into their familiar mountain-like shapes (figure 20.17). The Teton Mountains of Wyoming and the Sierra Nevadas of California are classic examples of fault block mountains that rise abruptly from the surrounding land. The various mountain ranges of Nevada, Arizona, Utah, and southeastern California have large numbers of fault block mountains that generally trend north and south.

Lava and other materials from volcanic vents can pile up to mountainous heights on the surface. These accumulations can form local volcano-formed mountains near mountains produced by folding or faulting. Such mixed-origin mountains are common in northern Arizona, New Mexico, and in western

Texas. The Cascade Mountains of Washington and Oregon are a series of towering volcanic peaks, most of which are not active today. As a source of mountains, volcanic activity has an overall limited impact on the continents. The major mountains built by volcanic activity are the mid-oceanic ridges formed at diverging plate boundaries.

Deep within the earth previously solid rock melts at high temperatures to become *magma,* a pocket of molten rock. Magma is not just melted rock alone, however, as the melt contains variable mixtures of minerals (resulting in different types of lava flows). It also includes gases such as water vapor, sulfur dioxide, hydrogen sulfide, carbon dioxide, and hydrochloric acid. You can often smell some of these gases around volcanic vents and hot springs. Hydrogen sulfide smells like rotten eggs or sewer gas. The sulfur smells like a wooden match that has just been struck.

The gases dissolved in magma play a major role in forcing magma out of the ground. As magma nears the surface it comes under less pressure, and this releases some of the dissolved gases from the magma. The gases help push the magma out of the ground. This process is similar to releasing the pressure on a can of warm soda, which releases dissolved carbon dioxide.

Magma works its way upward from its source below to the earth's surface, here to erupt into a lava flow or a volcano. A **volcano** is a hill or mountain formed by the extrusion of lava or rock fragments from magma below. Some lavas have a lower viscosity than others, are more fluid, and flow out over the land rather than forming a volcano. Such *lava flows* can accumulate into a plateau of basalt, the rock that the lava formed as it cooled and solidified. The Columbia Plateau of the states of Washington, Idaho, and Oregon is made up of layer after layer of basalt that accumulated from lava flows. Individual flows of lava

FIGURE 20.18

This is the top of Mount St. Helens several years after the 1980 explosive eruption.

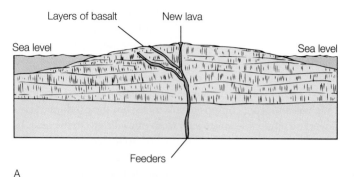

A

B

FIGURE 20.19

(*A*) A schematic cross section of an idealized shield volcano. (*B*) A photo of a shield volcano, Mauna Loa in Hawaii.

formed basalt layers up to 100 m (about 330 ft) thick, covering an area of hundreds of square kilometers. In places, the Columbia Plateau is up to 3 km (about 2 mi) thick from the accumulation of many individual lava flows.

The hill or mountain of a volcano is formed by ejected material that is deposited in a conical shape. The materials are deposited around a central *vent,* an opening through which an eruption takes place. The *crater* of a volcano is a basin-like depression over a vent at the summit of the cone. Figure 20.18 is an air view of Mount St. Helens, looking down into the crater at a volcanic dome that formed as magma periodically welled upward into the floor of the crater. This photo was taken several years after Mount St. Helens erupted in May 1980. Since that time the volcano has been quiet. It is not known when Mount St. Helens will erupt again. Geologists can tell when an eruption is near, since they are preceded by thousands of small earthquakes. Mount Garibaldi, Mount Rainier, Mount Hood, Mount Baker, and about ten other volcanic cones of the Cascade Range in Washington and Oregon could erupt next year, in the next decade, or in the next century. There are strong reasons that make it very likely that any one of the volcanic cones of the Cascade Range will indeed erupt again.

Volcanic materials do not always come out from a central vent. In a *flank eruption,* lava pours from a vent on the side of a volcano. The crater on Mount St. Helens was formed by a flank explosion. Magma was working its way toward the summit of Mount St. Helens, along with the localized earthquakes that usually accompany the movement of magma beneath the surface. One of the earthquakes was fairly strong and caused a landslide, removing the rock layers from over the slope. This reduced the pressure on the magma, and gases were suddenly released in a huge explosion, which ripped away the north flank of the volcano. The exploding gases continued to propel volcanic ash into the atmosphere for the next thirty hours.

There are three major types of volcanoes, (1) shield, (2) cinder cone, and (3) composite. The *shield volcanoes* are broad, gently sloping cones constructed of solidified lava flows. The lava that forms this type of volcano has a low viscosity, spreading widely from a vent. The islands of Hawaii are essentially a series of shield volcanoes built upward from the ocean floor. These enormous Hawaiian volcanoes range from elevations of up to 8.9 km (5.5 mi) above the ocean floor, typically with slopes that range from 2° to 20° (figure 20.19).

The Hawaiian volcanoes form from a magma about 60 km (about 40 mi) deep, probably formed by a mantle plume. The magma rises buoyantly to the volcanic cones, where it erupts by pouring from vents. Hawaiian eruptions are rarely explosive. Shield volcanoes also occur on the oceanic ridge (Iceland) and occasionally on the continents (Columbia Plateau).

A *cinder cone volcano* is constructed of, as the name states, cinders. The cinders are rock fragments, usually with sharp edges since they formed from frothy blobs of lava that cooled as they were thrown into the air. The cinder cone volcano is a pile or piles of loose cinders, usually red or black in color, that have been ejected from a vent. Cinder cones can have steep sides, with slope angles of 35° to 40°. This is the maximum steepness at which the sharp-edged, unconsolidated cinders will stay without falling. The cinder cone volcanoes are much steeper than the gentle slopes of the shield volcanoes. They also tend to be much smaller, with cones that are usually no more than about 500 m (1,600 ft) high.

A *composite volcano* (also called stratovolcano) is built up of alternating layers of cinders, ash, and lava flows (figure 20.20), forming what many people believe is the most imposing and majestic of Earth's mountains. The steepness of the sides, as you might expect, is somewhere between the steepness of the low shield volcanoes and the steep cinder cone volcanoes. The Cascade volcanoes are composite volcanoes, but the mixture of lava flows and cinders seems to vary from one volcano to the next. In addition to the alternating layers of solid basaltic rock and deposits of cinders and ash, a third type of layer is formed by volcanic mudflow. The volcanic mud can be hot or cold and it rolls down the volcano cone like wet concrete, depositing layers of volcanic conglomerate. Volcanic mudflows often do as much or more damage to the countryside as the other volcanic hazards.

The shield, cinder cone, and composite volcanoes form when magma breaks out at the earth's surface. Only a small fraction of all the magma generated actually reaches the earth's surface. Most of it remains below the ground, cooling and solidifying to form igneous rocks that were described in chapter 19. A large amount of magma that has crystallized below the surface is known as a *batholith*. A small protrusion from a batholith is called a *stock*. By definition a stock has less than 100 km² (40 mi²) of exposed surface area, and a batholith is larger. Both batholiths and stocks become exposed at the surface through erosion of the overlying rocks and rock materials, but not much is known about their shape below. The sides seem to angle away with depth, suggesting that they become larger with depth. The intrusion of a batholith sometimes tilts rock layers upward, forming a *hogback*, a ridge with equal slopes on its sides (figure 20.21). Other forms of intruded rock were formed as moving magma took the paths of least resistance, flowing into joints, faults, and planes between sedimentary bodies of rock. An intrusion that has flowed into a joint or fault that cuts across rock bodies is called a *dike*. A dike is usually tabular in shape, sometimes appearing as a great wall when exposed at the surface. One dike can occur by itself, but frequently dikes occur in great numbers, sometimes radiating out from a batholith like spokes around a wheel. If the intrusion flowed into the plane of contact between sedimentary rock layers, it is called a *sill*. A *laccolith* is similar to a sill, but has an arched top where the intrusion has raised the overlying rock into a blister-like uplift (figure 20.21).

Where does the magma that forms volcanoes and other volcanic features come from? It is produced within the outer 100 km (about 60 mi) or so of the earth's surface, presumably from a partial melting of the rocks within the crust or the uppermost part of the mantle. Basalt, the same rock type that makes up the oceanic crust, is the most abundant extrusive rock, both on the continents and along the mid-oceanic ridges. Volcanoes that rim the Pacific Ocean and Mediterranean extrude lava with a slightly different chemistry. This may be from a partial melting of the oceanic crust as it is subducted beneath the continental crust. The Cascade volcanoes are typical of those that rim the Pacific Ocean. The source of magma for Mount St. Helens and the other Cascade volcanoes is the Juan de Fuca Plate, a small plate whose spreading center is in the Pacific Ocean a little west of the Washington and Oregon coastline. The Juan de Fuca Plate is subducted beneath the continental

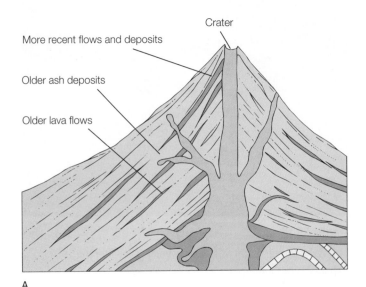

A

B

FIGURE 20.20

(*A*) A schematic cross section of an idealized composite volcano, which is built up of alternating layers of cinders, ash, and lava flows. (*B*) A photo of Mount Shasta, a composite volcano in California. You can still see the shapes of former lava flows from Mount Shasta.

A CLOSER LOOK | Volcanoes Change the World

A volcanic eruption changes the local landscape, that much is obvious. What is not so obvious are the worldwide changes that can happen just because of the eruption of a single volcano. Perhaps the most discussed change brought about by a volcano occurred back in 1815–16 after the eruption of Tambora in Indonesia. The Tambora eruption was massive, blasting huge amounts of volcanic dust, ash, and gas high into the atmosphere. Most of the ash and dust fell back to the earth around the volcano, but some dust particles and sulfur dioxide gas were pushed high into the stratosphere. It is known today that the sulfur dioxide from such explosive volcanic eruptions reacts with water vapor in the stratosphere, forming tiny droplets of diluted sulfuric acid. In the stratosphere there is no convection, so the droplets of acid and dust from the volcano eventually form a layer around the entire globe. This forms a haze that remains in the stratosphere for years, reflecting and scattering sunlight.

What were the effects of a volcanic haze around the entire world? There were fantastic, brightly colored sunsets from the added haze in the stratosphere. On the other hand, it was also cooler than usual, presumably because of the reflected sunlight that did not reach the earth's surface. It snowed in New England in June of 1816, and the cold continued into July. Crops failed, and 1816 became known as the "year without summer."

More information is available about the worldwide effects of present-day volcanic eruptions because there are now instruments to make more observations in more places. However, it is still necessary to do a great deal of estimating because of the relative inaccessibility of the worldwide stratosphere. It was estimated, for example, that the 1982 eruption of El Chichon in Mexico created enough haze in the stratosphere to reflect 5 percent of the solar radiation away from the earth.

Researchers also estimated that the El Chichon cooled the global temperatures by a few tenths of a degree for two or three years. The cooling did take place, but the actual El Chichon contribution to the cooling is not clear because of other interactions. The earth may have been undergoing global warming from the greenhouse effect, for example, so the El Chichon cooling effect could have actually been much greater. Other complicating factors such as the effects of El Niño (see chapter 21) make changes difficult to predict.

In June 1991, the Philippine volcano Mount Pinatubo erupted, blasting twice as much gas and dust into the stratosphere as El Chichon had about a decade earlier. The Philippine volcano is expected to continue erupting for years. The haze from such eruptions has the potential to cool the climate about 0.5°C (1°F). The overall result, however, will always depend on a possible greenhouse effect, a possible El Niño effect, and other complications.

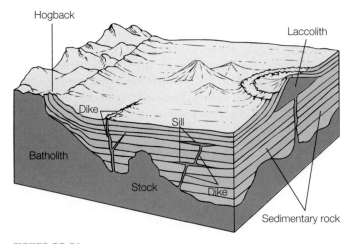

FIGURE 20.21

Here are the basic intrusive igneous bodies that form from volcanic activity.

lithosphere, and as it descends, it comes under higher and higher temperatures. Partial melting takes place, forming magma. The magma is less dense than the surrounding rock and is buoyed toward the earth's surface, erupting as a volcano.

Overall, the origin of mountain systems and belts of mountains such as the Cascades involves a complex mixture of volcanic activity as well as folding and faulting. An individual mountain, such as Mount St. Helens, can be identified as having a volcanic origin. The overall picture is best seen, however, from generalizations about how the mountains have grown along the edge of plates that are converging. Such converging boundaries are the places of folding, faulting, and associated earthquakes. They are also the places of volcanic activities, events that build and thicken the earth's crust. Thus, plate tectonics explains that mountains are built as the crust thickens at a convergent boundary between two plates. These mountains are slowly weathered and worn down as the next belt of mountains begins to build at the new continental edge.

A CLOSER LOOK | Geothermal Energy

Geothermal energy means earth (geo) heat (thermal) energy, or energy in the form of heat from the earth. Beneath the surface of the earth is a very large energy resource in the form of hot water and steam that can be used directly for heat or converted to electricity.

Most of the U.S. geothermal resources are located in the western part of the nation where the Juan de Fuca Plate is subducted beneath the continental lithosphere. The Juan de Fuca partially melts, forming magma that is buoyed toward the earth's surface, erupting as volcanoes (see figure 19.27). This subduction is the source of heating for the hot water and steam resources in ten western U.S. states. There are also geothermal resources in Hawaii from a different geothermal source.

One use of geothermal energy is to generate electricity. U.S. geothermal power plants are located in California (ten sites), Hawaii, Nevada (ten sites), Oregon, and Utah (two sites). These sites have a total generating capacity of 2,700 megawatts, which is enough electricity to supply the needs of 3.5 million people. The world's largest geothermal power plant is located at The Geysers Power Plant in northern California. Dry steam provides the energy for 23 units at this site, which generate more than 1,700 megawatts of electrical power. All of the other sites use hot water rather than steam to generate electricity, with a total generating capacity of 1,000 megawatts.

In addition to producing electricity, geothermal hot water is also used directly for space heating. Space heating in individual houses is accomplished by piping hot water from one geothermal well. District systems, on the other hand, pipe hot water from one or more geothermal wells to several buildings, houses, or blocks of houses. Currently, geothermal hot water is used in individual and in district space-heating systems at more than 120 locations. There are more than 1,200 potential geothermal sites that could be developed to provide hot water to more than 370 cities in eight states. The creation of such geothermal districts could result in a savings of up to 50 percent over the cost of natural gas heating.

Geothermal hot water is also used directly in greenhouses and aquaculture facilities. There are more than 35 large geothermal-energized greenhouses raising vegetables and flowers and more than 25 geothermal-energized aquaculture facilities raising fish in Arizona, California, Colorado, Idaho, Montana, Nevada, New York, Oregon, South Dakota, Utah, and Wyoming. A food dehydration facility in Nevada, for example, uses geothermal energy to process more than 15 million pounds of dried onions and garlic per year. Other uses of geothermal energy include laundries, swimming pools, spas, and resorts. There are over 200 resorts using geothermal hot water in the United States.

Geothermal energy is considered to be one of the renewable energy resources since the energy supply is maintained by plate tectonics. Currently, geothermal energy production is ranked third behind hydroelectricity and biomass, but ahead of solar and wind. It has been estimated that known geothermal resources could supply thousands of megawatts more power beyond current production, and development of the potential direct-use applications could displace the use—and greenhouse gas emissions—of 18 million barrels of oil per year.

PROCESSES THAT TEAR DOWN THE SURFACE

Sculpturing of the earth's surface takes place through agents and processes acting so gradually that humans are usually not aware that it is happening. Sure, some events such as a landslide or the movement of a big part of a beach by a storm are noticed. But the continual, slow, downhill drift of all the soil on a slope or the constant shift of grains of sand along a beach are outside the awareness of most people. People do notice the muddy water moving rapidly downstream in the swollen river after a storm, but few are conscious of the slow, steady dissolution of limestone by acid rain percolating through it. Yet, it is the processes of slow moving, shifting grains and bits of rocks, and slow dissolving that will wear down the mountains, removing all the features of the landscape that you can see.

A mountain of solid granite on the surface of the earth might appear to be a very solid, substantial structure, but it is always undergoing slow and gradual changes. Granite on the earth's surface is exposed to and constantly altered by air, water, and other agents of change. It is altered both in appearance and in composition, slowly crumbling, and then dissolving in water. Smaller rocks and rock fragments are moved downhill by gravity or streams, exposing more granite that was previously deeply buried. The process continues, and ultimately—over much time—a mountain of solid granite is reduced to a mass of loose rock fragments and dissolved materials. The photograph in figure 20.22 is a snapshot of a mountain-sized rock mass in a stage somewhere between its formation and its eventual destruction to rock fragments. Can you imagine the length of time that such a process requires?

Weathering

The slow changes that result in the breaking up, crumbling, and destruction of any kind of solid rock are called **weathering.** The term implies changes in rocks from the action of the weather, but it actually includes chemical, physical, and biological processes. These weathering processes are important and necessary in (1) the rock cycle, (2) the formation of soils, and (3) the movement of rock materials over the earth's surface. Weathering is important in the

FIGURE 20.22
The piles of rocks and rock fragments around a mass of solid rock is evidence that the solid rock is slowly crumbling away. This solid rock that is crumbling to rock fragments is in the Grand Canyon, Arizona.

rock cycle (see chapter 19) because it produces sediment, the raw materials for new rocks. It is important in the formation of soils because soil is an accumulation of rock fragments and organic matter. Weathering is also important in the rock cycle and in the further weathering of a rock structure because it prepares the rock materials for movement. Before the process of weathering, the rock is mostly confined to one location as a solid mass.

Weathering breaks down rocks physically and chemically, and this breaking down can occur while the rocks are stationary or while they are moving. The process of weathering is straightforward enough, but there is a fine line of distinction between the next two processes that occur in nature. The process of physically removing weathered materials is called **erosion.** *Weathering* prepares the way for erosion by breaking solid rock into fragments. The fragments are then *eroded,* physically picked up by an agent

such as a stream or a glacier. After they are eroded, the materials are then removed by *transportation.* Transportation is the movement of eroded materials by agents such as rivers, glaciers, wind, or waves. The weathering process continues during transportation. A rock being tumbled downstream, for example, is physically worn down as it bounces from rock to rock. It may be chemically altered as well as it is bounced along by the moving water. Overall, the combined action of weathering and erosion wears away and lowers the elevated parts of the earth and sculpts their surfaces.

There are two basic kinds of weathering that act to break down rocks: *mechanical weathering* and *chemical weathering. Mechanical weathering* is the physical breaking up of rocks without any changes in their chemical composition. Mechanical weathering results in the breaking up of rocks into smaller and smaller pieces, so it is also called *disintegration.* If you smash a sample of granite into smaller and smaller pieces you are mechanically weathering the granite. *Chemical weathering* is the alteration of minerals by chemical reactions with water, gases of the atmosphere, or solutions. Chemical weathering results in the dissolving or breaking down of the minerals in rocks, so it is also called *decomposition.* If you dissolve a sample of limestone in a container of acid, you are chemically weathering the limestone.

Examples of mechanical weathering in nature include the disintegration of rocks caused by (1) *wedging effects,* and (2) the *effects of reduced pressure.* Wedging effects are often caused by the repeated freezing and thawing of water in the pores and small cracks of otherwise solid rock. If you have ever seen what happens when water in a container freezes, you know that freezing water expands and exerts a pressure on the sides of its container. As water in a pore or a crack of a rock freezes, it also expands, exerting a pressure on the walls of the pore or crack, making it slightly larger. The ice melts and the enlarged pore or crack again becomes filled with water for another cycle of freezing and thawing. As the process is repeated many times, small pores and cracks become larger and larger, eventually forcing pieces of rock to break off. This process is called **frost wedging** (figure 20.23A). It is an important cause of mechanical weathering in mountains and other locations where repeated cycles of water freezing and thawing occur. The roots of trees and shrubs can also mechanically wedge rocks apart as they grow into cracks. You may have noticed the work of roots when trees or shrubs have grown next to a sidewalk for some period of time (figure 20.24).

The other example of mechanical weathering is believed to be caused by the reduction of pressure on rocks. As more and more weathered materials are removed from the surface, the downward pressure from the weight of the material on the rock below becomes less and less. The rock below begins to expand upward, fracturing into concentric sheets from the effect of reduced pressure. These curved, sheet-like plates fall away later in the mechanical weathering process called *exfoliation* (figure 20.23B). Exfoliation is the term given to the process of spalling off of layers of rock, somewhat analogous to peeling layers from an onion. Granite commonly weathers by exfoliation, producing characteristic dome-shaped hills and rounded boulders. Stone Mountain, Georgia is a well-known example of an exfoliation-shaped dome. The onion-like structure of exfoliated granite is a

Water freezing in crack expands, widening crack.

Eventually rock is broken free by progressive fracturing.

A

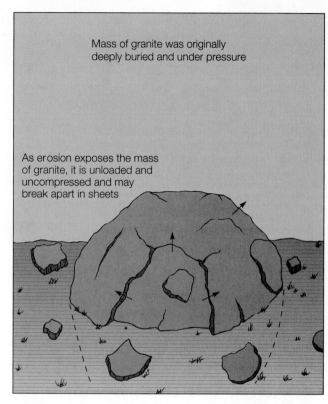

Mass of granite was originally deeply buried and under pressure

As erosion exposes the mass of granite, it is unloaded and uncompressed and may break apart in sheets

B

FIGURE 20.23

(*A*) Frost wedging and (*B*) exfoliation are two examples of mechanical weathering, or disintegration, of solid rock.

A

B

FIGURE 20.24

Growing trees can break, separate, and move solid rock. (*A*) Note how this tree has raised the sidewalk. (*B*) This tree is surviving by growing roots into tiny joints and cracks, which become larger as the tree grows.

FIGURE 20.25

Spheroidal weathering of granite. The edges and corners of an angular rock are attacked by weathering from more than one side and retreat faster than flat rock faces. The result is rounded granite boulders, which often shed partially weathered minerals in onion-like layers.

common sight in the Sierras, Adirondacks, and any mountain range where older exposed granite is at the surface (figure 20.25).

Examples of chemical weathering include (1) oxidation, (2) carbonation, and (3) hydration. *Oxidation* is a reaction between oxygen and the minerals making up rocks. The ferromagnesian minerals contain iron, magnesium, and other metal ions in a silicate structure. Iron can react with oxygen to produce several different iron oxides, each with its own characteristic color. The most common iron oxide (hematite) has a deep red color. Other oxides of iron are brownish to yellow-brownish. It is the presence of such iron oxides that color many sedimentary rocks and soils. The red soils of Oklahoma, Georgia, and many other places are colored by the presence of iron oxides produced by chemical weathering.

Carbonation is a reaction between carbonic acid and the minerals making up rocks. Rainwater is naturally somewhat acidic because it dissolves carbon dioxide from the air. This forms a weak acid known as carbonic acid (H_2CO_3), the same acid that is found in your carbonated soda pop. Carbonic acid rain falls on the land, seeping into cracks and crevices where it reacts with minerals. Limestone, for example, is easily weathered to a soluble form by carbonic acid. The limestone caves of Missouri, Kentucky, New Mexico, and elsewhere were produced by the chemical weathering of limestone by carbonation (figure 20.26). Minerals

A

B

FIGURE 20.26

Limestone caves develop when slightly acidic groundwater dissolves limestone along joints and bedding planes, carrying away rock components in solution. (*A*) Joints and bedding planes in a limestone bluff. (*B*) This stream has carried away less-resistant rock components, forming a cave under the ledge.

containing calcium, magnesium, sodium, potassium, and iron are chemically weathered by carbonation to produce salts that are soluble in water.

Hydration is a reaction between water and the minerals of rocks. The process of hydration includes (1) the dissolving of a mineral and (2) the combining of water directly with a mineral. Some minerals, for example halite (which is sodium chloride), dissolve in water to form a solution. The carbonates formed from carbonation are mostly soluble, so they are easily leached from a rock by dissolving. Water also combines directly with some minerals to form new, different minerals. The feldspars, for example, undergo hydration and carbonation to produce (1) water-soluble potassium carbonate, (2) a chemical product that combines with water to produce a clay mineral, and (3) silica. The silica, which is silicon dioxide (SiO_2), may appear as a suspension of finely divided particles or in solution.

Mechanical and chemical weathering are interrelated, working together in breaking up and decomposing solid rocks of the earth's surface. In general, mechanical weathering results in cracks in solid rocks and broken-off coarse fragments. Chemical weathering results in finely pulverized materials and ions in solution, the ultimate decomposition of a solid rock. Consider, for example, a mountain of solid granite, the most common rock found on continents. In general, granite is made up of 65 percent feldspars, 25 percent quartz, and about 10 percent ferromagnesian minerals. Mechanical weathering begins the destruction process as exfoliation and frost wedging create cracks in the solid mass of granite. Rainwater, with dissolved oxygen and carbon dioxide, flows and seeps into the cracks and reacts with ferromagnesian minerals to form soluble carbonates and metal oxides. Feldspars undergo carbonation and hydration, forming clay minerals and soluble salts, which are washed away. Quartz is less susceptible to chemical weathering and remains mostly unchanged to form sand grains. The end products of the complete weathering of granite are quartz sand, clay minerals, metal oxides, and soluble salts.

Erosion

Weathering has prepared the way for erosion to pick up and for some agent of transportation to move or carry away the fragments, clays, and solutions that have been produced from solid rock. The weathered materials can be moved to a lower elevation by the direct result of gravity acting alone. They can also be moved to a lower elevation by gravity acting through some intermediate agent, such as running water, wind, or glaciers. The erosion of weathered materials as a result of gravity alone will be considered first.

Mass Movement

Gravity constantly acts on every mass of materials on the surface of the earth, pulling parts of elevated regions toward lower levels. Rocks in the elevated regions are able to temporarily resist this constant pull through their cohesiveness with a main rock mass or by the friction of the rock on a slope. Whenever anything happens to reduce the cohesiveness or to reduce the fric-

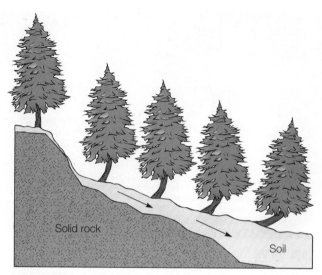

FIGURE 20.27

The slow creep of soil is evidenced by the strange growth pattern of these trees.

tion, gravity pulls the freed material to a lower elevation. Thus, gravity acts directly on individual rock fragments and on large amounts of surface materials as a mass, pulling all to a lower elevation. Erosion caused by gravity acting directly is called **mass movement** (also called mass wasting). Mass movement can be so slow that it is practically imperceptible. *Creep,* the slow downhill movement of soil down a steep slope, for example, is detectable only from the peculiar curved growth patterns of trees growing in the slowly moving soil (figure 20.27). At the other extreme, mass movement can be as sudden and swift as a rock bounding and clattering down a slope below a cliff. A *landslide* is a generic term used to describe any slow to rapid movement of any type or mass of materials, from the short slump of a hillside to the slide of a whole mountainside. Either slow or sudden, mass movement is a small victory for gravity in the ongoing process of leveling the landmass of the earth.

Running Water

Running water is the most important of all the erosional agents of gravity that remove rock materials to lower levels. Streams and major rivers are at work, for the most part, 24 hours a day every day of the year moving rock fragments and dissolved materials from elevated landmasses to the oceans. Any time you see mud, clay, and sand being transported by a river, you know that the river is at work moving mountains, bit by bit, to the

FIGURE 20.28

Moving streams of water carry away dissolved materials and sediments as they slowly erode the land.

FIGURE 20.29

A river usually stays in its channel, but during a flood it spills over and onto the adjacent flat land called the *floodplain.*

ocean. It has been estimated that rivers remove enough dissolved materials and sediments to lower the whole surface of the United States flat in little over 20 million years, a very short time compared to the 4.6-billion-year age of the earth.

Erosion by running water begins with rainfall. Each raindrop impacting the soil moves small rock fragments about, but also begins to dissolve some of the soluble products of weathering. If the rainfall is heavy enough, a shallow layer, or sheet of water forms on the surface, transporting small fragments and dissolved materials across the surface. This *sheet erosion* picks up fragments and dissolved material, then transports them to small streams at lower levels (figure 20.28). The small streams move to larger channels and the running water transports materials three different ways, (1) as dissolved rock materials carried in solution, (2) as clay minerals and small grains carried in suspension, and (3) as sand and larger rock fragments that are rolled, bounced, and slid along the bottom of the stream bed. Just how much material is eroded and transported by the stream depends on the volume of water, its velocity, and the load that it is already carrying.

In addition to transporting materials that were weathered and eroded by other agents of erosion, streams do their own erosive work. Streams can dissolve soluble materials directly from rocks and sediments. They also quarry and pluck fragments and pieces of rocks from beds of solid rock by hydraulic action. Most of the erosion accomplished directly by streams, however, is

done by the more massive fragments that are rolled, bounced, and slid along the stream bed and against each other. This results in a grinding and filing action on the fragments and a wearing away of the stream bed.

As a stream cuts downward into its bed, other agents of erosion such as mass movement begin to widen the channel as materials slump into the moving water. The load that the stream carries is increased by this slumping, which slows the stream. As the stream slows, it begins to develop bends, or *meanders* along the channel. Meanders have a dramatic effect on stream erosion because the water moves faster around an outside bank than it does around the inside bank downstream. This difference in stream velocity means that the stream has a greater erosion ability on the outside, downstream side and less on the sheltered area inside of curves. The stream begins to widen the floor of the valley through which it runs by eroding on the outside of the meander, then depositing the eroded material on the inside of another bend downstream. The stream thus begins to erode laterally, slowly working its way across the land. Sometimes two bends in the stream meet, forming a cut-off meander called an *oxbow lake.*

A stream, along with mass movement, develops a valley on a widening floodplain. A **floodplain** is the wide, level floor of a valley built by a stream (figure 20.29). It is called a floodplain because this is where the stream floods when it spills out of its channel. The development of a stream channel into a widening floodplain seems to follow a general, idealized aging pattern (figure 20.30). When a stream is on a recently uplifted landmass, it has a steep gradient, a vigorous, energetic ability to erode the land, and characteristic features known as the stage of youth. *Youth* is characterized by a steep gradient, a V-shaped valley without a floodplain, and the presence of features that interrupt its smooth flow such as boulders in the stream bed, rapids, and waterfalls (figure 20.31). Stream erosion during youth is predominantly downward. The stream eventually erodes its way into *maturity* by eroding away the boulders, rapids, and waterfalls, and in general smoothing and lowering the stream gradient.

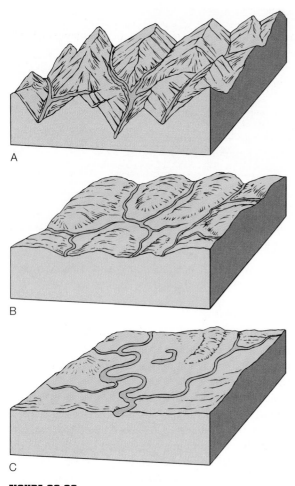

FIGURE 20.30

Three stages in the aging and development of a stream valley, (*A*) youth, (*B*) maturity, and (*C*) old age.

FIGURE 20.31

The waterfall and rapids on the Yellowstone River in Wyoming indicate that the river is actively down cutting. Note the V-shaped cross-profile and lack of floodplain, characteristics of a young stream valley.

During maturity, meanders form over a wide floodplain that now occupies the valley floor. The higher elevations are now more sloping hills at the edge of the wide floodplain rather than steep-sided walls close to the river channel. *Old age* is marked by a very low gradient in extremely broad, gently sloping valleys. The stream now flows slowly in broad meanders over the wide floodplain. Floods are more common in old age since the stream is carrying a full load of sediments and flows sluggishly.

Many assumptions are made in any generalized scheme of the erosional aging of a stream. Streams and rivers are dynamic systems that respond to local conditions, so it is possible to find an "old age feature" such as meanders in an otherwise youthful valley. This is not unlike finding a gray hair on an 18-year-old youth, and in this case the presence of the gray hair does not mean old age. In general, old age characteristics are observed near the *mouth* of a stream where it flows into an ocean, lake, or another stream. Youthful characteristics are observed at the *source,* where the water collects to first form the stream channel. As the stream slowly lowers the land, the old age characteristics will move slowly but surely toward the source.

When the stream flows into the ocean or a lake, it loses all of its sediment-carrying ability. It drops the sediments, forming a deposit at the mouth called a **delta** (figure 20.32). Large rivers such as the Mississippi River have large and extensive deltas that actually extend the landmass more and more over time. In a way, you could think of the Mississippi River delta as being formed from pieces and parts of the Rocky Mountains, the Ozark Mountains, and other elevated landmasses that the Mississippi has carried there over time.

Activities

Measure and record the speed of a river or stream every day for a month. The speed can be calculated from the time required for a floating object to cover a measured distance. Define the clearness of the water by shining a beam of light through a sample, then comparing to a beam of light through clear water. Use a scale, such as 1 for perfectly clear to 10 for no light coming through, to indicate clearness. Graph your findings to see if there is a relationship between clearness and speed of flow of the stream.

A

FIGURE 20.32

(*A*) Delta of Nooksack River, Washington. Note the sediment-laden water and how the land is being built outward by river sedimentation. (*B*) Cross section showing how a small delta might form. Large deltas are more complicated than this.

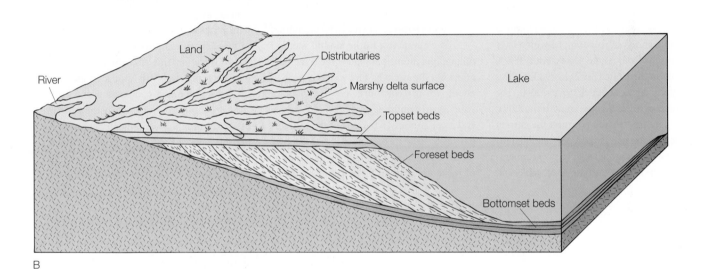

B

Glaciers

Glaciers presently cover only about 10 percent of the earth's continental land area, and as much of this is at higher latitudes, so it might seem that glaciers would not have much of an overall effect in eroding the land. However, ice has sculptured much of the present landscape, and features attributed to glacial episodes are found over about three-quarters of the continental surface. Only a few tens of thousands of years ago, sheets of ice covered major portions of North America, Europe, and Asia. Today, the most extensive glaciers in the United States are those of Alaska, which cover about 3 percent of the state's land area. Less extensive glacier ice is found in the mountainous regions of Washington, Montana, California, Colorado, and Wyoming.

A **glacier** is a mass of ice on land that moves under its own weight. Glacier ice forms gradually from snow, but the quantity of snow needed to form a glacier does not fall in a single winter.

Glaciers form in cold climates where some snow and ice persists throughout the year. The amount of winter snowfall must exceed the summer melting to accumulate a sufficient mass of snow to form a glacier. As the snow accumulates, it is gradually transformed into ice. The weight of the overlying snow packs it down, driving out much of the air, and causing it to recrystallize into a coarser, denser mass of interlocking ice crystals that appears to have a blue to deep blue color. Complete conversion of snow into glacial ice may take from 5 to 3,500 years, depending on such factors as climate and rate of snow accumulation at the top of the pile. Eventually the mass of ice will become large enough that it begins to flow, spreading out from the accumulated mass. Glaciers that form at high elevations in mountainous regions, which are called *alpine glaciers,* tend to flow downhill through a valley so they are also called *valley glaciers* (figure 20.33). Glaciers that cover a large area of a continent are called *continental glaciers.*

A CLOSER LOOK | Municipal Solid Waste Problems

Municipal solid waste is becoming a major problem for many large cities with the filling of old landfills and lack of suitable land for new ones. The people of the United States were producing close to an estimated 200 million tons of municipal solid waste per year by the year 2000. In general, this throwaway stuff was made up of discarded newspapers and other paper (39%), yard trimmings (17%), rubber, textiles, and wood (12%), metal (9%), glass (8%), plastic (8%), and food waste (7%).

About 80 percent of the municipal solid waste of North America goes into landfills where it is compacted by bulldozers, mixed with soil, and eventually capped with a layer of clay and soil. There were about 18,500 landfills in the United States in 1980, but this had declined to about 3,250 by 2000. The numbers were less because landfills were filled up or because they did not meet the current environmental standards. A new landfill must have complex bottom layers to stop contaminants from leaking into the water table, and

they must have monitoring systems to detect groundwater contamination in addition to methane gas created by rotting garbage.

New landfills are difficult to start because of a lack of suitable land and because of public concern over groundwater contamination, odors, blowing papers, and the traffic created by trash haulers. Local opposition to such a new landfill is sometimes called a NIMBY, or "not in my backyard" reaction. Sweden, Switzerland, and Japan have already moved away from relying on landfills, disposing of 15 percent or less of their waste this way. Compare this to the current North American practice of disposing of 80 percent of the waste in landfills.

One alternative to the landfill method of handling municipal solid waste is incineration. On the average, solid waste consists of about 75 percent burnable materials in paper, yard trimmings, wood, food scraps, and so on. About 15 percent of the municipal solid waste in the United States is currently incinerated. This reduces the amount of waste up to about 90 percent by volume

and 75 percent by weight. This does create possible air pollution problems and the ash must be disposed of, but the heat from the burning can be used to make steam and generate electricity.

Other alternatives to the landfill method of handling waste include recycling and composting. Thousands of cities across the United States and Canada have comprehensive recycling programs for aluminum, paper, glass, and some plastic items. Side benefits from recycling programs include the saving of natural resources and reduction of air pollution. Composting of yard trimmings and food scraps is another form of recycling since it turns waste into a useful soil additive, and this can reduce the amount of waste materials by about 20 percent. Intensive combined composting and recycling programs can easily reduce the amount of waste going to a landfill site by up to 50 percent. You can reduce waste even further by purchasing only reusable or recyclable materials and avoiding throwaway products.

FIGURE 20.33
Valley glacier on Mount Logan, Yukon Territory.

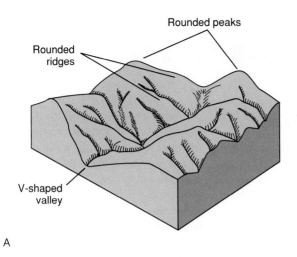

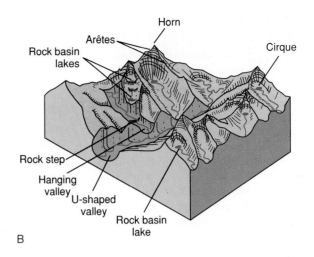

A

B

FIGURE 20.34

(*A*) A stream-carved mountainside before glaciation. (*B*) The same area after glaciation, with some of the main features of mountain glaciation labeled.

Continental glaciers can cover whole continents and reach a thickness of 1 km or more. Today, the remaining continental glaciers are found on Greenland and the Antarctic.

Glaciers move slowly and unpredictably, spreading like a huge blob of putty under the influence of gravity. As an alpine glacier moves downhill through a V-shaped valley, the sides and bottom of the valley are eroded wider and deeper. When the glacier later melts, the V-shaped valley is now a U-shaped valley that has been straightened and deepened by the glacial erosion. The glacier does its erosional work using three different techniques, (1) by bulldozing, (2) by plucking, and (3) by abrasion. *Bulldozing,* as the term implies, is the pushing along of rocks, soil, and sediments by the leading edge of an advancing glacier. Deposits of bulldozed rocks and other materials that remain after the ice melts are called *moraines. Plucking* occurs as water seeps into cracked rocks and freezes, becoming a part of the solid glacial mass. As the glacier moves on, it pulls the fractured rock apart and plucks away chunks of it. The process is accelerated by the frost-wedging action of the freezing water. Plucking at the uppermost level of an alpine glacier, combined with weathering of the surrounding rocks, produces a rounded or bowl-like depression known as a *cirque* (figure 20.34). *Abrasion* occurs as the rock fragments frozen into the moving glacial ice scratch, polish, and grind against surrounding rocks at the base and along the valley walls. The result of this abrasion is the pulverizing of rock into ever finer fragments, eventually producing a powdery, silt-sized sediment called *rock flour.* Suspended rock flour in meltwater from a glacier gives the water a distinctive gray to blue-gray color.

Glaciation is continuously at work eroding the landscape in Alaska and many mountainous regions today. The glaciation that formed the landscape features in the Rockies, the Sierras, and across the northeastern United States took place thousands of years ago.

Wind

Like running water and ice, wind also acts as an agent shaping the surface of the land. It can erode, transport, and deposit materials. However, wind is considerably less efficient than ice or water in modifying the surface. Wind is much less dense and does not have the eroding or carrying power of water or ice. In addition, a stream generally flows most of the time but the wind blows only occasionally in most locations. Thus, on a worldwide average, winds move only a few percent as much material as do streams. Wind also lacks the ability to attack rocks chemically as water does through carbonation and other processes, and the wind cannot carry dissolved sediments in solution. Even in many deserts, more sediment is moved during the brief periods of intense surface runoff following the occasional rainstorms than is moved by wind during the prolonged dry periods.

Flowing air and moving water do have much in common as agents of erosion since both are fluids. Both can move larger particles by rolling them along the surface and can move finer particles by carrying them in suspension. Both can move larger and more massive particles with increased velocities. Water is denser and more viscous than air, so it is more efficient at transporting quantities of material than is the wind, but the processes are quite similar.

Two major processes of wind erosion are called (1) abrasion and (2) deflation. *Wind abrasion* is a natural sandblasting process that occurs when the particles carried along by the wind break off small particles and polish what they strike. Generally, the harder mineral grains such as quartz sand accomplish this best near the ground where the wind is bouncing them along. Wind abrasion can strip paint from a car exposed to the moving particles of a dust storm, eroding the paint along with rocks on the surface. Rocks and boulders exposed to repeated wind storms where the wind blows consistently from one or a few directions may be planed off from

A CLOSER LOOK | Acid Rain and Chemical Weathering

Acid rain is a generic term used to describe any acidic substances, wet or dry, that fall from the atmosphere. Wet acidic deposition could be in the form of rain, but snow, sleet, and fog could also be involved. Dry acidic deposition could include gases, dust, or any solid particles that settle out of the atmosphere to produce an acid condition.

The acid or alkaline strength of a solution is measured on the pH scale, which is a powers of ten scale with values from 0 to 14. A pH of 0 is the strongest acid reading on the scale, 7 is neutral (neither acidic nor alkaline), and a pH of 14 is the strongest alkaline reading. Remember that the pH is not a simple linear scale, and each lower number between 7 and 0 is *ten times* more acidic than the next higher number. For example, a pH of 3 is ten times more acidic than a pH of 4; a pH of 2 is ten times more acidic than a pH of 3. A pH of 2 is ten times ten, or one-hundred times more acidic than a pH of 4. Values above 7 are alkaline, again increasing by powers of ten for each number on the pH scale.

Pure, unpolluted rain is naturally acidic. Carbon dioxide in the atmosphere is absorbed by rainfall, forming carbonic acid (H_2CO_3). Carbonic acid lowers the pH of pure rainfall to a range of 5.6–6.2. Decaying vegetation in local areas can provide more CO_2, making the pH even lower. A pH range of 4.5–5.0, for example, has been measured in remote areas of the Amazon jungle. Human-produced exhaust emissions of sulfur and nitrogen oxides can lower the pH of rainfall even more, to a 4.0–5.5 range. This is the pH range of acid rain (box table 20.1).

BOX TABLE 20.1

The approximate pH of some common acidic substances

Substance	pH (or pH Range)
Hydrochloric acid (4%)	0
Gastric (stomach) solution	1.6–1.8
Lemon juice	2.2–2.4
Vinegar	2.4–3.4
Carbonated soft drinks	2.0–4.0
Grapefruit	3.0–3.2
Oranges	3.2–3.6
Acid rain	**4.0–5.5**
Tomatoes	4.2–4.4
Potatoes	5.7–5.8
Natural rainwater	**5.6–6.2**
Milk	6.3–6.7
Pure water	7.0

The sulfur and nitrogen oxides that produce acid rain come from exhaust emissions of industries and electric utilities that burn fossil fuels and from the exhaust of cars, trucks, and buses. The oxides are the raw materials of acid rain and are not themselves acidic. They react with other atmospheric chemicals to form sulfates and nitrates, which combine with water vapor to form sulfuric acid and nitric acid. These are the chemicals of concern in acid rain.

Acid rain falls on the land, bodies of water, forests, crops, buildings, and people, so the concerns about acid rain center on its environmental impact on lakes, forests, crops, materials, and human health. It is believed, for example, that acid rain accelerates chemical weathering and also leaches essential nutrients from the soil. It can also cause more rapid chemical weathering of rocks and buildings and acidify lakes and streams, sometimes to the detriment of wildlife.

The type of rocks making up the local landscape can either moderate or aggravate the problems of acid rain. Limestone, and the soils of arid climates, tend to neutralize acid, while waters in granitic rocks and soils already tend to be somewhat acidic.

Natural phenomena, such as volcanoes, and human-produced emissions of sulfur and nitrogen oxides from burning fossil fuels are the cause of acid rain. The heavily industrialized northeastern part of the United States, from the midwest through New England, releases sulfur and nitrogen emissions that result in a precipitation pH of 4.0 to 4.5. Unfortunately, the area downwind of major sulfur sources has a granitic bedrock, which means that the effects of acid rain will not be moderated as they would in the west or midwest. This region, the northeast section of North America, is the geographic center of the acid rain problem. The solution to the problem is being sought by (1) using fuels other than fossil fuels when possible and (2) reducing the thousands of tons of sulfur and nitrogen oxides that are dumped into the atmosphere per day when fossil fuels are used.

---

the repeated action of this natural sandblasting. Rocks sculptured by wind abrasion are called *ventifacts,* after the Latin meaning "wind-made" (figure 20.35).

Deflation, named after the Latin meaning "to blow away," is the widespread picking up of loose materials from the surface. Deflation is naturally most active where winds are unobstructed and the materials are exposed and not protected by vegetation. These conditions are often found on deserts, beaches, and on unplanted farmland between crops. During the 1930s several

years of drought killed the native vegetation in the Plains states during a period of increased farming activity. Unusually strong winds eroded the unprotected surface, removing and transporting hundreds of millions of tons of soil. This period of prolonged drought, dust storms, and general economic disaster for farmers in the area is known as the Dust Bowl episode.

The most common wind-blown deposits are (1) dunes and (2) loess. A **dune** is a low mound or ridge of sand or other sediments. Dunes form when sediment-bearing wind encounters an

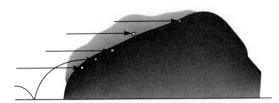

If wind is predominantly from one direction, rocks will be planed off or flattened on the upwind side.

With a persistent shift in wind direction, additional facets are cut in the rock.

FIGURE 20.35
Ventifact formation by abrasion from one or several directions.

FIGURE 20.36
The fossil record of the hard parts is beautifully preserved, along with a carbon film, showing a detailed outline of the fish and some of its internal structure.

obstacle that reduces the wind velocity. With a slower velocity the wind cannot carry as large a load, so sediments are deposited on the surface. This creates a windbreak, which results in a growing obstacle, a dune. Once formed, a dune tends to migrate, particularly if the winds blow predominantly from one direction. Dunes are commonly found in semiarid areas or near beaches.

Another common wind deposit is called *loess.* Loess is a very fine dust, or silt, that has been deposited over a large area. One such area is located in the central part of the United States, particularly to the east sides of the major rivers of the Mississippi basin. Apparently this deposit originated from the rock flour produced during the last great ice age. The rock flour was probably deposited along the major river valleys and later moved eastward by the prevailing westerly winds. Since rock flour is produced by the mechanical grinding action of glaciers, it has not been chemically broken down. Thus, the loess deposit contains many minerals that were not leached out of the deposit as typically occurs with chemical weathering. It also has an open, porous structure since it does not have as much of the chemically-produced clay minerals. The good moisture-holding capacity from this open structure, together with the presence of minerals that serve as plant nutrients, make farming on the deposit particularly productive.

GEOLOGIC TIME

Geology is the study of the earth and the processes that shape it. *Physical geology* is a branch of geology concerned with the materials of the earth, processes that bring about changes in the materials and structures they make up, and the physical features of the earth formed as a result. *Historical geology,* on the other hand, is a branch of geology concerned with the development of

the earth and the organisms on it over time. Physical and historical geology together provide a basis for understanding much about the earth and how it has developed.

One reason to study geology is to satisfy an intellectual curiosity about how the earth works. Piecing together the history of how a mountain range formed or inferring the history of an individual rock can be exciting as well as satisfying. As a result, you appreciate the beauty of our earth from a different perspective. Distinctive features such as the granite domes of Yosemite, the geysers of Yellowstone, and the rocks exposed in a roadcut take on a whole new meaning. Often, part of the new meaning is a story that tells the history and how that distinctive feature came to be.

Altogether, the story of the individual rock and the landscape features describes the history of the region and how it came to be what it is today. The resulting knowledge of geologic processes and events can also have a practical aspect. Certain earth materials are used for energy or for raw materials used in the manufacture of technological devices. Knowing how, where, and when such resources are formed can be very useful information to modern society.

Fossils

What would you think if you were on the top of a mountain, broke open a rock, and discovered the fossil fish pictured in figure 20.36? How could you explain what you found? There are several ways that a fish could end up on a mountaintop as a fossil. For example, perhaps the ocean was once much deeper and covered the mountaintop. On the other hand, maybe the mountaintop was once below sea level and pushed its way up to its present high altitude. Another explanation might be that someone left the fossil on the mountain as a practical joke. What would you look for to help you figure out what actually happened? In every rock and fossil there are fascinating clues that help you read what happened in the past, including clues that tell you if an ocean had covered the area or if a mountain pushed its way up from lower levels. There are even clues that tell you if a rock has been brought in from another place. This section is about some of the clues found in fossils and rocks, and what the clues mean.

Early Ideas about Fossils

The story about finding a fossil fish on a mountain is not as far-fetched as it might seem, and in fact one of the first recorded evidences of understanding the meaning of fossils took place in a similar setting. The ancient Greek historian Herodotus was among the first to realize that fossil shells found in rocks far from any ocean were remnants of organisms left by a bygone sea. Other Greek philosophers were not convinced that this conclusion was as obvious as it might seem today. Aristotle, for example, could see no connection between the shells of organisms of his time and the fossils, which he believed to have formed inside the rocks. Note that he believed that living organisms could arise by spontaneous generation from mud. A belief that the fossils must have "grown" in place in rocks would seem to be consistent with a belief in spontaneous generation.

It was a long time before it was generally recognized that fossils had anything to do with living things. Even when people started to recognize some similarities between living organisms and certain fossils, they did not make a connection. Fossils were considered to be the same as quartz crystals, or any other mineral crystals, meaning they were either formed with the earth, or grew there later (depending on the philosophical view of the interpreter). Fossils of marine organisms that were well preserved and very similar to living organisms were finally accepted as remains of once-living organisms—that were buried in Noah's flood. By the time of the Renaissance some people were starting to think of other fossils, too, as the remains of former life-forms. Leonardo da Vinci, like other Renaissance scholars, argued that fossils were the remains of organisms that had lived in the past.

By the early 1800s the true nature of fossils was becoming widely accepted. William "Strata" Smith, an English surveyor, discovered at this time that sedimentary rock strata could be identified by the fossils they contain. Smith grew up in a region of England where fossils were particularly plentiful. He became a collector, keeping careful notes on where he found each fossil and in which type of sedimentary rock layers. During his travels he discovered that the succession of rock layers on the south coast of England was the same as the succession of rock layers on the east coast. Through his keen observations, Smith found that each kind of sedimentary rock had a distinctive group of fossils that was unlike the group of fossils in other rock layers. Smith amazed his friends by telling them where and in what type of rock they had found their fossils.

Today, the science of discovering fossils, studying the fossil record, and deciphering the history of life from fossils is known as *paleontology*. The word "paleontology" was invented in 1838

by the British geologist Charles Lyell to describe this newly established branch of geology. It is derived from classic Greek roots and means "study of ancient life," and this means a study of fossils. A **fossil** is *any* evidence of former life, so the term means more than fossilized remains such as those pictured in figure 20.36. Evidence can include actual or altered remains of plants and animals. It could also be just simple evidence of former life such as the imprint of a leaf, the footprint of a dinosaur, or droppings from bats in a cave.

People sometimes blur the distinction between *paleontology* and *archaeology*. Archaeology is the study of past human life and culture from material evidence of artifacts, such as graves, buildings, tools, pottery, landfills, and so on ("artifact" literally means "something made"). The artifacts studied in archaeology can be of any age, from the garbage added to the city landfill yesterday to the pot shards of an ancient tribe that disappeared hundreds of years ago. The word "fossil" originally meant "anything dug up," but today carries the meaning of any *evidence of ancient organisms in the history of life*. Artifacts are therefore not fossils, as you can see from the definitions.

Types of Fossilization

Considering all the different things that can happen to the remains of an organism, and considering the conditions needed to form a fossil, it seems amazing that any fossils are formed and then found. Consider, for example, that you have probably never seen many dead birds, if any, in your neighborhood. This is probably true even if you live in an area where birds are exceptionally plentiful. The reason that you do not see dead birds is that scavengers have most likely destroyed the remains. Earth is teeming with many types, sizes, and forms of life, all searching for something to eat. As a result, very little digestible organic matter escapes destruction, but indigestible skeletal material, such as shells, bones, and teeth, has a much better chance of not being destroyed.

When an organism dies, scavengers, insects, and bacteria consume the soft parts of plants and animals, both on land and underwater. Thus a fossil is not likely to form unless there is rapid burial of a recently deceased organism. The presence of hard parts, such as a shell or a skeleton, will also favor the formation of a fossil if there is rapid burial.

There are three broad ways in which fossils are commonly formed (table 20.3):

1. preservation or alteration of hard parts.
2. preservation of the shape.
3. preservation of signs of activity.

Only rarely are the unaltered remains of an organism's *soft parts* found. The best examples of this uncommon method of fossilization include protection by freezing and entombing in tree resin. Mammoths, for example, have been found frozen and preserved by natural refrigeration in the ices of Alaska and Siberia. Insects and spiders, complete with delicate appendages, have been found preserved in amber, which is fossilized tree resin. In each case—ice and resin—the soft parts were protected from scavengers, insects, and bacteria.

TABLE 20.3	Summary of the types of fossil preservation

I. Preservation of all or part of the organism
   A. Unaltered
      1. Soft parts
      2. Hard parts
   B. Altered
      1. Mineralization
      2. Replacement
      3. Carbon films
II. Preservation of the organism's shape
   A. Cast
   B. Mold
III. Preservation of signs of activity
   A. Tracks
   B. Trails
   C. Burrows
   D. Borings
   E. Coprolites

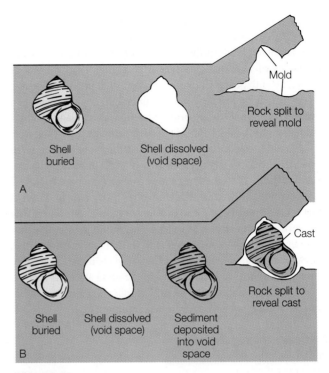

FIGURE 20.37

Origin of molds and casts. (*A*) Formation of a mold. (*B*) Formation of a cast.

Fossils are more commonly formed from *remains of hard parts* such as shells, bones, and teeth of animals or the pollen and spores of plants. Such parts are composed of calcium carbonate, calcium phosphate, silica, chitin, or other tough organic coverings. The fish fossil pictured in figure 20.36 is from Wyoming's Green River Formation. This freshwater fish died in its natural environment, and was soon covered by fine-grained sediment. The sediment preserved the complete articulated skeleton, along with some carbon traces of soft tissue. Plant fossils are often found as carbon traces, sometimes looking like a photograph of a leaf on a slab of shale or limestone.

Shells and other hard parts of invertebrates are sometimes preserved without alteration, or with changes in the chemical composition. Most corals, mollusks and other shelled invertebrates have calcium carbonate shells, but some do have calcium phosphate shells. Silica makes up the hard shells of protozoans, sponges, and diatoms, one of the most abundant marine plants. Silica is the most resistant common substance found in fossils. Chitin is the tough material that makes up the exoskeletons of insects, crabs, and lobsters.

Calcium carbonate shell material may be dissolved by groundwater in certain buried environments, leaving an empty *mold* in the rock. Sediment, or groundwater deposits may fill the mold and make a *cast* of the organism (figure 20.37).

Figure 20.38 is a photograph of part of the Petrified Forest National Park in Arizona. Petrified trees are not trees that have been "turned to stone." They are stony *replicas* that were formed as the wood was altered by circulating groundwater carrying elements in solution. There are two processes involved in the making of petrified fossils, and they are not restricted to wood. The

FIGURE 20.38

These logs of petrified wood are in the Petrified Forest National Park, Arizona.

processes involve (1) *mineralization,* which is the filling of pore spaces with deposits of calcium carbonate, silica, or pyrite, and/or (2) *replacement,* which is the dissolving of the original material and depositing of new material an ion at a time. Petrified wood is formed by both processes over a long period of time. As it decayed, the original wood was replaced by mineral matter an ion at a time. Over time the "mix" of minerals being deposited changed and the various resulting colors appear to preserve the texture of the wood.

Finally, a fossil must be protected, formed, and then *found and studied* to reveal its part in the history of life. This means the rocks in which the fossil formed must now somehow make it back to the surface of the earth. This usually involves movement and uplift of the rock, and weathering and erosion of the surrounding rock to release or reveal the fossil. Most fossils are found in recently eroded sedimentary rocks—sometimes atop mountains that were under the ocean a long, long time ago. The complete record of what has happened in the past is not found in the fossil alone, as there is a wealth of information in the rocks. The following section describes how the rocks can be read for understanding.

Activities

Make a collection of fossils from road cuts and old quarries in your area. Make an exhibit showing the fossils with sketches of what the animals or plants were like. What do these particular animals and plants mean about the history of your area?

Reading Rocks

Reading history from the rocks of the earth's crust requires both a feel for the immensity of geologic time and an understanding of geologic processes. By "geologic time" we mean the age of the earth; the very long span of the earth's history. This span of time is difficult for most of us to comprehend. Human history is measured in units of hundreds and thousands of years, and even the events that can take place in a thousand years are hard to imagine. Geologic time, on the other hand, is measured in units of millions and billions of years.

The understanding of geologic processes has been made possible through the development of various means of measuring ages and time spans in geologic systems. An understanding of geologic time leads to an understanding of geologic processes, which then leads to an understanding of the environmental conditions that must have existed in the past. Thus, the mineral composition, texture, and sedimentary structure of rocks are clues about past events, events that make up the history of earth.

Arranging Events in Order

The clues provided by thinking about geologic processes that must have occurred in the past are interpreted within a logical frame of reference that can be described by several basic principles. The following is a summary of these basic guiding principles that are used to read a story of geologic events from the rocks.

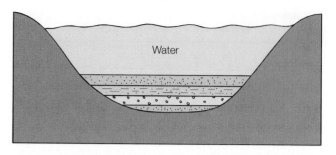

A

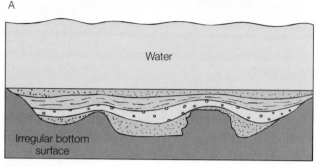

B

FIGURE 20.39

The principle of original horizontality. (*A*) Sediments tend to be deposited in horizontal layers. (*B*) Even where the sediments are draped over an irregular surface, they tend toward the horizontal.

The cornerstone of the logic used to guide thinking about geologic time is the *principle of uniformity.* As described earlier, this principle is sometimes stated as "the present is the key to the past." This means that the geologic features that you see today have been formed in the past by the same processes of crustal movement, erosion, and deposition that are observed today. By studying the processes now shaping the earth, you can understand how it has evolved through time. This principle establishes the understanding that the surface of the earth has been continuously and gradually modified over the immense span of geologic time.

The *principle of original horizontality* is a basic principle that is applied to sedimentary rocks. It is based on the observation that, on a large scale, sediments are commonly deposited in flat-lying layers. Old rocks are continually being changed to new ones in the continuous processes of crustal movement, erosion, and deposition. As sediments are deposited in a basin of deposition, such as a lake or ocean, they accumulate in essentially flat-lying, approximately horizontal layers (figure 20.39). Thus, any layer of sedimentary rocks that is not horizontal has been subjected to forces that have deformed the earth's surface.

The *principle of superposition* is another logical and obvious principle that is applied to sedimentary rocks. Layers of sediments are usually deposited in succession in horizontal layers, which later are compacted and cemented into layers of sedimentary rock. An undisturbed sequence of horizontal layers is thus arranged in chronological order with the oldest layers at the bottom. Each consecutive layer will be younger than the one below it (figure 20.40). This is true, of course, only if the layers have not been turned over by deforming forces (figure 20.41).

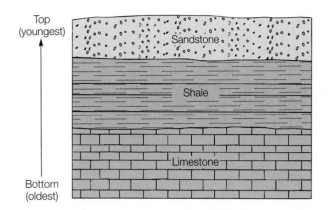

FIGURE 20.40

The principle of superposition. In an undisturbed sedimentary sequence, the rocks on the bottom were deposited first, and the depositional ages decrease as you progress to the top of the pile.

The *principle of crosscutting relationships* is concerned with igneous and metamorphic rock, in addition to sedimentary rock layers. Any geologic feature that cuts across or is intruded into a rock mass must be younger than the rock mass. Thus, if a fault cuts across a layer of sedimentary rocks, the fault is the youngest feature. Faults, folds, and igneous intrusions are always younger than the rocks they originally occur in. Often, there is a further clue to the correct sequence: The hotter igneous rock may have "baked," or metamorphosed, the surrounding rock immediately adjacent to it, so again the igneous rock must have come second (figure 20.42).

Shifting sites of erosion and deposition: The principle of uniformity states that the earth processes going on today have always been occurring. This does not mean, however, that they always occur in the same place. As erosion wears away the rock layers at a site, the sediments produced are deposited someplace else. Later, the sites of erosion and deposition may shift, and the sediments are

FIGURE 20.41

The Grand Canyon, Arizona, provides a majestic cross section of horizontal sedimentary rocks. According to the principle of superposition, traveling deeper and deeper into the Grand Canyon means that you are moving into older and older rocks.

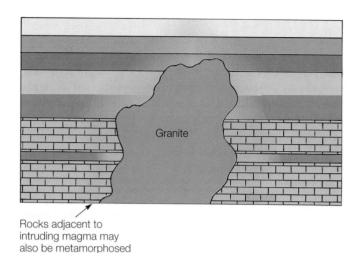

Rocks adjacent to intruding magma may also be metamorphosed by its heat.

FIGURE 20.42

A granite intrusion cutting across older rocks.

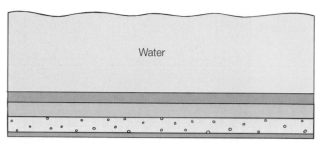

Deposition

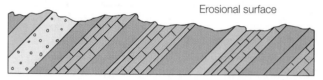

Erosional surface

Rocks tilted, eroded

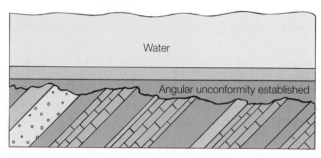

Water

Angular unconformity established

Subsequent deposition

FIGURE 20.43

Angular unconformity. Development involves some deformation and erosion before sedimentation is resumed.

deposited on top of the eroded area. When the new sediments later are formed into new sedimentary rocks, there will be a time lapse between the top of the eroded layer and the new layers. A time break in the rock record is called an **unconformity.** The unconformity is usually shown by a surface within a sedimentary sequence on which there was a lack of sediment deposition, or where active erosion may even have occurred for some period of time. When the rocks are later examined, that time span will not be represented in the record, and if the unconformity is erosional, some of the record once present will have been lost. An unconformity may occur within a sedimentary sequence of the same kind or between different kinds of rocks. The most obvious kind of unconformity to spot is an *angular unconformity.* An angular unconformity, as illustrated in figure 20.43, is one in which the bedding planes above and below the unconformity are not parallel. An angular unconformity usually implies some kind of tilting or folding, followed by a significant period of erosion, which in turn was followed by a period of deposition (figure 20.44).

Correlation

The principle of superposition, the principle of crosscutting relationships, and the presence of an unconformity all have meaning about the order of geologic events that have occurred in the past. This order can be used to unravel a complex sequence of events such as the one shown in figure 20.45. The presence of fossils can help too. The *principle of faunal succession* recognizes that life forms have changed through time. Old life forms disappear from the fossil record and new ones appear, but the same form is never exactly duplicated independently at two different times in history. This principle implies that the same type of fossil organisms that lived only a brief geologic time should occur only in rocks that are the same age. According to the principle of faunal succession, then, once the basic sequence of fossil forms in the rock record is determined, rocks can be placed in their correct relative chronological position on the basis of the fossils contained in them. The

FIGURE 20.44

A time break in the rock record in the Grand Canyon, Arizona. The horizontal sedimentary rock layers overlie almost vertically foliated metamorphic rocks. Metamorphic rocks form deep in the earth, so they must have been uplifted, and the overlying material eroded away before being buried again.

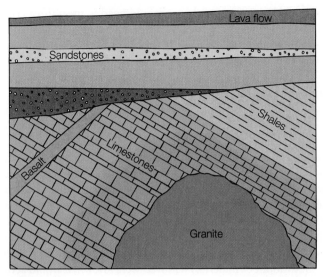

FIGURE 20.45

Deciphering a complex rock sequence. The limestones must be oldest (law of superposition), followed by the shales. The granite and basalt must both be younger than the limestone they crosscut (note the metamorphosed zone around the granite). It is not possible to tell whether the igneous rocks predate or postdate the shales or to determine whether the sedimentary rocks were tilted before or after the igneous rocks were emplaced. After the limestones and shales were tilted, they were eroded, and then the sandstones were deposited on top. Finally, the lava flow covered the entire sequence.

principle also means that if the same type of fossil organism is preserved in two different rocks, the rocks should be the same age. This is logical even if the two rocks have very different compositions and are from places far, far apart (figure 20.46).

Distinctive fossils of plant or animal species that were distributed widely over the earth but lived only a brief time are called **index fossils.** Index fossils, together with the other principles used in reading rocks, make it possible to compare the ages of rocks exposed in two different locations. This is called age *correlation* between rock units. Correlations of exposed rock units separated by a few kilometers are easier to do, but correlations have been done with exposed rock units that are separated by an ocean. Correlation allows the ordering of geologic events according to age. Since this process is only able to determine the age of a rock unit or geologic event relative to some other unit or event, it is called *relative dating* (figure 20.47). Dates with numerical ages are determined by means different from correlation.

The usefulness of correlation and relative dating through the concept of faunal succession is limited because the principles can be applied only to rocks in which fossils are well preserved, which are almost exclusively sedimentary rocks. Correlation also can be based on the occurrence of unusual rock types, distinctive rock sequences, or other geologic similarities. All this is useful in clarifying relative age relationships among rock units. It is not useful in answering questions about the age of rocks or the time required for certain events, such as the eruption of a volcano, to occur. Questions requiring numerical answers went unanswered until the twentieth century.

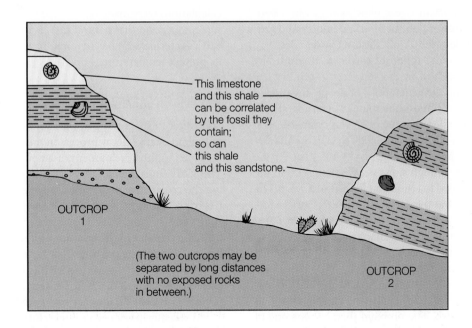

FIGURE 20.46

Similarity of fossils suggests similarity of ages, even in different rocks widely separated in space.

FIGURE 20.47

This dinosaur footprint is in shale near Tuba City, Arizona. It tells you something about the relative age of the shale, since it must have been soft mud when the dinosaur stepped here.

Early Attempts at Earth Dating

Most human activities are organized by time intervals of minutes, hours, days, months, and years. These time intervals are based on the movements of Earth and the Moon, which were discussed in chapter 18. Short time intervals are measured in minutes and hours, which are typically tracked by watches and clocks. Longer time intervals are measured in days, months, and years, which are typically tracked by calendars. How do you measure and track time intervals for something as old as the earth? First, you would need to know the age of the earth; then you would need some consistent, measurable events to divide the overall age into intervals. Questions about the age of the earth have puzzled people for thousands of years, dating back at least to the time of the ancient Greek philosophers. Many people have attempted to answer this question and understand geologic time, but with little success until the last few decades.

One early estimate of the age of the earth was attempted by Archbishop Ussher of Ireland in the seventeenth century. He painstakingly counted up the generations of people mentioned in biblical history, added some numerological considerations, and arrived at the conclusion that the earth was created at 9:00 A.M. on Tuesday, October 26, in the year 4004 B.C. On the authority of biblical scholars this date was generally accepted for the next century or so, even though some people thought that the geology of the earth seemed to require far longer to develop. The date of 4004 B.C. meant that the earth and all of the surface features on the earth had formed over a period of about 6,000 years. This required a model of great cataclysmic catastrophes to explain how all the earth's features could possibly have formed over a span of 6,000 years.

Near the end of the eighteenth century, James Hutton reasoned out the principle of uniformity and people began to assume a much older earth. The problem then became one of finding some uniform change or process that could serve as a geologic clock to measure the age of the earth. To serve as a geo-logic clock a process or change would need to meet three criteria: (1) the process must have been operating since the earth began, (2) the process must be uniform or at least subject to averaging, and (3) the process must be measurable.

During the nineteenth century many attempts were made to find earth processes that would meet the criteria to serve as a geologic clock. Among others, the processes explored were (1) the rate that salt is being added to the ocean, (2) the rate that sediments are being deposited, and (3) the rate that the earth is cooling. Comparing the load of salts being delivered to the ocean by all the rivers, and assuming the ocean was initially pure water, it was calculated that about 100 million years would be required for the present salinity to be reached. The calculations did not consider the amount of materials being removed from the ocean by organisms and by chemical sedimentation, however, so this technique was considered to be unacceptable. Even if the amount of materials removed were known, it would actually result in the age of the ocean, not the age of the earth.

A number of separate and independent attempts were made to measure the rate of sediment deposition, then compare that rate to the thickness of sedimentary rocks found on the earth. Dividing the total thickness by the rate of deposition resulted in estimates of an earth age that ranged from about 20 to 1,500 million years. The wide differences occurred because there are gaps in many sedimentary rock sequences, periods when sedimentary rocks were being eroded away to be deposited elsewhere as sediments again. There were just too many unknowns for this technique to be considered as acceptable.

The idea of measuring the rate that the earth is cooling for use as a geologic clock assumed that the earth was initially a molten mass that has been cooling ever since. Calculations estimating the temperature that the earth must have been to be molten were compared to the earth's present rate of cooling. This resulted in an estimated age of 20 to 40 million years. These calculations were made back in the nineteenth century before it was understood that natural radioactivity is adding heat to the earth's interior, so it has required much longer to cool down to its present temperature.

Modern Techniques

Soon after the beginning of the twentieth century the discovery of the radioactive decay process in the elements of minerals and rocks led to the development of a new, accurate geologic clock. This clock finds the *radiometric age* of rocks in years by measuring the radioactive decay of unstable elements within the crystals of certain minerals. Since radioactive decay occurs at a constant, known rate, the ratio of the remaining amount of an unstable element to the amount of decay products present can be used to calculate the time that the unstable element has been a part of that crystal (see chapter 15). Potassium, uranium, and thorium are radioactive isotopes that are often included in the minerals of rocks, so they are often used as radioactive clocks.

A recently developed geologic clock is based on the magnetic orientation of magnetic minerals. These minerals become aligned with the earth's magnetic field when the igneous rock crystallizes,

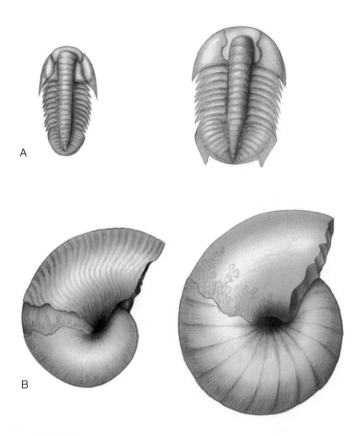

FIGURE 20.48

(A) Fossil trilobites from Cambrian rocks. (B) Fossil ammonites from Permian rocks.

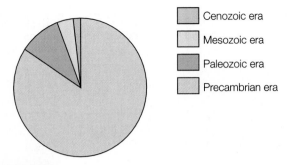

FIGURE 20.49

Geologic history is divided into four main eras. The Precambrian era was first, lasting the first 4 billion years, or about 85 percent of the total 4.6 billion years of geologic time. The Paleozoic lasted about 10 percent of geologic time, the Mesozoic about 4 percent, and the Cenozoic only about 1.5 percent of all geologic time.

making a record of the field at that time. The earth's magnetic field is global and has undergone a number of reversals in the past. A *geomagnetic time scale* has been established from the number and duration of magnetic field reversals occurring during the past 6 million years. Combined with radiometric age dating, the geomagnetic time scale is making possible a worldwide geologic clock that can be used to determine local chronologies.

The Geologic Time Scale

A yearly calendar helps you keep track of events over periods of time by dividing the year into months, weeks, and days. In a similar way, the **geologic time scale** helps you keep track of events that have occurred in the earth's geologic history. The first development of this scale came from the work of William "Strata" Smith, the English surveyor described in the section on fossils earlier in this chapter. Recall that Smith discovered that certain rock layers in England occurred in the same order, top to bottom, wherever they were located. He also found that he could correlate and identify each layer by the kinds of fossils in the rocks of the layers. In 1815, he published a geologic map of England, identifying the rock layers in a sequence from oldest to youngest. Smith's work was followed by extensive geological studies of the rock layers in other countries. Soon it was realized that similar, distinctive index fossils appeared in rocks of the

same age when the principle of superposition was applied. For example, the layers at the bottom contained fossils of trilobites (figure 20.48A), but trilobites were not found in the upper levels. (Trilobites are extinct marine arthropods that may be closely related to living crustaceans, spiders, and horseshoe crabs.) On the other hand, fossil shells of ammonites (figure 20.48B) appeared in the middle levels, but not the lower nor upper levels of the rocks. The topmost layer was found to contain the fossils of animals identified as still living today. The early appearance and later disappearance of fossils in progressively younger rocks is explained by organic evolution and extinction, events that could be used to mark the time boundaries of the earth's geologic history.

The major blocks of time in the earth's geologic history are called *eras,* and each era is identified by the appearance and disappearance of particular fossils in the sedimentary rock record. There are four main eras, and the pie chart in figure 20.49 compares how long each era lasted. The eras are: (1) *Cenozoic,* which refers to the time of recent life. Recent life means that the fossils for this time period are similar to the life found on earth today. (2) *Mesozoic,* which refers to the time of middle life. Middle life means that some of the fossils for this time period are similar to the life found on earth today, but many are different from anything living today. (3) *Paleozoic,* which refers to the time of ancient life. Ancient life means that the fossils for this time period are very different from anything living on the earth today. (4) *Precambrian,* which refers to the time before the time of ancient life. This means that the rocks for this time period contain very few fossils. The eras were divided into blocks of time called *periods,* and the periods were further subdivided into smaller blocks of time called *epochs* (figure 20.50).

The geologic time scale developed in the 1800s was without dates because geologists did not yet have a way to measure the length of the eras, periods, or epochs. This was a *relative time scale,* based on the superposition of rock layers and fossil records of organic evolution. With the development of radiometric dating, it became possible to attach numbers to the time scale.

Era	Period	Epoch	Time of start (millions of years ago)
Cenozoic	Quaternary	Holocene	0.1
		Pleistocene	2
	Tertiary	Pliocene	5
		Miocene	24
		Oligocene	37
		Eocene	58
		Paleocene	66
Mesozoic	Cretaceous		144
	Jurassic		208
	Triassic		245
Paleozoic	Permian		286
	Carboniferous:		
	Pennsylvanian		320
	Mississippian		360
	Devonian		408
	Silurian		438
	Ordovician		505
	Cambrian		534
Precambrian			4,600

FIGURE 20.50

The divisions of the geologic time scale.

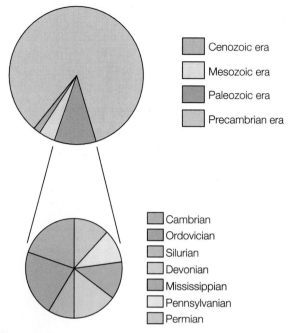

FIGURE 20.51

The periods of the Paleozoic era, which refers to the time of ancient life. Ancient life means that the fossils for this time period are very different from anything living on the earth today. Each period represents a dominant life-form of a different species of animal. This pie chart compares the relative time that each period lasted.

When this was accomplished it became apparent that geologic history spanned very long periods of time. In addition, the part of the time scale with evidence of life makes up about 15 percent of all of Earth's history. This does not mean that life appeared suddenly in the Cambrian period. The fossil record is incomplete since it is the hard parts of animals or plants that form fossils, usually after rapid burial. Thus, the soft-bodied life forms that existed during the Precambrian era would make Precambrian fossils exceedingly rare. The Precambrian fossils that have been found are chiefly those of deposits from algae, a few fungi, and the burrow holes of worms. Another problem in finding ancient fossils of soft-bodied life forms is that heat and pressure have altered many of the ancient rocks over time, destroying any fossil evidence that may have been present. Figure 20.51 shows the periods of the Paleozoic era in a pie chart, each slice representing the length of time that each period lasted.

In general, the earliest abundant fossils are found in rocks from the Cambrian period at the beginning of the Paleozoic era. They represent an abundance of oceanic life with over a thousand different species of animals. There is no fossil evidence of life of any kind living on the land during the Cambrian. The dominant life forms of the Cambrian ocean were trilobites and brachiopods (figure 20.48). The trilobites, now extinct, made up more than half the life population during the Cambrian.

More and more species of brachiopods appear in the rocks of the Ordovician and Silurian periods, reaching their climax in the Devonian. Primitive forms of ammonites, shelled organisms

that later become more important, first appear in the Devonian. Fossils of fish first appear in the Ordovician, then become abundant and diversified by the Devonian. Sharks were common, as was a primitive form of an air-breathing fish. Primitive evergreen and fernlike trees appeared on the land at this time, according to the Devonian fossils.

The Carboniferous age, which is made up of the Mississippian and Pennsylvanian periods in the United States, was a time of fast-growing, soft-wooded forests. These were not the modern trees of today, but were spore-bearing giants that later would form great coal deposits. Fossils of the first reptiles and the first winged insects are found in rocks from this age. The Paleozoic era closed with the extinction of many types of plant and animal life of that time.

The dinosaurs first appeared in the Triassic, outnumbering all the other reptiles until the close of the Mesozoic. Fossils of the first birds, the first mammals, the first flowering plants, and the first deciduous trees appeared in the rocks of this era. Like the close of the Paleozoic, the Mesozoic era ended with a great dying of land and marine life that resulted in the extinction of many species, including the dinosaurs.

Throughout the Mesozoic era the reptiles dominated life on the earth. As the Cenozoic era opened, the dinosaurs were extinct and the mammals became the dominant life form. The Cenozoic is thus called The Age of the Mammals.

You may wonder about the major extinctions of life on the earth during the close of the Paleozoic and Mesozoic eras. There are several theories about what could possibly cause such world-wide extinctions. The terminal Paleozoic extinctions are thought to be the result of plate tectonics: The joining of all the continents into one supercontinent, Pangaea, destroyed much of the suitable shallow water, maritime, continental shelf area. The cause of the terminal Cretaceous extinctions is debated by geologists. Based on evidence of a thin clay layer marking the boundary between the Cretaceous and Tertiary periods, one theory proposes that a huge (16 km diameter and 10^{15} kg mass) meteorite struck the earth. The impact would have thrown a tremendous amount of dust into the atmosphere, obscuring the sun and significantly changing the climate and thus the conditions of life on the earth. The resulting colder climate may have led to the extinction of many plant and animal species, including the dinosaurs. This theory is based on the clay layer, which theoretically formed as the dust settled, and its location in the rock record at the time of the extinctions. The layer is enriched with a rare metal, iridium, which is not found on the earth in abundance, but occurs in certain meteorites in greater abundance. Perhaps the great extinctions of the past have occurred when heavy bombardments occur every 100 million years or so.

In 1994 the people of the earth were treated to a once-in-a-lifetime spectacle of how cosmic bombardment can disturb a planet. The comet Shoemaker-Levy broke apart into twenty-two fragments, which lined up to impact Jupiter one after the other. The fireballs of many of the impacts were visible on Earth through telescopes, and the explosion from the seventh fragment amounted to the equivalent of an explosion of 6 billion tons of TNT. Plumes arising from the successive strikes cumulated across Jupiter's face and gradually flattened into spots, some as large as the Great Red Spot. Jupiter does not have the same composition as Earth, where plumes of dust, dirt, and rock debris would form rather than the ammonia and frozen hydrogen plumes of Jupiter. There are few who would argue that such a bombardment of Earth could raise enough dust and smoke to block sunlight. In turn, this could result in drastic changes of climate and other factors that influence the whole ecosystem.

SUMMARY

The *principle of uniformity* is the frame of reference that the same geologic processes you see changing rocks today are the same processes that changed them in the past.

Diastrophism is the process of deformation that changes the earth's surface, and the movement of magma is called *vulcanism. Diastrophism, vulcanism,* and *earthquakes* are closely related, and their occurrence can be explained most of the time by events involving *plate tectonics.*

Stress is a force that tends to compress, pull apart, or deform a rock, and the adjustment to stress is called *strain.* Rocks respond to stress by (1) withstanding the stress without change, (2) undergoing *elastic strain,* (3) undergoing *plastic strain,* or (4) by *breaking* in *fracture strain.* Exactly how a particular rock responds to stress depends on (1) the nature of the rock, (2) the temperature, and (3) how quickly the stress is applied over time.

Deeply buried rocks are at a higher temperature and tend to undergo *plastic deformation,* resulting in a wrinkling of the layers into *folds.* The most common are an arch-shaped fold called an *anticline* and a trough-shaped fold called a *syncline.* Anticlines and synclines are most easily observed in sedimentary rocks because they have bedding planes, or layers.

Rocks near the surface tend to break from a sudden stress. A break with movement on one side of the break relative to the other side is called a *fault.* The vibrations that move out as waves from the movement of rocks is called an *earthquake.* The actual place where an earthquake originates is called its *focus.* The place on the surface directly above a focus is called an *epicenter.* There are three kinds of waves that travel from the focus, *S-, P-,* and *surface waves.* The magnitude of earthquake waves is measured on the *Richter scale.*

Folding and faulting produce prominent features on the surface called mountains. *Mountains* can be classified as having an origin of *folding, faulting,* or *volcanic.* In general, mountains that occur in long narrow belts called *ranges* have an origin that can be explained by *plate tectonics.*

Weathering is the breaking up, crumbling, and destruction of any kind of solid rock. The process of physically picking up weathered rock materials is called *erosion.* After the eroded materials are picked up, they are removed by *transportation* agents. The combined action of weathering, erosion, and transportation wears away and lowers the surface of the earth.

The physical breaking up of rocks is called *mechanical weathering.* Mechanical weathering occurs by *wedging effects* and the *effects of reduced pressure. Frost wedging* is a wedging effect that occurs from repeated cycles of water freezing and thawing. The process of spalling off of curved layers of rock from reduced pressure is called *exfoliation.*

The breakdown of minerals by chemical reactions is called *chemical weathering.* Examples include *oxidation,* a reaction between oxygen and the minerals making up rocks; *carbonation,* a reaction between carbonic acid (carbon dioxide dissolved in water) and minerals making up rocks; and *hydration,* the dissolving or combining of a mineral with water. When the end products of complete weathering of rocks are removed directly by gravity, the erosion is called *mass movement.* Erosion and transportation also occur through the agents of *running water, glaciers,* or *wind.* Each create their own characteristic features of erosion and deposition.

A *fossil* is any evidence of former life. This evidence could be in the form of *actual remains, altered remains, preservation of the shape of an organism,* or *any sign of activity (trace fossils).* Actual remains of former organisms are rare, occurring usually from protection of remains by freezing, entombing in tree resin, or embalming in tar. Remains of organisms are sometimes altered by groundwater in the process of *mineralization,* deposition of mineral matter in the pore spaces of an object. *Replacement* of original materials occurs by dissolution and deposition of mineral matter. *Petrified wood* is an example of mineralization and replacement of wood. Another alteration occurs when volatile and gaseous constituents are distilled away, leaving a *carbon film.* Removal of an organism may leave a *mold,* a void where an organism was buried. A *cast* is formed if the void becomes filled with mineral matter.

Clues provided by geologic processes are interpreted within a logical framework of references to read the story of geologic events from the rocks. These clues are interpreted within a frame of reference based on (1) the *principle of uniformity,* (2) the *principle of original horizontality,* (3) the *principle of superposition,* (4) the *principle of crosscutting relationships,* (5) that *sites of past erosion and deposition* have shifted over time (shifting sites produce an unconformity, or break in the rock record when erosion removes part of the rocks), and (6) the *principle of faunal succession.*

Geologic time is measured through the radioactive decay process, determining the *radiometric age* of rocks in years. A *geomagnetic time scale* has been established from the number and duration of reversals in the magnetic field of the earth's past.

Correlation and the determination of the numerical ages of rocks and events have led to the development of a *geologic time scale.* The major blocks of time on this calendar are called *eras.* The eras are the (1) *Cenozoic,* the time of recent life, (2) *Mesozoic,* the time of middle life, (3) *Paleozoic,* the time of ancient life, and (4) *Precambrian,* the time before the time of ancient life. The eras are divided into smaller blocks of time called *periods,* and the periods are further subdivided into *epochs.* The fossil record is seen to change during each era, ending with *great extinctions* of plant and animal life.

KEY TERMS

anticline (p. **491**)	mass movement (p. **506**)
delta (p. **508**)	mountains (p. **497**)
dune (p. **512**)	normal fault (p. **492**)
earthquake (p. **494**)	principle of uniformity (p. **489**)
elastic rebound (p. **494**)	reverse fault (p. **492**)
erosion (p. **503**)	seismic waves (p. **494**)
fault (p. **492**)	strain (p. **490**)
floodplain (p. **507**)	stress (p. **490**)
folds (p. **490**)	syncline (p. **491**)
fossil (p. **514**)	thrust fault (p. **492**)
frost wedging (p. **503**)	tsunami (p. **496**)
geologic time scale (p. **521**)	unconformity (p. **518**)
glacier (p. **509**)	volcano (p. **498**)
index fossils (p. **519**)	weathering (p. **502**)

APPLYING THE CONCEPTS

1. The basic difference in the frame of reference called the principle of uniformity and the catastrophic frame of reference used by previous thinkers is
   a. the energy for catastrophic changes is much less.
   b. the principle of uniformity requires more time.
   c. catastrophic changes have a greater probability of occurring.
   d. none of the above.

2. The difference between elastic deformation and plastic deformation of rocks is that plastic deformation (is)
   a. permanently alters the shape of a rock layer.
   b. always occurs just before a rock layer breaks.
   c. returns to its original shape after the pressure is removed.
   d. all of the above.

3. Whether a rock layer subjected to stress undergoes elastic deformation, plastic deformation, or rupture depends on
   a. the temperature of the rock.
   b. the confining pressure on the rock.
   c. how quickly or slowly the stress is applied over time.
   d. all of the above.

4. When subjected to stress, rocks buried at great depths are under great pressure at high temperatures, so they tend to undergo
   a. no change because of the pressure.
   b. elastic deformation because of the high temperature.
   c. plastic deformation.
   d. breaking or rupture.

5. A fault where the footwall has moved upward relative to the hanging wall is called (a)
   a. normal fault.
   b. reverse fault.
   c. thrust fault.
   d. none of the above.

6. Reverse faulting probably resulted from which stress?
   a. compressional stress
   b. pulling-apart stress
   c. a twisting stress
   d. stress associated with diverging tectonic plates

7. Earthquakes that occur at the boundary between two tectonic plates moving against each other occur along
   a. the entire length of the boundary at once.
   b. short segments of the boundary at different times.
   c. the entire length of the boundary at different times.
   d. none of the above.

8. Each higher number of the Richter scale
   a. increases with the magnitude of an earthquake.
   b. means 10 times more ground movement.
   c. indicates about 30 times more energy released.
   d. means all of the above.

9. Other than igneous activity, all mountain ranges have an origin resulting from
   a. folding.
   b. faulting.
   c. stresses.
   d. sedimentation.

10. The preferred name for the very large ocean waves that are generated by an earthquake, landslide, or volcanic explosion is
    a. tidal wave.
    b. tsunami.
    c. tidal bore.
    d. Richter wave.

11. Freezing water exerts pressure on the wall of a crack in a rock mass, making the crack larger. This is an example of
    a. mechanical weathering.
    b. chemical weathering.
    c. exfoliation.
    d. hydration.

12. Of the following rock weathering events, the last one to occur would probably be
    a. exfoliation.
    b. frost wedging.
    c. carbonation.
    d. disintegration.

13. Which of the following would have the greatest overall effect in lowering the elevation of a continent such as North America?
    a. continental glaciers
    b. alpine glaciers
    c. wind
    d. running water

14. Broad meanders on a very wide, gently sloping floodplain with oxbow lakes are characteristics you would expect to find in a river valley during what stage?
    a. newborn
    b. youth
    c. maturity
    d. old age

15. A glacier forms when
    a. the temperature does not rise above freezing.
    b. snow accumulates to form ice, which begins to flow.
    c. a summer climate does not occur.
    d. a solid mass of snow moves downhill under the influence of gravity.

16. A moraine is (a)
    a. wind deposit.
    b. glacier deposit.
    c. river deposit.
    d. any of the above.

17. A fossil is
    a. remains of plants or animals such as shells or wood.
    b. remains that have been replaced by mineral matter.
    c. any sign of former life older than 10,000 years.
    d. any of the above.

18. Some of the oldest fossils are about how many years old?
    a. 4.55 billion
    b. 3.5 billion
    c. 250 million
    d. 10 thousand

19. According to the evidence, a human footprint found preserved in baked clay is about 5,000 years old. According to the definitions used by paleontologists, this footprint is a (an)
    a. fossil since it is an indication of former life.
    b. mold since it preserves the shape.
    c. actual fossil since it is preserved.
    d. none of the above.

20. A fossil of a jellyfish, if found, would most likely be
    a. a preserved fossil.
    b. one formed by mineralization.
    c. a carbon film.
    d. any of the above.

21. Which of the basic guiding principles used to read a story of geologic events tells you that layers of undisturbed sedimentary rocks have progressively older layers as you move toward the bottom?
    a. superposition
    b. horizontality
    c. crosscutting relationships
    d. None of the above are correct.

22. In any sequence of sedimentary rock layers that has not been subjected to stresses, you would expect to find
    a. essentially horizontal stratified layers.
    b. the oldest layers at the bottom and the youngest at the top.
    c. younger faults, folds, and intrusions in the rock layers.
    d. all of the above.

23. An unconformity is (a)
    a. rock bed that is not horizontal.
    b. rock bed that has been folded.
    c. rock sequence with rocks missing from the sequence.
    d. all of the above.

24. Correlation and relative dating of rock units is made possible by application of (the)
    a. principle of crosscutting relationships.
    b. principle of faunal succession.
    c. principle of superposition.
    d. all of the above.

25. The geologic time scale identified major blocks of time in the earth's past by
    a. major worldwide extinctions of life on the earth.
    b. the beginning of radioactive decay of certain unstable elements.
    c. measuring reversals in the earth's magnetic field.
    d. none of the above.

26. You would expect to find the least number of fossils in rocks from which era?
    a. Cenozoic
    b. Mesozoic
    c. Paleozoic
    d. Precambrian

27. You would expect to find relatively more fossils, but fossils of life very different from anything living today, in rocks from which era?
    a. Cenozoic
    b. Mesozoic
    c. Paleozoic
    d. Precambrian

28. The numerical dates associated with events on the geologic time scale were determined by
    a. relative dating of the rate of sediment deposition.
    b. radiometric dating using radioactive decay.
    c. the temperature of the earth.
    d. the rate that salt is being added to the ocean.

Answers

1. b **2.** a **3.** d **4.** c **5.** a **6.** a **7.** b **8.** d **9.** c **10.** b **11.** a **12.** c **13.** d **14.** d **15.** b **16.** b **17.** d **18.** b **19.** a **20.** c **21.** a **22.** d **23.** c **24.** d **25.** a **26.** d **27.** c **28.** b

QUESTIONS FOR THOUGHT

1. What is the principle of uniformity? What are the underlying assumptions of this principle?
2. Describe the responses of rock layers to increasing compressional stress when it increases slowly on deeply buried, warm layers, it increases slowly on cold rock layers, and it is applied quickly to rock layers of any temperature.
3. Describe the difference between a syncline and an anticline, using sketches as necessary.
4. What does the presence of folded sedimentary rock layers mean about the geologic history of an area?

5. Describe the conditions that would lead to faulting as opposed to folding of rock layers.

6. How would plate tectonics explain the occurrence of normal faulting? Reverse faulting?

7. What is an earthquake? What produces an earthquake?

8. Where would the theory of plate tectonics predict that earthquakes would occur?

9. Describe how a seismic recording station identifies the location of an earthquake.

10. Briefly explain how and where folded mountains form. Explain how fault block mountains form.

11. The magnitude of an earthquake is measured on the Richter scale. What does each higher number mean about an earthquake?

12. Identify some areas of probable active volcanic activity today in the United States. Explain your reasoning for selecting these areas.

13. Discuss the basic source of energy that produces the earthquakes in southern California.

14. Describe any possible relationships between volcanic activity and changes in the weather.

15. What is the source of magma that forms volcanoes? Explain how the magma is generated.

16. Describe how the nature of the lava produced results in the three major classification types of volcanoes.

17. What are mountains? Why do they tend to form in long, thin belts?

18. Compare and contrast mechanical and chemical weathering.

19. Granite is the most common rock found on continents. What are the end products after granite has been completely weathered? What happens to these weathering products?

20. What other erosion processes are important as a stream of running water carves a valley in the mountains? Explain.

21. Describe three ways in which a river erodes its channel.

22. What is a floodplain?

23. Describe the characteristic features associated with stream erosion as the stream valley passes through the stages of youth, maturity, and old age.

24. What is a glacier? How does a glacier erode the land?

25. What is rock flour and how is it produced?

26. Could a stream erode the land lower than sea level? Explain.

27. Explain why glacial erosion produces a U-shaped valley but stream erosion produces a V-shaped valley.

28. Name and describe as many ways as you can think of that mechanical weathering occurs in nature. Do not restrict your thinking to those discussed in this chapter.

29. What essential condition must be met before mass wasting can occur?

30. Compare the features caused by stream erosion, wind erosion, and glacial erosion.

31. Compare the materials deposited by streams, wind, and glaciers.

32. Why do certain stone buildings tend to weather more rapidly in cities than they do in rural areas?

33. What is the geologic time scale? What is the meaning of the eras?

34. Why does the rock record go back only 3.8 million years? If this missing record were available, what do you think it would show? Explain.

35. Do igneous, metamorphic, or sedimentary rocks provide the most information about Earth's history? Explain.

36. What major event marked the end of the Paleozoic and Mesozoic eras according to the fossil record? Describe one theory that proposes to account for this.

37. Briefly describe the principles and assumptions that form the basis of interpreting Earth's history from the rocks.

38. Describe the sequence of geologic events represented by an angular unconformity. Begin with the deposition of the sediments that formed the oldest rocks represented.

39. Describe how the principles of superposition, horizontality, and faunal succession are used in the relative dating of sedimentary rock layers.

40. How are the numbers of the ages of eras and other divisions of the geologic time scale determined?

41. Describe the three basic categories of fossilization methods.

42. Describe some of the things that fossils can tell you about Earth's history.

This cloud forms a thin covering over the mountaintop. Likewise, the earth's atmosphere forms a thin shell around the earth, with 99 percent of the mass within 32 km, or about 20 mi of the surface.

CHAPTER | Twenty-One

Earth's Weather

Almost all planets in our solar system have an *atmosphere,* a shell of gases that surrounds the solid mass of the planet. In general, the planets located in the outer part of the solar system have an atmosphere consisting mostly of hydrogen and helium, or a mixture of these two gases with methane. Earth is located between two planets, Venus and Mars, that have atmospheres of mostly carbon dioxide. Mercury, which is closest to the Sun, is the only planet without an atmosphere. Earth is also unusual since it is the only planet with an atmosphere of mostly nitrogen and oxygen. An explanation of why these differences exist between the planets was discussed in chapter 17.

The earth's atmosphere has a unique composition because of the cyclic flow of materials that takes place between different parts of the earth. These cycles, which do not exist on the other planets, involve the movement of materials between the surface and the interior and the building-up and tearing-down cycles on the surface. Materials also cycle in and out of the earth's atmosphere. Carbon dioxide, for example, is the major component of the atmospheres around Venus and Mars, and the early Earth had a similar atmosphere. Today, carbon dioxide is a very minor part of Earth's atmosphere. It has been maintained as a minor component in a mostly balanced state for about the past 570 million years, cycling into and out of the atmosphere.

Water is also involved in a global cyclic flow between the atmosphere and the surface. Water on the surface is mostly in the ocean, with lesser amounts in lakes, streams, and under ground. Not much water is found in the atmosphere at any one time on a worldwide basis, but billions of tons are constantly evaporating into the atmosphere each year and returning as precipitation in an ongoing cycle.

The cycling of carbon dioxide and water to and from the atmosphere takes place in a dynamic system that is energized by the sun. Radiant energy from the sun heats some parts of the earth more than others. Winds redistribute this energy with temperature changes, rain, snow, and other changes that are generally referred to as the *weather.*

Understanding and predicting the weather is the subject of **meteorology.** Meteorology is the science of the atmosphere and weather phenomena, from understanding everyday rain and snow to predicting not-so-common storms and tornadoes (figure 21.1). Understanding weather phenomena depends on a knowledge of the atmosphere and the role of radiant energy on a rotating Earth that is revolving around the Sun. This chapter is concerned with understanding the atmosphere of the earth, its cycles, and the influence of radiant energy on the atmosphere.

THE ATMOSPHERE

The atmosphere is a relatively thin shell of gases that surrounds the solid earth. If you could see the molecules making up the atmosphere, you would see countless numbers of rapidly moving particles, all undergoing a terrific, chaotic jostling from the billions and billions of collisions occurring every second. Since this jostling mass of tiny particles is pulled toward the earth by gravity, more are found near the surface than higher up. Thus, the atmosphere thins rapidly with increasing distance above the surface, gradually merging with the very diffuse medium of outer space.

To understand how rapidly the atmosphere thins with altitude, imagine a very tall stack of open boxes. At any given instant each consecutively higher box would contain fewer of the jostling molecules than the box below it. Molecules in the lowest box on the surface, at sea level, might be able to move a distance of only 1×10^{-8} m (about 3×10^{-6} in) before colliding with another molecule. A box moved to an altitude of 80 km (about 50 mi) above sea level would have molecules that could move perhaps 10^{-2} m (about 1/2 in) before colliding with another molecule. At 160 km (about 100 mi) the distance

FIGURE 21.1

The probability of a storm can be predicted, but nothing can be done to stop or slow a storm. Understanding the atmosphere may help in predicting weather changes, but it is doubtful that weather will ever be controlled on a large scale.

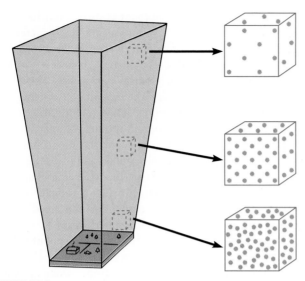

FIGURE 21.2

At greater altitudes, the same volume contains fewer molecules of the gases that make up the air. This means that the density of air decreases with increasing altitude.

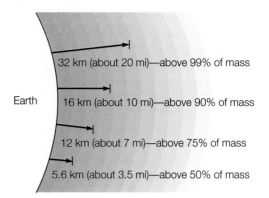

32 km (about 20 mi)—above 99% of mass

16 km (about 10 mi)—above 90% of mass

Earth

12 km (about 7 mi)—above 75% of mass

5.6 km (about 3.5 mi)—above 50% of mass

FIGURE 21.3

The earth's atmosphere thins rapidly with increasing altitude and is much closer to the earth than most people realize.

traveled would be about 2 m (about 7 ft). As you can see, the distance between molecules increases geometrically with increasing altitude. Since air density is defined by the number of molecules in a unit volume, the density of the atmosphere decreases geometrically with increasing altitude (figure 21.2).

It is often difficult to imagine a distance above the surface because there is nothing visible in the atmosphere for comparison. Imagine a stack of boxes from the surface upward, each with progressively fewer molecules per unit volume. Imagine that this stack of boxes is so tall that it reaches from the surface past the top of the atmosphere. Now imagine that this tremendously tall stack of boxes is tipped over and carefully laid out horizontally on the surface of the earth. How far would you have to move along these boxes to reach the box that was in outer space, outside of the atmosphere? From the bottom box, you would cover a distance of only 5.6 km (about 3.5 mi) to reach the box that was above 50 percent of the mass of the earth's atmosphere. At 12 km (about 7 mi), you would reach the box that was above 75 percent of the earth's atmosphere. At 16 km (about 10 mi), you will reach the box that was above about 90 percent of the atmosphere. And, after only 32 km (about 20 mi), you will reach the box that was above 99 percent of the earth's atmosphere. The significance of these distances might be better appreciated if you can imagine the distances to some familiar locations; for example, from your campus to a store 16 km (about 10 mi) away would place you above 90 percent of the atmosphere if you were to travel this same distance straight up.

Since the average radius of the solid earth is about 6,373 km (3,960 mi), you can see that the atmosphere is a very thin shell with 99 percent of the mass within 32 km (about 20 mi) by comparison. The atmosphere is much closer to the earth than most people realize (figure 21.3).

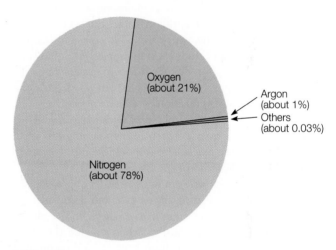

FIGURE 21.4

The earth's atmosphere has a unique composition of gases when compared to that of the other planets in the solar system.

Composition of the Atmosphere

A sample of pure, dry air is colorless, odorless, and composed mostly of the molecules of just three gases, nitrogen (N_2), oxygen (O_2), and argon (Ar). Nitrogen is the most abundant (about 78 percent of the total volume), followed by oxygen (about 21 percent), then argon (about 1 percent). The molecules of these three gases are well mixed, and this composition is nearly constant everywhere near the earth's surface (figure 21.4). Nitrogen does not readily enter into chemical reactions with rocks, so it has accumulated in the atmosphere. Some nitrogen is involved in a nitrogen cycle, however, as it is removed from the atmosphere by certain bacteria in the soil and by lightning. Both form nitrogen compounds that are essential to the growth of plants. These nitrogen compounds are absorbed by plants and consequently utilized throughout the food chain. Eventually the nitrogen is returned to the atmosphere through the decay of plant and animal matter. Overall, these processes of nitrogen removal and release must be in balance since the amount of nitrogen in the atmosphere is essentially constant over time.

Oxygen gas also cycles into and out of the atmosphere in balanced processes of removal and release. Oxygen is removed (1) by living organisms as food units are oxidized to carbon dioxide and water and (2) by chemical weathering of rocks as metals and other elements combine with oxygen to form oxides. Oxygen is released by green plants as a result of photosynthesis and the amount released balances the amount removed by organisms and weathering. So oxygen, as well as nitrogen, is maintained in a state of constant composition through balanced chemical reactions.

The third major component of the atmosphere, argon, is inert and does not enter into any chemical reactions or cycles. It is produced as a product of radioactive decay and once released, remains in the atmosphere as an inactive filler.

In addition to the relatively fixed amounts of nitrogen, oxygen, and argon, the atmosphere contains variable amounts of water vapor. Water vapor is the invisible, molecular form of water in the gaseous state, which should not be confused with fog or clouds. Fog and clouds are tiny droplets of liquid water, not water in the single molecular form of water vapor. The amount of water vapor in the atmosphere can vary from a small fraction of a percent composition by volume in cold, dry air to about 4 percent in warm, humid air. This small, variable percentage of water vapor is essential in maintaining life on the earth. It enters the atmosphere by evaporation, mostly from the ocean, and leaves the atmosphere as rain or snow.

Apart from the variable amounts of water vapor, the relatively fixed amounts of nitrogen, oxygen, and argon make up about 99.97 percent of the volume of a sample of dry air. The remaining 0.03 percent is mostly carbon dioxide (CO_2) and traces of the inert gases neon, helium, krypton, and xenon, along with less than 5 parts per million of free hydrogen, methane, and nitrous oxide. The carbon dioxide content varies locally near cities from the combustion of fossil fuels and from the respiration and decay of organisms and materials produced by organisms. The overall atmospheric concentration of carbon dioxide is regulated (1) by removal from the atmosphere through the photosynthesis process of green plants, (2) by massive exchanges of carbon dioxide between the ocean and the atmosphere, and (3) by chemical reactions between the atmosphere and rocks of the surface, primarily limestone.

The ocean contains some fifty times more carbon dioxide than the atmosphere in the form of carbonate ions and in the form of a dissolved gas. The ocean seems to serve as an equilibrium buffer, absorbing more if the atmospheric concentration increases and releasing more if the atmospheric concentration decreases. Limestone rocks contain an amount of carbon dioxide that is equal to about twenty times the mass of all of Earth's present atmosphere. If all this chemically locked-up carbon dioxide were released, the atmosphere would have a concentration of carbon dioxide similar to the present atmosphere of Venus. This amount of carbon dioxide would result in a tremendous increase in the atmospheric pressure and temperatures on Earth. Overall, however, equilibrium exchange processes with the ocean, rocks,

and living things regulate the amount of carbon dioxide in the atmosphere. Measurements have indicated a yearly increase of about 1 part per million of carbon dioxide in the atmosphere over the last several decades. This increase is believed to be a result of the destruction of tropical rain forests along with increased fossil fuel combustion. The possible consequences of a continuing increase in the carbon dioxide composition of the atmosphere will be discussed soon in another section.

In addition to gases and water vapor, the atmosphere contains particles of dust, smoke, salt crystals, and tiny solid or liquid particles called *aerosols*. The particles of an aerosol are typically 0.005 micrometers in diameter (a micrometer is 10^{-6} meter). These particles become suspended and are dispersed throughout the molecules of the atmospheric gases. Aerosols are produced by combustion, often resulting in air pollution (see the feature in this chapter). Aerosols are also produced by volcanoes and forest fires. Volcanoes, smoke from combustion, and the force of the wind lifting soil and mineral particles into the air all contribute to dust particles larger than aerosols in the atmosphere. These larger particles, which range in size up to 500 micrometers, are not suspended as the aerosols are, and they soon settle out of the atmosphere as dust and soot.

Tiny particles of salt crystals that are suspended in the atmosphere come from the mist created by ocean waves and the surf. This mist forms an atmospheric aerosol of seawater that evaporates, leaving the solid salt crystals suspended in the air. The aerosol of salt crystals and dust becomes well mixed in the lower atmosphere around the globe, playing a large and important role in the formation of clouds.

Atmospheric Pressure

At the earth's surface (sea level), the atmosphere exerts a force of about 10.0 newtons on each square centimeter (14.7 lb/sq in). As you go to higher altitudes above sea level the pressure geometrically decreases with increasing altitude. At an altitude of about 5.6 km (about 3.5 mi) the air pressure is about half of what it is at sea level, about 5.0 newtons/cm² (7.4 lb/in²). At 12 km (about 7 mi) the air pressure is about 2.5 newtons/cm² (3.7 lb/in²). Compare this decreasing air pressure at greater elevations to figure 21.3. Again, you can see that most of the atmosphere is very close to the earth, and it thins rapidly with increasing altitude. Even a short elevator ride takes you high enough that the atmospheric pressure on your eardrum is reduced. You equalize the pressure by opening your mouth, allowing the air under greater pressure inside the eardrum to move through the eustachian tube. This makes a "pop" sound that most people associate with changes in air pressure.

Atmospheric pressure is measured by an instrument called a **barometer.** The mercury barometer was invented in 1643 by an Italian named Torricelli. He closed one end of a glass tube and then filled it with mercury. The tube was then placed, open end down, in a bowl of mercury while holding the mercury in the tube with a finger. When Torricelli removed his finger with the open end below the surface in the bowl, a small amount of mercury moved into the bowl leaving a vacuum at the top end

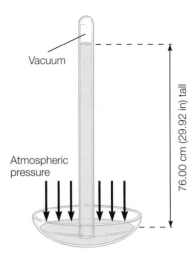

FIGURE 21.5

The mercury barometer measures the atmospheric pressure from the balance between the pressure exerted by the weight of the mercury in a tube and the pressure exerted by the atmosphere. As the atmospheric pressure increases and decreases, the mercury rises and falls. This sketch shows the average height of the column at sea level.

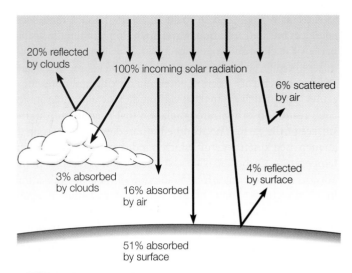

FIGURE 21.6

On the average, the earth's surface absorbs only 51 percent of the incoming solar radiation after it is filtered, absorbed, and reflected. This does not include the radiation emitted back to the surface from the greenhouse effect, which is equivalent to 93 units if the percentages in this figure are considered as units of energy.

of the tube. The mercury remaining in the tube was supported by the atmospheric pressure on the surface of the mercury in the bowl. The pressure exerted by the weight of the mercury in the tube thus balanced the pressure exerted by the atmosphere. At sea level, Torricelli found that atmospheric pressure balanced a column of mercury about 76.00 cm (29.92 in) tall (figure 21.5).

As the atmospheric pressure increases and decreases, the height of the supported mercury column moves up and down. Atmospheric pressure can be expressed in terms of the height of such a column of mercury. Public weather reports give the pressure by referring to such a mercury column, for example, "The pressure is 30 inches (about 76 cm) and rising." Meteorologists use a measure of atmospheric pressure called a **millibar.** A millibar of pressure is defined as 1,000 dynes per cm^2 (a dyne is a unit of force equal to 10^{-5} newton; see chapter 3).

If the atmospheric pressure at sea level is measured many times over long periods of time, an average value of 76.00 cm (29.92 in) of mercury is obtained. This average measurement is called the **standard atmospheric pressure** and is sometimes referred to as the *normal pressure.* It is also called *one atmosphere of pressure.* In meteorology, the normal sea level pressure is known as 1,013.25 millibars (or 760.0 mm) of mercury.

Warming the Atmosphere

Of the total energy radiated by the sun, about 40 percent is in the visible light (violet to red) part of the electromagnetic spectrum, about 9 percent is in the invisible ultraviolet part, and about 51 percent is in the invisible infrared part of the spectrum. This is the radiation from the sun that reaches the outermost part of the earth's atmosphere. Here, when the sunlight is perpendicular to

the outer edge and the earth is at an average distance from the sun, it produces about 1,370 watts per square meter. This amount has been called the **solar constant** because the quantity was believed to remain essentially constant. Recent measurements by spacecraft and satellites outside the earth's atmosphere have found, however, that the amount of solar energy received outside the earth's atmosphere decreases slightly, up to a tenth of a percent, when large numbers of sunspots are present on the surface of the sun. As more measurements are taken over longer periods of time, the solar constant may be found to be not so constant after all.

Radiation from the sun must first pass through the atmosphere before reaching the earth's surface. The atmosphere filters, absorbs, and reflects incoming solar radiation as shown in figure 21.6. On the average, the earth as a whole reflects about 30 percent of the total radiation back into space, with two-thirds of the reflection occurring from clouds. The amount reflected at any one time depends on the extent of cloud cover, the amount of dust in the atmosphere, and the extent of snow and vegetation on the surface. Substantial changes in any of these influencing variables could increase or decrease the reflectivity, leading to increased heating or cooling of the atmosphere.

As figure 21.6 shows, only about one-half of the incoming solar radiation reaches the earth's surface. The reflection and selective filtering by the atmosphere allow a global average of about 240 watts per square meter to reach the surface. The value of solar radiation received by the surface is less than half of the solar constant because it is a worldwide average that has been selectively filtered. Wide variations from the average occur with latitude as well as with the season.

The incoming solar radiation that does reach the earth's surface is absorbed. Rocks, soil, water, and the ground become warmer as a result. These materials emit the absorbed solar energy as infrared radiation, wavelengths longer than the visible part of the electromagnetic spectrum. This longer-wavelength infrared radiation has a frequency that matches some of the natural frequencies of vibration of carbon dioxide and water molecules. This match means that carbon dioxide and water molecules readily absorb infrared radiation that is emitted from the surface of the earth. The absorbed infrared energy shows up as an increased kinetic energy, which is indicated by an increase in temperature. Carbon dioxide and water vapor molecules in the atmosphere now emit infrared radiation of their own, this time in all directions. Some of this reemitted radiation is again absorbed by other molecules in the atmosphere, some is emitted to space, and significantly, some is absorbed by the surface to start the process all over again. The net result is that less of the energy from the sun escapes immediately to space after being absorbed and emitted as infrared. It is retained through the process of being redirected to the surface, increasing the surface temperature more than it would have otherwise been. The more carbon dioxide that is present in the atmosphere, the more energy that will be bounced around and redirected back toward the surface, increasing the temperature near the surface. The process of heating the atmosphere in the lower parts by the absorption of solar radiation and reemission of infrared radiation is called the **greenhouse effect.** It is called the greenhouse effect because greenhouse glass allows the short wavelengths of solar radiation to pass into the greenhouse but does not allow the longer infrared radiation to leave. This analogy is misleading, however, because carbon dioxide and water vapor molecules do not "trap" infrared radiation, but they are involved in a dynamic absorption and downward reemission process that increases the surface temperature. The more carbon dioxide molecules that are involved in this dynamic process, the more the temperature will increase. More layers of glass on a greenhouse will not increase the temperature significantly. The significant heating factor in a real greenhouse is the blockage of convection by the glass, a process that does not occur from the presence of carbon dioxide and water vapor in the atmosphere.

Structure of the Atmosphere

Convection currents and the repeating absorption and reemission processes of the greenhouse effect tend to heat the atmosphere from the ground up. In addition, the higher-altitude parts of the atmosphere lose radiation to space more readily than the lower-altitude parts. Thus, the lowest part of the atmosphere is warmer, and the temperature decreases with increasing altitude. On the average, the temperature decreases about 6.5°C for each kilometer of altitude (3.5°F/1,000 ft). This change of temperature with altitude is called the *observed lapse rate.* The observed lapse rate applies only to air that is not rising or sinking, and the actual change with altitude can be very different from this average value. For example, a stagnant mass of very cold air may settle over an area, producing colder temperatures near the surface than in the air layers above. Such a layer where the temperature increases with height is called an **inversion** (figure 21.7). Inver-

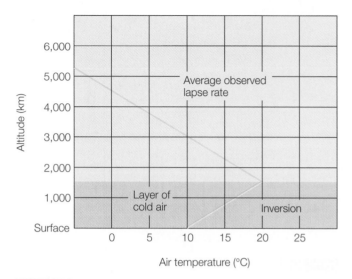

FIGURE 21.7

On the average, the temperature decreases about 6.5°C/1,000 km, which is known as the *observed lapse rate.* An inversion is a layer of air in which the temperature increases with height.

sions often occur on calm winter days after the arrival of a cold front. They also occur on calm, clear, and cool nights ("C" nights) when the surface rapidly loses radiant energy to space. In either case, the situation results in a "cap" of cooler, more dense air overlying the warmer air beneath. This often leads to an increase of air pollution because the inversion prevents dispersion of the pollutants.

Temperature decreases with height at the observed lapse rate until an average altitude of about 11 km (about 6.7 mi), where it then begins to remain more or less constant with increasing altitude. The layer of the atmosphere from the surface up to where the temperature stops decreasing with height is called the *troposphere.* Almost all weather occurs in the troposphere, which is named after the Greek meaning "turning layer." The upper boundary of the troposphere is called the *tropopause.* The tropopause is identified by the altitude where the temperature stops decreasing and remains constant with increasing altitude. This altitude varies with latitude and with the season. In general, the tropopause is nearly one and one-half times higher than the average over the equator and about half the average altitude over the poles. It is also higher in the summer than in the winter at a given latitude. Whatever its altitude, the tropopause marks the upper boundary of the atmospheric turbulence and the weather that occurs in the troposphere. The average temperature at the tropopause is about −60°C (about −80°F).

Above the tropopause is the second layer of the atmosphere called the *stratosphere.* This layer is named after the Greek for "stratified layer." It is stratified, or layered, because the temperature increases with height. Cooler air below means that consecutive layers of air are denser on the bottom, which leads to a stable situation rather than the turning turbulence of the troposphere below. The stratosphere contains little moisture or dust and lacks convective turbulence, making it a desirable

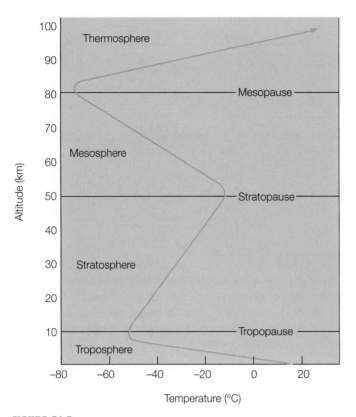

FIGURE 21.8

The structure of the atmosphere based on temperature differences. Note that the "pauses" are actually not lines, but are broad regions that merge.

altitude to fly. Temperature in the lower stratosphere increases gradually with increasing altitude to a height of about 48 km (about 30 mi), where it reaches a maximum of about 10°C (about 50°F). This altitude marks the upper boundary of the stratosphere, the *stratopause* (figure 21.8).

The temperature increases in the stratosphere as a result of interactions between high-energy ultraviolet radiation and ozone. Ozone is triatomic oxygen (O_3) that is concentrated mainly in the upper portions of the stratosphere. Diatomic molecules of oxygen (O_2) are concentrated in the troposphere and monatomic molecules of oxygen (O) are found in the outer edges of the atmosphere. Although the amount of ozone present in the stratosphere is not great, its presence is vital to life on the earth's surface. Without the **ozone shield** ultraviolet radiation would reach the surface of the earth at a level sufficient to damage both skin and eyes even more readily than it now does. It is believed that the incidence of skin cancer would rise dramatically without the protection offered by the ozone.

Here is how the ozone shield works. The ozone concentration is not static because there is an ongoing process of ozone formation and destruction. For ozone to form, diatomic oxygen must first be broken down into the monatomic form. Short ultraviolet (UV) radiation is absorbed by ordinary oxygen, which breaks it down into the single-atom (O) form. This reac-

tion is significant because of (1) the high-energy ultraviolet that is removed from the sunlight and (2) the monatomic oxygen that is formed, which will combine to make ozone that will absorb even more ultraviolet radiation. This initial reaction is

$$O_2 + UV \rightarrow O + O$$

When the O molecule collides with an O_2 molecule and any third, neutral molecule (NM), the following reaction takes place:

$$O_2 + O + NM \rightarrow O_3 + NM$$

When O_3 is exposed to ultraviolet, the ozone absorbs the UV radiation and breaks down to two forms of oxygen in the following reaction:

$$O_3 + UV \rightarrow O_2 + O$$

The monatomic molecule that is produced combines with an ozone molecule to produce two diatomic molecules,

$$O + O_3 \rightarrow 2\,O_2$$

and the process starts all over again. There is much concern about Freon and other similar chemicals that make their way to the stratosphere. These chemicals are broken down by UV radiation, releasing chlorine, which reacts with the oxygen. This forms a compound that is stable, so it effectively removes the oxygen from the ongoing ozone-forming and UV-radiation-absorbing process.

Above the stratopause, the temperature decreases again, just as in the stratosphere, then increases with altitude. The rising temperature is caused by the absorption of solar radiation by molecular fragments present at this altitude.

Layers above the stratopause are the *mesosphere* (Greek for "middle layer") and the *thermosphere* (Greek for "warm layer"). The name "thermosphere" and the high temperature readings of the thermosphere would seem to indicate an environment that is actually not found at this altitude. The gas molecules here do have a high kinetic energy, but the air here is very thin and the molecules are far apart. Thus, the average kinetic energy is very high, but the few molecules do not transfer much energy to a thermometer. A thermometer here would show a temperature far below zero for this reason, even though the same average kinetic energy back at the surface would result in a temperature beyond any temperature ever recorded in the hottest climates.

The *exosphere* (Greek for "outer layer") is the outermost layer where the molecules merge with the diffuse vacuum of space. Molecules of this layer that have sufficient kinetic energy are able to escape and move off into space. The thermosphere and upper mesosphere are sometimes called the *ionosphere* because of the free electrons and ions at this altitude. The electrons and ions here are responsible for reflecting radio waves around the earth and for the northern lights.

THE WINDS

The troposphere is heated from the bottom up as the surface of the earth absorbs incoming solar radiation and emits infrared radiation. The infrared radiation is absorbed and reemitted numerous times by carbon dioxide and water molecules as the energy works its way back to space. The overall result of this ongoing absorption and reemission process is the observed lapse rate, the decrease of temperature upward in the troposphere. The observed lapse rate is an *average* value, which means that the actual condition at a given place and time is probably higher or lower than this value. The composition of the surface varies from place to place, consisting of many different types and forms of rock, soil, water, ice, snow, and so forth. These various materials absorb and emit energy at various rates, which results in an uneven heating of the surface. You may have noticed that different materials vary in their abilities to absorb and emit energy if you have ever walked barefooted across some combination of grass, concrete, asphalt, and dry sand on a hot sunny day.

Uneven heating of the earth's surface sets the stage for *convection* (see chapter 5). As a local region of air becomes heated, the increased kinetic energy of the molecules expands the mass of air, reducing its density. This less dense air is buoyant and is pushed upward by nearby cooler, more dense air. This results in three general motions of air: (1) the upward movement of air over a region of greater heating, (2) the sinking of air over a cooler region, and (3) a horizontal air movement between the cooler and warmer regions. In general, a horizontal movement of air is called **wind,** and the direction of a wind is defined as the direction from which it blows.

Air in the troposphere rises, moves as the wind, and sinks. All three of these movements are related, and all occur at the same time in different places. During a day with gentle breezes on the surface, the individual, fluffy clouds you see are forming over areas where the air is moving upward. The clear air between the clouds is over areas where the air is moving downward. On a smaller scale, air can be observed moving from a field of cool grass toward an adjacent asphalt parking lot on a calm, sunlit day. Soap bubbles or smoke will often reveal the gentle air movement of this localized convection.

Depending on local surface conditions, which are discussed in the next section, the wind usually averages about 16 km/hr (about 10 mi/hr) and has an average rising and sinking velocity of about 2 km/hr (about 1 mi/hr). These normal, average values are greatly exceeded during storms and severe weather events. A hurricane has winds that exceed 120 km/hr (about 75 mi/hr), and a thunderstorm can have updrafts and downdrafts between 50 and 100 km/hr (about 30 to 60 mi/hr). The force exerted by such winds can be very destructive to structures on the surface. An airplane unfortunate enough to be caught in a thunderstorm can be severely damaged as it is tossed about by the updrafts and downdrafts.

Local Wind Patterns

Considering average conditions, there are two factors that are important for a generalized model to help you understand local wind patterns. These factors are (1) the relationship between air temperature and air density and (2) the relationship between air pressure and the movement of air. The relationship between air temperature and air density was discussed in the introduction to this section. This relationship is that cool air

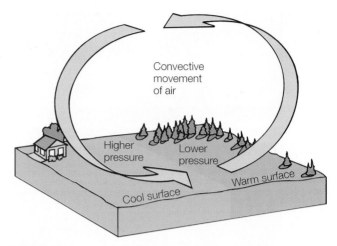

FIGURE 21.9

A model of the relationships between differential heating, the movement of air, and pressure difference in a convective cell. Cool air pushes the less dense, warm air upward, reducing the surface pressure. As the uplifted air cools and becomes more dense, it sinks, increasing the surface pressure.

has a greater density than warm air because a mass of warming air expands and a mass of cooling air contracts. Warm, less dense air is buoyed upward by cooler, more dense air, which results in the upward, downward, and horizontal movement of air called a convection cell.

The upward and downward movement of air leads to the second part of the generalized model, that (1) the upward movement produces a "lifting" effect on the surface that results in an area of lower atmospheric pressure and (2) the downward movement produces a "piling up" effect on the surface that results in an area of higher atmospheric pressure. On the surface, air is seen to move from the "piled up" area of higher pressure horizontally to the "lifted" area of lower pressure (figure 21.9). In other words, air generally moves from an area of higher pressure to an area of lower pressure. The movement of air and the pressure differences occur together, and neither is the cause of the other. This is an important relationship in a working model of air movement that can be observed and measured on a very small scale, such as between an asphalt parking lot and a grass field. It can also be observed and measured for local, regional wind patterns and for worldwide wind systems.

Adjacent areas of the surface can have different temperatures because of different heating or cooling rates. The difference is very pronounced between adjacent areas of land and water. Under identical conditions of incoming solar radiation, the temperature changes experienced by the water will be much less than the changes experienced by the adjacent land. There are three principal reasons for this difference: (1) The specific heat of water is about twice the specific heat of soil (see chapter 5). This means that it takes more energy to increase the temperature of water than it does for soil. Equal masses of soil and water exposed to sunlight will result in the soil heating about 1°C while the water heats 1/2°C from absorbing the same amount of

A CLOSER LOOK | The Wind Chill Factor

The term *wind chill* is attributed to the Antarctic explorer Paul A. Siple. During the 1940s Siple and Charles F. Passel conducted experiments on how long it took a can of water to freeze at various temperatures and wind speeds. They found that the time depended on the air temperature and the wind speed. From this data an equation was developed to calculate the wind chill factor for humans.

Here is the reason why the wind chill factor is an important consideration. The human body constantly produces heat to maintain a core temperature, and some of this heat is radiated to the surroundings. When the wind is not blowing (and you are not moving), your body heat is also able to warm some of the air next to your body. This warm blanket of air provides some insulation, protecting your skin from the colder air farther away. If the wind blows, however, it moves this air away from your body and you will feel cooler. How much cooler will depend on how fast the air is moving and upon the outside temperature—which is what the wind chill factor tells you. Thus, wind chill is an attempt to measure the effect of combinations of low temperature and wind on humans. It is just one of the many factors that can affect winter comfort. Others include the type of clothes, level of physical exertion, amount of sunshine, humidity, age, and body type. Finally, note that the use of numbers does not necessarily make a wind chill factor precise or exact. Wind chill is an approximation or estimate of a cooling effect, and other indices of "physiological" or "perceptual" temperature do exist.

Here is how to figure a wind chill factor: Suppose the air temperature is 10°F and the wind is blowing at 10 miles per hour. According to the chart (box figure 21.1), the wind chill is −9°. A wind chill of −9° is not saying that a chilled object will be cooled to −9°. It is saying that is *how cold you will feel* because of the low temperature and the wind, which blew away your warm blanket of air. If the air temperature is 10°F then any object in the air—such as the radiator in your car—will be no cooler than 10°F, no matter what the wind velocity is or if the car is driven or not.

An equation to calculate a Celsius wind chill:

$$T_{wc} = (0.045)\left(5.27\sqrt{v} + 10.45 - 0.28v\right)(T - 33) + 33$$

where

T_{wc} is the wind chill temperature.

v is the wind speed in kilometers/hour.

T is the present temperature in °C.

Wind Chill Chart

Wind (mph)	Temperature (°F)												
	35	30	25	20	15	10	5	0	-5	-10	-15	-20	-25
5	32	27	22	16	11	6	0	-5	-10	-15	-21	-26	-31
10	22	16	10	3	-3	-9	-15	-22	-27	-34	-40	-46	-52
15	16	9	2	-5	-11	-18	-25	-31	-38	-45	-51	-58	-65
20	12	4	-3	-10	-17	-24	-31	-39	-46	-53	-60	-67	-74
25	8	1	-7	-15	-22	-29	-36	-44	-51	-59	-66	-74	-81
30	6	-2	-10	-18	-25	-33	-41	-49	-56	-64	-71	-79	-86
35	4	-4	-12	-20	-27	-35	-43	-52	-58	-67	-74	-82	-89
40	3	-5	-13	-21	-29	-37	-45	-53	-60	-69	-76	-84	-92

BOX FIGURE 21.1

solar radiation. (2) Water is a transparent fluid that is easily mixed, so the incoming solar radiation warms a body of water throughout, spreading out the heating effect. Incoming solar radiation on land, on the other hand, warms a relatively thin layer on the top, concentrating the heating effect. (3) The water is cooled by evaporation, which helps keep a body of water at a lower temperature than an adjacent landmass under identical conditions of incoming solar radiation.

A local wind pattern may result from the resulting temperature differences between a body of water and adjacent landmasses. If you have ever spent some time along a coast, you may have observed that a cool, refreshing gentle breeze blows from the water

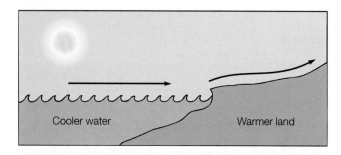

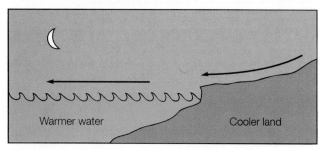

FIGURE 21.10

The land warms and cools more rapidly than an adjacent large body of water. During the day, the land is warmer, and air over the land expands and is buoyed up by cooler, more dense air from over the water. During the night, the land cools more rapidly than the water, and the direction of the breeze is reversed.

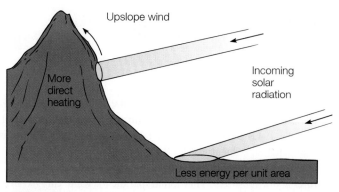

FIGURE 21.11

Incoming solar radiation falls more directly on the side of a mountain, which results in differential heating. The same amount of sunlight falls on the areas shown in this illustration, with the valley floor receiving a more spread-out distribution of energy per unit area. The overall result is an upslope mountain breeze during the day. During the night, dense cool air flows downslope for a reverse wind pattern.

toward the land during the summer. During the day the temperature of the land increases more rapidly than the water temperature. The air over the land is therefore heated more, expands, and becomes less dense. Cool, dense air from over the water moves inland under the air over the land, buoying it up. The air moving from the sea to the land is called a *sea breeze*. The sea breeze along a coast may extend inland several miles during the hottest part of the day in the summer. The same pattern is sometimes observed around the Great Lakes during the summer, but this breeze usually does not reach more than several city blocks inland. During the night the land surface cools more rapidly than the water, and a breeze blows from the land to the sea (figure 21.10).

Another pattern of local winds develops in mountainous regions. If you have ever visited a mountain in the summer, you may have noticed that there is usually a breeze or wind blowing up the mountain slope during the afternoon. This wind pattern develops because the air over the mountain slope is heated more than the air in a valley. As shown in figure 21.11, the air over the slope becomes warmer because it receives more direct sunlight than the valley floor. Sometimes this air movement is so gentle that it would be unknown except for the evidence of clouds that form over the peaks during the day and evaporate at night. During the night the air on the slope cools as the land loses radiant energy to space. As the air cools, it becomes denser and flows downslope, forming a reverse wind pattern to the one observed during the day.

During cooler seasons cold, dense air may collect in valleys or over plateaus, forming a layer or "puddle" of cold air. Such an accumulation of cold air often results in some very cold nighttime temperatures for cities located in valleys, temperatures that

are much colder than anywhere in the surrounding region. Some weather disturbance, such as an approaching front, can disturb such an accumulation of cold air and cause it to pour out of its resting place and through canyons or lower valleys. Air moving from a higher altitude like this becomes compressed as it moves to lower elevations under increasing atmospheric pressure. Compression of air, or any gas for that matter, increases the temperature by increasing the kinetic energy of the molecules. This creates a wind called a *Chinook,* which is common to mountainous and adjacent regions. A Chinook is a wind of compressed air with sharp temperature increases that can melt away any existing snow cover in a single day. The *Santa Ana* is a well-known compressional wind that occurs in southern California.

Global Wind Patterns

Local wind patterns tend to mask the existence of the overall global wind pattern that is also present. The global wind pattern is not apparent if the winds are observed and measured for a particular day, week, or month. It does become apparent when the records for a long period of time are analyzed. These records show that the earth has a large-scale pattern of atmospheric circulation that varies with latitude. There are belts in which the winds average an overall circulation in one direction, belts of higher atmospheric pressure averages, and belts of lower atmospheric pressure averages. This has led to a generalized pattern of atmospheric circulation and a global atmospheric model. This model, as you will see, today provides the basis for the daily weather forecast for local and regional areas.

As with local wind patterns, it is temperature imbalances that drive the global circulation of the atmosphere. The earth receives more direct solar radiation in the equatorial region than it does at higher latitudes (figure 21.12). As a result, the temperatures of the lower troposphere are generally higher in the equatorial region, decreasing with latitude toward both poles. The

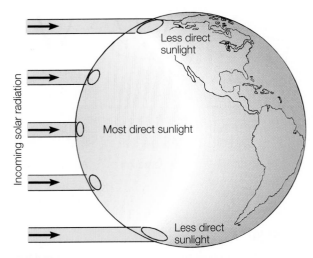

FIGURE 21.12

On a global, yearly basis, the equatorial region of the earth receives more direct incoming solar radiation than the higher latitudes. As a result, average temperatures are higher in the equatorial region and decrease with latitude toward both poles. This sets the stage for worldwide patterns of prevailing winds, high and low areas of atmospheric pressure, and climatic patterns.

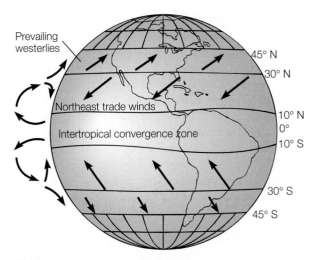

FIGURE 21.13

Part of the generalized global circulation pattern of the earth's atmosphere. The scale of upward movement of air above the intertropical convergence zone is exaggerated for clarity. The troposphere over the equator is thicker than elsewhere, reaching a height of about 20 km (about 12 mi).

lower troposphere from 10°N to 10°S of the equator is heated, expands, and becomes less dense. Hot air rises in this belt around the equator, known as the *intertropical convergence zone.* The rising air cools because it expands as it rises, resulting in heavy average precipitation. The tropical rain forests of the earth occur in this zone of high temperatures and heavy rainfall. As the now dry, rising air reaches the upper parts of the troposphere it begins to spread toward the north and toward the south, sinking back toward the earth's surface (figure 21.13). The descending air reaches the surface to form a high-pressure belt that is centered about 30°N and 30°S of the equator. Air moving on the surface away from this high-pressure belt produces the prevailing northeast trade winds and the prevailing westerly winds of the Northern Hemisphere. The great deserts of the earth are also located in this high-pressure belt of descending dry air.

Poleward of the belt of high pressure the atmospheric circulation is controlled by a powerful belt of wind near the top of the troposphere called a **jet stream.** Jet streams are sinuous, meandering loops of winds that tend to extend all the way around the earth, moving generally from the west in both hemispheres at speeds of 160 km/hr (about 100 mi/hr) or more. A jet stream may occur as a single belt, or loop, of wind but sometimes it divides into two or more parts. The jet stream develops north and south loops of waves much like the waves you might make on a very long rope. These waves vary in size, sometimes beginning as a small ripple but then growing slowly as the wave moves eastward. Waves that form on the jet stream bulge toward the poles (called a crest) or toward the equator (called a trough). Warm air masses move toward the poles ahead of a trough and cool air masses move toward the equator behind a trough as it moves eastward. The development of a wave in the jet stream is

understood to be one of the factors that influences the movement of warm and cool air masses, a movement that results in weather changes on the surface.

The intertropical convergence zone, the 30° belt of high pressure, and the northward and southward migration of a meandering jet stream all shift toward or away from the equator during the different seasons of the year. The troughs of the jet stream influence the movement of alternating cool and warm air masses over the belt of the prevailing westerlies, resulting in frequent shifts of fair weather to stormy weather, then back again. The average shift during the year is about 6° of latitude, which is sufficient to control the overall climate in some locations. The influence of this shift of the global circulation of the earth's atmosphere will be considered as a climatic influence after considering the roles of water and air masses in frequent weather changes.

WATER AND THE ATMOSPHERE

Water exists on the earth in all three states: (1) as a liquid when the temperature is generally above the freezing point of 0°C (32°F), (2) as a solid in the form of ice, snow, or hail when the temperature is generally below the freezing point, and (3) as the invisible, molecular form of water in the gaseous state, which is called *water vapor.*

Over 98 percent of all the water on the earth exists in the liquid state, mostly in the ocean, and only a small, variable amount of water vapor is in the atmosphere at any given time. Since so much water seems to fall as rain or snow at times, it may be a surprise that the overall atmosphere really does not contain very much water vapor. If the average amount of water vapor in the earth's atmosphere were condensed to liquid form, the vapor *and* all the droplets present in clouds would form a uniform layer around the earth only 3 cm (about 1 in) thick. Nonetheless, it is this small amount of water vapor that is eventually responsible for (1) contributing to the greenhouse effect, which helps make the earth a warmer planet, (2) serving as one of the principal agents in the weathering and erosion of the land, which creates soils and sculptures the landscape, and (3) maintains life, for almost all organisms (bacteria, protozoa, algae, fungi, plants, and animals) cannot survive without water. It is the ongoing cycling of water vapor into and out of the atmosphere that makes all this possible. Understanding this cycling process and the energy exchanges involved is also closely related to understanding the earth's weather patterns.

Evaporation and Condensation

Water tends to undergo a liquid-to-gas or a gas-to-liquid phase change at any temperature. The phase change can occur in either direction at any temperature. The temperature of liquid water and the temperature of water vapor are associated with the *average* kinetic energy of the water molecules. The word *average* implies that some of the molecules have a greater kinetic energy and some have less. If a molecule of water that has an exceptionally high kinetic energy is near the surface, and is headed in the right direction, it may overcome the attractive forces of the other water molecules and escape the liquid to become a gas. This is the process of evaporation. A supply of energy must be present to maintain the process of evaporation, and the water robs this energy from the surroundings. This explains why water at a higher temperature evaporates more rapidly than water at a lower temperature. More energy is available at higher temperatures to maintain the process at a faster rate.

Water molecules that evaporate move about in all directions, and some will strike the liquid surface. The same forces that it escaped from earlier now capture the molecule, returning it to the liquid state. This is called the process of condensation. Condensation is the opposite of evaporation. In *evaporation,* more molecules are leaving the liquid state than are returning. In *condensation,* more molecules are returning to the liquid state than are leaving. This is a dynamic, ongoing process with molecules leaving and returning continuously (figure 21.14). If the air were perfectly dry and still, more molecules would leave (evaporate) the liquid state than would return (condense). Eventually, however, an equilibrium would be reached with as many molecules returning to the liquid state per unit of time as are leaving. An equilibrium condition between evaporation and condensation occurs in **saturated air.** Saturated air occurs when the processes of evaporation and condensation are in balance.

Air will remain saturated as long as (1) the temperature remains constant and (2) the processes of evaporation and condensation remain balanced. Temperature influences the equilibrium condition of saturated air because increases or decreases in the temperature mean increases or decreases in the kinetic energy of water vapor molecules. Water vapor molecules usually undergo condensation when attractive forces between the molecules can pull them together into the liquid state. Lower temperature means lower kinetic energies, and slow-moving water

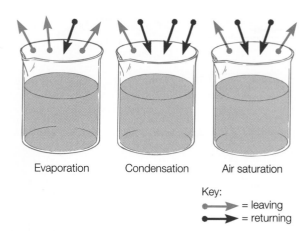

Key:
⟶ = leaving
⟶ = returning

FIGURE 21.14

Evaporation and condensation are occurring all the time. If the number of molecules leaving the liquid state exceeds the number returning, the water is evaporating. If the number of molecules returning to the liquid state exceeds the number leaving, the water vapor is condensing. If both rates are equal, the air is saturated; that is, the relative humidity is 100 percent.

vapor molecules spend more time close to one another and close to the surface of liquid water. Spending more time close together means an increased likelihood of attractive forces pulling the molecules together. On the other hand, higher temperature means higher kinetic energies, and molecules with higher kinetic energy are less likely to be pulled together. As the temperature increases, there is therefore less tendency for water molecules to return to the liquid state. If the temperature is increased in an equilibrium condition, more water vapor must be added to the air to maintain the saturated condition. Warm air can therefore hold more water vapor than cooler air. In fact, warm air on a typical summer day can hold five times as much water vapor as cold air on a cold winter day.

Humidity

The amount of water vapor in the air is referred to generally as **humidity.** Damp, moist air is more likely to have condensation than evaporation, and this air is said to have a *high humidity.* Dry air is more likely to have evaporation than condensation, on the other hand, and this air is said to have a *low humidity.* A measurement of the amount of water vapor in the air at a particular time is called the **absolute humidity** (figure 21.15). At room temperature, for example, humid air might contain 15 grams of water vapor in each cubic meter of air. At the same temperature, air of low humidity might have an absolute humidity of only 2 grams per cubic meter. The absolute humidity can range from near zero up to a maximum that is determined by the temperature at the time. Since the temperature of the water vapor present in the air is the same as the temperature of the air, the maximum absolute humidity is usually said to be determined by the air temperature. What this really means is that the maximum absolute humidity is determined by the temperature of the water vapor, that is, the average kinetic energy of the water vapor.

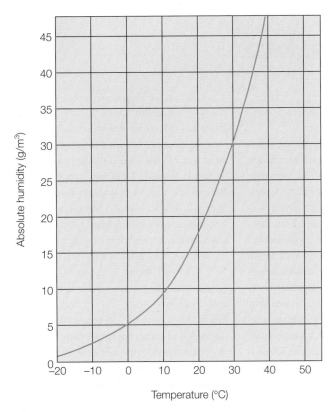

FIGURE 21.15

The maximum amount of water vapor that can be in the air at different temperatures. The amount of water vapor in the air at a particular temperature is called the *absolute humidity*.

The relationship between the *actual* absolute humidity at a particular temperature and the *maximum* absolute humidity that can occur at that temperature is called the **relative humidity**. Relative humidity is a ratio between (1) the amount of water vapor in the air and (2) the amount of water vapor needed to saturate the air at that temperature. The relationship is

$$\frac{\text{actual absolute humidity at present temperature}}{\text{maximum absolute humidity at present temperature}} \times 100\%$$

$$= \frac{\text{relative}}{\text{humidity}}$$

For example, suppose a measurement of the water vapor in the air at 10°C (50°F) finds an absolute humidity of 5.0 g/m³. According to figure 21.15, the maximum amount of water vapor that can be in the air when the temperature is 10°C is about 10 g/m³. The relative humidity is then

$$\frac{5.0 \text{ g/m}^3}{10 \text{ g/m}^3} \times 100\% = 50\%$$

If the absolute humidity had been 10 g/m³, then the air would have all the water vapor it could hold, and the relative humidity would be 100 percent. A humidity of 100 percent means that the air is saturated at the present temperature.

The important thing to understand about relative humidity is that the capacity of air to hold water vapor changes with the temperature. Cold air cannot hold as much water vapor, and warming the air will increase its capacity. With the capacity *increased* and the same amount of water in the air, the relative humidity *decreases* because you can now add more water vapor to the air than you could before. Just warming the air, for example, can reduce the relative humidity from 50 percent to 3 percent. Lower relative humidity results because warming the air increases the capacity of air to hold water vapor. This explains the need to humidify a home in the winter. Evaporation occurs very rapidly when the humidity is low. Evaporation is a cooling process since the molecules with higher kinetic energy are the ones to escape, lowering the average kinetic energy as they evaporate. Dry air will therefore cause you to feel cool even though the air temperature is fairly high. Adding moisture to the air will enable you to feel warmer at lower air temperatures, and thus lower your fuel bill. The relationship between the capacity of air to hold water vapor and temperature also explains why the relative humidity increases in the evening after the sun goes down. A cooler temperature means less capacity of air to hold water vapor. With the same amount of vapor in the air, a reduced capacity means a higher relative humidity.

Relative humidity is closely associated with your comfort. When the wind blows evaporation is increased, because the moving air also moves evaporated water vapor molecules away from your body, reducing the possibility that they could condense and release the latent heat of vaporization. Thus, wind makes the temperature *seem* much lower. In the winter, the cooling power of the wind is called the **wind chill factor**. For example, a temperature of 2°C (about 35°F) in air moving at 16 km/hr (about 10 mi/hr) will have the cooling equivalent of still air with a temperature of about −7°C (about 20°F).

The Condensation Process

In still air under a constant pressure, the rate of evaporation depends primarily on three factors: (1) the surface area of the liquid that is exposed to the atmosphere, (2) the air and water temperature, and (3) the amount of water vapor in the air at the time, that is, the relative humidity. The opposite process, condensation, depends primarily on two factors: (1) the relative humidity and (2) the temperature of the air, or more directly, the kinetic energy of the water vapor molecules. During condensation, molecules of water vapor join together to produce liquid water on the surface as dew or in the air as the droplets of water making up fog or clouds. Water molecules may also join together to produce solid water in the form of frost or snow. Before condensation can occur, however, the air must be saturated, which means that the relative humidity must be 100 percent with the air containing all the vapor it

A CLOSER LOOK | Environment Wind Power

Four nonrenewable sources of energy—coal, oil, natural gas, and nuclear—provide more than 97 percent of the world's energy supply. The remainder is supplied by renewable energy sources, including solar energy, tidal power, and geothermal. The term *solar energy* describes a number of technologies that directly or indirectly utilize sunlight as an alternative energy source. In part, these include solar cells, power towers, biomass, and wind energy.

The earth's surface is warmed by the radiant energy from the sun, and some parts are warmed more than other parts. This differential heating sets up convection currents in the atmosphere, resulting in horizontal movements of air called wind. Millions of windmills were installed in the United States between the late 1800s and the late 1940s in rural areas. These windmills used wind energy to pump water, grind grain, or generate electricity. Some are still in use today, but most were no longer needed after inexpen-

sive electric power became generally available in rural areas of the country.

In the 1970s wind energy made a comeback as a clean renewable energy alternative to fossil fuels. The windmills of the past were replaced by wind turbines of today. A wind turbine is usually mounted on a tower, with blades that are rotated by the wind. This rotary motion drives a generator that produces electricity. An area should have yearly wind speeds of about 12 miles per hour to provide enough wind energy for a turbine, and a greater yearly average means more energy is available. Farms, homes, and businesses in these areas can use smaller turbines, which are generally 50 kilowatts or less. Large turbines of 500 kilowatts or more are used in "wind farms," which are large clusters of interconnected wind turbines connected to a utility power grid.

Many areas of the United States have a high potential for wind-power use. The states of North Dakota, South Dakota, and Texas have enough wind resources to provide elec-

tricity for the entire nation. Today only California has extensively developed wind farms, with more than 13,000 wind turbines on three wind farms in the Altamont Pass region (east of San Francisco), Techachapi (southeast of Bakersfield), and San Gorgonio (near Palm Springs). With a total of 1,700 megawatts of installed capacity in California, wind energy generates enough electricity to more than meet the needs of a city the size of San Francisco.

The wind farms in California have a rated capacity that is comparable to two large coal-fired power plants, but without the pollution and limits of this nonrenewable energy source. Wind energy makes economic as well as environmental sense and new wind farms are being developed in the states of Minnesota, Oregon, and Wyoming. Other states with a strong wind-power potential include Kansas, Montana, Nebraska, Oklahoma, Iowa, Colorado, Michigan, and New York. All of these states, in fact, have a greater wind-energy potential than the state of California.

can hold at the present temperature. A parcel of air can become saturated as a result of (1) water vapor being added to the air from evaporation, (2) cooling, which reduces the capacity of the air to hold water vapor and therefore increases the relative humidity, or (3) a combination of additional water vapor with cooling.

The process of condensation of water vapor explains a number of common observations. You are able to "see your breath" on a cold day, for example, because the high moisture content of your exhaled breath is condensed into tiny water droplets by cold air. The small fog of water droplets evaporates as it spreads into the surrounding air with a lower moisture content. The white trail behind a high-flying jet aircraft is also a result of condensation of water vapor. Water is one of the products of combustion, and the white trail is condensed water vapor, a trail of tiny droplets of water in the cold upper atmosphere. The trail of water droplets is called a *contrail* after "condensation trail." Back on the surface, a cold glass of beverage seems to "sweat" as water vapor molecules near the outside of the glass are cooled, moving more slowly. Slowly moving water vapor molecules spend more time closer together, and the molecular forces between the molecules pull them together, forming a thin layer of liquid water on the outside of the cold glass. This same condensation

process sometimes results in a small stream of water from the cold air conditioning coils of an automobile or home mechanical air conditioner.

As air is cooled, its capacity to hold water vapor is reduced to lower and lower levels. Even without water vapor being added to the air, a temperature will eventually be reached at which saturation, 100 percent humidity, occurs. Further cooling below this temperature will result in condensation. The temperature at which condensation begins is called the **dew point temperature.** If the dew point is above 0°C (32°F) the water vapor will condense on surfaces as a liquid called **dew.** If the temperature is at or below 0°C the vapor will condense on surfaces as a solid called **frost.** Note that dew and frost form on the tops, sides, and bottoms of objects. Dew and frost condense directly on objects and do not "fall out" of the air. Note also that the temperature that determines if dew or frost forms is the temperature of the object where they condense. This temperature near the open surface can be very different from the reported air temperature, which is measured more at eye level in a sheltered instrument enclosure.

Observations of where and when dew and frost form can lead to some interesting things to think about. Dew and frost, for example, seem to form on "C" nights, nights that can be described by the three "C" words of clear, calm, and cool. Dew and frost also seem to form more (1) in open areas rather than under trees or other shelters, (2) on objects such as grass rather than on the flat, bare ground, and (3) in low-lying areas before they form on slopes or the sides of hills. What is the meaning of these observations?

Dew and frost are related to clear nights and open areas because these are the conditions best suited for the loss of infrared radiation. Air near the surface becomes cooler as infrared radiation is emitted from the grass, buildings, streets, and everything else that absorbed the shorter-wavelength radiation of incoming solar radiation during the day. Clouds serve as a blanket, keeping the radiation from escaping to space so readily. So a clear night is more conducive to the loss of infrared radiation and therefore to cooling. On a smaller scale, a tree serves the same purpose, holding in radiation and therefore retarding the cooling effect. Thus, an open area on a clear, calm night would have cooler air near the surface than would be the case on a cloudy night or under the shelter of a tree.

The observation that dew and frost form on objects such as grass before forming on flat, bare ground is also related to loss of infrared radiation. Grass has a greater exposed surface area than the flat, bare ground. A greater surface area means a greater area from which infrared radiation can escape, so grass blades cool more rapidly than the flat ground. Other variables, such as specific heat, may be involved, but overall frost and dew are more likely to form on grass and low-lying shrubs before they form on the flat, bare ground.

Dew and frost form in low-lying areas before forming on slopes and the sides of hills because of the density differences of cool and warm air. Cool air is more dense than warm air and is moved downhill by gravity, pooling in low-lying areas. You may have noticed the different temperatures of low-lying areas if you have ever driven across hills and valleys on a clear, calm, and cool evening. Citrus and other orchards are often located on slopes of hills rather than on valley floors because of the gravity drainage of cold air.

It is air near the surface that is cooled first by the loss of radiation from the surface. Calm nights favor dew or frost formation because the wind mixes the air near the surface that is being cooled with warmer air above the surface. If you have ever driven near a citrus orchard, you may have noticed the huge, airplanelike fans situated throughout the orchard on poles. These fans are used on "C" nights when frost is likely to form to mix the warmer, upper layers of air with the cooling air in the orchard (figure 21.16).

Condensation occurs on the surface as frost or dew when the dew point is reached. When does condensation occur in the air? Water vapor molecules in the air are constantly colliding and banging into each other, but they do not just join together to form water droplets, even if the air is saturated. The water molecules need something to condense upon. Condensation of water vapor into fog or cloud droplets takes place on tiny particles present in the air. The particles are called **condensation nuclei.** There are hundreds of tiny dust, smoke, soot, and salt crystals suspended in each cubic centimeter of the air that serve as condensation nuclei. Tiny salt crystals, however, are

FIGURE 21.16

Fans like this one are used to mix the warmer, upper layers of air with the cooling air in the orchard on nights when frost is likely to form.

particularly effective condensation nuclei because salt crystals attract water molecules. You may have noticed that salt in a salt shaker becomes moist on a humid day because of the way it attracts water molecules. Tiny salt crystals suspended in the air act the same way, serving as nuclei that attract water vapor into tiny droplets of liquid water.

After water vapor molecules begin to condense on a condensation nucleus, other water molecules will join the liquid water already formed, and the tiny droplet begins to increase in volume. The water droplets that make up a cloud are about 1,500 times larger than a condensation nuclei, and these droplets can condense out of the air in a matter of minutes. As the volume increases, however, the process slows, and hours and days are required to form the even larger droplets and drops. For comparison to the sizes shown in figure 21.17, consider that the average human hair is about 100 microns in diameter. This is about the same diameter as the large cloud droplet of water. Large raindrops have been observed falling from clouds that formed only a few hours previously, so it must be some process or processes other than the direct condensation of raindrops that form precipitation. These processes are discussed in the next chapter.

Fog and Clouds

Fog and clouds are both accumulations of tiny droplets of water that have been condensed from the air. These water droplets are very small, on the order of 0.02 to 0.1 mm (about 8×10^{-4} to 4×10^{-3} inches) in diameter, and a very slight upward movement of the air will keep them from falling. If they do fall they usually evaporate. Fog is sometimes described as a cloud that forms at or near the surface. A fog, as a cloud, forms because air containing

- • Condensation nucleus (0.2 micron)
- ● Average cloud droplet (20 microns)
- ◯ Large cloud droplet (100 microns)
- ◯ Drizzle droplet (300 microns)
- Average raindrop (2,000 microns)

FIGURE 21.17

This figure compares the size of the condensation nuclei to the size of typical condensation droplets. Note that 1 micron is 1/1,000 mm.

water vapor and condensation nuclei has been cooled to the dew point. Some types of fog form under the same "C" night conditions favorable for dew or frost to form, that is, on clear, cool, and calm nights when the relative humidity is high. Sometimes this type of fog forms only in valleys and low-lying areas where cool air accumulates (figure 21.18). This type of fog is typical of inland fogs, those that form away from bodies of water. Other types of fog may form somewhere else, such as in the humid air over an ocean, and then move inland. Many fogs that occur along coastal regions were formed over the ocean and then carried inland by breezes. A third type of fog looks much like steam rising from melting snow on a street, steam rising over a body of water into cold air, or steam rising over streets after a summer rain shower. These are examples of a temporary fog that forms as a lot of water vapor is added to cool air. This is a cool fog, like other fogs, and is not hot as the steamlike appearance may lead you to believe.

Sometimes a news report states something about the sun "burning off" a fog. A fog does not burn, of course, because it is made up of droplets of water. What the reporter really means is that the sun's radiation will increase the temperature, which increases the air capacity to hold water vapor. With an increased capacity to hold water the relative humidity drops, and the fog simply evaporates back to the state of invisible water vapor molecules.

Clouds, like fogs, are made up of tiny droplets of water that have been condensed from the air. Luke Howard, an English weather observer, made one of the first cloud classification schemes in 1803. He used the Latin terms *cirrus* (curly), *cumulus* (piled up), and *stratus* (spread out) to identify the basic shapes of clouds (figure 21.19). The clouds usually do not occur just in these basic cloud shapes but in combinations of the different shapes. Later, Howard's system was modified by expanding the different shapes of clouds into ten classes by using the basic cloud

A

B

FIGURE 21.18

(*A*) An early morning aerial view of fog between mountain at top and river below that developed close to the ground in cool, moist air on a clear, calm night. (*B*) Fog forms over the ocean where air moves from a warm current over a cool current, and the fog often moves inland.

shapes and altitude as criteria. Clouds give practical hints about the approaching weather. The relationship between the different cloud shapes and atmospheric conditions and what clouds can mean about the coming weather are discussed in the next section.

Precipitation

Water that returns to the surface of the earth, in either the liquid or solid form, is called **precipitation** (figure 21.20). Note that dew and frost are not classified as precipitation because they form directly on the surface and do not fall through the air. Precipitation

FIGURE 21.19

(A) Cumulus clouds. (B) Stratus and stratocumulus. Note the small stratocumulus clouds forming from increased convection over each of the three small islands. (C) An aerial view between the patchy cumulus clouds below and the cirrus and cirrostratus above (the patches on the ground are clear-cut forests). (D) Altocumulus. (E) A rain shower at the base of a cumulonimbus. (F) Stratocumulus.

A CLOSER LOOK | Pollution of Earth's Atmosphere

The earth's atmosphere is never "pure" in the sense of being just a mixture of gases, because the atmosphere naturally contains a mixture of suspended particles of dust, soot, and salt crystals. These are not pollutants, however, since they occur naturally as part of the atmosphere. An *air pollutant* is defined as a human-produced foreign substance that reduces visibility, irritates the senses, or is in some way detrimental to humans or their surroundings. The primary source of air pollutants today is the waste products released into the air from the exhaust of (1) internal combustion engines and (2) boilers and furnaces of industries, power plants, and homes.

Engines, boilers, and furnaces do not always produce air pollution just because they are operating. The atmosphere can usually absorb a limited amount of waste products from exhaust gases without detrimentally changing the environment. Whether the waste products become pollutants or not depends on a number of factors, such as the population density, the number of automobiles, the concentration of industries, and the immediate weather conditions. The right combination of population density, automobiles, and industries can produce air pollution under certain weather conditions, such as an inversion. Under other weather conditions the same amount of waste products can be released into the air without producing air pollution. Many people realize, however, that the limit of the atmosphere to absorb waste products is being rapidly reached in many locations. This reading is about what the waste products are, how they can be harmful to people and their environment, and how they can be limited.

All fuels used today are basically made up of hydrocarbon molecules (hydrogen and carbon), mixtures of other chemicals, and impurities. Coal also contains carbon in an uncombined state. If the fuels were just pure hydrocarbons or uncombined carbon burned completely at "ordinary" temperatures, the reaction would yield just water and carbon dioxide with the energy released. Of these products, carbon dioxide can be a pollutant when produced in sufficient quantities. The complete reaction rarely happens, however, because (1) incomplete combustion often occurs, (2) some boilers and furnaces often operate at some very high temperatures, and (3) fuels often contain impurities, some more than others. Incomplete combustion can yield carbon monoxide, unburned particles of carbon, and mixtures of various hydrocarbon chemical by-products. All of these can be toxic, irritating, and vision-obscuring pollutants under the right conditions. High combustion temperatures can eliminate these products of incomplete combustion, but high temperatures produce their own set of possible pollutants. High temperatures can oxidize nitrogen in the air to form different nitrogen oxide compounds with the general formula of NO_x. Compounds with this general formula are called "knox." The subscript x of knox means that several oxides are possible, nitrogen oxide (NO) and nitrogen dioxide (NO_2), for example. The general formula NO_x implies the presence of any or all of the nitrogen oxides. The NO_x compounds are associated with the dirty brown haze of air pollution and are partly responsible for the acid rain problem. So far, there are three separate sets of pollutants that could possibly be produced by (1) complete combustion at normal temperatures, (2) incomplete combustion, and (3) combustion at high temperatures. To each of these possibilities you can add the pollutants produced by the ever-present fuel impurities. Troublesome fuel impurities include (1) nitrogen compounds, which can produce NO_x compounds at lower combustion tempera-

tures, (2) sulfur compounds, which can produce sulfur dioxide (SO_2), another compound implicated in the acid rain problem, and (3) various solid impurities, which can become tiny particles called *particulates*.

After water vapor, carbon dioxide is the second most abundant waste product from burning and combustion of fuels. The use of fossil fuels (petroleum and coal) releases about twenty billion tons of carbon dioxide into the atmosphere each year. The total amount of carbon dioxide has been slowly increasing over the years, apparently from the combined effect of increased fossil fuel consumption and the clearing and burning of the world's tropical rain forests. Scientists are concerned that increases in the amount of carbon dioxide will eventually lead to increases in the average temperatures of the world through the greenhouse effect. Opinions differ about when or if the increased warming will happen and about what the consequences will be if it does happen. Some believe that an average temperature increase will melt some of the polar ice cap, flooding coastal cities and changing the global climate patterns. Others believe that increased carbon dioxide and warmer climate patterns would produce lush vegetation, therefore benefiting humans through increased agricultural production. Still other scientists have the opinion that no predictions can be made because of the number of other variables involved. Increased water vapor and dust in the atmosphere, for example, could reflect more and more sunlight. This could cancel any warming from the greenhouse effect or, in fact, could lead to an overall cooling trend. Scientists do seem to agree on at least two points about carbon dioxide, water vapor, and the other products of combustion. That agreement is that human activity has now reached the

—Continued top of next page

Continued—

level that (1) pollution is becoming an influencing part of the atmosphere and (2) any resulting changes in the earth's climate will last a long time.

Carbon monoxide is an odorless, colorless, and poisonous gas. Exposure to concentrations of 100 parts per million produces symptoms of breathing difficulties and headaches. Exposure to concentrations of 1,000 parts per million can be fatal. Carbon monoxide is the most abundant poisonous pollutant in the air of most large cities. It is mostly produced by automobile engines in need of a tune-up, but all engines produce some carbon monoxide. Even if a tuned engine only produces between 0.01 and 0.1 percent carbon monoxide in its exhaust, too many cars in too little space during certain types of weather and atmospheric conditions can lead to unacceptable levels of carbon monoxide.

Sulfur is released to the atmosphere from natural sources, such as decaying organic matter, and from metal smelting, petroleum refining, and the burning of fuels such as petroleum and coal, which contain sulfur impurities. There are different forms of sulfur oxides, but the most common is sulfur dioxide (SO_2). A study of all sources of SO_2 released in the eastern parts of Canada and the United States found that an average of 85,000 tons of sulfur emissions were released per day. About 60 percent of this came from coal-fired power plants.

Sulfur dioxide gas can injure plants and plant materials. If inhaled in sufficient quantities, it can make breathing difficult because it causes constriction of the air tubes in human lungs. Sulfur dioxide has also been implicated as an acid rain precursor, one of the substances that is later converted to acid by atmospheric processes.

Nitrogen oxides in the atmosphere are oxidized to nitrogen dioxide, which absorbs ultraviolet radiation from the sun. This breaks the compound down into nitric oxide and monatomic oxygen (O). Monatomic oxygen combines with diatomic oxygen (O_2) to produce ozone (O_3). Ozone is strongly reactive and causes the cracking and deterioration of rubber and other materials. In sufficient quantities, it is also a health hazard. Note the seemingly odd situation that ozone is considered beneficial when it is high in the stratosphere, but is considered a pollutant when it is near the earth's surface.

Ozone and nitrogen oxide combine with the hydrocarbons from incomplete fossil fuel combustion in a photochemical reaction. In the presence of sunlight they combine to form peroxyacyl nitrate (PAN), a principal component of smog that irritates eyes and damages mucous membranes. Photochemical smog usually occurs in large cities, but it can occur anywhere that a mixture of hydrocarbons and nitrogen oxides is exposed to sunlight.

Particulates include the larger dust, ash, and smoke particles and the much smaller suspended aerosols. About 80 percent of the particulates in the atmosphere are from windblown dust, sea mist, volcanoes, and other natural sources. Automobiles, incinerators, and a variety of industrial and power plant burning of fuels contribute to the human-added 20 percent. Particulates are of concern because of potential health problems and aesthetics. Smaller particles can lodge in the lower respiratory tract of humans, leading to asthma, emphysema, or lung cancer. In the past, cities were covered with black soot from the particulates released by coal burning. The increased use of cleaner-burning petroleum and the control technologies required with the use of coal have reduced this problem to one of reduced visibility.

The readily definable sources of air pollution, in order of greatest to least emissions, are (1) automobiles, (2) major industries, (3) power plants, and (4) home space heating. The automobile is the prime polluter, contributing almost all of the carbon monoxide and, along with industries, most of the hydrocarbons. Automobiles and power plants contribute most of the nitrogen oxides, and power plants and industries contribute most of the sulfur oxides.

The problem of air pollution is more acute in cities, where most people live. Large numbers of people use the same air, which receives the waste products from automobiles and other sources. Stringent state and federal automobile emission standards have reduced the total automobile emissions other than CO_2 and water vapor. The problem continues, however, because of too many cars in too little space. So, we will continue to hear about "acceptable levels" of air pollution and warnings about dangerous levels when they occur. Many newspaper weather reports include information about carbon monoxide, ozone, and particulates and the "acceptable" or "health hazard" levels. The Clean Air Act of 1971 established tight federal standards on the combustion emissions of industries and power plants. All are required to employ the "latest technology" of pollution controls, no matter what the expense. The cost of pollution control equipment is now often up to one-fourth of the total cost of building a new coal-fired power plant. This increased construction cost, and the associated operational costs, has resulted in a 10 to 15 percent increase in a total monthly electric bill. As new, modern pollution control equipment is developed through research, more options for emission control will become available. The question may change from "What can we do about air pollution?" to "How clean do you want the air . . . and how much are you willing to pay for it?"

FIGURE 21.20

Precipitation is water in the liquid or solid form that returns to the surface of the earth. The precipitation you see here is liquid, and each raindrop is made from billions of the tiny droplets that make up the clouds. The tiny droplets of clouds become precipitation by merging to form larger droplets or by the growth of ice crystals that melt while falling.

seems to form in clouds by one of two processes: (1) the *coalescence of cloud droplets* or (2) the *growth of ice crystals.* It would appear difficult for cloud droplets to merge, or coalesce, with one another since any air movement would seem to move them all at the same time, not bringing them together. Condensation nuclei come in different sizes, however, and cloud droplets of many different sizes form on these different-sized nuclei. Larger cloud droplets are slowed less by air friction as they drift downward, and they collide and merge with smaller droplets as they fall. They may merge, or coalesce, with a million other droplets before they fall from the cloud as raindrops. This *coalescence process* of forming precipitation is thought to take place in warm cumulus clouds that form near the ocean in the tropics. These clouds contain giant salt condensation nuclei and have been observed to produce rain within about twenty minutes after forming.

Clouds at middle latitudes, away from the ocean, also produce precipitation, so there must be a second way that precipitation forms. The *ice-crystal process* of forming precipitation is important in clouds that extend high enough in the atmosphere to be above the freezing point of water. Water molecules are more strongly bonded to each other in an ice crystal than in liquid water. Thus, an ice crystal can capture water molecules and grow to a larger size while neighboring water droplets are evaporating. As they grow larger and begin to drift toward the surface, they may coalesce with other ice crystals or droplets of water, soon falling from the cloud. During the summer, they fall through warmer air below and reach the ground as raindrops. During the winter, they fall through cooler air below and reach the ground as snow.

Tiny water droplets do not freeze as readily as a larger mass of liquid water, and many droplets do not freeze until the temperature is below about $-40°C$ ($-40°F$). Water that is still in the liquid state when the temperature is below the freezing temperature is said to be *supercooled.* Supercooled clouds of water droplets are common between the temperatures of $-40°C$ and $0°C$ ($-40°F$

and $32°F$), a range of temperatures that is often found in the upper atmosphere. The liquid droplets at these temperatures need solid particles called **ice-forming nuclei** to freeze upon. Generally, dust from the ground serves as ice-forming nuclei that start the ice-crystal process of forming precipitation. Artificial rainmaking has been successful by (1) dropping crushed dry ice, which is cooler than $-40°C$, on top of a supercooled cloud and (2) by introducing "seeds" of ice-forming nuclei in supercooled clouds. Tiny crystals from the burning of silver iodide are effective ice-forming nuclei, producing ice crystals at temperatures as high as $-4.0°C$ (about $25°F$). Attempts at ground-based cloud seeding with silver iodide in the mountains of the western United States have suggested up to 15 percent more snowfall, but it is difficult to know how much snowfall would have resulted without the seeding.

In general, the basic form of a cloud has meaning about the general type of precipitation that can occur as well as the coming weather. Cumulus clouds usually produce showers or thunderstorms that last only brief periods of time. Longer periods of drizzle, rain, or snow usually occur from stratus clouds. Cirrus clouds do not produce precipitation of any kind, but they may have meaning about the coming weather, which is discussed in the following section.

WEATHER PRODUCERS

The general circulation of the earth's atmosphere is useful as a model to help you understand why the weather changes as it does on the earth. The general circulation of the atmosphere described is a simplified, idealized model based on averages of a featureless, uniform earth. A more detailed model would require an application of physics that cannot be developed in a short chapter in a science textbook. Also, note that the winds and the belts of high and low pressure described in this model are generalized, average values that are different from the everyday changes of winds and pressures that are shown on daily weather maps.

The idealized model of the general atmospheric circulation starts with the poleward movement of warm air from the tropics. The region between 10°N and 10°S of the equator receives more direct radiation, on the average, than other regions of the earth's surface. The air over this region is heated more, expands, and becomes less dense as a consequence of the heating. This less dense air is buoyed up by convection to heights up to 20 km (about 12 mi) as it is cooled by radiation to less than $-73°C$ (about $-110°F$). This accumulating mass of cool, dry air spreads north and south toward both poles (see figure 21.13), then sinks back toward the surface at about 30°N and 30°S. The descending air is warmed by compression and is warm and dry by the time it reaches the surface. Part of the sinking air then moves back toward the equator across the surface, completing a large convective cell. This giant cell has a low-pressure belt over the equator and high-pressure belts over the subtropics near latitudes of 30°N and 30°S. The other part of the sinking air moves poleward across the surface, producing belts of westerly winds in both hemispheres to latitudes of about 60°. On an earth without landmasses next to bodies of water, a belt of low pressure would probably form around 60° in both hemispheres, and a high-pressure region would form at both poles.

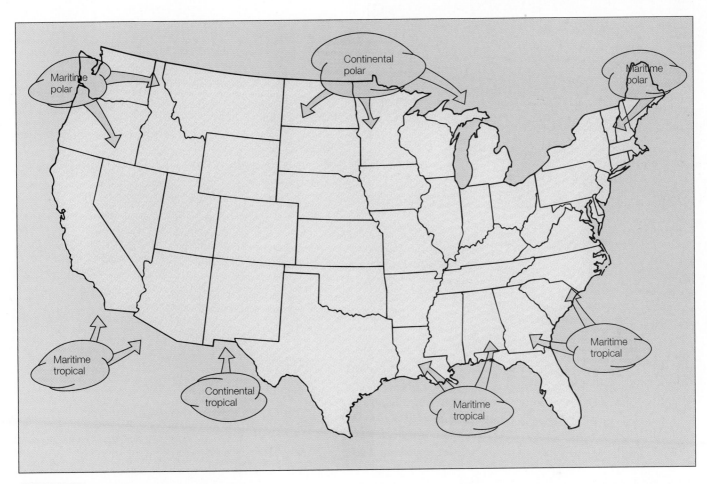

FIGURE 21.21

The general movement of the four main types of air masses that influence the weather over the contiguous United States. The tropical air masses visit most often in the summer, and the polar air masses visit most often during the winter. During other times, the polar and tropical air masses battle back and forth over the land.

The overall pattern of pressure belts and belts of prevailing winds is seen to shift north and south with the seasons, resulting in a seasonal shift in the types of weather experienced at a location. This shift of weather is related to three related weather producers: (1) the movement of large bodies of air, called *air masses,* that have acquired the temperature and moisture conditions where they have been located, (2) the leading *fronts* of air masses when they move, and (3) the local *high- and low-pressure* patterns that are associated with air masses and fronts. These are the features shown on almost all daily weather maps, and they are the topics of this section.

Air Masses

An **air mass** is defined as a large, more or less uniform body of air with nearly the same temperature and moisture conditions. An air mass forms when a large body of air, perhaps covering millions of square kilometers, remains over a large area of land or water for an extended period of time. While it is stationary it acquires the temperature and moisture characteristics of the land or water through the heat transfer processes of conduction, convection, and radiation and through the moisture transfer processes of

evaporation and condensation. For example, a large body of air that remains over the cold, dry, snow-covered surface of Siberia for some time will become cold and dry. A large body of air that remains over a warm tropical ocean, on the other hand, will become warm and moist. Knowledge about the condition of air masses is important because they tend to retain the acquired temperature and moisture characteristics when they finally break away, sometimes moving long distances. An air mass that formed over Siberia can bring cold, dry air to your location, while an air mass that formed over a tropical ocean will bring warm, moist air.

Air masses are classified according to the temperature and moisture conditions where they originate. There are two temperature extreme possibilities, a *polar air mass* from a cold region and a *tropical air mass* from a warm region. There are also two moisture extreme possibilities, a moist *maritime air mass* from over the ocean and a generally dry *continental air mass* from over the land. Thus, there are four main types of air masses that can influence the weather where you live: (1) continental polar, (2) maritime polar, (3) continental tropical, and (4) maritime tropical. Figure 21.21 shows the general direction in which these air masses usually move over the mainland United States.

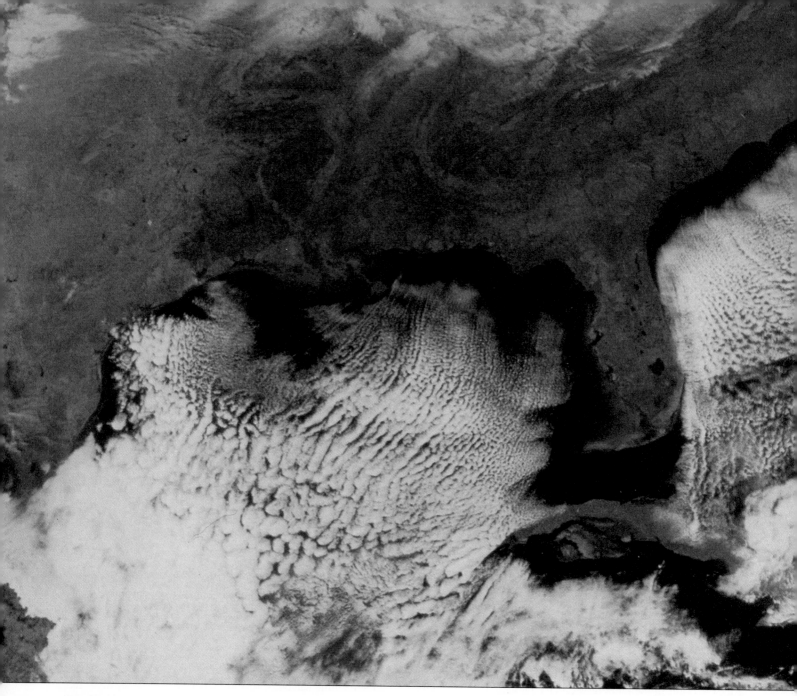

FIGURE 21.22

This satellite photograph shows the result of a polar air mass moving southeast over the southern United States. Clouds form over the warmer waters of the Gulf of Mexico and the Atlantic Ocean, showing the state of atmospheric instability from the temperature differences.

Once an air mass leaves its source region it can move at speeds of up to 800 km (about 500 mi) per day while mostly retaining the temperature and moisture characteristics of the source region (figure 21.22). If it slows and stagnates over a new location, however, the air may again begin to acquire a new temperature and moisture equilibrium with the surface. When a location is under the influence of an air mass, the location is having a period of *air mass weather.* This means that the weather conditions will generally remain the same from day to day with slow, gradual changes. Air mass weather will remain the same until a new air mass moves in or until the air mass acquires the conditions of the new location. This process may take days or

several weeks, and the weather conditions during this time depend on the conditions of the air mass and conditions at the new location. For example, a polar continental air mass arriving over a cool, dry land area may produce a temperature inversion with the air colder near the surface than higher up. When the temperature increases with height the air is stable and cloudless, and cold weather continues with slow, gradual warming. The temperature inversion may also result in hazy periods of air pollution in some locations. A continental air mass arriving over a generally warmer land area, on the other hand, results in a condition of instability. In this situation each day will start clear and cold, but differential heating during the day develops cumulus

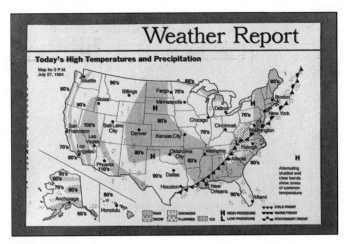

FIGURE 21.23

This weather map of the United States shows a cold front running from Houston, Texas to near Raleigh, North Carolina, where it becomes a stationary front that runs in a northeasterly direction. Note the areas of showers and the temperature predictions.

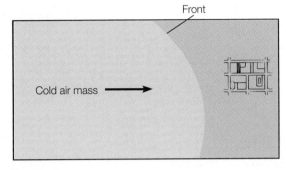

FIGURE 21.24

(A) A cold air mass is similar to a huge, flattened bubble of cold air that moves across the land. The front is the boundary between two air masses, a narrow transition zone of mixing. (B) A front is represented by a line on a weather map, which shows the location of the front at ground level.

clouds in the unstable air. After sunset the clouds evaporate, and a clear night results because the thermals during the day carried away the dust and air pollution. Thus, a dry, cold air mass can bring different weather conditions, each depending on the properties of the air mass and the land it moves over.

Weather Fronts

The boundary between air masses of different temperatures is called a **front.** A front is actually a thin transition zone between two air masses that ranges from about 5 to 30 km thick (about 3 to 20 mi), and the air masses do not mix other than in this narrow zone. The density differences between the two air masses prevent any general mixing since the warm, less-dense air mass is forced upward by the cooler, more dense air moving under it. You may have noticed on a daily weather map that fronts are usually represented with a line bulging outward in the direction of cold air mass movement (figure 21.23). A cold air mass is much like a huge, flattened bubble of air that moves across the land (figure 21.24). The line on a weather map represents the place where the leading edge of this huge, flattened bubble of air touches the surface of the earth.

A **cold front** is formed when a cold air mass moves into warmer air, displacing it in the process. A cold front is generally steep, and when it runs into the warmer air it forces it to rise quickly. If the warm air is moist, it is quickly cooled adiabatically to the dew point temperature, resulting in large, towering cumulus clouds and thunderclouds along the front (figure 21.25). You may have observed that thunderstorms created by an advancing cold front often form in a line along the front. These thunderstorms can be intense but are usually over quickly, soon followed by a rapid drop in temperature from the cold air mass moving past your location. The passage of the cold front is also marked by a rapid shift in the wind direction and a rapid increase in the barometric pressure. Before the cold front arrives, the wind is generally moving toward

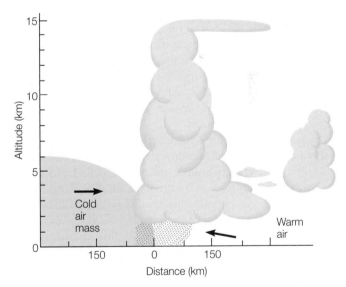

FIGURE 21.25

An idealized cold front, showing the types of clouds that might occur when an unstable cold air mass moves through unstable warm air. Stable air would result in more stratus clouds rather than cumulus clouds.

the front as warm, less dense air is forced upward by the cold, more dense air. The lowest barometric pressure reading is associated with the lifting of the warm air at the front. After the front passes your location, you are in the cooler, more dense air that is settling outward, so the barometric pressure increases and the wind shifts with the movement of the cold air mass.

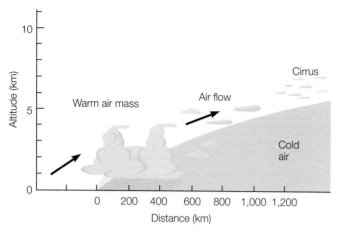

FIGURE 21.26

An idealized warm front, showing a warm air mass overriding and pushing cold air in front of it. Notice that the overriding warm air produces a predictable sequence of clouds far in advance of the moving front.

A **warm front** forms when a warm air mass advances over a mass of cooler air. Since the advancing warm air is less dense than the cooler air it is displacing, it generally overrides the cooler air, forming a long, gently sloping front. Because of this, the overriding warm air may form clouds far in advance of the ground-level base of the front (figure 21.26). This may produce high cirrus clouds a day or more in advance of the front, which are followed by thicker and lower stratus clouds as the front advances. Usually these clouds result in a broad band of drizzle, fog, and the continuous light rain usually associated with stratus clouds. This light rain (and snow in the winter) may last for days as the warm front passes.

Sometimes the forces influencing the movement of a cold or warm air mass lessen or become balanced, and the front stops advancing. When this happens a stream of cold air moves along the north side of the front, and a stream of warm air moves along the south side in an opposite direction. This is called a **stationary front** because the edge of the front is not advancing. A stationary front may sound as if it is a mild frontal weather maker because it is not moving. Actually, a stationary front represents an unstable situation that can result in a major atmospheric storm. This type of storm is discussed in the following section.

Waves and Cyclones

A slowly advancing cold front and a stationary front often develop a bulge, or *wave*, in the boundary between cool and warm air moving in opposite directions (figure 21.27B). The wave grows as the moving air is deflected, forming a warm front moving northward on the right side and a cold front moving southward on the left side. Cold air is more dense than warm air, and the cold air moves faster than the slowly moving warm front. As the faster moving cold air catches up with the slower moving warm air, the cold air underrides the warm air, lifting it upward. This lifting action produces a low-pressure area at the point where the two fronts come together (fig-

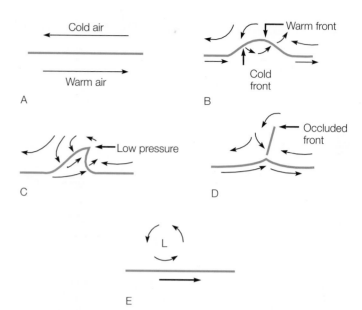

FIGURE 21.27

The development of a low-pressure center, or cyclonic storm, along a stationary front as seen from above. (*A*) A stationary front with cold air on the north side and warm air on the south side. (*B*) A wave develops, producing a warm front moving northward on the right side and a cold front moving southward on the left side. (*C*) The cold front lifts the warm front off the surface at the apex, forming a low-pressure center. (*D*) When the warm front is completely lifted off the surface, an occluded front is formed. (*E*) The cyclonic storm is now a fully developed low-pressure center.

ure 21.27C). The lifted air expands, cools adiabatically, and reaches the dew point. Clouds form and precipitation begins from the lifting and cooling action. Within days after the wave first appears the cold front completely overtakes the warm front, forming an occlusion (figure 21.27D). An **occluded front** is one that has been lifted completely off the ground into the atmosphere. The disturbance is now a *cyclonic storm* with a fully developed low-pressure center. Since its formation, this low-pressure cyclonic storm has been moving, taking its associated stormy weather with it in a generally easterly direction. Such cyclonic storms usually follow principal tracks along a front. Since they are observed generally to follow these same tracks, it is possible to predict where the storm might move next.

A *cyclone* is defined as a low-pressure center where the winds move into the low-pressure center and are forced upward. As air moves in toward the center, the Coriolis effect and friction with the ground cause the moving air to veer to the right of the direction of motion. In the Northern Hemisphere this veering of moving air produces a counterclockwise circulation pattern of winds around the low-pressure center (figure 21.28). The upward movement associated with the low-pressure center of a cyclone cools the air adiabatically, resulting in clouds, precipitation, and stormy conditions.

Air is sinking in the center of a region of high pressure, producing winds that move outward. In the Northern Hemisphere, the Coriolis effect and frictional forces deflect this wind to the

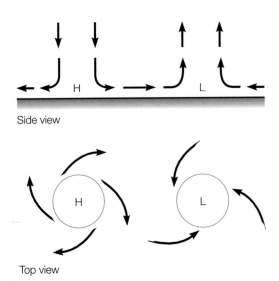

FIGURE 21.28

Air sinks over a high-pressure center and moves away from the center on the surface, veering to the right in the Northern Hemisphere to create a clockwise circulation pattern. Air moves toward a low-pressure center on the surface, rising over the center. As air moves toward the low-pressure center on the surface, it veers to the left in the Northern Hemisphere to create a counterclockwise circulation pattern.

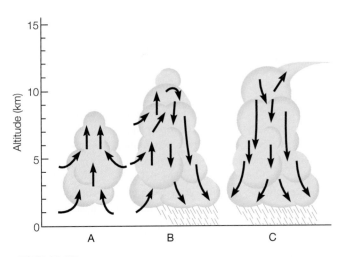

FIGURE 21.29

Three stages in the life of a thunderstorm cell. (A) The cumulus stage begins as warm, moist air is lifted in an unstable atmosphere. All the air movement is upward in this stage. (B) The mature stage begins when precipitation reaches the ground. This stage has updrafts and downdrafts side by side, which create violent turbulence. (C) The final stage begins when all the updrafts have been cut off, and only downdrafts exist. This cuts off the supply of moisture, and the rain decreases as the thunderstorm dissipates. The anvil-shaped top is a characteristic sign of this stage.

right, producing a clockwise circulation (figure 21.28). A high-pressure center is called an *anticyclone,* or simply a *high.* Since air in a high-pressure zone sinks, it is warmed adiabatically, and the relative humidity is lowered. Thus, clear, fair weather is usually associated with a high. By observing the barometric pressure, you can watch for decreasing pressure, which can mean the coming of a cyclone and its associated stormy weather. You can also watch for increasing pressure, which means a high and its associated fair weather are coming. Consulting a daily weather map makes such projections a much easier job, however.

Major Storms

A wide range of weather changes can take place as a front passes, because there is a wide range of possible temperature, moisture, stability, and other conditions between the new air mass and the air mass that it is displacing. The changes that accompany some fronts may be so mild that they go unnoticed. Others are noticed only as a day with breezes or gusty winds. Still other fronts are accompanied by a rapid and violent weather change called a **storm.** A snowstorm, for example, is a rapid weather change that may happen as a cyclonic storm moves over a location. The most rapid and violent changes occur with three kinds of major storms: (1) thunderstorms, (2) tornadoes, and (3) hurricanes.

Thunderstorms

A **thunderstorm** is a brief but intense storm with rain, lightning and thunder, gusty and often strong winds, and sometimes hail. Thunderstorms usually develop in warm, very moist, and unsta-

ble air. These conditions set the stage for a thunderstorm to develop when something lifts a parcel of air, starting it moving upward. This is usually accomplished by the same three general causes that produce cumulus clouds: (1) differential heating, (2) mountain barriers, or (3) along an occluded or cold front. Thunderstorms that occur from differential heating usually occur during warm, humid afternoons after the sun has had time to establish convective thermals. In the Northern Hemisphere most of these convective thunderstorms occur during the month of July. Frontal thunderstorms, on the other hand, can occur any month and any time of the day or night that a front moves through warm, moist, and unstable air.

Frontal thunderstorms generally move with the front that produced them. Thunderstorms that developed in mountains or over flat lands from differential heating can move miles after they form, sometimes appearing to wander aimlessly across the land. These storms are not just one big rain cloud but are sometimes made up of cells that are born, grow to maturity, then die out in less than an hour. The thunderstorm, however, may last longer than an hour because new cells are formed as old ones die out. Each cell is about 2 to 8 km (about 1 to 5 mi) in diameter and goes through three main stages in its life: (1) cumulus, (2) mature, and (3) final (figure 21.29).

Damage from a thunderstorm is usually caused by the associated lightning, strong winds, or hail. As illustrated in figure 21.29, the first stage of a thunderstorm begins as convection, mountains, or a dense air mass slightly lifts a mass of warm, moist air in an unstable atmosphere. The lifted air mass expands and cools adiabatically to the dew point temperature, and a

A CLOSER LOOK | Hydroelectric Energy Sources

ossil fuels—oil, coal, and natural gas— supply about 90 percent of world's supply of energy today. Fossil fuels were formed over millions of years ago as organisms were slowly transformed into the carbon-rich substances of today. Fossil fuels are being used much faster than they can form, so they are understood to be nonrenewable energy sources.

Renewable energy sources are fuel resources that are quickly replaced by natural processes. Renewable energy sources will not run out, as the fossil fuels must. Renewable sources include solar energy, and the technologies that use solar cells, power towers, biomass, and wind energy, among others. Geothermal and tidal energy round out the renewable alternative energy sources.

Flowing water is another energy source that is renewable as long as the rain continues to fall. At first, flowing water was used as an energy source to grind grain and run machinery and this meant the flour mill and the textile industry had to be located near a large source of running water. Then hydroelectric power plants were built, and this had a major impact on everyday life and helped spur industrial development. Now wires could carry the electrical energy far from the hydroelectric source and people could live and work anywhere. Today, 2,400 hydroelectric power plants produce a significant amount of energy, providing 9 percent of the national electrical needs and about 49 percent of all the renewable energy used in the United States. The nation's largest hydroelectric power plant is the 10,080-megawatt capacity Grand Coulee plant on the Columbia River in the state of Washington.

Hydroelectric power plants are usually located on artificial reservoirs, and the higher the dam the greater the amount of electricity that can be produced. Penstocks carry water from the reservoir to turbines inside the powerhouse, which drives generators that produce electricity. Larger dams with greater water flow will have more penstocks, turbines, and generators and will produce more power. It is also possible to construct a small hydroelectric plant on streams without the use of stored water as long as the stream does not experience seasonal water-flow changes.

Hydroelectric power plants are renewable sources of energy that produce no pollution. However, the construction of a reservoir can cause environmental damage with the destruction of the natural ecosystem, loss of fertile farmland, and relocation of communities. Without considering potential environmental damage, damage to scenic areas, and other concerns, it is estimated that only about 20 percent of the potential U.S. hydroelectric power plants have been developed. The remaining sites are unsuitable for new hydroelectric power plants because of environmental concerns. Instead of building new plants, existing dams can be retrofitted with advanced technology for increased energy production.

cumulus cloud forms. The latent heat of vaporization released by the condensation process accelerates the upward air motion, called an *updraft*, and the cumulus cloud continues to grow to towering heights. Soon the upward-moving, saturated air reaches the freezing level and ice crystals and snowflakes begin to form. When they become too large to be supported by the updraft, they begin to fall toward the surface, melting into raindrops in the warmer air they fall through. When they reach the surface, this marks the beginning of the mature stage. As the raindrops fall through the air, friction between the falling drops and the cool air produces a downdraft in the region of the precipitation. The cool air accelerates toward the surface at speeds up to 90 km/hr (about 55 mi/hr), spreading out on the ground when it reaches the surface. In regions where dust is raised by the winds, this spreading mass of cold air from the thunderstorm has the appearance of a small cold front with a steep, bulging leading edge. This miniature cold front may play a role in lifting other masses of warm, moist air in front of the thunderstorm, leading to the development of new cells. This stage in the life of a thunderstorm has the most intense rainfall, winds, and possibly hail. As the downdraft spreads throughout the cloud, the supply of new moisture from the updrafts is cut off and the thunderstorm enters the final, dissipating stage. The entire life

cycle, from cumulus cloud to the final stage, lasts for about an hour as the thunderstorm moves across the surface. During the mature stage of powerful updrafts, the top of the thunderstorm may reach all the way to the top of the troposphere, forming a cirrus cloud that is spread into an anvil shape by the strong winds at this high altitude.

The updrafts, downdrafts, and falling precipitation separate tremendous amounts of electric charges that accumulate in different parts of the thundercloud. Large drops of water tend to carry negative charges, and cloud droplets tend to lose them. The upper part of the thunderstorm develops an accumulation of positive charges as cloud droplets are uplifted, and the middle portion develops an accumulation of negative charges from larger drops that fall. There are many other charging processes at work, such as induction (see chapter 7), and the lower part of the thundercloud develops both negative and positive charges. The voltage of these charge centers builds to the point that the electrical insulating ability of the air between them is overcome and a giant electrical discharge called *lightning* occurs (figure 21.30). Lightning discharges occur from the cloud to the ground, from the ground to a cloud, from one part of the cloud to another part, or between two different clouds. The discharge takes place in a fraction of a second and may actually consist of

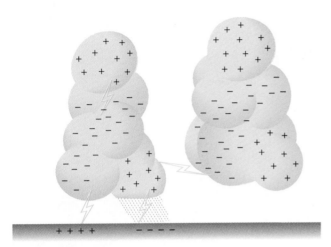

FIGURE 21.30

Different parts of a thunderstorm cloud develop centers of electric charge. Lightning is a giant electric spark that discharges the accumulated charges.

FIGURE 21.31

These hailstones fell from a thunderstorm in Iowa, damaging automobiles, structures, and crops.

© Telegraph Herald/Photo by Patti Carr

a number of strokes rather than one big discharge. The discharge produces an extremely high temperature around the channel, which may be only 6 cm or so wide (about 2 in). The air it travels through is heated quickly, expanding into a sudden pressure wave that you hear as *thunder.* A nearby lightning strike produces a single, loud crack. Farther away strikes sound more like a rumbling boom as the sound from the separate strokes become separated over distance. Echoing of the thunder produced at farther distances also adds to the rumbling sounds. The technique of estimating the distance to a lightning stroke by measuring the interval between the flash of the lightning and the boom of the thunder is discussed in chapter 6. Lightning can present a risk for people in the open, near bodies of water, or under a single, isolated tree during a thunderstorm. The safest place to be during a thunderstorm is inside a car or a building with a metal frame.

Updrafts are also responsible for **hail,** a frozen form of precipitation that can be very destructive to crops, automobiles, and other property. Hailstones can be irregular, somewhat spherical, or flattened forms of ice that range from the size of a BB to the size of a softball (figure 21.31). Most hailstones, however, are less than 2 cm (about 1 in) in diameter. The larger hailstones have alternating layers of clear and opaque, cloudy ice. These layers are believed to form as the hailstone goes through cycles of falling then being returned to the upper parts of the thundercloud by updrafts. The clear layers are believed to form as the hailstone moves through heavy layers of supercooled water droplets, which accumulate quickly on the hailstone but freeze slowly because of the release of the latent heat of fusion. The cloudy layers are believed to form as the hailstone accumulates snow crystals or moves through a part of the cloud with less supercooled water droplets. In either case, rapid freezing traps air bubbles, which result in the opaque, cloudy layer. Thunderstorms with hail are most common during the month of May in the states of Colorado, Kansas, and Nebraska.

Tornadoes

A **tornado** is the smallest, most violent weather disturbance that occurs on the earth (figure 21.32). Tornadoes occur with intense thunderstorms, resembling a long, narrow funnel or ropelike structure that drops down from a thundercloud and may or may not touch the ground. This ropelike structure is a rapidly whirling column of air, usually 100 to 400 m (about 330 to 1,300 ft) in diameter. The bottom of the column moves across the surface, sometimes skipping into the air, then back down again at speeds that average about 50 km/hr (about 30 mi/hr). The speed of the whirling air in the column has been estimated to be well over 300 km/hr (about 200 mi/hr), leaving a path of destruction wherever the column touches the surface. The destruction is produced by the powerful winds, the sudden drop in atmospheric pressure that occurs at the center of the funnel, and the debris that is flung through the air like projectiles. A passing tornado sounds like very loud, continuous rumbling thunder with cracking and hissing noises that are punctuated by the crashing of debris projectiles.

On the average, several hundred tornadoes are reported in the United States every year. These occur mostly during spring and early summer afternoons over the Great Plains states. The states of Texas, Oklahoma, Kansas, and Iowa have such a high occurrence of tornadoes that the region is called "tornado alley." During the spring and early summer, this region has maritime tropical air from the Gulf of Mexico at the surface. Above this warm, moist layer is a layer of dry, unstable air that has just crossed the Rocky Mountains, moved along rapidly by the jet stream. The stage is now set for some event, such as a cold air mass moving in from the north, to shove the warm, moist air upward, and the result will be violent thunderstorms with tornadoes.

FIGURE 21.32

A tornado might be small, but it is the most violent storm that occurs on the earth. This tornado, moving across an open road, eventually struck Dallas, Texas.

FIGURE 21.33

This is a satellite photo of hurricane John, showing the eye and counterclockwise motion.

Hurricanes

A **hurricane** is a large, violent circular storm that is born over the warm, tropical ocean near the equator. A hurricane is a *tropical cyclone* with winds that exceed 120 km/hr (75 mi/hr). It is a cyclone because it is a low-pressure center with counterclockwise winds in the Northern Hemisphere (figure 21.33). It is called a tropical cyclone because it is born in the tropics. The same type of tropical cyclone in the western Pacific is called a **typhoon.** Both typhoons and hurricanes, however, are tropical cyclones. The wind speed of 120 km/hr (75 mi/hr) is the defined difference between a tropical storm and a tropical cyclone. A storm with winds less than this defined value is a tropical storm. A storm with winds at or above 120 km/hr is a tropical cyclone, which is called a hurricane in some places and a typhoon in other places.

A tropical cyclone is similar to the wave cyclone of the mid-latitudes because both have low-pressure centers with a counter-clockwise circulation in the Northern Hemisphere. They are different because a wave cyclone is usually about 2,500 km (about 1,500 mi) wide, has moderate winds, and receives its energy from the temperature differences between two air masses. A tropical cyclone, on the other hand, is often less than 200 km (about 125 mi) wide, has very strong winds, and receives its energy from the latent heat of vaporization released during condensation.

A fully developed hurricane has heavy bands of clouds, showers, and thunderstorms that rapidly rotate around a relatively clear, calm eye (figure 21.34). As a hurricane approaches a location the air seems unusually calm as a few clouds appear, then thicken as the wind begins to gust. Over the next six hours or so the overall wind speed increases as strong gusts and intense rain showers occur. Thunderstorms, perhaps with tornadoes, and the strongest winds occur just before the winds suddenly die down and the sky clears with the arrival of the eye of the hurricane. The eye is an average of 10 to 15 km (about 6 to 9 mi) across, and it takes about an hour or so to cross a location. When the eye passes the intense rain showers, thunderstorms, and hurricane-speed winds begin again, this time blowing from the opposite direction. The whole sequence of events may be over in a day or two, but hurricanes are unpredictable and sometimes stall in one location for days. In general, they move at a rate of 15 to 50 km/hr (about 10 to 30 mi/hr).

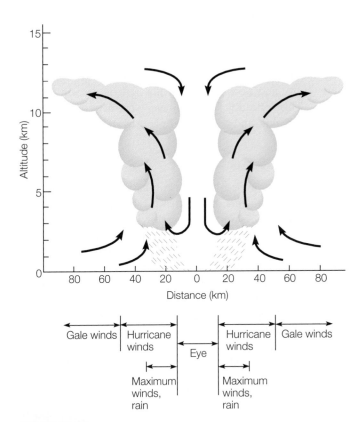

FIGURE 21.34

Cross section of a hurricane.

Most of the damage from hurricanes results from strong winds, flooding, and the occasional tornado. Flooding occurs from the intense, heavy rainfall but also from the increased sea level that results from the strong, constant winds blowing seawater toward the shore. The sea level can be raised some 5 m (about 16 ft) above normal, with storm waves up to 15 m (about 50 ft) high on top of this elevated sea level. Overall, large inland areas can be flooded with extensive property damage. A single hurricane moving into a populated coastal region has caused billions of dollars of damage and the loss of hundreds of lives in the past. Today, the National Weather Service tracks hurricanes by weather satellites. Warnings of hurricanes, tornadoes, and severe thunderstorms are broadcast locally over special weather alert stations located across the country.

WEATHER FORECASTING

Today, weather predictions are based on information about the characteristics, location, and rate of movement of air masses and associated fronts and pressure systems. This information is summarized as average values, then fed into a computer model of the atmosphere. The model is a scaled-down replica of the real atmosphere, and changes in one part of the model result in changes in another part of the model just as they do in the real atmosphere. Underlying the computer model are the basic scientific laws concerning solar radiation, heat, motion, and the gas laws. All these laws are written as a series of mathematical equations, which are

FIGURE 21.35

Supercomputers make routine weather forecasts possible by solving mathematical equations that describe changes in a mathematical model of the atmosphere. This "fish-eye" view was necessary to show all of this Cray supercomputer at CERN, the European Center of Particle Physics.

applied to thousands of data points in a three-dimensional grid that represents the atmosphere. The computer is given instructions about the starting conditions at each data point, that is, the average values of temperature, atmospheric pressure, humidity, wind speed, and so forth. The computer is then instructed to calculate the changes that will take place at each data point, according to the scientific laws, within a very short period of time. This requires billions of mathematical calculations when the program is run on a worldwide basis. The new calculated values are then used to start the process all over again, and it is repeated some 150 times to obtain a one-day forecast (figure 21.35).

A problem with the computer model of the atmosphere is that small-scale events are inadequately treated, and this introduces errors that grow when predictions are attempted for farther and farther into the future. Small eddies of air, for example, or gusts of wind in a region have an impact on larger-scale atmospheric motions such as those larger than a cumulus cloud. But all of the small eddies and gusts cannot be observed without filling the atmosphere with measuring instruments. This lack of ability to observe small events that can change the large-scale events introduces uncertainties in the data, which, over time, will increasingly affect the validity of a forecast.

To find information about the accuracy of a forecast the computer model can be run several different times, with each run having slightly different initial conditions. If the results of all the runs are close to each other the forecasters can feel confident that the atmosphere is in a predictable condition, and this means the forecast is probably accurate. In addition, multiple computer runs can provide forecasts in the form of probabilities. For example, if 8 out of 10 forecasts indicate rain, the "averaged" forecast might call for an 80 percent chance of rain.

FIGURE 21.36

The climate determines what types of plants and animals live in a location, the types of houses that people build, and the lifestyles of people. This orange tree, for example, requires a climate that is relatively frost-free, yet it requires some cool winter nights to produce a sweet fruit.

The use of new computer technology has improved the accuracy of next-day forecasts tremendously, and the forecasts up to three days are fairly accurate, too. For forecasts of more than five days, however, the number of calculations and the effect of uncertainties increase greatly. It has been estimated that the reductions of observational errors could increase the range of accurate forecasting up to two weeks. The ultimate range of accurate forecasting will require a better understanding—and thus an improved model—of patterns of changes that occur in the ocean as well as in the atmosphere. All of this increased understanding and reduction of errors leads to an estimated ultimate future forecast of three weeks, beyond which any pinpoint forecast would be only slightly better than a wild guess. In the meantime, regional and local daily weather forecasts are fairly accurate, and computer models of the atmosphere now provide the basis for extending the forecasts for up to about a week.

CLIMATE

Changes in the atmospheric condition over a brief period of time, such as a day or a week, are referred to as changes in the *weather.* Weather changes follow a yearly pattern of seasons that are referred to as winter weather, summer weather, and so on. All of these changes are part of a composite, larger pattern called **climate.** Climate is the general pattern of the weather that occurs for a region over a number of years. Among other things, the climate determines what types of vegetation grow in a particular region, resulting in characteristic groups of plants associated with the region (figure 21.36). For example, orange, grapefruit, and palm trees grow in a region that has a climate with warm monthly temperatures throughout the year. On the other hand, blueberries, aspen, and birch trees grow in a region that has cool temperature patterns throughout the year. Climate determines

what types of plants and animals live in a location, the types of houses that people build, and the lifestyles of people. Climate also influences the processes that shape the landscape, the type of soils that form, the suitability of the region for different types of agriculture, and how productive the agriculture will be in a region. This section is about climate, what determines the climate of a region, and how climate patterns are classified.

Major Climate Groups

Climate is determined by the same basic three elements that determine the weather, that is, temperature, moisture, and the movement of air. As you learned earlier, incoming solar radiation determines the air temperature, provides the energy that drives the hydrologic cycle, and is the ultimate cause of nearly all motion of the atmosphere. It is the uneven distribution of the incoming solar radiation that results in the great variety of temperature conditions, moisture patterns, and general circulation of the atmosphere at different latitudes of the earth. Since the climate varies so much from one location to the next, even at the same latitude, it is evident that other climatic influences are also at work. These climatic influences will be considered after first looking at the most important factor in different climates, incoming solar radiation.

The earth's atmosphere is heated directly by incoming solar radiation and by absorption of infrared radiation from the surface. The amount of heating at any particular latitude on the surface depends primarily on two factors: (1) the *intensity* of the incoming solar radiation, which is determined by the angle at which the radiation strikes the surface, and (2) the *time* that the radiation is received at the surface, that is, the number of daylight hours compared to the number of hours of night.

The earth is so far from the sun that all rays of incoming solar radiation reaching the earth are essentially parallel. The earth, however, has a mostly constant orientation of its axis with respect to the stars as it moves around the sun in its orbit. Since the inclined axis points toward the sun part of the year and away from the sun the other part, radiation reaches different latitudes at different angles during different parts of the year. The orientation of the earth's axis to the sun during different parts of the year also results in days and nights of nearly equal length in the equatorial region but increasing differences at increasing latitudes to the poles. During the polar winter months, the night is twenty-four hours long, which means no solar radiation is received at all. The equatorial region receives more solar radiation during a year, and the amount received decreases toward the poles as a result of (1) yearly changes in intensity and (2) yearly changes in the number of daylight hours (see chapter 18).

In order to generalize about the amount of radiation received at different latitudes, some means of organizing, or grouping, the latitudes is needed (figure 21.37). For this purpose the latitudes are organized into three groups: (1) the *low latitudes,* those that some time of the year receive *vertical* solar radiation at noon, (2) the *high latitudes,* those that some time of the year receive *no* solar radiation at noon, and (3) the *middle latitudes,* which are between the low and high latitudes. This definition of low, middle, and high latitudes means that the low latitudes are between the tropics of

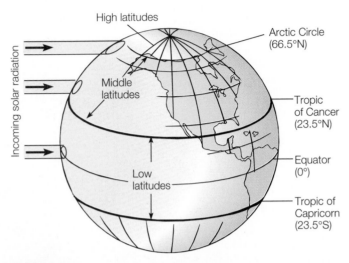

FIGURE 21.37

Latitude groups based on incoming solar radiation. The low latitudes receive vertical solar radiation at noon some time of the year, the high latitudes receive no solar radiation at noon during some time of the year, and the middle latitudes are in between.

Cancer and Capricorn (between 23 1/2°N and 23 1/2°S latitudes) and that the high latitudes are above the Arctic and Antarctic circles (above 66 1/2°N and above 66 1/2°S latitudes).

In general, (1) the low latitudes receive a high amount of incoming solar radiation that varies little during a year. Temperatures are high throughout the year, varying little from month to month. (2) The middle latitudes receive a higher amount of incoming radiation during one part of the year and a lower amount during the other part. Overall temperatures are cooler than in the low latitudes and have a wide seasonal variation. (3) The high latitudes receive a maximum amount of radiation during one part of the year and none during another part. Overall temperatures are low, with the highest range of annual temperatures.

The low, middle, and high latitudes provide a basic framework for describing the earth's climates. These climates are associated with the low, middle, and high latitudes illustrated in figure 21.37, but they are defined in terms of yearly temperature averages. It is necessary to define the basic climates in terms of temperature because land and water surfaces react differently to incoming solar radiation, creating a different temperature. Temperature and moisture are the two most important climate factors, and temperature will be considered first.

The principal climate zones are defined in terms of yearly temperature averages, which occur in broad regions (figure 21.38). They are (1) the *tropical climate zone* of the low latitudes

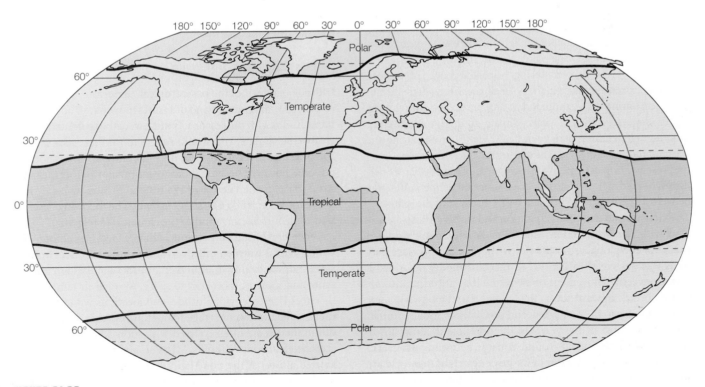

FIGURE 21.38

The principal climate zones are defined in terms of yearly temperature averages, which are determined by the amount of solar radiation received at the different latitude groups.

FIGURE 21.39

A wide variety of plant life can grow in a tropical climate, as you can see here.

FIGURE 21.40

Polar climates occur at high elevations as well as high latitudes. This mountain location has a highland polar climate and tundra vegetation, but little else.

(figure 21.39), (2) the *polar climate zone* of the high latitudes (figure 21.40), and (3) the *temperate climate zone* of the middle latitudes (figure 21.41). The tropical climate zone is near the equator and receives the greatest amount of sunlight throughout the year. Overall, the tropical climate zone is hot. Average monthly temperatures stay above 18°C (64°F), even during the coldest month of the year. The other extreme is found in the polar climate zone, where the sun never sets during some summer days and never rises during some winter days. Overall, the polar climate zone is cold. Average monthly temperatures stay below 10°C (50°F), even during the warmest month of the year. The temperate climate zone is between the polar and tropical zones, with average temperatures that are neither very cold nor very hot. Average monthly temperatures stay between 10°C and 18°C (50°F and 64°F) throughout the year.

General patterns of precipitation and winds are also associated with the low, middle, and high latitudes. An idealized model of the global atmospheric circulation and pressure patterns was described earlier. Recall that this model described a huge convective movement of air in the low latitudes, with air being forced upward over the equatorial region. This air expands, cools to the dew point, and produces abundant rainfall throughout most of the year. On the other hand, air is slowly sinking over 30°N and 30°S of the equator, becoming warm and dry as it is compressed. Most of the great deserts of the world are near 30°N or 30°S latitude for this reason. There is another wet zone near 60° latitudes and another dry zone near the poles. These wet and dry zones are shifted north and south during the year with the changing seasons. This results in different precip-

558

itation patterns in each season. Figure 21.42 shows where the wet and dry zones are in winter and in summer seasons.

Regional Climatic Influence

Latitude determines the basic tropical, temperate, and polar climatic zones, and the wet and dry zones move back and forth over the latitudes with the seasons. If these were the only factors influencing the climate, you would expect to find the same climatic conditions at all locations with the same latitude. This is not what is found, however, because there are four major factors that affect a regional climate. These are (1) altitude, (2) mountains, (3) large bodies of water, and (4) ocean currents. The following describes how these four factors modify the climate of a region.

The first of the four regional climate factors is *altitude*. The atmosphere is warmed mostly by the greenhouse effect from the surface upward, and air at higher altitudes increasingly radiates more and more of its energy to space. Average air temperatures therefore decrease with altitude, and locations with higher altitudes will have lower average temperatures. This is why the tops of mountains are often covered with snow when none is found at lower elevations. St. Louis, Missouri and Denver, Colorado are located almost at the same latitude (within 1° of 39°N), so you might expect the two cities to have about the same average temperature. Denver, however, has an altitude of 1,609 m (5,280 ft), and the altitude of St. Louis is 141 m (465 ft). The yearly average temperature for Denver is about 10°C (about 50°F) and for

FIGURE 21.41

This temperate-climate deciduous forest responds to seasonable changes in autumn with a show of color.

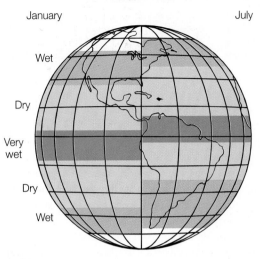

FIGURE 21.42

The idealized general rainfall patterns over the earth shift with seasonal shifts in the wind and pressure areas of earth's general atmospheric circulation patterns.

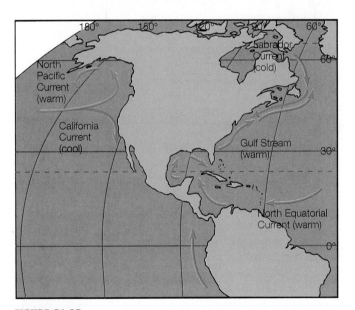

FIGURE 21.43

Ocean currents can move large quantities of warm or cool water to influence the air temperatures of nearby landmasses.

St. Louis it is about 14°C (about 57°F). In general, higher altitude means lower average temperature.

The second of the regional climate factors is *mountains*. In addition to the temperature change caused by the altitude of the mountain, mountains also affect the conditions of a passing air mass. The western United States has mountainous regions along the coast. When a moist air mass from the Pacific meets these mountains it is forced upward and cools adiabatically. Water vapor in the moist air mass condenses, clouds form, and the air mass loses much of its moisture as precipitation falls on the western side of the mountains. Air moving down the eastern slope is compressed and becomes warm and dry. As a result the western slopes of these mountains are moist and have forests of spruce, redwood, and fir trees. The eastern slopes are dry and have grassland or desert vegetation.

The third of the regional climate factors is a large body of *water*. Water, as discussed previously, has a higher specific heat than land material, it is transparent, and it loses energy through evaporation. All of these affect the temperature of a landmass located near a large body of water, making the temperatures more even from day to night and from summer to winter. San Diego, California and Dallas, Texas, for example, are at about the same latitude (both almost 33°N), but San Diego is at a seacoast and Dallas is inland. Because of its nearness to water, San Diego has an average summer temperature about 7°C (about 13°F) cooler and an average winter temperature about 5°C (about 9°F) warmer than the average temperatures in Dallas. Nearness to a large body of water keeps the temperature more even at San Diego. This relationship is observed to occur for the earth as a whole as well as locally. The Northern Hemisphere is about 39 percent land and 61 percent water and has an average yearly temperature range of about 14°C (about 25°F). On the other hand the Southern Hemisphere is about 19 percent land and 81 percent water and has an average yearly temperature range of about

7°C (about 13°F). There is little doubt about the extent to which nearness to a large body of water influences the climate.

The fourth of the regional climate factors is *ocean currents*. In addition to the evenness brought about by being near the ocean, currents in the ocean can bring water that has a different temperature than the land. For example, currents can move warm water northward or they can move cool water southward (figure 21.43). This can influence the temperatures of air masses that move from the water to the land and, thus, the temperatures of the land. For example, the North Pacific current brings warm waters to the western coast of North America, which results in warmer temperatures for cities near the coast.

A CLOSER LOOK | El Niño and La Niña

The term "El Niño" was originally used to describe an occurrence of warm, above-normal ocean temperatures off the South American coast. Fishermen along this coast learned long ago to expect associated changes in fishing patterns about every 3 to 7 years, which usually lasted for about 18 months. They called this event El Niño, which is Spanish for "the boy child" or "Christ child," because it typically occurred near Christmas. The El Niño event occurs when the trade winds along the equatorial Pacific become reduced or calm, allowing sea surface temperatures to increase much above normal. The warm water drives the fish to deeper waters or farther away from usual fishing locations.

Today "El Niño" is understood to be much more involved than just a warm ocean in the Pacific. It is more than a local event, and the bigger picture is sometimes called the "El Niño–Southern Oscillation," or ENSO. In addition to the warmer of "El Niño," the boy, the term "La Niña," the girl, has been used to refer to the times when the water of the tropical Pacific is *colder* than normal. The "Southern Oscillation" part of the name comes from observations that atmospheric pressure around Australia seems to be inversely linked to the atmospheric pressure in Tahiti. They seem to be linked because when the pressure is low in Australia, it is high in Tahiti. Conversely, when the atmospheric pressure is high in Australia, it is low in Tahiti. The strength of this Southern Oscillation is measured by the Southern Oscillation Index (SOI), which is defined as the pressure at Darwin, Australia, subtracted from that at Tahiti. Negative values of SOI are usually associated with El Niño events, so the Southern Oscillation and the El Niño are obviously linked. How ENSO can impact the weather in other parts of the world has only recently become better understood.

The atmosphere is a system that responds to incoming solar radiation, the spinning earth, and other factors, such as the amount of water vapor present. The ocean and atmospheric systems undergo changes by interacting with each other, most visibly in the tropical cyclone. The ocean system supplies water vapor, latent heat, and condensation nuclei, which are the essential elements of a tropical cyclone as well as everyday weather changes and climate. The atmosphere, on the other hand, drives the ocean with prevailing winds, moving warm or cool water to locations where they affect the climate on the land. There is a complex, interdependent relationship between the ocean and the atmosphere, and it is probable that even small changes in one system can lead to bigger changes in the other.

Normally, during non–El Niño times, the Pacific Ocean along the equator has established systems of prevailing wind belts, pressure systems, and ocean currents. In July, these systems push the surface seawater offshore from South America, westward along the equator and toward Indonesia. During El Niño times the trade winds weaken and the warm water moves back eastward, across the Pacific to South America, where it then spreads north and south along the coast. Why the trade winds weaken and become calm is unknown, the subject of ongoing research.

Warmer waters along the coast of South America bring warmer, more humid air and the increased possibility of thunderstorms. Thus the possibility of towering thunderstorms, tropical storms, or hurricanes increases along the Pacific coast of South America as the warmer waters move north and south. This creates the possibility of weather changes not only along the western coast, but elsewhere, too. The towering thunderstorms reach high into the atmosphere, adding tropical moisture and creating changes in prevailing wind belts. These wind belts carry or steer weather systems across the middle latitudes of North America, so typical storm paths are shifted. This shifting can result in

- increased precipitation in California during the fall to spring season;
- a wet winter with more and stronger storms along regions of the southern United States and Mexico;

- a warmer and drier than average winter across the northern regions of Canada and the United States;
- a variable effect on central regions of the United States, ranging from reduced snowfall to no effect at all; and
- other changes in the worldwide global complex of ocean and weather events, such as droughts in normally wet climates and heavy rainfall in normally dry climates.

One major problem in these predictions is a lack of understanding of what causes many of the links and a lack of consistency in the links themselves. For example, southern California did not always have an unusually wet season every time an El Niño occurred and in fact experienced a drought during one event.

Scientists have continued to study the El Niño since the mid-1980s, searching for patterns that will reveal consistent cause-and-effect links. Part of the problem may be that other factors, such as a volcanic eruption, may influence part of the linkage but not another part. Another part of the problem may be the influence of unknown factors, such as the circulation of water deep beneath the ocean surface, the track taken by tropical cyclones, or the energy released by tropical cyclones one year compared to the next.

The results so far have indicated that atmosphere-ocean interactions are much more complex than early theoretical models had predicted. Sometimes a new model will predict some weather changes that occur with El Niño, but no model is yet consistently correct in predicting the conditions that lead to the event and the weather patterns that result. All this may someday lead to a better understanding of how the ocean and the atmosphere interact on this dynamic planet.

Recent years in which El Niño events have occurred are 1951, 1953, 1957–1958, 1965, 1969, 1972–1973, 1976, 1982–1983, 1986–1987, 1991–1992, 1994, and 1997–1998.

Describing the earth's climates presents a problem because there are no sharp boundaries that exist naturally between two adjacent regions with different climates. Even if two adjacent climates are very different, one still blends gradually into the other. For example, if you are driving from one climate zone to another you might drive for miles before becoming aware that the vegetation is now different than it was an hour ago. Since the vegetation is very different from what it was before, you know that you have driven from one regional climate zone to another.

Actually, no two places on the earth have exactly the same climate. Some plants will grow on the north or south side of a building, for example, but not on the other side. The two sides of the building could be considered as small, local climate zones within a larger, major climate zone.

SUMMARY

The earth's *atmosphere* thins rapidly with increasing altitude. Pure, dry air is mostly *nitrogen, oxygen,* and *argon,* with traces of *carbon dioxide* and other gases. Atmospheric air also contains a variable amount of *water vapor.* Water vapor cycles into and out of the atmosphere through the *hydrologic cycle* of evaporation and precipitation.

Atmospheric pressure is measured with a *mercury barometer.* At sea level, the atmospheric pressure will support a column of mercury about 76.00 cm (about 29.92 in) tall. This is the average pressure at sea level, and it is called the *standard atmospheric pressure, normal pressure,* or *one atmosphere of pressure.*

Materials on the earth's surface absorb sunlight, emitting more and more *infrared radiation* as they are warmed. *Carbon dioxide* and *molecules* of water vapor in the atmosphere absorb infrared radiation, which then reemit the energy many times before it reaches outer space again. The overall effect warms the lower atmosphere from the bottom up in a process called the *greenhouse effect.*

The layer of the atmosphere from the surface up to where the temperature stops decreasing with height is called the *troposphere.* The *stratosphere* is the layer above the troposphere. Temperatures in the stratosphere increase because of the interaction between ozone (O_3) and ultraviolet radiation from the sun. Other layers of the atmosphere are the *mesosphere, thermosphere, exosphere,* and the *ionosphere.*

The surface of the earth is not heated uniformly by sunlight. This results in a *differential heating,* which sets the stage for *convection.* The horizontal movement of air on the surface from convection is called *wind.* A generalized model for understanding why the wind blows involves (1) the relationship between *air temperature and air density* and (2) the relationship between *air pressure and the movement of air.* This model explains local wind patterns and wind patterns observed on a global scale.

The amount of water vapor in the air at a particular time is called the *absolute humidity.* The *relative humidity* is a ratio between the amount of water vapor that is in the air and the amount needed to saturate the air at the present temperature.

When the air is saturated, condensation can take place. The temperature at which this occurs is called the *dew point temperature.* If the dew point temperature is above freezing, *dew* will form. If the temperature is below freezing, *frost* will form. Both dew and frost form directly on objects and do not fall from the air.

Water vapor condenses in the air on *condensation nuclei.* If this happens near the ground, the accumulation of tiny water droplets is called a *fog. Clouds* are accumulations of tiny water droplets in the air

above the ground. In general, there are three basic shapes of clouds, *cirrus, cumulus,* and *stratus.* These basic cloud shapes have meaning about the atmospheric conditions and about the coming weather conditions.

Water that returns to the earth in liquid or solid form falls from the clouds as *precipitation.* Precipitation forms in clouds through two processes: (1) the *coalescence* of cloud droplets or (2) the *growth of ice crystals* at the expense of water droplets.

Weather changes are associated with the movement of large bodies of air called *air masses,* the leading *fronts* of air masses when they move, and local *high- and low-pressure* patterns that accompany air masses or fronts. Examples of air masses include (1) *continental polar,* (2) *maritime polar,* (3) *continental tropical,* and (4) *maritime tropical.*

When a location is under the influence of an air mass the location is having *air mass weather* with slow, gradual changes. More rapid changes take place when the *front,* a thin transition zone between two air masses, passes a location.

A *stationary front* often develops a bulge, or *wave,* that forms into a moving cold front and a moving warm front. The faster moving cold front overtakes the warm front, lifting it into the air to form an *occluded front.* The lifting process forms a low-pressure center called a *cyclone.* Cyclones are associated with heavy clouds, precipitation, and stormy conditions because of the lifting action.

A *thunderstorm* is a brief, intense storm with rain, lightning and thunder, gusty and strong winds, and sometimes hail. A *tornado* is the smallest, most violent weather disturbance that occurs on the earth. A *hurricane* is a *tropical cyclone,* a large, violent circular storm that is born over warm tropical waters near the equator.

The general pattern of the weather that occurs for a region over a number of years is called *climate.* The three principal climate zones are (1) the *tropical climate zone,* (2) the *polar climate zone,* and (3) the *temperate climate zone.* The climate in these zones is influenced by four factors that determine the local climate: (1) *altitude,* (2) *mountains,* (3) *large bodies of water,* and (4) *ocean currents.* The climate for a given location is described by first considering the principal climate zone, then looking at subdivisions within each that result from local influences.

KEY TERMS

absolute humidity (p. **538**)	meteorology (p. **528**)
air mass (p. **547**)	millibar (p. **531**)
barometer (p. **530**)	occluded front (p. **550**)
climate (p. **556**)	ozone shield (p. **533**)
cold front (p. **549**)	precipitation (p. **542**)
condensation nuclei (p. **541**)	relative humidity (p. **539**)
dew (p. **540**)	saturated air (p. **538**)
dew point temperature (p. **540**)	solar constant (p. **531**)
front (p. **549**)	standard atmospheric pressure (p. **531**)
frost (p. **540**)	stationary front (p. **550**)
greenhouse effect (p. **532**)	storm (p. **551**)
hail (p. **553**)	thunderstorm (p. **551**)
humidity (p. **538**)	tornado (p. **553**)
hurricane (p. **554**)	typhoon (p. **554**)
ice-forming nuclei (p. **546**)	warm front (p. **550**)
inversion (p. **532**)	wind (p. **534**)
jet stream (p. **537**)	wind chill factor (p. **539**)

APPLYING THE CONCEPTS

1. An airplane flying at about 6 km (20,000 feet) is above how much of the earth's atmosphere?
   a. 99 percent
   b. 90 percent
   c. 75 percent
   d. 50 percent

2. The earth's atmosphere is mostly composed of
   a. oxygen, carbon dioxide, and water vapor.
   b. nitrogen, oxygen, and argon.
   c. oxygen, carbon dioxide, and nitrogen.
   d. oxygen, nitrogen, and water vapor.

3. Which of the following gases cycle into and out of the atmosphere?
   a. nitrogen
   b. carbon dioxide
   c. oxygen
   d. All of the above are correct.

4. If it were not for the ocean, the earth's atmosphere would probably be mostly
   a. nitrogen.
   b. carbon dioxide.
   c. oxygen.
   d. argon.

5. Your ear makes a "pop" sound as you descend in an elevator because
   a. air is moving from the atmosphere into your eardrum.
   b. air is moving from your eardrum to the atmosphere.
   c. air is not moving in or out of your eardrum.
   d. none of the above.

6. Most of the total energy radiated by the sun is
   a. visible light.
   b. ultraviolet radiation.
   c. infrared radiation.
   d. gamma radiation.

7. Of the total amount of solar radiation reaching the outermost part of the earth's atmosphere, how much reaches the surface?
   a. all of it
   b. about 99 percent
   c. about 75 percent
   d. about half

8. The solar radiation that does reach the earth's surface
   a. is eventually radiated back to space.
   b. shows up as an increase in temperature.
   c. is reradiated at different wavelengths.
   d. is all of the above.

9. The greenhouse effect results in warmer temperatures near the surface because
   a. clouds trap infrared radiation near the surface.
   b. some of the energy is reradiated back toward the surface.
   c. carbon dioxide molecules do not permit the radiation to leave.
   d. carbon dioxide and water vapor both trap infrared radiation.

10. The temperature increases with altitude in the stratosphere because
    a. it is closer to the sun than the troposphere.
    b. heated air rises to the stratosphere.
    c. of a concentration of ozone.
    d. the air is less dense in the stratosphere.

11. Ozone is able to protect the earth from harmful amounts of ultraviolet radiation by
    a. reflecting it back to space.
    b. absorbing it and decomposing, then reforming.
    c. refracting it to a lower altitude.
    d. all of the above.

12. Summertime breezes would not blow if the earth did not experience
    a. cumulus clouds.
    b. differential heating.
    c. the ozone layer.
    d. a lapse rate in the troposphere.

13. In a sea breeze the wind blows from an area of high pressure to an area of low pressure. Comparing the high- and low-pressure areas to the movement of air, the pressure areas are
    a. the cause of the movement of air.
    b. an effect of the air movement.
    c. not related to the air movement.

14. On a clear, calm, cool night you would expect the air temperature over a valley floor to be what temperature compared to the air temperature over a slope to the valley?
    a. cooler
    b. warmer
    c. the same temperature
    d. sometimes warmer and sometimes cooler

15. Air moving down a mountain slope is often warm because
    a. it has been closer to the sun.
    b. cool air is more dense and settles to lower elevations.
    c. it is compressed as it moves to lower elevations.
    d. this only occurs during the summertime.

16. Considering the overall earth's atmosphere, you would expect more rainfall to occur in a zone of
    a. high atmospheric pressure.
    b. low atmospheric pressure.
    c. prevailing westerly winds.
    d. prevailing trade winds.

17. Considering the overall earth's atmosphere, you would expect to find a desert located in a zone of
    a. high atmospheric pressure.
    b. low atmospheric pressure.
    c. prevailing westerly winds.
    d. prevailing trade winds.

18. Water molecules can go (1) from the liquid state to the vapor state and (2) from the vapor state to the liquid state. When is the movement from the liquid to the vapor state only?
    a. evaporation
    b. condensation
    c. saturation
    d. Usually, none of the above are correct.

19. What condition means a balance between the number of water molecules moving to and from the liquid state?
    a. evaporation
    b. condensation
    c. saturation
    d. None of the above are correct.

20. Without adding or removing any water vapor, a sample of air experiencing an increase in temperature will have
    a. a higher relative humidity.
    b. a lower relative humidity.
    c. the same relative humidity.
    d. a changed absolute humidity.

21. Cooling a sample of air results in a (an)
    a. increased capacity to hold water vapor.
    b. decreased capacity to hold water vapor.
    c. unchanged capacity to hold water vapor.

22. On a clear, calm, and cool night, dew or frost is most likely to form
    a. under trees or other shelters.
    b. on bare ground on the side of a hill.
    c. under a tree on the side of a hill.
    d. on grass in an open, low-lying area.

23. When water vapor in the atmosphere condenses to liquid water,
    a. dew falls to the ground.
    b. rain or snow falls to the ground.
    c. a cloud forms.
    d. All of the above are correct.

24. In order for liquid cloud droplets at the freezing point to freeze into ice crystals,
    a. condensation nuclei are needed.
    b. further cooling is required.
    c. ice-forming nuclei are needed.
    d. nothing more is required.

25. Longer periods of drizzle, rain, or snow usually occur from which basic form of a cloud?
    a. stratus
    b. cumulus
    c. cirrus
    d. None of the above are correct.

26. Brief periods of showers are usually associated with which type of cloud?
    a. stratus
    b. cumulus
    c. cirrus
    d. None of the above are correct.

27. The type of air mass weather that results from the arrival of polar continental air is
    a. frequent snowstorms with rapid changes.
    b. clear and cold with gradual changes.
    c. unpredictable, but with frequent and rapid changes.
    d. much the same from day to day, the conditions depending on the air mass and the local conditions.

28. The appearance of high cirrus clouds followed by thicker, lower stratus clouds, and then continuous light rain over several days probably means which of the following air masses has moved to your area?
    a. continental polar
    b. maritime tropical
    c. continental tropical
    d. maritime polar

29. A fully developed cyclonic storm is most likely to form
    a. on a stationary front.
    b. in a high-pressure center.
    c. from differential heating.
    d. over a cool ocean.

30. The basic difference between a tropical storm and a hurricane is
    a. size.
    b. location.
    c. wind speed.
    d. amount of precipitation.

31. Most of the great deserts of the world are located
    a. near the equator.
    b. 30° north or south latitude.
    c. 60° north or south latitude.
    d. anywhere, as there is no pattern to their location.

32. The average temperature of a location is made more even by the influence of
    a. a large body of water.
    b. elevation.
    c. nearby mountains.
    d. dry air.

33. The climate of a specific location is determined by
    a. its latitude.
    b. how much sunlight it receives.
    c. its altitude and nearby mountains and bodies of water.
    d. all of the above.

Answers

1. d 2. b 3. d 4. b 5. a 6. c 7. d 8. d 9. b 10. c 11. b 12. b 13. b 14. a 15. c 16. b 17. a 18. d 19. c 20. b 21. b 22. d 23. c 24. c 25. a 26. b 27. d 28. b 29. a 30. c 31. b 32. a 33. d

QUESTIONS FOR THOUGHT

1. What is the meaning of "normal" atmospheric pressure?
2. Explain the "greenhouse effect." Is a greenhouse a good analogy for the earth's atmosphere? Explain.
3. What is a temperature inversion? Why does it increase air pollution?
4. Describe how the ozone shield protects living things on the earth's surface. Why is there some concern about this shield?
5. What is wind? What is the energy source for wind? Explain.
6. Explain the relationship between air temperature and air density.
7. Why does heated air rise?

8. Provide an explanation for the observation that an airplane flying at the top of the troposphere takes several hours longer to fly from the east coast to the west coast than it does to make the return trip.

9. If evaporation cools the surroundings, does condensation warm the surroundings? Explain.

10. Explain why warm air can hold more water vapor than cool air.

11. Explain why a cooler temperature can result in a higher relative humidity when no water vapor was added to the air.

12. What is the meaning of the expression, "It's not the heat, it's the humidity"?

13. What is the meaning of the dew point temperature?

14. Explain why frost is more likely to form on a clear, calm, and cool night than on nights with other conditions.

15. What is a cloud? Describe how a cloud forms.

16. Describe two ways that precipitation may form from the water droplets of a cloud.

17. What is an air mass?

18. What kinds of clouds and weather changes are usually associated with the passing of (a) a warm front? (b) a cold front?

19. Describe the wind direction, pressure, and weather conditions that are usually associated with (a) low-pressure centers and (b) high-pressure centers.

20. In which of the four basic types of air masses would you expect to find afternoon thunderstorms? Explain.

21. Describe the three main stages in the life of a thunderstorm cell, identifying the events that mark the beginning and end of each stage.

22. What is a tornado? When and where do tornadoes usually form?

23. What is a hurricane? Describe how the weather conditions change as a hurricane approaches, passes directly over, then moves away from a location.

24. How is climate different from the weather?

25. Describe the average conditions found in the three principal climate zones.

26. Identify the four major factors that influence the climate of a region and explain how each does its influencing.

The earth's vast oceans cover more than 70 percent of the surface of the earth. Freshwater is generally abundant on the land because the supply is replenished from ocean waters through the hydrologic cycle.

CHAPTER | Twenty-Two

Earth's Waters

T hroughout history humans have diverted rivers and reshaped the land to ensure a supply of freshwater. There is evidence, for example, that ancient civilizations along the Nile River diverted water for storage and irrigation some five thousand years ago. The ancient Greeks and Romans built systems of aqueducts to divert streams to their cities some two thousand years ago. Some of these aqueducts are still standing today. More recent water diversion activities were responsible for the name of Phoenix, Arizona. Phoenix was named after a mythical bird that arose from its ashes after being consumed by fire. The city was given this name because it is built on a system of canals that were first designed and constructed by ancient Native Americans, then abandoned hundreds of years before settlers reconstructed the ancient canal system (figure 22.1). Water is and always has been an essential resource. Where water is in short supply, humans have historically turned to extensive diversion and supply projects to meet their needs.

Precipitation is the basic source of the water supply found today in streams, lakes, and beneath the earth's surface. Much of the precipitation that falls on the land, however, evaporates back into the atmosphere before it has a chance to become a part of this supply. The water that does not evaporate mostly moves directly to rivers and streams, flowing back to the ocean, but some soaks into the land. The evaporation of water, condensation of water vapor, and the precipitation-making processes were introduced in chapter 21 as important weather elements. They are also part of the generalized *hydrologic cycle* of evaporation from the ocean, transport through the atmosphere by moving air masses, precipitation on the land, and movement of water back to the ocean. Only part of this cycle was considered previously, however, and this was the part from evaporation through precipitation. This chapter is concerned with the other parts of the hydrologic cycle, that is, what happens to the water that falls on the land and makes it back to the ocean. The chapter begins with a discussion of how water is distributed on the earth and a more detailed look at the hydrologic cycle. Then the travels of water across and into the land will be considered as streams, wells, springs, and other sources of usable water are discussed as limited resources. The tracing of the hydrologic cycle will be completed as the water finally makes it back to the ocean. This last part of the cycle will consider the nature of the ocean floor, the properties of seawater, and how waves and currents are generated. The water is now ready to evaporate, starting another one of earth's never-ending cycles.

WATER ON THE EARTH

Some water is tied up in chemical bonds deep in the earth's interior, but free water is the most abundant chemical compound near the surface. Water is five or six times more abundant than the most abundant mineral in the outer 6 km (about 4 mi) of the earth, so it should be no surprise that water covers about 70 percent of the surface. On the average, about 98 percent of this water exists in the liquid state in depressions on the surface and in sediments. Of the remainder, about 2 percent exists in the solid state as snow and ice on the surface in colder locations. Only a fraction of a percent exists as a variable amount of water vapor in the atmosphere at a given time. Water is continually moving back and forth between these "reservoirs," but the percentage found in each is assumed to be essentially constant.

As shown in figure 22.2, over 97 percent of the earth's water is stored in the earth's oceans. This water contains a relatively high level of dissolved salts, which will be discussed in a later section. This dissolved salt makes ocean water unfit for human consumption and unfit for most agricultural purposes. All other water, which is fit for human consumption and agriculture, is

called **freshwater.** About two-thirds of the earth's freshwater supply is locked up in the ice caps of Greenland and the Antarctic and in glaciers. This leaves less than 1 percent of all the water found on the earth as available freshwater. There is a generally abundant supply, however, because the freshwater supply is continually replenished by the hydrologic cycle.

Evaporation of water from the ocean is an important process of the hydrologic cycle because (1) water vapor leaves the dissolved salts behind, forming precipitation that is freshwater, and (2) the gaseous water vapor is easily transported in the atmosphere from one part of the earth to another. Over a year this natural desalination process produces and transports enough freshwater to cover the entire earth with a layer about 85 cm (about 33 in) deep. Precipitation is not evenly distributed like this, of course, and some places receive much more, while other places receive almost none. Considering global averages, more water is evaporated from the ocean than returns by precipitation. On the other hand, more water is precipitated over the land than evaporates back to the atmosphere. The net amount evaporated and precipitated over the land and over the ocean is balanced by the return of water to

FIGURE 22.1

This is one of the water canals of the present-day system in Phoenix, Arizona. These canals were reconstructed from a system that was built by American Indians, then abandoned. Phoenix is named after a mythical bird that was consumed by fire and then arose from its ashes.

the ocean by streams and rivers (figure 22.3). This freshwater returning on and under the land is the source of freshwater. What happens to the freshwater during its return to the ocean is discussed in the following section.

Freshwater

The basic source of freshwater is precipitation, but not all precipitation ends up as part of the freshwater supply. Liquid water is always evaporating, even as it falls. In arid climates, rain sometimes evaporates completely before reaching the surface, even from a fully developed thunderstorm. Evaporation continues from the water that does reach the surface. Puddles and standing water on the hard surface of city parking lots and streets, for example, gradually evaporate back to the atmosphere after a rain and the surface is soon dry. There are many factors that determine how much of a particular rainfall evaporates, but in general more than two-thirds of the rain eventually returns to the atmosphere. The remaining amount either (1) flows downhill across the surface of the land toward a lower place or (2) soaks into the ground. Water moving across the surface is called **runoff.** Runoff begins as rain accumulates in thin sheets of water that move across the surface of the land. These sheets collect into a small body of running water called a *stream.* A stream is defined as any body of water that is moving across the land, from one so small that you could step across it to the widest river. Water that soaks into the ground moves down to a saturated zone where it is called *groundwater.* Groundwater moves through sediments and rocks beneath the surface, slowly moving downhill. Streams carry the runoff of a recent rainfall or melting snow, but otherwise most of the flow comes from

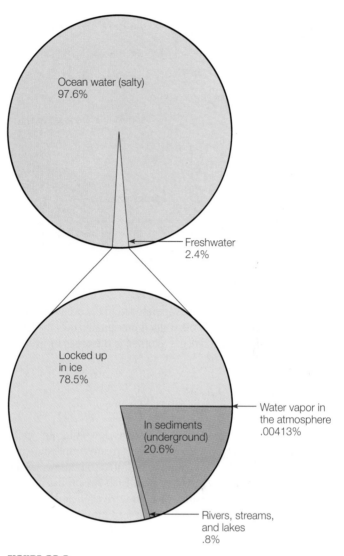

FIGURE 22.2

The distribution of all the water found on the earth's surface.

groundwater that has seeped into the stream channel. This explains how a permanent stream is able to continue flowing when it is not being fed by runoff or melting snow (figure 22.4). Where or when the source of groundwater is in low supply a stream may flow only part of the time, and it is designated as an *intermittent stream.*

The amount of a rainfall that becomes runoff or groundwater depends on a number of factors, including (1) the type of soil on the surface, (2) how dry the soil is, (3) the amount and type of vegetation, (4) the steepness of the slope, and (5) if the rainfall is a long, gentle one or a cloudburst. Different combinations of these factors can result in from 5 percent to almost 100 percent of given rainfall running off, with the rest evaporating or soaking into the ground. On the average, however, about 70 percent of all precipitation evaporates back into the atmosphere, about 30 percent becomes runoff, and less than 1 percent soaks into the ground.

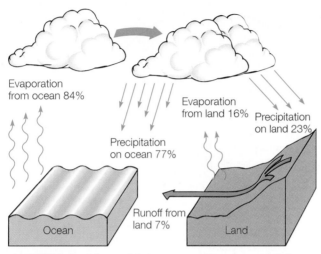

100% is based on a global average of 85 cm/yr precipitation.

FIGURE 22.3

On the average, more water is evaporated from the ocean than is returned by precipitation. More water is precipitated over the land than evaporates. The difference is returned to the ocean by rivers and streams.

Surface Water

If you could follow the smallest of streams downhill, you would find that it eventually merges with other streams until you reach a major river. The land area drained by a stream is known as the stream's drainage basin, or **watershed.** Each stream has its own watershed, but the watershed of a large river includes all the watersheds of the smaller streams that feed into the larger river. Figure 22.5 shows the watersheds of the Columbia River, the Colorado River, and the Mississippi River. Note that the water from the Columbia River and the Colorado River watersheds empties into the Pacific Ocean. The Mississippi River watershed drains into the Gulf of Mexico, which is part of the Atlantic Ocean.

Two adjacent watersheds are separated by a line called a *divide.* Rain that falls on one side of a divide flows into one watershed and rain that falls on the other side flows into the other watershed. A *continental divide* separates river systems that drain into opposite sides of a continent. The North American continental divide trends northwestward through the Rocky Mountains. Imagine standing over this line with a glass of water in each hand, then pouring the water to the ground. The water from one glass will eventually end up in the Atlantic Ocean, and the water from the other glass will end up in the Pacific Ocean. Sometimes the Appalachian Mountains are considered to be an eastern continental divide, but water from both sides of this divide ends up on the same side of the continent, in the Atlantic Ocean.

Water moving downhill is sometimes stopped by a depression in a watershed, a depression where water temporarily collects as a standing body of freshwater. A smaller body of standing water is usually called a *pond,* and one of much larger size is called a *lake.* A lake is supposed to be deep enough in some place or places that sunlight does not reach the bottom, but this defi-

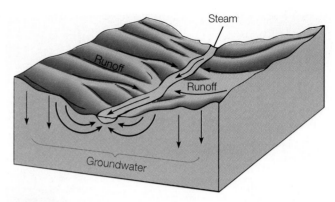

FIGURE 22.4

Some of the precipitation soaks into the ground to become groundwater. Groundwater slowly moves underground, and some of it emerges in streambeds keeping the streams running during dry spells.

nition is not always followed when lakes and ponds are named. Part of the problem is the interpretation of what is meant by "sunlight." In general, only blue-green light remains at a depth of about 10 m (about 33 ft), and there is insufficient light for plant life at a depth of about 80 m (about 260 ft).

A pond or lake can occur naturally in a depression, or it can be created by building a dam on a stream. A natural pond, a natural lake, or a pond or lake created by building a dam is called a *reservoir* if it is used for (1) water storage, (2) flood control, or (3) generating electricity. A reservoir can be used for one or two of these purposes but not generally for all three. A reservoir built for water storage, for example, is kept as full as possible to store water. This use is incompatible with use for flood control, which would require a low water level in the reservoir in order to catch runoff, preventing waters from flooding the land. In addition, extensive use of reservoir water to generate electricity requires the release of water, which could be incompatible with water storage. The water of streams, ponds, lakes, and reservoirs is collectively called *surface water,* and all serve as sources of freshwater. The management of surface water, as you can see, can present some complicated problems.

Groundwater

Precipitation soaks into the ground, or *percolates* slowly downward until it reaches an area, or zone, where the open spaces between rock and soil particles are completely filled with water. Water from such a saturated zone is called **groundwater.** There is a tremendous amount of water stored as groundwater, which makes up a supply about thirty times larger than all the surface water on the earth. Groundwater is an important source of freshwater for human consumption and for agriculture. Groundwater is often found within 100 m (about 330 ft) of the surface, even in arid regions where little surface water is found. Groundwater is the source of water for wells in addition to being the source that keeps streams flowing during dry periods.

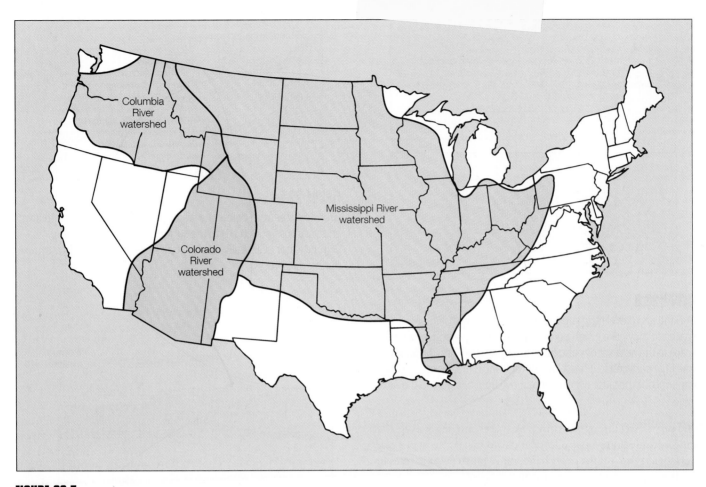

FIGURE 22.5

The approximate watersheds of the Columbia River, the Colorado River, and the Mississippi River.

Water is able to percolate down to a zone of saturation because sediments contain open spaces between the particles called *pore spaces.* The more pore space a sediment has, the more water it will hold. The total amount of pore spaces in a given sample of sediment is a measure of its *porosity.* Sand and gravel sediments, for example, have grains that have large pore spaces, so these sediments have a high porosity. In order for water to move through a sediment, however, the pore spaces must be connected. The ability of a given sample of sediment to transmit water is a measure of its *permeability.* Sand and gravel have a high permeability because the grains do not fit tightly together, allowing water to move from one pore space to the next. Sand and gravel sediments thus have a high porosity as well as a high permeability. Clay sediments, on the other hand, have small, flattened particles that fit tightly together. Clay thus has a low porosity, and when saturated or compressed, clay becomes *impermeable,* meaning water cannot move through it at all (figure 22.6).

The amount of groundwater available in a given location depends on a number of factors, such as the present and past climate, the slope of the land, and the porosity and permeability of the sediments beneath the surface. Generally sand and gravel

sediments, along with solid sandstone, have the best porosity and permeability for transmitting groundwater. Other solid rocks, such as granite, can also transmit groundwater if they are sufficiently fractured by joints and cracks. In any case, groundwater will percolate downward until it reaches an area where pressure and other conditions have eliminated all pores, cracks, and joints. Above this downward limit it collects in all available spaces to form a *zone of saturation.* Water from the zone of saturation is considered to be groundwater. Water from the zone above is not considered to be groundwater. The surface of the boundary between the zone of saturation and the zone above is called the **water table.** The surface of a water table is not necessarily horizontal, but it tends to follow the topography of the surface in a humid climate. A hole that is dug or drilled through the earth to the water table is called a well. The part of the well that is below the water table will fill with groundwater, and the surface of the water in the well is generally at the same level as the water table.

Precipitation falls on the land and percolates down to the zone of saturation, then begins to move laterally, or sideways, to lower and lower elevations until it finds its way back to the surface. This surface outflowing could take place at a stream, pond, lake, swamp, or spring (figure 22.7). Groundwater flows gradually

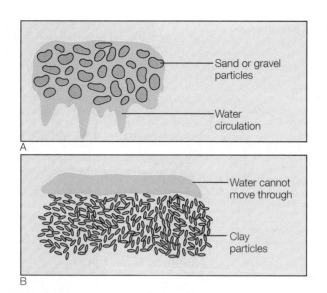

FIGURE 22.6

(*A*) Sand and gravel have large, irregular particles with large pore spaces, so they have a high porosity. Water can move from one pore space to the next, so they also have a high permeability. (*B*) Clay has small, flat particles, so it has a low porosity and is practically impermeable because water cannot move from one pore to the next.

and very slowly through the tiny pore spaces, moving at a rate that ranges from kilometers (miles) per day to meters (feet) per year. Surface streams, on the other hand, move much faster at rates up to about 30 km per hour (about 20 mi/hr).

An **aquifer** is a layer of sand, gravel, sandstone, or other highly permeable material beneath the surface that is capable of producing water. In some places an aquifer carries water from a higher elevation, resulting in a pressure on water trapped by impermeable layers at lower elevations. Groundwater that is under such a confining pressure is in an *artesian* aquifer. "Artesian" refers to the pressure, and groundwater from an artesian well rises above the top of the aquifer but not necessarily to the surface. Some artesian wells are under sufficient pressure to produce a fountainlike flow or spring (figure 22.8). Some people call groundwater from any deep well "artesian water," which is technically incorrect.

A c t i v i t i e s

After doing some research, make a drawing to show the depth of the water table in your area, and its relationship to local streams, lakes, swamps, and water wells.

Freshwater as a Resource

Water is an essential resource, not only because it is required for life processes but also because of its role in a modern industrialized society. Water is used in the home for drinking, cooking, and cleaning, as a carrier to remove wastes, and for maintaining lawns and gardens. These domestic uses lead to an equivalent consumption

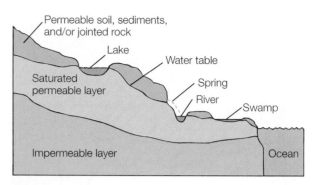

FIGURE 22.7

Groundwater from below the water table seeps into lakes, streams, and swamps and returns to the surface naturally at a spring. Groundwater eventually returns to the ocean, but the trip may take hundreds of years.

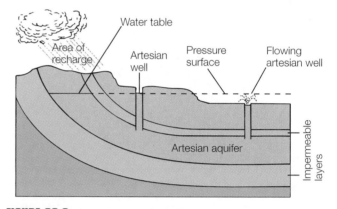

FIGURE 22.8

An artesian aquifer has groundwater that is under pressure from the water at higher elevations. The pressure will cause the water to rise above the water table in a well drilled into the aquifer, becoming a flowing well if the pressure is sufficiently high.

of about 570 liters per person each day (about 150 gal/person/day), but this is only about 10 percent of the total consumed. Average daily use of water in the United States amounts to some 5,700 liters per person each day (about 1,500 gal/person/day), or about enough water to fill a small swimming pool once a week. The bulk of the water is used by agriculture (about 40 percent), for the production of electricity (about 40 percent), and for industrial purposes (about 10 percent). These overall percentages of use vary from one region of the country to another, depending on (1) the relative proportions of industry, agriculture, and population, (2) the climate of the region, (3) the nature of the industrial or agricultural use, and (4) other variables. In an arid climate with a high proportion of farming and fruit growing, for example, up to two-thirds of the available water might be used for agriculture.

Most of the water supply is obtained from the surface water resources of streams, lakes, and reservoirs. Surface water contains more sediments, bacteria, and possible pollutants than groundwater because it is more active and is directly exposed to the

FIGURE 22.9

The filtering beds of a city water treatment facility. Surface water contains more sediments, bacteria, and other suspended materials because it is active on the surface and is exposed to the atmosphere. This means that surface water must be filtered and treated when used as a domestic resource. Such processing is not required when groundwater is used as the resource.

FIGURE 22.10

This is groundwater pumped from the ground for irrigation. In some areas, groundwater is being removed from the ground like this faster than it is being replaced by precipitation, resulting in a water table that is falling deeper and deeper. It is thus possible that the groundwater resource will soon become depleted in some areas.

atmosphere. This means that surface water requires filtering to remove suspended particles, treatment to kill bacteria, and sometimes processing to remove pollution. In spite of the additional processing and treatment costs, surface water is less costly as a resource than groundwater. Groundwater is naturally filtered as it moves through the pore spaces of an aquifer, so it is usually relatively free of suspended particles and bacteria. Thus the processing or treatment of groundwater is usually not necessary (figure 22.9). But groundwater, on the other hand, will cost more to use as a resource because it must be pumped to the surface. The energy required for this pumping can be very expensive. In addition, groundwater generally contains more dissolved minerals (hard water), which may require additional processing or chemical treatment to remove the troublesome minerals.

The use of surface water as a source of freshwater means that the supply depends on precipitation. When a drought occurs, low river and lake resources may require curtailing water consumption. The curtailing of consumption occurs more often when a drought lasts for a longer period of time and when smaller lakes and reservoirs make the supply sensitive to rainfall amounts. In some parts of the western United States, such as the Colorado River watershed, *all* of the surface water is already being used, with certain percentages allotted for domestic, industrial, and irrigation uses. Groundwater is also used in this watershed, and in some locations it is being pumped from the ground faster than it is being replenished by precipitation (figure 22.10). As the population grows and new industries develop, more and more demands are placed on the surface water supply, which has already been committed to other uses, and on the diminishing supply of ground-

water. This raises some very controversial issues about how freshwater should be divided among agriculture, industries, and city domestic use. Agricultural interests claim they should have the water because they produce the food and fibers that people must have. Industrial interests claim they should have the water because they create the jobs and the products that people must have. Cities, on the other hand, claim that domestic consumption is the most important because people cannot survive without water. Who should have the first priority for water use in such cases?

Some have suggested that people should not try to live and grow food in areas that have a short water supply, that plenty of freshwater is available elsewhere. Others have suggested that humans have historically moved rivers and reshaped the land to obtain water, so perhaps one answer to the problem is to find new sources of freshwater. Possible sources include the recycling of waste water and turning to the largest supply of water in the world, the ocean. About 90 percent of the water used by industries is presently dumped as a waste product. In some areas city waste water is already being recycled for use in power plants and for watering parks. A practically limitless supply of freshwater could be available by desalting ocean water, something which occurs naturally in the hydrologic cycle. The ocean, and the nature of seawater, is the topic of the following section. The treatment of seawater to obtain a new supply of freshwater is presently too expensive because of the cost of energy to accomplish the task. New technologies, perhaps ones that use solar energy, may make this more practical in the future. In the meantime, the best sources of extending the supply of freshwater appear to be the control of pollution, the recycling of waste water, and conservation of the existing supply.

A CLOSER LOOK | Water Quality and Wastewater Treatment

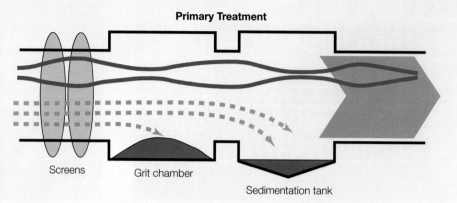

Primary Treatment

Screens Grit chamber Sedimentation tank

BOX FIGURE 22.1

What do you think of when you see a stream? Do you think about the water and how deep it is? Do you think of something to do with the water such as swimming, fishing, or having a good time? As you might imagine, not all people look at a stream and think about the same thing. A city engineer, for example, might wonder if the stream has enough water to serve as a source to supplement the city water supply. A rancher or farmer might wonder how the stream could be easily diverted to serve as a source of water for irrigation. An electric utility planner, on the other hand, might wonder if the stream could serve as a source of power.

Water in a stream is a resource that can be used many different ways, but using it requires knowing about the water quality as well as quantity. We need to know if the quality of the water is good enough for the intended use—and different uses have different requirements. Water fit for use in an electric power plant, for example, might not be suitable for use as a city water supply. Indeed, water fit for use in a power plant might not be suitable for irrigation. Water quality is determined by the kinds and amounts of substances dissolved and suspended in the water and the consequences to users. If a source of water can be used for drinking water or not, for example, is regulated by stringent rules and guidelines about what cannot be in the water. These rules do not call for pure water, but rather are designed to protect human health.

The water of even the healthiest stream is not absolutely pure. All water contains many naturally occurring substances such as ions of bicarbonates, calcium, and magnesium. A *pollutant* is not naturally occurring, usually a waste material that contaminates air, soil, or water. There are basically two types of water pollutants: degradable and persistent. Examples of degradable pollutants include domestic sewage, fertilizers, and some industrial wastes. As the term implies, degradable pollutants can be broken down into simple, nonpolluting substances such as carbon dioxide and nitrogen. The chemical reactions or natural bacteria processes that break down the pollutants can

lead to low oxygen levels if the pollution load is high, but this is usually reversible.

Examples of persistent pollutants include some pesticides (such as DDT), petroleum and petroleum products, plastic materials, leached chemicals from landfill sites, oil-based paints, heavy metals and metal compounds such as lead, mercury, and cadmium, and certain radioactive materials. The damage they cause is either irreversible or reparable only over long periods of time.

One of the most common forms of degradable pollution control in the United States is wastewater treatment. The United States has a vast system of sewer pipes, pumping stations, and treatment plants, and about 74 percent of all Americans are served by such wastewater systems. Sewer pipelines collect the wastewater from homes, businesses, and many industries, and deliver it to treatment plants. Most of these plants were designed to make wastewater fit for discharge into streams or other receiving waters.

Years ago raw sewage was dumped directly into waterways. The natural process of purification began as the larger volume of clean water in the stream diluted the wastes. How does water clean itself? Water is purified in large part by the normal actions of organisms. Energy from the sun drives the process of photosynthesis in aquatic plants, which produces oxygen as a by-product. Bacteria use oxygen to break down organic material, including sewage that may have been dumped

into the water. This decomposition produces the carbon dioxide, nutrients, and other substances needed by plants living in the water and the cycle continues. Thus the water can purify itself biologically but only to a degree since the water can absorb only so much. There is a point where the natural cleaning processes can no longer cope. The greater populations of today and greater volume of domestic and industrial wastewater require that communities provide treatment to give nature a hand.

The basic function of a waste treatment plant is to speed up the natural process of purifying the water. There are two basic stages in the treatment, called the *primary stage* and the *secondary stage*. The primary stage physically removes solids from the wastewater. The secondary stage uses biological processes to further purify wastewater. Sometimes, these stages are combined into one operation.

As raw sewage enters a treatment plant it first flows through a screen to remove large floating objects such as rags and sticks that might cause clogs. After this initial screening, it passes into a grit chamber where cinders, sand, and small stones settle to the bottom (box figure 22.1). A grit chamber is particularly important in communities with combined sewer systems where sand or gravel may wash into the system along with rain, mud, and other stuff, all with the storm water.

—Continued next page

Continued—

After screening and grit removal the sewage is basically a mixture of organic and inorganic matter along with other suspended solids. The solids are minute particles that can be removed in a sedimentation tank. The speed of the flow through the larger sedimentation tank is slower and suspended solids gradually sink to the bottom of the tank. They form a mass of solids called *raw primary sludge,* which is usually removed from the tank by pumping. The sludge may be further treated for use as a fertilizer or disposed of through incineration if necessary. If the treatment is for a primary stage only, the effluent from the sedimentation tank is usually disinfected with chlorine, then discharged into receiving waters. Chlorine is fed into the water to kill pathogenic bacteria and to reduce unpleasant odors.

Once it was enough, but today primary stage treatment alone has been increasingly unable to meet the requirements of many communities for higher water quality. Many cities and industries have added a secondary stage treatment to meet higher water quality requirements. Some cities also use advanced treatment processes beyond the secondary stage to remove nutrients and other contaminants.

The secondary stage of treatment removes about 85 percent of the organic matter in sewage by making use of the bacteria that are naturally a part of the sewage. There are two principal techniques used in secondary treatment, and these involve (1) trickling filters and (2) activated sludge. A trickling filter is simply a bed of stones from three to six feet deep through which the effluent from the sedimentation tank flows. Interlocking pieces of corrugated plastic or other synthetic media have also been used in trickling beds, but the important part is that it provides a place for bacteria to live and grow. Bacteria gather and multiply on the stones or synthetic media until they can consume most of the organic matter flowing by in the effluent. The now cleaner water trickles out through pipes to another sedimentation tank to remove excess bacteria. Disinfection of the effluent with chlorine is generally used to complete this secondary stage of basic treatment.

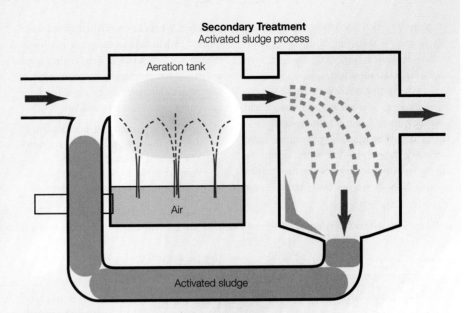

Secondary Treatment
Activated sludge process

Aeration tank

Air

Activated sludge

BOX FIGURE 22.2

The trend today is toward the use of an activated sludge process instead of trickling filters. The activated sludge process speeds up the work of the bacteria by bringing air and sludge heavily laden with bacteria into close contact with the effluent (box figure 22.2). After the effluent leaves the sedimentation tank in the primary stage, it is pumped into an aeration tank, where it is mixed with air and sludge loaded with bacteria and allowed to remain for several hours. During this time, the bacteria break down the organic matter into harmless by-products.

The sludge, now activated with additional millions of bacteria, can be used again by returning it to the aeration tank for mixing with new effluent and ample amounts of air. As with trickling, the final step is generally the addition of chlorine to the effluent which kills more than 99 percent of the harmful bacteria. Some municipalities are now manufacturing chlorine solution on site to avoid the necessity of transporting and storing large amounts of chlorine, sometimes in a gaseous form. Alternatives to chlorine disinfection, such as ultraviolet light or ozone, are also being used in situations where chlorine in sewage effluents can be harmful to fish and other aquatic life.

New pollution problems have placed additional burdens on wastewater treatment systems. Today's pollutants may be more difficult to remove from water. Increased demands on the water supply only aggravate the problem. These challenges are being met through better and more complete methods of removing pollutants at treatment plants, or through prevention of pollution at the source. Pretreatment of industrial waste, for example, removes many troublesome pollutants at the beginning, rather than at the end, of the pipeline.

The increasing need to reuse water calls for better and better wastewater treatment. Every use of water—whether at home, in the factory, or on a farm—results in some change in its quality. New methods for removing pollutants are being developed to return water of more usable quality to receiving lakes and streams. Advanced waste treatment techniques in use or under development range from biological treatment capable of removing nitrogen and phosphorus to physical-chemical separation techniques such as filtration, carbon adsorption, distillation, and reverse osmosis.

These wastewater treatment processes, alone or in combination, can achieve almost any degree of pollution control desired. As waste effluents are purified to higher

—Continued top of next page

Continued—

degrees by such treatment, the effluent water can be used for industrial, agricultural, or recreational purposes, or even drinking water supplies.

When pollution makes water unsuitable for drinking, recreation, agriculture, and industry, it eventually also diminishes the aesthetic quality of lakes and rivers. Even more seriously, when polluted water is able to destroy aquatic life it is able to menace human health. Nobody escapes the effects of water pollution. What can you do to improve water quality? Each individual effort can make a difference to water quality and the environment as a whole. You can start by not misusing your sewage system, especially by dumping toxic household products. Also consider:

- tossing items such as dental floss, disposable diapers, and plastic items into the wastebasket, not the toilet;
- using completely the contents of oven, toilet bowl, and sink drain cleaners, bleaches, paints, rust removers, and solvents;
- saving food scraps and composting them, not dumping them down the drain; and
- choosing water-based latex paint instead of oil-based paint, and using it up instead of storing or dumping it.

There is more you can do, and this begins by reading up on environmental issues for your local area.

Drawings, and some text, from *How Wastewater Treatment Works ... The Basics*, The Environmental Protection Agency, Office of Water, http://www.epa.gov/owmitnet/basics.htm

---

Activities

Find out how much water is used in the industrial processes in your location. Compare this to the amount of water used in the home for drinking, cooking, cleaning, and so on.

SEAWATER

More than 70 percent of the surface of the earth is covered by seawater, with an average depth of 3,800 m (about 12,500 ft). The land areas cover 30 percent, less than a third of the surface, with an average elevation of only about 830 m (about 2,700 ft). With this comparison, you can see that humans live on and fulfill most of their needs by drawing from a small part of the total earth. As populations continue to grow and as resources of the land continue to diminish, the ocean will be looked at more as a resource rather than a convenient place for dumping wastes. The ocean already provides some food and is a source of some minerals, but it can possibly provide freshwater, new sources of food, new sources of important minerals, and new energy sources in the future. There are vast deposits of phosphorite and manganese nodules on the ocean bottom, for example, that can provide valuable minerals. Phosphate is an important fertilizer needed in agriculture, and the land supplies are becoming depleted. Manganese nodules, which occur in great abundance on the ocean bottom, can be a source of manganese, iron, copper, cobalt, and nickel. Seawater contains enough deuterium to make it a feasible source of energy. One gallon of seawater contains about a spoonful of deuterium, with the energy equivalent of three hundred gallons of gasoline. It has been estimated there is sufficient deuterium in the oceans to supply power at one hundred times the present consumption for the next ten billion years. The development of controlled nuclear fusion is needed, however, to utilize this potential energy source. The sea may provide new sources of food through *aquaculture*, the farming of the sea the way that the land is presently farmed. Some aquaculture projects have already started with the farming of oysters, clams, and certain fishes, but these projects have barely begun to utilize the full resources that are possible.

Part of the problem of utilizing the ocean is that the ocean has remained mostly unexplored and a mystery until recent times. Only now are scientists beginning to understand the complex patterns of the circulation of ocean waters, the nature of the chemical processes at work in the ocean, and the interactions of the ocean and the atmosphere and to chart the topography of the ocean floor. The following section will briefly describe some of these findings about the nature of seawater, how the oceans move, and what lies beneath the vast, watery surface of the earth.

Oceans and Seas

The vast body of salt water that covers more than 70 percent of the earth's surface is usually called the *ocean* or the *sea*. Although there is really only one big ocean on the earth, specific regions have been given names for convenience in describing locations. For this purpose, three principal regions are recognized, the (1) Atlantic Ocean, (2) Indian Ocean, and (3) Pacific Ocean. As shown in figure 22.11, these are not separate, independent bodies of salt water but are actually different parts of earth's single, continuous ocean. In general, the **ocean** is a single, continuous body of salt water on the surface of the earth. Specific regions (Atlantic, Indian, and Pacific) are often subdivided further into North Atlantic Ocean, South Atlantic Ocean, and so on.

A *sea* is usually a smaller part of the ocean, a region with different characteristics that distinguish it from the larger ocean of which it is a part. However, the term "sea" is also used in the name of certain inland bodies of salty water, but not always. The ancient term "Seven Seas" is today synonymous with "ocean," meaning all parts of earth's huge body of salt water.

The Pacific Ocean is the largest of the three principal ocean regions. It has the largest surface area, covering 180 million km^2 (about 70 million mi^2), and has the greatest average depth of 3.9 km (about 2.4 mi). The Pacific is circled by active converging

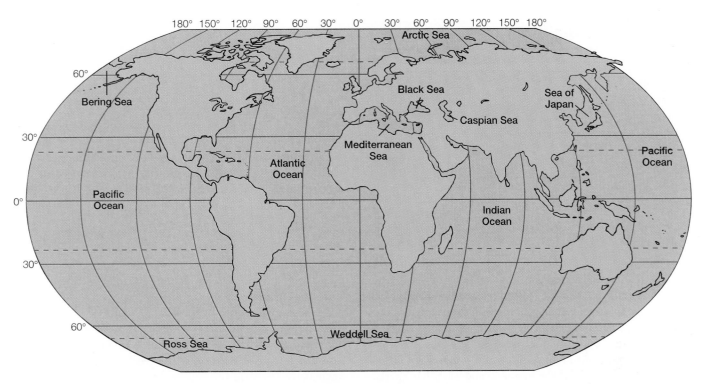

FIGURE 22.11

Distribution of the oceans and major seas on the earth's surface. There is really only one ocean; for example, where is the boundary between the Pacific, Atlantic, and Indian oceans in the Southern Hemisphere?

plate boundaries, so it is sometimes described as being circled by a "rim of fire." It is called this because of the volcanoes associated with the converging plates. The "rim" also has the other associated features of converging plate boundaries such as oceanic trenches, island arcs, and earthquakes. The Atlantic Ocean is second in size, with a surface area of 107 million km² (about 41 million mi²), and the shallowest average depth of only 3.3 km (about 2.1 mi). The Atlantic Ocean is bounded by nearly parallel continental margins with a diverging plate boundary between. It lacks the trench and island arc features of the Pacific, but it does have islands, such as Iceland, that are a part of the Mid-Atlantic Ridge at the plate boundary. The shallow seas of the Atlantic, such as the Mediterranean, Caribbean, and Gulf of Mexico, contribute to the shallow average depth of the Atlantic. The Indian Ocean has the smallest surface area, with 74 million km² (about 29 million mi²), and an average depth of 3.8 km (about 2.4 mi).

As mentioned earlier, a "sea" is usually a part of an ocean that is identified because some characteristic sets it apart. For example, the Mediterranean, Gulf of Mexico, and Caribbean seas are bounded by land, and they are located in a warm, dry climate. Evaporation of seawater is greater than usual at these locations, which results in the seawater being saltier. Being bounded by land and having saltier seawater characterizes these locations as being different from the rest of the Atlantic. The Sargasso Sea, on the other hand, is a part of the Atlantic that is not bounded by land and has a normal concentration of sea salts. This sea is characterized by having an abundance of floating brown seaweeds that accumulate in this region because of the global wind and ocean current patterns. The Arctic Sea, which is also sometimes called the Arctic Ocean, is a part of the North Atlantic Ocean that is less salty. Thus, the terms "ocean" and "sea" are really arbitrary terms that are used to describe different parts of the earth's one continuous ocean.

The Nature of Seawater

According to one theory, the ocean is an ancient feature of the earth's surface, forming at least three billion years ago as the earth cooled from its early molten state. The seawater, and much of the dissolved materials, are believed to have formed from the degassing of water vapor and other gases from molten rock materials. The degassed water vapor soon condensed, and over a period of time it began collecting as a liquid in the depression of the early ocean basin. Ever since, seawater has continuously cycled through the hydrologic cycle, returning water to the ocean through the world's rivers. For millions of years, these rivers have carried large amounts of suspended and dissolved materials to the ocean. These dissolved materials, including salts, stay behind in the seawater as the water again evaporates, condenses, falls on the land, and then brings more dissolved materials much like a continuous conveyor belt.

You might wonder why the ocean basin has not become filled in by the continuous supply of sediments and dissolved

TABLE 22.1	Major dissolved materials in seawater
Ion	**Percent (by weight)**
Chloride (Cl^-)	55.05
Sodium (Na^+)	30.61
Sulfate (SO_4^{-2})	7.68
Magnesium (Mg^{+2})	3.69
Calcium (Ca^{+2})	1.16
Potassium (K^+)	1.10
Bicarbonate (HCO_3^-)	0.41
Bromine (Br^-)	0.19
Total	99.89

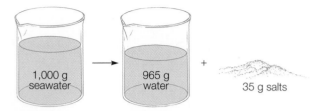

FIGURE 22.12

Salinity is defined as the mass of salts dissolved in 1.0 kg of seawater. Thus, if a sample of seawater has a salinity of 35 ‰, a 1,000 g sample would evaporate 965 g of water and leave 35 g of sea salts behind.

materials that would accumulate over millions of years. The basin has not filled in because (1) accumulated sediments have been recycled to the earth's interior through plate tectonics and (2) dissolved materials are removed by natural processes just as fast as they are supplied by the rivers. Some of the dissolved materials, such as calcium and silicon, are removed by plants and animals to make solid shells, bones, and other hard parts. Other dissolved materials, such as iron, magnesium, and phosphorous, form solid deposits directly and also make sediments that settle to the ocean floor. Hard parts of plants and animals and solid deposits are cycled to the earth's interior along with suspended sediments that have settled out of the seawater. Studies of fossils and rocks indicate that the composition of seawater has changed little over the past 600 million years.

The dissolved materials of seawater are present in the form of ions because of the strong dissolving ability of water molecules. Almost all of the chemical elements are present, but only six ions make up more than 99 percent of any given sample of seawater. As shown in table 22.1, chlorine and sodium are the most abundant ions. These are the elements of sodium chloride, or common table salt. As a sample of seawater evaporates, the positive metal ions join with the different negative ions to form a complex mixture of ionic compounds known as *sea salt*. Sea salt is mostly sodium chloride, but it also contains salts of the four metal ions (sodium, magnesium, calcium, and potassium) combined with the different negative ions of chlorine, sulfate, bicarbonate, and so on. Note that this mixture also includes magnesium sulfate, a strong laxative known as Epsom salt.

The amount of dissolved salts in seawater is measured as **salinity.** Salinity is defined as the mass of salts dissolved in 1.0 kg, or 1,000 g of seawater. Since the salt content is reported in parts per thousand, the symbol ‰ is used (% means parts per hundred). Thus 35 ‰ means that 1,000 g of seawater contains 35 g of dissolved salts (and 965 g of water). This is the same concentration as a 3.5 percent salt solution (figure 22.12). Oceanographers use the salinity measure because the mass of a sample of seawater does not change with changes in the water temperature. Other measures of concentration are based on the volume of a sample, and the volume of a liquid does vary as it expands and contracts with changes in the temperature. Thus, by using

the salinity measure, any corrections due to temperature differences are eliminated.

The average salinity of seawater is about 35 ‰, but the concentration varies from a low of about 32 ‰ in some locations up to a high of about 36 ‰ in other locations. The salinity of seawater in a given location is affected by factors that tend to increase or decrease the concentration. The concentration is increased by two factors, evaporation and the formation of sea ice. Evaporation increases the concentration because it is water vapor only that evaporates, leaving the dissolved salts behind in a greater concentration. Ice that forms from freezing seawater increases the concentration because when ice forms the salts are excluded from the crystal structure. Thus, sea ice is freshwater, and the removal of this water leaves the dissolved salts behind in a greater concentration. The salinity of seawater is decreased by three factors: heavy precipitation, the melting of ice, and the addition of freshwater by a large river. All three of these factors tend to dilute seawater with freshwater, which lowers the concentration of salts.

Note that increases or decreases in the salinity of seawater are brought about by the addition or removal of freshwater. This changes only the amount of water present in the solution. The *kind* or *proportion* of the ions present (table 22.1) in seawater does not change with increased or decreased amounts of freshwater. The same proportion, meaning the same chemical composition, is found in seawater of any salinity of any sample taken from any location anywhere in the world, from any depth of the ocean, or taken any time of the year. Seawater has a remarkably uniform composition that varies only in concentration. This means that the ocean is well mixed and thoroughly stirred around the entire earth. How seawater becomes so well mixed and stirred on a worldwide basis is discussed in the next section.

If you have ever allowed a glass of tap water to stand for a period of time, you may have noticed tiny bubbles collecting as the water warms. These bubbles are atmospheric gases, such as nitrogen and oxygen, that were dissolved in the water (figure 22.13). Seawater also contains dissolved gases in addition to the dissolved salts. Near the surface, seawater contains mostly nitrogen and oxygen, in similar proportions to the mixture that is found in the atmosphere. There is more carbon dioxide than

A CLOSER LOOK | Estuary Pollution

Pollution is usually understood to mean something that is not naturally occurring and contaminates air, soil, or water to interfere with human health, well being, or quality of the environment. An important factor in understanding pollution is the size of the human population and the amount of material that might become a pollutant. When the human population was small and produced few biological wastes, there was not a pollution problem. The decomposers broke down the material into simpler nonpolluting substances such as water and carbon dioxide and no harm was done. For example, suppose one person empties the tea leaves remaining from a cup of tea into a nearby river once a week. In this case decomposer organisms in the water would break down the tea leaves almost as fast as they are added to the river. But imagine 100,000 people doing this every day. In this case the tea leaves are released faster than they decompose and the leaves become pollutants.

The part of the wide lower course of a river where the freshwater of the river mixes with the saline water from the oceans is called a coastal *estuary.* Estuary waters include bays and tidal rivers that serve as nursery areas for many fish and shellfish populations, including shrimp, oysters, crabs, and scallops. Unfortunately, the rivers carry pollution from their watersheds and adjacent wetlands to the estuary, where it impacts the fish and shellfish industry, swimming, and recreation.

In 1996 the U.S. Environmental Protection Agency asked the coastal states to rate the general water quality in their estuaries. The states reported that pollutants impact aquatic life in 31 percent of the area surveyed, violate shellfish harvesting criteria in 27 percent of the area surveyed, and violate swimming-use criteria in 16 percent of the area surveyed.

The most common pollutants affecting the surveyed estuaries were excessive *nutrients,* which were found in 22 percent of all the estuaries surveyed. Excessive nutrients stimulate population explosions of algae.

Fast-growing masses of algae block light from the habitat below, stressing the aquatic life. The algae die and eventually decompose and this depletes the available oxygen supply, leading to further fish and shellfish kills.

The second most common pollutant was the presence of *bacteria,* which pollute 16 percent of all the estuary waters surveyed. The presence of *E. coli* is evidence that sewage is polluting the water. Bacteria interfere with recreation activities of people and can contaminate fish and shellfish.

The states also reported that *toxic organic chemicals* pollute 15 percent of the surveyed waters, *oxygen-depleting chemicals* pollute 12 percent, and *petroleum products* pollute another 8 percent of the surveyed waters. These pollutants impact the fish and shellfish industry, swimming, and recreation activities that require contact with the water.

The leading sources of the pollutants were identified as industrial discharges, urban runoff, municipal wastewater, agriculture runoff, and wastes from landfills.

FIGURE 22.13

Air will dissolve in water, and cooler water will dissolve more air than warmer water. The bubbles you see here are bubbles of carbon dioxide that came out of solution as the soda became warmer.

you would expect, however, as seawater contains a large amount of this gas. More carbon dioxide can dissolve in seawater because it reacts with water to form carbonic acid, H_2CO_3, the same acid that is found in a bubbly cola. In seawater, carbonic acid breaks down into bicarbonate and carbonate ions, which tend to remain in solution. Water temperature and the salinity have an influence on how much gas can be dissolved in seawater, and increasing either, or both, will reduce the amount of gases that can be dissolved. Cold, lower salinity seawater in colder regions will dissolve more gases than the warm, higher salinity seawater in tropical locations. Abundant plant life in the upper, sunlit water tends to reduce the concentration of carbon dioxide and increase the concentration of dissolved oxygen through the process of photosynthesis. With increasing depth, less light penetrates the water, and below about 80 m (about 260 ft) there is insufficient light for photosynthesis. Thus, more plant life and more dissolved oxygen are found above this depth. Below this depth there are no plants, more dissolved carbon dioxide, and less dissolved oxygen. The oxygen-poor, deep ocean water does eventually circulate back to the surface, but the complete process may take several thousand years.

Movement of Seawater

Consider the enormity of the earth's ocean, which has a surface area of some 361 million km² (about 139 million mi²) and a volume of 1,370 million km³ (about 328 million mi³) of seawater. There must be a terrific amount of stirring in such an enormous amount of seawater to produce the well-mixed, uniform chemical composition that is found in seawater throughout the world. The amount of mixing required is more easily imagined if you consider the long history of the ocean, the very long period of time over which the mixing has occurred. Based on investigations of the movement of seawater, it has been estimated that there is a complete mixing of all the earth's seawater about every 2,000 years or so. With an assumed age of 3 billion years, this means that the earth's seawater has been mixed 3,000,000,000 ÷ 2,000, or 1.5 million times. With this much mixing, you would be surprised if seawater were *not* identical all around the earth.

How does seawater move to accomplish such a complete mixing? Seawater is in a constant state of motion, both on the surface and deep within. The surface has two types of motion: (1) *waves*, which have been produced by some disturbance, such as the wind, and (2) *currents*, which move water from one place to another. Waves travel across the surface as a series of wrinkles that range from a few centimeters high to more than 30 m (100 ft) high. Waves crash on the shore as booming breakers and make the surf. This produces local currents as water moves along the shore and back out to sea. There are also permanent, worldwide currents that move ten thousand times more water across the ocean than all the water moving in all the large rivers on the land. Beneath the surface there are currents that move water up in some places and move it down in other places. Finally, there are enormous deep ocean currents that move tremendous volumes of seawater. The overall movement of many of the currents on the surface and their relationship to the deep ocean currents are not yet fully mapped or understood. The surface waves are better understood. The general trend and cause of permanent, worldwide currents in the ocean can also be explained. The following is a brief description and explanation of waves, currents, and the deep ocean movements of seawater.

Waves

Any slight disturbance will create ripples that move across a water surface. For example, if you gently blow on the surface of water in a glass you will see a regular succession of small ripples moving across the surface. These ripples, which look like small, moving wrinkles, are produced by the friction of the air moving across the water surface. The surface of the ocean is much larger, but a gentle wind produces patches of ripples in a similar way. These patches appear, then disappear as the wind begins to blow over calm water. If the wind continues to blow, larger and longer-lasting ripples are made, and the moving air can now push directly on the side of the ripples. A ripple may eventually grow into an **ocean wave,** a moving disturbance that travels across the surface of the ocean. In its simplest form, each wave has a ridge, or mound, of water called a *crest,* which is followed by a depression called a *trough.* Ocean waves are basically repeating series of these crests and troughs that move across the surface like wrinkles (figure 22.14).

FIGURE 22.14
The surface of the ocean is rarely, if ever, still. Any disturbance can produce a wave, but most waves on the open ocean are formed by a local wind.

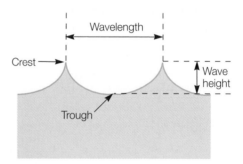

FIGURE 22.15
The simplest form of ocean waves, showing some basic characteristics. Most waves do not look like this representation because most are complicated mixtures of superimposed waves with a wide range of sizes and speeds.

The simplest form of an ocean wave can be described by measurements of three distinct characteristics: (1) the *wave height,* which is the vertical distance between the top of a crest and the bottom of the next trough, (2) the *wavelength,* which is the horizontal distance between two successive crests (or other successive parts of the wave), and (3) the *wave period,* which is the time required for two successive crests (or other successive parts) of the wave to pass a given point (figure 22.15).

The characteristics of an ocean wave formed by the wind depend on three factors: (1) the wind speed, (2) the length of time that the wind blows, and (3) the *fetch,* which is the distance the wind blows across the open ocean. As you can imagine, larger waves are produced by strong winds that blow for a longer time over a long fetch. In general, longer-blowing, stronger winds produce waves with greater wave heights, longer wavelengths, and longer periods, but a given wind produces waves with a wide range of sizes and speeds. In addition, the wind does not blow in just one direction, and shifting winds produce a chaotic pattern of waves of many dif-

FIGURE 22.16

Water particles are moved in a circular motion by a wave passing in the open ocean. On the surface, a water particle traces out a circle with a diameter that is equal to the wave height. The diameters of the circles traced out by water particles decrease with depth to a depth that is equal to one-half the wavelength of the ocean wave.

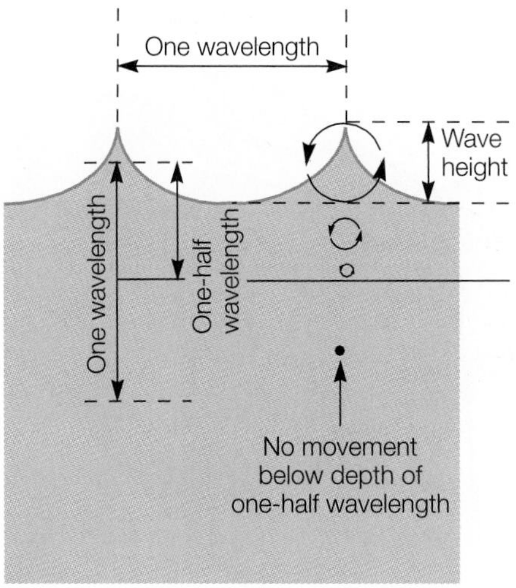

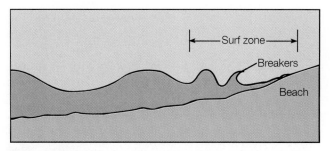

FIGURE 22.17

As a pattern of swell approaches a gently sloping beach, friction between the circular motion of the water particles and the bottom slows the wave, and the wave front becomes steeper and steeper. When the depth is about one and one-third times the wave height, the wave breaks forward, moving water toward the beach.

ferent heights and wavelengths. Thus, the surface of the ocean in the area of a storm or strong wind has a complicated mixture of many sizes and speeds of superimposed waves. The smaller waves soon die out from friction within the water, and the larger ones grow as the wind pushes against their crests. Ocean waves range in height from a few centimeters up to more than 30 m (about 100 ft), but giant waves more than 15 m (about 50 ft) are extremely rare.

The larger waves of the chaotic, superimposed mixture of waves in a storm area last longer than the winds that formed them, and they may travel for hundreds or thousands of kilometers from their place of origin. The longer wavelength waves travel faster and last longer than the shorter wavelength waves, so the longer wavelength waves tend to outrun the shorter wavelength waves as they die out from energy losses to water friction. Thus, the irregular, superimposed waves created in the area of a storm become transformed as they travel away from the area. They become regular groups of low-profile, long-wavelength waves that are called **swell.** The regular waves of swell that you might observe near a shore may have been produced by a storm that occurred days before, thousands of kilometers across the ocean.

The regular, low-profile crests and troughs of swell carry energy across the ocean, but they do not transport water across the open ocean. If you have ever been in a boat that is floating in swell, you know that you move in a regular pattern of up and forward on each crest, then backward and down on the following trough. The boat does not move along with the waves unless it is moved along by a wind or by some current. Likewise, a particle of water on the surface moves upward and forward with each wave crest, then backward and down on the following trough, tracing out a nearly circular path through this motion. The particle returns to its initial position, without any forward movement while tracing out the

small circle. Note that the diameter of the circular path is equal to the wave height (figure 22.16). Water particles farther below the surface also trace out circular paths as a wave passes. The diameters of these circular paths below the surface are progressively smaller with increasing depth. Below a depth equal to about half the wavelength there is no circular movement of the particles. Thus, you can tell how deeply the passage of a wave disturbs the water below if you measure the wavelength.

As swell moves from the deep ocean to the shore the waves pass over shallower and shallower water depths. When a depth is reached that is equal to about half the wavelength, the circular motion of the water particles begins to reach the ocean bottom. The water particles now move across the ocean bottom, and the friction between the two results in the waves moving slower as the wave height increases. These important modifications result in a change in the direction of travel and in an increasingly unstable situation as the wave height increases.

Most waves move toward the shore at some angle. As the wave crest nearest the shore starts to slow, the part still over deep water continues on at the same velocity. The slowing at the shoreward side *refracts*, or bends, the wave so it is more parallel to the shore. Thus, waves always appear to approach the shore head-on, arriving at the same time on all parts of the shore.

After the waves reach water that is less than one-half the wavelength, friction between the bottom and the circular motion of the water particles progressively slow the bottom part of the wave. The wave front becomes steeper and steeper as the top overruns the bottom part of the wave. When the wave front becomes too steep the top part breaks forward and the wave is now called a *breaker* (figure 22.17). In general, this occurs where the water depth is about one and one-third times the wave height. The zone where the breakers occur is called **surf** (figure 22.18).

Waves break in the foamy surf, sometimes forming smaller waves that then proceed to break in progressively shallower water. The surf may have several sets of breakers before the water is finally thrown on the shore as a surging sheet of seawater. The turbulence of the breakers in the surf zone and the final surge expend all the energy that the waves may have brought from

FIGURE 22.18

The white foam is in the surf zone, which is where the waves grow taller and taller, then break forward into a froth of turbulence. Do you see any evidence of rip currents in this picture?

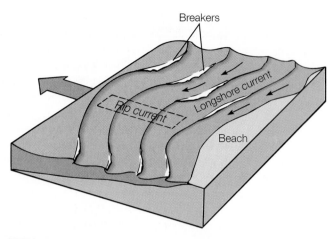

FIGURE 22.19

Breakers result in a buildup of water along the beach that moves as a longshore current. Where it finds a shore bottom that allows it to return to the sea, it surges out in a strong flow called a *rip current.*

thousands of kilometers away. Some of the energy does work in eroding the shoreline, breaking up rock masses into the sands that are carried by local currents back to the ocean. The rest of the energy goes into the kinetic energy of water molecules, which appears as a temperature increase.

Swell does not transport water with the waves over a distance, but small volumes of water are moved as a growing wave is pushed to greater heights by the wind over the open ocean. A strong wind can topple such a wave on the open ocean, producing a foam-topped wave known as a *whitecap.* In general, whitecaps form when the wind is blowing at 30 km/hr (about 20 mi/hr) or more.

Waves do transport water where breakers occur in the surf zone. When a wave breaks, it tosses water toward the shore, where the water begins to accumulate. This buildup of water tends to move away in currents, or streams, as the water returns to a lower level. Some of the water might return directly to the sea by moving beneath the breakers. This direct return of water forms a weak current known as *undertow.* Other parts of the accumulated water might be pushed along by the waves, producing a *longshore current* that moves parallel to the shore in the surf zone. This current moves parallel to the shore until it finds a lower place or a channel that is deeper than the adjacent bottom. Where the current finds such a channel, it produces a *rip current,* a strong stream of water that bursts out against the waves and returns water through the surf to the sea (figure 22.19). The rip current usually extends beyond the surf zone, and then diminishes. A rip current, or where rip currents are occurring, can usually be located by looking for the combination of (1) a lack of surf, (2) darker looking water, which means a deeper channel, and (3) a turbid, or muddy, streak of water that extends seaward from the channel indicated by the darker water that lacks surf.

In addition to waves created by winds, important waves are created by earthquakes and by tides. Movement of the earth's crust or underwater landslides can produce giant destructive waves called **tsunamis.** Tsunamis are sometimes incorrectly called "tidal waves" (they are in no way associated with tides or

tide-making processes). Tsunamis can be the largest of all ocean waves, but their tremendous, destructive energy is often not apparent until they reach the shore. A tsunami traveling across the deep ocean can have wavelengths up to 200 km (about 120 mi) and wave heights up to 0.5 m (about 1.5 ft) while traveling in an impulse-produced series of five to ten or so waves. Because of the small wave height, a tsunami often moves unnoticed across the deep ocean. But the tsunami forms taller waves as it slows in the shallow water near the shore, waves that can reach over 8 m (about 25 ft) tall. Waves of this size can result in tremendous property damage and loss of life along low coastal areas, especially if the waves make giant breakers. Many low coastal areas of Japan, Alaska, Hawaii, Chile, and the Philippines have experienced the destructive effects of tsunamis.

Tides are basically produced by the gravitational pull of the Moon and the rotation of the Earth-Moon system as discussed in chapter 18. The basic pattern of two high tides and two low tides every twenty-four hours and fifty minutes can produce strong *tidal currents* in narrow bays. These are reversing currents, moving up the bay with the rising tide and out to sea with the falling tide. If the bay is long and narrow, such as some in Alaska and Nova Scotia, a *tidal bore* may be produced. A tidal bore is a wave that moves rapidly up the bay as the tide rises. Sometimes the tidal bore is easily observed, and when the conditions are right, the bore is large enough to carry a person on a surfboard for a great distance.

Ocean Currents

Waves generated by the winds, earthquakes, and tidal forces keep the surface of the ocean in a state of constant motion. Local, temporary currents associated with this motion, such as rip currents or tidal currents, move seawater over a short distance. Seawater also moves in continuous **ocean currents,** streams of water that stay in about the same path as they move through other seawater over large distances. Ocean currents can be difficult to observe directly since they are surrounded by water that looks just like the water in the current. Wind is likewise difficult to observe directly since the moving air looks just like the rest of the atmosphere. Unlike the wind, an ocean current moves *continuously* in about the same path, often carrying water with different chemical and physical properties than the water it is moving through. Thus, an ocean current can be identified and tracked by measuring the physical and chemical characteristics of the current and the surrounding water. This shows where the current is coming from and where in the world it is going. In general, ocean currents are produced by (1) density differences in seawater and (2) winds that blow persistently in the same direction.

Density Currents. The density of seawater is influenced by three factors: (1) the water temperature, (2) salinity, and (3) suspended sediments. Cold water is generally more dense than warm water, thus sinking and displacing warmer water. Seawater of a high salinity has a higher relative density than less salty water, so it sinks and displaces water of less salinity. Likewise, seawater with a larger amount of suspended sediments has a higher relative density than clear water, so it sinks and displaces clear water. The following describes how these three ways of changing the density of seawater result in the ocean current known as a *density current,* which is an ocean current that flows because of density differences.

The earth receives more incoming solar radiation in the tropics than it does at the poles, which establishes a temperature difference between the tropical and polar oceans. The surface water in the polar ocean is often at or below the freezing point of freshwater, while the surface water in the tropical ocean averages about 26°C (about 79°F). Seawater freezes at a temperature below that of freshwater because the salt content lowers the freezing point. Seawater does not have a set freezing point, however, because as it freezes the salinity is increased as salt is excluded from the ice structure. Increased salinity lowers the freezing point more, so the more ice that freezes from seawater, the lower the freezing point for the remaining seawater. Cold seawater near the poles is therefore the densest, sinking and creeping slowly as a current across the ocean floor toward the equator. Where and how such a cold, dense bottom current moves is influenced by the shape of the ocean floor, the rotation of the earth, and other factors. The size and the distance that cold bottom currents move can be a surprise. Cold, dense water from the Arctic, for example, moves in a 200 m (about 660 ft) diameter current on the ocean bottom between Greenland and Iceland. This current carries an estimated 5 million cubic meters per second (about 177 million cubic ft/sec) of seawater to the 3.5 km (about 2.1 mi) deep water of the North Atlantic Ocean. This is a flow rate about 250 times larger than that of the Mississippi River. At about 30°N, the cold Arctic waters meet even denser water that has moved in currents all the way from the Antarctic to the deepest part of the North Atlantic Basin (figure 22.20).

A second type of density current results because of differences in salinity. The water in the Mediterranean, for example, has a high salinity because it is mostly surrounded by land in a warm, dry climate. The Mediterranean seawater, with its higher salinity, is more dense than the seawater in the open Atlantic Ocean. This density difference results in two separate currents that flow in opposite directions between the Mediterranean and the Atlantic.

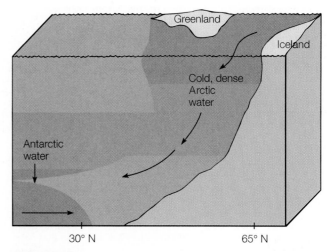

FIGURE 22.20

A cold density current carries about 250 times more water than the Mississippi River from the Arctic and between Greenland and Iceland to the deep Atlantic Ocean. At about 30°N latitude, it meets water that has moved by cold density currents all the way from the Antarctic.

The greater density seawater flows from the bottom of the Mediterranean into the Atlantic, while the less dense Atlantic water flows into the Mediterranean near the surface. The dense Mediterranean seawater sinks to a depth of about 1,000 m (about 3,300 ft) in the Atlantic, where it spreads over a large part of the North Atlantic Ocean. This increases the salinity of this part of the ocean, making it one of the more saline areas in the world.

The third type of density current occurs when underwater sediments on a slope slide toward the ocean bottom, producing a current of muddy or turbid water called a *turbidity current.* Turbidity currents are believed to be a major mechanism that moves sediments from the continents to the ocean basin. They may also be responsible for some undersea features, such as submarine canyons. Turbidity currents are believed to occur only occasionally, however, and none has ever been directly observed or studied. There is thus no data or direct evidence of how they form or what effects they have on the ocean floor.

Surface Currents. There are broad and deep-running ocean currents that slowly move tremendous volumes of water relatively near the surface. As shown in figure 22.21, each current is actually part of a worldwide system, or circuit, of currents. This system of ocean currents is very similar to the worldwide system of prevailing winds. This similarity exists because it is the friction of the prevailing winds on the seawater surface that drives the ocean currents. The currents are modified by other factors, such as the rotation of the earth and the shape of the ocean basins, but they are basically maintained by the wind systems.

Each ocean has a great system of moving water called a **gyre** that is centered in the mid-latitudes. The gyres rotate to the right in the Northern Hemisphere and to the left in the Southern Hemisphere because of the Coriolis effect (see chapter 18). The movement of water around these systems, or gyres, plus some smaller systems, form the surface circulation system

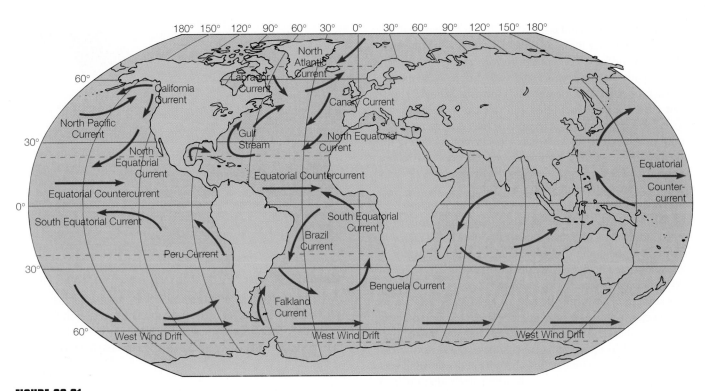

FIGURE 22.21

The earth's system of ocean currents.

of the world ocean. Each part of the system has a separate name, usually based on its direction of flow. All are called "currents" except one that is called a "stream" (the Gulf Stream) and those that are called "drifts." Both the Gulf Stream and the drifts are currents that are part of the connected system.

The major surface currents are like giant rivers of seawater that move through the ocean near the surface. You know that all the currents are connected, for a giant river of water cannot just start moving in one place, then stop in another. The Gulf Stream, for example, is a current about 100 km (about 60 mi) wide that may extend to a depth of 1 km (about 0.6 mi) below the surface, moving more than 75 million cubic meters of water per second (about 2.6 billion cubic ft/sec). The Gulf Stream carries more than 370 times more water than the Mississippi River. The California Current is weaker and broader, carrying cool water southward at a relatively slow rate. The flow rate of all the currents must be equal, however, since all the ocean basins are connected and the sea level is changing very little, if at all, over long periods of time.

THE OCEAN FLOOR

Some of the features of the ocean floor were discussed earlier because they were important in developing the theory of plate tectonics. Many features of the present ocean basins were created from the movement of large crustal plates, according to plate tectonics theory, and in fact some ocean basins are thought to have originated with the movement of these plates. There is also evidence that some features of the ocean floor were modi-

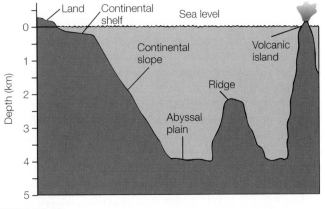

FIGURE 22.22

Some features of the ocean floor. Note that the inclination of the features is not as steep as this vertical exaggeration would suggest.

fied during the ice ages of the past. During an ice age, much water becomes locked up in glacial ice, which lowers the sea level. The sea level dropped as much as 140 m (about 460 ft) during the most recent major ice age, exposing the margins of the continents to erosion. Today, these continental margins are flooded with seawater, forming a zone of relatively shallow water called the **continental shelf** (figure 22.22). The continental shelf is considered to be a part of the continent and not the ocean, even though it is covered with an average depth of about 130 m (about 425 ft) of seawater. The shelf slopes gently away from the

A CLOSER LOOK | Deep Ocean Exploration and the Dynamic Earth

According to the theory of plate tectonics, the earth's surface is made up of a dozen or so rigid plates that are in motion, moving about 5 to 15 cm (about 2 to 6 in) per year with respect to one another. The plates are moving apart at the mid-ocean ridges where molten rock moves up from the mantle below, cools, and forms new crust. The size of the earth does not increase from the addition of the new crust, however, as old crust is incorporated back into the mantle at subduction zones. The subduction zone is where one plate is being forced under another plate. The spreading and the subduction boundaries have associated volcanic activity, faulting, and earthquakes, but the type of geologic activity associated with each boundary is very different.

The continents are generally viewed as less dense materials that "float" higher than the more dense materials making up the ocean crust. Thus, continents are moved away on plates as new ocean crust forms at ridges, then moves steadily away from the ridges and eventually sinks in a subduction zone. The continental plates sometimes collide and merge with other continental plates in the process. In addition, continental plates have split apart with a new ocean basin forming between the parts. In fact, the geologic evidence indicates that the continents were all together in one huge supercontinent at least twice in the past. The original supercontinent split apart to form new ocean basins and continents, then moved together to close the ocean basins and form a supercontinent for a second time. The second supercontinent split into separate continents, then moved apart for the second time to form the continents and ocean basins of today.

More evidence indicates that the opening and closing of the Atlantic Ocean basin might be periodic, operating on a 500-million-year cycle. Around the Pacific Ocean, on the other hand, there is no evidence of opening and closing. Another difference observed is that the ocean crust of the Pacific is being subducted under the continents around it, while the ocean crust of the Atlantic generally moves with the continents. This could mean that the Pacific basin is the original, permanent feature and the Atlantic basin is a temporary feature that opens and closes with movements of the continents.

If there is a pattern to the observations and if there is regularity to the opening and closing of the Atlantic Ocean, and not the Pacific, it could have meaning about changes in the climate of the entire earth. Do such patterns of change exist, and if they do, what does it mean about the future of the earth? One of the problems of seeking answers to such questions has been that the mid-ocean ridges and the subduction zones are at the bottom of the ocean, far beyond the reach of scientists. Only since the 1970s have advances in technology opened the way for study and direct observation of this important part of the earth's crust.

The investigation of the features of the ocean floor was made possible by the development of highly sensitive but durable cameras, high-resolution sonar scanners, and sensing devices to measure the magnetization, electrical, and seismic wave properties of crustal rocks. In addition to these instruments, which could be lowered and towed underwater by a surface ship, deep-diving submarine vehicles were developed to take research scientists directly to the deepest parts of the ocean. Both the instrument packages and a submarine vehicle named *Alvin* were used to first study the Atlantic and Pacific Ocean ridges in the 1970s. Some amazing discoveries were made during these expeditions.

The *Alvin* made several explorations in different locations of the ridge in the Pacific, in addition to an exploration of the Atlantic Ocean ridge and other studies. During the Pacific studies, which lowered three scientists to a depth of 2.6 km (about 1.6 mi), something unexpected was found on the ridge.

There were young flows of basalt lava, as was expected, but on the lava were mounts of chemical precipitates and organic debris. The mounds had funnel-shaped structures that were up to 10 m (about 33 ft) tall and 0.5 m (about 2 ft) wide. Hot, black fluids were billowing from these structures, vents that released water with temperatures up to 350°C (about 660°F) into the near-freezing seawater. The black color was from tiny particles of iron, copper, and zinc sulfides. Dense colonies of white crabs, large clams, and clusters of giant white tube worms with bright red plumes occupied an area about the size of a football field around the hydrothermal vents. These animals, previously unknown, were discovered in 1977 living in total darkness under a pressure that is 275 times greater than the pressure at the surface. Later studies found that these animals represented a whole new ecosystem, one based on sulfide-oxidizing bacteria rather than on green plants that utilize sunlight. The energy source for this ecosystem is chemical rather than sunlight. The hydrothermal vents released hot seawater, which seeped downward to the magma under the ridge, where it became superheated, dissolved sulfide minerals, then discharged back into the ocean. Such hydrothermal vents, and the sulfide-oxidizing ecosystems, were found in two separate locations some 3 km (about 2 mi) apart in the Pacific.

The mid-ocean ridge in the Pacific was found to be very different from the mid-ocean ridge in the Atlantic, indicating that perhaps the Pacific and the Atlantic ridges are created by different mechanisms. How or why these mechanisms are different is not yet known. Further deep ocean exploration will be needed to understand the processes that are involved in creating deep ocean ridges and what the differences might mean about the future of the earth. Who knows what other deep ocean surprises await the scientists who are trying to understand this complex earth and its cycles!

A CLOSER LOOK | Key Forecasting Tool for the Chesapeake Bay

Submerged aquatic vegetation—SAV for short—is vital to the Chesapeake Bay ecosystem and a key measuring stick for the Chesapeake's overall health. Eel and wigeon grasses grow in the shallows around the bay and across the bottom, providing food and shelter for baby blue crabs and fish, filtering pollutants, and providing life-sustaining oxygen for the water. So the health of the grasses provides an indication of the overall health of the Chesapeake.

The Chesapeake Bay once had an estimated 600,000 acres of grasses and provided an abundance of oysters, blue crab, shad, haddock, sturgeon, rockfish, and other prized sport fish and seafood. At this time the bay water was clear, and watermen reported they could clearly see grasses on the bottom some 6 meters (about 20 feet) below their boats. Then the water clouded, the grasses began to die, and the ecosystem of bay began to decline. The low point was reached in 1984 when less than 40,000 acres of grasses could be found in the now murky waters of the Chesapeake.

With the decline of the grasses came the decline of the aquatic species living in the bay. Shad, haddock, and sturgeon once supported large fisheries but are now scarce. The rockfish (striped bass) as well as the prized blue crab were once abundant but seem to decline and recover over the years. The decline of the blue crab population has shocked watermen because the crab is amazingly fertile. The crabs reach sexual maturity in about a year and one female bears millions of eggs. Part of the problem is the loss of the underwater grasses, which shelter the baby crabs from predators. In addition to the loss of habitat, some believe the crab is being overfished.

What happened to the underwater grasses of the Chesapeake? Scientists believe it is a combination of natural erosion and pollutants that wash from farms in the watershed. The pollutants include nutrient-rich fertilizers, chemical residues, and overflow from sewage treatment plants. There are some 6,000 chicken houses around the Chesapeake, raising more than 600 million chickens a year and producing 750,000 tons of manure. The

manure is used as fertilizer and significant amounts of nitrogen and phosphates wash from these fields. These nutrients accelerate algae growth in the bay, which blocks sunlight. The grass dies and the loss results in muddying of the water, making it impossible for new grasses to begin growing.

There is some evidence to support this idea since the 600,000 acres of underwater grasses died back to a low of 40,000 acres in 1984, then began to rebound when sewage treatment plants were modernized and there were less pollutants from industry and farms. The grasses recovered to some 63,500 acres by 1999, but are evidently very sensitive to even slight changes in water quality. Thus they may be expected to die back with unusual weather conditions that might bring more pollutants or muddy conditions, but continue to rebound when the conditions are right. The abundance of blue crabs and other aquatic species can be expected to fluctuate with the health of the grasses. The trends of underwater grass growth does indeed provide a key measure for the health of the Chesapeake Bay.

---

shore for an average of 75 km (about 47 mi), but it is much wider on the edge of some parts of continents than other parts.

The continental shelf is a part of the continent that happens to be flooded by seawater at the present time. It still retains some of the general features of the adjacent land that is above water, such as hills, valleys, and mountains, but these features were smoothed off by the eroding action of waves when the sea level was lower. Today, a thin layer of sediments from the adjacent land covers these smoothed-off features.

Beyond the gently sloping continental shelf is a steeper feature called the **continental slope.** The continental slope is the transition between the continent and the deep ocean basin. The water depth at the top of the continental slope is about 120 m (about 390 ft), then plunges to a depth of about 3,000 m (about 10,000 ft) or more. The continental slope is generally 20 to 40 km (about 12 to 25 mi) wide, so the inclination is similar to that encountered driving down a steep mountain road on an interstate highway. At various places around the world the continental slopes are cut by long, deep, and steep-sided *submarine canyons.* Some of these canyons extend from the top of the slope and down the slope to the ocean basin. Such a submarine canyon can be similar in size and depth to the Grand Canyon on the Colorado River

of Arizona. Submarine canyons are believed to have been eroded by turbidity currents, which were discussed earlier.

Beyond the continental slope is the bottom of the ocean floor, the **ocean basin.** Ocean basins are the deepest part of the ocean, covered by about 4 to 6 km (about 2 to 4 mi) of seawater. The basin is mostly a practically level plain called the **abyssal plain** and long, rugged mountain chains called *ridges* that rise thousands of meters above the abyssal plain. The Atlantic Ocean and Indian Ocean basins have ridges that trend north and south near the center of the basin. The Pacific Ocean basin has its ridge running north and south near the eastern edge. The Pacific Ocean basin also has more *trenches* than the Atlantic Ocean or the Indian Ocean basins. A trench is a long, relatively narrow, steep-sided trough that occurs along the edges of the ocean basins. Trenches range in depth from about 8 to 11 km (about 5 to 7 mi) deep below sea level.

The ocean basin and ridges of the ocean cover more than half of the earth's surface, accounting for more of the total surface than all the land of the continents. The plain of the ocean basin alone, in fact, covers an area about equal to the area of the land. Scattered over the basin are more than ten thousand steep volcanic peaks called *seamounts.* By definition, seamounts rise more than 1 km (about 0.6 mi) above the ocean floor, sometimes higher than the sea-level surface of the ocean. A seamount that sticks above the water level makes an island. The Hawaiian Islands are examples of such giant volcanoes that have formed islands. Most seamount-formed islands are in the Pacific Ocean. Most islands in the Atlantic, on the other hand, are the tops of volcanoes of the Mid-Atlantic Ridge.

SUMMARY

Precipitation that falls on the land either evaporates, flows across the surface, or soaks into the ground. Water moving across the surface is called *runoff*. Water that moves across the land as a small body of running water is called a *stream*. A stream drains an area of land known as the stream drainage basin or *watershed*. The watershed of one stream is separated from the watershed of another by a line called a *divide*. Water that collects as a small body of standing water is called a *pond*, and a larger body is called a *lake*. A *reservoir* is a natural pond, a natural lake, or a lake or pond created by building a dam for water management or control. The water of streams, ponds, lakes, and reservoirs is collectively called *surface water*.

Precipitation that soaks into the ground *percolates* downward until it reaches a *zone of saturation*. Water from the saturated zone is called *groundwater*. The amount of water that a material will hold depends on its *porosity*, and how well the water can move through the material depends on its *permeability*. The surface of the zone of saturation is called the *water table*.

The *ocean* is the single, continuous body of salt water on the surface of the earth. A *sea* is a smaller part of the ocean with different characteristics. The dissolved materials in seawater are mostly the ions of six substances, but sodium ions and chlorine ions are the most abundant. *Salinity* is a measure of the mass of salts dissolved in 1,000 grams of seawater.

An *ocean wave* is a moving disturbance that travels across the surface of the ocean. In its simplest form a wave has a ridge called a *crest* and a depression called a *trough*. Waves have a characteristic *wave height, wavelength,* and *wave period*. The characteristics of waves made by the wind depend on the wind *speed,* the *time* the wind blows, and the *fetch*. Regular groups of low-profile, long-wavelength waves are called *swell*. When swell approaches a shore, the wave slows and increases in wave height. This slowing *refracts*, or bends, the waves so they approach the shore head-on. When the wave height becomes too steep, the top part breaks forward, forming *breakers* in the *surf zone*. Water accumulates at the shore from the breakers and returns to the sea as *undertow*, as *longshore currents*, or in *rip currents*.

Earthquakes or undersea landslides produce large, destructive waves called *tsunamis*. Tsunamis do not have a large wave height at open sea, but they can do tremendous damage when they reach a low coastal area. A tide moving in or out of a narrow bay can produce a wave called a *tidal bore*.

Ocean currents are streams of water that move through other seawater over large distances. Some ocean currents are *density currents*, which are caused by differences in *water temperature, salinity,* or *suspended sediments*. Each ocean has a great system of moving water called a *gyre* that is centered in mid-latitudes. Different parts of a gyre are given different names such as the *Gulf Stream* or the *California Current*.

The ocean floor is made up of the *continental shelf,* the *continental slope,* and the *ocean basin*. The ocean basin has two main parts, the *abyssal plain* and mountain chains called *ridges*.

KEY TERMS

abyssal plain (p. **584**)

aquifer (p. **570**)

continental shelf (p. **582**)

continental slope (p. **584**)

freshwater (p. **566**)

groundwater (p. **568**)

gyre (p. **581**)

ocean (p. **574**)

ocean basin (p. **584**)

ocean currents (p. **580**)

ocean wave (p. **578**)

runoff (p. **567**)

salinity (p. **576**)

surf (p. **579**)

surface current (p. **581**)

swell (p. **579**)

tsunamis (p. **580**)

watershed (p. **568**)

water table (p. **569**)

APPLYING THE CONCEPTS

1. The most abundant chemical compound at the surface of the earth is
   a. silicon dioxide.
   b. nitrogen gas.
   c. water.
   d. minerals of iron, magnesium, and silicon.

2. Of the total supply, the amount of water that is available for human consumption and agriculture is
   a. 97 percent.
   b. about two-thirds.
   c. about 3 percent.
   d. less than 1 percent.

3. Considering yearly global averages of precipitation that fall on and evaporate from the land,
   a. more is precipitated than evaporates.
   b. more evaporates than is precipitated.
   c. there is a balance between the amount precipitated and the amount evaporated.
   d. there is no pattern that can be generalized.

4. In general, how much of all the precipitation that falls ends up as runoff and groundwater?
   a. 97 percent
   b. about half
   c. about one-third
   d. less than 1 percent

5. Groundwater is
   a. any water beneath the earth's surface.
   b. water beneath the earth's surface from a saturated zone.
   c. water that soaks into the ground.
   d. any of the above.

6. In a region of abundant rainfall a layer of extensively cracked, but otherwise solid, granite could serve as a limited source of groundwater because it has
   a. limited permeability and no porosity.
   b. average porosity and average permeability.
   c. no permeability and no porosity.
   d. limited porosity and no permeability.

7. How many different oceans are actually on the earth's surface?
   a. 14
   b. 7
   c. 3
   d. 1

8. The largest of the three principal ocean regions of the earth is the
   a. Atlantic Ocean.
   b. Pacific Ocean.
   c. Indian Ocean.
   d. South American Ocean.

9. The Gulf of Mexico is a shallow sea of the
   a. Atlantic Ocean.
   b. Pacific Ocean.
   c. Indian Ocean.
   d. South American Ocean.

10. Measurement of the salts dissolved in seawater taken from various locations throughout the world show that seawater has a
    a. uniform chemical composition and a variable concentration.
    b. variable chemical composition and a variable concentration.
    c. uniform chemical composition and a uniform concentration.
    d. variable chemical composition and a uniform concentration.

11. The percentage of dissolved salts in seawater averages about
    a. 35%.
    b. 3.5%.
    c. 0.35%.
    d. 0.035%.

12. The salinity of seawater is *increased* locally by
    a. the addition of water from a large river.
    b. heavy precipitation.
    c. the formation of sea ice.
    d. none of the above.

13. Considering only the available light and the dissolving ability of gases in seawater, more abundant life should be found in a
    a. cool, relatively shallow ocean.
    b. warm, very deep ocean.
    c. warm, relatively shallow ocean.
    d. cool, very deep ocean.

14. The regular, low profile waves called swell are produced from
    a. constant, prevailing winds.
    b. small, irregular waves becoming superimposed.
    c. longer wavelengths outrunning and outlasting shorter wavelengths.
    d. all wavelengths becoming transformed by gravity as they travel any great distance.

15. If the wavelength of swell is 10.0 m, then you know that the fish below the surface feel the waves to a depth of
    a. 5.0 m.
    b. 10.0 m.
    c. 20.0 m.
    d. however deep it is to the bottom.

16. In general, a breaker forms where the water depth is about 1 1/3 times the wave
    a. period.
    b. length.
    c. height.
    d. width.

17. The largest of all ocean waves is the
    a. tidal bore.
    b. swell.
    c. storm wave.
    d. tsunami.

18. Ocean currents are generally driven by
    a. the rotation of the earth.
    b. the prevailing winds.
    c. rivers from the land.
    d. all of the above.

19. The greatest volume of water is moved by the
    a. Mississippi River.
    b. California Current.
    c. Gulf Stream.
    d. Colorado River.

20. The continental shelf, which is covered with an average depth of 130 m of seawater, is part of (the)
    a. continent, which is why it is called the continental shelf.
    b. abyssal plain.
    c. ocean basin.
    d. none of the above.

Answers

1. c **2.** d **3.** a **4.** c **5.** b **6.** a **7.** d **8.** b **9.** a **10.** a **11.** b **12.** c **13.** a **14.** c **15.** a **16.** c **17.** d **18.** b **19.** c **20.** a

QUESTIONS FOR THOUGHT

1. How are the waters of the earth distributed as a solid, a liquid, and a gas at a given time? How much of the water is salt water and how much is freshwater?
2. Describe the hydrologic cycle. Why is the hydrologic cycle important in maintaining a supply of freshwater? Why is the hydrologic cycle called a cycle?
3. Describe in general all the things that happen to the water that falls on the land.
4. Explain how a stream can continue to flow even during a dry spell.
5. What is the water table? What is the relationship between the depth to the water table and the depth that a well must be drilled? Explain.
6. Compare the advantages and disadvantages of using (a) surface water and (b) groundwater as a source of freshwater.
7. Prepare arguments for (a) agriculture, (b) industries, and (c) cities each having first priority in the use of a limited water supply. Identify one of these arguments as being the "best case" for first priority, then justify your choice.
8. Discuss some possible ways of extending the supply of freshwater.
9. The world's rivers and streams carry millions of tons of dissolved materials to the ocean each year. Explain why this does not increase the salinity of the ocean.
10. What is swell and how does it form?
11. Why do waves always seem to approach the shore head-on?
12. What factors determine the size of an ocean wave made by the wind?
13. Describe how a breaker forms from swell. What is surf?
14. Describe what you would look for to avoid where rip current occurs at a beach.

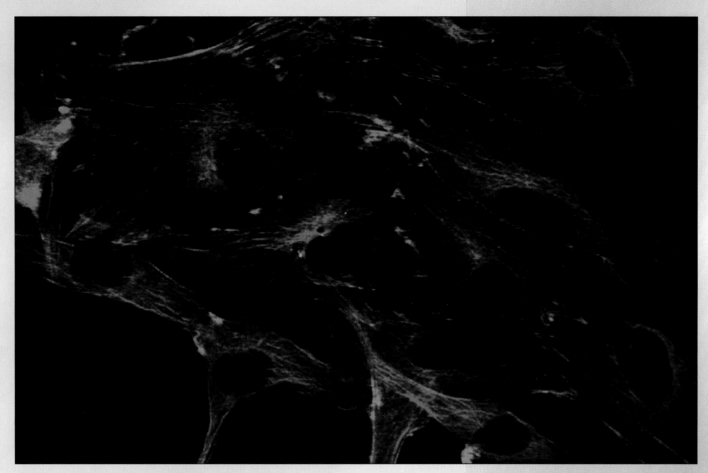

The cell is the simplest structure capable of existing as an individual living unit. Within this unit, many chemical reactions are required for maintaining life. These reactions do not occur at random, but are associated with specific parts of the many kinds of cells.

CHAPTER | Twenty-Three

What Is Life?

The science of **biology** is, broadly speaking, the study of living things. It draws on chemistry and physics for its foundation and applies these basic physical laws to living things. Because there are many kinds of living things, there are many special areas of study in biology. Practical biology—such as medicine, crop science, plant breeding, and wildlife management—is balanced by basic research in biology—such as medical microbiological physiology, photosynthetic biochemistry, plant taxonomy, and animal behavior (ethology). There is also just plain fun biology like insect collecting and bird watching. Specifically, biology is a science that deals with living things and how they interact with their surroundings.

What does it mean to be alive? You would think that a science textbook could answer this question easily. However, this is more than just a theoretical question because in recent years it has been necessary to construct legal definitions of what life is and especially of when it begins and ends. The legal definition of death is important because it may determine whether a person will receive life insurance benefits or if body parts may be used in transplants. In the case of heart transplants, the person donating the heart may be legally "dead," but the heart certainly isn't since it can be removed while it still has "life." In other words, there are different kinds of death. There is the death of the whole living unit and the death of each cell within the living unit. A person actually "dies" before every cell has died. Death, then, is the absence of life, but that still doesn't tell us what life is. At this point, we won't try to define life but we will describe some of the basic characteristics of living things.

CHARACTERISTICS OF LIFE

The ability to manipulate energy and matter is unique to living things. Understanding just how this is accomplished will help you understand how living things differ from nonliving objects. Living things show five characteristics that the nonliving do not display: (1) metabolic processes, (2) generative processes, (3) responsive processes, (4) control processes, and (5) a unique structural organization. This section gives a brief introduction to the basic characteristics of living things that will be expanded upon later.

Metabolic processes are all the chemical reactions and associated energy changes taking place within an organism. Energy is necessary for movement, growth, and many other activities. The energy that organisms use comes from energy stored in the chemical bonds of complex molecules. This energy becomes available through a controlled sequence of chemical reactions.

There are three essential aspects of metabolism: (1) *nutrient uptake,* (2) *nutrient processing,* and (3) *waste elimination.* All living things expend energy to take in nutrients (raw materials) from their environment. Many animals take in these materials by eating or swallowing other organisms. Microorganisms and plants absorb raw materials into their cells to maintain their lives. Once inside, raw materials are used in a series of chemical reactions to manufacture new parts, make repairs, reproduce, and provide energy for essential activities. However, not all the raw materials entering a living thing are valuable to it. There may be portions that are useless or even harmful. Organisms eliminate these portions as waste. Metabolic processes also produce unusable heat energy, which may also be considered a waste product.

The second group of characteristics of life, **generative processes,** are reactions that result in an increase in the size of an individual organism—*growth*—or an increase in the number of individuals in a population of organisms—*reproduction.*

During growth, living things add to their structure, repair parts, and store nutrients for later use. Growth and reproduction are directly related to metabolism, since neither can occur without the acquisition and processing of nutrients.

Reproduction is one of the most important life functions because all organisms eventually die. Thus reproduction is the only way that living things can perpetuate themselves. There are a number of different ways that various kinds of organisms reproduce and guarantee their continued existence. Some reproductive processes known as sexual reproduction involve two organisms contributing to the creation of a unique, new organism. Asexual reproduction occurs when organisms make identical copies of themselves.

Organisms also respond to changes within their bodies and in their surroundings in a meaningful way. These **responsive processes** have been organized into three categories: *irritability, individual adaptation,* and *population adaptation* or *evolution.* Irritability is an individual's rapid response to a stimulus, such as your response to a loud noise, beautiful sunset, or noxious odor. This type of response occurs only in the individual receiving the stimulus and is rapid because the mechanism that allows the response to occur (i.e., muscles, bones, and nerves) is already in place. Individual adaptation is also an individual response but is slower since it requires a growth or some other fundamental change in an organism. For example, a weasel's fur color will change from its brown summer coat to its white winter coat when genes responsible for the production of brown pigment are "turned off." Or the response of our body to disease organisms requires a change in the way cells work that eventually gets control of the organism causing the disease. Population adaptation involves changes in the kinds of characteristics displayed by individuals within the population. It is also known as *evolution,* which is a change in the genetic makeup of a *population* of organisms. This process occurs over long periods of time and enables a species (specific kind of organism) to adapt and better survive long-term changes in its environment over many generations. For example, the structures that give birds the ability to fly long distances allow them to respond to a world in which the winter season presents severe conditions that would threaten survival. Similarly, the ability of humans to think and use tools allows them to survive and be successful in a great variety of environmental conditions.

The **control processes** of *coordination* and *regulation* constitute the fourth characteristic of life. Control processes are mechanisms that ensure that an organism will carry out all metabolic activities in the proper sequence (coordination) and at the proper rate (regulation). All the chemical reactions of an organism are coordinated and linked together in specific pathways. The orchestration of all the reactions ensures that there will be specific stepwise handling of the nutrients needed to maintain life. The molecules responsible for coordinating these reactions are known as *enzymes.* **Enzymes** are molecules produced by organisms, molecules that are able to increase and control the rate at which life's chemical reactions occur. Enzymes also regulate the amount of nutrients processed into other forms.

In addition to these four basic processes that are typical of living things, living things also share some basic **structural similarities.** All living things are organized into complex structural units called *cells.* The cell units have an outer limiting membrane and internal structural units that have specific functions. Some living things, like you, consist of trillions of cells with specialized abilities that interact to provide the independently functioning unit called an *organism* (figure 23.1). Typically, in such large, multicellular organisms as humans, cells cooperate with one another in units called *tissues* (e.g., muscle, nervous). Groups of tissues are organized into larger units known as *organs* (e.g., heart), and in turn, into *organ systems* (e.g., circulatory system). Other organisms, such as bacteria or yeast, carry out all four of the life processes within a single cell. Nonliving materials, such as rocks, water, or gases, do not share a structurally complex common subunit. Figure 23.2 summarizes the characteristics of living things.

THE CELL THEORY

The concept of a *cell* is one of the most important ideas in biology because it applies to all living things. It did not emerge all at once, but has been developed and modified over many years. It is still being modified today.

Several individuals made key contributions to the cell concept. Anton van Leeuwenhoek (1632–1723) was one of the first to make use of a *microscope* to examine biological specimens. When van Leeuwenhoek discovered that he could see things moving in pond water using his microscope, his curiosity stimulated him to look at a variety of other things. He studied blood, semen, feces, pepper, and tartar, for example. He was the first to see individual cells and recognize them as living units, but he did not call them cells. The name he gave to these "little animals" that he saw moving around in the pond water was *animalcules.*

The first person to use the term "cell" was Robert Hooke (1635–1703) of England, who was also interested in how things looked when magnified. He chose to study thin slices of cork from the bark of a cork oak tree. He saw a mass of cubicles fitting neatly together, which reminded him of the barren rooms in a monastery. Hence, he called them *cells.* As it is currently used, the term **cell** refers to the basic structural unit that makes up all living things. When Hooke looked at cork, the tiny boxes he saw were, in fact, only the cell walls that surrounded the living portions of plant cells. We now know that the cell wall is composed of the complex carbohydrate cellulose, which provides strength and protection to the living contents of the cell. The cell wall appears to be a rigid, solid layer of material, but in reality it is composed of many interwoven strands of cellulose molecules. Its structure allows certain very large molecules to pass through it readily, but it acts as a screen to other molecules.

Hooke's use of the term *cell* in 1666 in his publication *Micrographia* was only the beginning, for nearly two hundred years passed before it was generally recognized that all living things are made of cells and that these cells can reproduce themselves. In 1838, Mathias Jakob Schleiden stated that all plants are made up of smaller cellular units. In 1839, Theodor Schwann published the idea that all animals are composed of cells.

Soon after the term *cell* caught on, it was recognized that the cell's vitally important portion is inside the cell wall. This living material was termed **protoplasm,** which means *first-formed substance.* The term *protoplasm* allowed scientists to distinguish between the living portion of the cell and the nonliving cell wall. Very soon microscopists were able to distinguish two different regions of protoplasm. One type of protoplasm was more viscous and darker than the other. This region, called the **nucleus** or core, is a central body within a more fluid material surrounding it. **Cytoplasm** is the name given to the colloidal fluid portion of the protoplasm (figure 23.3). While the term *protoplasm* is seldom used today, the term *cytoplasm* is still very common in the vocabulary of cell biologists.

The development of better microscopes and better staining techniques revealed that protoplasm contains many tiny structures called **organelles.** It has been determined that certain functions are performed in certain organelles. The essential job an organelle does is related to its structure. Each organelle is dynamic in its operation, changing shape and size as it works. Organelles move throughout the cell, and some even self-duplicate.

To date, most biologists recognize two major cell types, *prokaryotes* and *eukaryotes. Prokaryotic cells* are structurally more simple because they do not have the typical nucleus and do not have as great a variety of organelles in comparison to eukaryotes. Most single-celled organisms we commonly refer to as *bacteria* are prokaryotic cells. Other less well-known prokaryotes display significantly different traits that have caused biologists to believe that they are more ancient than the bacteria. This has resulted in some biologists creating a second category of prokaryotes, the archaebacteria. All other living things are based on the *eukaryotic cell* plan, with a true nucleus and many organelles. Eukaryotic cells are found in algae, fungi, plants, and animals.

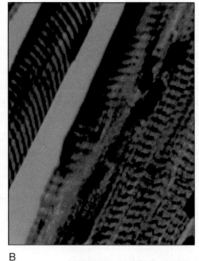

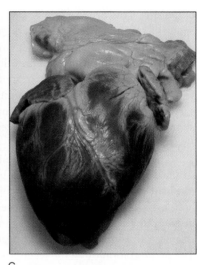

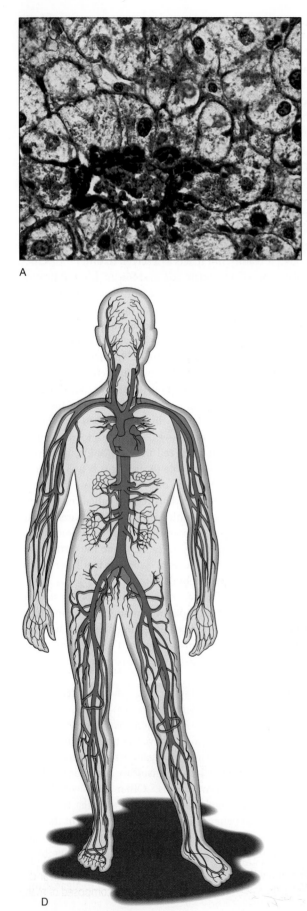

A

B

C

D

E

FIGURE 23.1

The basic building block of all living things is the cell. (*A*) These cells are typical human liver cells. Although the components of these cells are not "alive," they function together as a "living" thing. (*B*) When cells with similar structure and function work together to serve a common purpose, they are called a tissue. This microscopic photograph shows the cells of human skeletal muscle. When various tissues work together they form organs, such as (*C*) a human heart. Organs in turn can work together as organ systems, such as (*D*) the human circulatory system. An organism (*E*) consists of several organ systems that are interconnected and function together as a unit.

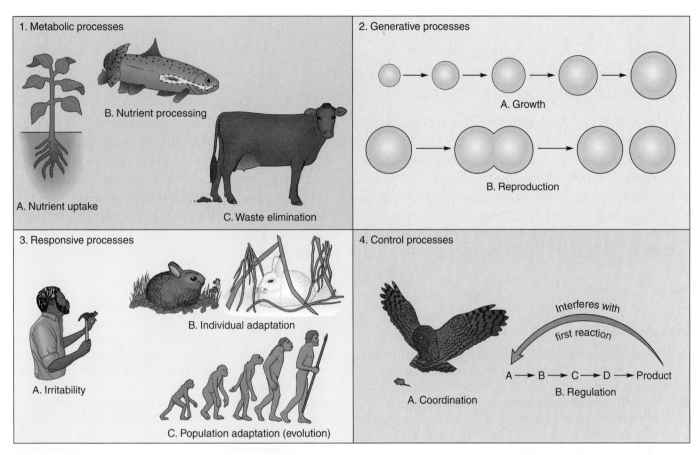

FIGURE 23.2

Living things demonstrate many common characteristics.

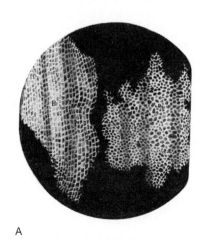

A

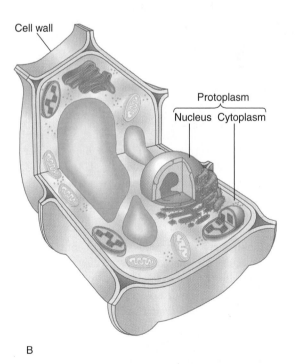

B

FIGURE 23.3

The cell concept has changed considerably over the last 300 years. Robert Hooke's idea of a cell (*A*) was based on his observation of slices of cork (cell walls of the bark of the cork oak tree). One of the first subcellular differentiations was to divide the protoplasm into cytoplasm and nucleus as shown in this plant cell (*B*). We now know that cells are much more complex than even this illustration; they are composed of many kinds of subcellular structures, some components numbering in the thousands.

CONNECTIONS

Casual Contact, Catching AIDS/HIV, and Cell Receptor Sites

Cell membranes have molecules on their surface known as *clusters of differentiation* or *CD markers*. These are designated by numbered groups, that is, CD1, CD2, CD3, CD4, etc. Different cell types have different markers on their surfaces. Those with CD4 markers include macrophages, nerve cells, and certain cells of the immune system known as T-helper (T_H) and T-delayed-type (T_D) hypersensitivity cells. CD4 cells are typically found deep inside the body (e.g., in blood) and not on the skin. The CD4 marker acts as the attachment site for human immunodeficiency virus (HIV) which is responsible for *Acquired ImmunoDeficiency Syndrome* (AIDS). In order to cause AIDS, the virus must enter the CD4 cell and take command of its metabolism. HIV cannot attach to cells that do not have CD4 cell surface receptors. Since CD4 cells do not normally occur on the surface of the skin, casual contact (e.g., shaking hands, touching) is not likely to be the way this virus is transmitted from one person to another.

CELL MEMBRANES

One feature common to all cells and many of the organelles they contain is a thin layer of material called *membrane*. Membrane can be folded and twisted into many different structures, shapes, and forms. The particular arrangement of membrane of an organelle is related to the functions that it is capable of performing. This is similar to the way a piece of fabric can be fashioned into a pair of pants, a shirt, sheets, pillowcases, or a rag doll. All cellular membranes have a fundamental molecular structure that allows them to be fashioned into a variety of different organelles.

Cellular membranes are thin sheets composed primarily of phospholipids and proteins. The current hypothesis of how membranes are constructed is known as the **fluid-mosaic model,** which proposes that the various molecules of the membrane are able to flow and move about. The membrane maintains its form because of the physical interaction of its molecules with its surroundings. The phospholipid molecules of the membrane have one end (the glycerol portion) that is soluble in water and is therefore called **hydrophilic** (water loving). The other end that is not water soluble, called **hydrophobic** (water hating), is comprised of fatty acid. We commonly represent this molecule like a balloon with two strings. The inflated balloon represents the glycerol and negatively charged phosphate, while the two strings represent the uncharged fatty acids. Consequently, when phospholipid molecules are placed in water, they form a double-layered sheet, with the water-soluble (hydrophilic) portions of the molecules facing away from each other. This is commonly referred to as a *phospholipid bilayer*. If phospholipid molecules are shaken in a glass of water, the molecules will automatically form double-layered membranes. It is important to understand that the membrane formed is not rigid or stiff but resembles a heavy olive oil in consistency. The component phospholipids are in constant motion as they move with the surrounding water molecules and slide past one another.

The protein component of cellular membranes can be found on either surface of the membrane or in the membrane, among the phospholipid molecules. Many of the protein molecules are capable of moving from one side to the other. Some of these proteins help with the chemical activities of the cell. Others aid in the movement of molecules across the membrane by forming channels through which substances may travel or by acting as transport molecules (figure 23.4). In addition to phospholipid and protein, some protein molecules found on the outside surfaces of cellular membranes have carbohydrates or fats attached to them. These combination molecules are important in determining the "sidedness" (inside-outside) of the membrane and also help organisms recognize differences between types of cells. Your body can recognize disease-causing organisms because their surface proteins are different from those of its own cellular membranes. Some of these molecules also serve as attachment sites for specific chemicals, bacteria, protozoa, white blood cells, and viruses. Many dangerous agents cannot stick to the surface of cells and therefore cannot cause harm. For this reason cell biologists explore the exact structure and function of these molecules. They are also attempting to identify molecules that can interfere with the binding of such agents as viruses and bacteria in the hope of controlling infections.

Other molecules found in cell membranes are cholesterol and carbohydrates. Cholesterol is found in the middle of the membrane, in the hydrophobic region, because cholesterol is not water soluble. It appears to play a role in stabilizing the membrane and keeping it flexible. Carbohydrates are usually found on the outside of the membrane, where they are bound to proteins or lipids. They appear to play a role in cell-to-cell interactions and are involved in binding with regulatory molecules.

GETTING THROUGH MEMBRANES

If a cell is to stay alive it must meet the characteristics of life outlined earlier. This includes taking in nutrients and eliminating wastes and other by-products of metabolism. Several mechanisms allow cells to carry out the processes characteristic of life. They include diffusion, osmosis, dialysis, facilitated diffusion, active transport, and phagocytosis.

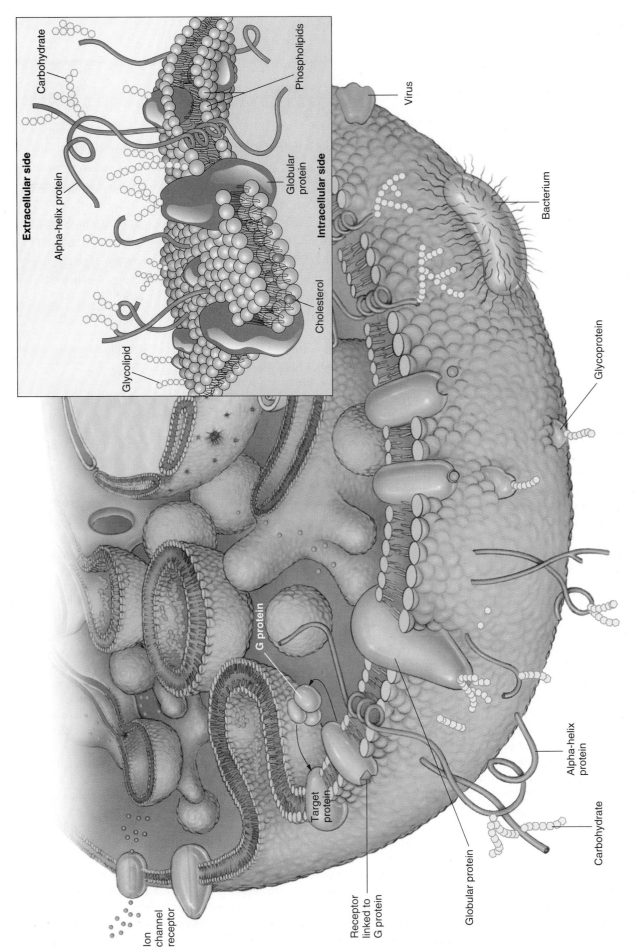

Extracellular side

Carbohydrate

Alpha-helix protein

Glycolipid

Phospholipids

Globular protein

Cholesterol

Intracellular side

Virus

Bacterium

Glycoprotein

Alpha-helix protein

Carbohydrate

Globular protein

Receptor linked to G protein

G protein

Target protein

Ion channel receptor

FIGURE 23.4

Notice in this section of a generalized human cell that there is no surrounding cell wall as pictured in Hooke's cell, figure 23.3. Membranes in all cells are composed of protein and phospholipids. Two layers of phospholipid are oriented so that the hydrophobic fatty ends extend toward each other and the hydrophilic glycerol portions are on the outside. The phosphate-containing chain of the phospholipid is coiled near the glycerol portion. Buried within the phospholipid layer and/or floating on it are the globular proteins. Some of these proteins accumulate materials from outside the cell; others act as sites of chemical activity. Carbohydrates are often attached to one surface of the membrane.

CONNECTIONS

The Other Outer Layer—The Cell Wall

The *cell walls* of microorganisms, plants, and fungi appear to be rigid, solid layers of material but are really loosely woven layers. And like water pouring through a leaky straw basket, many types of molecules easily pass through a cell wall. The cell wall lends strength and protection to the contents of the cell but hampers flexibility and movement. There are three kinds of materials typically used for cell walls. All the higher plants and most algae have *cellulose* as their wall material. When found in large amounts, it is known as *wood*. Another cell wall material, *chitin,* is found in the fungi. Chitin also constitutes the exoskeleton material in insects, and the outer shell of a beetle or a shrimp is chitin. But in those animals, the chitin surrounds masses of tissue instead of individual cells. The fungi are not crunchy or brittle like shrimp skeletons because the chitin is very thin. Many bacteria have their walls composed of *peptidoglycan*. It is composed of both amino acids and carbohydrates. The lengths of peptidoglycan chains and how they are interlinked may determine the shape of a bacterium.

Diffusion

There is a natural tendency in gases and liquids for molecules of different types to completely mix with each other. This is because they are constantly moving about with various levels of kinetic energy. Consider two types of molecules. As the molecules of one type move about, they tend to scatter from a central location. The other type of molecule also tends to disperse. The result of this random motion is that the two types of molecules are eventually mixed.

Remember that the motion of the molecules is completely random. If you follow the paths of molecules from a sugar cube placed in a glass of water, you will find that some of the sugar molecules move away from the cube while others move in the opposite direction. However, more sugar molecules would move away from the original cube because there were more there to start with.

We generally are not interested in the individual movement but rather in the overall movement, which is called the *net movement.* The direction of greatest movement (net movement) is determined by the relative concentration of the molecules. **Diffusion** is defined as the net movement of a kind of molecule from a place where that molecule is in higher concentration to a place where that molecule is scarcer. When a kind of molecule is completely dispersed, and movement is equal in all directions, we say that the system has reached a state of **dynamic equilibrium.** There is no longer a net movement because movement in one direction equals movement in the other. It is dynamic, however, because the system still has energy, and the molecules are still moving.

Because the cell membrane is composed of phospholipid and protein molecules that are in constant motion, temporary openings are formed that allow small molecules to cross from one side of the membrane to the other. Molecules close to the membrane are in constant motion as well. They are able to move into and out of a cell by passing through these openings in the membrane.

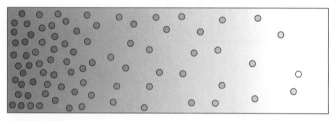

FIGURE 23.5

Gradual changes in concentrations of molecules over distance are called concentration gradients. This bar shows a color gradient of molecules with full color (concentrated molecules) at one end and no color (few molecules) at the other end. A concentration gradient is necessary for diffusion to occur. Diffusion results in net movement of molecules from an area of higher concentration to an area of lower concentration.

The rate of diffusion is related to the kinetic energy and size of the molecules. Since diffusion occurs only when molecules are unevenly distributed, the relative concentration of the molecules is important in determining how fast diffusion occurs. The difference in concentration of the molecules is known as a **concentration gradient** or **diffusion gradient.** When the molecules are equally distributed, no such gradient exists (figure 23.5).

Diffusion can take place only as long as there are no barriers to the free movement of molecules. In the case of a cell, the membrane permits some molecules to pass through, while others are not allowed to pass or are allowed to pass more slowly. This permeability is based on size, ionic charge, and solubility of the molecules involved. The membrane does not, however, distinguish direction of movement of molecules; therefore, the membrane does not influence the direction of diffusion. The direction of diffusion is determined by the relative concentration of specific molecules on the two sides of the membrane, and the energy that causes diffusion to occur is supplied by the kinetic energy of the molecules themselves (figure 23.6).

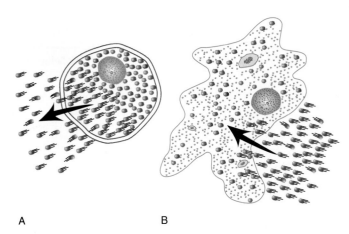

A   B

FIGURE 23.6

As a result of molecular motion, molecules move from areas where they are concentrated to areas where they are less concentrated. This figure shows (A) molecules leaving an animal cell by diffusion and (B) molecules entering a cell by diffusion. The direction is controlled by concentration (always high-to-low concentration), and the energy necessary is supplied by the kinetic energy of the molecules themselves.

Diffusion is an important means by which materials are exchanged between a cell and its environment. Since the movement of the molecules is random, the cell has little control over the process; thus, diffusion is considered a passive process, that is, chemical bond energy does not have to be expended. For example, animals are constantly using oxygen in various chemical reactions. Consequently, the oxygen concentration in cells always remains low. The cells, then, contain a lower concentration of oxygen than the oxygen level outside of the cells. This creates a diffusion gradient, and the oxygen molecules diffuse from the outside of the cell to the inside of the cell.

In large animals, many cells are buried deep within the body; if it were not for the animals' circulatory systems, cells would have little opportunity to exchange gases directly with their surroundings. The circulatory system is a transportation system within a body composed of blood vessels of various sizes. These vessels carry many different molecules from one place to another. Oxygen may diffuse into blood through the membranes of the lungs, gills, or other moist surfaces of the animal's body. The circulatory system then transports the oxygen-rich blood throughout the body. The oxygen automatically diffuses into cells. This occurs since the inside of cells is always low in oxygen because the oxygen combines with other molecules as soon as it enters. The opposite is true of carbon dioxide. Animal cells constantly produce carbon dioxide as a waste product, and so there is always a high concentration of it within the cells. These molecules diffuse from the cells into the blood, where the concentration of carbon dioxide is kept constantly low since the blood is pumped to the moist surface (gills, lungs, etc.), and the carbon dioxide again diffuses into the surrounding environment. In a similar manner, many other types of molecules constantly enter and leave cells.

Dialysis and Osmosis

Another characteristic of all membranes is that they are selectively permeable. **Selectively permeable** means that a membrane will allow certain molecules to pass across it and will prevent others from doing so. Molecules that are able to dissolve in phospholipids, such as vitamins A and D, can pass through the membrane rather easily; however, many molecules cannot pass through at all. In certain cases, the membrane differentiates on the basis of molecular size; that is, the membrane allows small molecules, such as water, to pass through and prevents the passage of larger molecules. The membrane may also regulate the passage of ions. If a particular portion of the membrane has a large number of positive ions on its surface, positively charged ions in the environment will be repelled and prevented from crossing the membrane.

We make use of diffusion across a selectively permeable membrane when we use a *dialysis machine* to remove wastes from the blood. If a kidney is unable to function normally, blood from a patient is diverted to a series of tubes composed of selectively permeable membrane. The toxins that have concentrated in the blood diffuse into the surrounding fluids in the dialysis machine, and the cleansed blood is returned to the patient. Thus, the machine functions in place of the kidney.

Water molecules easily diffuse through cell membranes. The net movement (diffusion) of water molecules through a selectively permeable membrane is known as **osmosis.** In any osmotic situation, there must be a selectively permeable membrane separating two solutions. For example, a solution of 90 percent water and 10 percent sugar separated by a selectively permeable membrane from a different sugar solution, such as one of 80 percent water and 20 percent sugar, demonstrates osmosis. The membrane allows water molecules to pass freely but prevents the larger sugar molecules from crossing. There is a higher concentration of water molecules in one solution compared to the concentration of water molecules in the other, so more of the water molecules move from the solution with 90 percent water to the other solution with 80 percent water. Be sure that you recognize that osmosis is really diffusion in which the diffusing substance is water, and that the regions of different concentrations are separated by a membrane that is more permeable to water.

A proper amount of water is required if a cell is to function efficiently. Too much water in a cell may dilute the cell contents and interfere with the chemical reactions necessary to keep the cell alive. Too little water in the cell may result in a buildup of poisonous waste products. As with the diffusion of other molecules, osmosis is a passive process because the cell has no control over the diffusion of water molecules. This means that the cell can remain in balance with an environment only if that environment does not cause the cell to lose or gain too much water.

If cells contain a concentration of water and dissolved materials equal to that of their surroundings, the cells are said to be **isotonic** to their surroundings. For example, the ocean contains many kinds of dissolved salts. Organisms such as sponges,

TABLE 23.1	The effects of osmosis on different cell types	
Cell Type	**What Happens When Cell Is Placed in Hypotonic Solution**	**What Happens When Cell Is Placed in Hypertonic Solution**
With cell wall; e.g., bacteria, fungi cell wall	Swells; does not burst due to presence of protective cell wall. Cells will become swollen (*turgid*) under these conditions.	Shrinks; cell membrane pulls away from inside of cell wall and forms compressed mass of protoplasm, a process known as *plasmolysis*. Cells will shrink under these conditions. Placing cells in salt water causes certain types of bacterial cells to tear their cell membranes away from the cell wall and results in their death.
Without cell wall; e.g., human	Swells and may burst. In red blood cells this process is called *hemolysis*.	Red blood cells shrink into compact mass, a process known as *crenation*.

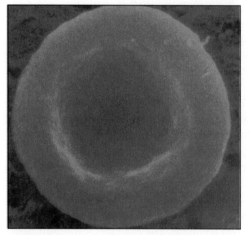

A Isotonic

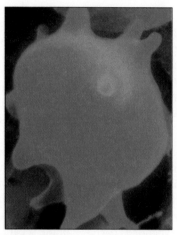

B Cell in hypertonic solution

C Cell in hypotonic solution

FIGURE 23.7

The cells in these three photographs were subjected to three different environments. (*A*) The cell is isotonic to its surroundings. The water concentration inside the red blood cell and the water concentration in the environment are in balance with each other, so movement of water into the cell equals movement of water out of the cell, and the cell has its normal shape. (*B*) The cell is in a hypertonic solution. Water has diffused from the cell to the environment because a high concentration of water was in the cell and the cell has shrunk. (*C*) A cell has accumulated water from the environment because a higher concentration of water was outside the cell than in its protoplasm. The cell is in a hypotonic solution so it is swollen.

jellyfishes, and protozoa are isotonic because the amount of material dissolved in their cellular water is equal to the amount of salt dissolved in the ocean's water.

If an organism is going to survive in an environment that has a different concentration of water than does its cells, it must expend energy to maintain this difference. Organisms that live in freshwater have a lower concentration of water (higher concentration of dissolved materials) than their surroundings and tend to gain water by osmosis very rapidly. They are said to be *hypertonic* to their surroundings, and the surroundings are *hypotonic*. These two terms are always used to compare two different solutions. The **hypertonic** solution is a solution with more dissolved material and less water; the **hypotonic** solution has less dissolved material and more water. It may help to remember that the water goes where the salt is (table 23.1). Organisms whose cells gain water by osmosis must expend energy to eliminate any excess if they are to keep from swelling and bursting (figure 23.7).

Under normal conditions, when we drink small amounts of water the cells of the brain swell a little, and signals are sent to the kidneys to rid the body of excess water. By contrast, marathon runners may drink large quantities of water in a very short time following a race. This rapid addition of water to the body may cause abnormal swelling of brain cells because the excess water cannot be gotten rid of rapidly enough. If this happens, the person may lose consciousness or even die because the brain cells have swollen too much.

Plant cells also experience osmosis. If the water concentration outside the plant cell is higher than the water concentration inside, more water molecules enter the cell than leave. This creates internal pressure within the cell. But plant cells do not burst because they are surrounded by a rigid cell wall. Lettuce cells that are crisp are ones that have gained water so that there is high internal pressure. Wilted lettuce has lost some of its water to its surroundings so that it has only slight internal cellular water pressure. Osmosis occurs when you put salad dressing on a

salad. Because the dressing has a very low water concentration, water from the lettuce diffuses from the cells into the surroundings. Salad that has been "dressed" too long becomes limp and unappetizing.

So far, we have considered only situations in which cells have no control over the movement of molecules. Cells cannot rely solely on diffusion and osmosis, however, because many of the molecules they require either cannot pass through the cell membranes or occur in relatively low concentrations in the cells' surroundings.

Controlled Methods of Transporting Molecules

Some molecules move across the membrane by combining with specific carrier proteins. When the rate of diffusion of a substance is increased in the presence of a carrier, we call this **facilitated diffusion.** Since this is diffusion, the net direction of movement is in accordance with the concentration gradient. Therefore, this is considered a passive transport method, although it can occur only in living organisms with the necessary carrier proteins. One example of facilitated diffusion is the movement of glucose molecules across the membranes of certain cells. In order for the glucose molecules to pass into these cells, specific proteins are required to carry them across the membrane. The action of the carrier does not require an input of energy other than the kinetic energy of the molecules (figure 23.8).

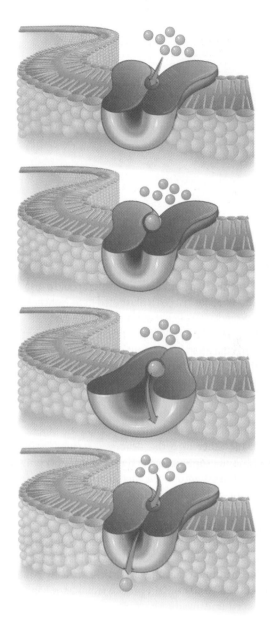

FIGURE 23.8

This method of transporting materials across membranes is a diffusion process; i.e., a movement of molecules from a high to a low concentration. However, the process is helped (facilitated) by a particular membrane protein. No chemical-bond energy in the form of ATP is required for this process. The molecules being moved through the membrane attach to a specific transport carrier protein in the membrane. This causes a change in its shape that propels the molecule or ion through to the other side.

When molecules are moved across the membrane from an area of *low* concentration to an area of *high* concentration, the cell must expend energy. The process of using a carrier protein to move molecules up a concentration gradient is called **active transport** (figure 23.9). Active transport is very specific: Only certain molecules or ions can be moved in this way, and specific proteins in the membrane must carry them. The action of the

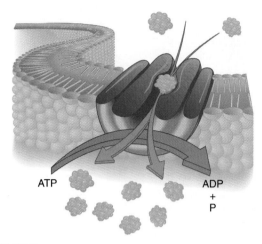

FIGURE 23.9

One possible method whereby active transport could cause materials to accumulate in a cell is illustrated here. Notice that the concentration gradient is such that if simple diffusion were operating, the molecules would leave the cell. The action of the carrier protein requires an active input of energy (the compound ATP) other than the kinetic energy of the molecules; therefore, this process is termed *active* transport.

carrier requires an input of energy other than the kinetic energy of the molecules; therefore, this process is termed *active* transport. For example, some ions, such as sodium and potassium, are actively pumped across cell membranes. Sodium ions are pumped out of cells up a concentration gradient. Potassium ions are pumped into cells up a concentration gradient.

In addition to active transport, materials can be transported into a cell by *endocytosis* and out by *exocytosis*. **Phagocytosis** is another name for one kind of endocytosis that is the process cells use to wrap membrane around a particle (usually food) and engulf it (figure 23.10). This is the process leukocytes (white blood cells) use to surround invading bacteria, viruses, and other foreign materials. Because of this, these kinds of cells are called *phagocytes*. When phagocytosis occurs, the material to be engulfed touches the surface of the phagocyte and causes a portion of the outer cell membrane to be indented. The indented cell membrane is pinched off inside the cell to form a sac containing the engulfed material. This sac, composed of a single membrane, is called a **vacuole.** Once inside the cell, the membrane of the vacuole is broken down, releasing its contents inside the cell, or it may combine with another vacuole containing destructive enzymes.

Many types of cells use phagocytosis to acquire large amounts of material from their environment. If a cell is not surrounding a large quantity of material but is merely engulfing some molecules dissolved in water, the process is termed **pinocytosis.** In this form of endocytosis, the sacs that are formed are very small in comparison to those formed during phagocytosis. Because of this size difference, they are called **vesicles.** In fact, an electron microscope is needed in order to see them. The processes of phagocytosis and pinocytosis differ from active transport in that the cell surrounds large amounts of

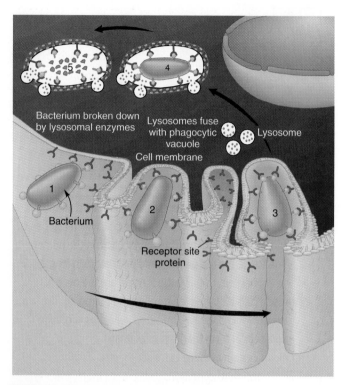

FIGURE 23.10

The sequence 1–5 illustrates a cell engulfing a large amount of material at one time and surrounding it with a membrane (phagocytosis). Once encased in a portion of the cell membrane (now called a phagocytic vacuole) a lysosome adds its digestive enzymes to it which speeds the breakdown of the material. Finally, the hydrolyzed (digested) material moves from the vacuole into the cytoplasm of the cell for further metabolism.

material with a membrane rather than taking the material in molecule by molecule through the membrane.

CELL SIZE

The size of a cell is directly related to its level of activity and the rate that molecules move across its membranes. In order to stay alive, a cell must have a constant supply of nutrients, oxygen, and other molecules. It must also be able to get rid of carbon dioxide and other waste products that are harmful to it. The larger a cell becomes, the more difficult it is to satisfy these requirements; consequently, most cells are very small. There are a few exceptions to this general rule, but they are easily explained. Egg cells, like the yolk of a hen's egg, are very large cells. However, the only part of an egg cell that is metabolically active is a small spot on its surface. The central portion of the egg is simply inactive stored food called *yolk.* Similarly, some plant cells are very large but consist of a large, centrally located region filled with water. Again, the metabolically active portion of the cell is at the surface (outer face), where exchange by diffusion or active transport is possible. The living material is the volume (amount of space) within the cell.

There is a mathematical relationship between the surface area and volume of a cell referred to as the *surface-to-volume ratio*. As cells grow, the amount of surface area increases by the square (A^2) but volume increases by the cube (V^3). They do not increase at the same rate. The surface area increases at a slower rate than the volume. Thus, the surface area-to-volume ratio changes as the cell grows. As a cell gets larger, cells have a problem with transporting materials across the plasma membrane. For example, diffusion of molecules is quite rapid over a short distance, but becomes slower over a longer distance. If a cell were to get too large, the center of the cell would die because transport mechanisms such as diffusion would not be rapid enough to allow for the exchange of materials. When the surface area is not large enough to permit sufficient exchange between the cell volume and the outside environment, cell growth stops. For example, the endoplasmic reticulum of eukaryotic cells provides an increase in surface area for taking up or releasing molecules. Cells lining the intestinal tract of humans have fingerlike extensions that also help in solving this problem.

ORGANELLES COMPOSED OF MEMBRANES

Now that you have some background concerning the structure and the function of membranes, let's turn our attention to the way cells use membranes to build the structural components of their protoplasm. The outer boundary of the cell is termed the **cell membrane** or **plasma membrane.** It is associated with a great variety of metabolic activities including taking up and releasing molecules, sensing stimuli in the environment, recognizing other cell types, and attaching to other cells and nonliving objects. In addition to the cell membrane, many other organelles are composed of membranes. Each of these membranous organelles has a unique shape or structure that is associated with particular functions.

One of the most common organelles found in cells is the *endoplasmic reticulum.* The **endoplasmic reticulum,** or *ER,* is a set of folded membranes and tubes throughout the cell. This system of membranes provides a large surface upon which chemical activities take place (figure 23.11). Since the ER has an enormous surface area, many chemical reactions can be carried out in an extremely small space.

Another organelle composed of membrane is the **Golgi apparatus.** Even though this organelle is also composed of membrane, the way in which it is structured enables it to perform jobs different from those performed by the ER. The typical Golgi is composed of from five to twenty flattened, smooth, membranous sacs, which resemble a stack of pancakes. The Golgi apparatus is the site of the synthesis and packaging of certain molecules produced in the cell. It is also the place where particular chemicals are concentrated prior to their release from the cell or distribution within the cell. Some Golgi vesicles are used to transport such molecules as mucus, carbohydrates, glycoproteins, insulin, and enzymes to the outside of the cell. The molecules are concentrated inside the Golgi, and tiny vesicles are pinched off or budded off the outside surfaces of the Golgi sacs. The vesicles move to and merge with the endoplasmic reticulum or cell membrane. In so doing, the contents are placed in the ER where they can be utilized or transported from the cell.

An important group of molecules that is necessary to the cell includes the hydrolytic enzymes. This group of enzymes is capable of destroying carbohydrates, nucleic acids, proteins, and lipids. Since cells contain large amounts of these molecules, these enzymes must be controlled in order to prevent the destruction of the cell. The Golgi apparatus is the site where these enzymes are converted from their inactive to their active forms and packaged in membranous sacs. These vesicles are pinched off from the outside surfaces of the Golgi sacs and given the special name **lysosomes,** or "bursting body." Cells use the lysosomes in four major ways:

1. When a cell is damaged, the membranes of the lysosomes break and the enzymes are released. These enzymes then begin to break down the contents of the damaged cell so that surrounding cells can use the component parts.
2. Lysosomes also play a part in the normal development of an organism. For example, as a tadpole slowly changes into a frog, the cells of the tail are destroyed by the action of lysosomes. In humans, the developing embryo has paddle-shaped hands and feet. At a prescribed point in the development, the cells between the bones of the fingers and toes release the enzymes that had been stored in the lysosomes. As these cells begin to disintegrate, individual fingers or toes begin to take shape. Occasionally this process does not take place, and infants are born with "webbed" fingers or toes. This developmental defect, called *syndactylism,* may be surgically corrected soon after birth (figure 23.12).
3. In many kinds of cells, the lysosomes are known to combine with food vacuoles. When this occurs, the enzymes of the lysosome break down the food particles into smaller and smaller molecular units. This process is common in one-celled organisms such as *Paramecium.*
4. Lysosomes are also used in the destruction of engulfed, disease-causing microorganisms such as bacteria, viruses, and fungi. As these invaders are taken into the cell by phagocytosis, lysosomes fuse with the phagocytic vacuole. When this occurs, the hydrolytic enzymes and proteins called *defensins* move from the lysosome into the vacuole to destroy the microorganisms.

Another submicroscopic vesicle is the **peroxisome.** In human cells, peroxisomes are responsible for producing hydrogen peroxide, H_2O_2. The peroxisome enzymes are able to manufacture H_2O_2 that is used in destroying invading microbes. The activity of H_2O_2 is easily demonstrated by mixing the enzyme catalase with H_2O_2. The enzyme converts the hydrogen peroxide to water and oxygen, which forms bubbles. It is the O_2 that is responsible for oxidizing potentially harmful microbes and other dangerous materials. Peroxisomes are also important since they contain enzymes that are responsible for the breakdown of long-chain fatty acids and the synthesis of cholesterol.

The many kinds of vacuoles and vesicles contained in cells are frequently described by their function. Thus, food vacuoles hold food, and water vacuoles store water. Specialized water vacuoles called *contractile vacuoles* are able to forcefully expel excess water that has accumulated in the cytoplasm as a result of osmosis. The contractile vacuole is a necessary organelle in cells that live in freshwater. The water constantly diffuses into the cell since the environment contains a higher concentration than inside the cell and therefore must be actively pumped out. The special containers that hold the contents resulting from pinocytosis are called *pinocytic vesicles.*

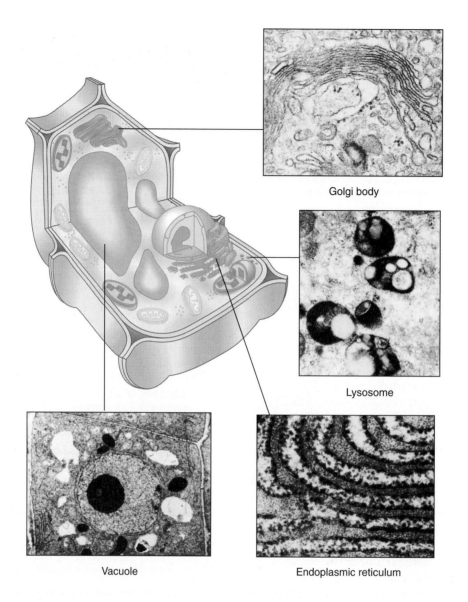

Golgi body

Lysosome

Vacuole

Endoplasmic reticulum

FIGURE 23.11

Certain structures in the cytoplasm are constructed of membranes. Membranes are composed of protein and phospholipids. The four structures here—the Golgi body, lysosomes, endoplasmic reticulum, and vacuoles—are constructed of simple membranes.

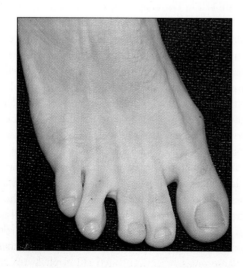

FIGURE 23.12

This person displays the trait known as syndactylism (*syn* = connected; *dactyl* = finger or toe). In most people, enzymes break down the connecting tissue, allowing the toes to separate. In this genetic abnormality, these enzymes fail to do their job.

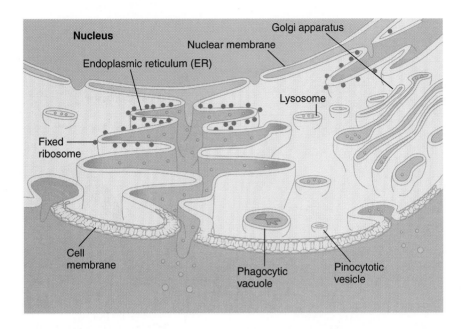

FIGURE 23.13

Eukaryotic cells contain a variety of organelles composed of phospholipids and proteins. Each has a unique shape and function. Many of these organelles are interconverted from one to another as they perform their essential functions. Cell membranes can become vacuolar membrane or endoplasmic reticulum, which can become vesicular membrane, which in turn can become Golgi or nuclear membrane. However, mitochondria cannot exchange membrane parts with other membranous organelles.

In all cases, these simple containers are constructed of a surrounding membrane. In most plants, there is one huge, centrally located vacuole in which water, food, wastes, and minerals are stored.

A nucleus is a place in a cell—not a solid mass. Just as a room is a place created by walls, a floor, and a ceiling, the nucleus is a place in the cell created by the **nuclear membrane.** This membrane separates the *nucleoplasm,* liquid material in the nucleus, from the cytoplasm. Because they are separated, the cytoplasm and nucleoplasm can maintain different chemical compositions. If the membrane were not formed around the genetic material, the organelle we call the nucleus would not exist. The nuclear membrane is formed from many flattened sacs fashioned into a hollow sphere around the genetic material, DNA (*d*eoxyribo*n*ucleic *a*cid). It also has large openings called nuclear pores that allow thousands of relatively large molecules to pass into and out of the nucleus each minute. These pores are held open by donut-shaped molecules that resemble the "eyes" in shoes through which the shoelace is strung.

All of the membranous organelles just described can be converted from one form to another (figure 23.13). For example, phagocytosis results in the formation of vacuolar membrane from cell membrane that fuses with lysosomal membrane, which in turn came from Golgi membrane. Two other organelles composed of membranes are chemically different and are incapable of interconversion. Both types of organelles are associated with energy conversion reactions in the cell. These organelles are the *mitochondrion* and the *chloroplast* (figure 23.14).

The **mitochondrion** is an organelle resembling a small bag with a larger bag inside that is folded back on itself. These inner folded surfaces are known as the **cristae.** Located on the surface of the cristae are particular proteins and enzymes involved in *aerobic cellular respiration.* **Aerobic cellular respiration** is the series of reactions involved in the release of usable energy from food molecules, which requires the participation of oxygen molecules. Enzymes that speed the breakdown of simple nutrients are arranged in a sequence on the mitochondrial membrane. The average human cell contains upwards of ten thousand mitochondria. Cells that are involved in activities that require large amounts of energy, such as muscle cells, contain many more mitochondria. When properly stained, they can be seen with a compound light microscope. When cells are functioning aerobically, the mitochondria swell with activity. But when this activity diminishes, they shrink and appear as threadlike structures.

A second energy-converting organelle is the **chloroplast.** This membranous, saclike organelle is found only in plants—organisms that carry out photosynthesis—and contains the green pigment *chlorophyll.* Some cells contain only one large chloroplast, while others contain hundreds of smaller chloroplasts. In this organelle light energy is converted to chemical-bond energy in a process known as *photosynthesis.* Chemical-bond energy is found in food molecules. A study of the ultrastructure—that is, the structures seen with an electron microscope—of a chloroplast shows that the entire organelle is enclosed by a membrane, while other membranes are folded and interwoven throughout. As shown in figure 23.14A, in some areas concentrations of these membranes are stacked up or folded back on themselves. Chlorophyll molecules are attached to these membranes. These areas of concentrated chlorophyll are called the

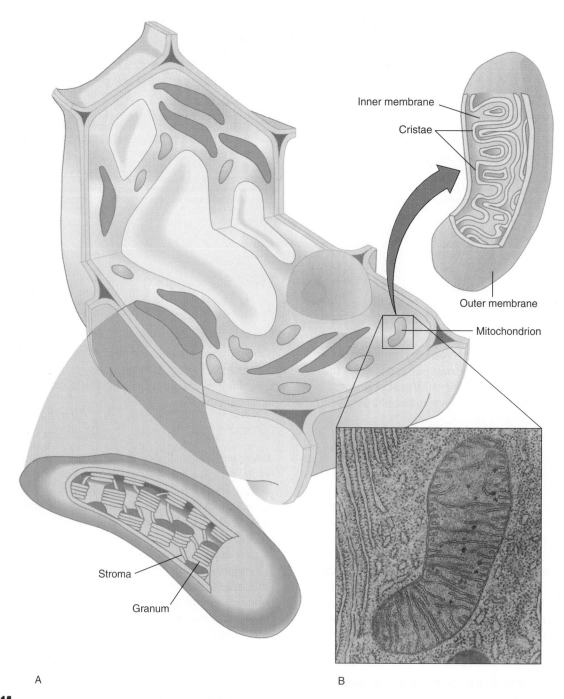

A

B

FIGURE 23.14

(*A*) The chloroplast, the container of the pigment chlorophyll, is the site of photosynthesis. The chlorophyll, located in the grana, captures light energy that is used to construct organic, sugarlike molecules in the stroma. (*B*) The mitochondria with their inner folds, called cristae, are the site of aerobic cellular respiration, where food energy is converted to usable cellular energy. Both organelles are composed of phospholipid and protein membranes.

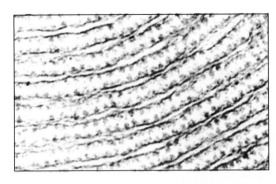

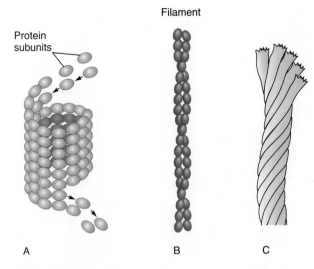

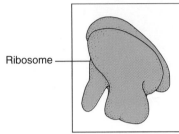

Ribosome

FIGURE 23.15

Each ribosome is constructed of two subunits of protein and ribonucleic acid. These globular organelles are associated with the construction of protein molecules from individual amino acids. They are sometimes located individually in the cytoplasm where protein is being assembled, or they may be attached to endoplasmic reticulum (ER). They are so obvious on the ER when using electron micrograph techniques that, when they are present, we label this ER *rough ER*.

grana of the chloroplast. The space between the grana, which has no chlorophyll, is known as the *stroma*.

Mitochondria and chloroplasts are different from other kinds of membranous structures in several ways. First, their membranes are chemically different from those of other membranous organelles. Second, they are composed of double layers of membrane—an inner and an outer membrane. Third, both of these structures have ribosomes and DNA that are similar to those of bacteria. Finally, these two structures have a certain degree of independence from the rest of the cell—they have a limited ability to reproduce themselves but must rely on nuclear DNA for assistance.

All of the organelles just described are composed of membranes. Many of these membranes are modified for particular functions. Each membrane is composed of the double phospholipid layer with protein molecules associated with it.

FIGURE 23.16

(*A*) Microtubules are hollow tubes constructed of the protein spheres called tubulin. The dynamic nature of the microtubule is useful in the construction of certain organelles in a cell, such as centrioles, spindle fibers, and cilia or flagella. (*B*) Microfilaments are composed of the contractile protein, actin. This is the same contractile protein found in human muscle cells. (*C*) Intermediate filaments are also protein but resemble a multistranded wire cable. These link microtubules and microfilaments.

NONMEMBRANOUS ORGANELLES

Suspended in the cytoplasm and associated with the membranous organelles are various kinds of structures that are not composed of phospholipids and proteins arranged in sheets.

In the cytoplasm are many very small structures called **ribosomes** that are composed of ribonucleic acid (RNA) and protein. Ribosomes function in the manufacture of protein. Many ribosomes are also found floating freely in the cytoplasm (figure 23.15), wherever proteins are being assembled. Cells that are actively producing protein (e.g., liver cells) have great numbers of free and attached ribosomes.

Among the many types of nonmembranous organelles found there are elongated protein structures that are known as *microtubules, microfilaments,* and *intermediate filaments* (figure 23.16). Their various functions are as complex as those provided by the structural framework and cables of a high-rise office building, geodesic dome, or skeletal and muscular systems of a large animal. All three types of organelles

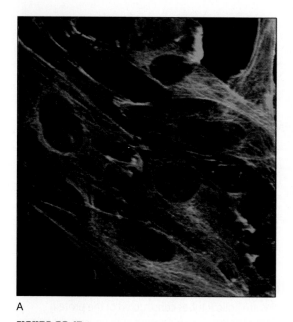

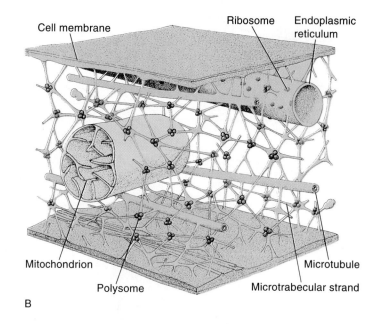

A

B

FIGURE 23.17

A complex array of microfilaments, microtubules, and intermediate filaments provides an internal framework structure for the cell. The cellular skeleton is not a rigid, fixed in place structure, but is dynamic and changes as the microfilament and microtubule component parts assemble and disassemble. (A) Elements of the cytoskeleton have been labeled with a fluorescent dye to make them visible. The microtubules have fluorescent red dye and actin filaments are green. Part (B) shows how the various parts of the cytoskeleton are interconnected.

Source for (B) from "The Ground Substance of the Living Cell" by Keith R. Porter and Jonathan B. Tucker, in *Scientific American,* March 1981. Copyright © 1981 by Scientific American, Inc. All Rights Reserved.

interconnect and some are attached to the inside of the cell membrane, forming what is known as the *cytoskeleton* of the cell (figure 23.17). These cellular components provide the cell with shape, support, and ability to move about the environment. They also serve to transport materials from place-to-place within the cytoplasm. Think of the cytoskeleton components as the internal supports and cables required to construct a circus tent. The shape of the flexible canvas cover (i.e., cell membrane) is determined by the location of internal tent poles (i.e., microtubules) and the tension placed on them by attached wire or rope cables (i.e., contractile microfilaments). The poles are made light and strong by being tubular (*tubulin* protein) and are attached to the inner surface of the canvas cover at specific points. The comparable cell membrane attachment points for microtubules are cell membrane molecules known as *integrins*. The reason the poles stay in place is because of their attachment to the canvas and the tension placed on them by the cables. As the cables are adjusted, the shape of the canvas (i.e., cell) changes. The intermediate filaments serve as cables that connect microfilaments and microtubules, thus providing additional strength and support. Just as in the tent analogy, when one of the microfilaments or intermediate filaments is adjusted, the shape of the entire cell changes. For example, when a cell is placed on a surface to which it cannot stick, the

internal tensions created by the cytoskeleton components can pull together and cause the cell to form a sphere.

An arrangement of two sets of microtubules at right angles to each other makes up a structure known as the *centriole*. The centrioles of many cells are located in a region known as the *centrosome,* or a structure called a *basal body*. The centrosome is usually located close to the nuclear membrane, while basal bodies are at the bases of cilia and flagella. They operate by organizing microtubules into a complex of strings called *spindle fibers*. The *spindle* is the structure upon which chromosomes are attached in order that they may be properly separated during cell division. Each set is composed of nine groups of short microtubules arranged in a cylinder (figure 23.18). One curious fact about centrioles is that they are present in most animal cells but not in many types of plant cells.

Many cells have microscopic, hairlike structures projecting from their surfaces; these are *cilia* or *flagella* (figure 23.19). In general, we call them flagella if they are long and few in number, and cilia if they are short and more numerous. They are similar in structure, and each functions to move the cell through its environment or to move the environment past the cell. They are constructed of a cylinder of nine sets of microtubules similar to those in the centriole, but they have an additional two microtubules in the center. These long strands of microtubules project

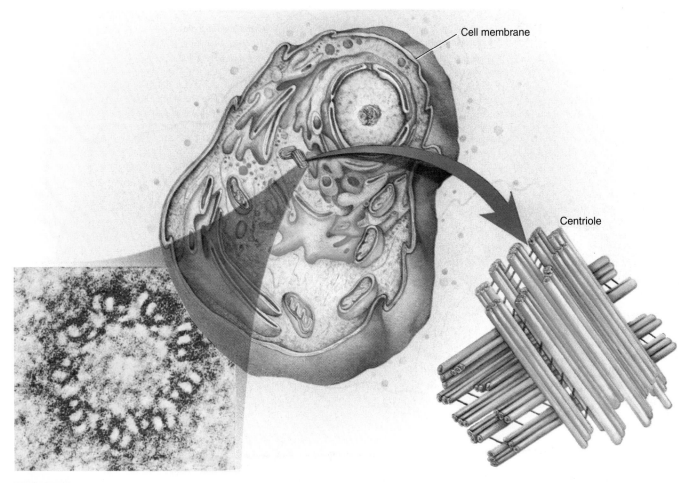

Cell membrane

Centriole

FIGURE 23.18

These two sets of short microtubules that constitute the centriole are located just outside the nuclear membrane in many types of cells. The micrograph shows an end view of one of these sets. Magnification is about 160,000 times.

from the cell surface and are covered by cell membrane. When cilia and flagella are sliced crosswise, their cut ends show what is referred to as the *9 + 2 arrangement* of microtubules. The cell has the ability to control the action of these microtubular structures, enabling them to be moved in a variety of different ways. Their coordinated actions either propel the cell through the environment or the environment past the cell surface. The protozoan *Paramecium* is covered with thousands of cilia that actively beat a rhythmic motion to move the cell through the water. The cilia on the cells that line your trachea move mucous-containing particles from deep within your lungs.

NUCLEAR COMPONENTS

As stated earlier, one of the first structures to be identified in cells was the nucleus. The nucleus was referred to as the cell center. If the nucleus is removed from a cell, the cell can live only a

short time. For example, human red blood cells begin life in bone marrow, where they have nuclei. Before they are released into the bloodstream to serve as oxygen and carbon dioxide carriers, they lose their nuclei. As a consequence, red blood cells are able to function only for about 120 days before they disintegrate.

When nuclear structures were first identified, it was noted that certain dyes stained some parts more than others. The parts that stained more heavily were called **chromatin,** which means colored material. Chromatin is composed of long molecules of deoxyribonucleic acid (DNA) in association with proteins. These DNA molecules contain the genetic information for the cell, the blueprints for its construction and maintenance. Chromatin is loosely organized DNA in the nucleus. When the chromatin is tightly coiled into shorter, denser structures, we call them **chromosomes.** Chromatin and chromosomes are really the same molecules but differ in structural arrangement. In addition to chromosomes, the nucleus may also contain one, two, or several *nucleoli*. A **nucleolus** is the site of ribosome manufacture. Nucleoli are composed of specific

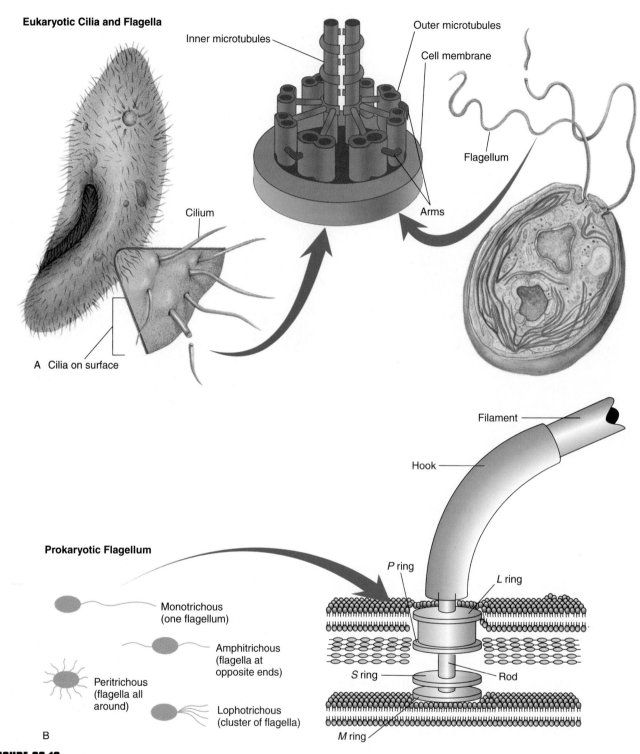

Eukaryotic Cilia and Flagella

Inner microtubules

Outer microtubules

Cell membrane

Cilium

Flagellum

A Cilia on surface

Arms

Filament

Hook

Prokaryotic Flagellum

P ring

L ring

Monotrichous
(one flagellum)

Amphitrichous
(flagella at
opposite ends)

Peritrichous
(flagella all
around)

Lophotrichous
(cluster of flagella)

S ring

Rod

M ring

B

FIGURE 23.19

(*A*) Cilia and flagella function like oars or propellers that move the cell through its environment or move the environment past the cell. Cilia and flagella are constructed of groups of microtubules as in the ciliated protozoan shown on the left and the flagellated alga on the right. Flagella are usually less numerous and longer than cilia. (*B*) The flagella of prokaryotes are composed of a single type of protein arranged in a fiber that is anchored into the cell wall and membrane. Bacterial flagella move the cell by rotating.

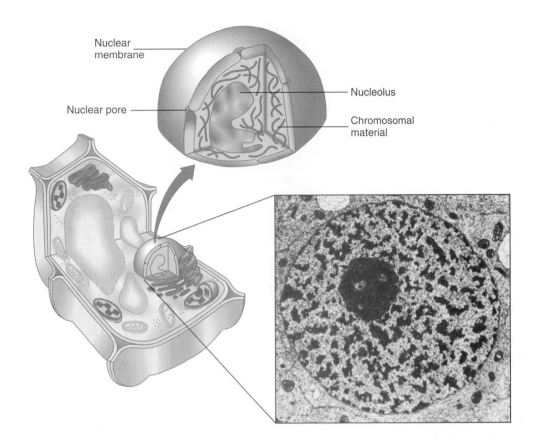

Nuclear membrane

Nuclear pore

Nucleolus

Chromosomal material

FIGURE 23.20

One of the two major regions of protoplasm, the nucleus has its own complex structure. It is bounded by two layers of membrane that separate it from the cytoplasm. Inside the nucleus are the nucleoli, chromosomes or chromatin material composed of DNA and protein, and the liquid matrix (nucleoplasm). Magnification is about 20,000 times.

granules and fibers in association with the cell's DNA used in the manufacture of ribosomes. These regions, together with the completed or partially completed ribosomes, are called nucleoli.

The final component of the nucleus is its liquid matrix called the **nucleoplasm.** It is a colloidal mixture composed of water and the molecules used in the construction of ribosomes, nucleic acids, and other nuclear material (figure 23.20). Table 23.2 compares the structure and function of cellular organelles.

MAJOR CELL TYPES

Not all of the cellular organelles we have just described are located in every cell. Some cells typically have combinations of organelles that differ from others. For example, some cells have nuclear membrane, mitochondria, chloroplasts, ER, and Golgi, while others have mitochondria, centrioles, Golgi, ER, and nuclear membrane. Other cells are even simpler and lack the

complex membranous organelles described in this chapter. Because of this fact, biologists have been able to classify cells into two major types: *prokaryotic* and *eukaryotic* (figure 23.21).

The Prokaryotic Cell Structure

Prokaryotic cells, the *Bacteria* and *Archaea,* do not have a typical nucleus bound by a nuclear membrane, nor do they contain mitochondria, chloroplasts, Golgi, or extensive networks of ER. However, prokaryotic cells contain DNA and enzymes and are able to reproduce and engage in metabolism. They perform all of the basic functions of living things with fewer and simpler organelles. As yet, members of the Archaea are of little concern to the medical profession since none have been identified as disease-causing. They are typically found growing in extreme environments where the pH, salt concentration, or temperatures make it impossible for most other organisms to survive. The other prokaryotic cells are called bacteria and about 5 percent cause diseases such as tuberculosis, strep throat, gonorrhea, and acne.

TABLE 23.2	Comparison of the structure and function of the cellular organelles		
Organelle	**Type of Cell in Which Located**	**Structure**	**Function**
Plasma membrane	Prokaryotic and eukaryotic	Membranous; typical membrane structure; phospholipid and protein present	Controls passage of some materials to and from the environment of the cell
Inclusions (granules)	Prokaryotic and eukaryotic	Nonmembranous; variable	May have a variety of functions
Chromatin material	Prokaryotic and eukaryotic	Nonmembranous; composed of DNA and proteins (histones in eukaryotes and HU proteins in prokaryotes)	Contains the hereditary information that the cell uses in its day-to-day life and passes it on to the next generation of cells
Ribosomes	Prokaryotic and eukaryotic	Nonmembranous; protein and RNA structure	Site of protein synthesis
Microtubules, microfilaments, and intermediate filaments	Eukaryotic	Nonmembranous; solid or hollow tubes composed of protein	Provide structural support and allow for movement
Lysosomes	Eukaryotic	Membranous; submicroscopic membrane-enclosed vesicle	Isolates very strong enzymes from the rest of the cell
Mitochondria	Eukaryotic	Membranous; double membranous organelle; large membrane folded inside of a smaller membrane	Associated with the release of energy from food; site of aerobic cellular respiration
Centriole	Eukaryotic	Two clusters of nine microtubules	Associated with cell division
Contractile vacuole	Eukaryotic	Membranous; single-membrane container	Expels excess water
Cilia and flagella	Eukaryotic	Nonmembranous; prokaryotes composed of single type of protein arranged in a fiber that is anchored into the cell wall and membrane; 9 + 2 tubulin protein in eukaryotes	Flagellar movement in prokaryotic type rotate; ciliary and flagellar movement in eukaryotic type seen as waving or twisting
Nuclear membrane	Eukaryotic	Membranous; sacs of typical membrane formed into a single container of nucleoplasm and nucleic acids	Separates the nucleus from the cytoplasm
Nucleolus	Eukaryotic	Nonmembranous; group of RNA molecules and DNA located in the nucleus	Site of ribosome manufacture and storage
Endoplasmic reticulum	Eukaryotic	Membranous; folds of membrane forming sheets and canals	Surface for chemical reactions and intracellular transport system
Golgi apparatus	Eukaryotic	Membranous stack of single membrane sacs	Associated with the production of secretions and enzyme activation
Vacuoles	Eukaryotic	Membranous; microscopic single membranous sacs	Containers of materials
Peroxisomes	Eukaryotic	Membranous; submicroscopic membrane-enclosed vesicle	Releases enzymes to break down hydrogen peroxide

Other prokaryotic cells are responsible for the decay and decomposition of dead organisms. While some bacteria have a type of green photosynthetic pigment and carry on photosynthesis, they do so without chloroplasts and use different chemical reactions.

One significant difference between prokaryotic and eukaryotic cells is in the chemical makeup of their ribosomes. The ribosomes of prokaryotic cells contain different proteins than those found in eukaryotic cells. Prokaryotic ribosomes are also smaller. This discovery was important to medicine because many cellular forms of life that cause common diseases are bacterial. As soon as differences in the ribosomes were noted, researchers began to look for ways in which to interfere with the prokaryotic ribosome's function but *not* interfere with the ribosomes of eukaryotic cells. *Antibiotics,* such as streptomycin, are the result of this research.

This drug combines with prokaryotic ribosomes and causes the death of the prokaryote by preventing the production of proteins essential to its survival. Since eukaryotic ribosomes differ from prokaryotic ribosomes, streptomycin does not interfere with the normal function of ribosomes in human cells.

The Eukaryotic Cell Structure

Eukaryotic cells contain a true nucleus and most of the membranous organelles described earlier. Eukaryotic organisms can be further divided into several categories based on the specific combination of organelles they contain. The cells of plants, fungi, protozoa and algae, and animals are all eukaryotic. The most obvious characteristic that sets the plants and

Domain Bacteria	Domain Archaea	Domain Eukarya			
Prokaryotic Cells Characterized by few membranous organelles; DNA not separated from the cytoplasm by a membrane		**Eukaryotic Cells** Cells larger than prokaryotic cells; DNA found within nucleus with a membrane separating it from the cytoplasm; many complex organelles composed of many structures including phospholipid bilayer membranes			
Kingdom Prokaryotae		Kingdom Protista	Kingdom Mycetae	Kingdom Plantae	Kingdom Animalia
Unicellular microbes; typically associated with bacterial "diseases," but 90%–95% are ecologically important and not pathogens	Unicellular microbes; typically associated with extreme environments including low pH, high salinity, and extreme temperatures	Unicellular microbes; some in colonies; both photosynthetic and heterotrophic nutrition; a few are parasites	Multicellular organisms or loose colonial arrangements of cells; organism is a row or filament of cells; decay fungi and parasites	Multicellular organisms; cells supported by a rigid cell wall of cellulose; some cells have chloroplasts; complex arrangement into tissues	Multicellular organisms with division of labor into complex tissues; no cell wall present; acquire food from the environment; some are parasites
Examples: gram-positive bacteria such as *Streptococcus pneumonia* and gram-negative bacteria such as *E. coli*	Examples: *Methanococcus,* halophiles, and *Thermococcus*	Examples: protozoans such as *Amoeba* and *Paramecium* and algae such as *Chlamydomonas* and *Euglena*	Examples: yeast such as bakers yeast, molds such as *Penicillium;* morels, mushrooms, and rusts	Examples: moss, ferns, cone-bearing trees, and flowering plants	Examples: worms, insects, starfish, frogs, reptiles, birds, mammals

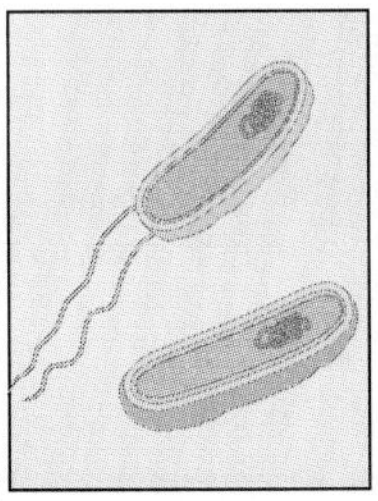

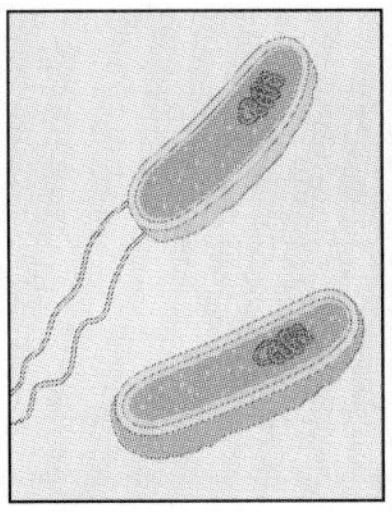

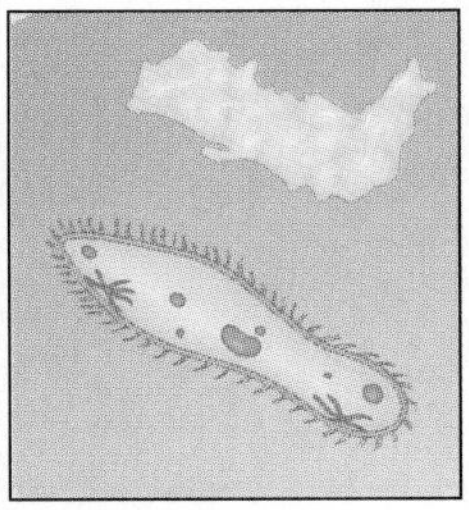

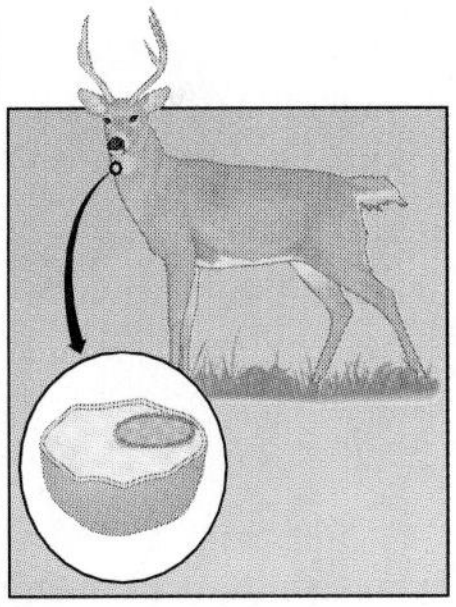

FIGURE 23.21

The two types of cells (prokaryotic and eukaryotic) are described in relationship to the major patterns found in all living things, the five kingdoms of life. Note the similarities of all five kingdoms and the subtle differences among them.

algae apart from other organisms is their green color, which indicates that the cells contain chlorophyll. Chlorophyll is necessary for the process of photosynthesis—the conversion of light energy into chemical-bond energy in food molecules. These cells, then, are different from the other cells in that they contain chloroplasts in their cytoplasm. Another distinguishing characteristic of plants and algae is the presence of cellulose in their cell walls.

The group of organisms that have a cell wall but lack chlorophyll in chloroplasts is collectively known as *fungi.* They were previously thought to be either plants that had lost their ability to make their own food or animals that had developed cell walls. Organisms that belong in this category of eukaryotic cells include yeasts, molds, mushrooms, and the fungi that cause such human diseases as athlete's foot, jungle rot, and ringworm. Now we have come to recognize this group as different enough from plants and animals to place them in a separate kingdom.

Eukaryotic organisms that lack cell walls and cannot photosynthesize are placed in separate groups. Organisms that consist of only one cell are called protozoans—examples are *Amoeba* and *Paramecium.* They have all the cellular organelles described in this chapter except the chloroplast; therefore, protozoans must consume food as do the fungi and the multicellular animals.

While the differences in these groups of organisms may seem to set them worlds apart, their similarity in cellular structure is one of the central themes unifying the field of biology (figure 23.22). One can obtain a better understanding of how

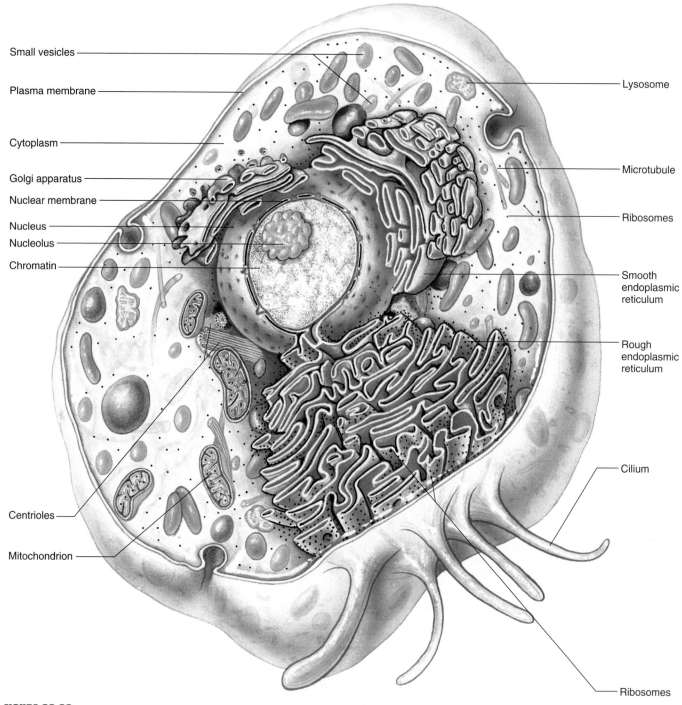

Small vesicles

Plasma membrane

Cytoplasm

Golgi apparatus

Nuclear membrane

Nucleus

Nucleolus

Chromatin

Centrioles

Mitochondrion

Lysosome

Microtubule

Ribosomes

Smooth endoplasmic reticulum

Rough endoplasmic reticulum

Cilium

Ribosomes

FIGURE 23.22

The major organelles of a typical animal cell are shown here. Use this illustration to identify structures when you are reviewing table 23.2.

cells operate in general by studying specific examples. Since the organelles have the same general structure and function regardless of the kind of cell in which they are found, we can learn more about how mitochondria function in plants by studying how mitochondria function in animals. There is a commonality among all living things with regard to their cellular structure and function.

THE IMPORTANCE OF CELL DIVISION

The process of cell division replaces dead cells with new ones, repairs damaged tissues, and allows living organisms to grow. For example, you began as a single cell that resulted from the union of a sperm and an egg. One of the *first* activities of this single cell was to divide. As this process continued, the number

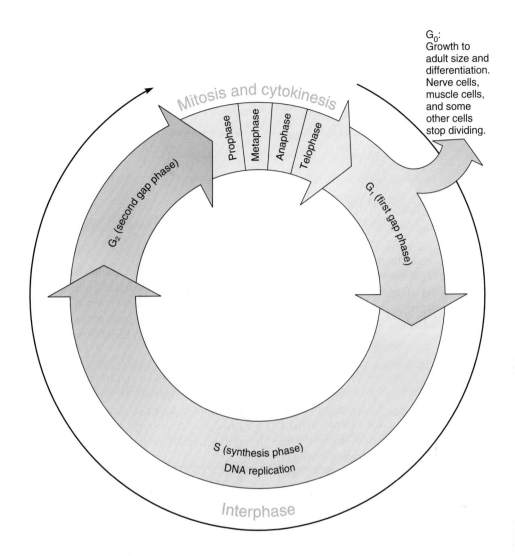

G$_0$:
Growth to adult size and differentiation. Nerve cells, muscle cells, and some other cells stop dividing.

Mitosis and cytokinesis

Prophase
Metaphase
Anaphase
Telophase

G$_2$ (second gap phase)

G$_1$ (first gap phase)

S (synthesis phase)
DNA replication

Interphase

FIGURE 23.23

During the cell cycle, tRNA, mRNA, ribosomes, and enzymes are produced in the G$_1$ stage. DNA replication occurs in the S stage. Proteins required for the spindles are synthesized in the G$_2$ stage. The nucleus is replicated in mitosis and two cells are formed by cytokinesis. Once some organs, such as the brain, have completely developed, certain types of cells, such as nerve cells, enter the G$_0$ stage. The time periods indicated are relative and vary depending on the type of cell and the age of the organism.

of cells in your body increased, so that as an adult your body consists of several trillion cells.

The *second* function of cell division is to maintain the body. Certain cells in your body, such as red blood cells and cells of the gut lining and skin, wear out. As they do, they must be replaced with new cells. Altogether, you lose about 50 million cells per second; this means that millions of cells are dividing in your body at any given time.

A *third* purpose of cell division is repair. When a bone is broken, the break heals because cells divide, increasing the number of cells available to knit the broken pieces together. If some skin cells are destroyed by a cut or abrasion, cell division produces new cells to repair the damage.

During cell division, two events occur. The replicated genetic information of a cell is equally distributed to two daughter nuclei in a process called **mitosis.** As the nucleus goes

through its division, the cytoplasm also divides into two new cells. This division of the cell's cytoplasm is called **cytokinesis**—cell splitting. Each new cell gets one of the two daughter nuclei so that both have a complete set of genetic information.

THE CELL CYCLE

All cells go through the same basic life cycle, but they vary in the amount of time they spend in the different stages. A generalized picture of a cell's life cycle may help you to understand it better (figure 23.23). Once begun, cell division is without a beginning or an end. It is a cycle in which cells continue to grow and divide. There are five stages to the life cycle of a eukaryotic cell: (1) G$_1$, gap (growth)—phase one; (2) S, synthesis; (3) G$_2$, gap (growth)—phase two; (4) cell division

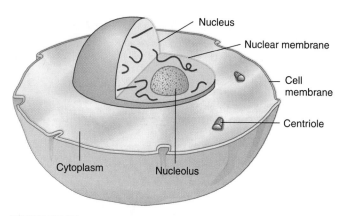

FIGURE 23.24

Growth and the production of necessary organic compounds occur during interphase. If the cell is going to divide, DNA replication also occurs during this phase. The individual chromosomes are not visible, but a distinct nuclear membrane and nucleolus are present. (Some cells have more than one nucleolus.)

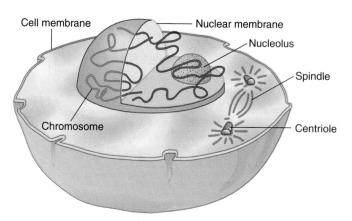

FIGURE 23.25

In early prophase, chromosomes begin to appear as thin tangled threads and the nucleolus and nuclear membrane are present. The two sets of microtubules known as the centrioles begin to separate and move to opposite poles of the cell. A series of fibers known as the spindle will shortly begin to form.

(mitosis and cytokinesis); and G_0, gap (growth)—mitotic dormancy or differentiated.

During the G_0 phase cells are not considered to be in the cycle of division, but become differentiated or specialized in their function. It is at this time that they "mature" to play the role specified by their genetic makeup. While some cells entering the G_0 phase remain there more-or-less permanently (e.g., nerve cells), others have the ability to move back into the cell cycle of mitosis, G_1, S, and G_2 with ease (e.g., skin cells).

The first three phases of the cell cycle—G_1, S, and G_2—occur during a period of time known as *interphase*. **Interphase** is the stage between cell divisions. During the G_1 stage, the cell grows in volume as it produces tRNA, mRNA, ribosomes, enzymes, and other cell components. During the S stage, DNA replication occurs in preparation for the distribution of genes to daughter cells. During the G_2 stage that follows, final preparations are made for mitosis with the synthesis of spindle-fiber proteins.

During interphase, the cell is not dividing but is engaged in metabolic activities such as muscle-cell contractions, photosynthesis, or glandular-cell secretion. During interphase, the nuclear membrane is intact and the individual chromosomes are not visible (figure 23.24). The individual chromatin strands are too thin and tangled to be seen. Remember that chromosomes include various kinds of histone proteins as well as DNA, the cell's genetic information. The double helix of DNA and the nucleosomes are arranged as a chromatid, and there are two attached chromatids for each replicated chromosome. It is these chromatids (chromosomes) that will be distributed during mitosis.

THE STAGES OF MITOSIS

All stages in the life cycle of a cell are continuous; there is no precise point when the G_1 stage ends and the S stage begins, or when the interphase period ends and mitosis begins. Likewise, in the individual stages of mitosis there is a gradual transition from one stage to the next. However, for purposes of study and communication, scientists have divided the process into four stages based on recognizable events. These four phases are prophase, metaphase, anaphase, and telophase.

Prophase

As the G_2 stage of interphase ends, mitosis begins. **Prophase** is the first stage of mitosis. One of the first noticeable changes is that the individual chromosomes become visible (figure 23.25). The thin, tangled chromatin present during interphase gradually coils and thickens, becoming visible as separate chromosomes. The DNA portion of the chromosome has genes that are arranged in a specific order. Each chromosome carries its own set of genes that is different from the sets of genes on other chromosomes.

As prophase proceeds, and as the chromosomes become more visible, we recognize that each chromosome is made of two parallel, threadlike parts lying side by side. Each parallel thread is called a **chromatid** (figure 23.26). These chromatids were formed during the S stage of interphase, when DNA synthesis occurred. The two identical chromatids are attached at a genetic region called the *centromere*. This portion of the DNA is not replicated during prophase, but remains base-paired as in the original duplex DNA. The centromere is vital to the cell division process. Without the centromere, cells will not complete mitosis and die.

In the diagrams in this text, a few genes are shown as they might occur on human chromosomes. The diagrams show fewer chromosomes and fewer genes on each chromosome than are actually present. Normal human cells have 10 billion nucleotides arranged into 46 chromosomes, each chromosome with thousands of genes. In this book, smaller numbers of genes and chromosomes are used to make it easier to follow the events that happen in mitosis.

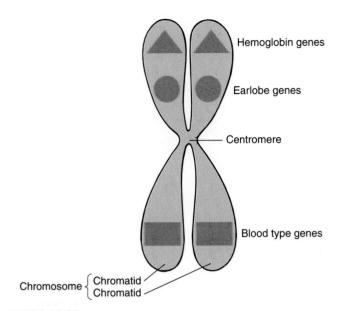

FIGURE 23.26

During interphase, when chromosome replication occurs, the original duplex DNA unzips to form two identical double strands that are attached at the centromere. Each of these double strands is a chromatid. The two identical chromatids of the chromosome are sometimes termed a dyad, to reflect that there are two duplex DNA molecules, one in each chromatid. The DNA contains the genetic data. (The examples presented here are for illustrative purposes only. Do not assume that the traits listed are actually located in the positions shown on these hypothetical chromosomes.)

Several other events occur as the cell proceeds to the late prophase stage (figure 23.27). One of these events is the duplication of the *centrioles*. Remember that human and many other eukaryotic cells contain centrioles, microtubule-containing organelles located just outside the nucleus. As they duplicate, they move to the poles of the cell. As the centrioles move to the poles, the microtubules are assembled into the *spindle*. Recall the spindle is an array of microtubules extending from pole to pole that is used in the movement of chromosomes.

In most eukaryotic cells, as prophase is occurring, the nuclear membrane is gradually disassembled. It is present at the beginning of prophase but disappears by the time this stage is completed. In addition to the disassembled nuclear membrane, the nucleoli within the nucleus disappear. Because of the disassembly of the nuclear membrane, the chromosomes are free to move anywhere within the cytoplasm of the cell. As prophase progresses, the chromosomes become attached to the spindle fibers at their centromeres. Initially they are distributed randomly throughout the cytoplasm. As this movement occurs, the cell enters the next stage of mitosis.

Metaphase

During **metaphase,** the second stage in mitosis, the chromosomes align at the equatorial plane. There is no nucleus present during metaphase, and the spindle, which started to form during

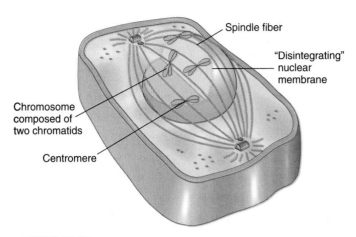

FIGURE 23.27

In late prophase, the chromosomes appear as two chromatids (a dyad) connected at a centromere. The nucleolus and the nuclear membrane have disassembled. The centrioles have moved farther apart, and the spindle is produced.

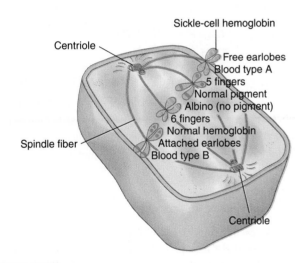

FIGURE 23.28

During metaphase the chromosomes travel along the spindle and align at the equatorial plane. Notice that each chromosome still consists of two chromatids.

prophase, is completed. The centrioles are at the poles, and the microtubules extend between them to form the spindle. Then the chromosomes are their most tightly coiled and move until all their centromeres align themselves along the equatorial plane at the equator of the cell (figure 23.28). At this stage in mitosis, each chromosome still consists of two chromatids attached at a centromere. In a human cell, there are 46 chromosomes, or 92 chromatids, aligned at the cell's equatorial plane during metaphase.

If we view a cell in the metaphase stage from the side (figure 23.29), it is an equatorial view. In this view, the chromosomes appear as if they were in a line. If we view the cell from the pole, it is a polar view. The chromosomes are seen on the equatorial plane (see figure 23.29). Chromosomes viewed from this

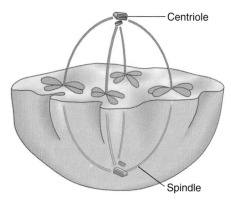

FIGURE 23.29

This view shows how the chromosomes spread out on the equatorial plane.

CONNECTIONS

Keep Your Lawn Crabgrass Free with a Mitosis Inhibitor

The crabgrass control agent dithiopyr appears to bind to a cellular protein other than tubulin. It is thought that dithiopyr acts on the chemical Microtubule Associated Protein (MAP), altering microtubule stability. When this occurs, spindle formation does not occur and the cell dies in metaphase.

direction look like hot dogs scattered on a plate. In late metaphase, each chromosome splits as the centromeres replicate as the cell enters the next phase, anaphase.

Anaphase

Anaphase is the third stage of mitosis. The nuclear membrane is still absent and the spindle extends from pole to pole. The two chromatids within the chromosome separate as they move along the spindle fibers toward opposite ends of the poles (figure 23.30). Although this movement has been observed repeatedly, no one knows the exact mechanism of its action. As this separation of chromatids occurs, the chromatids are called *daughter chromosomes.* Daughter chromosomes contain identical genetic information.

Examine figure 23.30 closely and notice that the four chromosomes moving to one pole have exactly the same genetic information as the four moving to the opposite pole. It is the alignment of the chromosomes in metaphase, and their separation in anaphase, that causes this type of distribution. It is during anaphase that the second important event occurs, cytokinesis. Cytokinesis divides the cytoplasm of the original cell so that two smaller, separate daughter cells result. Daughter cells are two cells formed by cell division that have identical genetic information. At the end of anaphase, there are

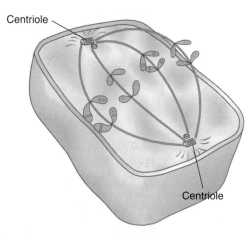

FIGURE 23.30

In anaphase, the pairs of chromatids separate after the centromeres replicate. The chromatids, now called daughter chromosomes, are separating and moving toward the poles and the cell will begin cytokinesis.

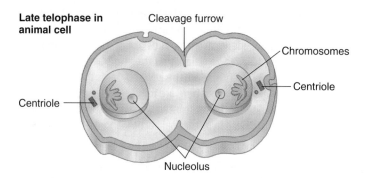

FIGURE 23.31

During telophase the spindle disassembles and the nucleolus and nuclear membrane form. Daughter cells are formed as a result of the division of the cytoplasm.

two identical groups of chromosomes, one group at each pole. The next stage completes the mitosis process.

Telophase

Telophase is the last stage in mitosis. It is during telophase that *daughter nuclei* are reformed. Each set of chromosomes becomes enclosed by a nuclear membrane and the nucleoli reappear. Now the cell has two identical daughter nuclei (figure 23.31). In addition, the microtubules are disassembled, so the spindle disappears. With the formation of the daughter nuclei, mitosis, the first process in cell division, is completed, and the second process, cytokinesis, can occur. Cytokinesis (cytoplasm splitting) divides the cytoplasm of the original cell and forms two smaller daughter cells. *Daughter cells* are two cells formed by cell division that have identical genetic information. Each of the newly formed daughter cells then enters the G_1 stage of interphase. These cells can grow, replicate their DNA, and enter another round of mitosis and cytokinesis to continue the cell cycle.

SUMMARY

Living things show the *characteristics* of (1) *metabolic processes,* (2) *generative processes,* (3) *responsive processes,* (4) *control processes,* and (5) *a unique structural organization.* The concept of the *cell* has developed over a number of years. Initially, only two regions, the *cytoplasm* and the *nucleus,* could be identified. At present, numerous *organelles* are recognized as essential components of both *prokaryotic* and *eukaryotic* cell types. The *structure* and *function* of some *organelles* are compared in table 23.2.

The *cell* is the common unit of life. We study individual cells and their structures to understand how they function as *individual living organisms* and as parts of *many-celled beings.* Knowing how *prokaryotic* and *eukaryotic cell* types resemble or differ from each other helps physicians control some organisms dangerous to humans.

Cell division is necessary for *growth, repair,* and *reproduction.* Cells go through a *cell cycle* that includes *cell division* (*mitosis* and *cytokinesis*) and *interphase. Interphase* is the period of growth and preparation for division. Mitosis is divided into four stages: *prophase, metaphase, anaphase,* and *telophase.* During mitosis, *two daughter nuclei* are formed from *one parent nucleus.* These nuclei have *identical sets of chromosomes and genes* that are *exact copies* of those of the parent. Although the process of *mitosis* has been presented as a series of phases, you should realize that it is a *continuous, flowing process* from *prophase* through *telophase.* Following mitosis, *cytokinesis* divides the *cytoplasm,* and the cell returns to *interphase.*

KEY TERMS

active transport (p. **597**)

aerobic cellular respiration (p. **601**)

anaphase (p. **614**)

biology (p. **588**)

cell (p. **589**)

cell membrane (p. **599**)

cellular membranes (p. **592**)

chloroplast (p. **601**)

chromatid (p. **612**)

chromatin (p. **603**)

chromosomes (p. **603**)

concentration gradient (p. **594**)

control processes (p. **589**)

cristae (p. **601**)

cytokinesis (p. **611**)

cytoplasm (p. **589**)

diffusion (p. **594**)

diffusion gradient (p. **594**)

dynamic equilibrium (p. **594**)

endoplasmic reticulum (p. **599**)

enzymes (p. **589**)

eukaryotic cells (p. **608**)

facilitated diffusion (p. **597**)

fluid-mosaic model (p. **592**)

generative processes (p. **588**)

Golgi apparatus (p. **599**)

hydrophilic (p. **592**)

hydrophobic (p. **592**)

hypertonic (p. **596**)

hypotonic (p. **596**)

interphase (p. **612**)

isotonic (p. **595**)

lysosomes (p. **599**)

metabolic processes (p. **588**)

metaphase (p. **613**)

mitochondrion (p. **601**)

mitosis (p. **611**)

nuclear membrane (p. **601**)

nucleolus (p. **603**)

nucleoplasm (p. **607**)

nucleus (p. **589**)

organelles (p. **589**)

osmosis (p. **595**)

peroxisome (p. **599**)

phagocytosis (p. **598**)

pinocytosis (p. **598**)

plasma membrane (p. **599**)

prokaryotic cells (p. **607**)

prophase (p. **612**)

protoplasm (p. **589**)

responsive processes (p. **588**)

ribosomes (p. **603**)

selectively permeable (p. **595**)

structural similarities (p. **589**)

telophase (p. **614**)

vacuole (p. **598**)

vesicles (p. **598**)

APPLYING THE CONCEPTS

1. Which one of the following represents a generative process?
   a. enzymes
   b. individual adaptation
   c. nutrient uptake
   d. cell division

2. Which of these is a biological problem?
   a. famine in Rhodesia
   b. hurricanes kill thousands in Honduras
   c. pesticide research leads to toxic waste
   d. All of these have a biological component.

3. When dealing with responsive processes:
   a. metabolic processes decrease.
   b. populations evolve through time.
   c. organisms grow.
   d. individuals coordinate activities.

4. Metabolic processes include
   a. nutrient processing.
   b. waste elimination.
   c. nutrient uptake.
   d. all of these.

5. The centromeres split during
   a. anaphase.
   b. prophase.
   c. metaphase.
   d. interphase.

6. What would happen if microtubules were prevented from forming during mitosis?
   a. The cell plate would not form.
   b. Replication would not occur.
   c. Centromeres would not split.
   d. Anaphase cannot occur.

7. The normal state of chromosomes in prophase of mitosis is as
   a. daughter chromosomes.
   b. chromosomes composed of two chromatids.
   c. chromatids composed of two chromosomes.
   d. chromosomes consisting of single chromatids.

8. The presence of cell walls in plants is associated with _____ in telophase.
   a. cleavage furrows
   b. spindle fiber formation
   c. differentiation
   d. cytokinesis

9. During which stage of the cell cycle does DNA replication occur?

    a. the S stage of interphase

    b. anaphase of mitosis

    c. G_2 stage of metaphase

    d. prophase

10. Nerve cells do not normally undergo mitosis. This means that

    a. the brain is unimportant.

    b. your brain cannot grow.

    c. cytokinesis will be common in nerve tissue.

    d. transcription of DNA will not occur.

11. Radiation is able to successfully control cancer because

    a. cancer cells do not grow rapidly.

    b. cancer cells spend most of their time in the S stage of prophase.

    c. it stimulates programmed cell death.

    d. it affects only diseased cells.

12. The correct order for the stages of mitosis is

    a. prophase, anaphase, metaphase, telophase.

    b. metaphase, prophase, anaphase, telophase.

    c. prophase, metaphase, anaphase, telophase.

    d. prophase, metaphase, telophase, anaphase.

13. Compared to the mother cell, daughter cells at the end of mitosis have _____ number of chromosomes.

    a. half the

    b. the same

    c. twice the

    d. none of the above

14. Which one of the following is *not* an event of telophase?

    a. nucleoli reappear

    b. spindle disappears

    c. daughter nuclei form

    d. centrioles duplicate

15. During anaphase

    a. chromosomes become visible.

    b. daughter chromosomes migrate to the poles.

    c. chromosomes line up at the equatorial plane.

    d. cytokinesis is completed.

16. Cell division is needed for

    a. growth.

    b. replacement of worn-out cells.

    c. healing of damaged tissue.

    d. all of these.

17. Chromosomes are composed of sister chromatids during

    a. interphase.

    b. prophase only.

    c. prophase and metaphase.

    d. prophase, metaphase, and anaphase.

18. Sister chromatids contain

    a. identical strands of DNA.

    b. half of a duplex DNA molecule.

    c. different genetic information.

    d. one gene.

Answers

1. d 2. d 3.b 4. d 5.a 6.d 7.b 8. d 9. a 10. b 11. c 12. c 13. b 14. d 15. b 16. d 17.c 18. a

QUESTIONS FOR THOUGHT

1. Make a list of the membranous organelles of a eukaryotic cell and describe the function of each.
2. Describe how the concept of the cell has changed over the past two hundred years.
3. What three methods allow the exchange of molecules between cells and their surroundings?
4. How do diffusion, facilitated diffusion, osmosis, and active transport differ?
5. What are the differences between the cell wall and the cell membrane?
6. Diagram a cell and show where proteins, nucleic acids, carbohydrates, and lipids are located.
7. Make a list of the nonmembranous organelles of the cell and describe their functions.
8. Define the following terms: cytoplasm, stroma, grana, cristae, chromatin, and chromosome.
9. Why does putting salt on meat preserve it from spoilage by bacteria?
10. In what ways do mitochondria and chloroplasts resemble one another?
11. Name the four stages of mitosis and describe what occurs in each stage.
12. What is meant by cell cycle?
13. During which stage of a cell's cycle does DNA replication occur?
14. At what phase of mitosis does the DNA become most visible?
15. What are the differences between plant and animal mitosis?
16. Why can radiation treatment be used to control cancer?
17. What is the purpose of mitosis?
18. What is the difference between cytokinesis in plants and animals?
19. What types of activities occur during interphase?
20. List five differences between an interphase cell and a cell in mitosis.

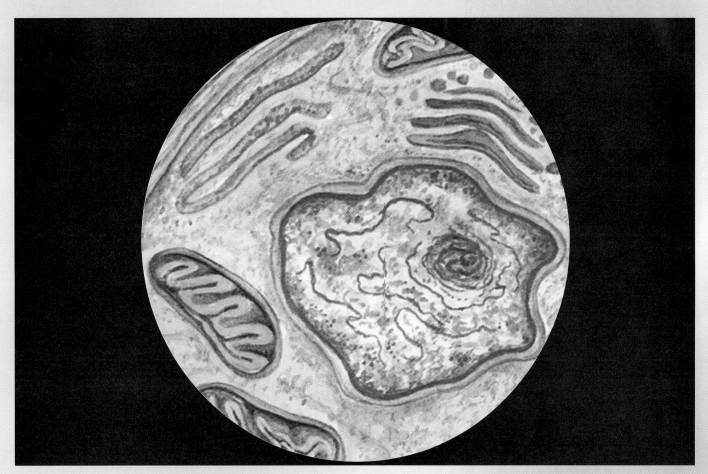

Early in Earth's history, inorganic materials were converted to organic materials. These organic materials have become combined into living units called cells. Cells are life. Scientists always ask these two questions: What is the nature of life? And how did life originate?

CHAPTER | Twenty-Four

The Origin and Evolution of Life

For centuries humans have studied the basic nature of their environment. The vast amount of information presented in previous chapters is evidence of our ability to gather and analyze information. These efforts have resulted in solutions to many problems and have simultaneously revealed new and more challenging topics to study. Despite the growth in scientific knowledge, two questions have continued to be subjects of speculation: What is the nature of life? And how did life originate?

In earlier times, no one ever doubted that life originated from nonliving things. The Greeks, Romans, Chinese, and many other ancient peoples believed that maggots arose from decaying meat; mice developed from wheat stored in dark, damp places; lice formed from sweat; and frogs originated from damp mud. The concept of **spontaneous generation**—the theory that living organisms arise from nonliving material—was widely believed until the seventeenth century (figure 24.1). However, there were some who doubted this theory. These people subscribed to an opposing theory, called *biogenesis*. **Biogenesis** is the concept that life originates only from preexisting life.

One of the earliest challenges to the theory of spontaneous generation came in 1668. Francesco Redi, an Italian physician, set up a controlled experiment designed to disprove the theory of spontaneous generation (figure 24.2). He used two sets of jars that were identical except for one aspect. Both sets of jars contained decaying meat, and both were exposed to the atmosphere; however, one set of jars was covered by gauze, and the other was uncovered. Redi observed that flies settled on the meat in the open jar, but the gauze blocked their access to the covered jars. When maggots appeared on the meat in the uncovered jars but not on the meat in the covered ones, Redi concluded that the maggots arose from the eggs of the flies and not from spontaneous generation in the meat.

CURRENT THINKING ABOUT THE ORIGIN OF LIFE

Today, we still have basically two major scientific theories regarding the origin of life. One holds that life arrived on Earth from some extraterrestrial source and the other maintains that life was created on Earth from nonliving material. It is important to recognize that we will probably never know for sure how life on Earth came to be, but it is interesting to speculate and examine the evidence related to this fundamental question.

The concept that life arrived from extraterrestrial sources received some support in 1996 when a meteorite from Antarctica was analyzed. It has been known for many years that meteorites often contain organic molecules and this suggested that life might have existed elsewhere in the solar system. The chemical makeup of the Antarctic meteorite suggests that it was a portion of the planet Mars which was ejected from Mars as a result of a collision between the planet and an asteroid. Analysis of the meteorite shows the presence of complex organic molecules and small structures that resemble those found on Earth that are thought to be the result of the activity of ancient microorganisms. Since Mars currently has some water as ice and shows features that resemble dried-up river systems, Mars may have had much more water in the past. For these reasons many feel it is reasonable to consider that life of a nature similar to that presently found on Earth could have existed on Mars.

The alternative view, that life originated on Earth, has also received support. Let us look at several lines of evidence.

1. Earth is the only planet in our solar system that has a temperature range that allows for water to exist as a liquid on its surface and water is the most common compound in most kinds of living things.

2. Analysis of the atmospheres of other planets shows that they all lack oxygen. The oxygen in Earth's atmosphere is the result of current biological activity. Therefore, before life on Earth the atmosphere probably lacked oxygen. This could have led to a buildup of organic molecules in the oceans.

3. Experiments demonstrate that organic molecules can be generated in an atmosphere that lacks oxygen.

4. Since it is assumed that all of the planets have been cooling off as they age, it is very likely that Earth was much hotter in the past. The large portions of Earth's surface that are of volcanic origin strongly suggest a hotter past. There is also the likelihood that various large bodies collided with Earth early in its history and that they could have led to increased temperatures at least in the site of the collision.

5. Recognition that there are distinct prokaryotic organisms that live in extreme environments of high temperature, high salinity, low pH, or the absence of oxygen suggests that they may have been adapted to life in a world that is very different from today's Earth. These kinds of organisms are found today in unusual locations such as hot springs and around thermal vents in the ocean floor and may be descendants of the first organisms formed on the primitive Earth.

MEETING METABOLIC NEEDS— HETEROTROPHS OR AUTOTROPHS

Fossil evidence indicates that there were primitive forms of life on Earth at least 3.5 billion years ago. Regardless of how they developed, these first primitive cells would have needed a way to add new organic molecules to their structure as previously existing molecules were lost or destroyed. There are two ways to accomplish this. **Heterotrophs** capture organic molecules such as sugars, amino acids, or organic acids from their surroundings, which they use to make new molecules and to provide themselves with a source of energy. **Autotrophs** use some external energy source such as sunlight or the energy from inorganic chemical reactions to allow them to combine together simple inorganic molecules like water and carbon dioxide to make new organic molecules. These new organic molecules can then be used as building materials for new cells or can be broken down at a later date to provide a source of energy.

Many scientists support the idea that the first living things produced on Earth were heterotrophs that lived off the organic molecules that would have been found in the oceans. Since the early heterotrophs are thought to have developed in a reducing atmosphere that lacked oxygen, they would have been of necessity anaerobic organisms; therefore, they did not obtain the maximum amount of energy from the organic molecules they obtained from their environment. At first, this would not have been a problem. The organic molecules that had been accumulating in the ocean for millions of years served as an ample source of organic material for the heterotrophs. However, as the population of heterotrophs increased through reproduction, the supply of organic material would have been consumed faster than it was being spontaneously produced in the atmosphere. If there was no other source of organic compounds, the heterotrophs would have eventually exhausted their nutrient supply, and they would have become extinct.

Even though the early heterotrophs probably contained nucleic acids and were capable of producing enzymes that could regulate chemical reactions, they probably carried out a minimum of biochemical activity. There is evidence to suggest that a wide variety of compounds were present in the early oceans, some of which could have been used unchanged by the heterotrophs. There was no need for the heterotrophs to modify the compounds to meet their needs.

Those compounds that could be easily used by heterotrophs would have been the first to become depleted from the early environment. However, some of the heterotrophs may have contained a mutated form of nucleic acid, which allowed them to convert material that was not directly usable into a compound that

FIGURE 24.1

Many works of art explore the idea that living things could originate from very different types of organisms or even from nonliving matter. M. C. Escher's work entitled "The Reptiles, 1943" shows the life cycle of a little alligator. Amid all kinds of objects, a drawing book lies open at a drawing of a mosaic of reptilian figures in three contrasting shades. Evidently, one of them is tired of lying flat and rigid among its fellows, so it puts one plastic-looking leg over the edge of the book, wrenches itself free, and launches out into "real" life. It climbs up the back of the zoology book and works its way laboriously up the slippery slope of the set square to the highest point of its existence. Then after a quick snort, tired but fulfilled, it goes downhill again, via an ashtray, to the level surface, to that flat drawing paper, and meekly rejoins its erstwhile friends, taking up once more its function as one element of surface division.

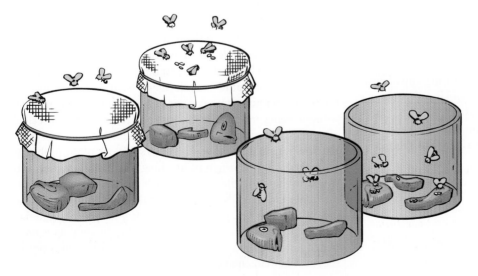

FIGURE 24.2

The two sets of jars here are identical in every way except one—the gauze covering. The set on the left is called the control group; the set on the right is the experimental group. Any differences seen between the control and the experimental groups are the result of a single variable. In this manner, Redi concluded that the presence of maggots in meat was due to flies laying their eggs on the meat and not spontaneous generation.

could be used. Generic mutations changes may have been common because the amount of ultraviolet light, one cause of mutations, would have been high. The absence of ozone in the upper atmosphere of the early Earth would have allowed high amounts of ultraviolet light to reach Earth's surface. Heterotrophs with such mutations could have survived, while those without it would have become extinct as the compounds they used for food became scarce. It has been suggested that through a series of mutations in the early heterotrophs, a more complex series of biochemical reactions originated within some of the cells. Such cells could use chemical reactions to convert ingestable chemicals into usable organic compounds.

As with many areas of science there are often differences of opinion. Although this heterotroph hypothesis for the origin of living things was the prevailing theory for many years, recent discoveries have caused many scientists to consider an alternative; that the first organism was an autotroph. Several kinds of information support this theory. Many kinds of very primitive prokaryotic organisms known as the Archaea are autotrophic and live in extremely hostile environments. These organisms are found in hot springs like those found in Yellowstone National Park or near hot thermal vents—areas where hot mineral-rich water enters seawater from the ocean floor. They use inorganic chemical reactions as a source of energy to allow them to synthesize organic molecules from inorganic components. The fact that many of these organisms live in very hot environments suggests that they may have originated on an Earth that was much hotter than it is currently. There is much evidence that Earth was a much hotter place in the past. If the first organism was an autotroph there could have been subsequent evolution of a variety of kinds of cells, both autotrophic and heterotrophic, that could have led to the variety of different prokaryotic cells seen today in the Bacteria and the Archaea.

REPRODUCTION AND THE ORIGIN OF GENETIC MATERIAL

Two important *nucleic acid* molecules store and transfer information within a cell. Deoxyribonucleic acid (DNA) is found in the cell nucleus and contains the genetic information for the cell, the blueprints for its construction and maintenance. Ribonucleic acid (RNA) is found in small structures in the cytoplasm (ribosomes) and functions in the manufacture of protein.

As scientists began to understand the chemical makeup of the nucleic acids, an attempt was made to understand how DNA and RNA relate to inheritance, cell structure, and cell activities. The concept that resulted is known as the *central dogma*, main belief, or "source of all information." It is most easily written in this form:

DNA ← (replication) ← $\boxed{\text{DNA}}$ → (transcription) → RNA →

(translation) → Proteins $<$ structural / regulatory

What this concept map says is that at the center of it all is DNA, the genetic material of the cell and (going to the left) it is capable of reproducing itself, a process called DNA *replication*. Going to the right, DNA is capable of supervising the manufacture of RNA (a process known as *transcription*), which in turn is involved in the production of protein molecules, a process known as *translation*.

DNA replication occurs in cells in preparation for the cell division processes of mitosis and meiosis. Without replication, daughter cells would not receive the library of information required to sustain life. The transcription process results in the formation of a strand of RNA that is a copy of a segment of the DNA on which it is formed. Some of the RNA molecules become involved in various biochemical processes, while others are used in the translation of the RNA information into proteins. Structural proteins are used by the cell as building materials (feathers, hair-collagen), while regulatory proteins are used to direct and control chemical reactions (enzymes, some hormones).

The reproduction of most current organisms involves the replication of DNA and the distribution of the copied DNA or RNA to subsequent cells. It is difficult to see how this complicated sequence of events, which involves many steps and the assistance of several enzymes, could have been generated spontaneously, so scientists have looked for simpler systems that could have led to the DNA system we see today.

Science works simultaneously on several fronts. Scientists involved in studying the structure and function of viruses discovered that many viruses do not contain DNA but store their genetic information in the structure of RNA. In order for these RNA-viruses to reproduce, they must enter a cell and have their RNA reverse-transcribed into DNA, which the host cell translates to manufacture new virus protein and RNA.

Other scientists who study viral diseases find that it is difficult to develop vaccines for many viral diseases, because their genetic material easily mutates. Because of this, researchers have been studying the nature of viral DNA or RNA to see what causes the high rate of mutation. This has led others to explore the RNA viruses and ask the question, Can RNA replicate itself without DNA? This is an important question, because if RNA can replicate itself it would have all of the properties necessary to serve as genetic material. It could store information, translate information into protein structure, mutate, and make copies of itself.

Other research about the nature of RNA provides interesting food for thought. RNA can be assembled from simpler subunits that could have been present on the early Earth. Scientists have also shown that RNA molecules are able to make copies of themselves without the need for enzymes, and they can do so without being inside cells. These molecules have been called *ribozymes*. This new evidence suggests that RNA may have been the first genetic material and helps to solve one of the problems associated with the origin of life: how genetic information was stored in these primitive life-forms. Since RNA is a much simpler molecule than DNA and it can make copies of itself without the aid of enzymes, perhaps it was the first genetic material. Once a primitive life-form had the ability to copy its genetic material it would be able to reproduce. Reproduction is one of the most fundamental characteristics of living things.

As a result of this discussion you should understand that we do not know how life on Earth originated. Scientists look at many kinds of evidence and continue to explore new avenues of research. So we currently have three competing theories for the origin of life on Earth;

1. Life arrived from some extraterrestrial source (biogenesis).
2. Life originated on Earth as a heterotroph (spontaneous generation).
3. Life originated on Earth as an autotroph (spontaneous generation).

MAJOR EVOLUTIONARY CHANGES IN THE NATURE OF LIVING THINGS

Once living things existed and had a genetic material that stored information but was changeable (mutations), living things could have proliferated into a variety of kinds that were adapted to specific environmental conditions. Earth has not been static but has been changing as a result of its cooling, volcanic activity, and encounters with asteroids. In addition, the organisms have had an impact on the way in which Earth has developed. Regardless of the way in which life originated on Earth, there have been several major events in the subsequent evolution of living things.

The Development of an Oxidizing Atmosphere

Ever since its formation, Earth has undergone constant change. In the beginning, it was too hot to support an atmosphere. Later, as it cooled and as gases escaped from volcanoes, a reducing atmosphere (lacking oxygen) was likely to have been formed. The early life-forms would have lived in this reducing atmosphere. However, today we have an oxidizing atmosphere and most organisms use this oxygen as a way to extract energy from organic molecules through a process of aerobic respiration. But what caused the atmosphere to change? Today it is clear that the oxygen in our atmosphere is the result of the process of photosynthesis. Prokaryotic Cyanobacteria are the simplest organisms that are able to photosynthesized, so it seemed logical that the first organisms could have accumulated many mutations over time that could have resulted in photosynthetic autotrophs. One of the waste products of the process of photosynthesis is molecular oxygen (O_2). This would have been a significant change because it would have led to the development of an **oxidizing atmosphere,** which contains molecular oxygen. The development of an oxidizing atmosphere created an environment unsuitable for the formation of organic molecules. Organic molecules tend to break down (oxidize) when oxygen is present. The presence of oxygen in the atmosphere would make it impossible for life to spontaneously originate in the manner described earlier in this chapter, because an oxidizing atmosphere would not allow for the accumulation of organic molecules in the seas. However, new life is generated through reproduction, and new kinds of life are generated through mutation and evolution. The presence of oxygen in the atmosphere had one other important outcome: It opened the door for the evolution of aerobic organisms.

It appears that an oxidizing atmosphere began to develop about 2 billion years ago. Although various chemical reactions released small amounts of molecular oxygen into the atmosphere, it was photosynthesis that generated most of the oxygen. The oxygen molecules also reacted with one another to form ozone (O_3). Ozone collected in the upper atmosphere and acted as a screen to prevent most of the ultraviolet light from reaching Earth's surface. The reduction of ultraviolet light diminished the spontaneous formation of complex organic molecules. It also reduced the number of mutations in cells. In an oxidizing atmosphere, it was no longer possible for organic molecules to accumulate over millions of years to be later incorporated into living material.

The appearance of oxygen in the atmosphere also allowed for the evolution of aerobic respiration. Since the first heterotrophs were of necessity anaerobic organisms, they did not derive large amounts of energy from the organic materials available as food. With the evolution of aerobic heterotrophs, there could be a much more efficient conversion of food into usable energy. Aerobic organisms would have a significant advantage over anaerobic organisms: They could use the newly generated oxygen as a final hydrogen acceptor and, therefore, generate many more energy-rich ATP molecules from the food molecules they consumed.

The Establishment of Three Major Domains of Life

In 1977 Carl Woese published the idea that the "bacteria" (organisms that lack a nucleus), which had been considered a group of similar organisms, were really made up of two very different kinds of organisms, the Bacteria and Archaea. Furthermore the Archaea shared some characteristics with eukaryotic organisms. Subsequent investigations have supported these ideas and led to an entirely different way of looking at the classification and evolution of living things. Although biologists have traditionally divided organisms into kingdoms based on their structure and function, it was very difficult to do this with microscopic organisms. With the newly developed ability to decode the sequence of nucleic acids, it became possible to look at the genetic nature of organisms without being confused by their external structures. Woese studied the sequences of ribosomal RNA and compared similarities and differences. As a result of his studies and those of many others a new concept of the relationships between various kinds of organisms has emerged.

The three main kinds of living things, Bacteria, Archaea, and Eucarya, have been labeled "domains." Within each domain there are several kingdoms. In the Eucarya there are four kingdoms that we already recognize: Animalia, Plantae, Mycetae, and Protista. However, previously all of the Bacteria and Archaea have been lumped into the same kingdom: Prokaryotae. It has become clear that there are great differences between the Bacteria and Archaea and within each of these groups there are greater differences than are found between the other four kingdoms (Animalia, Plantae, Mycetae, and Protista).

This new picture of living things requires us to reorganize our thinking. It appears that the oldest organisms may have been

TABLE 24.1   Major domains of life

Bacteria	Archaea	Eucarya
No nuclear membrane.	No nuclear membrane.	Nuclear membrane present.
Chlorophyll-based photosynthesis is a bacterial invention but most can function anaerobically.	None have been identified as pathogenic.	Many kinds of membranous organelles present in cells.
Oxygen-generating photosynthesis was an invention of the Cyanobacteria.	Probably have a common ancestor with Eucarya.	Chloroplasts are probably derived from Cyanobacteria.
Large number of heterotrophs oxidize organic molecules for energy.	Many use energy from inorganic reactions to make organic matter: 1. methanogens $H_2 + CO_2 \rightarrow CH_4$ 2. sulfur $H_2 + S \rightarrow H_2S$	Mitochondria probably derived from Proteobacteria. Probably have a common ancestor with Archaea.
Some use energy from inorganic chemical reactions to produce organic molecules.	Typically found in extreme environments.	Common evolutionary theme is the development of complex cells through symbiosis with other organisms.
Some live at high temperatures and may be ancestral of Archaea.	Few heterotrophs.	
Much metabolic diversity among closely related organisms.		

bacteria that were able to live in hot situations and that they gave rise to the Archaea, many of which still require extreme environments. Perhaps most interesting is the idea that the Archaea and Eucarya share many characteristics suggesting that they are more closely related to each other than either is to the Bacteria.

It appears that each domain developed specific abilities. The Archaea are primarily organisms that use inorganic chemical reactions to generate the energy they need to make organic matter. Often these reactions result in the production of methane. These organisms are known as methanogens. Others use sulfur and produce H_2S. Most of these organisms are found in extreme environments such as hot springs or in extremely salty or acidic environments.

The Bacteria developed many different metabolic abilities. Today many are able to use organic molecules as a source of energy, some are able to carry on photosynthesis, and still others are able to get energy from inorganic chemical reactions similar to Archaea.

The Eucarya are the most familiar and appear to have exploited the metabolic abilities of other organisms by incorporating them into their own structure. Chloroplasts and mitochondria are both bacteria-like structures found inside eukaryotic cells. Table 24.1 summarizes the major characteristics of these three domains.

THE ORIGIN OF EUKARYOTIC CELLS

The earliest fossils appear to be similar in structure to that of present-day bacteria. Therefore, it is likely that the early heterotrophs and autotrophs were probably simple one-celled organisms like bacteria. They were **prokaryotes** that lacked nuclear mem-

branes and other membranous organelles, such as mitochondria, an endoplasmic reticulum, chloroplasts, and a Golgi apparatus. Present-day bacteria and blue-green bacteria and archaea are prokaryotes. All other forms of life are **eukaryotes,** which possess a nuclear membrane and other membranous organelles.

Biologists generally believe that the eukaryotes evolved from the prokaryotes. The **endosymbiotic theory** attempts to explain this evolution. This theory suggests that present-day eukaryotic cells evolved from the combining of several different types of primitive prokaryotic cells. It is thought that some organelles found in eukaryotic cells may have originated as free-living prokaryotes. Since mitochondria and chloroplasts contain bacteria-like DNA and ribosomes, control their own reproduction, and synthesize their own enzymes, it has been suggested that they originated as free-living prokaryotic bacteria. These bacterial cells could have established a symbiotic relationship with another primitive nuclear-membrane-containing cell type (figure 24.3). When this theory was first suggested it met with a great deal of criticism. However, continuing research has uncovered several other instances of the probable joining of two different prokaryotic cells to form one.

If these cells adapted to one another and were able to survive and reproduce better as a team, it is possible that this relationship may have evolved into present-day eukaryotic cells. If this relationship had included only a nuclear-membrane-containing cell and aerobic bacteria, the newly evolved cell would have been similar to present-day heterotrophic protozoa, fungi, and animal cells. If this relationship had included both aerobic bacteria and photosynthetic bacteria, the newly formed cell would have been similar to present-day autotrophic algae and plant cells. In addition it is likely that endosymbiosis occurred among eukaryotic

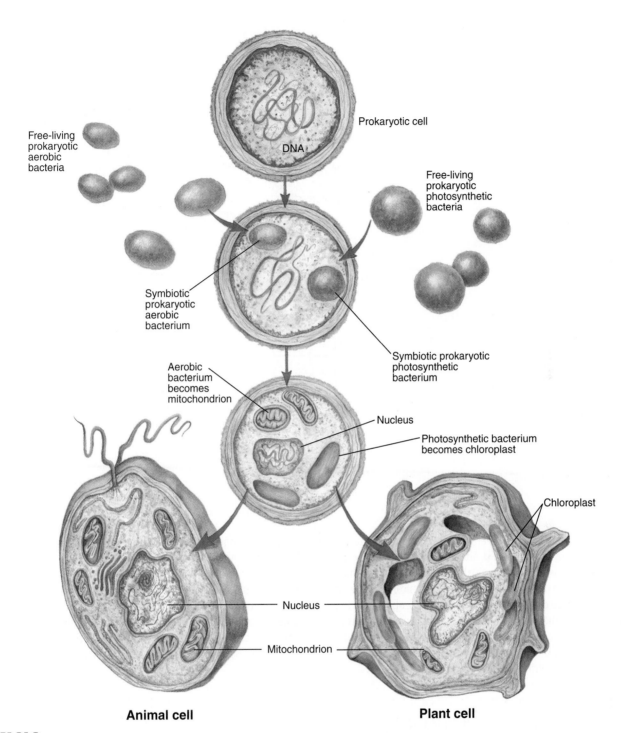

Prokaryotic cell

DNA

Free-living
prokaryotic
aerobic
bacteria

Free-living
prokaryotic
photosynthetic
bacteria

Symbiotic
prokaryotic
aerobic
bacterium

Symbiotic prokaryotic
photosynthetic
bacterium

Aerobic
bacterium
becomes
mitochondrion

Nucleus

Photosynthetic bacterium
becomes chloroplast

Chloroplast

Nucleus

Mitochondrion

Animal cell

Plant cell

FIGURE 24.3

The endosymbiotic theory proposes that some free-living prokaryotic bacteria developed symbiotic relationships with a host cell. When some aerobic bacteria developed into mitochondria and photosynthetic bacteria developed into chloroplasts, a eukaryotic cell evolved. These cells evolved into eukaryotic plant and animal cells.

organisms as well. Several kinds of eukaryotic red and brown algae contain chloroplast-like structures that appear to have originated as free-living eukaryotic cells.

Regardless of the type of cell (prokaryotic or eukaryotic), or whether the organisms are heterotrophic or autotrophic, all organisms have a common basis. DNA is the universal genetic material; protein serves as structural material and enzymes; and ATP is the source of energy. Although there is a wide variety of organisms, they all are built from the same basic molecular building blocks. Therefore, it is probable that all life derived from a single origin and that the variety of living things seen today evolved from the first protocells.

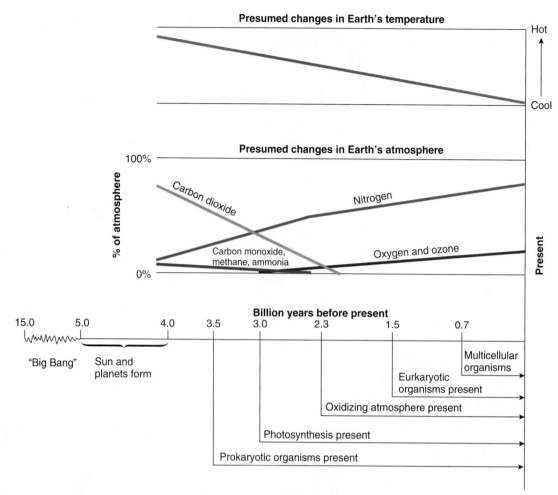

FIGURE 24.4

Current thinking suggests that several factors shaped the nature of Earth and its atmosphere. It is probable that as Earth cooled, water remained on the surface as a liquid. Additional water probably arrived in the form of comets that collided with Earth. Conceivably, the initial atmosphere was primarily carbon dioxide (CO_2) with smaller amounts of nitrogen (N_2), ammonia (NH_3), methane (CH_4), and carbon monoxide (CO). Carbon dioxide decreased as a result of chemical reactions at the surface of Earth. Living organisms have had a significant effect on the atmosphere because once photosynthesis began carbon dioxide was further reduced and oxygen (O_2) began to increase. Carbon dioxide was further reduced by the accumulation of carbonates in the shells of many kinds of marine protozoa and animals. The presence of oxygen allowed for the increase in ozone (O_3), which reduces the amount of ultraviolet light penetrating the atmosphere to reach the surface of Earth. Because ultraviolet light causes mutations, a protective layer of ozone probably has reduced the frequency of mutations from that which was present on primitive Earth.

Let us return for a moment to the question that perplexed early scientists and caused the controversies surrounding the opposing theories of spontaneous generation and biogenesis. From our modern perspective we can see that all life we experience comes into being as a result of reproduction. Life is generated from other living things, the process of biogenesis. However, reproduction does not answer the question: Where did life come from in the first place? We can speculate, test hypotheses, and discuss various possibilities, but we will probably never know for sure. Life either always was or it started at some point in the past. If it started, then spontaneous generation of some type had to occur at least once, but it is not happening today.

In this chapter we have discussed several ideas about how cells may have originated. It is thought that from these cells evolved the great diversity we see in living organisms today. Figure 24.4 summarizes several of the major events in the formation of Earth and the development of life on it.

THE ROLE OF NATURAL SELECTION IN EVOLUTION

In many cultural contexts, the word *evolution* means progressive change. We talk about the evolution of economies, fashion, or musical tastes. From a biological perspective, the word has a more specific meaning. **Evolution** is the continuous genetic adaptation of a population of organisms to its environment over time. Evolution involves changes in the genes that are present in a population. By definition individual organisms are not able to evolve—only populations can. The organism's surroundings determine which characteristics favor survival and reproduction (i.e., which characteristics best fit the organism to its environment). The mechanism by which evolution occurs involves the selective passage of genes from one generation to the next through sexual reproduction. The various processes that encourage the passage of beneficial genes to future generations and discourage the passage of harmful or less valuable genes are collectively known as **natural selection.**

The idea that some individuals whose gene combinations favor life in their surroundings will be most likely to survive, reproduce, and pass their genes on to the next generation is known as the **theory of natural selection.** The *theory of evolution,* however, states that populations of organisms become genetically adapted to their surroundings over time. Natural selection is the process that brings about evolution by "selecting" which genes will be passed to the next generation.

Recall that a theory is a well-established generalization supported by many different kinds of evidence. The theory of natural selection was first proposed by Charles Darwin and Alfred Wallace and was clearly set forth in 1859 by Darwin in his book *On the Origin of Species by Means of Natural Selection, or the Preservation of Favored Races in the Struggle for Life.* Since the time it was first proposed, the theory of natural selection has been subjected to countless tests and remains the core concept for explaining how evolution occurs.

There are two common misinterpretations associated with the process of natural selection. The first involves the phrase "survival of the fittest." Individual survival is certainly important because those that do not survive will not reproduce. But the more important factor is the number of descendants an organism leaves. An organism that has survived for hundreds of years but has not reproduced has not contributed any of its genes to the next generation and so has been selected against. The key, therefore, is not survival alone but survival and reproduction of the more fit organisms.

Second, the phrase "struggle for life" does not necessarily refer to open conflict and fighting. It is usually much more subtle than that. When a resource such as nesting material, water, sunlight, or food is in short supply, some individuals survive and reproduce more effectively than others. For example, many kinds of birds require holes in trees as nesting places (see figure 24.5). If these are in short supply, some birds will be fortunate and find a top-quality nesting site, others will occupy less suitable holes, and some may not find any. There may or may not be fighting for possession of a site. If a site is already occupied, a bird may not necessarily try to dislodge its occupant but may just continue to search for suitable but less valuable sites. Those that successfully occupy good nesting sites will be much more successful in raising young than will those that must occupy poor sites or those that do not find any.

Similarly, on a forest floor where there is little sunlight, some small plants may grow fast and obtain light while shading

FIGURE 24.5

Many kinds of birds, like this red-bellied woodpecker, nest in holes in trees. If old and dead trees are not available they may not be able to breed. Many people build birdhouses that provide artificial "tree holes" to encourage such birds to nest near their homes.

out plants that grow more slowly. The struggle for life in this instance involves a subtle difference in the rate at which the plants grow. But the plants are indeed engaged in a struggle, and a superior growth rate is the weapon for survival.

WHAT INFLUENCES NATURAL SELECTION?

Now that we have a basic understanding of how natural selection works, we can look in more detail at factors that influence it. Genetic variety within a species, genetic recombination as a result of sexual reproduction, the degree to which genes are expressed, and the ability of most species to reproduce excess offspring all exert an influence on the process of natural selection.

Genetic Variety Resulting from Mutation

In order for natural selection to occur, there must be genetic differences among the many individuals of an interbreeding population of organisms. If all individuals are identical genetically, it does not matter which ones reproduce—the same genes will be passed to the next generation and natural selection cannot occur. Genetic variety is generated in two ways. First of all,

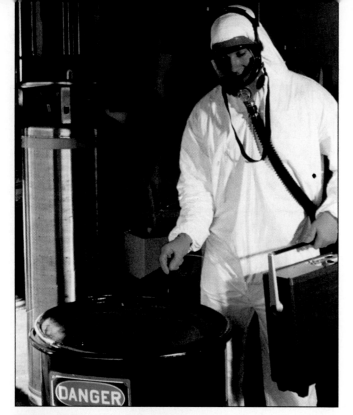

FIGURE 24.6

Because radiation and certain chemicals increase the likelihood of mutations, people who work in hazardous environments receive special training and use protective measures to reduce their exposure to mutagenic agents.

mutations may alter existing genes, resulting in the introduction of entirely new genetic information into a species' gene pool. **Spontaneous mutations** are changes in DNA that cannot be tied to a particular causative agent. It is suspected that cosmic radiation or naturally occurring mutagenic chemicals might be the cause of many of these mutations. It is known that subjecting organisms to high levels of radiation or to certain chemicals increases the rate at which mutations occur. It is for this reason that people who work with radioactive materials or other mutagenic agents take special safety precautions (figure 24.6).

Naturally occurring mutation rates are low (perhaps one chance in one hundred thousand that a gene will be altered), and mutations usually result in a gene that is harmful. However, in populations of millions of individuals, each of whom has thousands of genes, over thousands of generations it is quite possible that a new beneficial piece of genetic information could come about as a result of mutation. When we look at the various genes that exist in humans or in any other organism, we should remember that every gene originated as a modification of a previously existing gene. For example, the gene for blue eyes may be a mutated brown-eye gene, or blond hair may have originated as a mutated brown-hair gene. Thus, mutations have been very important for introducing new genetic material into species over time.

In order for mutations to be important in the evolution of organisms, they must be in cells that will become sex cells. Mutations to the cells of the skin or liver will affect only those specific cells and will not be passed to the next generation.

Genetic Variety Resulting from Sexual Reproduction

A second very important process involved in generating genetic variety is sexual reproduction. While sexual reproduction does not generate new genetic information the way mutation does, it allows for the recombination of genes into mixtures that did not occur previously. Each individual entering a population by sexual reproduction carries a unique combination of genes; approximately half are donated by the mother and half are donated by the father. During the formation of sex cells the manipulation of chromosomes results in new combinations of genes. This means that there are millions of possible combinations of genes in the sex cells of any individual. When fertilization occurs, one of the millions of possible sperm unites with one of the millions of possible eggs, resulting in a genetically unique individual. The gene mixing that occurs during sexual reproduction is known as **genetic recombination.** The new individual has a complete set of genes that is different from that of any other organism that ever existed.

Acquired Characteristics Do Not Influence Natural Selection

Many organisms survive because they have characteristics that are not genetically determined. These **acquired characteristics** are gained during the life of the organism; they are not genetically determined and, therefore, cannot be passed on to future generations through sexual reproduction. Consider an excellent tennis player's skill. Although this person may have inherited characteristics that are beneficial to a tennis player, the ability to play a good game of tennis is acquired through practice, not through genes. An excellent tennis player's offspring will not automatically be excellent tennis players. They may inherit some of the genetically determined physical characteristics necessary to become excellent tennis players, but the skills are still acquired through practice.

We often desire a specific set of characteristics in our domesticated animals. For example, the breed of dog known as boxers are "supposed" to have short tails. However, the genes for short tails are rare in this breed. Consequently, the tails of these dogs are amputated—a procedure called docking. Similarly, the tails of lambs are also usually amputated. These acquired characteristics are not passed on to the next generation. Removing the tails of these animals does not remove the genes for tail production from their genetic information and each generation of puppies is born with tails.

PROCESSES THAT DRIVE NATURAL SELECTION

Several mechanisms allow for selection of certain individuals for successful reproduction. The specific environmental factors that favor certain characteristics are called **selecting agents.** If predators must pursue swift prey organisms, then the faster predators will be selected for, and the selecting agent is the swiftness of available prey. If predators must find prey that are slow but hard to see, then the selecting agent is the camouflage coloration of the prey, and keen eyesight is selected for. If plants are eaten by

A CLOSER LOOK | Human-Designed Organisms

Humans have designed several kinds of plants and animals for their own purposes through the process of domestication. Most modern cereal grains are special plants that rely on human activity for their survival; most would not live without fertilizer, cultivation, and other helps. These grains are the descendants of wild plants. Initially the process of domestication of plants by humans probably was not a conscious effort to develop crops, but the unconscious selection of individual plants that had particularly useful characteristics from among all the plants of that species. For example, the seeds may have been larger than those of other plants of the same species or they may have tasted better or they may have been easier to chew. The unconscious selection of these seeds could have led to the dispersal of these particular seeds along traditional travel routes and eventually trading with other groups of people. Wheat, barley, rice, and corn are all plants that have been domesticated and currently supply a major part of the food for the world's population. However, modern genetic techniques have resulted in domesticated plants that only superficially resemble their wild ancestors. Most domesticated animals are mammals or birds that are herbivores that grow rapidly, have a decent disposition, and will reproduce in captivity. Examples are cattle, sheep, pigs, goats, horses, camels, chickens, turkeys, ducks, and geese. (The carnivorous dog is an obvious exception to the general herbivore rule.) Obviously animals that do not reproduce in captivity would not be good candidates for domestication and those that had behaviors that made them dangerous to be around would not be chosen to be domesticated. Again, the choices by humans would have been "unconscious." They would not have planned to domesticate an animal for a particular purpose (milk, eggs, power, meat) and modern science has greatly modified domesticated animals to something that is quite different from their wild ancestors. In fact, many of the wild ancestors of modern domesticated plants and animals are extinct. Thousands of generations of selection have, in effect, caused the development of species that are known only by their domesticated remnants.

---

insects, then the production of toxic materials in the leaves is selected for. All selecting agents influence the likelihood that certain characteristics will be passed to subsequent generations.

Differential Survival

As stated previously, the phrase "survival of the fittest" is often associated with the theory of natural selection. Although this is recognized as an oversimplification of the concept, survival is an important factor in influencing the flow of genes to subsequent generations. If a population consists of a large number of genetically and structurally different individuals it is likely that some of them will possess characteristics that make their survival difficult. Therefore, they are likely to die early in life and not have an opportunity to pass their genes on to the next generation.

The English peppered moth provides a classic example. Two color types are found in the species: One form is light-colored and one is dark-colored. These moths rest on the bark of trees during the day, where they may be spotted and eaten by birds. The birds are the selecting agents. About 150 years ago, the light-colored moths were most common. However, with the advance of the Industrial Revolution in England, which involved an increase in the use of coal, air pollution increased. The fly ash in the air settled on the trees, changing the bark to a darker color. Because the light moths were more easily seen against a dark background, the birds ate them (figure 24.7). The darker ones were less conspicuous; therefore, they were less frequently eaten and more likely to reproduce successfully. The light-colored moth, which was originally the more common type, became much less common. This change in the frequency of light- and dark-colored forms occurred within the short span of 50 years. Scientists who have studied this situation have estimated that the dark-colored moths had a 20 percent better chance of reproducing than did the light-colored moths. This study is continuing today. As England has reduced its air pollution and tree bark has become lighter in color, the light-colored form of the moth increased in frequency again.

As another example of how differential survival can lead to changed gene frequencies, consider what has happened to many insect populations as we have subjected them to a variety of insecticides. Since there is genetic variety within all species of insects, an insecticide that is used for the first time on a particular species kills all those that are genetically susceptible. However, the insecticide may not kill individuals with slightly different genetic compositions.

Suppose that in a population of a particular species of insect, 5 percent of the individuals have genes that make them resistant to a specific insecticide. The first application of the insecticide could, therefore, kill 95 percent of the population. However, tolerant individuals would then constitute the majority of the breeding population that survived. This would mean that many insects in the second generation would be tolerant. The second use of the insecticide on this population would not be as effective as the first. With continued use of the same insecticide, each generation would become more tolerant.

Many species of insects produce a new generation each month. In organisms with a short generation time, 99 percent of the population could become resistant to the insecticide in just

FIGURE 24.7

This photo of the two variations of the peppered moth shows that the light-colored moth is much more conspicuous against the dark tree trunk. (The two dark moths are indicated by arrows.) The trees are dark because of an accumulation of pollutants from the burning of coal. The more conspicuous light-colored moths are more likely to be eaten by bird predators, and the genes for light color should become more rare in the population.

five years. As a result, the insecticide would no longer be useful in controlling the species. As a new factor (the insecticide) was introduced into the environment of the insect, natural selection resulted in a population that was tolerant of the insecticide. Figure 24.8 indicates that more than five hundred species of insects have populations that are resistant to many kinds of insecticides.

Differential Reproductive Rates

Survival alone does not always ensure reproductive success. For a variety of reasons, some organisms may be better able to utilize available resources to produce offspring. If one individual leaves 100 offspring and another leaves only 2, the first organism has passed more copies of its genetic information to the next generation than has the second. If we assume that all 102 individual offspring have similar survival rates, the first organism has been selected for and its genes have become more common in the subsequent population.

Scientists have conducted studies of the frequencies of genes for the height of clover plants (figure 24.9). Two identical fields of clover were planted and cows were allowed to graze in one of them. Cows acted as a selecting agent by eating the taller plants first. These tall plants rarely got a chance to reproduce. Only the shorter plants flowered and produced seeds. After some time, seeds were collected from both the grazed and ungrazed fields and grown in a greenhouse under identical conditions. The average height of the plants from the ungrazed field was compared to that of the plants from the grazed field. The seeds from the ungrazed field produced

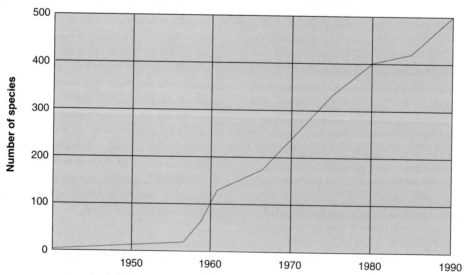

Pest species resistant to insecticides

FIGURE 24.8

The continued use of insecticides has constantly selected for genes that give resistance to a particular insecticide. As a result, many species of insects and other arthropods are now resistant to many kinds of insecticides, and the number continues to increase.

Source: Data from Georghiou, University of California at Riverside.

FIGURE 24.9

The clover field to the left of the fence is undergoing natural selection: the grazing cattle are eating the tall plants and causing them to reproduce less than do the short plants. The other field is not subjected to this selection pressure, so its clover population has more genes for tallness.

some tall, some short, but mostly medium-sized plants. However, the seeds from the grazed field produced many more shorter plants than medium or tall ones. The cows had selectively eaten the plants that had the genes for tallness. Since the flowers are at the tip of the plant, tall plants were less likely to successfully reproduce, even though they might have been able to survive grazing by cows.

Differential Mate Selection

Within animal populations, some individuals may be chosen as mates more frequently than others. Obviously, those that are frequently chosen have an opportunity to pass on more copies of their genes than those that are rarely chosen. Characteristics of the more frequently chosen individuals may involve general characteristics, such as body size or aggressiveness, or specific conspicuous characteristics attractive to the opposite sex.

For example, male red-winged blackbirds establish territories in cattail marshes where females build their nests. A male will chase out all other males but not females. Some males have large territories, some have small territories, and

some are unable to establish territories. Although it is possible for any male to mate, it has been demonstrated that those who have no territory are least likely to mate. Those who defend large territories may have two or more females nesting in their territories and are very likely to mate with those females. It is unclear exactly why females choose one male's territory over another, but the fact is that some males are chosen as mates and others are not.

In other cases, it appears that the females select males that display conspicuous characteristics. Male peacocks have very conspicuous tail feathers. Those with spectacular tails are more likely to mate and have offspring (figure 24.10). Darwin was puzzled by such cases as the peacock, in which the large and conspicuous tail should have been a disadvantage to the bird. Long tails require energy to produce, make it more difficult to fly, and make it more likely that predators will capture the individual. The current theory that seeks to explain this paradox involves female choice. If the females have an innate (genetic) tendency to choose the most elaborately decorated males, genes that favor such plumage will be regularly passed to the next generation. Such special cases in which females choose males with specific characteristics have been called sexual selection.

FIGURE 24.10

In many animal species the males display very conspicuous characteristics that are attractive to females. Because the females choose the males they will mate with, those males with the most attractive characteristics will have more offspring and, in future generations, there will be a tendency to enhance the characteristic. With peacocks, those individuals with large colorful displays are more likely to mate.

A SUMMARY OF THE CAUSES OF EVOLUTIONARY CHANGE

Earlier in this chapter, evolution was described as the change in gene frequency over time. We can now see that several different mechanisms operate to bring about this change. Mutations can either change one genetic message into another or introduce an entirely new piece of genetic information into the population. Immigration can introduce new genetic information if the entering organisms have unique genes. Emigration and death remove genes from the gene pool. Natural selection systematically filters some genes from the population, allowing other genes to remain and become more common. The primary mechanisms involved in natural selection are differences in death rates, reproductive rates, and the rate at which individuals are selected as mates (figure 24.11). In addition, gene frequencies are more easily changed in small populations since events such as death, immigration, emigration, and mutation can have a greater impact on a small population than on a large population.

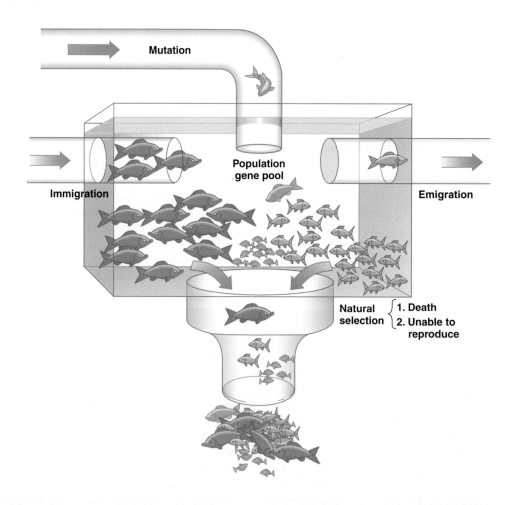

FIGURE 24.11

Several different processes cause gene frequencies to change. Genes enter populations through immigration and mutation. Genes leave populations through death and emigration. Natural selection operates within populations through death and rates of reproduction.

A

B

C

FIGURE 24.12

Even though they don't do so in nature, (*A*) horses and (*B*) donkeys can be mated. The offspring is called a (*C*) mule and is sterile. Because the mule is sterile, the horse and the donkey are considered to be of different species.

SPECIES: A WORKING DEFINITION

The smallest irreversible step in the evolutionary process is the invention of a new species from previous species. Before we consider how new species are produced, let's establish how one species is distinguished from another. A **species** is commonly defined as a population of organisms whose members have the potential to interbreed naturally to produce fertile offspring but do not interbreed with other groups. This is a working definition since it applies in most cases but must be interpreted to encompass some exceptions. There are two key ideas within this definition. First, a species is a population of organisms. An individual—you, for example—is not a species. You can only be a member of a group that is recognized as a species. The human species, *Homo sapiens,* consists of more than 6 billion individuals, whereas the endangered California condor species, *Gymnogyps californianus,* consists of about 100 individuals.

Second, the definition involves the ability of individuals within the group to produce fertile offspring. Obviously, we cannot check every individual to see if it is capable of mating with any other individual that is similar to it, so we must make some judgment calls. Do most individuals within the group potentially have

the capability of interbreeding to produce fertile offspring? In the case of humans we know that some individuals are sterile and cannot reproduce, but we don't exclude them from the human species because of this. If they were not sterile, they would have the potential to interbreed. We recognize that, although humans normally choose mating partners from their subpopulation, humans from all parts of the world are potentially capable of interbreeding. We know this to be true because of the large number of instances of reproduction involving people of different ethnic and racial backgrounds. The same is true for many other species that have local subpopulations but have a wide geographic distribution.

Another way to look at this question is to think about **gene flow.** Gene flow is the movement of genes from one generation to the next or from one region to another. Two or more populations that demonstrate gene flow between them constitute a single species. Conversely, two or more populations that do not demonstrate gene flow between them are generally considered to be different species. Some examples will clarify this working definition.

The mating of a male donkey and a female horse produces young that grow to be adult mules, incapable of reproduction (figure 24.12). Since mules are nearly always sterile, there can be no gene flow between horses and donkeys and they are considered

A CLOSER LOOK | Is the Red Wolf a Species?

The red wolf (*Canis rufus*) is listed as an endangered species, so the U.S. Fish and Wildlife Service has instituted a captive breeding program to preserve the animal and reintroduce it to a suitable habitat in the southeastern United States, where it was common into the 1800s. Biologists have long known that red wolves will hybridize with both the coyote, *Canis latrans*, and the gray wolf, *Canis lupus*, and many suspect that the red wolf is really a hybrid between the gray wolf and the coyote. Gray wolf–coyote hybrids are common in nature where one or the other species is rare. Some have argued that the red wolf does not meet the definition of a species and should not be protected under the Endangered Species Act.

Museums have helped shed light on this situation by providing skulls of all three kinds of animals preserved in the early 1900s. It is known that during the early 1900s as the number of red wolves in the southeastern United States declined, they readily interbred with coyotes, which were very common. The gray wolf had been exterminated by the early 1900s. Some scientists believe that the skulls of the few remaining "red wolves" might not represent the true red wolf but a "red wolf" with many coyote characteristics. Studies of the structure of the skulls of red wolves, coyotes, and gray wolves show that the red wolves were recognizably different and intermediate in structure between coyotes and gray wolves. This supports the hypothesis that the red wolf is a distinct species.

DNA studies were performed using DNA from preserved red wolf pelts. The red wolf DNA was compared to coyote and gray wolf DNA. These studies show that red wolves contain DNA sequences typical of both gray wolves and coyotes but do not appear to have distinct base sequences found only in the red wolf. These studies support the hypothesis that the red wolf is not a species but a population that resulted from hybridization between gray wolves and coyotes.

There is still no consensus on the status of the red wolf. Independent researchers disagree with one another and with Fish and Wildlife Service scientists, who have been responsible for developing and administering a captive breeding program and planning reintroductions of the red wolf.

---

to be separate species. Similarly, lions and tigers can be mated in zoos to produce offspring. However, this does not happen in nature and so gene flow does not occur naturally; thus they are considered to be two separate species.

Still another way to try to determine if two organisms belong to different species is to determine their genetic similarity. The recent advances in molecular genetics allow scientists to examine the structure of the genes present in individuals from a variety of different populations. Those that have a great deal of similarity are assumed to have resulted from populations that have exchanged genes through sexual reproduction in the recent past. If there are significant differences in the genes present in individuals from two populations they have not exchanged genes recently and are more likely to be members of separate species. Interpretation of the results obtained by examining genetic differences still requires the judgment of experts. It will not unequivocally settle every dispute related to the identification of species, but it is another tool that helps to clarify troublesome situations.

HOW NEW SPECIES ORIGINATE

The geographic area over which a species can be found is known as its **range.** The range of the human species is the entire world, while that of a bird known as a snail kite is a small region of southern Florida. As a species expands its range or environmental conditions change in some parts of the range, portions of the population can become separated from the rest. Thus, many species consist of partially isolated populations that display characteristics that differ significantly from other local populations. Many of the differences observed may be directly related to adaptations to local environmental conditions. This means that new colonies or isolated populations may have infrequent gene exchange with their geographically distant relatives. These genetically distinct populations are known as subspecies or demes.

A portion of a species can become totally isolated from the rest of the gene pool by some geographic change, such as the formation of a mountain range, river valley, desert, or ocean. When this happens the portion of the species is said to be in **geographic isolation** from the rest of the species. If two populations of a species are geographically isolated they are also reproductively isolated, and gene exchange is not occurring between them. The geographic features that keep the different portions of the species from exchanging genes are called **geographic barriers.** The uplifting of mountains, the rerouting of rivers, and the formation of deserts all may separate one portion of a gene pool from another. For example, two kinds of squirrels are found on opposite sides of the Grand Canyon. Some people consider them to be separate species, while others consider them to be different isolated subpopulations of the same species (figure 24.13). Even small changes may cause geographic isolation in species that have little ability to move. A fallen tree, a plowed field, or even a new freeway may effectively isolate populations within such species. Snails in two valleys separated by a high ridge have been found to be closely related but different species. The

Kaibab squirrel

Aberts squirrel

FIGURE 24.13

These two squirrels are found on opposite sides of the Grand Canyon. Some people consider them to be different species; others consider them to be distinct populations of the same species.

snails cannot get from one valley to the next because of the height and climatic differences presented by the ridge (figure 24.14).

The separation of a species into two or more isolated subpopulations is not enough to generate new species. Even after many generations of geographic isolation, these separate groups may still be able to exchange genes (mate and produce fertile offspring) if they overcome the geographic barrier, because they have not accumulated enough genetic differences to prevent reproductive success. Differences in environments and natural selection play very important roles in the process of forming new species. Following separation from the main portion of the gene pool by geographic isolation, the organisms within the small, local population are likely to experience different environmental conditions. If, for example, a mountain range has separated a species into two populations, one of them may

receive more rain or more sunlight than the other (figure 24.15). These environmental differences act as natural selecting agents on the two gene pools and acting over a long period of time account for different genetic combinations in the two places. Furthermore, different mutations may occur in the two isolated populations, and each may generate different random combinations of genes as a result of sexual reproduction. This would be particularly true if one of the populations was very small. As a result, the two populations may show differences in color, height, enzyme production, time of seed germination, or many other characteristics.

Over a long period of time, the genetic differences that accumulate may result in regional populations called **subspecies** that are significantly modified structurally, physiologically, or behaviorally. The differences among some subspecies

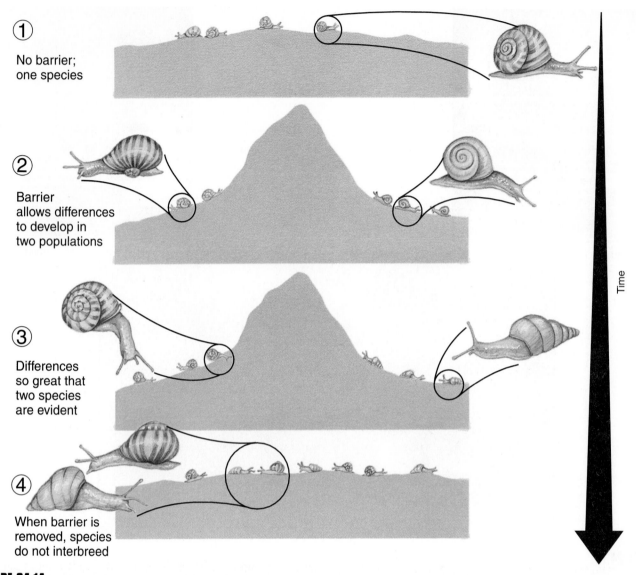

① No barrier; one species

② Barrier allows differences to develop in two populations

③ Differences so great that two species are evident

④ When barrier is removed, species do not interbreed

Time

FIGURE 24.14

If a single species of snail was to be divided into two different populations by the development of a ridge between them, the two populations could be subjected to different environmental conditions. This could result in a slow accumulation of changes that could ultimately result in two populations that would not be able to interbreed even if the ridge between them were to erode. They would be different species.

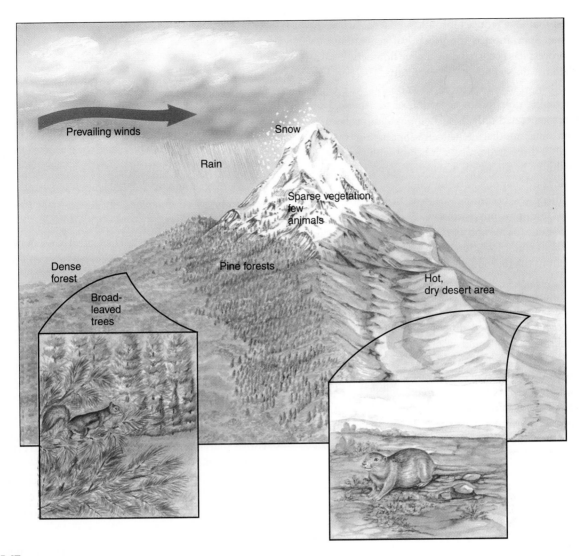

FIGURE 24.15

Most mountain ranges affect the local environment. Because of the prevailing winds, most rain falls on the windward side of the mountain. This supports abundant vegetation. The other side of the mountain receives much less rain and is drier. Often a desert may exist. Both plants and animals must be adapted to the kind of climate typical for their regions. Cactus and ground squirrels would be typical of the desert and pine trees and tree squirrels would be typical of the windward side of the mountain.

may be so great that they reduce reproductive success when the subspecies mate. **Speciation** is the process of generating new species. This process has occurred only if gene flow between isolated populations does not occur even after barriers are removed. In other words, the process of speciation can begin with the geographic isolation of a portion of the species, but new species are generated only if isolated populations become separate from one another genetically. Speciation by this method is really a three-step process. It begins with geographic isolation, is followed by the action of selective agents that choose specific genetic combinations as being valuable, and ends with the genetic differences becoming so great that reproduction between the two groups is impossible.

It is also possible to envision ways in which speciation could occur without geographic isolation being necessary. Any process that could result in the reproductive isolation of a portion of a species could lead to the possibility of speciation. For example, within populations, some individuals may breed or flower at a somewhat different time of the year. If the difference in reproductive time is genetically based, different breeding populations could be established, which could eventually lead to speciation. Among animals, variations in the genetically determined behaviors related to courtship and mating could effectively separate one species into two or more separate breeding populations. In plants, genetically determined incompatibility of the pollen of one population of flowering plants with the flowers of other populations of the same species could lead to separate species.

THE DEVELOPMENT OF EVOLUTIONARY THOUGHT

Today, most scientists consider speciation an important first step in the process of evolution. However, this was not always the case. For centuries people believed that the various species of plants and animals were fixed and unchanging—that is, they were thought to have remained unchanged from the time of their creation. This was a reasonable assumption because people knew nothing about DNA, meiosis, or population genetics. Furthermore, the process of evolution is so slow that the results of evolution were usually not evident during a human lifetime. It is even difficult for modern scientists to recognize this slow change in many kinds of organisms. In the mid-1700s, Georges-Louis Buffon, a French naturalist, expressed some curiosity about the possibilities of change (evolution) in animals, but he did not suggest any mechanism that would result in evolution.

In 1809, Jean-Baptiste de Lamarck, a student of Buffon's, suggested a process by which evolution could occur. He proposed that acquired characteristics could be transmitted to offspring. For example, he postulated that giraffes originally had short necks. Since giraffes constantly stretched their necks to obtain food, their necks got slightly longer. This slightly longer neck acquired through stretching could be passed to the offspring, who were themselves stretching their necks, and over time, the necks of giraffes would get longer and longer. Although we now know Lamarck's theory was wrong (because acquired characteristics are not inherited), it stimulated further thought

CONNECTIONS

Fossil Organisms Found Alive

At one time a primitive kind of conifer known as *Ginkgo biloba* was known only from fossils. However, in the 1700s it was discovered in China. It was considered to be a sacred tree and was cultivated on the grounds of religious temples. Since their rediscovery they have been planted around the world. They grow well in urban environments, but female trees produce foul-smelling seeds that are toxic. People have developed a skin rash simply by handling the seeds. Therefore, it is recommended that only male trees be planted in urban settings. Extracts from the tree have become a common herbal medicine with claims that it increases intelligence and protects from cancer.

as to how evolution could occur. All during this period, from the mid-1700s to the mid-1800s, lively arguments continued about the possibility of evolutionary change. Some, like Lamarck and others, thought that change did take place, while many others said that it was not possible. It was the thinking of two English scientists that finally provided a mechanism that explained how evolution could occur.

In 1858, Charles Darwin and Alfred Wallace suggested the theory of natural selection as a mechanism for evolution. They based their theory on the following assumptions about the nature of living things:

1. All organisms produce more offspring than can survive.
2. No two organisms are exactly alike.
3. Among organisms, there is a constant struggle for survival.
4. Individuals that possess favorable characteristics for their environment have a higher rate of survival and produce more offspring.
5. Favorable characteristics become more common in the species, and unfavorable characteristics are lost.

Using these assumptions, the Darwin-Wallace theory of evolution by natural selection offers a different explanation for the development of long necks in giraffes (figure 24.16):

1. In each generation, more giraffes would be born than the food supply could support.
2. In each generation, some giraffes would inherit longer necks, and some would inherit shorter necks.
3. All giraffes would compete for the same food sources.
4. Giraffes with longer necks would obtain more food, have a higher survival rate, and produce more offspring.
5. As a result, succeeding generations would show an increase in the neck length of the giraffe species.

This logic seems simple and obvious today, but remember that at the time Darwin and Wallace proposed their theory, the processes of meiosis and fertilization were poorly understood,

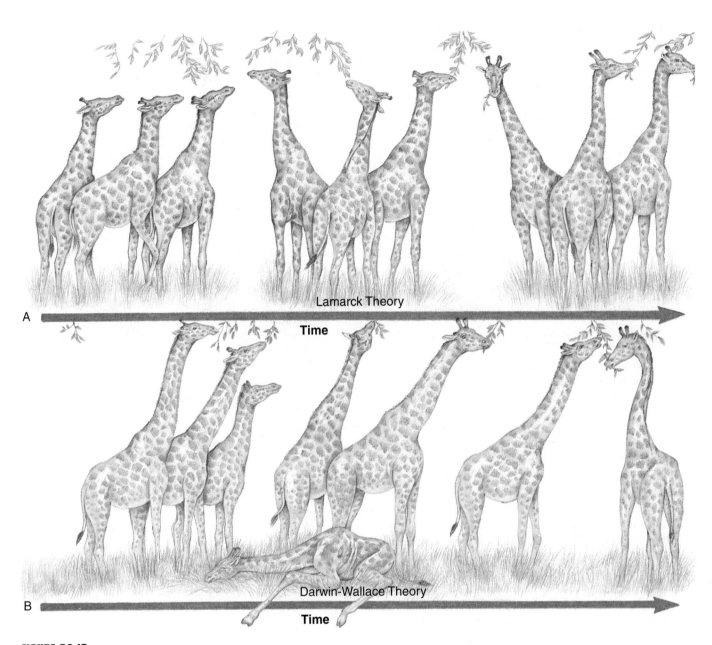

FIGURE 24.16

(*A*) Lamarck thought that acquired characteristics could be passed on to the next generation. Therefore, he postulated that as giraffes stretched their necks to get food, their necks got slightly longer. This characteristic was passed on to the next generation, which would have longer necks. (*B*) The Darwin-Wallace theory states that there is variation within the population and that those with longer necks would be more likely to survive and reproduce and pass on their genes for long necks to the next generation.

and the concept of the gene was only beginning to be discussed. Nearly 50 years after Darwin and Wallace suggested their theory, the rediscovery of the work of Gregor Mendel provided an explanation for how characteristics could be transmitted from one generation to the next. Not only did Mendel's idea of the gene provide a means of passing traits from one generation to the next, it also provided the first step in understanding mutations, gene flow, and the significance of reproductive isolation. All of these ideas are interwoven into the modern concept of

evolution. If we look at the same five ideas from the thinking of Darwin and Wallace and update them with modern information, they might look something like this:

1. An organism's capacity to overproduce results in surplus organisms.
2. Because of mutation, new genes enter the gene pool. Because of sexual reproduction, involving meiosis and fertilization, new combinations of genes are present in every generation. These processes are so powerful that each individual in a sexually

reproducing population is genetically unique. The genes present are expressed as the phenotype of the organism.

3. Resources such as food, soil nutrients, water, mates, and nest materials are in short supply, so some individuals will do without. Other environmental factors such as disease organisms, predators, or helpful partnerships with other species also affect survival. All of these factors that affect survival are called selecting agents.

4. Selecting agents favor individuals with the best combination of genes. They will be more likely to survive and reproduce, passing more of their genes on to the next generation. An organism is selected against if it has fewer offspring than other individuals that have a more favorable combination of genes. It does not need to die to be selected against.

5. Therefore, genes or gene combinations that produce characteristics favorable to survival will become more common, and the species will become better adapted to its environment.

THE TENTATIVE NATURE OF THE EVOLUTIONARY HISTORY OF ORGANISMS

It is important to understand that thinking about the concept of evolution can take us in several different directions. First, it is clear that genetic changes do occur. Mutations introduce new genes into a species. This has been demonstrated repeatedly with chemicals and radiation. Our recognition of this danger is evident by the ways we protect ourselves against excessive exposure to mutagenic agents. We also recognize that species can change. We purposely manipulate the genetic constitution of our domesticated plants and animals and change their characteristics to suit our needs. We also recognize that different populations of the same species show genetic differences. Examination of fossils shows that species of organisms that once existed are no longer in existence. We even have historical examples of plants and animals that are now extinct. We can also demonstrate that new species come into existence, and this is easily done with plants. It is clear from this evidence that species are not fixed, unchanging entities.

However, when we try to piece together the evolutionary history of organisms over long periods of time, we must use much indirect evidence, and it becomes difficult to state definitively the specific sequence of steps that the evolution of a species followed. Although there is little question that evolution occurs, it is not possible to state unconditionally that evolution of a particular group of organisms has followed a specific path. There will always be new information that will require changes in thinking, and equally reputable scientists will disagree on the evolutionary processes or the sequence of events that led to a specific group of organisms. But there can be no question that evolution occurred in the past and continues to occur today.

HUMAN EVOLUTION

There is intense curiosity about the how our species (*Homo sapiens*) came to be and the evolution of the human species remains an interesting and controversial topic. We recognize that humans show genetic diversity, experience mutations, and are subject to the same evolutionary forces as other organisms. We also recognize that some individuals have genes that make them subject to early death or unable to reproduce. On the other hand, since all of our close evolutionary relatives are extinct, it is difficult for us to visualize our evolutionary development and we tend to think we are unique and not subject to the laws of nature.

We use several kinds of evidence to try to sort out our evolutionary history. Fossils of various kinds of human and prehuman ancestors have been found, but these are often fragmentary and hard to date. Stone tools of various kinds have also been found associated with human and prehuman sites. Finally, other aspects of the culture of our human ancestors have been found in burial sites, cave paintings, and the creation of ceremonial objects. Various methods have been used to age these findings. Some can be dated quite accurately, while others are more difficult to pinpoint.

When fossils are examined, anthropologists can identify differences in the structures of bones that are consistent with changes in species. Based on the amount of change they see and the ages of the fossils, these scientists make judgments about the species to which the fossil belongs. As new discoveries are made, opinions of experts will change and our evolutionary history may become clearer as old ideas are replaced. It is also clear from the fossil record that humans are relatively recent additions to the forms of life. Assembling all of these bits of information into a clear picture is not possible at this point, but several points are well accepted.

1. Several species of hominids of the genus *Australopithecus* are the oldest hominid fossils. They were herbivores and walked upright. Their fossils are found only in Africa.
2. Several species of the genus *Homo* became prominent in Africa and appear to have made a change from a primarily herbivorous diet to a carnivorous diet.
3. Fossils of *Homo erectus* are found throughout Africa, Europe, and Asia, but not in Australia or the Americas.
4. Based on fossil evidence it appears that the climate of Africa was becoming dryer during the time that hominid evolution was occurring.
5. Various kinds of hominids appear to have been sexually dimorphic, with the males being larger. This suggests that there may have been differences in activities as well.

When we try to put all of these bits of information together we can construct the following scenario for the evolution of our species. Monkeys, apes, and other primates are adapted to living in forested areas where their grasping hands, opposable thumbs and big toes, and wide range of movement of the shoulders allow them to move freely in the trees. As the climate became dryer grasslands replaced the forests and, as is always the case, some organisms became extinct while others adapted to the change.

Australopithecus is the earliest humanlike organism known from the fossil record. Australopithecines were present in Africa from about 4.4 million years ago until about 1 million years ago. It is important to recognize that there are few fossils of these early humanlike organisms and that often they are fragments of the whole organism. This has led to much speculation and argument among experts about the specific position each fossil has in the evolutionary history of humans. However, from examin-

A CLOSER LOOK | Neandertals Were Probably a Different Species

An ongoing controversy surrounds the relationship between Neandertal and other forms of prehistoric humans. One position is that Neandertals were a small separate race or subspecies of human that could have interbred with other humans and may have disappeared because their subspecies was eliminated by interbreeding with more populous, successful groups. (Many small, remote tribes have been eliminated in the same way in recent history.) Others maintain that the Neandertals show such great difference from other early humans that they must be of different species and became extinct because they could not compete with more successful immigrants from Africa. (The names of these ancient people typically are derived from the place where the fossils were first discovered. For example, Neandertals were first found in the Neander Valley of Germany and Cro-Magnons were initially found in the Cro-Magnon caves in France.)

The use of molecular genetic technology has shed some light on the relationship of Neandertals to other kinds of humans. Examination of the mitochondrial DNA obtained from the bones of a Neandertal individual reveals that there are significant differences between Neandertals and other kinds of early humans. This greatly strengthens the argument that Neandertals were a separate species.

---

ing the fossil bones of the leg, pelvis, and foot it is apparent that *Australopithecus* was relatively short and stocky and walked upright like humans. Anthropologists recognize three different species of *Australopithecus.*

An upright posture has several advantages. It allows for more rapid movement over long distances, the ability to see longer distances, and reduces the amount of heat gained from the sun. In addition, upright posture frees the arms for other uses such as carrying things, manipulating objects, and using tools. The various species of *Australopithecus* shared these characteristics and based on the structure of their skull, jaws, and teeth appeared to have been herbivores with relatively small brains. About 2.5 million years ago the first members of the genus *Homo* appeared on the scene. *Homo habilis* had a larger brain and smaller teeth than australopithecenes and made much more use of stone tools. Some people feel that it was a direct descendant of *Australopithecus africanus.* Many experts feel that *Homo habilis* was a scavenger that made use of group activities, tools, and higher intelligence to commandeer the kills made by other carnivores. The higher-quality diet would have supported the metabolic needs of the larger brain.

About 1.8 million years ago *Homo erectus* appeared on the scene. It was much larger (up to 6 ft tall) than *H. habilis* (about 4 ft tall) and also had a much larger brain. The larger brain appears to be associated with extensive use of stone tools. Hand axes were manufactured and used to cut the flesh of prey and crush the bones for marrow. *Homo erectus* appears to have been a predator while *Homo habilis* was a scavenger. The use of meat as food allows animals to move about more freely, because appropriate food is available almost everywhere. By contrast, herbivores are often confined to places that have foods appropriate to their use; fruits for fruit eaters, grass for grazers, forests for browsers, etc. In fact, *H. erectus* has been found throughout the world with the exception of Australia and the Americas. Most experts feel that *H. erectus* originated in Africa and migrated to Asia and Europe.

Homo erectus disappears from the fossil record at about 100,000 years ago. During the period of about 1.7 million years, *H. erectus* shows a progressive increase in the size of the cranial capacity and reduction in the size of the jaw. Later fossils of *H. erectus* are thus difficult to distinguish from early *Homo sapiens* and some scientists feel that *H. erectus* evolved into the early forms of *H. sapiens.* However, at about 800,000 years ago archaic forms classified as *Homo sapiens* are identified in the fossil record.

Two different theories seek to explain what happened to *Homo erectus.* One theory, known as the out-of-Africa hypothesis, states that modern humans migrated from Africa to Asia and Europe and displaced *H. erectus* that had migrated into these areas previously. The other theory states that *H. erectus* evolved into *H. sapiens* throughout Africa, Asia, and Europe and that interbreeding among the various groups has given rise to the various races of humans we see today.

Another continuing puzzle is the relationship of humans that clearly belong to the species *Homo sapiens* with a contemporary group known as Neandertals. Some people consider Neandertals to be a race of humans specially adapted to life in the harsh conditions found in postglacial Europe. Others consider them to be a separate species. The Neandertals were muscular, had a larger brain capacity than modern humans, and had many elements of culture, including burials. The cause of their disappearance from the fossil record at about 25,000 years ago remains a mystery. Perhaps climate change to a warmer climate was responsible. Perhaps contact with *Homo sapiens* resulted in their elimination either through hostile interactions or, if they were able to interbreed with *H. sapiens,* they could have been absorbed into the larger *H. sapiens* population.

Large numbers of fossils of prehistoric humans have been found in all parts of the world. Many of these show evidence of a collective group memory we call *culture.* Cave paintings, carvings in wood and bone, tools of various kinds, and burials are examples. These are also evidence of a capacity to think and invent, and "free time" to devote to things other than gathering

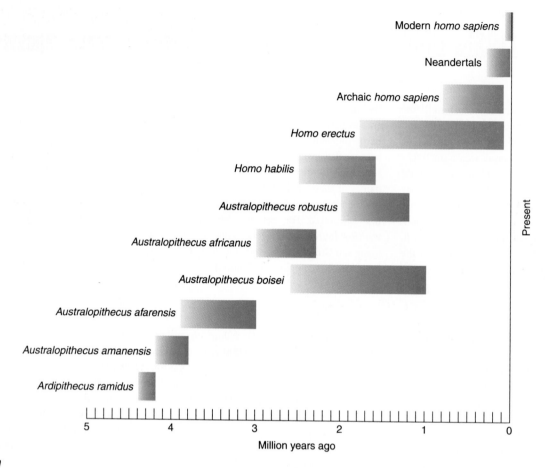

FIGURE 24.17

This diagram shows the various organisms thought to be ancestors of humans. All are extinct today except for modern humans. Notice that several of the different species of organisms coexisted for extensive periods.

food and other necessities of life. We may never know how we came to be but we will always be curious and will continue to search and speculate about our beginnings. Figure 24.17 summarizes the current knowledge of the historical record of humans and their relatives.

THE CLASSIFICATION OF ORGANISMS

Every day you see a great variety of living things. Just think of how many different species of plants and animals you have observed. Biologists at the Smithsonian Institution estimated that there are over 30 million species in the world; approximately 1.5 million of these have been named. What names do you assign to each? Is the name you use the same as that used in other sections of the country or regions of the world? In much of the United States and Canada, the fish pictured in figure 24.18A is known as a largemouth black bass, but in sections of the southern United States it is called a trout. This use of local names can lead to confusion. If a student in Mississippi writes to a friend in Wisconsin about catching a 6-pound trout, the person in Wisconsin thinks that the friend caught the kind of fish

pictured in figure 24.18B. In the scientific community, accuracy is essential; local names cannot be used. When a biologist is writing about a species, all biologists in the world who read that article must know exactly what that species is.

Taxonomy is the science of naming organisms and grouping them into logical categories. Various approaches have been used to classify organisms. The Greek philosopher Aristotle (384–322 B.C.) had an interest in nature and was the first person to attempt a logical classification system. The root word for *taxonomy* is the Greek word *taxis*, which means *arrangement*. Aristotle used the size of plants to divide them into the categories of trees, shrubs, and herbs.

During the Middle Ages, Latin was widely used as the scientific language. As new species were identified, they were given Latin names, often using as many as fifteen words. Although using Latin meant that most biologists, regardless of their native language, could understand a species name, it did not completely do away with duplicate names. Because many of the organisms could be found over wide geographic areas and communication was slow, there could be two or more Latin names for a species. To make the situation even more confusing, ordinary people still called organisms by their common local names.

A

B

FIGURE 24.18

Using the scientific name *Micropterus salmoides* for largemouth black bass (*A*) and *Salmo trutta* for brown trout (*B*) correctly indicates which of these two species of fish a biologist is talking about. Both fish are called trout in some parts of the world.

FIGURE 24.19

Carolus Linnaeus (1707–1778), a Swedish doctor and botanist, originated the modern system of taxonomy.

The modern system of classification began in 1758 when Carolus Linnaeus (1707–1778), a Swedish doctor and botanist, published his tenth edition of *Systema Naturae* (figure 24.19). (Linnaeus's original name was Karl von Linne, which he "latinized" to Carolus Linnaeus.) In the previous editions, Linnaeus had used a polynomial (many-names) Latin system. However, in the tenth edition he introduced the **binomial** (two-name) **system of nomenclature.** This system used two Latin names, genus and specific epithet (epithet = descriptive word), for each species of organism.

Recall that a species is a population of organisms capable of interbreeding and producing fertile offspring. Individual organisms are members of a species. A **genus** (plural, *genera*) is a group of closely related organisms; the specific epithet is a word added to the genus name to identify which one of several species within the genus we are discussing. It is similar to the naming system we use with people. When you look in the phone book you look for the last name (surname), which gets you in the correct general category. Then you look for the first name (given name) to identify the individual you wish to call. The unique name given to a particular type of organism is called its species name or scientific name. In order to clearly identify the scientific name, binomial names are either *italicized* or *underlined*. The first letter of the genus name is capitalized. The specific epithet is always written in lowercase. *Micropterus salmoides* is the binomial name for the largemouth black bass.

When biologists adopted Linnaeus's binomial method, they eliminated the confusion that was the result of using common local names. For example, with the binomial system the white water lily is known as *Nymphaea odorata.* Regardless of which of the 245 common names is used in a botanist's local area, when botanists read *Nymphaea odorata,* they know exactly which plant is being referred to. The binomial name cannot be changed unless there is compelling evidence to justify doing so. The rules that govern the worldwide classification and naming of species are expressed in the International Rules for Botanical Nomenclature, the International Rules for Zoological Nomenclature, and the International Bacteriological Code of Nomenclature.

In addition to assigning a specific name to each species, Linnaeus recognized a need for placing organisms into groups. This system divides all forms of life into **kingdoms,** the largest grouping used in the classification of organisms. Originally there were two kingdoms, Plantae and Animalia. Today most biologists recognize five kingdoms of life: Plantae, Animalia, Mycetae (fungi), Protista (protozoa and algae), and Prokaryotae (bacteria) (figure 24.20). Each of these kingdoms is divided into smaller units and given specific names. The taxonomic subdivision under each kingdom is usually called a **phylum,** although microbiologists and botanists replace this term with the word *division.* All kingdoms have more than one phylum. For example, the kingdom Plantae contains several phyla, including flowering plants, conifer trees, mosses, ferns, and several other groups. Organisms are placed in phyla based on careful investigation of the specific nature of their structure, metabolism, and biochemistry. An attempt is made to identify natural groups rather than artificial or haphazard arrangements. For example, while nearly all plants are green and carry on photosynthesis, only flowering plants

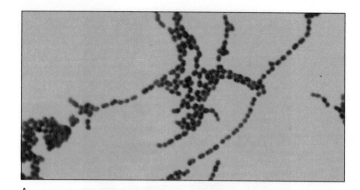

A

B

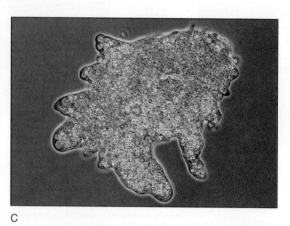

C

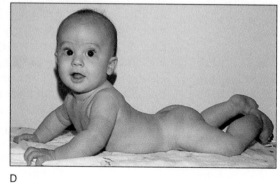

D

E

FIGURE 24.20

(A) Prokaryotae is represented by the bacterium *Streptococcus pyogenes;* (B) Mycetae, by the ascomycete *Morchella esculenta;* (C) Protista, by the one-celled *Amoeba proteus;* (D) Animalia, by the animal *Homo sapiens;* and (E) Plantae, by the tree *Acer saccharum.*

have flowers and produce seeds; conifers lack flowers but have seeds in cones; ferns lack flowers, cones, and seeds; and mosses lack tissues for transporting water.

A **class** is a subdivision within a phylum. For example, within the phylum Chordata there are seven classes: mammals, birds, reptiles, amphibians, and three classes of fishes. An **order** is a category within a class. Carnivora is an order of meat-eating animals within the class Mammalia. There are several other orders of mammals, including horses and their relatives, cattle and their relatives, rodents, rabbits, bats, seals, whales, and many others. A **family** subdivision of an order consists of a group of closely related genera, which in turn are composed of groups of closely related species. The cat family, Felidae, is a subgrouping

of the order Carnivora and includes many species in several genera, including the Canada lynx and bobcat (genus *Lynx*), the cougar (genus *Puma*), the leopard, tiger, jaguar, and lion (genus *Panthera*), the house cat (genus *Felis*), and several other genera. Thus, in the present-day science of taxonomy, each organism that has been classified has its own unique binomial name. In turn, it is assigned to larger groupings that are thought to have a common evolutionary history. Table 24.2 uses the classification of humans to show how the various categories are used.

Phylogeny is the science that explores the evolutionary relationships among organisms and seeks to reconstruct evolutionary history. Taxonomists and phylogenists work together so that the products of their work are compatible. A taxonomic

CONNECTIONS

What Is Carbon-14 Dating?

Carbon is an element that occurs naturally in several forms. The most common form is carbon-12. A second, heavier radioactive form, carbon-14, is constantly being produced in the atmosphere by cosmic rays. Radioactive elements are unstable and break down into other forms of matter. Hence, radioactive carbon-14 naturally decays. The rate at which carbon-14 is formed and the rate at which it decays is about the same, therefore, the concentration of carbon-14 on the earth stays relatively constant. All living things contain large quantities of the element carbon. Plants take in carbon in the form of carbon dioxide from the atmosphere, and animals obtain carbon from the food they eat. While an organism is alive, the proportion of carbon-14 to carbon-12 within its body is equal to its surroundings. When an organism

dies, the carbon-14 within its tissues disintegrates, but no new carbon-14 is added. Therefore, the age of plant and animal remains can be determined by the ratio of carbon-14 to carbon-12 in the tissues. The less carbon-14 present, the older the specimen. Radioactive decay rates are measured in half-life. One half-life is the amount of time it takes for one-half of a radioactive sample to decay.

The half-life of carbon-14 is 5,730 years. Therefore, a bone containing one-half the normal proportion of carbon-14 is 5,730 years old. If the bone contains one-quarter of the normal proportion of carbon-14, it is $2 \times 5{,}730 = 11{,}460$ years old, and if it contains one-eighth of the naturally occurring proportion of carbon-14, it is $3 \times 5{,}730 = 17{,}190$ years old.

TABLE 24.2   Classification of humans

Taxonomic Category	Human Classification	Representative Organisms in the Same Category
Kingdom	Animalia	Heterotrophic organisms with specialized tissues that are usually mobile; insects, snails, starfish, worms, snakes, fish, dogs
Phylum	Chordata	Animals with stiffening rod in the back; reptiles, amphibians, birds, fish
Class	Mammalia	Animals with hair and mammary glands; dogs, whales, mice
Order	Primates	Animals with large brains and opposable thumbs; apes, squirrel monkeys, chimpanzees, baboons
Family	Hominidae	Lack tail and have upright posture; humans and extinct relatives (Neandertal)
Genus	*Homo*	Humans are the only surviving member of the genus, although other members of this genus existed in the past (*H. erectus*)
Species	*Homo sapiens*	Humans

ranking should reflect the evolutionary relationships among the organisms being classified. Although taxonomy and phylogeny are sciences, there is no complete agreement as to how organisms are classified or how they are related. Just as there was dissension two hundred years ago when biologists disagreed on the theories of spontaneous generation and biogenesis, there are differences in opinion about the evolutionary relationships of organisms. People arrive at different conclusions because they use different kinds of evidence or interpret this evidence differently. Phylogenists use several lines of evidence to develop evolutionary histories: fossils, comparative anatomy, life-cycle information, and biochemical/molecular evidence.

Fossils are physical evidence of previously existing life and are found in several different forms. Some fossils may be preserved whole and relatively undamaged. For example, mam-

moths and humans have been found frozen in glaciers, and bacteria and insects have been preserved after becoming embedded in plant resins. Other fossils are only parts of once-living organisms. The outlines or shapes of extinct plant leaves are often found in coal deposits, and individual animal bones that have been chemically altered over time are often dug up (figure 24.21). Animal tracks have also been discovered in the dried mud of ancient riverbeds. It is important to understand that some organisms are more easily fossilized than others. Those that have hard parts like cell walls, skeletons, and shells are more likely to be preserved than are tiny, soft-bodied organisms. Aquatic organisms are much more likely to be buried in the sediments at the bottom of the oceans or lakes than are their terrestrial counterparts. Later, when these sediments are pushed up by geologic forces, aquatic fossils are found in their layers of sediments on dry land.

A   B

FIGURE 24.21

Fossils are either the remains of prehistoric organisms or evidence of their existence. (*A*) The remains of an ancient fly preserved in amber. (*B*) A bony fish specimen. The skeletons of fish make good fossils.

FIGURE 24.22

Because new layers of sedimentary rock are formed on top of older layers of sedimentary rock, it is possible to determine the relative ages of fossils found in various layers. The layers of rock shown here represent on the order of hundreds of millions of years of formation. The fossils of the lower layers are millions of years older than the fossils in the upper layers.

Evidence obtained from the discovery and study of fossils allows biologists to place organisms in a time sequence. This can be accomplished by comparing one type of fossil with another. As geological time passes and new layers of sediment are laid down, the older organisms should be in deeper layers, providing the sequence of layers has not been disturbed (figure 24.22). In addition, it is possible to age-date rocks by comparing the amounts of certain radioactive isotopes they contain. The older sediment layers have less of these specific radioactive isotopes than do younger layers. A comparison of the layers gives an indication of the relative age of the fossils found in the rocks. Therefore, fossils found in the same layer must have been alive during the same geological period.

It is also possible to compare subtle changes in particular kinds of fossils over time. For example, the size of the leaf of a specific fossil plant has been found to change extensively through long geological periods. A comparison of the extremes, the oldest with the newest, would lead to their classification into different categories. However, the fossil links between the extremes clearly show that the younger plant is a descendant of the older.

The comparative anatomy of fossils or currently living organisms can be very useful in developing a phylogeny. Since the structures of an organism are determined by its genes and developmental processes, those organisms having similar structures are thought to be related. Plants can be divided into several categories: All plants that have flowers are thought to be more closely related to one another than to plants like ferns, which do not have flowers. In the animal kingdom, all organisms that nurse their young from mammary glands are grouped together, and all animals in the bird category have feathers and beaks and lay eggs with shells. Reptiles also have shelled eggs but differ from birds in that reptiles lack feathers and have scales covering their bodies. The fact that these two groups share this fundamental eggshell characteristic implies that they are more closely related to each other than they are to other groups.

Another line of evidence useful to phylogenists and taxonomists comes from the field of developmental biology. Many organisms have complex life cycles that include many completely different stages. After fertilization, some organisms grow into free-living developmental stages that do not resemble the adults of their species. These are called *larvae* (singular, *larva*). Larval stages often provide clues to the relatedness of organisms. For example, barnacles live attached to rocks and other solid marine objects and look like small, hard cones. Their outward appearance does not suggest that they are related to shrimp; however, the larval stages of barnacles and shrimp are very similar. Detailed anatomical studies of barnacles confirm that they share many structures with shrimp; their outward appearance tends to be misleading (figure 24.23). This same kind of evidence is available in the plant kingdom. Many kinds of plants, such as peas, peanuts, and lima beans, produce large, two-parted seeds in pods (you can easily split the seeds into two parts). Even though peas grow as vines, lima beans grow as bushes, and peanuts have their seeds underground, all these plants are considered to be related.

Like all aspects of biology, the science of taxonomy is constantly changing as new techniques develop. Recent advances in

A

B

FIGURE 24.23

The adult barnacle (A) and shrimp (B) are very different from each other, but the early larval stages look very much alike.

DNA analysis are being used to determine genetic similarities among species. In the field of ornithology, which deals with the study of birds, there are those who believe that storks and flamingos are closely related; others believe that flamingos are more closely related to geese. An analysis of the DNA points to a higher degree of affinity between flamingos and storks than between flamingos and geese. This is interpreted to mean that the closest relationship is between flamingos and storks. Algae and plants have several different kinds of chlorophyll: chlorophyll *a, b, c, d,* and *e.* Most photosynthetic organisms contain a combination of two of these chlorophyll molecules. Members of the kingdom Plantae have chlorophyll *a* and *b.* The large seaweeds, like kelp, superficially resemble terrestrial plants like trees and shrubs. However, a comparison of the chlorophylls present shows that kelp has chlorophyll *a* and *d.* When another group of algae, called the *green algae,* are examined, they are found to have chlorophyll *a* and *b.* Along with other anatomical and developmental evidence, this biochemical information has helped to establish an evolutionary link between the green algae and plants. All of these kinds of evidence (fossils, comparative anatomy, developmental stages, and biochemical evidence) have been used to develop the various taxonomic categories, including kingdoms.

CONNECTIONS

New Discoveries Lead to Changes in the Classification System

The classification of organisms requires constant rethinking and reorganizing as new information is discovered. The recent discovery of a dinosaur fossil shows an adult apparently sitting on a nest of eggs. This suggests that dinosaurs may have incubated eggs in a manner similar to birds and implies that dinosaurs may have greater similarity to birds than we had previously thought. Bacteria discovered at the openings of deep ocean volcanoes have such different structural and metabolic abilities from those of other bacteria that some microbiologists feel that they should be in a different kingdom from the more typical bacteria. A recently discovered fossil of a whale-like animal with well-developed front legs and functional hind limbs strengthens the belief that whales are descendants of four-legged terrestrial carnivores.

Given all these sources of evidence, biologists have developed a hypothetical picture of how all organisms are related (figure 24.24). At the base of this evolutionary scheme is the biochemical evolution of cells. These first cells are thought to be the origin of the five kingdoms. While protocells no longer exist, their descendants have diversified over millions of years. Of these groups, the Prokaryotae have the simplest structure and are probably most similar to some of the first cellular organisms on Earth.

Kingdom Prokaryotae

Members of the kingdom Prokaryotae are commonly known as bacteria. Some are disease-causing, such as *Streptococcus pneumoniae,* but most are not. In addition, many are able to photosynthesize. Members of the Prokaryotae are grouped together because they all have the same cellular structure. They are small, single-celled organisms ranging from 1 to 10 micrometers (μm). Their cell walls typically contain complex organic molecules, such as peptidoglycan, polymers of unique sugars, and unusual amino acids not found in other kinds of organisms.

However, studies of the various kinds of Prokaryotes have led to the conclusion that there are two very different groups of Prokaryotes; the Bacteria and the Archaea. Eventually the Prokaryotes will be reclassified into two or more kingdoms.

Prokaryotes have no nucleus, and the genetic material is a single loop of DNA. Some have as few as 5,000 genes. The cells reproduce by binary fission. This is a type of asexual cell division that does not involve the more complex structures used by eukaryotes

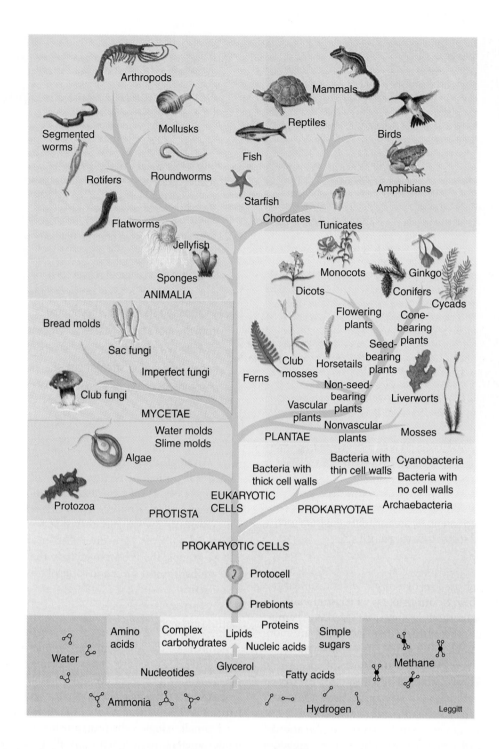

FIGURE 24.24

The theory of chemical evolution proposes that the molecules in the early atmosphere and early oceans accumulated to form prebionts—nonliving structures composed of carbohydrates, proteins, lipids, and nucleic acids. The prebionts are believed to have been the forerunners of the protocells—the first living cells. These protocells probably evolved into prokaryotic cells, on which the kingdom Prokaryotae is based. Some prokaryotic cells probably gave rise to eukaryotic cells. The organisms formed from these early eukaryotic cells were probably similar to members of the kingdom Protista. Members of this kingdom are thought to have given rise to the kingdoms Animalia, Plantae, and Mycetae. Thus, all present-day organisms evolved from the protocells.

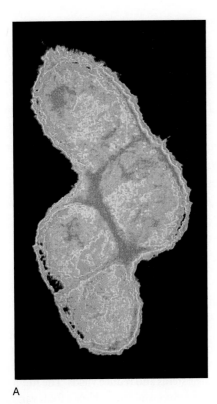

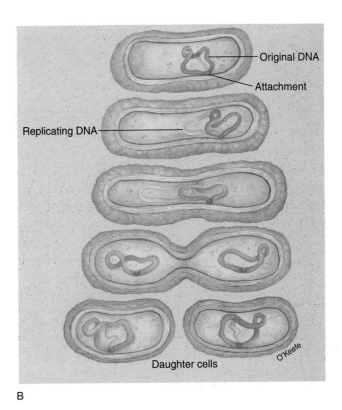

A   B

FIGURE 24.25

Two cells of the bacterium *Bacillus megaterium* formed by binary fission (*A*). (*B*) Binary fission consists of DNA replication and cytoplasmic division.

in mitosis or meiosis. As a result, the daughter cells produced have a single copy of the parental DNA loop (figure 24.25). Some cells move by secreting a slime that glides over the cell's surface, causing it to move through the environment. Others move by means of a kind of flagellum. The structure of the flagellum is different from the flagellum found in eukaryotic organisms.

Since the early atmosphere is thought to have been a reducing atmosphere, the first Prokaryotae were probably anaerobic organisms. Today there are both anaerobic and aerobic Prokaryotae.

Some prokaryotic heterotrophs are **saprophytes,** organisms that obtain energy by the decomposition of dead organic material and cause decomposition; others are parasites that obtain energy and nutrients from living hosts and cause disease; still others are mutualistic and have a mutually beneficial relationship with their host; finally, some are commensalistic and derive benefit from a host without harming it. Several kinds of Prokaryotae are autotrophic. Many are called cyanobacteria because they contain a blue-green pigment, which allows them to capture sunlight and carry on photosynthesis. They can become extremely numerous in some polluted waters where nutrients are abundant. Others use inorganic chemical reactions for their energy source and are called chemosynthetic.

Some biologists hypothesize that eukaryotic cells evolved from prokaryotic cells by a process of endosymbiosis. This hypothesis proposes that structures like mitochondria, chloroplasts, and other membranous organelles originated from separate cells that were ingested by larger, more primitive cells. Once

inside, these structures and their functions became integrated with the host cell and ultimately became essential to its survival. This new type of cell was the forerunner of present-day eukaryotic cells. Single-celled eukaryotic organisms are members of the kingdom Protista (see figure 24.24).

Kingdom Protista

The changes in cell structure that led to eukaryotic organisms most probably gave rise to single-celled organisms similar to those currently grouped in the kingdom Protista. Most members of this kingdom are one-celled organisms, although there are some colonial forms. Eukaryotic cells are usually much larger than the prokaryotes, typically having more than a thousand times the volume of prokaryotic cells. Their larger size was made possible by the presence of specialized membranous organelles, such as mitochondria, the endoplasmic reticulum, chloroplasts, and nuclei.

There is a great deal of diversity within the 60,000 known species of Protista. Many species live in freshwater; others are found in marine or terrestrial habitats, and some are parasitic, commensalistic, or mutualistic. All species can undergo mitosis, resulting in asexual reproduction. Some species can also undergo meiosis and reproduce sexually. Many contain chlorophyll in chloroplasts and are autotrophic; others require organic molecules as a source of energy and are heterotrophic. Both autotrophs and heterotrophs have mitochondria and respire aerobically.

CONNECTIONS

Dogs and Wolves Are the Same Species

Scientists compared the mitochondrial DNA sequences from 67 breeds of dogs and 162 wolves and coyotes and jackals. Mitochondrial DNA passes only from the mother to the offspring. All the wolf and dog sequences were similar and both differed significantly from coyotes and jackals. Careful analysis of the differences in mitochondrial DNA suggests that there were two major domestication events from different parent wolf populations and that there have been incidences of interbreeding of wolves and dogs several times. Furthermore, the various "breeds" of dogs could not be distinguished by the sequence of DNA. Dogs that superficially resemble one another may have more differences in mitochondrial DNA than those that appear quite different. The breeds of dogs are really rather superficial differences in the basic animal. The unmistakable conclusion is that dogs are simply domesticated wolves.

Because members of this kingdom are so diverse with respect to form, metabolism, and reproductive methods, most biologists do not feel that the Protista form a valid phylogenetic unit. However, it is still a convenient taxonomic grouping. By placing these organisms together in this group it is possible to gain a useful perspective on how they relate to other kinds of organisms. After the origin of eukaryotic organisms, evolution proceeded along several different pathways. Three major lines of evolution can be seen today in the plantlike autotrophs (algae), animal-like heterotrophs (protozoa), and the funguslike heterotrophs (slime molds). *Amoeba* and *Paramecium* are commonly encountered examples of protozoa. Many seaweeds and pond scums are collections of large numbers of algal cells. Slime molds are less frequently seen because they live in and on the soil in moist habitats; they are most often encountered as slimy masses on decaying logs.

Through the process of evolution, the plantlike autotrophs probably gave rise to the kingdom Plantae, the animal-like heterotrophs probably gave rise to the kingdom Animalia, and the funguslike heterotrophs were probably the forerunners of the kingdom Mycetae (see figure 24.24).

Kingdom Mycetae

Fungus is the common name for members of the kingdom Mycetae. The majority of fungi are nonmotile. They have a rigid, thin cell wall, which in most species is composed of chitin, a complex carbohydrate containing nitrogen. Members of the kingdom Mycetae are nonphotosynthetic, eukaryotic organisms. The majority (mushrooms and molds) are multicellular, but a few, like yeasts, are single-celled. In the multicellular fungi the basic structural unit is a network of multicellular filaments. Because all of these organisms are heterotrophs, they must obtain nutrients from organic sources. Most are saprophytes and secrete enzymes that digest large molecules into smaller units that are absorbed. They are very important as decomposers in all ecosystems. They feed on a variety of nutrients ranging from dead organisms to such products as shoes, foodstuffs, and clothing. Most synthetic organic molecules are not attacked as readily by fungi; this is why plastic bags, foam cups, and organic pesticides are slow to decompose.

Some fungi are parasitic and others are mutualistic. Many of the parasitic fungi are important plant pests. Some attack and kill plants (chestnut blight, Dutch elm disease); others injure the fruit, leaves, roots, or stems and reduce yields. The fungi that are human parasites are responsible for athlete's foot, vaginal yeast infections, valley fever, "ringworm," and other diseases. Mutualistic fungi are important in lichens and in combination with the roots of certain kinds of plants.

Kingdom Plantae

Another major group that has its roots in the kingdom Protista is the green, photosynthetic plants. The ancestors of plants were most likely specific kinds of algae commonly called *green algae.* Members of the kingdom Plantae are nonmotile, terrestrial, multicellular organisms that contain chlorophyll and produce their own organic compounds. All plant cells have a cellulose cell wall. Over 300,000 species of plants have been classified; about 85 percent are flowering plants, 14 percent are mosses and ferns, and the remaining 1 percent are cone-bearers and several other small groups within the kingdom.

A wide variety of plants exist on the earth today. Members of the kingdom Plantae range from simple mosses to vascular plants with stems, roots, leaves, and flowers. Most biologists feel that the evolution of this kingdom began about 400 million years ago when the green algae of the kingdom Protista gave rise to two lines: The nonvascular plants like the mosses evolved as one type of plant and the vascular plants like the ferns evolved as a second type (figure 24.26). Some of the vascular plants evolved into seed-producing plants, which today are the cone-bearing and flowering plants, while the ferns lack seeds. The development of vascular plants was a major step in the evolution of plants from an aquatic to a terrestrial environment.

Plants have a unique life cycle. There is a gametophyte stage that produces sex cells by mitosis. There is also a sporophyte stage that produces spores (figure 24.27). In addition to sexual reproduction, plants are able to reproduce asexually.

A CLOSER LOOK | The World's Oldest and Largest Living Organisms

Several organisms have been suggested as record holders for the title of oldest and largest organisms. Several of these are plants. The bristlecone pine (*Pinus longaeva*) in the White Mountains of California have been aged to over 5,000 years. Creosote bush (*Larrea divaricata*) forms clones that grow out from the center as the central portion of the plant dies. Several clones of creosote bush in the Mojave Desert have been estimated to have an age of 12,000 years. The title of the largest organism can be determined in several ways. The Giant Redwood (*Sequoiadendron giganteum*) is the tree with the single largest stem. The General Sherman tree probably weighs about 1,400 tons. However, a clone of trembling aspen (*Populus tremuloides*) consists of many individual stems that are probably all joined together by roots. One such clone in the Rocky Mountains covers about 0.4 square kilometer and probably weighs about 6,000 tons. However, it may be a clone of a fungus (*Armillaria*) that lives in the soil that holds the record for the largest organism. A clone in the state of Washington is estimated to cover about 3 square kilometers.

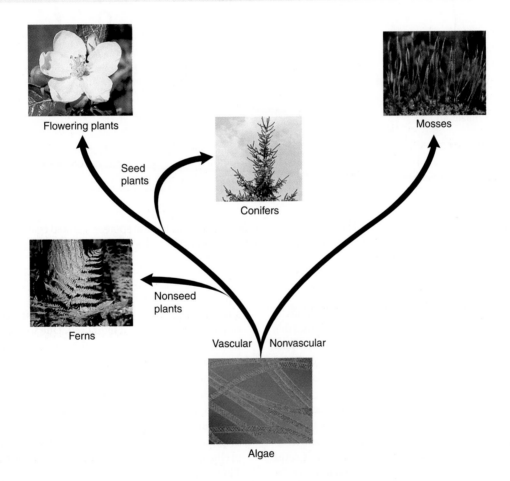

FIGURE 24.26

Two lines of plants are thought to have evolved from the plantlike Protista, the algae. The nonvascular mosses evolved as one type of plant. The second type, the vascular plants, evolved into the seed and nonseed plants.

Kingdom Animalia

Like fungi and plants, animals are thought to have evolved from the Protista. Over a million species of animals have been classified. These range from microscopic types, like mites or aquatic larvae of marine animals, to huge animals like elephants or whales. Regardless of their type, all animals have some common traits. They all are composed of eukaryotic cells, and all species are heterotrophic and multicellular. Most animals are motile; however, some, like the sponges, barnacles, mussels, and corals are sessile (not able to move). All animals are capable of sexual reproduction, but many of the less complex animals are also able to reproduce asexually.

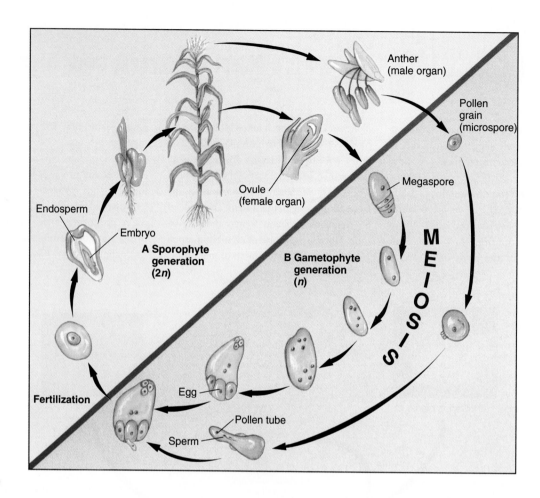

FIGURE 24.27

(A) Plants have a multicellular structure (sporophyte generation) that undergoes meiosis to form spores—haploid cells. (B) These spores give rise to haploid organisms (gametophyte generation) that form the gametes—sperm cells and egg cells. The union of these gametes forms the diploid stage, which develops into the sporophyte.

It is thought that animals originated from certain kinds of Protista that had flagella (see figure 24.24). This idea proposes that colonies of flagellated Protista gave rise to simple multicellular forms of animals like the ancestors of present-day sponges. These first animals lacked specialized tissues and organs. As cells became more specialized, organisms developed special organs and systems of organs and the variety of kinds of animals increased.

Although taxonomists have grouped organisms into five kingdoms, some organisms do not easily fit into these categories. Viruses, which lack all cellular structures, still show some characteristics of life. In fact, some scientists consider them to be highly specialized parasites that have lost some of their complexity as they developed as parasites. Others consider them to be the simplest of living organisms. Some even consider them to be nonliving. For these reasons viruses are considered separately from the five kingdoms.

Viruses

A **virus** consists of a nucleic acid core surrounded by a coat of protein (figure 24.28). Viruses are **obligate intracellular para-**

sites, which means they are infectious particles that can function only when inside a living cell. Due to their unusual characteristics, viruses are not a member of any kingdom. Biologists do not consider them to be living because they are not capable of living and reproducing by themselves and show the characteristics of life only when inside living cells.

Soon after viruses were discovered in the late part of the nineteenth century, biologists began speculating on how they originated. One early hypothesis was that they were either prebionts or parts of prebionts that did not evolve into cells. This idea was discarded as biologists learned more about the complex relationship between viruses and host cells. A second hypothesis was that viruses developed from intracellular parasites that became so specialized that they needed only the nucleic acid to continue their existence. Once inside a cell, this nucleic acid can take over and direct the host cell to provide for all of the virus's needs. A third hypothesis is that viruses are runaway genes that have escaped from cells and must return to a host cell to replicate. Regardless of how the viruses came into being, today they are important as parasites in all forms of life.

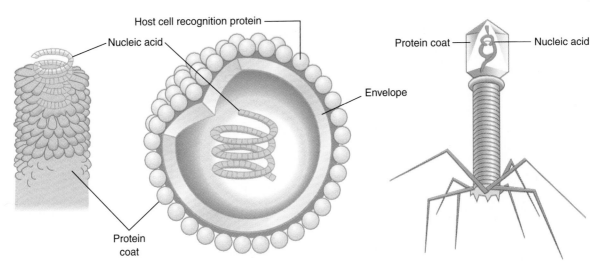

FIGURE 24.28

Viruses consist of a core of nucleic acid, either DNA or RNA depending on the kind of virus. Some are surrounded by an envelope.

Viruses are typically host-specific, which means that they usually attack only one kind of cell. The host is a specific kind of cell that provides what the virus needs to function. Viruses can infect only those cells that have the proper receptor sites to which the virus can attach. This site is usually a glycoprotein molecule on the surface of the cell membrane. For example, the virus responsible for measles attaches to the membranes of skin cells, hepatitis viruses attach to liver cells, and mumps viruses attach to cells in the salivary glands. Host cells for the HIV virus include some types of human brain cells and several types belonging to the immune system.

Once it has attached to the host cell, the virus either enters the cell or injects its nucleic acid into the cell. If it enters the cell, the virus loses its protein coat, releasing the nucleic acid. Once released into the cell, the nucleic acid of the virus may remain free in the cell or it may link with the host's genetic material. Some viruses contain as few as three genes, others contain as many as 500. A typical eukaryotic cell contains tens of thousands of genes. Most viruses need only a small number of genes since they rely on the host to perform most of the activities necessary for viral reproduction.

Some viruses have DNA as their genetic material but many have RNA. The RNA must first be reverse transcribed to DNA before the virus can reproduce. Reverse transcriptase, the enzyme that accomplishes this, has become very important in the new field of molecular genetics because its use allows scientists to make large numbers of copies of a specific molecule of DNA from RNA molecules.

Viral genes are able to take command of the host's metabolic pathways and direct it to carry out the work of making new copies of the original virus. The virus makes use of the host's available enzymes and ATP for this purpose. When enough new viral nucleic acid and protein coat are produced, complete virus particles are assembled and released from the host (figure 24.29). The number of viruses released ranges from ten to thousands. The virus that causes polio releases about 10,000 new virus particles from each human host cell. Some viruses remain in cells and are only occasionally triggered to reproduce, causing symptoms of disease. Herpes viruses, which cause cold sores, genital herpes, and shingles, reside in nerve cells.

Viruses vary in size and shape, which helps in classifying them. Some are rod-shaped, others are round, and still others are in the shape of a coil or helix. Viruses are some of the smallest infecting agents known to humans. Only a few can be seen with a standard laboratory microscope; most require an electron microscope to make them visible. A great deal of work is necessary to isolate viruses from the environment and prepare them for observation with an electron microscope. For this reason, most viruses are more quickly identified by their activities in host cells. Almost all of the species in the five kingdoms serve as hosts to some form of virus (table 24.3).

TABLE 24.3	Viral diseases
Type of Virus	**Disease**
Papovaviruses	Warts in humans
Paramyxoviruses	Mumps and measles in humans; distemper in dogs
Adenoviruses	Respiratory infections in most mammals
Poxviruses	Smallpox
Wound-tumor viruses	Diseases in corn and rice
Potexviruses	Potato diseases
Bacteriophage	Infections in many types of bacteria

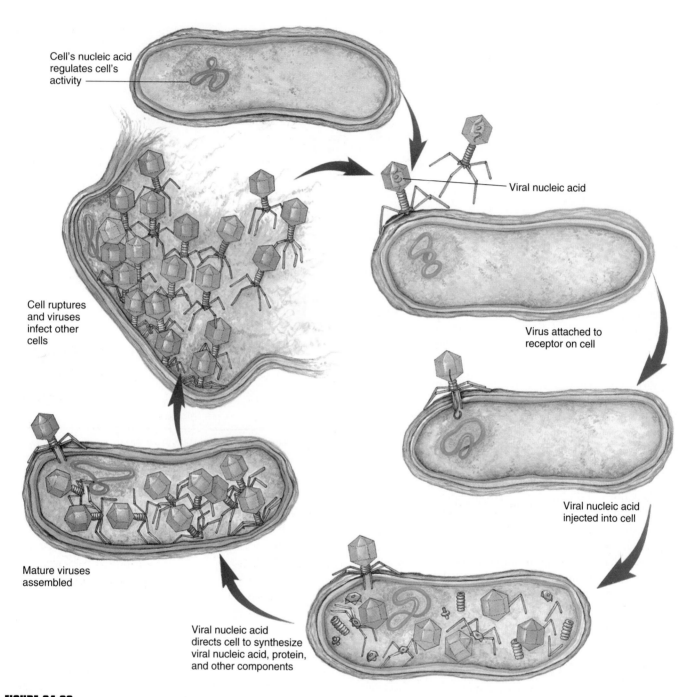

Cell's nucleic acid regulates cell's activity

Viral nucleic acid

Cell ruptures and viruses infect other cells

Virus attached to receptor on cell

Viral nucleic acid injected into cell

Mature viruses assembled

Viral nucleic acid directs cell to synthesize viral nucleic acid, protein, and other components

FIGURE 24.29

The viral nucleic acid takes control of the activities of the host cell. Because the virus has no functional organelles of its own, it can become metabolically active only while it is within a host cell.

SUMMARY

Current theories about the *origin of life* include the two ideas (1) that the primitive Earth environment led to the *spontaneous organization of organic chemicals* into primitive cells and (2) that primitive forms of life *arrived on Earth from space*. Basic units of life were probably similar to present-day *prokaryotes*. These primitive cells could have *changed through time* as a result of *mutation* and *in response to a changing environment*.

All sexually reproducing organisms naturally exhibit genetic variety among the individuals in the population as a result of *mutations* and the *genetic recombination* resulting from *meiosis* and *fertilization*. The *genetic differences* are reflected in *phenotypic differences* among individuals. These *genetic differences* are important for the survival of the species because *natural selection* must have *genetic variety* to select from. *Natural selection* by the *environment* results in better-suited individual organisms that have greater numbers of offspring than those that are less well off genetically. Not all genes are equally expressed. Some express themselves only during specific periods in the life of an organism and some may be *recessive genes* that show themselves only when in the *homozygous state*. Characteristics that are acquired during the life of the individual and are not determined by genes cannot be raw material for natural selection.

Organisms with wide *geographic distribution* often show different *gene frequencies* in different parts of their *range*. A *species* is a group of *organisms* that can interbreed to produce fertile offspring. The process of *speciation* usually involves the *geographic separation* of the species into two or more isolated *populations*. While they are separated, natural selection operates to *adapt each population to its environment*.

At one time, people thought that all organisms had remained *unchanged* from the time of their creation. Lamarck suggested that *change did occur* and thought that *acquired characteristics could be passed from generation to generation*. Darwin and Wallace proposed the *theory of natural selection* as the mechanism that drives *evolution*. *Evolution* is basically a *divergent process* upon which other patterns can be superimposed. *Adaptive radiation* is a very rapid *divergent evolution*, while *convergent evolution* involves the development of *superficial similarities* among widely different organisms. The rate at which evolution has occurred probably varies. The *fossil record* shows *periods of rapid change* interspersed with *periods of little change*. This has caused some to look for *mechanisms* that could cause the sudden appearance of *large numbers of new species* in the fossil record and to challenge the traditional idea of slow, steady change accumulating enough differences to cause a new species to be formed.

The early *evolution of humans* has been difficult to piece together because of the *fragmentary evidence*. Beginning about 4.4 million years ago the earliest forms of *Australopithecus* showed upright posture and other humanlike characteristics. The structure of the jaw and teeth indicates that the various kinds of *Australopithecus* were herbivores. *Homo habilis* had a larger brain and appears to have been a scavenger. *Homo erectus* had a larger cranium, was a carnivore, and may have been the direct ancestor of modern humans, *Homo sapiens*.

To facilitate accurate *communication*, biologists assign a *specific name* to each *species* that is cataloged. The various species are cataloged into larger groups on the basis of similar traits.

Taxonomy is the science of classifying and naming organisms. *Phylogeny* is the science of trying to figure out the evolutionary history of a particular organism. The *taxonomic ranking of organisms* reflects their *evolutionary relationships*. *Fossil evidence, comparative anatomy, developmental stages*, and *biochemical evidence* are employed in the sciences of taxonomy and phylogeny.

The first organisms thought to have evolved were single-celled organisms of the kingdom *Prokaryotae*. From this simple beginning, more complex, many-celled organisms evolved, creating members of the kingdoms *Protista, Mycetae, Plantae*, and *Animalia*.

KEY TERMS

acquired characteristics (p. **626**)

autotroph (p. **619**)

binomial system of nomenclature (p. **641**)

biogenesis (p. **618**)

class (p. **642**)

endosymbiotic theory (p. **622**)

eukaryotes (p. **622**)

evolution (p. **624**)

family (p. **642**)

fungus (p. **648**)

gene flow (p. **631**)

genetic recombination (p. **626**)

genus (p. **641**)

geographic barriers (p. **632**)

geographic isolation (p. **632**)

heterotroph (p. **618**)

kingdoms (p. **641**)

natural selection (p. **625**)

obligate intracellular parasite (p. **650**)

order (p. **642**)

oxidizing atmosphere (p. **621**)

phylogeny (p. **642**)

phylum (p. **641**)

prokaryotes (p. **622**)

range (p. **632**)

saprophytes (p. **647**)

selecting agents (p. **626**)

speciation (p. **636**)

species (p. **631**)

spontaneous generation (p. **618**)

spontaneous mutation (p. **626**)

subspecies (p. **633**)

taxonomy (p. **640**)

theory of natural selection (p. **625**)

virus (p. **650**)

APPLYING THE CONCEPTS

1. Which was *not* a major component of the early Earth's reducing atmosphere?
   a. H_2
   b. CO_2
   c. NH_3
   d. ozone

2. Hypothetically, which came first in the evolution of cells?
   a. autotrophs
   b. heterotrophs
   c. aerobic cells
   d. eukaryotes

3. Prokaryotic cells hypothetically came into existence approximately _____ years ago?
   a. 20 billion
   b. 4–5 billion
   c. 3.5 billion
   d. 1.5 billion

4. Which one of the following researchers found evidence not to support spontaneous generation?
   a. Redi
   b. Needham
   c. Escher
   d. Woese

5. The three main kinds of living things, Bacteria, Archaea, and Eucarya, have been labeled as
   a. species.
   b. kingdoms.
   c. domains.
   d. families.

6. Hybrid animals like mules are not considered to be a species because they
   a. do not reproduce through many generations.
   b. are not common enough.
   c. can be maintained only by humans.
   d. are the result of convergent evolution.

7. Two closely related organisms are not considered to be separate species unless they
   a. look different.
   b. are reproductively isolated.
   c. are able to interbreed.
   d. are in different geographic parts of the world.

8. The Darwin-Wallace theory of natural selection differs from Lamarck's ideas in that
   a. Lamarck understood the role of genes and Darwin and Wallace did not.
   b. Lamarck assumed that characteristics obtained during an organism's lifetime could be passed to the next generation; Darwin and Wallace did not.
   c. Lamarck did not think that evolution took place; Darwin and Wallace did.
   d. Lamarck developed the basic ideas of speciation, which Darwin and Wallace refined.

9. What percent of the number of species that have ever been present on the face of the earth are extinct?
   a. 10 percent
   b. 50 percent
   c. 75 percent
   d. 99 percent

10. Which of the following is *not* necessary for speciation?
    a. genetic isolation from other species
    b. genetic variety within a species
    c. hundreds of millions of years
    d. reproduction

11. Although several species of firefly may live in the same area, they do not interbreed. Each species of firefly has a unique flash that is used to signal potential mates of the same species. In fireflies, cross-species matings are prevented by
    a. mechanical isolation.
    b. ecological isolation.
    c. seasonal isolation.
    d. behavioral isolation.

12. Two groups of organisms belong to different species if
    a. gene flow between the two groups is not possible, even in the absence of physical barriers.
    b. physical barriers separate the two groups, thereby preventing cross matings.
    c. the two groups of organisms have a different physical appearance.
    d. individuals from the two groups, when mated, produce hybrid offspring.

13. Which of the following scientific names is written correctly?
    a. *Micropterus* Salmoides
    b. *Treponema pallidum*
    c. Nymphaea odorata
    d. salmo *Trutta*

14. Which is the proper sequence of taxa?
    a. kingdom, phylum, class, order, family
    b. phylum, kingdom, family, order, class
    c. phylum, class, order, family, species
    d. kingdom, phylum, family, class, order

15. Which is commonly used as sources of information when developing a phylogeny?
    a. fossil evidence
    b. biochemical information
    c. DNA analysis
    d. all the above

16. Which type of organism is *not* included in the five kingdom system of classification?
    a. fungi
    b. marine red algae
    c. viruses
    d. larva

17. The single-cell Mycetae are commonly called
    a. algae.
    b. molds.
    c. protozoa.
    d. yeast.

18. Which of the following kingdoms contain members that are autotrophs?
    a. Mycetae
    b. Animalia
    c. viruses
    d. Protista

19. Infectious diseases that occur throughout the world at unacceptably high rates are referred to as
    a. endemic.
    b. epidemic.
    c. pandemic.
    d. international.

20. The cause of AIDS is a kind of
    a. bacterium.
    b. fungus.
    c. protozoan.
    d. virus.

21. Of the following, two organisms which belong to the same _____ are most closely related.
    a. family
    b. order
    c. class
    d. kingdom

22. Alternation of generations is a characteristic of
    a. Prokaryotae.
    b. Protista.
    c. Mycetae.
    d. Plantae.

23. Viruses
    a. can be free-living or parasitic.
    b. belong to the kingdom Prokaryotae.
    c. can function and reproduce only inside a host cell.
    d. all of the above.

24. An organism that is multicellular, eukaryotic, and heterotrophic is a(n)
    a. plant.
    b. animal.
    c. protist.
    d. animal or fungus.

25. Organisms that obtain energy from the decomposition of other organisms are
    a. saprophytes.
    b. gametophytes.
    c. sporophytes.
    d. obligate intracellular parasites.

26. Most Plantae are
    a. cone-bearing plants.
    b. ferns.
    c. flowering plants.
    d. mosses.

27. Algae, amoeba, and paramecia belong to the kingdom
    a. Mycetae.
    b. Prokaryotae.
    c. Protista.
    d. virus.

28. Viruses are composed of
    a. prokaryotic cells.
    b. eukaryotic cells.
    c. protein and nucleic acid.
    d. membranous organelles.

Answers

1. d 2. b 3. c 4. a 5. d 6. a 7. b 8. b 9. d 10. c 11. d 12. a 13. b 14. a 15. d
16. c 17. d 18. d 19. c 20. d 21. a 22. d 23. c 24. d 25. a 26. c 27. c 28. c

QUESTIONS FOR THOUGHT

1. In what sequence did the following things happen: living cell, oxidizing atmosphere, autotrophy, heterotrophy, reducing atmosphere, first organic molecule?
2. What is meant by *spontaneous generation?* What is meant by *biogenesis?*
3. What were the circumstances on primitive Earth that favored the survival of anaerobic heterotrophs?
4. The current theory of the origin of life as a result of nonbiologic manufacture of organic molecules depends on our knowing something of Earth's history. Why is this so?
5. List two important effects caused by the increase of oxygen in the atmosphere.
6. What evidence supports the theory that eukaryotic cells arose from the development of a symbiotic relationship between primitive prokaryotic cells?
7. Why are acquired characteristics of little interest to evolutionary biologists?
8. What factors can contribute to variety in the gene pool?
9. What is natural selection? How does it work?
10. Give two examples of selecting agents and explain how they operate.
11. The smaller the population, the more likely it is that random changes will influence gene frequencies. Why is this true?
12. List three factors that can lead to changed gene frequencies from one generation to the next.
13. Why is geographic isolation important in the process of speciation?
14. How does speciation differ from the formation of subspecies or races?
15. Why aren't mules considered a species?
16. Describe three kinds of genetic isolating mechanisms that prevent interbreeding between different species.
17. Can you always tell by looking at two organisms whether or not they belong to the same species?
18. Why has Lamarck's theory been rejected?
19. Describe two differences between convergent evolution and adaptive radiation.
20. Give an example of seasonal isolation, ecological isolation, and behavioral isolation.
21. List the series of events necessary for speciation to occur.
22. "Evolution is a fact." "Evolution is a theory." Explain how both of these statements can be true.
23. What are some of the major steps thought to have been involved in the evolution of humans?
24. What are the five kingdoms of living things?
25. What is the difference between the kingdom Prokaryotae and the kingdom Plantae?
26. What is the value of taxonomy?
27. An order is a collection of what similar groupings?
28. How are viruses thought to have originated?
29. Eukaryotic cells are found in which kingdoms?
30. How do viruses reproduce?
31. What characteristics are there in common between the members of the kingdoms Mycetae and Plantae?
32. Why are Latin names used for genus and species?
33. Who designed the present-day system of classification? How does this system differ from previous systems?

Even organisms of the same species affect one another in the course of their normal daily activities.

CHAPTER | Twenty-Five

Ecology and Environment

Today we hear people from all walks of life using the terms *ecology* and *environment*. Students, homeowners, politicians, planners, and union leaders speak of "environmental issues" and "ecological concerns." Often these terms are interpreted in different ways, so we need to establish some basic definitions.

Ecology is the branch of biology that studies the relationships between organisms and their environments. This is a very simple definition for a very complex branch of science. Most ecologists define the word **environment** very broadly as anything that affects an organism during its lifetime. These environmental influences can be divided into two categories. Other living things that affect an organism are called **biotic factors,** and nonliving influences are called **abiotic factors** (figure 25.1). If we consider a fish in a stream, we can identify many environmental factors that are important to its life. The temperature of the water is extremely important as an abiotic factor, but it may be influenced by the presence of trees (biotic factor) along the stream bank that shade the stream and prevent the sun from heating it. Obviously, the kind and number of food organisms in the stream are important biotic factors as well. The type of material that makes up the stream bottom and the amount of oxygen dissolved in the water are other important abiotic factors, both of which are related to how rapidly the water is flowing.

As you can see, characterizing the environment of an organism is a complex and challenging process; everything seems to be influenced or modified by other factors. A plant is influenced by many different factors during its lifetime: the types and amounts of minerals in the soil, the amount of sunlight hitting the plant, the animals that eat the plant, and the wind, water, and temperature. Each item on this list can be further subdivided into other areas of study. For instance, water is important in the life of plants, so rainfall is studied in plant ecology. But even the study of rainfall is not simple. The rain could come during one part of the year, or it could be evenly distributed throughout the year. The rainfall could be hard and driving, or it could come as gentle, misty showers of long duration. The water could soak into the soil for later use or it could run off into streams and be carried away.

Temperature is also very important to the life of a plant. For example, two areas of the world can have the same average daily temperature of 10°C (50°F) but not have the same plants because of different temperature extremes. In one area, the temperature may be 13°C (about 55°F) during the day and 7°C (about 45°F) at night, for a 10°C (50°F) average. In another area, the temperature may be 20°C (68°F) in the day and only 0°C (32°F) at night, for a 10°C (about 45°F) average. Plants react to extremes in temperature as well as to the daily average. Furthermore, different parts of a plant may respond differently to temperature. Tomato plants will grow at temperatures below 13°C (about 55°F) but will not begin to develop fruit below 13°C (about 55°F).

The animals in an area are influenced as much by abiotic factors as are the plants. If nonliving factors do not favor the growth of plants, there will be little food and few hiding places for animal life. Two types of areas that support only small numbers of living animals are deserts and polar regions. Near the polar regions of the earth, the low temperature and short growing season inhibits growth; therefore, there are relatively few species of animals with relatively small numbers of individuals. Deserts receive little rainfall and therefore have poor plant growth and low concentrations of animals. On the other hand, tropical rainforests have high rates of plant growth and large numbers of animals of many kinds.

As you can see, living things are themselves part of the environment of other living things. If there are too many animals in an area, they could demand such large amounts of food that they would destroy the plant life, and the animals themselves would die. So far we have discussed how organisms interact with their environment in rather general terms. Ecologists have developed several concepts that help us understand how biotic and abiotic factors interrelate in a complex system.

A CLOSER LOOK | What Is Environmental Science?

Environmental science is a recently developed field of study that integrates the traditional scientific disciplines, such as chemistry, physics, ecology, and agriculture, with an understanding of humans as social and political beings. It is an interdisciplinary study that focuses on how humans use and affect the natural world. Environmental scientists seek remedies for the negative impacts we have on the world. Since environmental science deals with human activity, there are many different ways to solve problems. Sometimes the solutions are political—a new environmental law is passed. Sometimes the problems are dealt with by encouraging social change—birth control is made available to people who cannot afford it. Sometimes environmental progress involves the use of technology—new methods for bleaching paper are substituted for methods that generate toxic dioxins or energy-conserving lighting is substituted for wasteful incandescent lighting.

Who Is an Environmental Scientist?

An environmental scientist is a chemist who studies water quality, a biologist who manages fish populations, a political scientist who lobbies congress, a lawyer who prosecutes environmental offenders, a physician who provides nutritional and family-planning services to impoverished people, a hunter or fisher who supports sound game and fish management practices, an activist who seeks to protect parkland from development, or a writer who creates word pictures that inspire people to look at nature differently.

THE ORGANIZATION OF LIVING SYSTEMS

Living systems can be examined from several different levels of organization. The smallest living unit is the individual *organism.* Groups of organisms of the same species are called *populations.*

Interacting groups of species are called *communities.* And an **ecosystem** consists of all the interacting organisms in an area and their interactions with their abiotic surroundings. Figure 25.2 shows how these different levels of organization are related to one another.

All living things require a continuous supply of energy to maintain life. Therefore, many people like to organize living systems by the energy relationships that exist among the different kinds of organisms present. An ecosystem contains several different kinds of organisms. Those that trap sunlight for photosynthesis, resulting in the production of organic material from inorganic material, are called **producers.** Green plants and other photosynthetic organisms such as algae and cyanobacteria are, in effect, converting sunlight energy into the energy contained within the chemical bonds of organic compounds. There is a flow of energy from the sun into the living matter of plants.

The energy that plants trap can be transferred through a number of other organisms in the ecosystem. Since all of these organisms must obtain energy in the form of organic matter, they are called **consumers.** Consumers cannot capture energy from the sun as plants do. All animals are consumers. They either eat plants directly or eat other sources of organic matter derived from plants. Each time the energy enters a different organism, it is said to enter a different **trophic level,** which is a step, or stage, in the flow of energy through an ecosystem

A   B

FIGURE 25.1

(*A*) The woodpecker feeding its young in the hole in this tree is influenced by several biotic factors. The tree itself is a biotic factor as is the disease that weakened it, causing conditions that allowed the woodpecker to make a hole in the rotting wood. (*B*) The irregular shape of the trees is the result of wind and snow, both abiotic factors. Snow driven by the prevailing winds tends to "sandblast" one side of the tree and prevent limb growth.

FIGURE 25.2

The same organism can be looked at from several different perspectives. We can study it as an individual, as a member of a population, or as a participant in a community or ecosystem.

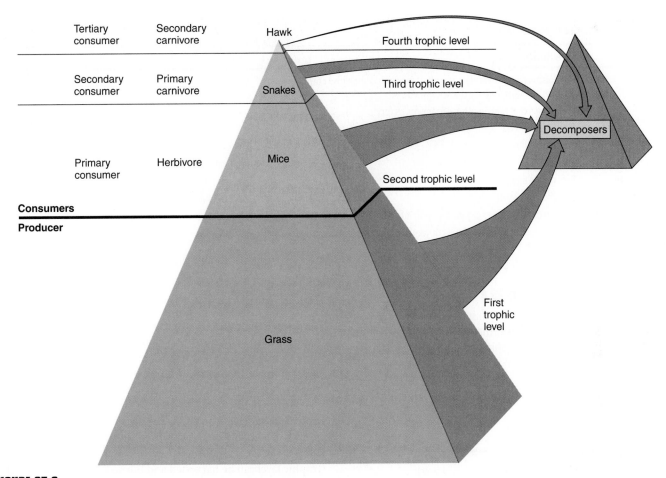

FIGURE 25.3

Organisms within ecosystems can be divided into several different trophic levels on the basis of how they obtain energy. Several different sets of terminology are used to identify these different roles. This illustration shows how the different sets of terminology are related to one another.

(figure 25.3). The plants (producers) receive their energy directly from the sun and are said to occupy the *first trophic level.*

Various kinds of consumers can be divided into several categories, depending on how they fit into the flow of energy through an ecosystem. Animals that feed directly on plants are called **herbivores,** or **primary consumers,** and occupy the *second trophic level.* Animals that eat other animals are called **carnivores,** or **secondary consumers,** and can be subdivided into different trophic levels depending on what animals they eat. Animals that feed on herbivores occupy the *third trophic level* and are known as **primary carnivores.** Animals that feed on the primary carnivores are known as **secondary carnivores** and occupy the *fourth trophic level.* For example, a human may eat a fish that ate a frog that ate a spider that ate an insect that consumed plants for food.

This sequence of organisms feeding on one another is known as a **food chain.** Figure 25.4 shows the six different trophic levels in this food chain. Obviously, there can be higher categories, and some organisms don't fit neatly into this theoretical scheme. Some animals are carnivores at some times and herbivores at others; they are called **omnivores.** They are classified into different trophic levels depending on what they happen to be eating at the moment. If an organism dies, the energy contained within the organic compounds of its body is finally released to the environment as heat by organisms that decompose the dead body into carbon dioxide, water, ammonia, and other simple inorganic molecules. Organisms of decay, called **decomposers,** are things such as bacteria, fungi, and other organisms that use dead organisms as sources of energy.

This group of organisms efficiently converts nonliving organic matter into simple inorganic molecules that can be used by producers in the process of trapping energy. Decomposers are thus very important components of ecosystems that cause materials to be recycled. As long as the sun supplies the energy, elements are cycled through ecosystems repeatedly. Table 25.1 summarizes the various categories of organisms within an ecosystem. Now that we have a better idea of how ecosystems are organized, we can look more closely at energy flow through ecosystems.

TABLE 25.1	Roles in an ecosystem	
Classification	**Description**	**Examples**
Producers	Organisms that convert simple inorganic compounds into complex organic compounds by photosynthesis	Trees, flowers, grasses, ferns, mosses, algae, cyanobacteria
Consumers	Organisms that rely on other organisms as food; animals that eat plants or other animals	
Herbivore	Eats plants directly	Deer, goose, cricket, vegetarian human, many snails
Carnivore	Eats meat	Wolf, pike, dragonfly
Omnivore	Eats plants and meat	Rat, most humans
Scavenger	Eats food left by others	Coyote, skunk, vulture, crayfish
Parasite	Lives in or on another organism, using it for food	Tick, tapeworm, many insects
Decomposers	Organisms that return organic compounds to inorganic compounds; important components in recycling	Bacteria, fungi

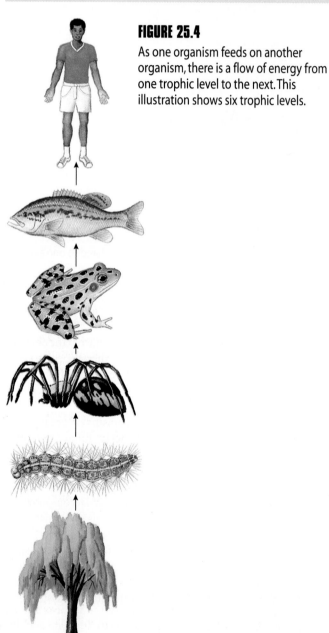

FIGURE 25.4

As one organism feeds on another organism, there is a flow of energy from one trophic level to the next. This illustration shows six trophic levels.

THE GREAT PYRAMID OF ENERGY

The ancient Egyptians constructed elaborate tombs we call *pyramids*. The broad base of the pyramid is necessary to support the upper levels of the structure, which narrows to a point at the top. This same kind of relationship exists when we look at how the various trophic levels of ecosystems are related to one another.

At the base of the pyramid is the producer trophic level, which contains the largest amount of energy of any of the trophic levels within an ecosystem. In an ecosystem, the total energy can be measured in several ways. The total producer trophic level can be harvested and burned. The number of calories of heat energy produced by burning is equivalent to the energy content of the organic material of the plants. Another way of determining the energy present is to measure the rate of photosynthesis and respiration and calculate the amount of energy being trapped in the living material of the plants.

Since only the plants, algae, and cyanobacteria in the producer trophic level are capable of capturing energy from the sun, all other organisms are directly or indirectly dependent on the producer trophic level. The second trophic level consists of herbivores that eat the producers. This trophic level has significantly less energy in it for several reasons. *In general, there is about a 90 percent loss of energy as we proceed from one trophic level to the next higher level.* Actual measurements will vary from one ecosystem to another, but 90 percent is a good rule of thumb. This loss in energy content at the second and subsequent trophic levels is primarily due to the second law of thermodynamics. This law states that whenever energy is converted from one form to another, some energy is converted to useless heat. Think of any energy-converting machine; it probably releases a great deal of heat energy. For example, an automobile engine must have a cooling system to get rid of the heat energy produced. An incandescent lightbulb also produces large amounts of heat. Although living systems are somewhat different, they must follow the same energy rules.

In addition to the loss of energy as a result of the second law of thermodynamics, there is an additional loss involved in the capture and processing of food material by herbivores. Although

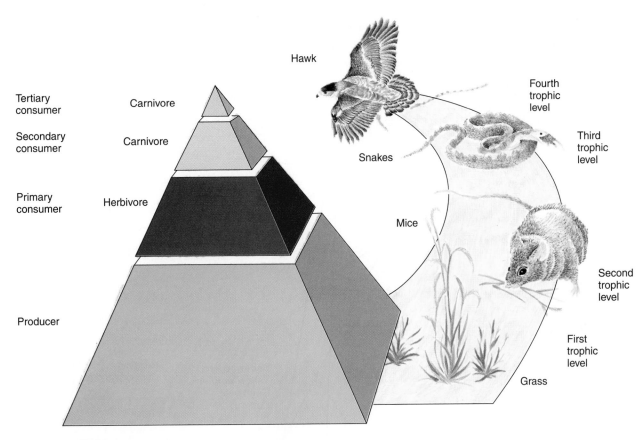

FIGURE 25.5

As energy flows from one trophic level to the next, approximately 90 percent of it is lost. This means that the amount of energy at the producer level must be 10 times larger than the amount of energy at the herbivore level.

herbivores don't need to chase their food, they do need to travel to where food is available, then gather, chew, digest, and metabolize it. All these processes require energy.

Just as the herbivore trophic level experiences a 90 percent loss in energy content, the higher trophic levels of primary carnivores, secondary carnivores, and tertiary carnivores also experience a reduction in the energy available to them. Figure 25.5 shows an energy pyramid in which the energy content decreases by 90 percent as we pass from one trophic level to the next.

ECOLOGICAL COMMUNITIES

In the previous section we looked at things from the point of view of entire ecosystems and the way energy flows through them. But we can also study interactions from other perspectives. For example, we can distinguish between the ecosystem and the interacting organisms that are a part of it. An ecosystem is a unit that consists of the physical environment and all the interacting organisms within that area. The collection of interacting organisms within an ecosystem is called a **community** and consists of many kinds of organisms. The num-

ber of individuals of a particular species in an area is called a **population.** Therefore, we can look at the same organism from several points of view. We can look at it as an individual, as a part of a population of similar individuals, as a part of a community that includes other populations, and as a part of an ecosystem, which includes abiotic factors as well as living organisms.

As you know from the discussion in the previous section, one of the ways that organisms interact is by feeding on one another. A community includes many different food chains and many organisms may be involved in several of the food chains at the same time, so the food chains become interwoven into a **food web** (figure 25.6). In a community, the interacting food chains usually result in a relatively stable combination of populations. Although communities are relatively stable, we need to recognize that they are also dynamic collections of organisms: As one population increases, another decreases. This might occur over several years, or even in the period of one year. This happens because most ecosystems are not constant. There may be differences in rainfall throughout the year or changes in the amount of sunlight and in the average temperature. We should expect populations to fluctuate as abiotic factors change. A change in the size of one population will

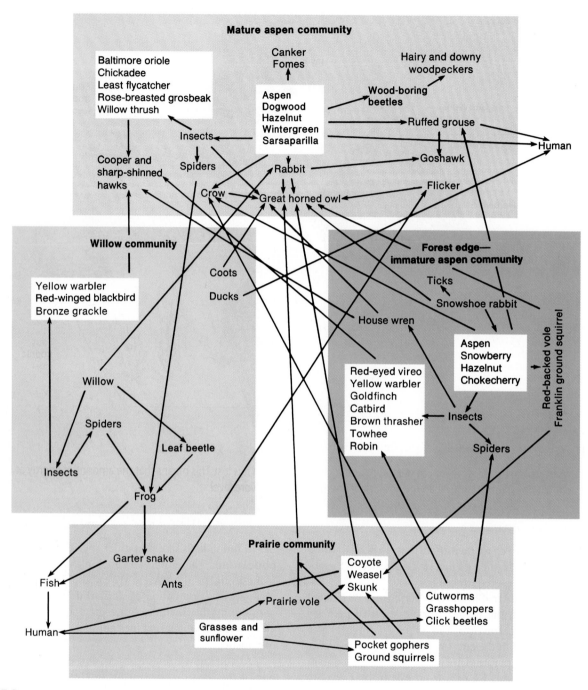

FIGURE 25.6

When many different food chains are interlocked with one another, a food web results. The arrows indicate the direction of energy flow. Notice that some organisms are a part of several food chains—the great horned owl in particular. Because of the interlocking nature of the food web, changing conditions may shift the way in which food flows through this system.

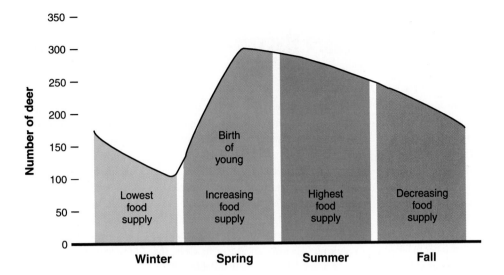

FIGURE 25.7

The number of organisms living in an area varies during the year. The availability of food is the primary factor determining the size of the population of deer in this illustration, but water availability, availability of soil nutrients, and other factors could also be important.

trigger changes in other populations as well. Figure 25.7 shows what happens to the size of a population of deer as the seasons change. The area can support 100 deer from January through February, when plant food for deer is least available. As spring arrives, plant growth increases. It is no accident that deer breed in the fall and give birth in the spring. During the spring producers are increasing and the area has more available food to support a large deer population. It is also no accident that wolves and other carnivores that feed on deer give birth in the spring. The increased available energy in the form of plants (producers) means more food for deer (herbivores), which, in turn, means more energy for the wolves (carnivores) at the next trophic level.

If numbers of a particular kind of organism in a community increase or decrease significantly, some adjustment usually occurs in the populations of other organisms within the community. For example, the populations of many kinds of small mammals fluctuate from year to year. This results in changes in the numbers of their predators or the predators must switch to other prey species and impact other parts of the community. As another example, humans have used insecticides to control the populations of many kinds of insects. Reduced insect populations may result in lower numbers of insect-eating birds and affect the predators that use these birds as food. Furthermore, the indiscriminate use of insecticides often increases the populations of herbivorous, pest insects because insecticides kill many beneficial predator insects that normally feed on the pest, rather than just the one or two target pest species.

Since communities are complex and interrelated, it is helpful if we set artificial boundaries that allow us to focus our study on a definite collection of organisms. An example of a community with easily determined natural boundaries is a small pond (figure 25.8). The water's edge naturally defines the limits of this community. You would expect to find certain animals and plants living in the pond, such as fish, frogs, snails, insects, algae, pondweeds, bacteria, and fungi. But you might ask at this point, what about the plants and animals that live right at the water's edge? That leads us to think about the animals that spend only part of their lives in the water. That awkward-looking, long-legged bird wading in the shallows and darting its long beak down to spear a fish has its nest atop some tall trees away from the water. Should it be considered part of the pond community? Should we also include the deer that comes to drink at dusk and then wanders away? Small parasites could enter the body of the deer as it drinks. The immature parasite would develop into an adult within the deer's body. That same parasite must spend part of its life cycle in the body of a certain snail. Are these parasites part of the pond community? Several animals are members of more than one community. What originally seemed to be a clear example of a community has become less clear-cut. Although the general outlines of a community can be arbitrarily set for the purposes of a study, we must realize that the boundaries of a community, or any ecosystem for that matter, must be considered somewhat artificial.

TYPES OF COMMUNITIES

Ponds and other small communities are parts of biomes, large regional communities primarily determined by climate. The primary climatic factors are the amount of precipitation and

A CLOSER LOOK | Zebra Mussels—Invaders from Europe

In the mid-1980s a clamlike organism called the zebra mussel, *Dressenia polymorpha*, was introduced into the waters of the Great Lakes. It probably arrived in the ballast water of a ship from Europe. Ballast water is pumped into empty ships to make them more stable when crossing the ocean. Immature stages of the zebra mussel were probably emptied into Lake St. Clair, near Detroit, Michigan, when the ship discharged its ballast water to take on cargo. This organism has since spread to many areas of the Great Lakes and smaller inland lakes. It has also been discovered in other parts of the United States, including the mouth of the Mississippi River. Zebra mussels attach to any hard surface and reproduce rapidly. Densities of more than 20,000 individuals per square meter have been documented in Lake Erie.

These invaders are of concern for several reasons. First, they coat the intake pipes of municipal water plants and other facilities, requiring expensive measures to clean the pipes. Second, they coat any suitable surface, preventing native organisms from using the space. Third, they introduce a new organism into the food chain. Zebra mussels filter small aquatic organisms from the water very efficiently and may remove food organisms required by native species. Their filtering activity has significantly increased the clarity of the water in several areas in the Great Lakes. This can affect the kinds of fish present, because greater clarity allows predator fish to more easily find prey. There is concern that they will significantly change the ecological organization of the Great Lakes.

FIGURE 25.8

Although a pond community would seem to be an easy community to characterize, it interacts extensively with the surrounding land-based communities. Some of the organisms associated with a pond community are always present in the water (fish, pondweeds, clams); others occasionally venture from the water to the surrounding land (frogs, dragonflies, turtles, muskrats); still others are occasional or rare visitors (minks, heron, ducks).

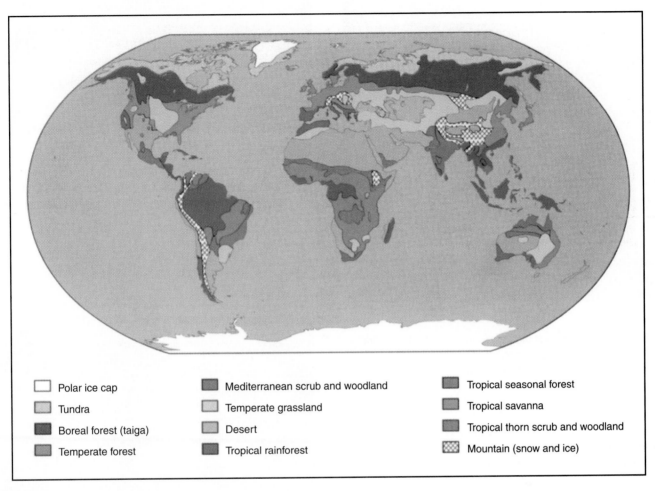

FIGURE 25.9

Major climatic differences determine the kind of vegetation that can live in a region of the world. Associated with specialized groups of plants are particular kinds of animals. These regional ecosystems are called biomes.

Legend:
- Polar ice cap
- Tundra
- Boreal forest (taiga)
- Temperate forest
- Mediterranean scrub and woodland
- Temperate grassland
- Desert
- Tropical rainforest
- Tropical seasonal forest
- Tropical savanna
- Tropical thorn scrub and woodland
- Mountain (snow and ice)

typical temperature ranges. The map in figure 25.9 shows the distribution of the major biomes of the world. One of the large, land-based biomes in the eastern part of North America is the *temperate deciduous forest*. This biome, like other land-based biomes, is named for a major feature of the ecosystem, which in this case happens to be the dominant vegetation. The predominant plants are large trees that lose their leaves more or less completely during the fall of the year and are therefore called *deciduous* (figure 25.10). Most of the trees require a considerable amount of rainfall.

This naming system works fairly well, since the major type of plant determines the other kinds of plants and animals that can occur. Of course, since the region is so large and has different climatic conditions in different areas, we can find some differences in the particular species of trees (and other organisms) in this biome. For instance, in Maryland the tulip tree is one of the common large trees, whereas in Michigan it is so unusual that people plant it in lawns and parks as a decorative tree. Aspen, birch, cottonwood, oak, hickory, beech, and maple are

typical trees found in this geographic region. Typical animals of this biome are skunks, porcupines, deer, frogs, opossums, owls, mosquitoes, and beetles.

The temperate deciduous forest covers a large area from the Mississippi River to the Atlantic Coast, and from Florida to southern Canada. This type of biome is also found in parts of Europe and Asia. Many local spots within this biome are quite different from one another. Many of them have no trees at all. For example, the tops of some of the mountains along the Appalachian Trail, the sand dunes of Lake Michigan, and the scattered grassy areas in Illinois are natural areas within this biome that lack trees. In much of this region, the natural vegetation has been removed to allow for agriculture, so the original character of the biome is gone except where farming is not practical or the original forest has been preserved.

The biome located to the west of the temperate deciduous forest in North America is the grassland or *prairie* biome (figure 25.11). This kind of biome is also common in parts of Eurasia, Africa, Australia, and South America. The dominant vegetation

FIGURE 25.10

Temperate deciduous forest biome is found in parts of the world that have significant rainfall (100 centimeters or more) and cold weather for a significant part of the year when the trees are without leaves.

FIGURE 25.11

This typical short-grass prairie biome of western North America is associated with an annual rainfall of 25 to 50 centimeters. This community contains a unique grouping of plant and animal species.

FIGURE 25.12

A savanna biome is likely to develop in areas that have a rainy season and a dry season. During the dry season, fires are frequent. The fires kill tree seedlings and prevent the establishment of forests.

in this region is made up of various species of grasses. The rainfall in this grassland is not adequate to support the growth of trees, which are common in this biome only along streams where they can obtain sufficient water. The common plants in this area are those that can grow in drier conditions. Animals found in this area include the prairie dog, pronghorn antelope, prairie chicken, grasshopper, rattlesnake, and meadowlark. Most of the original grasslands, like the temperate deciduous forest, have been converted to agricultural uses. Breaking the sod (the thick layer of grass roots) so that wheat, corn, and other grains can be grown exposes the soil to the wind, which may cause excess drying and result in soil erosion that depletes the fertility of the soil. Grasslands that are too dry to allow for farming typically have been used as grazing land for cattle and sheep. The grazing of these domesticated animals has modified the natural vegetation, as has farming in the moister grassland regions.

A biome that is similar to a prairie is a *savanna* (figure 25.12). Savannas are typical biomes of central Africa and parts of South America and typically consist of grasses with scattered trees. Such areas generally have wet and dry seasons and experience fires during the dry part of the year.

Very dry areas are known as *deserts* and are found throughout the world wherever rainfall is low and irregular. Some deserts are extremely hot, while others can be quite cold during much of the year. The distinguishing characteristic of desert biomes is low rainfall, not high temperature. Furthermore, deserts show large daily fluctuations in air temperature. When the sun goes down at night, the land cools off very rapidly, because there is no insulating blanket of clouds to keep the heat from radiating into space.

A desert biome is characterized by scattered, thorny plants that lack leaves or have reduced leaves (figure 25.13). Since leaves tend to lose water rapidly, the lack of leaves is an adaptation to dry conditions. Under these conditions the stems are green and carry on photosynthesis. Many of the plants, like cacti, are capable of storing water in their fleshy stems. Others store water in their roots. Although this is a very harsh environment, many kinds of flowering plants, insects, reptiles, and mammals can live in this biome. The animals usually avoid the hottest part of the day by staying in burrows or other shaded, cool areas. Staying underground or in the shade also allows the animals to conserve water.

Through parts of southern Canada, extending southward along the mountains of the United States, and in much of

FIGURE 25.13

The desert biome gets less than 25 centimeters of precipitation per year, but it contains many kinds of living things. Cacti, sagebrush, lichens, snakes, small mammals, birds, and insects inhabit the desert. All deserts are dry, and the plants and animals show adaptations that allow them to survive under these extreme conditions. In hot deserts where daytime temperatures are high, most animals are active only at night when the air temperature drops significantly.

FIGURE 25.14

Conifers are the dominant vegetation. They are found in most of Canada, in a major part of Russia, and at high altitudes in sections of western North America. The boreal forest biome is characterized by cold winters with abundant snowfall.

Northern Asia we find communities that are dominated by evergreen trees. This is the *taiga, boreal forest,* or *coniferous forest* biome (figure 25.14). The evergreen trees are especially adapted to withstand long, cold winters with abundant snowfall. Most of the trees in the wetter, colder areas are spruces and firs, but some drier, warmer areas have pines. The wetter areas generally have dense stands of small trees intermingled with many other kinds of vegetation and many small lakes and bogs. In the mountains of the western United States, pine trees are often widely scattered and very large, with few branches near the ground. The area has a parklike appearance because there is very little vegetation on the forest floor. Characteristic animals in this biome include mice, bears, wolves, squirrels, moose, midges, and flies.

The coastal areas of Northern California, Oregon, Washington, British Columbia, and southern Alaska contain an unusual set of environmental conditions that supports a *temperate rainforest.* The prevailing winds from the west bring moisture-laden air to the coast. As the air meets the coastal mountains and is forced to rise, it cools and the moisture falls as rain or snow. Most of these areas receive 200 centimeters (80 inches) or more of precipitation per year. This abundance of water, along with fertile soil and mild temperatures, results in a lush growth of plants.

Sitka spruce, Douglas fir, and western hemlock are typical evergreen coniferous trees in the temperate rainforest. Undisturbed (old growth) forests of this region have trees as old as 800 years that are nearly 100 meters tall. Deciduous trees of various kinds (red alder, big leaf maple, black cottonwood) also exist in open areas where they can get enough light. All trees are covered with mosses, ferns, and other plants that grow on the surface of the trees. The dominant color is green, since most surfaces have something photosynthetic growing on them.

When a tree dies and falls to the ground it rots in place and often serves as a site for the establishment of new trees. This is such a common feature of the forest that the fallen, rotting trees are called nurse trees. The fallen tree also serves as a food source for a variety of insects, which are food for a variety of larger animals.

Because of the rich resource of trees, 90 percent of the original temperate rainforest has already been logged. What remains may be protected because it serves as the home to the endangered northern spotted owl and marbled murrelet (a seabird).

North of the coniferous forest biome is an area known as the *tundra* (figure 25.15). It is characterized by extremely long, severe winters and short, cool summers. The deeper layers of the soil remain permanently frozen, forming a layer called the permafrost. Because the deeper layers of the soil are frozen, when the surface thaws the water forms puddles on the surface. Under these conditions of low temperature and short growing season, very few kinds of animals and plants can survive. No trees can live in this region. Typical plants and animals of the area are dwarf willow and some other shrubs, reindeer moss (actually a lichen), some flowering plants, caribou, wolves, musk oxen, fox, snowy owls, mice, and many kinds of insects. Many kinds of birds are summer residents only. The tundra community is relatively simple, so any changes may have drastic and long-lasting effects. The tundra is easy to injure and slow to heal; therefore, we must treat it gently. The construction of the Alaskan pipeline has left scars that could still be there a hundred years from now.

The *tropical rainforest* is at the other end of the climate spectrum from the tundra. Tropical rainforests are found primarily near the equator in Central and South America, Africa, parts of

FIGURE 25.15

The tundra biome is located in northern parts of North America and Eurasia. It is characterized by short, cool summers and long, extremely cold winters. There is a layer of soil below the surface that remains permanently frozen; consequently, no large trees exist in this biome. Relatively few kinds of plants and animals can survive this harsh environment.

FIGURE 25.16

The tropical rainforest biome is a moist, warm region of the world located near the equator. The growth of vegetation is extremely rapid. There are more kinds of plants and animals in this biome than in any other.

southern Asia, and some Pacific Islands (figure 25.16). The temperature is high, rain falls nearly every day, and there are thousands of species of plants in a small area. Balsa (a very light wood), teak (used in furniture), and ferns the size of trees are examples of plants from the tropical rainforest. Typically, every plant has other plants growing on it. Tree trunks are likely to be covered with orchids, many kinds of vines, and mosses. Tree frogs, bats, lizards, birds, monkeys, and an almost infinite variety of insects inhabit the rainforest. These forests are very dense and little sunlight reaches the forest floor. When the forest is opened up (by a hurricane or the death of a large tree) and sunlight reaches the forest floor, the opened area is rapidly overgrown with vegetation.

Since plants grow so quickly in these forests, people assume the soils are fertile, and many attempts have been made to bring this land under cultivation. In reality, the soils are poor in nutrients. The nutrients are in the organisms, and as soon as an organism dies and decomposes its nutrients are reabsorbed by other organisms. Typical North American agricultural methods, which require the clearing of large areas, cannot be used with the soil and rainfall conditions of the tropical rainforest. The constant rain falling on these fields quickly removes the soil's nutrients so that heavy applications of fertilizer are required. Often these soils become hardened when exposed in this way. Although most of

these forests are not suitable for agriculture, large expanses of tropical rainforest are being cleared yearly because of the pressure for more farmland in the highly populated tropical countries and the desire for high-quality lumber from many of the forest trees.

The distribution of terrestrial ecosystems is primarily related to temperature and precipitation. Air temperatures are warmest near the equator and become cooler as the poles are approached. Similarly, air temperature decreases as altitude increases. This means that even at the equator it is possible to have cold temperatures on the peaks of tall mountains. Therefore, as one proceeds from sea level to the tops of mountains, it is possible to pass through a series of biomes that are similar to what one would encounter traveling from the equator to the North Pole (figure 25.17).

COMMUNITY, HABITAT, AND NICHE

People approach the study of organism interactions in two major ways. Many people look at interrelationships from the broad ecosystem point of view, while others focus on individual

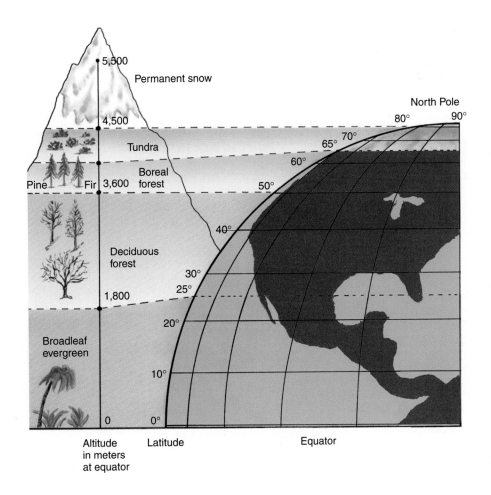

FIGURE 25.17

As one travels up a mountain, the climate changes. The higher the elevation, the cooler the climate. Even in the tropics tall mountains can have snow on the top. Thus it is possible to experience the same change in vegetation by traveling up a mountain as one would experience traveling from the equator to the North Pole.

organisms and the specific things that affect them in their daily lives. The first approach involves the study of all the organisms that interact with one another—the community—and usually looks at general relationships among them. The previous section described categories of organisms—producers, consumers, and decomposers—that perform different functions in a community.

Another way of looking at interrelationships is to study in detail the ecological relationships of certain species of organisms. Each organism has particular requirements for life and lives where the environment provides what it needs. The environmental requirements of a whale include large expanses of ocean, but with seasonally important feeding areas and protected locations used for giving birth. The kind of place, or part of an ecosystem, occupied by an organism is known as its **habitat.** Habitats are usually described in terms of conspicuous or particularly significant features in the area where the organism lives. For example, the habitat of a prairie dog is usually described as a grassland, while the habi-

tat of a tuna is described as the open ocean. The habitat of the fiddler crab is sandy ocean shores and the habitat of various kinds of cacti is the desert. The key thing to keep in mind when you think of habitat is the *place* in which a particular kind of organism lives. When describing the habitats of organisms, we sometimes use the terminology of the major biomes of the world, such as desert, grassland, or savanna. It is also possible to describe the habitat of the bacterium *Escherichia coli* as the gut of humans and other mammals, or the habitat of a fungus as a rotting log. Organisms that have very specific places in which they live simply have more restricted habitats.

Each species has particular requirements for life and places specific demands on the habitat in which it lives. The specific functional role of an organism is its **niche.** Its niche is the way it goes about living its life. Just as the word *place* is the key to understanding the concept of habitat, the word *function* is the key to understanding the concept of a niche. Understanding the niche of an organism involves a detailed

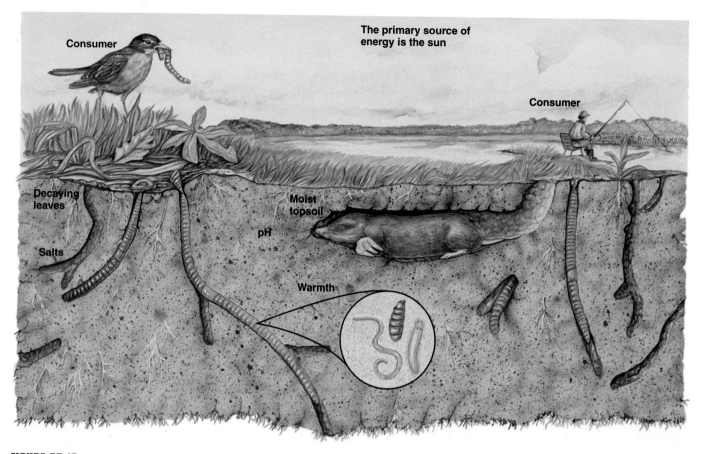

The primary source of
energy is the sun

Consumer

Consumer

Decaying
leaves

Moist
topsoil

pH

Salts

Warmth

FIGURE 25.18

The niche of an earthworm involves a great many factors. It includes the fact that the earthworm is a consumer of dead organic matter, a source of food for other animals, a host to parasites, and bait for an angler. Furthermore, that the earthworm loosens the soil by its burrowing and "plows" the soil when it deposits materials on the surface are other factors. Additionally, the pH, texture, and moisture content of the soil have an impact on the earthworm. Keep in mind that this is but a small part of what the niche of the earthworm includes.

understanding of the impacts an organism has on its biotic and abiotic surroundings as well as all of the factors that affect the organism. For example, the niche of an earthworm includes abiotic items such as soil particle size, soil texture, and the moisture, pH, and temperature of the soil. The earthworm's niche also includes biotic impacts such as serving as food for birds, moles, and shrews; as bait for anglers; or as a consumer of dead plant organic matter (figure 25.18). In addition, an earthworm serves as a host for a variety of parasites, transports minerals and nutrients from deeper soil layers to the surface, incorporates organic matter into the soil, and creates burrows that allow air and water to penetrate the soil more easily. And this is only a limited sample of all of the aspects of its niche.

Some organisms have rather broad niches; others, with very specialized requirements and limited roles to play, have niches that are quite narrow. The opossum (figure 25.19A) is an animal with a very broad niche. It eats a wide variety of plant and animal foods, can adjust to a wide variety of climates, is used as

food by many kinds of carnivores (including humans), and produces large numbers of offspring. By contrast, the koala of Australia (figure 25.19B) has a very narrow niche. It can live only in areas of Australia with specific species of *Eucalyptus* trees, because it eats the leaves of only a few kinds of these trees. Furthermore, it cannot tolerate low temperatures and does not produce large numbers of offspring. As you might guess, the opossum is expanding its range and the koala is endangered in much of its range.

The complete description of an organism's niche involves a very detailed inventory of influences, activities, and impacts. It involves what the organism does and what is done to the organism. Some of the impacts are abiotic, others are biotic. Since the niche of an organism is a complex set of items, it is often easy to overlook important roles played by some organisms.

For example, when Europeans introduced cattle into Australia—a continent where there had previously been no large, hoofed mammals—they did not think about the impact of cow manure or the significance of a group of beetles called *dung bee-*

A CLOSER LOOK | The Importance of Habitat Size

Many people interested in songbird populations have documented a significant decrease in the numbers of certain songbird species. Species particularly affected are those that migrate between North America and South America and require relatively large areas of undisturbed forest in both their northern and southern homes. Many of these species are being hurt by human activities that fragment large patches of forest into many smaller patches, creating more edges between different habitat types. Bird and other animal species that thrive in edge habitats replace the songbirds, which require large patches of undisturbed forest. Cowbirds that normally live in open areas are a particular problem. Cowbirds do not build nests but lay their eggs in the nests of other birds after removing the eggs of the host species. When forests are broken into small patches, cowbirds reach a larger percentage of forest nesting birds because the forest birds must nest closer to the edge.

The species most severely affected are those that have both their northern and southern habitats disturbed. A study of migrating sharp-shinned hawks indicated that their numbers have also been greatly reduced in recent years. It is thought that since sharp-shinned hawks use small songbirds as their primary source of food, the reduction in hawks is directly related to the reduction in migratory songbirds.

A

B

FIGURE 25.19

(*A*) The opossum has a very broad niche. It eats a variety of foods, is able to live in a variety of habitats, and has a large reproductive capacity. It is generally extending its range in the United States.
(*B*) The koala has a narrow niche. It feeds on the leaves of only a few species of *Eucalyptus* trees, is restricted to relatively warm, forested areas, and is generally endangered in much of its habitat.

tles. These beetles rapidly colonize fresh dung and cause it to be broken down. No such beetles existed in Australia; therefore, in areas where cattle were raised, a significant amount of land became covered with accumulated cow dung. This reduced the area where grass could grow and reduced productivity. The problem was eventually solved by the importation of several species of dung beetles from Africa, where large, hoofed mammals are common. The dung beetles made use of what the cattle did not digest, returning it to a form that plants could more easily recycle into plant biomass.

A

B

FIGURE 25.20

(*A*) Many predators capture prey by making use of speed. The cheetah can reach estimated speeds of 100 kilometers per hour (about 70 mph) during sprints to capture its prey. (*B*) Other predators, like this veiled chameleon, blend in with their surroundings, lie in wait, and ambush their prey. Because strength is needed to kill the prey, the predator is generally larger than the prey. Obviously, predators benefit from the food they obtain to the detriment of the prey organism.

KINDS OF ORGANISM INTERACTIONS

One of the important components of an organism's niche is the other living things with which it interacts. When organisms encounter one another in their habitats, they can influence one another in numerous ways. Some interactions are harmful to one or both of the organisms. Others are beneficial. Ecologists have classified kinds of interactions between organisms into several broad categories, which we will discuss here.

Predation

Predation occurs when one animal captures, kills, and eats another animal. The organism that is killed is called the **prey** and the one that does the killing is called the **predator.** The predator obviously benefits from the relationship, while the prey organism is harmed. Most predators are relatively large compared to their prey and have specific adaptations that aid them in catching prey. Many spiders build webs that serve as nets to catch flying insects. The prey are quickly paralyzed by the spider's bite and wrapped in a tangle of silk threads. Other rapidly moving spiders, like wolf spiders and jumping spiders, have large eyes that help them find prey without using webs. Dragonflies patrol areas where they can capture flying insects. Hawks and owls have excellent eyesight that allows them to find their prey. Many predators, like leopards, lions, and cheetahs, use speed to run down their prey, while others such as frogs, toads, and many kinds of lizards blend in with their surroundings and strike quickly when a prey organism happens by (figure 25.20).

Many kinds of predators are useful to us because they control the populations of organisms that do us harm. For example, snakes eat many kinds of rodents that eat stored grain and other agricultural products. Many birds and bats eat insects that are agricultural pests. It is even possible to think of a predator as having a beneficial effect on the prey species. Certainly the *individual* organism that is killed is harmed, but the *population* can benefit. Predators can prevent large populations of prey organisms from destroying their habitat by preventing overpopulation of prey species or they can reduce the likelihood of epidemic disease by eating sick or diseased individuals. Furthermore, predators act as selecting agents. The individuals who fall to them as prey are likely to be less well adapted than the ones that escape predation. Predators usually kill slow, unwary, sick, or injured individuals. Thus, the genes that may have contributed to slowness, inattention, illness, or the likelihood of being injured are removed from the gene pool and a better-adapted population remains. Because predators eliminate poorly adapted *individuals,* the *species* benefits. What is bad for the individual can be good for the species.

Parasitism

Another kind of interaction in which one organism is harmed and the other aided is the relationship of parasitism. In fact, there are more species of parasites in the world than there are nonparasites, making this a very common kind of relationship. **Parasitism** involves one organism living in or on another living organism from which it derives nourishment. The **parasite** derives the benefit and harms the **host,** the organism it lives in or on (figure 25.21). Many kinds of fungi live on trees and other kinds of plants, including those that are commercially valuable. Dutch elm disease is caused by a fungus that infects the living, sap-carrying parts of the tree. Mistletoe is a common plant that is a parasite on other plants. The mistletoe plant invades the tissues of the tree it is living on and derives nourishment from the tree.

A

B

C

FIGURE 25.21

Parasites benefit from the relationship because they obtain nourishment from the host. Tapeworms (*A*) are internal parasites in the guts of their host, where they absorb food from the host's gut. The lamprey (*B*) is an external parasite that sucks body fluids from its host. Mistletoe (*C*) is a photosynthesizing plant that also absorbs nutrients from the tissues of its host tree. The host in any of these three situations may not be killed directly by the relationship, but it is often weakened, thus becoming more vulnerable to predators or diseases. There are more species of parasites in the world than species of organisms that are not parasites.

Many kinds of worms, protozoa, bacteria, and viruses are important parasites. Parasites that live on the outside of their hosts are called **external parasites.** For example, fleas live on the outside of the bodies of mammals like rats, dogs, cats, and humans, where they suck blood and do harm to their host. At the same time, the host could also have a tapeworm in its intestine. Since the tapeworm lives inside the host, it is called an **internal parasite.** Another kind of parasite that may be found in the blood of rats is the bacterium *Yersinia pestis.* It does little harm to the rat but causes a disease known as *plague* or *Black Death* if it is transmitted to humans. Because fleas can suck the blood of rats and also live on and bite humans, they can serve as a carrier of the bacterium between rats and humans. An organism that can carry a disease from one individual to another is called a **vector.** During the mid-1300s, when living conditions were poor and rats and fleas were common, epidemics of plague killed millions of people. In some

countries in Western Europe, 50 percent of the population was killed by this disease. Plague is still a problem today when living conditions are poor and sanitation is lacking. Cases of plague are even found in developed countries like the United States on occasion.

Lyme disease is also a vector-borne disease caused by the bacterium, *Borrelia burgdorferi,* that is spread by certain species of ticks (figure 25.22). Over 90 percent of the cases are centered in the Northeast (New York, Pennsylvania, Maryland, Delaware, Connecticut, Rhode Island, and New Jersey).

Predation and parasitism are both relationships in which one member of the pair is helped and the other is harmed. But there are many kinds of interactions in which one is harmed and the other aided that don't fit neatly into the categories of interactions dreamed up by scientists. For example, when a cow eats grass, it is certainly harming the grass while deriving benefit from it. We could call cows *grass predators,* but we usually refer to them

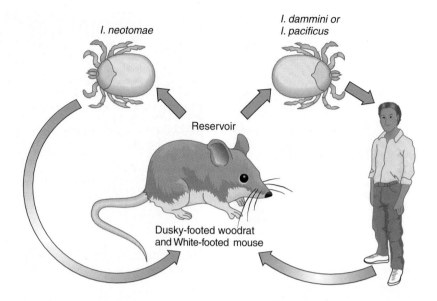

FIGURE 25.22

Lyme disease is a bacterial disease originally identified in a small number of individuals in the Old Lyme, Connecticut, area. Today it is found throughout the United States and Canada. Once the parasite, *Borrelia burgdorferi,* has been transferred into a suitable susceptible host (e.g., humans, mice, horses, cattle, domestic cats, and dogs), it causes symptoms that have been categorized into three stages. The first-stage symptoms may appear 3 to 32 days after an individual is bitten by an infected tick (various members of the genus *Ixodes*) and include a spreading red rash, headache, nausea, fever, aching joints and muscles, and fatigue. Stage two may not appear for weeks or months after the infection and may affect the heart and nervous system. The third stage may occur months or years later and typically appears as severe arthritis attacks. The main reservoir of the disease is the white-footed mouse and dusky-footed woodrat.

as *herbivores.* Likewise, such animals as mosquitoes, biting flies, vampire bats, and ticks take blood meals but don't usually live permanently on the host, nor do they kill it. Are they temporary parasites or specialized predators? Finally, birds like cowbirds and some species of European cuckoos lay their eggs in the nests of other species of birds, who raise these foster young rather than their own. The adult cowbird and cuckoo often remove eggs from the host nest or their offspring eject the eggs or the young of the host-bird species, so that usually only the cowbird or cuckoo is raised by the foster parents. This kind of relationship has been called nest parasitism, because the host parent birds are not killed and aid the cowbird or cuckoo by raising their young.

Commensalism

Predation and parasitism both involve one organism benefiting while the other is harmed. Another common relationship is one in which one organism benefits while the other is not affected. This is known as **commensalism.** For example, sharks often have another fish, the remora, attached to them. The remora has a sucker on the top side of its head that allows it to attach to the shark and get a free ride (figure 25.23A). While the remora benefits from the free ride and by eating leftovers from the shark's meals, the shark does not appear to be troubled by this uninvited guest, nor does it benefit from the presence of the remora.

Another example of commensalism is the relationship between trees and epiphytic plants. **Epiphytes** are plants that

Mites in Your Bed

Your pillow and mattress are inhabited by extremely small mites that live on the tiny flakes of skin that you shed constantly. (You probably didn't know that you shared your bed with these creatures.) The relationship they have with you is one of commensalism. You are not affected by them, although some people may have an allergic reaction to the mites when they are inhaled. However, your skin fragments are essential to their survival.

live on the surface of other plants but do not derive nourishment from them (figure 25.23B). Many kinds of plants (e.g., orchids, ferns, and mosses) use the surface of trees as a place to live. These kinds of organisms are particularly common in tropical rainforests. Many epiphytes derive benefit from the relationship because they are able to be located in the top of the tree, where they receive more sunlight and moisture. The trees derive no benefit from the relationship, nor are they harmed; they simply serve as a support surface for epiphytes.

A

B

FIGURE 25.23

In the relationship called commensalism, one organism benefits and the other is not affected. (*A*) The remora fish shown here hitchhike a ride on the shark. They eat scraps of food left over by the messy eating habits of the shark. The shark does not seem to be hindered in any way. (*B*) The epiphytic plants growing on this tree do not harm the tree but are aided by using the tree surface as a place to grow.

Mutualism

So far in our examples, only one species has benefited from the association of two species. There are also many situations in which two species live in close association with one another, and both benefit. This is called **mutualism.** One interesting example of mutualism involves digestion in rabbits. Rabbits eat plant material that is high in cellulose, even though they do not produce the enzymes capable of breaking down cellulose molecules into simple sugars. They manage to get energy out of these cellulose molecules with the help of special bacteria living in their digestive tracts. The bacteria produce cellulose-digesting enzymes, called *cellulases,* which break down cellulose into smaller carbohydrate molecules that the rabbit's digestive enzymes can break down into smaller glucose molecules. The bacteria benefit because the gut of the rabbit provides them with a moist, warm, nourishing environment in which to live. The rabbit benefits because the bacteria provide them with a source of food. Termites, cattle, buffalo, and antelope also have collections of bacteria and protozoa living in their digestive tracts that help them to digest cellulose.

Another kind of mutualistic relationship exists between flowering plants and bees. Undoubtedly you have observed bees and other insects visiting flowers to obtain nectar from the blossoms. Usually the flowers are constructed in such a manner that the bees pick up pollen (sperm-containing packages) on their hairy bodies and transfer it to the female part of the next flower they visit (figure 25.24). Because bees normally visit many individual flowers of the same species for several minutes and ignore other species of flowers, they can serve as pollen carriers between two flowers of the same species. Plants pollinated in this manner produce less pollen than do plants that rely on the wind to transfer pollen. This saves the plant energy because it doesn't need to produce huge quantities of pollen. It does, however, need to transfer some of its energy savings into the production of showy flowers and nectar to attract the bees. The bees benefit from both the nectar and pollen; they use both for food.

FIGURE 25.24

Mutualism is an interaction between two organisms in which both benefit. The plant benefits because cross-fertilization (exchange of gametes from a different plant) is more probable; the butterfly benefits by acquiring nectar for food.

Lichens and corals exhibit a more intimate kind of mutualism. In both cases the organisms consist of the cells of two different organisms intermingled with one another. Lichens consist of fungal cells and algal cells in a partnership; corals consist of the cells of the coral organism intermingled with algal cells. In both cases the algae carry on photosynthesis and provide nutrients, while the fungus or coral provides a moist, fixed structure for the algae to live in.

FIGURE 25.25

Whenever a needed resource is in limited supply, organisms compete for it. This competition may be between members of the same species (*intraspecific*), illustrated by the vultures shown in the photograph, or may involve different species (*interspecific*).

One other term that relates to parasitism, commensalism, and mutualism is *symbiosis*. **Symbiosis** literally means "living together." Unfortunately, this word is used in several ways, none of which is very precise. It is often used as a synonym for mutualism, but it is also often used to refer to commensalistic relationships and parasitism. The emphasis, however, is on interactions that involve a close physical relationship between the two kinds of organisms.

Competition

So far in our discussion of organism interactions we have left out the most common one. It is reasonable to envision every organism on the face of the earth being involved in *competitive* interactions. **Competition** is a kind of interaction between organisms in which both organisms are harmed to some extent. Competition occurs whenever two organisms both need a vital resource that is in short supply (figure 25.25). The vital resource could be food, shelter, nesting sites, water, mates, or space. It can be a snarling tug-of-war between two dogs over a scrap of food, or it can be a silent struggle between plants for access to available light. If you have ever started tomato seeds (or other garden plants) in a garden and failed to eliminate the weeds, you have witnessed competition. If the weeds are not removed, they compete with the garden plants for available sunlight, water, and nutrients, resulting in poor growth of both the garden plants and the weeds.

It is important to recognize that although competition results in harm to both organisms there can still be winners and losers. The two organisms may not be harmed to the same extent, which results in one having greater access to the limited resource. Furthermore, even the loser can continue to survive if it migrates to an area where competition is less intense, or evolves to exploit a different niche. Thus, competition provides a major mechanism for natural selection. With the development of slight differences between niches, the intensity of competition is reduced. For example, many birds catch flying insects as food. However, they do not compete directly with each other because some feed at night, some feed high in the air, some feed only near the ground, and still others perch on branches and wait for insects to fly past. The insect-eating niche can be further subdivided by specializing on particular sizes or kinds of insects.

Many of the relationships just described involve the transfer of nutrients from one organism to another (predation, parasitism, mutualism). Another important way scientists look at ecosystems is to look at how materials are cycled from organism to organism.

THE CYCLING OF MATERIALS IN ECOSYSTEMS

Although some new atoms are being added to the earth from cosmic dust and meteorites, this amount is not significant in relation to the entire biomass of the earth. Therefore, the earth can be considered to be a closed ecosystem as far as matter is concerned. Only sunlight energy comes to the earth in a continuous stream, and even this is ultimately returned to space as heat energy. However, it is this flow of energy that drives all biological processes. Living systems have evolved ways of using this energy to continue life through growth and reproduction and the continual reuse of existing atoms. In this recycling process, inorganic molecules are combined to form the organic compounds of living things. If there were no way of recycling this organic matter back into its inorganic forms, organic material would build up as the bodies of dead organisms. This is thought to have occurred millions of years ago when the present deposits of coal, oil, and natural gas were formed. Under most conditions decomposers are available to break down organic material to inorganic material that can then be reused by other organisms to rebuild organic material. One way to get an appreciation of how various kinds of organisms interact to cycle materials is to look at a specific kind of atom and follow its progress through an ecosystem.

The Carbon Cycle

Living systems contain many kinds of atoms, but some are more common than others. Carbon, nitrogen, oxygen, hydrogen, and phosphorus are found in all living things and must be recycled when an organism dies. Let's look at some examples of this recycling process. Carbon and oxygen atoms combine to form the molecule carbon dioxide (CO_2), which is a gas

A CLOSER LOOK | Carbon Dioxide and Global Warming

Humans have significantly altered the carbon cycle. As we burn fossil fuels, the amount of carbon dioxide in the atmosphere continually increases. Carbon dioxide allows light to enter the atmosphere but does not allow heat to exit. Since this is similar to what happens in a greenhouse, carbon dioxide and other gases that have similar effects are called greenhouse gases. Therefore, increased amounts of carbon dioxide in the atmosphere could lead to a warming of the earth. Many are concerned that increased carbon dioxide levels will lead to a warming of the planet that would cause major changes in weather and climate, leading to the flooding of coastal cities and major changes in agricultural production. The Intergovernmental Panel on Climate Change (IPCC)

established by the United Nations concluded that there has been an increase in average temperature of the earth and humans are the cause because of the burning of fossil fuels and the destruction of forests. There is no doubt that the amount of carbon dioxide in the atmosphere has been increasing. Despite this fact and the conclusions of the IPCC, there is still controversy about this topic and some doubt that warming is occurring. At a meeting in Kyoto, Japan, in 1998, many countries agreed to reduce the amount of carbon dioxide and other greenhouse gases they release into the atmosphere. However, emissions of carbon dioxide are directly related to economic activity and the energy usage that fuels it. Therefore, it remains to be seen if countries will meet their goals or will suc-

cumb to economic pressures to allow continued use of large amounts of fossil fuels. However, some countries have sought to control the change in the amount of carbon dioxide in the atmosphere by planting millions of trees or preventing the destruction of forests. The thought is that the trees carry on photosynthesis, grow, and store carbon in their bodies, leading to reduced carbon dioxide levels. At the same time, people in other parts of the world continue to destroy forests at a rapid rate. Tree planting does not offset deforestation. In addition, the trees that have been planted will ultimately die and decompose, releasing carbon dioxide back into the atmosphere, so it is not clear that this is an effective means of reducing atmospheric carbon dioxide over the long term.

found in small quantities in the atmosphere. During photosynthesis, carbon dioxide (CO_2) combines with water (H_2O) to form complex organic molecules like sugar ($C_6H_{12}O_6$). At the same time, oxygen molecules (O_2) are released into the atmosphere.

The organic matter in the bodies of plants may be used by herbivores as food. When a herbivore eats a plant, it breaks down the complex organic molecules into more simple molecules, like simple sugars, amino acids, glycerol, and fatty acids. These can be used as building blocks in the construction of its own body. Thus, the atoms in the body of the herbivore can be traced back to the plants that were eaten. Similarly, when carnivores eat herbivores these same atoms are transferred to them. Finally, the waste products of plants and animals and the remains of dead organisms are used by decomposer organisms as sources of carbon and other atoms they need for survival. In addition, all the organisms in this cycle—plants, herbivores, carnivores, and decomposers—obtain energy (ATP) from the process of respiration, in which oxygen (O_2) is used to break down organic compounds into carbon dioxide (CO_2) and water (H_2O). Thus, the carbon atoms that started out as components of carbon dioxide

(CO_2) molecules have passed through the bodies of living organisms as parts of organic molecules and returned to the atmosphere as carbon dioxide, ready to be cycled again. Similarly, the oxygen atoms (O) released as oxygen molecules (O_2) during photosynthesis have been used during the process of respiration (figure 25.26).

The Nitrogen Cycle

Another important element for living things is nitrogen (N). Nitrogen is essential in the formation of amino acids, which are needed to form proteins, and in the formation of nitrogenous bases, which are a part of ATP and the nucleic acids DNA and RNA. Nitrogen (N) is found as molecules of nitrogen gas (N_2) in the atmosphere. Although nitrogen gas (N_2) makes up approximately 80 percent of the earth's atmosphere, only a few kinds of bacteria are able to convert it into nitrogen compounds that other organisms can use. Therefore, in most terrestrial ecosystems, the amount of nitrogen available limits the amount of plant biomass that can be produced. (Most aquatic ecosystems are limited by the amount of phosphorus rather than the amount of nitrogen.) Plants utilize several different nitrogen-containing

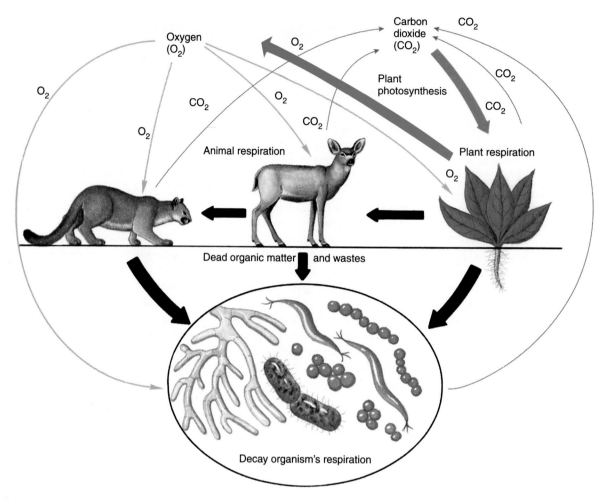

FIGURE 25.26

Carbon atoms are cycled through ecosystems. Carbon dioxide (green arrows) produced by respiration is the source of carbon that plants incorporate into organic molecules when they carry on photosynthesis. These carbon-containing organic molecules (black arrows) are passed to animals when they eat plants and other animals. Organic molecules in waste or dead organisms are consumed by decay organisms in the soil when they break down organic molecules into inorganic molecules. All organisms (plants, animals, and decomposers) return carbon atoms to the atmosphere as carbon dioxide when they carry on cellular respiration. Oxygen (blue arrows) is being cycled at the same time that carbon is. The oxygen is released into the atmosphere and into the water during photosynthesis and taken up during cellular respiration.

compounds to obtain the nitrogen atoms they need to make amino acids and other compounds (figure 25.27).

Symbiotic nitrogen-fixing bacteria live in the roots of certain kinds of plants, where they convert nitrogen gas molecules into compounds that the plants can use to make amino acids and nucleic acids. The most common plants that enter into this mutualistic relationship with bacteria are the legumes, such as beans, clover, peas, alfalfa, and locust trees. Some other organisms, such as alder trees and even a kind of aquatic fern, can also participate in this relationship. There are also **free-living nitrogen-fixing bacteria** in the soil that provide nitrogen compounds that can be taken up through the roots, but the bacteria do not live in a close physical union with plants.

Another way plants get usable nitrogen compounds involves a series of different bacteria. Decomposer bacteria convert organic nitrogen-containing compounds into ammonia (NH_3). **Nitrifying bacteria** can convert ammonia (NH_3) into nitrite-containing (NO_2^-) compounds, which in turn can be converted into nitrate-containing (NO_3^-) compounds. Many kinds of plants can use either ammonia (NH_3) or nitrate (NO_3^-) from the soil as building blocks for amino acids and nucleic acids.

All animals obtain their nitrogen from the food they eat. The ingested proteins are broken down into their component amino acids during digestion. These amino acids can then be reassembled into new proteins characteristic of the animal. All dead organic matter and waste products of plants and animals

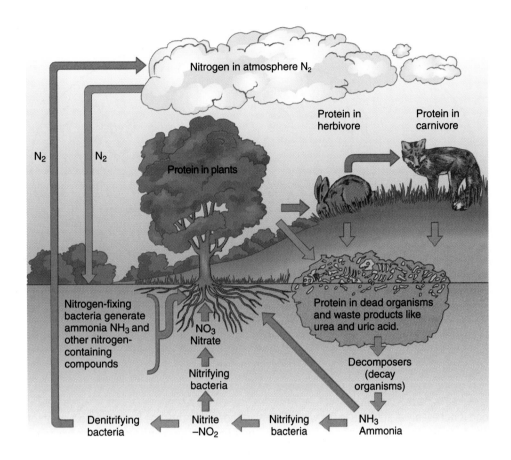

FIGURE 25.27

Nitrogen atoms are cycled through ecosystems. Atmospheric nitrogen is converted by nitrogen-fixing bacteria to nitrogen-containing compounds that plants can use to make proteins and other compounds. Proteins are passed to other organisms when one organism is eaten by another. Dead organisms and their waste products are acted upon by decay organisms to form ammonia, which may be reused by plants and converted to other nitrogen compounds by nitrifying bacteria. Denitrifying bacteria return nitrogen as a gas to the atmosphere.

are acted upon by decomposer organisms, and the nitrogen is released as ammonia (NH_3), which can be taken up by plants or acted upon by nitrifying bacteria to make nitrate (NO_3^-).

Finally, other kinds of bacteria called **denitrifying bacteria** are capable of converting nitrite (NO_2^-) to nitrogen gas (N_2), which is released into the atmosphere. Thus, in the nitrogen cycle, nitrogen from the atmosphere is passed through a series of organisms, many of which are bacteria, and ultimately returns to the atmosphere to be cycled again. However, there is also a secondary cycle in which nitrogen compounds are recycled without returning to the atmosphere.

Since nitrogen is in short supply in most ecosystems, farmers usually find it necessary to supplement the natural nitrogen sources in the soil to obtain maximum plant growth. This can be done in a number of ways. Alternating nitrogen-producing crops with nitrogen-demanding crops helps to maintain high levels of usable nitrogen in the soil. One year, a crop such as beans or clover that has symbiotic nitrogen-fixing bacteria in its roots can be planted. The following year, the farmer can plant a nitrogen-

demanding crop, such as corn. The use of manure is another way of improving nitrogen levels. The waste products of animals are broken down by decomposer bacteria and nitrifying bacteria, resulting in enhanced levels of ammonia and nitrate. Finally, the farmer can use industrially produced fertilizers containing ammonia or nitrate. These compounds can be used directly by plants or converted into other useful forms by nitrifying bacteria.

Fertilizers usually contain more than just nitrogen compounds. The numbers on a fertilizer bag tell you the percentages of nitrogen, phosphorus, and potassium in the fertilizer. For example, a 6-24-24 fertilizer would have 6 percent nitrogen compounds, 24 percent phosphorus compounds, and 24 percent potassium-containing compounds. These other elements (phosphorus and potassium) are also cycled through ecosystems. In natural, nonagricultural ecosystems these elements are released by decomposers and enter the soil, where they are available for plant uptake through the roots. However, when crops are removed from fields, these elements are removed with them and must be replaced by adding more fertilizer.

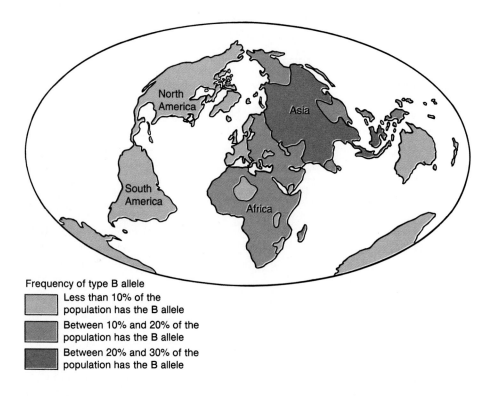

Frequency of type B allele

Less than 10% of the population has the B allele

Between 10% and 20% of the population has the B allele

Between 20% and 30% of the population has the B allele

FIGURE 25.28

The allele for type B blood is not evenly distributed in the world. This map shows that the type B allele is most common in parts of Asia and has been dispersed to the Middle East and parts of Europe and Africa. There has been very little flow of the allele to the Americas.

POPULATION CHARACTERISTICS

A **population** is a group of organisms of the same species located in the same place at the same time. Examples are the number of dandelions in your front yard, the rat population in the sewers of your city, or the number of people in your biology class. On a larger scale, all the people of the world constitute the human population. The terms *species* and *population* are interrelated because a species is a population—the largest possible population of a particular kind of organism. The term *population,* however, is often used to refer to portions of a species by specifying a space and time. For example, the size of the human population in a city changes from hour to hour during the day and varies according to where you set the boundaries of the city.

Since each local population is a small portion of a species, we should expect distinct populations of the same species to show differences. One of the ways in which they can differ is in *gene frequency.* **Gene frequency** is a measure of how often a specific gene shows up in the gametes (sex cells) of a population. Two populations of the same species often have quite different gene frequencies. For example, many populations of mosquitoes have high frequencies of insecticide-resistant genes, whereas others do not. The frequency of the genes for tallness in humans is greater in certain African tribes than in any other human population. Figure 25.28 shows that the frequency of the allele for type B blood differs significantly from one human population to another.

Since members of a population are of the same species, sexual reproduction can occur, and genes can flow from one gener-

ation to the next. Genes can also flow from one place to another as organisms migrate or are carried from one geographic location to another. *Gene flow* is used to refer to both the movement of genes within a species due to migration and the movement from one generation to the next as a result of gene replication and sexual reproduction. Typically both happen together as individuals migrate to new regions and subsequently reproduce, passing their genes to the next generation in the new area.

Another feature of a population is its *age distribution,* which is the number of organisms of each age in the population. In addition, organisms are often grouped into the following categories:

- prereproductive juveniles—insect larvae, plant seedlings, or babies;
- reproductive adults—mature insects, plants producing seeds, or humans in early adulthood; or
- postreproductive adults no longer capable of reproduction— annual plants that have shed their seeds, salmon that have spawned, and many elderly humans.

A population is not necessarily divided into equal thirds (figure 25.29). In some situations, a population may be made up of a majority of one age group. If the majority of the population is prereproductive, then a "baby boom" should be anticipated in the future. If a majority of the population is reproductive, the population should be growing rapidly. If the majority of the population is postreproductive, a population decline should be anticipated. Many organisms that only live a short time and have high reproductive rates can have age distributions that change significantly

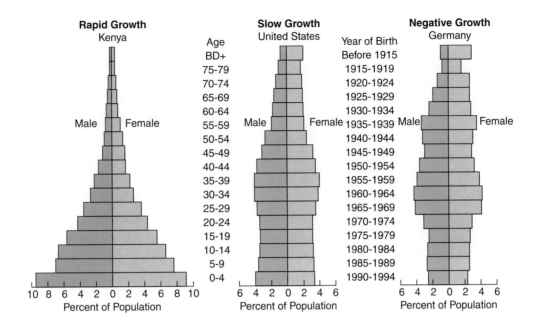

FIGURE 25.29

The relative number of individuals in each of the three categories (prereproductive, reproductive, and postreproductive) can give a good clue to the future of the population. Kenya has a large number of young individuals who will become reproducing adults. Therefore this population will grow rapidly and will double in about 27 years. The United States has a declining proportion of prereproductive individuals but a relatively large reproductive population. Therefore it will continue to grow for a time but will probably stabilize in the future. Germany's population has a large proportion of postreproductive individuals and a small proportion of prereproductive individuals. Its population is beginning to fall.

Source: U.S. Bureau of the Census and the United Nations, as reported in Joseph A. McFalls, Jr., "Population: A Lively Introduction," *Population Bulletin,* vol. 46, no. 2. Washington, D.C.: Population Reference Bureau, Inc., October 1991.

in a matter of weeks or months. For example, many birds have a flurry of reproductive activity during the summer months. Therefore, if you sample the population of a specific species of bird at different times during the summer you would have widely different proportions of reproductive and prereproductive individuals.

Populations can also differ in their sex ratios. The *sex ratio* is the number of males in a population compared to the number of females. In bird and mammal species where strong pair-bonding occurs, the sex ratio may be nearly one to one (1:1). Among mammals and birds that do not have strong pair-bonding, sex ratios may show a larger number of females than males. This is particularly true among game species, where more males than females are shot, resulting in a higher proportion of females surviving. Since one male can fertilize several females, the population can remain large even though the females outnumber the males. However, if the population of these managed game species becomes large enough to cause a problem, it becomes necessary to harvest some of the females as well, since their number determines how much reproduction can take place. In addition to these examples, many species of animals like bison, horses, and elk have mating systems in which one male maintains a harem of females. The sex ratio in these small groups is quite different from a 1:1 ratio (figure 25.30).

FIGURE 25.30

Some male animals defend a harem of females; therefore the sex ratio in these groups is several females per male.

A

FIGURE 25.31

This population of lodgepole pine seedlings consists of a large number of individuals very close to one another (*A*). As the trees grow, many of the weaker trees will die, the distance between individuals will increase, and the population density will be reduced (*B*).

B

There are very few situations in which the number of males exceeds the number of females. In some human and other populations, there may be sex ratios in which the males dominate if female mortality is unusually high or if some special mechanism separates most of one sex from the other.

Regardless of the sex ratio of a population, most species can generate large numbers of offspring, producing a concentration of organisms in an area. Population density is the number of organisms of a species per unit area. Some populations are extremely concentrated into a limited space, while others are well dispersed. As the population density increases, competition among members of the population for the necessities of life increases. This increases the likelihood that some individuals will explore new habitats and migrate to new areas. Increases in the intensity of competition that cause changes in the environment and lead to dispersal are often referred to as *population pressure*. The dispersal of individuals to new areas can relieve the pressure on the home area and lead to the establishment of new populations. Among animals, it is often the juveniles who participate in this dispersal process. Female bears generally mate every other year and abandon their nearly grown young the summer before the next set of cubs is to be born. The abandoned young bears tend to wander and disperse to new areas. Similarly, young turtles, snakes, rabbits, and many other common animals

disperse during certain times of the year. That is one of the reasons you see so many road-killed animals in the spring and fall.

If dispersal cannot relieve population pressure, there is usually an increase in the rate at which individuals die due to predation, parasitism, starvation, and accidents. In plant populations, dispersal is not very useful for relieving population density; instead, the death of weaker individuals usually results in reduced population density. In the lodgepole pine, seedlings become established in areas following fire and dense thickets of young trees are established. As the stand ages, many small trees die and the remaining trees grow larger as the population density drops (figure 25.31).

THE POPULATION GROWTH CURVE

Because most species of organisms have a high reproductive capacity, there is a tendency for populations to grow if environmental conditions permit. For example, if the usual litter size for a pair of mice is four, the four would produce eight, which in turn would produce sixteen, and so forth. Figure 25.32 shows a graph of change in population size over time known as a **population growth curve.** This kind of curve is typical for situations where a species is introduced into a previously unutilized area.

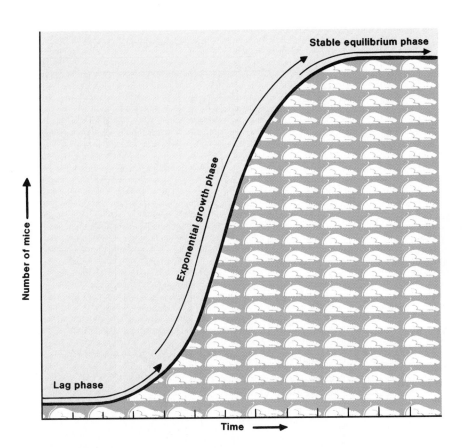

Stable equilibrium phase

Exponential growth phase

Lag phase

Number of mice

Time

The change in the size of a population depends on the rate at which new organisms enter the population compared to the rate at which they leave. The number of new individuals added to the population by reproduction per thousand individuals is called **natality.** The number of individuals leaving the population by death per thousand individuals is called **mortality.** When a small number of organisms (two mice) first invades an area, there is a period of time before reproduction takes place when the population remains small and relatively constant. This part of the population growth curve is known as the **lag phase.** During the lag phase both natality and mortality are low. The lag phase occurs because reproduction is not an instantaneous event. Even after animals enter an area they must mate and produce young. This may take days or years depending on the animal. Similarly, new plant introductions must grow to maturity, produce flowers, and set seed. Some annual plants may do this in less than a year, while some large trees may take several years of growth before they produce flowers.

In organisms that take a long time to mature and produce young, such as elephants, deer, and many kinds of plants, the lag phase may be measured in years. With the mice in our example, it will be measured in weeks. The first litter of young will be able to reproduce in a matter of weeks. Furthermore, the original parents will probably produce an additional litter or two during this time period. Now we have several pairs of mice reproducing more than just once. With several pairs of mice reproducing, natality increases while mortality remains low; therefore, the population begins to grow at an ever-increasing (accelerating) rate. This portion of the population growth curve is known as the **exponential growth phase.**

The number of mice (or any other organism) cannot continue to increase at a faster and faster rate because, eventually, something in the environment will become limiting and cause an increase in the number of deaths. For animals, food, water, or nesting sites may be in short supply, or predators or disease may kill many individuals. Plants may lack water, soil nutrients, or sunlight. Eventually, the number of individuals entering the population will come to equal the number of individuals leaving it by death or migration, and the population size becomes stable. Often there is both a decrease in natality and an increase in mortality at this point. This portion of the population growth curve is known as the **stable equilibrium phase.** It is important to recognize that this is still a population with births, deaths, migration, and a changing mix of individuals; however, the size of the population is stable.

POPULATION-SIZE LIMITATIONS

Populations cannot continue to increase indefinitely; eventually, some factor or set of factors acts to limit the size of a population, leading to the development of a stable equilibrium phase or even to a reduction in population size. The identifiable factors that prevent unlimited population growth are known as **limiting factors.** All of the different limiting factors that act on a population

FIGURE 25.33

A number of factors in the environment, such as food, oxygen supply, diseases, predators, and space determine the number of organisms that can survive in a given area—the carrying capacity of that area. The environmental factors that limit populations are collectively known as environmental resistance.

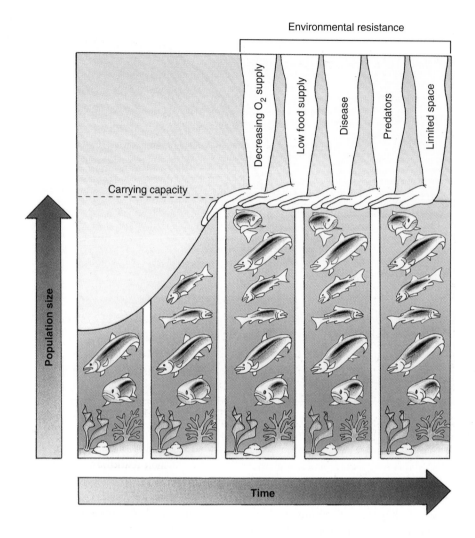

are collectively known as **environmental resistance,** and the maximum population that an area can support is known as the carrying capacity of the area. In general, organisms that are small and have short life spans tend to have fluctuating populations and do not reach a **carrying capacity,** whereas large organisms that live a long time tend to reach an optimum population size that can be sustained over an extended period (figure 25.33). A forest ecosystem contains populations of many insect species that fluctuate widely and rarely reach a carrying capacity, but the number of specific tree species or large animals such as owls or deer is relatively constant. Each is at the carrying capacity of the ecosystem for its species.

Carrying capacity is not an inflexible number, however. Often such environmental differences as successional changes, climate variations, disease epidemics, forest fires, or floods can change the carrying capacity of an area for specific species. In aquatic ecosystems one of the major factors that determine the carrying capacity is the amount of nutrients in the water. In areas where nutrients are abundant, the numbers of various kinds of organisms are high. Often nutrient levels fluctuate with

changes in current or runoff from the land, and plant and animal populations fluctuate as well. In addition, a change that negatively affects the carrying capacity for one species may increase the carrying capacity for another. For example, the cutting down of a mature forest followed by the growth of young trees increases the carrying capacity for deer and rabbits, which use the new growth for food, but decreases the carrying capacity for squirrels, which need mature, fruit-producing trees as a source of food and old, hollow trees for shelter.

Wildlife management practices often encourage modifications to the environment that will increase the carrying capacity for the designated game species. The goal of wildlife managers is to have the highest sustainable population available for harvest by hunters. Typical habitat modifications include creating water holes, cutting forests to provide young growth, and encouraging the building of artificial nesting sites.

In some cases the size of the organisms in a population also affects the carrying capacity. For example, an aquarium of a certain size can support only a limited number of fish, but the size of the fish makes a difference. If all the fish are tiny, a large num-

A CLOSER LOOK | Government Policy and Population Control in China

The actions of government can have a significant impact on the population growth pattern of a nation. Some countries have policies that encourage couples to have children. The U.S. tax code indirectly encourages births by providing tax advantages to the parents of children. Some countries in Europe are concerned about the lack of working-age people in the future and are considering ways to encourage births.

China has long been the most populated country in the world. Its current population is over 1.2 billion people, more than 21 percent of the world's people. This number would be much larger if the Chinese government had not instituted stringent policies of population control. When the People's Republic of China was established in 1949, the population was about 540 million. The official policy of the government at that time was to encourage births because more Chinese would be able to produce more goods and services, and production was the key to economic prosperity.

Because of its high birthrate and falling deathrate, the population increased to 614 million by 1955. However, this rapid increase was coupled with a lack of economic growth, which led government officials to change the policy toward population growth. Abortions became legal in 1953, and the first family planning program began in 1955 as a means of improving maternal and child health. The failure of several government development and agricultural programs during the Great Leap Forward resulted in widespread famine and increased deathrates and low birthrates in the late 1950s and early 1960s.

The present family planning policy began in 1971 with the launching of the *wan xi shao* campaign. Translated, this phrase means "later" (marriages), "longer" (intervals between births), and "fewer" (children). As part of this program the legal ages for marriage were raised. For women and men in rural areas, the ages were raised to 23 and 25, respectively; for women and men in urban areas, the ages were raised to 25 and 28, respectively. These policies resulted in a reduction of birthrates by nearly 50 percent between 1970 and 1979.

An even more restrictive one-child campaign was begun in 1978–1979. The program offered incentives for couples to restrict their family size to one child. Couples enrolled in the program would receive free medical care, cash bonuses for their work, special housing treatment, and extra old-age benefits. Those who broke their pledge were penalized by the loss of these benefits as well as other economic penalties. By the mid-1980s less than 20 percent of the eligible couples were signing up for the program. Rural couples particularly desired more than one child. In fact, in a country where about 70 percent of the population is rural, the rural total fertility rate was 2.5 children per woman. (The total fertility rate is the number of children born per woman per lifetime.) In 1988 a second child was sanctioned for rural couples if their first child was a girl, which legalized what had been happening anyway.

However, the programs appear to have had an effect because the current total fertility rate has fallen to 1.8 children per woman. Replacement fertility, the total fertility rate at which the population would eventually stabilize, is 2.1 children per woman per lifetime. Furthermore, over 80 percent of couples use contraception. Abortion is also an important aspect of this program, with a ratio of more than 600 abortions per 1,000 live births. However, because a large proportion of the population is under 30 years of age, the Chinese population is still expected to double in about 67 years.

---

ber can be supported, and the carrying capacity is high; however, the same aquarium may be able to support only one large fish. In other words, the biomass of the population makes a difference (figure 25.34). Similarly, when an area is planted with small trees, the population size is high. But as the trees get larger, competition for nutrients and sunlight becomes more intense, and the number of trees declines while the biomass increases.

HUMAN POPULATION GROWTH

Today we hear differing opinions about the state of the world's human population. On one hand we hear that the population is growing rapidly. By contrast we hear that some countries are afraid that their populations are shrinking. Other countries are concerned about the aging of their populations because birthrates and deathrates are low. In magazines and on television we see that there are starving people in the world. At the same time we hear discussions about the problem of food sur-

pluses and obesity in many countries. Some have even said that the most important problem in the world today is the rate at which the human population is growing; others maintain that the growing population will provide markets for goods and be an economic boon. How do we reconcile this mass of conflicting information?

It is important to realize that human populations follow the same patterns of growth and are acted upon by the same kinds of limiting factors as are populations of other organisms. When we look at the curve of population growth over the past several thousand years, estimates are that the human population remained low and constant for thousands of years but has increased rapidly in the past few hundred years (figure 25.35). For example, it has been estimated that when Columbus discovered America, the Native American population was about 1 million and was at or near the carrying capacity. Today, the population of North America is about 300 million people. Does this mean that humans are different from other animal species? Can the human population continue to grow forever?

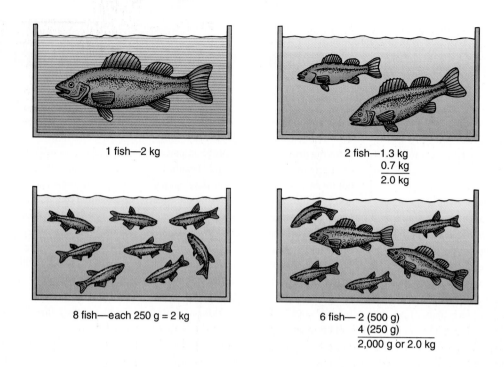

FIGURE 25.34

Each aquarium can support a biomass of 2 kilograms of fish. The size of the population is influenced by the body size of the fish in the population.

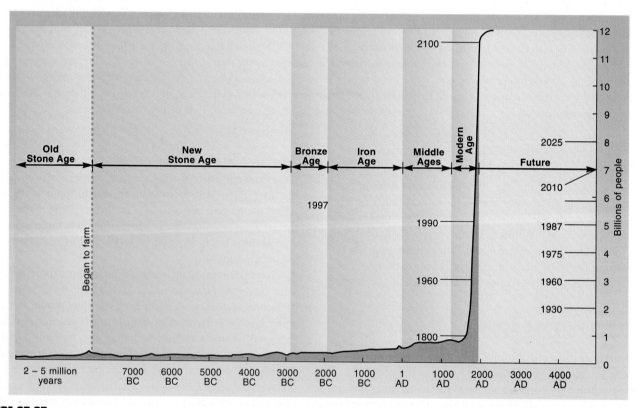

FIGURE 25.35

The number of humans doubled from A.D. 1800 to 1930 (from 1 billion to 2 billion), doubled again by 1975 (4 billion), and is projected to double again by the year 2025. How long can the human population continue to double before the earth's ultimate carrying capacity is reached?

Source: Data from Jean Van Der Tak et al., "Our Population Predicament: A New Look," in *Population Bulletin,* vol. 34, no. 5, December 1979, Population Reference Bureau, Washington, D.C.

The human species is no different from other animals. It has an upper limit set by the carrying capacity of the environment, but the human population has been able to increase astronomically because technological changes and displacement of other species have allowed us to shift the carrying capacity upward. Much of the exponential growth phase of the human population can be attributed to the removal of diseases, improvement in agricultural methods, and replacement of natural ecosystems with artificial agricultural ecosystems. But even these conditions have their limits. There will be some limiting factors that eventually cause a leveling off of our population growth curve. We cannot increase beyond our ability to get raw materials and energy, nor can we ignore the waste products we produce or the other organisms with which we interact.

To many of us, raw materials consist simply of the amount of food available, but we should not forget that in a technological society, iron ore, lumber, irrigation water, and silicon chips are also raw materials. However, most people of the world have much more basic needs. For the past several decades, large portions of the world's population have not had enough food. Although it is biologically accurate to say that the world can currently produce enough food to feed all the people of the world, there are many reasons why people can't get food or won't eat it. Many cultures have food taboos or traditions that prevent the use of some available food sources. For example, pork is forbidden in some cultures. Certain groups of people find it almost impossible to digest milk. Some African cultures use a mixture of cow's milk and cow's blood as food, which people of other cultures might be unable to eat.

In addition, there are complex political, economic, and social issues related to the production and distribution of food. In some cultures, farming is a low-status job, which means that people would rather buy their food from someone else than grow it themselves. This can result in underutilization of agricultural resources. Food is sometimes used as a political weapon when governments want to control certain groups of people. But probably most important is the fact that transportation of food from centers of excess to centers of need is often very difficult and expensive.

A more fundamental question is whether the world can continue to produce enough food. In 1999 the world population was growing at a rate of 1.4 percent per year. This amounts to more than 149 new people added to the world population every minute, which will result in a doubling of the world population in 49 years. With a continuing increase in the number of mouths to feed, it is unlikely that food production will be able to keep pace with the growth in human population. A primary indicator of the status of the world food situation is the amount of grain produced for each person in the world (per capita grain production). World per capita grain production peaked in 1984. The less-developed nations of the world have a disproportionately large increase in population and a decline in grain production because they are less able to afford costly fertilizer, machinery, and the energy necessary to run the machines and irrigate the land to produce their own grain.

The availability of energy is the second broad limiting factor that affects human populations as well as other kinds of organisms. All species on the earth ultimately depend on sunlight for energy—including the human species. Whether one produces electrical power from a hydroelectric dam, burns fossil fuels, or uses a solar cell, the energy is derived from the sun. Energy is needed for transportation, building and maintaining homes, and food production. It is very difficult to develop unbiased, reasonably accurate estimates of global energy "reserves" in the form of petroleum, natural gas, and coal. Therefore, it is difficult to predict how long these "reserves" might last. We do know, however, that the quantities are limited and that the rate of use has been increasing, particularly in the developed and developing countries.

If the less-developed countries were to attain a standard of living equal to that of the developed nations, the global energy "reserves" would disappear overnight. Since the United States constitutes approximately 4.6 percent of the world's population and consumes approximately 25 percent of the world's energy resources, raising the standard of living of the entire world population to that of the United States would result in a tremendous increase in the rate of consumption of energy and reduce theoretical reserves drastically. Humans should realize there is a limit to our energy resources; we are living on solar energy that was stored over millions of years, and we are using it at a rate that could deplete it in hundreds of years. Will energy availability be the limiting factor that determines the ultimate carrying capacity for humans, or will problems of waste disposal predominate?

One of the most talked-about aspects of human activity is the problem of waste disposal. Not only do we have normal biological wastes, which can be dealt with by decomposer organisms, but we generate a variety of technological wastes and by-products that cannot be efficiently degraded by decomposers. Most of what we call pollution results from the waste products of technology. The biological wastes usually can be dealt with fairly efficiently by building waste-water treatment plants and other sewage facilities. Certainly these facilities take energy to run, but they rely on decomposers to degrade unwanted organic matter to carbon dioxide and water. Earlier in this chapter we discussed the problem that bacteria and yeasts face when their metabolic waste products accumulate. In this situation, the organisms so "befoul their nest" that their wastes poison them. Are humans in a similar situation on a much larger scale? Are we dumping so much technological waste, much of which is toxic, into the environment that we are being poisoned? Some people believe that disregard for the quality of our environment will be a major factor in decreasing our population growth rate. In any case, it makes good sense to do everything possible to stop pollution and work toward cleaning our nest.

The fourth category of limiting factors that determine carrying capacity is interaction among organisms. Humans interact with other organisms in as many ways as other animals do. We have parasites and occasionally predators. We are predators in relation to a variety of animals, both domesticated and wild. We have mutualistic relationships with many of our domesticated plants and animals, since they could not survive without our agricultural practices and we would not survive without the food they provide. Competition is also very important. Insects and rodents compete for the food we raise, and we compete directly with many kinds of animals for the use of ecosystems.

CONNECTIONS

Problems in Predicting Future Population Size

Predicting the course of future population growth of a country is extremely difficult, because many factors influence population growth rates. All other things being equal, if a country has a total fertility rate of 2.1 children per woman the population of the country should stabilize. The total fertility rate is the number of children born per woman over her entire reproductive lifetime. However, there are several factors that may cause a population to grow even though the total fertility rate is 2.1 or less.

1. If the deathrate of a country falls dramatically, then people are living longer and are not leaving the population. There-

fore, even low numbers of births will still result in an increasing population.

2. If a very large proportion of the population is young, and if many of these young people have 1 or 2 children, the population will likely grow rapidly. This is because the proportion of the population approaching death is small and the number who are in their reproductive years is large.

3. In addition to natural reproduction, immigration of new people into a country adds to the population.

CONNECTIONS

Ecologist versus Economist

Both economists and ecologists look at human populations and the resources required to satisfy the population. However, their views are often quite different from one another. The chart below compares several of the major differences in their ways of think-

ing. Although there are major differences of opinion between these two fields of study, it is possible to understand and appreciate the significance of both points of view.

Ecologist	Economist	Example
Resources are finite and we will eventually run out of a key resource, which will limit the population.	When resources become limiting, the price rises and a less costly substitute is found as a replacement.	1. Native American populations were limited by the amount of game animals available. 2. Agricultural practices allowed for more food to be grown on the same land.
Increasing population is a problem that will lead to widespread famine and disease.	Increasing population provides both increased amounts of labor and increased markets for goods and services.	1. The Chinese population is the largest in the world and has regularly suffered from famine. 2. The most rapidly growing portion of the world economy is China and other Asian countries.
Food is in short supply in many parts of the world and this will become a greater problem.	There is adequate food in the world but it is not being distributed to the people who need it because they lack the money to buy it. Furthermore, there will be increases in production of food as new technology is deployed.	1. Many people in Africa are malnourished because land is often not suitable for raising crops. 2. Irrigation and the use of fertilizer and pesticides can be used to increase crop yield.
Natural ecosystems are being destroyed with the loss of many species.	Natural ecosystems are resources that should be exploited for the benefit of the human population.	1. Elephants are endangered in many areas because their habitat has been modified by human activity. 2. Elephants are resources that should be used to provide benefit to people. Selling ivory would be an example.

As humans convert more and more land to agricultural and other purposes, many other organisms are displaced. Many of these displaced organisms are not able to compete successfully and must leave the area, have their populations reduced, or become extinct. The American bison (buffalo), African and Asian elephants, the panda, and the grizzly bear are a few species that are much reduced in number because they were not able to compete successfully with the human species. The passenger pigeon, Carolina parakeet, and great auk are a few that have become extinct. Our parks and natural areas have become tiny refuges for plants and animals that once occupied vast expanses of the world. If these refuges are lost, many organisms will become extinct. What today might seem to be an insignificant organism that we can easily do without may tomorrow be seen as a link to our very survival. We humans have been extremely successful in our efforts to convert ecosystems to our own uses at the expense of other species.

Competition with one another (intraspecific competition), however, is a different matter. Since competition is negative to both organisms, competition between humans harms humans. We are not displacing another species, we are displacing some of our own kind. Certainly, when resources are in short supply, there is competition. Unfortunately, it is usually the young that are least able to compete, and high infant mortality is the result.

Humans are different from most other organisms in a fundamental way: We are able to predict the outcome of a specific course of action. Current technology and medical knowledge are available to control human population and improve the health and well-being of the people of the world. Why then does the human population continue to grow, resulting in human suffering and stressing the environment in which we live? Because we are social animals that have freedom of choice, we frequently do not do what is considered "best" from an unemotional, unselfish, biological point of view. People make decisions based on historical, social, cultural, ethical, and personal considerations. What is best for the population as a whole may be bad for you as an individual. The biggest problems associated with control of the human population are not biological problems but rather require the efforts of philosophers, theologians, politicians, and sociologists. As population increases, so will political, social, and biological problems; there will be less individual freedom, there will be intense competition for resources, and famine and starvation will become even more common. The knowledge and technology necessary to control the human population are available, but the will is not. What will eventually limit the size of our population? Will it be lack of resources, lack of energy, accumulated waste products, competition among ourselves, or rational planning of family size?

SUMMARY

Ecology is the study of how organisms interact with their environment. The *environment* consists of *biotic* and *abiotic* components that are interrelated in an *ecosystem.* All *ecosystems* must have a constant *input of energy* from the sun. *Producer organisms* are capable of trapping the sun's energy and converting it into *biomass.* *Herbivores* feed on producers and are in turn eaten by *carnivores,* which may be eaten by *other carnivores.* Each

level in the *food chain* is known as a *trophic level.* Other kinds of organisms involved in food chains are *omnivores,* which eat both plant and animal food, and *decomposers,* which break down dead organic matter and waste products. All *ecosystems* have a large *producer* base with successively smaller amounts of energy at the *herbivore, primary carnivore,* and *secondary carnivore trophic levels.* This is because each time *energy* passes from *one trophic level* to the next, about 90 percent of the energy is lost from the *ecosystem.* A *community* consists of the interacting *populations* of *organisms* in an area. The organisms are interrelated in many ways in food chains that interlock to create *food webs.* Because of this interlocking, changes in one part of the *community* can have effects elsewhere.

Major land-based regional *ecosystems* are known as *biomes.* The *temperate deciduous forest, coniferous forest, tropical rainforest, desert, savanna,* and *tundra* are examples of biomes. *Ecosystems* go through a series of predictable changes that lead to a relatively stable collection of plants and animals. This stable unit is called a *climax community,* and the process of change is called *succession.*

Humans use *ecosystems* to provide themselves with necessary food and raw materials. As the *human population* increases, most people will be living as *herbivores* at the *second trophic level* because we cannot afford to lose 90 percent of the energy by first feeding it to a herbivore, which we then eat. Humans have converted most *productive ecosystems* to *agricultural production* and continue to seek more agricultural land as population increases.

Each organism in a community occupies a specific space known as its *habitat* and has a specific functional role to play known as its *niche.* An *organism's habitat* is usually described in terms of some conspicuous element of its surroundings. The *niche* is difficult to describe because it involves so many interactions with the physical environment and other living things.

Interactions between organisms fit into several categories. *Predation* is one organism benefiting (*predator*) at the expense of the organism killed and eaten (*prey*). *Parasitism* is one organism benefiting (*parasite*) by living in or on another organism (*host*) and deriving nourishment from it. Organisms that carry *parasites* from one host to another are called *vectors.* *Commensal* relationships exist when one organism is helped but the other is not affected. *Mutualistic* relationships benefit both organisms. *Symbiosis* is any interaction in which two organisms live together in a close physical relationship. *Competition* causes harm to both of the organisms involved, although one may be harmed more than the other and may become extinct, evolve into a different niche, or be forced to migrate.

Many atoms are cycled through ecosystems. The *carbon atoms* of living things are trapped by *photosynthesis,* passed from organism to organism as food, and released to the atmosphere by *respiration. Nitrogen* originates in the atmosphere, is trapped by *nitrogen-fixing bacteria,* passes through a series of organisms, and is ultimately released to the atmosphere by *denitrifying bacteria.*

A *population* is a group of organisms of the same species in a particular place at a particular time. *Populations* differ from one another in *gene frequency, age distribution, sex ratio,* and *population density.* A typical population growth curve consists of a *lag phase* in which population rises very slowly, followed by an *exponential growth phase* in which the population increases at an accelerating rate, followed by a leveling off of the population in a *stable equilibrium phase* as the *carrying capacity* of the environment is reached. In some *populations,* a fourth phase may occur, known as the *death phase.*

Humans as a species have the same limits and influences that other organisms do. Our current problems of *food production, energy needs, pollution,* and *habitat destruction* are outcomes of *uncontrolled population growth.* However, humans can reason and predict, thus offering the possibility of population control through *conscious population limitation.*

KEY TERMS

abiotic factors (p. **658**)

biotic factors (p. **658**)

carnivore (p. **661**)

carrying capacity (p. **686**)

commensalism (p. **676**)

community (p. **663**)

competition (p. **678**)

consumer (p. **659**)

decomposers (p. **661**)

denitrifying bacteria (p. **680**)

ecology (p. **658**)

ecosystem (p. **659**)

environment (p. **658**)

environmental resistance
(p. **686**)

epiphytes (p. **676**)

exponential growth phase
(p. **685**)

external parasite (p. **675**)

food chain (p. **661**)

food web (p. **663**)

free-living nitrogen-fixing
bacteria (p. **680**)

gene frequency (p. **682**)

habitat (p. **671**)

herbivore (p. **661**)

host (p. **674**)

internal parasite (p. **675**)

lag phase (p. **685**)

limiting factors (p. **685**)

mortality (p. **685**)

mutualism (p. **677**)

natality (p. **685**)

niche (p. **671**)

nitrifying bacteria (p. **680**)

omnivore (p. **661**)

parasite (p. **674**)

parasitism (p. **674**)

population (p. **663, 682**)

population growth curve
(p. **684**)

predation (p. **674**)

predator (p. **674**)

prey (p. **674**)

primary carnivore (p. **661**)

primary consumer (p. **661**)

producer (p. **659**)

secondary carnivore (p. **661**)

secondary consumer (p. **661**)

stable equilibrium phase
(p. **685**)

symbiosis (p. **678**)

symbiotic nitrogen-fixing
bacteria (p. **680**)

trophic level (p. **659**)

vector (p. **675**)

APPLYING THE CONCEPTS

1. As energy is passed from one trophic level to the next in a food chain, about _____ percent is lost at each transfer.
   a. 10
   b. 50
   c. 75
   d. 90

2. The primary factor that determines whether a geographic area will support temperate deciduous forest or prairie is
   a. the amount of rainfall.
   b. the severity of the winters.
   c. the depth of the soil.
   d. the kinds of animals present.

3. The concept of a community differs from that of an ecosystem in that
   a. a community includes plants and animals but not decomposers, whereas an ecosystem includes all of these.
   b. a community does not include abiotic factors and an ecosystem does.
   c. a community is larger than an ecosystem.
   d. a community cannot be organized in a pyramid and an ecosystem can.

4. Which one of the following organisms is most likely to be at the second trophic level?
   a. a maple tree
   b. a snake
   c. a fungus
   d. an elephant

5. Which one of the following is typical for organisms in a food chain?
   a. Most organisms will use only one species of organism as a source of food.
   b. Most organisms occupy 3–4 different trophic levels depending on what they happen to be eating at the time.
   c. Most organisms will be involved in several different food chains.
   d. Most organisms will always feed at the producer level.

6. Tree → insect → spider → frog → fish → human
   In the food chain above, the spider is
   a. a tertiary consumer.
   b. a herbivore.
   c. at the third trophic level.
   d. a producer.

7. If the first trophic level contains 10,000 units of energy, the _____ trophic level will contain _____ units of energy.
   a. fifth; 10
   b. second; 9,000
   c. third; 100
   d. second; 5,000

8. Which of the following is biotic?
   a. water
   b. bacteria
   c. pH
   d. energy

9. The biome you would typically find located between temperate deciduous forest and tundra is
   a. prairie.
   b. savanna.
   c. desert.
   d. boreal forest.

10. Which one of the following describes, in part, the niche of a rabbit?
    a. the wind in the area where it lives
    b. the golf course on which it lives
    c. rabbits are eaten by coyotes
    d. sunlight

11. An epiphyte is in a _____ relationship.
    a. commensal
    b. parasitic
    c. competitive
    d. mutualistic

12. If two species of organisms occupy the same niche,
    a. mutualism will result.
    b. competition will be very intense.
    c. both organisms will become extinct.
    d. both will need to enlarge their habitat.

13. Mutualism, parasitism, and commensalism are all examples of
    a. nitrogen-fixing bacteria.
    b. symbiosis.
    c. habitats.
    d. competition.

14. If nitrogen-fixing bacteria were to become extinct
    a. life on earth would stop immediately since there would be no source of nitrogen.
    b. life on earth would continue indefinitely.
    c. life on earth would slowly dwindle as useful nitrogen became less available.
    d. life on earth would be unchanged except that proteins would be less common.

15. Nitrogen of the organic molecules in your body comes from
    a. the air you breath.
    b. the water you drink.
    c. carbohydrates in the food you eat.
    d. proteins in the food you eat.

16. Nitrogen is returned to the atmosphere as N_2 by
    a. plants.
    b. nitrogen-fixing bacteria.
    c. denitrifying bacteria.
    d. nitrifying bacteria.

17. The habitat of an earthworm is
    a. the planet Earth.
    b. the forest ecosystem.
    c. topsoil.
    d. eating dead organic matter.

18. Many plants (flowers) provide nectar for insects. The insects, in turn, pollinate the flower. This relationship between the insect and plant represents
    a. parasitism.
    b. commensalism.
    c. mutualism.
    d. predation.

19. Mosquitoes do not cause malaria, but carry and transfer the organism that does cause malaria. Mosquitoes in this instance are playing the role of a(n)
    a. vector.
    b. predator.
    c. epiphyte.
    d. competitor.

20. As population density increases, which one of the following is likely to occur?
    a. Natality will increase.
    b. Mortality will decrease.
    c. The population will experience exponential growth.
    d. Individuals will migrate from the area of highest density.

21. Density-independent limiting factors
    a. increase in intensity as the population increases.
    b. are unrelated to population size.
    c. usually influence the size of populations of large animals.
    d. never affect population size.

22. Which one of the following is *not* the direct result of increasing human population?
    a. extinction of some kinds of animals
    b. increased standard of living
    c. decreased availability of energy
    d. pollution

23. Populations of organisms that are in small, confined situations often have their population limited by
    a. the production of their own wastes.
    b. an inability to reproduce.
    c. reduced biotic potential.
    d. increased energy input.

24. A population made up primarily of prereproductive individuals will
    a. increase rapidly in the future.
    b. become extinct.
    c. rarely occur.
    d. remain stable for several generations.

25. Some organisms have a low reproductive capacity but are very successful because
    a. they have a short life span.
    b. they have high mortality.
    c. most of their offspring live.
    d. they do not compete with other organisms.

26. Which one of the following is an extrinsic limiting factor?
    a. fights between males over females
    b. competition over food
    c. death due to unusual weather
    d. increased sexual activity

27. Listed below are the sex ratios for four populations. All other things being equal (including current population size), which population should experience the greatest future growth?
    a. 1 male: 1 female
    b. 2 male: 1 female
    c. 1 male: 2 female
    d. 3 male: 1 female

28. Listed below are the natality and mortality numbers for four populations. All other things being equal, which population will experience the greatest growth?
    a. natality = 26/1,000; mortality = 17/1,000
    b. natality = 19/1,000; mortality = 8/1,000
    c. natality = 13/1,000; mortality = 25/1,000
    d. natality = 11/1,000; mortality = 5/1,000

29. Mortality exceeds natality during the _____ phase.
    a. death
    b. lag
    c. stable equilibrium
    d. exponential growth

30. Which one of the following is an example of an exponential increase:
    a. 100, 200, 300, 400, 500
    b. 1, 2, 3, 4, 5
    c. 2, 4, 8, 16, 32
    d. 5, 10, 15, 20, 25

31. The current human population of the world is experiencing
    a. a population decline.
    b. slow steady growth.
    c. stable equilibrium.
    d. exponential growth.

32. Increasing a population's food supply may directly increase
    a. environmental resistance on the population.
    b. the carrying capacity for the population.
    c. the population's mortality.
    d. population pressure.

33. When yeasts ferment the sugar in grape juice, they produce ethyl alcohol. When alcohol concentration reaches a certain level, the yeast population declines and eventually dies. In this example, population growth of the yeast is stopped by
    a. limited space.
    b. limited food supply.
    c. accumulation of waste.
    d. disease.

Answers

1. d 2. a 3. b 4. d 5. c 6. c 7. c 8. c 9. d 10. c 11. a 12. b 13. b 14. c 15. d 16. c 17. c 18. c 19. a 20. d 21. b 22. b 23. a 24. a 25. c 26. c 27. c 28. b 29. a 30. c 31. d 32. b 33. c

QUESTIONS FOR THOUGHT

1. Why are rainfall and temperature important in an ecosystem?
2. Describe the flow of energy through an ecosystem.
3. What is the difference between the terms *ecosystem* and *environment*?

4. What role does each of the following play in an ecosystem: sunlight, plants, the second law of thermodynamics, consumers, decomposers, herbivores, carnivores, and omnivores?
5. Give an example of a food chain.
6. What is meant by the term *trophic level*?
7. Why is there usually a larger herbivore biomass than a carnivore biomass?
8. List a predominant abiotic factor in each of the following biomes: temperate deciduous forest, coniferous forest, desert, tundra, tropical rainforest, and savanna.
9. Can energy be recycled through an ecosystem? Explain why or why not.
10. What is the difference between an ecosystem and a community?
11. How does primary succession differ from secondary succession?
12. How does a climax community differ from a successional community?
13. Describe your niche.
14. What is the difference between a habitat and a niche?
15. What do parasites, commensal organisms, and mutualistic organisms have in common? How are they different?
16. Describe two situations in which competition may involve combat and two that do not involve combat.
17. Trace the flow of carbon atoms through a community that contains plants, herbivores, decomposers, and parasites.
18. Describe four different roles played by bacteria in the nitrogen cycle.
19. How have past practices of predator control and habitat destruction negatively altered biological communities?
20. List three ways the carbon and nitrogen cycles are similar and three ways they differ.
21. Draw the population growth curve of a yeast culture during the wine-making process. Label the lag, exponential growth, stable equilibrium, and death phases.
22. List four ways in which two populations of the same species could be different.
23. Why do populations grow?
24. List four kinds of limiting factors that help to set the carrying capacity for a species.
25. How do the concepts of biomass and population size differ?
26. Differentiate between density-dependent and density-independent limiting factors. Give an example of each.
27. Differentiate between intrinsic and extrinsic limiting factors. Give an example of each.
28. As the human population continues to grow, what should we expect to happen to other species?
29. How does the population growth curve of humans compare with that of other kinds of animals?

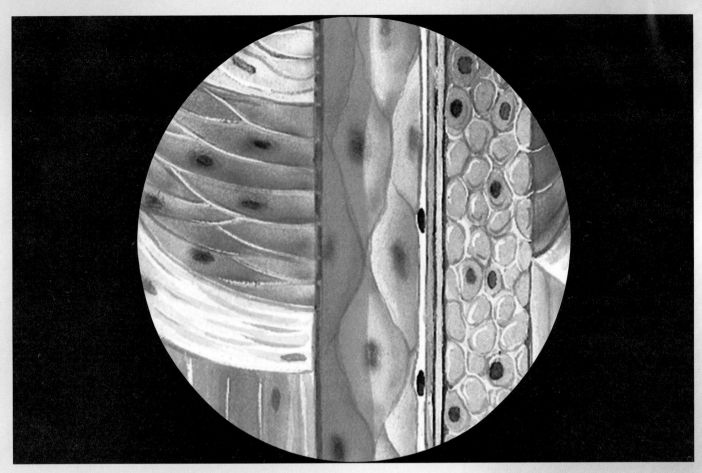

Because many different systems each perform a specific set of tasks in the human body, it is necessary to coordinate these various activities with one another.

CHAPTER | Twenty-Six

Anatomy and Physiology

More and more, maintaining good health has become the responsibility of individuals rather than the duty of health care professionals. This chapter covers some of the major systems of the human organism and how to provide for the health and the well-being of the whole organism.

An understanding of basic **anatomy** (the study of the structure of a body and the relation of its parts) and **physiology** (the study of the function of an organ or organism) allows one to make better decisions about their own health and the health of others. For example, what nutrients we put into our bodies determines how well our various parts will function. Digestion and processing of food materials to release energy are covered here along with muscle structure and function, sexual function and dysfunction, and the nervous and endocrine systems.

SEXUAL REPRODUCTION

The most successful kinds of plants and animals are those that have developed a method of shuffling and exchanging genetic information. In many cases this involves organisms that have two sets of genetic data, one inherited from each parent. **Sexual reproduction** is the formation of a new individual by the union of two sex cells. Before sexual reproduction can occur, the two sets of genetic information must be reduced to one set. This is somewhat similar to shuffling a deck of cards and dealing out hands; the shuffling and dealing assure that each hand will be different. An organism with two sets of chromosomes can produce many combinations of chromosomes when it produces sex cells, just as many different hands can be dealt from one pack of cards. When one of these sex cells unites with another, a new organism containing two sets of genetic information is formed. This new organism's genetic information might have survival advantages over the information found in either parent; this is the value of sexual reproduction.

The cell cycle is a continuous process, without a beginning or an end. The process of mitosis followed by growth is important in the life cycle of any organism. Thus, the cell cycle is part of an organism's *life cycle* (figure 26.1).

The sex cells produced by male organisms are called **sperm** and those produced by females are called **eggs.** A general term sometimes used to refer to either eggs or sperm is the term **gamete** (sex cell). The cellular process that is responsible for generating gametes is called *gametogenesis.* The uniting of an egg and sperm (gametes) is known as *fertilization.*

In many organisms the *zygote,* which results from the union of an egg and a sperm, divides repeatedly by mitosis to form the complete organism. Notice in figure 26.1 that the zygote and its descendants have two sets of chromosomes. However, the male gamete and the female gamete each contain only one set of chromosomes. These sex cells are said to be **haploid.** The haploid number of chromosomes is noted as *n.* A zygote contains two sets and is said to be **diploid.** The diploid number of chromosomes is noted as $2n$ ($n + n = 2n$). Diploid cells have two sets of chromosomes, one set from each parent. Remember, a chromosome is composed of two chromatids, each containing duplex DNA. These two chromatids are attached to each other at a point called the *centromere.* In a diploid nucleus, the chromosomes occur as *homologous chromosomes*—a pair of chromosomes in a diploid cell that contain similar genes throughout their length. One of the chromosomes of a homologous pair was donated by the father, the other by the mother (figure 26.2).

It is necessary for organisms that reproduce sexually to form gametes having only one set of chromosomes. If gametes contained two sets of chromosomes, the zygote resulting from their union would have four sets of chromosomes. The number of chromosomes would continue to double with each new generation, which could result in the extinction of the species. However, this does not usually happen; the number of chromosomes remains constant generation after generation. Since cell division by mitosis and cytokinesis results in cells that have the same number of chromosomes as the parent cell, two questions arise: How are sperm and egg cells formed, and how do they get only half the chromosomes of the diploid cell? The answers lie in the process of **meiosis,** the specialized pair of cell divisions that reduce the chromosome number from diploid ($2n$) to haploid (n). One of the major functions of meiosis is to produce cells that have one set of genetic information. Therefore, when fertilization occurs the zygote will have two sets of chromosomes, as did each parent.

Not every cell goes through the process of meiosis. Only specialized organs are capable of producing haploid cells (figure 26.3). In animals, the organs in which meiosis occurs are called **gonads.** The female gonads that produce eggs are called **ovaries.** The male gonads that produce sperm are called **testes.** Organs that produce gametes are also found in algae and plants. Some of these are very simple. In algae such as *Spirogyra,* individual cells become specialized for gamete production. In plants, the structures are very complex. In flowering plants, the **pistil** produces eggs or ova, and the **anther** produces pollen, which contains sperm.

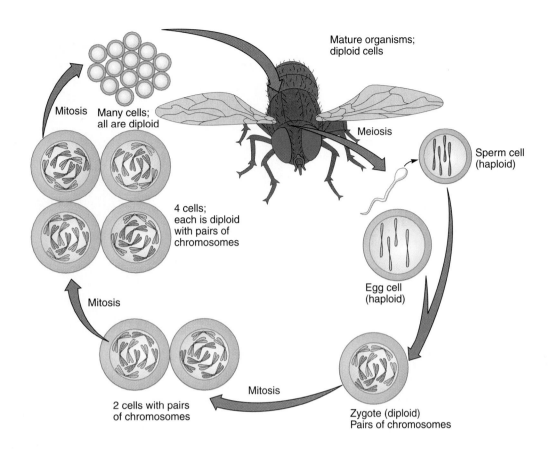

FIGURE 26.1

The cells of this adult fruit fly have eight chromosomes in their nuclei. In preparation for sexual reproduction, the number of chromosomes must be reduced by half so that fertilization will result in the original number of eight chromosomes in the new individual. The offspring will grow and produce new cells by mitosis, completing the life cycle.

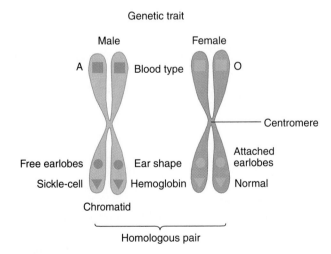

FIGURE 26.2

A pair of chromosomes of similar size and shape that have genes for the same traits are said to be homologous. Notice that the genes may not be identical but code for the same type of information. Homologous chromosomes are of the same length, have the same types of genes in the same sequence, and have their centromeres in the same location—one came from the male parent and the other was contributed by the female parent.

To illustrate meiosis we have chosen to show a cell that has only eight chromosomes (figure 26.4). (In reality, humans have 46 chromosomes, or 23 pairs.) The haploid number of chromosomes in this cell is four, and these haploid cells contain only one complete set of four chromosomes. You can see that there are eight chromosomes in this cell—four from the mother and four from the father. A closer look at figure 26.4 shows you that there are only four types of chromosomes, but two of each type:

1. long chromosomes consisting of chromatids attached at centromeres near the center;
2. long chromosomes consisting of chromatids attached near one end;
3. short chromosomes consisting of chromatids attached near one end; and
4. short chromosomes consisting of chromatids attached near the center.

We can talk about the number of chromosomes in two ways. We can say that our hypothetical diploid cell has eight replicated chromosomes, or we can say that it has four pairs of homologous chromosomes.

Haploid cells, on the other hand, do not have homologous chromosomes. They have one of each type of chromosome. The whole point of meiosis is to distribute the chromosomes and the

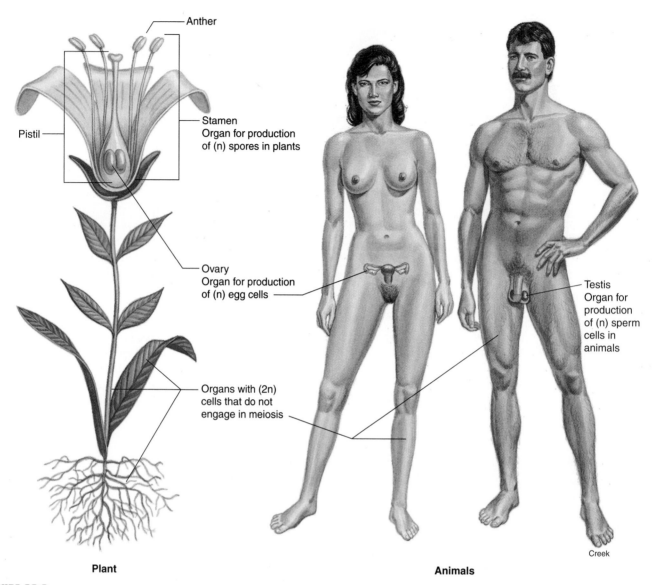

Anther

Pistil

Stamen
Organ for production
of (n) spores in plants

Ovary
Organ for production
of (n) egg cells

Testis
Organ for
production
of (n) sperm
cells in
animals

Organs with (2n)
cells that do not
engage in meiosis

Creek

Plant

Animals

FIGURE 26.3

Both plants and animals produce cells with a haploid number of chromosomes. The male anther in plants and the testes in animals produce haploid male cells, sperm. In both plants and animals, while the ovaries produce haploid female cells, eggs, most other cells are diploid.

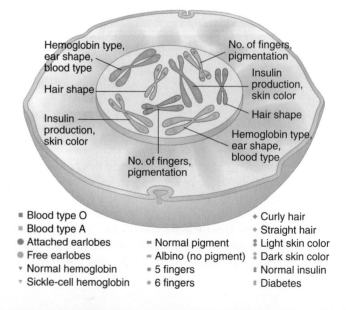

Hemoglobin type,
ear shape,
blood type

No. of fingers,
pigmentation

Hair shape

Insulin
production,
skin color

Insulin
production,
skin color

Hair shape

No. of fingers,
pigmentation

Hemoglobin type,
ear shape,
blood type

FIGURE 26.4

In this diagram of a cell, the eight chromosomes are scattered in the nucleus. Even though they are not arranged in pairs, note that there are four pairs of replicated (each pair consisting of one green and one purple chromosome) homologous chromosomes. Check to be certain you can pair them up using the list of characteristics.

- Blood type O
- Blood type A
- Attached earlobes
- Free earlobes
- Normal hemoglobin
- Sickle-cell hemoglobin

- Normal pigment
- Albino (no pigment)
- 5 fingers
- 6 fingers

- Curly hair
- Straight hair
- Light skin color
- Dark skin color
- Normal insulin
- Diabetes

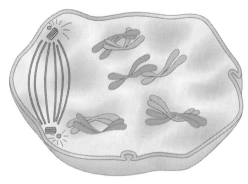

FIGURE 26.5

During prophase I, the cell is preparing for division. A unique event that occurs in prophase I is the synapsis of the chromosomes. Notice that the nuclear membrane is no longer apparent and that the paired homologs are free to move about the cell.

genes they carry so that each daughter cell gets one member of each homologous pair. In this way, each daughter cell gets one complete set of genetic information.

The Mechanics of Meiosis: Meiosis I

Meiosis is preceded by an interphase stage during which DNA replication occurs. In a sequence of events called meiosis I, members of homologous pairs of chromosomes divide into two complete sets. This is sometimes called a **reduction division,** a type of cell division in which daughter cells get only half the chromosomes from the parent cell. The division begins with replicated chromosomes composed of two chromatids. The sequence of events in meiosis I is artificially divided into four phases: prophase I, metaphase I, anaphase I, and telophase I.

Prophase I

During prophase I, the cell is preparing itself for division (figure 26.5). The chromatin material coils and thickens into chromosomes, the nucleoli disappear, the nuclear membrane is disassembled, and the spindle begins to form. The spindle is formed in animals when the centrioles move to the poles. There are no centrioles in plant cells, but the spindle does form. However, there is an important difference between the prophase stage of mitosis and prophase I of meiosis. During prophase I, homologous chromosomes recognize one another by their centromeres, move through the cell toward one another, and come to lie next to each other in a process called *synapsis*. While the chromosomes are synapsed, a unique event called *crossing-over* may occur. **Crossing-over** is the exchange of equivalent sections of DNA on homologous chromosomes. We will fit crossing-over into the whole picture of meiosis later.

Metaphase I

The synapsed pair of homologous chromosomes now moves into position on the equatorial plane of the cell. In this stage, the centromere of each chromosome attaches to the spindle. The synapsed homologous chromosomes move to the equator of the cell as single units. How they are arranged on the equator (which one is on

the left and which one is on the right) is determined by chance (figure 26.6). In the cell in figure 26.6, three green chromosomes from the father and one purple chromosome from the mother are lined up on the left. Similarly, one green chromosome from the father and three purple chromosomes from the mother are on the right. They could have aligned themselves in several other ways. For instance, they could have lined up as shown to in figure 26.6.

Anaphase I

Anaphase I is the stage during which homologous chromosomes separate (figure 26.7). During this stage, the chromosome number is reduced from diploid to haploid. The two members of each pair of homologous chromosomes move away from each other toward opposite poles. The centromeres do not replicate during this phase. The direction each takes is determined by how each pair was originally arranged on the spindle. Each chromosome is independently attached to a spindle fiber at its centromere. Unlike the anaphase stage of mitosis, *the centromeres that hold the chromatids together do not divide during anaphase I of meiosis* (the chromosomes are still in their replicated form). Each chromosome still consists of two chromatids. Because the homologous chromosomes and the genes they carry are being separated from one another, this process is called *segregation*. The way in which a single pair of homologous chromosomes segregates does not influence how other pairs of homologous chromosomes segregate. That is, each pair segregates independently of other pairs. This is known as *independent assortment* of chromosomes.

Telophase I

Telophase I consists of changes that return the cell to an interphase-like condition (figure 26.8). The chromosomes uncoil and become long, thin threads, the nuclear membrane reforms around them, and nucleoli reappear. During this activity, cytokinesis divides the cytoplasm into two separate cells.

Because of meiosis I, the total number of chromosomes is divided equally, and each daughter cell has one member of each homologous chromosome pair. This means that the genetic data each cell receives is one-half of the total, but each cell still has a complete set of the genetic information. Each individual chromosome is still composed of two chromatids joined at the centromere, and the chromosome number is reduced from diploid $(2n)$ to haploid (n). In the cell we have been using as our example, the number of chromosomes is reduced from eight to four. The four pairs of chromosomes have been distributed to the two daughter cells. Depending on the type of cell, there may be a time following telophase I when a cell engages in normal metabolic activity that corresponds to an interphase stage. However, the chromosomes do not replicate before the cell enters meiosis II. Figure 26.9 shows the events in meiosis I.

The Mechanics of Meiosis: Meiosis II

Meiosis II includes four phases: prophase II, metaphase II, anaphase II, and telophase II. The two daughter cells formed during meiosis I continue through meiosis II, so that usually four cells result from the two divisions.

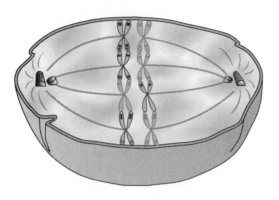

■ Blood type A
■ Blood type O
● Free earlobes
● Attached earlobes
▼ Sickle-cell hemoglobin
▼ Normal hemoglobin
━ Albino (no pigment)
━ Normal pigment
● 6 fingers
● 5 fingers
◆ Straight hair
◆ Curly hair
⸎ Dark skin color
⸎ Light skin color
□ Diabetes
□ Normal insulin

FIGURE 26.6

During metaphase I, notice that the homologous chromosome pairs are arranged on the equatorial plane in the synapsed condition. The cell shows one way the chromosomes could be lined up. A second possible arrangement is shown on the right. How many other ways can you diagram metaphase I?

■ Blood type O
■ Blood type A
● Attached earlobes
● Free earlobes
▼ Normal hemoglobin
▼ Sickle-cell hemoglobin
━ Normal pigment
━ Albino (no pigment)
● 5 fingers
● 6 fingers
◆ Curly hair
◆ Straight hair
⸎ Light skin color
⸎ Dark skin color

FIGURE 26.7

During anaphase I, one member of each homologous pair is segregated from the other member of the pair. Notice that the centromeres of the chromosomes do not replicate.

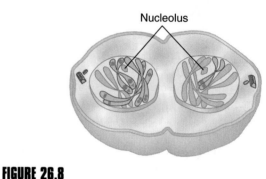

Nucleolus

FIGURE 26.8

What activities would you expect during the telophase I stage of cell division? What term is used to describe the fact that the cytoplasm is beginning to split the parent cell into two daughter cells?

Prophase I   Metaphase I   Anaphase I   Telophase I

FIGURE 26.9

The stages in meiosis I result in reduction division. This reduces the number of chromosomes in the parental cell from the diploid number to the haploid number in each of the two daughter cells.

Prophase II

Prophase II is similar to prophase in mitosis; the nuclear membrane is disassembled, nucleoli disappear, and the spindle apparatus begins to form. However, it differs from prophase I because these cells are haploid, not diploid (figure 26.10). Also, synapsis, crossing-over, segregation, and independent assortment do not occur during meiosis II.

Metaphase II

The metaphase II stage is typical of any metaphase stage because the chromosomes attach by their centromeres to the spindle at the equatorial plane of the cell. Since pairs of chromosomes are no longer together in the same cell, each chromosome moves as a separate unit (figure 26.11).

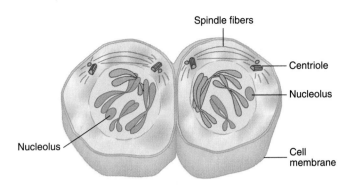

Spindle fibers

Centriole

Nucleolus

Nucleolus

Cell
membrane

FIGURE 26.10

During prophase II, the two daughter cells are preparing for the second division of meiosis. Study this diagram carefully. Can you list the events of this stage?

FIGURE 26.12

Anaphase II is very similar to the anaphase of mitosis. The centromere of each chromosome divides and one chromatid separates from the other. As soon as this happens, we no longer refer to them as chromatids; we now call each strand of nucleoprotein a chromosome.

FIGURE 26.11

During metaphase II, each chromosome lines up on the equatorial plane. Each chromosome is composed of two chromatids (a replicated chromosome) joined at a centromere. How does metaphase II of meiosis compare to metaphase I of meiosis?

FIGURE 26.13

During telophase II, what events would you expect?

Prophase II   Metaphase II   Anaphase II   Telophase II

FIGURE 26.14

During meiosis II, the centromere of each chromosome replicates and each chromosome divides into separate chromatids. Four haploid cells are produced, each having one chromatid of each kind.

Anaphase II

Anaphase II differs from anaphase I because during anaphase II the centromere of each chromosome divides, and the chromatids, now called *daughter chromosomes,* move to the poles as in mitosis (figure 26.12). Remember, there are no paired homologs in this stage; therefore, segregation and independent assortment cannot occur.

Telophase II

During telophase II, the cell returns to a nondividing condition. As cytokinesis occurs, new nuclear membranes form, chromosomes uncoil, nucleoli re-form, and the spindles disappear (fig-

ure 26.13). This stage is followed by differentiation; the four cells mature into gametes—either sperm or eggs. The events of meiosis II are summarized in figure 26.14.

The formation of a haploid cell by meiosis and the combination of two haploid cells to form a diploid cell by sexual reproduction results in variety in the offspring. Five factors influence genetic variation in offspring: mutations, crossing-over, segregation, independent assortment, and fertilization.

In many organisms, egg cells are produced in such a manner that three of the four cells resulting from meiosis in a female disintegrate. However, since the one that survives is randomly chosen, the likelihood of any one particular combination of genes being formed is not affected. The whole point of learning the mechanism of meiosis is to see how variation happens.

Chromosomes and Sex Determination

You already know that there are several different kinds of chromosomes, that each chromosome carries genes unique to it, and that these genes are found at specific places. Furthermore, diploid organisms have homologous pairs of chromosomes. Genes determine sexual characteristics in the same manner as other types of characteristics. In many organisms, sex-determining genes are located on specific chromosomes known as **sex chromosomes.** All other chromosomes not involved in determining the sex of an individual are known as **autosomes.** In humans and all other mammals, and in some other organisms (e.g., fruit flies), the sex of an individual is determined by the presence of a certain chromosome combination. The genes that determine maleness are located on a small chromosome known as the *Y chromosome.* This Y chromosome behaves as if it and another larger chromosome, known as the *X chromosome,* were homologs. Most males have one X and one Y chromosome. Most females have two X chromosomes. Some animals have their sex determined in a completely different way. In bees, for example, the females are diploid and the males are haploid. Other plants and animals have still other chromosomal mechanisms for determining their sex.

Spermatogenesis

One of the biological reasons for sexual activity is the production of offspring. The process of producing gametes includes meiosis and is called **gametogenesis** (gamete formation) (figure 26.15). The term **spermatogenesis** is used to describe gametogenesis that takes place in the testes of males. The two bean-shaped testes are composed of many small sperm-producing tubes, or **seminiferous tubules,** and collecting ducts that store sperm. These are held together by a thin covering membrane. The seminiferous tubules join together and eventually become the epididymis, a long, narrow convoluted tube in which sperm cells are stored and mature before ejaculation (figure 26.16).

Leading from the epididymis is the vas deferens, or sperm duct; this empties into the urethra, which conducts the sperm out of the body through the **penis** (figure 26.17). Before puberty, the seminiferous tubules are packed solid with diploid cells called spermatogonia. These cells, which are found just inside the tubule wall, undergo *mitosis* and produce more spermatogonia. Beginning about age eleven, some of the spermatogonia specialize and begin the process of *meiosis,* while others continue to divide by mitosis, assuring a constant and continuous supply of spermatogonia. Once spermatogenesis begins, the seminiferous tubules become hollow and can transport the mature sperm.

Spermatogenesis involves several steps. Some of the spermatogonia in the walls of the seminiferous tubules differentiate and enlarge to become *primary spermatocytes.* These

diploid cells undergo the first meiotic division, which produces two haploid *secondary spermatocytes.* The secondary spermatocytes go through the second meiotic division, resulting in four haploid *spermatids,* which lose much of their cytoplasm and develop long tails. These cells are then known as **sperm** (figure 26.18). The sperm have only a small amount of food reserves. Therefore, once they are released and become active swimmers, they live no more than 72 hours. However, if the temperature is lowered drastically, the sperm freeze, become deactivated, and can live for years outside the testes. This has led to the development of sperm banks. Artificial insemination of cattle, horses, and other domesticated animals with sperm from sperm banks is common. Human artificial insemination is much less common and is usually related to couples with fertility problems.

Spermatogenesis in human males takes place continuously throughout a male's reproductive life, although the number of sperm produced decreases as a man ages. Sperm counts can be taken and used to determine the probability of successful fertilization. For reasons not totally understood, a man must be able to release at least 100 million sperm at one insemination to be fertile. It appears that enzymes in the head of sperm are needed to digest through mucus and protein found in the female reproductive tract. Millions of sperm contribute in this way to the process of fertilization, but only one is involved in fertilizing the egg. A healthy male probably releases about 300 million sperm during each act of **sexual intercourse,** also known as **coitus** or **copulation.**

Oogenesis

The term **oogenesis** refers to the production of egg cells. This process starts during prenatal development of the ovary, when diploid oogonia cease dividing by mitosis and enlarge to become *primary oocytes.* All of the primary oocytes that a woman will ever have are already formed prior to her birth. At this time they number approximately 2 million, but that number is reduced by cell death to between 300,000 to 400,000 cells by the time of puberty. Oogenesis halts at this point, and all the primary oocytes remain just under the surface of the ovary.

At puberty and on a regular basis thereafter, the sex hormones stimulate a primary oocyte to continue its maturation process, and it goes through the first meiotic division. But in telophase I, the two cells that form receive unequal portions of cytoplasm. You might think of it as a lopsided division (see figure 26.15). The smaller of the two cells is called a *polar body,* and the larger haploid cell is the *secondary oocyte.* The other primary oocytes remain in the ovary. Ovulation begins when the soon-to-be released secondary oocyte, encased in a saclike structure known as a **follicle,** grows and moves near the surface of the ovary. When this maturation is complete, the follicle erupts and the secondary oocyte is released. It is swept into the oviduct (fallopian tube) by ciliated cells and travels toward the uterus (figure 26.19). Because of the action of the luteinizing hormone, the rest of the follicle develops into a glandlike structure, the **corpus luteum,** which produces hormones (progesterone and estrogen) that prevent the release of other secondary oocytes.

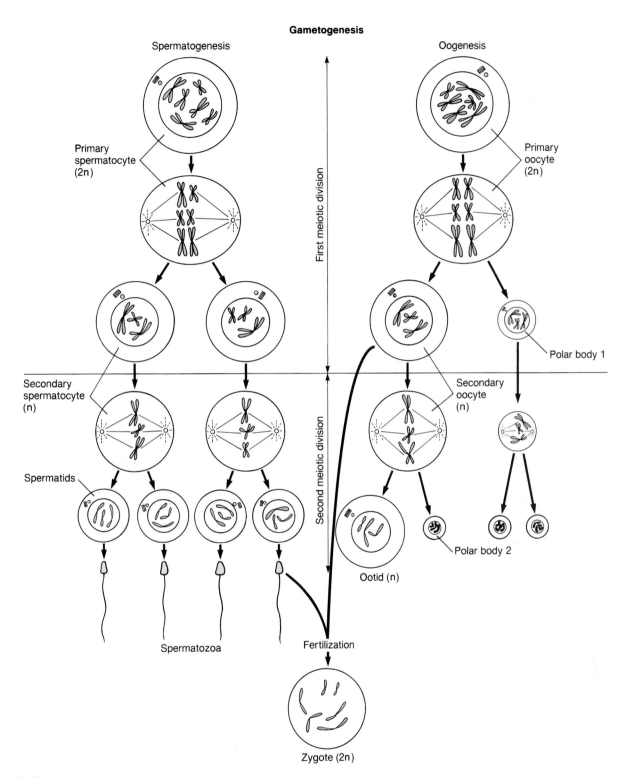

FIGURE 26.15

This diagram illustrates the process of gametogenesis in human males and females. Not all of the 46 chromosomes are shown. Carefully follow the chromosomes as they segregate, recalling the details of the process of meiosis explained previously.

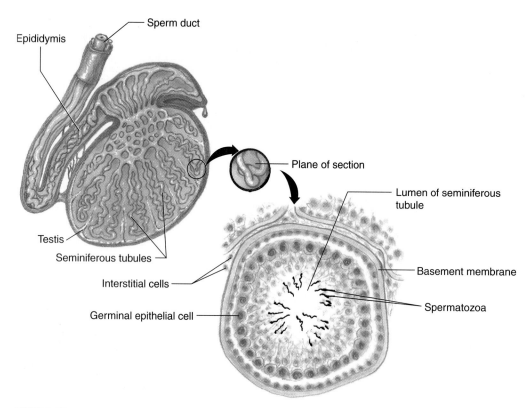

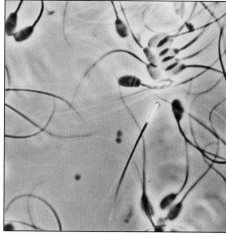

FIGURE 26.16

The testis consists of many tiny tubes called seminiferous tubules. The walls of the tubes consist of cells that continually divide, producing large numbers of sperm. The sperm leave the seminiferous tubules and enter the epididymis, where they are stored prior to ejaculation through the sperm duct.

FIGURE 26.18

These human sperm cells are primarily DNA-containing packages produced by the male.

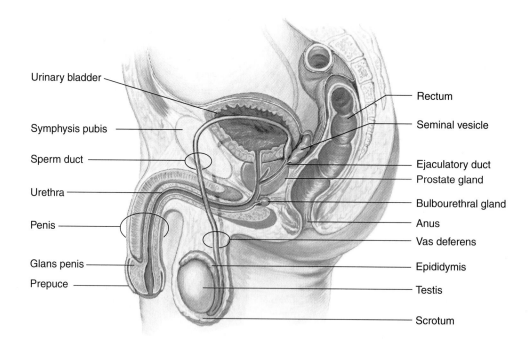

FIGURE 26.17

The male reproductive system consists of two testes that produce sperm, ducts that carry the sperm, and various glands. Muscular contractions propel the sperm through the vas deferens past the seminal vesicles, prostate gland, and bulbourethral gland, where most of the liquid of the semen is added. The semen passes through the urethra of the penis to the outside of the body.

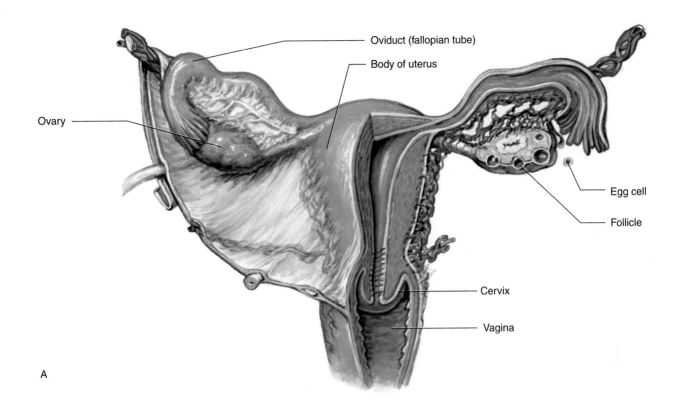

Oviduct (fallopian tube)

Body of uterus

Ovary

Egg cell

Follicle

Cervix

Vagina

A

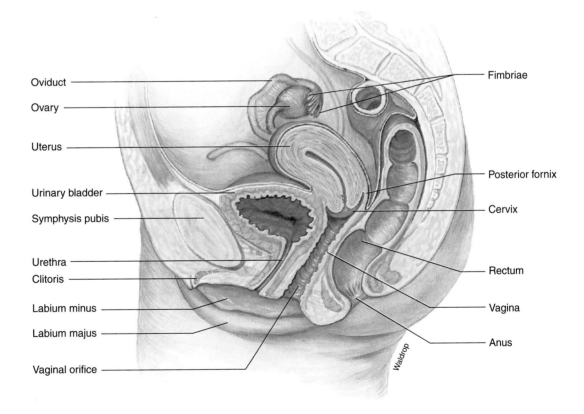

Oviduct

Ovary

Uterus

Urinary bladder

Symphysis pubis

Urethra

Clitoris

Labium minus

Labium majus

Vaginal orifice

Fimbriae

Posterior fornix

Cervix

Rectum

Vagina

Anus

Waldrop

B

FIGURE 26.19

In the human female reproductive system (A) After ovulation, the cell travels down the oviduct to the uterus. If it is not fertilized, it is shed when the uterine lining is lost during menstruation. (B) The human female reproductive system, side view.

A CLOSER LOOK | Sexually Transmitted Diseases

Diseases currently referred to as sexually transmitted diseases (STDs) were formerly called venereal diseases (VDs). The term venereal is derived from the name of the Roman goddess for love, Venus. Although these kinds of illnesses are most frequently transmitted by sexual activity, many can also be spread by other methods of direct contact, by hypodermic needles, by blood transfusions, and by blood-contaminated materials. Currently, the Centers for Disease Control and Prevention in Atlanta, Georgia, recognizes nineteen diseases as being sexually transmitted and a twentieth, gay bowel syndrome, which is actually caused by a great variety of different microorganisms (box table 26.1).

The United States has the highest rate of sexually transmitted diseases among industrially developed countries. The Centers for Disease Control and Prevention estimates there are 12 million new cases of sexually transmitted diseases each year, 3 million among teenagers. Box table 26.2 lists the most common STDs and estimates by the Centers for Disease Control and Prevention of the number of new cases each year. The portions of the public that are most at risk are teenagers, minorities, and women. Some of the most important STDs are described here because of their high incidence in the population and our inability to bring some of them under control. For example, there is no known cure for the HIV virus that is responsible for AIDS. There has also been a sharp rise in the number of gonorrhea cases in the United States caused by a form of the bacterium *Neisseria gonorrhoeae,* which has become resistant to the drug penicillin by producing an enzyme that actually destroys the antibiotic. However, most of the infectious agents can be controlled if diagnosis occurs early and the patient carefully follows treatment programs.

The spread of STDs during sexual intercourse is significantly diminished by the use of condoms. Other types of sexual contact

BOX TABLE 26.1

Sexually transmitted diseases

Disease	Agent
Genital herpes	Virus
Gonorrhea	Bacterium
Syphilis	Bacterium
Acquired immunodeficiency syndrome (AIDS)	Virus
Candidiasis	Yeast
Chancroid	Bacterium
Condyloma acuminatum (venereal warts)	Virus
Gardnerella vaginalis	Bacterium
Genital chlamydia infection	Bacterium
Genital cytomegalovirus infection	Virus
Genital mycoplasma infection	Bacterium
Group B streptococcus infection	Bacterium
Nongonococcal urethritis	Bacterium
Pelvic inflammatory disease (PID)	Bacterium
Reiter's syndrome	Bacterium
Crabs	Body lice
Scabies	Mite
Trichomoniasis	Protozoan
Viral hepatitis (HBV)	Virus
Gay bowel syndrome	Variety of agents

BOX TABLE 26.2

Yearly estimates of the number of new cases of sexually transmitted diseases

Sexually Transmitted Disease	New Cases Each Year (Estimate)
Chlamydia	4 million
Gonorrhea	800,000
Human papilloma virus	500,000 to 1 million
Genital herpes (people who show symptoms)	200,000 to 500,000
Syphilis	101,000
Congenital syphilis	3,400
Human immunodeficiency virus (HIV) (AIDS)	40,000 to 80,000

—*Continued top of next page*

Continued—

(i.e., hand, oral, anal) and congenital transmission (i.e., from the mother to the fetus during pregnancy) help to maintain some of these diseases in the population. These are at high enough levels to warrant attention by public health officials, the United States Public Health Service, the Centers for Disease Control and Prevention, and state and local public health agencies. All of these agencies are involved in attempts to raise the general public health to a higher level. Their investigations have resulted in the successful control of many diseases and the identification of special problems, such as those associated with the STDs. Since the United States has an incidence rate of STDs that is 50 to 100 times higher than other industrially developed countries, there is still much that needs to be done.

All public health agencies are responsible for warning members of the public about things that may be dangerous to them. In order to meet these obligations when dealing with sexually transmitted diseases, such as AIDS and syphilis, they encourage the use of one of their most potent weapons, sex education. Individuals must know about their own sexuality if they are to understand the transmission and nature of STDs. Then it will be possible to alter their behavior in ways that will prevent the spread of these diseases. The intent is to present people with biological facts, not to scare them. Public health officials do not have the luxury of advancing their personal opinions when it comes to their jobs. The biological nature of sexual behavior is not a moral issue, but biological facts are needed if people are to make intelligent decisions relating to their sexual behavior. It is hoped that through education, people will alter their high-risk sexual behaviors and avoid situations where they could become infected with one of the STDs. As one health official stated, we should be knowledgeable enough about our own sexuality and the STDs to answer the question, Is what I'm about to do worth dying for?

If the secondary oocyte is fertilized, it completes meiosis by proceeding through meiosis II with the sperm DNA inside. During the second meiotic division, the secondary oocyte again divides unevenly, so that a second polar body forms. None of the polar bodies survive; therefore, only one large secondary oocyte is produced from each primary oocyte that begins oogenesis. If the cell is not fertilized, the secondary oocyte passes through the **vagina** to the outside during menstruation. During her lifetime, a female releases about 300 to 500 secondary oocytes. Obviously, few of these cells are fertilized.

One of the characteristics to note here is the relative age of the sex cells. In males, sperm production is continuous throughout life. Sperm do not remain in the tubes of the male reproductive system for very long. They are either released shortly after they form or die and are harmlessly absorbed. In females, meiosis begins before birth, but the oogenesis process is not completed, and the cell is not released for many years. A secondary oocyte released when a woman is 37 years old began meiosis 37 years before! During that time, the cell was exposed to many influences, a number of which may have damaged the DNA or interfered with the meiotic process. This has been postulated as a possible reason for the increased incidence of nondisjunction (abnormal meiosis) in older women. Such alterations are less likely to occur in males because new gametes are being produced continuously. Also, defective sperm appear to be much less likely to be involved in fertilization.

Hormones control the cycle of changes in breast tissue, in the ovaries, and in the uterus. In particular, estrogen and progesterone stimulate milk production by the breasts and cause the lining of the uterus to become thicker and more vascularized prior to the release of the secondary oocyte. This ensures that if the secondary oocyte becomes fertilized, the resultant embryo will be able to attach itself to the wall of the uterus and receive nourishment. If the cell is not fertilized, the lining of the uterus is shed. This is known as *menstruation, menstrual flow,* the *menses,* or a *period.* Once the wall of the uterus has been shed, it begins to build up again. This continual building up and shedding of the wall of the uterus is known as the menstrual cycle.

At the same time that hormones are regulating the release of the secondary oocyte and the menstrual cycle, changes are taking place in the breasts. The same hormones that prepare the uterus to receive the embryo also prepare the breasts to produce milk. These changes in the breasts, however, are relatively minor unless pregnancy occurs.

Sexual Maturation of Young Adults

Following birth, sexuality plays only a small part in physical development for several years. Culture and environment shape the responses that the individual will come to recognize as normal behavior. During **puberty,** normally between 12 and 14 years of age, increasing production of sex hormones causes major changes as the individual reaches sexual maturity. Generally females reach puberty six months to a year before males. After puberty, humans are sexually mature and have the capacity to produce offspring.

The Maturation of Females

Female children typically begin to produce quantities of sex hormones from the hypothalamus, pituitary gland, ovaries, and adrenal glands at 8 to 13 years of age. This marks the onset of puberty. The **hypothalamus** controls the functioning of many other glands throughout the body, including the **pituitary gland.** At puberty, the pituitary gland in girls begins to produce **follicle-stimulating hormone (FSH).** The increasing amount of this hormone circulating in the blood causes the ovaries to grow and, under the influence of another hormone, luteinizing

TABLE 26.1   Human reproductive hormones

Hormone	Production Site	Target Organ	Function
1. Prolactin (lactogenic or luteotropic hormone)	Pituitary gland	Breasts, ovary	Stimulates milk production; also helps maintain normal ovarian cycle
2. Follicle-stimulating hormone	Pituitary gland	Ovary, testis	Stimulates ovary and testis development; stimulates egg production in females and sperm production in males
3. Luteinizing hormone (interstitial cell-stimulating hormone)	Pituitary gland	Ovary, testis	Stimulates ovulation in females and sex-hormone (estrogen and testosterone) production in both males and females
4. Estrogen	Follicle of the ovary	Entire body	Stimulates development of female reproductive tract and secondary sexual characteristics
5. Testosterone	Testes	Entire body	Stimulates development of male reproductive tract and secondary sexual characteristics
6. Progesterone	Ovaries	Uterus, breasts	Causes uterine thickening and maturation; maintains pregnancy
7. Oxytocin	Pituitary gland	Breasts, uterus	Causes uterus to contract and breasts to release milk
8. Androgens	Testes, adrenal glands	Entire body	Stimulate development of male reproductive tract and secondary sexual characteristics in males and females

hormone, begin producing larger quantities of **estrogen.** The increasing supply of estrogen is responsible for the many changes in sexual development that can be noted at this time. These changes include breast growth, changes in the walls of the uterus and vagina, increased blood supply to the clitoris, and changes in the pelvic bone structure.

Estrogen also stimulates the female adrenal gland to produce **androgens,** male sex hormones. The androgens are responsible for the production of pubic hair and they seem to have an influence on the female sex drive. The adrenal gland secretions may also be involved in the development of acne. Those features that are not primarily involved in sexual reproduction but are characteristic of a sex are called **secondary sex characteristics.** In women, the distribution of body hair, patterns of fat deposits, and a higher voice are examples.

A major development during this time is the establishment of the **menstrual cycle.** This involves the periodic growth and shedding of the lining of the uterus. These changes are under the control of a number of hormones produced by the ovaries. The ovaries are stimulated to release their hormones by the pituitary gland, which is in turn influenced by the ovarian hormones. Follicle-stimulating hormone (FSH) and luteinizing hormone (LH) are both produced by the pituitary gland. FSH causes the maturation and development of the ovaries, and LH is important in causing ovulation and in maintaining the menstrual cycle. In addition, secretions of the hypothalamus at puberty bring about the development of a menstrual cycle (table 26.1). Associated with the menstrual cycle is the periodic release of sex cells from the surface of the ovary, called **ovulation** (figure 26.20). Initially, these two cycles, menstruation and ovulation, may be irregular, which is normal during puberty. Eventually, hormone production becomes regulated so that ovulation and menstruation take

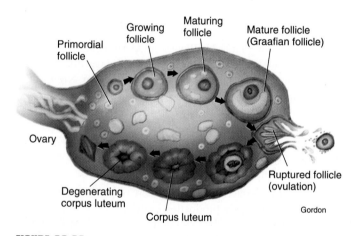

FIGURE 26.20

In the ovary, the egg begins development inside a sac of cells known as a follicle. Each month, one of these follicles develops and releases its product. This release through the wall of the ovary is known as ovulation.

place on a regular monthly basis in most women, although normal cycles may vary from 21 to 45 days.

As girls progress through puberty, curiosity about the changing female body form and new feelings leads to self-investigation. Studies have shown that sexual activity such as manipulation of the clitoris, which causes pleasurable sensations, is performed by a large percentage of young women. Self-stimulation, frequently to orgasm, is a common result. This stimulation is termed **masturbation,** and it should be stressed that it is considered a normal part of sexual development. **Orgasm** is a complex response to mental and physical stimulation that causes rhythmic contractions of the muscles of the reproductive organs and an intense frenzy of excitement.

A CLOSER LOOK | Speculation on the Evolution of Human Sexual Behavior

There has been much speculation about how human sexual behavior evolved. It is important to recognize that this speculation is not fact but an attempt to evaluate human sexual behaviors from an evolutionary perspective.

When we compare human sexuality with that of other mammals, there are several ways in which human sexuality is different from that of most other mammals. While most mammals are sexually active during specific periods of the year, humans may engage in sexual intercourse at any time throughout the year. The sex act appears to be important as a pleasurable activity rather than a purely reproductive act. Associated with this difference is the fact that human females do not display changes that indicate they are releasing eggs (ovulating).

All other female mammals display changes in odor, appearance, or behavior that clearly indicate to the males of the species that the female is ovulating and sexually responsive. This is referred to as "being in heat." This is not true for humans. Human males are unable to differentiate ovulating females from those that are not ovulating.

With few exceptions in other mammals, infants grow to sexual maturity in a year or less. Although extremely long-lived mammals (elephants or whales) do not reach sexual maturity in a year, their young have well-developed muscles that allow them to move about on their own. While not sexually mature, they display a high degree of independence. This is not true for human infants, which are extremely immature when born, develop walking skills slowly, and require an extensive period of training before they are able to function independently. Perhaps the extremely immature condition in which human infants are born is related to human brain size. The size of the head is very large and just fits through the birth canal in the pelvis. One way to accommodate a large brain size and not need to redesign the basic anatomy of the pelvis would be to have the young be born in a very immature condition while the brain is still in the process of grow-ing. The immature condition of human infants is associated with a need to provide extensive care for them.

Raising young requires a considerable investment of time and resources. Females invest considerable resources in the pregnancy itself. Fat stores provide energy necessary to a successful pregnancy. Female mammals, including humans, that have little stored fat often have difficulty becoming pregnant in the first place and also are more likely to die of complications resulting from the pregnancy. Nutritional counseling is an important part of modern prenatal care, because it protects the health of both the mother and developing fetus. The long period of pregnancy in humans requires good nutrition over a long period. Once the child is born the mother continues to require good nutrition, since she provides the majority of food for the infant through her breast milk. As the child grows, other food items are added to its diet. These also require an expenditure of energy on the part of the mother or father or both parents.

With these ideas in mind we can speculate about how human sexual behavior may have evolved. Imagine a primitive stone-age human culture. Females have a great deal invested in each child produced. They will be able to produce only a few children during their lifetime, and many children will die due to malnutrition, disease, and accidents. Those females that have genes that will allow more of their offspring to survive will be selected for. Human males, on the other hand, have very little invested in each child and can impregnate many different females. Males that have many children that survive are selected for. How might these different male and female goals fit together to explain the sexual behaviors we see in humans today?

The males of most mammals provide little toward the raising of young. In many species males only meet with females for mating (deer). Some form short-term pair bonds for one season and may share in raising young. Only a few (wolves) form longer-term bonds lasting for years, in which males and females cooperate in the raising of young. However, pair bonding in humans is usually a long-term relationship. The significance of this relationship can be evaluated from an evolutionary perspective. This long-term pair bond can serve the interests of both males and females. When males form long-term relationships with females, the females gain an additional source of nutrition for their offspring, who will be completely dependent on their parents for food and protection for several years, thus increasing the likelihood that the young will survive. Human males benefit from the long-term pair bond as well. Since human females do not display the fact that they are ovulating, the only way a male can be assured that the child he is raising is his is to have exclusive mating rights with a specific female. The establishment of bonding involves a great deal of sexual activity, much more than is necessary to just bring about reproduction. It is interesting to speculate that sexual behavior in humans is as much involved in maintaining pair bonds as it is in creating new humans.

Much has been written about the differences in sexual behavior between men and women and that men and women look for different things when assessing individuals as potential mates. It is very difficult to distinguish behaviors that are truly biological and those that are culturally determined. However, some differences may have biological roots. Females benefit from bonding with males that have access to resources that are shared in the raising of young. Do women look for financial security and a willingness to share in a mate? Since pregnancy requires a great deal of nutrition, it is in the male's interest to choose a mate that is healthy, young, and in good nutritional condition. Do men look for youth and appropriate amounts of nutritionally important fat stored in the breasts and buttocks? If these differences between men and women really exist, are they purely cultural, or is there an evolutionary input from our primitive ancestors?

The Maturation of Males

Males typically reach puberty about two years later than females, (ages 10 to 15), but puberty in males also begins with a change in hormone levels. At puberty, follicle-stimulating hormone (FSH), the same pituitary-gland secretion produced by females, increases. This is the primary growth stimulator of the testes and ovaries. FSH produced by the male during embryology was responsible for the embryonic development of testes. FSH also stimulates the production of luteinizing hormone, which is often known as **interstitial cell-stimulating hormone (ICSH)** in males, which stimulates the testes to produce **testosterone,** the primary sex hormone in males. The testosterone produced by the embryonic testes caused the differentiation of internal and external genital anatomy in the male embryo. Testosterone is also important in the maturation and production of sperm.

The major changes during puberty include growth of the testes and scrotum, pubic-hair development, and increased size of the penis. Secondary sex characteristics begin to become apparent at age 13 or 14. Facial hair, underarm hair, and chest hair are some of the most obvious. The male voice changes as the larynx (voice box) begins to change shape. Body contours also change and a growth spurt increases height. In addition, the proportion of the body that is muscle increases, while the proportion of body fat decreases. At this time, a boy's body begins to take on the characteristic adult male shape, with broader shoulders and heavier muscles.

In addition to these external changes, the FSH released from the pituitary causes the production of seminal fluid by the **seminal vesicles,** prostate gland, and the bulbourethral glands. Later, FSH stimulates the production of sperm cells. The release of sperm cells and seminal fluid also begins during puberty and is termed **ejaculation.** This release is generally accompanied by the pleasurable sensations of orgasm. The sensations associated with ejaculation may lead to self-stimulation, or masturbation. Masturbation is a common and normal activity as a boy goes through puberty. Studies of sexual behavior have shown that nearly all men masturbate at some time during their lives.

MATERIALS EXCHANGE IN THE BODY

Living things are complex machines with many parts that must function in a coordinated fashion. All systems are integrated and affect one another in many ways. For example, when you run up a hill, your leg and arm muscles move in a coordinated way to provide power. They burn fuel (glucose) for energy and produce carbon dioxide and lactic acid as waste products, which tend to lower the pH of the blood. Your heart beats faster to provide oxygen and nutrients to the muscles, you breathe faster to supply the muscles with oxygen and get rid of carbon dioxide, and the blood vessels in the muscles dilate to allow more blood to flow to them. As you run you generate excess heat. As a result, more blood flows to the skin to get rid of the heat and sweat glands begin to secrete, thus cooling the skin. All of these automatic internal adjustments help the body maintain a constant level of oxygen, carbon dioxide, and glucose in the blood; constant pH; and constant body temperature.

They can be summed up in the concept of *homeostasis.* **Homeostasis** is the process of maintaining a constant internal environment as a result of monitoring and modifying the functioning of various systems. To explore the various mechanisms that help organisms maintain homeostasis, we will begin at the cellular level.

Exchanging Materials: Basic Principles

Cells are highly organized units that require a constant flow of energy in order to maintain themselves. The energy they require is provided in the form of nutrient molecules that enter the cell. Oxygen is required for the efficient release of energy from the large organic molecules that serve as fuel. Inevitably, as oxidation takes place, waste products form that are useless or toxic. These must be removed from the cell. All these exchanges of food, oxygen, and waste products must take place through the cell surface.

The ability to transport materials into or out of a cell is determined by its surface area, while its metabolic demands are determined by its volume. So, the larger a cell becomes, the more difficult it is to satisfy its needs. Some cells overcome this handicap by having highly folded cell membranes that substantially increase their surface area. This is particularly true of cells that line the intestine or are involved in the transport of large numbers of nutrient molecules. These cells have many tiny, folded extensions of the cell membrane called **microvilli** (figure 26.21).

In addition to the limitation that surface area presents to the transport of materials, large cells also have a problem with diffusion. The molecular process of diffusion is quite rapid over short distances but becomes very slow over longer distances. Diffusion is generally insufficient to handle the needs of cells if it must take place over a distance of more than 1 millimeter. The center of the cell would die before it received the molecules it needed if the distance were greater. Because of this, it is understandable that the basic unit of life, the cell, must remain small.

All single-celled organisms are limited to a small body size because they handle the exchange of molecules through their cell membranes. Large, multicellular organisms consist of a multitude of cells, many of which are located far from the surface of the organism. Each cell within a multicellular organism must solve the same materials-exchange problems as single-celled organisms. Large organisms have several interrelated systems that are involved in the exchange and transport of materials so that each cell can meet its metabolic needs.

Circulation

Large, multicellular organisms like humans consist of trillions of cells. Since many of these cells are buried within the organism far from the body surface, there must be some sort of distribution system to assist them in solving their materials-exchange problems. The primary mechanism used is the circulatory system.

The circulatory system consists of several fundamental parts. **Blood** is the fluid medium that assists in the transport of materials and heat. The **heart** is a pump that forces the fluid blood from one part of the body to another. The heart pumps blood into **arteries,** which distribute blood to organs. It flows

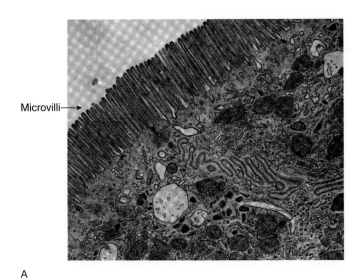

Microvilli→

A

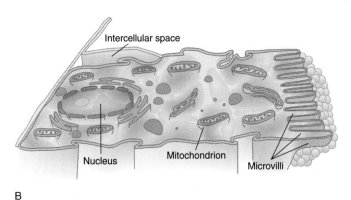

Intercellular space

Nucleus   Mitochondrion   Microvilli

B

FIGURE 26.21

Intestinal cells that are in contact with the food in the gut have highly folded surfaces. The tiny projections of these cells are called microvilli. These can be clearly seen in the photomicrograph in (*A*). The drawing in (*B*) shows that only one surface has these projections.

into successively smaller arteries until it reaches tiny vessels called **capillaries,** where materials are exchanged between the blood and tissues through the walls of the capillaries. The blood flows from the capillaries into **veins** that combine into larger veins that ultimately return the blood to the heart from the tissues.

The Nature of Blood

Blood is a fluid that consists of several kinds of cells suspended in a watery matrix called **plasma.** This fluid plasma also contains many kinds of dissolved molecules. The primary function of the blood is to transport molecules, cells, and heat from one part of the body to another. The major kinds of molecules that are distributed by the blood are respiratory gases (oxygen and carbon dioxide), nutrients of various kinds, waste products, disease-fighting antibodies, and chemical messengers (hormones). Blood has special characteristics that allow it to distribute respiratory gases very efficiently. Although little oxygen is carried as free, dissolved oxygen in the plasma, *red blood cells* (*RBCs*) contain **hemoglobin,** an iron-containing molecule, to which oxygen molecules readily bind. This allows for much more oxygen to be carried than could be possible if it were simply dissolved in the blood.

Table 26.2 lists the variety of cells found in blood. While the red, hemoglobin-containing erythrocytes serve in the transport of oxygen and carbon dioxide, the *white blood cells* (*WBCs*) carried in the blood are involved in defending against harmful agents. These cells help the body resist many diseases. They constitute the core of the **immune system.** The various WBCs participate in providing immunity in several ways. First of all, immune system cells are able to recognize cells and molecules that are dangerous to the body. If a molecule is recognized as dangerous, certain WBCs produce *antibodies* (*immunoglobulins*) that attach to the dangerous materials. The dangerous molecules that stimulate the

production of antibodies are called *antigens* (*immunogens*). When harmful microorganisms (e.g., bacteria, viruses, fungi), cancer cells, or toxic molecules enter the body, other WBCs (1) recognize, (2) boost their abilities to respond to, (3) move toward, and (4) destroy the problem causers. *Neutrophils, eosinophils, basophils,* and *monocytes* are specific kinds of WBCs capable of engulfing dangerous material, a process called phagocytosis. Thus they are often called *phagocytes.* While most can move from the bloodstream into the surrounding tissue, monocytes undergo such a striking increase in size that they are given a different name—*macrophages.* Macrophages can be found throughout the body and are the most active of the phagocytes.

Another kind of cellular particle in the blood is the platelet. These are fragments of specific kinds of white blood cells and are important in blood clotting. They collect at the site of a wound where they break down, releasing molecules. This begins a series of reactions that results in the formation of fibers that trap bloods cells and form a plug in the opening of the wound.

The Heart

Blood can perform its transportation function only if it moves. The organ responsible for providing the energy to pump the blood is the heart. The heart is a muscular pump that provides the pressure necessary to propel the blood throughout the body. It must continue its cycle of contraction and relaxation, or blood stops flowing and body cells are unable to obtain nutrients or get rid of wastes. Some cells, such as brain cells, are extremely sensitive to having their flow of blood interrupted because they require a constant supply of glucose and oxygen. Others, such as muscle cells or skin cells, are much better able to withstand temporary interruptions of blood flow.

The hearts of humans, other mammals, and birds consist of four chambers and four sets of valves that work together to

TABLE 26.2	The composition of blood	
Component	**Quantity Present**	**Function**
Plasma	55%	Maintain fluid nature of blood
Water	91.5%	
Protein	7.0%	
Other materials	1.5%	
Cellular material	45%	Carry oxygen and carbon dioxide
Red blood cells (erythrocytes)	4.3–5.8 million/mm^3	
White blood cells (leukocytes)	5–9 thousand/mm^3	Immunity
Lymphocytes	25%–30% of white cells present	
Monocytes	3%–7% of white cells present	
Neutrophils	57%–67% of white cells present	
Eosinophils	1%–3% of white cells present	
Basophils	Less than 1% of white cells present	
Platelets	250–400 thousand/mm^3	Clotting

Neutrophils   Eosinophils   Basophils

Lymphocytes   Monocytes   Platelets   Erythrocytes

ensure that blood flows in one direction only. Two of these chambers, the right and left **atria** (singular, *atrium*), are relatively thin-walled structures that collect blood from the major veins and empty it into the larger, more muscular ventricles (figure 26.22). Most of the flow of blood from the atria to the ventricles is caused by the lowered pressure produced within the ventricles as they relax. The contraction of the thin-walled atria assists in emptying them more completely.

The right and left **ventricles** are chambers that have powerful muscular walls whose contraction forces blood to flow through the arteries to all parts of the body. The valves between the atria and ventricles, known as **atrioventricular valves,** are important one-way valves that allow the blood to flow from the atria to the ventricles but prevent flow in the opposite direction. Similarly, there are valves in the aorta and pulmonary artery, known as **semilunar valves.** The **aorta** is the large artery that

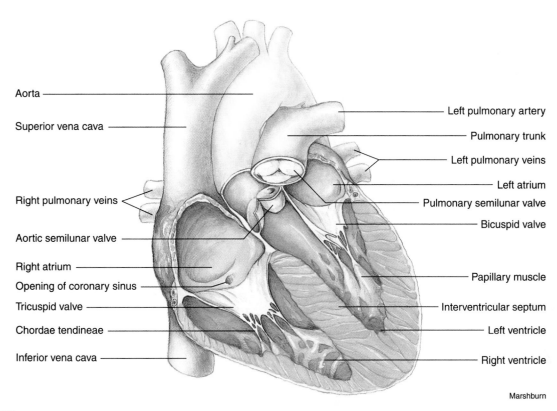

Aorta

Superior vena cava

Right pulmonary veins

Aortic semilunar valve

Right atrium

Opening of coronary sinus

Tricuspid valve

Chordae tendineae

Inferior vena cava

Left pulmonary artery

Pulmonary trunk

Left pulmonary veins

Left atrium

Pulmonary semilunar valve

Bicuspid valve

Papillary muscle

Interventricular septum

Left ventricle

Right ventricle

Marshburn

FIGURE 26.22

The heart consists of two thin-walled chambers called atria that contract to force blood into the two ventricles. When the ventricles contract, the atrioventricular valves (bicuspid and tricuspid) close, and blood is forced into the aorta and pulmonary artery. Semilunar valves in the aorta and pulmonary artery prevent the blood from flowing back into the ventricles when they relax.

carries blood from the left ventricle to the body, and the **pulmonary artery** carries blood from the right ventricle to the lungs. The semilunar valves prevent blood from flowing back into the ventricles. If the atrioventricular or semilunar valves are damaged or function improperly, the efficiency of the heart as a pump is diminished, and the person may develop an enlarged heart or other symptoms.

The right and left sides of the heart have slightly different jobs, because they pump blood to different parts of the body. The right side of the heart receives blood from the general body and pumps it through the pulmonary arteries to the lungs, where exchange of oxygen and carbon dioxide takes place and the blood returns from the lungs to the left atrium. This is called **pulmonary circulation.** The larger, more powerful left side of the heart receives blood from the lungs, delivers it through the aorta to all parts of the body, and returns it to the right atrium by way of veins. This is known as **systemic circulation.** Both circulatory pathways are shown in figure 26.23. The systemic circulation is responsible for gas, nutrient, and waste exchange in all parts of the body except the lungs.

Arteries and Veins

Arteries and veins are the tubes that transport blood from one place to another within the body. Figure 26.24 compares the structure and function of arteries and veins. Arteries carry blood away from the heart, because it is under considerable pressure from the contraction of the ventricles. The contraction of the walls of the ventricles increases the pressure in the arteries. A typical pressure recorded in a large artery while the heart is contracting is about 120 millimeters of mercury. This is known as the **systolic blood pressure.** The pressure recorded while the heart is relaxing is about 80 millimeters of mercury. This is known as the **diastolic blood pressure.** A blood pressure reading includes both of these numbers and is recorded as 120/80.

The walls of arteries are relatively thick and muscular, yet elastic. Healthy arteries have the ability to expand as blood is pumped into them and return to normal as the pressure drops. This ability to expand absorbs some of the pressure and reduces the peak pressure within the arteries, thus reducing the likelihood that they will burst. If arteries become hardened and less resilient, the peak blood pressure rises and they are more likely

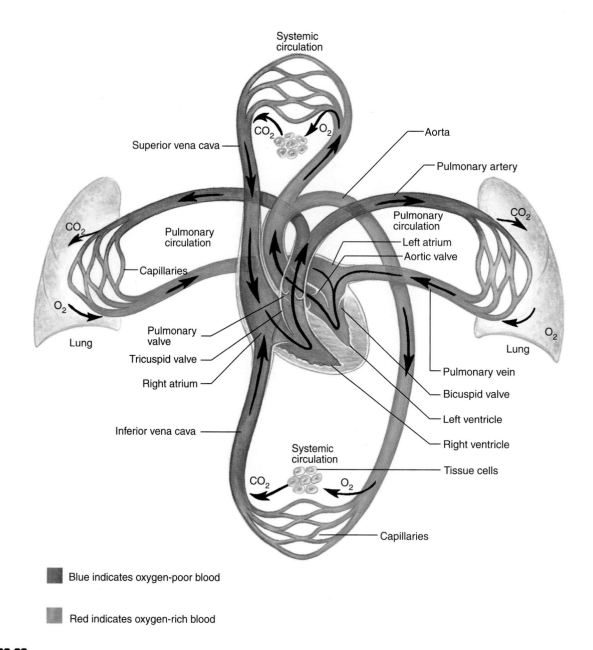

FIGURE 26.23

The right ventricle pumps blood that is poor in oxygen to the two lungs by way of the pulmonary arteries, where it receives oxygen and turns bright red. The blood is then returned to the left atrium by way of four pulmonary veins. This part of the circulatory system is known as pulmonary circulation. The left ventricle pumps oxygen-rich blood by way of the aorta to all parts of the body except the lungs. This blood returns to the right atrium, depleted of its oxygen, by way of the superior vena cava from the head region and the inferior vena cava from the rest of the body. This portion of the circulatory system is known as systemic circulation.

Artery

Vein

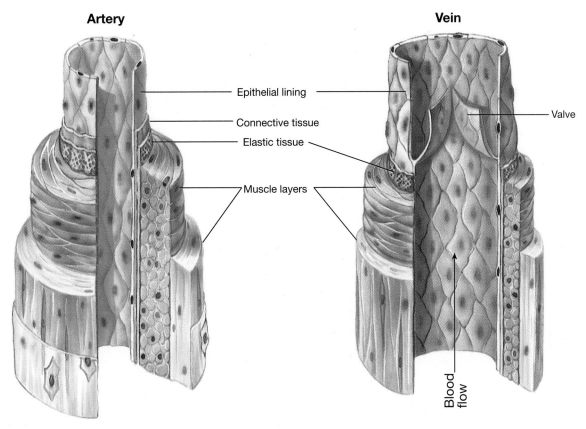

Epithelial lining

Connective tissue

Elastic tissue

Muscle layers

Valve

Blood flow

FIGURE 26.24

The walls of arteries are much thicker than the walls of veins. (The pressure in arteries is much higher than the pressure in veins.) The pressure generated by the ventricles of the heart forces blood through the arteries. Veins often have very low pressure. The valves in the veins prevent the blood from flowing backward, away from the heart.

to rupture. The elastic nature of the arteries is also responsible for assisting the flow of blood. When they return to normal from their stretched condition they give a little push to the blood that is flowing through them.

Blood is distributed from the large aorta through smaller and smaller blood vessels to millions of tiny capillaries. Some of the smaller arteries, called **arterioles,** may contract or relax to regulate the flow of blood to specific parts of the body. Major parts of the body that receive differing amounts of blood, depending on need, are the digestive system, muscles, and skin.

Veins collect blood from the capillaries and return it to the heart. The pressure in these blood vessels is very low. Some of the largest veins may have a blood pressure of 0.0 mmHg for brief periods. Since pressure in veins is so low, muscular movements of the body are important in helping to return blood to the heart. When muscles of the body contract, they compress nearby veins, and this pressure pushes blood along in the veins. Because valves in the veins allow blood to flow only toward the heart, this activity acts as an additional pump to help return blood to the heart.

Although the arteries are responsible for distributing blood to various parts of the body and arterioles regulate where blood goes, it is the function of capillaries to assist in the exchange of materials between the blood and cells.

Capillaries

Capillaries are tiny, thin-walled tubes that receive blood from arterioles. They are so small that red blood cells must go through them in single file. They are so numerous that each cell in the body has a capillary located near it. It is estimated that there are about 1,000 square meters of surface area represented by the capillary surface in a typical human. Each capillary wall consists of a single layer of cells and therefore presents only a thin barrier to the diffusion of materials between the blood and cells. It is also possible for liquid to flow through tiny spaces between the individual cells of most capillaries (figure 26.25). The flow of blood through these smallest blood vessels is relatively slow. This allows time for the diffusion of such materials as oxygen, glucose, and water from the blood to surrounding cells, and for the movement of such materials as carbon dioxide, lactic acid, and ammonia from the cells into the blood.

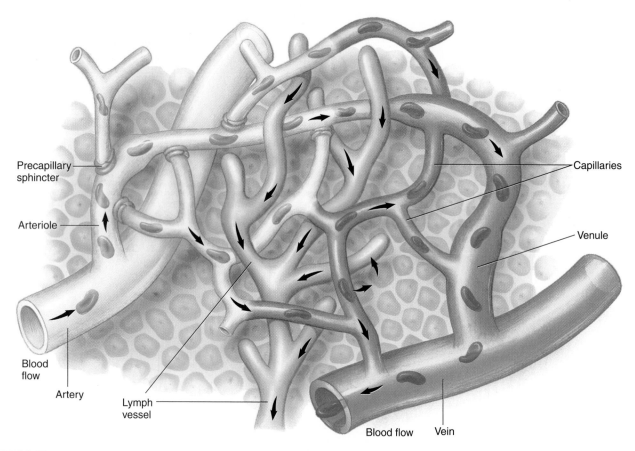

FIGURE 26.25

Capillaries are tiny blood vessels. Exchange of cells and molecules can occur between blood and tissues through their thin walls. Molecules diffuse in and out of the blood, and cells such as monocytes can move from the blood through the thin walls into the surrounding tissue. There is also a flow of liquid through holes in the capillary walls. This liquid, called lymph, bathes the cells and eventually enters small lymph vessels that return lymph to the circulatory system near the heart.

In addition to molecular exchange, considerable amounts of water and dissolved materials leak through the small holes in the capillaries. This liquid is known as **lymph.** Lymph is produced when the blood pressure forces water and some small dissolved molecules through the walls of the capillaries. Lymph bathes the cells but must eventually be returned to the circulatory system by lymph vessels or swelling will occur. Return is accomplished by the **lymphatic system,** a collection of thin-walled tubes that branch throughout the body. These tubes collect lymph that is filtered from the circulatory system and ultimately empty it into major blood vessels near the heart. Figure 26.26 shows the structure of the lymphatic system.

Respiratory Anatomy

The **lungs** are organs of the body that allow gas exchange to take place between the air and blood. Associated with the lungs is a set of tubes that conducts air from outside the body to the lungs. The single large-diameter **trachea** is supported by rings of cartilage that prevent its collapse. It branches into two major **bronchi** (sin-gular, *bronchus*) that deliver air to smaller and smaller branches. Bronchi are also supported by cartilage. The smallest tubes, known as **bronchioles,** contain smooth muscle and are therefore capable of constricting. Finally, the bronchioles deliver air to clusters of tiny sacs, known as **alveoli** (singular, *alveolus*), where the exchange of gases takes place between the air and blood.

The nose, mouth, and throat are also important parts of the air-transport pathway because they modify the humidity and temperature of the air and clean the air as it passes. The lining of the trachea contains cells that have cilia that beat in a direction that moves mucus and foreign materials from the lungs. The foreign matter may then be expelled by swallowing, coughing, or other means. Figure 26.27 illustrates the various parts of the respiratory system.

Breathing-System Regulation

Breathing is the process of moving air in and out of the lungs. It is accomplished by the movement of a muscular organ known as the **diaphragm,** which separates the chest cavity and the lungs from the abdominal cavity. In addition, muscles located between

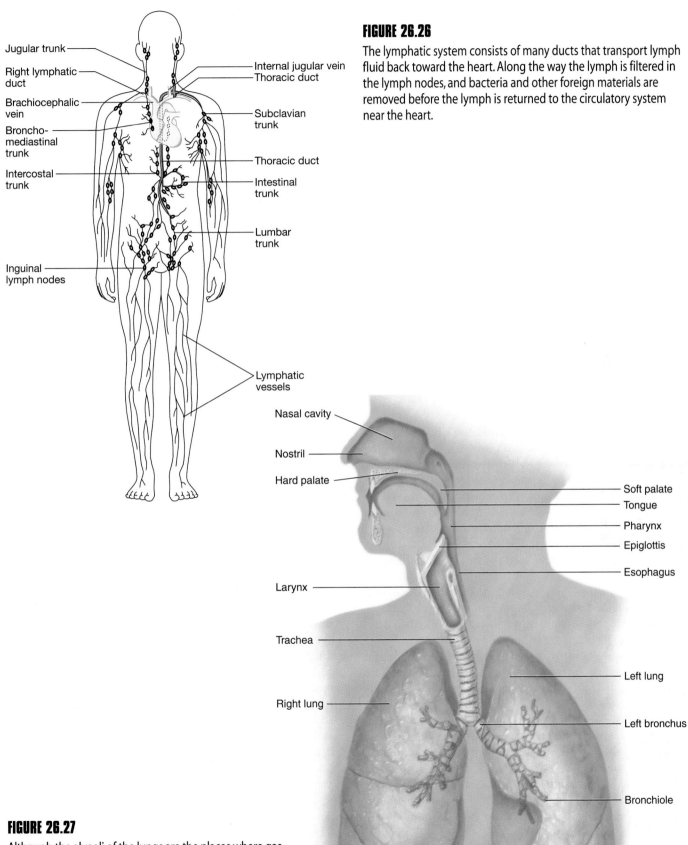

Jugular trunk

Right lymphatic duct

Brachiocephalic vein

Broncho-mediastinal trunk

Intercostal trunk

Inguinal lymph nodes

Internal jugular vein

Thoracic duct

Subclavian trunk

Thoracic duct

Intestinal trunk

Lumbar trunk

Lymphatic vessels

FIGURE 26.26

The lymphatic system consists of many ducts that transport lymph fluid back toward the heart. Along the way the lymph is filtered in the lymph nodes, and bacteria and other foreign materials are removed before the lymph is returned to the circulatory system near the heart.

Nasal cavity

Nostril

Hard palate

Larynx

Trachea

Right lung

Soft palate

Tongue

Pharynx

Epiglottis

Esophagus

Left lung

Left bronchus

Bronchiole

FIGURE 26.27

Although the alveoli of the lungs are the places where gas exchange takes place, there are many other important parts of the respiratory system. The nasal cavity cleans, warms, and humidifies the air entering the lungs. The trachea is also important in cleaning the air going to the lungs.

Lew

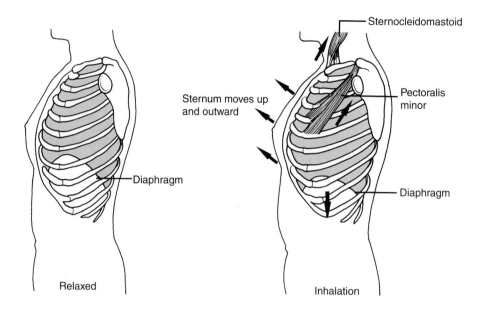

Sternocleidomastoid

Pectoralis minor

Sternum moves up and outward

Diaphragm

Diaphragm

Relaxed

Inhalation

FIGURE 26.28

During inhalation, the diaphragm and external intercostal muscles between the ribs contract, causing the volume of the chest cavity to increase. During a normal exhalation, these muscles relax, and the chest volume returns to normal.

the ribs (*intercostal* muscles) are attached to the ribs in such a way that their contraction causes the chest wall to move outward and upward, which increases the size of the chest cavity. During inhalation, the diaphragm moves downward and the external intercostal muscles of the chest wall contract, causing the volume of the chest cavity to increase. This results in a lower pressure in the chest cavity compared to the outside air pressure. Consequently, air flows from the outside high-pressure area through the trachea, bronchi, and bronchioles to the alveoli. During normal relaxed breathing, exhalation is accomplished by the chest wall and diaphragm simply returning to their normal, relaxed position. Muscular contraction is not involved (figure 26.28).

Exercising causes an increase in the amount of carbon dioxide in the blood because muscles are oxidizing glucose more rapidly. This lowers the pH of the blood. Certain brain cells and specialized cells in the aortic arch and carotid arteries are sensitive to changes in blood pH. When they sense a lower blood pH, nerve impulses are sent more frequently to the diaphragm and intercostal muscles. These muscles contract more rapidly and more forcefully, resulting in more rapid, deeper breathing. Because more air is being exchanged per minute, carbon dioxide is lost from the lungs more rapidly. When exercise stops, blood pH rises, and breathing eventually returns to normal. Bear in mind, however, that moving air in and out of the lungs is of no value unless oxygen is diffusing into the blood and carbon dioxide is diffusing out.

Obtaining Nutrients

All cells must have a continuous supply of nutrients that provides the energy they require and the building blocks needed to construct the macromolecules typical of living things. This sec-

tion will deal with the processing and distribution of different kinds of nutrients.

The digestive system consists of a muscular tube with several specialized segments. In addition, there are glands that secrete digestive juices into the tube. Four different kinds of activities are involved in getting nutrients to the cells that need them: mechanical processing, chemical processing, nutrient uptake, and chemical alteration.

Mechanical and Chemical Processing

The digestive system is designed as a disassembly system. Its purpose is to take large chunks of food and break them down to smaller molecules that can be taken up by the circulatory system and distributed to cells. The first step in this process is mechanical processing.

It is important to grind large particles into small pieces by chewing in order to increase their surface area and allow for more efficient chemical reactions. It is also important to add water to the food, which further disperses the particles and provides the watery environment needed for these chemical reactions. Materials must also be mixed, so that all the molecules that need to interact with one another have a good chance of doing so. The mouth and the stomach are the major body regions involved in reducing the size of food particles. The teeth are involved in cutting and grinding food to increase its surface area. The watery mixture that is added to the food in the mouth is known as **saliva,** and the three pairs of glands that produce saliva are known as **salivary glands.** Saliva contains the enzyme *salivary amylase,* which initiates the chemical breakdown of starch. Saliva also lubricates the mouth and helps to bind food before

A CLOSER LOOK | Cigarette Smoking and Your Health

Cigarette smoking is becoming less and less acceptable in our society. The banning of smoking in public buildings and on domestic air flights attests to this fact. Yet in spite of social pressure to quit smoking, research linking smoking with lung and heart disease, and evidence that even secondhand smoke can be harmful, over one-fifth of American adults are smokers (a smoker in this case is defined as someone who has smoked 100 or more cigarettes in his or her lifetime). Among Americans with a high school education or less, smoking is even more prevalent.

Hazards of Cigarette Smoking

Bronchitis—Cigarette smoking is the leading cause of chronic bronchitis, which involves the inflammation of the bronchi. A common symptom of bronchitis is a harsh cough that expels a greenish-yellow mucus.

Emphysema—Emphysema is a progressive disease in which some of the alveoli are lost. People afflicted with this disease have less and less respiratory surface area and experience greater difficulty getting adequate oxygen, even though they may be breathing more rapidly. It may be caused by cigarette smoke and other air pollutants that damage alveoli. This damage reduces the capacity of the lungs to exchange gases with the bloodstream. A common symptom of emphysema is difficulty exhaling. Several years of an emphysema sufferer's forced breathing can increase the size of the chest and give it a barrel appearance.

Asthma—Cigarette smoke is one of many environmental factors that may trigger an asthma attack. Asthma is an allergic reaction that results in the narrowing of the lungs' air passages and the excess production of fluids that limit the amount of air that can enter the lungs. Symptoms of asthma include coughing, wheezing, and difficulty breathing.

Lung Cancer—Lung cancer develops twenty times more frequently in heavy smokers than in nonsmokers. Typically, lung cancer starts in the bronchi. Cigarette smoke and other pollutants cause cells below the surface of the bronchi to divide at an abnormally high rate. This malignant growth may spread through the lung and move into other parts of the body. Occurrence of cancers of the mouth, larynx, esophagus, pancreas, and bladder are also significantly greater in smokers than in nonsmokers.

Pneumonia—Cigarette smokers have an increased risk of developing pneumonia. Pneumonia involves the infection or inflammation of alveoli, which leads to fluid filling the alveolar sacs. Pneumonia is typically caused by the bacterium *Streptococcus pneumoniae* but in some cases is caused by other bacteria, fungi, protists, or viruses.

Smoking during Pregnancy—Cigarette smoking during pregnancy has been linked to low birth weight and higher rates of fetal and infant death. Children of mothers who smoke during pregnancy have a higher incidence of heart abnormalities, cleft palate, and sudden infant death syndrome (SIDS). Nursing infants of smoking mothers have higher than normal rates of intestinal problems, and infants exposed to secondhand smoke have an increased incidence of respiratory disorders.

Heart Disease—Smoking is a major contributor to heart disease. The action of nicotine from cigarette smoke results in constriction of blood vessels and the reduction of blood flow.

---

swallowing. Figure 26.29 describes and summarizes the function of the structures of the digestive system.

Once the food has been chewed, it is swallowed and passes down the esophagus to the stomach. The process of swallowing involves a complex series of events. First, a ball of food, known as a *bolus,* is formed by the tongue and moved to the back of the mouth cavity. Here it stimulates the walls of the throat, also known as the **pharynx.** Nerve endings in the lining of the pharynx are stimulated, causing a reflex contraction of the walls of the esophagus, which transports the bolus to the stomach. Since both food and air pass through the pharynx it is important to prevent food from getting into the lungs. During swallowing, the larynx is pulled upward. This causes a flap of tissue called the *epiglottis* to cover the opening to the trachea and prevent food from entering the trachea. In the stomach, additional liquid, called **gastric juice,** is added to the food. Gastric juice contains enzymes and hydrochloric acid. The major enzyme of the stomach is *pepsin,* which initiates the chemical breakdown of protein. The pH of gastric juice is very low, generally around pH 2. The entire mixture is churned by the contractions of the three layers of muscle in the stomach wall. The combined activities of enzymatic breakdown, chemical breakdown by hydrochloric acid, and mechanical processing by muscular movement result in a thoroughly mixed liquid called *chyme.* Chyme eventually leaves the stomach through a valve known as the pyloric sphincter and enters the small intestine.

The first part of the **small intestine** is known as the **duodenum.** In addition to producing enzymes, the duodenum secretes several kinds of hormones that regulate the release of food from the stomach and the release of secretions from the pancreas and liver. The **pancreas** produces a number of different digestive enzymes and also secretes large amounts of bicarbonate ions, which neutralize stomach acid so that the pH of the duodenum is about pH 8. The **liver** is a large organ in the upper abdomen

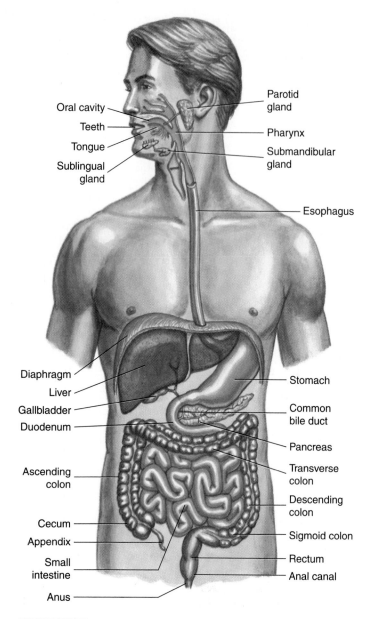

Oral cavity

Teeth

Tongue

Sublingual gland

Parotid gland

Pharynx

Submandibular gland

Esophagus

Diaphragm

Liver

Gallbladder

Duodenum

Ascending colon

Cecum

Appendix

Small intestine

Anus

Stomach

Common bile duct

Pancreas

Transverse colon

Descending colon

Sigmoid colon

Rectum

Anal canal

FIGURE 26.29

In the digestive system, the teeth, tongue, and enzymes from the salivary glands modify the food before it is swallowed. The stomach adds acid and enzymes and further changes the texture of the food. The food is eventually emptied into the duodenum, where the liver and pancreas add their secretions. The small intestine also adds enzymes and is involved in absorbing nutrients. The large intestine is primarily involved in removing water.

that performs several functions. One of its functions is the secretion of *bile*. When bile leaves the liver it is stored in the *gallbladder* prior to being released into the duodenum. When bile is released from the gallbladder, it assists mechanical mixing by breaking large fat globules into smaller particles. This process is called *emulsification*.

Along the length of the intestine, additional watery juices are added until the mixture reaches the **large intestine.** The large intestine is primarily involved in reabsorbing the water that has been added to the food tube along with saliva, gastric juice, bile, pancreatic secretions, and intestinal juices. The large intestine is also home to a variety of different kinds of bacteria. Most live on the undigested food that makes it through the small intestine. Some provide additional benefit by producing vitamins that can be absorbed from the large intestine. A few kinds may cause disease.

Nutrient Uptake

The process of digestion has resulted in a variety of simple organic molecules that are available for absorption from the tube of the gut into the circulatory system. The small intestine is a very long tube; the longer the tube, the greater the internal surface area. In a typical adult human it is about 3 meters long. In addition to length, the lining of the intestine consists of millions of fingerlike projections called **villi,** which increase the surface area. When we examine the cells that make up the villi, we find that they also have folds in their surface membranes. All of these characteristics increase the surface area available for the transport of materials from the gut into the circulatory system (figure 26.30). Scientists estimate that the cumulative effect of all of these features produces a total intestinal surface area of about 250 square meters. That is equivalent to about half the area of a football field.

The surface area by itself would be of little value if it were not for the intimate contact of the circulatory system with this lining. Each villus contains several capillaries and a branch of the lymphatic system called a *lacteal*. The close association between the intestinal surface and the circulatory and lymphatic systems allows for the efficient uptake of nutrients from the cavity of the gut into the circulatory system.

Several different kinds of processes are involved in the transport of materials from the intestine to the circulatory system. Some molecules, such as water and many ions, simply diffuse through the wall of the intestine into the circulatory system. Other materials, such as amino acids and simple sugars, are assisted across the membrane by carrier molecules. Fatty acids and glycerol are absorbed into the intestinal lining cells where

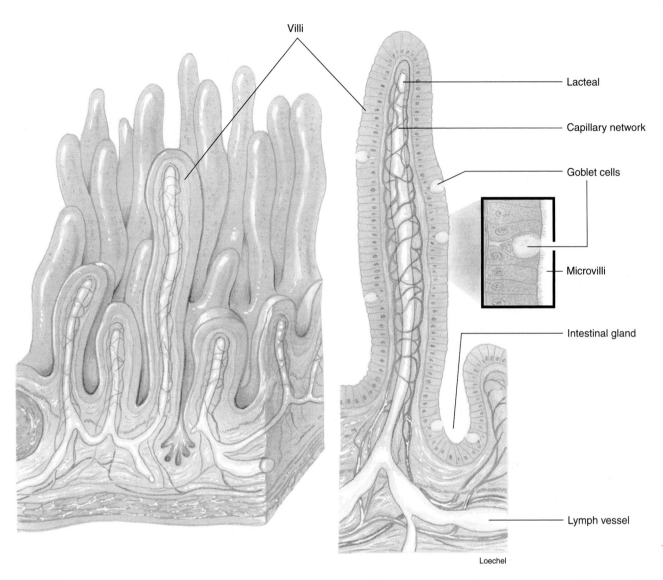

Villi

Lacteal

Capillary network

Goblet cells

Microvilli

Intestinal gland

Lymph vessel

Loechel

FIGURE 26.30

The surface area of the intestinal lining is increased by the many fingerlike projections known as villi. Within each villus are capillaries and lacteals. Most kinds of materials enter the capillaries, but most fat-soluble substances enter the lacteals, giving them a milky appearance. Lacteals are part of the lymphatic system. Because the lymphatic system empties into the circulatory system, fat-soluble materials also eventually enter the circulatory system. The close relationship between these vessels and the epithelial lining of the villus allows for efficient exchange of materials from the intestinal cavity to the circulatory system.

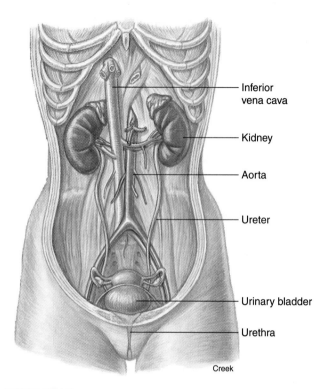

Inferior
vena cava

Kidney

Aorta

Ureter

Urinary bladder

Urethra

Creek

FIGURE 26.31

The primary organs involved in removing materials from the blood are the kidneys. The urine produced by the kidneys is transported by the ureters to the urinary bladder. From the bladder, the urine is emptied to the outside of the body by way of the urethra.

they are resynthesized into fats and enter lacteals in the villi. Since the lacteals are part of the lymphatic system, which eventually empties its contents into the circulatory system, fats are also transported by the blood. They just reach the blood by a different route.

Waste Disposal

Because cells are modifying molecules during metabolic processes, harmful waste products are constantly being formed. Urea is a common waste; many other toxic materials must be eliminated as well. Among these are large numbers of hydrogen ions produced by metabolism. This excess of hydrogen ions must be removed from the bloodstream. Other molecules, such as water and salts, may be consumed in excessive amounts and must be removed. The primary organs involved in regulating the level of toxic or unnecessary molecules are the **kidneys** (figure 26.31).

The kidneys consist of about 2.4 million tiny units called **nephrons.** At one end of a nephron is a cup-shaped structure called *Bowman's capsule,* which surrounds a knot of capillaries known as a *glomerulus* (figure 26.32). In addition to Bowman's capsule, a nephron consists of three distinctly different regions: the *proximal convoluted tubule,* the *loop of Henle,* and the *distal convoluted tubule.* The distal convoluted tubule of a nephron is

connected to a collecting duct that transports fluid to the ureters, and ultimately to the urinary bladder, where it is stored until it can be eliminated.

As in the other systems, the excretory system involves a close connection between the circulatory system and a surface. In this case the large surface is provided by the walls of the millions of nephrons, which are surrounded by capillaries. Three major activities occur at these surfaces: filtration, reabsorption, and secretion. The glomerulus presents a large surface for the filtering of material from the blood to Bowman's capsule. Blood that enters the glomerulus is under pressure from the muscular contraction of the heart. The capillaries of the glomerulus are quite porous and provide a large surface area for the movement of water and small dissolved molecules from the blood into Bowman's capsule. Normally, only the smaller molecules, such as glucose, amino acids, and ions, are able to pass through the glomerulus into Bowman's capsule at the end of the nephron. The various kinds of blood cells and larger molecules like proteins do not pass out of the blood into the nephron. This physical filtration process allows many kinds of molecules to leave the blood and enter the nephron. The volume of material filtered in this way through the approximately 2.4 million nephrons of our kidneys is about 7.5 liters per hour. Since your entire blood supply is about 5 to 6 liters, there must be some method of recovering much of this fluid.

Some molecules that pass through the nephron are relatively unaffected by the various activities going on in the kidney. One of these is urea, which is filtered through the glomerulus into Bowman's capsule. As it passes through the nephron, much of it stays in the tubule and is eliminated in the urine. Many other kinds of molecules, such as minor metabolic waste products and some drugs, are also treated in this manner.

NUTRITION

Organisms maintain themselves by constantly processing molecules to provide building blocks for new living material and energy to sustain themselves. Autotrophs can manufacture organic molecules from inorganic molecules, but heterotrophs must consume organic molecules to get what they need. All molecules required to support living things are called **nutrients.** Some nutrients are inorganic molecules such as calcium, iron, or potassium, others are organic molecules such as carbohydrates, proteins, fats, and vitamins. All heterotrophs obtain the nutrients they need from food and each kind of heterotroph has particular nutritional requirements. This section deals with the nutritional requirements of humans.

Living Things as Chemical Factories: Matter and Energy Manipulators

The word **nutrition** is used in two related contexts. First of all nutrition is a branch of science that seeks to understand food, its nutrients, how the nutrients are used by the body, and how inappropriate combinations of quantities of nutrients lead to ill

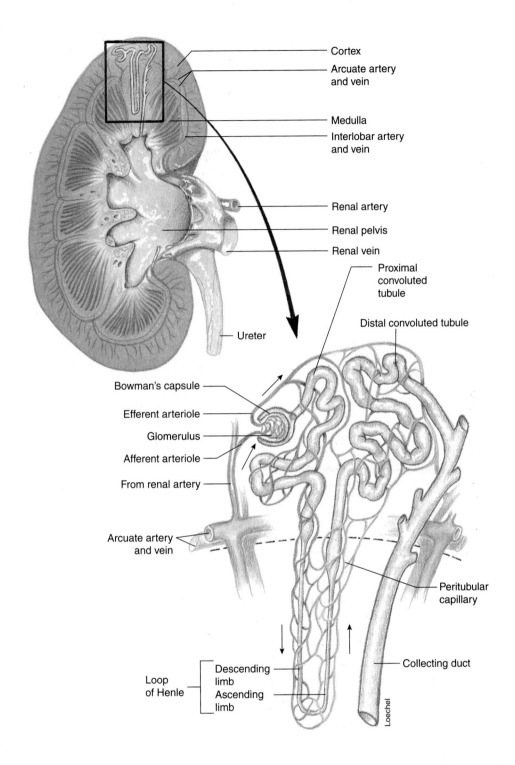

FIGURE 26.32

The nephron and the closely associated blood vessels create a system that allows for the passage of materials from the circulatory system to the nephron by way of the glomerulus and Bowman's capsule. Materials are added to and removed from the fluid in the nephron via the tubular portions of the nephron.

TABLE 26.3 Typical energy requirements for common activities			
Light Activities (120–150 kcal/hr)	**Light to Moderate Activities** (150–300 kcal/hr)	**Moderate Activities** (300–400 kcal/hr)	**Heavy Activities** (420–600 kcal/hr)
Dressing	Sweeping floors	Pulling weeds	Chopping wood
Typing	Painting	Walking behind lawnmower	Shoveling snow
Studying	Store clerking	Walking 3.5–4 mph on level surface	Walking 5 mph
Standing	Bowling	Calisthenics	Walking up hills
Slow walking	Walking 2–3 mph	Canoeing 4 mph	Cross-country skiing
Sitting activities	Canoeing 2.5–3 mph	Doubles tennis	Swimming
	Bicycling on level surface at 5.5 mph	Volleyball	Jogging 5 mph
		Golf (no cart)	Bicycling 11–12 mph or in hills

health. The word nutrition is also used in a slightly different context to refer to all the processes by which we take in food and utilize it, including *ingestion, digestion, absorption,* and *assimilation. Ingestion* involves the process of taking food into the body through eating. *Digestion* involves the breakdown of complex food molecules to simpler molecules. *Absorption* involves the movement of simple molecules from the digestive system to the circulatory system for dispersal throughout the body. *Assimilation* involves the modification and incorporation of absorbed molecules into the structure of the organism.

Many of the nutrients that enter living cells undergo chemical changes before they are incorporated into the body. These interconversion processes are ultimately under the control of the genetic material, DNA. It is DNA that codes the information necessary to manufacture the enzymes required to extract energy from chemical bonds and to convert raw materials (nutrients) into the structure (anatomy) of the organism.

The food and drink consumed from day to day constitutes a person's **diet.** It must contain the minimal nutrients necessary to manufacture and maintain the body's structure (bones, skin, tendons, muscles, etc.) and regulatory molecules (enzymes and hormones), and to supply the energy (ATP) needed to run the body's machinery. If the diet is deficient in nutrients, or if a person's body cannot process nutrients efficiently, a dietary deficiency and ill health may result. A good understanding of nutrition can promote good health and involves an understanding of the energy and nutrient content in various foods.

Kilocalories, Basal Metabolism, and Weight Control

The unit used to measure the amount of energy in foods is the **kilocalorie (kcal).** One kilocalorie is the amount of heat needed to raise the temperature of one *kilogram* of water one degree Celsius. Remember that the prefix *kilo-* means "1,000 times" the value listed. Therefore, a kilocalorie is 1,000 times more heat energy than a **calorie,** which is the amount of heat needed to raise the temperature of one *gram* of water one degree Celsius. However, the amount of energy contained in food is usually called a Calorie with a capital C. This is unfortunate because it

is easy to confuse a Calorie, which is really a kilocalorie, with a calorie. Most books on nutrition and dieting use the term Calorie to refer to *food calories.* The energy requirements in kilocalories for a variety of activities are listed in table 26.3.

Significant energy expenditure is required for muscular activity. However, even at rest, energy is required to maintain breathing, heart rate, and other normal body functions. The rate at which the body uses energy when at rest is known as the **basal metabolic rate (BMR).** The basal metabolism of most people requires more energy than their voluntary muscular activity. Much of this energy is used to keep your body temperature constant. A true measurement of basal metabolic rate requires a measurement of oxygen used over a specific period under controlled conditions.

Since few of us rest for twenty-four hours a day, we normally require more than the energy needed for basal metabolism. One of these requirements is the amount of energy needed to process the food that we eat. This is called **specific dynamic action (SDA)** and is equal to approximately 10 percent of your total daily kilocalorie intake. In addition to basal metabolism and specific dynamic action, the activity level of a person determines the number of kilocalories needed.

Depending on how obesity is defined, obesity is a problem for 30 percent to 40 percent of the population of the United States. Why is weight control a problem for such a large portion of the population? There are several metabolic pathways that convert carbohydrates (glucose) or proteins to fat. Stored body fat was very important for our prehistoric ancestors, because it allowed them to survive periods of food scarcity. In periods of food scarcity the stored body fat can be used to supply energy. The glycerol portion of the fat molecule can be converted to a small amount of glucose, which can supply energy for red blood cells and nervous tissue that must have glucose. The fatty acid portion of the molecule can be metabolized by most other tissues directly to produce ATP. Today, however, for most of us food scarcity is not a problem, and even small amounts of excess food consumed daily tend to add to our fat stores.

Although energy doesn't weigh anything, the nutrients that contain the energy do. Weight control is a matter of balancing the kilocalories ingested as a result of dietary intake with the kilocalories of energy expended by normal daily activities and exercise. There is a limit to the rate at which a moderately active human body can use fat as an energy source. At the most, one or two pounds of fat tissue per week are lost by an average person when dieting. Since one pound of fatty tissue contains about 3,500 kilocalories, decreasing your kilocalorie intake by 500 to 1,000 kilocalories per day while maintaining a balanced diet (including proteins, carbohydrates, and fats) will result in fat loss of one to two pounds per week. (A pound of pure fat contains about 4,100 kilocalories, but fat tissue contains other materials besides fat, such as water.)

For those who need to gain weight, increasing kilocalorie intake by 500 to 1,000 kilocalories per day will result in an increase of one to two pounds per week, provided the low weight is not the result of a health problem.

If, like millions of others, you feel that you are overweight, you have probably tried numerous diet plans. Not all of these plans are the same, and not all are suitable to your particular situation. If a diet plan is to be valuable in promoting good health, it must satisfy your needs in several ways. It must provide you with needed kilocalories, protein, fat, and carbohydrates. It should also contain readily available foods from all the basic food groups, and it should provide enough variety to prevent you from becoming bored with the plan and going off the diet too soon. A diet should not be something you follow only for a while, then abandon and regain the lost weight.

Amounts and Sources of Nutrients

In order to give people some guidelines for planning a diet that provides adequate amounts of the six classes of nutrients, nutritional scientists in the United States and many other countries have developed nutrient standards. In the United States, these guidelines are known as the **Recommended Dietary Allowances,** or **RDAs.** RDAs are dietary recommendations, not requirements or minimum standards. They are based on the needs of a healthy person already eating an adequate diet. RDAs do not apply to persons with medical problems who are under stress or suffering from malnutrition. The amount of each nutrient specified by the RDAs has been set relatively high so that most of the population eating those quantities will be meeting their nutritional needs. Keep in mind that since everybody is different, eating the RDA amounts may not meet your personal needs. You may have a special need for additional amounts of a particular nutrient if you have an unusual metabolic condition.

The U.S. RDAs are used when preparing product labels. The federal government requires by law that labels list ingredients from the greatest to the least in quantity. The volume in the package must be stated along with the weight, and the name of the manufacturer or distributor. If any nutritional claim is made, it must be supported by factual information.

A product label that proclaims, for example, that a serving of cereal provides 25 percent of the RDA for vitamin A means that you are getting at least one-fourth of your RDA of vitamin A from a single serving of that cereal. To figure your total RDA

of vitamin A, consult a published RDA table for adults. It tells you that an adult male requires 1,000 and a female 800 micrograms (μg) of vitamin A per day. Of this, 25 percent is 250 and 200 μg, respectively—the amount you are getting in a serving of that cereal. You will need to get the additional amounts (750 for men and 600 for women) by having more of that cereal or eating other foods that contain vitamin A. If a product claims to have 100 percent of the RDA of a particular nutrient, that amount must be present in the product. However, restricting yourself to that one product will surely deprive you of many of the other nutrients necessary for good health. Ideally, you should eat a variety of complex foods containing a variety of nutrients to ensure that all your health requirements are met.

The Food Guide Pyramid with Five Food Groups

Using RDAs and product labels is a pretty complicated way for a person to plan a diet. Planning a diet around basic food groups is generally easier. The four basic food groups first developed and introduced in 1953 have been modified and updated several times to serve as guidelines in maintaining a balanced diet (figure 26.33). In May 1992, the U.S. Department of Agriculture released the results of its most recent study on how best to educate the public about daily nutrition. The federal government adopted the **Food Guide Pyramid** of the Department of Agriculture as one of its primary tools to help the general public plan for good nutrition. The Food Guide Pyramid contains five basic groups of foods with guidelines for the amounts one needs daily from each group for ideal nutritional planning. The Food Guide Pyramid differs from previous federal government information in than it encourages a reduction in the amount of fats and sugars in the diet while increasing our daily servings of fruits and vegetables. In addition, the new guidelines suggest significantly increasing the amount of grain products we eat each day.

Grain Products Group

Grains include vitamin-enriched or whole-grain cereals and products such as breads, bagels, buns, crackers, dry and cooked cereals, pancakes, pasta, and tortillas. Items in this group are typically dry and seldom need refrigeration. They should provide most of your kilocalorie requirements in the form of complex carbohydrates like starch, which is the main ingredient in most grain products. These foods help you feel you have satisfied your appetite, and many of them are very low in fat.

Fruits Group

You probably can remember discussions of whether a tomato is a vegetable or a fruit. This controversy arises from the fact that the term *vegetable* is not scientifically precise but means a plant material eaten during the main part of the meal. *Fruit,* on the other hand, is a botanical term for the structure that is produced from the female part of the flower to protect and nourish the ripening seeds. Although, botanically, green beans, peas, and corn are all fruits, nutritionally speaking, they are placed in the vegetable category because they are generally eaten during the main part of the meal. Nutritionally speaking, fruits include

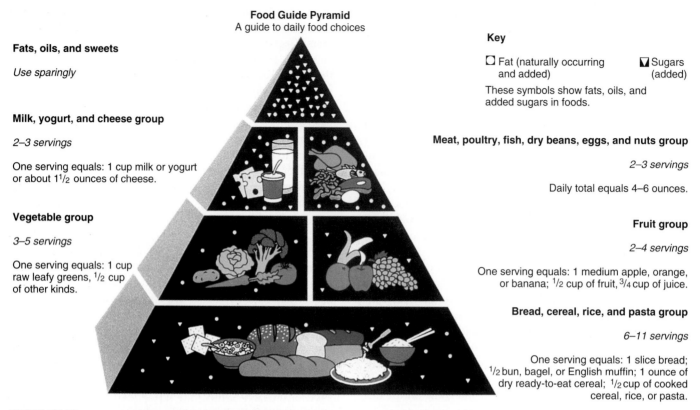

Food Guide Pyramid
A guide to daily food choices

Fats, oils, and sweets

Use sparingly

Milk, yogurt, and cheese group

2–3 servings

One serving equals: 1 cup milk or yogurt
or about 1¹/₂ ounces of cheese.

Vegetable group

3–5 servings

One serving equals: 1 cup
raw leafy greens, ¹/₂ cup
of other kinds.

Key

☐ Fat (naturally occurring   ▼ Sugars
   and added)   (added)

These symbols show fats, oils, and
added sugars in foods.

Meat, poultry, fish, dry beans, eggs, and nuts group

2–3 servings

Daily total equals 4–6 ounces.

Fruit group

2–4 servings

One serving equals: 1 medium apple, orange,
or banana; ¹/₂ cup of fruit, ³/₄ cup of juice.

Bread, cereal, rice, and pasta group

6–11 servings

One serving equals: 1 slice bread;
¹/₂ bun, bagel, or English muffin; 1 ounce of
dry ready-to-eat cereal; ¹/₂ cup of cooked
cereal, rice, or pasta.

FIGURE 26.33

In May 1992, the Department of Agriculture released a new guide to good eating. This Food Guide Pyramid suggests that we eat particular amounts of five different food groups while decreasing our intake of fats and sugars. This guide should simplify our menu planning and help ensure that we get all the recommended amounts of basic nutrients.

Source: U.S. Department of Agriculture, 1992.

such sweet plant products as melons, berries, apples, oranges, and bananas. The Food Guide Pyramid suggests two to four servings of fruit per day. However, since these foods tend to be high in natural sugars, consumption of large amounts of fruits can add significant amounts of kilocalories to your diet. A small apple, half a grapefruit, a half cup of grapes, or six ounces of fruit juice is considered a serving. Fruits provide fiber, carbohydrate, water, and certain of the vitamins, particularly vitamin C.

Vegetables Group

The Food Guide Pyramid suggests three to five servings from this group each day. Items in this group include nonsweet plant materials, such as broccoli, carrots, cabbage, corn, green beans, tomatoes, potatoes, lettuce, and spinach. A serving is considered one cup of raw leafy vegetables or a half cup of other types. It is wise to include as much variety as possible in this group. If you eat only carrots, several cups each day can become very boring. There is increasing evidence indicating that cabbage, broccoli, and cauliflower can provide some protection from certain types of cancers. This is a good reason to include these foods in your diet.

Foods in this group provide vitamins A and C as well as water and minerals. They also provide fiber, which assists in the proper functioning of the digestive tract.

Dairy Products Group

All of the cheeses, ice cream, yogurt, and milk are in this group. Two servings from this group are recommended each day. Each of these servings should be about 1 cup, or 2 ounces of cheese. Using product labels will help you determine the appropriate serving size of individual items. This group provides minerals, such as calcium, in your diet, but also provides water, vitamins, carbohydrates, and protein. You must also remember that many cheeses contain large amounts of cholesterol and fat for each serving. Low-fat dairy products are increasingly common as manufacturers seek to match their products with the desire of the public for less fat in the diet.

Meat, Poultry, Fish, and Dry Beans Group

This group contains most of the things we eat as a source of protein; for example, nuts, peas, tofu, and eggs are considered a part of this group. It is recommended that we include 4 to 6 ounces (100–170 grams) of these items in our daily diet. In the past, people in the economically developed world ate at least this quantity and frequently much more. Since many sources of protein also include significant fat, and health recommendations suggest reduced fat, more attention is being paid to the quantity of the protein-rich foods in the diet. Not only have we decreased our intake of items from this group, but we have also shifted from the high-fat-content foods, such as beef and pork, to foods that are high in protein but lower in fat content, such as fish and poultry. Beans (except for the oil-rich soybean) are also excellent ways to get needed protein without unwanted fat. Modern food preparers tend to use smaller portions and cook foods in ways that reduce the fat content. Broiled fish, rather than fried, and baked, skinless portions of chicken or turkey (the fat is attached to the skin) are seen more and more often on restaurant menus and on dining-room tables at home.

Remember that the recommended daily portion from this group is only 4 to 6 ounces. This means that one double cheeseburger greatly exceeds this recommendation for the daily intake. Eating excessive amounts of protein can stress the kidneys by causing higher concentrations of calcium in the urine, increase the demand for water to remove toxic keto acids produced from the breakdown of amino acids, and lead to weight gain because of the fat normally associated with many sources of protein. It should be noted, however, that vegetarians must pay particular attention to acquiring adequate sources of protein because they have eliminated a major source from their diet. Although nuts and soybeans are high in protein, they should not be consumed in large quantities because they are also high in fats.

Eating Disorders

The three most common eating disorders are obesity, bulimia, and anorexia nervosa. All three disorders are related to the prevailing perceptions and values of the culture in which we live. In many cases there is a strong psychological component as well.

Obesity

People who gain a great deal of unnecessary weight and are 15 percent to 20 percent above their ideal weight are obese, and suffer from a disease condition known as obesity. **Obesity** is the condition of being overweight to the extent that a person's health and life span are adversely affected. Obesity occurs when people consistently take in more food energy than is necessary to meet their daily requirements.

On the surface it would appear that obesity is a simple problem to solve. To lose weight all that people must do is consume fewer kilocalories, exercise more, or do both. While all obese people have an imbalance between their need for kilocalories and the amount of food they eat, the reasons for this imbalance are complex and varied.

Bulimia

Bulimia ("hunger of an ox" in Greek) is a disease condition in which the person has a cycle of eating binges followed by purging the body of the food by inducing vomiting or the use of laxatives. Many bulimics also use diuretics that cause the body to lose water and therefore reduce weight. It is often called the silent killer because it is difficult to detect. Bulimics are usually of normal body weight or overweight. The cause is thought to be psychological, stemming from depression, low self-esteem, displaced anger, a need to be in control of one's body, or a person-

ality disorder. Many bulimics have other compulsive behaviors such as drug abuse and are often involved in incidences of theft and suicide.

Vomiting may be induced physically or by the use of some nonprescription drugs. Case studies have shown that bulimics may take forty to sixty laxatives a day to rid themselves of food. For some, the laxative becomes addictive. The binge-purge cycle and associated use of diuretics results in a variety of symptoms that can be deadly.

Anorexia Nervosa

Anorexia nervosa is a nutritional deficiency disease characterized by severe, prolonged weight loss as a result of a voluntary severe restriction in food intake. An anorexic person's fear of becoming overweight is so intense that even though weight loss occurs, it does not lessen the fear of obesity, and the person continues to diet, often even refusing to maintain the optimum body weight for his or her age, sex, and height. This nutritional deficiency disease is thought to stem from sociocultural factors. Our society's preoccupation with weight loss and the desirability of being thin strongly influences this disorder.

Just turn on your television or radio, or look at newspapers, magazines, or billboards, and you can see how our culture encourages people to be thin. Male and female models are thin. Muscular bodies are considered healthy and any stored body fat unhealthy. Unless you are thin, so the advertisements imply, you will never be popular, get a date, or even marry. Our culture's constant emphasis on being thin has influenced many people to become anorexic and lose too much weight. Anorexic individuals frequently starve themselves to death. Individuals with anorexia are mostly adolescent and preadolescent females, although the disease does occur in males.

Deficiency Diseases

Without minimal levels of the essential amino acids in the diet, a person may develop health problems that could ultimately lead to death. In many parts of the world, large populations of people live on diets that are very high in carbohydrates and fats but low in complete protein. This is easy to understand, since carbohydrates and fats are inexpensive to grow and process in comparison to proteins. For example, corn, rice, wheat, and barley are all high-carbohydrate foods. Corn and its products (meal, flour) contain protein, but it is an incomplete protein that has very low amounts of the amino acids tryptophan and lysine. Without enough of these amino acids, many necessary enzymes cannot be made in sufficient amounts to keep a person healthy. One protein-deficiency disease is called **kwashiorkor,** and the symptoms are easily seen (figure 26.34). A person with this deficiency has a distended belly, slow growth, slow movement, and is emotionally depressed. If the disease is caught in time, brain damage may be prevented and death averted. This requires a change in diet that includes expensive protein, such as poultry, fish, beef, shrimp, or milk. As the world food problem increases, these expensive foods will be in even shorter supply and will become more and more costly.

FIGURE 26.34

This starving child shows the symptoms of kwashiorkor, a protein-deficiency disease. If treated with a proper diet containing all amino acids, the symptoms can be reduced.

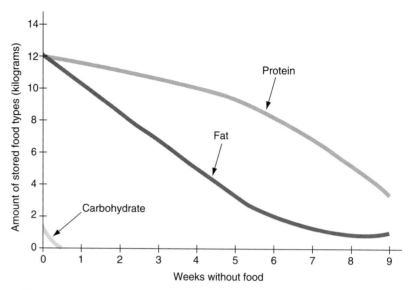

FIGURE 26.35

Starving yourself results in a very selective loss of the kinds of nutrients stored in the body. Notice how the protein level in the body has the slowest decrease of the three nutrients. This protein-conservation mechanism enables the body to preserve essential amounts of enzymes and other vital proteins.

Starvation is also a common problem in many parts of the world. Very little carbohydrate is stored in the body. If you starve yourself, this small amount will last as a stored form of energy only for about two days. Even after a few hours of fasting, the body begins to use its stored fat deposits as a source of energy; and as soon as the carbohydrates are gone proteins will begin to be used to provide a source of glucose. Some of the keto acids produced during the breakdown of fats and amino acids are released in the breath and can be detected as an unusual odor. People who are fasting, anorexic, diabetic, or have other metabolic problems often have this "ketone breath." During the early stages of starvation, the amount of fat in the body steadily decreases, but the amount of protein drops only slightly (20–30 grams per day) (figure 26.35). This can continue only up to a certain point. After several weeks of fasting, so much fat has been lost from the body that proteins are no longer protected, and cells begin to use them as a primary source of energy (as much as 125 grams per day). This results in a loss of proteins from the cells that prevents them from carrying out their normal functions. When starvation gets to this point it is usually fatal.

The lack of a particular vitamin in the diet can result in a **vitamin-deficiency disease.** A great deal has been said about the need for vitamin and mineral supplements in diets. Some people claim that supplements are essential, while others claim that a well-balanced diet provides adequate amounts of vitamins and miner-als. Supporters of vitamin supplements have even claimed that extremely high doses of certain vitamins can prevent poor health or even create "supermen." It is very difficult to substantiate many of these claims, however. Since the function of vitamins and minerals and their regulation in the body is not completely understood, the RDAs are best estimates by experts who have looked at the data from a variety of studies. In fact, the minimum daily requirement of a number of vitamins has not been determined. Vitamin-deficiency diseases that show recognizable symptoms are extremely rare except in cases of extremely poor nutrition.

CONTROL MECHANISMS

A large, multicellular organism, which consists of many different kinds of systems, must have some way of integrating various functions so that it can survive. The various systems must be coordinated to maintain a reasonably constant internal environment. Recall that the condition of maintaining a constant internal environment is called *homeostasis.* To allow for homeostasis there must be constant monitoring and modification of the way specific parts of the organism function. If the organism does not respond appropriately, it will die. There are many kinds of sense organs located within organs and on the surface that respond to specific kinds of stimuli. A **stimulus** is any change in the

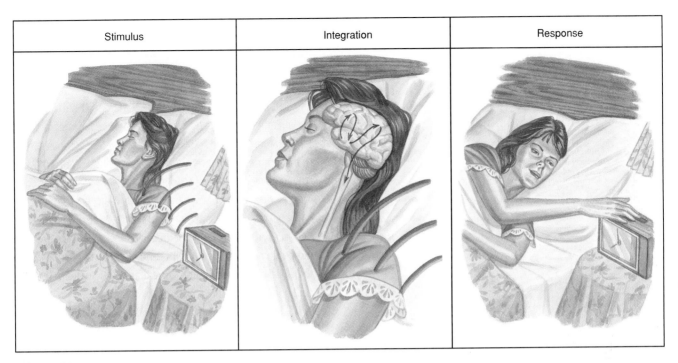

Stimulus	Integration	Response

FIGURE 26.36

A stimulus is any detectable change in the surroundings of an organism. When an organism receives a stimulus, it processes the information and may ignore the stimulus or generate a response to it.

environment that the organism can detect. Some stimuli, like light or sound, are typically external to the organism; others, like the pain generated by an infection, are internal. The reaction of the organism to a stimulus is known as a **response** (figure 26.36).

Integration of Input

The nervous and endocrine systems are the major systems of the body that integrate stimuli and generate appropriate responses necessary to maintain homeostasis. The **nervous system** consists of a network of cells with fibrous extensions that carry information throughout the body. The **endocrine system** consists of a number of glands that communicate with one another and with other tissues through chemicals distributed throughout the organism. **Glands** are organs that manufacture molecules that are secreted either through ducts or into surrounding tissue, where they are picked up by the circulatory system. *Endocrine glands* have no ducts and secrete their products, called **hormones,** into the circulatory system; other glands, such as the digestive glands and sweat glands, empty their contents through ducts. These kinds of glands are called *exocrine glands.*

Although the functions of the nervous and endocrine systems can overlap and be interrelated, these two systems have quite different methods of action. The nervous system functions very much like a computer. A message is sent along established pathways from a specific initiating point to a specific end point, and the transmission is very rapid. The endocrine system functions in a manner analogous to a radio broadcast system. Radio

stations send their signals in all directions, but only those radio receivers that are tuned to the correct frequency can receive the message. Messenger molecules (hormones) are typically distributed throughout the body by the circulatory system, but only those cells that have the proper receptor sites can receive and respond to the molecules.

The Structure of the Nervous System

The basic unit of the nervous system is a specialized cell called a **neuron,** or *nerve cell.* A typical neuron consists of a central body called the **soma,** or *cell body,* that contains the nucleus and several long, protoplasmic extensions called nerve fibers. There are two kinds of fibers: **axons,** which carry information away from the cell body, and **dendrites,** which carry information toward the cell body (figure 26.37). Most nerve cells have one axon and several dendrites.

Neurons are arranged into two major systems. The **central nervous system,** which consists of the brain and spinal cord, is surrounded by the skull and the vertebrae of the spinal column. It receives input from sense organs, interprets information, and generates responses. The **peripheral nervous system** is located outside the skull and spinal column and consists of bundles of long axons and dendrites called nerves. There are two different sets of neurons in the peripheral nervous system. *Motor neurons* carry messages from the central nervous system to muscles and glands, and *sensory neurons* carry input from sense organs to the central nervous system. Motor neurons typically have one long

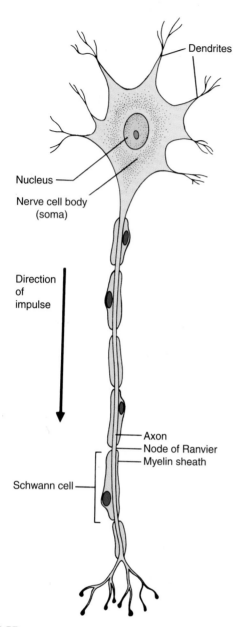

FIGURE 26.37

Nerve cells consist of a nerve-cell body that contains the nucleus and several fibrous extensions. The shorter, more numerous fibers that carry impulses to the nerve cell body are dendrites. The long fiber that carries the impulse away from the cell body is the axon.

axon that runs from the spinal cord to a muscle or gland, while sensory neurons have long dendrites that carry input from the sense organs to the central nervous system.

The Nature of the Nerve Impulse

Since most nerve cells have long fibrous extensions, it is possible for information to be passed along the nerve cell from one end to the other. The message that travels along a neuron is known as a **nerve impulse.** A nerve impulse is not a simple electric cur-

Human Comfort

Calculating the average temperature of a large sample of people, we find that the human body maintains an average internal body temperature of about 37.0°C (98.6°F). This average temperature is of the internal body core, not the skin and not the arms or legs, which can be considerably cooler or warmer. In addition, the core temperature is not *always* the average value, even if you are healthy. It is slightly less at night and slightly more during the day, but usually not more than a degree or so. In addition, the average temperature varies slightly with certain hormonal changes. Thus there are daily and monthly temperature fluctuations, but the core body temperature is always maintained independent of the temperature of the immediate environment. It is maintained by a section of the brain called the hypothalamus, which acts as a thermostat by conserving body heat at low environmental temperatures and getting rid of body heat at high environmental temperatures. This built-in thermostat is generally activated when the environmental temperature is below 20°C (68°F) or above 28°C (about 82°F). Between these two temperatures, most people generally have a feeling of comfort. The level of activity, age, gender, the relative humidity, air movement, and combinations of these factors influence the actual comfort range. Temperature, however, is usually the most important variable influencing a sense of comfort. We use heat technology to control our living environment, increasing or lowering the temperature to somewhere between 20°C and 28°C for comfort of the people.

rent but involves a specific sequence of chemical events involving activities at the cell membrane.

Since all cell membranes are differentially permeable, it is difficult for some ions to pass through the membrane, and the combination of ions inside the membrane is different from that on the outside. Cell membranes also contain proteins that actively transport specific ions from one side of the membrane to the other. Active transport involves the cell's use of ATP to move materials from one side of the cell membrane to the other. Because ATP is required, this is an ability that cells lose when they die. One of the ions that is actively transported from cells is the sodium ion (Na^+). At the same time sodium ions (Na^+) are being transported out of cells, potassium ions (K^+) are being transported into the normal resting cells. However, there are more sodium ions (Na^+) transported out than potassium ions (K^+) transported in.

Because a normal resting cell has more positively charged Na^+ ions on the outside of the cell than on the inside, a small but measurable voltage exists across the membrane of the cell. The

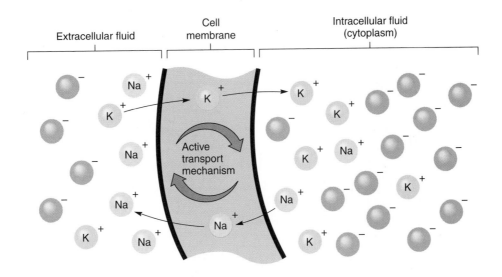

Extracellular fluid   Cell membrane   Intracellular fluid (cytoplasm)

FIGURE 26.38

All cells, including nerve cells, have an active transport mechanism that pumps Na$^+$ out of cells and simultaneously pumps K$^+$ into them. The end result is that there are more Na$^+$ ions outside the cell and more K$^+$ ions inside the cell. In addition, negative ions such as Cl$^-$ are more numerous inside the cell. Consequently, the outside of the cell is positive (+) compared to the inside, which is negative (−).

voltage difference between the inside and outside of a cell membrane is about 70 millivolts (.07 volt). The two sides of the cell membrane are, therefore, polarized in the same sense that a battery is polarized, with a positive and negative pole. A resting neuron has its positive pole on the outside of the cell membrane and its negative pole on the inside of the membrane (figure 26.38).

When a cell is stimulated at a specific point on the cell membrane, the cell membrane changes its permeability and lets sodium ions (Na$^+$) pass through it from the outside to the inside and potassium ions (K$^+$) diffuse from the inside to the outside. The membrane is thus *depolarized;* it loses its difference in charge as sodium ions (Na$^+$) diffuse into the cell from the outside. Sodium ions diffuse into the cell because initially they are in greater concentration outside the cell than inside. When the membrane becomes more permeable, they are able to diffuse into the cell, toward the area of lower concentration. The depolarization of one point on the cell membrane causes the adjacent portion of the cell membrane to change its permeability as well, and it also depolarizes. Thus, a wave of depolarization passes along the length of the neuron from one end to the other (figure 26.39). The depolarization and passage of an impulse along any portion of the neuron is a momentary event. As soon as a section of the membrane has been depolarized, potassium ions diffuse back into the cell. This reestablishes the original polarized state, and the membrane is said to be *repolarized.* When the nerve impulse reaches the end of the axon, it stimulates the release of a molecule that stimulates depolarization of the next neuron in the chain.

Activities at the Synapse

Between the fibers of adjacent neurons in a chain is a space called the **synapse.** Many chemical events occur in the synapse that are important in the function of the nervous system. When a neuron is stimulated, an impulse passes along its length from one end to the other. When the impulse reaches a synapse, a molecule called a **neurotransmitter** is released into the synapse from the axon. It diffuses across the synapse and binds to specific receptor sites on the dendrite of the next neuron. When enough neurotransmitter molecules have bound to the second neuron, an impulse is initiated in it as well. Several kinds of neurotransmitters are produced by specific neurons. These include dopamine, epinephrine, acetylcholine, and several other molecules. The first neurotransmitter identified was *acetylcholine.* Acetylcholine molecules are manufactured in the soma and migrate down the axon where they are stored until needed.

As long as a neurotransmitter is bound to its receptor it continues to stimulate the nerve cell. Thus if acetylcholine continues to occupy receptors, the neuron continues to be stimulated again and again. An enzyme called *acetylcholinesterase* destroys acetylcholine and prevents this from happening. The breakdown products of the acetylcholine can be used to remanufacture new acetylcholine molecules at the end of the axon. The destruction of acetylcholine allows the second neuron in the chain to return to normal. Thus it will be ready to accept another burst of acetylcholine from the first neuron a short time later. Neurons must also constantly manufacture new acetylcholine molecules or they will exhaust their supply and be unable to conduct an impulse across a synapse.

Caffeine and nicotine are considered stimulants because they make it easier for impulses to pass through the synapse. The nerve cells are sensitized and will respond to smaller amounts of acetylcholine than normal. Drinking coffee, taking caffeine pills, or smoking will increase the sensitivity of the nervous system to stimulation. However, as with most kinds of drugs, continual use or abuse tends to lead to a loss of the effect as the nervous system adapts to the constant presence of the drugs.

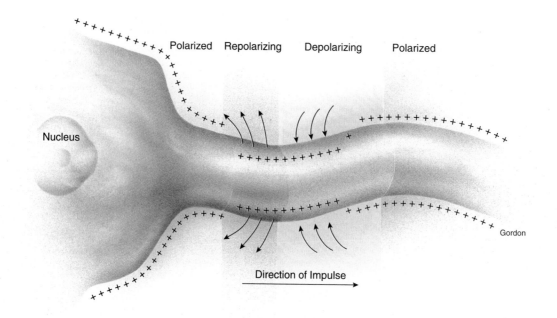

Polarized   Repolarizing   Depolarizing   Polarized

Nucleus

Gordon

Direction of Impulse

FIGURE 26.39

When a nerve cell is stimulated, a small portion of the cell membrane depolarizes as Na^+ flows into the cell through the membrane. This encourages the depolarization of an adjacent portion of the membrane, and it depolarizes a short time later. In this way a wave of depolarization passes down the length of the nerve cell. Shortly after a portion of the membrane is depolarized, the ionic balance is reestablished. It is repolarized and ready to be stimulated again.

Because of the way the synapse works, impulses can go in only one direction: Only axons secrete acetylcholine, and only dendrites have receptors. This explains why there are sensory and motor neurons to carry messages to and from the central nervous system.

Endocrine System Function

As mentioned previously, the endocrine system is basically a broadcasting system in which glands secrete messenger molecules, called hormones, which are distributed throughout the body by the circulatory system (figure 26.40). However, each kind of hormone attaches only to appropriate receptor molecules on the surfaces of certain cells. The cells that receive the message typically respond in one of three ways: (1) Some cells release products that have been previously manufactured, (2) other cells are stimulated to synthesize molecules or to begin metabolic activities, and (3) some are stimulated to divide and grow.

These different kinds of responses mean that some endocrine responses are relatively rapid, while others are very slow. For example, the release of the hormones *epinephrine* and *norepinephrine* from the adrenal medulla, located near the kidney, causes a rapid change in the behavior of an organism. The heart rate increases, blood pressure rises, blood is shunted to muscles, and the breathing rate increases. You have certainly experienced this reaction many times in your lifetime as when you nearly have an automobile accident or slip and nearly fall.

Another hormone, called *antidiuretic hormone (ADH)*, acts more slowly. It is released from the posterior pituitary gland at the base of the brain and regulates the rate at which the body loses water through the kidneys. It does this by encouraging the reabsorption of water from their collecting ducts. The effects of this hormone can be noticed in a matter of minutes to hours. Insulin is another hormone whose effects are quite rapid. Insulin is produced by the pancreas, located near the stomach, and stimulates cells—particularly muscle, liver, and fat cells—to take up glucose from the blood. After a meal that is high in carbohydrates, the level of glucose in the blood begins to rise, stimulating the pancreas to release insulin. The increased insulin causes glucose levels to fall as the sugar is taken up by cells. People with diabetes have insufficient or improperly acting insulin or lack the receptors to respond to the insulin, and therefore have difficulty regulating glucose levels in their blood.

The responses that result from the growth of cells may take weeks or years to occur. For example, *growth-stimulating hormone (GSH)* is produced by the anterior pituitary gland over a period of years and results in typical human growth. After sexual maturity, the amount of this hormone generally drops to very low levels and body growth stops. Sexual development is also largely the result of the growth of specific tissues and organs. The male sex hormone *testosterone*, produced by the testes, causes the growth of male sex organs and a change to the adult body form. The female counterpart, *estrogen*, results in the development of female sex organs and body form. In all of these cases, it is the

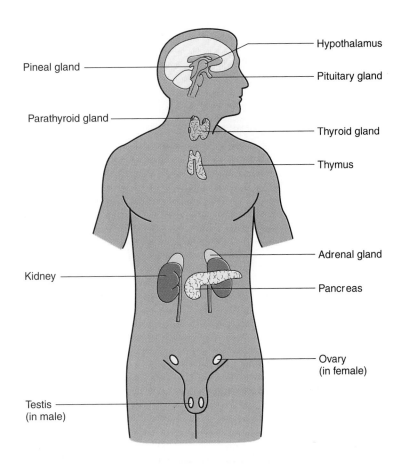

Pineal gland

Hypothalamus

Pituitary gland

Parathyroid gland

Thyroid gland

Thymus

Adrenal gland

Kidney

Pancreas

Ovary
(in female)

Testis
(in male)

FIGURE 26.40

The endocrine glands are located at various locations within the body and cause their effects by secreting hormones.

release of hormones over long periods, continually stimulating the growth of sensitive tissues, that results in the normal developmental pattern. The absence or inhibition of any of these hormones early in life changes the normal growth process.

Because the pituitary is constantly receiving information from the hypothalamus of the brain, many kinds of sensory stimuli to the body can affect the functioning of the endocrine system. One example is the way in which the nervous system and endocrine system interact to influence the menstrual cycle. At least three different hormones are involved in the cycle of changes that affect the ovary and the lining of the uterus. It is well documented that stress caused by tension or worry can interfere with the normal cycle of hormones and delay or stop menstrual cycles. In addition, young women living in groups, such as in college dormitories, often find that their menstrual cycles become synchronized. Although the exact mechanism involved in this phenomenon is unknown, it is suspected that input from the nervous system causes this synchronization. (Odors and sympathetic feelings have been suggested as causes.)

Although we still tend to think of the nervous and endocrine systems as being separate and different, it is becoming clear that they are interconnected. These two systems cooperate to bring about appropriate responses to environmental challenges. The nervous system is specialized for receiving and sending short-

term messages, whereas activities that require long-term, growth-related actions are handled by the endocrine system.

Sensory Input

The activities of the nervous and endocrine systems are often responses to some kind of input received from the sense organs. Sense organs of various types are located throughout the body. Many of them are located on the surface, where environmental changes can be easily detected. Hearing, sight, and touch are good examples of such senses. Other sense organs are located within the body and indicate to the organism how its various parts are changing. For example, pain and pressure are often used to monitor internal conditions. The sense organs detect changes, but the brain is responsible for perception—the recognition that a stimulus has been received. Sensory abilities involve many different kinds of mechanisms, including chemical recognition, the detection of energy changes, and the monitoring of forces.

Chemical Detection

All cells have receptors on their surfaces that can bind selectively to molecules they encounter. This binding process can cause changes in the cell in several ways. In some cells it causes depolarization. When this happens, the cells can stimulate neurons

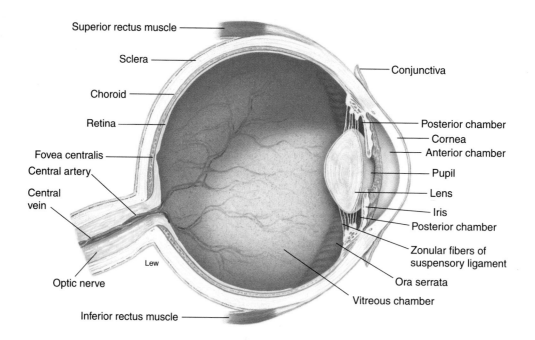

Superior rectus muscle
Sclera
Choroid
Retina
Fovea centralis
Central artery
Central vein
Optic nerve
Inferior rectus muscle
Lew
Conjunctiva
Posterior chamber
Cornea
Anterior chamber
Pupil
Lens
Iris
Posterior chamber
Zonular fibers of suspensory ligament
Ora serrata
Vitreous chamber

FIGURE 26.41

The eye contains a cornea and lens that focus the light on the retina of the eye. The light causes pigments in the rods and cones of the retina to decompose. This leads to the depolarization of these cells and the stimulation of neurons that send messages to the brain.

and cause messages to be sent to the central nervous system, informing it of some change in the surroundings. In other cases, a molecule binding to the cell surface may cause certain genes to be expressed, and the cell responds by changing the molecules it produces. This is typical of the way the endocrine system receives and delivers messages.

Most cells have specific binding sites for particular molecules. Others, such as the taste buds on the tongue, appear to respond to classes of molecules. They are able to distinguish four kinds of tastes: sweet, sour, salt, and bitter. These different kinds of taste buds are found at specific locations on the tongue's surface. The taste buds that give us the sour sensation appear to respond to the presence of hydrogen ions (H^+), because acids taste sour. Sodium chloride stimulates the taste buds that give us the sensation of a salty taste. However, the sensation of sweetness can be stimulated by many kinds of organic molecules, including sugars and artificial sweeteners, and also by inorganic lead compounds. The sweet taste of lead salts in old paints partly explains why children may eat paint chips. Since the lead interferes with normal brain development, this can have disastrous results. Many other kinds of compounds of diverse structures give the bitter sensation.

It is also important to understand that much of what we often refer to as taste involves such inputs as temperature, texture, and smell. Cold coffee has a different taste than hot coffee even though they are chemically the same. Lumpy, cooked cereal and smooth cereal have different tastes. If you are unable to smell food, it doesn't taste as it should, which is why you sometimes lose your appetite when you have a stuffy nose. We still have much to learn about how the tongue detects chemicals and the role that other associated senses play in modifying taste.

The other major chemical sense, the sense of smell, is much more versatile; it can detect thousands of different molecules at very low concentrations. The cells that make up the **olfactory epithelium,** the cells that line the nasal cavity and respond to smells, apparently bind molecules to receptors on their surface. Exactly how this can account for the large number of recognizably different odors is unknown, but the receptor cells are extremely sensitive. In some cases a single molecule of a substance is sufficient to cause a receptor cell to send a message to the brain, where the sensation of odor is perceived. These sensory cells also fatigue rapidly. You have probably noticed that when you first walk into a room, specific odors are readily detected, but after a few minutes you are unable to detect them. Most perfumes and aftershaves are undetectable after 15 minutes of continuous stimulation.

Many internal sense organs also respond to specific molecules. For example, the brain and aorta contain cells that respond to concentrations of hydrogen ions, carbon dioxide, and oxygen in the blood. Remember, too, that the endocrine system relies on the detection of specific messenger molecules to trigger its activities.

Light Detection

The eyes primarily respond to changes in the flow of light energy. The structure of the eye is designed to focus light on a light-sensitive layer of the back of the eye known as the **retina** (figure 26.41). There are two kinds of receptors in the retina of the eye. The cells called *rods* respond to a broad range of wavelengths of light and are responsible for black-and-white vision.

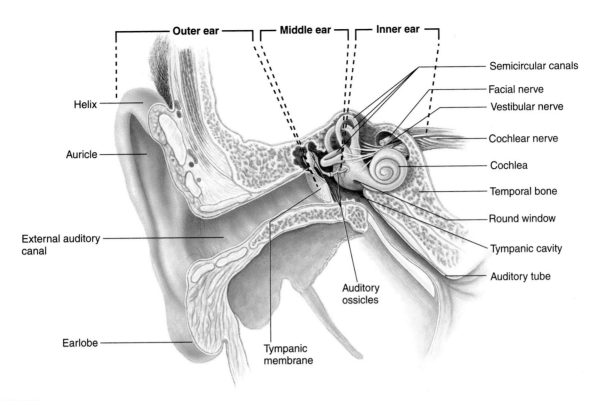

FIGURE 26.42

The ear consists of an external cone that directs sound waves to the tympanum. Vibrations of the tympanum move the ear bones and vibrate the oval window of the cochlea, where the sound is detected. The semicircular canals monitor changes in the position of the head, helping us maintain balance.

Since rods are very sensitive to light, they are particularly useful in dim light. Rods are located over most of the retinal surface except for the area of most acute vision known as the *fovea centralis*. The other receptor cells, called *cones*, are found throughout the retina but are particularly concentrated in the fovea centralis. Cones are not as sensitive to light, but they can detect different wavelengths of light. This combination of receptors gives us the ability to detect color when light levels are high, but we rely on black-and-white vision at night. There are three different varieties of cones: One type responds best to red light, another responds best to green light, and the third responds best to blue light. Stimulation of various combinations of these three kinds of cones allows us to detect different shades of color.

Rods and the three different kinds of cones each contain a pigment that decomposes when struck by light of the proper wavelength and sufficient strength. The pigment found in rods is called *rhodopsin*. This change in the structure of rhodopsin causes the rod to depolarize. Cone cells have a similar mechanism of action, and each of the three kinds of cones has a different pigment. Since rods and cones synapse with neurons, they stimulate a neuron when depolarized and cause a message to be sent to the brain. Thus, the pattern of color and light intensity recorded on the retina is detected by rods and cones and converted into a series of nerve impulses that are received and interpreted by the brain.

Sound Detection

The ears respond to changes in sound waves. Sound is produced by the vibration of molecules. Consequently, the ears are detecting changes in the quantity of energy and the quality of sound waves. Sound has several characteristics. Loudness, or volume, is a measure of the intensity of sound energy that arrives at the ear. Very loud sounds will literally vibrate your body and can cause hearing loss if they are too intense. Pitch is a quality of sound that is determined by the frequency of the sound vibrations. High-pitched sounds have short wavelengths; low-pitched sounds have long wavelengths.

Figure 26.42 shows the anatomy of the ear. The sound that arrives at the ear is first funneled by the external ear to the **tympanum**, also known as the *eardrum*. The cone-shaped nature of the external ear focuses sound on the tympanum and causes it to vibrate at the same frequency as the sound waves reaching it. Attached to the tympanum are three tiny bones known as the *malleus* (hammer), *incus* (anvil), and *stapes* (stirrup). The malleus is attached to the tympanum, the incus is attached to the malleus and stapes, and the stapes is attached to a small, membrane-covered opening called the oval window in a snail-shaped structure known as the cochlea. The vibration of the tympanum causes the tiny bones (malleus, incus, and stapes) to vibrate, and they in turn cause a corresponding vibration in the membrane of the *oval window.*

The *cochlea* of the ear is the structure that detects sound and consists of a snail-shaped set of fluid-filled tubes. When the oval window vibrates, the fluid in the cochlea begins to move, causing a membrane in the cochlea, called the *basilar membrane,* to vibrate. High-pitched, short-wavelength sounds cause the basilar membrane to vibrate at the base of the cochlea near the oval window. Low-pitched, long-wavelength sounds vibrate the basilar membrane far from the oval window. Loud sounds cause the basilar membrane to vibrate more vigorously than do faint sounds. Cells on this membrane depolarize when they are stimulated by its vibrations. Because they synapse with neurons, messages can be sent to the brain.

Because sounds of different wavelengths stimulate different portions of the cochlea, the brain is able to determine the pitch of a sound. Most sounds consist of a mixture of pitches that are heard. Louder sounds stimulate the membrane more forcefully, causing the sensory cells in the cochlea to send more nerve impulses per second. Thus, the brain is able to perceive the loudness of various sounds, as well as the pitch.

Associated with the cochlea are two fluid-filled chambers and a set of fluid-filled tubes called the *semicircular canals.* These structures are not involved in hearing but are involved in maintaining balance and posture. In the walls of these canals and chambers are cells similar to those found on the basilar membrane. These cells are stimulated by movements of the head and by the position of the head with respect to the force of gravity. The constantly changing position of the head results in sensory input that is important in maintaining balance.

Touch

What we normally call the sense of *touch* consists of a variety of different kinds of input. Some receptors respond to pressure, others to temperature, and others, which we call *pain receptors,* usually respond to cell damage. When these receptors are appropriately stimulated, they send a message to the brain. Because receptors are stimulated in particular parts of the body, the brain is able to localize the sensation. However, not all parts of the body are equally supplied with these receptors. The tips of the fingers, lips, and external genitals have the highest density of these nerve endings, while the back, legs, and arms have far fewer receptors.

Some internal receptors, such as pain and pressure receptors, are important in allowing us to monitor our internal activities. Many pains generated by the internal organs are often perceived as if they were somewhere else. For example, the pain associated with heart attack is often perceived to be in the left arm. Pressure receptors in joints and muscles are important in providing information about the degree of stress being placed on a portion of the body. This is also important information to send back to the brain so that adjustments can be made in movements to maintain posture. If you have ever had your foot "go to sleep" because the nerve stopped functioning, you have experienced what it is like to lose this constant input of nerve messages from the pressure

sensors to assist in guiding the movements you make. Your movements become uncoordinated until the nerve function returns to normal.

Output Coordination

The nervous system and endocrine system cause changes in several ways. Both systems can stimulate muscles to contract and glands to secrete. The endocrine system is also able to change metabolism of cells and regulate the growth of tissues. The nervous system acts upon two kinds of organs: muscles and glands. The actions of muscles and glands are simple and direct: muscles contract and glands secrete.

Muscles

The ability to move is one of the fundamental characteristics of animals. Through the coordinated contraction of many muscles, the intricate, precise movements of a dancer, basketball player, or writer are accomplished. It is important to recognize that muscles can pull only by contracting; they are unable to push by lengthening. The work of any muscle is done during its contraction. Relaxation is the passive state of the muscle. There must always be some force available that will stretch a muscle after it has stopped contracting and relaxes. Therefore, the muscles that control the movements of the skeleton are present in antagonistic sets—for every muscle's action there is another muscle that has the opposite action. For example, the biceps muscle causes the arm to flex (bend) as the muscle shortens. The contraction of its antagonist, the triceps muscle, causes the arm to extend (straighten) and at the same time stretches the relaxed biceps muscle (figure 26.43).

There are three major types of muscle: *skeletal, smooth,* and *cardiac.* These differ from one another in several ways. *Skeletal muscle* is voluntary muscle; it is under the control of the nervous system. The brain or spinal cord sends a message to skeletal muscles, and they contract to move the legs, fingers, and other parts of the body. This does not mean that you must make a conscious decision every time you want to move a muscle. Many of the movements we make are learned initially but become automatic as a result of practice. For example, walking, swimming, or riding a bicycle required a great amount of practice originally, but now you probably perform these movements without thinking about them. They are, however, still considered voluntary actions.

Smooth muscles make up the walls of muscular internal organs, such as the gut, blood vessels, and reproductive organs. They have the property of contracting as a response to being stretched. Since much of the digestive system is being stretched constantly, the responsive contractions contribute to the normal rhythmic movements associated with the digestive system. These are involuntary muscles; they can contract on their own without receiving direct messages from the nervous system.

Cardiac muscle is the muscle that makes up the heart. It has the ability to contract rapidly like skeletal muscle, but does not require nervous stimulation to do so. Nervous stimulation can, however,

A CLOSER LOOK | Which Type of Exercise Do You Do?

Aerobic exercise occurs when the muscles being contracted are supplied with sufficient oxygen to continue aerobic cellular respiration:

$$C_6H_{12}O_6 + 6O_2 \rightarrow 6CO_2 + 6H_2O + energy$$

This type of exercise involves long periods of activity with elevated breathing and heart rate. It results in strengthened chest muscles, which enable more complete exchange of air during breathing. It also improves the strength of the heart, enabling it to pump more efficiently. Flow of blood to the muscles also improves. All these changes increase endurance.

Anaerobic exercise takes place when insufficient amounts of oxygen reach the contracting muscle cells and they shift to anaerobic cellular respiration:

$$C_6H_{12}O_6 \rightarrow lactic\ acid + less\ energy$$

The buildup of lactic acid results in muscle pain and eventually prevents further contraction. Anaerobic exercise involves explosive bouts of activity, as in sprints or jumping. This kind of exercise increases muscle strength but does little to improve endurance.

Resistance exercise occurs when muscles contract against an object that does not allow the muscles to move that object. This type of exercise does not improve the ability of your body to deliver oxygen to your muscles, nor does it increase your endurance. However, it does stimulate your muscle cells to manufacture more contractile protein fibers.

Make a list of your activities that would be considered aerobic, anaerobic, or resistance exercise. Which activities would result in weight reduction? Weight gain? Improved cardiovascular fitness?

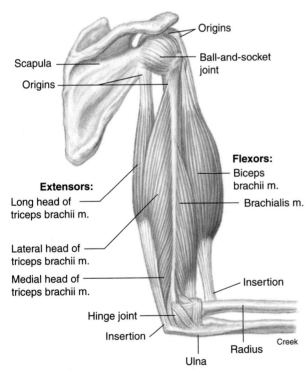

FIGURE 26.43

Because muscles cannot actively lengthen, it is necessary to have sets of muscles that oppose one another. The contraction and shortening of one muscle cause the stretching of a relaxed muscle.

cause the heart to speed or slow its rate of contraction. Hormones, such as epinephrine and norepinephrine, also influence the heart by increasing its rate and strength of contraction. Cardiac muscle also has the characteristic of being unable to stay contracted. It will contract quickly but must have a short period of relaxation before it will be able to contract a second time. This makes sense in light of its continuous, rhythmic, pumping function.

Glands

The glands of the body are of two different kinds. Those that secrete into the bloodstream are called **endocrine glands.** We have already talked about several of these: the pituitary, thyroid, ovary, and testis are examples. The **exocrine glands** are those that secrete to the surface of the body or into one of the tubular organs of the body, such as the gut or reproductive tract. Examples are the salivary glands, intestinal mucous glands, and sweat glands. Some of these glands, such as salivary glands and sweat glands, are under nervous control. When stimulated by the nervous system, they secrete their contents.

The Russian physiologist Ivan Petrovich Pavlov showed that salivary glands were under the control of the nervous system when he trained dogs to salivate in response to hearing a bell. You may recall that initially the animals were presented with food at the same time the bell was rung. Eventually they would salivate when the bell was rung even if food was not present. This demonstrated that saliva production was under the control of the central nervous system.

Many other exocrine glands are under hormonal control. Many of the digestive enzymes of the stomach and intestine are secreted in response to local hormones produced in the gut. These are circulated through the blood to the digestive glands, which respond by secreting the appropriate digestive enzymes and other molecules.

Growth Responses

The hormones produced by the endocrine system can have a variety of effects. As mentioned earlier, hormones can stimulate smooth muscle to contract and can influence the contraction of cardiac muscle as well. Many kinds of glands, both endocrine and exocrine, are caused to secrete as a result of a hormonal stimulus. However, the endocrine system has one major effect that is not equaled by the nervous system: Hormones regulate growth. Several examples of the many kinds of long-term growth changes that are caused by the endocrine system were given earlier in the chapter. *Growth-stimulating hormone (GSH)* is produced over a period of years to bring about the increase in size of most of the structures of the body. A low level of this hormone results in a person with small body size. It is important to recognize that the amount of growth-stimulating hormone (GSH) present varies from time to time. It is present in fairly high amounts throughout childhood and results in steady growth. It also appears to be present at higher levels at certain times, resulting in growth spurts. Finally, as adulthood is reached, the level of this hormone falls, and growth stops.

Similarly, testosterone produced during adolescence influences the growth of bone and muscle to provide men with larger, more muscular bodies than those of women. In addition, there is growth of the penis, growth of the larynx, and increased growth of hair on the face and body. The primary female hormone, estrogen, causes growth of reproductive organs and development of breast tissue. It is also involved, along with other hormones, in the cyclic growth and sloughing of the wall of the uterus.

SUMMARY

Meiosis is a specialized process of cell division resulting in the production of four cells, each of which has the *haploid number* of *chromosomes*. The total process involves two sequential divisions during which one *diploid cell* reduces to four *haploid cells*. Since the *chromosomes* act as carriers for *genetic information*, *genes* separate into different sets during meiosis.

The sex of many kinds of organisms is determined by specific chromosome combinations. In humans, *females* have *two X chromosomes*, while *males* have *an X and a Y chromosome*. At *puberty*, *hormones* influence the development of *secondary sex characteristics* and the functioning of *gonads*.

Sexual reproduction involves the production of *gametes* by *meiosis* in the *ovaries and testes*. In males, each cell that undergoes *spermatogenesis* results in four *sperm*; in females, each cell that undergoes *oogenesis* results in one *oocyte*.

The body's various *systems* must be integrated in such a way that the *internal environment* stays relatively constant. This concept is called *homeostasis*. The *circulatory, respiratory, digestive,* and *excretory systems* are involved in the exchange of materials across cell membranes. All of these systems have special features that provide *large surface* areas to allow for necessary *exchanges*.

The *circulatory system* consists of a pump, the *heart*, and *blood vessels* that distribute the blood to all parts of the body. The *blood* is a carrier fluid that transports molecules and heat. The exchange of materials between the *blood* and body *cells* takes place through the walls of the *capillaries*. Hemoglobin in *red blood cells* is very important in the transport of *oxygen*.

The *respiratory system* consists of the *lungs* and associated tubes that allow air to enter and leave the lungs. The *diaphragm* and *muscles* of the chest wall are important in the process of *breathing*. In the lungs, tiny sacs called *alveoli* provide a large surface area in association with capillaries, which allows for rapid exchange of oxygen and carbon dioxide.

The *digestive system* is involved in disassembling food molecules. This involves several processes: *grinding* by the teeth and stomach, *emulsification* of fats by bile from the liver, *addition of water* to dissolve molecules, and *enzymatic action* to break complex molecules into simpler molecules for *absorption*. The intestine provides a *large surface area* for the *absorption of nutrients* because it is long and its wall contains many tiny projections that increase surface area. Once absorbed, the materials are carried to the *liver*, where molecules can be modified.

The *excretory system* is a filtering system of the body. The *kidneys* consist of *nephrons* into which the *circulatory system* filters fluid. Most of this fluid is useful and is reclaimed by the cells that make up the walls of these *tubules*. Materials that are present in excess or those that are harmful are allowed to escape.

To maintain good health, people must receive *nutrient molecules* that can enter the cells and function in the metabolic processes. The *proper quantity and quality* of *nutrients* are essential to good health. An important measure of the amount of energy required to sustain a human at rest is the *basal metabolic rate*. To meet this and all additional requirements, the United States has established the *RDAs, Recommended Dietary Allowances,* for each nutrient. Should there be metabolic or psychological problems associated with a person's *normal metabolism*, a variety of disorders may occur, including *obesity, anorexia nervosa, bulimia, kwashiorkor,* and *vitamin-deficiency diseases*.

A *nerve impulse* is caused by sodium ions entering the cell as a result of a change in the permeability of the cell membrane. Thus, a *wave of depolarization* passes down the length of a *neuron* to the *synapse*. The *axon* of a neuron secretes a *neurotransmitter*, such as *acetylcholine,* into the *synapse,* where these molecules bind to the dendrite of the next cell in the chain, resulting in an impulse in it as well. The *acetylcholinesterase* present in the synapse destroys acetylcholine so that it does not repeatedly stimulate the dendrite.

Several kinds of *sensory inputs* are possible. Many kinds of chemicals can bind to cell surfaces and be recognized. This is probably how the *sense of taste* and the *sense of smell* function. Light energy can be detected because light causes certain molecules in the *retina* of the eye to decompose and stimulate neurons. *Sound* can be detected because fluid in the *cochlea* of the *ear* is caused to vibrate, and special cells detect this movement and stimulate neurons. The sense of *touch* consists of a variety of receptors that respond to *pressure, cell damage,* and *temperature*.

Glands are of two types: *exocrine glands,* which secrete through ducts into the cavity of an organ or to the surface of the skin, and *endocrine glands,* which release their secretions into the circulatory system. It is becoming clear that the endocrine system and the nervous system are interrelated. Actions of the endocrine system can change how the nervous system functions and the reverse is also true.

KEY TERMS

alveoli (p. **716**)

anatomy (p. **696**)

androgens (p. **708**)

anorexia nervosa (p. **727**)

anther (p. **696**)

aorta (p. **712**)

arteries (p. **710**)

arterioles (p. **715**)

atria (p. **712**)

atrioventricular valves (p. **712**)

autosomes (p. **702**)

axons (p. **729**)

basal metabolic rate (p. **724**)

blood (p. **710**)

breathing (p. **716**)

bronchi (p. **716**)

bronchioles (p. **716**)

bulimia (p. **727**)

calorie (p. **724**)

capillaries (p. **711**)

central nervous system (p. **729**)

coitus (p. **702**)

copulation (p. **702**)

corpus luteum (p. **702**)

crossing-over (p. **699**)

dendrites (p. **729**)

diaphragm (p. **716**)

diastolic blood pressure (p. **713**)

diet (p. **724**)

diploid (p. **696**)

duodenum (p. **719**)

egg (p. **696**)

ejaculation (p. **710**)

endocrine glands (p. **737**)

endocrine system (p. **729**)

estrogen (p. **708**)

exocrine glands (p. **737**)

follicle (p. **702**)

follicle stimulating hormone (p. **707**)

Food Guide Pyramid (p. **725**)

gamete (p. **696**)

gametogenesis (p. **702**)

gastric juice (p. **719**)

glands (p. **729**)

gonad (p. **696**)

haploid (p. **696**)

heart (p. **710**)

hemoglobin (p. **711**)

homeostasis (p. **710**)

hormones (p. **729**)

hypothalamus (p. **707**)

immune system (p. **711**)

interstitial cell-stimulating hormone (p. **710**)

kidney (p. **722**)

kilocalorie (p. **724**)

kwashiorkor (p. **727**)

large intestine (p. **720**)

liver (p. **719**)

lung (p. **716**)

lymph (p. **716**)

lymphatic system (p. **716**)

masturbation (p. **708**)

meiosis (p. **696**)

menstrual cycle (p. **708**)

microvilli (p. **710**)

nephrons (p. **722**)

nerve impulse (p. **730**)

nervous system (p. **729**)

neuron (p. **729**)

neurotransmitter (p. **731**)

nutrients (p. **722**)

nutrition (p. **722**)

obesity (p. **727**)

olfactory epithelium (p. **734**)

oogenesis (p. **702**)

orgasm (p. **708**)

ovary (p. **696**)

ovulation (p. **708**)

pancreas (p. **719**)

penis (p. **702**)

peripheral nervous system (p. **729**)

pharynx (p. **719**)

physiology (p. **696**)

pistil (p. **696**)

pituitary gland (p. **707**)

plasma (p. **711**)

puberty (p. **707**)

pulmonary artery (p. **713**)

pulmonary circulation (p. **713**)

Recommended Dietary Allowances (p. **725**)

reduction division (p. **699**)

response (p. **729**)

retina (p. **734**)

saliva (p. **718**)

salivary glands (p. **718**)

secondary sexual characteristics (p. **708**)

semilunar valves (p. **712**)

seminal vesicles (p. **710**)

seminiferous tubules (p. **702**)

sex chromosomes (p. **702**)

sexual intercourse (p. **702**)

sexual reproduction (p. **696**)

small intestine (p. **719**)

soma (p. **729**)

specific dynamic action (p. **724**)

sperm (p. **696, 702**)

spermatogenesis (p. **702**)

stimulus (p. **728**)

synapse (p. **731**)

systemic circulation (p. **713**)

systolic blood pressure (p. **713**)

testes (p. **696**)

testosterone (p. **710**)

trachea (p. **716**)

tympanum (p. **735**)

vagina (p. **707**)

veins (p. **711**)

ventricles (p. **712**)

villi (p. **720**)

vitamin-deficiency disease (p. **728**

APPLYING THE CONCEPTS

1. The exchange of genetic material (genes) between segments of homologous chromosomes results in

   a. new gene combinations.

   b. zygotes.

   c. diploid cells.

   d. segregation of genes.

2. A process that occurs during prophase I is

   a. segregation.

   b. synapsis.

   c. reduction division.

   d. independent assortment.

3. The diploid number of chromosomes is found in cells during

   a. prophase II.

   b. telophase I.

   c. anaphase II.

   d. prophase I.

4. The fact that each homologous pair of chromosomes in humans separates and moves to the poles without being influenced by the other pairs is

   a. segregation.

   b. disintegration.

   c. independent assortment.

   d. fertilization.

5. A new nuclear membrane is formed in

   a. anaphase I.

   b. prophase II.

   c. telophase I.

   d. anaphase II.

6. In mitosis the centromeres split during anaphase; in meiosis they split during
   a. anaphase I.
   b. telophase I.
   c. prophase II.
   d. anaphase II.

7. Diploid cells are formed by
   a. synapsis.
   b. reduction division.
   c. fertilization.
   d. independent assortment.

8. An organism having a diploid number of 12 forms gametes having
   a. 6 chromosomes.
   b. 12 chromosomes.
   c. 18 chromosomes.
   d. 24 chromosomes.

9. Segregation of homologous chromosomes occurs during
   a. mitosis.
   b. meiosis I.
   c. meiosis II.
   d. all of the above.

10. Gametogenesis produces
    a. sex cells.
    b. gonads.
    c. zygotes.
    d. testes.

11. Which of the following represents chromosome number before and after the process of meiosis?
    a. $n \rightarrow n$
    b. $n \rightarrow 2n$
    c. $2n \rightarrow n$
    d. $2n \rightarrow 2n$

12. You could identify human cells as being either male or female by
    a. looking at the size of a cell.
    b. examining the chromosomes.
    c. looking for differentiation.
    d. weighing them.

13. Ovaries and testes differ from other kinds of organs because
    a. their cells divide.
    b. their cells live for only a short time.
    c. meiosis occurs in them.
    d. they grow continuously.

14. The sex of an individual is set
    a. at conception.
    b. at birth.
    c. at puberty.
    d. during pregnancy.

15. Which one of the following would be a secondary sex characteristic?
    a. ovaries in a female
    b. the presence of a penis in a male
    c. a menstrual cycle in a female
    d. pubic hair in males and females

16. Which one of the following hormones is responsible for stimulating the testes to produce testosterone?
    a. interstitial cell-stimulating hormone
    b. estrogen
    c. androgens
    d. oxytocin

17. The primary structures involved in pumping blood are the
    a. veins.
    b. atria.
    c. capillaries.
    d. ventricles.

18. The fluid portion of the blood that leaves the capillaries and surrounds the cells is
    a. hemoglobin.
    b. edema.
    c. lymph.
    d. lacteal.

19. Blood is carried through vessels to all parts of the body except the lungs by
    a. pulmonary circulation.
    b. the pulmonary artery.
    c. systemic circulation.
    d. the lymphatic system.

20. As air passes through the lungs it follows the path:
    a. trachea $\rightarrow$ bronchi $\rightarrow$ bronchioles $\rightarrow$ alveoli
    b. trachea $\rightarrow$ bronchioles $\rightarrow$ bronchi $\rightarrow$ alveoli
    c. bronchi $\rightarrow$ trachea $\rightarrow$ alveoli $\rightarrow$ bronchioles
    d. bronchioles $\rightarrow$ alveoli $\rightarrow$ bronchi $\rightarrow$ trachea

21. The levels of water, hydrogen ions, salts, and urea in the blood are regulated by the
    a. liver.
    b. kidneys.
    c. bladder.
    d. rectum.

22. Persons of normal weight who overeat and force the food out of their digestive tract
    a. are obese.
    b. are bulimic.
    c. have anorexia nervosa.
    d. have beriberi.

23. A complete protein contains
    a. all the recommended dietary allowances.
    b. essential fatty acids.
    c. the suggested kilocalorie amount.
    d. all the essential amino acids.

24. A person with kwashiorkor suffers from a lack of
    a. proteins.
    b. minerals.
    c. vitamins.
    d. kilocalories.

25. Converting a nutrient into a molecule in your body's cells is
    a. digestion.
    b. assimilation.
    c. carbohydrate loading.
    d. protein-sparing.

26. RDAs are easily met by
    a. eating breakfast cereal.
    b. using vitamin supplements when you are not feeling well.
    c. eating a well-balance, varied diet.
    d. consuming more fiber.

27. Carbohydrates that are not digested in the human digestive tract are
    a. fibers.
    b. minerals.
    c. vitamins.
    d. fatty acids.

28. The **C**alories listed on package labels are equal to _____ calories.
    a. 1
    b. 100
    c. 1,000
    d. 10,000

29. Osteoporosis results from a _____ deficiency.
    a. vitamin
    b. calcium
    c. protein
    d. energy

30. An organism's reaction to a change in the environment is a(n)
    a. stimulus.
    b. impulse.
    c. response.
    d. perception.

31. Acetylcholine is destroyed by
    a. cholinesterase.
    b. endocrine glands.
    c. exocrine glands.
    d. axons.

32. A light stimulus is received by the nervous system which results in growth. This is the result of
    a. release of hormones from the thyroid.
    b. activating muscles.
    c. stimulating the endocrine system.
    d. increasing nervous activity.

33. When a nerve cell is stimulated,
    a. acetylcholine is destroyed.
    b. potassium ions enter the neuron.
    c. sodium ions enter the neuron.
    d. calcium attaches to the dendrites.

34. Which one of the following is necessary for muscle contraction?
    a. calcium ions
    b. oxygen
    c. testosterone
    d. fat

35. When the temperature of a home falls below a set point, the furnace produces heat. Once the home has reached the desired temperature, the furnace shuts off. As the home cools and again falls below the set point the furnace will again produce heat. This cycle is similar to _____ that occurs in the human body.
    a. depolarization
    b. negative feedback control
    c. muscle contraction
    d. a synapse

36. Which one of the following is *not* a type of muscle?
    a. skeletal muscle
    b. cardiac muscle
    c. neuron muscle
    d. smooth muscle

37. The central nervous system consists of the
    a. brain only.
    b. brain and spinal cord.
    c. brain, spinal cord, and nerves.
    d. motor neurons and sensory neurons.

38. Olfactory senses detect
    a. light.
    b. sound.
    c. chemicals.
    d. pain.

39. The ear bones are the
    a. malleus, incus, and stapes.
    b. tympanum and cochlea.
    c. rods and cones.
    d. fovea centralis and olfactory epithelium.

40. The chemical messengers secreted by endocrine glands are called
    a. enzymes.
    b. hormones.
    c. neurotransmitters.
    d. impulses.

41. Which of the following is *not* part of a nerve cell?
    a. dendrites
    b. soma
    c. axon
    d. myosin

Answers

1. a **2.** b **3.** d **4.** c **5.** c **6.** d **7.** c **8.** a **9.** b **10.** a **11.** c **12.** b **13.** c **14.** a **15.** d **16.** a **17.** d **18.** c **19.** c **20.** a **21.** b **22.** b **23.** d **24.** a **25.** b **26.** c **27.** a **28.** c **29.** b **30.** c **31.** a **32.** c **33.** c **34.** a **35.** b **36.** c **37.** b **38.** c **39.** a **40.** b **41.** d

QUESTIONS FOR THOUGHT

1. How do haploid cells differ from diploid cells?
2. What is unique about prophase I?
3. Why is meiosis necessary in organisms that reproduce sexually?
4. Diagram the metaphase I stage of a cell with the diploid number of 8.
5. Diagram fertilization as it would occur between a sperm and an egg with the haploid number of 3.
6. Describe the processes that cause about 50 percent of babies to be born male and 50 percent female.
7. What structures are associated with the human female reproductive system? What are their functions?
8. What structures are associated with the human male reproductive system? What are their functions?
9. What are the differences between oogenesis and spermatogenesis in humans?
10. How are ovulation and menses related to each other?
11. What are the functions of the heart, arteries, veins, arterioles, the blood, and capillaries?
12. Describe three ways in which the digestive system increases its ability to absorb nutrients.
13. Describe the mechanics of breathing.
14. Name the five basic food groups and give two examples of each.
15. What do the initials RDA stand for?
16. Describe how changing permeability of the cell membrane and the movement of sodium ions cause a nerve impulse.
17. What is the role of acetylcholine in a synapse? What is the role of acetylcholinesterase?
18. Give an example of the interaction between the endocrine system and the nervous system.
19. What is actually detected by the nasal epithelium, taste buds, cochlea of the ear, and retina of the eye?
20. List the differences between the central and peripheral nervous systems.

The expression of many genes is influenced by the environment. The gene for dark hair in the cat is sensitive to temperature and expresses itself only in the parts of the body that stay cool

CHAPTER | Twenty-Seven

Mendelian and Molecular Genetics

Why do you have a particular blood type or hair color? Why do some people have the same skin color as their parents, while others have a skin color different from that of their parents? Why do flowers show such a wide variety of colors? Why is it that generation after generation of plants, animals, and microbes look so much like members of their own kind? These questions and many others can be better answered if you have an understanding of genetics.

GENETICS, MEIOSIS, AND CELLS

A **gene** is a portion of DNA that determines a characteristic. Through meiosis and reproduction, genes can be transmitted from one generation to another. The study of genes, how genes produce characteristics, and how the characteristics are inherited is the field of biology called **genetics.** The first person to systematically study inheritance and formulate laws about how characteristics are passed from one generation to the next was an Augustinian monk named Gregor Mendel (1822–1884). Mendel's work was not generally accepted until 1900, when three men, working independently, rediscovered some of the ideas that Mendel had formulated more than thirty years earlier. Because of his early work, the study of the pattern of inheritance that follows the laws formulated by Gregor Mendel is often called **Mendelian genetics.**

To understand this chapter, you need to know some basic terminology. One term that you have already encountered is *gene.* Mendel thought of a gene as a particle that could be passed from the parents to the *offspring* (children, descendants, or progeny). Today we know that genes are actually composed of specific sequences of DNA nucleotides. The particle concept is not entirely inaccurate, because a particular gene is located at a specific place on a chromosome called its *locus* (loci = pl.; location).

Another important idea to remember is that all sexually reproducing organisms have a diploid (2*n*) stage. Since gametes are haploid (*n*) and most organisms are diploid, the conversion of diploid to haploid cells during meiosis is an important process.

$$2 \, (n) \rightarrow \text{meiosis} \rightarrow (n) \text{ gametes}$$

The diploid cells have two sets of chromosomes—one set inherited from each parent.

$$n + n \text{ gametes} \rightarrow \text{fertilization} \rightarrow 2n$$

Therefore, they have two chromosomes of each kind and have two genes for each characteristic. When sex cells are produced by meiosis, reduction division occurs, and the diploid number is reduced to haploid. Therefore, the sex cells produced by meiosis have only one chromosome of each of the homologous pairs that were in the diploid cell that began meiosis. Diploid organisms usually result from the fertilization of a haploid egg by a haploid sperm. Therefore, they inherit one gene of each type from each parent. For example, each of us has two genes for earlobe shape: one came with our father's sperm, the other with our mother's egg (figure 27.1).

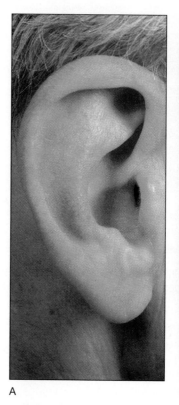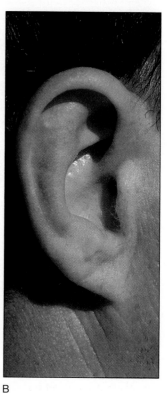

A   B

FIGURE 27.1

Whether your earlobe is free (*A*) or attached (*B*) depends on the genes you have inherited. As genes express themselves, their actions affect the development of various tissues and organs. Some people's earlobes do not separate from the sides of their heads in the same manner as do those of others. How genes control this complex growth pattern and why certain genes function differently than others is yet to be clarified.

SINGLE-GENE INHERITANCE PATTERNS

In diploid organisms there may be two different forms of the gene. In fact, there may be *several* alternative forms or **alleles** of each gene within a population. In people, for example, there are two alleles for earlobe shape. One allele produces an earlobe that is fleshy and hangs free, while the other allele produces a lobe that is attached to the side of the face and does not hang free. The type of earlobe that is present is determined by the type of allele (gene) received from each parent and the way in which these

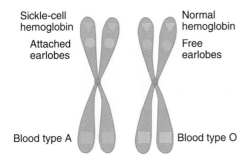

Sickle-cell hemoglobin

Attached earlobes

Normal hemoglobin

Free earlobes

Blood type A

Blood type O

FIGURE 27.2

Homologous chromosomes contain genes for the same characteristics at the same place. Note that the attached-earlobe allele is located at the ear-shape locus on one chromosome, and the free-earlobe allele is located at the ear-shape locus on the other member of the homologous pair of chromosomes. The other two genes are for hemoglobin structure (alleles for normal and sickled cells) and blood type (alleles for blood types A and O). The examples presented here are for illustrative purposes only. We do not really know if these particular genes are on these chromosomes.

alleles interact with one another. Alleles are located on the same pair of homologous chromosomes—one allele on each chromosome. These alleles are also at the same specific location, or locus (figure 27.2).

The **genome** is a set of all the genes necessary to specify an organism's complete list of characteristics. The term genome is used in two ways. It may refer to the diploid ($2n$) or haploid (n) number of chromosomes in a cell. Be sure to clarify how your instructor uses this term. The **genotype** of an organism is a listing of the genes present in that organism. It consists of the cell's DNA code; therefore, you cannot see the genotype of an organism. It is not yet possible to know the complete genotype of most organisms, but it is often possible to figure out the genes present that determine a particular characteristic. For example, there are three possible genotypic combinations of the two alleles for earlobe shape. Genotypes are typically represented by upper- and lowercase letters. In the case of the earlobe trait, the allele for free earlobes is designated "E" while that for attached earlobes is "e." A person's genotype could be (1) two alleles for attached earlobes, (2) one allele for attached earlobes and one allele for free earlobes, or (3) two alleles for free earlobes.

How would individuals with each of these three genotypes look? The way each combination of alleles *expresses* (shows) itself is known as the **phenotype** of the organism. The phrase *gene expression* refers to the degree to which a gene goes through transcription and translation to show itself as a physical feature of the individual.

A person with two alleles for attached earlobes will have earlobes that do not hang free. A person with one allele for attached earlobes and one allele for free earlobes will have a phenotype that exhibits free earlobes. An individual with two alleles for free earlobes will also have free earlobes. Notice that there are three genotypes, but only two phenotypes. The indi-

viduals with the free-earlobe phenotype can have different genotypes.

Alleles	Genotypes	Phenotypes
E = free earlobes	EE	free earlobes
e = attached earlobes	Ee	free earlobes
	ee	attached earlobes

The expression of some genes is directly influenced by the presence of other alleles. For any particular pair of alleles in an individual, the two alleles from the two parents are either identical or not identical. A person is **homozygous** for a trait when they have the combination of two identical alleles for that particular characteristic, for example, EE and ee. A person with two alleles for freckles is said to be homozygous for that trait. A person with two alleles for no freckles is also homozygous. If an organism is homozygous, the characteristic expresses itself in a specific manner. A person homozygous for free earlobes has free earlobes, and a person homozygous for attached earlobes has attached earlobes.

An individual is designated as **heterozygous** when they have two different allelic forms of a particular gene, for example, Ee. The heterozygous individual received one form of the gene from one parent and a different allele from the other parent. For instance, a person with one allele for freckles and one allele for no freckles is heterozygous. If an organism is heterozygous, these two different alleles interact to determine a characteristic. A **carrier** is any person who is heterozygous for a trait. In this situation, the recessive allele is hidden, that is, does not expresses itself enough to be a phenotype.

Dominant and Recessive Alleles

Often, one allele in the pair expresses itself more than the other. A **dominant allele** masks the effect of other alleles for the trait. For example, if a person has one allele for free earlobes and one allele for attached earlobes, that person has a phenotype of free earlobes. We say the allele for free earlobes is dominant. A **recessive allele** is one that, when present with another allele, has its actions overshadowed by the other; it is masked by the effect of the other allele. Having attached earlobes is the result of having a combination of two recessive characteristics. A person with one allele for free earlobes and one allele for attached earlobes has a phenotype of free earlobes. The expression of recessive alleles is only noted when the organism is homozygous for the recessive alleles. If you have attached earlobes, you have two alleles for that trait. Don't think that recessive genes are necessarily bad. *The term recessive has nothing to do with the significance or value of the gene— it simply describes how it can be expressed. Recessive alleles are not less likely to be inherited but must be present in a homozygous condition to express themselves. Also, recessive alleles are not necessarily less frequent in the population.* Sometimes, the physical environment determines whether or not dominant or recessive genes function. For example, in humans genes for freckles do not show themselves fully unless the person's skin is exposed to sunlight (figure 27.3).

FIGURE 27.3

The expression of many genes is influenced by the environment. The allele for dark hair in the cat is sensitive to temperature and expresses itself only in the parts of the body that stay cool. The allele for freckles expresses itself more fully when a person is exposed to sunlight.

Codominance

In cases of dominance and recessiveness, one allele of the pair clearly overpowers the other. Although this is common, it is not always the case. In some combinations of alleles, there is a **codominance.** This is a situation in which both alleles in a heterozygous person express themselves. An example involves the genetic abnormality, *sickle-cell disease.* Having the two recessive alleles for sickle-cell hemoglobin ($Hb^S Hb^S$) can result in abnormally shaped red blood cells. This occurs because the hemoglobin molecules are synthesized with the wrong amino acid sequence. These abnormal hemoglobin molecules tend to attach to one another in long, rodlike chains when oxygen is in short supply, that is, with exercise, pneumonia, or emphysema. These rodlike chains distort the shape of the red blood cells into a sickle shape. When these abnormal red blood cells change shape, they clog small blood vessels. The sickled red cells are also destroyed more rapidly than normal cells. This results in a shortage of red blood cells, a condition known as anemia, and an oxygen deficiency in the tissues that have become clogged. People with sickle-cell anemia may experience pain, swelling, and damage to organs such as the heart, lungs, brain, and kidneys.

Sickle-cell anemia can be lethal in the homozygous recessive condition. In the homozygous dominant condition ($Hb^A Hb^A$), the person has normal red blood cells. In the heterozygous condition ($Hb^A Hb^S$), patients produce both kinds of red blood cells. When the amount of oxygen in the blood falls below a certain level, those able to sickle will distort. However, when this occurs, most people heterozygous for the trait do not show severe symptoms. Therefore, these alleles related to one another in a codominant fashion. However, under the right circumstances, being heterozygous can be beneficial. A person with a single sickle-cell allele is more resistant to malaria than a person without this allele.

Genotype	Phenotype
$Hb^A Hb^A$	Normal hemoglobin and nonresistance to malaria
$Hb^A Hb^S$	Normal hemoglobin and resistance to malaria
$Hb^S Hb^S$	Resistance to malaria but death from sickle-cell anemia

Originally, sickle-cell anemia was found at a high frequency in parts of the world where malaria was common, such as tropical regions of Africa and South America. Today, however, this genetic disease can be found anywhere in the world. In the United States, it is most common among black populations whose ancestors came from equatorial Africa.

A classic example of codominance in plants involves the color of the petals of snapdragons. There are two alleles for the color of these flowers. Because neither allele is recessive, we cannot use the traditional capital and small letters as symbols for these alleles. Instead, the allele for white petals is given the symbol F^W, and the one for red petals is given the symbol F^R (figure 27.4).

There are three possible combinations of these two alleles:

Genotype	Phenotype
$F^W F^W =$	White flower
$F^R F^R =$	Red flower
$F^R F^W =$	Pink flower

Notice that there are only two different alleles, red and white, but there are three phenotypes, red, white, and pink. Both the red-flower allele and the white-flower allele partially express themselves when both are present, and this results in pink.

X-Linked Genes

Pairs of alleles located on nonhomologous chromosomes separate independently of one another during meiosis when the chromosomes separate into sex cells. Since each chromosome has many genes on it, these genes tend to be inherited as a group. Genes located on the same chromosome that tend to be inherited together are called a **linkage group.** The process of crossing-over, which occurs during prophase I of meiosis I, may split up these linkage groups. Crossing-over happens between homologous chromosomes donated by the mother and the father, and results in a mixing of genes. The closer two genes are to each other on a chromosome, the more probable it is that they will be inherited together.

FIGURE 27.4

The colors of these snapdragons are determined by two alleles for petal color, F^W and F^R. There are three different phenotypes because of the way in which the alleles interact with one another. In the heterozygous condition, neither of the alleles dominates the other.

People and many other organisms have two types of chromosomes. Autosomes (22 pairs) are not involved in sex determination and have the same genes on both members of the homologous pair of chromosomes. *Sex chromosomes* are a pair of chromosomes that control the sex of an organism. In humans, and some other animals, there are two types of sex chromosomes—the X chromosome and the Y chromosome. The Y chromosome is much shorter than the X chromosome and has few genes for traits found on the X chromosome (figure 27.5). One genetic trait that is located on the Y chromosome contains the testis-determining gene—SRY. Females are normally produced when two X chromosomes are present. Males are usually produced when one X chromosome and one Y chromosome are present.

Genes found together on the X chromosomes are said to be **X-linked.** Because the Y chromosome is shorter than the X chromosome, it does not have many of the alleles that are found on the comparable portion of the X chromosome. Therefore, in a man, the presence of a single allele on his only X chromosome

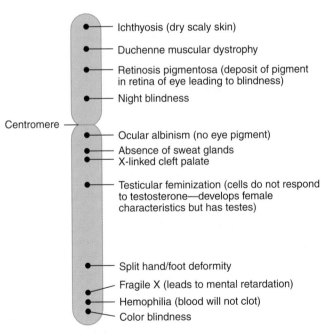

X chromosome

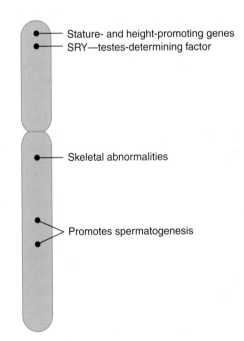

Y chromosome

FIGURE 27.5

Why is the Y chromosome so small? Is there an advantage to a species in having one sex chromosome deficient in genes? One hypothesis answers yes! Consider the idea that, with genes for supposedly "female" characteristics eliminated from the Y chromosome, crossing-over and recombining with "female" genes on the X chromosome during meiosis could help keep sex traits separated. Males would be males and females would stay females. The chances of "male-determining" and "female-determining" genes getting mixed onto the same chromosome would be next to impossible because they would not even exist on the Y chromosome.

TABLE 27.1	Dominant and recessive traits in pea plants	
Characteristic	**Dominant Allele**	**Recessive Allele**
Plant height	Tall	Dwarf
Pod shape	Full	Constricted
Pod color	Green	Yellow
Seed surface	Round	Wrinkled
Seed color	Yellow	Green
Flower color	Purple	White

will be expressed, regardless of whether it is dominant or recessive. A Y-linked trait in humans is the SRY gene. This gene controls the differentiation of the embryonic gonad to a male testis. By contrast, more than 100 genes are on the X chromosome. Some of these X-linked genes can result in abnormal traits such as *color blindness, hemophilia, brown teeth,* and at least two forms of *muscular dystrophy* (Becker's and Duchenne's).

MENDEL'S LAWS OF HEREDITY

Heredity problems are concerned with determining which alleles are passed from the parents to the offspring and how likely it is that various types of offspring will be produced. The first person to develop a method of predicting the outcome of inheritance patterns was Mendel, who performed experiments concerning the inheritance of certain characteristics in sweet pea plants. From his work, Mendel concluded which traits were dominant and which were recessive. Some of his results are shown in table 27.1.

What made Mendel's work unique was that he studied only one trait at a time. Previous investigators had tried to follow numerous traits at the same time. When this was attempted, the total set of characteristics was so cumbersome to work with that no clear idea could be formed of how the offspring inherited traits. Mendel used traits with clear-cut alternatives, such as purple or white flower color, yellow or green seed pods, and tall or dwarf pea plants. He was very lucky to have chosen pea plants in his study because they naturally self-pollinate. When self-pollination occurs in pea plants over many generations, it is possible to develop a population of plants that is homozygous for a number of characteristics. Such a population is known as a *pure line.*

Mendel took a pure line of pea plants having purple flower color, removed the male parts (anthers), and discarded them so that the plants could not self-pollinate. He then took anthers from a pure-breeding white-flowered plant and pollinated the antherless purple flower. When the pollinated flowers produced seeds, Mendel collected, labeled, and planted them. When these seeds germinated and grew, they eventually produced flowers.

You might be surprised to learn that all of the plants resulting from this cross had purple flowers. One of the prevailing hypotheses of Mendel's day would have predicted that the purple and white colors would have blended, resulting in flowers

that were lighter than the parental purple flowers. Another hypothesis would have predicted that the offspring would have had a mixture of white and purple flowers. The unexpected result—all of the offspring produced flowers like those of one parent and no flowers like those of the other—caused Mendel to examine other traits as well and formed the basis for much of the rest of his work. He repeated his experiments using pure strains for other traits. Pure-breeding tall plants were crossed with pure-breeding dwarf plants. Pure-breeding plants with yellow pods were crossed with pure-breeding plants with green pods. The results were all the same: the offspring showed the characteristic of one parent and not the other.

Next, Mendel crossed the offspring of the white-purple cross (all of which had purple flowers) with each other to see what the third generation would be like. Had the characteristic of the original white-flowered parent been lost completely? This second-generation cross was made by pollinating these purple flowers that had one white parent among themselves. The seeds produced from this cross were collected and grown. When these plants flowered, three-fourths of them produced purple flowers and one-fourth produced white flowers.

After analyzing his data, Mendel formulated several genetic laws to describe how characteristics are passed from one generation to the next and how they are expressed in an individual.

Mendel's **law of dominance**—When an organism has two different alleles for a given trait, the allele that is expressed, overshadowing the expression of the other allele, is said to be dominant. The gene whose expression is overshadowed is said to be recessive.

Mendel's **law of segregation**—When gametes are formed by a diploid organism, the alleles that control a trait separate from one another into different gametes, retaining their individuality.

Mendel's **law of independent assortment**—Members of one gene pair separate from each other independently of the members of other gene pairs.

At the time of Mendel's research, biologists knew nothing of chromosomes or DNA or of the processes of mitosis and meiosis. Mendel assumed that each gene was separate from other genes. It was fortunate for him that most of the characteristics he picked to study were found on separate chromosomes. If two or more of these genes had been located on the same chromosome (*linked genes*), he probably would not have been able to formulate his laws. The discovery of chromosomes and DNA has led to modifications in Mendel's laws, but it was Mendel's work that formed the foundation for the science of genetics.

PROBABILITY VERSUS POSSIBILITY

In order to solve heredity problems, you must have an understanding of probability. **Probability** is the chance that an event will happen, and is often expressed as a percent or a fraction. Probability is not the same as possibility. It is possible to toss a coin and have it come up heads. But the probability of getting a head is more precise than just saying it is possible to get a head. The probability of getting a head is one out of two (1/2 or 0.5 or

A CLOSER LOOK | Cystic Fibrosis—What's the Probability?

One in every twenty persons has a defective allele that causes cystic fibrosis. *Cystic fibrosis (CF)* is among the most common lethal genetic disorders affecting most whites in the USA and affects nearly 30,000 people in North America. About 5 percent of American whites are asymptomatic carriers of the recessive allele. About 1,000 new cases are identified in the United States each year. The gene for CF occurs on chromosome 7 and is responsible for the manufacture of cystic fibrosis transmembrane regulator (CFTR) protein. A deletion mutation in this gene is responsible for the loss of a single amino acid in CFTR. As a result, CFTR is unable to control the movement of chloride ions through chloride-

permeable cell membrane channels. This results in the malfunction of sweat glands in the skin and secretion of excess chloride ions, which causes a salty taste of the skin of children with CF. It is also responsible for other symptoms including:

1. Mucus filling bronchioles that results in blocked breathing and frequent upper-respiratory infections
2. Bile duct clogging that interferes with digestion and liver function
3. Pancreas ducts clogged with mucus that prevents the flow of digestive enzymes into the intestinal tract
4. Bowel obstructions caused by thickened stools

5. Sterility in males due to absence of vas deferens development and, on occasion, female sterility due to dense mucus that blocks sperm from reaching eggs

Until recently, the test used to detect this gene in people with a history of cystic fibrosis was only 50 percent accurate and could not detect the gene in persons with no family history of the disease. A new test is 75 percent accurate in detecting people with the gene regardless of their family history. This new test enables physicians to screen prospective parents to better determine the probability that they will have an offspring with cystic fibrosis. Confirmation of CF requires additional testing.

50%), because there are two sides to the coin, only one of which is a head. Probability can be expressed as a fraction:

$$\text{Probability} = \frac{\text{number of events that can produce a given outcome}}{\text{total number of possible outcomes}}$$

What is the probability of cutting a deck of cards and getting the ace of hearts? The number of times that the ace of hearts can occur is 1. The total number of possible outcomes (number of cards in the deck) is 52. Therefore, the probability of cutting an ace of hearts is 1/52.

What is the probability of cutting an ace? The total number of aces in the deck is 4, and the total number of cards is 52. Therefore, the probability of cutting an ace is 4/52 or 1/13.

It is also possible to determine the probability of two independent events occurring together. *The probability of two or more events occurring simultaneously is the product of their individual probabilities.* If you throw a pair of dice, it is possible that both will be fours. What is the probability that both will be fours? The probability of one die being a four is 1/6. The probability of the other die being a four is also 1/6. Therefore, the probability of throwing two fours is

$$1/6 \times 1/6 = 1/36$$

STEPS IN SOLVING HEREDITY PROBLEMS: SINGLE-FACTOR CROSSES

The first type of problem we will consider is the easiest type, a single-factor cross. A **single-factor cross** (sometimes called a monohybrid cross: *mono* = one; *hybrid* = combination) is a genetic

cross or mating in which a single characteristic is followed from one generation to the next.

For example, in humans, the allele for Tourette syndrome (TS) is inherited as an autosomal dominant allele. For centuries, people displaying this genetic disorder were thought to be possessed by the devil since they displayed such unusual behaviors. These motor and verbal behaviors or tics are involuntary and range from mild (e.g., leg tapping, eye blinking, face twitching) to the more violent forms such as the shouting of profanities, head jerking, spitting, compulsive repetition of words, or even barking like a dog. The symptoms result from an excess production of the brain messenger, dopamine. The most widely used drug to control elevated dopamine levels in Tourette syndrome patients is haloperidol. The penetrance of this dominant allele is different for males and females. Penetrance refers to the all-or-none expression of a given genotype in heterozygous individuals in a population. Understanding penetrance helps answer the question, "If all these people have this gene, why don't I see more people with the condition?" For example, heterozygous males show a 90 percent penetrance of TS, that is, 90 percent of men with TS express the trait enough to be diagnosed with TS. There is only 60 to 70 percent penetrance in the female population. If both parents are heterozygous (have one allele for Tourette syndrome and one allele for no Tourette syndrome) what is the probability that they can have a child without Tourette syndrome? with Tourette syndrome?

In solving a heredity problem there are five basic steps:

Step 1: Assign a symbol for each allele.
Usually a capital letter is used for a dominant allele and a small letter for a recessive allele. Use the symbol *T* for Tourette and *t* for no Tourette.

A CLOSER LOOK | Blame That Trait on Your Mother!

Within a eukaryotic cell, the bulk of DNA is located in the nucleus. It is this genetic material that controls the majority of biochemical processes of the cell. It has long been recognized that the mitochondria also contain DNA, mtDNA. This genetic material works in conjunction with that located in the nucleus. However, an interesting thing happens when a sperm is formed as a result of meiosis. All the mitochondria are packed into the "necks" and not the "heads" of the sperm. When a sperm penetrates a secondary oocyte, the head enters but the majority of the mitochondria in the neck remain outside. Therefore, fathers rarely if ever contribute mitochondrial genetic information to their children. Should a mutation occur in the mother's mitochondrial DNA, she will pass the abnormality on to her children. Should a mutation occur in the father's mitochondrial DNA, he will not transmit the abnormality. This unusual method of transmission is called *mitochondrial* (or *maternal*) *inheritance*. Most abnormalities transmitted through mitochondrial genes are associated with muscular weakness because the mitochondria are the major source of ATP in eukaryotic cells. People with one form of this rare type of genetic abnormality, *Leber's hereditary optic neuropathy,* show symptoms of sudden loss of central vision due to optic nerve death in young adults with onset at about 20 years of age. Both males and females may be affected.

Allele	Genotype	Phenotype
T = Tourette	*TT*	Tourette syndrome
t = normal	*Tt*	Tourette syndrome
	tt	Normal

Step 2:   Determine the genotype of each parent and indicate a mating.

Because both parents are heterozygous, the male genotype is *Tt*. The female genotype is also *Tt*. The × between them is used to indicate a mating.

$$Tt \times Tt$$

Step 3:   Determine all the possible kinds of gametes each parent can produce.

Remember that gametes are haploid; therefore, they can have only one allele instead of the two present in the diploid cell. Because the male has both the Tourette syndrome allele and the normal allele, half of his gametes will contain the Tourette syndrome allele and the other half will contain the normal allele. Because the female has the same genotype, her gametes will be the same as his.

For genetic problems, a *Punnett square* is used. A **Punnett square** is a box figure that allows you to determine the probability of genotypes and phenotypes of the progeny of a particular cross. Remember, because of the process of meiosis, each gamete receives only one allele for each characteristic listed. Therefore, the male will produce sperm with either a *T* or a *t*; the female will produce ova with either a *T* or a *t*. The possible gametes produced by the male parent are listed on the left side of the square and the female gametes are listed on the top. In our example, the Punnett square would show a single dominant allele and a single recessive allele from the male on the left side. The alleles from the female would appear on the top.

Male genotype
Tt

Possible male gametes
T and *t*

Female genotype
Tt

Possible female gametes
T and *t*

	T	*t*
T		
t		

Step 4:   Determine all the gene combinations that can result when these gametes unite.

To determine the possible combinations of alleles that could occur as a result of this mating, simply fill in each of the empty squares with the alleles that can be donated from each parent. Determine all the gene combinations that can result when these gametes unite.

	T	*t*
T	*TT*	*Tt*
t	*Tt*	*tt*

Step 5:   Determine the phenotype of each possible gene combination.

In this problem, three of the offspring, *TT*, *Tt*, and *Tt*, have Tourette syndrome. One progeny, *tt*, is normal. Therefore, the answer to the problem is that the probability of having offspring with Tourette syndrome is 3/4; for no Tourette syndrome, it is 1/4.

Take the time to learn these five steps. All single-factor problems can be solved using this method; the only variation in the problems will be the types of alleles and the number of possible types of gametes the parents can produce.

A CLOSER LOOK | Muscular Dystrophy and Genetics

Because most genes are comprised of thousands of nucleotide base pairs, there are many different changes in these sequences that can result in the formation of multiple "bad" gene products and therefore, an abnormal phenotype. The same gene can have many different "bad" forms. The fact that a phenotypic characteristic can be determined by many different alleles for a particular characteristic is called *genetic heterogeneity*. For example, two of the best-known forms of *muscular dystrophy* (*MD*) are Duchenne's and Becker's. Duchenne's (DMD) is characterized by a severe, progressive weakening of the muscles, the appearance of a false muscle atrophy (degeneration) in the calves of the legs, onset in early childhood, and a high likelihood of death in the thirties. DMD is caused by a mutation in the dystrophin gene located on the X chromosome and is a recessive trait. Becker's (BMD) is caused by a mutation in the same dystrophin gene but at a different location. BMD is a more mild form of MD.

REAL-WORLD SITUATIONS

So far we have considered a characteristic that is determined by simple dominance and recessiveness between two alleles. Other situations, however, may not fit this pattern. Some genetic characteristics are determined by more than two alleles; moreover, some traits are influenced by gene interactions and some traits are inherited differently, depending on the sex of the offspring.

Multiple Alleles

So far we have discussed only traits that are determined by two alleles, for example, *A, a*. However, there can be more than two different alleles for a single trait. All the various forms of the same gene (alleles) that control a particular trait are referred to as **multiple alleles.** However, one person can have only a maximum of two of the alleles for the characteristic. A good example of a characteristic that is determined by multiple alleles is the ABO blood type. There are three alleles for blood type:

Allele*

I^A = blood has type A antigens on red blood cell surface

I^B = blood has type B antigens on red blood cell surface

i = blood type O has neither type A nor type B antigens on surface of red blood cell

In the ABO system, A and B show *codominance* when they are together in the same individual, but both are dominant over the O allele. These three alleles can be combined as pairs in six different ways, resulting in four different phenotypes:

Genotype		Phenotype
$I^A I^A$	=	Blood type A
$I^A i$	=	Blood type A
$I^B I^B$	=	Blood type B
$I^B i$	=	Blood type B
$I^A I^B$	=	Blood type AB
ii	=	Blood type O

Multiple-allele problems are worked as single-factor problems. Some examples are in the practice problems at the end of this chapter.

Polygenic Inheritance

Thus far we have considered phenotypic characteristics that are determined by alleles at a specific, single place on homologous chromosomes. However, some characteristics are determined by the interaction of genes at several different loci (on different chromosomes or at different places on a single chromosome). This is called **polygenic inheritance.** A number of different pairs of alleles may combine their efforts to determine a characteristic. Skin color in humans is a good example of this inheritance pattern. According to some experts, genes for skin color are located at a minimum of three loci. At each of these loci, the allele for dark skin is dominant over the allele for light skin. Therefore, a wide variety of skin colors is possible depending on how many dark-skin alleles are present (figure 27.6).

Polygenic inheritance is very common in determining characteristics that are quantitative in nature. In the skin-color example, and in many others as well, the characteristics cannot be categorized in terms of *either/or*, but the variation in phenotypes can be classified as *how much* or *what amount*.

For instance, people show great variations in height. There are not just tall and short people—there is a wide range. Some people are as short as one meter, and others are taller than two meters. This quantitative trait is probably determined by a number of different genes. Intelligence also varies significantly, from those who are severely retarded to those who are geniuses. Many

---

*The symbols I^A and I^B stand for the technical term for the antigenic carbohydrates attached to red blood cells, the *Immunogens*. These alleles are located on human chromosome 9. The ABO system is not the only one used to type blood. Others include the Rh, MNS, and Xg systems.

Locus 1	d^1d^1	d^1D^1	d^1D^1	D^1D^1	D^1d^1	D^1d^1	D^1D^1
Locus 2	d^2d^2	d^2d^2	d^2D^2	D^2d^2	D^2d^2	D^2D^2	D^2D^2
Locus 3	d^3d^3	d^3d^3	d^3d^3	d^3d^3	D^3D^3	D^3D^3	D^3D^3
Total number of dark-skin genes	0	1	2	3	4	5	6

Very light   Medium   Very dark

FIGURE 27.6

Skin color in humans is an example of polygenic inheritance. The darkness of the skin is determined by the number of dark-skin genes a person inherits from his or her parents.

A

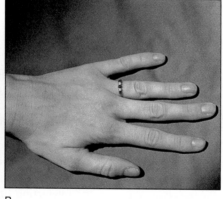

B

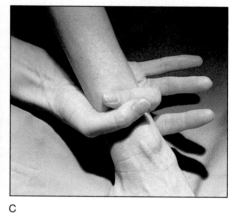

C

FIGURE 27.7

It is estimated that about 40,000 (the incidence is 1 out of 10,000) people in the United States have Marfan syndrome, an autosomal dominant abnormality. Notice the lanky appearance to the body and face of this person with Marfan syndrome (A). Photos (B) and (C) illustrate their unusually long fingers.

of these traits may be influenced by outside environmental factors such as diet, disease, accidents, and social factors. These are just a few examples of polygenic inheritance patterns.

Pleiotropy

Even though a single gene produces only one type of mRNA during transcription, it often has a variety of effects on the phenotype of the person. This is called *pleiotropy*. **Pleiotropy** (*pleio-* = changeable) is a term used to describe the multiple effects that a gene may have on the phenotype. A good example of pleiotropy is *Marfan syndrome* (figure 27.7), a disease suspected to have occurred in former U.S. President Abraham Lincoln. Marfan is a disorder of the body's connective tissue but can also have effects in many other organs, including the eyes, heart, blood, skeleton, and lungs. Symptoms generally appear as a tall, lanky body with long arms and spider-fingers, scoliosis (curvature of the spine), and depression of the chest

CONNECTIONS

The Inheritance of Eye Color

It is commonly thought that eye color is inherited in a simple dominant/recessive manner. Brown eyes are considered to be dominant over blue eyes. The real pattern of inheritance, however, is considerably more complicated than this. Eye color is determined by the amount of a brown pigment, known as melanin, present in the iris of the eye. If there is a large quantity of melanin present on the anterior surface of the iris, the eyes are dark. Black eyes have a greater quantity of melanin than brown eyes.

If there is not a large amount of melanin present on the anterior surface of the iris, the eyes will appear blue, not because of a blue pigment but because blue light is returned from the iris (see box figure 27.1). The iris appears blue for the same reason that deep bodies of water tend to appear blue. There is no blue pigment in the water, but blue wavelengths of light are returned to the eye from the water. People appear to have blue eyes because the blue wavelengths of light are reflected from the iris.

Just as black and brown eyes are determined by the amount of pigment present, colors such as green, gray, and hazel are produced by the various amounts of melanin in the iris. If a very small amount of brown melanin is present in the iris, the eye tends to appear green, whereas relatively large amounts of melanin produce hazel eyes.

Several different genes are probably involved in determining the quantity and placement of the melanin and, therefore, in determining eye color. These genes interact in such a way that a wide range of eye color is possible. Eye color is probably determined by polygenic inheritance, just as skin color and height are. Some newborn babies have blue eyes that later become brown. This is because they have not yet begun to produce melanin in their irises at the time of birth.

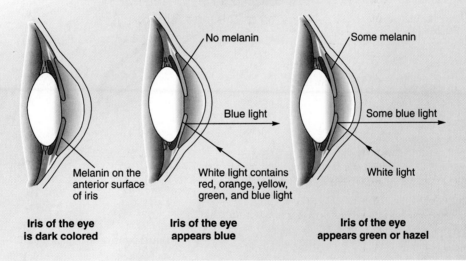

Iris of the eye is dark colored

Melanin on the anterior surface of iris

Iris of the eye appears blue

No melanin

Blue light

White light contains red, orange, yellow, green, and blue light

Iris of the eye appears green or hazel

Some melanin

Some blue light

White light

---

wall (funnel chest/pectus excavatum or pigeon chest/pectus carinatum). In many cases these nearsighted people also show dislocation of the lens of the eye. The white of the eye (sclera) may appear bluish. Heart problems include dilation of the aorta and prolapse of the heart's mitral valve. Death may be caused by a dissection (tear) in the aorta from the rupture in a weakened and dilated area of the aorta called an aortic *aneurysm.*

Another example of pleiotropy is the genetic abnormality PKU. In this example, a single gene affects many different chemical reactions that depend on the way a cell metabolizes the amino acid phenylalanine, commonly found in many foods (refer to figure 27.8).

ENVIRONMENTAL INFLUENCES ON GENE EXPRESSION

Maybe you assumed that the dominant allele would always be expressed in a heterozygous individual. It is not so simple! Here, as in other areas of biology, there are exceptions. For example, the allele for six fingers (*polydactylism*) is dominant over the allele for five fingers in humans. Some people who have received the allele for six fingers have a fairly complete sixth finger; in others, it may appear as a little stub. In another case, a dominant allele causes the formation of a little finger that cannot be bent as a normal little finger. However, not all people who are believed to have inherited that allele will have a stiff little finger. In some cases, this dominant characteristic is not

CONNECTIONS

Inheritance Pattern of Spina Bifida

Many polygenic traits are influenced by outside environmental factors, such as diet, drugs, disease, accidents, and social factors. The degree to which the environment impacts the genes can determine the nature of phenotype. Patterns of inheritance that are the result of polygenic traits influenced by environmental factors are called multifactorial inheritance. In fact, most polygenic traits are multifactorial. *Multifactorial inheritance* is not yet completely understood and is hard to trace since it does not follow the typical single-gene inheritance pattern. The first generation (children) to follow individuals with such diseases (parents) show a markedly high risk of acquiring the gene, whereas the second generation (grandchildren) has a much lower risk. Examples of multifactorial diseases include neural tube defects (e.g., spina bifida), insulin-dependent diabetes mellitus, cleft palate, different fingerprints in identical twins, and congenital heart disease. In the case of spina bifida the arches of the vertebrae fail to enclose the spinal cord (typically in the lower back region) and in some cases the cord and nerves may be exposed at birth. This abnormality is a leading cause of still births and death during early infancy, and if unable to be surgically corrected, is a handicap in survivors. One factor that appears to influence this disease is the amount of the vitamin folic acid the mother has in her system prior to conception. Other factors include chromosome disorders, single-gene mutations, and teratogens (chemicals that cause developmental abnormalities).

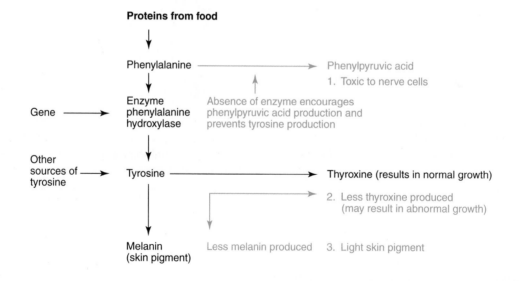

FIGURE 27.8

Phenylketonuria (PKU) is an autosomal recessive disorder located on chromosome 12. This diagram shows how the normal pathways work (these are shown in black). If the enzyme phenylalanine hydroxylase is not produced because of a mutated gene, the amino acid phenylalanine cannot be broken down, and is converted into phenylpyruvic acid, which accumulates in body fluids. There are three major results: (1) mental retardation because phenylpyruvic acid kills nerve cells, (2) abnormal body growth because less of the growth hormone thyroxine is produced, and (3) pale skin pigmentation because less melanin is produced (abnormalities are shown in color). It should also be noted that if a woman who has PKU becomes pregnant, her baby is likely to be born retarded. While the embryo may not have the genetic disorder, the phenylpyruvic acid produced by the pregnant mother will damage the developing brain cells. This is called maternal PKU.

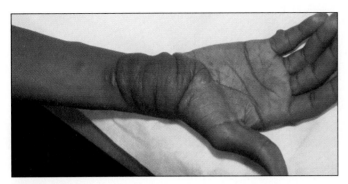

FIGURE 27.9

Neurofibromatosis I is seen in many forms including benign fibromatous skin tumors, "Café-au-lait" spots, nodules in the iris, and possible malignant tumors. It is extremely variable in its expressivity, i.e., the traits may be almost unnoticeable or extensive. An autosomal dominant trait, it is the result of a mutation and the production of a protein (neurofibromin) that normally would suppress the activity of a gene that causes tumor formation.

expressed or perhaps only shows on one hand. Thus, there may be variation in the degree to which an allele expresses itself in an *individual*. Geneticists refer to this as *variable expressivity*. A good example of this occurs in the genetic abnormality *neurofibromatosis type I* (NF1) (figure 27.9). In some cases it may not be expressed in the *population* at all. This is referred to as a *lack of penetrance*. Other genes may be interacting with these dominant alleles, causing the variation in expression.

Both internal and external environmental factors can influence the expression of genes. For example, at conception, a male receives genes that will eventually determine the pitch of his voice. However, these genes are expressed differently after puberty. At puberty, male sex hormones are released. This internal environmental change results in the deeper male voice. A male who does not produce these hormones retains a higher-pitched voice in later life. Similarly, the gene for baldness expresses itself when it is exposed to testosterone. Therefore, baldness is more common in males than females (figure 27.10). Comparable changes can occur in females when an abnormally functioning adrenal gland causes the release of large amounts of male hormones. This results in a female with a deeper voice or may result in hair loss. Also recall the genetic disease PKU. If children with *Phenylketonuria* (*PKU*) are allowed to eat foods containing the amino acid phenylalanine, they will become mentally retarded. However, if the amino acid phenylalanine is excluded from the diet, and certain other dietary adjustments are made, the person will develop normally. NutraSweet is a phenylalanine-based sweetener, so people with this genetic disorder must use caution when buying products that contain it.

FIGURE 27.10

It is a common misconception that males have genes for *baldness* and females do not. Male-pattern baldness is a sex-influenced trait in which both males and females may possess alleles coding for baldness. These genes are turned on by high levels of the hormone testosterone. This is another example of an internal gene-regulating factor.

Diet is an external environmental factor that can influence the phenotype of an individual. *Diabetes mellitus*, a metabolic disorder in which glucose is not properly metabolized and is passed out of the body in the urine, has a genetic basis. Some people who have a family history of diabetes are thought to have inherited the trait for this disease. Evidence indicates that they can delay the onset of the disease by reducing the amount of sugar in their diet. This change in the external environment influences gene expression in much the same way that sunlight affects the expression of freckles in humans (see figure 27.3).

WHAT IS A GENE AND HOW DOES IT WORK?

As scientists began to understand the chemical makeup of the **nucleic acids,** an attempt was made to understand how DNA and RNA relate to inheritance, cell structure, and cell activities. The concept that resulted is known as the *central dogma*. The belief can be written in this form:

$$\text{DNA} \Longleftarrow \text{(replication)} \Longleftarrow \boxed{\text{DNA}} \Longrightarrow \text{(transcription)} \Longrightarrow \text{RNA} \Longrightarrow \text{(translation)} \Longrightarrow \text{Proteins} \begin{array}{c} \nearrow \text{structural} \\ \searrow \text{regulatory} \end{array}$$

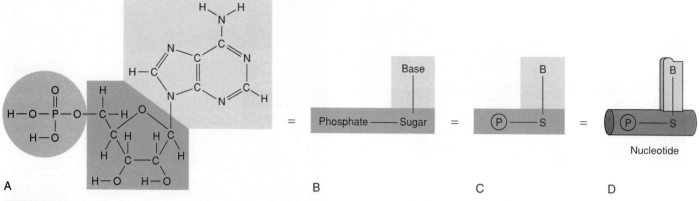

FIGURE 27.11

(*A*) All nucleotides are constructed in the basic way shown. The nucleotide is the basic structural unit of all nucleic acid molecules. Notice in (*C*) that the phosphate group is written in "shorthand" form as a P inside a circle. Part (*D*) is a stylized version of a nucleotide and will be used throughout the chapter. Remember that this style is only representative of the kind of complex organic molecule shown in (*A*).

What this concept map says is that at the center of it all is DNA, the genetic material of the cell and (going to the left) it is capable of reproducing itself, a process called **DNA replication.** Going to the right, DNA is capable of supervising the manufacture of RNA (a process known as **transcription**), which in turn is involved in the production of protein molecules, a process known as **translation.**

DNA replication occurs in cells in preparation for the cell division processes of mitosis and meiosis. Without replication, daughter cells would not receive the library of information required to sustain life. The transcription process results in the formation of a strand of RNA that is a copy of a segment of the DNA on which it is formed. Some of the RNA molecules become involved in various biochemical processes, while others are used in the translation of the RNA information into proteins. Structural proteins are used by the cell as building materials (feathers, hair); regulatory proteins are used to direct and control chemical reactions (enzymes, some hormones).

It is the processes of transcription and translation that result in the manufacture of all enzymes. Each unique enzyme molecule is made from a blueprint in the form of a DNA nucleotide sequence, or *gene.* Some of the thousands of enzymes manufactured in the cell are the tools required for transcription and translation to take place. *The process of making enzymes is carried out by the enzymes made by the process!* Tools are made to make more tools! The same is true for DNA replication.

Enzymes made from the DNA blueprints by transcription and translation are used as tools to make exact copies of the genetic material! More blueprints are made so that future generations of cells will have the genetic materials necessary for them to manufacture their own regulatory and structural proteins. Without DNA, RNA, and enzymes functioning in the proper manner, life as we know it would not occur.

DNA has four properties that enable it to function as genetic material. It is able to (1) *replicate* by directing the manufacture of copies of itself; (2) *mutate,* or chemically change, and transmit these changes to future generations; (3) *store* information that determines the characteristics of cells and organisms; and (4) use this information to *direct* the synthesis of structural and regulatory proteins essential to the operation of the cell or organism.

The Structure of DNA and RNA

Nucleic acids are enormous and complex polymers made up of monomers called **nucleotides.** Each nucleotide is composed of a sugar molecule (S) containing five carbon atoms, a phosphate group (P), and a molecule containing nitrogen that will be referred to as a *nitrogenous base* (B) (figure 27.11).

There are eight common types of nucleotides available in a cell for building nucleic acids. These eight nucleotides differ from one another in the kind of sugar- and nitrogen-containing parts they possess (figure 27.12). Because of these differences, it

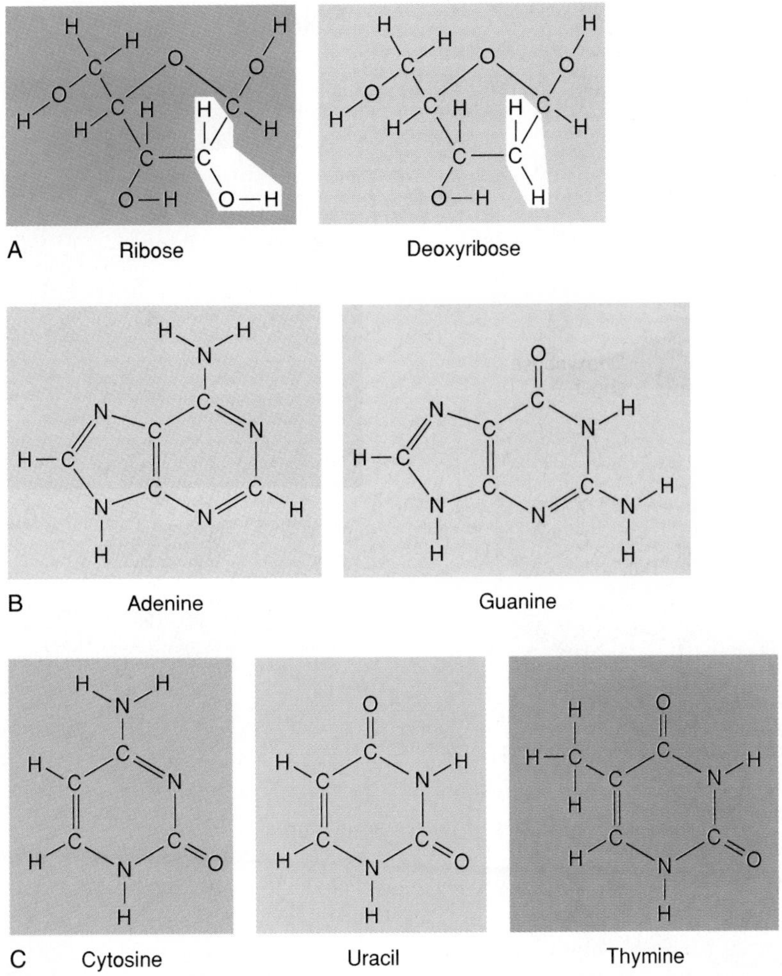

FIGURE 27.12

All nucleic acids are composed of two organic components: a 5-carbon sugar molecule and a nitrogenous base. (A) Notice the difference (highlighted in white) between the two sugar molecules. The nitrogenous bases are divided into two groups according to their size. The large *purines* in (B)—adenine and guanine molecules—differ from each other in their attached groups, as do the three smaller *pyrimidine* nitrogenous bases—cytosine, uracil, and thymine—in (C). The two types of nucleic acids—DNA and RNA—are composed of these eight building blocks. (D) Note that each building block has a sugar, base, and phosphate component. These nucleotides are color coded throughout the chapter so that you can recognize the difference between DNA and RNA. *(continued)*

is possible to classify nucleic acids into two main groups: *ribonucleic acid* (RNA) and *deoxyribonucleic acid* (DNA). The nucleotides can contain two sizes of nitrogenous bases. The larger purine molecules are *adenine* (A) and *guanine* (G), which differ in the kinds of atoms attached to their double-ring structure (figure 27.12B). The smaller nitrogen-containing pyrimidine molecules are *cytosine* (C), *uracil* (U), and *thymine* (T). Each of these differs from the others in the kinds of atoms attached to its single-ring structure (figure 27.12C). These differences in size are important, as you will see later.

In cells, DNA is the nucleic acid that functions as the original blueprint for the synthesis of proteins. It contains the sugar

deoxyribose; phosphates; and adenine, guanine, cytosine, and thymine (A, G, C, T). RNA is a type of nucleic acid that is directly involved in the synthesis of protein. It contains the sugar *ribose;* phosphates; and adenine, guanine, cytosine, and uracil (A, G, C, U). There is no thymine in RNA and no uracil in DNA. The construction of a nucleotide involves the bonding of a nitrogenous base to a 5-carbon sugar. For example, when adenine is chemically bonded to ribose sugar, the result is a molecule called *adenosine.* When a phosphate is added to the ribose of the adenosine, a new molecule called *adenosine monophosphate (AMP)* is formed. This is a complete nucleotide. Monophosphate nucleotides are energetically stable, and they are the building blocks of RNA.

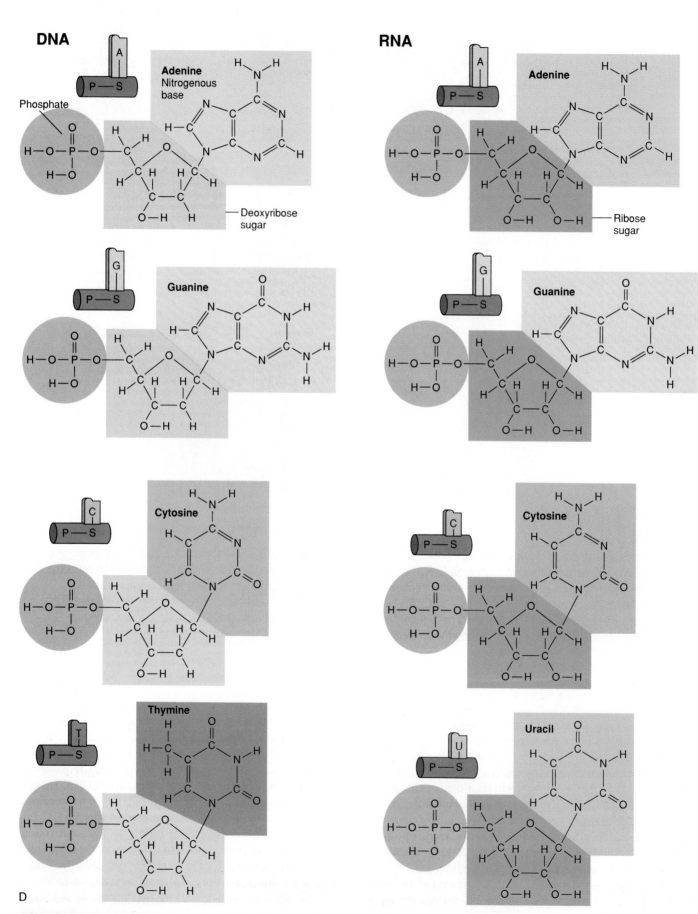

FIGURE 27.12 *(continued)*

A CLOSER LOOK | Of Men (and Women), Microbes, and Molecules

Microorganisms were very important in the research that led to our understanding of DNA, its structure and function. The better understanding of the microbe ushered in a period of rapid advancement in biology. A major contribution came in 1952, when Alfred Hershey and Martha Chase demonstrated by using bacteria and viruses that DNA is the controlling molecule of cells. Their work with the viruses that infect bacterial cells, bacteriophages, was so significant that the phage became a standard laboratory research organism. In 1953, just one year later, James

D. Watson and Frances Crick used the information, and that of other researchers, to propose a double-helix molecular structure for DNA. Ten years later, Watson, Crick, and co-worker Maurice Wilkins shared a Nobel Prize for their work. In 1958, George Beadle and Edward Tatum won a Nobel Prize for their discovery that genes operate by regulating specific chemical reactions in the cell, their "one gene–one enzyme" concept. The chemical reactions of the cell are controlled by the action of enzymes and it is the DNA that chemically codes the structure of those special protein molecules.

At first glance, some research by microbiologists may seem irrelevant or unrelated to everyday life. But it is a rare occasion when the results of such research do not make their way into our lives in some practical, beneficial form. The work of Watson, Crick, Beadle, and Tatum has been applied in hospitals and doctor's offices. Their basic research into DNA provided the information necessary to develop medicines that control disease-causing organisms and medicines that regulate basic metabolic processes in our bodies.

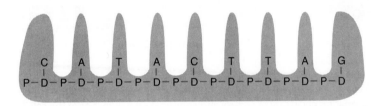

FIGURE 27.13

A single strand of DNA resembles a comb. The molecule is much longer than pictured here and is composed of a sequence of linked nucleotides.

The molecule adenosine triphosphate (ATP) is a source of chemical-bond energy. Keep in mind that all nucleotides exist in mono-, di-, and triphosphate forms. As in ATP, the other triphosphates contain high-energy phosphate bonds used in the dehydration synthesis reaction that binds nucleotides together as RNA or DNA. The result of this nucleic acid synthesis is a polymer that can be compared to a comb. The protruding "teeth" (A, T, G, C, or U) are connected to a common "backbone" (sugar and phosphate molecules). This is the basic structure of both RNA and DNA (figure 27.13).

DNA and RNA differ in one other respect. DNA is actually a double molecule. It consists of two flexible comblike strands held together between their protruding teeth. The two strands are twisted about each other in a coil or double helix (plural helices) also called **duplex DNA** (figure 27.14). The two combs of the molecule are held together because they "fit" each other like two jigsaw puzzle pieces that interlock with one another and by weak chemical forces, hydrogen bonds. The four kinds of teeth always pair in a definite way: adenine (A) with thymine (T), and guanine (G) with cytosine (C). Notice that the large molecules (A and G) pair with the small ones (T and C), thus

keeping the two complementary (matched) strands parallel. The bases that pair are said to be **complementary bases.** Three hydrogen bonds are formed between guanine and cytosine:

$$G : : : C$$

and two between adenine and thymine:

$$A : : : T$$

You can "write" a message in the form of a stable DNA molecule by combining the four different DNA nucleotides (A, T, G, C) in particular sequences. Notice in figure 27.13 that it is possible to make sense out of the sequence of nitrogen-containing portions of the molecule. If you "read" them from left to right in groups of three, you can read three words: CAT, ACT, and TAG. In this case, the four DNA nucleotides are being used as an alphabet to construct three-letter words. In order to make sense out of such a code, it is necessary to read in one direction. Reading the sequence in reverse does not always make sense, just as reading this paragraph in reverse would not make sense.

The genetic material of humans and other eukaryotic organisms are *strands* of coiled duplex DNA, which has histone

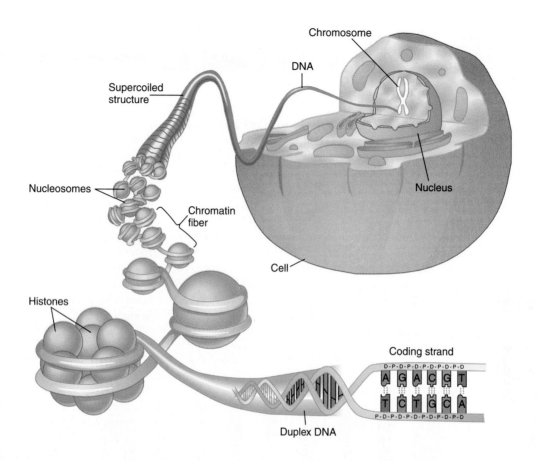

FIGURE 27.14

Eukaryotic cells contain duplex DNA in their nuclei, which takes the form of a three-dimensional helix. One strand is a chemical code (the coding strand) that contains the information necessary to control and coordinate the activities of the cell. The two strands fit together and are bound by weak hydrogen bonds formed between the complementary, protruding nitrogenous bases according to the base-pairing rule: A pairs with T, and C pairs with G. The length of a DNA molecule is measured in numbers of "base pairs"—the number of rungs on the ladder.

proteins attached along its length. These coiled DNA strands with attached proteins, which become visible during mitosis and meiosis, are called **nucleoproteins,** or *chromatin fibers*. The histone protein and DNA are not arranged randomly, but come together in a highly organized pattern. The duplex DNA spirals around repeating clusters of eight histone spheres. Histone clusters with their encircling DNA are called **nucleosomes** (figure 27.15A). When eukaryotic chromatin fibers coil into condensed, highly knotted bodies, they are seen easily through a microscope after staining with dye. Condensed like this, a chromatin fiber is referred to as a **chromosome** (figure 27.15C). The genetic material in bacteria is also double-stranded (duplex) DNA, but the ends of the molecule are connected to form a *loop* and they do not form condensed chromosomes (figure 27.15B). However, prokaryotic cells have an attached protein called *HU protein*.

Each chromatin strand is different because each strand has a different chemical code. Coded DNA serves as a central cell library. Tens of thousands of messages are in this storehouse of information. This information tells the cell such things as (1) how to produce enzymes required for the digestion of nutrients, (2) how to manufacture enzymes that will metabolize the nutrients and eliminate harmful wastes, (3) how to repair and assemble cell parts, (4) how to reproduce healthy offspring, (5) when and how to react to favorable and unfavorable changes in the environment, and (6) how to coordinate and regulate all of life's essential functions. If any of these functions are not performed properly, the cell may die. The importance of maintaining essential DNA in a cell becomes clear when we consider cells that have lost it. For example, human red blood cells lose their nuclei as they become specialized for carrying oxygen and carbon dioxide throughout the body. Without DNA they are unable to manufacture the essential cell components needed to sustain themselves. They continue to exist for about 120 days, functioning only on enzymes manufactured earlier in their lives. When these enzymes are gone, the cells die. Because these specialized cells begin to die the moment they lose their DNA, they are more accurately called *red blood corpuscles* (*RBCs*): "little dying red bodies."

DNA Replication

Since all cells must maintain a complete set of genetic material, there must be a doubling of DNA in order to have enough to pass on to the offspring. DNA replication is the process of duplicating the genetic material prior to its distribution to daughter

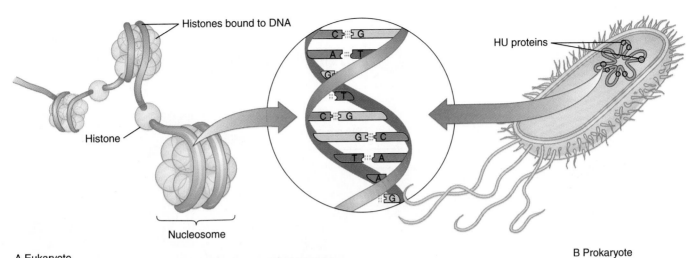

Histones bound to DNA

Histone

Nucleosome

A Eukaryote

HU proteins

B Prokaryote

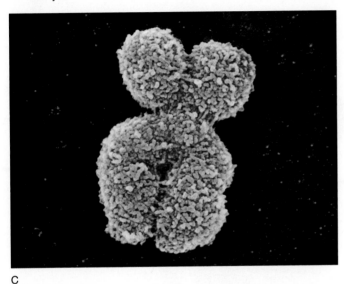

C

FIGURE 27.15

The term *nucleoprotein* is used to describe the combination of DNA and its associated protein in eukaryotic cells. When these giant nucleoprotein molecules are found loose inside a cell's nucleus, they are called chromatin. (A) Upon close examination of a portion of a eukaryotic nucleoprotein, it is possible to see how the DNA and protein are arranged. The protein component histone is found along the DNA in globular masses; histones may be individual or arranged in groups, the nucleosomes. The nucleic acid of prokaryotic cells (the bacteria) does not have the histone protein; rather, it has proteins called HU proteins. In addition, the ends of the giant nucleoprotein molecule overlap and bind with one another to form a loop (B). (C) During certain stages in the reproduction of a eukaryotic cell, the nucleoprotein coils and "supercoils," forming tightly bound masses. When stained, these are easily seen through the microscope. In their supercoiled form, they are called chromosomes, meaning colored bodies.

cells. When a cell divides into two daughter cells, each new cell must receive a complete copy of the parent cell's genetic information, or it will not be able to manufacture all the proteins vital to its existence. Accuracy of duplication is essential in order to guarantee the continued existence of that type of cell. Should the daughters not receive exact copies, they would most likely die.

1. The DNA replication process begins as an enzyme breaks the attachments between the two strands of DNA. In eukaryotic cells, this occurs in hundreds of different spots along the length of the DNA (figure 27.16A).

2. Moving along the DNA, the enzyme "unzips" the halves of the DNA (figure 27.16B and C). Hydrogen bonds hold each in position (A ::: T, G ::: C) while the new nucleotide is covalently bonded between the sugar and phosphate of the new backbone.

3. Proceeding in opposite directions on each side, the enzyme **DNA polymerase** moves down the length of the DNA, attaching new DNA nucleotides into position (figure 27.16C).

4. The enzyme that speeds the addition of new nucleotides to the growing chain works along with another enzyme to make sure that no mistakes are made. If the wrong nucleotide appears to be headed for a match, the enzyme will reject it in favor of the correct nucleotide. If a mistake is made and a wrong nucleotide is paired into position, specific enzymes have the ability to replace it with the correct one (figure 27.16D).

5. Replication proceeds in both directions, appearing as "bubbles" (figure 27.16E).

6. The complementary molecules (A ::: T, G ::: C) pair with the exposed nitrogenous bases of both DNA strands by forming new hydrogen bonds (figure 27.16F).

7. Once properly aligned, a bond is formed between the sugars and phosphates of the newly positioned nucleotides. A strong sugar and phosphate backbone is formed in the process (figure 27.16G).

8. This process continues until all the replication "bubbles" join (figure 27.16H). Figure 27.17 summarizes this process.

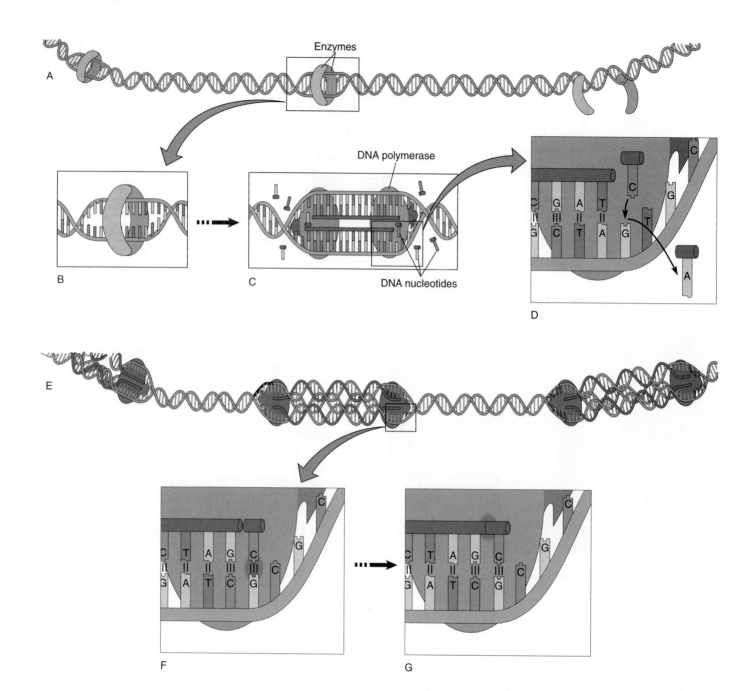

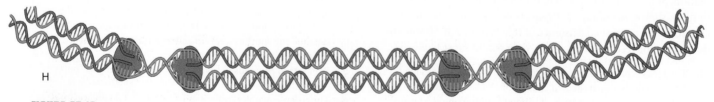

FIGURE 27.16

These illustrations summarize the basic events that occur during the replication of duplex DNA.

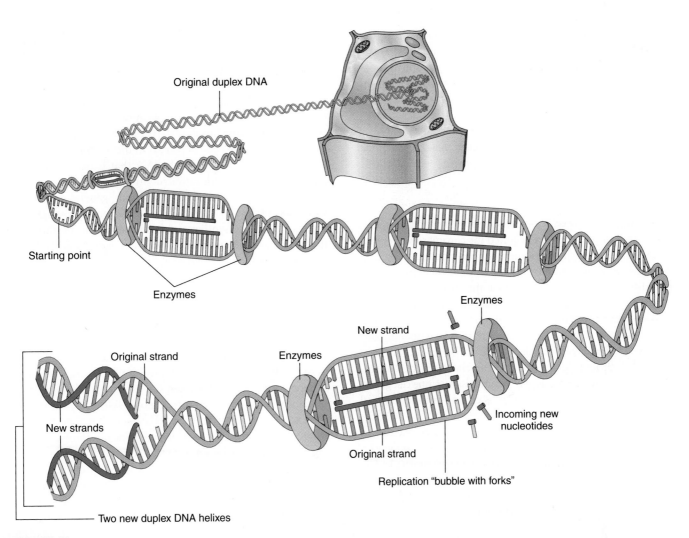

Original duplex DNA

Starting point

Enzymes

New strands

Original strand

Enzymes

Two new duplex DNA helixes

Enzymes

New strand

Enzymes

Incoming new nucleotides

Original strand

Replication "bubble with forks"

FIGURE 27.17

In eukaryotic cells, the "unzipping" enzymes attach to the DNA at numerous points, breaking the bonds that bind the complementary strands. As the DNA replicates, numerous replication "bubbles" and "forks" appear along the length of the DNA. Eventually all the forks come together, completing the replication process.

C O N N E C T I O N S

Antibiotics and DNA Replication

The DNA of most bacteria is in the form of a loop that is super-coiled. Supercoiling occurs when, like a rubber band, double-stranded DNA is very tightly twisted around itself. The unique bacterial enzyme, DNA gyrase, fashions supercoils. Without this enzyme, the DNA would not be compact enough to fit into the tiny bacterial cell or function properly. Two antibiotics, nalidixic acid and novobiocin, interfere with the action of DNA gyrase, thus contributing to bacterial cell death. Nalidixic acid can be used to treat urinary tract infections caused by bacteria such as E. coli, and novobiocin has been used to treat infections caused by Staphylococci.

A new complementary strand of DNA forms on each of the old DNA strands, resulting in the formation of two double-stranded duplex DNA molecules. In this way, the exposed nitrogenous bases of the original DNA serve as a *template,* or pattern, for the formation of the new DNA. As the new DNA is completed, it twists into its double-helix shape.

The ends of a chromosome contain a special sequence of nucleotides called **telomeres.** *In humans these chromosome "caps" contain the nucleotide base pair sequence*

TTAGGG
AATCCC

repeated many times over. Telomeres are very important segments of the chromosome. They are required for chromosome replication, they protect the chromosome from being destroyed by dangerous DNAase enzymes, and they keep chromosomes from bonding end-to-end. Evidence shows that the loss of telomeres is associated with cell "aging," whereas their maintenance has been linked to cancer. Every time a cell reproduces itself, it loses telomeres because in "young" cells the enzyme *telomerase* is not normally produced. However, cancer cells appear to be "immortal" as a result of their producing this enzyme. This enables them to maintain, if not increase, the number of telomeres from one cell generation to the next. Telomerase activity is critical to the continued reproduction of tumor cells.

The completion of the DNA replication process yields two double helices that are identical in their nucleotide sequences. Half of each is new, half is the original parent DNA molecule. The DNA replication process is highly accurate. It has been estimated that there is only one error made for every 2×10^9 nucleotides. A human cell contains 46 chromosomes consisting of about 5,000,000,000 (5 billion) base pairs. This averages to about five errors per cell! Don't forget that this figure is an estimate. While some cells may have five errors per replication, others may have more, and some may have no errors at all. It is also important to note that some errors may be major and deadly, while others are insignificant. Because this error rate is so small, DNA replication is considered by most to be essentially error-free. Following DNA replication, the cell now contains twice the amount of genetic information and is ready to begin the process of distributing one set of genetic information to each of its two daughter cells.

The distribution of DNA involves splitting the cell and distributing a set of genetic information to the two new daughter cells. In this way, each new cell has the necessary information to control its activities. The mother cell ceases to exist when it divides its contents between the two smaller daughter cells (figure 27.18).

A cell does not really die when it reproduces itself; it merely starts over again. This is called the *life cycle* of a cell. A cell may divide and redistribute its genetic information to the next generation in a number of ways.

DNA Transcription

DNA functions in the manner of a reference library that does not allow its books to circulate. Information from the originals must be copied for use outside the library. The second major function of DNA is the making of these single-stranded, complementary RNA copies of DNA. This operation is called transcription (*scribe* = to write), which means to transfer data from one form to another. In this case, the data are copied from DNA language to RNA language. The same base-pairing rules that control the accuracy of DNA replication apply to the process of transcription. Using this process, the genetic information stored as a DNA chemical code is carried in the form of an RNA copy to other parts of the cell. It is RNA that is used to guide the assembly of amino acids into structural and regulatory proteins. Without the process of transcription, genetic information would be useless in directing cell functions. Although many types of RNA are synthesized from the genes, the three most important are *messenger RNA (mRNA), transfer RNA (tRNA),* and *robosomal RNA (rRNA).*

Transcription begins in a way that is similar to DNA replication. The duplex DNA is separated by an enzyme, exposing the nitrogenous-base sequences of the two strands. However, unlike DNA replication, transcription occurs only on one of the two DNA strands, which serves as a template, or pattern, for the synthesis of RNA (see figure 27.19). This side is also referred to as the **coding strand** of the DNA. But which strand is copied? Where does it start and when does it stop? Where along the sequence of thousands of nitrogenous bases does the chemical code for the manufacture of a particular enzyme begin and where does it end? If transcription begins randomly, the resulting RNA may not be an accurate copy of the code, and the enzyme product may be useless or deadly to the cell. To answer these questions, it is necessary to explore the nature of the genetic code itself.

We know that genetic information is in chemical-code form in the DNA molecule. When the coded information is used or *expressed,* it guides the assembly of particular amino acids into structural and regulatory polypeptides and proteins. If DNA is molecular language, then each nucleotide in this language can be thought of as a letter within a four-letter alphabet. Each word, or code, is always three letters (nucleotides) long, and only three-letter words can be written. A **DNA code** is a triplet nucleotide sequence that codes for one of the twenty common amino acids. The number of codes in this language is limited because there are only four different nucleotides, which are used only in groups of three. The order of these three letters is just as important in DNA language as it is in our language. We recognize that CAT is not the same as TAC. If all the possible three-letter codes were written using only the four DNA nucleotides for letters, there would be a total of 64 combinations.

$$4^3 = 4 \times 4 \times 4 = 64$$

When codes are found at a particular place along a coding strand of DNA, and the sequence has meaning, the sequence is a gene. "Meaning" in this case refers to the fact that the gene can be transcribed into an RNA molecule, which in turn may control the assembly of individual amino acids into a polypeptide.

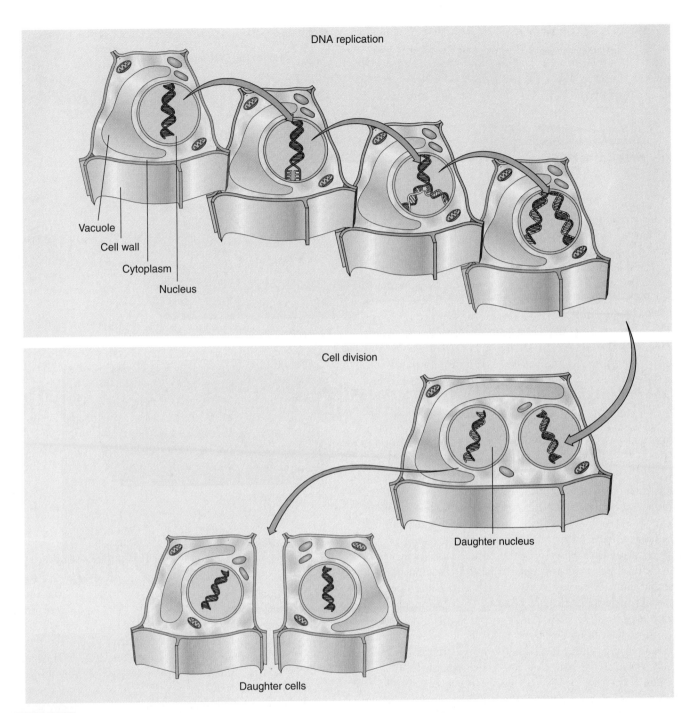

DNA replication

Vacuole

Cell wall

Cytoplasm

Nucleus

Cell division

Daughter nucleus

Daughter cells

FIGURE 27.18

These are the generalized events in the nucleus of a eukaryotic cell during the process of DNA replication. Notice that the daughter cells each have double helices; they are identical to each other and identical to the original duplex strands of the parent cell.

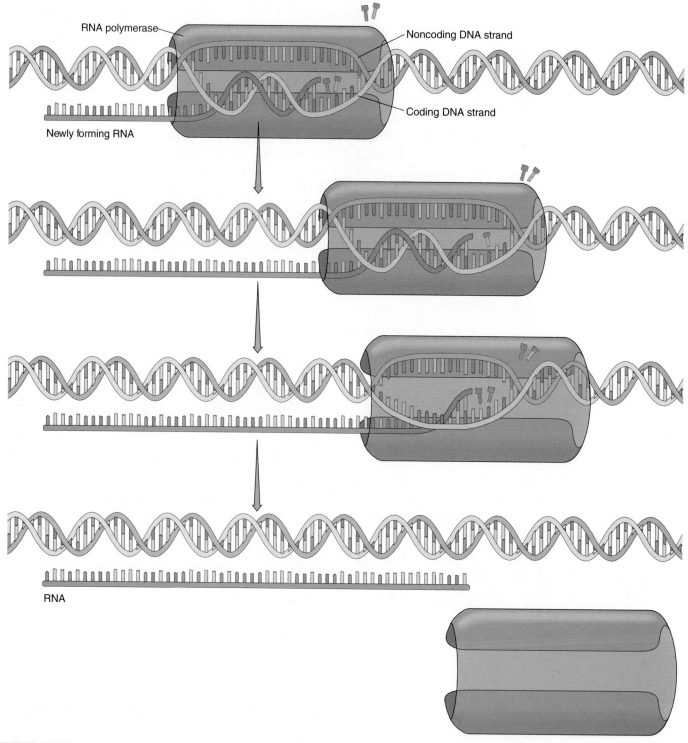

FIGURE 27.19

This summary illustrates the basic events that occur during the transcription of one side (the coding strand) of double-stranded DNA. The enzyme attaches to the DNA at a point that allows it to separate the complementary strands. As this enzyme, RNA polymerase, moves down the DNA, new complementary RNA nucleotides are base-paired on one of the exposed strands and linked together, forming a new strand that is complementary to the nucleotide sequence of the DNA. The newly formed (transcribed) RNA is then separated from its DNA complement. Depending on the DNA segment that has been transcribed, this RNA molecule may be a messenger RNA (mRNA), a transfer RNA (tRNA), a ribosomal RNA (rRNA), or an RNA molecule used for other purposes within the cell.

Prokaryotic Transcription

Each bacterial gene is made of attached nucleotides that are transcribed in order into a single strand of RNA. This RNA molecule is used to direct the assembly of a specific sequence of amino acids to form a polypeptide. This system follows the pattern of:

one DNA gene → one RNA → one polypeptide

The beginning of each gene on a DNA strand is identified by the presence of a region known as the *promoter,* just ahead of an *initiation code* that has the base sequence TAC. The gene ends with a terminator region, just in back of one of three possible *termination codes*—ATT, ATC, or ACT. These are the "start reading here" and "stop reading here" signals. The actual genetic information is located between initiation and termination codes:

promoter::initiator code:::::gene:::::terminator code::terminator region

When a bacterial gene is transcribed into RNA, the duplex DNA is "unzipped," and an enzyme known as **RNA polymerase** attaches to the DNA at the promoter region. It is from this region that the enzymes will begin to assemble RNA nucleotides into a complete, single-stranded copy of the gene, including initiation and termination codes. Triplet RNA nucleotide sequences complementary to DNA codes are called **codons.** Remember that there is no thymine in RNA molecules; it is replaced with uracil. Therefore, the initiation code in DNA (TAC) would be base-paired by RNA polymerase to form the RNA codon AUG. When transcription is complete, the newly assembled RNA is separated from its DNA template and made available for use in the cell; the DNA recoils into its original double-helix form.

In summary (figure 27.19):

1. The process begins as one portion of the enzyme RNA polymerase breaks the attachments between the two strands of DNA; the enzyme "unzips" the two strands of the DNA.
2. A second portion of the enzyme RNA polymerase attaches at a particular spot on the DNA called the start codon. It proceeds in one direction along one of the two DNA strands, attaching new RNA nucleotides into position until it reaches a stop codon. The enzymes then assemble RNA nucleotides into a complete, single-stranded RNA copy of the gene. There is no thymine in RNA molecules; it is replaced by uracil. Therefore, the start codon in DNA (TAC) would be paired by RNA polymerase to form the RNA codon AUG.
3. The enzyme that speeds the addition of new nucleotides to the growing chain works along with another enzyme to make sure that no mistakes are made.
4. When transcription is complete, the newly assembled RNA is separated from its DNA template and made available for use in the cell; the DNA recoils into its original double-helix form.

As previously mentioned, three general types of RNA are produced by transcription: messenger RNA, transfer RNA, and ribosomal RNA. Each kind of RNA is made from a specific gene and performs a specific function in the synthesis of polypeptides from individual amino acids at ribosomes. **Messenger RNA (mRNA)** is a mature, straight-chain copy of a gene that describes the exact sequence in which amino acids should be bonded together to form a polypeptide.

Transfer RNA (tRNA) molecules are responsible for picking up particular amino acids and transferring them to the ribosome for assembly into the polypeptide. All tRNA molecules are shaped like a cloverleaf. This shape is formed when they fold and some of the bases form hydrogen bonds that hold the molecule together. One end of the tRNA is able to attach to a specific amino acid. Toward the midsection of the molecule, a triplet nucleotide sequence can base-pair with a codon on mRNA. This triplet nucleotide sequence on tRNA that is complementary to a codon of mRNA is called an **anticodon. Ribosomal RNA (rRNA)** is a highly coiled molecule and is used, along with protein molecules, in the manufacture of all ribosomes, the cytoplasmic organelles where tRNA, mRNA, and rRNA come together to help in the synthesis of proteins.

Eukaryotic Transcription

The transcription system is different in eukaryotic cells. A eukaryotic gene begins with a promoter region and an initiation code and ends with a termination code and region. However, the intervening gene sequence contains patches of nucleotides that apparently have no meaning but do serve important roles in maintaining the cell. If they were used in protein synthesis, the resulting proteins would be worthless. To remedy this problem, eukaryotic cells prune these segments from the mRNA after transcription. When such *split genes* are transcribed, RNA polymerase synthesizes a strand of pre-mRNA that initially includes copies of both *exons* (meaningful mRNA coding sequences) and *introns* (meaningless mRNA coding sequences). Soon after its manufacture, this pre-mRNA molecule has the meaningless introns clipped out and the exons spliced together into the final version, or *mature mRNA,* which is used by the cell (figure 27.20).

Translation or Protein Synthesis

The mRNA molecule is a coded message written in the biological world's universal nucleic acid language. The code is read in one direction starting at the initiator. The information is used to assemble amino acids into protein by a process called translation. The word *translation* refers to the fact that nucleic acid language is being changed to protein language. To translate mRNA language into protein language, a dictionary is necessary. Remember, the four letters in the nucleic acid alphabet yield 64 possible three-letter words. The protein language has 20 words in the form of 20 common amino acids (table 27.2).

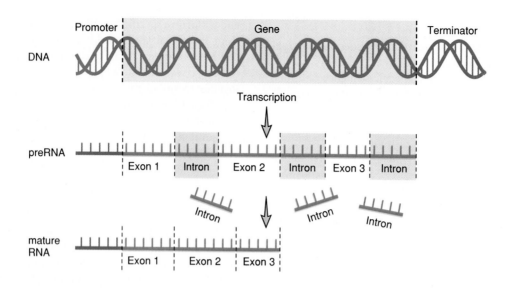

DNA — Promoter — Gene — Terminator

Transcription

preRNA — Exon 1 | Intron | Exon 2 | Intron | Exon 3 | Intron

Intron   Intron   Intron

mature RNA — Exon 1 | Exon 2 | Exon 3

FIGURE 27.20

This is a summary of the events that occur in the nucleus during the manufacture of mRNA in a eukaryotic cell. Notice that the original nucleotide sequence is first transcribed into an RNA molecule that is later "clipped" and then rebonded to form a shorter version of the original. It is during this time that the introns are removed.

TABLE 27.2	The 20 common amino acids and their codes	
Amino Acid	**Three-Letter Code**	**Single-Letter Code**
glycine	Gly	G
alanine	Ala	A
valine	Val	V
leucine	Leu	L
isoleucine	Ile	I
methionine	Met	M
phenylalanine	Phe	F
tryptophan	Trp	W
proline	Pro	P
serine	Ser	S
threonine	Thr	T
cysteine	Cys	C
tyrosine	Tyr	Y
asparagine	ASN	N
glutamine	Gln	Q
aspartic acid	Asp	D
glutamic acid	Glu	E
lysine	Lys	K
arginine	Arg	R
histidine	His	H

These are the 20 common amino acids used in the protein synthesis operation of a cell. Each has a known chemical structure.

Thus, there are more than enough nucleotide words for the 20 amino acid molecules because each nucleotide triplet codes for an amino acid.

Table 27.3 is an amino acid–nucleic acid dictionary. Notice that more than one codon may code for the same amino acid. Some would contend that this is needless repetition, but such "synonyms" can have survival value. If, for example, the gene or the mRNA becomes damaged in a way that causes a particular nucleotide base to change to another type, the chances are still good that the proper amino acid will be read into its proper position. But not all such changes can be compensated for by the codon system, and an altered protein may be produced (figure 27.21). Changes can occur that cause great harm. Some damage is so extensive that the entire strand of DNA is broken, resulting in improper protein synthesis or a total lack of synthesis. Any change in DNA is called a **mutation.**

The construction site of the protein molecules (i.e., the translation site) is on the ribosome, a cellular organelle that serves as the meeting place for mRNA and the tRNA that is carrying amino acid building blocks. Ribosomes can be found free in the cytoplasm or attached to the endoplasmic reticulum. Proteins destined to be part of the cell membrane or packaged for export from the cell are synthesized on ribosomes attached to the endoplasmic reticulum. Proteins that are to perform their function in the cytoplasm are synthesized on unattached or free ribosomes (figure 27.22).

Thus, the mRNA moves through the ribosomes, its specific codon sequence allowing for the chemical bonding of a specific sequence of amino acids. Remember that the sequence was orig-

TABLE 27.3	Amino acid–mRNA nucleic acid dictionary								

Second Letter

First Letter		U		C		A		G		Third Letter
U	UUU UUC	Phe	UCU UCC	Ser	UAU UAC	Tyr	UGU UGC	Cys	U C
	UUA UUG	Leu	UCA UCG		UAA UAG	Stop Stop	UGA UGG	Stop Try	A G
C	CUU CUC	Leu	CCU CCC	Pro	CAU CAC	His	CGU CGC	Arg	U C
	CUA CUG		CCA CCG		CAA CAG	Gln	CGA CGG		A G
A	AUU AUC	Leu	ACU ACC	Thr	AAU AAC	Asn	AGU AGC	Ser	U C
	AUA		ACA		AAA	Lys	AGA	Arg	A
	AUG	Met Met Start	ACG		AAG		AGG		G
G	GUU GUC	Val	GCU GCC	Ala	GAU GAC	Asp	GGU GGC	Gly	U C
	GUA GUG		GCA GCG		GAA GAG	Glu	GGA GGG		A G

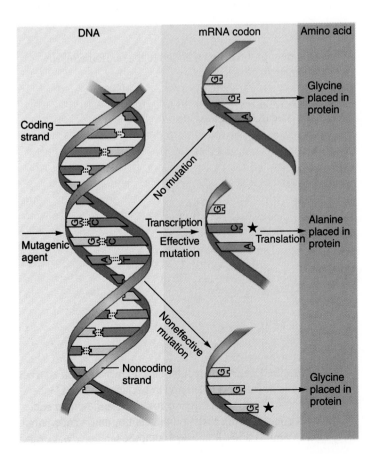

DNA   mRNA codon   Amino acid

Coding strand

No mutation

Glycine placed in protein

Mutagenic agent

Transcription
Effective mutation

Translation

Alanine placed in protein

Noneffective mutation

Noncoding strand

Glycine placed in protein

FIGURE 27.21

A nucleotide substitution changes the genetic information only if the changed codon results in a different amino acid being substituted into a protein chain. This feature of DNA serves to better ensure that the synthesized protein will be functional.

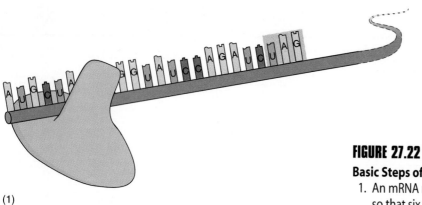

(1)

FIGURE 27.22
Basic Steps of Translation

1. An mRNA molecule is placed in the small portion of a ribosome so that six nucleotides (two codons) are locked into position.

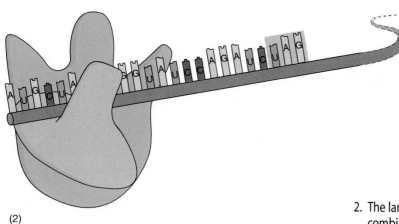

(2)

2. The larger ribosomal unit is added to the ribosome/mRNA combination.

(3)

3. A tRNA with bases that match the second mRNA codon attaches to the mRNA. The tRNA is carrying a specific amino acid. Once attached, a second tRNA carrying another specific amino acid moves in and attaches to its complementary mRNA codon right next to the first tRNA/amino acid complex.

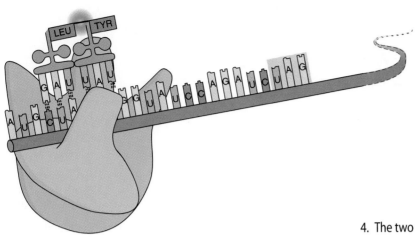

(4)

4. The two tRNAs properly align their two amino acids so that they may be chemically attached to one another.

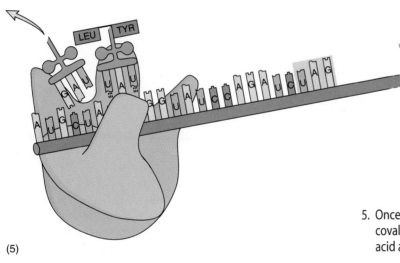

(5)

5. Once the two amino acids are connected to one another by a covalent peptide bond, the first tRNA detaches from its amino acid and mRNA codon and leaves.

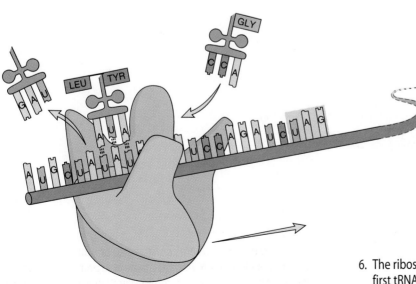

(6)

6. The ribosome moves along the mRNA to the next codon (the first tRNA is set free to move through the cytoplasm to attach to and transfer another amino acid).

(7)

7. The next tRNA/amino acid unit enters the ribosome and attaches to its codon next to the first set of amino acids.

(8)

8. The tRNAs properly align their amino acids so that they may be chemically attached to one another, forming a chain of three amino acids.

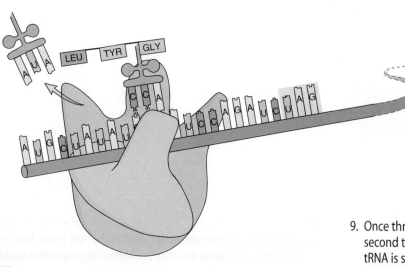

(9)

9. Once three amino acids are connected to one another, the second tRNA is released from its amino acid and mRNA (this tRNA is set free to move through the cytoplasm to attach to and transfer another amino acid).

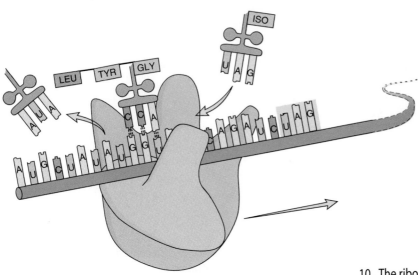

(10)

10. The ribosome moves along the mRNA to the next codon and the fourth tRNA arrives.

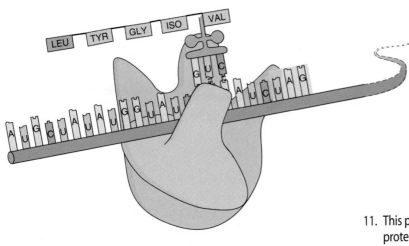

(11)

11. This process repeats until all the amino acids needed to form the protein have attached to one another in the proper sequence. This amino acid sequence was encoded by the DNA gene.

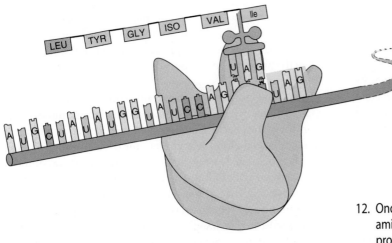

(12)

12. Once the final amino acid is attached to the growing chain of amino acids, all the molecules (mRNA, tRNA, and newly formed protein) are released from the ribosome. The stop mRNA codon signals this action.

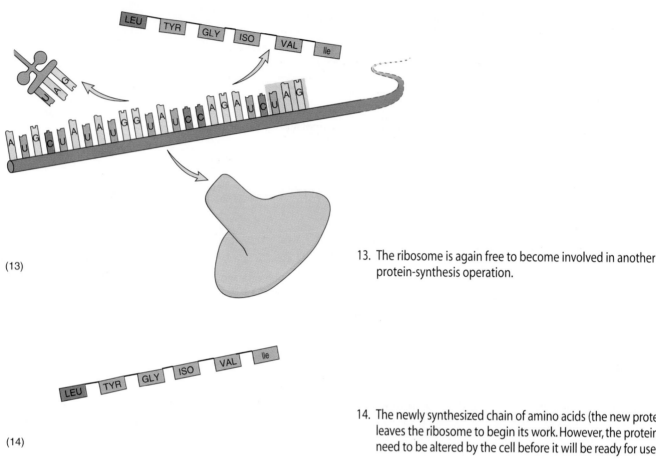

(13)

13. The ribosome is again free to become involved in another protein-synthesis operation.

(14)

14. The newly synthesized chain of amino acids (the new protein) leaves the ribosome to begin its work. However, the protein may need to be altered by the cell before it will be ready for use.

inally determined by the DNA. Figure 27.23 shows a possible result of protein synthesis. After transcribing the DNA code, the mRNA delivers its message to a ribosome, where a protein is made. In the case of figure 27.23, the amino acid sequence contains phenylalanine, serine, lysine, and arginine.

Each protein has a specific sequence of amino acids that determines its three-dimensional shape. This shape determines the activity of the protein molecule. The protein may be a structural component of a cell or a regulatory protein, such as an enzyme. Any changes in amino acids or their order changes the action of the protein molecule. The protein insulin, for example, has a different amino acid sequence than the digestive enzyme trypsin. Both proteins are essential to human life and must be produced constantly and accurately. The amino acid sequence of each is determined by a different gene. Each gene is a particular sequence of DNA nucleotides. Any alteration of that sequence can directly alter the protein structure and, therefore, the survival of the organism.

Alterations of DNA

Several kinds of changes to DNA may result in mutations. Phenomena that are either known or suspected causes of DNA damage are called **mutagenic agents.** Agents known to cause damage to DNA are certain viruses (e.g., papillomavirus), weak or "fragile" spots in the DNA, X radiation (X rays) and chemi-

cals found in foods and other products, such as nicotine in tobacco. All have been studied extensively, and there is little doubt that they cause mutations. **Chromosomal aberrations** is the term used to describe major changes in DNA. Four types of aberrations include inversions, translocations, duplications, and deletions. An *inversion* occurs when a chromosome is broken and this piece becomes reattached to its original chromosome but in reverse order. It has been cut out and flipped around. A *translocation* occurs when one broken segment of DNA becomes integrated into a different chromosome. *Duplications* occur when a portion of a chromosome is replicated and attached to the original section in sequence. *Deletion* aberrations result when the broken piece becomes lost or is destroyed before it can be reattached.

In some individuals, a single nucleotide of the gene may be changed. This type of mutation is called a **point mutation.** An example of the effects of altered DNA may be seen in human red blood cells. Red blood cells contain the oxygen-transport molecule, hemoglobin. Normal hemoglobin molecules are composed of 150 amino acids in four chains—two alpha and two beta. The nucleotide sequence of the gene for the beta chain is known, as is the amino acid sequence for this chain. In normal individuals, the sequence begins like this:

Val-His-Leu-Thr-Pro-Glu-Glu-Lys . . .

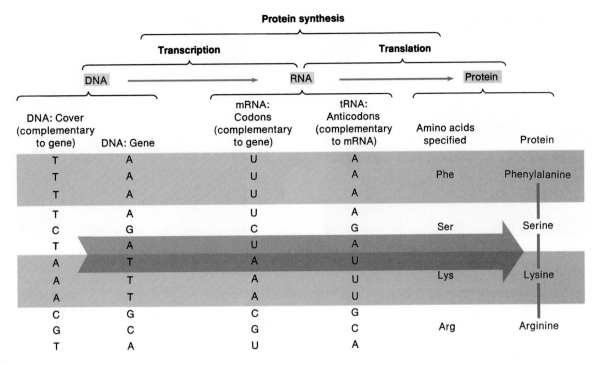

FIGURE 27.23

There are several steps involved in protein synthesis. (1) mRNA is manufactured from a DNA molecule in the transcription operation; (2) the mRNA enters the cytoplasm and attaches to ribosomes; (3) the tRNA carries amino acids to the ribosome and positions them in the order specified by the mRNA in the translation operation; (4) the amino acids are combined chemically to form a protein.

The result is a new amino acid sequence in all the red blood cells:

Val-His-Leu-Thr-Pro-Val-Glu-Lys . . .

This single nucleotide change (known as a *missense point muta-tion*), which causes a single amino acid to change, may seem minor. However, it is the cause of **sickle-cell anemia,** a disease that affects the red blood cells by changing them from a circular to a sickle shape when oxygen levels are low (figure 27.24). When this sickling occurs, the red blood cells do not flow smoothly through capillaries. Their irregular shapes cause them to clump, clogging the blood vessels. This prevents them from delivering their oxygen load to the oxygen-demanding tissues. A number of physical disabilities may result, including physical weakness, brain damage, pain and stiffness of the joints, kidney damage, rheumatism, and, in severe cases, death.

Other mutations occur as a result of changing the number of nucleotide bases in a gene. *Transposons* or "jumping genes" are segments of DNA capable of moving from one chromosome to another. When the jumping gene is spliced into its new location, it alters the normal nucleotide sequence, causing normally stable genes to be misread during transcription. The result may be a mutant gene. It is estimated that 10 percent of all human genes are transposons. Transposons can alter the genetic activity of a cell when they leave their original location, stop transcription of the gene they "jump" into, or change the reading of codons from their normal sequence. For example, one person who developed hemophilia ("bleeders disease") did so as a result of a transposon "jumping" into

the gene that was responsible for producing a specific clotting factor, factor VIII.

Changes in the structure of DNA may have harmful effects on the next generation if they occur in the sex cells. Some damage to DNA is so extensive that the entire strand of DNA is broken, resulting in the synthesis of abnormal proteins or a total lack of protein synthesis. A number of experiments indicate that many street drugs, such as LSD (lysergic acid diethylamide), are mutagenic agents and cause DNA to break. Abnormalities have also been identified that are the result of changes in the number or sequence of bases. One way to illustrate these various kinds of mutations is seen in table 27.4.

A powerful new science of gene manipulation, **biotechnology,** suggests that in the future, genetic diseases may be controlled or cured. Since 1953, when the structure of the DNA molecule was first described, there has been a rapid succession of advances in the field of genetics. It is now possible to transfer the DNA from one organism to another. This has made possible the manufacture of human genes and gene products by bacteria.

Manipulating DNA to Our Advantage

Biotechnology includes the use of a method of splicing genes from one organism into another, resulting in a new form of DNA called **recombinant DNA.** This process is accomplished using enzymes that are naturally involved in the DNA-replication process and others naturally produced by bacteria. When genes are spliced from different organisms into host cells, the host cell replicates these new,

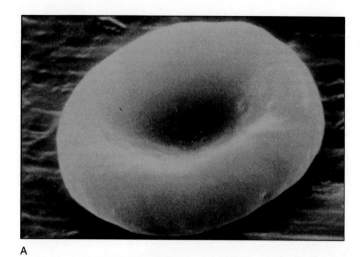

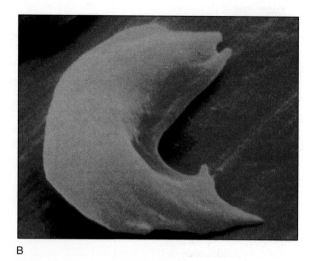

A

B

FIGURE 27.24

(*A*) A normal red blood cell is shown in comparison with (*B*) a cell having the sickle shape. This sickling is the result of a single amino acid change in the hemoglobin molecule.

TABLE 27.4	Types of chromosomal mutations
Normal Sequence	**THE ONE BIG FLY HAD ONE RED EYE**
Kind of Mutation	*Sequence Change*
Missense	THQ ONE BIG FLY HAD ONE RED EYE
Nonsense	THE ONE BIG
Frameshift:	THE ONE QBI GFL YHA DON ERE DEY
Deletion	THE ONE BIG HAD ONE RED EYE
Duplication	THE ONE BIG FLY FLY HAD ONE RED EYE
Insertion	THE ONE BIG WET FLY HAD ONE RED EYE
Expanding mutation:	
Parents	THE ONE BIG FLY HAD ONE RED EYE
Children	THE ONE BIG FLY FLY FLY HAD ONE RED EYE
Grand children	THE ONE BIG FLY FLY FLY FLY FLY FLY HAD ONE RED EYE

A sentence composed of three-letter words can provide an analogy to the effect of mutations on a gene's nucleotide sequence.
Based on R. Lewis, *Human Genetics, Concepts and Applications,* 2nd ed., Dubuque, IA: Wm. C. Brown Publishers, 1997.

CONNECTIONS

Tobacco and Genes

For decades cancer has been empirically (based on commonplace experiences) linked to the use of tobacco, but only recently have researchers showed a cause-and-effect relationship between the two. Studies of patients with lung cancer revealed that 60 percent exhibited mutations in a gene known as p53. This gene is responsible for the synthesis of a protein that keeps cells growing in a controlled fashion and, when necessary, causes cells to commit suicide, a process called apoptosis. To be more specific, one of the compounds in tobacco (benzo[a]pyrene) has the ability to attach and mutate three spots in this gene, the same spots where the p53 genes have been changed in lung cancer patients. Studies of this nature give strong scientific evidence that smoking causes lung cancer.

"foreign" genes and synthesizes proteins encoded by them. Gene splicing begins with the laboratory isolation of DNA from an organism that contains the desired gene; for example, from human cells that contain the gene for the manufacture of insulin. If the gene is short enough and its base sequence is known, it may be synthesized in the laboratory from separate nucleotides. If the gene is too long and complex, it is cut from the chromosome with enzymes called *restriction endonucleases*. They are given this name because these enzymes (*-ases*) only cut DNA (*nucle-*) at certain base sequences (restricted in their action) and work inside (*endo-*) the DNA. These particular enzymes do not cut the DNA straight across, but in a zig-zag pattern that leaves one strand slightly longer than its complement. The short nucleotide sequence that sticks out and remains unpaired is called a *sticky end* because it can be reattached

to another complementary strand. DNA segments have been successfully cut from rats, frogs, bacteria, and humans.

This isolated gene with its "sticky end" is spliced into microbial DNA. The host DNA is opened up with the proper restriction endonuclease, and ligase (i.e., tie together) enzymes are used to attach the sticky ends into the host DNA. This gene-splicing procedure may be performed with small loops of bacterial DNA that are not part of the main chromosome. These small DNA loops are called *plasmids*. Once the splicing is completed, the *plasmids* can be inserted into the bacterial host by treating the cell with special chemicals that encourage it to take in these large chunks of DNA. A more efficient alternative is to splice the desired gene into the DNA of a bacterial virus so that it can carry the new gene into the bacterium as it infects the host cell. Once inside the host cell, the genes

A CLOSER LOOK | The PCR and Genetic Fingerprinting

In 1989, the American Association for the Advancement of Science named DNA polymerase Molecule of the Year. The value of this enzyme in the polymerase chain reaction (PCR) is so great that it could not be ignored. Just what is the PCR, how does it work, and what can you do with it?

The PCR is a laboratory procedure for copying selected segments of DNA. A single cell can provide enough DNA for analysis and identification! Having a large number of copies of a "target sequence" of nucleotides enables biochemists to more easily work with DNA. This is like increasing the one "needle in the haystack" to such large numbers (100 billion in only a matter of hours) that they're not hard to find, recognize, and work with. The types of specimens that can be used include semen, hair, blood, bacteria, protozoa, viruses, mummified tissue, and frozen cells. The process requires the DNA specimen, free DNA nucleotides, synthetic "primer" DNA, DNA polymerase, and simple lab equipment, such as a test tube and a source of heat.

Having decided which target sequence of nucleotides (which "needle") is to be replicated, scientists heat the specimen of DNA to separate the coding and noncoding strands. Molecules of synthetic "primer" DNA are added to the specimen. These primer molecules are specifically designed to attach to the ends of the target sequence. Next, a mixture of triphosphorylated nucleotides is added so that they can become the newly replicated DNA. The presence of the primer, attached to the DNA and added nucleotides, serves as the substrate for the DNA polymerase. Once added, the polymerase begins making its way down the length of the DNA from one attached primer end to the other. The enzyme bonds the new DNA nucleotides to the strand, replicating the molecule as it goes. It stops when it reaches the other end, having produced a new copy of the target sequence. Because the enzyme will continue to operate as long as enzymes and substrates are available, the process continues, and in a short time there are billions of small pieces of DNA, all replicas of the target sequence.

So what, you say? Well, consider the following. Using the PCR, scientists have been able to:

1. more accurately diagnose such diseases as sickle-cell anemia, cancer, Lyme disease, AIDS, and Legionnaires disease;
2. perform highly accurate tissue typing for matching organ-transplant donors and recipients;
3. help resolve criminal cases of rape, murder, assault, and robbery by matching suspect DNA to that found at the crime scene;
4. detect specific bacteria in environmental samples;
5. monitor the spread of genetically engineered microorganisms in the environment;
6. check water quality by detecting bacterial contamination from feces;
7. identify viruses in water samples;
8. identify disease-causing protozoa in water;
9. determine specific metabolic pathways and activities occurring in microorganisms;
10. determine races, distribution patterns, kinships, migration patterns, evolutionary relationships, and rates of evolution of long-extinct species;
11. accurately settle paternity suits;
12. confirm identity in amnesia cases;
13. identify a person as a relative for immigration purposes;
14. provide the basis for making human antibodies in specific bacteria;
15. possibly provide the basis for replicating genes that could be transplanted into individuals suffering from genetic diseases; and
16. identify nucleotide sequences peculiar to the human genome (an application currently underway as part of the Human Genome Project).

may be replicated, along with the rest of the DNA, to clone the "foreign" gene, or they may begin to synthesize the encoded protein.

As this highly sophisticated procedure has been refined, it has become possible to quickly and accurately splice genes from a variety of species into host bacteria, making possible the synthesis of large quantities of medically important products. For example, recombinant DNA procedures are responsible for the production of human insulin, used in the control of diabetes; interferon, used as an antiviral agent; human growth hormone, used to stimulate growth in children lacking this hormone; and somatostatin, a brain hormone also implicated in growth. Over 200 such products have been manufactured using these methods.

The possibilities that open up with the manipulation of DNA are revolutionary. These methods enable cells to produce molecules that they would not normally make. Some research laboratories have even spliced genes into laboratory-cultured human cells. Should such a venture prove to be practical, genetic diseases such as sickle-cell anemia could be controlled. The process of recombinant DNA gene splicing also enables cells to be more efficient at producing molecules that they normally synthesize. Some of the likely rewards are

1. production of additional, medically useful proteins;
2. mapping of the locations of genes on human chromosomes;
3. more complete understanding of how genes are regulated;
4. production of crop plants with increased yields; and
5. development of new species of garden plants.

The discovery of the structure of DNA over forty years ago seemed very far removed from the practical world. The importance of this "pure" or "basic" research is just now being realized. Many companies are involved in recombinant DNA research with the aim of alleviating or curing disease.

SUMMARY

Genes are *units of heredity* composed of specific *lengths of DNA* that determine the *characteristics* an organism displays. *Specific genes* are at *specific loci* on *specific chromosomes*. The *phenotype* displayed by an organism is the result of the effect of the environment on the ability of the genes to express themselves.

Diploid organisms have *two genes* for each characteristic. The *alternative forms of genes* for a characteristic are called *alleles*. There may be many *different* alleles for a *particular characteristic*. Organisms with *two identical alleles* are *homozygous* for a characteristic; those with *different alleles* are *heterozygous*. Some alleles are *dominant* over other alleles that are said to be *recessive*.

Sometimes *two alleles both express themselves*, and often a gene has more than one recognizable effect on the phenotype of the organism. Some characteristics may be determined by several different *pairs of alleles*.

The successful operation of a *living cell* depends on its ability to *accurately reproduce genes and control chemical reactions. DNA replication* results in an exact *doubling* of the *genetic material*. The process virtually guarantees that *identical* strands of DNA will be passed on to the next generation of cells.

The *enzymes* are responsible for the *efficient control of a cell's metabolism*. However, the *production of protein molecules* is under the control of the *nucleic acids*, the *primary control molecules* of the cell. The structure of the nucleic acids *DNA and RNA* determine the *structure of the proteins*, whereas the *structure of the proteins* determines their *function* in the cell's life cycle. *Protein synthesis* involves the *decoding* of the DNA into specific protein molecules and the use of the *intermediate molecules*, mRNA and tRNA, at the *ribosome*. Errors in any of the *codons* of these molecules may produce observable *changes* in the cell's functioning and lead to cell *death*.

Methods of *manipulating DNA* have led to the *controlled transfer of genes* from one kind of organism to another. This has made it possible for *bacteria* to produce a number of *human* gene products.

KEY TERMS

alleles (p. 744)

anticodon (p. 767)

biotechnology (p. 775)

carrier (p. 745)

chromosomal aberrations (p. 774)

chromosome (p. 760)

coding strand (p. 764)

codominance (p. 746)

codon (p. 767)

complementary base (p. 759)

DNA code (p. 764)

DNA polymerase (p. 761)

DNA replication (p. 756)

dominant allele (p. 745)

duplex DNA (p. 759)

gene (p. 744)

genetics (p. 744)

genome (p. 745)

genotype (p. 745)

heterozygous (p. 745)

homozygous (p. 745)

law of dominance (p. 748)

law of independent assortment (p. 748)

law of segregation (p. 748)

linkage group (p. 746)

Mendelian genetics (p. 744)

messenger RNA (mRNA) (p. 767)

multiple alleles (p. 751)

mutagenic agent (p. 774)

mutation (p. 768)

nucleic acids (p. 755)

nucleoproteins (p. 760)

nucleosomes (p. 760)

nucleotides (p. 755)

phenotype (p. 745)

pleiotropy (p. 752)

point mutation (p. 774)

polygenic inheritance (p. 751)

probability (p. 748)

Punnett square (p. 750)

recessive allele (p. 745)

recombinant DNA (p. 775)

ribosomal RNA (rRNA) (p. 767)

RNA polymerase (p. 767)

sickle-cell anemia (p. 775)

single-factor cross (p. 749)

telomeres (p. 764)

transcription (p. 756)

transfer RNA (tRNA) (p. 767)

translation (p. 756)

X-linked gene (p. 747)

APPLYING THE CONCEPTS

1. BbCCDd is the genotype of an organism. How many different types of gametes can this organism produce?
   a. 1
   b. 2
   c. 3
   d. 4

2. An example of a phenotype is
   a. AB type blood.
   b. an allele for type A and B blood.
   c. a sperm with an allele for A blood.
   d. lack of iron in the diet causing anemia.

3. Body size is determined by the interaction of numerous alleles. This is an example of
   a. autosomes.
   b. pleiotropy.
   c. single-factor crosses.
   d. polygenic inheritance.

4. The probability of guessing the correct answer to this question is
   a. four.
   b. one-fourth.
   c. sixteen.
   d. 1/4 times 1/4.

5. If both parents are heterozygous, the probability that the recessive trait will appear in the offspring is
   a. 1/2.
   b. 1/4.
   c. 3/4.
   d. zero.

6. In order for a recessive X-linked trait to appear in a female, she must inherit a recessive allele from
   a. neither parent.
   b. both parents.
   c. her father only.
   d. her mother only.

7. It was noticed that in certain flowers, if a flower had red petals they were usually small petals. If the petals were white, pink, or orange they were usually large petals. This could be the result of
   a. lack of dominance.
   b. pleiotropy.
   c. linkage groups.
   d. polygenic inheritance.

8. In pea plants, tall plant height (*T*) dominates dwarf (*t*). If a plant heterozygous for the height trait is crossed with a dwarf plant, what will be the phenotypic ratios of the offspring?

   a. all *Tt*

   b. 1/2 *Tt*, 1/4 *TT*, 1/4 *tt*

   c. 1/2 tall, 1/2 dwarf

   d. 3/4 tall, 1/4 dwarf

9. A man with normal color vision marries a woman with normal vision, but who is a carrier for color blindness. The probability that their first child will be color-blind is _____; the probability that their first daughter will be color-blind is _____; and the probability that their first son will be color-blind is _____.

   a. 1/2; 1/2; 1/2

   b. 1/4; 0; 1/2

   c. 1/2; 0; 1/2

   d. 1/4; 1/2; 1/2

10. A woman with blood type O and a man with blood type AB have a child together. What are the possible bloodtypes of this child?

    a. AB or O

    b. A, B, or O

    c. A or B

    d. AB, A, B, or O

11. A radish plant that produces round radishes is crossed with a radish plant that produces long radishes. All of the offspring have oval radishes. Radish shape may be inherited by

    a. polygenic inheritance.

    b. pleiotropy.

    c. multiple alleles.

    d. lack of dominance.

12. In the following cross, what is the probability that the offspring will exhibit both dominant traits?

    Aabb × aaBB

    a. 1/2

    b. 3/8

    c. 4/16

    d. 1

13. "She has really long fingers and toes and is exceptionally tall." This is a statement of

    a. genotype.

    b. phenotype.

    c. monohybridization.

    d. locus placement.

14. One difference between preRNA and mature RNA is:

    a. kinds of nucleotide components.

    b. presence of extra RNA in preRNA.

    c. the part of DNA from which they were coded.

    d. only the age of the molecules.

15. Which of the following is *not* a chromosomal aberration?

    a. inversion

    b. duplication

    c. deletion

    d. X ray

16. The site of protein synthesis is:

    a. at the nuclear membrane.

    b. at ribosomes.

    c. near microfilaments.

    d. always at a Golgi body.

17. While one strand of duplex DNA is being transcribed to mRNA:

    a. the complementary strand makes tRNA.

    b. the complementary strand is inactive.

    c. the complementary strand at this point is replicating.

    d. mutations are impossible during this short period.

18. One way to introduce new DNA into an organism is

    a. gene splicing.

    b. replication.

    c. removing introns.

    d. transcription.

19. Removing only one base in a DNA sequence:

    a. usually has no effect on the organism.

    b. could result in a chromosomal mutation.

    c. cannot occur without extremes of heat and pressure.

    d. can result in a significant change in the information about a protein.

20. A major difference between the genetic data of prokaryotic and eukaryotic cells is that in prokaryotes the

    a. genes are RNA not DNA.

    b. histones are arranged differently.

    c. duplex DNA is circular.

    d. duplex DNA is absent in bacteria.

21. If the DNA gene strand has the base sequence CCA-TAT-TCG, the complimentary DNA strand will have the sequence:

    a. CCA-TAT-TCG

    b. GGU-AUA-AGC

    c. CCA-UAU-UCG

    d. GGT-ATA-AGC

22. A DNA gene strand with the base sequence CCA-TAT-TCG will be transcribed into RNA with the base sequence:

    a. CCA-TAT-TCG.

    b. GGU-AUA-AGC.

    c. CCA-UAU-UCG.

    d. GGT-ATA-AGC.

23. A DNA gene strand with the base sequence CCA-TAT-TCG codes for the amino acid sequence: (Consult the amino acid–nucleic acid dictionary in your text.)

    a. proline-tyrosine-serine

    b. glycine-isoleucine-threonine

    c. proline-tyrosine-threonine

    d. glycine-isoleucine-serine

24. The mRNA codon CAU will form temporary bonds with the

    a. mRNA anticodon CAU.

    b. mRNA codon GUA.

    c. tRNA anticodon GUA.

    d. tRNA codon CAU.

25. If the DNA base sequence GAG is mutated to GAC: (Consult the amino acid–nucleic acid dictionary in your text.)

    a. aspartic acid will substitute for glutamic acid in the resulting polypeptide.

    b. the resulting protein will be unable to function.

    c. there will be no change in the amino acid sequence of the resulting polypeptide.

    d. a chromosomal mutation has occurred.

26. In eukaryotic cells, mature RNA is formed by the

    a. removal of introns.

    b. removal of exons.

    c. addition of introns.

    d. addition of exons.

Answers

1. d 2. a 3. d 4. b 5. b 6. b 7. c 8. c 9. c 10. c 11. d 12. a 13. b 14. b 15. d
16. b 17. b 18. a 19. d 20. c 21. d 22. b 23. d 24. c 25. c 26. a

QUESTIONS FOR THOUGHT

1. What is the difference between a nucleotide, a nitrogenous base, and a codon?
2. What are the differences between DNA and RNA?
3. List the sequence of events that takes place when a DNA message is translated into protein.
4. Chromosomal and point mutations both occur in DNA. In what ways do they differ? How is this related to recombinant DNA?
5. Why is DNA replication necessary?
6. What is polymerase and how does it function?
7. How does DNA replication differ from the manufacture of an RNA molecule?
8. If a DNA nucleotide sequence is CATAAAGCA, what is the mRNA nucleotide sequence that would base-pair with it?
9. What amino acids would occur in the protein chemically coded by the sequence of nucleotides in question 8?
10. How do tRNA, rRNA, and mRNA differ in function?
11. How many kinds of gametes are possible with each of the following genotypes?
    a. Aa
    b. AaBB
    c. AaBb
    d. AaBbCc
12. What is the probability of getting the gamete ab from each of the following genotypes?
    a. aabb
    b. Aabb
    c. AaBb
    d. AABb
13. What is the probability of each of the following sets of parents producing the given genotypes in their offspring?

	Parents		Offspring	Genotype
a.	AA	×	aa	Aa
b.	Aa	×	Aa	Aa
c.	Aa	×	Aa	aa
d.	AaBb	×	AaBB	AABB
e.	AaBb	×	AaBB	AaBb
f.	AaBb	×	AaBb	AABB

14. If an offspring has the genotype Aa, what possible combinations of parental genotypes can exist?
15. In humans, the allele for albinism is recessive to the allele for normal skin pigmentation.
    a. What is the probability that a child of a heterozygous mother and father will be an albino?
    b. If a child is normal, what is the probability that it is a carrier of the recessive albino allele?
16. In certain pea plants, the allele T for tallness is dominant over t for shortness.
    a. If a homozygous tall and homozygous short plant are crossed, what will be the phenotype and genotype of the offspring?
    b. If both individuals are heterozygous, what will be the phenotypic and genotypic ratios of the offspring?
17. Dwarfism (achondroplasia) is the result of a mutation in the gene for the production of fibroblast growth factor. It is inherited as an autosomal dominant disorder and often occurs as a new mutation. What might be the explanation if two people of normal height and body stature have five children, one who displays achondroplasia?
18. What is the probability of a child having type AB blood if one of the parents is heterozygous for A blood and the other is heterozygous for B? What other genotypes are possible in these children?
19. A color-blind woman marries a man with normal vision. They have ten children—six boys and four girls.
    a. How many are expected to have normal vision?
    b. How many are expected to be color-blind?
20. A light-haired man has blood type O. His wife has dark hair and blood type AB, but her father had light hair.
    a. What is the probability that this couple will have a child with dark hair and blood type A?
    b. What is the probability that they will have a light-haired child with blood type B?
    c. How many different phenotypes could their children show?
21. Certain kinds of cattle have two alleles for coat color: R = red, and r = white. When an individual cow is heterozygous, it is spotted with red and white (roan). When two red alleles are present, it is red. When two white alleles are present, it is white. The allele L, for lack of horns, is dominant over l, for the presence of horns.
    a. If a bull and a cow both have the genotype $RrLl$, how many possible phenotypes of offspring can they have?
    b. How probable is each phenotype?
22. Hemophilia is a disease that prevents the blood from clotting normally. It is caused by a recessive allele located on the X chromosome. A boy has the disease; neither his parents nor his grandparents have the disease. What are the genotypes of his parents and grandparents?

APPENDIX A | Mathematical Review

WORKING WITH EQUATIONS

Many of the problems of science involve an equation, a shorthand way of describing patterns and relationships that are observed in nature. Equations are also used to identify properties and to define certain concepts, but all uses have well-established meanings, symbols that are used by convention, and allowed mathematical operations. This appendix will assist you in better understanding equations and the reasoning that goes with the manipulation of equations in problem-solving activities.

Background

In addition to a knowledge of rules for carrying out mathematical operations, an understanding of certain quantitative ideas and concepts can be very helpful when working with equations. Among these helpful concepts are (1) the meaning of inverse and reciprocal, (2) the concept of a ratio, and (3) fractions.

The term *inverse* means the opposite, or reverse, of something. For example, addition is the opposite, or inverse, of subtraction, and division is the inverse of multiplication. A *reciprocal* is defined as an inverse multiplication relationship between two numbers. For example, if the symbol n represents any number (except zero), then the reciprocal of n is $1/n$. The reciprocal of a number ($1/n$) multiplied by that number (n) always gives a product of 1. Thus, the number multiplied by 5 to give 1 is $1/5$ ($5 \times 1/5 = 5/5 = 1$). So $1/5$ is the reciprocal of 5, and 5 is the reciprocal of $1/5$. Each number is the *inverse* of the other.

The fraction $1/5$ means 1 divided by 5, and if you carry out the division it gives the decimal 0.2. Calculators that have a $1/x$ key will do the operation automatically. If you enter 5, then press the $1/x$ key, the answer of 0.2 is given. If you press the $1/x$ key again, the answer of 5 is given. Each of these numbers is a reciprocal of the other.

A *ratio* is a comparison between two numbers. If the symbols m and n are used to represent any two numbers, then the ratio of the number m to the number n is the fraction m/n. This expression means to divide m by n. For example, if m is 10 and n is 5, the ratio of 10 to 5 is $10/5$, or $2:1$.

Working with *fractions* is sometimes necessary in problem-solving exercises, and an understanding of these operations is needed to carry out unit calculations. It is helpful in many of these operations to remember that a number (or a unit) divided by itself is equal to 1, for example,

$$\frac{5}{5} = 1 \qquad \frac{\text{inch}}{\text{inch}} = 1 \qquad \frac{5 \text{ inches}}{5 \text{ inches}} = 1$$

When one fraction is divided by another fraction, the operation commonly applied is to "invert the denominator and multiply." For example, 2/5 divided by 1/2 is

$$\frac{\dfrac{2}{5}}{\dfrac{1}{2}} = \frac{2}{5} \times \frac{2}{1} = \frac{4}{5}$$

What you are really doing when you invert the denominator of the larger fraction and multiply is making the denominator (1/2) equal to 1. Both the numerator (2/5) and the denominator (1/2) are multiplied by 2/1, which does not change the value of the overall expression. The complete operation is

$$\frac{\dfrac{2}{5}}{\dfrac{1}{2}} \times \frac{\dfrac{1}{2}}{\dfrac{2}{1}} = \frac{\dfrac{2}{5} \times \dfrac{2}{1}}{\dfrac{1}{2} \times \dfrac{2}{1}} = \frac{\dfrac{4}{5}}{\dfrac{2}{2}} = \frac{\dfrac{4}{5}}{1} = \frac{4}{5}$$

Symbols and Operations

The use of symbols seems to cause confusion for some students because it seems different from their ordinary experiences with arithmetic. The rules are the same for symbols as they are for numbers, but you cannot do the operations with the symbols until you know what values they represent. The operation signs, such as $+$, $\div$, $\times$, and $-$, are used with symbols to indicate the operation that you *would* do if you knew the values. Some of the mathematical operations are indicated several ways. For example, $a \times b$, $a \cdot b$, and ab all indicate the same thing, that a is to be multiplied by b. Likewise, $a \div b$, a/b, and $a \times 1/b$ all indicate that a is to be divided by b. Since it is not possible to carry out the operations on symbols alone, they are called *indicated operations*.

Operations in Equations

An equation is a shorthand way of expressing a simple sentence with symbols. The equation has three parts: (1) a left side, (2) an equal sign ($=$), which indicates the equivalence of the two sides, and (3) a right side. The left side has the same value and units as the right side, but the two sides may have a very different appearance. The two sides may also have the symbols that indicate mathematical operations ($+$, $-$, $\times$, and so forth) and may be in certain forms that indicate operations (a/b, ab, and so forth). In any case, the equation is a complete expression that states the left side has the same value and units as the right side.

Equations may contain different symbols, each representing some unknown quantity. In science, the expression "solve the equation" means to perform certain operations with one symbol (which represents some variable) by itself on one side of the equation. This single symbol is usually, but not necessarily, on the left side and is not present on the other side. For example, the equation $F = ma$ has the symbol F on the left side. In science, you would say that this equation is solved for F. It could also be solved for m or for a, which will be considered shortly. The equation $F = ma$ is solved for F, and the *indicated operation* is to multiply m by a because they are in the form ma, which means the same thing as $m \times a$. This is the only indicated operation in this equation.

A solved equation is a set of instructions that has an order of indicated operations. For example, the equation for the relationship between a Fahrenheit and Celsius temperature, solved for °C, is $C = 5/9(F - 32)$. A list of indicated operations in this equation is as follows:

1. Subtract 32° from the given Fahrenheit temperature.
2. Multiply the result of (1) by 5.
3. Divide the result of (2) by 9.

Why are the operations indicated in this order? Because the bracket means 5/9 of the *quantity* $(F - 32°)$. In its expanded form, you can see that $5/9(F - 32°)$ actually means $5/9(F) - 5/9(32°)$. Thus, you cannot multiply by 5 or divide by 9 until you have found the quantity of $(F - 32°)$. Once you have figured out the order of operations, finding the answer to a problem becomes almost routine as you complete the needed operations on both the numbers and the units.

Solving Equations

Sometimes it is necessary to rearrange an equation to move a different symbol to one side by itself. This is known as solving an equation for an unknown quantity. But you cannot simply move a symbol to one side of an equation. Since an equation is a statement of equivalence, the right side has the same value as the left side. If you move a symbol, you must perform the operation in a way that the two sides remain equivalent. This is accomplished by "canceling out" symbols until you have the unknown on one side by itself. One key to understanding the canceling operation is to remember that a fraction with the same number (or unit) over itself is equal to 1. For example, consider the equation $F = ma$, which is solved for F. Suppose you are considering a problem in which F and m are given, and the unknown is a. You need to solve the equation for a so it is on one side by itself. To eliminate the m, you do the *inverse* of the indicated operation on m, dividing both sides by m. Thus,

$$F = ma$$

$$\frac{F}{m} = \frac{ma}{m}$$

$$\frac{F}{m} = a$$

Since m/m is equal to 1, the a remains by itself on the right side. For convenience, the whole equation may be flipped to move the unknown to the left side,

$$a = \frac{F}{m}$$

Thus, a quantity that indicated a multiplication (ma) was removed from one side by an inverse operation of dividing by m.

Consider the following inverse operations to "cancel" a quantity from one side of an equation, moving it to the other side:

If the Indicated Operation of the Symbol You Wish to Remove Is:	Perform This Inverse Operation on Both Sides of the Equation
multiplication	division
division	multiplication
addition	subtraction
subtraction	addition
squared	square root
square root	square

Q & A

The equation for finding the kinetic energy of a moving body is $KE = 1/2mv^2$. You need to solve this equation for the velocity, v.

Answer

The order of indicated operations in the equation is as follows:

1. Square v.
2. Multiply v^2 by m.
3. Divide the result of (2) by 2.

To solve for v, this order is *reversed* as the "canceling operations" are used:

Step 1: Multiply both sides by 2

$$KE = \frac{1}{2}mv^2$$

$$2KE = \frac{2}{2}mv^2$$

$$2KE = mv^2$$

Step 2: Divide both sides by m

$$\frac{2KE}{m} = \frac{mv^2}{m}$$

$$\frac{2KE}{m} = v^2$$

Step 3: Take the square root of both sides

$$\sqrt{\frac{2KE}{m}} = \sqrt{v^2}$$

$$\sqrt{\frac{2KE}{m}} = v$$

or

$$v = \sqrt{\frac{2KE}{m}}$$

The equation has been solved for v, and you are now ready to substitute quantities and perform the needed operations.

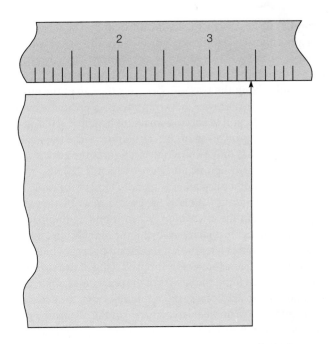

SIGNIFICANT FIGURES

The numerical value of any measurement will always contain some uncertainty. Suppose, for example, that you are measuring one side of a square piece of paper as shown in figure A.1. You could say that the paper is *about* 3.5 cm wide and you would be correct. This measurement, however, would be unsatisfactory for many purposes. It does not approach the true value of the length and contains too much uncertainty. It seems clear that the paper width is larger than 3.4 cm but shorter than 3.5 cm. But how much larger than 3.4 cm? You cannot be certain if the paper is 3.44, 3.45, or 3.46 cm wide. As your best estimate, you might say that the paper is 3.45 cm wide. Everyone would agree that you can be certain about the first two numbers (3.4) and they should be recorded. The last number (0.05) has been estimated and is not certain. The two certain numbers, together with one uncertain number, represent the greatest accuracy possible with the ruler being used. The paper is said to be 3.45 cm wide.

A *significant figure* is a number that is believed to be correct with some uncertainty only in the last digit. The value of the width of the paper, 3.45 cm, represents three significant figures. As you can see, the number of significant figures can be determined by the degree of accuracy of the measuring instrument being used. But suppose you need to calculate the area of the paper. You would multiply 3.45 cm $\times$ 3.45 cm and the product for the area would be 11.9025 cm^2. This is a greater accuracy than you were able to obtain with your measuring instrument. The result of a calculation can be no more accurate than the values being treated. Because the measurement had only three significant figures (two certain, one uncertain), then the answer can have only three significant figures. The area is correctly expressed as 11.9 cm^2.

There are a few simple rules that will help you determine how many significant figures are contained in a reported measurement:

1. All digits reported as a direct result of a measurement are significant.

2. Zero is significant when it occurs between nonzero digits. For example, 607 has three significant figures, and the zero is one of the significant figures.

3. In figures reported as *larger than the digit one*, the digit zero is not significant when it follows a nonzero digit to indicate place. For example, in a report that "23,000 people attended the rock concert," the digits 2 and 3 are significant but the zeros are not significant. In this situation the 23 is the measured part of the figure, and the three zeros tell you an estimate of how many attended the concert, that is, 23 thousand. If the figure is a measurement rather than an estimate, then it is written *with a decimal point after the last zero* to indicate that the zeros *are* significant. Thus 23,000 has *two* significant figures (2 and 3), but 23,000. has *five* significant figures. The figure 23,000 means "about 23 thousand," but 23,000. means 23,000. and not 22,999 or 23,001.

4. In figures reported as *smaller than the digit 1*, zeros after a decimal point that come before nonzero digits *are not* significant and serve only as place holders. For example, 0.0023 has two significant figures, 2 and 3. Zeros alone after a decimal point or zeros after a nonzero digit indicate a measurement, however, so these zeros *are* significant. The figure 0.00230, for example, has three significant figures since the 230 means 230 and not 229 or 231. Likewise, the figure 3.000 cm has four significant figures because the presence of the three zeros means that the measurement was actually 3.000 and not 2.999 or 3.001.

Multiplication and Division

When multiplying or dividing measurement figures, the answer may have no more significant figures than the *least* number of significant figures in the figures being multiplied or divided. This simply means that an answer can be no more accurate than the least accurate measurement entering into the calculation, and that you cannot improve the accuracy of a measurement by doing a calculation. For example, in multiplying 54.2 mi/hr $\times$ 4.0 hr to find out the total distance traveled, the first figure (54.2) has three significant figures but the second (4.0) has only two significant figures. The answer can contain only two significant figures since this is the weakest number of those involved in the calculation. The correct answer is therefore 220 mi, not 216.8 mi. This may seem strange since multiplying the two numbers together gives the answer of 216.8 mi. This answer, however, means a greater accuracy than is possible, and the accuracy cannot be improved over the weakest number involved in the calculation. Since the weakest number (4.0) has only two significant figures the answer must also have only two significant figures, which is 220 mi.

The result of a calculation is *rounded* to have the same least number of significant figures as the least number of a measurement involved in the calculation. When rounding numbers the last significant figure is increased by 1 if the number after it is 5 or larger. If the number after the last significant figure is 4 or less, the nonsignificant figures are simply dropped. Thus, if two significant figures are called for in the answer of the previous example, 216.8 is rounded up to 220 because the last number after the two significant figures is 6 (a number larger than 5). If the calculation result had been 214.8, the rounded number would be 210 miles.

Note that *measurement figures* are the only figures involved in the number of significant figures in the answer. Numbers that are counted or **defined** are not included in the determination of significant figures in an answer. For example, when dividing by 2 to find an average, the 2 is ignored when considering the number of significant figures. Defined numbers are defined exactly and are not used in significant figures. Since 1 kilogram is *defined* to be exactly 1,000 grams, such a conversion is not a measurement.

Addition and Subtraction

Addition and subtraction operations involving measurements, as multiplication and division, cannot result in an answer that implies greater accuracy than the measurements had before the calculation. Recall that the last digit to the right in a measurement is uncertain, that is, it is the result of an estimate. The answer to an addition or subtraction calculation can have this uncertain number *no farther from the decimal place than it was in the weakest number involved in the calculation.* Thus, when 8.4 is added to 4.926, the weakest number is 8.4, and the uncertain number is .4, one place to the right of the decimal. The sum of 13.326 is therefore rounded to 13.3, reflecting the placement of this weakest doubtful figure.

The rules for counting zeros tell us that the numbers 203 and 0.200 both have three significant figures. Likewise, the numbers 230 and 0.23 only have two significant figures. Once you remember the rules, the counting of significant figures is straightforward. On the other hand, sometimes you find a number that seems to make it impossible to follow the rules. For example, how would you write 3,000 with two significant figures? There are several special systems in use for taking care of problems such as this, including the placing of a little bar over the last significant digit. One of the convenient ways of showing significant figures for difficult numbers is to use scientific notation, which is also discussed elsewhere in this appendix. The convention for writing significant figures is to display one digit to the left of the decimal. The exponents are not considered when showing the number of

significant figures in scientific notation. Thus if you want to write three thousand showing one significant figure, you would write 3×10^3. To show two significant figures it is 3.0×10^3 and for three significant figures it becomes 3.00×10^3. As you can see, the correct use of scientific notation leaves little room for doubt about how many significant figures are intended.

In an example concerning percentage error, an experimental result of 511 Hz was found for a tuning fork with an accepted frequency value of 522 Hz. The error calculation is

$$\frac{(522\ Hz - 511\ Hz)}{522\ Hz} \times 100\% = 2.1\%$$

Since $522 - 511$ is 11, the least number of significant figures of measurements involved in this calculation is *two*. Note that the "100" does not enter into the determination since it is not a measurement number. The calculated result (from a calculator) is 2.1072797, which is rounded off to have only two significant figures, so the answer is recorded as 2.1%.

CONVERSION OF UNITS

The measurement of most properties results in both a numerical value and a unit. The statement that a glass contains 50 cm³ of a liquid conveys two important concepts—the numerical value of 50 and the referent unit of cubic centimeters. Both the numerical value and the unit are necessary to communicate correctly the volume of the liquid.

When working with calculations involving measurement units, *both* the numerical value and the units are treated mathematically. As in other mathematical operations, there are general rules to follow.

1. Only properties with *like units* may be added or subtracted. It should be obvious that adding quantities such as 5 dollars and 10 dimes is meaningless. You must first convert to like units before adding or subtracting.
2. Like or unlike units may be multiplied or divided and treated in the same manner as numbers. You have used this rule when dealing with area (length × length = length², for example, cm × cm = cm²) and when dealing with volume (length × length × length = length³, for example, cm × cm × cm = cm³).

You can use these two rules to create a *conversion ratio* that will help you change one unit to another. Suppose you need to convert 2.3 kg to grams. First, write the relationship between kilograms and grams:

$$1,000\ g = 1\ kg$$

Next, divide both sides by what you wish to convert *from* (kilograms in this example):

$$\frac{1,000\ \text{g}}{1\ \text{kg}} = \frac{1\ \text{kg}}{1\ \text{kg}}$$

One kilogram divided by 1 kg equals 1, just as 10 divided by 10 equals 1. Therefore, the right side of the relationship becomes 1 and the equation is:

$$\frac{1,000\ \text{g}}{1\ \text{kg}} = 1$$

The 1 is usually understood, that is, not stated, and the operation is called *canceling*. Canceling leaves you with the fraction 1,000 g/1 kg, which is a conversion ratio that can be used to convert from kilograms to grams. You simply multiply the conversion ratio by the numerical value and unit you wish to convert:

$$= 2.3\ \text{kg} \times \frac{1,000\ \text{g}}{1\ \text{kg}}$$

$$= \frac{2.3 \times 1,000}{1}\ \frac{\text{kg} \times \text{g}}{\text{kg}}$$

$$= \boxed{2,300\ \text{g}}$$

The kilogram units cancel. Showing the whole operation with units only, you can see how you end up with the correct unit of grams:

$$\text{kg} \times \frac{\text{g}}{\text{kg}} = \frac{\text{kg} \cdot \text{g}}{\text{kg}} = \text{g}$$

Since you did obtain the correct unit, you know that you used the correct conversion ratio. If you had blundered and used an inverted conversion ratio, you would obtain

$$2.3\ \text{kg} \times \frac{1\ \text{kg}}{1,000\ \text{g}} = .0023\ \frac{\text{kg}^2}{\text{g}}$$

which yields the meaningless, incorrect units of kg^2/g. Carrying out the mathematical operations on the numbers and the units will always tell you whether or not you used the correct conversion ratio.

A distance is reported as 100.0 km, and you want to know how far this is in miles.

Solution

First, you need to obtain a *conversion factor* from a textbook or reference book, which usually lists the conversion factors by properties in a table. Such a table will show two conversion factors for kilometers and miles: (1) 1 km = 0.621 mi and (2) 1 mi = 1.609 km. You select the factor that is in the same form as your problem; for example, your problem is 100.0 km = ? mi. The conversion factor in this form is 1 km = 0.621 mi.

Second, you convert this conversion factor into a *conversion ratio* by dividing the factor by what you wish to convert *from:*

conversion factor:	1 km = 0.621 mi
divide factor by what you want to convert from:	$\dfrac{1\ \text{km}}{1\ \text{km}} = \dfrac{0.621\ \text{mi}}{1\ \text{km}}$
resulting conversion rate:	$\dfrac{0.621\ \text{mi}}{\text{km}}$

Note that if you had used the 1 mi = 1.609 km factor, the resulting units would be meaningless. The conversion ratio is now multiplied by the numerical value *and unit* you wish to convert:

$$100.0\ \text{km} \times \frac{0.621\ \text{mi}}{\text{km}}$$

$$(100.0)(0.621)\ \frac{\text{km} \cdot \text{mi}}{\text{km}}$$

$$62.1\ \text{mi}$$

Q & A

A service station sells gasoline by the liter, and you fill your tank with 72 liters. How many gallons is this? (Answer: 19 gal)

SCIENTIFIC NOTATION

Most of the properties of things that you might measure in your everyday world can be expressed with a small range of numerical values together with some standard unit of measure. The range of numerical values for most everyday things can be dealt with by using units (1s), tens (10s), hundreds (100s), or perhaps thousands (1,000s). But the actual universe contains some objects of incredibly large size that require some very big numbers to describe. The sun, for example, has a mass of about 1,970,000,000,000,000,000,000,000,000,000 kg. On the other hand, very small numbers are needed to measure the size and parts of an atom. The radius of a hydrogen atom, for example, is about 0.00000000005 m. Such extremely large and small numbers are cumbersome and awkward since there are so many zeros to keep track of, even if you are successful in carefully counting all the zeros. A method does exist to deal with extremely large or small numbers in a more condensed form. The method is called *scientific notation*, but it is also sometimes called *powers of ten* or *exponential notation*, since it is based on exponents of 10. Whatever it is called, the method is a compact way of dealing with numbers that not only helps you keep track of zeros but provides a simplified way to make calculations as well.

In algebra you save a lot of time (as well as paper) by writing $(a \times a \times a \times a \times a)$ as a^5. The small number written to the right and above a letter or number is a superscript called an *exponent.* The exponent means that the letter or number is to be multiplied by itself that many times, for example, a^5 means a multiplied by itself five times, or $a \times a \times a \times a \times a$. As you can see, it is much easier to write the exponential form of this operation than it is to write it out in the long form. Scientific notation uses an exponent to indicate the power of the base 10. The exponent tells how many times the base, 10, is multiplied by itself. For example,

$$10,000 = 10^4$$
$$1,000 = 10^3$$
$$100 = 10^2$$
$$10 = 10^1$$
$$1 = 10^0$$
$$0.1 = 10^{-1}$$
$$0.01 = 10^{-2}$$
$$0.001 = 10^{-3}$$
$$0.0001 = 10^{-4}$$

This table could be extended indefinitely, but this somewhat shorter version will give you an idea of how the method works. The symbol 10^4 is read as "ten to the fourth power" and means $10 \times 10 \times 10 \times 10$. Ten times itself four times is 10,000, so 10^4 is the scientific notation for 10,000. It is also equal to the number of zeros between the 1 and the decimal point, that is, to write the longer form of 10^4 you simply write 1, then move the decimal point four places to the *right;* 10 to the fourth power is 10,000.

The power of ten table also shows that numbers smaller than 1 have negative exponents. A negative exponent means a reciprocal:

$$10^{-1} = \frac{1}{10} = 0.1$$

$$10^{-2} = \frac{1}{100} = 0.01$$

$$10^{-3} = \frac{1}{1000} = 0.001$$

To write the longer form of 10^{-4}, you simply write 1 then move the decimal point four places to the *left;* 10 to the negative fourth power is 0.0001.

Scientific notation usually, but not always, is expressed as the product of two numbers: (1) a number between 1 and 10 that is called the *coefficient* and (2) a power of ten that is called the *exponent.* For example, the mass of the sun that was given in long form earlier is expressed in scientific notation as

$$1.97 \times 10^{30}\,\text{kg}$$

and the radius of a hydrogen atom is

$$5.0 \times 10^{-11}\,\text{m}$$

In these expressions, the coefficients are 1.97 and 5.0, and the power of ten notations are the exponents. Note that in both of these examples, the exponent tells you where to place the deci-

mal point if you wish to write the number all the way out in the long form. Sometimes scientific notation is written without a coefficient, showing only the exponent. In these cases the coefficient of 1.0 is understood, that is, not stated. If you try to enter a scientific notation in your calculator, however, you will need to enter the understood 1.0, or the calculator will not be able to function correctly. Note also that 1.97×10^{30} kg and the expressions 0.197×10^{31} kg and 19.7×10^{29} kg are all correct expressions of the mass of the sun. By convention, however, you will use the form that has one digit to the left of the decimal.

What is 26,000,000 in scientific notation?

Answer

Count how many times you must shift the decimal point until one digit remains to the left of the decimal point. For numbers larger than the digit 1, the number of shifts tells you how much the exponent is increased, so the answer is

$$2.6 \times 10^7$$

which means the coefficient 2.6 is multiplied by 10 seven times.

What is 0.000732 in scientific notation? (Answer: 7.32×10^{-4})

It was stated earlier that scientific notation provides a compact way of dealing with very large or very small numbers, but it provides a simplified way to make calculations as well. There are a few mathematical rules that will describe how the use of scientific notation simplifies these calculations.

To *multiply* two scientific notation numbers, the coefficients are multiplied as usual, and the exponents are *added* algebraically. For example, to multiply (2×10^2) by (3×10^3), first separate the coefficients from the exponents,

$$(2 \times 3) \times (10^2 \times 10^3),$$

then multiply the coefficients and add the exponents,

$$6 \times 10^{(2 + 3)} = 6 \times 10^5$$

Adding the exponents is possible because $10^2 \times 10^3$ means the same thing as $(10 \times 10) \times (10 \times 10 \times 10)$, which equals $(100) \times (1,000)$, or 100,000, which is expressed as 10^5 in scientific notation. Note that two negative exponents add algebraically, for example $10^{-2} \times 10^{-3} = 10^{[(-2) + (-3)]} = 10^{-5}$. A negative and a positive exponent also add algebraically, as in $10^5 \times 10^{-3} = 10^{[(+5) + (-3)]} = 10^2$.

If the result of a calculation involving two scientific notation numbers does not have the conventional one digit to the left of the decimal, move the decimal point so it does, changing the exponent according to which way and how much the decimal point is moved. Note that the exponent increases by one number for each decimal point moved to the left. Likewise, the exponent decreases by one number for each decimal point moved to the right. For example, $938. \times 10^3$ becomes 9.38×10^5 when the decimal point is moved two places to the left.

To *divide* two scientific notation numbers, the coefficients are divided as usual and the exponents are *subtracted*. For example, to divide (6×10^6) by (3×10^2), first separate the coefficients from the exponents,

$$(6 \div 3) \times (10^6 \div 10^2)$$

then divide the coefficients and subtract the exponents,

$$2 \times 10^{(6-2)} = 2 \times 10^4$$

Note that when you subtract a negative exponent, for example, $10^{[(3) - (-2)]}$, you change the sign and add, $10^{(3+2)} = 10^5$.

Solve the following problem concerning scientific notation:

$$\frac{(2 \times 10^4) \times (8 \times 10^{-6})}{8 \times 10^4}$$

Answer

First, separate the coefficients from the exponents,

$$\frac{2 \times 8}{8} \times \frac{10^4 \times 10^{-6}}{10^4}$$

then multiply and divide the coefficients and add and subtract the exponents as the problem requires,

$$2 \times 10^{\{[(4) + (-6)] - [4]\}}$$

Solving the remaining additions and subtractions of the coefficients gives

$$2 \times 10^{-6}$$

APPENDIX B | Solubilities Chart

	Acetate	Bromide	Carbonate	Chloride	Fluoride	Hydroxide	Iodide	Nitrate	Oxide	Phosphate	Sulfate	Sulfide
Aluminum	S	S	—	S	s	i	S	S	i	i	S	d
Ammonium	S	S	S	S	S	S	S	S	—	S	S	S
Barium	S	S	i	S	s	S	S	S	S	i	i	d
Calcium	S	S	i	S	i	s	S	S	s	i	s	d
Copper(I)	—	s	i	s	i	—	i	—	i	—	d	i
Copper(II)	S	S	i	S	S	i	S	S	i	i	S	i
Iron(II)	S	S	i	S	s	i	S	S	i	i	S	i
Iron(III)	S	S	i	S	s	i	S	S	i	i	S	d
Lead	S	s	i	s	i	i	s	S	i	i	i	i
Magnesium	S	S	i	S	i	i	S	S	i	i	S	d
Mercury(I)	s	i	i	i	d	d	i	S	i	i	i	i
Mercury(II)	S	s	i	S	d	i	i	S	i	i	i	i
Potassium	S	S	S	S	S	S	S	S	S	S	S	i
Silver	s	i	i	i	S	—	i	S	i	i	i	i
Sodium	S	S	S	S	S	S	S	S	d	S	S	S
Strontium	S	S	s	S	i	s	S	S	—	i	i	i
Zinc	S	S	i	S	S	i	S	S	i	i	S	i

S—soluble
i—insoluble
s—slightly soluble
d—decomposes

APPENDIX C | Relative Humidity Table

Dry-Bulb Temperature (°C)	Difference between Wet-Bulb and Dry-Bulb Temperatures (°C)																			
	1	2	3	4	5	6	7	8	9	10	11	12	13	14	15	16	17	18	19	20
0	81	64	46	29	13															
1	83	66	49	33	17															
2	84	68	52	37	22	7														
3	84	70	55	40	26	12														
4	86	71	57	43	29	16														
5	86	72	58	45	33	20	7													
6	86	73	60	48	35	24	11													
7	87	74	62	50	38	26	15													
8	87	75	63	51	40	29	19	8												
9	88	76	64	53	42	32	22	12												
10	88	77	66	55	44	34	24	15	6											
11	89	78	67	56	46	36	27	18	9											
12	89	78	68	58	48	39	29	21	12											
13	89	79	69	59	50	41	32	23	15	7										
14	90	79	70	60	51	42	34	26	18	10										
15	90	80	71	61	53	44	36	27	20	13	6									
16	90	81	71	63	54	46	38	30	23	15	8									
17	90	81	72	64	55	47	40	32	25	18	11									
18	91	82	73	65	57	49	41	34	27	20	14	7								
19	91	82	74	65	58	50	43	36	29	22	16	10								
20	91	83	74	66	59	51	44	37	31	24	18	12	6							
21	91	83	75	67	60	53	46	39	32	26	20	14	9							
22	92	83	76	68	61	54	47	40	34	28	22	17	11	6						
23	92	84	76	69	62	55	48	42	36	30	24	19	13	8						
24	92	84	77	69	62	56	49	43	37	31	26	20	15	10	5					
25	92	84	77	70	63	57	50	44	39	33	28	22	17	12	8					
26	92	85	78	71	64	58	51	46	40	34	29	24	19	14	10	5				
27	92	85	78	71	65	58	52	47	41	36	31	26	21	16	12	7				
28	93	85	78	72	65	59	53	48	42	37	32	27	22	18	13	9	5			
29	93	86	79	72	66	60	54	49	43	38	33	28	24	19	15	11	7			
30	93	86	79	73	67	61	55	50	44	39	35	30	25	21	17	13	9	5		
31	93	86	80	73	67	61	56	51	45	40	36	31	27	22	18	14	11	7		
32	93	86	80	74	68	62	57	51	46	41	37	32	28	24	20	16	12	9	5	
33	93	87	80	74	68	63	57	52	47	42	38	33	29	25	21	17	14	10	7	
34	93	87	81	75	69	63	58	53	48	43	39	35	30	28	23	19	15	12	8	5
35	94	87	81	75	69	64	59	54	49	44	40	36	32	28	24	20	17	13	10	7

APPENDIX D | Solutions for Group A Parallel Exercises

Exercises CHAPTER 1

1.1. Answers will vary but should have the relationship of 100 cm in 1 m, for example, 178 cm = 1.78 m.

1.2. Since mass density is given by the relationship $\rho = m/V$, then

$$\rho = \frac{m}{V} = \frac{272 \text{ g}}{20.0 \text{ cm}^3}$$

$$= \frac{272}{20.0} \frac{\text{g}}{\text{cm}^3}$$

$$= \boxed{13.6 \frac{\text{g}}{\text{cm}^3}}$$

1.3. The volume of a sample of lead is given and the problem asks for the mass. From the relationship of $\rho = m/V$, solving for the mass (m) tells you that the mass density (ρ) times the volume (V), or $m = \rho V$. The mass density of lead, 11.4 g/cm³, can be obtained from table 1.4, so

$$\rho = \frac{m}{V}$$

$$V\rho = \frac{m\cancel{V}}{\cancel{V}}$$

$$m = \rho V$$

$$m = \left(11.4 \frac{\text{g}}{\text{cm}^3}\right)(10.0 \text{ cm}^3)$$

$$11.4 \times 10.0 \frac{\text{g}}{\text{cm}^3} \times \text{cm}^3$$

$$114 \frac{\text{g} \cdot \cancel{\text{cm}^3}}{\cancel{\text{cm}^3}}$$

$$= \boxed{114 \text{ g}}$$

1.4. Solving the relationship $\rho = m/V$ for volume gives $V = m/\rho$, and

$$\rho = \frac{m}{V}$$

$$V\rho = \frac{m\cancel{V}}{\cancel{V}}$$

$$\frac{V\cancel{\rho}}{\cancel{\rho}} = \frac{m}{\rho}$$

$$V = \frac{m}{\rho}$$

$$V = \frac{600 \text{ g}}{3.00 \dfrac{\text{g}}{\text{cm}^3}}$$

$$= \frac{600}{3.00} \frac{\text{g}}{1} \times \frac{\text{cm}^3}{\text{g}}$$

$$= 200 \frac{\text{g} \cdot \text{cm}^3}{\cancel{\text{g}}}$$

$$= \boxed{200 \text{ cm}^3}$$

1.5. A 50.0 cm³ sample with a mass of 34.0 grams has a density of

$$\rho = \frac{m}{V} = \frac{34.0 \text{ g}}{50.0 \text{ cm}^3}$$

$$= \frac{34.0}{50.0} \frac{\text{g}}{\text{cm}^3}$$

$$= \boxed{0.680 \frac{\text{g}}{\text{cm}^3}}$$

According to table 1.4, 0.680 g/cm³ is the mass density of gasoline, so the substance must be gasoline.

1.6. The problem asks for a mass and gives a volume, so you need a relationship between mass and volume. Table 1.4 gives the mass density of water as 1.00 g/cm³, which is a density that is easily remembered. The volume is given in liters (L), which should first be converted to cm³ because this is the unit in which density is expressed. The relationship of $\rho = m/V$ solved for mass is ρV, so the solution is

$$\rho = \frac{m}{V} \therefore m = \rho V$$

$$m = \left(1.00\frac{g}{cm^3}\right)(40,000 \text{ cm}^3)$$

$$= 1.00 \times 40,000 \frac{g}{cm^3} \times cm^3$$

$$= 40,000 \frac{g \cdot cm^3}{cm^3}$$

$$= 40,000 \text{ g}$$

$$= \boxed{40 \text{ kg}}$$

1.7. From table 1.4, the mass density of aluminum is given as 2.70 g/cm³. Converting 2.1 kg to the same units as the density gives 2,100 g. Solving $\rho = m/V$ for the volume gives

$$V = \frac{m}{\rho} = \frac{2,100 \text{ g}}{2.70 \dfrac{g}{cm^3}}$$

$$= \frac{2,100}{2.70} \frac{g}{1} \times \frac{cm^3}{g}$$

$$= 777.78 \frac{g \cdot cm^3}{g}$$

$$= \boxed{780 \text{ cm}^3}$$

1.8. The length of one side of the box is 0.1 m. Reasoning: Since the density of water is 1.00 g/cm³, then the volume of 1,000 g of water is 1,000 cm³. A cubic box with a volume of 1,000 cm³ is 10 cm (since 10 × 10 × 10 = 1,000). Converting 10 cm to m units, the cube is 0.1 m on each edge.

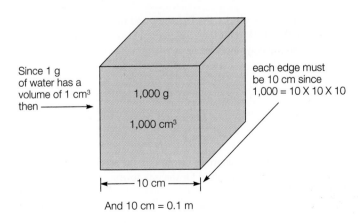

Since 1 g of water has a volume of 1 cm³ then →

1,000 g

1,000 cm³

each edge must be 10 cm since 1,000 = 10 × 10 × 10

|← 10 cm →|

And 10 cm = 0.1 m

1.9. The relationship between mass, volume, and density is $\rho = m/V$. The problem gives a volume, but not a mass. The mass, however, can be assumed to remain constant during the compression of the bread so the mass can be obtained from the original volume and density, or

$$\rho = \frac{m}{V} \therefore m = \rho V$$

$$m = \left(0.2\frac{g}{cm^3}\right)(3,000 \text{ cm}^3)$$

$$= 0.2 \times 3,000 \frac{g}{cm^3} \times cm^3$$

$$= 600 \frac{g \cdot cm^3}{cm^3}$$

$$= 600 \text{ g}$$

A mass of 600 g and the new volume of 1,500 cm³ means that the new density of the crushed bread is

$$\rho = \frac{m}{V}$$

$$= \frac{600 \text{ g}}{1,500 \text{ cm}^3}$$

$$= \frac{600}{1,500} \frac{g}{cm^3}$$

$$= \boxed{0.4 \frac{g}{cm^3}}$$

1.10. According to table 1.4, lead has a density of 11.4 g/cm³. Therefore, a 1.00 cm³ sample of lead would have a mass of

$$\rho = \frac{m}{V} \therefore m = \rho V$$

$$m = \left(11.4\frac{g}{cm^3}\right)(1.00 \text{ cm}^3)$$

$$= 11.4 \times 1.00 \frac{g}{cm^3} \times cm^3$$

$$= 11.4 \frac{g \cdot cm^3}{cm^3}$$

$$= 11.4 \text{ g}$$

Also according to table 1.4, copper has a density of 8.96 g/cm³. To balance a mass of 11.4 g of lead, a volume of this much copper would be required:

$$\rho = \frac{m}{V} \therefore V = \frac{m}{\rho}$$

$$V = \frac{11.4 \text{ g}}{8.96 \dfrac{g}{cm^3}}$$

$$= \frac{11.4}{8.96} \frac{g}{1} \times \frac{cm^3}{g}$$

$$= 1.27 \frac{g \cdot cm^3}{g}$$

$$= \boxed{1.27 \text{ cm}^3}$$

Exercises CHAPTER 2

2.1. Listing these quantities given in this problem, with their symbols, we have

$$d = 22 \text{ km}$$

$$t = 15 \text{ min}$$

$$\bar{v} = ?$$

The usual units for a speed problem are km/hr or m/s, and the problem specifies that the answer should be in km/hr. We see that 15 minutes is 15/60, or 1/4, or 0.25 of an hour. We will now make a new list of the quantities with the appropriate units:

$$d = 22 \text{ km}$$

$$t = 0.25 \text{ hr}$$

$$\bar{v} = ?$$

These quantities are related in the average speed equation, which is already solved for the unknown average velocity:

$$\bar{v} = \frac{d}{t}$$

Substituting the known quantities, we have

$$\bar{v} = \frac{22.0 \text{ km}}{0.25 \text{ hr}}$$

$$= \boxed{88 \frac{\text{km}}{\text{hr}}}$$

2.2. Listing the quantities with their symbols:

$$\bar{v} = 3.0 \times 10^8 \text{ m/s}$$

$$t = 20.0 \text{ min}$$

$$d = ?$$

We see that the velocity units are meters per second, but the time units are minutes. We need to convert minutes to seconds, and:

$$\bar{v} = 3.0 \times 10^8 \text{ m/s}$$

$$t = 1.20 \times 10^3 \text{ s}$$

$$d = ?$$

These relationships can be found in the average speed equation, which can be solved for the unknown:

$$\bar{v} = \frac{d}{t} \quad \therefore \quad d = \bar{v}t$$

$$d = \left(3.0 \times 10^8 \frac{\text{m}}{\text{s}}\right)(1.20 \times 10^3 \text{ s})$$

$$= (3.0)(1.20) \times 10^{8+3} \frac{\text{m}}{\text{s}} \times \text{s}$$

$$= 3.6 \times 10^{11} \text{ m}$$

$$= \boxed{3.6 \times 10^8 \text{ km}}$$

2.3. Listing the quantities with their symbols, we can see the problem involves the quantities found in the definition of average speed:

$$\bar{v} = 350.0 \text{ m/s}$$

$$t = 5.00 \text{ s}$$

$$d = ?$$

$$\bar{v} = \frac{d}{t} \quad \therefore \quad d = \bar{v}t$$

$$d = \left(350.0 \frac{\text{m}}{\text{s}}\right)(5.00\text{s})$$

$$= (350.0)(5.00) \frac{\text{m}}{\text{s}} \times \text{s}$$

$$= \boxed{1,750 \text{ m}}$$

2.4. Note that the two speeds given (100.0 km/hr and 50.0 km/hr) are *average* speeds for two different legs of a trip. They are not initial and final speeds of an accelerating object, so you cannot add them together and divide by 2. The average speed for the total (entire) trip can be found from definition of average speed, that is, average speed is the *total* distance covered divided by the *total* time elapsed. So, we start by finding the distance covered for each of the two legs of the trip:

$$\bar{v} = \frac{d}{t} \quad \therefore d = \bar{v}t$$

$$\text{Leg 1 distance} = \left(100.0 \frac{\text{km}}{\text{hr}}\right)(2.00 \text{ hr})$$

$$= 200.0 \text{ km}$$

$$\text{Leg 2 distance} = \left(50.0 \frac{\text{km}}{\text{hr}}\right)(1.00 \text{ hr})$$

$$= 50.0 \text{ km}$$

Total distance (leg 1 plus leg 2) = 250.0 km
Total time = 3.00 hr

$$\bar{v} = \frac{d}{t} = \frac{250.0 \text{ km}}{3.00 \text{ hr}} = \boxed{83.3 \text{ km/hr}}$$

2.5. The initial velocity, final velocity, and time are known and the problem asked for the acceleration. Listing these quantities with their symbols, we have

$$v_i = 0 \text{ m/s}$$

$$v_f = 15.0 \text{ m/s}$$

$$t = 10.0 \text{ s}$$

$$a = ?$$

These are the quantities involved in the acceleration equation, which is already solved for the unknown:

$$a = \frac{v_f - v_i}{t}$$

$$a = \frac{15.0 \text{ m/s} - 0 \text{ m/s}}{10.0 \text{ s}}$$

$$= \frac{15.0}{10.0} \frac{m}{s} \times \frac{1}{s}$$

$$= \boxed{1.50 \frac{m}{s^2}}$$

2.6. The initial velocity, final velocity, and acceleration are known and the problem asked for the time. Listing these quantities with their symbols, we have

$$v_i = 20.0 \text{ m/s}$$

$$v_f = 25.0 \text{ m/s}$$

$$a = 3.0 \text{ m/s}^2$$

$$t = ?$$

These are the quantities involved in the acceleration equation, which must first be solved for the unknown time:

$$a = \frac{v_f - v_i}{t} \quad \therefore \quad t = \frac{v_f - v_i}{a}$$

$$t = \frac{25.0 \frac{m}{s} - 20.0 \frac{m}{s}}{3.0 \frac{m}{s^2}}$$

$$= \frac{5.00 \frac{m}{s}}{3.0 \frac{m}{s^2}}$$

$$= \frac{5.00}{3.0} \frac{m}{s} \times \frac{s \cdot s}{m}$$

$$= \boxed{1.7s}$$

2.7. The initial velocity, final velocity, and time are known and the problem asked for the distance. Listing the quantities with their symbols, we have

$$v_i = 80.0 \text{ m/s}$$

$$v_f = 0 \text{ m/s}$$

$$t = 5.00 \text{ s}$$

$$d = ?$$

Looking over each of the equations in the equation list at the end of the chapter, we find that only $\bar{v} = \frac{d}{t}$ and $d = 1/2 \, at^2$ have a distance quantity. Either equation could be used to solve this problem, but the first seems to be less involved.

First, $\bar{v}$ can be obtained from $\bar{v} = \frac{v_f + v_i}{2}$ since the initial and final velocities are known. Thus,

$$\bar{v} = \frac{0 + 80.0 \frac{km}{hr}}{2} = 40.0 \frac{km}{hr}$$

Second, km/hr must be converted to m/s because the time unit is in seconds (please see appendix A for information on converting units):

$$\bar{v} = \frac{0.2778 \frac{m}{s}}{\frac{km}{hr}} \times 40.0 \frac{km}{hr}$$

$$= (0.2778)(40.0) \frac{m}{s} \times \frac{hr}{km} \times \frac{km}{hr}$$

$$= 11.1 \frac{m}{s}$$

Finally, the distance traveled can be found by solving the equation for average speed:

$$\bar{v} = \frac{d}{t} \quad \therefore \quad d = \bar{v}t$$

$$= \left(11.1 \frac{m}{s}\right)(5.00s)$$

$$= \boxed{55.5 \text{ m}}$$

2.8. The distance (d) and the time (t) quantities are given in the problem, and

$$\bar{v} = \frac{d}{t}$$

$$= \frac{285 \text{ mi}}{5.0 \text{ hr}}$$

$$= \frac{285}{5.0} \frac{mi}{hr}$$

$$= \boxed{57 \text{ mi/hr}}$$

The units cannot be simplified further. Note two significant figures in the answer, which is the least number of significant figures involved in the division operation.

2.9. **(a)** The sprinter's average speed is

$$\bar{v} = \frac{d}{t}$$

$$= \frac{200.0 \text{ m}}{21.4 \text{ s}}$$

$$= \frac{200.0}{21.4} \frac{m}{s}$$

$$= \boxed{9.35 \text{ m/s}}$$

(Note the *calculator answer* of 9.3457944 m/s is incorrect. The least number of significant figures involved in the division is three, so *your answer* should have three significant figures. This speed is approximately equivalent to 30 ft/s or 20 mi/hr).

(b) To find the time involved in maintaining a speed for a certain distance the relationship between average speed ($\bar{v}$), distance (d), and time (t) can be solved for t:

$$\bar{v} = \frac{d}{t}$$

$$\bar{v}t = d$$

$$t = \frac{d}{\bar{v}}$$

$$t = \frac{420{,}000 \text{ m}}{9.35 \dfrac{\text{m}}{\text{s}}}$$

$$= \frac{420{,}000}{9.35} \times \frac{\text{m}\cdot\text{s}}{\text{m}}$$

$$= 44{,}900 \text{ s}$$

The seconds can be converted to hours by dividing by 3,600 s/hr (60 seconds in a minute times 60 minutes in an hour):

$$\frac{44{,}900 \text{ s}}{3{,}600 \dfrac{\text{s}}{\text{hr}}} = \frac{44{,}900}{3{,}600} \frac{\text{s}}{1} \times \frac{\text{hr}}{\text{s}}$$

$$= 12.5 \frac{\text{s}\cdot\text{hr}}{\text{s}}$$

$$= \boxed{13 \text{ hr}}$$

2.10. Average speed is represented by the symbol $\bar{v}$, and the relationship between the quantities is

(a) $\bar{v} = \dfrac{d}{t} = \dfrac{400.0 \text{ mi}}{8.00 \text{ hr}} = \boxed{50.0 \text{ mi/hr}}$

(b) $\bar{v} = \dfrac{d}{t} = \dfrac{400.0 \text{ mi}}{7.00 \text{ hr}} = \boxed{57.1 \text{ mi/hr}}$

2.11. Again, the relationship between average velocity ($\bar{v}$), distance (d), and time (t) can be solved for time:

$$\bar{v} = \frac{d}{t}$$

$$\bar{v}t = d$$

$$t = \frac{d}{\bar{v}}$$

$$t = \frac{5{,}280 \text{ ft}}{2{,}360 \dfrac{\text{ft}}{\text{s}}}$$

$$= \frac{5{,}280}{2{,}360} \frac{\text{ft}}{1} \times \frac{\text{s}}{\text{ft}}$$

$$= 2.24 \frac{\text{ft}\cdot\text{s}}{\text{ft}}$$

$$= \boxed{2.24 \text{ s}}$$

2.12. The relationship between average velocity ($\bar{v}$), distance (d), and time (t) can be solved for distance:

$$\bar{v} = \frac{d}{t} \therefore d = \bar{v}t$$

$$d = \left(40.0 \frac{\text{m}}{\text{s}}\right)(0.4625 \text{ s})$$

$$= 40.0 \times 0.4625 \frac{\text{m}\cdot\text{s}}{\text{s}}$$

$$= \boxed{18.5 \text{ m}}$$

A distance of 18.5 m is equivalent to about 20 yards (60 feet) in the English system of measurement.

2.13. "How many minutes ...," is a question about time and the distance is given. Since the distance is given in km and the speed in m/s, a unit conversion is needed. The easiest thing to do is to convert km to m. There are 1,000 m in a km, and

$$(1.50 \times 10^8 \text{ km}) \times (1 \times 10^3 \text{ m/km}) = 1.50 \times 10^{11} \text{ m}$$

The relationship between average velocity ($\bar{v}$), distance (d), and time (t) can be solved for time:

$$\bar{v} = \frac{d}{t} \therefore t = \frac{d}{\bar{v}}$$

$$t = \frac{1.50 \times 10^{11} \text{ m}}{3.00 \times 10^8 \dfrac{\text{m}}{\text{s}}}$$

$$= \frac{1.50}{3.00} \times 10^{11-8} \frac{\text{m}}{1} \times \frac{\text{s}}{\text{m}}$$

$$= 0.500 \times 10^3 \frac{\text{m}\cdot\text{s}}{\text{m}}$$

$$= 5.00 \times 10^2 \text{ s}$$

$$\frac{500 \text{ s}}{60 \dfrac{\text{s}}{\text{min}}} = \frac{500}{60} \frac{\text{s}}{1} \times \frac{\text{min}}{\text{s}}$$

$$= 8.33 \frac{\text{s}\cdot\text{min}}{\text{s}}$$

$$= \boxed{8.33 \text{ min}}$$

[Information on how to use scientific notation (also called powers of ten or exponential notation) is located in the Mathematical Review of appendix A.]
All significant figures are retained here because the units are defined exactly, without uncertainty.

2.14. The initial velocity (v_i) is given as 100.0 m/s, the final velocity (v_f) is given as 51.0 m/s, and the time is given as 5.00 s. Acceleration, including a deceleration or negative acceleration, is found from a change of velocity during a given time. Thus,

$$a = \frac{v_f - v_i}{t}$$

$$= \frac{\left(51.0\,\dfrac{m}{s}\right) - \left(100.0\,\dfrac{m}{s}\right)}{5.00\ s}$$

$$= \frac{-49.0\,\dfrac{m}{s}}{5.00\ s}$$

$$= -9.80\,\frac{m}{s} \times \frac{1}{s}$$

$$= \boxed{-9.80\,\frac{m}{s^2}}$$

(The negative sign means a negative acceleration, or deceleration.)

2.15. The initial velocity (v_i) is given as 0 ft/s ("from rest"), the final velocity (v_f) is given as 235 ft/s, and the time is given as 5.0 s. Acceleration is a change of velocity during a given time period, so

$$a = \frac{v_f - v_i}{t}$$

$$= \frac{\left(235\dfrac{ft}{s}\right) - \left(0\dfrac{ft}{s}\right)}{5.0\ s}$$

$$= \frac{235\,\dfrac{ft}{s}}{5.0\ s}$$

$$= 47\,\frac{ft}{s} \times \frac{1}{s}$$

$$= \boxed{47\,\frac{ft}{s^2}}$$

(An acceleration of 47 ft/s² is equivalent to increasing the speed about 32 mph every second.)

2.16. In this problem the initial speed is something other than zero. The change of velocity is the difference between the final velocity and the initial velocity, or

$$a = \frac{v_f - v_i}{t}$$

$$= \frac{\left(220\dfrac{ft}{s}\right) - \left(145\dfrac{ft}{s}\right)}{11\ s}$$

$$= \frac{75\,\dfrac{ft}{s}}{11\ s}$$

$$= 6.8\,\frac{ft}{s} \times \frac{1}{s}$$

$$= \boxed{6.8\,\frac{ft}{s^2}}$$

(A velocity of 145 ft/s is equivalent to about 100 mph, and 220 ft/s is equivalent to about 150 mph. An acceleration of 6.8 ft/s² is a change of speed of about 5 mph every second.)

2.17. The question is asking for a final velocity (v_f) of an accelerating car. The acceleration ($a = 9.0$ ft/s²) and the time ($t = 8.0$ s) are given. The initial velocity (v_i) is zero since the car accelerates from "rest." Therefore, the question involves a relationship between acceleration (a), final velocity (v_f), and time (t), and

$$a = \frac{v_f - v_i}{t}$$

$$at = v_f - v_i$$

$$v_f = at + v_i$$

$$v_f = \left(9.0\,\frac{ft}{s^2}\right)(8.0\ s) + 0\,\frac{ft}{s}$$

$$= 9.0 \times 8.0\,\frac{ft \cdot \cancel{s}}{\cancel{s} \cdot s}$$

$$= \boxed{72\ \text{ft/s}}$$

(A velocity of 72 ft/s is about 50 mph. An acceleration of 9.0 ft/s² means that you are increasing your speed about 6 mph every second.)

2.18. For this type of problem, note that the velocity, distance, and time relationship calls for $\bar{v}$, or *average* velocity. The 88.0 ft/s is the *initial* velocity (v_i) that the car had before stopping ($v_f = 0$). The average velocity must be obtained before solving for the time required:

(a) $\bar{v} = \dfrac{d}{t}$

$$\bar{v}t = d$$

$$t = \frac{d}{\bar{v}}$$

$$\bar{v} = \frac{v_f + v_i}{2} = \frac{0 + 88.0\ \text{ft/s}}{2} = 44.0\ \text{ft/s}$$

$$t = \frac{100.0\ \text{ft}}{44.0\ \text{ft/s}}$$

$$= \frac{100.0}{44.0}\,\text{ft} \times \frac{s}{ft}$$

$$= 2.27\,\frac{\cancel{ft} \cdot s}{\cancel{ft}}$$

$$= \boxed{2.27\ s}$$

(The 2 is not considered in significant figures since it is not the result of a measurement. A velocity of 88.0 ft/s is exactly 60 mph.)

(b) $a = \dfrac{v_f - v_i}{t} = \dfrac{0 - 88.0\ \text{ft/s}}{2.27\ s}$

$$= \frac{-88.0}{2.27}\,\frac{ft}{s} \times \frac{1}{s}$$

$$= -38.8\,\frac{ft}{s \cdot s}$$

$$= \boxed{-38.8\,\frac{ft}{s^2}}$$

(The negative sign simply means that the car is slowing, or decelerating, from some given speed. The deceleration is equivalent to about 26 mph every second.)

One g is 32 ft/s^2, so the deceleration was

$$\frac{38.8 \text{ ft/s}^2}{32 \text{ ft/s}^2} = \frac{38.8}{32} \frac{\text{ft/s}^2}{\text{ft/s}^2} = 1.2 \, g$$

2.19. A ball thrown straight up decelerates to a velocity of zero, then accelerates back to the surface, just as a dropped ball would do from the height reached. Thus, the time required to decelerate upwards is the same as the time required to accelerate downwards. The ball returns to the surface with the same velocity with which it was thrown (neglecting friction). Therefore:

$$a = \frac{v_f - v_i}{t}$$

$$at = v_f - v_i$$

$$v_f = at + v_i$$

$$= \left(9.8 \frac{\text{m}}{\text{s}^2}\right)(3.0 \text{ s})$$

$$= (9.8)(3.0) \frac{\text{m}}{\text{s}^2} \times \text{s}$$

$$= 29 \frac{\text{m} \cdot \text{s}}{\text{s} \cdot \text{s}}$$

$$= \boxed{29 \text{ m/s}}$$

(or 96 ft/s in English units)

(The velocity of the ball—29 m/s or 96 ft/s—is about 65 mph.)

2.20. These three questions are easily answered by using the three sets of relationships, or equations, that were presented in this chapter:

(a) $v_f = at + v$, and when v_i is zero,

$$v_f = at$$

$$v_f = \left(9.8 \frac{\text{m}}{\text{s}^2}\right)(4.00 \text{ s})$$

$$= 9.8 \times 4.00 \frac{\text{m}}{\text{s}^2} \times \text{s}$$

$$= 39 \frac{\text{m} \cdot \text{s}}{\text{s} \cdot \text{s}}$$

$$= \boxed{39 \text{ m/s}}$$

(b) $\bar{v} = \dfrac{v_f + v_i}{2} = \dfrac{39 \text{ m/s} + 0}{2} = 19.5 \text{ m/s}$

(c) $\bar{v} = \dfrac{d}{t} \therefore d = \bar{v}t = \left(19.5 \dfrac{\text{m}}{\text{s}}\right)(4.00 \text{ s})$

$$= 19.5 \times 4.00 \frac{\text{m}}{\text{s}} \times \text{s}$$

$$= 78 \frac{\text{m} \cdot \text{s}}{\text{s}}$$

$$= \boxed{78 \text{ m}}$$

(A velocity of 39 m/s is equivalent to about 87 mph.)

2.21. Note that this problem can be solved with a series of three steps as in the previous problem. It can also be solved by the equation that combines all the relationships into one step. Either method is acceptable, but the following example of a one-step solution reduces the possibilities of error since fewer calculations are involved:

$$d = \frac{1}{2}gt^2 = \frac{1}{2}\left(9.8 \frac{\text{m}}{\text{s}^2}\right)(5.00 \text{ s})^2$$

$$= \frac{1}{2}\left(9.8 \frac{\text{m}}{\text{s}^2}\right)(25.0 \text{ s}^2)$$

$$= \left(\frac{1}{2}\right)(9.8)(25.0) \frac{\text{m}}{\text{s}^2} \times \text{s}^2$$

$$= 4.90 \times 25.0 \frac{\text{m} \cdot \text{s}^2}{\text{s}^2}$$

$$= 122.5 \text{ m}$$

$$= \boxed{120 \text{ m}}$$

2.22. Note that "how long must a car accelerate?" is asking for a unit of time and that the acceleration (a), initial velocity (v_i), and the final velocity (v_f) are given. The *first step*, before anything else is done, is to write the equation expressing a relationship between these quantities and then solve the equation for the unknown (time in this problem). Since unit conversions are not necessary, quantities are then substituted for the symbols and mathematical operations are performed on both the numbers and units:

$$a = \frac{v_f - v_i}{t} \therefore t = \frac{v_f - v_i}{a} = \frac{50.0 \dfrac{\text{m}}{\text{s}} - 10.0 \dfrac{\text{m}}{\text{s}}}{4.00 \dfrac{\text{m}}{\text{s}^2}}$$

$$= \frac{40.0}{4.00} \frac{\text{m}}{\text{s}} \times \frac{\text{s}^2}{\text{m}}$$

$$= 10.0 \frac{\text{m} \cdot \text{s} \cdot \text{s}}{\text{m} \cdot \text{s}}$$

$$= \boxed{10.0 \text{ s}}$$

2.23. $a = g \therefore g = \dfrac{v_f - v_i}{t} \therefore v_f = gt$ (when v_i is zero)

$$= \left(9.8 \frac{\text{m}}{\text{s}^2}\right)(3.0 \text{ s})$$

$$= (9.8)(3.0) \frac{\text{m}}{\text{s}^2} \times \text{s}$$

$$= 29.4 \frac{\text{m} \cdot \text{s}}{\text{s} \cdot \text{s}}$$

$$= 29.4 \frac{\text{m}}{\text{s}} = \boxed{29 \frac{\text{m}}{\text{s}}}$$

Exercises CHAPTER 3

3.1. Listing the known and unknown quantities:

$$m = 40.0 \text{ kg}$$
$$a = 2.4 \text{ m/s}^2$$
$$F = ?$$

These are the quantities found in Newton's second law of motion, $F = ma$, which is already solved for force (F). Thus,

$$F = (40.0 \text{ kg})\left(2.4 \frac{m}{s^2}\right)$$
$$= 40.0 \times 2.4 \frac{kg \cdot m}{s^2}$$
$$= \boxed{96 \text{ N}}$$

3.2. Listing the known and unknown quantities:

$$F = 100 \text{ N}$$
$$m = 5 \text{ kg}$$
$$a = ?$$

These are the quantities of Newton's second law of motion, $F = ma$, and

$$F = ma \therefore a = \frac{F}{m}$$
$$= \frac{100 \frac{kg \cdot m}{s^2}}{5 \text{ kg}}$$
$$= \frac{100}{5} \frac{kg \cdot m}{s^2} \times \frac{1}{kg}$$
$$= \boxed{20 \frac{m}{s^2}}$$

3.3. Listing the known and unknown quantities:

$$v_i = 72 \text{ km/hr}$$
$$v_f = 108 \text{ km/hr}$$
$$t = 5 \text{ s}$$
$$m = 1,000 \text{ kg}$$
$$F = ?$$

First, unit conversions are needed (see appendix A for help with this). We are looking for a force, which means that the answer unit will be in newtons (N). A newton of force is defined as the force needed to give a one kilogram mass an acceleration of one meter per second, so we know that the units used in the problem should be kilograms, meters, and seconds. Speeds of km/hr must first be converted to m/s to have the correct units in the problem, and

$$\left(\frac{0.2778 \frac{m}{s}}{\frac{km}{hr}}\right)\left(72 \frac{km}{hr}\right)$$
$$= 0.2778 \times 72 \frac{m}{s} \times \frac{hr}{km} \times \frac{km}{hr}$$
$$= 20 \frac{m}{s}$$

$$\left(\frac{0.2778 \frac{m}{s}}{\frac{km}{hr}}\right)\left(108 \frac{km}{hr}\right)$$
$$= 0.2778 \times 108 \frac{m}{s} \times \frac{hr}{km} \times \frac{km}{hr}$$
$$= 30 \frac{m}{s}$$

We need a mass and an acceleration to find a force using Newton's second law of motion, $F = ma$. The mass of the car is given, but not the acceleration. The acceleration can be calculated, however, from the initial velocity, the final velocity, and the time, $a = \frac{v_f - v_i}{t}$, which are given, and this can be substituted for the a in $F = ma$.

$$F = ma \text{ and } a = \frac{v_f - v_i}{t}, \text{ so}$$
$$F = m\left(\frac{v_f - v_i}{t}\right)$$
$$= (1,000 \text{ kg})\left(\frac{30 \text{ m/s} - 20 \text{ m/s}}{5 \text{ s}}\right)$$
$$= (1,000 \text{ kg})\left(\frac{10 \text{ m/s}}{5 \text{ s}}\right)$$
$$= (1,000)\left(\frac{10}{5}\right)\frac{kg}{1} \times \frac{m}{s} \times \frac{1}{s}$$
$$= 1,000 \times 2 \frac{kg \cdot m}{s^2}$$
$$= \boxed{2,000 \text{ N}}$$

3.4. List the known and unknown quantities for the first situation, using an unbalanced force of 18.0 N to give the object an acceleration of 3 m/s²:

$$F_1 = 18 \text{ N}$$
$$a_1 = 3 \text{ m/s}^2$$

For the second situation, we are asked to find the force needed for an acceleration of 10 m/s²:

$$a_2 = 10 \text{ m/s}^2$$
$$F = ?$$

These are the quantities of Newton's second law of motion,
$F = ma$, but the mass appears to be missing. The mass can be
found from

$$F = ma \therefore m_1 = \frac{F_1}{a_1}$$

$$= \frac{18 \frac{kg \cdot m}{s^2}}{3 \frac{m}{s^2}}$$

$$= \frac{18}{3} \frac{kg \cdot m}{s^2} \times \frac{s^2}{m}$$

$$= 6 \text{ kg}$$

Now that we have the mass, we can easily find the force
needed for an acceleration of 10 m/s²:

$$F_2 = m_2 a_2$$

$$= (6 \text{ kg})\left(10 \frac{m}{s^2}\right)$$

$$= 6 \times 10 \frac{kg \cdot m}{s^2}$$

$$= \boxed{60 \text{ N}}$$

3.5.   Listing the known and unknown quantities:

$$F = 100 \text{ N}$$

$$a = 0.5 \text{ m/s}^2$$

$$m = ?$$

These are the quantities found in Newton's second law of motion, $F = ma$, which must be first solved for m. Thus,

$$F = ma \therefore m = \frac{F}{a}$$

$$= \frac{100 \frac{kg \cdot m}{s^2}}{0.5 \frac{m}{s^2}}$$

$$= \frac{100}{0.5} \frac{kg \cdot m}{s^2} \times \frac{s^2}{m}$$

$$= \boxed{200 \text{ kg}}$$

3.6.   Listing the known and unknown quantities:

$$m = 1,500 \text{ kg}$$

$$a = 2 \text{ m/s}^2$$

$$F = ?$$

These are the quantities found in Newton's second law of motion, $F = ma$, which is already solved for force (F). Thus,

$$F = ma$$

$$= (1,500 \text{ kg})\left(2 \frac{m}{s^2}\right)$$

$$= 1,500 \times 2 \frac{kg \cdot m}{s^2}$$

$$= \boxed{3,000 \text{ N}}$$

3.7.   Listing the known and unknown quantities:

mass of the earth $m_e = 5.98 \times 10^{24} \text{ kg}$

mass of the moon $m_m = 7.36 \times 10^{22} \text{ kg}$

distance between m_e and m_m $d = 3.84 \times 10^8 \text{ m}$

$$G = 6.67 \times 10^{-11} \text{ N} \cdot \text{m}^2/\text{kg}^2$$

$$F = ?$$

These are the quantities found in the equation for Newton's
universal law of gravitation, $F = \frac{G m_e m_m}{d^2}$, where m_e and m_m
are the mass of the earth and the moon, d is the distance
between their centers, and G is a constant of proportionality
with a value of 6.67×10^{-11} N·m²/kg². The equation is already solved for the gravitational force, F. Thus,

$$F = \frac{G m_e m_m}{d^2}$$

$$= \frac{\left(6.67 \times 10^{-11} \frac{Nm^2}{kg^2}\right)(5.98 \times 10^{24} \text{ kg})(7.36 \times 10^{22} \text{ kg})}{(3.84 \times 10^8 \text{ m})^2}$$

$$= \frac{(6.67 \times 10^{-11})(5.98 \times 10^{24})(7.36 \times 10^{22})}{1.47 \times 10^{17}} \frac{Nm^2}{kg^2} \times \frac{kg^2}{1} \times \frac{1}{m^2}$$

$$= \frac{2.94 \times 10^{37}}{1.47 \times 10^{17}} \text{ N}$$

$$= \boxed{1.99 \times 10^{20} \text{ N}}$$

3.8.   As usual, there is more than one way to solve this problem,
including the use of a ratio of the distances involved
(squared) to g on the surface. The equation $g = G \frac{m_e}{d^2}$ was
derived from Newton's second law of motion ($F = ma$) and
Newton's universal law of gravitation, so using this equation
might seem more conceptually satisfying. The symbol m_e is
the mass of the earth, d is the distance from the center of the
earth to the 500 km altitude (6,400 km + 500 km = 6,900
km), and G is a constant with a value of 6.67×10^{-11}
N·m²/kg². Listing the known and unknown quantities:

$$m_e = 5.98 \times 10^{24} \text{ kg}$$

$$d = 6900 \text{ km}$$

$$G = 6.67 \times 10^{-11} \text{ N} \cdot \text{m}^2/\text{kg}^2$$

$$g = ?$$

Thus,

$$g = G \frac{m_e}{d^2}$$

$$= \frac{\left(6.67 \times 10^{-11} \frac{Nm^2}{kg^2}\right)(5.98 \times 10^{24} \text{ kg})}{(6.9 \times 10^6 \text{ m})^2}$$

$$= \frac{(6.67 \times 10^{-11})(5.98 \times 10^{24})}{4.8 \times 10^{13}} \frac{kg \cdot m}{s^2} \times \frac{m^2}{kg^2} \times \frac{1}{m^2} \times \frac{kg}{1}$$

$$= \frac{3.99 \times 10^{14}}{4.8 \times 10^{13}} \frac{m}{s^2}$$

$$= \boxed{8.3 \frac{m}{s^2}}$$

3.9. A student who weighs 591 N on the surface of the earth will have a mass of

$$w = mg \therefore m = \frac{w}{g}$$

$$= \frac{591 \frac{kg \cdot m}{s^2}}{9.8 \frac{m}{s^2}}$$

$$= \frac{591}{9.8} \frac{kg \cdot m}{s^2} \times \frac{s^2}{m}$$

$$= 60.3 \text{ kg}$$

At an altitude of 1,100 km, which is 7,500 km from the center of the earth, the value of g would be

$$g = G \frac{m_e}{d^2}$$

$$= \frac{\left(6.67 \times 10^{-11} \frac{Nm^2}{kg^2}\right)(5.98 \times 10^{24} \text{ kg})}{(7.5 \times 10^6 \text{ m})^2}$$

$$= \frac{(6.67 \times 10^{-11})(5.98 \times 10^{24})}{5.6 \times 10^{13}} \frac{kg \cdot m}{s^2} \frac{m^2}{kg^2} \times \frac{1}{m^2} \times \frac{kg}{1}$$

$$= \frac{3.99 \times 10^{14}}{5.6 \times 10^{13}} \frac{m}{s^2}$$

$$= 7.1 \frac{m}{s^2}$$

So a 60.3 kg person at this altitude would weigh

$$w = mg$$

$$= (60.3 \text{ kg})\left(7.1 \frac{m}{s^2}\right)$$

$$= 60.3 \times 7.1 \frac{kg \cdot m}{s^2}$$

$$= 438 \text{ N}$$

$$= \boxed{440 \text{ N}}$$

3.10. Listing the known and unknown quantities:

$$m = 100 \text{ kg}$$

$$v = 6 \text{ m/s}$$

$$p = ?$$

These are the quantities found in the equation for momentum, $p = mv$, which is already solved for momentum (p). Thus,

$$p = mv$$

$$= (100 \text{ kg})\left(6 \frac{m}{s}\right)$$

$$= \boxed{600 \frac{kg \cdot m}{s}}$$

(Note the lowercase p is the symbol used for momentum. This is one of the few cases where the English letter does not provide a clue about what it stands for. The units for momentum are also somewhat unusual for metric units since they do not have a name or single symbol to represent them.)

3.11. Listing the known and unknown quantities:

$$w = 13,720 \text{ N}$$

$$v = 91 \text{ km/hr}$$

$$p = ?$$

The equation for momentum is $p = mv$, which is already solved for momentum (p). The weight unit must be first converted to a mass unit:

$$w = mg \therefore m = \frac{w}{g}$$

$$= \frac{13,720 \frac{kg \cdot m}{s^2}}{9.8 \frac{m}{s^2}}$$

$$= \frac{13,720}{9.8} \frac{kg \cdot m}{s^2} \times \frac{s^2}{m}$$

$$= 1,400 \text{ kg}$$

The km/hr unit should next be converted to m/s. Using the conversion factor from inside the front cover:

$$\frac{0.2778 \frac{m}{s}}{1 \frac{km}{hr}} \times 91 \frac{km}{hr}$$

$$0.2778 \times 91 \frac{m}{s} \times \frac{hr}{km} \times \frac{km}{hr}$$

$$25 \frac{m}{s}$$

Now, listing the converted known and unknown quantities:

$$m = 1,400 \text{ kg}$$

$$v = 25 \text{ m/s}$$

$$p = ?$$

and solving for momentum (p),

$$p = mv$$

$$= (1{,}400 \text{ kg})\left(25 \frac{\text{m}}{\text{s}}\right)$$

$$= \boxed{35{,}000 \frac{\text{kg·m}}{\text{s}}}$$

3.12.   Listing the known and unknown quantities:

Car A $\rightarrow m = 1{,}200$ kg   Car B $\rightarrow m = 2{,}200$ kg

Car A $\rightarrow v = 25$ m/s   Car B $\rightarrow v = 15$ m/s

What is the result of a head-on collision?

This question is easily answered by finding which car had the greater momentum before the collision:

Car A: $p = mv = (1{,}200 \text{ kg})(25 \text{ m/s}) = 30{,}000$ kg·m/s

Car B: $p = mv = (2{,}200 \text{ kg})(15 \text{ m/s}) = 33{,}000$ kg·m/s

Car B had the greatest momentum, so the wreckage should move south (Car A was driving north and collided head-on with car B, so car B must have been heading south).

3.13.   Listing the known and unknown quantities:

Bullet $\rightarrow m = 0.015$ kg   Rifle $\rightarrow m = 6$ kg

Bullet $\rightarrow v = 200$ m/s   Rifle $\rightarrow v = ?$ m/s

Note the mass of the bullet was converted to kg. This is a conservation of momentum question, where the bullet and rifle can be considered as a system of interacting objects:

Bullet momentum $= -$rifle momentum

$$(mv)_b = -(mv)_r$$

$$(mv)_b - (mv)_r = 0$$

$$(0.015 \text{ kg})\left(200 \frac{\text{m}}{\text{s}}\right) - (6 \text{ kg})v_r = 0$$

$$\left(3 \text{ kg·} \frac{\text{m}}{\text{s}}\right) - (6 \text{ kg·}v_r) = 0$$

$$\left(3 \text{ kg·} \frac{\text{m}}{\text{s}}\right) = (6 \text{ kg·}v_r)$$

$$v_r = \frac{3 \text{ kg·} \dfrac{\text{m}}{\text{s}}}{6 \text{ kg}}$$

$$= \frac{3}{6} \frac{\text{kg}}{1} \times \frac{1}{\text{kg}} \times \frac{\text{m}}{\text{s}}$$

$$= \boxed{0.5 \frac{\text{m}}{\text{s}}}$$

The rifle recoils with a velocity of 0.5 m/s.

3.14.   Listing the known and unknown quantities:

Astronaut $\rightarrow w = 2{,}156$ N   Wrench $\rightarrow m = 5.0$ kg

Astronaut $\rightarrow v = ?$ m/s   Wrench $\rightarrow v = 5.0$ m/s

Note the astronaut's weight is given, but we need mass for the conservation of momentum equation. Mass can be found be-

cause the weight on the earth was given, where we know $g = 9.8$ m/s². Thus the mass is

$$w = mg \therefore m = \frac{w}{g}$$

$$= \frac{2{,}156 \dfrac{\text{kg·m}}{\text{s}^2}}{9.8 \dfrac{\text{m}}{\text{s}^2}}$$

$$= \frac{2{,}156}{9.8} \frac{\text{kg·m}}{\text{s}^2} \times \frac{\text{s}^2}{\text{m}}$$

$$= 220 \text{ kg}$$

So the converted known and unknown quantities are:

Astronaut $\rightarrow m = 220$ kg   Wrench $\rightarrow m = 5.0$ kg

Astronaut $\rightarrow v = ?$ m/s   Wrench $\rightarrow v = 5.0$ m/s

This is a conservation of momentum question, where the astronaut and wrench can be considered as a system of interacting objects:

Wrench momentum $= -$astronaut momentum

$$(mv)_w = -(mv)_a$$

$$(mv)_w - (mv)_a = 0$$

$$(5.0 \text{ kg})\left(5.0 \frac{\text{m}}{\text{s}}\right) - (220 \text{ kg})v_a = 0$$

$$\left(25 \text{ kg·} \frac{\text{m}}{\text{s}}\right) - (220 \text{ kg·}v_a) = 0$$

$$v_a = \frac{25 \text{ kg·} \dfrac{\text{m}}{\text{s}}}{220 \text{ kg}}$$

$$= \frac{25}{220} \frac{\text{kg}}{1} \times \frac{1}{\text{kg}} \times \frac{\text{m}}{\text{s}}$$

$$= \boxed{0.11 \frac{\text{m}}{\text{s}}}$$

The astronaut moves away with a velocity of 0.11 m/s.

3.15.   Listing the known and unknown quantities:

Student and Boat $\rightarrow m = 100.0$ kg   Rock $\rightarrow m = 5.0$ kg

Student and Boat $\rightarrow v = ?$ m/s   Rock $\rightarrow v = 5.0$ m/s

This is a conservation of momentum question, where the student-boat and rock can be considered as a system of interacting objects:

Rock momentum $= -$student and boat momentum

$$(mv)_r = -(mv)_{S\&B}$$

$$(mv)_r - (mv)_{S\&B} = 0$$

$$(5.0 \text{ kg})\left(5.0 \frac{\text{m}}{\text{s}}\right) - (100.0 \text{ kg})v_{S\&B} = 0$$

$$\left(25 \text{ kg·} \frac{\text{m}}{\text{s}}\right) - (100.0 \text{ kg·}v_{S\&B}) = 0$$

$$\left(25 \text{ kg} \cdot \frac{\text{m}}{\text{s}}\right) = (100.0 \text{ kg} \cdot v_{\text{S\&B}})$$

$$v_{\text{S\&B}} = \frac{25 \text{ kg} \cdot \dfrac{\text{m}}{\text{s}}}{100.0 \text{ kg}}$$

$$= \frac{25}{100.0} \frac{\text{kg}}{1} \times \frac{1}{\text{kg}} \times \frac{\text{m}}{\text{s}}$$

$$= \boxed{0.25 \frac{\text{m}}{\text{s}}}$$

The student and boat move ahead with a velocity of 0.25 m/s.

3.16. **(a)** Weight (w) is a downward force from the acceleration of gravity (g) on the mass (m) of an object. This relationship is the same as Newton's second law of motion, $F = ma$, and

$$\text{a) } w = mg = (1.25 \text{ kg})\left(9.8 \frac{\text{m}}{\text{s}^2}\right)$$

$$= (1.25)(9.8)\text{kg} \times \frac{\text{m}}{\text{s}^2}$$

$$= 12.25 \frac{\text{kg} \cdot \text{m}}{\text{s}^2}$$

$$= \boxed{12 \text{ N}}$$

(b) First, recall that a force (F) is measured in newtons (N) and a newton has units of $\text{N} = \dfrac{\text{kg} \cdot \text{m}}{\text{s}^2}$. Second, the relationship between force (F), mass (m), and acceleration (a) is given by Newton's second law of motion, force = mass times acceleration, or $F = ma$. Thus,

$$\text{b) } F = ma \therefore a = \frac{F}{m} = \frac{10.0 \dfrac{\text{kg} \cdot \text{m}}{\text{s}^2}}{1.25 \text{ kg}}$$

$$= \frac{10.0}{1.25} \frac{\text{kg} \cdot \text{m}}{\text{s}^2} \times \frac{1}{\text{kg}}$$

$$= 8.00 \frac{\text{kg} \cdot \text{m}}{\text{kg} \cdot \text{s}^2}$$

$$= \boxed{8.00 \frac{\text{m}}{\text{s}^2}}$$

(Note how the units were treated mathematically in this solution and why it is necessary to show the units for a newton of force. The resulting unit in the answer *is* a unit of acceleration, which provides a check that the problem was solved correctly. For your information, 8.00 m/s² is equivalent to a change of velocity of about 29 km/hr each s, or 18 mph each s.)

3.17.
$$F = ma = (1.25 \text{ kg})\left(5.00 \frac{\text{m}}{\text{s}^2}\right)$$

$$= (1.25)(5.00)\text{kg} \times \frac{\text{m}}{\text{s}^2}$$

$$= 6.25 \frac{\text{kg} \cdot \text{m}}{\text{s}^2}$$

$$= \boxed{6.25 \text{ N}}$$

(Note that the solution is correctly reported in *newton* units of force rather than kg·m/s².)

3.18. The bicycle tire exerts a backwards force on the road, and the equal and opposite reaction force of the road on the bicycle produces the forward motion. (The motion is always in the direction of the applied force.) Therefore,

$$F = ma = (70.0 \text{ kg})\left(2.0 \frac{\text{m}}{\text{s}^2}\right)$$

$$= (70.0)(2.0)\text{kg} \times \frac{\text{m}}{\text{s}^2}$$

$$= 140 \frac{\text{kg} \cdot \text{m}}{\text{s}^2}$$

$$= \boxed{140 \text{ N}}$$

3.19. The question requires finding a force in the metric system, which is measured in newtons of force. Since newtons of force are defined in kg, m, and s, unit conversions are necessary, and these should be done first.

$$1 \frac{\text{km}}{\text{hr}} = \frac{1,000 \text{ m}}{3,600 \text{ s}} = 0.2778 \frac{\text{m}}{\text{s}}$$

Dividing both sides of this conversion factor by what you are converting *from* gives the conversion ratio of

$$\frac{0.2778 \dfrac{\text{m}}{\text{s}}}{\dfrac{\text{km}}{\text{hr}}}$$

Multiplying this conversion ratio times the two velocities in km/hr will convert them to m/s as follows:

$$\left(\frac{0.2778 \dfrac{\text{m}}{\text{s}}}{\dfrac{\text{km}}{\text{hr}}}\right)\left(80.0 \frac{\text{km}}{\text{hr}}\right)$$

$$= [0.2778][80.0] \frac{\text{m}}{\text{s}} \times \frac{\text{hr}}{\text{km}} \times \frac{\text{km}}{\text{hr}}$$

$$= 22.2 \frac{\text{m}}{\text{s}}$$

$$\left(\frac{0.2778 \dfrac{\text{m}}{\text{s}}}{\dfrac{\text{km}}{\text{hr}}}\right)\left(44.0 \frac{\text{km}}{\text{hr}}\right)$$

$$= [0.2778][44.0] \frac{\text{m}}{\text{s}} \times \frac{\text{hr}}{\text{km}} \times \frac{\text{km}}{\text{hr}}$$

$$= 12.2 \frac{\text{m}}{\text{s}}$$

Now you are ready to find the appropriate relationship between the quantities involved. This involves two separate equations: Newton's second law of motion and the relationship of quantities involved in acceleration. These may be combined as follows:

$$F = ma \text{ and } a = \frac{v_f - v_i}{t} \therefore F = m\left(\frac{v_f - v_i}{t}\right)$$

Now you are ready to substitute quantities for the symbols and perform the necessary mathematical operations:

$$= (1,500 \text{ kg})\left(\frac{22.2 \text{ m/s} - 12.2 \text{ m/s}}{10.0 \text{ s}}\right)$$

$$= (1,500 \text{ kg})\left(\frac{10.0 \text{ m/s}}{10.0 \text{ s}}\right)$$

$$= 1,500 \times 1.00 \frac{\text{kg} \cdot \frac{\text{m}}{\text{s}}}{\text{s}}$$

$$= 1,500 \frac{\text{kg} \cdot \text{m}}{\text{s}} \times \frac{1}{\text{s}}$$

$$= 1,500 \frac{\text{kg} \cdot \text{m}}{\text{s} \cdot \text{s}}$$

$$= 1,500 \frac{\text{kg} \cdot \text{m}}{\text{s}^2}$$

$$= 1,500 \text{ N} = \boxed{1.5 \times 10^3 \text{ N}}$$

3.20. A unit conversion is needed as in the previous problem:

$$\left(90.0 \frac{\text{km}}{\text{hr}}\right)\left(0.2778 \frac{\frac{\text{m}}{\text{s}}}{\frac{\text{km}}{\text{hr}}}\right) = 25.0 \text{ m/s}$$

(a) $F = ma \therefore m = \frac{F}{a}$ and $a = \frac{v_f - v_i}{t}$, so

$$m = \frac{F}{\frac{v_f - v_i}{t}} = \frac{5,000.0 \frac{\text{kg} \cdot \text{m}}{\text{s}^2}}{\frac{25.0 \text{ m/s} - 0}{5.0 \text{ s}}}$$

$$= \frac{5,000.0 \frac{\text{kg} \cdot \text{m}}{\text{s}^2}}{5.0 \frac{\text{m}}{\text{s}^2}}$$

$$= \frac{5,000.0}{5.0} \frac{\text{kg} \cdot \text{m}}{\text{s}^2} \times \frac{\text{s}^2}{\text{m}}$$

$$= 1,000 \frac{\text{kg} \cdot \text{m} \cdot \text{s}^2}{\text{m} \cdot \text{s}^2}$$

$$= \boxed{1.0 \times 10^3 \text{ kg}}$$

(b)

$$w = mg$$

$$= (1.0 \times 10^3 \text{ kg})\left(9.8 \frac{\text{m}}{\text{s}^2}\right)$$

$$= (1.0 \times 10^3)(9.8) \text{ kg} \times \frac{\text{m}}{\text{s}^2}$$

$$= 9.8 \times 10^3 \frac{\text{kg} \cdot \text{m}}{\text{s}^2}$$

$$= \boxed{9.8 \times 10^3 \text{ N}}$$

3.21.

$$w = mg$$

$$= (70.0 \text{ kg})\left(9.8 \frac{\text{m}}{\text{s}^2}\right)$$

$$= 70.0 \times 9.8 \text{ kg} \times \frac{\text{m}}{\text{s}^2}$$

$$= 686 \frac{\text{kg} \cdot \text{m}}{\text{s}^2}$$

$$= \boxed{690 \text{ N}}$$

3.22. You were given a mass (m), a force (F), and a time (t) and asked to find a final speed (v_f). These relationships are found in Newton's second law of motion and in the definition of acceleration. The two equations can be combined and solved for v_f:

$$F = ma \text{ and } a = \frac{v_f - v_i}{t}$$

$$\therefore$$

$$F = m\left(\frac{v_f - v_i}{t}\right)$$

$$Ft = m(v_f - v_i)$$

$$\frac{Ft}{m} = v_f - v_i$$

$$v_f = \frac{Ft}{m} + v_i$$

Now the solution becomes a matter of substituting values and carrying out the mathematical operations:

$$v_f = \frac{\left(1,000.0 \frac{\text{kg} \cdot \text{m}}{\text{s}^2}\right)(10.0 \text{ s})}{1,000.0 \text{ kg}} + 0 \frac{\text{m}}{\text{s}}$$

$$= \frac{(1,000.0)(10.0)}{1,000.0} \frac{\text{kg} \cdot \text{m}}{\text{s}^2} \times \frac{\text{s}}{1} \times \frac{1}{\text{kg}}$$

$$= 10.0 \frac{\text{kg} \cdot \text{m} \cdot \text{s}}{\text{kg} \cdot \text{s} \cdot \text{s}}$$

$$= \boxed{10.0 \frac{\text{m}}{\text{s}}}$$

3.23.

$$p = mv$$

$$= (50 \text{ kg})\left(2\frac{\text{m}}{\text{s}}\right)$$

$$= \boxed{100 \frac{\text{kg} \cdot \text{m}}{\text{s}}}$$

3.24.

$$F = \frac{mv^2}{r}$$

$$= \frac{(0.20 \text{ kg})\left(3.0\frac{\text{m}}{\text{s}}\right)^2}{1.5 \text{ m}}$$

$$= \frac{(0.20 \text{ kg})\left(9.0\frac{\text{m}^2}{\text{s}^2}\right)}{1.5 \text{ m}}$$

$$= \frac{0.20 \times 9.0}{1.5} \frac{\text{kg} \cdot \text{m}^2}{\text{s}^2} \times \frac{1}{\text{m}}$$

$$= 1.2 \frac{\text{kg} \cdot \text{m} \cdot \text{m}}{\text{s}^2 \cdot \text{m}}$$

$$= \boxed{1.2 \text{ N}}$$

3.25. Note that unit conversions are necessary since the definition of the newton unit of force involves different units:

$$F = \frac{mv^2}{r}$$

$$\frac{Fr}{m} = v^2$$

$$v = \sqrt{\frac{Fr}{m}}$$

$$v = \sqrt{\frac{\left(1.0\frac{\text{kg} \cdot \text{m}}{\text{s}^2}\right)(0.50 \text{ m})}{0.10 \text{ kg}}}$$

$$= \sqrt{\frac{(1.0)(0.50)}{0.10} \frac{\text{kg} \cdot \text{m}}{\text{s}^2} \times \frac{\text{m}}{1} \times \frac{1}{\text{kg}}}$$

$$= \sqrt{5.0 \text{ m}^2/\text{s}^2}$$

$$= \boxed{2.2 \text{ m/s}}$$

3.26. **(a)**

$$F = \frac{mv^2}{r} = \frac{(1,000.0 \text{ kg})(10.0 \text{ m/s})^2}{20.0 \text{ m}}$$

$$= \frac{(1,000.0 \text{ kg})(100.0 \text{ m}^2/\text{s}^2)}{20.0 \text{ m}}$$

$$= \frac{(1,000.0)(100.0)}{20.0} \frac{\text{kg} \cdot \text{m}^2}{\text{s}^2} \times \frac{1}{\text{m}}$$

$$= 5,000 \frac{\text{kg} \cdot \text{m} \cdot \text{m}}{\text{s}^2 \cdot \text{m}}$$

$$= \boxed{5.00 \times 10^3 \text{ N}}$$

(b) The centripetal force that keeps a car on the road while moving around a curve is the frictional force between the tires and the road.

3.27. **(a)** Newton's laws of motion consider the resistance to a change of motion, or mass, and not weight. The astronaut's mass is

$$w = mg \therefore m = \frac{w}{g} = \frac{1,960.0 \frac{\text{kg} \cdot \text{m}}{\text{s}^2}}{9.8 \frac{\text{m}}{\text{s}^2}}$$

$$= \frac{1,960.0}{9.8} \frac{\text{kg} \cdot \text{m}}{\text{s}^2} \times \frac{\text{s}^2}{\text{m}} = 200 \text{ kg}$$

(b) From Newton's second law of motion, you can see that the 100 N rocket gives the 200 kg astronaut an acceleration of:

$$F = ma \therefore a = \frac{F}{m} = \frac{100 \frac{\text{kg} \cdot \text{m}}{\text{s}^2}}{200 \text{ kg}}$$

$$= \frac{100 \text{ kg} \cdot \text{m}}{200 \text{ s}^2} \times \frac{1}{\text{kg}} = 0.5 \text{ m/s}^2$$

(c) An acceleration of 0.5 m/s² for 2.0 s will result in a final velocity of

$$a = \frac{v_f - v_i}{t} \therefore v_f = at + v_i$$

$$= (0.5 \text{ m/s}^2)(2.0 \text{ s}) + 0 \text{ m/s}$$

$$= \boxed{1 \text{ m/s}}$$

Exercises CHAPTER 4

4.1. Listing the known and unknown quantities:

$$F = 200 \text{ N}$$

$$d = 3 \text{ m}$$

$$W = ?$$

These are the quantities found in the equation for work, $W = Fd$, which is already solved for work (W). Thus,

$$W = Fd$$

$$= \left(200 \frac{\text{kg} \cdot \text{m}}{\text{s}^2}\right)(3 \text{ m})$$

$$= (200)(3) \text{ N} \cdot \text{m}$$

$$= \boxed{600 \text{ J}}$$

4.2.   Listing the known and unknown quantities:

$$F = 440 \text{ N}$$

$$d = 5.0 \text{ m}$$

$$w = 880 \text{ N}$$

$$W = ?$$

These are the quantities found in the equation for work, $W = Fd$, which is already solved for work (W). As you can see in the equation, the force exerted and the distance the box was moved are the quantities used in determining the work accomplished. The weight of the box is a different variable, and one that is not used in this equation. Thus,

$$W = Fd$$

$$= \left(440 \frac{\text{kg·m}}{\text{s}^2}\right)(5.0 \text{ m})$$

$$= 2,200 \text{ N·m}$$

$$= \boxed{2,200 \text{ J}}$$

4.3.   Note that 10.0 kg is a mass quantity, and not a weight quantity. Weight is found from $w = mg$, a form of Newton's second law of motion. Thus the force that must be exerted to lift the backpack is its weight, or $(10.0 \text{ kg}) \times (9.8 \text{ m/s}^2)$, which is 98 N. Therefore, a force of 98 N was exerted on the backpack through a distance of 1.5 m, and

$$W = Fd$$

$$= \left(98 \frac{\text{kg·m}}{\text{s}^2}\right)(1.5 \text{ m})$$

$$= 147 \text{ N·m}$$

$$= \boxed{150 \text{ J}}$$

4.4.   Weight is defined as the force of gravity acting on an object, and the greater the force of gravity the harder it is to lift the object. The force is proportional to the mass of the object, as the equation $w = mg$ tells you. Thus the force you exert when lifting is $F = w = mg$, so the work you do on an object you lift must be $W = mgh$.

You know the mass of the box and you know the work accomplished. You also know the value of the acceleration due to gravity, g, so the list of known and unknown quantities is:

$$m = 102 \text{ kg}$$

$$g = 9.8 \text{ m/s}^2$$

$$W = 5,000 \text{ J}$$

$$h = ?$$

The equation $W = mgh$ is solved for work, so the first thing to do is to solve it for h, the unknown height in this problem (note that height is also a distance):

$$W = mgh \therefore h = \frac{W}{mg}$$

$$= \frac{5,000 \frac{\text{kg·m}}{\text{s}^2} \times \text{m}}{(102 \text{ kg})\left(9.8 \frac{\text{m}}{\text{s}^2}\right)}$$

$$= \frac{5,000.0}{102 \times 9.8} \frac{\text{kg·m}}{\text{s}^2} \times \frac{\text{m}}{1} \times \frac{1}{\text{kg}} \times \frac{\text{s}^2}{\text{m}}$$

$$= \frac{5,000}{999.6} \text{ m}$$

$$= \boxed{5 \text{ m}}$$

4.5.   A student running up the stairs has to lift herself, so her weight is the required force needed. Thus the force exerted is $F = w = mg$, and the work done is $W = mgh$. You know the mass of the student, the height, and the time. You also know the value of the acceleration due to gravity, g, so the list of known and unknown quantities is:

$$m = 60.0 \text{ kg}$$

$$g = 9.8 \text{ m/s}^2$$

$$h = 5.00 \text{ m}$$

$$t = 3.94 \text{ s}$$

$$P = ?$$

The equation $P = \dfrac{mgh}{t}$ is already solved for power, so:

(a) $P = \dfrac{mgh}{t}$

$$= \frac{(60.0 \text{ kg})\left(9.8 \frac{\text{m}}{\text{s}^2}\right)(5.00 \text{ m})}{3.92 \text{ s}}$$

$$= \frac{(60.0)(9.8)(5.00)}{(3.94)} \frac{\left(\frac{\text{kg·m}}{\text{s}^2}\right) \times \text{m}}{\text{s}}$$

$$= \frac{2940}{3.92} \frac{\text{N·m}}{\text{s}}$$

$$= 750 \frac{\text{J}}{\text{s}}$$

$$= \boxed{750 \text{ W}}$$

(b) A power of 750 watts is almost one horsepower.

4.6.

(a)

$$\frac{1.00 \text{ hp}}{746 \text{ W}} \times 1,400 \text{ W}$$

$$\frac{1,400}{746} \frac{\text{hp·W}}{\text{W}}$$

$$1.9 \text{ hp}$$

(b)

$$\frac{746 \text{ W}}{1 \text{ hp}} \times 3.5 \text{ hp}$$

$$746 \times 3.5 \frac{\text{W·hp}}{\text{hp}}$$

$$\boxed{2,611 \text{ W}}$$

4.7. Listing the known and unknown quantities:

$$m = 2,000 \text{ kg}$$

$$v = 72 \text{ km/hr}$$

$$KE = \, ?$$

These are the quantities found in the equation for kinetic energy, $KE = 1/2mv^2$, which is already solved. However, note that the velocity is in units of km/hr, which must be changed to m/s before doing anything else (it must be m/s because all energy and work units are in units of the joule (J). A joule is a newton-meter, and a newton is a kg·m/s²). Using the conversion factor from inside the front cover of your text,

$$\frac{0.2778 \, \dfrac{m}{s}}{1.0 \, \dfrac{km}{hr}} \times 72 \, \frac{km}{hr}$$

$$(0.2778)\,(72) \, \frac{m}{s} \times \frac{hr}{km} \times \frac{km}{hr}$$

$$20 \, \frac{m}{s}$$

and

$$KE = \frac{1}{2} mv^2$$

$$= \frac{1}{2}\,(2,000 \text{ kg})\left(20 \, \frac{m}{s}\right)^2$$

$$= \frac{1}{2}\,(2,000 \text{ kg})\left(400 \, \frac{m^2}{s^2}\right)$$

$$= \frac{1}{2} \times 2,000 \times 400 \, \frac{kg\cdot m^2}{s^2}$$

$$= 400,000 \, \frac{kg\cdot m}{s^2} \times m$$

$$= 400,000 \text{ N·m}$$

$$= \boxed{4 \times 10^5 \text{ J}}$$

Scientific notation is used here to simplify a large number and to show one significant figure.

4.8. Recall the relationship between work and energy—that you do work on an object when you throw it, giving it kinetic energy, and the kinetic energy it has will do work on something else when stopping. Because of the relationship between work and energy you can calculate (1) the work you do, (2) the kinetic energy a moving object has as a result of your work, and (3) the work it will do when coming to a stop, and all three answers should be the same. Thus you do not have a force or a distance to calculate the work needed to stop a moving car, but you can simply calculate the kinetic energy of the car. Both answers should be the same.

Before you start, note that the velocity is in units of km/hr, which must be changed to m/s before doing anything else (it must be m/s because all energy and work units are in units of the joule (J). A joule is a newton-meter, and a newton is a kg·m/s²). Using the conversion factor from inside the front cover,

$$\frac{0.2778 \, \dfrac{m}{s}}{1.0 \, \dfrac{km}{hr}} \times 54.0 \, \frac{km}{hr}$$

$$0.2778 \times 54.0 \, \frac{m}{s} \times \frac{hr}{km} \times \frac{km}{hr}$$

$$15.0 \, \frac{m}{s}$$

and

$$KE = \frac{1}{2} mv^2$$

$$= \frac{1}{2}\,(1,000.0 \text{ kg})\left(15.0 \, \frac{m}{s}\right)^2$$

$$= \frac{1}{2}\,(1,000.0 \text{ kg})\left(225 \, \frac{m^2}{s^2}\right)$$

$$= \frac{1}{2} \times 1,000.0 \times 225 \, \frac{kg\cdot m^2}{s^2}$$

$$= 112,500 \, \frac{kg\cdot m}{s^2} \times m$$

$$= 112,500 \text{ N·m}$$

$$= \boxed{1.13 \times 10^5 \text{ J}}$$

Scientific notation is used here to simplify a large number and to easily show three significant figures. The answer could likewise be expressed as 113 kJ.

4.9. **(a)** How much energy was used by a 1,000 kg car climbing a hill 50.02 m high is answered by how much work the car did. In this case $W = Fd$ and the force exerted is the weight of the car, $w = mgh$. Thus,

$$w = mgh$$

$$= (1,000 \text{ kg})\left(9.8 \, \frac{m}{s^2}\right)(51.02 \text{ m})$$

$$= 1,000 \times 9.8 \times 51.02 \text{ kg} \times \frac{m}{s^2} \times m$$

$$= 499,996 \, \frac{kg\cdot m}{s^2} \times m$$

$$= 500,000 \text{ N·m}$$

$$= 5 \times 10^5 \text{ J (or 500 kJ)}$$

(b) How much potential energy the car has is found in the potential energy equation, $PE = mgh$. Note the potential energy is path independent, that is, depends only on the vertical height of the hill. Thus,

$$PE = mgh$$

$$= (1{,}000 \text{ kg})\left(9.8 \, \frac{\text{m}}{\text{s}^2}\right)(51.02 \text{ m})$$

$$= 1{,}000 \times 9.8 \times 51.02 \text{ kg} \times \frac{\text{m}}{\text{s}^2} \times \text{m}$$

$$= 499{,}996 \, \frac{\text{kg·m}}{\text{s}^2} \times \text{m}$$

$$= 500{,}000 \text{ N·m}$$

$$= \boxed{5 \times 10^5 \text{ J (or 500 kJ)}}$$

As you can see, the potential energy of the car is exactly the same as the amount of energy used to climb the hill.

4.10. The velocity of any object in free fall can be found from the velocity of falling object equation, $v = \sqrt{2gh}$. Thus a "bundle" of falling water, or anything else that falls 100.0 feet, will have a velocity of

$$v = \sqrt{2gh}$$

$$= \sqrt{2\left(9.8 \, \frac{\text{m}}{\text{s}^2}\right)(100.0 \text{ m})}$$

$$= \sqrt{(2)(9.8)(100.0) \frac{\text{m}}{\text{s}^2} \times \text{m}}$$

$$= \sqrt{1{,}960 \, \frac{\text{m}^2}{\text{s}^2}}$$

$$= 44.27 \, \frac{\text{m}}{\text{s}}$$

$$= 44 \, \frac{\text{m}}{\text{s}}$$

4.11. Listing the known and unknown quantities for the four trucks:

truck 1: $m = 1{,}000$ kg

truck 2: $m = 2{,}000$ kg

truck 3: $m = 3{,}000$ kg

truck 4: $m = 4{,}000$ kg

v of all trucks = 72 km/hr

KE of trucks 1, 2, 3, and 4 = ?

We are supposed to find the kinetic energy for each truck, $KE = 1/2mv^2$, then see if we can generalize a relationship concerning mass and kinetic energy. First, however, note that the velocity is in units of km/hr which must be changed to m/s before doing anything else (it must be m/s because all energy and work units are in units of the joule (J). A joule is a newton-meter, and a newton is a kg·m/s²). Using the conversion factor from inside the front cover,

$$\frac{0.2778 \, \dfrac{\text{m}}{\text{s}}}{1 \, \dfrac{\text{km}}{\text{hr}}} \times 72 \, \frac{\text{km}}{\text{hr}}$$

$$0.2778 \times 72 \, \frac{\text{m}}{\text{s}} \times \frac{\text{hr}}{\text{km}} \times \frac{\text{km}}{\text{hr}}$$

$$20 \, \frac{\text{m}}{\text{s}}$$

Truck 1:

$$KE = \frac{1}{2} \, mv^2$$

$$= \frac{1}{2} \, (1{,}000 \text{ kg})\left(20 \, \frac{\text{m}}{\text{s}}\right)^2$$

$$= \frac{1}{2} \, (1{,}000 \text{ kg})\left(400 \, \frac{\text{m}^2}{\text{s}^2}\right)$$

$$= \frac{1}{2} \times 1{,}000 \times 400 \, \frac{\text{kg·m}^2}{\text{s}^2}$$

$$= 200{,}000 \, \frac{\text{kg·m}}{\text{s}^2} \times \text{m}$$

$$= 200{,}000 \text{ N·m}$$

$$= \boxed{2 \times 10^5 \text{ J}}$$

Truck 2:

$$KE = \frac{1}{2} \, mv^2$$

$$= \frac{1}{2} \, (2{,}000 \text{ kg})\left(20 \, \frac{\text{m}}{\text{s}}\right)^2$$

$$= \frac{1}{2} \, (2{,}000 \text{ kg})\left(400 \, \frac{\text{m}^2}{\text{s}^2}\right)$$

$$= \frac{1}{2} \times 2{,}000 \times 400 \, \frac{\text{kg·m}^2}{\text{s}^2}$$

$$= 400{,}000 \, \frac{\text{kg·m}}{\text{s}^2} \times \text{m}$$

$$= 400{,}000 \text{ N·m}$$

$$= \boxed{4 \times 10^5 \text{ J}}$$

Truck 3:

$$KE = \frac{1}{2} \, mv^2$$

$$= \frac{1}{2} \, (3{,}000 \text{ kg})\left(20 \, \frac{\text{m}}{\text{s}}\right)^2$$

$$= \frac{1}{2} \, (3{,}000 \text{ kg})\left(400 \, \frac{\text{m}^2}{\text{s}^2}\right)$$

$$= \frac{1}{2} \times 3{,}000 \times 400 \, \frac{\text{kg·m}^2}{\text{s}^2}$$

$$= 600,000 \, \frac{\text{kg·m}}{\text{s}^2} \times \text{m}$$

$$= 600,000 \, \text{N·m}$$

$$= \boxed{6 \times 10^5 \, \text{J}}$$

Truck 4:

$$KE = \frac{1}{2} mv^2$$

$$= \frac{1}{2} (4,000 \, \text{kg}) \left(20 \, \frac{\text{m}}{\text{s}} \right)^2$$

$$= \frac{1}{2} (4,000 \, \text{kg}) \left(400 \, \frac{\text{m}^2}{\text{s}^2} \right)$$

$$= \frac{1}{2} \times 4,000 \times 400 \, \frac{\text{kg·m}^2}{\text{s}^2}$$

$$= 800,000 \, \frac{\text{kg·m}}{\text{s}^2} \times \text{m}$$

$$= 800,000 \, \text{N·m}$$

$$= \boxed{8 \times 10^5 \, \text{J}}$$

Analysis: Doubling the mass of the truck from 1,000 to 2,000 kg at the same velocity doubles the KE from 200,000 J to 400,000 J. The same relationship is observed between 2,000 and 4,000 kg trucks, so the KE appears to vary directly with the mass of the truck, $KE \propto m$.

4.12. Listing the known and unknown quantities for the three trucks:

truck 1: $v = 10.0$ m/s

truck 2: $v = 20.0$ m/s

truck 3: $v = 30.0$ m/s

m of all trucks: $= 3,000$ kg

KE of trucks 1, 2, and 3 $= ?$

We are supposed to find the kinetic energy for each truck, $KE = 1/2mv^2$, then see if we can generalize a relationship concerning velocity and kinetic energy.

Truck 1:

$$KE = \frac{1}{2} mv^2$$

$$= \frac{1}{2} (3,000.0 \, \text{kg}) \left(10.0 \, \frac{\text{m}}{\text{s}} \right)^2$$

$$= \frac{1}{2} (3,000.0 \, \text{kg}) \left(100 \, \frac{\text{m}^2}{\text{s}^2} \right)$$

$$= \frac{1}{2} \times 3,000.0 \times 100 \, \frac{\text{kg·m}^2}{\text{s}^2}$$

$$= 150,000 \, \frac{\text{kg·m}}{\text{s}^2} \times \text{m}$$

$$= 150,000 \, \text{N·m}$$

$$= \boxed{1.50 \times 10^5 \, \text{J}}$$

Truck 2:

$$KE = \frac{1}{2} mv^2$$

$$= \frac{1}{2} (3,000.0 \, \text{kg}) \left(20.0 \, \frac{\text{m}}{\text{s}} \right)^2$$

$$= \frac{1}{2} (3,000.0 \, \text{kg}) \left(400 \, \frac{\text{m}^2}{\text{s}^2} \right)$$

$$= \frac{1}{2} \times 3,000.0 \times 400 \, \frac{\text{kg·m}^2}{\text{s}^2}$$

$$= 600,000 \, \frac{\text{kg·m}}{\text{s}^2} \times \text{m}$$

$$= 600,000 \, \text{N·m}$$

$$= \boxed{6.00 \times 10^5 \, \text{J}}$$

Truck 3:

$$KE = \frac{1}{2} mv^2$$

$$= \frac{1}{2} (3,000.0 \, \text{kg}) \left(30.0 \, \frac{\text{m}}{\text{s}} \right)^2$$

$$= \frac{1}{2} (3,000.0 \, \text{kg}) \left(900 \, \frac{\text{m}^2}{\text{s}^2} \right)$$

$$= \frac{1}{2} \times 3,000.0 \times 900 \, \frac{\text{kg·m}^2}{\text{s}^2}$$

$$= 1,350,000 \, \frac{\text{kg·m}}{\text{s}^2} \times \text{m}$$

$$= 1,350,000 \, \text{N·m}$$

$$= \boxed{1.35 \times 10^6 \, \text{J}}$$

Analysis: If you double the speed from 10 to 20 m/s the energy is increased four times (150,000 J times 4 = 600,000 J). If you triple the speed from 10 to 30 m/s the energy is increased nine times (150,000 J times 9 = 1,350,000). It appears that $KE \propto v^2$. Note that doubling and tripling the amount of energy requires you to double and triple the amount of work needed to bring a truck to rest.

4.13. **(a)**

$$W = Fd$$

$$= (10 \, \text{lb})(5 \, \text{ft})$$

$$= (10)(5) \, \text{ft} \times \text{lb}$$

$$= 50 \, \text{ft·lb}$$

(b) The distance of the bookcase from some horizontal reference level did not change, so the gravitational potential energy does not change.

4.14. The force (F) needed to lift the book is equal to the weight (w) of the book, or $F = w$. Since $w = mg$, then $F = mg$. Work is defined as the product of a force moved through a distance, or $W = Fd$. The work done in lifting the book is therefore $W = mgd$, and:

(a)
$$W = mgd$$
$$= (2.0 \text{ kg})(9.8 \text{ m/s}^2)(2.00 \text{ m})$$
$$= (2.0)(9.8)(2.00) \frac{\text{kg·m}}{\text{s}^2} \times \text{m}$$
$$= 39.2 \frac{\text{kg·m}^2}{\text{s}^2}$$
$$= 39.2 \text{ J} = \boxed{39 \text{ J}}$$

(b)
$$PE = mgh = \boxed{39 \text{ J}}$$

(c)
$$PE_{\text{lost}} = KE_{\text{gained}} = mgh = \boxed{39 \text{ J}}$$

(or)
$$v = \sqrt{2gh} = \sqrt{(2)(9.8 \text{ m/s}^2)(2.00 \text{ m})}$$
$$= \sqrt{39.2 \text{ m}^2/\text{s}^2} \quad \text{(Note)}$$
$$= 6.26 \text{ m/s}$$
$$KE = \frac{1}{2}mv^2 = \left(\frac{1}{2}\right)(2.0 \text{ kg})(6.26 \text{ m/s})^2$$
$$= \left(\frac{1}{2}\right)(2.0 \text{ kg})(39.2 \text{ m}^2/\text{s}^2)$$
$$= (1.0)(39.2)\frac{\text{kg·m}^2}{\text{s}^2}$$
$$= \boxed{39 \text{ J}}$$

4.15. Note that the gram unit must be converted to kg to be consistent with the definition of a newton-meter, or joule unit of energy:

$$KE = \frac{1}{2}mv^2 = \left(\frac{1}{2}\right)(0.15 \text{ kg})(30.0 \text{ m/s})^2$$
$$= \left(\frac{1}{2}\right)(0.15 \text{ kg})(900 \text{ m}^2/\text{s}^2)$$
$$= \left(\frac{1}{2}\right)(0.15)(900)\frac{\text{kg·m}^2}{\text{s}^2}$$
$$= 67.5 \text{ J} = \boxed{68 \text{ J}}$$

4.16. The km/hr unit must first be converted to m/s before finding the kinetic energy. Note also that the work done to put an object in motion is equal to the energy of motion, or kinetic energy that it has as a result of the work. The work needed to bring the object to a stop is also equal to the kinetic energy of the moving object:

Unit conversion:

$$1\frac{\text{km}}{\text{hr}} = 0.2778\frac{\frac{\text{m}}{\text{s}}}{\frac{\text{km}}{\text{hr}}} = \left(90.0\frac{\text{km}}{\text{hr}}\right)\left(0.2778\frac{\frac{\text{m}}{\text{s}}}{\frac{\text{km}}{\text{hr}}}\right) = 25.0 \text{ m/s}$$

(a)
$$KE = \frac{1}{2}mv^2 = \frac{1}{2}(1{,}000.0 \text{ kg})\left(25.0\frac{\text{m}}{\text{s}}\right)^2$$
$$= \frac{1}{2}(1{,}000.0 \text{ kg})\left(625\frac{\text{m}^2}{\text{s}^2}\right)$$
$$= \frac{1}{2}(1{,}000.0)(625)\frac{\text{kg·m}^2}{\text{s}^2}$$
$$= 312.5 \text{ kJ} = \boxed{313 \text{ kJ}}$$

(b)
$$W = Fd = KE = \boxed{313 \text{ kJ}}$$

(c)
$$KE = W = Fd = \boxed{313 \text{ kJ}}$$

4.17.
$$KE = \frac{1}{2}mv^2$$
$$= \frac{1}{2}(60.0 \text{ kg})\left(2.0\frac{\text{m}}{\text{s}}\right)^2$$
$$= \frac{1}{2}(60.0 \text{ kg})\left(4.0\frac{\text{m}^2}{\text{s}^2}\right)$$
$$= 30.0 \times 4.0 \text{ kg} \times \left(\frac{\text{m}^2}{\text{s}^2}\right)$$
$$= \boxed{120 \text{ J}}$$

$$KE = \frac{1}{2}mv^2$$
$$= \frac{1}{2}(60.0 \text{ kg})\left(4.0\frac{\text{m}}{\text{s}}\right)^2$$
$$= \frac{1}{2}(60.0 \text{ kg})\left(16\frac{\text{m}^2}{\text{s}^2}\right)$$
$$= 30.0 \times 16 \text{ kg} \times \left(\frac{\text{m}^2}{\text{s}^2}\right)$$
$$= \boxed{480 \text{ J}}$$

Thus, doubling the speed results in a four-fold increase in kinetic energy.

4.18.
$$KE = \frac{1}{2}mv^2$$
$$= \frac{1}{2}(70.0 \text{ kg})(6.00 \text{ m/s})^2$$
$$= (35.0 \text{ kg})(36.0 \text{ m}^2/\text{s}^2)$$
$$= 35.0 \times 36.0 \text{ kg} \times \frac{\text{m}^2}{\text{s}^2}$$
$$= \boxed{1{,}260 \text{ J}}$$

$$KE = \frac{1}{2}mv^2$$
$$= \frac{1}{2}(140.0 \text{ kg})(6.00 \text{ m/s})^2$$
$$= (70.0 \text{ kg})(36.0 \text{ m}^2/\text{s}^2)$$
$$= 70.0 \times 36.0 \text{ kg} \times \frac{\text{m}^2}{\text{s}^2}$$
$$= \boxed{2{,}520 \text{ J}}$$

Thus, doubling the mass results in a doubling of the kinetic energy.

4.19. **(a)** The force needed is equal to the weight of the student. The English unit of a pound is a force unit, so

$$W = Fd$$

$$= (170.0 \text{ lb})(25.0 \text{ ft})$$

$$= \boxed{4,250 \text{ ft·lb}}$$

(b) Work (W) is defined as a force (F) moved through a distance (d), or $W = Fd$. Power (P) is defined as work (W) per unit of time (t), or $P = W/t$. Therefore,

$$P = \frac{Fd}{t}$$

$$= \frac{(170.0 \text{ lb})(25.0 \text{ ft})}{10.0 \text{ s}}$$

$$= \frac{(170.0)(25.0)}{10.0} \frac{\text{ft·lb}}{\text{s}}$$

$$= 425 \frac{\text{ft·lb}}{\text{s}}$$

One hp is defined as $550 \frac{\text{ft·lb}}{\text{s}}$ and

$$\frac{425 \text{ ft·lb/s}}{550 \frac{\text{ft·lb/s}}{\text{hp}}} = \boxed{0.77 \text{ hp}}$$

(Note that the student's power rating (425 ft·lb/s) is less than the power rating defined as 1 horsepower (550 ft·lb/s). Thus, the student's horsepower must be *less* than 1 horsepower. A simple analysis such as this will let you know if you inverted the ratio or not.)

4.20. **(a)** The force (F) needed to lift the elevator is equal to the weight of the elevator. Since the work (W) is = to Fd and power (P) is = to W/t, then

$$P = \frac{Fd}{t} \therefore t = \frac{Fd}{P}$$

$$= \frac{[2,000.0 \text{ lb}][20.0 \text{ ft}]}{\left(550 \frac{\text{ft·lb}}{\text{s}}\right)[20.0 \text{ hp}]}$$

$$= \frac{40,000}{11,000} \frac{\text{ft·lb}}{\frac{\text{ft·lb}}{\text{s}}} \times \frac{1}{\text{hp}} \times \text{hp}$$

$$= \frac{40,000}{11,000} \frac{\text{ft·lb}}{1} \times \frac{\text{s}}{\text{ft·lb}}$$

$$= 3.64 \frac{\text{ft·lb·s}}{\text{ft·lb}}$$

$$= \boxed{3.64 \text{ s}}$$

(b)

$$\bar{v} = \frac{d}{t}$$

$$= \frac{20.0 \text{ ft}}{3.64 \text{ s}}$$

$$= \boxed{5.49 \text{ ft/s}}$$

4.21. Since $PE_{\text{lost}} = KE_{\text{gained}}$, then $mgh = \frac{1}{2} mv^2$. Solving for v,

$$v = \sqrt{2gh} = \sqrt{(2)(32.0 \text{ ft/s}^2)(9.8 \text{ ft})}$$

$$= \sqrt{(2)(32.0)(9.8) \text{ ft}^2/\text{s}^2}$$

$$= \sqrt{627 \text{ ft}^2/\text{s}^2}$$

$$= 25 \text{ ft/s}$$

4.22. $KE = \frac{1}{2} mv^2 \therefore v = \sqrt{\dfrac{2KE}{m}}$

$$= \sqrt{\frac{(2)\left(200,000 \dfrac{\text{kg·m}^2}{\text{s}^2}\right)}{1,000.0 \text{ kg}}}$$

$$= \sqrt{\frac{400,000}{1,000.0} \frac{\text{kg·m}^2}{\text{s}^2} \times \frac{1}{\text{kg}}}$$

$$= \sqrt{\frac{400,000}{1,000.0} \frac{\text{kg·m}^2}{\text{kg·s}^2}}$$

$$= \sqrt{400 \text{ m}^2/\text{s}^2}$$

$$= \boxed{20 \text{ m/s}}$$

4.23. The maximum velocity occurs at the lowest point with a gain of kinetic energy equivalent to the loss of potential energy in falling 3.0 in (which is 0.25 ft), so

$$KE_{\text{gained}} = PE_{\text{lost}}$$

$$\frac{1}{2} mv^2 = mgh$$

$$v = \sqrt{2gh}$$

$$= \sqrt{(2)(32 \text{ ft/s}^2)(0.25 \text{ ft})}$$

$$= \sqrt{(2)(32)(0.25) \text{ft/s}^2 \times \text{ft}}$$

$$= \sqrt{16 \text{ ft}^2/\text{s}^2}$$

$$= \boxed{4.0 \text{ ft/s}}$$

4.24. **(a)** $W = Fd$ and the force F that is needed to lift the load upward is mg, so $W = mgh$. Power is W/t, so

$$P = \frac{mgh}{t}$$

$$= \frac{(250.0 \text{ kg})(9.8 \text{ m/s}^2)(80.0 \text{ m})}{39.2 \text{ s}}$$

$$= \frac{(250.0)(9.8)(80.0)}{39.2} \frac{\text{kg}}{1} \times \frac{\text{m}}{\text{s}^2} \times \frac{\text{m}}{1} \times \frac{1}{\text{s}}$$

$$= \frac{196,000}{39.2} \frac{\text{kg·m}^2}{\text{s}^2} \times \frac{1}{\text{s}}$$

$$= 5,000 \frac{\text{J}}{\text{s}}$$

$$= \boxed{5.0 \text{ kW}}$$

(b) There are 746 watts per horsepower, so

$$\frac{5{,}000 \text{ W}}{746 \dfrac{\text{W}}{\text{hp}}} = \frac{5{,}000}{746} \frac{\text{W}}{1} \times \frac{\text{hp}}{\text{W}}$$

$$= 6.70 \frac{\text{W}\cdot\text{hp}}{\text{W}}$$

$$= \boxed{6.70 \text{ hp}}$$

Exercises **CHAPTER 5**

5.1. Listing the known and unknown quantities:

body temperature $\quad T_F = 98.6°$

$$T_C = ?$$

These are the quantities found in the equation for conversion of Fahrenheit to Celsius, $T_C = \dfrac{5}{9}(T_F - 32°)$, where T_F is the temperature in Fahrenheit and T_C is the temperature in Celsius. This equation describes a relationship between the two temperature scales and is used to convert a Fahrenheit temperature to Celsius. The equation is already solved for the Celsius temperature, T_C. Thus,

$$T_C = \frac{5}{9}(T_F - 32°)$$

$$= \frac{5}{9}(98.6° - 32°)$$

$$= \frac{333°}{9}$$

$$= \boxed{37° \text{ C}}$$

5.2. Listing the known and unknown quantities:

bank temperature $\quad T_C = 20°$

$$T_F = ?$$

These are the quantities found in the equation for conversion of Celsius to Fahrenheit, $T_F = \dfrac{9}{5}T_C + 32°$ where T_C is the temperature in Celsius and T_F is the temperature in Fahrenheit. This equation describes a relationship between the two temperature scales and is used to convert a Celsius temperature to Fahrenheit. The equation is already solved for the Fahrenheit temperature, T_F. Thus,

$$T_F = \frac{9}{5}T_C + 32°$$

$$= \frac{9}{5}20° + 32°$$

$$= \frac{180°}{5} + 32°$$

$$= 36° + 32°$$

$$= \boxed{68°F}$$

5.3. Listing the known and unknown quantities:

$$T_K = 300.0$$

$$T_C = ?$$

$$T_F = ?$$

(a) What is the equivalent Celsius temperature?

There are the quantities found in the equation for conversion of Celsius to Kelvin, $T_K = T_C + 273$, where T_K is the temperature in Kelvins and T_C is the temperature in Celsius. This equation describes a relationship between the two temperature scales and is used to convert Celsius temperature degrees to Kelvins. The equation must be first solved for the Celsius temperature, T_C. Thus,

$$T_K = T_C + 273 \therefore T_C = T_K - 273°$$

$$= 300.0 - 273°$$

$$= \boxed{27.0°C}$$

5.4. $\quad Q = mc\Delta T$

$$= (221 \text{ g})\left(0.093 \frac{\text{cal}}{\text{gC°}}\right)(38.0°C - 20.0°C)$$

$$= (221)(0.093)(18.0)\text{g} \times \frac{\text{cal}}{\text{gC°}} \times °C$$

$$= 370 \frac{\text{g}\cdot\text{cal}\cdot°C}{\text{gC°}}$$

$$= \boxed{370 \text{ cal}}$$

5.5. First, you need to know the energy of the moving bike and rider. Since the speed is given as 36.0 km/hr, convert to m/s by multiplying times 0.2778 m/s per km/hr:

$$\left(36.0 \frac{\text{km}}{\text{hr}}\right)\left(0.2778 \frac{\text{m/s}}{\text{km/hr}}\right)$$

$$= (36.0)(0.2778) \frac{\text{km}}{\text{hr}} \times \frac{\text{hr}}{\text{km}} \times \frac{\text{m}}{\text{s}}$$

$$= 10.0 \text{ m/s}$$

Then,

$$KE = \frac{1}{2}mv^2$$

$$= \frac{1}{2}(100.0 \text{ kg})(10.0 \text{ m/s})^2$$

$$= \frac{1}{2}(100.0 \text{ kg})(100 \text{ m}^2/\text{s}^2)$$

$$= \frac{1}{2}(100.0)(100) \frac{\text{kg}\cdot\text{m}^2}{\text{s}^2}$$

$$= \boxed{5.00 \times 10^3 \text{J}}$$

Second, this energy is converted to the calorie heat unit through the mechanical equivalent of heat relationship, that 1.0 kcal = 4,184 J, or that 1.0 cal = 4.184 J. Thus,

$$\frac{1.0 \text{ cal}}{4.184 \text{ J}} = 0.2390 \frac{\text{cal}}{\text{J}}$$

$$= (0.2390)(5,000) \frac{\text{cal}}{\text{J}} \times \text{J}$$

$$= 1,195 \frac{\text{cal} \cdot \cancel{J}}{\cancel{J}}$$

$$= \boxed{1 \text{ kcal}}$$

5.6.   First, you need to find the energy of the falling bag. Since the potential energy lost equals the kinetic energy gained, the energy of the bag just as it hits the ground can be found from

$$PE = mgh$$

$$= (15.53 \text{ kg})(9.8 \text{ m/s}^2)(5.50 \text{ m})$$

$$= (15.53)(9.8)(5.50) \frac{\text{kg} \cdot \text{m}}{\text{s}^2} \times \text{m}$$

$$= 837 \text{ J}$$

In calories, this energy is equivalent to

$$(0.2390 \text{ cal/J})(837 \text{ J}) = 200 \text{ cal}$$

Second, the temperature change can be calculated from the equation giving the relationship between a quantity of heat (Q), mass (m), specific heat of the substance (c), and the change of temperature:

$$Q = mc\Delta T \therefore \Delta T = \frac{Q}{mc}$$

$$= \frac{0.200 \text{ kcal}}{(15.53 \text{ kg})\left(0.200 \frac{\text{kcal}}{\text{kgC}^\circ}\right)}$$

$$= \frac{0.200}{(15.53)(0.200)} \frac{\text{kcal}}{1} \times \frac{1}{\text{kg}} \times \frac{\text{kgC}^\circ}{\text{kcal}}$$

$$= 0.0644 \frac{\cancel{\text{kcal}} \cdot \text{kgC}^\circ}{\cancel{\text{kcal}} \cdot \cancel{\text{kg}}}$$

$$= \boxed{6.44 \times 10^{-2} \, ^\circ\text{C}}$$

5.7.   The Calorie used by dietitians is a kilocalorie; thus 250.0 Cal is 250.0 kcal. The mechanical energy equivalent is 1 kcal = 4,184 J, so (250.0 kcal)(4,184 J/kcal) = 1,046,250 J.
Since $W = Fd$ and the force needed is equal to the weight (mg) of the person, $W = mgh = (75.0 \text{ kg})(9.8 \text{ m/s}^2)(10.0 \text{ m}) = 7,350 \text{ J}$ for each stairway climb.
A total of 1,046,250 J of energy from the French fries would require 1,046,250 J/7,350 J per climb, or 142.3 trips up the stairs.

5.8.   For unit consistency,

$$T_C = \frac{5}{9}(T_F - 32^\circ) = \frac{5}{9}(68^\circ - 32^\circ) = \frac{5}{9}(36^\circ) = 20^\circ\text{C}$$

$$= \frac{5}{9}(32^\circ - 32^\circ) = \frac{5}{9}(0^\circ) = 0^\circ\text{C}$$

Glass Bowl:

$$Q = mc\Delta T$$

$$= (0.5 \text{ kg})\left(0.2 \frac{\text{kcal}}{\text{kg}^\circ\text{C}}\right)(20^\circ\text{C})$$

$$= (0.5)(0.2)(20) \frac{\text{kg}}{1} \times \frac{\text{kcal}}{\text{kgC}^\circ} \times \frac{^\circ\text{C}}{1}$$

$$= \boxed{2 \text{ kcal}}$$

Iron Pan:

$$Q = mc\Delta T$$

$$= (0.5 \text{ kg})\left(0.11 \frac{\text{kcal}}{\text{kgC}^\circ}\right)(20^\circ\text{C})$$

$$= (0.5)(0.11)(20) \text{ kg} \times \frac{\text{kcal}}{\text{kgC}^\circ} \times {^\circ\text{C}}$$

$$= \boxed{1 \text{ kcal}}$$

5.9.   Note that a specific heat expressed in cal/g has the same numerical value as a specific heat expressed in kcal/kg because you can cancel the k units. You could convert 896 cal to 0.896 kcal, but one of the two conversion methods is needed for consistency with other units in the problem.

$$Q = mc\Delta T \therefore m = \frac{Q}{c\Delta T}$$

$$= \frac{896 \text{ cal}}{\left(0.056 \frac{\text{cal}}{\text{gC}^\circ}\right)(80.0^\circ\text{C})}$$

$$= \frac{896}{(0.056)(80.0)} \frac{\text{cal}}{1} \times \frac{\text{gC}^\circ}{\text{cal}} \times \frac{1}{\text{C}}$$

$$= 200 \text{ g}$$

$$= \boxed{0.20 \text{ kg}}$$

5.10.   Since a watt is defined as a joule/s, finding the total energy in joules will tell the time:

$$Q = mc\Delta T$$

$$= (250.0 \text{ g})\left(1.00 \frac{\text{cal}}{\text{gC}^\circ}\right)(60.0^\circ\text{C})$$

$$= (250.0)(1.00)(60.0) \text{ g} \times \frac{\text{cal}}{\text{gC}^\circ} \times {^\circ\text{C}}$$

$$= 1.50 \times 10^4 \text{ cal}$$

This energy in joules is $(1.50 \times 10^4 \text{ cal})\left(4.184 \frac{\text{J}}{\text{cal}}\right) = 62,800 \text{ J}$

A 300 watt heater uses energy at a rate of $300 \frac{\text{J}}{\text{s}}$, so $\frac{62,800 \text{ J}}{300 \text{ J/s}}$

$$= 209 \text{ s is required, which is } \frac{209 \text{ s}}{60 \frac{\text{s}}{\text{min}}} = 3.48 \text{ min, or}$$

$$\boxed{\text{about } 3\frac{1}{2} \text{ min}}$$

5.11.

$$Q = mc\Delta T \therefore c = \frac{Q}{m\Delta T}$$

$$= \frac{60.0 \text{ cal}}{(100.0 \text{ g})(20.0°C)}$$

$$= \frac{60.0}{(100.0)(20.0)} \frac{\text{cal}}{\text{gC°}}$$

$$= \boxed{0.0300 \frac{\text{cal}}{\text{gC°}}}$$

5.12. Since the problem specified a solid changing to a liquid without a temperature change, you should recognize that this is a question about a phase change only. The phase change from solid to liquid (or liquid to solid) is concerned with the latent heat of fusion. For water, the latent heat of fusion is given as 80.0 cal/g, and

$$m = 250.0 \text{ g} \qquad Q = mL_f$$

$$L_{f \text{ (water)}} = 80.0 \text{ cal/g} \qquad = (250.0 \text{ g})\left(80.0 \frac{\text{cal}}{\text{g}}\right)$$

$$Q = ? \qquad = 250.0 \times 80.0 \frac{\text{g·cal}}{\text{g}}$$

$$= 20,000 \text{ cal} = \boxed{20.0 \text{ kcal}}$$

5.13. To change water at 80.0°C to steam at 100.0°C requires two separate quantities of heat that can be called Q_1 and Q_2. The quantity Q_1 is the amount of heat needed to warm the water from 80.0°C to the boiling point, which is 100.0°C at sea level pressure ($\Delta T = 20.0°C$). The relationship between the variable involved is $Q = mc\Delta T$. The quantity Q_2 is the amount of heat needed to take 100.0° water through the phase change to steam (water vapor) at 100.0°C. The phase change from a liquid to a gas (or gas to liquid) is concerned with the latent heat of vaporization. For water, the latent heat of vaporization is given as 540.0 cal/g.

$$m = 250.0 \text{ g} \qquad Q_1 = mc\Delta T$$

$$L_{v \text{ (water)}} = 540.0 \text{ cal/g} \qquad = (250.0 \text{ g})\left(1.00 \frac{\text{cal}}{\text{gC°}}\right)(20.0°C)$$

$$Q = ? \qquad = (250.0)(1.00)(20.0) \text{ g} \times \frac{\text{cal}}{\text{gC°}} \times °C$$

$$= 5,000 \frac{\text{g·cal·°C}}{\text{gC°}}$$

$$= 5,000 \text{ cal}$$

$$= 5.00 \text{ kcal}$$

$$Q_2 = mL_v$$

$$= (250.0 \text{ g})\left(540.0 \frac{\text{cal}}{\text{g}}\right)$$

$$= 250.0 \times 540.0 \frac{\text{g·cal}}{\text{g}}$$

$$= 135,000 \text{ cal}$$

$$= 135.0 \text{ kcal}$$

$$Q_{\text{Total}} = Q_1 + Q_2$$

$$= 5.00 \text{ kcal} + 135.0 \text{ kcal}$$

$$= \boxed{140.0 \text{ kcal}}$$

5.14. To change 20.0°C water to steam at 125.0°C requires three separate quantities of heat. First, the quantity Q_1 is the amount of heat needed to warm the water from 20.0°C to 100.0°C ($\Delta T = 80.0°C$). The quantity Q_2 is the amount of heat needed to take 100.0°C water to steam at 100.0°C. Finally, the quantity Q_3 is the amount of heat needed to warm the steam from 100.0° to 125.0°C. According to table 5.4, the c for steam is 0.48 cal/g°C.

$$m = 100.0 \text{ g} \qquad Q_1 = mc\Delta T$$

$$\Delta T_{\text{water}} = 80.0°C \qquad = (100.0 \text{ g})\left(1.00 \frac{\text{cal}}{\text{gC°}}\right)(80.0°C)$$

$$\Delta T_{\text{steam}} = 25.0°C$$

$$L_{v(\text{water})} = 540.0 \text{ cal/g} \qquad = (100.0)(1.00)(80.0) \text{ g} \times \frac{\text{cal}}{\text{gC°}} \times °C$$

$$c_{\text{steam}} = 0.48 \text{ cal/gC°} \qquad = 8,000 \frac{\text{g·cal·°C}}{\text{gC°}}$$

$$= 8,000 \text{ cal}$$

$$= 8.00 \text{ kcal}$$

$$Q_2 = mL_v$$

$$= (100.0 \text{ g})\left(540.0 \frac{\text{cal}}{\text{g}}\right)$$

$$= 100.0 \times 540.0 \frac{\text{g·cal}}{\text{g}}$$

$$= 54,000 \text{ cal}$$

$$= 54.00 \text{ kcal}$$

$$Q_3 = mc\Delta T$$

$$= (100.0 \text{ g})\left(0.48 \frac{\text{cal}}{\text{gC°}}\right)(25.0°C)$$

$$= (100.0)(0.48)(25.0) \text{ g} \times \frac{\text{cal}}{\text{gC°}} \times °C$$

$$= 1,200 \frac{\text{g·cal·°C}}{\text{g·C°}}$$

$$= 1,200 \text{ cal}$$

$$= 1.2 \text{ kcal}$$

$$Q_{\text{total}} = Q_1 + Q_2 + Q_3$$

$$= 8.00 \text{ kcal} + 54.00 \text{ kcal} + 1.2 \text{ kcal}$$

$$= \boxed{63.2 \text{ kcal}}$$

5.15. **(a)** **Step 1:** Cool water from 18.0°C to 0°C.

$$Q_1 = mc\Delta T$$

$$= (400.0 \text{ g})\left(1.00 \frac{\text{cal}}{\text{gC°}}\right)(18.0°C)$$

$$= (400.0)(1.00)(18.0)\text{g} \times \frac{\text{cal}}{\text{gC°}} \times °C$$

$$= 7,200 \frac{\text{g} \cdot \text{cal} \cdot °C}{\text{gC°}}$$

$$= 7,200 \text{ cal}$$

$$= 7.20 \text{ kcal}$$

Step 2: Find the energy needed for the phase change of water at 0°C to ice at 0°C.

$$Q_2 = mL_f$$

$$= (400.0 \text{ g})\left(80.0 \frac{\text{cal}}{\text{g}}\right)$$

$$= 400.0 \times 80.0 \frac{\text{g} \cdot \text{cal}}{\text{g}}$$

$$= 32,000 \text{ cal}$$

$$= 32.0 \text{ kcal}$$

Step 3: Cool the ice from 0°C to ice at −5.00°C.

$$Q_3 = mc\Delta T$$

$$= (400.0 \text{ g})\left(0.50 \frac{\text{cal}}{\text{gC°}}\right)(5.00°C)$$

$$= 400.0 \times 0.50 \times 5.00 \text{ g} \times \frac{\text{cal}}{\text{gC°}} \times °C$$

$$= 1,000 \frac{\text{g} \cdot \text{cal} \cdot °C}{\text{gC°}}$$

$$= 1,000 \text{ cal}$$

$$= 1.0 \text{ kcal}$$

$$Q_{total} = Q_1 + Q_2 + Q_3$$

$$= 7.20 \text{ kcal} + 32.0 \text{ kcal} + 1.0 \text{ kcal}$$

$$= \boxed{40.2 \text{ kcal}}$$

(b) $$Q = mL_v \therefore m = \frac{Q}{L_v}$$

$$= \frac{40,200 \text{ cal}}{40.0 \frac{\text{cal}}{\text{g}}}$$

$$= \frac{40,200}{40.0} \frac{\text{cal}}{1} \times \frac{\text{g}}{\text{cal}}$$

$$= 1,005 \frac{\text{cal} \cdot \text{g}}{\text{cal}}$$

$$= \boxed{1.01 \times 10^3 \text{ g}}$$

5.16. First, note that the heat values (or the value of J) must be converted for unit consistency.

$$W = J(Q_H - Q_L)$$

$$= 4,184 \frac{\text{J}}{\text{kcal}} (0.3000 \text{ kcal} - 0.2000 \text{ kcal})$$

$$= 4,184 \frac{\text{J}}{\text{kcal}} (0.1000 \text{ kcal})$$

$$= 4,184 \times 0.1000 \frac{\text{J} \cdot \text{kcal}}{\text{kcal}}$$

$$= \boxed{418.4 \text{ J}}$$

5.17. $$W = J(Q_H - Q_L)$$

$$= 4,184 \frac{\text{J}}{\text{kcal}} (55.0 \text{ kcal} - 40.0 \text{ kcal})$$

$$= 4,184 \frac{\text{J}}{\text{kcal}} (15.0 \text{ kcal})$$

$$= 4,184 \times 15.0 \frac{\text{J} \cdot \text{kcal}}{\text{kcal}}$$

$$= 62,760 \text{ J}$$

$$= \boxed{62.8 \text{ kJ}}$$

Exercises CHAPTER 6

6.1. $$v = f\lambda$$

$$= \left(10\frac{1}{\text{s}}\right)(0.50 \text{ m})$$

$$= 5 \frac{\text{m}}{\text{s}}$$

6.2. The distance between two *consecutive* condensations (or rarefactions) is one wavelength, so $\lambda = 3.00$ m and

$$v = f\lambda$$

$$= \left(112.0 \frac{1}{\text{s}}\right)(3.00 \text{ m})$$

$$= 336 \frac{\text{m}}{\text{s}}$$

6.3. **(a)** One complete wave every 4 s means that $T = 4.00$ s. (Note that the symbol for the *time of a cycle* is T. Do not confuse this symbol with the symbol for temperature.)

(b)
$$f = \frac{1}{T}$$

$$= \frac{1}{4.0 \text{ s}}$$

$$= \frac{1}{4.0} \frac{1}{\text{s}}$$

$$= 0.25 \frac{1}{\text{s}}$$

$$= \boxed{0.25 \text{ Hz}}$$

6.4. The distance from one condensation to the next is one wavelength, so

$$v = f\lambda \therefore \lambda = \frac{v}{f}$$

$$= \frac{330 \frac{\text{m}}{\text{s}}}{260 \frac{1}{\text{s}}}$$

$$= \frac{330}{260} \frac{\text{m}}{\text{s}} \times \frac{\text{s}}{1}$$

$$= \boxed{1.3 \text{ m}}$$

6.5. **(a)** $\quad v = f\lambda = \left(256 \frac{1}{\text{s}}\right)(1.34 \text{ m}) \qquad = \boxed{343 \text{ m/s}}$

(b) $\qquad = \left(440.0 \frac{1}{\text{s}}\right)(0.780 \text{ m}) \qquad = \boxed{343 \text{ m/s}}$

(c) $\qquad = \left(750.0 \frac{1}{\text{s}}\right)(0.457 \text{ m}) \qquad = \boxed{343 \text{ m/s}}$

(d) $\qquad = \left(2,500.0 \frac{1}{\text{s}}\right)(0.1372 \text{ m}) \qquad = \boxed{343 \text{ m/s}}$

6.6. The speed of sound at 0.0°C is 1,087 ft/s, and

(a) $V_{T_F} = V_0 + \left[\frac{2.0 \text{ ft/s}}{\text{°C}}\right][T_P]$

$$= 1,087 \text{ ft/sec} + \left[\frac{2.0 \text{ ft/s}}{\text{°C}}\right][0.0\text{°C}]$$

$$= 1,087 + (2.0)(0.0) \text{ ft/s} + \frac{\text{ft/s}}{\text{°C}} \times \text{°C}$$

$$= 1,087 \text{ ft/s} + 0.0 \text{ ft/s}$$

$$= \boxed{1,087 \text{ ft/s}}$$

(b) $\quad V_{20°} = 1,087 \text{ ft/s} + \left[\frac{2.0 \text{ ft/s}}{\text{°C}}\right][20.0\text{°C}]$

$$= 1,087 \text{ ft/s} + 40 \text{ ft/s}$$

$$= \boxed{1,127 \text{ ft/s}}$$

(c) $\quad V_{40°} = 1,087 \text{ ft/s} + \left[\frac{2.0 \text{ ft/s}}{\text{°C}}\right][40.0\text{°C}]$

$$= 1,087 \text{ ft/s} + 80 \text{ ft/s}$$

$$= \boxed{1,167 \text{ ft/s}}$$

(d) $\quad V_{80°} = 1,087 \text{ ft/s} + \left[\frac{2.0 \text{ ft/s}}{\text{°C}}\right][80.0\text{°C}]$

$$= 1,087 \text{ ft/s} + 160 \text{ ft/s}$$

$$= \boxed{1,247 \text{ ft/s}}$$

6.7. For consistency with the units of the equation given, 43.7°F is first converted to 6.50°C. The velocity of sound in this air is:

$$V_{T_F} = V_0 + \left[\frac{2.0 \text{ ft/s}}{\text{°C}}\right][T_P]$$

$$= 1,087 \text{ ft/s} + \left[\frac{2.0 \text{ ft/s}}{\text{°C}}\right][6.50\text{°C}]$$

$$= 1,087 \text{ ft/s} + 13 \text{ ft/s}$$

$$= 1,100 \text{ ft/s}$$

The distance that a sound with this velocity travels in the given time is

$$v = \frac{d}{t} \therefore d = vt$$

$$= (1,100 \text{ ft/s})(4.80 \text{ s})$$

$$= (1,100)(4.80) \frac{\text{ft} \cdot \text{s}}{\text{s}}$$

$$= 5,280 \text{ ft}$$

$$\frac{5,280 \text{ ft}}{2}$$

$$= 2,640 \text{ ft}$$

$$= \boxed{2,600}$$

Since the sound traveled from the rifle to the cliff and then back, the cliff must be about one-half mile away.

6.8. This problem requires three steps, (1) conversion of the °F temperature value to °C, (2) calculating the velocity of sound in air at this temperature, and (3) calculating the distance from the calculated velocity and the given time:

$$V_{T_F} = V_0 + \left[\frac{2.0 \text{ ft/s}}{\text{°C}}\right][T_P]$$

$$= 1,087 \text{ ft/s} + \left[\frac{2.0 \text{ ft/s}}{\text{°C}}\right][26.67\text{°C}]$$

$$= 1,087 \text{ ft/s} + 53 \text{ ft/s} = 1,140 \text{ ft/s}$$

$$v = \frac{d}{t} \therefore d = vt$$

$$= (1,140 \text{ ft/s})(4.63 \text{ s})$$

$$= \boxed{5,280 \text{ ft (one mile)}}$$

6.9. **(a)**

$$v = f\lambda \therefore \lambda = \frac{v}{f}$$

$$= \frac{1{,}125 \frac{\text{ft}}{\text{s}}}{440 \frac{1}{\text{s}}}$$

$$= \frac{1{,}125}{440} \frac{\text{ft}}{\text{s}} \times \frac{\text{s}}{1}$$

$$= 2.56 \frac{\text{ft} \cdot \cancel{s}}{\cancel{s}}$$

$$= \boxed{2.6 \text{ ft}}$$

(b)

$$v = f\lambda \therefore \lambda = \frac{v}{f}$$

$$= \frac{5{,}020}{440} \frac{\text{ft}}{\text{s}} \times \frac{\text{s}}{1}$$

$$= 11.4 \text{ ft} = \boxed{11 \text{ ft}}$$

Exercises CHAPTER 7

7.1. First, recall that a negative charge means an excess of electrons. Second, the relationship between the total charge (q), the number of electrons (n), and the charge of a single electron (e) is $q = ne$. The fundamental charge of a single ($n = 1$) electron (e) is 1.60×10^{-19} C. Thus

$$q = ne \therefore n = \frac{q}{e}$$

$$= \frac{1.00 \times 10^{-14} \text{ C}}{1.60 \times 10^{-19} \frac{\text{C}}{\text{electron}}}$$

$$= \frac{1.00 \times 10^{-14}}{1.60 \times 10^{-19}} \frac{\text{C}}{1} \times \frac{\text{electron}}{\text{C}}$$

$$= 6.25 \times 10^{4} \frac{\cancel{\text{C}} \cdot \text{electron}}{\cancel{\text{C}}}$$

$$= \boxed{6.25 \times 10^{4} \text{ electron}}$$

7.2. **(a)** Both balloons have negative charges so the force is repulsive, pushing the balloons away from each other.
(b) The magnitude of the force can be found from Coulomb's law:

$$F = \frac{kq_1 q_2}{d^2}$$

$$= \frac{(9.00 \times 10^9 \text{ N} \cdot \text{m}^2/\text{C}^2)(3.00 \times 10^{-14} \text{ C})(2.00 \times 10^{-12} \text{ C})}{(2.00 \times 10^{-2} \text{ m})^2}$$

$$= \frac{(9.00 \times 10^9)(3.00 \times 10^{-14})(2.00 \times 10^{-12}) \frac{\text{N} \cdot \text{m}^2}{\text{C}^2} \times \text{C} \times \text{C}}{4.00 \times 10^{-4} \text{ m}^2}$$

$$= \frac{5.40 \times 10^{-16}}{4.00 \times 10^{-4}} \frac{\text{N} \cdot \cancel{\text{m}^2}}{\cancel{\text{C}^2}} \times \cancel{\text{C}^2} \times \frac{1}{\cancel{\text{m}^2}}$$

$$= \boxed{1.35 \times 10^{-12} \text{ N}}$$

7.3.

$$\frac{\text{potential}}{\text{difference}} = \frac{\text{work}}{\text{charge}}$$

or

$$V = \frac{W}{q}$$

$$= \frac{7.50 \text{ J}}{5.00 \text{ C}}$$

$$= 1.50 \frac{\text{J}}{\text{C}}$$

$$= \boxed{1.50 \text{ V}}$$

7.4.

$$\frac{\text{electric}}{\text{current}} = \frac{\text{charge}}{\text{time}}$$

or

$$I = \frac{q}{t}$$

$$= \frac{6.00 \text{ C}}{2.00 \text{ s}}$$

$$= 3.00 \frac{\text{C}}{\text{s}}$$

$$= \boxed{3.00 \text{ A}}$$

7.5. A current of 1.00 amp is defined as 1.00 coulomb/s. Since the fundamental charge of the electron is 1.60×10^{-19} C/electron,

$$\frac{1.00 \frac{\text{C}}{\text{s}}}{1.60 \times 10^{-19} \frac{\text{C}}{\text{electron}}}$$

$$= 6.25 \times 10^{18} \frac{\cancel{\text{C}}}{\text{s}} \times \frac{\text{electron}}{\cancel{\text{C}}}$$

$$= \boxed{6.25 \times 10^{18} \frac{\text{electrons}}{\text{s}}}$$

7.6.

$$R = \frac{V}{I}$$

$$= \frac{120.0 \text{ V}}{4.00 \text{ A}}$$

$$= 30.0 \frac{\text{V}}{\text{A}}$$

$$= \boxed{30.0 \text{ }\Omega}$$

7.7.

$$R = \frac{V}{I} \therefore I = \frac{V}{R}$$

$$= \frac{120.0 \text{ V}}{60.0 \frac{\text{V}}{\text{A}}}$$

$$= \frac{120.0}{60.0} \cancel{\text{V}} \times \frac{\text{A}}{\cancel{\text{V}}}$$

$$= \boxed{2.00 \text{ A}}$$

7.8. (a) $R = \dfrac{V}{I} \therefore V = IR$

$$= (1.20 \text{ A})\left(10.0 \dfrac{V}{A}\right)$$

$$= \boxed{12.0 \text{ V}}$$

(b)   Power = (current)(potential difference)

or

$$P = IV$$

$$= \left(1.20 \dfrac{C}{s}\right)\left(12.0 \dfrac{J}{C}\right)$$

$$= (1.20)(12.0)\dfrac{C}{s} \times \dfrac{J}{C}$$

$$= 14.4 \dfrac{J}{s}$$

$$= \boxed{14.4 \text{ W}}$$

7.9.   Note that there are two separate electrical units that are rates: (1) the amp (coulomb/s) and (2) the watt (joule/s). The question asked for a rate of using energy. Energy is measured in joules, so you are looking for the power of the radio in watts. To find watts ($P = IV$), you will need to calculate the current (I) since it is not given. The current can be obtained from the relationship of Ohm's law:

$$I = \dfrac{V}{R}$$

$$= \dfrac{3.00 \text{ V}}{15.0 \dfrac{V}{A}}$$

$$= 0.200 \text{ A}$$

$$P = IV$$

$$= (0.200 \text{ C/s})(3.00 \text{ J/C})$$

$$= \boxed{0.600 \text{ W}}$$

7.10. $\dfrac{(1,200 \text{ W})(0.25 \text{ hr})}{1,000 \dfrac{W}{kW}} = \dfrac{(1,200)(0.25)}{1,000} \dfrac{W \cdot hr}{1} \times \dfrac{kW}{W}$

$$= 0.30 \text{ kWhr}$$

$$(0.30 \text{ kWhr})\left(\dfrac{\$0.10}{\text{kWhr}}\right) = \boxed{\$0.03} \quad (3 \text{ cents})$$

7.11.   The relationship between power (P), current (I), and volts (V) will provide a solution. Since the relationship considers

power in watts, the first step is to convert horsepower to watts. One horsepower is equivalent to 746 watts, so:

$$(746 \text{ W/hp})(2.00 \text{ hp}) = 1,492 \text{ W}$$

$$P = IV \therefore I = \dfrac{P}{V}$$

$$= \dfrac{1,492 \dfrac{J}{s}}{12.0 \dfrac{J}{C}}$$

$$= \dfrac{1,492}{12.0} \dfrac{J}{s} \times \dfrac{C}{J}$$

$$= 124.3 \dfrac{C}{s}$$

$$= \boxed{124 \text{ A}}$$

7.12.   (a)   The rate of using energy is joule/s, or the watt. Since 1.00 hp = 746 W,

inside motor: (746 W/hp)(1/3 hp) = 249 W

outside motor: (746 W/hp)(1/3 hp) = 249 W

compressor motor: (746 W/hp)(3.70 hp) = 2,760 W

249 W + 249 W + 2,760 W = $\boxed{3,258 \text{ W}}$

(b) $\dfrac{(3,258 \text{ W})(1.00 \text{ hr})}{1,000 \text{ W/kW}} = 3.26 \text{ kWhr}$

$$(3.26 \text{ kWhr})(\$0.10/\text{kWhr}) = \$0.33 \text{ per hour}$$

(c) $(\$0.33/\text{hr})(12 \text{ hr/day})(30 \text{ day/mo}) = \boxed{\$118.80}$

7.13.   The solution is to find how much current each device draws and then to see if the total current is less or greater than the breaker rating:

Toaster: $I = \dfrac{V}{R} = \dfrac{120 \text{ V}}{15 \text{ V/A}} = 8.0 \text{ A}$

Motor: (0.20 hp)(746 W/hp) = 150 W

$$I = \dfrac{P}{V} = \dfrac{150 \text{ J/s}}{102 \text{ J/C}} = 1.3 \text{ A}$$

Three 100 W bulbs: 3 × 100 W = 300 W

$$I = \dfrac{P}{V} = \dfrac{300 \text{ J/s}}{120 \text{ J/C}} = 2.5 \text{ A}$$

Iron: $I = \dfrac{P}{V} = \dfrac{600 \text{ J/s}}{120 \text{ J/C}} = 5.0 \text{ A}$

The sum of the currents is 8.0A + 1.3A + 2.5A + 5.0A = 16.8A, so the total current is greater than 15.0 amp and the circuit breaker will trip.

7.14. (a) $V_p = 1{,}200$ V

$N_p = 1$ loop

$N_s = 200$ loops

$V_s = ?$

$$\frac{V_p}{N_p} = \frac{V_s}{N_s} \therefore V_s = \frac{V_p N_s}{N_p}$$

$$V_s = \frac{(1{,}200 \text{ V})(200 \text{ loop})}{1 \text{ loop}}$$

$$= \boxed{240{,}000 \text{ V}}$$

(b) $I_p = 40$ A $V_p I_p = V_s I_s \therefore I_s = \dfrac{V_p I_p}{V_s}$

$I_s = ?$ $I_s = \dfrac{1{,}200 \text{ V} \times 40 \text{ A}}{240{,}000 \text{ V}}$

$$= \frac{1{,}200 \times 40}{240{,}000} \frac{\text{V·A}}{\text{V}}$$

$$= \boxed{0.2 \text{ A}}$$

7.15. (a) $V_s = 12$ V $\dfrac{V_p}{N_p} = \dfrac{V_s}{N_s} \therefore \dfrac{N_p}{N_s} = \dfrac{V_p}{V_s}$

$I_s = 0.5$ A

$V_p = 120$ V $\dfrac{N_p}{N_s} = \dfrac{120 \text{ V}}{12 \text{ V}} = \dfrac{10}{1}$

$\dfrac{N_p}{N_s} = ?$ or

$$\boxed{10 \text{ primary to 1 secondary}}$$

(b) $I_p = ?$ $V_p I_p = V_s I_s \therefore I_p = \dfrac{V_s I_s}{V_p}$

$$I_p = \frac{(12 \text{ V})(0.5 \text{ A})}{120 \text{ V}}$$

$$= \frac{12 \times 0.5}{120} \frac{\text{V·A}}{\text{V}}$$

$$= \boxed{0.05 \text{ A}}$$

(c) $P_s = ?$ $P_s = I_s V_s$

$$= (0.5 \text{ A})(12 \text{ V})$$

$$= 0.5 \times 12 \frac{\text{C}}{\text{s}} \times \frac{\text{J}}{\text{C}}$$

$$= 6 \frac{\text{J}}{\text{s}}$$

$$= \boxed{6 \text{ W}}$$

7.16. (a) $V_p = 120$ V $\dfrac{V_p}{N_p} = \dfrac{V_s}{N_s} \therefore V_s = \dfrac{V_p N_s}{N_p}$

$N_p = 50$ loops

$N_s = 150$ loops $V_s = \dfrac{120 \text{ V} \times 150 \text{ loops}}{50 \text{ loops}}$

$I_p = 5.0$ A

$V_s = ?$ $= \dfrac{120 \times 150}{50} \dfrac{\text{V·loops}}{\text{loops}}$

$$= \boxed{360 \text{ V}}$$

(b) $I_s = ?$ $V_p I_p = V_s I_s \therefore I_s = \dfrac{V_p I_p}{V_s}$

$$I_s = \frac{(120 \text{ V})(5.0 \text{ A})}{360 \text{ V}}$$

$$= \frac{120 \times 5.0}{360} \frac{\text{V·A}}{\text{V}}$$

$$= \boxed{1.7 \text{ A}}$$

(c) $P_s = ?$ $P_s = I_s V_s$

$$= \left(1.7 \frac{\text{C}}{\text{s}}\right)\left(360 \frac{\text{J}}{\text{C}}\right)$$

$$= 1.7 \times 360 \frac{\text{C}}{\text{s}} \times \frac{\text{J}}{\text{C}}$$

$$= 612 \frac{\text{J}}{\text{s}}$$

$$= \boxed{600 \text{ W}}$$

Exercises CHAPTER 8

8.1. The relationship between the speed of light in a transparent material (v), the speed of light in a vacuum ($c = 3.00 \times 10^8$ m/s), and the index of refraction (n) is $n = c/v$. According to table 8.1, the index of refraction for water is $n = 1.33$ and for ice is $n = 1.31$.

(a) $c = 3.00 \times 10^8$ m/s

$n = 1.33$

$v = ?$

$$n = \frac{c}{v} \therefore v = \frac{c}{n}$$

$$v = \frac{3.00 \times 10^8 \text{ m/s}}{1.33}$$

$$= \boxed{2.26 \times 10^8 \text{ m/s}}$$

(b) $c = 3.00 \times 10^8$ m/s

$n = 1.31$

$v = ?$

$$v = \frac{3.00 \times 10^8 \text{ m/s}}{1.31}$$

$$= \boxed{2.29 \times 10^8 \text{ m/s}}$$

8.2.
$$d = 1.50 \times 10^8 \text{ km}$$
$$= 1.50 \times 10^{11} \text{ m}$$
$$c = 3.00 \times 10^8 \text{ m/s}$$
$$t = ?$$
$$v = \frac{d}{t} \therefore t = \frac{d}{v}$$
$$t = \frac{1.50 \times 10^{11} \text{ m}}{3.00 \times 10^8 \frac{\text{m}}{\text{s}}}$$
$$= \frac{1.50 \times 10^{11}}{3.00 \times 10^8} \text{ m} \times \frac{\text{s}}{\text{m}}$$
$$= 5.00 \times 10^2 \frac{\text{m} \cdot \text{s}}{\text{m}}$$
$$= \frac{5.00 \times 10^2 \text{ s}}{60.0 \frac{\text{s}}{\text{min}}}$$
$$= \frac{5.00 \times 10^2}{60.0} \text{ s} \times \frac{\text{min}}{\text{s}}$$
$$= \boxed{8.33 \text{ min}}$$

8.3.
$$d = 6.00 \times 10^9 \text{ km}$$
$$= 6.00 \times 10^{12} \text{ m}$$
$$c = 3.00 \times 10^8 \text{ m/s}$$
$$t = ?$$
$$v = \frac{d}{t} \therefore t = \frac{d}{v}$$
$$t = \frac{6.00 \times 10^{12} \text{ m}}{3.00 \times 10^8 \frac{\text{m}}{\text{s}}}$$
$$= \frac{6.00 \times 10^{12}}{3.00 \times 10^8} \text{ m} \times \frac{\text{s}}{\text{m}}$$
$$= 2.00 \times 10^4 \text{ s}$$
$$= \frac{2.00 \times 10^4 \text{ s}}{3,600 \frac{\text{s}}{\text{hr}}}$$
$$= \frac{2.00 \times 10^4}{3.600 \times 10^3} \text{ s} \times \frac{\text{hr}}{\text{s}}$$
$$= \boxed{5.56 \text{ hr}}$$

8.4. From equation 8.1, note that both angles are measured from the normal and that the angle of incidence (θi) equals the angle of reflection (θr), or
$$\theta_i = \theta_r \therefore \boxed{\theta_i = 10°}$$

8.5.
$$v = 2.20 \times 10^8 \text{ m/s}$$
$$c = 3.00 \times 10^8 \text{ m/s}$$
$$n = ?$$
$$n = \frac{c}{v}$$
$$= \frac{3.00 \times 10^8 \frac{\text{m}}{\text{s}}}{2.20 \times 10^8 \frac{\text{m}}{\text{s}}}$$
$$= 1.36$$

According to table 8.1, the substance with an index of refraction of 1.36 is $\boxed{\text{ethyl alcohol.}}$

8.6. **(a)** From equation 8.3:
$$\lambda = 6.00 \times 10^{-7} \text{ m} \qquad c = \lambda f \therefore f = \frac{c}{\lambda}$$
$$c = 3.00 \times 10^8 \text{ m/s}$$
$$f = ?$$
$$f = \frac{3.00 \times 10^8 \frac{\text{m}}{\text{s}}}{6.00 \times 10^{-7} \text{m}}$$
$$= \frac{3.00 \times 10^8}{6.00 \times 10^{-7}} \frac{\text{m}}{\text{s}} \times \frac{1}{\text{m}}$$
$$= 5.00 \times 10^{14} \frac{1}{\text{s}}$$
$$= \boxed{5.00 \times 10^{14} \text{ Hz}}$$

(b) From equation 8.5:
$$f = 5.00 \times 10^{14} \text{ Hz}$$
$$h = 6.63 \times 10^{-34} \text{ J} \cdot \text{s}$$
$$E = ?$$
$$E = hf$$
$$= (6.63 \times 10^{-34} \text{ J} \cdot \text{s})\left(5.00 \times 10^{14} \frac{1}{\text{s}}\right)$$
$$= (6.63 \times 10^{-34})(5.00 \times 10^{14}) \text{J} \cdot \text{s} \times \frac{1}{\text{s}}$$
$$= \boxed{3.32 \times 10^{-19} \text{ J}}$$

8.7. First, you can find the energy of one photon of the peak intensity wavelength (5.60×10^{-7} m) by using equation 8.3 to find the frequency, then equation 8.5 to find the energy:

Step 1: $c = \lambda f \therefore f = \frac{c}{\lambda}$
$$= \frac{3.00 \times 10^8 \frac{\text{m}}{\text{s}}}{5.60 \times 10^{-7} \text{ m}}$$
$$= 5.36 \times 10^{14} \text{ Hz}$$

Step 2: $E = hf$
$$= (6.63 \times 10^{-34} \text{ J} \cdot \text{s})(5.36 \times 10^{14} \text{ Hz})$$
$$= 3.55 \times 10^{-19} \text{J}$$

Step 3: Since one photon carries an energy of 3.55×10^{-19} J and the overall intensity is 1,000.0 W, each square meter must receive an average of

$$\frac{1,000.0 \, \dfrac{J}{s}}{3.55 \times 10^{-19} \, \dfrac{J}{photon}}$$

$$\frac{1.000 \times 10^{3} \, J}{3.55 \times 10^{-19} \, s} \times \frac{photon}{J}$$

$$\boxed{2.82 \times 10^{21} \, \frac{photon}{s}}$$

8.8. **(a)** $f = 4.90 \times 10^{14}$ Hz $\qquad c = \lambda f \therefore \lambda = \dfrac{c}{f}$

$c = 3.00 \times 10^8$ m/s

$\lambda = ?$

$$\lambda = \frac{3.00 \times 10^8 \, \dfrac{m}{s}}{4.90 \times 10^{14} \, \dfrac{1}{s}}$$

$$= \frac{3.00 \times 10^8}{4.90 \times 10^{14}} \, \frac{m}{s} \times \frac{s}{1}$$

$$= \boxed{6.12 \times 10^{-7} \, m}$$

(b) According to table 8.2, this is the frequency and wavelength of orange light.

8.9. $f = 5.00 \times 10^{20}$ Hz

$h = 6.63 \times 10^{-34}$ J·s

$E = ?$

$E = hf$

$$= (6.63 \times 10^{-34} \, J \cdot s)\left(5.00 \times 10^{20} \, \frac{1}{s}\right)$$

$$= (6.63 \times 10^{-34})(5.00 \times 10^{20}) \, J \cdot s \times \frac{1}{s}$$

$$= \boxed{3.32 \times 10^{-13} \, J}$$

8.10. $\lambda = 1.00$ mm

$\quad = 0.001$ m

$f = ?$

$c = 3.00 \times 10^8$ m/s

$h = 6.63 \times 10^{-34}$ J·s

$E = ?$

Step 1: $c = \lambda f \therefore f = \dfrac{v}{\lambda}$

$$f = \frac{3.00 \times 10^8 \, \dfrac{m}{s}}{1.00 \times 10^{-3} \, m}$$

$$= \frac{3.00 \times 10^8}{1.00 \times 10^{-3}} \, \frac{m}{s} \times \frac{1}{m}$$

$$= 3.00 \times 10^{11} \, Hz$$

Step 2: $E = hf$

$$= (6.63 \times 10^{-34} \, J \cdot s)\left(3.00 \times 10^{11} \, \frac{1}{s}\right)$$

$$= (6.63 \times 10^{-34})(3.00 \times 10^{11}) \, J \cdot s \times \frac{1}{s}$$

$$= \boxed{1.99 \times 10^{-22} \, J}$$

Exercises CHAPTER 9

9.1. $m = 1.68 \times 10^{-27}$ kg

$v = 3.22 \times 10^3$ m/s

$h = 6.63 \times 10^{-34}$ J·s

$\lambda = ?$

$$\lambda = \frac{h}{mv}$$

$$= \frac{6.63 \times 10^{-34} \, J \cdot s}{(1.68 \times 10^{-27} \, kg)\left(3.22 \times 10^3 \, \dfrac{m}{s}\right)}$$

$$= \frac{6.63 \times 10^{-34}}{(1.68 \times 10^{-27})(3.22 \times 10^3)} \, \frac{J \cdot s}{kg \times \dfrac{m}{s}}$$

$$= \frac{6.63 \times 10^{-34}}{5.41 \times 10^{-24}} \, \frac{\dfrac{kg \cdot m^2}{s \cdot s} \times s}{kg \times \dfrac{m}{s}}$$

$$= 1.23 \times 10^{-10} \, \frac{kg \cdot m \cdot m}{s} \times \frac{1}{kg} \times \frac{s}{m}$$

$$= \boxed{1.23 \times 10^{-10} \, m}$$

9.2. **(a)** $\qquad n = 6$

$E_1 = -13.6$ eV

$E_6 = ?$

$$E_n = \frac{E_1}{n^2}$$

$$E_6 = \frac{-13.6 \, eV}{6^2}$$

$$= \frac{-13.6 \, eV}{36}$$

$$= \boxed{-0.378 \, eV}$$

(b) $\qquad = (-0.378 \, eV)\left(1.60 \times 10^{-19} \, \dfrac{J}{eV}\right)$

$$= (-0.378)(1.60 \times 10^{-19}) \, eV \times \frac{J}{eV}$$

$$= \boxed{-6.05 \times 10^{-20} \, J}$$

9.3. **(a)** Energy is related to the frequency and Planck's constant in equation 9.1, $E = hf$. From equation 9.6,

$$hf = E_H - E_L \therefore E = E_H - E_L$$

For $n = 6$, $E_H = 6.05 \times 10^{-20}$ J

For $n = 2$, $E_L = 5.44 \times 10^{-19}$ J

$$E = ? \text{ J}$$

$$E = E_H - E_L$$

$$= (-6.05 \times 10^{-20} \text{ J}) - (-5.44 \times 10^{-19} \text{ J})$$

$$= \boxed{4.84 \times 10^{-19} \text{ J}}$$

(b) $E_H = -0.377$ eV*

$E_L = -3.40$ eV*

$E = ?$ eV

$E = E_H - E_L$

$$= (-0.377 \text{ eV}) - (-3.40 \text{ eV})$$

$$= \boxed{3.02 \text{ eV}}$$

*From figure 9.14

9.4. For $n = 6$, $E_H = -6.05 \times 10^{-20}$ J

For $n = 2$, $E_L = -5.44 \times 10^{-19}$ J

$$h = 6.63 \times 10^{-34} \text{ J·s}$$

$$f = ?$$

$$hf = E_H - E_L \therefore f = \frac{E_H - E_L}{h}$$

$$f = \frac{(-6.05 \times 10^{-20} \text{ J}) - (-5.44 \times 10^{-19} \text{ J})}{6.63 \times 10^{-34} \text{ J·s}}$$

$$= \frac{4.84 \times 10^{-19}}{6.63 \times 10^{-34}} \frac{\cancel{J}}{\cancel{J}\text{·s}}$$

$$= 7.29 \times 10^{14} \frac{1}{s}$$

$$= \boxed{7.29 \times 10^{14} \text{ Hz}}$$

9.5. $(n = 1) = -13.6$ eV $E_n = \frac{E_1}{n^2}$

$E = ?$

$$= \frac{-13.6 \text{ eV}}{1^2}$$

$$= -13.6 \text{ eV}$$

Since the energy of the electron is -13.6 eV, it will require 13.6 eV (or 2.17×10^{-18} J) to remove the electron.

9.6. $q/m = -1.76 \times 10^{11}$ C/kg

$q = -1.60 \times 10^{-19}$ C

$m = ?$

$$\text{mass} = \frac{\text{charge}}{\text{charge/mass}}$$

$$= \frac{-1.60 \times 10^{-19} \text{ C}}{-1.76 \times 10^{11} \frac{\text{C}}{\text{kg}}}$$

$$= \frac{-1.60 \times 10^{-19}}{-1.76 \times 10^{11}} \cancel{C} \times \frac{\text{kg}}{\cancel{C}}$$

$$= \boxed{9.09 \times 10^{-31} \text{ kg}}$$

9.7. $\lambda = -1.67 \times 10^{-10}$ m

$m = 9.11 \times 10^{-31}$ kg

$v = ?$

$$\lambda = \frac{h}{mv} \therefore v = \frac{h}{m\lambda}$$

$$v = \frac{6.63 \times 10^{-34} \text{ J·s}}{(9.11 \times 10^{-31} \text{ kg})(1.67 \times 10^{-10} \text{ m})}$$

$$= \frac{6.63 \times 10^{-34}}{(9.11 \times 10^{-31})(1.67 \times 10^{-10})} \frac{\text{J·s}}{\text{kg·m}}$$

$$= \frac{6.63 \times 10^{-34} \frac{\text{kg·m}^2}{s\cancel{·s}} \times \cancel{s}}{1.52 \times 10^{-40} \text{ kg·m}}$$

$$= 4.36 \times 10^6 \frac{\text{kg·m·}\cancel{m}}{s} \times \frac{1}{\cancel{kg}} \times \frac{1}{\cancel{m}}$$

$$= \boxed{4.36 \times 10^6 \frac{\text{m}}{s}}$$

9.8. **(a)** Boron: $1s^2 2s^2 2p^1$

(b) Aluminum: $1s^2 2s^2 2p^6 3s^2 3p^1$

(c) Potassium: $1s^2 2s^2 2p^6 3s^2 3p^6 4s^1$

9.9. **(a)** Boron is atomic number 5 and there are 5 electrons.

(b) Aluminum is atomic number 13 and there are 13 electrons.

(c) Potassium is atomic number 19 and there are 19 electrons.

9.10. **(a)** Argon: $1s^2 2s^2 2p^6 3s^2 3p^6$

(b) Zinc: $1s^2 2s^2 2p^6 3s^2 3p^6 4s^2 3d^{10}$

(c) Bromine: $1s^2 2s^2 2p^6 3s^2 3p^6 4s^2 3d^{10} 4p^5$

Exercises CHAPTER 10

10.1. The answers to this question are found in the list of elements and their symbols, which is located on the inside back cover of this text.

(a) Silicon: Si

(b) Silver: Ag

(c) Helium: He

(d) Potassium: K

(e) Magnesium: Mg

(f) Iron: Fe

10.2. Atomic weight is the weighted average of the isotopes as they occur in nature. Thus

Lithium-6: 6.01512 u × 0.0742 = 0.446 u

Lithium-7: 7.016 u × 0.9258 = 6.4954 u

Lithium-6 contributes 0.446 u of the weighted average and lithium-7 contributes 6.4954 u. The atomic weight of lithium is therefore

$$
\begin{array}{rl}
0.446 & u \\
+\ 6.4954 & u \\
\hline
6.941 & u
\end{array}
$$

10.3. Recall that the subscript is the atomic number, which identifies the number of protons. In a neutral atom, the number of protons equals the number of electrons, so the atomic number tells you the number of electrons, too. The superscript is the mass number, which identifies the number of neutrons and the number of protons in the nucleus. The number of neutrons is therefore the mass number minus the atomic number.

	Protons	Neutrons	Electrons
(a)	6	6	6
(b)	1	0	1
(c)	18	22	18
(d)	1	1	1
(e)	79	118	79
(f)	92	143	92

10.4.

		Period	Family
(a)	Radon (Rn)	6	VIIIA
(b)	Sodium (Na)	3	IA
(c)	Copper (Cu)	4	IB
(d)	Neon (Ne)	2	VIIIA
(e)	Iodine (I)	5	VIIA
(f)	Lead (Pb)	6	IVA

10.5. Recall that the number of outer-shell electrons is the same as the family number for the representative elements:

(a) Li: 1   (d) Cl: 7

(b) N: 5   (e) Ra: 2

(c) F: 7   (f) Be: 2

10.6. The same information that was used in question 5 can be used to draw the dot notation:

(a) $\overset{\bullet}{\underset{\bullet}{B}}\bullet$ (c) Ca: (e) $\cdot\overset{\bullet}{\underset{\bullet}{O}}\cdot$

(b) $\cdot\overset{\bullet}{\underset{\bullet\bullet}{Br}}:$ (d) K·   (f) $\overset{\bullet\bullet}{\underset{\bullet}{S}}\cdot$

10.7. The charge is found by identifying how many electrons are lost or gained in achieving the noble gas structure:

(a) Boron 3+

(b) Bromine 1−

(c) Calcium 2+

(d) Potassium 1+

(e) Oxygen 2−

(f) Nitrogen 3−

10.8. Metals have one, two, or three outer electrons and are located in the left two-thirds of the periodic table. Semiconductors are adjacent to the line that separates the metals and nonmetals. Look at the periodic table on the inside back cover and you will see:

(a) Krypton—nonmetal

(b) Cesium—metal

(c) Silicon—semiconductor

(d) Sulfur—nonmetal

(e) Molybdenum—metal

(f) Plutonium—metal

10.9. (a) Bromine gained an electron to acquire a 1− charge, so it must be in family VIIA (the members of this family have seven electrons and need one more to acquire the noble gas structure).

(b) Potassium must have lost one electron, so it is in IA.

(c) Aluminum lost three electrons, so it is in IIIA.

(d) Sulfur gained two electrons, so it is in VIA.

(e) Barium lost two electrons, so it is in IIA.

(f) Oxygen gained two electrons, so it is in VIA.

10.10. (a) $^{16}_{8}O$ (c) $^{3}_{1}H$

       (b) $^{23}_{11}Na$ (d) $^{35}_{17}Cl$

Exercises CHAPTER 11

11.1.

11.2. **(a)** Sulfur is in family VIA, so sulfur has six valence electrons and will need two more to achieve a stable outer structure like the noble gases. Two more outer shell electrons will give the sulfur atom a charge of $2-$. Copper^{2+} will balance the $2-$ charge of sulfur, so the name is copper(II) sulfide. Note the "-ide" ending for compounds that have only two different elements.

(b) Oxygen is in family VIA, so oxygen has six valence electrons and will have a charge of $2-$. Using the crossover technique in reverse, you can see that the charge on the oxygen is $2-$, and the charge on the iron is $3-$. Therefore the name is iron(III) oxide.

(c) From information in (a) and (b), you know that oxygen has a charge of $2-$. The chromium ion must have the same charge to make a neutral compound as it must be, so the name is chromium(II) oxide. Again, note the "-ide" ending for a compound with two different elements.

(d) Sulfur has a charge of $2-$, so the lead ion must have the same positive charge to make a neutral compound. The name is lead(II) sulfide.

11.3. The name of some common polyatomic ions are in table 11.4. Using this table as a reference, the names are

(a) hydroxide

(b) sulfite

(c) hypochlorite

(d) nitrate

(e) carbonate

(f) perchlorate

11.4. The Roman numeral tells you the charge on the variable-charge elements. The charges for the polyatomic ions are found in table 11.4. The charges for metallic elements can be found in tables 11.1 and 11.2. Using these resources and the crossover technique, the formulas are as follows:

(a) $Fe(OH)_3$

(b) $Pb_3(PO_4)_2$

(c) $ZnCO_3$

(d) NH_4NO_3

(e) $KHCO_3$

(f) K_2SO_3

11.5. Table 11.7 has information about the meaning of prefixes and stem names used in naming covalent compounds. (a), for example, asks for the formula of carbon tetrachloride. Carbon has no prefixes, so there is one carbon atom, and it comes first in the formula because it comes first in the name. The "tetra-" prefix means four, so there are four chlorine atoms. The name ends in "-ide," so you know there are only two elements in the compound. The symbols can be obtained from the list of elements on the inside back cover of this text. Using all this information from the name, you can think out the formula for carbon tetrachloride. The same process is used for the other compounds and formulas:

(a) CCl_4

(b) H_2O

(c) MnO_2

(d) SO_3

(e) N_2O_5

(f) As_2S_5

11.6. Again using information from table 11.7, this question requires you to reverse the thinking procedure you learned in question 5.

(a) carbon monoxide

(b) carbon dioxide

(c) carbon disulfide

(d) dinitrogen monoxide

(e) tetraphosphorus trisulfide

(f) dinitrogen trioxide

11.7. The types of bonds formed are predicted by using the electronegativity scale in table 11.5 and finding the absolute difference. On this basis:

(a) Difference $= 1.7$, which means ionic bond

(b) Difference $= 0$, which means nonpolar covalent

(c) Difference $= 0$, which means nonpolar covalent

(d) Difference $= 0.4$, which means nonpolar covalent

(e) Difference $= 3.0$, which means ionic

(f) Difference $= 1.6$, which means polar covalent and almost ionic

Exercises CHAPTER 12

12.1. (a) $MgCl_2$ is an ionic compound, so the formula has to be empirical.

(b) C_2H_2 is a covalent compound, so the formula might be molecular. Since it is not the simplest whole number ratio (which would be CH), then the formula is molecular.

(c) BaF_2 is ionic; the formula is empirical.

(d) C_8H_{18} is not the simplest whole number ratio of a covalent compound, so the formula is molecular.

(e) CH_4 is covalent, but the formula might or might not be molecular (?).

(f) S_8 is a nonmetal bonded to a nonmetal (itself); this is a molecular formula.

12.2. (a) $CuSO_4$

$$1 \text{ of Cu} = 1 \times 63.5 \text{ u} = 63.5 \text{ u}$$
$$1 \text{ of S} = 1 \times 32.1 \text{ u} = 32.1 \text{ u}$$
$$4 \text{ of O} = 4 \times 16.0 \text{ u} = \underline{64.0 \text{ u}}$$
$$159.6 \text{ u}$$

(b) CS_2

$$1 \text{ of C} = 1 \times 12.0 \text{ u} = 12.0 \text{ u}$$
$$2 \text{ of S} = 2 \times 32.0 \text{ u} = \underline{64.0 \text{ u}}$$
$$76.0 \text{ u}$$

(c) $CaSO_4$

$$1 \text{ of Ca} = 1 \times 40.1 \text{ u} = 40.1 \text{ u}$$
$$1 \text{ of S} = 1 \times 32.0 \text{ u} = 32.0 \text{ u}$$
$$4 \text{ of O} = 4 \times 16.0 \text{ u} = \underline{64.0 \text{ u}}$$
$$136.1 \text{ u}$$

(d) Na_2CO_3

$$2 \text{ of Na} = 2 \times 23.0 \text{ u} = 46.0 \text{ u}$$
$$1 \text{ of C} = 1 \times 12.0 \text{ u} = 12.0 \text{ u}$$
$$3 \text{ of O} = 3 \times 16.0 \text{ u} = \underline{48.0 \text{ u}}$$
$$106.0 \text{ u}$$

12.3. (a) FeS_2

For Fe: $\dfrac{(55.9 \text{ u Fe})(1)}{119.9 \text{ u FeS}_2} \times 100\% \text{ FeS}_2 = 46.6\% \text{ Fe}$

For S: $\dfrac{(32.0 \text{ u S})(2)}{119.9 \text{ u FeS}_2} \times 100\% \text{ FeS}_2 = 53.4\% \text{ S}$

or $(100\% \text{ FeS}_2) - (46.6\% \text{ Fe}) = 53.4\% \text{ S}$

(b) H_3BO_3

For H: $\dfrac{(1.0 \text{ u H})(3)}{61.8 \text{ u H}_3BO_3} \times 100\% \text{ H}_3BO_3 = 4.85\% \text{ H}$

For B: $\dfrac{(10.8 \text{ u B})(1)}{61.8 \text{ u H}_3BO_3} \times 100\% \text{ H}_3BO_3 = 17.5\% \text{ B}$

For O: $\dfrac{(16 \text{ u O})(3)}{61.8 \text{ u H}_3BO_3} \times 100\% \text{ H}_3BO_3 = 77.7\% \text{ O}$

(c) $NaHCO_3$

For Na: $\dfrac{(23.0 \text{ u Na})(1)}{84.0 \text{ u NaHCO}_3} \times 100\% \text{ NaHCO}_3 = 27.4\% \text{ Na}$

For H: $\dfrac{(1.0 \text{ u H})(1)}{84.0 \text{ u NaHCO}_3} \times 100\% \text{ NaHCO}_3 = 1.2\% \text{ H}$

For C: $\dfrac{(12.0 \text{ u C})(1)}{84.0 \text{ u NaHCO}_3} \times 100\% \text{ NaHCO}_3 = 14.3\% \text{ C}$

For O: $\dfrac{(16.0 \text{ u O})(3)}{84.0 \text{ u NaHCO}_3} \times 100\% \text{ NaHCO}_3 = 57.1\% \text{ O}$

(d) $C_9H_8O_4$

For C: $\dfrac{(12.0 \text{ u C})(9)}{180.0 \text{ u C}_9H_8O_4} \times 100\% \text{ C}_9H_8O_4 = 60.0\% \text{ C}$

For H: $\dfrac{(1.0 \text{ u H})(8)}{180.0 \text{ u C}_9H_8O_4} \times 100\% \text{ C}_9H_8O_4 = 4.4\% \text{ H}$

For O: $\dfrac{(16.0 \text{ u O})(4)}{180.0 \text{ u C}_9H_8O_4} \times 100\% \text{ C}_9H_8O_4 = 35.6\% \text{ O}$

12.4. (a) $2 SO_2 + O_2 \rightarrow 2 SO_3$

(b) $4 P + 5 O_2 \rightarrow 2 P_2O_5$

(c) $2 Al + 6 HCl \rightarrow 2 AlCl_3 + 3 H_2$

(d) $2 NaOH + H_2SO_4 \rightarrow Na_2SO_4 + 2 H_2O$

(e) $Fe_2O_3 + 3 CO \rightarrow 2 Fe + 3 CO_2$

(f) $3 Mg(OH)_2 + 2 H_3PO_4 \rightarrow Mg_3(PO_4)_2 + 6 H_2O$

12.5. (a) General form of $XY + AZ \rightarrow XZ + AY$ with precipitate formed: Ion exchange reaction.

(b) General form of $X + Y \rightarrow XY$: Combination reaction.

(c) General form of $XY \rightarrow X + Y + \ldots$: Decomposition reaction.

(d) General form of $X + Y \rightarrow XY$: Combination reaction.

(e) General form of $XY + A \rightarrow AY + X$: Replacement reaction.

(f) General form of $X + Y \rightarrow XY$: Combination reaction.

12.6. (a) $C_5H_{12}(g) + 8 O_2(g) \rightarrow 5 CO_2(g) + 6 H_2O(g)$

(b) $HCl(aq) + NaOH(aq) \rightarrow NaCl(aq) + H_2O(l)$

(c) $2 Al(s) + Fe_2O_3(s) \rightarrow Al_2O_3(s) + 2 Fe(l)$

(d) $Fe(s) + CuSO_4(aq) \rightarrow FeSO_4(aq) + Cu(s)$

(e) $MgCl(aq) + Fe(NO_3)_2(aq) \rightarrow$ No reaction (all possible compounds are soluble and no gas or water was formed).

(f) $C_6H_{10}O_5(s) + 6 O_2(g) \rightarrow 6 CO_2(g) + 5 H_2O(g)$

12.7. (a) $2 KClO_3 \xrightarrow{\Delta} 2 KCl(S) + 3 O_2 \uparrow$

(b) $2 Al_2O_3(l) \xrightarrow{elec} 4 Al(s) + 3 O_2 \uparrow$

(c) $CaCO_3(s) \xrightarrow{\Delta} CaO(s) + CO_2 \uparrow$

12.8. (a) $2 Na(s) + 2 H_2O(l) \rightarrow 2 NaOH(aq) + H_2 \uparrow$

(b) $Au(s) + HCl(aq) \rightarrow$ No reaction (gold is below hydrogen in the activity series).

(c) $Al(s) + FeCl_3(aq) \rightarrow AlCl_3(aq) + Fe(s)$

(d) $Zn(s) + CuCl_2(aq) \rightarrow ZnCl_2(aq) + Cu(s)$

12.9. (a) $NaOH(aq) + HNO_3(aq) \rightarrow NaNO_3(aq) + H_2O(l)$

(b) $CaCl_2(aq) + KNO_3(aq) \rightarrow$ No reaction

(c) $3 Ba(NO_3)_2(aq) + 2 Na_3PO_4(aq) \rightarrow 6 NaNO_3(aq) + Ba_3(PO_4)_2\downarrow$

(d) $2 KOH(aq) + ZnSO_4(aq) \rightarrow K_2SO_4(aq) + Zn(OH)_2\downarrow$

12.10. One mole of oxygen combines with 2 moles of acetylene, so 0.5 mole of oxygen would be needed for 1 mole of acetylene. Therefore, 1 L of C_2H_2 requires 0.5 L of O_2.

Exercises CHAPTER 13

13.1. $m_{solute} = 1.75\ g$

$m_{solution} = 50.0\ g$

% weight = ?

$$\%\ solute = \frac{m_{solute}}{m_{solution}} \times 100\%\ solution$$

$$= \frac{1.75\ g\ NaCl}{50.0\ g\ solution} \times 100\%\ solution$$

$$= \boxed{3.50\%\ NaCl}$$

13.2. $m_{solution} = 103.5\ g$

$m_{solute} = 3.50\ g$

% weight = ?

$$\%\ solute = \frac{m_{solute}}{m_{solution}} \times 100\%\ solution$$

$$= \frac{3.50\ g\ NaCl}{103.5\ g\ solution} \times 100\%\ solution$$

$$= \boxed{3.38\%\ NaCl}$$

13.3. Since ppm is defined as the weight unit of solute in 1,000,000 weight units of solution, the percent by weight can be calculated just like any other percent. The weight of the dissolved sodium and chlorine ions is the part, and the weight of the solution is the whole, so

$$\% = \frac{part}{whole} \times 100\%$$

$$= \frac{30{,}113\ g\ NaCl\ ions}{1{,}000{,}000\ g\ seawater} \times 100\%\ seawater$$

$$= \boxed{3.00\%\ NaCl\ ions}$$

13.4. $m_{solution} = 250\ g$

% solute = 3.0%

$m_{solute} = ?$

$$\%\ solute = \frac{m_{solute}}{m_{solution}} \times 100\%\ solution$$

$$\therefore$$

$$m_{solute} = \frac{(m_{solution})(\%\ solute)}{100\%\ solution}$$

$$= \frac{(250\ g)(3.0\%)}{100\%}$$

$$= \boxed{7.5\ g}$$

13.5. % solution = 12% solution

$V_{solution} = 200\ mL$

$V_{solute} = ?$

$$\%\ solution = \frac{V_{solute}}{V_{solution}} \times 100\%\ solution$$

$$\therefore$$

$$V_{solute} = \frac{(\%\ solution)(V_{solution})}{100\%\ solution}$$

$$= \frac{(12\%\ solution)(200\ mL)}{100\%\ solution}$$

$$= \boxed{24\ mL\ alcohol}$$

13.6. % solution = 40%

$V_{solution} = 50\ mL$

$V_{solute} = ?$

$$\%\ solution = \frac{V_{solute}}{V_{solution}} \times 100\%\ solution$$

$$\therefore$$

$$V_{solute} = \frac{(\%\ solution)(V_{solution})}{100\%\ solution}$$

$$= \frac{(40\%\ solution)(50\ mL)}{100\%\ solution}$$

$$= \boxed{20\ mL\ alcohol}$$

13.7. **(a)** $\%\ concentration = \dfrac{ppm}{1 \times 10^4}$

$$= \frac{5}{1 \times 10^4}$$

$$= \boxed{0.0005\%\ DDT}$$

(b) $\%\ part = \dfrac{part}{whole} \times 100\%\ whole$

$$\therefore$$

$$whole = \frac{(100\%)(part)}{\%\ part}$$

$$= \frac{(100\%)(17.0\ g)}{0.0005\%} = \boxed{3{,}400{,}000\ g\ or\ 3{,}400\ kg}$$

13.8.

(a) $\overset{acid}{(\mathrm{HC_2H_3O_2(aq)})}$ + $\overset{base}{\boxed{\mathrm{H_2O(l)}}}$ ⇌ $H_3O^+(aq) + C_2H_3O_2^-(aq)$

(b) $\underset{base}{\boxed{\mathrm{C_6H_6NH_2(l)}}}$ + $\underset{acid}{(\mathrm{H_2O(l)})}$ ⇌ $C_6H_6NH_3^+(aq) + OH^-(aq)$

(c) $\underset{acid}{(\mathrm{HClO_4(aq)})}$ + $\underset{base}{\boxed{\mathrm{HC_2H_3O_2(aq)}}}$ ⇌ $H_2C_2H_3O_2^+(aq)$
$+ ClO_4^-(aq)$

(d) $\underset{base}{\boxed{\mathrm{H_2O(l)}}}$ + $\underset{acid}{(\mathrm{H_2O(l)})}$ ⇌ $H_3O^+(aq) + OH^-(aq)$

Exercises **CHAPTER 14**

14.1.

(a)

```
    H   H   H   H   H
    |   |   |   |   |
H — C — C — C — C — C — H
    |   |   |   |   |
    H   H   H   H   H
```

(b)

```
        H
        |
    H — C — H
        |
    H       H
    |       |
H — C — C — C — H
    |   |   |
    H   |   H
        |
    H — C — H
        |
        H
```

(c) 2,2-dimethylpropane

14.2. *n*-hexane

```
    H   H   H   H   H   H
    |   |   |   |   |   |
H — C — C — C — C — C — C — H
    |   |   |   |   |   |
    H   H   H   H   H   H
```

3-methylpentane

```
    H   H   H   H   H
    |   |   |   |   |
H — C — C — C — C — C — H
    |   |   |   |   |
    H   H   |   H   H
            |
        H — C — H
            |
            H
```

2-methylpentane

```
    H   H   H   H   H
    |   |   |   |   |
H — C — C — C — C — C — H
    |   |   |   |   |
    H   |   H   H   H
        |
    H — C — H
        |
        H
```

2,2-dimethylbutane

```
    H   CH₃  H   H
    |   |    |   |
H — C — C — C — C — H
    |   |    |   |
    H   CH₃  H   H
```

14.3.

```
    H   H   CH3 CH3  H   H   H   H
    |   |   |   |    |   |   |   |
H — C — C — C — C — C — C — C — C — H
    |   |   |   |    |   |   |   |
    H   H   CH3  H    H   H   H   H
```

```
    H   CH3  H   H   H
    |   |    |   |   |
H — C = C — C — C — C — H
                 |   |   |
                 H   H   H
```

```
    H   H           CH3  H   H
    |   |           |    |   |
H — C — C — C ≡ C — C — C — C — H
    |   |                |    |   |
    H   H           CH3  H   H
```

14.4. (a) 2-chloro-4-methylpentane

(b) 2-methyl-l-pentene

(c) 3-ethyl-4-methyl-2-pentene

14.5. The 2,2,3-trimethylbutane is more highly branched, so it will have the higher octane rating.

2,2,3-trimethylbutane

```
    H   CH₃ CH₃  H
    |   |   |    |
H — C — C — C — C — H
    |   |   |    |
    H   CH₃  H    H
```

2,2-dimethylpentane

```
    H   CH₃  H   H   H
    |   |    |   |   |
H — C — C — C — C — C — H
    |   |    |   |   |
    H   CH₃  H   H   H
```

14.6. (a) alcohol
(b) amide
(c) ether
(d) ester
(e) organic acid

Exercises CHAPTER 15

15.1. (a) cobalt-60: 27 protons, 33 neutrons

(b) potassium-40: 19 protons, 21 neutrons

(c) neon-24: 10 protons, 14 neutrons

(d) lead-208: 82 protons, 126 neutrons

15.2. (a) $^{60}_{27}Co$ (c) $^{24}_{10}Ne$

(b) $^{40}_{19}K$ (d) $^{204}_{82}Pb$

15.3. (a) cobalt-60: Radioactive because odd numbers of protons (27) and odd numbers of neutrons (33) are usually unstable.

(b) potassium-40: Radioactive, again having an odd number of protons (19) and an odd number of neutrons (21).

(c) neon-24: Stable, because even numbers of protons and neutrons are usually stable.

(d) lead-208: Stable, because even numbers of protons and neutrons *and* because 82 is a particularly stable number of nucleons.

15.4. (a) $^{56}_{26}Fe \rightarrow {}^{0}_{-1}e + {}^{56}_{27}Co$

(b) $^{7}_{4}Be \rightarrow {}^{0}_{-1}e + {}^{7}_{5}B$

(c) $^{64}_{29}Cu \rightarrow {}^{0}_{-1}e + {}^{64}_{30}Zn$

(d) $^{24}_{11}Na \rightarrow {}^{0}_{-1}e + {}^{24}_{12}Mg$

(e) $^{214}_{82}Pb \rightarrow {}^{0}_{-1}e + {}^{214}_{83}Bi$

(f) $^{32}_{15}P \rightarrow {}^{0}_{-1}e + {}^{32}_{16}S$

15.5. (a) $^{235}_{92}U \rightarrow {}^{4}_{2}He + {}^{231}_{90}Th$

(b) $^{226}_{88}Ra \rightarrow {}^{4}_{2}He + {}^{222}_{86}Rn$

(c) $^{239}_{94}Pu \rightarrow {}^{4}_{2}He + {}^{235}_{92}U$

(d) $^{214}_{83}Bi \rightarrow {}^{4}_{2}He + {}^{210}_{81}Tl$

(e) $^{230}_{90}Th \rightarrow {}^{4}_{2}He + {}^{226}_{88}Ra$

(f) $^{210}_{84}Po \rightarrow {}^{4}_{2}He + {}^{206}_{82}Pb$

15.6. Thirty-two days is four half-lives. After the first half-life (8 days), 1/2 oz will remain. After the second half-life (8 + 8, or 16 days), 1/4 oz will remain. After the third half-life (8 + 8 + 8, or 24 days), 1/8 oz will remain. After the fourth half-life (8 + 8 + 8 + 8, or 32 days), 1/16 oz will remain, or 6.3×10^{-2} oz.

15.7. The relationship between half-life and the radioactive decay constant is found in equation 15.2, which is first solved for the decay constant:

$$t_{1/2} = \frac{0.693}{k} \therefore k = \frac{0.693}{t_{1/2}}$$

A decay constant calculation requires units of seconds, so 27.6 years is next converted to seconds:

$$(27.6 \text{ yr})(3.15 \times 10^7 \text{ s/yr}) = 8.69 \times 10^8 \text{ s}$$

This value in seconds is now used in equation 15.2 solved for *k*:

$$k = \frac{0.693}{8.69 \times 10^8 \text{ s}} = 7.97 \times 10^{-10}/\text{s}$$

15.8. The relationship between the radioactive decay constant, the decay rate, and the number of nuclei is found in equation 15.1, which can be solved for the rate:

$$k = \frac{\text{rate}}{n} \therefore \text{rate} = kn$$

The number of nuclei in a molar mass of strontium-90 (87.6 g) is Avogadro's number, 6.02×10^{23}, so

$$\text{rate} = (7.97 \times 10^{-10} \text{ s})(6.02 \times 10^{23} \text{ nuclei})$$

$$= 4.80 \times 10^{14} \text{ nuclei/s}$$

15.9. A curie (Ci) is defined as 3.70×10^{10} disintegrations/s, so a molar mass that is decaying at a rate of 4.80×10^{14} nuclei/s has a curie activity of

$$\frac{4.80 \times 10^{14}}{3.70 \times 10^{10}} = 1.30 \times 10^4 \text{ Ci}$$

15.10. The Fe-56 nucleus has a mass of 55.9206 u, but the individual masses of the nucleons are

$$26 \text{ protons} \times 1.00728 \text{ u} = 26.1893 \text{ u}$$

$$30 \text{ neutrons} \times 1.00867 \text{ u} = \underline{30.2601 \text{ u}}$$
$$56.4494 \text{ u}$$

The mass defect is thus

$$56.4494 \text{ u}$$
$$\underline{-55.9206 \text{ u}}$$
$$0.5288 \text{ u}$$

The atomic mass unit (u) is equal to the mass of a mole (g), therefore 0.5288 u = 0.5288 g. The mass defect is equivalent to the binding energy according to $E = mc^2$. For a molar mass of Fe-56, the mass defect is

$$E = (5.29 \times 10^{-4} \text{ kg})\left(3.00 \times 10^8 \frac{\text{m}}{\text{s}}\right)^2$$

$$= (5.29 \times 10^{-4} \text{ kg})\left(9.00 \times 10^{16} \frac{\text{m}^2}{\text{s}^2}\right)$$

$$= 4.76 \times 10^{13} \frac{\text{kg} \cdot \text{m}^2}{\text{s}^2}$$

$$= 4.76 \times 10^{13} \text{ J}$$

For a single nucleus,

$$\frac{4.76 \times 10^{13} \text{ J}}{6.02 \times 10^{23} \text{ nuclei}} = 7.90 \times 10^{-11} \text{ J / nuclei}$$

GLOSSARY

A

abiotic factors   nonliving parts of an organism's environment

absolute humidity   a measure of the actual amount of water vapor in the air at a given time—for example, in grams per cubic meter

absolute magnitude   a classification scheme to compensate for the distance differences to stars; calculations of the brightness that stars would appear to have if they were all at a defined, standard distance of 10 parsecs

absolute temperature scale   temperature scale set so that zero is at the theoretical lowest temperature possible, which would occur when all random motion of molecules has ceased

absolute zero   the theoretical lowest temperature possible, which occurs when all random motion of molecules has ceased

abyssal plain   the practically level plain of the ocean floor

acceleration   a change in velocity per change in time; by definition, this change in velocity can result from a change in speed, a change in direction, or a combination of changes in speed and direction

accretion disk   fat bulging disk of gas and dust from the remains of the gas cloud that forms around a protostar

acetylcholine   a neurotransmitter secreted into the synapse by many axons and received by dendrites

acetylcholinesterase   an enzyme present in the synapse that destroys acetylcholine

achondrites   homogeneously textured stony meteorites

acid   any substance that is a proton donor when dissolved in water; generally considered a solution of hydronium ions in water that can neutralize a base, forming a salt and water

acid-base indicator   a vegetable dye used to distinguish acid and base solutions by a color change

adenine   a double-ring nitrogenous-base molecule in DNA and RNA; the complementary base of thymine or uracil

air mass   a large, more or less uniform body of air with nearly the same temperature and moisture conditions throughout

air mass weather   the weather experienced within a given air mass; characterized by slow, gradual changes from day to day

alcohol   an organic compound with a general formula of ROH, where R is one of the hydrocarbon groups—for example, methyl or ethyl

aldehyde   an organic molecule with the general formula RCHO, where R is one of the hydrocarbon groups—for example, methyl or ethyl

alkali metals   members of family IA of the periodic table, having common properties of shiny, low-density metals that can be cut with a knife and that react violently with water to form an alkaline solution

alkaline earth metals   members of family IIA of the periodic table, having common properties of soft, reactive metals that are less reactive than alkali metals

alkanes   hydrocarbons with single covalent bonds between the carbon atoms

alkenes   hydrocarbons with a double covalent carbon-carbon bond

alkyne   hydrocarbon with a carbon-carbon triple bond

alleles   alternative forms of a gene for a particular characteristic (e.g., attached-earlobe and free-earlobe are alternative alleles for ear shape)

allotropic forms   elements that can have several different arrangements of atoms with different physical properties—for example, graphite and diamond are two allotropic forms of carbon

alpha particle   the nucleus of a helium atom (two protons and two neutrons) emitted as radiation from a decaying heavy nucleus; also known as an alpha ray

alpine glaciers   glaciers that form at high elevations in mountainous regions

alternating current   an electric current that first moves one direction, then the opposite direction with a regular frequency

alternation of generations   a term used to describe that aspect of the life cycle in which there are two distinctly different forms of an organism; each form is involved in the production of the other and only one form is involved in producing gametes

alveoli   tiny sacs that are part of the structure of the lungs where gas exchange takes place

amino acids   organic molecules that join to form polypeptides and proteins

amp   unit of electric current; equivalent to C/s

ampere   full name of the unit amp

amplitude   the extent of displacement from the equilibrium condition; the size of a wave from the rest (equilibrium) position

anaphase   the third stage of mitosis, characterized by dividing of the centromeres and movement of the chromosomes to the poles

androgens   male sex hormones produced by the testes that cause the differentiation of typical internal and external genital male anatomy

angle of incidence   angle of an incident (arriving) ray or particle to a surface; measured from a line perpendicular to the surface (the normal)

angle of reflection   angle of a reflected ray or particle from a surface; measured from a line perpendicular to the surface (the normal)

angular momentum quantum number   in the quantum mechanics model of the atom, one of four descriptions of the energy state of an electron wave; this quantum number describes the energy sublevels of electrons within the main energy levels of an atom

angular unconformity   a boundary in rock where the bedding planes above and below the time interruption unconformity are not parallel, meaning probable tilting or folding followed by a significant period of erosion, which in turn was followed by a period of deposition

annular eclipse   occurs when the penumbra reaches the surface of the earth; as seen from the earth, the sun forms a bright ring around the disk of the new moon

anorexia nervosa   a nutritional deficiency disease characterized by severe, prolonged weight loss for fear of becoming obese—thought to stem from sociocultural factors

Antarctic Circle   parallel identifying the limit toward the equator where the sun appears above the horizon all day for six months during the summer; located at 66.5°S latitude

anther   the sex organ in plants that produces the pollen that contains the sperm

anticline   an arch-shaped fold in layered bed rock

anticodon   a sequence of three nitrogenous bases on a tRNA molecule capable of forming hydrogen bonds with three complementary bases on an mRNA codon during translation

anticyclone   a high-pressure center with winds flowing away from the center; associated with clear, fair weather

antinode   region of maximum amplitude between adjacent nodes in a standing wave

aorta   the large blood vessel that carries blood from the left ventricle to the majority of the body

aphelion   the point at which an orbit is farthest from the sun

apogee   the point at which the moon's elliptical orbit takes the moon farthest from the earth

apoptosis   also known as "programmed cell death"; death of specific cells that has a genetic basis and is not the result of injury

apparent local noon   the instant when the sun crosses the celestial meridian at any particular longitude

apparent local solar time   the time found from the position of the sun in the sky; the shadow of the gnomon on a sundial

apparent magnitude   a classification scheme for different levels of brightness of stars that you see; brightness values range from one to six with the number one (first magnitude) assigned to the brightest star and the number six (sixth magnitude) assigned to the faintest star that can be seen

apparent solar day   the interval between two consecutive crossings of the celestial meridian by the sun

aquifer   a layer of sand, gravel, or other highly permeable material beneath the surface that is saturated with water and is capable of producing water in a well or spring

Arctic Circle   parallel identifying the limit toward the equator where the sun appears

above the horizon all day for one day up to six months during the summer; located at 66.5°N latitude

area   the extent of a surface; the surface bounded by three or more lines

arid   dry climate classification; receives less than 25 cm (10 in) precipitation per year

aromatic hydrocarbon   organic compound with at least one benzene ring structure; cyclic hydrocarbons and their derivatives

arteries   the blood vessels that carry blood away from the heart

arterioles   small arteries located just before capillaries that can expand and contract to regulate the flow of blood to parts of the body

artesian   term describing the condition where confining pressure forces groundwater from a well to rise above the aquifer

asbestos   the common name for any one of several incombustible fibrous minerals that will not melt or ignite and can be woven into a fireproof cloth or used directly in fireproof insulation; about six different commercial varieties of asbestos are used, one of which has been linked to cancer under heavy exposure

assimilation   the physiological process that takes place in a living cell as it converts nutrients in food into specific molecules required by the organism

asteroids   small rocky bodies left over from the formation of the solar system; most are accumulated in a zone between the orbits of Mars and Jupiter

asthenosphere   a plastic, mobile layer of the earth's structure that extends around the earth below the lithosphere; ranges in thickness from a depth of 130 km to 160 km

astronomical unit   the radius of the earth's orbit is defined as one astronomical unit (A.U.)

atmospheric stability   the condition of a parcel of air found by relating the parcel's actual lapse rate to the dry adiabatic lapse rate

atom   the smallest unit of an element that can exist alone or in combination with other elements

atomic mass unit   relative mass unit (u) of an isotope based on the standard of the carbon-12 isotope, which is defined as a mass of exactly 12.00 u; one atomic mass unit (1 u) is 1/12 the mass of a carbon-12 atom

atomic number   the number of protons in the nucleus of an atom

atomic weight   weighted average of the masses of stable isotopes of an element as they occur in nature, based on the abundance of each isotope of the element and the atomic mass of the isotope compared to C-12

atria   thin-walled sacs of the heart that receive blood from the veins of the body and empty into the ventricles; singular, atrium

atrioventricular valves   located between the atria and ventricles of the heart preventing blood from flowing backward from the ventricles into the atria

autosomes   chromosomes that typically carry genetic information used by the organism for characteristics other than the primary determination of sex

autotrophs   organisms that are able to make their food molecules from inorganic raw materials by using basic energy sources such as sunlight

autumnal equinox   one of two times a year that daylight and night are of equal length; occurs on or about September 23 and identifies the beginning of the fall season

Avogadro's number   the number of C-12 atoms in exactly 12.00 g of C; 6.02×10^{23} atoms or other chemical units; the number of chemical units in one mole of a substance

axis   the imaginary line about which a planet or other object rotates

axon   a neuronal fiber that carries information away from the nerve cell body

B

background radiation   ionizing radiation (alpha, beta, gamma, etc.) from natural sources; between 100 and 500 millirems/yr of exposure to natural radioactivity from the environment

Balmer series   a set of four line spectra, narrow lines of color emitted by hydrogen atom electrons as they drop from excited states to the ground state

band of stability   a region of a graph of the number of neutrons versus the number of protons in nuclei; nuclei that have the neutron to proton ratios located in this band do not undergo radioactive decay

barometer   an instrument that measures atmospheric pressure, used in weather forecasting and in determining elevation above sea level

basal metabolic rate (BMR)   the amount of energy required to maintain normal body activity while at rest

base   any substance that is a proton acceptor when dissolved in water; generally considered a solution that forms hydroxide ions in water that can neutralize an acid, forming a salt and water

basilar membrane   a membrane in the cochlea containing sensory cells that are stimulated by the vibrations caused by sound waves

basin   a large, bowl-shaped fold in the land into which streams drain; also a small enclosed or partly enclosed body of water

batholith   a large volume of magma that has cooled and solidified below the surface, forming a large mass of intrusive rock

beat   rhythmic increases and decreases of volume from constructive and destructive interference between two sound waves of slightly different frequencies

beta particle   high-energy electron emitted as ionizing radiation from a decaying nucleus; also known as a beta ray

big bang theory   current model of galactic evolution in which the universe was created from an intense and brilliant explosion from a primeval fireball

bile   a product of the liver, stored in the gallbladder, which is responsible for the emulsification of fats

binding energy   the energy required to break a nucleus into its constituent protons and neutrons; also the energy equivalent released when a nucleus is formed

binomial system of nomenclature   a naming system that uses two Latin names, genus and specific epithet, for each type of organism

biogenesis   the concept that life originates only from preexisting life

biology   the science that deals with the study of living things and how living things interact with things around them

biomes   large regional communities primarily determined by climate

biotechnology   the science of gene manipulation

biotic factors   living parts of an organism's environment

black hole   the theoretical remaining core of a supernova with an intense gravitational field

blackbody radiation   electromagnetic radiation emitted by an ideal material (the blackbody) that perfectly absorbs and perfectly emits radiation

blood   the fluid medium consisting of cells and plasma that assists in the transport of materials and heat

body wave   a seismic wave that travels through the earth's interior, spreading outward from a disturbance in all directions

Bohr model   model of the structure of the atom that attempted to correct the deficiencies of the solar system model and account for the Balmer series

boiling point   the temperature at which a phase change of liquid to gas takes place through boiling; the same temperature as the condensation point

boundary   the division between two regions of differing physical properties

breaker   a wave whose front has become so steep that the top part has broken forward of the wave, breaking into foam, especially against a shoreline

British thermal unit   the amount of energy or heat needed to increase the temperature of 1 pound of water 1 degree Fahrenheit (abbreviated Btu)

bronchi   major branches of the trachea that ultimately deliver air to bronchioles in the lungs

bronchioles   small tubes that deliver air to the alveoli in the lung; they are capable of contracting

buffer solution   a solution consisting of a weak acid and a salt that has the same negative ion as the acid; has the ability to resist changes in the pH when small amounts of an acid or a base are added to the solution

bulimia   a nutritional deficiency disease characterized by a binge-and-purge cycle of eating; thought to stem from psychological disorders

C

calorie   the amount of energy (or heat) needed to increase the temperature of 1 gram of water 1 degree Celsius

Calorie   the nutritionist's "calorie"; equivalent to 1 kilocalorie

cancer   a tumor that is malignant

capillaries   tiny blood vessels through the walls of which exchange between cells and the blood takes place

carbohydrates   organic compounds that include sugars, starches, and cellulose; carbohydrates are used by plants and animals for structure, protection, and food

carbon film   a type of fossil formed when the volatile and gaseous constituents of a buried organic structure are distilled away, leaving a carbon film as a record

carbonation   in chemical weathering, a reaction that occurs naturally between carbonic acid (H_2CO_3) and rock minerals

carcinogens   agents responsible for causing cancer

carnivores   animals that eat other animals

carrier (genetic)   any individual having a hidden, recessive allele

carrying capacity (ecosystem)   the *optimum* maximum population size an area can support over an extended period of time

cast   sediments deposited by groundwater in a mold, taking the shape and external features of the organism that was removed to form the mold, then gradually changing to sedimentary rock

cathode rays   negatively charged particles (electrons) that are emitted from a negative terminal in an evacuated glass tube

celestial equator   line of the equator of the earth directly above the earth; the equator of the earth projected on the celestial sphere

celestial meridian   an imaginary line in the sky directly above you that runs north through the north celestial pole, south through the south celestial pole, and back around the other side to make a big circle around the earth

celestial sphere   a coordinate system of lines used to locate objects in the sky by imagining a huge turning sphere surrounding the earth with the stars and other objects attached to the sphere; the latitude and longitude lines of the earth's surface are projected to the celestial sphere

cell   the basic structural unit that makes up all living things

cell plate   a plant-cell structure that begins to form in the center of the cell and proceeds to the cell membrane, resulting in cytokinesis

cellulose   a polysaccharide abundant in plants that forms the fibers in cell walls that provide structure for plant materials

Celsius scale   referent scale that defines numerical values for measuring hotness or coldness, defined as degrees of temperature; based on the reference points of the freezing point of water and the boiling point of water at sea-level pressure, with 100 degrees between the two points

cementation   process by which spaces between buried sediment particles under compaction are filled with binding chemical deposits, binding the particles into a rigid, cohesive mass of a sedimentary rock

Cenozoic   the most recent geologic era; the time of recent life, meaning the fossils of this era are identical to the life found on the earth today

central nervous system the portion of the nervous system consisting of the brain and spinal cord

centrifugal force an apparent outward force on an object following a circular path that is a consequence of the third law of motion

centrioles organelles containing microtubules located just outside the nucleus

centripetal force the force required to pull an object out of its natural straight-line path and into a circular path; centripetal means "center seeking"

centromere the unreplicated region where two chromatids are joined

cepheid variables a bright variable star that can be used to measure distance

cerebellum a region of the brain connected to the medulla oblongata that receives many kinds of sensory stimuli and coordinates muscle movement

cerebrum a region of the brain that surrounds most of the other parts of the brain and is involved in consciousness and thought

chain reaction a self-sustaining reaction where some of the products are able to produce more reactions of the same kind; in a nuclear chain reaction neutrons are the products that produce more nuclear reactions in a self-sustaining series

chemical bond an attractive force that holds atoms together in a compound

chemical change a change in which the identity of matter is altered and new substances are formed

chemical energy a form of energy involved in chemical reactions associated with changes in internal potential energy; a kind of potential energy that is stored and later released during a chemical reaction

chemical equation concise way of describing what happens in a chemical reaction

chemical equilibrium occurs when two opposing reactions happen at the same time and at the same rate

chemical reaction a change in matter where different chemical substances are created by forming or breaking chemical bonds

chemical sediments ions from rock materials that have removed from solution, for example, carbonate ions removed by crystallization or organisms to form calcium carbonate chemical sediments

chemical weathering the breakdown of minerals in rocks by chemical reactions with water, gases of the atmosphere, or solutions

chemistry the science concerned with the study of the composition, structure, and properties of substances and the transformations they undergo

Chinook a warm wind that has been warmed by compression; also called Santa Ana

chondrites subdivision of stony meteorites containing small spherical lumps of silicate minerals or glass

chondrules small spherical lumps of silicate minerals or glass found in some meteorites

chromatid one of two component parts of a chromosome formed by replication and attached at the centromere

chromatin fibers see nucleoproteins

chromosomal aberrations changes in the gene arrangement in chromosomes; for example, translocation and duplication mutations

chromosomes complex structures within the nucleus composed of various kinds of histone proteins and DNA that contains the cell's genetic information

cinder cone volcano a volcanic cone that formed from cinders, sharp-edged rock fragments that cooled from frothy blobs of lava as they were thrown into the air

cirque a bowl-like depression in the side of a mountain, usually at the upper end of a mountain valley, formed by glacial erosion

class a group of closely related orders found within a phylum

clastic sediments weathered rock fragments that are in various states of being broken down from solid bedrock; boulders, gravel, and silt

cleavage furrow an indentation of the cell membrane of an animal cell that pinches the cytoplasm into two parts

climate the general pattern of weather that occurs in a region over a number of years

coalescence process (meteorology) the process by which large raindrops form from the merging and uniting of millions of tiny water droplets

cochlea the part of the ear that converts sound into nerve impulses

codominance when an organism has two different alleles for a trait and they both express themselves

codon a sequence of three nucleotides of an mRNA molecule that directs the placement of a particular amino acid during translation

coitus see *sexual intercourse*

cold front the front that is formed as a cold air mass moves into warmer air

combination chemical reaction a synthesis reaction in which two or more substances combine to form a single compound

comets celestial objects originating from the outer edges of the solar system that move about the sun in highly elliptical orbits; solar heating and pressure from the solar wind form a tail on the comet that points away from the sun

commensalism a relationship between two organisms in which one organism is helped and the other is not affected

community all of the kinds of interacting organisms within an ecosystem

compaction the process of pressure from a depth of overlying sediments squeezing together the deeper sediments and squeezing out water

complementary base a nitrogenous base in nucleic acids that can form hydrogen bonds with another base of a specific nucleotide

composite volcano a volcanic cone that formed from a buildup of alternating layers of cinders, ash, and lava flows

compound a pure chemical substance that can be decomposed by a chemical change into simpler substances with a fixed mass ratio

compressive stress a force that tends to compress the surface as the earth's plates move into each other

concentration an arbitrary description of the relative amounts of solute and solvent in a solution; a larger amount of solute makes a concentrated solution, and a small amount of solute makes a dilute concentration

conception fertilization

condensation (sound) a compression of gas molecules; a pulse of increased density and pressure that moves through the air at the speed of sound

condensation (water vapor) when more vapor or gas molecules are returning to the liquid state than are evaporating

condensation nuclei tiny particles such as tiny dust, smoke, soot, and salt crystals that are suspended in the air on which water condenses

condensation point the temperature at which a gas or vapor changes back to a liquid

conduction the transfer of heat from a region of higher temperature to a region of lower temperature by increased kinetic energy moving from molecule to molecule

cones light-sensitive cells in the retina of the eye that respond to different colors of light

constellations patterns of 88 groups of stars imagined to resemble various mythological characters, inanimate objects, and animals

constructive interference the condition in which two waves arriving at the same place, at the same time and in phase, add amplitudes to create a new wave

consumers organisms that must obtain energy in the form of organic matter

continental air mass dry air masses that form over large land areas

continental climate a climate influenced by air masses from large land areas; hot summers and cold winters

continental drift a concept that continents shift positions on the earth's surface, moving across the surface rather than being fixed, stationary landmasses

continental glaciers glaciers that cover a large area of a continent; for example, Greenland and the Antarctic

continental shelf a feature of the ocean floor; the flooded margins of the continents that form a zone of relatively shallow water adjacent to the continents

continental slope a feature of the ocean floor; a steep slope forming the transition between the continental shelf and the deep ocean basin

control group the situation used as the basis for comparison in a controlled experiment

control processes mechanisms that ensure that an organism will carry out all metabolic activities in the proper sequence (coordination) and at the proper rate (regulation)

control rods rods inserted between fuel rods in a nuclear reactor to absorb neutrons and thus control the rate of the nuclear chain reaction

controlled experiment an experiment that allows for a comparison of two events that are identical in all but one respect

convection transfer of heat from a region of higher temperature to a region of lower temperature by the displacement of high-energy molecules—for example, the displacement of warmer, less dense air (higher kinetic energy) by cooler, more dense air (lower kinetic energy)

convection cell complete convective circulation pattern; also, slowly turning regions in the plastic asthenosphere that might drive the motion of plate tectonics

convection zone (of a star) part of the interior of a star according to a model; the region directly above the radiation zone where gases are heated by the radiation zone below and move upward by convection to the surface, where they emit energy in the form of visible light, ultraviolet radiation, and infrared radiation

conventional current opposite to electron current—that is, considers an electric current to consist of a drift of positive charges that flow from the positive terminal to the negative terminal of a battery

convergent boundaries boundaries that occur between two plates moving toward each other

coordinate covalent bond a "hole and plug" kind of covalent bond, formed when the shared electron pair is contributed by one atom

Copernican system heliocentric, or sun-centered solar system, model developed by Nicholas Copernicus in 1543

copulation see *sexual intercourse*

core (of a star) dense, very hot region of a star where nuclear fusion reactions release gamma and X-ray radiation

core (of the earth) the center part of the earth, which consists of a solid inner part and liquid outer part; makes up about 15 percent of the earth's total volume and about one-third of its mass

Coriolis effect the apparent deflection due to the rotation of the earth; it is to the right in the Northern Hemisphere

correlation the determination of the equivalence in geologic age by comparing the rocks in two separate locations

coulomb unit used to measure quantity of electric charge; equivalent to the charge resulting from the transfer of 6.24 billion particles such as the electron

Coulomb's law relationship between charge, distance, and magnitude of the electrical force between two bodies

covalent bond a chemical bond formed by the sharing of a pair of electrons

covalent compound chemical compound held together by a covalent bond or bonds

creep the slow downhill movement of soil down a steep slope

crest the high mound of water that is part of a wave; also refers to the condensation, or high-pressure part, of a sound wave

critical angle limit to the angle of incidence when all light rays are reflected internally

critical mass mass of fissionable material needed to sustain a chain reaction

crude oil petroleum pumped from the ground that has not yet been refined into usable products

crust the outermost part of the earth's interior structure; the thin, solid layer of rock that rests on top of the Mohorovicic discontinuity

curie unit of nuclear activity defined as 3.70×10^{10} nuclear disintegrations per second

cycle a complete vibration

cyclone a low-pressure center where the winds move into the low-pressure center and are forced upward; a low-pressure center with clouds, precipitation, and stormy conditions

cytokinesis division of the cytoplasm of one cell into two new cells

cytosine a single-ring nitrogenous-base molecule in DNA and RNA; complementary to guanine

data measurement information used to describe something

data points points that may be plotted on a graph to represent simultaneous measurements of two related variables

daughter cells two cells formed by cell division

daughter chromosomes chromosomes produced by DNA replication that contain identical genetic information; formed after chromosome division in anaphase

daughter nuclei two nuclei formed by mitosis

daylight saving time setting clocks ahead one hour during the summer to more effectively utilize the longer days of summer, then setting the clocks back in the fall

decibel scale a nonlinear scale of loudness based on the ratio of the intensity level of a sound to the intensity at the threshold of hearing

decomposers organisms that use dead organic matter as a source of energy

decomposition chemical reaction a chemical reaction in which a compound is broken down into the elements that make up the compound, into simpler compounds, or into elements and simpler compounds

deflation the widespread removal of base materials from the surface by the wind

degassing process where gases and water vapor were released from rocks heated to melting during the early stages of the formation of a planet

delta a somewhat triangular deposit at the mouth of a river formed where a stream flowing into a body of water slowed and lost its sediment-carrying ability

dendrites neuronal fibers that receive information from axons and carry it toward the nerve-cell body

density the compactness of matter described by a ratio of mass (or weight) per unit volume

density current an ocean current that flows because of density differences in seawater

deoxyribonucleic acid (DNA) a polymer of nucleotides that serves as genetic information; in prokaryotic cells, it is a duplex DNA (double-stranded) loop and contains attached HU proteins and in eukaryotic cells, it is found in strands with attached histone proteins—when tightly coiled, it is known as a chromosome

deoxyribose a 5-carbon sugar molecule; a component of DNA

depolarized having lost the electrical difference existing between two points or objects

destructive interference the condition in which two waves arriving at the same point at the same time out of phase add amplitudes to create zero total disturbance

dew condensation of water vapor into droplets of liquid on surfaces

dew point temperature the temperature at which condensation begins

diaphragm a muscle separating the lung cavity from the abdominal cavity that is involved in exchanging the air in the lungs

diastrophism all-inclusive term that means any and all possible movements of the earth's plates, including drift, isostatic adjustment, and any other process that deforms or changes the earth's surface by movement

diet the food and drink consumed by a person from day to day

differentiation the process of forming specialized cells within a multicellular organism

diffraction the bending of light around the edge of an opaque object

diffuse reflection light rays reflected in many random directions, as opposed to the parallel rays reflected from a perfectly smooth surface such as a mirror

dike a tabular-shaped intrusive rock that formed when magma moved into joints or faults that cut across other rock bodies

diploid having two sets of chromosomes: one set from the maternal parent and one set from the paternal parent

direct current an electrical current that always moves in one direction

direct proportion when two variables increase or decrease together in the same ratio (at the same rate)

disaccharides two monosaccharides joined together with the loss of a water molecule; examples of disaccharides are sucrose (table sugar), lactose, and maltose

dispersion the effect of spreading colors of light into a spectrum with a material that has an index of refraction that varies with wavelength

divergent boundaries boundaries that occur between two plates moving away from each other

divide line separating two adjacent watersheds

DNA code a sequence of three nucleotides of a DNA molecule

DNA polymerase an enzyme that bonds DNA nucleotides together when they base pair with an existing DNA strand

DNA replication the process by which the genetic material (DNA) of the cell reproduces itself prior to its distribution to the next generation of cells

dome a large, upwardly bulging, symmetrical fold that resembles a dome

dominant allele an allele that expresses itself and masks the effects of other alleles for the trait

Doppler effect an apparent shift in the frequency of sound or light due to relative motion between the source of the sound or light and the observer

double bond covalent bond formed when two pairs of electrons are shared by two atoms

dry adiabatic lapse rate the rate of adiabatic cooling or warming of air in the absence of condensation or evaporation in a parcel of air that is descending or ascending

dune a hill, low mound, or ridge of wind-blown sand or other sediments

duodenum the first part of the small intestine, which receives food from the stomach and secretions from the liver and pancreas

duplex DNA DNA in a double-helix shape

E

earthquake a quaking, shaking, vibrating, or upheaval of the earth's surface

earthquake epicenter point on the earth's surface directly above an earthquake focus

earthquake focus place where seismic waves originate beneath the surface of the earth

echo a reflected sound that can be distinguished from the original sound, which usually arrives 0.1 s or more after the original sound

eclipse when the shadow of a celestial body falls on the surface of another celestial body

ecology the branch of biology that studies the relationships between organisms and their environment

ecosystem an interacting unit consisting of a collection of organisms and abiotic factors

egg the haploid sex cell produced by sexually mature females

ejaculation the release of sperm cells and seminal fluid through the penis

El Niño changes in the atmospheric pressure systems, ocean currents, water temperatures, and wind patterns that seem to be linked to worldwide changes in the weather

elastic rebound the sudden snap of stressed rock into new positions; the recovery from elastic strain that results in an earthquake

elastic strain an adjustment to stress in which materials recover their original shape after a stress is released

electric circuit consists of a voltage source that maintains an electrical potential, a continuous conducting path for a current to follow, and a device where work is done by the electrical potential; a switch in the circuit is used to complete or interrupt the conducting path

electric current the flow of electric charge

electric field force field produced by an electrical charge

electric field lines a map of an electric field representing the direction of the force that a test charge would experience; the direction of an electric field shown by lines of force

electric generator a mechanical device that uses wire loops rotating in a magnetic field to produce electromagnetic induction in order to generate electricity

electric potential energy   potential energy due to the position of a charge near other charges

electrical conductors   materials that have electrons that are free to move throughout the material; for example, metals

electrical energy   a form of energy from electromagnetic interactions; one of five forms of energy—mechanical, chemical, radiant, electrical, and nuclear

electrical force   a fundamental force that results from the interaction of electrical charge and is billions and billions of times stronger than the gravitational force; sometimes called the "electromagnetic force" because of the strong association between electricity and magnetism

electrical insulators   electrical nonconductors, or materials that obstruct the flow of electric current

electrical nonconductors   materials that have electrons that are not moved easily within the material—for example, rubber; electrical nonconductors are also called electrical insulators

electrical resistance   the property of opposing or reducing electric current

electrolyte   water solution of ionic substances that conducts an electric current

electromagnet   a magnet formed by a solenoid that can be turned on and off by turning the current on and off

electromagnetic force   one of four fundamental forces; the force of attraction or repulsion between two charged particles

electromagnetic induction   process in which current is induced by moving a loop of wire in a magnetic field or by changing the magnetic field

electron   subatomic particle that has the smallest negative charge possible and usually found in an orbital of an atom, but gained or lost when atoms become ions

electron configuration   the arrangement of electrons in orbitals and suborbitals about the nucleus of an atom

electron current   opposite to conventional current; that is, considers electric current to consist of a drift of negative charges that flows from the negative terminal to the positive terminal of a battery

electron dot notation   notation made by writing the chemical symbol of an element with dots around the symbol to indicate the number of outer shell electrons

electron pair   a pair of electrons with different spin quantum numbers that may occupy an orbital

electron volt   the energy gained by an electron moving across a potential difference of one volt; equivalent to 1.60×10^{-19} J

electronegativity   the comparative ability of atoms of an element to attract bonding electrons

electrostatic charge   an accumulated electric charge on an object from a surplus or deficiency of electrons; also called "static electricity"

element   a pure chemical substance that cannot be broken down into anything simpler by chemical or physical means; there are over 100 known elements, the fundamental materials of which all matter is made

empirical formula   identifies the elements present in a compound and describes the simplest whole number ratio of atoms of these elements with subscripts

endocrine glands   secrete chemical messengers into the circulatory system

endocrine system   a number of glands that communicate with one another and other tissues through chemical messengers transported throughout the body by the circulatory system

endosymbiotic theory   a theory suggesting that some organelles found in eukaryotic cells may have originated as free-living prokaryotes

energy   the ability to do work

English system   a system of measurement that originally used sizes of parts of the human body as referents

environment   the surroundings; anything that affects an organism during its lifetime

environmental resistance   the collective set of factors that limit population growth

enzymes   molecules, produced by organisms, that are able to control the rate at which chemical reactions occur

epicycle   small secondary circular orbits of planets in the geocentric model that was invented to explain the occasional retrograde motion of the planets

epinephrine   a hormone produced by the adrenal medulla that increases heart rate, blood pressure, and breathing rate

epiphyte   a plant that lives on the surface of another plant without doing harm

epochs   subdivisions of geologic periods

equation   a statement that describes a relationship in which quantities on one side of the equal sign are identical to quantities on the other side

equation of time   the cumulative variation between the apparent local solar time and the mean solar time

equinoxes   Latin meaning "equal nights"; time when daylight and night are of equal length, which occurs during the spring equinox and the autumnal equinox

eras   the major blocks of time in the earth's geologic history; the Cenozoic, Mesozoic, Paleozoic, and Precambrian

erosion   the process of physically removing weathered materials, for example, rock fragments are physically picked up by an erosion agent such as a stream or a glacier

essential amino acids   those amino acids that cannot be synthesized by the human body and must be part of the diet (e.g., lysine, tryptophan, and valine)

esters   class of organic compounds with the general structure of RCOOR', where R is one of the hydrocarbon groups—for example, methyl or ethyl; esters make up fats, oils, and waxes and some give fruit and flowers their taste and odor

estrogen   one of the female sex hormones that cause the differentiation of typical female internal and external genital anatomy; responsible for the changes in breasts, vagina, uterus, clitoris, and pelvic bone structure at puberty

ether   class of organic compounds with the general formula ROR', where R is one of the hydrocarbon groups—for example, methyl or ethyl; mostly used as industrial and laboratory solvents

eukaryote   an organism composed of cells possessing a nuclear membrane and other membranous organelles

evaporation   process of more molecules leaving a liquid for the gaseous state than returning from the gas to the liquid; can occur at any given temperature from the surface of a liquid

excited states   as applied to an atom, describes the energy state of an atom that has electrons in a state above the minimum energy state for that atom; as applied to a nucleus, describes the energy state of a nucleus that has particles in a state above the minimum energy state for that nuclear configuration

exfoliation   the fracturing and breaking away of curved, sheet-like plates from bare rock surfaces via physical or chemical weathering, resulting in dome-shaped hills and rounded boulders

exocrine glands   use ducts to secrete to the surface of the body or into hollow organs of the body, for example, sweat glands or pancreas

exosphere   the outermost layer of the atmosphere where gas molecules merge with the diffuse vacuum of space

experiment a re-creation of an event in a way that enables a scientist to gain valid and reliable empirical evidence

experimental group the group in a controlled experiment that is identical to the control group in all respects but one

exponential growth phase a period of time during population growth when the population increases at an accelerating rate

F

Fahrenheit scale referent scale that defines numerical values for measuring hotness or coldness, defined as degrees of temperature; based on the reference points of the freezing point of water and the boiling point of water at sea-level pressure, with 180 degrees between the two points

family (elements) vertical columns of the periodic table consisting of elements that have similar properties

family (organisms) a group of closely related genera within an order

fats organic compounds of esters formed from glycerol and three long-chain carboxylic acids that are also called triglycerides; called fats in animals and oils in plants

fault a break in the continuity of a rock formation along which relative movement has occurred between the rocks on either side

fault plane the surface along which relative movement has occurred between the rocks on either side; the surface of the break in continuity of a rock formation

ferromagnesian silicates silicates that contain iron and magnesium; examples include the dark-colored minerals olivine, augite, hornblende, and biotite

fertilization the joining of haploid nuclei, usually from an egg and a sperm cell, resulting in a diploid cell called the zygote

fiber natural (plant) or industrially produced polysaccharides that are resistant to hydrolysis by human digestive enzymes

first law of motion every object remains at rest or in a state of uniform straight-line motion unless acted on by an unbalanced force

first quarter the moon phase between the new phase and the full phase when the moon is perpendicular to a line drawn through the earth and the sun; one half of the lighted moon can be seen from the earth, so this phase is called the first quarter

floodplain the wide, level floor of a valley built by a stream; the river valley where a stream floods when it spills out of its channel

fluids matter that has the ability to flow or be poured; the individual molecules of a fluid are able to move, rolling over or by one another

folds bends in layered bed rock as a result of stress or stresses that occurred when the rock layers were in a ductile condition, probably under considerable confining pressure from deep burial

foliation the alignment of flat crystal flakes of a rock into parallel sheets

follicle the saclike structure near the surface of the ovary that encases the soon-to-be-released secondary oocyte

follicle-stimulating hormone (FSH) the pituitary secretion that causes the ovaries to begin to produce larger quantities of estrogen and to develop the follicle and prepare the egg for ovulation

food chain a sequence of organisms that feed on one another, resulting in a flow of energy from a producer through a series of consumers

Food Guide Pyramid a tool developed by the U.S. Department of Agriculture to help the general public plan for good nutrition; guidelines for required daily intake from each of five food groups

food web a system of interlocking food chains

force a push or pull capable of changing the state of motion of an object; since a force has magnitude (strength) as well as direction, it is a vector quantity

force field a model describing action at a distance by giving the magnitude and direction of force on a unit particle; considers a charge or a mass to alter the space surrounding it and a second charge or mass to interact with the altered space with a force

formula describes what elements are in a compound and in what proportions

formula weight the sum of the atomic weights of all the atoms in a chemical formula

fossil any evidence of former prehistoric life

fossil fuels organic fuels that contain the stored radiant energy of the sun converted to chemical energy by plants or animals that lived millions of years ago; coal, petroleum, and natural gas are the common fossil fuels

Foucault pendulum a heavy mass swinging from a long wire that can be used to provide evidence about the rotation of the earth

fovea centralis the area of sharpest vision on the retina, containing only cones, where light is sharply focused

fracture strain an adjustment to stress in which materials crack or break as a result of the stress

free fall when objects fall toward the earth with no forces acting upward; air resistance is neglected when considering an object to be in free fall

freezing point the temperature at which a phase change of liquid to solid takes place; the same temperature as the melting point for a given substance

frequency the number of cycles of a vibration or of a wave occurring in one second, measured in units of cycles per second (hertz)

freshwater water that is not saline and is fit for human consumption

front the boundary, or thin transition zone, between air masses of different temperatures

frost ice crystals formed by water vapor condensing directly from the vapor phase; frozen water vapor that forms on objects

frost wedging the process of freezing and thawing water in small rock pores and cracks that become larger and larger, eventually forcing pieces of rock to break off

fuel rod long zirconium alloy tubes containing fissionable material for use in a nuclear reactor

full moon the moon phase when the earth is between the sun and the moon and the entire side of the moon facing the earth is illuminated by sunlight

functional group the atom or group of atoms in an organic molecule that is responsible for the chemical properties of a particular class or group of organic chemicals

fundamental charge smallest common charge known; the magnitude of the charge of an electron and a proton, which is 1.60×10^{-19} coulomb

fundamental frequency the lowest frequency (longest wavelength) that can set up standing waves in an air column or on a string

fundamental properties a property that cannot be defined in simpler terms other than to describe how it is measured; the fundamental properties are length, mass, time, and charge

fungus the common name for the kingdom Mycetae

G

g symbol representing the acceleration of an object in free fall due to the force of gravity; its magnitude is 9.8 m/s^2 (32 ft/s^2)

galactic clusters   gravitationally bound subgroups of as many as 1,000 stars that move together within the Milky Way galaxy

galaxy   group of billions and billions of stars that form the basic unit of the universe, for example, Earth is part of the solar system, which is located in the Milky Way galaxy

gallbladder   an organ attached to the liver that stores bile

gamete   a haploid sex cell

gametogenesis   the generating of gametes; the meiotic cell-division process that produces sex cells; oogenesis and spermatogenesis

gametophyte stage   a life-cycle stage in plants in which a haploid sex cell is produced by mitosis

gamma ray   very short wavelength electromagnetic radiation emitted by decaying nuclei

gases   a phase of matter composed of molecules that are relatively far apart moving freely in a constant, random motion and have weak cohesive forces acting between them, resulting in the characteristic indefinite shape and indefinite volume of a gas

gasohol   solution of ethanol and gasoline

gastric juice   the secretions of the stomach that contain enzymes and hydrochloric acid

Geiger counter   a device that indirectly measures ionizing radiation (beta and/or gamma) by detecting "avalanches" of electrons that are able to move because of the ions produced by the passage of ionizing radiation

gene   a unit of heredity located on a chromosome and composed of a sequence of DNA nucleotides

gene flow   the movement of genes from one generation to another or from one place to another

generative processes   actions that increase the size of an individual organism (growth) or increase the number of individuals in a population (reproduction)

genetics   the study of genes, how genes produce characteristics, and how the characteristics are inherited

genome   a set of all the genes necessary to specify an organism's complete list of characteristics

genotype   the catalog of genes of an organism, whether or not these genes are expressed

genus   (plural, genera) a group of closely related species within a family

geocentric   the idea that the earth is the center of the universe

geographic barriers   geographic features that keep different portions of a species from exchanging genes

geographic isolation   a condition in which part of the gene pool is separated by geographic barriers from the rest of the population

geologic time scale   a "calendar" of geologic history based on the appearance and disappearance of particular fossils in the sedimentary rock record

geomagnetic time scale   time scale established from the number and duration of magnetic field reversals during the past 6 million years

giant planets   the large outer planets of Jupiter, Saturn, Uranus, and Neptune that all have similar densities and compositions

glacier   a large mass of ice on land that formed from compacted snow and slowly moves under its own weight

gland   an organ that manufactures and secretes a material either through ducts or directly into the circulatory system

globular clusters   symmetrical and tightly packed clusters of as many as a million stars that move together as subgroups within the Milky Way galaxy

glycerol   an alcohol with three hydroxyl groups per molecule; for example, glycerin (1,2,3-propanetriol)

glycogen   a highly branched polysaccharide synthesized by the human body and stored in the muscles and liver; serves as a direct reserve source of energy

glycol   an alcohol with two hydroxyl groups per molecule; for example, ethylene glycol that is used as an antifreeze

gonad   a generalized term for organs in which meiosis occurs to produce gametes; ovary or testis

gram-atomic weight   the mass in grams of one mole of an element that is numerically equal to its atomic weight

gram-formula weight   the mass in grams of one mole of a compound that is numerically equal to its formula weight

gram-molecular weight   the gram-formula weight of a molecular compound

granite   light-colored, coarse-grained igneous rock common on continents; igneous rocks formed by blends of quartz and feldspars, with small amounts of micas, hornblende, and other minerals

greenhouse effect   the process of increasing the temperature of the lower parts of the atmosphere through redirecting energy back toward the surface; the absorption and reemission of infrared radiation by carbon dioxide, water vapor, and a few other gases in the atmosphere

ground state   energy state of an atom with electrons at the lowest energy state possible for that atom

groundwater   water from a saturated zone beneath the surface; water from beneath the surface that supplies wells and springs

growth-stimulating hormone (GSH)   a hormone produced by the anterior pituitary gland that stimulates tissues to grow

guanine   a double-ring nitrogenous-base molecule in DNA and RNA. It is the complementary base of cytosine

gyre   the great circular systems of moving water in each ocean

H

habitat   the place or part of an ecosystem occupied by an organism

hail   a frozen form of precipitation, sometimes with alternating layers of clear and opaque, cloudy ice

hair hygrometer   a device that measures relative humidity from changes in the length of hair

half-life   the time required for one-half of the unstable nuclei in a radioactive substance to decay into a new element

halogen   member of family VIIA of the periodic table, having common properties of very reactive nonmetallic elements common in salt compounds

haploid   having a single set of chromosomes resulting from the reduction division of meiosis

hard water   water that contains relatively high concentrations of dissolved salts of calcium and magnesium

heart   the muscular pump that forces the blood through the blood vessels of the body

heat   total internal energy of molecules, which is increased by gaining energy from a temperature difference (conduction, convection, radiation) or by gaining energy from a form conversion (mechanical, chemical, radiant, electrical, nuclear)

heat of formation   energy released in a chemical reaction

Heisenberg uncertainty principle you cannot measure both the exact momentum and the exact position of a subatomic particle at the same time—when the more exact of the two is known, the less certain you are of the value of the other

hemoglobin an iron-containing molecule found in red blood cells, to which oxygen molecules bind

herbivores animals that feed directly on plants

hertz unit of frequency; equivalent to one cycle per second

Hertzsprung-Russell diagram diagram to classify stars with a temperature-luminosity graph

heterotroph an organism that requires a source of organic material from its environment; it cannot produce its own food

heterozygous describes a diploid organism that has two different alleles for a particular characteristic

high short for high-pressure center (anticyclone), which is associated with clear, fair weather

high latitudes latitudes close to the poles; those that for a period of time during the winter months receive no solar radiation at noon

high-pressure center another term for anticyclone

homeostasis the process of maintaining a constant internal environment as a result of constant monitoring and modification of the functioning of various systems

homologous chromosomes a pair of chromosomes in a diploid cell that contain similar genes at corresponding loci throughout their length

homozygous describes a diploid organism that has two identical alleles for a particular characteristic

hormones chemical messengers secreted by endocrine glands to regulate other parts of the body

horsepower measurement of power defined as a power rating of 550 ft·lb/s

host an organism that a parasite lives in or on

hot spots sites on the earth's surface where plumes of hot rock materials rise from deep within the mantle

humid moist climate classification; receives more than 50 cm (20 in) precipitation per year

humidity the amount of water vapor in the air; see *relative humidity*

hurricane a tropical cyclone with heavy rains and winds exceeding 120 km/hr

hydration the attraction of water molecules for ions; a reaction that occurs between water and minerals that make up rocks

hydrocarbon an organic compound consisting of only the two elements hydrogen and carbon

hydrocarbon derivatives organic compounds that can be thought of as forming when one or more hydrogen atoms on a hydrocarbon have been replaced by an element or a group of elements other than hydrogen

hydrogen bond a weak to moderate bond between the hydrogen end ($+$) of a polar molecule and the negative end ($-$) of a second polar molecule

hydrologic cycle water vapor cycling into and out of the atmosphere through continuous evaporation of liquid water from the surface and precipitation of water back to the surface

hydronium ion a molecule of water with an attached hydrogen ion, H_3O^+

hypothalamus a region of the brain located in the floor of the thalamus and connected to the pituitary gland that is involved in sleep and arousal; emotions such as anger, fear, pleasure, hunger, sexual response, and pain; and automatic functions such as temperature, blood pressure, and water balance

hypothesis a tentative explanation of a phenomenon that is compatible with the data and provides a framework for understanding and describing that phenomenon

I

ice-crystal process a precipitation-forming process that brings water droplets of a cloud together through the formation of ice crystals

ice-forming nuclei small, solid particles suspended in air; ice can form on the suspended particles

igneous rocks rocks that formed from magma, which is a hot, molten mass of melted rock materials

immune system a system of white blood cells specialized to provide the body with resistance to disease; there are two types, antibody-mediated immunity and cell-mediated immunity

incandescent matter emitting visible light as a result of high temperature; for example, a light bulb, a flame from any burning source, and the sun are all incandescent sources because of high temperature

incident ray line representing the direction of motion of incoming light approaching a boundary

inclination of the earth axis tilt of the earth's axis measured from the plane of the ecliptic (23.5°); considered to be the same throughout the year

incus the ear bone that is located between the malleus and the stapes

index fossils distinctive fossils of organisms that lived only a brief time; used to compare the age of rocks exposed in two different locations

index of refraction the ratio of the speed of light in a vacuum to the speed of light in a material

inertia a property of matter describing the tendency of an object to resist a change in its state of motion; an object will remain in unchanging motion or at rest in the absence of an unbalanced force

infrasonic sound waves having too low a frequency to be heard by the human ear; sound having a frequency of less than 20 Hz

initiation code the code on DNA with the base sequence TAC that begins the process of transcription

inorganic chemistry the study of all compounds and elements in which carbon is not the principal element

insulators materials that are poor conductors of heat—for example, heat flows slowly through materials with air pockets because the molecules making up air are far apart; also, materials that are poor conductors of electricity, for example, glass or wood

intensity a measure of the energy carried by a wave

interference phenomenon of light where the relative phase difference between two light waves produces light or dark spots, a result of light's wavelike nature

intermediate-focus earthquakes earthquakes that occur in the upper part of the mantle, between 70 to 350 km below the surface of the earth

intermolecular forces forces of interaction between molecules

internal energy sum of all the potential energy and all the kinetic energy of all the molecules of an object

international date line the 180° meridian is arbitrarily called the international date line; used to compensate for cumulative time zone changes by adding or subtracting a day when the line is crossed

interphase the stage between cell divisions in which the cell is engaged in metabolic activities

interstitial cell-stimulating hormone(ICSH) the chemical messenger molecule released from the pituitary that causes the testes to produce testosterone, the primary male sex hormone; same as follicle-stimulating hormone in females

intertropical convergence zone a part of the lower troposphere in a belt from 10°N to 10°S of the equator where air is heated, expands, and becomes less dense and rises around the belt

intrusive igneous rocks coarse-grained igneous rocks formed as magma cools slowly deep below the surface

inverse proportion the relationship in which the value of one variable increases while the value of the second variable decreases at the same rate (in the same ratio)

inversion a condition of the troposphere when temperature increases with height rather than decreasing with height; a cap of cold air over warmer air that results in trapped air pollution

ion an atom or a particle that has a net charge because of the gain or loss of electrons; polyatomic ions are groups of bonded atoms that have a net charge

ion exchange reaction a reaction that takes place when the ions of one compound interact with the ions of another, forming a solid that comes out of solution, a gas, or water

ionic bond chemical bond of electrostatic attraction between negative and positive ions

ionic compounds chemical compounds that are held together by ionic bonds—that is, bonds of electrostatic attraction between negative and positive ions

ionization process of forming ions from molecules

ionization counter a device that measures ionizing radiation (alpha, beta, gamma, etc.) by indirectly counting the ions produced by the radiation

ionized an atom or a particle that has a net charge because it has gained or lost electrons

ionosphere refers to that part of the atmosphere—parts of the thermosphere and upper mesosphere—where free electrons and ions reflect radio waves around the earth and where the northern lights occur

iron meteorites meteorite classification group whose members are composed mainly of iron

island arcs curving chains of volcanic islands that occur over belts of deep-seated earthquakes; for example, the Japanese and Indonesian islands

isomers chemical compounds with the same molecular formula but different molecular structure; compounds that are made from the same numbers of the same elements but have different molecular arrangements

isostasy a balance or equilibrium between adjacent blocks of crust "floating" on the asthenosphere; the concept of less dense rock floating on more dense rock

isostatic adjustment a concept of crustal rocks thought of as tending to rise or sink gradually until they are balanced by the weight of displaced mantle rocks

isotope atoms of an element with identical chemical properties but with different masses; isotopes are atoms of the same element with different numbers of neutrons

J

jet stream a powerful, winding belt of wind near the top of the troposphere that tends to extend all the way around the earth, moving generally from the west in both hemispheres at speeds of 160 km/hr or more

joint a break in the continuity of a rock formation without a relative movement of the rock on either side of the break

joule metric unit used to measure work and energy; can also be used to measure heat; equivalent to newton-meter

K

Kelvin scale a temperature scale that does not have arbitrarily assigned referent points, and zero means nothing; the zero point on the Kelvin scale (also called absolute scale) is the lowest limit of temperature, where all random kinetic energy of molecules ceases

Kepler's first law relationship in planetary motion that each planet moves in an elliptical orbit, with the sun located at one focus

Kepler's laws of planetary motion the three laws describing the motion of the planets

Kepler's second law relationship in planetary motion that an imaginary line between the sun and a planet moves over equal areas of the ellipse during equal time intervals

Kepler's third law relationship in planetary motion that the square of the period of an orbit is directly proportional to the cube of the radius of the major axis of the orbit

ketone an organic compound with the general formula RCOR′, where R is one of the hydrocarbon groups; for example, methyl or ethyl

kidneys the primary organs involved in regulating blood levels of water, hydrogen ions, salts, and urea

kilocalorie the amount of energy required to increase the temperature of 1 kilogram of water 1 degree Celsius; equivalent to 1,000 calories

kilogram the fundamental unit of mass in the metric system of measurement

kinetic energy the energy of motion; can be measured from the work done to put an object in motion, from the mass and velocity of the object while in motion, or from the amount of work the object can do because of its motion

kinetic molecular theory the collection of assumptions that all matter is made up of tiny atoms and molecules that interact physically, that explain the various states of matter, and that have an average kinetic energy that defines the temperature of a substance

kingdom the largest grouping used in the classification of organisms

Kuiper Belt a disk-shaped region of small icy bodies some 30 to 100 A.U. from the sun; the source of short-period comets

kwashiorkor a protein-deficiency disease common in malnourished children caused by prolonged protein starvation leading to reduced body size, lethargy, and low mental ability

L

laccolith an intrusive rock feature that formed when magma flowed into the plane of contact between sedimentary rock layers, then raised the overlying rock into a blister-like uplift

lack of dominance the condition of two unlike alleles both expressing themselves, neither being dominant

lake a large inland body of standing water

large intestine the last portion of the food tube; primarily involved in reabsorbing water

last quarter the moon phase between the full phase and the new phase when the moon is perpendicular to a line drawn through the earth and the sun; one half of the lighted moon can be seen from the earth, so this phase is called the last quarter

latent heat refers to the heat "hidden" in phase changes

latent heat of fusion the heat absorbed when 1 gram of a substance changes from the solid to the liquid phase, or the heat released by 1 gram of a substance when changing from the liquid phase to the solid phase

latent heat of vaporization   the heat absorbed when 1 gram of a substance changes from the liquid phase to the gaseous phase, or the heat released when 1 gram of gas changes from the gaseous phase to the liquid phase

latitude   the angular distance from the equator to a point on a parallel that tells you how far north or south of the equator the point is located

lava   magma, or molten rock, that is forced to the surface from a volcano or a crack in the earth's surface

law of conservation of energy   energy is never created or destroyed; it can only be converted from one form to another as the total energy remains constant

law of conservation of mass   same as law of conservation of matter; mass, including single atoms, is neither created nor destroyed in a chemical reaction

law of conservation of matter   matter is neither created nor destroyed in a chemical reaction

law of conservation of momentum   the total momentum of a group of interacting objects remains constant in the absence of external forces

law of independent assortment   members of one allelic pair will separate from each other independently of the members of other allele pairs

law of segregation   when gametes are formed by a diploid organism, the alleles that control a trait separate from one another into different gametes, retaining their individuality

light ray   model using lines to show the direction of motion of light to describe the travels of light

light-year   the distance that light travels through empty space in one year, approximately 9.5×10^{12} km

line spectrum   narrow lines of color in an otherwise dark spectrum; these lines can be used as "fingerprints" to identify gases

linear scale   a scale, generally on a graph, where equal intervals represent equal changes in the value of a variable

lines of force   lines drawn to make an electric field strength map, with each line originating on a positive charge and ending on a negative charge; each line represents a path on which a charge would experience a constant force and lines closer together mean a stronger electric field

linkage group   genes located on the same chromosome that tend to be inherited together

liquids   a phase of matter composed of molecules that have interactions stronger than those found in a gas but not strong enough to keep the molecules near the equilibrium positions of a solid, resulting in the characteristic definite volume but indefinite shape of a liquid

liter   a metric system unit of volume usually used for liquids

lithosphere   solid layer of the earth's structure that is above the asthenosphere and includes the entire crust, the Moho, and the upper part of the mantle

liver   an organ of the body responsible for secreting bile, filtering the blood, detoxifying molecules, and modifying molecules absorbed from the gut

locus   the spot on a chromosome where an allele is located

loess   very fine dust or silt that has been deposited by the wind over a large area

longitude   angular distance of a point east or west from the prime meridian on a parallel

longitudinal wave   a mechanical disturbance that causes particles to move closer together and farther apart in the same direction that the wave is traveling

longshore current   a current that moves parallel to the shore, pushed along by waves that move accumulated water from breakers

loudness   a subjective interpretation of a sound that is related to the energy of the vibrating source, related to the condition of the transmitting medium, and related to the distance involved

low latitudes   latitudes close to the equator; those that receive vertical solar radiation at noon during some part of the year

luminosity   the total amount of energy radiated into space each second from the surface of a star

luminous   an object or objects that produce visible light; for example, the sun, stars, light bulbs, and burning materials are all luminous

lunar eclipse   occurs when the moon is full and the sun, the moon, and the earth are lined up so the shadow of the earth falls on the moon

lunar highlands   light-colored mountainous regions of the moon

lung   a respiratory organ in which air and blood are brought close to one another and gas exchange occurs

L-wave   seismic waves that move on the solid surface of the earth much as water waves move across the surface of a body of water

lymph   liquid material that leaves the circulatory system to surround cells

lymphatic system   a collection of thin-walled tubes that collects, filters, and returns lymph from the body to the circulatory system

macromolecule   very large molecule, with a molecular weight of thousands or millions of atomic mass units, that is made up of a combination of many smaller, similar molecules

magma   a mass of molten rock material either below or on the earth's crust from which igneous rock is formed by cooling and hardening

magnetic domain   tiny physical regions in permanent magnets, approximately 0.01 to 1 mm, that have magnetically aligned atoms, giving the domain an overall polarity

magnetic field   model used to describe how magnetic forces on moving charges act at a distance

magnetic poles   the ends, or sides, of a magnet about which the force of magnetic attraction seems to be concentrated

magnetic quantum number   from the quantum mechanics model of the atom, one of four descriptions of the energy state of an electron wave; this quantum number describes the energy of an electron orbital as the orbital is oriented in space by an external magnetic field, a kind of energy sub-sublevel

magnetic reversal   the flipping of polarity of the earth's magnetic field as the north magnetic pole and the south magnetic pole exchange positions

magnitude   the size of a measurement of a vector; scalar quantities that consist of a number and unit only, no direction, for example

main sequence stars   normal, mature stars that use their nuclear fuel at a steady rate; stars on the Hertzsprung-Russell diagram in a narrow band that runs from the top left to the lower right

malleus   the ear bone that is attached to the tympanum

manipulated variable   in an experiment, a quantity that can be controlled or manipulated; also known as the independent variable

mantle   middle part of the earth's interior; a 2,870 km (about 1,780 mile) thick shell between the core and the crust

maria   smooth dark areas on the moon

marine climate   a climate influenced by air masses from over an ocean, with mild winters and cool summers compared to areas farther inland

maritime air mass   a moist air mass that forms over the ocean

mass   a measure of inertia, which means a resistance to a change of motion

mass defect   the difference between the sum of the masses of the individual nucleons forming a nucleus and the actual mass of that nucleus

mass movement   erosion caused by the direct action of gravity

mass number   the sum of the number of protons and neutrons in a nucleus defines the mass number of an atom; used to identify isotopes; for example, uranium 238

masturbation   stimulation of one's own sex organs

matter   anything that occupies space and has mass

matter waves   any moving object has wave properties, but at ordinary velocities these properties are observed only for objects with a tiny mass; term for the wavelike properties of subatomic particles

mean solar day   a period of 24 hours, averaged from the mean solar time

mean solar time   a uniform time averaged from the apparent solar time

meanders   winding, circuitous turns or bends of a stream

measurement   the process of comparing a property of an object to a well-defined and agreed-upon referent

mechanical energy   the form of energy associated with machines, objects in motion, and objects having potential energy that results from gravity

mechanical weathering   the physical breaking up of rocks without any changes in their chemical composition

medulla oblongata   a region of the more primitive portion of the brain connected to the spinal cord that controls such automatic functions as blood pressure, breathing, and heart rate

meiosis   the specialized pair of cell divisions that reduces the chromosome number from diploid ($2n$) to haploid (n)

melting point   the temperature at which a phase change of solid to liquid takes place; the same temperature as the freezing point for a given substance

Mendelian genetics   the pattern of inheriting characteristics that follows the laws formulated by Gregor Mendel

menstrual cycle   (menses, menstrual flow, period) the repeated building up and shedding of the lining of the uterus

meridians   north-south running arcs that intersect at both poles and are perpendicular to the parallels

mesosphere   the term means "middle layer"—the solid, dense layer of the earth's structure below the asthenosphere but above the core; also the layer of the atmosphere below the thermosphere and above the stratosphere

Mesozoic   the third of four geologic eras; the time of middle life, meaning some of the fossils for this time period are similar to the life found on the earth today, but many are different from anything living today

messenger RNA (mRNA)   a molecule composed of ribonucleotides that functions as a copy of the gene and is used in the cytoplasm of the cell during protein synthesis

metabolic processes   the total of all chemical reactions within an organism; for example, nutrient uptake and processing, and waste elimination

metal   matter having the physical properties of conductivity, malleability, ductility, and luster

metamorphic rocks   previously existing rocks that have been changed into a distinctly different rock by heat, pressure, or hot solutions

metaphase   the second stage in mitosis, characterized by alignment of the chromosomes at the equatorial plane

meteor   the streak of light and smoke that appears in the sky when a meteoroid is made incandescent by friction with the earth's atmosphere

meteor shower   event when many meteorites fall in a short period of time

meteorite   the solid iron or stony material of a meteoroid that survives passage through the earth's atmosphere and reaches the surface

meteoroids   remnants of comets and asteroids in space

meteorology   the science of understanding and predicting weather

meter   the fundamental metric unit of length

metric system   a system of referent units based on invariable referents of nature that have been defined as standards

microclimate   a local, small-scale pattern of climate; for example, the north side of a house has a different microclimate than the south side

microvilli   tiny projections from the surfaces of cells that line the intestine

middle latitudes   latitudes equally far from the poles and equator; between the high and low latitudes

millibar   a measure of atmospheric pressure equivalent to 1.000 dynes per cm^2

mineral (chemistry)   a naturally occurring, inorganic solid element or chemical compound with a crystalline structure

minerals (biology)   inorganic elements that cannot be manufactured by the body but are required in low concentrations; essential to metabolism

miscible fluids   two or more kinds of fluids that can mix in any proportion

mitosis   a process that results in equal and identical distribution of replicated chromosomes into two newly formed nuclei

mixture   matter composed of two or more kinds of matter that has a variable composition and can be separated into its component parts by physical means

model   a mental or physical representation of something that cannot be observed directly that is usually used as an aid to understanding

moderator   a substance in a nuclear reactor that slows fast neutrons so the neutrons can participate in nuclear reactions

Mohorovicic discontinuity   boundary between the crust and mantle that is marked by a sharp increase in the velocity of seismic waves as they pass from the crust to the mantle

mold   the preservation of the shape of an organism by the dissolution of the remains of a buried organism, leaving an empty space where the remains were

mole   an amount of a substance that contains Avogadro's number of atoms, ions, molecules, or any other chemical unit; a mole is thus 6.02×10^{23} atoms, ions, or other chemical units

molecular formula   a chemical formula that identifies the actual numbers of atoms in a molecule

molecular weight   the formula weight of a molecular substance

molecule   from the chemical point of view: a particle composed of two or more atoms held together by an attractive force called a chemical bond; from the kinetic theory point of view: smallest particle of a compound or gaseous element that can exist and still retain the characteristic properties of a substance

momentum the product of the mass of an object times its velocity

monosaccharides simple sugars containing 3 to 6 carbon atoms; the most common kinds are 6-carbon molecules such as glucose and fructose

moraines deposits of rocks and other mounded materials bulldozed into position by a glacier and left behind when the glacier melted

mortality the number of individuals leaving the population by death per thousand individuals in the population

motor neurons those neurons that carry information from the central nervous system to muscles or glands

motor unit all of the muscle cells stimulated by a single neuron

mountain a natural elevation of the earth's crust that rises above the surrounding surface

multiple alleles a term used to refer to conditions in which there are several different alleles for a particular characteristic, not just two

mutagenic agent anything that causes permanent change in DNA

mutation any change in the genetic information of a cell

mutualism a relationship between two organisms in which both organisms benefit

N

natality the number of individuals entering the population by reproduction per thousand individuals in the population

natural frequency the frequency of vibration of an elastic object that depends on the size, composition, and shape of the object

neap tide period of less-pronounced high and low tides; occurs when the sun and the moon are at right angles to one another

nebulae a diffuse mass of interstellar clouds of hydrogen gas or dust that may develop into a star

negative electric charge one of the two types of electric charge; repels other negative charges and attracts positive charges

negative ion atom or particle that has a surplus, or imbalance, of electrons and, thus, a negative charge

nerve cell see *neuron*

nerve impulse a series of changes that take place in the neuron, resulting in a wave of depolarization that passes from one end of the neuron to the other

nerves bundles of neuronal fibers

nervous system a network of neurons that carry information from sense organs to the central nervous system and from the central nervous system to muscles and glands

net force the resulting force after all vector forces have been added; if a net force is zero, all the forces have canceled each other and there is not an unbalanced force

neuron the cellular unit consisting of a cell body and fibers that makes up the nervous system; also called nerve cell

neurotransmitter a molecule released by the axons of neurons that stimulates other cells

neutralized acid or base properties have been lost through a chemical reaction

neutron neutral subatomic particle usually found in the nucleus of an atom

neutron star very small superdense remains of a supernova with a center core of pure neutrons

new crust zone zone of a divergent boundary where new crust is formed by magma upwelling at the boundary

new moon the moon phase when the moon is between the earth and the sun and the entire side of the moon facing the earth is dark

newton a unit of force defined as $kg \cdot m/s^2$; that is, a 1 newton force is needed to accelerate a 1 kg mass 1 m/s^2

niche the functional role of an organism

nitrogenous base a category of organic molecules found as components of the nucleic acids. There are five common types: thymine, guanine, cytosine, adenine, and uracil

noble gas members of family VIII of the periodic table, having common properties of colorless, odorless, chemically inert gases; also known as rare gases or inert gases

node regions on a standing wave that do not oscillate

noise sounds made up of groups of waves of random frequency and intensity

nonelectrolytes water solutions that do not conduct an electric current; covalent compounds that form molecular solutions and cannot conduct an electric current

nonferromagnesian silicates silicates that do not contain iron or magnesium ions; examples include the minerals of muscovite (white mica), the feldspars, and quartz

nonmetal an element that is brittle (when a solid), does not have a metallic luster, is a poor conductor of heat and electricity, and is not malleable or ductile

nonsilicates minerals that do not have the silicon-oxygen tetrahedra in their crystal structure

noon the event of time when the sun moves across the celestial meridian

norepinephrine a hormone produced by the adrenal medulla that increases heart rate, blood pressure, and breathing rate

normal a line perpendicular to the surface of a boundary

normal fault a fault where the hanging wall has moved downward with respect to the foot wall

north celestial pole a point directly above the North Pole of the earth; the point above the north pole on the celestial sphere

north pole (of a magnet) short for "north seeking"; the pole of a magnet that points northward when it is free to turn

nova a star that explodes or suddenly erupts and increases in brightness

nuclear energy the form of energy from reactions involving the nucleus, the innermost part of an atom

nuclear fission nuclear reaction of splitting a massive nucleus into more stable, less massive nuclei with an accompanying release of energy

nuclear force one of four fundamental forces, a strong force of attraction that operates over very short distances between subatomic particles; this force overcomes the electric repulsion of protons in a nucleus and binds the nucleus together

nuclear fusion nuclear reaction of low-mass nuclei fusing together to form more stable and more massive nuclei with an accompanying release of energy

nuclear reactor steel vessel in which a controlled chain reaction of fissionable materials releases energy

nucleic acids complex molecules that store and transfer genetic information within a cell; constructed of fundamental monomers known as nucleotides

nucleons name used to refer to both the protons and neutrons in the nucleus of an atom

nucleoproteins the duplex DNA strands with attached proteins; also called chromatin fibers

nucleosomes histone clusters with their encircling DNA

nucleotide the building block of the nucleic acids, composed of a 5-carbon sugar, a phosphate, and a nitrogenous base

nucleus (atom) tiny, relatively massive and positively charged center of an atom containing protons and neutrons; the small, dense center of an atom

nucleus (cell) the central part of a cell that contains the genetic material

numerical constant a constant without units; a number

nutrients molecules required by the body for growth, reproduction, or repair

nutrition collectively, the processes involved in taking in, assimilating, and utilizing nutrients

O

obese a term describing a person who gains a great deal of unnecessary weight and is 15% to 20% above his or her ideal weight

oblate spheroid the shape of the earth—a somewhat squashed spherical shape

obligate intracellular parasites infectious particles (viruses) that can function only when inside a living cell

observed lapse rate the rate of change in temperature compared to change in altitude

occluded front a front that has been lifted completely off the ground into the atmosphere, forming a cyclonic storm

ocean the single, continuous body of salt water on the surface of the earth

ocean basin the deep bottom of the ocean floor, which starts beyond the continental slope

ocean currents streams of water within the ocean that stay in about the same path as they move over large distances; steady and continuous onward movement of a channel of water in the ocean

ocean wave a moving disturbance that travels across the surface of the ocean

oceanic ridges long, high, continuous, suboceanic mountain chains; for example, the Mid-Atlantic Ridge in the center of the Atlantic Ocean Basin

oceanic trenches long, narrow, deep troughs with steep sides that run parallel to the edge of continents

octet rule a generalization that helps keep track of the valence electrons in most representative elements; atoms of the representative elements (A families) attempt to acquire an outer orbital with eight electrons through chemical reactions

offspring descendants of a set of parents

ohm unit of resistance; equivalent to volts/amps

Ohm's law the electric potential difference is directly proportional to the product of the current times the resistance

oil field petroleum accumulated and trapped in extensive porous rock structure or structures

oils organic compounds of esters formed from glycerol and three long-chain carboxylic acids that are also called triglycerides; called fats in animals and oils in plants

olfactory epithelium the cells of the nasal cavity that respond to chemicals

omnivores animals that are carnivores at some times and herbivores at others

oogenesis the specific name given to the gametogenesis process that leads to the formation of eggs

Oort cloud a spherical "cloud" of small, icy bodies from 30,000 A.U. out to a light-year from the sun; the source of long-period comets

opaque materials that do not allow the transmission of any light

orbital the region of space around the nucleus of an atom where an electron is likely to be found

order a group of closely related families within a class

ore mineral mineral deposits with an economic value

organ a structure composed of two or more kinds of tissues

organ system a group of organs that performs a particular function

organic acids organic compounds with a general formula of RCOOH, where R is one of the hydrocarbon groups; for example, methyl or ethyl

organic chemistry the study of compounds in which carbon is the principal element

organism an independent living unit

orgasm a complex series of responses to sexual stimulation that results in intense frenzy of sexual excitement

orientation of the earth's axis direction that the earth's axis points; considered to be the same throughout the year

origin the only point on a graph where both the x and y variables have a value of zero at the same time

osteoporosis a disease condition resulting from the demineralization of the bone, resulting in pain, deformities, and fractures; related to a loss of calcium

oval window the membrane-covered opening of the cochlea, to which the stapes is attached

ovaries the female sex organs that produce haploid sex cells—the eggs or ova

overtones higher resonant frequencies that occur at the same time as the fundamental frequency, giving a musical instrument its characteristic sound quality

oviduct the tube (fallopian tube) that carries the oocyte to the uterus

ovulation the release of a secondary oocyte from the surface of the ovary

oxbow lake a small body of water, or lake, that formed when two bends of a stream came together and cut off a meander

oxidation the process of a substance losing electrons during a chemical reaction; a reaction between oxygen and the minerals making up rocks

oxidation-reduction reaction a chemical reaction in which electrons are transferred from one atom to another; sometimes called "redox" for short

oxidizing agents substances that take electrons from other substances

oxidizing atmosphere an atmosphere that contains molecular oxygen

oxytocin a hormone released from the posterior pituitary that causes contraction of the uterus

ozone shield concentration of ozone in the lower portions of the stratosphere that absorbs potentially damaging ultraviolet radiation, preventing it from reaching the surface of the earth

P

Paleozoic second of four geologic eras; time of ancient life, meaning the fossils from this time period are very different from anything living on the earth today

pancreas an organ of the body that secretes many kinds of digestive enzymes into the duodenum

parallels reference lines on the earth used to identify where in the world you are northward or southward from the equator; east- and west-running circles that are parallel to the equator on a globe with the distance from the equator called the latitude

parasite an organism that lives in or on another organism and derives nourishment from it

parsec astronomical unit of distance where the distance at which the angle made from 1 A.U. baseline is 1 arc second

parts per billion concentration ratio of parts of solute in every one billion parts of solution (ppb); could be expressed as ppb by volume or as ppb by weight

parts per million concentration ratio of parts of solute in every one million parts of solution (ppm); could be expressed as ppm by volume or as ppm by weight

Pauli exclusion principle no two electrons in an atom can have the same four quantum numbers; thus, a maximum of two electrons can occupy a given orbital

penis the portion of the male reproductive system that deposits sperm in the female reproductive tract

penumbra the zone of partial darkness in a shadow

pepsin an enzyme produced by the stomach that is responsible for beginning the digestion of proteins

percent by volume the volume of solute in 100 volumes of solution

percent by weight the weight of solute in 100 weight units of solution

perception recognition by the brain that a stimulus has been received

perigee when the moon's elliptical orbit brings the moon closest to the earth

perihelion the point at which an orbit comes closest to the sun

period (geologic time) subdivisions of geologic eras

period (periodic table) horizontal rows of elements with increasing atomic numbers; runs from left to right on the element table

period (wave) the time required for one complete cycle of a wave

periodic law similar physical and chemical properties recur periodically when the elements are listed in order of increasing atomic number

peripheral nervous system the fibers that communicate between the central nervous system and other parts of the body

permeability the ability to transmit fluids through openings, small passageways, or gaps

permineralization the process that forms a fossil by alteration of an organism's buried remains by circulating groundwater depositing calcium carbonate, silica, or pyrite

petroleum oil that comes from oil-bearing rock, a mixture of hydrocarbons that is believed to have formed from ancient accumulations of buried organic materials such as remains of algae

pH scale scale that measures the acidity of a solution with numbers below 7 representing acids, 7 representing neutral, and numbers above 7 representing bases

pharynx the region at the back of the mouth cavity; the throat

phase change the action of a substance changing from one state of matter to another; a phase change always absorbs or releases internal potential energy without a temperature change

phases of matter the different physical forms that matter can take as a result of different molecular arrangements, resulting in characteristics of the common phases of a solid, liquid, or gas

phenotype the physical, chemical, and behavioral expression of the genes possessed by an organism

photoelectric effect the movement of electrons in some materials as a result of energy acquired from absorbed light

photon a quantum of energy in a light wave; the particle associated with light

phylogeny the science that explores the evolutionary relationships among organisms and seeks to reconstruct evolutionary history

phylum a subdivision of a kingdom

physical change a change of the state of a substance but not the identity of the substance

pistil the sex organ in plants that produces eggs or ova

pitch the frequency of a sound wave

pituitary gland the gland at the base of the brain that controls the functioning of other glands throughout the organism

placenta an organ made up of tissues from the embryo and the uterus of the mother that allows for the exchange of materials between the mother's bloodstream and the embryo's bloodstream; it also produces hormones

Planck's constant proportionality constant in the relationship between the energy of vibrating molecules and their frequency of vibration; a value of 6.63×10^{-34} Js

plasma (biology) the watery matrix that contains the molecules and cells of the blood

plasma (physics) a phase of matter; a very hot gas consisting of electrons and atoms that have been stripped of their electrons because of high kinetic energies

plastic strain an adjustment to stress in which materials become molded or bent out of shape under stress and do not return to their original shape after the stress is released

plate tectonics the theory that the earth's crust is made of rigid plates that float on the asthenosphere

pleiotropy the multiple effects that a gene may have on the phenotype of an organism

point mutation a change in the DNA of a cell as a result of a loss or change in a nitrogenous-base sequence

polar air mass cold air mass that forms in cold regions

polar body the smaller of two cells formed by unequal meiotic division during oogenesis

polar climate zone climate zone of the high latitudes; average monthly temperatures stay below 10°C (50°F), even during the warmest month of the year

polar covalent bond a covalent bond in which there is an unequal sharing of bonding electrons

polarized light whose constituent transverse waves are all vibrating in the same plane; also known as plane-polarized light

Polaroid a film that transmits only polarized light

polyatomic ion ion made up of many atoms

polygenic inheritance the concept that a number of different pairs of alleles may combine their efforts to determine a characteristic

polymers huge, chainlike molecules made of hundreds or thousands of smaller repeating molecular units called monomers

polysaccharides polymers consisting of monosaccharide units joined together in straight or branched chains; starches, glycogen, or cellulose

pond a small body of standing water, smaller than a lake

pons a region of the brain immediately anterior to the medulla oblongata that connects to the cerebellum and higher regions of the brain and controls several sensory and motor functions of the head and face

population a group of organisms of the same species located in an area

population growth curve a graph of the change in population size over time

porosity the ratio of pore space to the total volume of a rock or soil sample, expressed as a percentage; freely admitting the passage of fluids through pores or small spaces between parts of the rock or soil

positive electric charge one of the two types of electric charge; repels other positive charges and attracts negative charges

positive ion   atom or particle that has a net positive charge due to an electron or electrons being torn away

potential energy   energy due to position; energy associated with changes in position (e.g., gravitational potential energy) or changes in shape (e.g., compressed or stretched spring)

power   the rate at which energy is transferred or the rate at which work is performed; defined as work per unit of time

Precambrian   first of four geologic eras; the time before the time of ancient life, meaning the rocks for this time period contain very few fossils

precession   the slow wobble of the axis of the earth similar to the wobble of a spinning top

precipitation   water that falls to the surface of the earth in the solid or liquid form

predation   a relationship between two organisms that involves the capturing, killing, and eating of one by the other

predator   an organism that captures, kills, and eats another animal

pressure   defined as force per unit area; for example, pounds per square inch (lb/in^2)

prey   an organism captured, killed, and eaten by a predator

primary carnivores   carnivores that eat herbivores and are therefore on the third trophic level

primary coil   part of a transformer; a coil of wire that is connected to a source of alternating current

primary consumers   organisms that feed directly on plants—herbivores

primary loop   part of the energy-converting system of a nuclear power plant; the closed pipe system that carries heated water from the nuclear reactor to a steam generator

primary oocyte   the diploid cell of the ovary that begins to undergo the first meiotic division in the process of oogenesis

primary spermatocyte   the diploid cell in the testes that undergoes the first meiotic division in the process of spermatogenesis

prime meridian   the referent meridian (0°) that passes through the Greenwich Observatory in England

principal quantum number   from the quantum mechanics model of the atom, one of four descriptions of the energy state of an electron wave; this quantum number describes the main energy level of an electron in terms of its most probable distance from the nucleus

probability   the chance that an event will happen, expressed as a percent or fraction

producers   organisms that produce new organic material from inorganic material with the aid of sunlight

prokaryote   an organism composed of cells that lack a nuclear membrane and other membranous organelles

promoter   a region of DNA at the beginning of each gene, just ahead of an initiator code

proof   a measure of ethanol concentration of an alcoholic beverage; proof is double the concentration by volume; for example, 50 percent by volume is 100 proof

properties   qualities or attributes that, taken together, are usually unique to an object; for example, color, texture, and size

prophase   the first phase of mitosis during which individual chromosomes become visible

proportionality constant   a constant applied to a proportionality statement that transforms the statement into an equation

protein synthesis   the process whereby the tRNA utilizes the mRNA as a guide to arrange the amino acids in their proper sequence according to the genetic information in the chemical code of DNA

proteins   macromolecular polymers made of smaller molecules of amino acids, with molecular weight from about six thousand to fifty million; proteins are amino acid polymers with roles in biological structures or functions; without such a function, they are known as polypeptides

protogalaxy   collection of gas, dust, and young stars in the process of forming a galaxy

proton   subatomic particle that has the smallest possible positive charge, usually found in the nucleus of an atom

protoplanet nebular model   a model of the formation of the solar system that states that the planets formed from gas and dust left over from the formation of the sun

protostar   an accumulation of gases that will become a star

pseudoscience   use of the appearance of science to mislead; the assertions made are not valid or reliable

psychrometer   a two-thermometer device used to measure the relative humidity

Ptolemaic system   geocentric model of the structure of the solar system that uses epicycles to explain retrograde motion

puberty   a time in the life of a developing individual characterized by the increasing production of sex hormones, which cause it to reach sexual maturity

pulmonary artery   the major blood vessel that carries blood from the right ventricle to the lungs

pulmonary circulation   the flow of blood through certain chambers of the heart and blood vessels to the lungs and back to the heart

pulsars   the source of regular, equally spaced pulsating radio signals believed to be the result of the magnetic field of a rotating neutron star

Punnett square   a method used to determine the probabilities of allele combinations in an offspring

pure substance   materials that are the same throughout and have a fixed definite composition

pure tone   sound made by very regular intensities and very regular frequencies from regular repeating vibrations

P-wave   a pressure, or compressional, wave in which a disturbance vibrates materials back and forth in the same direction as the direction of wave movement

P-wave shadow zone   a region on the earth between 103° and 142° of arc from an earthquake where no P-waves are received; believed to be explained by P-waves being refracted by the core

pyloric sphincter   a valve located at the end of the stomach that regulates the flow of food from the stomach to the duodenum

quad   one quadrillion Btu (10^{15} Btu); used to describe very large amounts of energy

quanta   fixed amounts; usually referring to fixed amounts of energy absorbed or emitted by matter ("quanta" is plural and "quantum" is singular)

quantities   measured properties; includes the numerical value of the measurement and the unit used in the measurement

quantum mechanics   model of the atom based on the wave nature of subatomic particles, the mechanics of electron waves; also called wave mechanics

quantum numbers   numbers that describe energy states of an electron; in the Bohr model of the atom, the orbit quantum numbers could be any whole number 1, 2, 3, and so on out from the nucleus; in the quantum mechanics model of the atom, four quantum numbers are used to describe the energy state of an electron wave

R

rad a measure of radiation received by a material (radiation absorbed dose)

radiant energy the form of energy that can travel through space; for example, visible light and other parts of the electromagnetic spectrum

radiation the transfer of heat from a region of higher temperature to a region of lower temperature by greater emission of radiant energy from the region of higher temperature

radiation zone part of the interior of a star according to a model; the region directly above the core where gamma and X rays from the core are absorbed and reemitted, with the radiation slowly working its way outward

radioactive decay the natural spontaneous disintegration or decomposition of a nucleus

radioactive decay constant a specific constant for a particular isotope that is the ratio of the rate of nuclear disintegration per unit of time to the total number of radioactive nuclei

radioactive decay series series of decay reactions that begins with one radioactive nucleus that decays to a second nucleus that decays to a third nucleus and so on until a stable nucleus is reached

radioactivity spontaneous emission of particles or energy from an atomic nucleus as it disintegrates

radiometric age age of rocks determined by measuring the radioactive decay of unstable elements within the crystals of certain minerals in the rocks

range the geographical distribution of a species

rarefaction a thinning or pulse of decreased density and pressure of gas molecules

ratio a relationship between two numbers, one divided by the other; the ratio of distance per time is speed

real image an image generated by a lens or mirror that can be projected onto a screen

recessive allele an allele that, when present with a dominant allele, does not express itself and is masked by the effect of the dominant allele

recombinant DNA DNA that has been constructed by inserting new pieces of DNA into the DNA of another organism, such as a bacterium

Recommended Dietary Allowances (RDA) U.S. dietary guidelines for a healthy person that focus on the amounts of foods desired from six classes of nutrients

red giant stars one of two groups of stars on the Hertzsprung-Russell diagram that have a different set of properties than the main sequence stars; bright, low-temperature giant stars that are enormously bright for their temperature

redox reaction short name for oxidation-reduction reaction

reducing agent supplies electrons to the substance being reduced in a chemical reaction

reduction division (also meiosis) a type of cell division in which daughter cells get only half the chromosomes from the parent cell

referent referring to or thinking of a property in terms of another, more familiar object

reflected ray a line representing direction of motion of light reflected from a boundary

reflection the change when light, sound, or other waves bounce backwards off a boundary

refraction a change in the direction of travel of light, sound, or other waves crossing a boundary

relative dating dating the age of a rock unit or geological event relative to some other unit or event

relative humidity ratio (times 100%) of how much water vapor is in the air to the maximum amount of water vapor that could be in the air at a given temperature

reliable a term used to describe results that remain consistent over successive trials

rem measure of radiation that considers the biological effects of different kinds of ionizing radiation

replacement (chemical reaction) reaction in which an atom or polyatomic ion is replaced in a compound by a different atom or polyatomic ion

replacement (fossil formation) process in which an organism's buried remains are altered by circulating groundwaters carrying elements in solution; the removal of original materials by dissolutions and the replacement of new materials an atom or molecule at a time

representative elements name given to the members of the A-group families of the periodic table; also called the main-group elements

reservoir a natural or artificial pond or lake used to store water, control floods, or generate electricity; a body of water stored for public use

resonance when the frequency of an external force matches the natural frequency of a material and standing waves are set up

responding variable the variable that responds to changes in the manipulated variable; also known as the dependent variable because its value depends on the value of the manipulated variable

response the reaction of an organism to a stimulus

responsive processes those abilities to react to external and internal changes in the environment; for example, irritability, individual adaptation, and evolution

retina the light-sensitive region of the eye

reverberation apparent increase in volume caused by reflections, usually arriving within 0.1 second after the original sound

reverse fault a fault where the hanging wall has moved upward with respect to the foot wall

revolution the motion of a planet as it orbits the sun

rhodopsin a light-sensitive pigment found in the rods of the retina

ribonucleic acid (RNA) a polymer of nucleotides formed on the template surface of DNA by transcription. Three forms that have been identified are mRNA, rRNA, and tRNA

ribose a 5-carbon sugar molecule that is a component of RNA

ribosomal RNA (rRNA) a globular form of RNA; a part of ribosomes

Richter scale expresses the intensity of an earthquake in terms of a scale with each higher number indicating 10 times more ground movement and about 30 times more energy released than the preceding number

ridges long, rugged mountain chains rising thousands of meters above the abyssal plains of the ocean basin

rift a split or fracture in a rock formation, land formation, or in the crust of the earth

rip current strong, brief current that runs against the surf and out to sea

RNA polymerase an enzyme that attaches to the DNA at the promoter region of a gene and assists in combining RNA nucleotides when the genetic information is transcribed into RNA

rock a solid aggregation of minerals or mineral materials that have been brought together into a cohesive solid

rock cycle understanding of igneous, sedimentary, or metamorphic rock as a temporary state in an ongoing transformation of rocks into new types; the process of rocks continually changing from one type to another

rock flour rock pulverized by a glacier into powdery, silt-sized sediment

rods light-sensitive cells in the retina of the eye that respond to low-intensity light but do not respond to different colors of light

rotation the spinning of a planet on its axis

runoff water moving across the surface of the earth as opposed to soaking into the ground

salinity a measure of dissolved salts in seawater, defined as the mass of salts dissolved in 1,000 g of solution

salivary amylase an enzyme present in saliva that breaks starch molecules into smaller molecules

salivary glands glands that produce saliva

salt any ionic compound except one with hydroxide or oxide ions

San Andreas fault in California, the boundary between the North American Plate and the Pacific Plate that runs north-south for some 1,300 km (800 miles) with the Pacific Plate moving northwest and the North American Plate moving southeast

saprophyte an organism that obtains energy by the decomposition of dead organic material

saturated air air in which an equilibrium exists between evaporation and condensation; the relative humidity is 100 percent

saturated molecule an organic molecule that has the maximum number of hydrogen atoms possible

saturated solution the apparent limit to dissolving a given solid in a specified amount of water at a given temperature; a state of equilibrium that exists between dissolving solute and solute coming out of solution

scalars measurements that have magnitude only, defined by the numerical value and a unit such as 30 miles per hour

scientific law a relationship between quantities, usually described by an equation in the physical sciences; is more important and describes a wider range of phenomena than a scientific principle

scientific principle a relationship between quantities concerned with a specific or narrow range of observations and behavior

scintillation counter a device that indirectly measures ionizing radiation (alpha, beta, gamma, etc.) by measuring the flashes of light produced when the radiation strikes a phosphor

sea a smaller part of the ocean with characteristics that distinguish it from the larger ocean

sea breeze cool, dense air from over water moving over land as part of convective circulation

seafloor spreading the process by which hot, molten rock moves up from the interior of the earth to emerge along mid-oceanic rifts, flowing out in both directions to create new rocks and spread apart the seafloor

seamounts steep, submerged volcanic peaks on the abyssal plain

second the standard unit of time in both the metric and English systems of measurement

second law of motion the acceleration of an object is directly proportional to the net force acting on that object and inversely proportional to the mass of the object

secondary carnivores carnivores that feed on primary carnivores and are therefore at the fourth trophic level

secondary coil part of a transformer, a coil of wire in which the voltage of the original alternating current in the primary coil is stepped up or down by way of electromagnetic induction

secondary consumers animals that eat other animals—carnivores

secondary loop part of a nuclear power plant; the closed pipe system that carries steam from a steam generator to the turbines, then back to the steam generator as feedwater

secondary oocyte the larger of the two cells resulting from the unequal cytoplasmic division of a primary oocyte in meiosis I of oogenesis

secondary sex characteristics characteristics of the adult male or female, including the typical shape that develops at puberty: broader shoulders, heavier long-bone muscles, development of facial hair, axillary hair, and chest hair, and changes in the shape of the larynx in the male; rounding of the pelvis and breasts and changes in deposition of fat in the female

secondary spermatocyte cells in the seminiferous tubules that go through the second meiotic division, resulting in four haploid spermatids

sedimentary rocks rocks formed from particles or dissolved minerals from previously existing rocks that were deposited from air or water

sediments accumulations of silt, sand, or gravel that settled out of the atmosphere or out of water

segregation the separation and movement of homologous chromosomes to the opposite poles of the cell

seismic waves vibrations that move as waves through any part of the earth, usually associated with earthquakes, volcanoes, or large explosions

seismograph an instrument that measures and records seismic wave data

semen the sperm-carrying fluid produced by the seminal vesicles, prostate glands, and bulbourethral glands of males

semiarid climate classification between arid and humid; receives between 25 and 50 cm (10 and 20 in) precipitation per year

semicircular canals a set of tubular organs associated with the cochlea that sense changes in the movement or position of the head

semiconductors elements that have properties between those of a metal and those of a nonmetal, sometimes conducting an electric current and sometimes acting like an electrical insulator depending on the conditions and their purity; also called metalloids

semilunar valves pulmonary artery and aorta valves that prevent the flow of blood backward into the ventricles

seminal vesicle a part of the male reproductive system that produces a portion of the semen

seminiferous tubules sperm-producing tubes in the testes

sensory neurons those neurons that send information from sense organs to the central nervous system

sex chromosomes a pair of chromosomes that determines the sex of an organism; X and Y chromosomes

sex-determining chromosome the chromosomes X and Y that are primarily responsible for determining if an individual will develop as a male or female

sexual intercourse the mating of male and female; the deposition of the male sex cells, or sperm cells, in the reproductive tract of the female; also known as coitus or copulation

sexuality a term used in reference to the totality of the aspects—physical, psychological, and cultural—of our sexual nature

shallow-focus earthquakes earthquakes that occur from the surface down to 70 km deep

shear stress produced when two plates slide past one another or by one plate sliding past another plate that is not moving

shell model of the nucleus model of the nucleus that has protons and neutrons moving in energy levels or shells in the nucleus (similar to the shell structure of electrons in an atom)

shells the layers that electrons occupy around the nucleus

shield volcano a broad, gently sloping volcanic cone constructed of solidified lava flows

shock wave a large, intense wave disturbance of very high pressure; the pressure wave created by an explosion, for example

sidereal day the interval between two consecutive crossings of the celestial meridian by a particular star

sidereal month the time interval between two consecutive crossings of the moon across any star

sidereal year the time interval required for the earth to move around its orbit so that the sun is again in the same position against the stars

silicates minerals that contain silicon-oxygen tetrahedra either isolated or joined together in a crystal structure

sill a tabular-shaped intrusive rock that formed when magma moved into the plane of contact between sedimentary rock layers

simple harmonic motion the vibratory motion that occurs when there is a restoring force opposite to and proportional to a displacement

single bond covalent bond in which a single pair of electrons is shared by two atoms

single-factor cross a genetic study in which a single characteristic is followed from the parental generation to the offspring

slope the ratio of changes in the y variable to changes in the x variable or how fast the y-value increases as the x-value increases

small intestine the portion of the digestive system immediately following the stomach; responsible for digestion and absorption

soil a mixture of unconsolidated weathered earth materials and humus, which is altered, decay-resistant organic matter

solar constant the average solar power received by the outermost part of the earth's atmosphere when the sunlight is perpendicular to the outer edge and the earth is at an average distance from the sun; about 1,370 watts per square meter

solenoid a cylindrical coil of wire that becomes electromagnetic when a current runs through it

solids a phase of matter with molecules that remain close to fixed equilibrium positions due to strong interactions between the molecules, resulting in the characteristic definite shape and definite volume of a solid

solstices time when the sun is at its maximum or minimum altitude in the sky, known as the summer solstice and the winter solstice

solubility dissolving ability of a given solute in a specified amount of solvent, the concentration that is reached as a saturated solution is achieved at a particular temperature

solute the component of a solution that dissolves in the other component; the solvent

solution a homogeneous mixture of ions or molecules of two or more substances

solvent the component of a solution present in the larger amount; the solute dissolves in the solvent to make a solution

soma the cell body of a neuron, which contains the nucleus

sonic boom sound waves that pile up into a shock wave when a source is traveling at or faster than the speed of sound

sound quality characteristic of the sound produced by a musical instrument; determined by the presence and relative strengths of the overtones produced by the instrument

south celestial pole a point directly above the South Pole of the earth; the point above the south pole on the celestial sphere

south pole (of a magnet) short for "south seeking"; the pole of a magnet that points southward when it is free to turn

speciation the process of generating new species

species a group of organisms that can interbreed naturally to produce fertile offspring

specific heat each substance has its own specific heat, which is defined as the amount of energy (or heat) needed to increase the temperature of 1 gram of a substance 1 degree Celsius

speed a measure of how fast an object is moving—the rate of change of position per change in time; speed has magnitude only and does not include the direction of change

sperm the haploid sex cells produced by sexually mature males

spermatids haploid cells produced by spermatogenesis that change into sperm

spermatogenesis the specific name given to the gametogenesis process that leads to the formation of sperm

spin quantum number from the quantum mechanics model of the atom, one of four descriptions of the energy state of an electron wave; this quantum number describes the spin orientation of an electron relative to an external magnetic field

spinal cord the portion of the central nervous system located within the vertebral column, which carries both sensory and motor information between the brain and the periphery of the body

spindle an array of microtubules extending from pole to pole; used in the movement of chromosomes

spontaneous generation the theory that living organisms arose from nonliving material

spring equinox one of two times a year that daylight and night are of equal length; occurs on or about March 21 and identifies the beginning of the spring season

spring tides unusually high or low tides that occur every two weeks because of the earth, the moon, and the sun

standard atmospheric pressure the average atmospheric pressure at sea level, which is also known as normal pressure; the standard pressure is 29.92 inches or 760.0 mm of mercury (1,013.25 millibar)

standard time zones 15° wide zones defined to have the same time throughout the zone, defined as the mean solar time at the middle of each zone

standard unit a measurement unit established as the standard upon which the value of the other referent units of the same type are based

standing waves condition where two waves of equal frequency traveling in opposite directions meet and form stationary regions of maximum displacement due to constructive interference and stationary regions of zero displacement due to destructive interference

stapes the ear bone that is attached to the oval window

starch a group of complex carbohydrates (polysaccharides) that plants use as a stored food source and that serves as an important source of food for animals

stationary front occurs when the edge of a weather front is not advancing

steam generator part of a nuclear power plant; the heat exchanger that heats feedwater from the secondary loop to steam with the very hot water from the primary loop

step-down transformer a transformer that decreases the voltage

step-up transformer a transformer that increases the voltage

stimulus any change in the internal or external environment of an organism that it can detect

stony meteorites   meteorites composed mostly of silicate minerals that usually make up rocks on the earth

stony-iron meteorites   meteorites composed of silicate minerals and metallic iron

storm   a rapid and violent weather change with strong winds, heavy rain, snow, or hail

strain   adjustment to stress; a rock unit might respond to stress by changes in volume, changes in shape, or by breaking

stratopause   the upper boundary of the stratosphere

stratosphere   the layer of the atmosphere above the troposphere where temperature increases with height

stream   body of running water

stress   a force that tends to compress, pull apart, or deform rock; stress on rocks in the earth's solid outer crust results as the earth's plates move into, away from, or alongside each other

strong acid   acid that ionizes completely in water, with all molecules dissociating into ions

strong base   base that is completely ionized in solution and has hydroxide ions

subduction zone   the region of a convergent boundary where the crust of one plate is forced under the crust of another plate into the interior of the earth

sublimation   the phase change of a solid directly into a vapor or gas

submarine canyons   a feature of the ocean basin; deep, steep-sided canyons that cut through the continental slopes

subspecies   regional groups within a species that are significantly different structurally, physiologically, or behaviorally, yet are capable of exchanging genes by interbreeding with other members of the species

summer solstice   in the Northern Hemisphere, the time when the sun reaches its maximum altitude in the sky, which occurs on or about June 22 and identifies the beginning of the summer season

superconductors   some materials in which, under certain conditions, the electrical resistance approaches zero

supercooled   water in the liquid phase when the temperature is below the freezing point

supernova   a rare catastrophic explosion of a star into an extremely bright, but short-lived phenomenon

supersaturated   containing more than the normal saturation amount of a solute at a given temperature

surf   the zone where breakers occur; the water zone between the shoreline and the outermost boundary of the breakers

surface wave   a seismic wave that moves across the earth's surface, spreading across the surface like water waves spread on the surface of a pond from a disturbance

S-wave   a sideways, or shear wave in which a disturbance vibrates materials from side to side, perpendicular to the direction of wave movement

S-wave shadow zone   a region of the earth more than 103° of arc away from the epicenter of an earthquake where S-waves are not recorded; believed to be the result of the core of the earth that is a liquid, or at least acts like a liquid

swell   regular groups of low-profile, long-wavelength waves that move continuously

symbiosis   a close physical relationship between two kinds of organisms—parasitism, commensalism, and mutualism may all be examples of symbiosis

synapse   the space between the axon of one neuron and the dendrite of the next, where chemicals are secreted to cause an impulse to be initiated in the second neuron

synapsis   the condition in which the two members of a pair of homologous chromosomes come to lie close to one another

syncline   a trough-shaped fold in layered bedrock

synodic month   the interval of time from new moon to new moon (or any two consecutive identical phases)

systemic circulation   the flow of blood through certain chambers of the heart and blood vessels to the general body and back to the heart

T

talus   steep, conical or apron-like accumulations of rock fragments at the base of a slope

taxonomy   the science of classifying and naming organisms

telophase   the last phase in mitosis characterized by the formation of daughter nuclei

temperate climate zone   climate zone of the middle latitudes; average monthly temperatures stay between 10°C and 18°C (50°F and 64°F) throughout the year

temperature   how hot or how cold something is; a measure of the average kinetic energy of the molecules making up a substance

tensional stress   the opposite of compressional stress; occurs when one part of a plate moves away from another part that does not move

termination code   the DNA nucleotide sequence at the end of a gene with the code ATT, ATC, or ACT that signals "stop here"

terrestrial planets   planets Mercury, Venus, Earth, and Mars that have similar densities and compositions as compared to the outer giant planets

testes   the male sex organs that produce haploid cells called sperm

testosterone   the male sex hormone produced in the testes that controls male sexual development

thalamus   a region of the brain that relays information between the cerebrum and lower portions of the brain, providing some level of awareness in that it determines pleasant and unpleasant stimuli and is involved in sleep and arousal

theory   a broad, detailed explanation that guides the development of hypotheses and interpretations of experiments in a field of study

thermometer   a device used to measure the hotness or coldness of a substance

thermosphere   thin, high, outer atmospheric layer of the earth where the molecules are far apart and have a high kinetic energy

third law of motion   whenever two objects interact, the force exerted on one object is equal in size and opposite in direction to the force exerted on the other object; forces always occur in matched pairs that are equal and opposite

thrust fault   a reverse fault with a low-angle fault plane

thunderstorm   a brief, intense electrical storm with rain, lightning, thunder, strong winds, and sometimes hail

thymine   a single-ring nitrogenous-base molecule in DNA but not in RNA; it is complementary to adenine

thyroid-stimulating hormone (TSH)   a hormone secreted by the pituitary gland that stimulates the thyroid to secrete thyroxine

thyroxine   a hormone produced by the thyroid gland that speeds up the metabolic rate

tidal bore   a strong tidal current, sometimes resembling a wave, produced in very long, very narrow bays as the tide rises

tidal currents   a steady and continuous onward movement of water produced in narrow bays by the tides

tides   periodic rise and fall of the level of the sea from the gravitational attraction of the moon and sun

tissue   a group of specialized cells that work together to perform a specific function

tornado   a long, narrow, funnel-shaped column of violently whirling air from a thundercloud that moves destructively over a narrow path when it touches the ground

total internal reflection   condition where all light is reflected back from a boundary between materials; occurs when light arrives at a boundary at the critical angle or beyond

total solar eclipse   eclipse that occurs when the earth, the moon, and the sun are lined up so the new moon completely covers the disk of the sun; the umbra of the moon's shadow falls on the surface of the earth

trachea   a major tube supported by cartilage that carries air to the bronchi; also known as the windpipe

transcription   the process of manufacturing RNA from the template surface of DNA; three forms of RNA that may be produced are mRNA, rRNA, and tRNA

transfer RNA (tRNA)   a molecule composed of ribonucleic acid. It is responsible for transporting a specific amino acid into a ribosome for assembly into a protein

transform boundaries   in plate tectonics, boundaries that occur between two plates sliding horizontally by each other along a long, vertical fault; sudden jerks along the boundary result in the vibrations of earthquakes

transformer   a device consisting of a primary coil of wire connected to a source of alternating current and a secondary coil of wire in which electromagnetic induction increases or decreases the voltage of the source

transition elements   members of the B-group families of the periodic table

translation   the assembly of individual amino acids into a polypeptide

transparent   term describing materials that allow the transmission of light; for example, glass and clear water are transparent materials

transportation   the movement of eroded materials by agents such as rivers, glaciers, wind, or waves

transverse wave   a mechanical disturbance that causes particles to move perpendicular to the direction that the wave is traveling

triglyceride   organic compound of esters formed from glycerol and three long-chain carboxylic acids; also called fats in animals and oils in plants

triple bond   covalent bond formed when three pairs of electrons are shared by two atoms

trophic level   a step in the flow of energy through an ecosystem

tropic of Cancer   parallel identifying the northern limit where the sun appears directly overhead; located at 23.5°N latitude

tropic of Capricorn   parallel identifying the southern limit where the sun appears directly overhead; located at 23.5°S latitude

tropical air mass   a warm air mass from warm regions

tropical climate zone   climate zone of the low latitudes; average monthly temperatures stay above 18°C (64°F), even during the coldest month of the year

tropical cyclone   a large, violent circular storm that is born over the warm, tropical ocean near the equator; also called hurricane (Atlantic and eastern Pacific) and typhoon (in western Pacific)

tropical year   the time interval between two consecutive spring equinoxes; used as standard for the common calendar year

tropopause   the upper boundary of the troposphere, identified by the altitude where the temperature stops decreasing and remains constant with increasing altitude

troposphere   layer of the atmosphere from the surface to where the temperature stops decreasing with height

trough   the low mound of water that is part of a wave; also refers to the rarefaction, or low-pressure part of a sound wave

tsunami   very large, fast, and destructive ocean wave created by an undersea earthquake, landslide, or volcanic explosion; a seismic sea wave

turbidity current   a muddy current produced by underwater landslides

tympanum   the eardrum

typhoon   the name for hurricanes in the western Pacific

U

ultrasonic   sound waves too high in frequency to be heard by the human ear; frequencies above 20,000 Hz

umbra   the inner core of a complete shadow

unconformity   a time break in the rock record

undertow   a current beneath the surface of the water produced by the return of water from the shore to the sea

unit   in measurement, a well-defined and agreed-upon referent

universal law of gravitation   every object in the universe is attracted to every other object with a force directly proportional to the product of their masses and inversely proportional to the square of the distance between the centers of the two masses

unpolarized light   light consisting of transverse waves vibrating in all conceivable random directions

unsaturated molecule   an organic molecule that does not contain the maximum number of hydrogen atoms; a molecule that can add more hydrogen atoms because of the presence of double or triple bonds

uracil   a single-ring nitrogenous-base molecule in RNA but not in DNA; it is complementary to adenine

uterus   the organ in female mammals in which the embryo develops

V

vagina   the passageway between the uterus and outside of the body; the birth canal

valence   the number of covalent bonds an atom can form

valence electrons   electrons of the outermost shell; the electrons that determine the chemical properties of an atom and the electrons that participate in chemical bonding

valid   a term used to describe meaningful data that fit into the framework of scientific knowledge

Van Allen belts   belts of radiation caused by cosmic-ray particles becoming trapped and following the earth's magnetic field lines between the poles

Van der Waals force   general term for weak attractive intermolecular forces

vapor   the gaseous state of a substance that is normally in the liquid state

variables   changing quantities usually represented by a letter or symbol

vector (biology)   an organism that carries a disease or parasite from one host to the next

vector (physics)   quantity that is described by both a magnitude and direction

veins   the blood vessels that return blood to the heart

velocity   describes both the speed and direction of a moving object; a change in velocity is a change in speed, in direction of travel, or both

ventifacts   rocks sculpted by wind abrasion

ventricles the powerful muscular chambers of the heart whose contractions force blood to flow through the arteries to all parts of the body

vernal equinox another name for the spring equinox, which occurs on or about March 21 and marks the beginning of the spring season

vibration a back-and-forth motion that repeats itself

villi tiny fingerlike projections in the lining of the intestine that increase the surface area for absorption

virtual image an image where light rays appear to originate from a mirror or lens; this image cannot be projected on a screen

virus a nucleic acid particle coated with protein that functions as an obligate intracellular parasite

vitamin-deficiency disease poor health caused by the lack of a certain vitamin in the diet; for example, scurvy for lack of vitamin C

vitamins organic molecules that cannot be manufactured by the body but are required in very low concentrations for good health

volcanism volcanic activity; the movement of magma

volcano a hill or mountain formed by the extrusion of lava or rock fragments from a mass of magma below

volt unit of potential difference equivalent to J/C

voltage drop the electric potential difference across a resistor or other part of a circuit that consumes power

voltage source source of electric power in an electric circuit that maintains a constant voltage supply to the circuit

volume how much space something occupies

vulcanism volcanic activity; the movement of magma

W

warm front the front that forms when a warm air mass advances against a cool air mass

water table the boundary below which the ground is saturated with water

watershed the region or land area drained by a stream; a stream drainage basin

watt metric unit for power; equivalent to J/s

wave a disturbance or oscillation that moves through a medium

wave equation the relationship of the velocity of a wave to the product of the wavelength and frequency of the wave

wave front a region of maximum displacement in a wave; a condensation in a sound wave

wave height the vertical distance of an ocean wave between the top of the wave crest and the bottom of the next trough

wave mechanics alternate name for quantum mechanics derived from the wavelike properties of subatomic particles

wave period the time required for two successive crests or other successive parts of the wave to pass a given point

wavelength the horizontal distance between successive wave crests or other successive parts of the wave

weak acid acids that only partially ionize because of an equilibrium reaction with water

weak base a base only partially ionized because of an equilibrium reaction with water

weathering slow changes that result in the breaking up, crumbling, and destruction of any kind of solid rock

white dwarf stars one of two groups of stars on the Hertzsprung-Russell diagram that have a different set of properties than the main sequence stars; faint, white-hot stars that are very small and dense

wind a horizontal movement of air that moves along or parallel to the ground, sometimes in currents or streams

wind abrasion the natural sand-blasting process that occurs when wind-driven particles break off small particles of rock and polish the rock they strike

wind chill factor a factor that compares heat loss from bodies in still air with those in moving air; moving air removes heat more rapidly and causes a person to feel that the air is colder than its actual temperature; the cooling power of wind

winter solstice in the Northern Hemisphere, the time when the sun reaches its minimum altitude, which occurs on or about December 22 and identifies the beginning of the winter season

work the magnitude of applied force times the distance through which the force acts; can be thought of as the process by which one form of energy is transformed to another

X

X chromosome the chromosome in a human female egg (and in one-half of sperm cells) that is associated with the determination of sexual characteristics

X-linked gene a gene located on one of the sex-determining X chromosomes

Y

Y chromosome the sex-determining chromosome in one-half of the sperm cells of human males responsible for determining maleness

Z

zone of saturation zone of sediments beneath the surface in which water has collected in all available spaces

zygote a diploid cell that results from the union of an egg and a sperm; a fertilized egg

CREDITS

Photographs

INDEX

Table of Atomic Weights (Based on Carbon-12)

Name	Symbol	Atomic Number	Atomic Weight	Name	Symbol	Atomic Number	Atomic Weight
Actinium	Ac	89	(227)	Mendelevium	Md	101	258.10
Aluminum	Al	13	26.9815	Mercury	Hg	80	200.59
Americium	Am	95	(243)	Molybdenum	Mo	42	95.94
Antimony	Sb	51	121.75	Neodymium	Nd	60	144.24
Argon	Ar	18	39.948	Neon	Ne	10	20.179
Arsenic	As	33	74.922	Neptunium	Np	93	(237)
Astatine	At	85	(210)	Nickel	Ni	28	58.71
Barium	Ba	56	137.34	Niobium	Nb	41	92.906
Berkelium	Bk	97	(247)	Nitrogen	N	7	14.0067
Beryllium	Be	4	9.0122	Nobelium	No	102	259.101
Bismuth	Bi	83	208.980	Osmium	Os	76	190.2
Bohrium	Bh	107	264	Oxygen	O	8	15.9994
Boron	B	5	10.811	Palladium	Pd	46	106.4
Bromine	Br	35	79.904	Phosphorus	P	15	30.9738
Cadmium	Cd	48	112.40	Platinum	Pt	78	195.09
Calcium	Ca	20	40.08	Plutonium	Pu	94	244.064
Californium	Cf	98	242.058	Polonium	Po	84	(209)
Carbon	C	6	12.0112	Potassium	K	19	39.098
Cerium	Ce	58	140.12	Praseodymium	Pr	59	140.907
Cesium	Cs	55	132.905	Promethium	Pm	61	144.913
Chlorine	Cl	17	35.453	Protactinium	Pa	91	(231)
Chromium	Cr	24	51.996	Radium	Ra	88	(226)
Cobalt	Co	27	58.933	Radon	Rn	86	(222)
Copper	Cu	29	63.546	Rhenium	Re	75	186.2
Curium	Cm	96	(247)	Rhodium	Rh	45	102.905
Dubnium	Db	105	(262)	Rubidium	Rb	37	85.468
Dysprosium	Dy	66	162.50	Ruthenium	Ru	44	101.07
Einsteinium	Es	99	(254)	Rutherfordium	Rf	104	(261)
Erbium	Er	68	167.26	Samarium	Sm	62	150.35
Europium	Eu	63	151.96	Scandium	Sc	21	44.956
Fermium	Fm	100	257.095	Seaborgium	Sg	106	(266)
Fluorine	F	9	18.9984	Selenium	Se	34	78.96
Francium	Fr	87	(223)	Silicon	Si	14	28.086
Gadolinium	Gd	64	157.25	Silver	Ag	47	107.868
Gallium	Ga	31	69.723	Sodium	Na	11	22.989
Germanium	Ge	32	72.59	Strontium	Sr	38	87.62
Gold	Au	79	196.967	Sulfur	S	16	32.064
Hafnium	Hf	72	178.49	Tantalum	Ta	73	180.948
Hassium	Hs	108	(269)	Technetium	Tc	43	(99)
Helium	He	2	4.0026	Tellurium	Te	52	127.60
Holmium	Ho	67	164.930	Terbium	Tb	65	158.925
Hydrogen	H	1	1.0079	Thallium	Tl	81	204.37
Indium	In	49	114.82	Thorium	Th	90	232.038
Iodine	I	53	126.904	Thulium	Tm	69	168.934
Iridium	Ir	77	192.2	Tin	Sn	50	118.69
Iron	Fe	26	55.847	Titanium	Ti	22	47.90
Krypton	Kr	36	83.80	Tungsten	W	74	183.85
Lanthanum	La	57	138.91	Uranium	U	92	238.03
Lawrencium	Lr	103	260.105	Vanadium	V	23	50.942
Lead	Pb	82	207.19	Xenon	Xe	54	131.30
Lithium	Li	3	6.941	Ytterbium	Yb	70	173.04
Lutetium	Lu	71	174.97	Yttrium	Y	39	88.905
Magnesium	Mg	12	24.305	Zinc	Zn	30	65.38
Manganese	Mn	25	54.938	Zirconium	Zr	40	91.22
Meitnerium	Mt	109	(268)				

Note: A value given in parentheses denotes the number of the longest-lived or best-known isotope.